fifth edition

Physical Geology Earth Revealed

David McGeary
Emeritus of California State University at Sacramento

Charles C. Plummer
California State University at Sacramento

Diane H. Carlson
California State University at Sacramento

Boston Burr Ridge, IL Dubuque, IA Madison, WI New York San Francisco St. Louis
Bangkok Bogotá Caracas Kuala Lumpur Lisbon London Madrid Mexico City
Milan Montreal New Delhi Santiago Seoul Singapore Sydney Taipei Toronto

The McGraw·Hill Companies

PHYSICAL GEOLOGY: EARTH REVEALED
FIFTH EDITION

Published by McGraw-Hill, a business unit of The McGraw-Hill Companies, Inc., 1221 Avenue of the Americas, New York, NY 10020.

 This book is printed on recycled, acid-free paper containing 10% postconsumer waste.

2 3 4 5 6 7 8 9 0 QPD/QPD 0 9 8 7 6 5 4

ISBN 0–07–246327–9

Publisher: *Margaret J. Kemp*
Sponsoring editor: *Thomas C. Lyon*
Developmental editor: *Lisa A. Leibold*
Executive marketing manager: *Lisa L. Gottschalk*
Lead project manager: *Joyce M. Berendes*
Production supervisor: *Sherry L. Kane*
Media technology producer: *Renee Russian*
Coordinator of freelance design: *Rick D. Noel*
Cover designer: *Jamie E. O'Neal*
Cover image: © *Getty Images, Hawaii, volcano at night, G. Brad Lewis*
Lead photo research coordinator: *Carrie K. Burger*
Compositor: *Carlisle Communications, Ltd.*
Typeface: *10.5/12 Times Roman*
Printer: *Quebecor World Dubuque, IA*

Library of Congress Cataloging-in-Publication Data

McGeary, David.
Physical geology : Earth revealed / David McGeary.— 5th ed.
p. cm.
Includes index.
ISBN 0–07–246327–9 (hard copy : alk. paper)
1. Physical geology. I. Plummer, Charles C., 1937– . II. Carlson, Diane H. III. Earth revealed (Television program). IV. Title.

QE28.2 .M34 2004
551–dc21

2002151425
CIP

www.mhhe.com

Contents in Brief

Contents

*These end-of-chapter sections appear in every chapter.

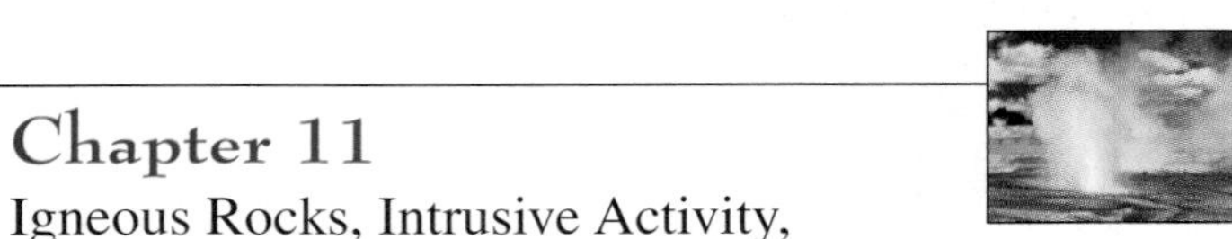

Chapter 14
Sediments and Sedimentary Rocks 335

Chapter 15
Metamorphism, Metamorphic Rocks, and Hydrothermal Rocks 365

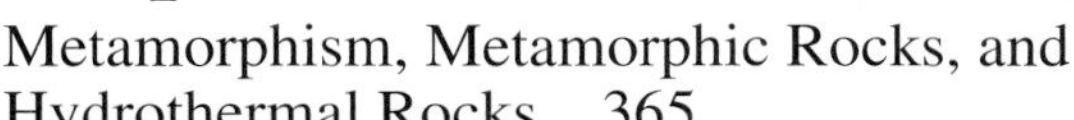

Chapter 16
Streams and Floods 387

Chapter 17
Ground Water 423

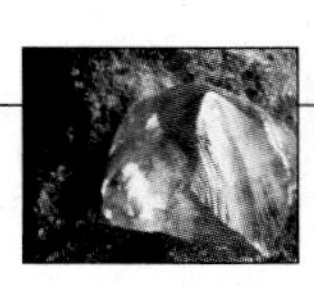

Chapter 18
Deserts and Wind Action 447

Chapter 19
Glaciers and Glaciation 469

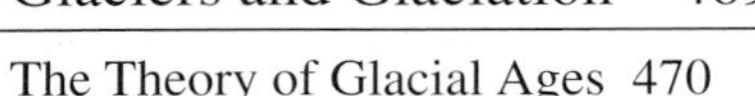

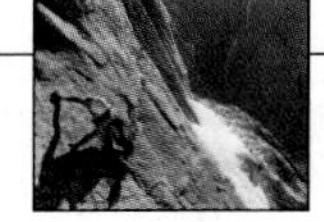

Chapter 20
Waves, Beaches, and Coasts 499

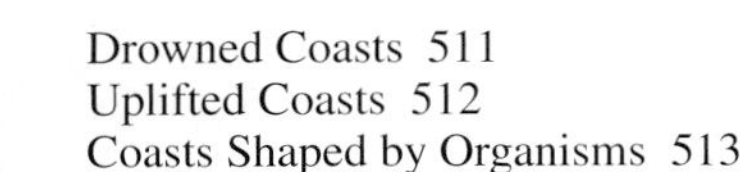

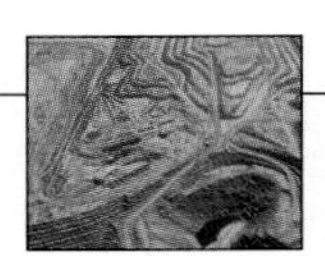

Chapter 21
Geologic Resources 519

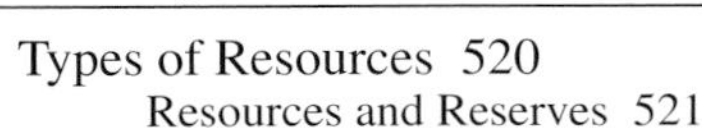

Preface

Physical Geology Earth has been updated to include the latest technology and most current information. *Physical Geology* is for both nonscience majors and students contemplating majoring in geology. The new art program and interactive writing style will grab your students' attention and further their interest in the subject.

This book contains the same text and illustrations as the ninth edition of *Physical Geology* by Plummer, McGeary, and Carlson. The chapter order has been changed so that internal processes (plate tectonics, earthquakes, etc.) are covered in the first part of the book and external processes (rivers, glacier, etc.) are described toward the end of the book. This ordering is favored by many geology instructors. *Physical Geology: Earth Revealed* is featured as the companion text to *Earth Revealed Introductory Geology,* a PBS television course and video resource produced in collaboration with the Annenberg/CPB project. *Earth Revealed* is a series of twenty-six half-hour video programs organized around the chapters of this text. The television programs document evidence of geologic principles at geographically diverse sites, often using a case study approach. Videocassettes can be purchased individually or as a thirteen-tape set. A *Study Guide* and *Faculty Guide* are also available to supplement the programs. For information regarding the use of *Earth Revealed Introduction Geology* as a television course, or to purchase videocassettes for institutional or classroom use, contact the Annenberg/CPB Multimedia Collection at 1-800-LEARNER.

What's New in This Edition

The Internet has revolutionized the way we learn. This edition expands upon the integration of the Internet and textbook. We have added boxes that have a brief summary in the book, while the complete boxes are accessible through this book's website. We have shortened some boxes from previous editions, but placed the full box on the website. When we have found excellent and appropriate websites, we have added URLs in the text and in figure captions. Our website has enjoyable and enlightening web exercises that we have tested with our students. An exciting addition to this edition is the figures that have been animated to more clearly illustrate processes active in geology.

We have added new and revised artwork and photos. Some of the changes we have made for this edition include the following items.

In chapter 1, we have added boxes on geology as a career and the origin of the solar system. We have added isostatic adjustment to the important concepts covered in the introductory chapter and have expanded the introduction to plate tectonics. In the minerals chapter, we have added a brief section on polarizing microscopy to the discussion of double refraction and referred the interested reader to a website for more information. The introduction to the rock cycle has been moved to the beginning of the chapter on igneous rocks. In the chapter on igneous rocks and processes, we have overhauled our presentation of Bowen's reaction series to present what students need to know to understand igneous processes and use the website for a more complete presentation of the reaction series. We give a thorough, illustrated explanation of how partial melting takes place in circulating asthenosphere above subducting crust. In the volcanoes chapter we have added a section on volcanoes and myths. We have also added a section that quantifies volcanic hazards. A new box looks at Mexico's Popocatepetl's recent eruptions and the potential for a disastrous eruption.

We have added a discussion of the twelve soil orders and updated the description and diagram of a soil profile to include the E Horizon. Abrasion has been removed as an agent of erosion. Chapter 14 has been expanded to include a discussion and diagram of the relation of plate tectonic settings and types of sedimentary rock. A new astrogeology box featuring the latest Mars Global Surveyor images discusses the importance of sedimentary rock for determining whether water and life once existed on Mars; the regression and transgression box has been moved to the website and now includes animated diagrams.

In chapter 15, we have enhanced the description of the role of water in metamorphism to include why retrograde metamorphism is uncommon. We tie in the dehydration of metamorphic minerals during subduction to supplying the water necessary to partial melting of asthenosphere as described in the chapter on igneous rocks. In the chapter on geologic time, we have greatly expanded our coverage of isotopic dating to include descriptions of the mechanisms of radioactive decay. The recently dated 4.004 billion-year-old zircon crystal and its implications regarding early Earth history are discussed in that chapter.

Chapter 16 includes new photos of Niagara Falls and braided streams. The stream piracy section has been removed. The astrogeology box has been updated to include a discussion and latest photos of streamlike features on Mars from Nanedi Vallis canyon. Website URLs provide easy access to additional images from the Mars Orbiter Camera. Chapter 17 includes a rewrite of the Darcy's Law box to address the influence of porosity on groundwater velocity through sediment or rock as well as revision of several diagrams showing the details of ground-water flow and fluctuation. The term *speleothem* has been added, and a discussion of thermophyllic bacteria around hot springs and the implication for early life is presented.

In the glaciers chapter, we have added a figure showing the extent of glaciation during the ice ages for the world (rather than just North America). We have pointed out that our present sea level is not permanent because of episodes of more extensive glaciation and global warming.

Chapter 18 has improved maps of deserts and photos of desert features and more realistic diagrams of blowouts and migration of sand dunes; an image of barchan dunes from Mars Proctor Crater has also been added. In Chapter 20, the box on rising sea level has been updated and many diagrams have been redrawn to look more realistic while retaining clarity for the beginning geology student.

In the structure chapter, text and diagrams have been rewritten and redrawn to improve clarity of difficult concepts. An exciting addition to this new edition is the animated diagrams of folding and faulting to show the mechanics of movement and accommodation of strain in the crust.

Chapter 7 has undergone a major revision to include information and spectacular photos of the recent major earthquakes that have struck around the world—Seattle, India, El Salvador, Turkey, and Taiwan. New boxes on earthquake engineering and lifesaving tips on what to do before, during, and after an earthquake have been added. The discussion of tsunamis has been revised and expanded to include new diagrams, photos, and a map of travel-time and early warning systems throughout the Pacific rim.

In the chapter on Earth's interior and its geophysical properties, we have updated and expanded our coverage of the core-mantle boundary to include a discussion of the D layer and ultra low velocity zone (ULVZ) as well as incorporating exciting new discoveries about the dynamics of the deep interior of Earth. Chapter 3 includes a new astrogeology box on the origin of the ocean. Maps of features on the sea floor have been revised.

The plate tectonics chapter has been partially rewritten and expanded to include an illustrated discussion of the paleontological evidence for continental drift. It also includes new information and an accompanying figure presenting the latest ideas about the dynamics of plates and mantle plumes at depth in the mantle.

In the chapter on mountains and the continental crust, we have expanded our coverage of the Appalachians by discussing their post-orogenic erosional and uplift history. Our geologic resources chapter now includes a box on frozen methane hydrates as a potential new energy resource along with its potential to contribute to global warming.

Features

The Internet has revolutionized the way we obtain knowledge and this book makes full use of its potential to help students learn. We have made the process student-friendly by having all websites that we mention in the book, a mouse-click away from this book's website. (We also include all URLs in the textbook for those who wish to go directly to a site.) Within our website we have Internet exercises to allow students to get the most out of appropriate sites as well as to raise interest for independent, further exploration on a topic. We expect to add more sites and exercises to our web pages as we discover new ones after the book has gone to press. Our website also features online quizzes and exercises to help a student succeed in a geology course.

Technology-Related Supplements

For Instructors:

- Online Learning Center at www.mhhe.com/plummer9e/ containing:
  - Access to PowerWeb—Geology
  - Password Protected Instructor's Manual
  - Web Links and more!
- Digital Content Manager CD-ROM with most of the line art and photographs from the text
- Interactive Plate Tectonics CD-ROM
- Geoscience Videotape Library (available to qualified adopters)
- Computerized testing software

For Students:

- Online Learning Center at www.mhhe.com/plummer9e/ containing:
  - 60 new Animations
  - Interactive Quizzing
  - Key Term Flashcards
  - Access to PowerWeb—Geology
  - Web Links and more!

Printed Supplements

- 200 Transparencies
- 477 Slides
- *Laboratory Manual for Physical Geology,* 11th edition, by Zumberge, Rutford, and Carter, ISBN 0-07-239195-2
- *Laboratory Manual for Physical Geology,* 4th edition, by Jones, ISBN 0-07-243655-7
- *Student Atlas of Environmental Issues,* by Allen, ISBN 0-697-36520-4
- *You Can Make a Difference: Be Environmentally Responsible,* by Getis, ISBN 0-07-292416-0

Acknowledgments

We have tried to write a book that will be useful to both students and instructors. We would be grateful for any comments by users, especially regarding mistakes within the text or sources of good geological photographs.

Diane Carlson would like to thank her husband, Reid Buell, for his support and technical assistance with several chapters. We thank Susan Slaymaker for writing the planetary geology material originally in early editions.

We are also very grateful to the following reviewers of the fifth edition for their careful evaluation and useful suggestions for improvement.

William W. Atkinson, Jr., *University of Colorado–Boulder*
J. Bret Bennington, *Hofstra University*
Stephen K. Boss, *University of Arkansas–Fayetteville*
Kevin Cornwell, *California State University, Sacramento*
P. Thompson Davis, *Bentley College*
Dave Evans, *California State University, Sacramento*
Tim Flood, *St. Norbert College*
Norm Harris, *Nassau Community College*
Tim Horner, *California State University, Sacramento*
Chris Hill, *Fullerton College*
Paul Hudak, *University of North Texas*
Leslie Kanat, *Johnson State College*
Alan Lester, *University of Colorado–Boulder*
Donald Lindsley, *SUNY at Stony Brook*
Jerry F. Magloughlin, *Colorado State University*
Penelope Morton, *University of Minnesota–Duluth*
Doug Oliver, *Tarrant County College*
Eugene Perry, *Northern Illinois University*
John D. Pigott, *University of Oklahoma*
Randye L. Rutberg, *Hunter College*
William E. Sanford, *Colorado State University*
Kevin J. Smart, *University of Oklahoma*
Mark Swanson, *University of Southern Maine*
Sarah Ulerick, *Lane Community College*
Andrew Warnock, *Colorado State University*
John Wickham, *University of Texas–Arlington*

Take a Closer Look *at* Today's WORLD...

The proven *Physical Geology* Learning System

Updated Art Program

The art program for this edition of *Physical Geology: Earth Revealed* has been carefully revised and developed for clarity and consistency to help students master fundamental concepts. In total, 150 new or revised art pieces have been introduced, including more three-dimensional diagrams and line art and photo combinations than ever before. We have digitized the art program which allowed us not only to improve the printed illustrations, but also to provide the artwork in electronic format, making it easier to include relevant figures and art in classroom presentations. We've also developed nearly 60 animations and 12 pieces of Active Art from this new art program. Active Art allows instructors to adapt figures for their lecture environment and illustrate complex processes and concepts more easily.

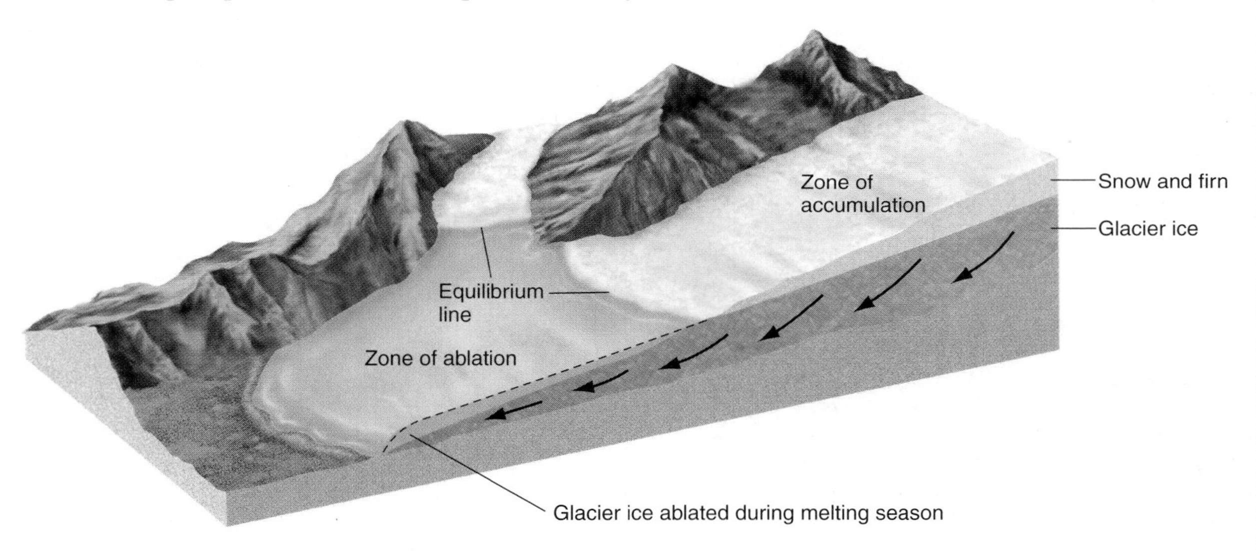

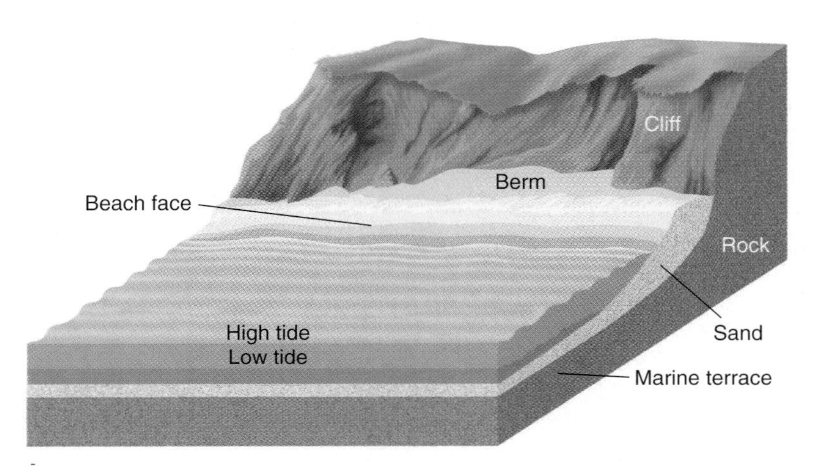

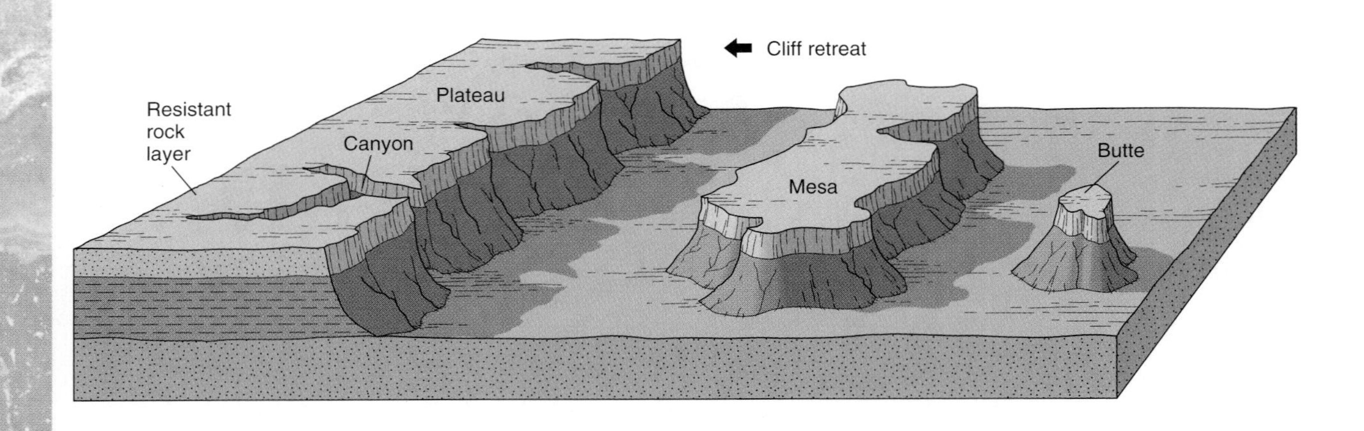
Cliff retreat
Resistant rock layer
Plateau
Canyon
Mesa
Butte

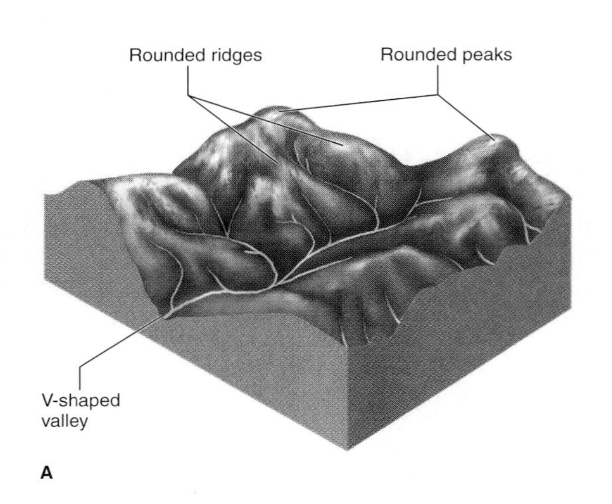
Rounded ridges
Rounded peaks
V-shaped valley
A

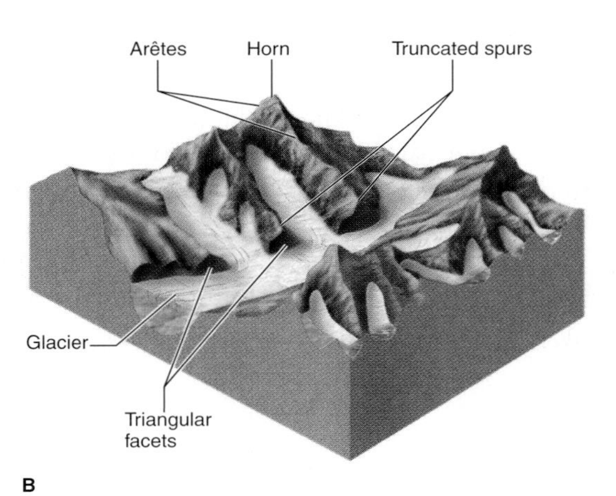
Arêtes
Horn
Truncated spurs
Glacier
Triangular facets
B

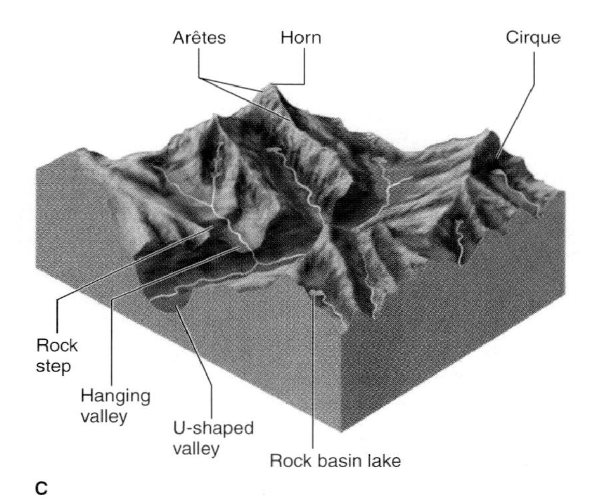
Arêtes
Horn
Cirque
Rock step
Hanging valley
U-shaped valley
Rock basin lake
C

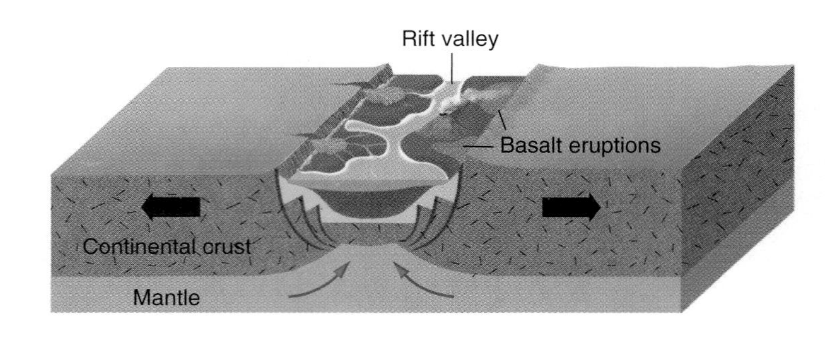
Rift valley
Basalt eruptions
Continental crust
Mantle

Boxed Readings

Each chapter contains at least one of 3 types of boxes—**In Greater Depth, Environmental Geology,** and **Astrogeology.** The *In Greater Depth* boxes cover interesting topics that are usually not an essential part of an introductory geology course. *Environmental Geology* boxes show how material pertaining to physical geology relates to environmental concerns such as oil spills, exploiting natural resources, and mitigating natural disasters. *Astrogeology* boxes relate topics discussed in the text to what has been discovered on other planets or the solar system.

The following satirical newspaper column was written by humorist Art Buchwald in 1978, a year, like the "El Niño" year of 1998, in which southern California had many landslides because of unusually wet weather.

Los Angeles—I came to Los Angeles last week for rest and recreation, only to discover that it had become a rain forest.

I didn't realize how bad it was until I went to dinner at a friend's house. I had the right address, but when I arrived there was nothing there. I went to a neighboring house where I found a man bailing out his swimming pool.

I beg your pardon, I said. Could you tell me where the Cables live?

"They used to live above us on the hill. Then, about two years ago, their house slid down in the mud, and they lived next door to us. I think it was last Monday, during the storm, that their house slid again, and now they live two streets below us, down there. We were sorry to see them go—they were really nice neighbors."

I thanked him and slid straight down the hill to the new location of the Cables' house. Cable was clearing out the mud from his car. He apologized for not giving me the new address and explained, "Frankly, I didn't know until this morning whether the house would stay here or continue sliding down a few more blocks."

Cable, I said, you and your wife are intelligent people, why do you build your house on the top of a canyon, when you know that during a rainstorm it has a good chance of sliding away?

"We did it for the view. It really was fantastic on a clear night up there. We could sit in our Jacuzzi and see all of Los Angeles, except of course when there were brush fires.

"Even when our house slid down two years ago, we still had a great sight of the airport. Now I'm not too sure what kind of view we'll have because of the house in front of us, which slid down with ours at the same time."

But why don't you move to safe ground so that you don't have to worry about rainstorms?

"We've thought about it. But once you live high in a canyon, it's hard to move to the plains. Besides, this house is built solid and has about three more good mudslides in it."

Still, it must be kind of hairy to sit in your home during a deluge and wonder where you'll wind up next. Don't you ever have the desire to just settle down in one place?

"It's hard for people who don't live in California to understand how we people out here think. Sure we have floods, and fire and drought, but that's the price you have to pay for living the good life. When Esther and I saw this house, we knew it was a dream come true. It was located right on the tippy top of the hill, way up there. We would wake up in the morning and listen to the birds, and eat breakfast out on the patio and look down on all the smog.

"Then, after the first mudslide, we found ourselves living next to people. It was an entirely different experience. But by that time we were ready for a change. Now we've slid again and we're in a whole new neighborhood. You can't do that if you live on solid ground. Once you move into a house below Sunset Boulevard, you're stuck there for the rest of your life.

"When you live on the side of a hill in Los Angeles, you at least know it's not going to last forever."

Then, in spite of what's happened, you don't plan to move out?

"Are you crazy? You couldn't replace a house like this in L.A. for $500,000."

What happens if it keeps raining and you slide down the hill again?

"It's no problem. Esther and I figure if we slide down too far, we'll just pick up and go back to the top of the hill, and start all over again; that is, if the hill is still there after the earthquake."

Reprinted by permission of the author.

Further Reading

John McPhee's *The control of nature* contains a factual, and highly readable, account of 1978 landslides in southern California.

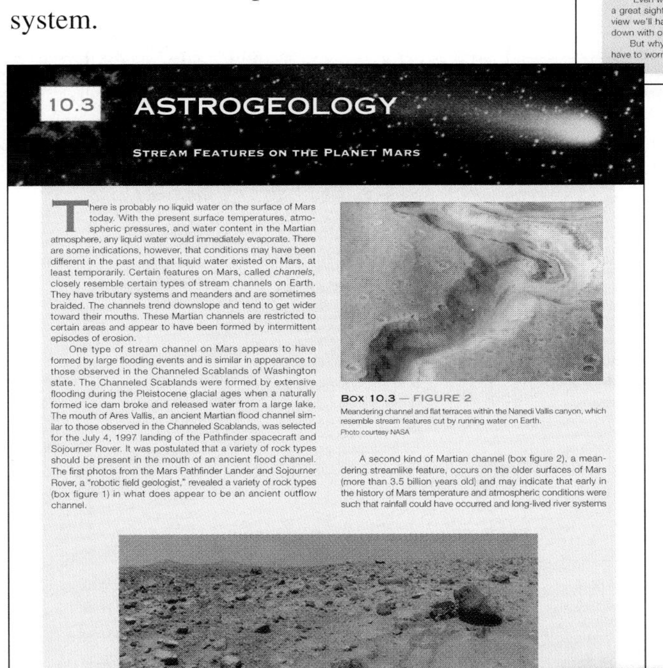

10.3 ASTROGEOLOGY

STREAM FEATURES ON THE PLANET MARS

There is probably no liquid water on the surface of Mars today. With the present surface temperatures, atmospheric pressures, and water content in the Martian atmosphere, any liquid water would immediately evaporate. There are some indications, however, that conditions may have been different in the past and that liquid water existed on Mars, at least temporarily. Certain features on Mars, called *channels*, closely resemble certain types of stream channels on Earth. They have tributary systems and meanders and are sometimes braided. The channels trend downslope and tend to get wider toward their mouths. These Martian channels are restricted to certain areas and appear to have been formed by intermittent episodes of erosion.

One type of stream channel on Mars appears to have formed by large flooding events and is similar in appearance to those observed in the Channeled Scablands of Washington state. The Channeled Scablands were formed by extensive flooding during the Pleistocene glacial ages when a naturally formed ice dam broke and released water from a large lake. The mouth of Ares Vallis, an ancient Martian flood channel similar to those observed in the Channeled Scablands, was selected for the July 4, 1997 landing of the Pathfinder spacecraft and Sojourner Rover. It was postulated that a variety of rock types should be present in the mouth of an ancient flood channel. The first photos from the Mars Pathfinder Lander and Sojourner Rover, a "robotic field geologist," revealed a variety of rock types (box figure 1) in what does appear to be an ancient outflow channel.

Box 10.3 — FIGURE 2

Meandering channel and flat terraces within the Nanedi Vallis canyon, which resemble stream features cut by running water on Earth.

Photo courtesy NASA

A second kind of Martian channel (box figure 2), a meandering streamlike feature, occurs on the older surfaces of Mars (more than 3.5 billion years old) and may indicate that early in the history of Mars temperature and atmospheric conditions were such that rainfall could have occurred and long-lived river systems

Box 10.3 — FIGURE 1

View from the Mars Pathfinder Lander showing the Sojourner Rover and a variety of rocks from Ares Vallis.

Photo courtesy of Jet Propulsion Laboratory/NASA

The oldest fossils found are prokaryotes—microscopic, single-celled organisms that lack a nucleus. These date back to around 3.5 billion years (b.y.) ago, so life on Earth is at least that old. It is likely that even more primitive organisms, similar to viruses, date back further in time but are not preserved in the fossil record. Fossils of much more complex, single-celled organisms that contained a nucleus (eukaryotes) are found in rocks as old as 1.4 b.y. These are the earliest living creatures to have reproduced sexually. Colonies of unicellular organisms likely evolved into multicellular organisms. Multicellular algae fossils date back at least a billion years.

Imprints of larger multicellular creatures appear in rocks of late Precambrian age, about 700 to 650 million years ago (m.y.). These resemble jellyfish and worms.

Sedimentary rocks from the Paleozoic, Mesozoic, and Cenozoic eras have abundant fossils. Large numbers of fossils appeared early in the Cambrian period. Trilobites (figure 8.21) evolved into many species and were particularly abundant during the Cambrian. They became less significant later in the Paleozoic and finally all trilobites became extinct by the end of the Paleozoic.

The most primitive fish, the first vertebrates, date back to late in the Cambrian. Fish similar to presently living species of fish (including sharks) flourished during the Devonian (named after Devonshire, England). The Devonian is often called the "age of fish." Amphibians evolved from fish that had developed lungs late in the Devonian. These were the first land animals. Land plants, however, date back to the Ordovician. Reptiles evolved from amphibians in Pennsylvanian time or perhaps earlier.

The Paleozoic ended with the greatest mass extinction ever to occur on earth. Over 95% of species that existed died out.

During the Mesozoic new creatures evolved to occupy ecological domains left vacant by extinct creatures. Dinosaurs and mammals evolved from reptiles that survived the great extinction. Dinosaurs became the dominant group of land animals. Birds likely evolved from dinosaurs in the Mesozoic. Large, now extinct, marine reptiles, such as ichthysaurs, lived in the Mesozoic seas while flying reptiles, pterosaurs, soared through the air.

The Cretaceous period (and Mesozoic era) ended with the second largest mass extinction (around 75% of species were wiped out).

The Cenozoic is often called the "age of mammals." Mammals, which were small, insignificant creatures during the Mesozoic, evolved into the many groups of mammals (whales, bats, canines, cats, elephants, primates, and so forth) that occupy the earth at present. Many species of mammals evolved and became extinct throughout the Cenozoic. Hominids (modern humans and our extinct ancestors) have a fossil record dating back 4 m.y. and likely evolved from a now extinct ancestor common to hominids, chimpanzees, and other apes.

We tend to think of mammals' evolution as being the great success story (because we are mammals), however, mammals pale in comparison to insects. Insects have been around far longer than mammals and now account for an estimated 1 million species.

Related Web Resource

University of California Museum of Paleontology

www.ucmp.berkeley.edu/

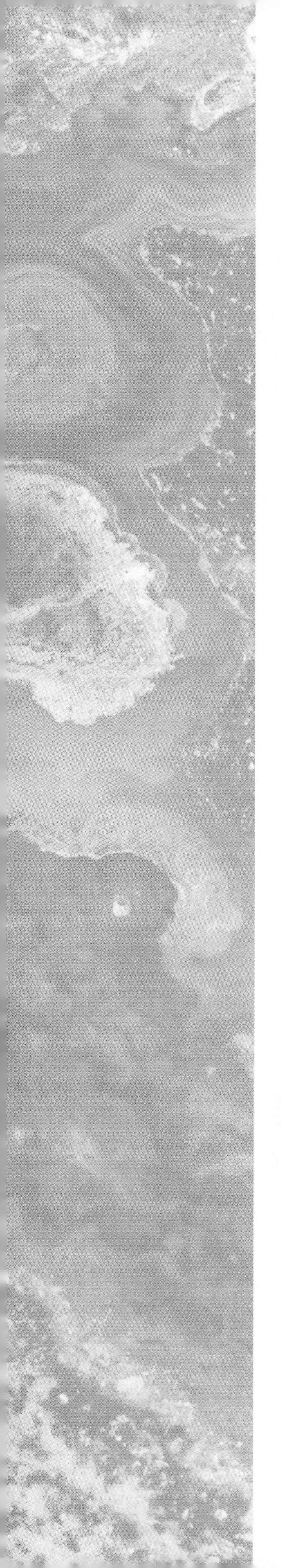

End-of-Chapter Learning Aids

Additional support helps you make the grade.

Use these helpful end-of-chapter learning aids to prepare for tests and quizzes:

Summary—overviews of chapter content.

Terms to Remember—important terms to review and understand.

Testing Your Knowledge—realistic sample tests you can use to prepare for exams and improve your grades.

Expanding Your Knowledge—questions that help you develop critical thinking skills.

Exploring Resources—supplemental references in a number of different media.

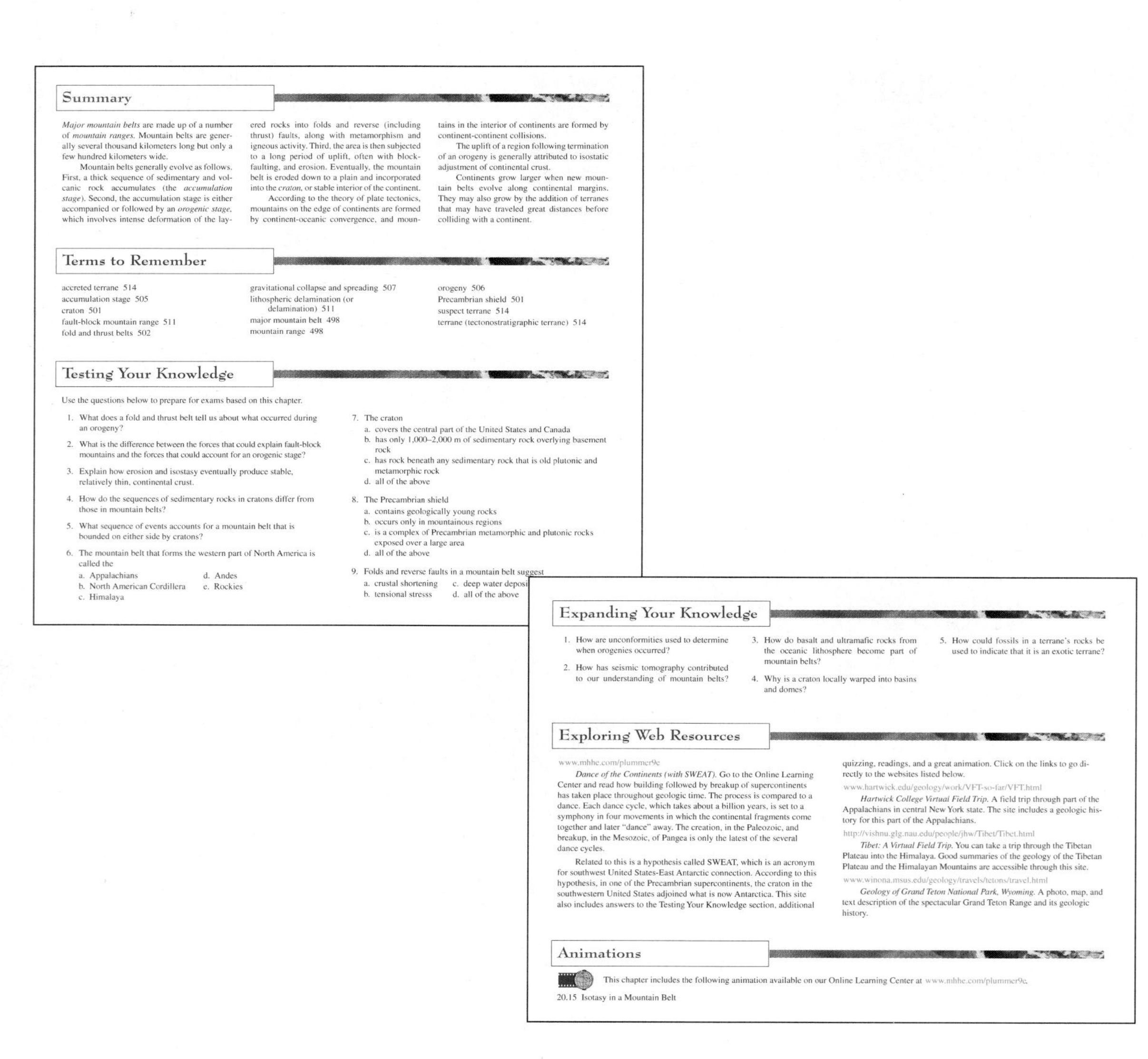

Summary

Major mountain belts are made up of a number of *mountain ranges.* Mountain belts are generally several thousand kilometers long but only a few hundred kilometers wide.

Mountain belts generally evolve as follows. First, a thick sequence of sedimentary and volcanic rock accumulates (the *accumulation stage*). Second, the accumulation stage is either accompanied or followed by an *orogenic stage,* which involves intense deformation of the layered rocks into folds and reverse (including thrust) faults, along with metamorphism and igneous activity. Third, the area is then subjected to a long period of uplift, often with block-faulting, and erosion. Eventually, the mountain belt is eroded down to a plain and incorporated into the *craton,* or stable interior of the continent.

According to the theory of plate tectonics, mountains on the edge of continents are formed by continent-oceanic convergence, and mountains in the interior of continents are formed by continent-continent collisions.

The uplift of a region following termination of an orogeny is generally attributed to isostatic adjustment of continental crust.

Continents grow larger when new mountain belts evolve along continental margins. They may also grow by the addition of terranes that may have traveled great distances before colliding with a continent.

Terms to Remember

accreted terrane 514
accumulation stage 505
craton 501
fault-block mountain range 511
fold and thrust belts 502
gravitational collapse and spreading 507
lithospheric delamination (or delamination) 511
major mountain belt 498
mountain range 498
orogeny 506
Precambrian shield 501
suspect terrane 514
terrane (tectonostratigraphic terrane) 514

Testing Your Knowledge

Use the questions below to prepare for exams based on this chapter.

1. What does a fold and thrust belt tell us about what occurred during an orogeny?
2. What is the difference between the forces that could explain fault-block mountains and the forces that could account for an orogenic stage?
3. Explain how erosion and isostasy eventually produce stable, relatively thin, continental crust.
4. How do the sequences of sedimentary rocks in cratons differ from those in mountain belts?
5. What sequence of events accounts for a mountain belt that is bounded on either side by cratons?
6. The mountain belt that forms the western part of North America is called the
   a. Appalachians
   b. North American Cordillera
   c. Himalaya
   d. Andes
   e. Rockies
7. The craton
   a. covers the central part of the United States and Canada
   b. has only 1,000–2,000 m of sedimentary rock overlying basement rock
   c. has rock beneath any sedimentary rock that is old plutonic and metamorphic rock
   d. all of the above
8. The Precambrian shield
   a. contains geologically young rocks
   b. occurs only in mountainous regions
   c. is a complex of Precambrian metamorphic and plutonic rocks exposed over a large area
   d. all of the above
9. Folds and reverse faults in a mountain belt suggest
   a. crustal shortening
   b. tensional stresss
   c. deep water deposi
   d. all of the above

Expanding Your Knowledge

1. How are unconformities used to determine when orogenies occurred?
2. How has seismic tomography contributed to our understanding of mountain belts?
3. How do basalt and ultramafic rocks from the oceanic lithosphere become part of mountain belts?
4. Why is a craton locally warped into basins and domes?
5. How could fossils in a terrane's rocks be used to indicate that it is an exotic terrane?

Exploring Web Resources

www.mhhe.com/plummer9e

Dance of the Continents (with SWEAT). Go to the Online Learning Center and read how building followed by breakup of supercontinents has taken place throughout geologic time. The process is compared to a dance. Each dance cycle, which takes about a billion years, is set to a symphony in four movements in which the continental fragments come together and later "dance" away. The creation, in the Paleozoic, and breakup, in the Mesozoic, of Pangea is only the latest of the several dance cycles.

Related to this is a hypothesis called SWEAT, which is an acronym for southwest United States-East Antarctic connection. According to this hypothesis, in one of the Precambrian supercontinents, the craton in the southwestern United States adjoined what is now Antarctica. This site also includes answers to the Testing Your Knowledge section, additional quizzing, readings, and a great animation. Click on the links to go directly to the websites listed below.

www.hartwick.edu/geology/work/VFT-so-far/VFT.html

Hartwick College Virtual Field Trip. A field trip through part of the Appalachians in central New York state. The site includes a geologic history for this part of the Appalachians.

http://vishnu.glg.nau.edu/people/jhw/Tibet/Tibet.html

Tibet: A Virtual Field Trip. You can take a trip through the Tibetan Plateau into the Himalaya. Good summaries of the geology of the Tibetan Plateau and the Himalayan Mountains are accessible through this site.

www.winona.msus.edu/geology/travels/tetons/travel.html

Geology of Grand Teton National Park, Wyoming. A photo, map, and text description of the spectacular Grand Teton Range and its geologic history.

Animations

This chapter includes the following animation available on our Online Learning Center at www.mhhe.com/plummer9e.

20.15 Isotasy in a Mountain Belt

Digital Content Manager CD-ROM

The Digital Content Manager CD-ROM is a multimedia collection of visual resources that allows instructors to create powerful presentations to help their students visualize difficult concepts.

The Digital Content Manager CD-ROM contains:

- 385 color images from the text.
- 284 photographs from the text.
- 58 animations using art from the text.
- 12 pieces of Active Art, which allow instructors to adapt figures to meet the needs of their lecture environment and illustrate complex processes and concepts more easily.
- all tables from the text.
- chapter-specific PowerPoint Lecture Outlines to help instructors create exciting lecture presentations

The digital assets on this cross-platform CD-ROM are grouped by chapter within easy-to-use folders. The Art Library contains full-color digital files of most of the illustrations in the text. The Photo Library contains digital files of instructionally significant photographs from the text. The Power Point Lectures consist of ready-made presentations that have lecture notes specifically written to cover each chapter. Every table that appears in the text is provided in electronic form in the Table Library.

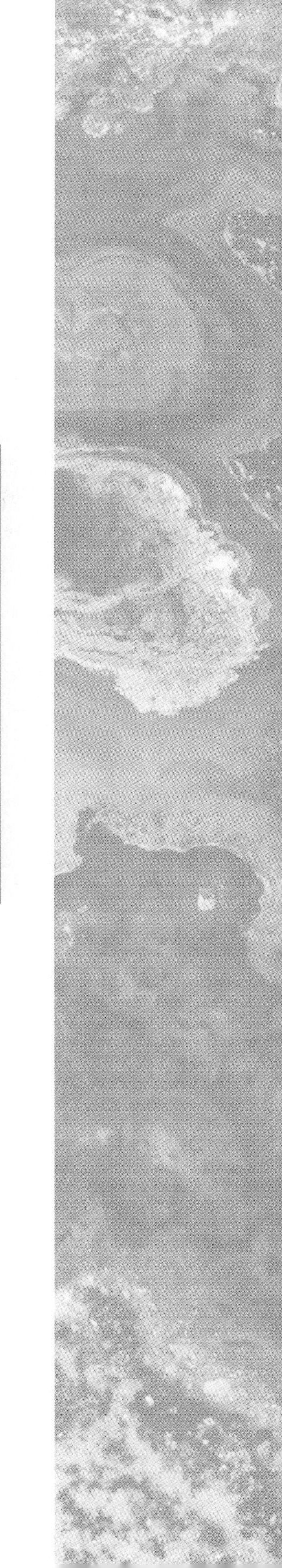

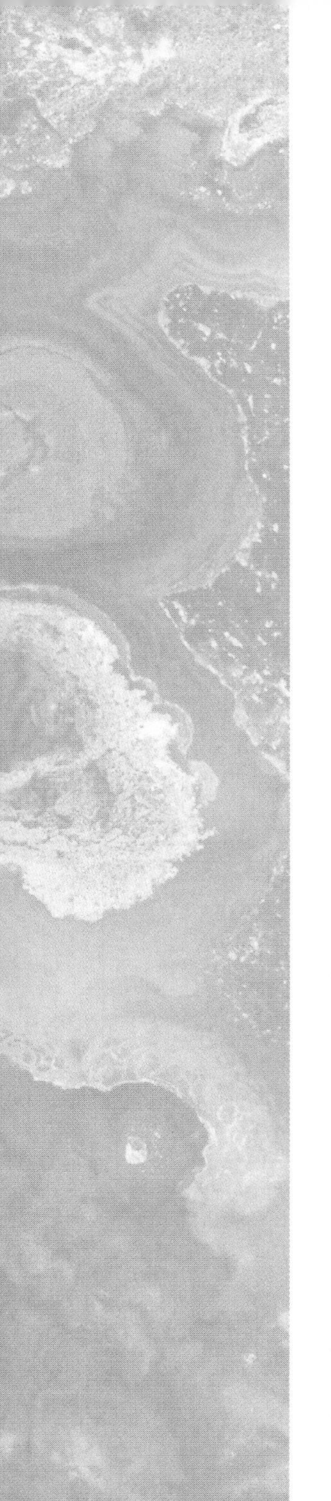

The Online Learning Center

Your Password to Success

www.mhhe.com/plummer9e/

The Online Learning Center offers a wealth of study aids for students as well as teaching aids for instructors. Take a look for yourself!

Students, learning geology can be easy and fun!

Visit the Online Learning Center for:

- nearly 60 animations of difficult concepts
- interactive quizzing
- visual and key term flashcards
- web links related to recent geologic events
- Internet exercises
- PowerWeb: Geology

Instructors, we've made your life just a little easier!

Just a sampling of the assets you will find:

- a password-protected Instructor's Manual
- a password-protected Test Item File
- web links related to recent geologic events
- professional resources
- PowerWeb: Geology

Access to PowerWeb: Geology

www.dushkin.com/powerweb

PowerWeb is a password-protected website developed by McGraw-Hill/Dushkin to give instructors and students:

- course-specific materials
- monitored course-specific web links and articles
- student study tools, including quizzing, review forms, time management tools, and web research
- interactive exercises
- weekly updates with assessment
- informative and timely world news
- access to Northern Light Research Engine (received multiple Editor's Choice awards for superior capabilities from *PC Magazine)*
- material on how to conduct web research
- daily news feed of topic-specific news

Imagine the advantages of having so many learning and teaching tools all in one place—all at your fingertips!

Course Management Systems

Course Management Systems expand the reach of your course. Online discussion and message boards can now complement your office hours. Also, because of a sophisticated tracking system, you will know which students need more attention—even if they don't ask for help. Online testing scores are recorded and automatically placed in your grade book. You can also create special alerts to show up if a student is struggling with coursework. Should you seek advice for your course, you have access to discussion and message boards where you can collaborate and communicate with other professors across campus or around the world. Our own specialists are also ready to answer your questions.

McGraw-Hill's Online Learning Center content is compatible with a variety of course management systems, including:

- **PageOut,** McGraw-Hill's exclusive course management system, which allows you to create a custom course website with links to book-specific course content. Simply select a template to post course information and an interactive syllabus links to Online Learning Center content.

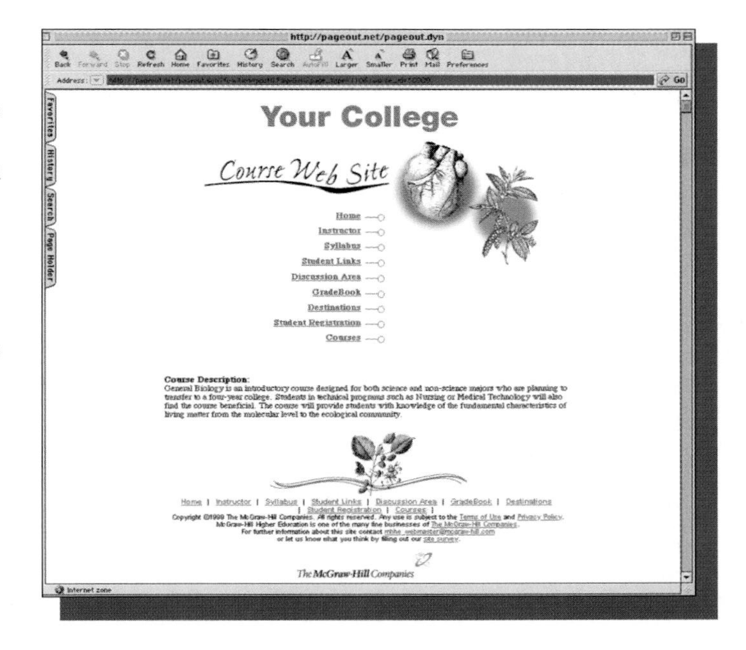

Online Learning Center content is also compatible with these commercially produced course management systems:

- WebCT
- Blackboard
- Topclass
- eCollege

Contact your McGraw-Hill sales representative for more information or visit *www.mhhe.com*.

Meet the Authors

David McGeary, Diane Carlson, and Charles Plummer at an outcrop of a Sierra Nevadan intrusive body.

Charles Plummer

Professor Charles "Carlos" Plummer grew up in the shadows of volcanoes in Mexico City. There, he developed a love for mountains and mountaineering that eventually led him into geology. He received his B.A. degree from Dartmouth College. After graduation, he served in the U.S. Army as an artillery officer. He resumed his geological education at the University of Washington where he received his M.S. and Ph.D. degrees. His geologic work has been in mountainous and polar regions, notably Antarctica (where a glacier is named in his honor). He taught at Olympic Community College in Washington before joining the faculty at California State University, Sacramento.

At CSUS he taught optical mineralogy, metamorphic petrology, and field courses before his semiretirement. He continues to teach introductory courses. He flies airplanes, skis, and recently became a certified open water SCUBA diver. (*plummercc@csus.edu*)

David McGeary

David McGeary retired from teaching in 1992 and from textbook writing in 1995. His activities today are nongeological—tending his house and land, traveling, carpentry, blacksmithing, and acting in community theatre.

Diane Carlson

Professor Diane Carlson grew up on the glaciated Precambrian shield of northern Wisconsin and received an A.A. degree at Nicolet College in Rhinelander and her B.S. in geology at the University of Wisconsin at Eau Claire. She continued her studies at the University of Minnesota–Duluth where she studied the structural complexities of high-grade metamorphic rocks along the margin of the Idaho batholith for her master's thesis. The lure of the West and an opportunity to work with the U.S. Geological Survey to map the Colville batholith in northeastern Washington led her to Washington State University for her Ph.D. Dr. Carlson accepted a position at California State University, Sacramento after her Ph.D. and teaches physical geology, structural geology, environmental geology, and field geology. Professor Carlson is a recipient of the Outstanding Teacher Award from the CSUS School of Arts and Sciences. She is also actively engaged in researching the structural and tectonic evolution of part of the Foothill Fault System in the northern Sierra Nevada of California. (*carlsondh@csus.edu*)

Physical Geology
Earth Revealed

Dedicated

to the memory of author

Dave McGeary

who passed away December 11, 2002.

CHAPTER

1

Introduction to Physical Geology

Geology uses the scientific method to explain natural aspects of the Earth—for example, how mountains form or why oil resources are concentrated in some rocks and not in others. This chapter briefly explains how and why Earth's surface, and its interior, are constantly changing. It relates this constant change to the major geological topics of interaction of the atmosphere, water and rock, the modern theory of plate tectonics, and geologic time. These concepts form a framework for the rest of the book. Understanding the "big picture" presented here will aid you in comprehending the chapters that follow.

Opposite: Natural rock sculpture in Paria Plateau, Arizona. Sandstone formed from ancient sand dunes. Running water has eroded the rock into the present distinctive shapes.
Photo © Kerrick James

STRATEGY for Using This Textbook

- As authors, we try to be thorough in our coverage of topics so the textbook can serve you as a resource. Your instructor may choose, however, to concentrate only on certain topics for *your* course. Find out which topics and chapters you should focus upon in your studying and concentrate your energies there.
- Your instructor may present additional material that is not in the textbook. Take good notes in class.
- Do not get overwhelmed by terms. (Every discipline has its own language.) Don't just memorize each term and its definition. If you associate a term with a concept or mental picture, remembering the term comes naturally when you understand the concept. (You remember names of people you know because you associate personality and physical characteristics with a name.) You may find it helpful to learn the meanings of frequently used prefixes and suffixes for geological terms. These can be found in appendix G.
- **Boldfaced** terms are ones you are likely to need to understand because they are important to the entire course.
- *Italicized* terms are not as important, but may be necessary to understand the material in a particular chapter.
- Pay particular attention to illustrations. Geology is a visually oriented science, and the photos and artwork are at least as important as the text. You should be able to sketch important concepts from memory.
- Find out to what extent your instructor expects you to learn the material in the boxes. They offer an interesting perspective on geology and how it is used, but much of the material might well be considered optional for an introductory course and not vital to your understanding of major topics. Many of the In Greater Depth boxes are meant to be challenging—do not be discouraged if you need your instructor's help in understanding them.
- Read through the appropriate chapter before going to class. Reread it after class, concentrating on the topics covered in the lecture or discussion. Especially concentrate on concepts that you do not fully understand. Return to previously covered chapters to refresh your memory on necessary background material.
- Use the end of chapter material for review. The Summary is just that, a summary. Don't expect to get through an exam by only reading the summary but not the rest of the chapter. Use the Terms to Remember to see if you can visually or verbally associate the appropriate concept with each term. Answer the Testing Your Knowledge questions in writing. Be honest with yourself. If you are fuzzy on an answer, return to that portion of the chapter and reread it. Remember that these are just a sampling of the kind of questions that might be on an exam.
- Geology, like most science, builds on previously acquired knowledge. You must retain what you learn from chapter to chapter. If you forget or did not learn significant concepts covered early in your course, you will find it frustrating later in the course. (To verify this, turn to chapter 20 and you will probably find it intimidating; but if you build on your knowledge as you progress through your course, the chapter material will fall nicely into place.)
- Get acquainted with the book's website at www.mhhe.com/plummer9e. You will find the online quizzes useful for review and the web exercises interesting.
- Be curious. Geologists are motivated by a sense of discovery. We hope you will be too.

WHO NEEDS GEOLOGY?

Imagine yourself as a student at California State University, Northridge (CSUN) in greater Los Angeles. At 4:30 A.M. on January 17, 1994, you are jolted awake when your apartment begins to sway violently. Dishes, bookcases, and ceiling plaster crash to the floor. The noise and shaking are terrifying as you struggle to stand up. In less than a minute the shaking stops and silence momentarily returns. You realize you survived an earthquake, but you are still scared and disoriented.

This seismic event took place because of a sudden shift of bedrock eighteen kilometers (For a metric conversion table, see appendix E.) beneath Northridge. Shock waves spread in all directions with damaging effects in many parts of Los Angeles. Northridge and vicinity suffered the heaviest damage. You feel fortunate—you are safe and your apartment building was not destroyed. Others nearby were not so fortunate. Two university students were among the sixteen killed in one of the forty apartment buildings that collapsed (figure 1.1*A*). (Emergency crews drove right past one apartment building because it appeared undamaged; but it had collapsed straight downward, completely crushing the first floor but leaving the upper floors standing.)

After the earthquake you and your neighbors do not have electricity (power was lost, temporarily, as far away as Edmonton, Canada). Water is in short supply because of broken water mains. Leaving the area is not a good option. Eleven major highway structures have been destroyed, including a segment of the Santa Monica Freeway, the nation's busiest highway.

A

B

FIGURE 1.1

Damage from the Northridge earthquake. (*A*) The Northridge Meadows apartments where 16 people, or roughly one-third of all those killed in the quake, died when the first floor was crushed by the weight of the top two floors. (*B*) The parking garage at California State University, Northridge that was destroyed by the earthquake.

Photo A © Michael Edwards/LA Times; Photo B by Frank M. Hanna

Despite the deprivations and the jitters caused by frequent aftershocks, you are comforted by how people pitch in to help each other. A tremendous sense of community grows from the shared adversity.

Weeks later, the spring semester begins at CSUN. You are among the 27,000 students who must cope with the fact that all the buildings on campus were damaged and some were destroyed. But at least the most severely mangled building was a large, prefabricated parking structure (figure 1.1*B*). Repairing your campus will take time and an estimated $350 million. Meanwhile, some of your classes are held outdoors, others in hastily erected temporary buildings. As the weeks pass, you share the pride of the university community and are determined that the quality of your education shall not be undermined by the earthquake and its damage. But your earthquake jitters continue even though the aftershocks now are infrequent and barely perceptible.

Sixty people were killed in the Northridge earthquake. The monetary cost was estimated at between $20 and $40 billion—financially the most costly natural disaster in North America's history. The loss in lives could have been far worse. For instance, 8,000 people died in the 1985 Mexico City earthquake and 5,000 were killed in 1995 at Kobe, Japan. The Northridge death toll was low due mainly to good planning. The February 2001 earthquake near Seattle (described in the earthquakes chapter), which was stronger than the Northridge earthquake, caused no fatalities but several billion dollars in damage. Again, being prepared for earthquakes saved lives.

Preparing for major earthquakes has been an ongoing process on the West Coast for decades. Earthquakes are expected because of what geologists have learned about how the Earth works. Skyscrapers in the Los Angeles and Seattle areas survived because they were built to meet high standards for seismic-resistance—the fruits of many decades of engineering studies and design. The apartment buildings that collapsed in Northridge probably would have survived if, during their construction, building codes had been enforced. It is likely that thousands of lives were saved because of adherence to building codes. Pillars for freeway bridges and underpasses were known to be vulnerable. The Santa Monica and other freeways were scheduled to have their pillars reinforced later in 1994, but the quake came first.

Luck also played a role in keeping the casualties down. Los Angeles' normally densely packed freeways were almost empty, because of the early hour and it being a holiday, Martin Luther King's birthday. (The San Francisco Bay area was equally lucky when, four years earlier, the Loma Prieta earthquake took place as the area's two baseball teams were about to begin a World Series game; the normally heavy rush hour traffic was the lightest in anyone's memory when a freeway and part of a bridge collapsed.)

The Northridge and Seattle earthquakes were reminders that our solid Earth does not sit still.

According to the theory of plate tectonics, the Earth's rigid outer shell is broken into a series of *plates.* Adjoining plates may slide past, move away from, or collide with one another. Plates generally move from 1 to 18 centimeters a year. But the motion may not be smooth and continuous. Plates may be "locked" against one another for many years and move suddenly.

Sudden motion along a fault caused the Northridge earthquake (figure 1.2). The mountain ranges that rise above Los Angeles are another product of relentless plate motion. It's as if a giant vise is slowly closing, forcing bedrock upward into mountains. The Santa Susana Mountains that border Northridge grew higher by 38 centimeters during the earthquake.

The awesome energy released by an earthquake is a product of forces within the Earth that move firm rock. Earthquakes

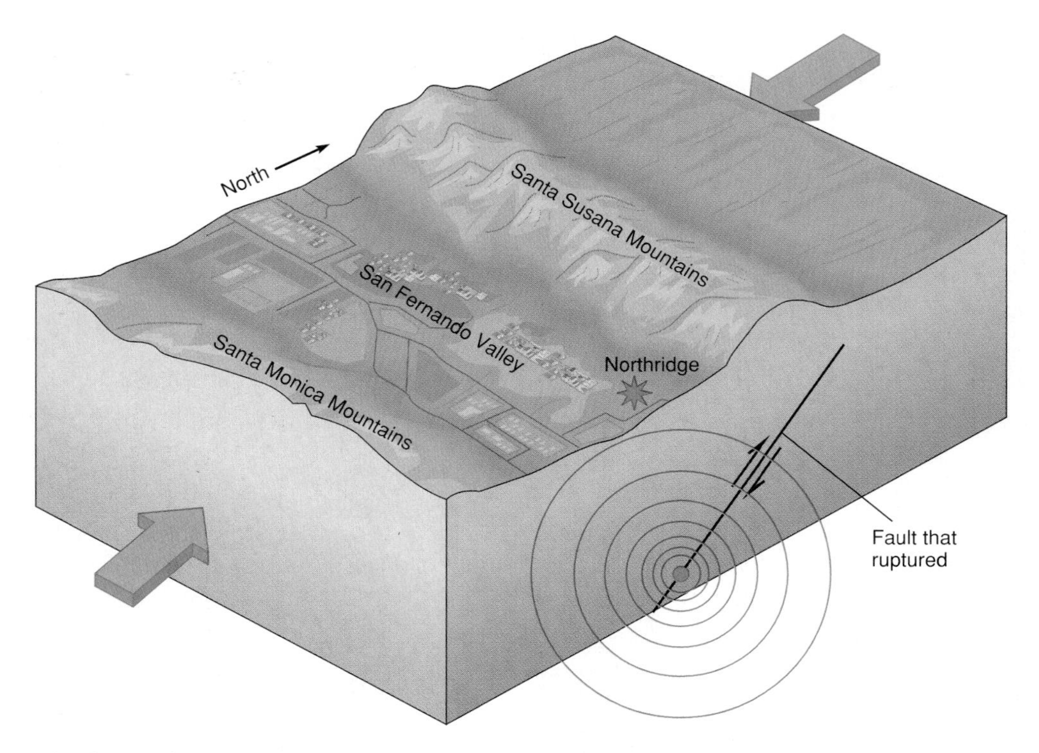

FIGURE 1.2
Compression (indicated by large arrows) in the Los Angeles area causes movement along a fault and the January 17, 1994, earthquake. The black line is the trace of the fault plane that underlies the San Fernando Valley and Santa Susana Mountains. The earthquake began at the red circle beneath Northridge.

are only one consequence of the ongoing changing of the Earth. Ocean basins open and close. Mountain ranges rise and are worn down to plains through slow, but very effective, processes. Studying how the Earth works can be as exciting as watching a great theatrical performance. Understanding the changes that take place in and on the Earth, and the reasons for those changes, is the challenging objective of *geology,* the scientific study of the Earth.

Physical geology is the division of geology concerned with Earth materials, changes in the surface and interior of the Earth, and the dynamic forces that cause those changes. Put another way, physical geology is about how the Earth works.

Earthquakes and other aspects of geology are interesting, but how does geology benefit you, as an inhabitant of this planet? Some of the ways are discussed next.

Avoiding Geologic Hazards

Geology can have a direct application in ensuring people's safety and well-being. For example, if you were building a house in an earthquake-prone area, you would want to know how to minimize danger to yourself and your home. You would want to build the house on a type of ground not likely to be shaken apart by an earthquake. You would want the house designed and built to absorb the kind of vibrations given off by earthquakes.

Volcanic eruptions, like earthquakes, are products of Earth's sudden release of energy. They can be dangerous; however, their biggest dangers are not what most people think. Neither falling volcanic debris nor lava flows are as big a killer as pyroclastic flows or volcanic mudflows. As described in the volcano chapter, a *pyroclastic flow* is a hot, turbulent mixture of exploding gases and volcanic ash that flows rapidly down the flank of a volcano. Pyroclastic flows often reach speeds of over 100 kilometers per hour and are extremely destructive. A *mudflow* is a slurry of water and rock debris that flows down a stream channel.

Mount Pinatubo's eruption in 1991 was the second-largest volcanic eruption of the twentieth century (box 1.1). Geologists successfully predicted the climactic eruption (figure 1.3) in time for Philippine officials to evacuate people living near the mountain. Tens of thousands of lives were saved from pyroclastic flows and mudflows.

By contrast, one of the worst volcanic disasters of the 1900s took place after a relatively small eruption of Nevado del Ruiz in Colombia in 1985. Hot volcanic debris blasted out of the volcano and caused part of the ice and snow capping the peak to melt. The water and loose debris turned into a mudflow. The mudflow overwhelmed the town of Armero at the base of the volcano, killing 23,000 people. Colombian geologists had previously predicted such a mudflow could occur and published maps showing the location and extent of

ENVIRONMENTAL GEOLOGY 1.1

The 1991 Eruption of Mount Pinatubo—Geologists Save Thousands of Lives

When minor steam eruptions began in April 1991, Mount Pinatubo was a vegetation-covered mountain that had last erupted 400 years earlier. As the eruptions intensified, Filipino geologists thought a major eruption might be developing. Geologic fieldwork completed in earlier years indicated that prehistoric eruptions of the volcano tended to be large and violent. Under a previous arrangement for cooperation, American geologists joined their Philippine colleagues and deployed portable seismographs to detect and locate small earthquakes within the volcano and tiltmeters for measuring the bulging of the volcano. These and other data were analyzed by state-of-the-art computer programs.

Fortunately, it took two months for the volcano to reach its climactic eruption, allowing time for the scientists to work with local officials and develop emergency evacuation plans. Geologists had to educate the officials about the principal hazards—mudflows and pyroclastic flows.

In June, explosions, ash eruptions, and minor pyroclastic flows indicated that magma (molten rock) was not far underground and a major eruption was imminent. Some 80,000 people were evacuated from the vicinity of the volcano. The U.S. military evacuated and later abandoned Clark Air Force Base, which was buried by ash. The climactic eruption occurred on June 15 when huge explosions blasted the top off the volcano and resulted in large pyroclastic flows. Volcanic debris was propelled high into the atmosphere. A typhoon 50 km away brought heavy rains, which mixed with the ash and resulted in numerous, large mudflows.

The estimated volume of magma that erupted from the climactic eruption is 5 km^3, making it the world's largest eruption since 1917. Its effects extended beyond the Philippines. Fine volcanic dust and gas blasted into the high atmosphere were carried around the world and would take years to settle out. For a while, we got more colorful sunsets worldwide. Because of the filtering effect for solar radiation, worldwide average temperature was estimated to drop by 0.5°C for two years, more than countering the long-term warming trend of the Earth's climate.

The death toll from the eruption was 374. Of these, 83 were killed in mudflows. Most of the rest died because roofs collapsed from the weight of ash. In addition, 358 people died from illness related to the eruptions. More than 108,000 homes were partly or totally destroyed. The death toll probably would have been in the tens of thousands had the prediction and warning system not been so successful. Although Mount Pinatubo is quiet now, lives and property are still being lost to mudflows, more than a decade after the big eruption.

Related Web Resource

Volcano World

The site contains a wealth of information on volcanoes, including Mount Pinatubo.

http://volcano.und.nodak.edu/

In the Path of a Killer Volcano is a first-rate videotape produced for the NOVA television series. Available from Films for the Humanities and Science, Princeton, New Jersey.

expected mudflows. The actual mudflow that wiped out the town matched that shown on the geologists' map almost exactly. Unfortunately, government officials had ignored the map and the geologists' report; otherwise the tragedy could have been averted.

Other geologic hazards (described in later chapters) that geologists investigate include floods, wave erosion at coastlines, collapsing ground surfaces, and landslides. (In the United States and Canada, far more property and lives have been lost due to landslides and floods than to earthquakes and volcanoes.)

Supplying Things We Need

We depend upon the Earth for energy resources and the raw materials we need for survival, comfort, and pleasure. The Earth, at work for billions of years, has distributed material into localized concentrations that humans can mine or extract. By learning how the Earth works and how different kinds of substances are distributed and why, we can intelligently search for metals, sources of energy, and gems. Even maintaining a supply of sand and gravel for construction purposes depends upon geology.

The economic systems of Western civilization currently depend on abundant and cheap energy sources. Nearly all our vehicles and machinery are powered by petroleum, coal, or nuclear power and depend on energy sources concentrated unevenly in the Earth. The United States economy in particular is geared to petroleum as a cheap source of energy. In a few decades, Americans have used up most of the country's known petroleum reserves, which took nature hundreds of millions of years to store in the Earth. Americans are now heavily dependent on imported oil. When fuel prices jump, people who are not aware that petroleum is a nonrenewable resource become upset and are quick to blame oil companies, politicians, and oil producing countries. (The Gulf War of 1991 was at least partially fought because of the industrialized nations' petroleum requirements.) To find more of this diminishing resource will require more money and increasingly sophisticated

FIGURE 1.3
The major eruption of Mount Pinatubo on June 15, 1991, as seen from Clark Air Force Base, Philippines.
Photo by Robert Lapointe, U.S. Air Force

knowledge of geology. Although many people are not aware of it, we face similar problems with diminishing resources of other materials, notably metals such as iron, aluminum, copper, and tin, each of which has been concentrated in a particular environment by the action of the Earth's geologic forces.

Just how much of our resources do we use? According to the Mineral Information Institute, for every person living in the United States we annually mine 18,000 kilograms (40,000 pounds) of resources, not including energy resources. The amount of each commodity mined is 4,400 kg stone, 3,500 kg sand and gravel, 325 kg limestone for cement, 160 kg clays, 165 kg salt, 760 kg other nonmetals, 545 kg iron, 19 kg aluminum, 9 kg copper, 5 kg each for lead and zinc, 3 kg manganese, and 11 kg other metals. Our yearly per capita consumption of energy resources is over 8,000 kg (17,000 pounds); of this, 3,500 kg is petroleum, 2,300 kg coal, 2,250 natural gas, and .02 kg uranium.

Protecting the Environment

Our demands for more energy and metals have, in the past, led us to extract them with little regard for effects on the balance of nature within the Earth, and therefore on us, Earth's residents. Mining of coal, if done carelessly, for example, can release acids into water supplies. Understanding geology can help us lessen or prevent damage to the environment—just as it can be used to find the resources in the first place.

The environment is further threatened because these are nonrenewable resources. Petroleum and metal deposits do not grow back after being harvested. As demands for these commodities increase, so does the pressure to disregard the ecological damage caused by the extraction of the remaining deposits.

Problems involving petroleum illustrate this. Oil companies employ geologists to discover new oil fields, while the public and government depend on other geologists to assess the potential environmental impact of petroleum's removal from the ground, the transportation of petroleum (see box 1.2), and disposal of any toxic wastes from petroleum products.

Understanding Our Surroundings

It is a uniquely human trait to want to understand the world around us. Most of us get satisfaction from understanding our cultural and family histories, how governments work or do not

ENVIRONMENTAL GEOLOGY 1.2

Delivering Alaskan Oil—The Environment vs. the Economy

In the 1960s, geologists discovered oil beneath the shores of the Arctic Ocean on Alaska's North Slope. It is now the United States' largest oil field. Thanks to the Alaska pipeline, completed in 1977, Alaska has supplied as much as 20% of the United States' domestic oil.

In the late 1970s before Alaskan oil began to flow, the United States was importing almost half its petroleum, at a loss of billions of dollars per year to the national economy. (By 1997, the United States was importing more than half of the petroleum it uses, despite Alaskan oil in the market.) The drain on the country's economy and the increasing cost of energy can be major causes of inflation, lower industrial productivity, unemployment, and the erosion of standards of living.

Despite its important role in the American economy, some considered the Alaska pipeline and the use of tankers as unacceptable threats to the area's ecology. Another concern was the risk of marine oil spills during shipping of oil in tankers from the southern end of the pipeline to the lower 48 states.

Geologists with the U.S. Geological Survey conducted the official environmental impact investigation of the proposed pipeline route. After an exhaustive study, they recommended against its construction, partly because of the hazards to oil tankers and partly because of the geologic hazards of the pipeline route. Their report was overruled. The Congress and the President of the United States exempted the pipeline from laws that require a favorable environmental impact statement before a major project can begin. The up to 2 million barrels of oil a day that flow from the Arctic oil fields mean that over $10 billion a year that would have been lost through the purchase of foreign oil instead remains in the American economy.

The 1,250-kilometer-long pipeline, through which 88,000 barrels of oil an hour flow, crosses regions of ice-saturated, frozen ground and major earthquake-prone mountain ranges that geologists regard as serious hazards to the structure.

Building anything on frozen ground creates problems. The pipeline presented enormous engineering problems. If the pipeline were placed on the ground, the hot oil flowing through it could melt the frozen ground. On a slope, mud could easily slide and rupture the pipeline. Careful (and costly) engineering minimized these hazards. Much of the pipeline is elevated above the ground (box figure 1). Radiators conduct heat out of the structure. In some places refrigeration equipment in the ground protects against melting.

Records indicate that a strong earthquake can be expected every few years in the earthquake belts crossed by the pipeline. An earthquake could rupture a pipeline—especially a conventional pipe as in the original design. When the Alaska pipeline was built, however, in several places sections were specially jointed to allow the pipe to shift as much as 6 meters without rupturing.

The original estimated cost of the pipeline was $900 million, but the final cost was $7.7 billion, making it the costliest privately financed construction project in history. The redesigning and construction that minimized the potential for an environmental disaster were among the reasons for the increased cost. There have been some minor spills from the pipeline. For instance, in January 1981, 5,000 barrels of oil were lost when a valve ruptured.

Box 1.2 — FIGURE 1
The Alaska pipeline.
Photo by David Applegate

When the tanker *Exxon Valdez* ran aground in 1989, over 240,000 barrels of crude oil were spilled into the waters of Alaska's Prince William Sound. It was the worst ever oil spill in U.S. waters. The spill, with its devastating effects on wildlife and the fishing industry, dramatically highlighted the conflicts between maintaining the energy demands of the American economy and conservation of the environment. The 1972 environmental impact statement had singled out marine oil spills as being the greatest threat to the environment. Based on statistical studies of tanker accidents worldwide, it gave the frequency with which large oil spills could be expected. The *Exxon Valdez* spill should not have been a surprise.

In the early 2000s, as the United States once again faces an energy crisis, one of the "fixes" being proposed is to allow exploitation of oil in the Arctic National Wildlife Refuge on Alaska's North Slope. The rhetoric in the early stages of the debate is more self-serving or emotional than scientific. At one extreme are those who feel that any significant, potential oil field should be developed without regard to environmental damage. At the other extreme are those who instinctively assume that any intrusion on an ecological environment is unacceptable. We can hope that the enormous amount of data from the Alaskan pipeline and the drilling of the North Slope oil field (which has been producing decreasing amounts of oil with ongoing pumping) will be used to help transcend the politics.

Related Web Resource

For a more in-depth look as well as links to related sites, go to the expanded box on the book's website (www.mhhe.com/plummer9e).

FIGURE 1.4
Mount Robson, the highest peak in the Canadian Rocky Mountains.
Photo © Superstock

work. Music and art help link our feelings to that which we have discovered through our life. The natural sciences involve understanding the physical and biological universe in which we live. Most scientists get great satisfaction from their work because, besides gaining greater knowledge from what has been discovered by scientists before them, they can find new truths about the world around them. Even after a basic geology course you can use what you learn to explain and be able to appreciate what you see around you, especially when you travel. If, for instance, you were traveling through the Canadian Rockies, you might see the scene in figure 1.4 and wonder how the landscape came to be.

You might wonder: (1) why there are layers in the rock exposed in the cliffs; (2) why the peaks are so jagged; (3) why there is a glacier in a valley carved into the mountain; (4) why this is part of a mountain belt that extends northward and southward for thousands of kilometers; (5) why there are mountain ranges here and not in the central part of the continent. After completing a course in physical geology, you should be able to answer these questions as well as understand how other kinds of landscapes formed.

AN OVERVIEW OF PHYSICAL GEOLOGY—IMPORTANT CONCEPTS

The remainder of this chapter is an overview of physical geology that should provide a framework for most of the material in this book. Although the concepts probably are totally new to you, it is important that you comprehend what follows. You may want to reread portions of this chapter while studying later chapters when you need to expand or reinforce your comprehension of this basic material.

The Earth can be visualized as a giant machine driven by two engines, one internal and the other external. Both are *heat engines,* devices that convert heat energy into mechanical energy. Two simple heat engines are shown in figure 1.5. An automobile is powered by a heat engine. When gasoline is ignited in the cylinders, the resulting hot gases expand, driving pistons to the far end of cylinders. In this way, the heat energy of the expanding gas has been converted to the mechanical energy of the moving pistons, then transferred to the wheels, where the energy is put to work moving the car.

Earth's *internal* heat engine is driven by heat moving from the hot interior of the Earth toward the cooler exterior. Moving plates and earthquakes are products of this heat engine.

Earth's *external* heat engine is driven by solar power. Heat from the sun provides the energy for circulating the atmosphere and oceans. Water, especially from the oceans, is evaporated due to solar heating. When moist air cools, rain or snow falls.

Over long periods of time, moisture at the Earth's surface helps rock disintegrate. Water washing down hillsides and flowing in streams loosens and carries away the rock particles. In this way mountains originally raised by the Earth's internal forces are worn away by processes driven by the external heat engine.

We will look at how the Earth's heat engines work and show how some of the major topics of physical geology are related to the *surficial* (on the Earth's surface) and *internal* processes powered by the heat engines.

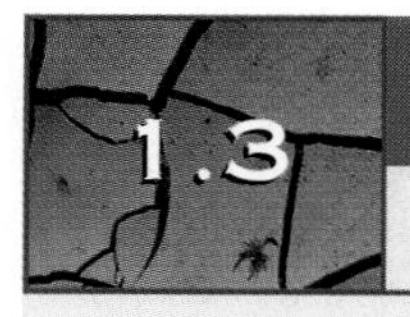

In Greater Depth

Geology as a Career

If someone tells you she or he is a geologist, it tells you almost nothing about what he or she does. This is because geology encompasses a broad spectrum of disciplines. Perhaps what most geologists have in common is that they were attracted to the outdoors. Most of us enjoyed hiking, skiing, climbing, or other outdoor activities before getting interested in geology. We like having one of our laboratories being Earth itself.

Geology is a collection of disciplines. When someone decides to become a geologist, she or he is selecting one of those disciplines. The choice is very large. Some are financially lucrative, others may be less so, but might be more satisfying. Below are a few of the areas in which geologists work.

Petroleum geologists work at trying to determine where existing oil fields might be expanded or where new oil fields might exist. Mining geologists might be concerned with trying to determine where to extend an existing mine to get more ore or trying to find new concentrations of ore that are potentially commercially viable. Environmental geologists might work at mitigating pollution or in preventing degradation of the environment. Marine geologists are concerned with understanding the sea floor. Hydrogeologists study surface and underground water and either assist in increasing our supply of clean water or in the isolation or clean-up of polluted water. Glaciologists and volcanologists obviously are specialists in glaciers and volcanoes, respectively. Geophysicists interpret earthquake waves or gravity measurements to determine the nature of Earth's interior. Seismologists are geophysicists who specialize in earthquakes. Teaching is an important field in which geologists work. Some teach at the college level and are usually involved in research as well. There is an increasing demand for geologists to teach Earth science (which includes meteorology, oceanography, astronomy as well as geology) in high schools. More and more secondary schools are adding Earth science to their curriculum and need qualified teachers. Engineering geologists determine whether rock or soil upon which structures (dams, bridges, buildings) are built can safely support those structures. Paleontologists study fossils and learn about when extinct creatures lived and the environment in which they existed.

Related Web Resource

For more information, go to the American Geological Institutes career site at

www.agiweb.org/careers.html

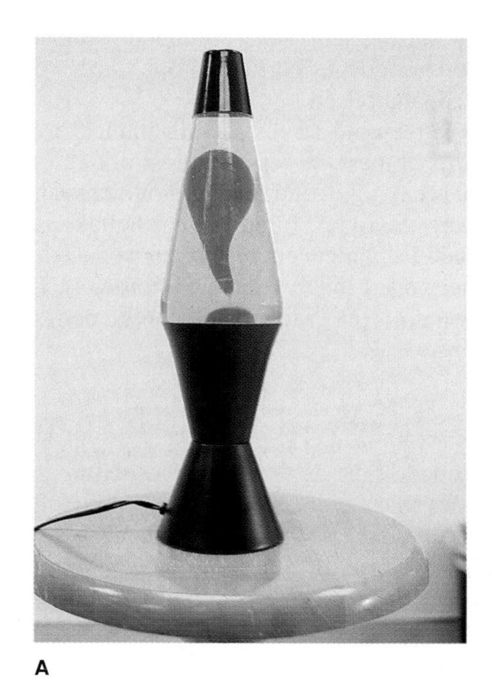

A

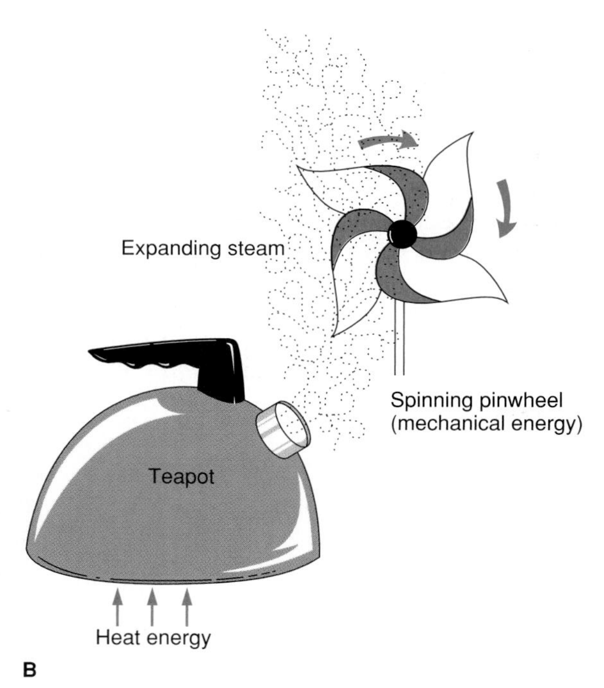

B

FIGURE 1.5

Two examples of simple heat engines. (*A*) A "lava lamp." Blobs are heated from below and rise. Blobs cool off at the top of the lamp and sink. (*B*) A pinwheel held over steam. Heat energy is converted to mechanical energy.

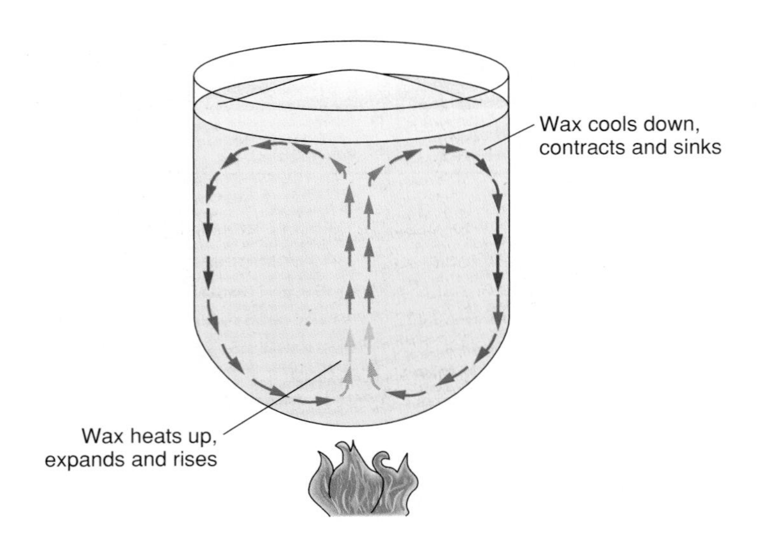

FIGURE 1.6
Movement of wax due to density differences caused by heating and cooling.

Internal Processes: How the Earth's Internal Heat Engine Works

The Earth's internal heat engine works because hot, buoyant material deep within the Earth moves slowly upward toward the cool surface and cold, denser material moves downward. Visualize a vat of hot wax, heated from below (figure 1.6). As the wax immediately above the fire gets hotter, it expands, becomes less dense (that is, a given volume of the material will weigh less) and will rise. Wax at the top of the vat loses heat to the air, cools, contracts, becomes denser, and will sink. A similar process takes place in the Earth's interior. Rock that is deep within the Earth and is very hot rises slowly toward the surface while rock that has cooled near the surface is denser and sinks downward. Instinctively, we don't want to believe that rock can flow like hot wax. Experiments have shown, however, that under the right conditions, rocks are capable of being molded (like wax or putty). Deeply buried rock that is hot and under high pressure can deform, like taffy or putty. But the deformation takes place very slowly. If we were somehow able to strike a rapid blow to the deeply buried rock with a hammer, it would fracture, just as rock at Earth's surface would.

Earth's Interior

As described in more detail in chapter 2, the **mantle** is the most voluminous of Earth's three major concentric zones (see figure 1.7). Although the mantle is solid rock, parts of it flow slowly, generally upward or downward, depending upon whether it is hotter or colder than adjacent mantle.

The other two zones are the **crust** and the **core.** The crust of the Earth is analogous to the skin on an apple. The thickness of the crust is insignificant compared to the whole Earth. We have direct access to only the crust, and not much of the crust at that. We are like microbes crawling on an apple, without the ability to penetrate its skin. Because it is our home and we depend on it for resources, we are concerned more with the crust than with the inaccessible mantle and core. The crust varies in thickness. Two major types of crust are *oceanic crust* and *continental crust.* The crust under the oceans is much thinner. It is made of rock that is somewhat denser than the rock that underlies the continents.

The lower parts of the crust and the entire mantle are inaccessible to direct observation. No mine or oil well has penetrated through the crust, so our concept of the Earth's interior is based on indirect evidence.

The crust and the uppermost part of the mantle are relatively rigid. Collectively they make up the **lithosphere.** (To help you remember terms, the meanings of commonly used prefixes and suffixes are given in appendix G. For example, *lith* means "rock" in Greek. You will find *lith* to be part of many geologic terms.) The uppermost mantle underlying the lithosphere, called the **asthenosphere,** is soft and therefore flows more readily than the underlying mantle. It provides a "lubricating" layer over which the lithosphere moves (*asthenos* means "weak" in Greek). Where hot mantle material wells upward, it will uplift the lithosphere. Where the lithosphere is coldest and densest, it will sink down through the asthenosphere and into the deeper mantle (figure 1.8). The effect of this internal heat engine on the crust is of great significance to geology. The forces generated inside the Earth, called **tectonic forces,** cause deformation of rock as well as vertical and horizontal movement of portions of the Earth's crust. Mountain ranges are evidence of tectonic forces strong enough to outdo gravitational forces. (Mount Everest, the world's highest peak, is made of rock that formed beneath an ancient sea.) Mountain ranges are built over extended periods, as portions of the Earth's crust are squeezed, stretched, and raised.

Most tectonic forces are mechanical forces. Some of the energy from these forces is put to work deforming rock, bending and breaking it, and raising mountain ranges. The mechanical energy may be stored (an earthquake is a sudden release of stored mechanical energy) or converted to heat energy (rock may melt, resulting in volcanic eruptions). The working of the machinery of the Earth is elegantly demonstrated by plate tectonics.

The Theory of Plate Tectonics

From time to time a theory emerges within a science that revolutionizes that field. (As explained in box 1.4 a *theory* in science is a concept that has been highly tested and in all likelihood is true. In common usage, "theory" is used for what scientists call a *hypothesis,* that is, a tentative answer to a question or solution to a problem.) The theory of plate tectonics is as important to geology as the theory of relativity is to physics, the atomic theory to chemistry, or evolution is to biology. The plate tectonic theory, currently accepted by virtually all geologists, is a unifying theory that accounts for many seemingly

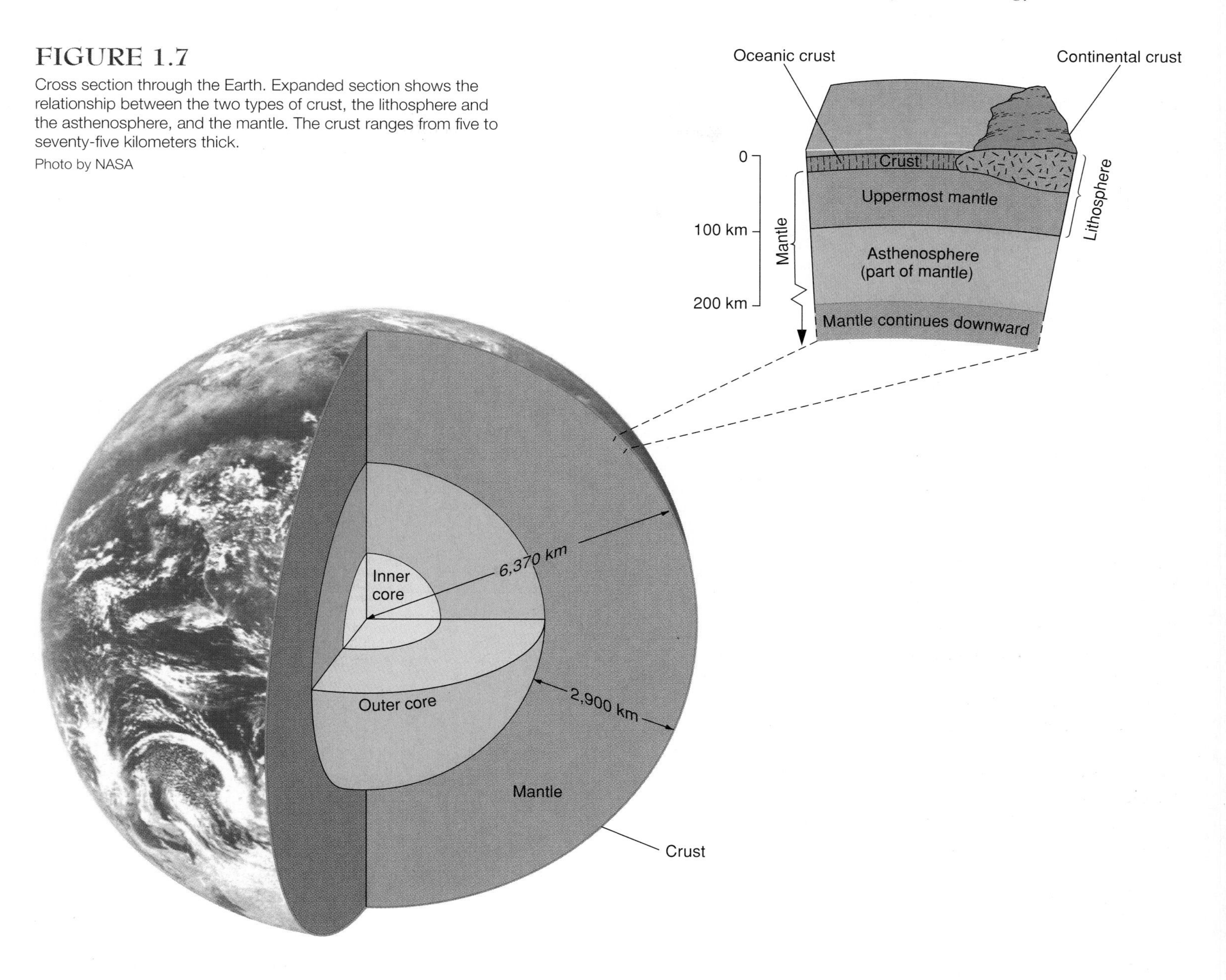

FIGURE 1.7

Cross section through the Earth. Expanded section shows the relationship between the two types of crust, the lithosphere and the asthenosphere, and the mantle. The crust ranges from five to seventy-five kilometers thick.

Photo by NASA

unrelated geological phenomena. Some of the disparate phenomena that plate tectonics explains are where and why we get earthquakes, volcanoes, mountain belts, deep ocean trenches, and mid-oceanic ridges.

Plate tectonics was seriously proposed as a hypothesis in the early 1960s, though the idea was based on earlier work—notably, the hypothesis of *continental drift.* In the chapters on igneous, sedimentary, and metamorphic rocks, as in the chapter on earthquakes, we will expand upon what you learn about the theory here to explain the origin of some rocks and why volcanoes and earthquakes occur. Chapter 4 is devoted to plate tectonics and will show that what you learn in other chapters is interrelated and explained by plate tectonic theory.

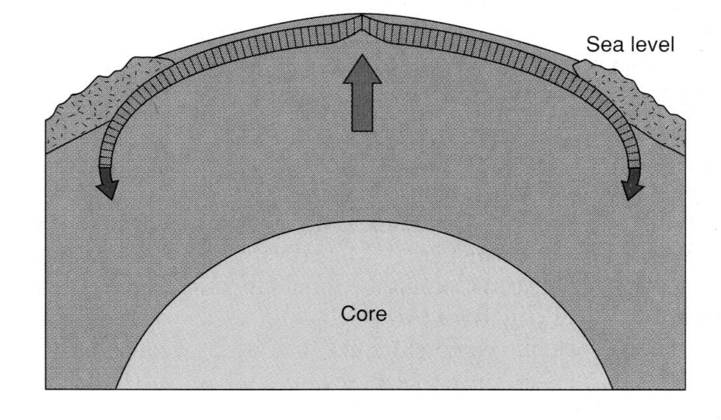

FIGURE 1.8

Hot mantle travels upward. Cold crust and mantle sink. Not drawn to scale.

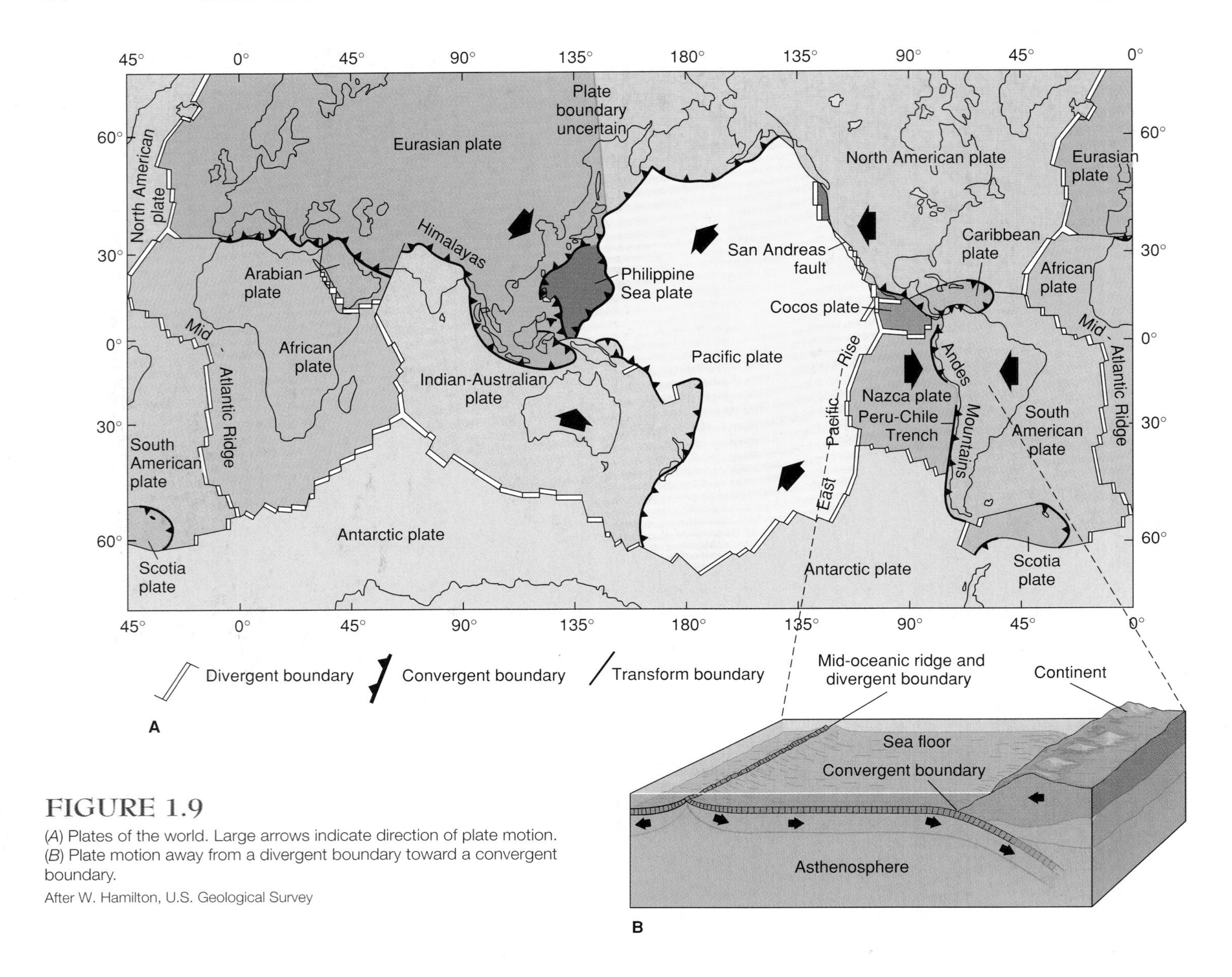

FIGURE 1.9

(*A*) Plates of the world. Large arrows indicate direction of plate motion. (*B*) Plate motion away from a divergent boundary toward a convergent boundary.

After W. Hamilton, U.S. Geological Survey

Plate tectonics regards the lithosphere as broken into *plates* that are in motion (figure 1.9). The plates, which are much like segments of the cracked shell on a boiled egg, move relative to one another, sliding on the underlying asthenosphere. Much of what we observe in the rock record can be explained by what takes place along *plate boundaries,* where two plates are pulling away from each other, sliding past each other, or moving toward each other.

According to plate tectonics, **divergent boundaries** exist where plates are moving apart. Most divergent boundaries coincide with the crests of submarine mountain ranges, called **mid-oceanic ridges** (figures 1.9*B* and 1.10).

A mid-oceanic ridge is higher than deep ocean floor (figure 1.10*A*) because the rocks, being hotter there, are less dense. Tensional cracks develop along the ridge crest (figure 1.10*B*). A *rift valley,* bounded by tensional cracks, runs along the crest of the ridge. Cracks extend downward and tap into localized **magma** (molten rock) chambers. The magma in the chamber comes from partial melting of the underlying asthenosphere. Magma from the magma chamber squeezes into fissures. Some magma erupts onto the floor of the rift valley and the rest solidifies in the fissure. Continued pulling apart of the ridge crest develops new cracks, and the process of filling and cracking continues indefinitely. Thus, new oceanic crust is continuously created at a divergent boundary. All of the mantle material does *not* melt; a solid residue remains under the newly created crust. New crust and underlying solid mantle make up the lithosphere that moves away from the ridge crest, traveling like the top of a conveyor belt. The rate of motion is generally 1 to 18 centimeters per year (approximately the growth rate of a fingernail), slow in human terms but quite fast by geologic standards.

As the lithosphere moves away from the divergent boundary, the material slowly cools. As it cools it contracts and becomes denser. The contraction of the lithosphere and its slow sinking (because of the increased density) cause the floors beneath oceans to deepen away from ridge crests.

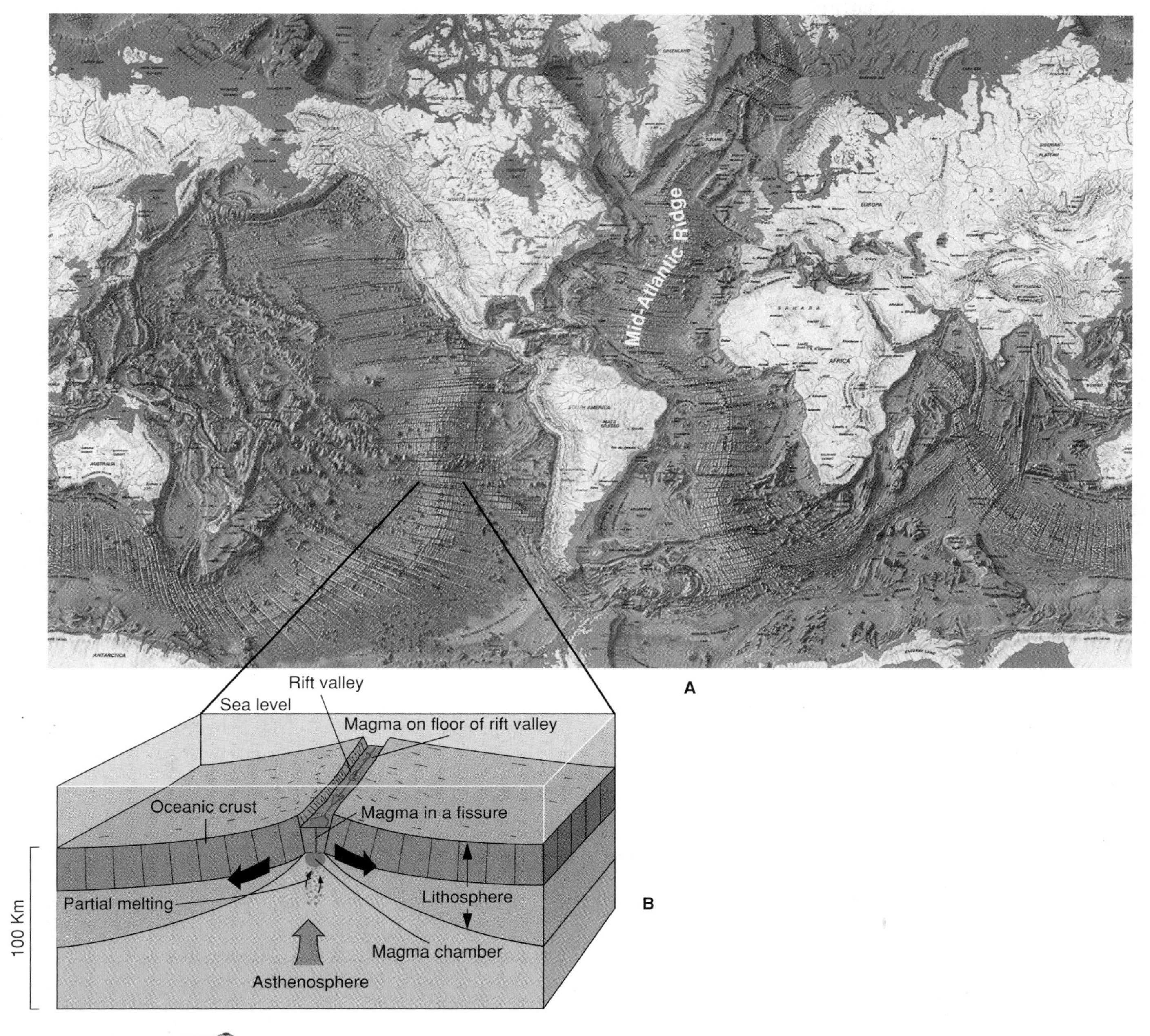

FIGURE 1.10

(*A*) The sea floors of the world. (*B*) A divergent boundary at a mid-oceanic ridge. Hot asthenosphere wells upward beneath the ridge crest. Magma forms and squirts into fissures. Solid material that does not melt remains as mantle in lower part of lithosphere. As lithosphere moves away from spreading axis, it cools, becomes denser, and sinks to a lower level.

The top of a plate may be composed exclusively of oceanic crust or include a continent or part of a continent. For example, if you live on the North American plate, you are riding westward relative to Europe because the plate's divergent boundary is along the mid-oceanic ridge in the North Atlantic Ocean (figures 1.9 and 1.10). The western half of the North Atlantic sea floor and North America are moving together in a westerly direction away from the mid-Atlantic ridge plate boundary.

A second type of boundary, a **transform boundary,** occurs where two plates slide past each other. The San Andreas fault in California is an example of this type of boundary, and the earthquakes along the fault are a result of plate motion.

The third type of boundary, one resulting in a wide range of geologic activities, is a **convergent boundary,** where plates move toward each other (figure 1.11). If one plate is capped by oceanic crust and the other by continental crust, the less dense, more buoyant continental plate will override the denser, oceanic plate. The oceanic plate sinks along what is known as a **subduction zone,** a zone where an oceanic plate descends into the mantle beneath an overriding plate.

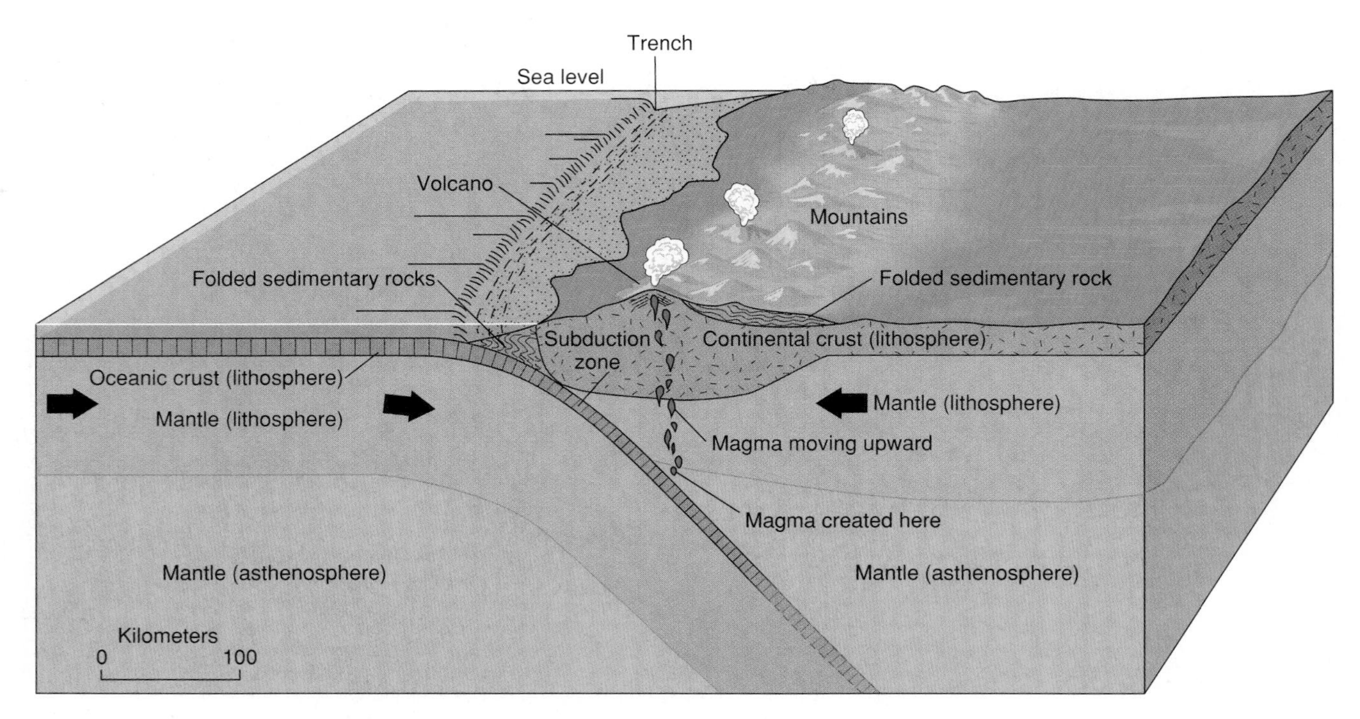

FIGURE 1.11 
A convergent boundary.

Deep *oceanic trenches* are found where oceanic lithosphere is starting to be subducted. These narrow, linear troughs are the deepest parts of the ocean floor.

In the region where the top of the subducting plate slides beneath the asthenosphere, melting takes place and magma is created. Magma is less dense than the overlying solid rock. Therefore, the magma created along the subduction zone works its way upward and either erupts on the Earth's surface to solidify as *extrusive* **igneous rock** or solidifies within the crust to become *intrusive* igneous rock. Hot rock, under high pressure, near the subduction zone that does not melt may change in the solid state to a new rock—**metamorphic rock.**

In addition to containing igneous and metamorphic rocks, major mountain belts show the effects of squeezing caused by plate convergence (for instance, the "folded sedimentary rocks" shown on figure 1.11). In the process, rock that may have been below sea level might be squeezed upward to become part of a mountain range.

Box 1.4 describes how plate tectonic theory was developed through the *scientific method.* If you do not have a thorough comprehension of how the scientific method works, be sure to study the box.

Surficial Processes: The Earth's External Heat Engine

Tectonic forces can squeeze formerly low-lying continental crustal rock along a convergent boundary and raise the upper part well above sea level. Portions of the crust also can rise because of **isostatic adjustment,** vertical movement of sections of Earth's crust to achieve balance. That is to say, lighter rock will "float" higher than denser rock on the underlying mantle. Isostatic adjustment is why an empty ship is higher above water than an identical one that is full of cargo. Continental crust, which is less dense than oceanic crust, will tend to float higher over the underlying mantle than oceanic crust (which is why the oceanic crust is below sea level and the continents are above sea level). After a portion of the continental crust is pulled downward by tectonic forces, it is out of isostatic balance. It will then rise slowly due to isostatic adjustment when tectonic forces are relaxed.

When a portion of crust rises above sea level, rocks are exposed to the atmosphere. Earth's external heat engine, driven by solar power, comes into play. Our weather patterns are largely a product of this heat engine. For instance, hot air rises near the equator and sinks in cooler zones to the north and south. Solar heating of air creates wind; ocean waves are, in turn, produced by wind. When moist air cools, it rains or snows. Rainfall on hillsides flows down slopes and into streams. Streams flow to lakes or seas. Glaciers grow where there is abundant snowfall at colder, high elevations and flow downhill because of gravity.

Where moving water, ice, or wind loosens and removes material, **erosion** is taking place. Streams flowing toward oceans remove some of the land over which they run. Crashing waves carve back a coastline. Glaciers grind and carry away underlying rock as they move. In each case, rock originally raised by the Earth's internal processes is worn down by surficial processes (figure 1.12).

IN GREATER DEPTH 1.4

Plate Tectonics and the Scientific Method

Although the hypothesis was proposed only a few decades ago, plate tectonics has been so widely accepted and disseminated that most people have at least a rough idea of what it is about. Most nonscientists can understand the television and newspaper reports (and occasional comic strip, such as that in box figure 1) that include plate tectonics in reports on earthquakes and volcanoes. Our description of plate tectonics implies little doubt about the existence of the process. The theory of plate tectonics has been accepted by nearly all geologists (this does not mean it is "true"). Plate tectonic theory, like all knowledge gained by science, has evolved through the processes of the **scientific method.** We will illustrate the scientific method by showing how plate tectonics has evolved from a vague idea into a plausible theory.

The basis for the scientific method is the belief that the universe is orderly and that by *objectively* analyzing phenomena, we can discover their workings. Science is a deeply human endeavor that involves creativity. A scientist's mind searches for connections and thinks of solutions to problems that might not have been considered by others. At the same time, a scientist must be aware of what work has been done by others, so that science can build on those works. Here, the scientific method is presented as a series of steps. A scientist is aware that his or her work must satisfy the requirements of the steps, but does not ordinarily go through a formal checklist.

1. A question is raised or a problem is presented.
2. Available information pertinent to the question or problem is analyzed. Facts, which scientists call **data,** are gathered.
3. After the data have been analyzed, tentative explanations or solutions, called **hypotheses,** are proposed.
4. One predicts what would occur in given situations if a hypothesis were correct.
5. Predictions are tested. Incorrect hypotheses are discarded.
6. A hypothesis that passes the testing becomes a **theory,** which is regarded as having an excellent chance of being true. In science, however, nothing is considered proven absolutely. All theories remain open to scrutiny, further testing, and refinement.

Like any human endeavor, the scientific method is not infallible. Objectivity is needed throughout. Someone can easily become attached to the hypothesis he or she has created and so tend subconsciously to find only supporting evidence. As in a court of law, every effort is made to have observers objectively examine the logic of both procedures and conclusions. Courts sometimes make wrong decisions; science, likewise, is not immune to error.

How the concept of plate tectonics evolved into a theory is outlined below.

Step 1: A question asked or problem raised. Actually, a number of questions were being asked about seemingly unrelated geological phenomena.

What caused the submarine ridge that extends through most of the oceans of the world? Why are rocks in mountain belts intensely deformed? What sets off earthquakes? What causes rock to melt underground and erupt as volcanoes?

Step 2: Gathering of data. Early in the 20th century, the amount of data was limited. But through the decades, the information gathered increased enormously. New data, most notably information gained from exploration of the sea floor in the mid-1900s, forced scientists to discard old hypotheses and come up with new ideas.

Step 3: Hypotheses proposed. Most of the questions being asked were treated as separate problems wanting separate hypotheses. Some appeared interrelated. One hypothesis, **continental drift,** did address several questions. It was advocated by Alfred Wegener, a German scientist, in a book published in the early 1900s.

Box 1.4 — FIGURE 1

Plate tectonics sometimes show up in comic strips.

FRANK & ERNEST reprinted by permission of Newspaper Enterprise Association, Inc.

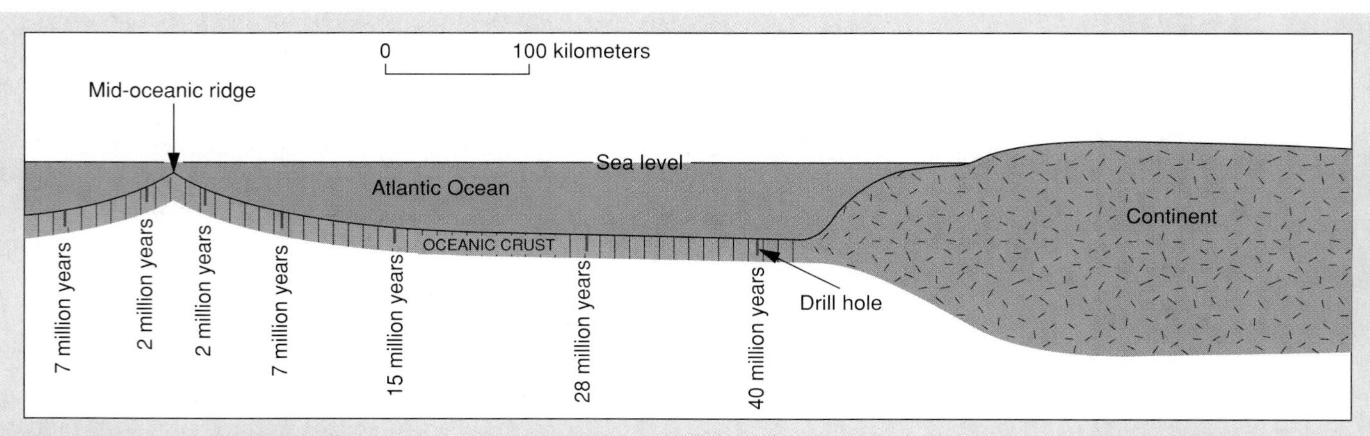

Box 1.4 — Figure 2

Ages of rocks from holes drilled into the oceanic crust. (Vertical scale of diagram is exaggerated).

Wegener postulated that the continents were all once part of a single supercontinent called Pangaea. The hypothesis explained why the coastlines of Africa and South America look like separated parts of a jigsaw puzzle. Some 200 million years ago this supercontinent broke up, and the various continents slowly drifted into their present positions. The hypothesis suggested that the rock within mountain belts becomes deformed as the leading edge of a continental crust moves against and over the stationary oceanic crust. Earthquakes were presumably caused by continuing movement of the continents.

Until the 1960s continental drift was not widely accepted. It was scoffed at by many geologists who couldn't conceive of how a continent could be plowing over oceanic crust. During the 1960s, after new data on the nature of the sea floor became available, the idea of continental drift was incorporated into the concept of plate tectonics. What was added in the plate tectonic hypothesis was the idea that oceanic crust, as well as continental crust, was shifting.

Step 4: Prediction. An obvious prediction, if plate tectonics is correct, is that if Europe and North America are moving away from each other, the distance measured between the two continents is greater from one year to the next. But we cannot stretch a tape measure across oceans, and until recently, we have not had the technology to accurately measure distances between continents. So in the 1960s, other testable predictions had to be made. Some of these predictions and results of their testing are described in the chapter on plate tectonics. One of these predictions was that the rocks of the oceanic crust will be progressively older the farther they are from the crest of a mid-oceanic ridge.

Step 5: Predictions are tested. Experiments were conducted in which holes were drilled in the deep sea floor from a specially designed ship. Rocks and sediment were collected from these holes, and the ages of these materials were determined. As the hypothesis predicted, the youngest sea floor (generally less than a million years old) is near the mid-oceanic ridges, whereas the oldest sea floor (up to about 200 million years old) is farthest from the ridges (box figure 2).

This test was only one of a series. Various other tests, described in some detail later in this book, tended to confirm the hypothesis of plate tectonics. Some tests did not work out exactly as predicted. Because of this, and more detailed study of data, the original concept was, and continues to be, modified. The basic premise, however, is generally regarded as valid.

Step 6: The hypothesis becomes a theory. Most geologists in the world considered the results of this and other tests as positive, indicating that the concept is not reasonably disputable and very probably true. It can now be called the plate tectonic theory.

During the last few years, plate tectonic theory has been further confirmed by the results of very accurate satellite surveys that determine where points on separate continents are relative to one another. The results indicate that the continents are indeed moving relative to one another. Europe and North America *are* moving farther apart.

Although it is unlikely that plate tectonic theory will be replaced by something we haven't thought of yet, aspects that fall under plate tectonics' umbrella continue to be analyzed and revised as new data become available.

Important Note

Words used by scientists do not always have the same meaning when used by the general public. A case in point is the word *theory.* To most people, a "theory" is what scientists regard as an "hypothesis." You may remember news reports about an airliner that exploded offshore from New York in 1996. A typical statement on television was: "One theory is that a bomb in the plane exploded; a second theory is that the plane was shot down by a missile fired from a ship at sea; a third theory is that a spark ignited in a fuel tank and the plane exploded." Clearly, each "theory" is an hypothesis in the scientific sense of the word. This has led to considerable confusion for nonscientists about science. You have probably heard the expression "It's just a theory." Statements such as "Evolution is just a theory" are used to imply that scientific support is weak. The reality is that theories such as evolution and plate tectonics have been so overwhelmingly verified that they come as close as possible to what scientists accept as being indisputable facts.

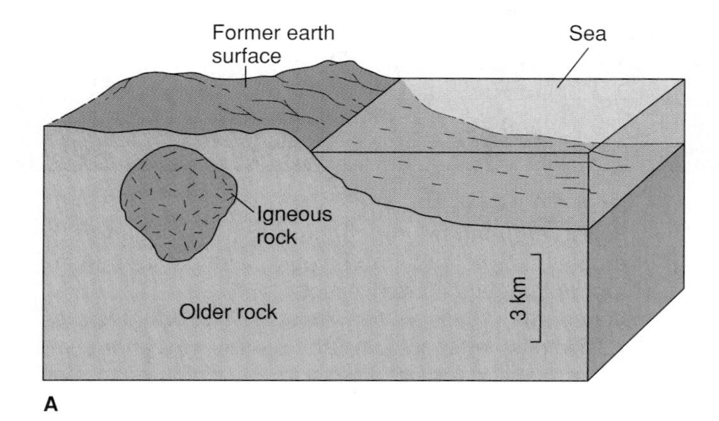

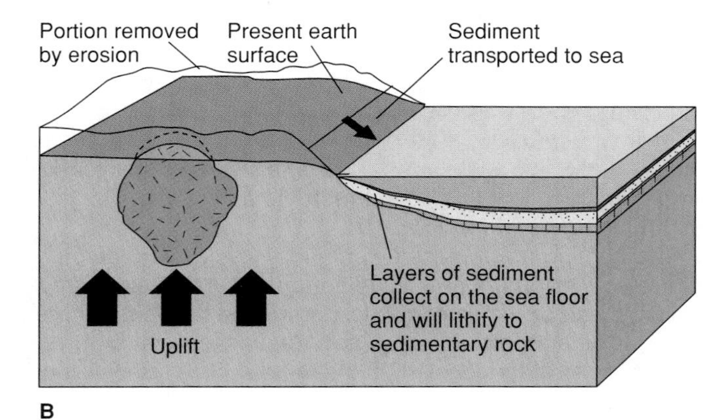

FIGURE 1.12
Uplift, erosion, and deposition. (*A*) Magma has solidified underground to become igneous rock. (*B*) Land is uplifted due to isostatic adjustment. Upper portion is eroded. Sediment is transported to the sea to become sedimentary rock.

Rocks formed at high temperature and under high pressure deep within the Earth and pushed upward by tectonic forces are unstable in their new environment. Air and water tend to cause the once deep-seated rocks to break down and form new materials. The new materials, stable under conditions at the Earth's surface, are said to be in **equilibrium,** that is, adjusted to the physical and chemical conditions of their environment so that they do not change or alter with time. For example, much of an igneous rock (such as granite) that formed at a high temperature tends to break down chemically to clay. Clay is in equilibrium, that is to say it is stable, at the Earth's surface.

The product of the breakdown of rock is **sediment,** loose material. Sediment may be transported by an agent of erosion, such as running water in a stream. Sediment is deposited when the transporting agent loses its carrying power. For example, when a river slows down as it meets the sea, the sand being transported by the stream is deposited as a layer of sediment.

In time a layer of sediment deposited on the sea floor becomes buried under another layer. This process may continue burying our original layer progressively deeper. The pressure from overlying layers compresses the sediment, helping to consolidate the loose material. With the cementation of the loose particles, the sediment becomes *lithified* (cemented or otherwise consolidated) into a **sedimentary rock.**

GEOLOGIC TIME

We have mentioned the great amount of time required for geological processes. As humans, we think in units of time related to personal experience—seconds, hours, years, a human lifetime. It stretches our imagination to contemplate ancient history that involves 1,000 or 2,000 years. Geology involves vastly greater amounts of time, often referred to as *deep time.*

To be sure, some geological processes occur quickly, such as a great landslide or a volcanic eruption. These events occur when stored energy (like the energy stored in a car battery) is suddenly released. Most geological processes, however, are slow but relentless, reflecting the pace at which the heat engines work. It is unlikely that a hill will visibly change in shape or height during your lifetime (unless through human activity). However, in a geologic time frame, the hill probably is eroding away quite rapidly. "Rapidly" to a geologist may mean that within a few million years the hill will be reduced nearly to a plain. Similarly, in the geologically "recent" past of several million years ago, a sea may have existed where the hill is now. Some processes are regarded by geologists as "fast" if they are begun and completed within a million years.

The rate of plate motion is relatively fast. If new magma erupts and solidifies along a mid-oceanic ridge, we can easily calculate how long it will take that igneous rock to move 1,000 kilometers away from the spreading center. At the rate of 1 centimeter per year, it will take 100 million years for the presently forming part of the crust to travel the 1,000 kilometers.

Although we will discuss geologic time in detail in chapter 8, table 1.1 shows some reference points to keep in mind. The Earth is estimated to be about 4.55 billion years old (4,550,000,000 years). Fossils in rocks indicate that complex forms of animal life have existed in abundance on the Earth for about the past 545 million years. Reptiles became abundant about 230 million years ago. Dinosaurs evolved from reptiles and became extinct about 65 million years ago. Humans have been here only about the last 3 million years. The eras and periods shown in table 1.1 comprise a kind of calendar for geologists into which geologic events are placed (as explained in the chapter on geologic time).

Not only are the immense spans of geologic time difficult to comprehend, but very slow processes are impossible to duplicate. A geologist who wants to study a certain process cannot repeat in a few hours a chemical reaction that takes a million years to occur in nature. As Mark Twain wrote in *Life on the Mississippi,* "Nothing hurries geology."

1.5 ASTROGEOLOGY

THE ORIGIN AND HISTORY OF THE SOLAR SYSTEM

From the perspective of the universe Earth is insignificant. Earth is one of nine planets that orbit the Sun. The Sun is an average star in the Milky Way galaxy. The Milky Way is just one of billions of galaxies. Despite its minuscule place in the universe, Earth is important to us—it's our planet. Among the planets, Earth is unique in many ways; notable is its abundance of liquid water. However, it shares many characteristics as well as an early history with its neighboring planets.

The solar system is the Sun and the planets and space debris that orbit the Sun. The solar system began with the condensation of a large volume of interstellar gas and dust called a *nebula.* Most of the material in the nebula collected in the center to form the Sun, but some of it remained farther out. Small groupings of particles formed randomly, and because of the combined gravitational pull of the particles in the group, they attracted other particles. The groupings continued to increase in size until, on the average, they were probably several kilometers in diameter.

These larger bodies sometimes joined together by collision and sometimes became fragmented. The largest bodies tended to increase in size by capturing smaller ones, and the smaller bodies tended to be fragmented or were captured. The bodies that grew larger became the nine planets and their satellites; those that were fragmented became interstellar dust or asteroids.

Variations in composition among the planets can be partly explained by differences in their sizes. Larger planets were able to attract and hold larger amounts of lighter gases, such as hydrogen and helium. Another important factor in determining planetary composition was the location within the nebula where the planet formed. Those bodies nearest the center (Mercury, Venus, Earth, the Moon, and Mars) would have experienced higher temperatures and so retained smaller percentages of materials that are easily vaporized. The outer planets and their satellites were formed at a great enough distance from the Sun for large quantities of gases and ices to collect around their rocky cores.

Heat from building of planets along with heat derived from the decay of radioactive elements caused partial melting of the planets' interiors. Heavier materials such as iron-nickel alloys would sink to the planets' centers, while lighter materials such as silicate magmas would rise to form crusts. On Earth, the original crust was mostly basalt, with some granitic rocks. The areas of granitic rocks would have formed the first continents. Convection in the mantle probably would have begun at the time of crustal differentiation and would have set in motion a form of rapid plate tectonic processes.

On Venus, which is about the same size as Earth, the presence of possible folded mountains (the parallel ridges that have been observed) and chains of volcanoes may indicate that differentiation and some kind of lithospheric motions have occurred. However, no surface features have been found yet that could be interpreted as diverging plate boundaries similar to Earth's oceanic ridges.

Space "debris" continued to be swept up by the planets throughout their early histories, producing heavily cratered surfaces and multi-ringed basins. The Moon and Mercury are still heavily cratered. Earth, too, was subjected to an early period of intensive bombardment. But erosion and tectonic activity have erased most of those early-formed craters.

The heating up of planetary interiors would produce great quantities of magma that would rise to the surface along fractures produced by meteorite impact or by some tectonic process. Vast areas of the Moon, Mercury, and Mars are covered with volcanic rocks, most formed early in the history of the solar system. Similar early extrusive activity must have occurred on Venus and on Earth. Volcanic activity on a much smaller scale still occurs on Earth and Venus.

Gases would escape from planetary interiors as a result of volcanic activity, later condensing to form atmospheres, oceans, ice shells, and polar caps. Lighter elements would have escaped quickly from objects as small as the Moon and Mercury but could have been partially retained by Earth, Venus, and Mars. The distances of these planets from the Sun might determine whether these light elements exist mostly as atmospheres (as on Venus); as atmospheres, oceans, underground water, and ice caps (as on Earth); or mostly as ice caps and underground ice (as on Mars). On planets with appropriate conditions and temperature ranges, life may have formed, as it did on Earth. The presence of life on Earth has certainly extensively modified Earth's atmosphere (by producing oxygen, for example).

The presence of atmospheres, oceans, underground ice, and so on would have determined the extent to which weathering and erosion modified a planet's surface.

Related Web Resource

The Nine Planets

www.nineplanets.org/

table 1.1 Some Important Ages in the Development of Life on Earth

Millions of Years Before Present	Noteworthy Life	Eras	Periods
4	Earliest hominids	Cenozoic	Quaternary Tertiary
65	First important mammals Extinction of dinosaurs		
		Mesozoic	Cretaceous Jurassic Triassic
	First dinosaurs		
245			
		Paleozoic	Permian Pennsylvanian Mississippian Devonian Silurian Ordovician Cambrian
300	First reptiles		
400	Fishes become abundant		
	First abundant fossils		
545			
3,500	Earliest single-celled fossils	Precambrian	(The Precambrian accounts for the vast majority of geologic time.)
4,500	Origin of the Earth		

Summary

Geology is the scientific study of Earth. Geological investigations indicate that Earth is changing because of internal and surficial processes. Internal processes are driven mostly by temperature differences within Earth's mantle. Surficial processes are driven by solar energy. Internal forces cause the crust of Earth to move. Plate tectonic theory visualizes the lithosphere (the crust and uppermost mantle) as broken into plates that move relative to each other over the asthenosphere. The plates are moving *away* from divergent boundaries usually located at the crests of mid-oceanic ridges where new crust is being created. Plates move *toward* convergent boundaries. Convergence results in subduction of one plate. Plates slide past one another at transform boundaries. Plate tectonics and isostatic adjustment cause parts of the crust to move up or down.

Erosion takes place at Earth's surface where rocks are exposed to air and water. Rocks that formed under high pressure and temperature inside Earth are out of equilibrium at the surface and tend to alter to substances that are stable at the surface. Sediment is transported to a lower elevation where it is deposited (commonly on a sea floor in layers). When sediment is cemented it becomes sedimentary rock.

Although Earth is changing constantly, the rates of change are generally extremely slow by human standards.

Terms to Remember

asthenosphere 12
continental drift 17
convergent boundary 15
core 12
crust 12
data 17
divergent boundary 14
equilibrium 19
erosion 16
hypothesis 17
igneous rock 16
isostatic adjustment 16
lithosphere 12
magma 14
mantle 12
metamorphic rock 16
mid-oceanic ridge 14
plate tectonics 14
scientific method 17
sediment 19
sedimentary rock 19
subduction zone 15
tectonic forces 12
theory 17
transform boundary 15

Testing Your Knowledge

Use the questions below to prepare for exams based on this chapter.

1. What is meant by equilibrium? What happens when rocks are forced out of equilibrium?
2. What tectonic plate are you presently on? Where is the nearest plate boundary and what kind of boundary is it?
3. What is the most likely geologic hazard in your part of the country?
4. What are the three major types of rocks?
5. What are the relationships among the mantle, the crust, the asthenosphere, and the lithosphere?
6. What would the surface of Earth be like if there were no tectonic activity?
7. Explain why cave dwellers never saw a dinosaur.
8. Plate tectonics is a result of Earth's internal heat engine, powered by (choose all that apply)
   a. the Sun b. gravity
   c. heat flowing from Earth's interior outward
9. A typical rate of plate motion is
   a. 3–4 meters per year b. 1 kilometer per year
   c. 1–10 centimeters per year d. 1,000 kilometers per year
10. Earthquakes may be caused by
    a. movement of plates b. motion along faults
    c. shifting of bedrock d. all of the above
11. The division of geology concerned with Earth materials, changes in the surface and interior of the Earth, and the dynamic forces that cause those changes is
    a. physical geology b. historical geology
    c. geophysics d. paleontology
12. Which is a geologic hazard?
    a. earthquake b. volcano
    c. mudflows d. floods
    e. wave erosion at coastlines f. landslides
    g. all of the above
13. The largest zone of Earth's interior by volume is the
    a. crust b. mantle
    c. outer core d. inner core
14. Oceanic and continental crust differ in
    a. composition b. density
    c. thickness d. all of the above
15. The forces generated inside Earth that cause deformation of rock as well as vertical and horizontal movement of portions of Earth's crust are called
    a. erosional forces b. gravitational forces
    c. tectonic forces d. all of the above
16. Plate tectonics is a
    a. conjecture b. opinion
    c. hypothesis d. theory
17. Which is the type of a plate boundary?
    a. divergent b. transform
    c. convergent d. all of the above
18. The lithosphere is
    a. the same as the crust b. the layer beneath the crust
    c. the crust and uppermost mantle d. only part of the mantle
19. Erosion is a result of Earth's external heat engine, powered by (choose all that apply)
    a. the Sun b. gravity
    c. heat flowing from Earth's interior outward

Expanding Your Knowledge

1. Why are some parts of the lower mantle hotter than other parts?
2. According to plate tectonic theory, where are crustal rocks created? Why doesn't Earth keep getting larger if rock is continually created?
3. What percentage of geologic time is accounted for by the last century?
4. What would Earth be like without solar heating?
5. What are some of the technical difficulties you would expect to encounter if you tried to drill a hole to the center of Earth?

Exploring Web Resources

www.mhhe.com/plummer9e
Visit our Online Learning Center for additional readings and media resources as well as answers to the Testing Your Knowledge section, more quizzing, and direct links to the sites listed below.

http://pubs.usgs.gov/publications/text/dynamic.html
This Dynamic Earth by the U.S. Geological Survey is an online, illustrated publication explaining plate tectonics. You may want to go to the section "Understanding plate motion." This will help reinforce what you read about plate tectonics in this chapter. It goes into plate tectonics in greater depth, however, covering material that is in chapter 4 of this textbook.

www.uh.edu/~jbutler/anon/anontrips.html
Virtual Field Trips. The site provides access to geologic sites throughout the world. Many are field trips taken by geology classes. Check the alphabetical listing and see if there are any sites near you. Or watch a video clip in one of the Quick Time field trips. One of the well-done trips in the alphabetical listing is the Oneonta to the Hudson River field trip in Central New York.

www.usgs.gov
The *U.S. Geological Survey's* home page. Use this as a gateway to a wide range of geologic information.

http://earthrise.sdsc.edu/
EarthRISE provides access to photographs taken from the space shuttle. You may retrieve photos by requesting a geologic feature (e.g., volcano) or of a specific geographic portion of Earth.

www.nrcan.gc.ca/gsc/
The *Geological Survey of Canada* home page.

Animations

This chapter includes the following animations available on our Online Learning Center at www.mhhe.com/plummer9e.

1.10 Divergent Boundary

1.11 Convergent Boundary

CHAPTER

2

Earth's Interior and Geophysical Properties

The only rocks that geologists can study directly in place are those of the crust; and Earth's crust is but a thin skin of rock, making up less than 1% of Earth's total volume. Mantle rocks brought to Earth's surface in basalt flows, in diamond-bearing kimberlite pipes, and also the tectonic attachment of lower parts of the oceanic lithosphere to the continental crust, give geologists a glimpse of what the underlying mantle might look like. Meteorites also give clues about the possible composition of the core of Earth. But to learn more about the deep interior of Earth, geologists must study it *indirectly,* largely by using the tools of geophysics—that is, seismic waves and the measurement of gravity, heat flow, and earth magnetism.

The evidence from geophysics suggests that Earth is divided into three major layers—the crust on Earth's surface, the rocky mantle beneath the crust, and the metallic core at the center of Earth. The study of plate tectonics has shown that the crust and uppermost mantle can be conveniently divided into the brittle lithosphere and the plastic asthenosphere.

You will learn in this chapter how gravity measurements can indicate where certain regions of the crust and upper mantle are being held up or held down out of their natural position of equilibrium. We will also discuss Earth's magnetic field and its history of reversals. We will show how magnetic anomalies can indicate hidden ore and geologic structures. We close with a discussion of the distribution and loss of Earth's heat.

Opposite: Because diamonds form in the mantle and are brought to the surface in kimberlite pipes, they give geologists a glimpse of Earth's interior.

Photo courtesy of the American Museum of Natural History

What *do* geologists know about Earth's interior? How do they obtain information about the parts of Earth beneath the surface? Geologists, in fact, are not able to sample rocks very far below Earth's surface. Some deep mines penetrate 3 kilometers into Earth, and a deep oil well may go as far as 8 kilometers beneath the surface; the deepest scientific well has reached 12 kilometers in Russia (see box 2.1). Rock samples can be brought up from a mine or a well for geologists to study.

A direct look at rocks from deeper levels can be gotten where mantle rocks have been brought up to the surface by basalt flows (see box 2.2), by the intrusion and erosion of diamond-bearing kimberlite pipes (see chapter 12), or where the lower part of the oceanic lithosphere (see chapter 3) has been tectonically attached to the continental crust at a convergent plate boundary. However, Earth has a radius of about 6,370 kilometers, so it is obvious that geologists can only scratch the surface when they try to study *directly* the rocks beneath their feet.

Deep parts of Earth are studied *indirectly,* however, largely through the branch of geology called **geophysics,** which is the application of physical laws and principles to a study of Earth. Geophysics includes the study of seismic waves and Earth's magnetic field, gravity, and heat. All these things tell us something about the nature of the deeper parts of Earth. Together they create a convincing picture of what makes up Earth's interior.

EVIDENCE FROM SEISMIC WAVES

Seismic waves from a large earthquake may pass through the entire Earth. A nuclear bomb explosion also generates seismic waves. Geologists obtain new information about Earth's interior after every large earthquake and bomb test.

One important way of learning about Earth's interior is the study of **seismic reflection,** the return of some of the energy of seismic waves to Earth's surface after the waves bounce off a rock boundary. If two rock layers of differing densities are separated by a fairly sharp boundary, seismic waves reflect off that boundary just as light reflects off a mirror (figure 2.1). These reflected waves are recorded on a seismogram, which shows the amount of time the waves took to travel down to the boundary, reflect off it, and return to the surface. From the amount of time necessary for the round trip, geologists calculate the depth of the boundary.

Another method used to locate rock boundaries is the study of **seismic refraction,** the bending of seismic waves as they pass from one material to another, which is similar to the way that light waves bend when they pass through the lenses of eyeglasses. As a seismic wave strikes a rock boundary, much of the energy of the wave passes across the boundary. As the wave crosses from one rock layer to another, it changes direction (figure 2.2). This change of direction, or refraction, occurs only if the velocity of seismic waves is different in each layer (which is generally true if the rock layers differ in density or strength).

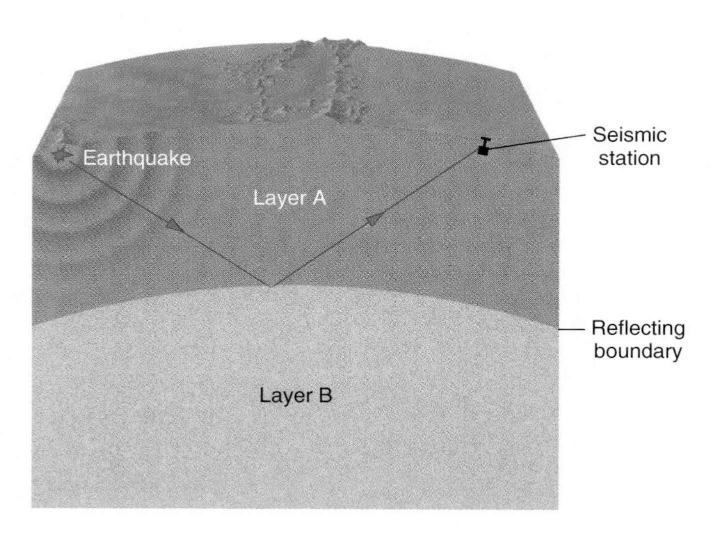

FIGURE 2.1
Seismic reflection. Seismic waves reflect from a rock boundary deep within the earth and return to a seismograph station on the surface.

The boundaries between such rock layers are usually distinct enough to be located by seismic refraction techniques, as shown in figure 2.3. Seismograph station 1 is receiving seismic waves that pass directly through the upper layer *A*. Stations farther from the epicenter, such as station 2, receive seismic waves from two pathways: (1) a direct path straight through layer *A* and (2) a refracted path through layer *A* to a higher-velocity layer *B* and back to layer *A*. Station 2 therefore receives the same wave twice.

Seismograph stations close to station 1 receive only the direct wave or possibly two waves, the direct (upper) wave arriving before the refracted (lower) wave. Stations near station 2 receive both the direct and the refracted waves. At some point between station 1 and station 2 there is a transformation from receiving the direct wave first to receiving the *refracted* wave first. Even though the refracted wave travels farther, it can arrive at a station first because most of its path is in the high-velocity layer *B*.

The distance between this point of transformation and the epicenter of the earthquake is a function of the depth to the rock boundary between layers *A* and *B*. A series of portable seismographs can be set up in a line away from an explosion (a *seismic shot*) to find this distance, and the depth to the boundary can then be calculated. The velocities of seismic waves within the layers can also be found.

Figure 2.2 shows how waves bend as they travel downward into higher velocity layers. But why do waves return to the surface, as shown in figure 2.3? The answer is that advancing waves give off energy in all directions. Much of this energy continues to travel horizontally within layer *B* (figure 2.3). This energy passes beneath station 2 and out of the figure toward the right. A small part of the energy "leaks" upward into layer *A,* and it is this pathway that is shown in the figure. There are many other pathways for this wave's energy that are not shown here.

A sharp rock boundary is not necessary for the refraction of seismic waves. Even in a thick layer of uniform rock, the

IN GREATER DEPTH 2.1

Deep Drilling on Continents

The structure and composition of most of the continental crust is unknown. Surface mapping and seismic reflection and refraction suggest that the continents are largely igneous and metamorphic rock, such as granite and gneiss, overlain by a veneer of sedimentary rocks. This sedimentary cover is generally thin, like icing on a cake, but it may thicken to 10 kilometers or more in giant sedimentary basins where the underlying "basement rock" has subsided. Although oil companies have drilled as deep as 8 kilometers on land, they drill in the sedimentary basins. The igneous and metamorphic basement, which averages 40 kilometers thick and makes up most of the continental crust, has rarely been sampled deeper than 2 or 3 kilometers (although uplift and erosion have exposed some rocks widely thought to have been formed much deeper in the crust).

Russia has drilled the world's deepest hole on the Kola Peninsula near Murmansk north of the Arctic Circle. The 12 kilometer-deep hole took 15 years to drill and penetrated ancient Precambrian basement rocks. The second deepest well drilled is the KTB hole in southeastern Germany, which reached a depth of 10 kilometers and cost more than a billion dollars (box figure 1). Deep drilling is as technically complex as space exploration. High pressures and 300°C temperatures require special equipment and techniques. The Russians used a turbodrill that rotated under the pressure of circulating drilling mud. Unlike normal drilling operations, the lightweight aluminum drill pipe does not turn. Because the Kola drilling operation resulted in a crooked hole, the Germans advanced deep-drilling technology by developing a system to keep the hole straight while being drilled.

The drilling at Kola shows that seismic models for this area are wrong. The Russians expected 4.7 kilometers of metamorphosed sedimentary and volcanic rock, then a granitic layer to a depth of 7 kilometers, and a "basaltic" layer below that. The granite, however, appeared at 6.8 kilometers and extends to more than 12 kilometers; the "basalt" has not yet been found. These results, and data from the other deep holes, show that seismic surveys of continental crust are being systematically misinterpreted.

The Russians and Germans unexpectedly found open fractures and circulating fluids throughout the borehole. The fluids include hydrogen, helium, and methane (natural gas), as well as mineralized waters forming ore bodies. Copper-nickel ore was found deeper than theory predicted, and gold mineralization was present from 9.5 to 11 kilometers down. These results will change geologists' models of ore formation and fluid circulation underground.

Box 2.1 — FIGURE 1

The KTB drilling operation in southeastern Germany reached a depth of 10 kilometers and has advanced the technology of deep drilling.

Photo courtesy of ICDP, GeoForschungsZentrum Potsdam

Additional Resources

Kerr, R. A. 1993. Looking — deeply — into the Earth's crust in Europe. *Science* 261:295.

Kozlovsky Y. A. 1987. The Superdeep well of the Kola Peninsula, Springer-Verlag, 558 p.

Related Web Resources

Scientific Information System for the world's deepest borehole, Kola SDB-3.

IGCP408: Rocks and minerals at great depth and on the surface.

http://icdp.gfz-potsdam.de/html/kola/IGCP408.html

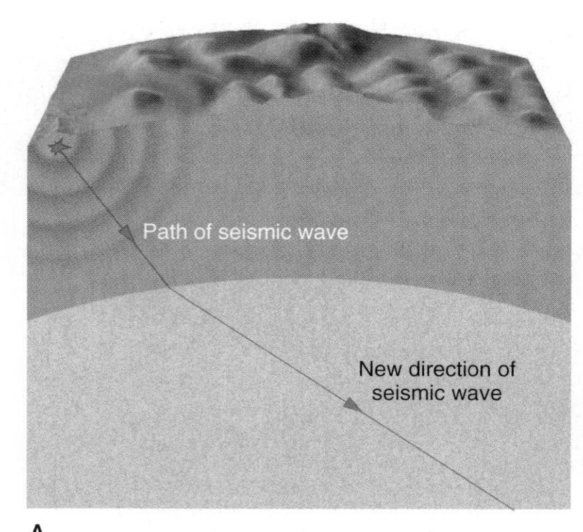

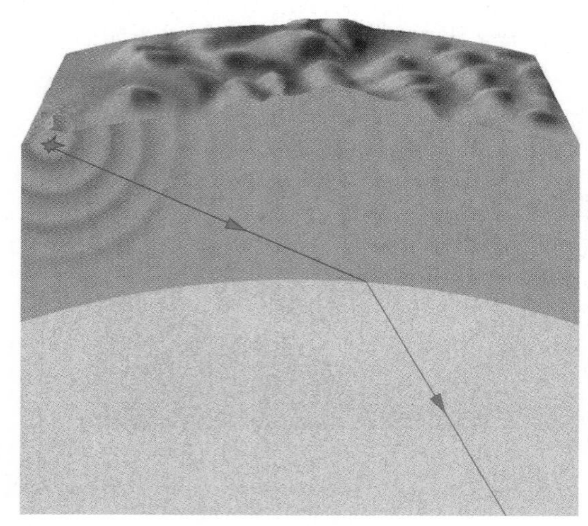

FIGURE 2.2

Seismic refraction occurs when seismic waves bend as they cross rock boundaries. At an interface, seismic (or sound or light) waves will bend toward the lower velocity material.

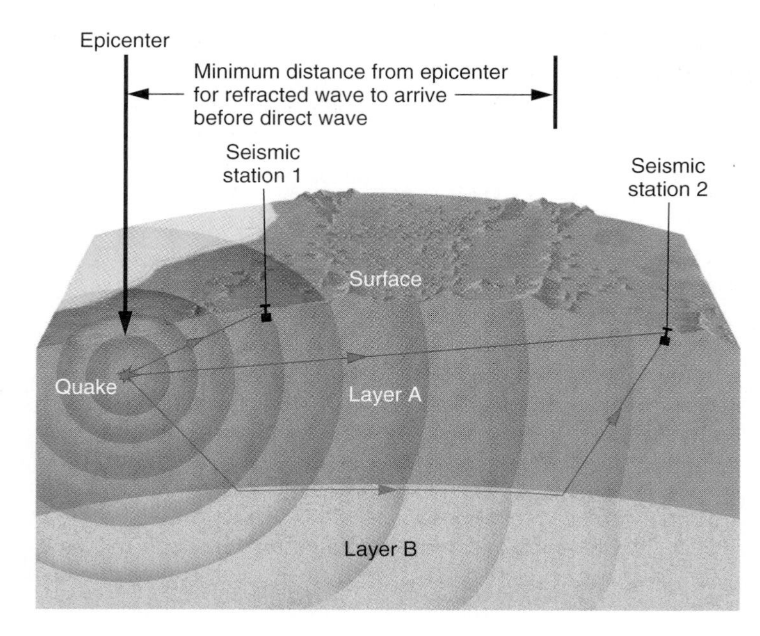

FIGURE 2.3

Seismic refraction can be used to detect boundaries between rock layers. See text for explanation.

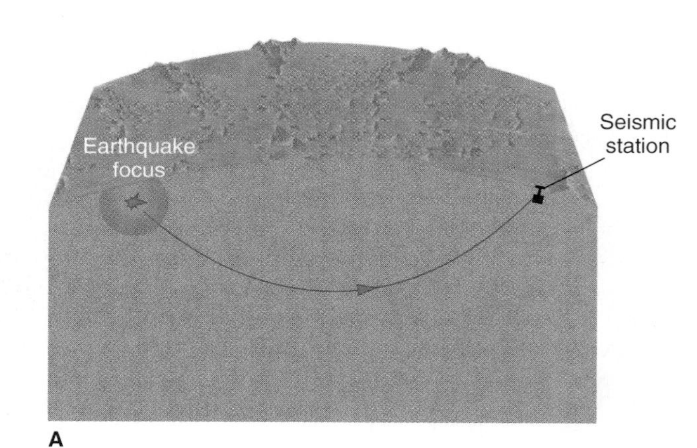

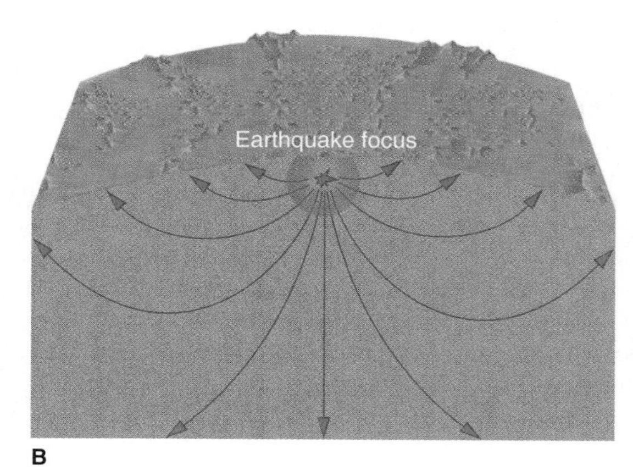

FIGURE 2.4

Curved paths of seismic waves caused by uniform rock with increasing seismic velocity with depth. (*A*) Path between earthquake and recording station. (*B*) Waves spreading out in all directions from earthquake focus.

increasing pressure with depth tends to increase the velocity of the waves. The waves follow curved paths through such a layer, as shown in figure 2.4. To understand the reason for the curving path, visualize the thick rock layer as a stack of very thin layers, each with a slightly higher velocity than the one above. The curved path results from many small changes in direction as the wave passes through the many layers.

EARTH'S INTERNAL STRUCTURE

It was the study of seismic refraction and seismic reflection that enabled scientists to plot the three main zones of Earth's interior (figure 2.5). The **crust** is the outer layer of rock, which forms a thin skin on Earth's surface. Below the crust lies the **mantle,** a thick shell of rock that separates the crust above from the core below. The **core** is the central zone of Earth. It is probably metallic and the source of Earth's magnetic field.

The Crust

Studies of seismic waves have shown (1) that the crust is thinner beneath the oceans than beneath the continents (figure 2.6) and (2) that seismic waves travel faster in oceanic crust than in continental crust. Because of this velocity difference, it is assumed that the two types of crust are made up of different kinds of rock.

Seismic P waves travel through oceanic crust at about 7 kilometers per second, which is also the speed at which they travel through basalt and gabbro (the coarse-grained equivalent of basalt). Samples of rocks taken from the sea floor by oceanographic ships verify that the upper part of the oceanic crust is basalt and suggest that the lower part is gabbro. The oceanic crust averages 7 kilometers in thickness, varying from 5 to 8 kilometers (table 2.1).

Seismic P waves travel more slowly through continental crust—about 6 kilometers per second, the same speed at which they travel through granite and gneiss. Continental crust is often called "granitic," but the term should be put in quotation marks because most of the rocks exposed on land are not granite. The continental crust is highly variable and complex, consisting of a crystalline basement composed of granite, other plutonic rocks, gneisses and schists, all capped by a layer of sedimentary rocks, like icing on a cake. Since a single rock term cannot accurately describe crust that varies so greatly in composition, some geologists use the term *felsic* (rocks high in *feldspar* and *silicon*) for continental crust and *mafic* (rocks high in magnesium and iron [ferric]) for oceanic crust.

Continental crust is much thicker than oceanic crust, averaging 30 to 50 kilometers in thickness, though it varies from 10 to 70 kilometers. Seismic waves show that the crust is thickest under geologically young mountain ranges, such as the Andes and

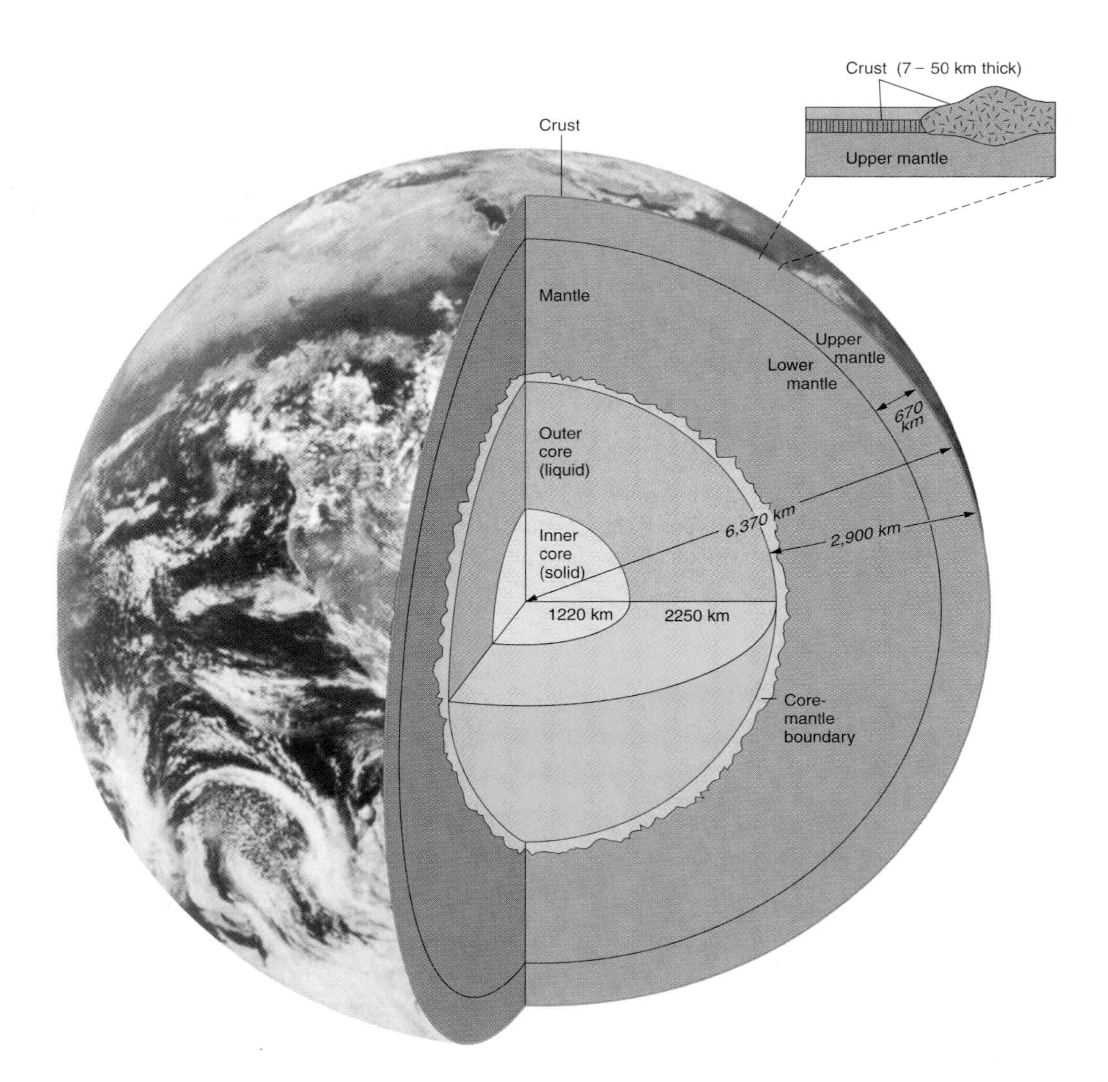

FIGURE 2.5
Earth's interior. Seismic waves show the three main divisions of Earth: the crust, the mantle, and the core.
Photo by NASA

table 2.1 Characteristics of Oceanic Crust and Continental Crust

	Oceanic Crust	Continental Crust
Average thickness	7 km	30 to 50 km (thickest under mountains)
Seismic P-wave velocity	7 km/second	6 km/second (higher in lower crust)
Density	3.0 gm/cm^3	2.7 gm/cm^3
Probable composition	Basalt underlain by gabbro	Granite, other plutonic rocks, schist, gneiss (with sedimentary rock cover)

Himalayas, bulging downward as a *mountain root* into the mantle (figure 2.6). The continental crust is also less dense than oceanic crust, a fact that is important in plate tectonics (table 2.1).

The boundary that separates the crust from the mantle beneath it is called the **Mohorovičić discontinuity** (**Moho** for short). Note from figure 2.6 that the mantle lies closer to Earth's surface beneath the ocean than it does beneath continents. The idea behind an ambitious program called Project Mohole (begun during the early 1960s) was to use specially equipped ships to drill through the oceanic crust and obtain samples from the mantle. Although the project was abandoned because of high costs, ocean-floor drilling has become routine since then, but not to the great depth necessary to sample the mantle. Perhaps in the future the original concept of drilling to the mantle through oceanic crust will be revived. (Ocean drilling is discussed in more detail in the following two chapters.)

The Mantle

Because of the way seismic waves pass through the mantle, geologists believe that it, like the crust, is made of solid rock. Localized magma chambers of melted rock may occur as isolated pockets of liquid in both the crust and the upper mantle, but most of the mantle seems to be solid. Because P waves travel at about 8 kilometers per second in the upper mantle, it appears that the mantle is a different type of rock from either oceanic crust or continental crust. The best hypothesis that geologists can make about the composition of the upper mantle is that it consists of ultramafic rock such as peridotite. *Ultramafic rock* is dense igneous rock made up chiefly of ferromagnesian minerals such as olivine and pyroxene (see box 2.2). Some ultramafic rocks contain garnet, and all of them lack feldspar.

The crust and uppermost mantle together form the **lithosphere,** the outer shell of Earth that is relatively strong and brittle. The lithosphere makes up the plates of plate tectonic theory. The lithosphere averages about 70 kilometers thick beneath oceans and may be 125 to 250 kilometers thick beneath continents. Its lower boundary is marked by a curious mantle layer in which seismic waves slow down (figure 2.6).

Generally, seismic waves increase in velocity with depth as increasing pressure alters the properties of the rock. Beginning at a depth of 70 to 125 kilometers, however, seismic waves travel more slowly than they do in shallower layers, and so this zone has been called the *low-velocity zone* (figure 2.6). This zone, extending to a depth of perhaps 200 kilometers, is also called, in plate tectonic theory, the **asthenosphere.** The rocks in this zone may be closer to their melting point than the rocks above or below the zone. (The rocks are probably not *hotter* than the rocks below—melting points are controlled by pressure as well as temperature.) Some geologists think that these rocks may actually be partially melted, forming a crystal-and-liquid slush; a very small percentage of liquid in the asthenosphere could help explain some of its physical properties.

If the rocks of the asthenosphere are close to their melting point, this zone may be important for two reasons: (1) it may represent a zone where magma is likely to be generated; and (2) the rocks here may have relatively little strength and therefore are likely to flow. If mantle rocks in the asthenosphere are weaker than they are in the overlying lithosphere, then the

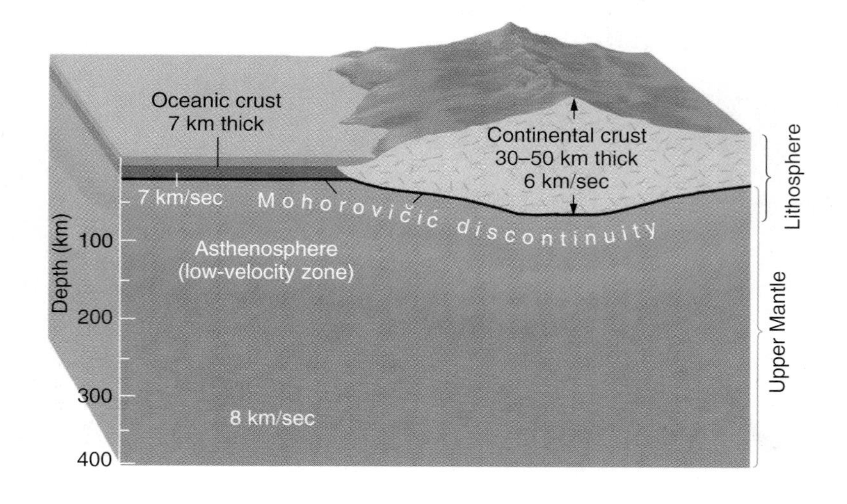

FIGURE 2.6

Thin oceanic crust has a P-wave velocity of 7 kilometers per second, whereas thick continental crust has a lower velocity. Mantle velocities are about 8 kilometers per second. The oceanic and continental crust, along with the upper rigid part of the upper mantle, form the lithosphere. The asthenosphere underlies the lithosphere and is defined by a decrease in P-wave velocities.

IN GREATER DEPTH 2.2

A CAT Scan of the Mantle

A new technique for looking at the mantle is similar to the medical technique of CAT scanning (CAT stands for computed axial tomography), which builds up a three-dimensional picture of soft body tissues such as the brain by taking a series of X-ray pictures along successive planes in the body.

Seismic tomography uses earthquake waves and powerful computers to study planar cross sections of the mantle following large earthquakes. Slight variations from expected arrival times at distant seismograph stations can be used to find temperature variations in the mantle. Hot rock slows down seismic waves, so a late arrival of a seismic wave shows that the wave went through hot rock. Cold rock is dense and strong, so it speeds up seismic waves, resulting in early arrivals. Sophisticated computer analysis of hundreds of sections through the mantle allows maps of seismic-wave velocity (and therefore mantle rock temperature) to be drawn for various depths.

Box figure 1, top shows mantle velocities at a depth of 100 kilometers. Red areas show low velocities (probably caused by hot rock) in generally expected positions—along the crest of the mid-oceanic ridge and beneath hot spots. Blue areas show high-velocity (probably cold) rock under continents and old sea floor such as the western Pacific. Box figure 1, bottom shows that these patterns are dramatically different at a depth of 300 kilometers. High-velocity rock extends to this depth below most continents, implying that continents have very deep roots. Some areas that appear hot at 100 kilometers are cold at 300 kilometers, such as the ridge crest just south of Australia. Areas such as the central Pacific and the Red Sea region appear cold at 100 kilometers and hot at 300 kilometers.

In box figure 2, vertical cross sections of seismic velocity are shown to a depth of 670 kilometers for two regions. Note in box figure 1 that high-velocity (cold) roots beneath North America, Asia, and Antarctica extend 400 to 600 kilometers downward. This finding casts doubt on our simple lithosphere-asthenosphere model of plate behavior—continental plates here seem to be hundreds of kilometers thick. Notice, too, how some low-velocity hot spots near Greenland (box figure 2, top) and in the south Atlantic and south Pacific (box figure 2, bottom) are underlain by apparently cold rock. This pattern suggests to some geologists that mantle plumes may be quite shallow and

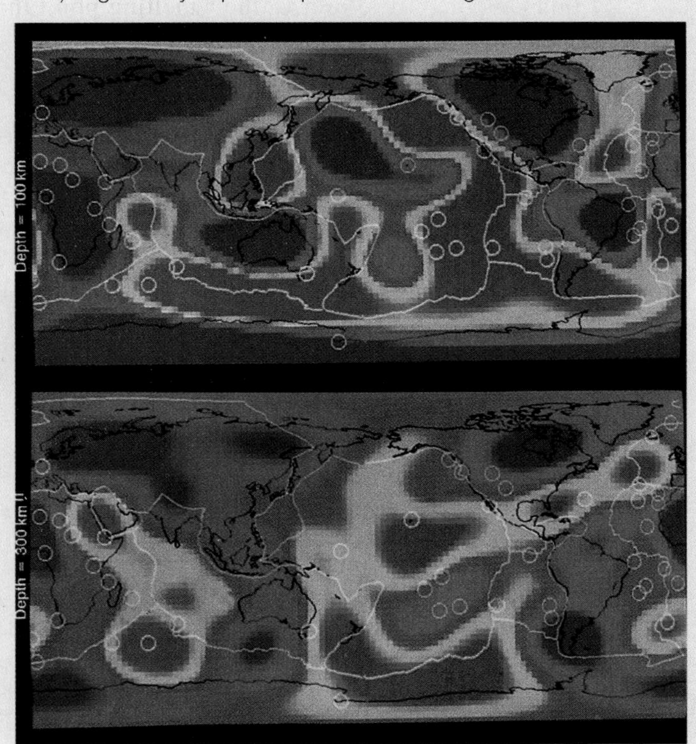

Box 2.2 — FIGURE 1

Map views of seismic-wave velocities in the mantle at depths of 100 and 300 kilometers, as determined by seismic tomography. Blue indicates high velocity (cold rock), red indicates low velocity (hot rock). White lines outline plates; white circles are major hot spots.

From Dziewonski and Anderson, *American Scientist,* 1984, 72:483–94

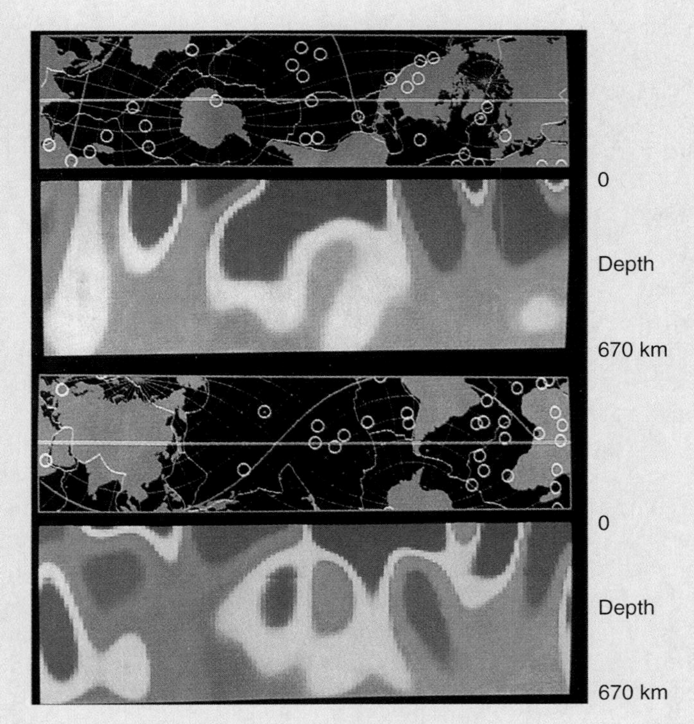

Box 2.2 — FIGURE 2

Vertical cross sections of seismic-wave velocities to a depth of 670 kilometers in the mantle. The orange lines show the locations of the cross sections.

From Dziewonski and Anderson, *American Scientist,* 1984, 72:483–94

lat.: 6.0 lon.: 125.0 lat.: −2.5 lon.: 35.0

Box 2.2 — FIGURE 3

Cross section of seismic-wave velocities from the earth's surface (upper curve) to core. Blue indicates fast seismic velocities (cold rock), and red indicates low velocities (hot rock). There is a presumed cold slab of rock, shown on the left side of the cross section, that is sinking into the lower mantle into other slabs that rest on the core-mantle boundary. Hot rocks, believed to represent mantle plumes, also emanate from the core-mantle boundary, on the right side of the cross section.

Photo courtesy of Stephen Grand, University of Texas at Austin

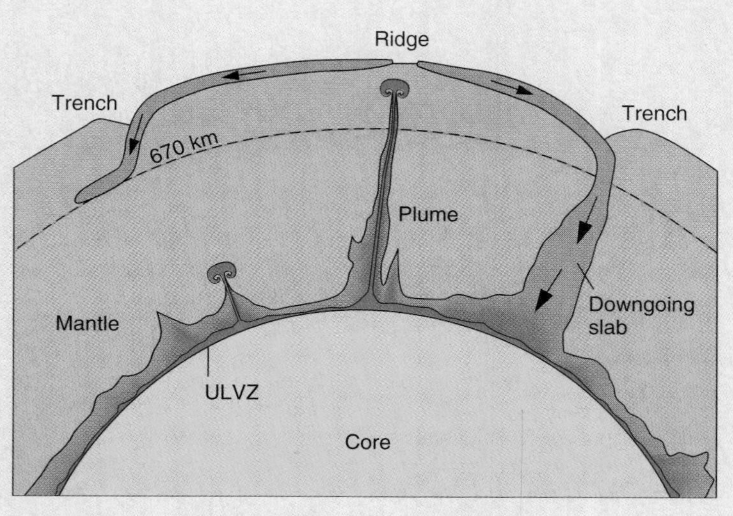

Box 2.2 — FIGURE 4

Seismic data suggest some plates sink to the base of the mantle, whereas other plates are impeded by the increase in density of the mantle at 670 km. Deep mantle plumes emanating from the core-mantle boundary are underlain by an ultra-low velocity zone (ULVZ).

may not extend vertically throughout the mantle. On the other hand, plume tails may be too narrow to be detected by this technique.

More recent, deeper CAT scans of the mantle (box figure 3) indicate that some mantle plumes emanate from the core-mantle boundary and are fed by heat loss from the core. The plume under the Hawaiian hot spot was recently found to contain material from the crust, mantle, and core. It is likely that the hot plumes originate from various depths in the mantle.

The new tomographic images also reveal high-velocity areas, which are interpreted as cold sinking slabs of subducted plates, that also extend all the way to the core-mantle boundary (box figures 3 and 4).

Other plates stop descending at the 670-kilometer boundary within the mantle. Perhaps the depth of sinking is controlled by plate density. The older the subducting rock is, the colder and denser it is. Old, dense plates may sink to the base of the mantle, while younger plates, being less dense, stop at a depth of 670 kilometers (box figure 4).

It is becoming increasingly apparent that the core-mantle boundary may play an important role in the overall mechanism of plate movement.

Additional Readings

Kerr, R. A. 1991. Do plumes stir earth's entire mantle? *Science* 252: 1068–69.

———. 1997. Deep-sinking slabs stir the mantle. *Science* 275: 613–15.

asthenosphere can deform easily by plastic flow. Plates of brittle lithosphere probably move easily over the asthenosphere, which may act as a lubricating layer below.

There is widespread agreement on the existence and depth of the asthenosphere under oceanic crust, but considerable disagreement about asthenosphere under continental crust. Figure 2.6 shows asthenosphere at a depth of 125 kilometers below the continents. Some geologists think that the lithosphere is much thicker beneath continents than shown in the figure, and that the asthenosphere begins at a depth of 250 kilometers (or even more). A few geologists say that there is *no* asthenosphere beneath continents at all. The reasons for this disagreement are the results of the rapidly developing field of seismic tomography, which is described in box 2.2.

Data from seismic reflection and refraction indicate several concentric layers in the mantle (figure 2.7), with prominent boundaries at 400 and 670 kilometers (670 kilometers is also the depth of the deepest earthquakes). It is doubtful that the layering is due to the presence of several different kinds of rock. Most geologists think that the chemical composition of the mantle rock is about the same throughout the mantle. Because pressure increases with depth into Earth, the boundaries between mantle layers possibly represent depths at which pressure collapses the internal structure of certain minerals into denser minerals. For example, at a pressure equivalent to a depth of about 670 kilometers, the mineral *olivine* should collapse into the denser structure of the mineral *perovskite.* If the boundaries between mantle layers represent pressure-caused transformations of minerals, the entire mantle may have the same *chemical* composition throughout, although not the same *mineral* composition. However, some geologists think that the 670-kilometer boundary represents a chemical change as well as a physical change and separates the *upper mantle* from the chemically different *lower mantle* below.

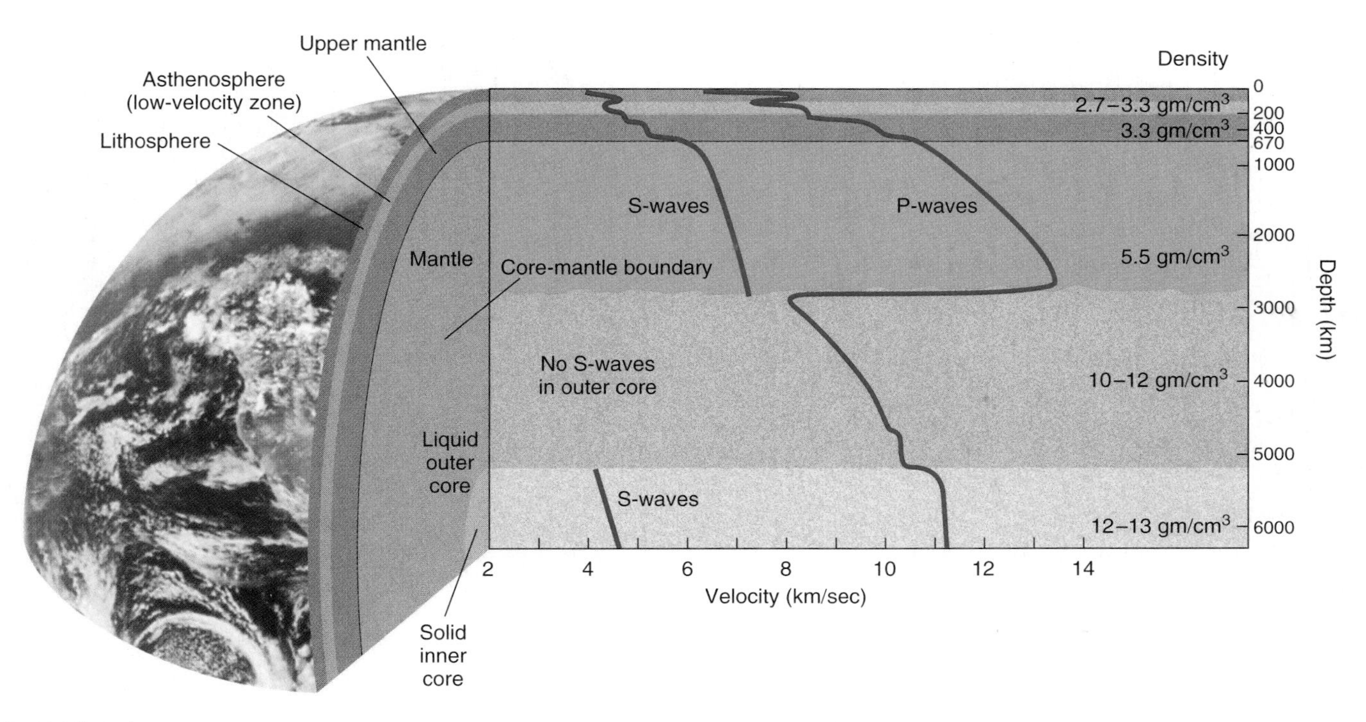

FIGURE 2.7
The concentric shell structure of Earth as defined by variation in S- and P-wave velocities and estimates of density. The velocity of seismic P and S waves generally increases with depth except in the low-velocity zone. The plastic asthenosphere slows down seismic waves. Velocity increases at 400 and 670 kilometers may be caused by mineral collapse. S waves do not pass through the outer core, but are thought to travel through the solid inner core.

The Core

Seismic-wave data provide the primary evidence for the existence of the core of Earth. (See chapter 7 for a discussion of seismic P and S waves.) Seismic waves do not reach certain areas on the opposite side of Earth from a large earthquake. Figure 2.8 shows how seismic P waves spread out from a quake until, at 103° of arc (11,500 km) from the epicenter, they suddenly disappear from seismograms. At more than 142° (15,500 km) from the epicenter, P waves reappear on seismograms. The region between 103° and 142°, which lacks P waves, is called the **P-wave shadow zone.**

The P-wave shadow zone can be explained by the refraction of P waves when they encounter the core boundary deep within Earth's interior. Because the paths of P waves can be accurately calculated, the size and shape of the core can be determined also. In figure 2.8, notice that Earth's core deflects the P waves and, in effect, "casts a shadow" where their energy does not reach the surface. In other words, P waves are missing within the shadow zone because they have been bent (refracted) by the core.

The chapter on earthquakes explains that while P waves can travel through solids and fluids, S waves can travel only through solids. As figure 2.9 shows, an **S-wave shadow zone** also exists and is larger than the P-wave shadow zone. Direct S waves are not recorded in the entire region more than 103° away from the epicenter. The S-wave shadow zone seems to indicate that S waves do not travel through the core at all. If this is true, it implies that the core of Earth is a liquid, or at least acts like a liquid.

The way in which P waves are refracted within Earth's core (as shown by careful analysis of seismograms) suggests that the core has two parts, a *liquid outer core* and a *solid inner core* (figure 2.7).

Composition of the Core

When evidence from astronomy and seismic-wave studies is combined with what we know about the properties of materials, it appears that Earth's core is made of metal—not silicate rock—and that this metal is probably iron (along with a minor amount of oxygen, silicon, sulfur, or nickel). How did geologists arrive at this conclusion?

The overall density of the earth is 5.5 gm/cm^3, based on calculations from Newton's law of gravitational attraction. The crustal rocks are relatively low density, from 2.7 gm/cm^3 for granite to 3.0 gm/cm^3 for basalt. The ultramafic rock thought to make up the mantle probably has a density of 3.3 gm/cm^3 in the upper mantle, although rock pressure should raise this value to about 5.5 gm/cm^3 at the base of the mantle (figure 2.7).

If the crust and the mantle, which have approximately 85% of Earth's volume, are at or below the average density of Earth, then the core must be very heavy to bring the average up to 5.5 gm/cm^3.

Calculations show that the core has to have a density of about 10 gm/cm^3 at the core-mantle boundary, increasing to 12 or 13 gm/cm^3 at the center of Earth (figure 2.7). This great density would be enough to give Earth an average density of 5.5 gm/cm^3.

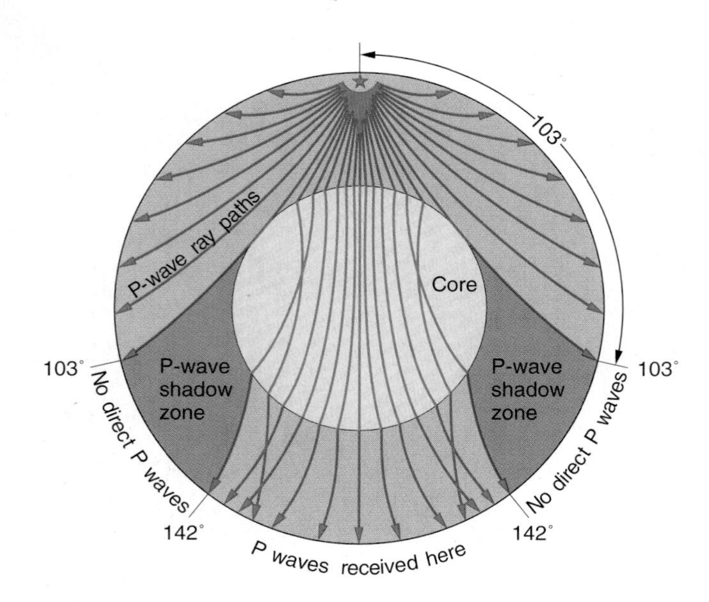

FIGURE 2.8

The P-wave shadow zone, caused by refraction of P waves within the Earth's core.

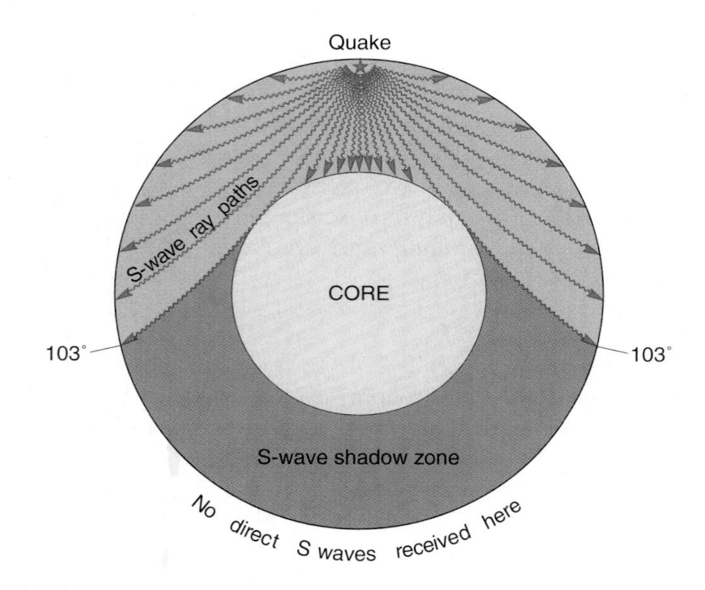

FIGURE 2.9

The S-wave shadow zone. Because no S waves pass through the core, the core is apparently a liquid (or acts like a liquid).

Under the great pressures existing in the core, iron would have a density slightly greater than that required in the core. Iron mixed with a small amount of a lighter element, such as oxygen, sulfur or silicon, would have the required density. Therefore, many geologists think that such a mixture makes up the core.

But a study of density by itself is hardly convincing evidence that the core is mostly iron, for many other heavy substances could be there instead. The choice of iron as the major

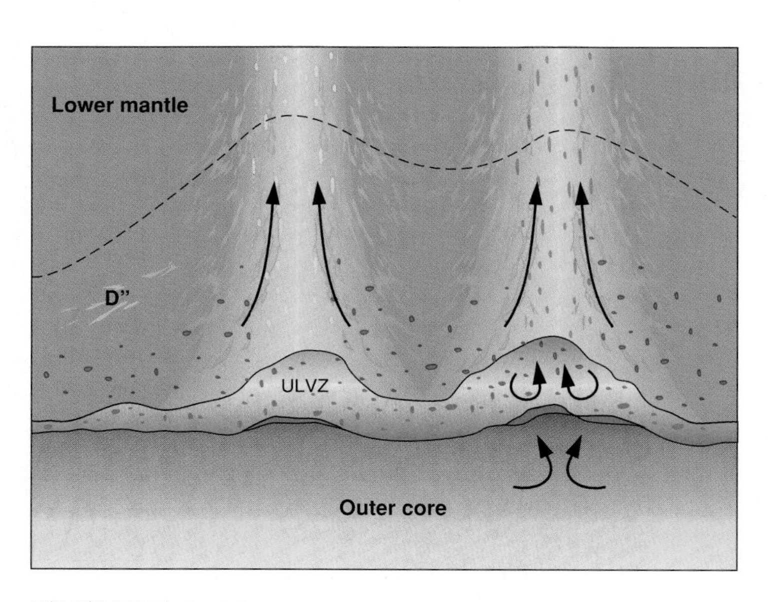

FIGURE 2.10

Recent seismic and geodetic studies are redefining the boundary between the lower 200 kilometers of the mantle (D layer) and the outer core. Iron silicate "sediments" (shown in brown) may rise from the underlying liquid core and fill pockets or inverted basins at the core-mantle boundary (CMB). Alternatively, the outer core material (shown in red) may be melting the lowermost mantle (shown in yellow) to form the ultra-low velocity zone (ULVZ). Modified from Garnero and Jeanloz, *Science*, 2000

component of the core comes from looking at meteorites (see box 2.3). Meteorites are thought by some scientists to be remnants of the basic material that created our own solar system. An estimated 10% of meteorites are composed of iron mixed with small amounts of nickel. Material similar to these meteorites may have helped create the earth, perhaps settling to the center of Earth because of metal's high density. The composition of these meteorites, then, may tell us what is in the core. Nickel is denser than iron, however, so a mixture of just iron and nickel would have a density greater than that required in the core. (The other 90% of meteorites are mostly ultramafic rock and perhaps represent material that formed the mantle.)

Seismic and density data, together with assumptions based on meteorite composition, point to a core that is largely iron, with at least the outer part being liquid. The existence of Earth's magnetic field, which is discussed later in this chapter, also suggests a metallic core. Of course, no geologist has seen the core, nor is anyone likely to in the foreseeable future. But since so many lines of indirect evidence point to a liquid metal outer core, most scientists accept this theory as the best conclusion that can be made about the core's composition.

The Core-Mantle Boundary

The boundary between the core and mantle is marked by great changes in seismic velocity density (figure 2.7), and temperature, as we see later in the chapter. Here there is a transition zone up to 200 kilometers thick, known as the *D layer,* at the base of the mantle where P-wave velocities decrease dramatically. The *ultra-low velocity zone* (ULVZ)

ASTROGEOLOGY 2.3

Meteorites

Small solid particles of rock, metal, and/or ice orbiting the Sun are called *meteoroids.* When these particles enter Earth's atmosphere, they are heated to incandescence by friction; these glowing particles are called *meteors* (or "shooting stars" or "falling stars"). Most meteors are small and burn up while still in the atmosphere, but about 150 per year are large enough to strike Earth's surface. Those that do are called *meteorites* (box figure 1). The largest fragment of a meteorite found (in South Africa) weighs 50 tons; much larger meteorites have hit Earth in the past.

Three basic types of meteorite are iron, stony-iron, and stony meteorites. Stony meteorites are by far the most common, but they look like Earth rocks, so they are hard to find. Iron meteorites are rare, but look so unique that they are commonly found; most museum meteorites are of the iron type.

Iron meteorites are mostly iron alloyed (mixed) with a small percentage of nickel. Small amounts of other metals or minerals may be present. Iron-nickel meteorites give an important source of information regarding the composition of Earth's core.

Stony-iron meteorites are made of iron-nickel alloy and silicate minerals in about equal parts.

Stony meteorites are made of silicate minerals such as plagioclase, olivine, and pyroxene; they may contain a small amount of iron-nickel alloy. About 90% of stony meteorites contain round silicate grains called *chondrules* and are called *chondrites.* The other 10% are *achondrites,* which lack chondrules.

Chondrules consist mostly of olivine and pyroxene, and range from distinct spheres to large bodies with fuzzy outlines. The composition of chondrite meteorites resembles the ultramafic rock peridotite, but peridotite lacks the chondritic texture and iron-nickel content of the meteorites.

One kind of chondrite is composed mostly of serpentine or pyroxene and contains up to 5% organic materials, including carbon, hydrocarbon compounds, and amino acids. These meteorites are called *carbonaceous chondrites.* All available evidence indicates that the organic compounds were in fact produced by inorganic processes. Carbonaceous chondrites are of particular interest to scientists because they are believed to have the same composition as the original material from which the solar system was formed.

Box 2.3 — FIGURE 1
A large meteorite from Meteor Crater, Arizona. Pocketknife for scale.
Photo by Frank M. Hanna

Achondrites are generally similar to terrestrial rocks in composition and texture. In composition they are most similar to basalts. Some have textures like ordinary igneous rocks, and others are breccias with fragments of different compositions and textures.

The origin of meteorites is controversial. Many meteorites have a coarse-grained texture, probably formed by slow cooling within a larger body, such as a planet. The similarity in iron-nickel composition among iron meteorites also suggests that they are fragments from a single, large body. The larger body may have differentiated into a heavy, iron-rich core and a lighter, rocky mantle before it fragmented into meteoroids. Isotopic dating shows that most meteorites have the same age, 4.6 billion years old. No terrestrial rocks have ages greater than 3.8–3.9 billion years; therefore, meteorites provide the best clue as to the age of the solar system and the formation of the planets.

(figure 2.10) that forms the undulating border at the core-mantle boundary may be due to hot core partially melting overlying mantle rock or could be due to part of the liquid outer core reacting chemically with the adjacent mantle. The latest seismic and geodetic studies have hinted that lighter iron alloys from the liquid outer core may react with silicates in the lower mantle to form iron silicates. The less dense iron silicate "sediment," along with liquid iron in pore spaces, rise and collect in uneven layers along the core-mantle boundary. The pressure of the accumulating "sediment" along the boundary causes some of the liquid iron to be squeezed out of the pore spaces to form an electrically conductive layer that connects the core and mantle and explains the decrease in seismic velocities at the ULVZ. It may be difficult to prove whether the lowermost mantle is being partially melted by the core or whether the core is instead chemically reacting with the mantle.

Both the mantle and the core are undergoing **convection,** a circulation pattern in which low density material rises and high density material sinks. Based on seismic tomography studies, heavy portions of the mantle (including subducted plates) sink to its base, but are unable to penetrate the denser core. Light portions of the core may rise to its top, and may be incorporated into the mantle above. This is suggested by recent isotopic studies of the mantle plume that feeds the Hawaiian hot spot. The resulting Hawaiian volcanic rocks (basalts) contain a light isotope signature that is characteristic of the core. Continent-sized blobs of liquid and liquid-crystal slush may accumulate at the core-mantle boundary, perhaps interfering with or helping cause heat loss from the core to help drive mantle convection and transfer of heat to the surface, and also causing changes in Earth's magnetic field. This boundary is an exciting frontier for geologic study, but data, of course, are sparse and hard to obtain.

ISOSTASY

Isostasy is a balance or *equilibrium* of adjacent blocks of brittle crust "floating" on the upper mantle. Since crustal rocks weigh less than mantle rocks, the crust can be thought of as floating on the denser mantle much as wood floats on water (figure 2.11).

Blocks of wood floating on water rise or sink until they displace an amount of water equal to their own mass weight. The weight of the displaced water buoys up the wood blocks, allowing them to float. The higher a wood block appears above the water surface, the deeper the block extends under water. Thus a tall block has a deep "root."

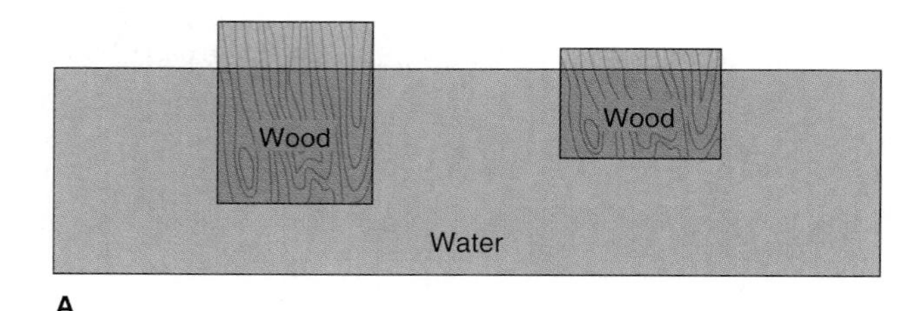

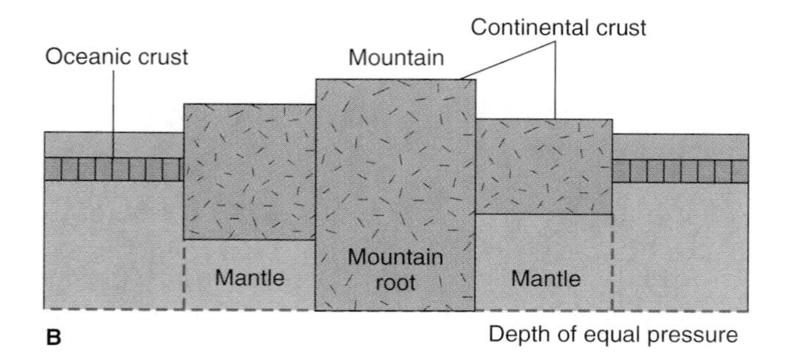

FIGURE 2.11

Isostatic balance. (*A*) Wood blocks float in water with most of their bulk submerged. (*B*) Crustal blocks "float" on mantle in approximately the same way. The thicker the block, the deeper it extends into the mantle.

In a greatly simplified way, crustal rocks can be thought of as tending to rise or sink gradually until they are balanced by the weight of displaced mantle rocks. This concept of vertical movement to reach equilibrium is called **isostatic adjustment.** Just as with the blocks of wood, once crustal blocks have come into isostatic balance, a tall block (a mountain range) extends deep into the mantle (a *mountain root,* as shown in figure 2.11).

Figure 2.11 shows both the blocks of wood and the blocks of crustal rock in isostatic balance. The weight of the wood is equal to the weight of the displaced water. Similarly, the weight of the crustal blocks is equal to the weight of the displaced mantle. As a result, the rocks (and overlying sea water) in figure 2.11 can be thought of as separated into vertical columns, each with the same pressure at its base. At some *depth of equal pressure* each column is in balance with the other columns, for each column has the same weight. A column of thick continental crust (a mountain and its root) has the same weight as a column containing thin continental crust and some of the upper mantle. A column containing sea water, thin oceanic crust, and a thick section of heavy mantle weighs the same as the other two columns.

Figure 2.11 shows the crust as isolated blocks free to move past each other along vertical faults, but this is not really a good picture of crustal structure. It is more accurate to think of the crust as bending in broad uplifts and downwarps without vertical faults, as shown in figures 2.12 and 2.13.

Let us look at some examples of isostatic balance (equilibrium) in crustal rocks. Suppose that two sections of crust of unequal thickness are next to each other, as in figure 2.12. Sediment from the higher part, which is subject to more rapid erosion, is deposited on the lower part. The decrease in weight from the high part causes it to rise, while the increase in weight on the low part causes it to sink. These vertical movements (isostatic adjustment) do in fact take place whenever large volumes of material are eroded from or deposited on parts of the crust.

Rising or sinking of the crust, of course, requires plastic flow of the mantle to accommodate the motion. By measuring the rate of rising or sinking, the viscosity of the mantle can be calculated. The plastic flow of the mantle probably takes place within the asthenosphere in the upper mantle.

Another example of isostatic adjustment, caused by plastic mantle flow, is the upward movement of large areas of the crust since the glacial ages. The weight of the thick continental ice sheets during the Pleistocene Epoch depressed the crust underneath the ice (figure 2.13). After the melting of the ice, the crust rose back upward, a process that is still going on in some areas (figure 2.14). This rise of the crust after the removal of the ice is known as **crustal rebound.**

Isostatic adjustment may occur at subduction zones. As explained in the chapter on igneous rocks, a subducting plate may generate molten magma, which sometimes rises all the way to the surface to erupt as lava. However, some geologists

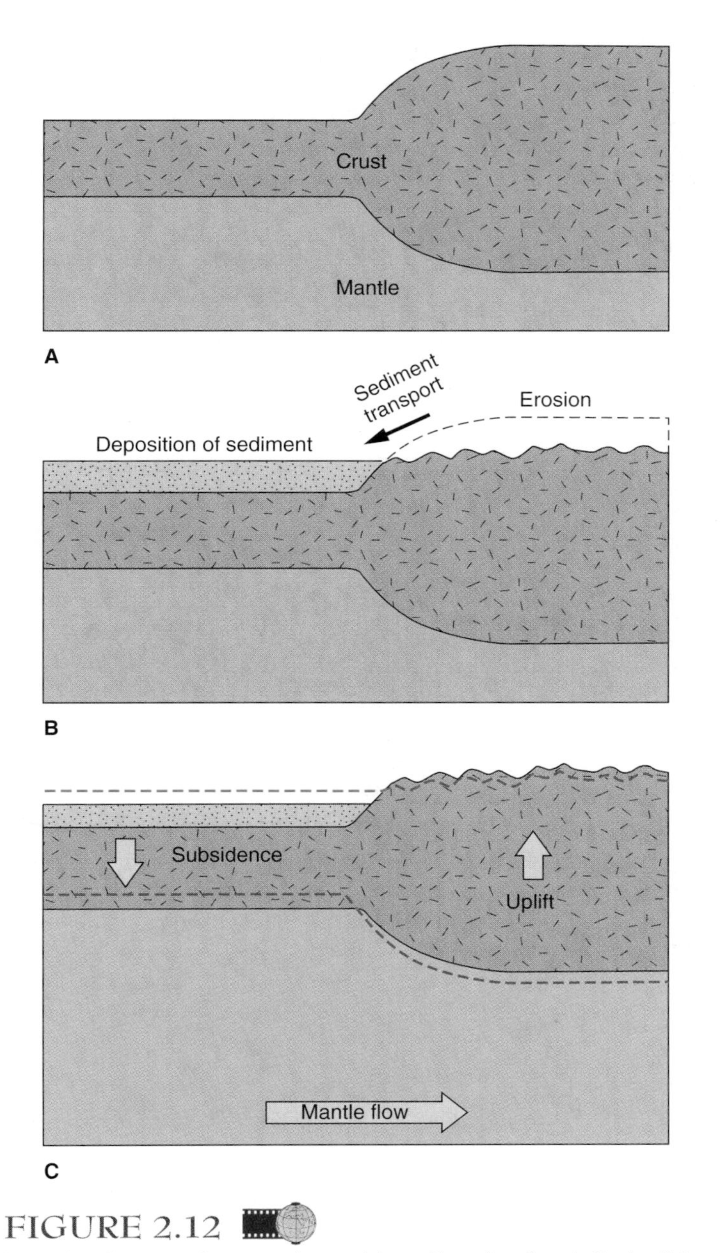

FIGURE 2.12

Isostatic adjustment due to erosion and deposition of sediment. Rock within the mantle must flow to accommodate vertical motion of crustal blocks. Mantle flow occurs in the asthenosphere, deeper than shown in *C*.

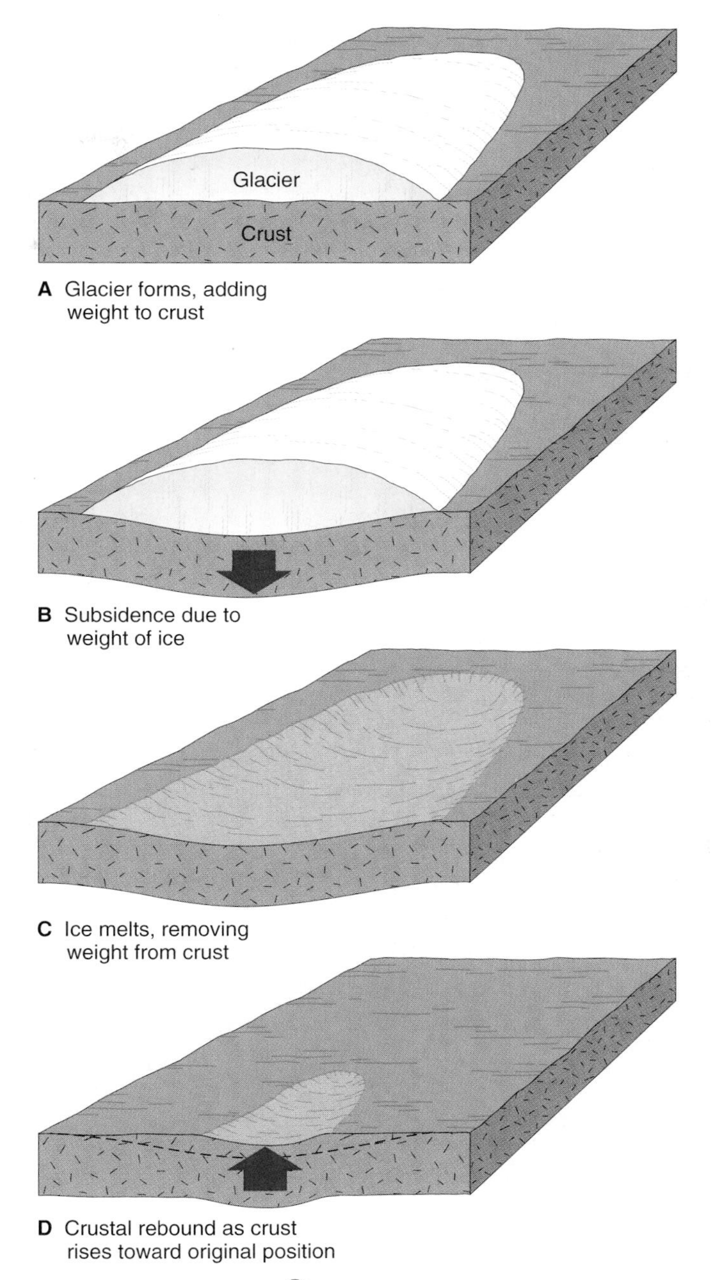

FIGURE 2.13

The weight of glaciers depresses the crust, and the crust rebounds when the ice melts.

believe that in certain cases the magma stops at the base of an overlying continent (or perhaps within the continent near its base). This accumulation of magma can locally thicken the continent when the magma cools (figure 2.15*A*). The thickening of continents from below (a theory not accepted by all geologists) causes the crust to be out of isostatic equilibrium. So it rises, reestablishing equilibrium and forming a mountain range (figure 2.15*B*).

Recent geophysical studies have shown that some mountains, such as the Rockies and southern Sierra Nevada, do not have thick roots and are instead buoyed by warm, less dense mantle. It appears that the upper mantle beneath some continents is not homogeneous, but has zones that are quite buoyant due to higher temperatures and less dense mineral phases.

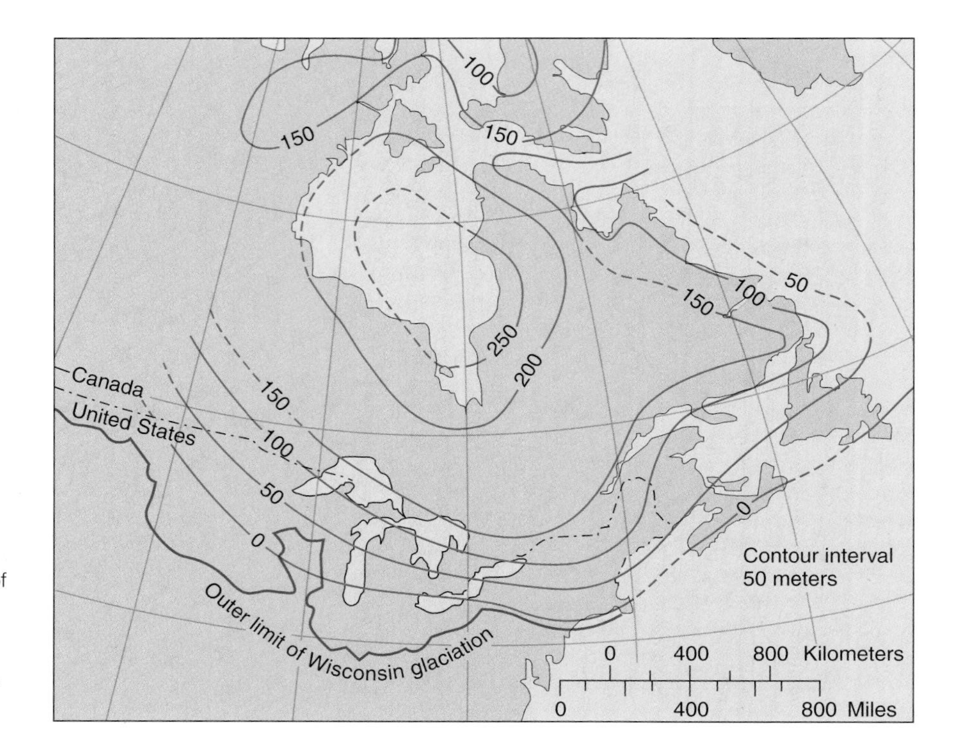

FIGURE 2.14

Uplift of land surface in Canada and the northern United States caused by crustal rebound after glaciers melted. Colored lines show the amount of uplift in meters since the ice disappeared.

From Phillip B. King, "Tectonics of Quaternary Time in Middle North America," in *The Quaternary of the United States,* H. D. Wright, Jr. and David G. Frey, eds., fig. 4A, p. 836. Reprinted by permission of Princeton University Press

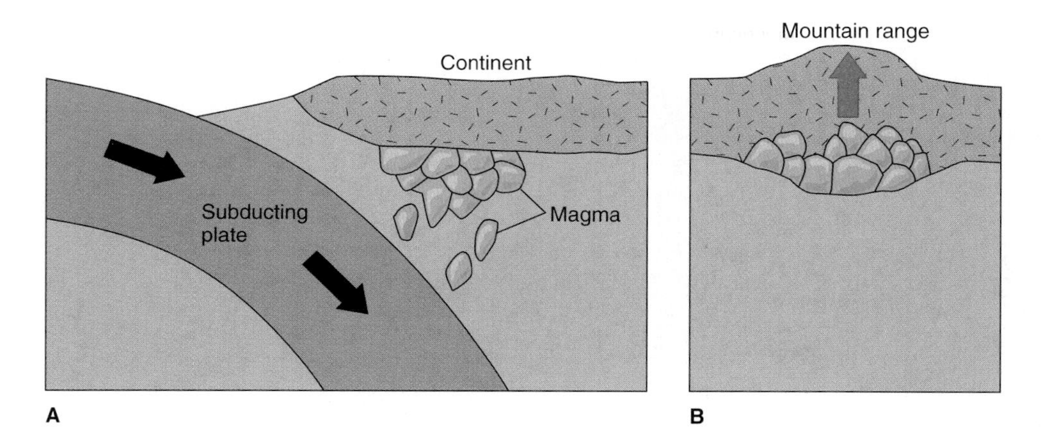

FIGURE 2.15

Isostatic uplift as a result of crustal thickening from below. (*A*) Rising blobs of magma accumulate at the base of a continent, thickening the crust (base of lithosphere below continent not shown). (*B*) Isostatic uplift of thickened continental crust produces a mountain range.

GRAVITY MEASUREMENTS

The force of gravity between two objects varies with the masses of the objects and the distance between them (figure 2.16):

$$\text{Force of gravity between } A \text{ and } B = \text{constant}\left(\frac{\text{mass}_A \times \text{mass}_B}{\text{distance}^2}\right)$$

The force increases with an increase in either mass. The gravitational attraction between Earth and the moon, for example, is vastly greater than the extremely small attraction that exists between two bowling balls. The equation also shows that force decreases with the square of the distance between the two objects. The farther two objects are apart, the less gravitational attraction there is between them.

A useful tool for studying the crust and upper mantle is the **gravity meter,** which measures the gravitational attraction between Earth and a mass within the instrument. One use of the gravity meter is to explore for local variations in rock density (mass = density × volume). Dense rock such as metal ores and ultramafic rock pulls strongly on the mass inside the meter (figure 2.17). The strong pull stretches a spring, and the amount of stretching can be very precisely determined. So a gravity meter can be used to explore for metallic ore deposits. A cavity or a

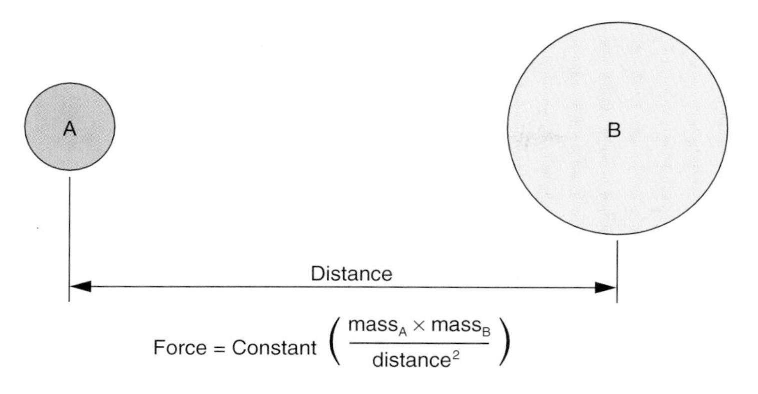

FIGURE 2.16

The force of gravitational attraction between two objects is a function of the masses of the objects and the distance between the centers of the objects.

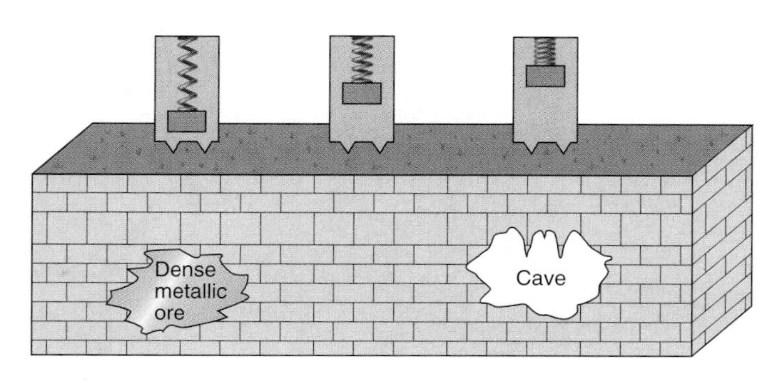

FIGURE 2.17

A gravity meter reading is affected by the density of the rocks beneath it. Dense rock pulls strongly on the mass within the meter, stretching a spring; a cavity exerts a weak pull on the mass. A gravity meter can be used to explore for hidden ore bodies, caves, and other features that have density contrasts with the surrounding rock.

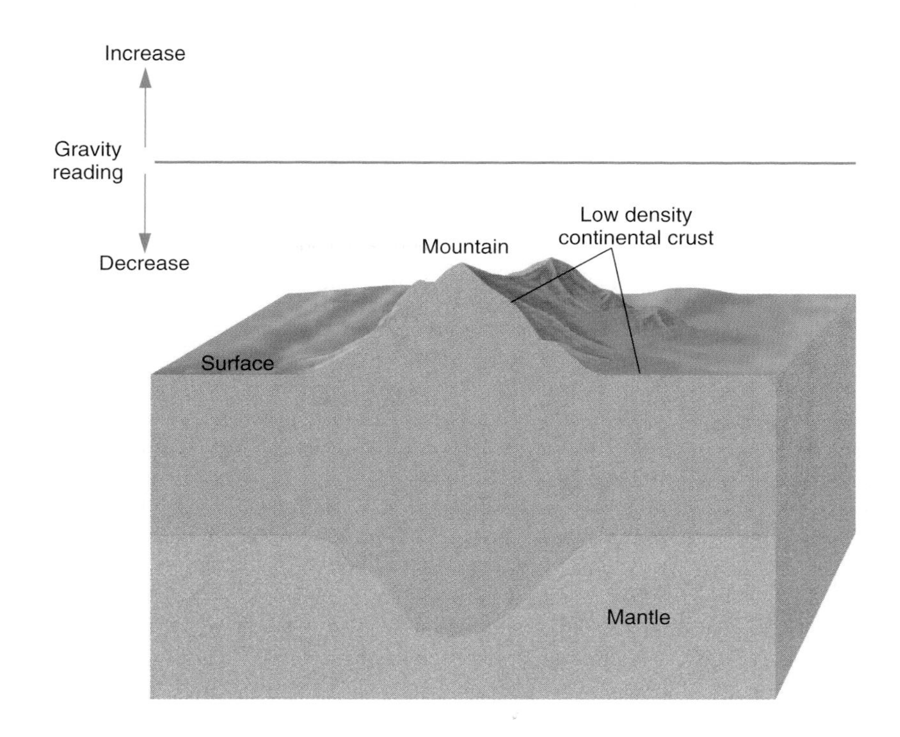

FIGURE 2.18

A region in isostatic balance gives a uniform gravity reading (no gravity anomalies).

body of low-density material such as sediment causes a much weaker pull on the meter's mass (figure 2.17). The use of a gravity meter to explore for salt domes and their associated traps for oil and gas is shown in box 21.1.

Another important use of a gravity meter is to discover whether regions are in isostatic equilibrium. If a region is in isostatic balance, as in figure 2.18, each column of rock has the same mass. If a gravity meter were carried across the rock columns, it would register the same amount of gravitational attraction for each column (after correcting for differences in elevation—gravitational attraction is less on a mountaintop than at sea level because the mountaintop is farther from the center of Earth).

Some regions, however, are held up out of isostatic equilibrium by deep tectonic forces. Figure 2.19 shows a region with uniformly thick crust. Tectonic forces are holding the center of the region up. This uplift creates a mountain range without a mountain root. There is a thicker section of heavy mantle rock under the mountain range than there is on either side of the mountain range. Therefore, the central "column" has more mass than the neighboring columns, and a gravity meter shows that the gravitational attraction is correspondingly greater over the central than over the side columns.

A gravity reading higher than the normal regional gravity is called a **positive gravity anomaly** (figure 2.19). It can indicate that tectonic forces are holding a region up out of isostatic

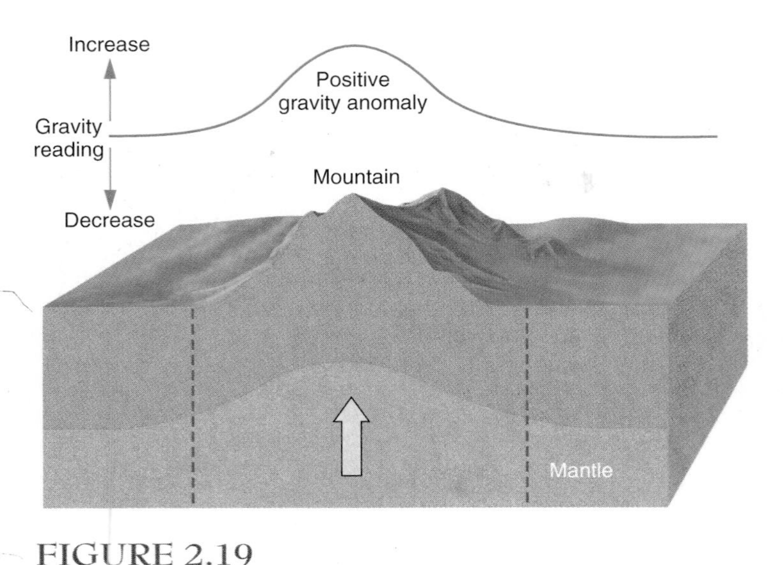

FIGURE 2.19

A region being held up out of isostatic equilibrium gives a positive gravity anomaly.

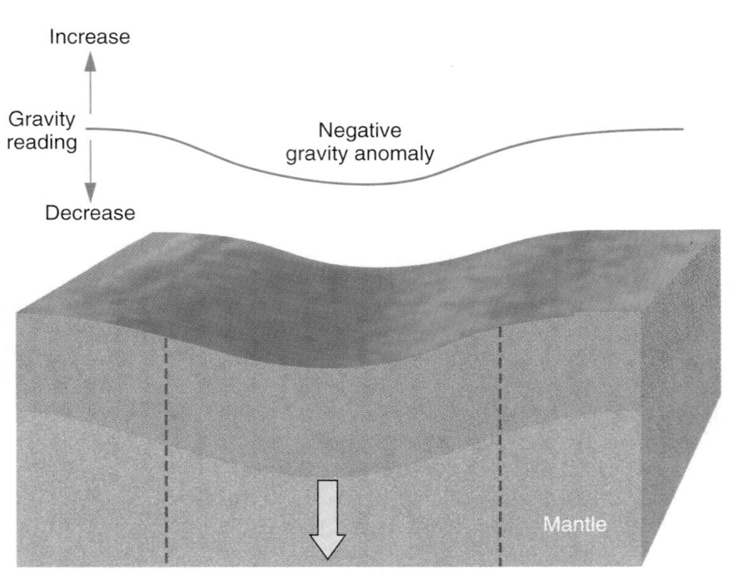

FIGURE 2.20

A region being held down out of isostatic equilibrium gives a negative gravity anomaly.

equilibrium, as shown in figure 2.19. When the forces stop acting, the land surface sinks until it reestablishes isostatic balance. The gravity anomaly then disappears. For the region shown in figure 2.19, equilibrium will be established when the land surface becomes level.

Positive gravity anomalies, particularly small ones, are also caused by local concentrations of dense rock such as metal ore. The gravity meter in figure 2.17 is registering a positive gravity anomaly over ore (the spring inside the meter is stretched). Since there can be more than one cause of a positive gravity anomaly, geologists may disagree about the interpretation of anomalies. Drilling into a region with a gravity anomaly usually discloses the reason for the anomaly.

A region can also be held down out of isostatic equilibrium, as shown in figure 2.20. The mass deficiency in such a region produces a **negative gravity anomaly**—a gravity reading lower than the normal regional gravity. Negative gravity anomalies indicate either that a region is being held down (figure 2.20) or that local mass deficiencies exist for other reasons (figure 2.17).

The greatest negative gravity anomalies in the world are found over oceanic trenches (see next two chapters). These negative anomalies are interpreted to mean that trenches are actively being held down and are out of isostatic balance.

THE EARTH'S MAGNETIC FIELD

A region of magnetic force—a **magnetic field**—surrounds Earth. The invisible lines of magnetic force surrounding Earth deflect magnetized objects, such as compass needles, that are free to move. The field has north and south **magnetic poles,** one near the geographic North Pole, the other near the geographic South Pole. (Because it has two poles, Earth's field is called *dipolar.*) The strength of the magnetic field is greatest at the magnetic poles where magnetic lines of force appear to leave and enter Earth vertically (figure 2.21).

Because the compass is important in navigation, Earth's magnetism has been observed for centuries. It has long been known that the magnetic poles are displaced about $11^1/_2°$ from the geographic poles (about which Earth rotates). Furthermore, changes in the position of the magnetic poles have been well documented, especially since the time of the great explorations of the globe. Because Earth's field is not 100% dipolar, the magnetic poles appear to be moving slowly around the geographic poles. The separation between the two types of poles has probably never been much greater than it is today.

More recently, geophysical studies have been directed toward the *source* of Earth's magnetism. The rate of the poles' changes in position, together with the strength of the magnetic field, strongly suggest that the magnetic field is generated within the liquid metal of the outer core rather than within the solid rock of the crust or the mantle.

How is Earth's magnetic field generated? A number of hypotheses have been put forth. One widely accepted hypothesis suggests that the magnetic field is created by electric currents within the liquid outer core. The outer core is extremely hot and flows at a rate of several kilometers per year in large convection currents, about one million times faster than mantle convection above it. Convecting metal creates electric currents, which in turn create a magnetic field. This hypothesis requires the core to be an electrical conductor. Metals are good conductors of electricity, whereas silicate rock is generally a poor electrical conductor. Indirectly, this is evidence that the core is metallic.

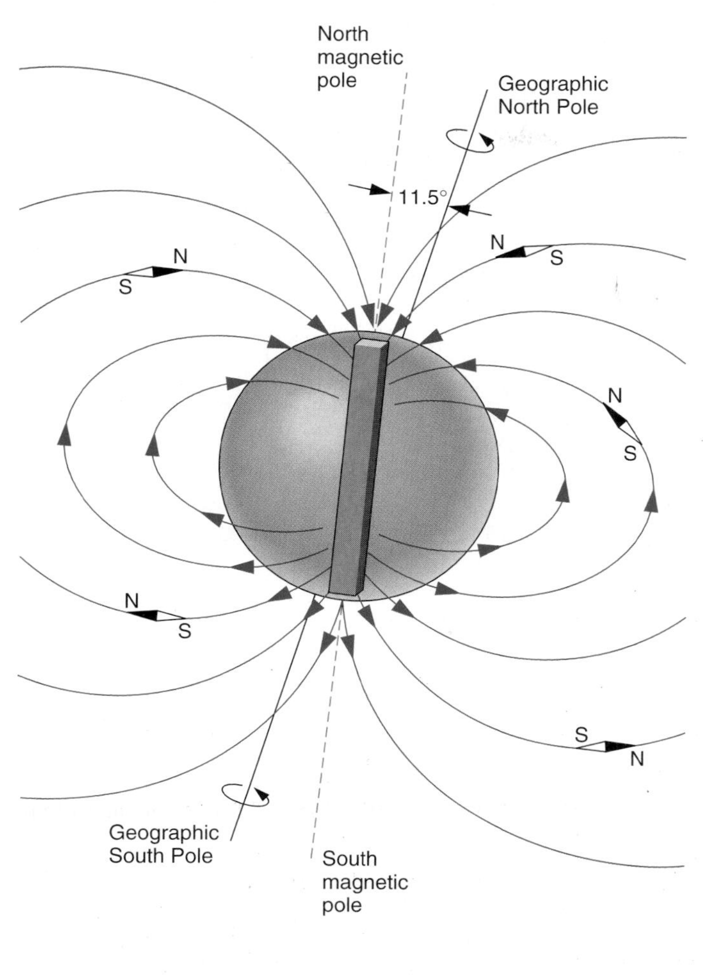

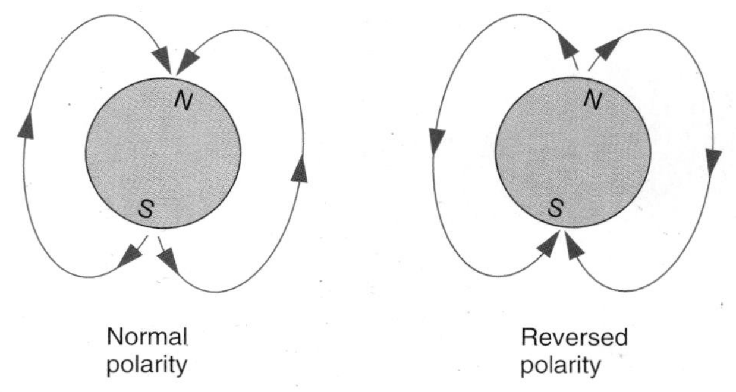

FIGURE 2.21

The Earth's magnetic field. The depiction of the internal field as a large bar magnet is a simplification of the real field, which is more complex. N and S in the two small figures indicate the *geographic* poles.

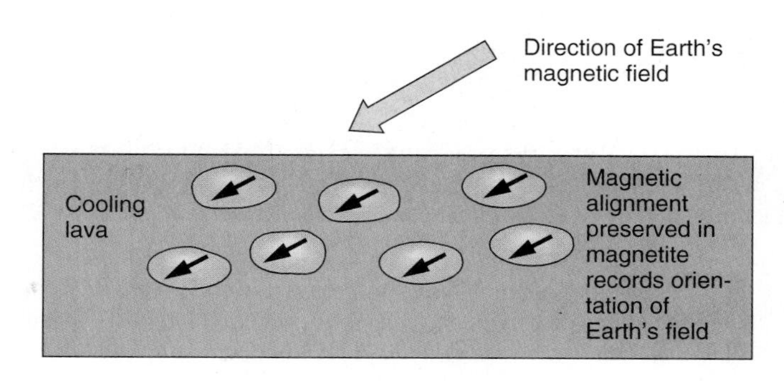

FIGURE 2.22

Some rocks preserve a record of Earth's magnetic field.

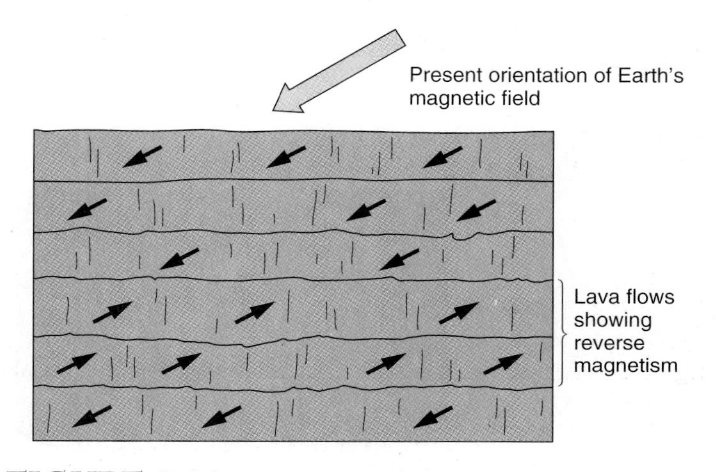

FIGURE 2.23

Cross section of stacked lava flows showing evidence of magnetic reversals.

Magnetic Reversals

In the 1950's evidence began to accumulate that Earth's magnetic field has periodically reversed its polarity in the past. Such a change in the polarity of the magnetic field is a **magnetic reversal.** During a time of *normal polarity,* magnetic lines of force leave Earth near the geographic South Pole and reenter near the geographic North Pole (figure 2.21). This orientation is called "normal" polarity because it is the same as the present polarity. During a time of *reversed polarity,* the magnetic lines of force run the other way, leaving Earth near the North Pole and entering near the South Pole (figure 2.21). In other words, during a magnetic reversal, the north magnetic pole and the south magnetic pole exchange positions.

Many rocks contain a record of the strength and direction of the magnetic field *at the time the rocks formed.* When the mineral magnetite, for example, is crystallizing in a cooling lava flow, the atoms within the crystals respond to Earth's magnetic field and form magnetic alignments that "point" toward the north magnetic pole. As the lava cools slowly below the **Curie point** (580°C for magnetite), this magnetic record is permanently trapped in the rock (figure 2.22). Unless the rock is heated again above the Curie point or temperature, this magnetic record is retained, and when studied reveals the direction of Earth's magnetic field at the time the lava cooled. Other rock types, including sedimentary rocks stained red by iron compounds, also record former magnetic field directions. The study of ancient magnetic fields is called **paleomagnetism.**

Most of the evidence for magnetic reversals comes from lava flows on the continents. Paleomagnetic studies of a series of stacked lava flows often show that some of the lava flows have a magnetic orientation directly opposite to Earth's present orientation (figure 2.23). That is, at the time these lava flows cooled, the

2.4

IN GREATER DEPTH

Earth's Spinning Inner Core

Recent studies have led to a new understanding of the dynamics of Earth's inner core and generation of Earth's magnetic field and periodic magnetic reversals. Gary A. Glatzmaier, of Los Alamos National Laboratory in New Mexico and Paul H. Roberts of the University of California, Los Angeles developed a very sophisticated computer model of convection in the outer core that has been successful in simulating a magnetic field very similar to that measured on Earth. The model utilizes circulating metallic fluids in the outer core, caused by cooling and heat loss, as the driving force of Earth's magnetic field. The circulation of metallic fluids in the outer core has been theorized for many years, and the computer model was successful in simulating and maintaining a magnetic field similar to that measured on Earth. The model also predicted that Earth's solid inner core spins faster than the rest of the planet, gaining a full lap on the rest of the planet every 150 years. Because the magnetic lines of force penetrate and connect both the inner and outer core, a faster rate of rotation of the inner core would play an important role in the generation of Earth's magnetic field and may also influence periodic magnetic reversals. Interestingly, Glatzmaier and Robert's model produced a magnetic reversal on its own without any additional input from the experimenters after about 35,000 years of simulated time (box figure 1).

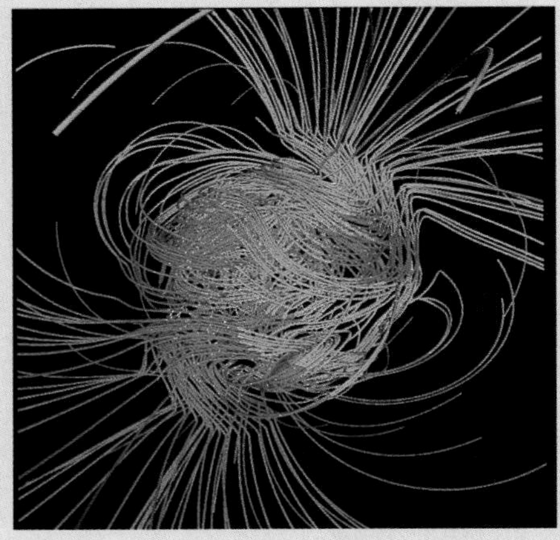

A

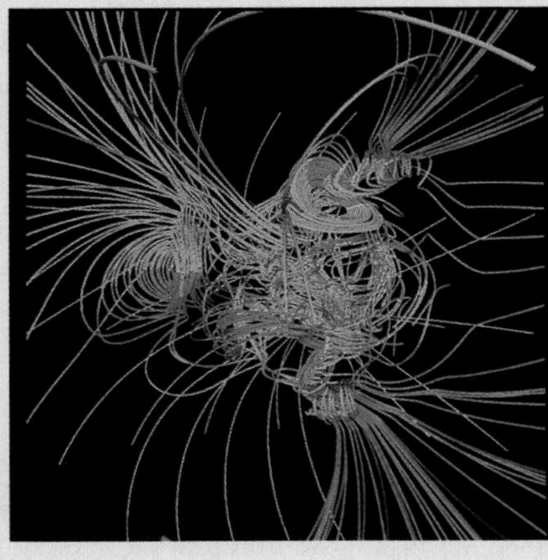

B

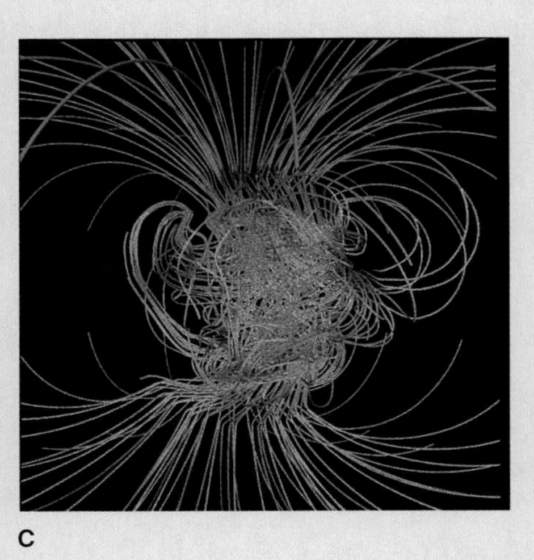

C

Box 2.4 — FIGURE 1

Computer simulation of Earth's magnetic field and magnetic reversal. (*A*) Reversed magnetic polarity with magnetic field lines leaving the north magnetic pole (orange) and reentering at the south pole. (*B*) Transitional magnetic field. (*C*) Normal magnetic field.

Photos from the Geodynamo Computer Simulation, courtesy of G. A. Glatzmaier, Los Alamos National Laboratory, and P. H. Roberts, University of California, Los Angeles

The results of this computer model inspired seismologists Xiao Dong Song and Paul Richards from Columbia's Lamont-Doherty Earth Observatory to look for evidence that the inner core actually spins at a more rapid rate than the rest of the planet. The seismologists knew that previous studies suggested seismic waves pass through the inner core faster along a nearly north-south route. This faster route, or high velocity pathway, is similar to the grain in a piece of wood. This pathway is not aligned directly with the inner core's spin axis but is tilted about 10 degrees from it (box figure 2). Seismic waves tend to travel slower along other paths, such as in an east-west direction parallel to the inner core's equator.

The seismologists studied seismic wave records from 38 separate, closely spaced earthquakes from 1967 to 1995 near the Sandwich Islands, off Argentina to determine how long it took them to reach a monitoring station in College, Alaska. The waves all took about the same amount of time to reach Alaska; however, the seismic waves in the 1990s arrived in Alaska about 0.3 seconds faster than the seismic waves in the 1960s. Since the seismic waves would have traveled through the inner core, the seismologists have explained the difference in travel time as indicating that the inner core had changed its position relative to the monitoring station in Alaska. That is, the inner core and the high velocity pathway had rotated slightly with respect to the rest of the planet.

Seismologists at Harvard looked at additional earthquake records and calculated that the inner core is rotating at approximately the same rate as Glatzmaier and Robert's model predicted. Future studies to examine earthquake records over a longer period of time are needed to confirm whether the inner core has been spinning faster than the rest of the earth. This is an exciting time for Earth scientists since we may now have a better idea about the inner motion of the core and the generation of Earth's magnetic field.

Additional Resources

Carlowicz, M. 1996. Spin control. *Earth* 12(21): 62–63.

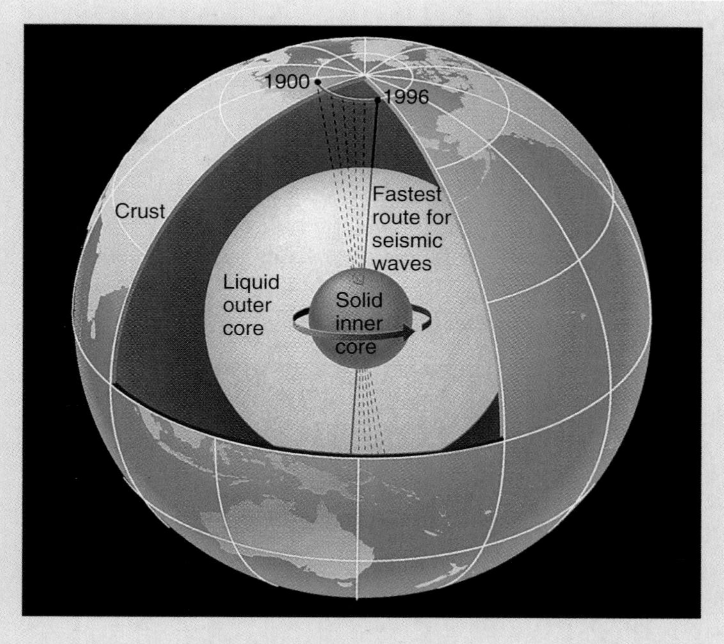

Box 2.4 — FIGURE 2

Seismic waves indicate that Earth's core rotates faster than the rest of the planet by about a degree per year. The solid line indicates the 1996 position of a point in the core relative to the surface of Earth, and the dashed line indicates where the point was in 1900.

Courtesy of Lamont-Doherty Earth Observatory, Columbia University. Data from Michael Carlowicz, *Earth Magazine,* p. 21, 1996

Related Web Resource

Core convection and the Geodynamo website discusses the recent model for reversals of Earth's magnetic field:
http://ees5-www.lanl.gov/IGPP/Geodynamo.html

magnetic poles had exchanged positions. During this time of magnetic reversal, a compass needle would have pointed south rather than north. Many periods of normal and reverse magnetization are recorded in continental lava flows. They are worldwide events. Since lava flows can be dated isotopically, the time of these reversals in Earth's past can be determined. Although reversals appear to occur randomly (figure 2.24), records for tens of millions of years suggest that Earth's field reverses on average about once every 500,000 years. The present normal orientation has lasted for the past 700,000 years. It takes time for one magnetic orientation to die out and the reverse orientation to build up. Most geologists think that it takes 10,000 years for a reversal to develop, although new evidence suggests that a reversal can occur much faster than that.

What causes magnetic reversals? The question is difficult to answer because no one knows how the magnetic field is generated in the first place. Recent computer modeling and seismological research support the theory that the magnetic field is generated by convection currents in the liquid outer core (see box 2.4). *If* the field is caused by convection currents within the liquid outer core, perhaps a reversal is caused when the currents change direction, or by a temporary current building up and then dying out. Some geologists think that reversals may be triggered by the impact of an asteroid or comet with the earth; other geologists disagree.

A magnetic reversal can have some profound effects on Earth. The strength of Earth's magnetic field probably declines to near zero before the orientation reverses; then the field strength increases to its usual values, but in the opposite orientation. This collapse of the magnetic field means that deadly cosmic radiation from the sun would be much more intense at the surface. When the magnetic field is at its usual strength, it shields Earth from these rays, but when the field collapses, this shielding is lost. Cosmic radiation affects organisms; the extinction of some species and the appearance of new species by mutation have been correlated with some magnetic reversals; however, there have been far more reversals than extinctions.

Magnetic Anomalies

A **magnetometer** is an instrument used to measure the strength of Earth's magnetic field. A magnetometer can be carried over the land surface or flown over land or sea. At sea, magnetometers can also be towed behind ships. They are also used as metal detectors in airports.

The strength of Earth's magnetic field varies from place to place. As with gravity, a deviation from average readings is called an *anomaly*. Very broad regional magnetic anomalies may be due to *circulation patterns in the liquid outer core* or to other deep-seated causes. Smaller anomalies generally reflect *variations in rock type*, for the magnetism of near-surface rocks adds to the main magnetic field generated in the core. Rocks differ in their magnetism, depending upon their content of iron-containing mineral, particularly magnetite.

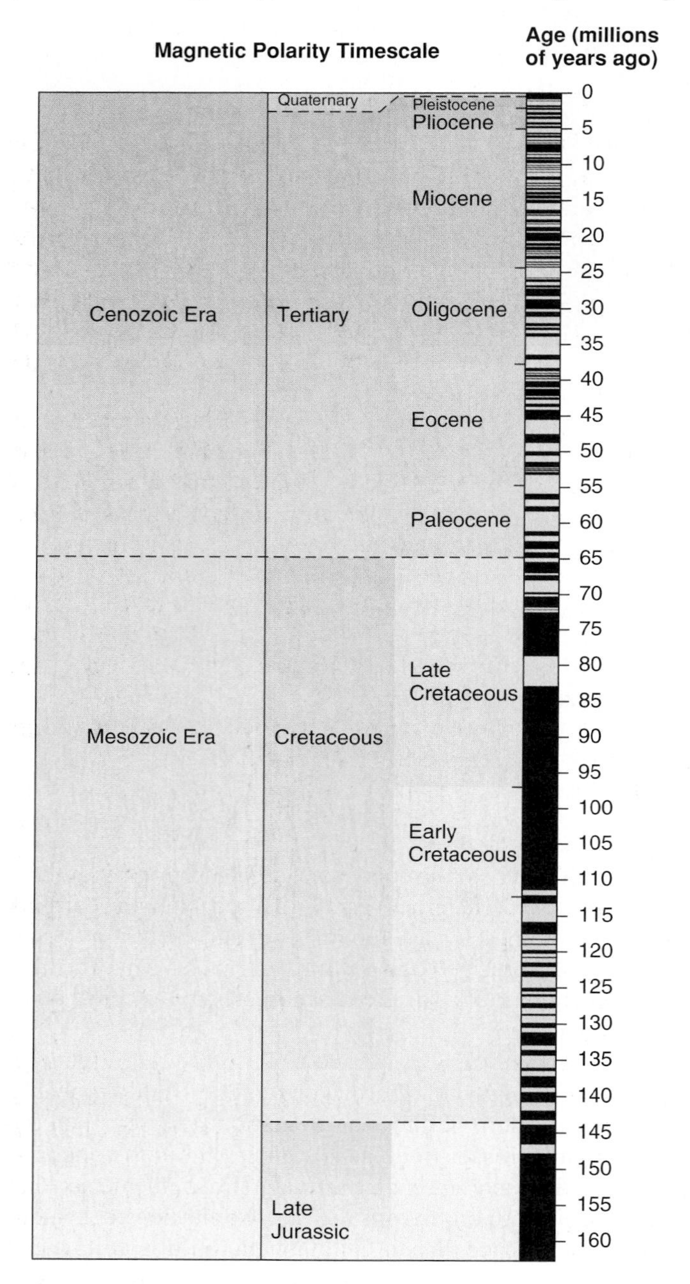

FIGURE 2.24

Worldwide magnetic polarity timescale for the Cenozoic and Mesozoic Eras. Black indicates positive anomalies (and therefore normal polarity). Blue indicates negative anomalies (reverse polarity).

Modified from R. L. Larson and W. C. Pitman, III, 1972, *Geological Society of America Bulletin*

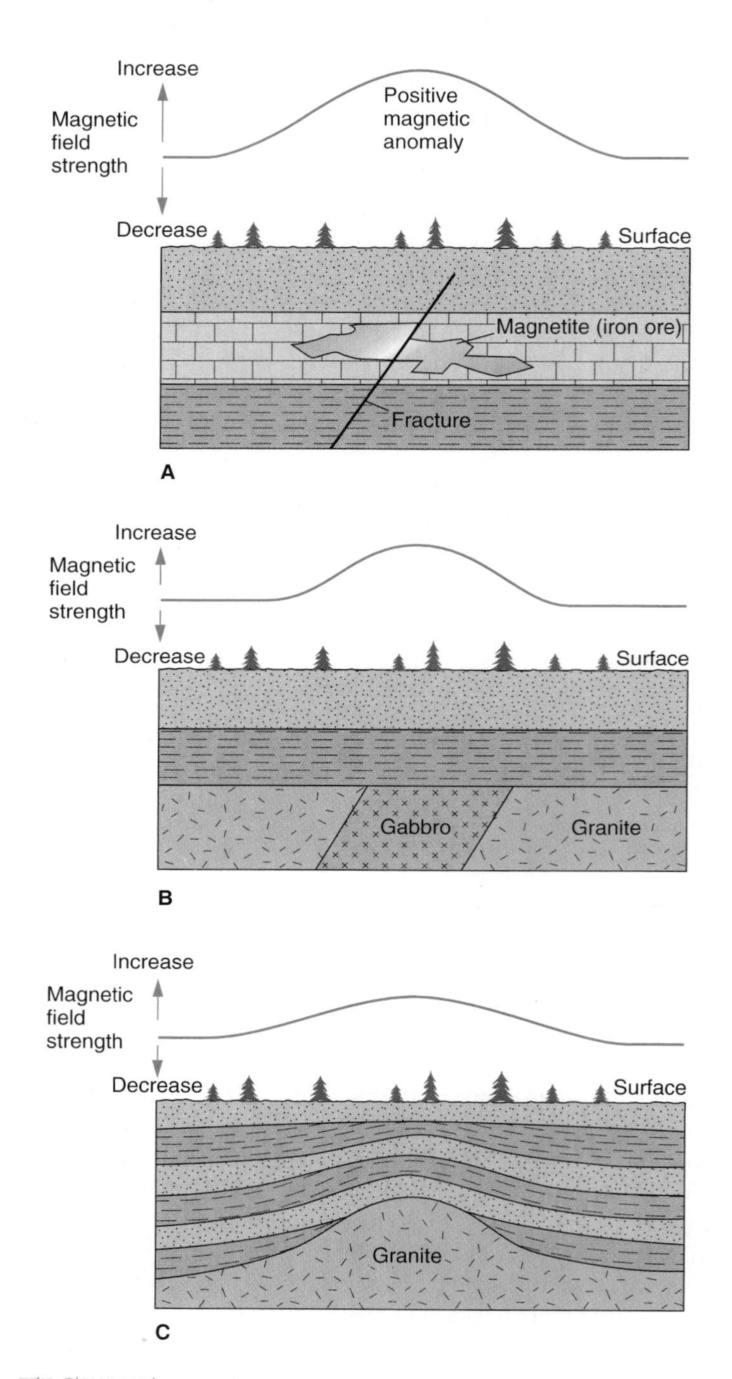

FIGURE 2.25

Positive magnetic anomalies can indicate hidden ore and geologic structures.

A **positive magnetic anomaly** is a reading of magnetic field strength that is higher than the regional average. Figure 2.25 shows three geologic situations that can cause positive magnetic anomalies. In figure 2.25*A*, a body of magnetite ore (a highly magnetic ore of the metal iron) has been emplaced in a bed of limestone by hot solutions rising along a fracture. The magnetism of the iron ore adds to the magnetic field of Earth, giving a stronger magnetic field measurement at the surface (a positive anomaly). In figure 2.25*B* a large dike of gabbro has intruded into granitic basement rock. Because gabbro contains more ferromagnesian minerals than granite, gabbro is more magnetic and causes a positive magnetic anomaly. Figure 2.25*C* shows a granitic basement high (perhaps originally a hill) that has influenced later sediment deposits, causing a draping of the layers as the sediments on the hilltop compacted less than the thicker sediments to the sides. Such a structure can form an *oil trap* (see chapter 21). The granite in the hill contains more iron in its ferromagnesian minerals than the surrounding sedimentary rocks, so a small positive magnetic anomaly occurs where the granite is closer to the surface. Note how each example shows horizontal sedimentary rocks at the surface, with no surface hint of the subsurface geology. The magnetometer helps find hidden ores and geologic structures.

A **negative magnetic anomaly** is a reading of magnetic field strength that is lower than the regional average. Figure 2.26 shows how a negative anomaly can be produced by a down-dropped fault block (a *graben*) in igneous rock. The thick sedimentary fill above the graben is less magnetic than is the igneous rock, so a weaker field (a negative magnetic anomaly) develops over the thick sediment.

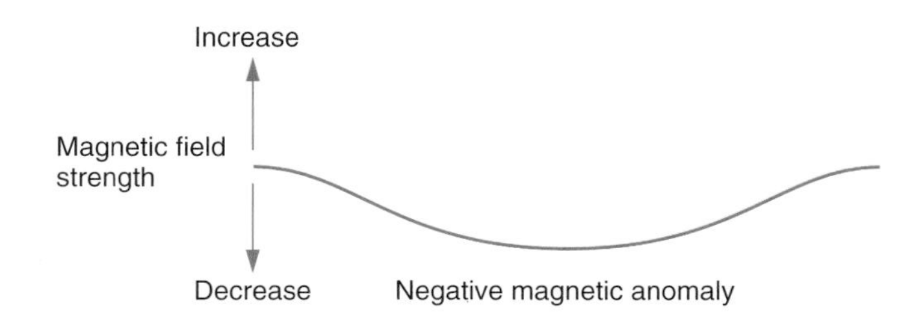

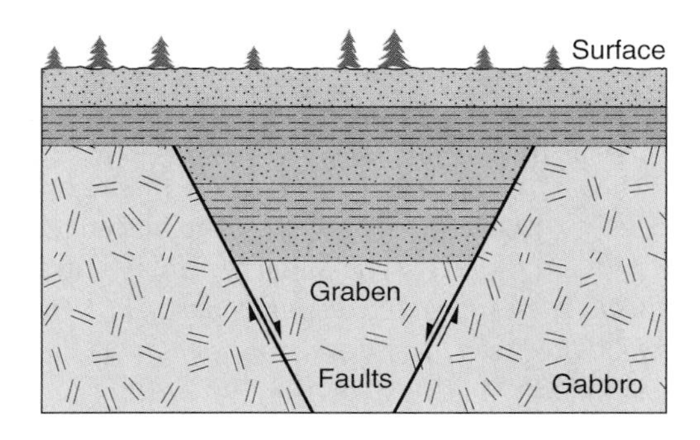

FIGURE 2.26

A graben filled with sediment can give a negative magnetic anomaly if the sediment contains fewer magnetic minerals than the rock beneath it.

Not all local magnetic anomalies are caused by variations in rock type. The linear magnetic anomalies found at sea are apparently caused by a *variation in the direction of magnetism,* as you will see in chapter 4.

HEAT WITHIN THE EARTH

Geothermal Gradient

The temperature increase with depth into Earth is called the **geothermal gradient.** The geothermal gradient can be measured on land in abandoned wells or on the sea floor by dropping specially designed probes into the mud. The average temperature increase is 25°C per kilometer (about 75°F per mile) of depth. Some regions have a much higher gradient, indicating concentrations of heat at shallow depths. Such regions have a potential for generating *geothermal energy* (discussed in chapter 17).

The temperature increase with depth creates a problem in deep mines, such as in a 3-kilometer-deep gold mine in South Africa, where the temperature is close to the boiling point of water. Deep mines must be cooled by air-conditioning for the miners to survive. High temperatures at depth also complicate the drilling of deep oil wells. A well drilled to a depth of 7 or 8 kilometers must pass through rock with a temperature of 200°C. At such high temperatures, a tough steel drilling pipe will become soft and flexible unless it is cooled with a special mud solution pumped down the hole.

Geologists believe that the geothermal gradient must taper off sharply a short distance into Earth. The high values of 25°C/kilometer recorded near Earth's surface could not continue very far into Earth. If they did, the temperature would be 2,500°C at the shallow depth of 100 kilometers. This temperature is above the melting point of all rocks at that depth—even though the increased pressure with depth into Earth increases the melting point of rocks. Seismic evidence seems to indicate a solid, not molten, mantle, so the geothermal gradient must drop to values as low as 1°C/kilometer within the mantle (figure 2.27*A*).

At the boundary between the inner core and the outer core, there would be some constraints on possible temperatures if the core is molten metal above the boundary and solid metal below. The weight of the thick rock layer of the mantle and the liquid metal of the outer core raises the pressure at this boundary (figure 2.27*B*) to about 3 million atmospheres. (An *atmosphere* of pressure is the force per unit area caused by the weight of the air in the atmosphere. It is about 1 kilogram per square centimeter, or 14.7 pounds per square inch.)

Using geophysical and geochemical data, in addition to computer modeling and high-pressure experiments, the internal temperature of Earth can be estimated. Recent laboratory

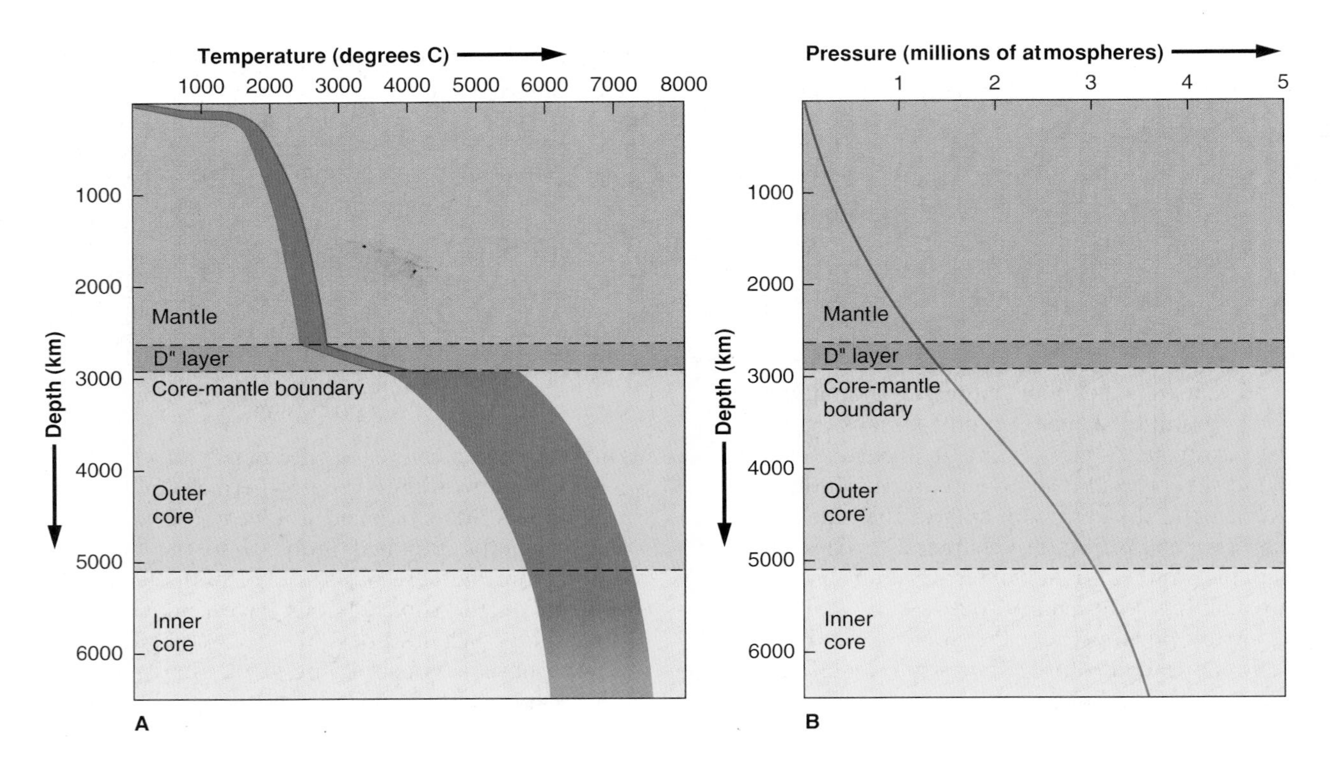

FIGURE 2.27
Estimated (*A*) temperature and (*B*) pressure with depth into Earth. The width of the red zone in graph (*A*) indicates the range of uncertainty of the estimate.

experiments with pressure anvils and giant guns have created (for a millionth of a second) the enormous pressures found at the center of Earth. The measured temperature at this pressure was far higher than expected. New estimates of Earth's internal temperatures have resulted: 3,800°C at the core-mantle boundary, 6,300°C ± 800°C at the inner-core/outer-core boundary, and 6,400°C ± 600°C at Earth's center (hotter than the surface of the sun!).

Heat Flow

A small but measurable amount of heat from Earth's interior is being lost gradually through the surface. This gradual loss of heat through Earth's surface is called **heat flow.** What is the origin of the heat? It could be "original" heat from the time that Earth formed, that is, *if* the earth formed as a mass of planetesimals that coalesced and compressed the inner material. Or the heat could be a by-product of the decay of radioactive isotopes inside Earth. Radioactive decay *may* actually be warming up the planet. Geologists are not sure whether Earth formed as a hot or cold mass, or whether the planet is now cooling off or warming up. Changes in Earth's internal temperature are extremely slow (on the order of 100 million years), and trying to work out its thermal history is a slow, often frustrating, job.

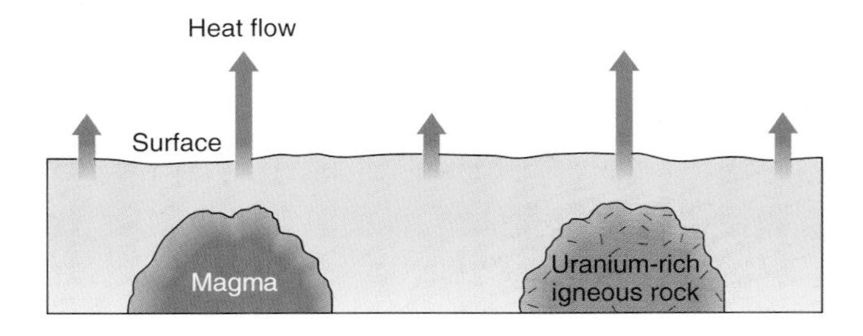

FIGURE 2.28
Some regions have higher heat flow than others; the amount of heat flow is indicated by the length of the arrow. Regions of high heat flow may be underlain by cooling magma or uranium-rich igneous rock.

Some regions on Earth have a high heat flow. More heat is being lost through the surface in these regions than is normal. High heat flow is usually caused by the presence of a magma body or still-cooling pluton near the surface (figure 2.28). An old body of igneous rock that is rich in uranium and other radioactive isotopes can cause a high heat flow, too, because radioactive decay produces heat as it occurs. High heat flow over an extensive area may be due to the rise of warm mantle rock beneath abnormally thin crust.

The average heat flow from continents is the same as the average heat flow from the sea floor, a surprising fact if you consider the greater concentration of radioactive material in continental rock (figure 2.29). The unexpectedly high average heat flow under the ocean may be due to hot mantle rock rising slowly by convection under parts of the ocean (see chapter 1). Regional patterns of high heat flow and low heat flow on the sea floor (heat flow decreases away from the crest of the mid-oceanic ridge) may also be explained by convection of mantle rock, as we discuss in chapter 4.

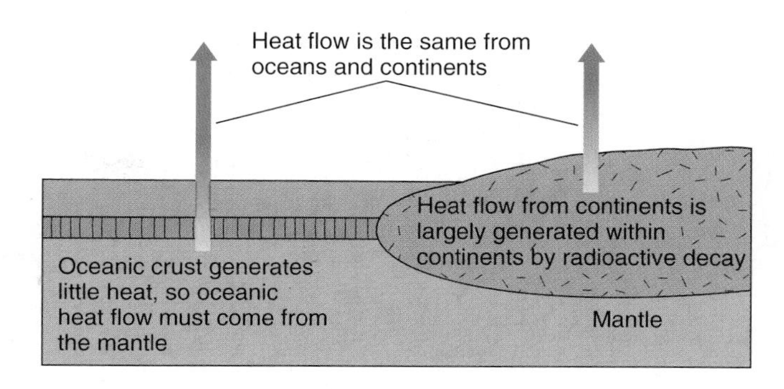

FIGURE 2.29
The average heat flow from oceans and continents is the same, but the origin of the heat differs from the ocean to continents.

Summary

The interior of Earth is studied indirectly by *geophysics*—a study of seismic waves, gravity, Earth magnetism, and Earth heat.

Seismic reflection and *seismic refraction* can indicate the presence of boundaries between rock layers.

Earth is divided into three major units—the *crust,* the *mantle,* and the *core.*

The crust beneath oceans is 7 kilometers thick and made of basalt on top of gabbro. Continental crust is 30 to 50 kilometers thick and consists of a crystalline basement of granite and gneiss (and other rocks) capped by sedimentary rocks.

The *Mohorovičić discontinuity* separates the crust from the mantle.

The mantle is a layer of solid rock 2,900 kilometers thick and is probably composed of an ultramafic rock such as peridotite. Seismic waves show the mantle has a structure of concentric shells, perhaps caused by pressure transformations of minerals.

The *lithosphere,* which forms plates, is made up of brittle crust and upper mantle. It is 70 to 125 (or more) kilometers thick and moves over the plastic asthenosphere.

The *asthenosphere* lies below the lithosphere and may represent rock close to its melting point (seismic waves slow down here). It is probably the region of most magma generation and isostatic adjustment.

Seismic-wave shadow zones show the core has a radius of 3,450 kilometers and is divided into a liquid outer core and a solid inner core. A core composition of mostly iron is suggested by Earth's density, the composition of meteorites, and the existence of Earth's magnetic field.

Isostasy is the equilibrium of crustal columns floating on plastic mantle. *Isostatic adjustment* occurs when weight is added to or subtracted from a column of rock. *Crustal rebound* is isostatic adjustment that occurs after the melting of glacial ice.

A *gravity meter* can be used to study variations in rock density or to find regions that are out of isostatic equilibrium.

A *positive gravity anomaly* forms over dense rock or over regions being held up out of isostatic balance. A *negative gravity anomaly* indicates low-density rock or a region being held down.

Earth's *magnetic field* has two *magnetic poles,* probably generated by convection circulation and electric currents in the outer core.

Some rocks record Earth's magnetism at the time they form. *Paleomagnetism* is the study of ancient magnetic fields.

Magnetic reversals of polarity occurred in the past, with the north magnetic pole and south magnetic pole exchanging positions. Isotopic dating of rocks shows the ages of the reversals.

A *magnetometer* measures the strength of the magnetic field.

A *positive magnetic anomaly* develops over rock that is more magnetic than neighboring rock. A *negative magnetic anomaly* indicates rock with low magnetism.

Magnetic anomalies can also be caused by circulation patterns in Earth's core and by variations in the direction of rock magnetism.

The *geothermal gradient* is about 25°C/kilometer near the surface but decreases rapidly at depth. The temperature at the center of Earth may be 6,900°C. *Heat flow* measurements show that heat loss per unit area from continents and oceans is about the same, perhaps because of convection of hot mantle rock beneath the oceans.

Terms to Remember

asthenosphere 30
convection 35
core 29
crust 28
crustal rebound 36
Curie point 41
geophysics 26
geothermal gradient 45
gravity meter 38
heat flow 46
isostasy 36
isostatic adjustment 36
lithosphere 30
magnetic field 40
magnetic pole 40
magnetic reversal 41
magnetometer 44
mantle 29
Mohorovičić discontinuity (Moho) 30
negative gravity anomaly 40
negative magnetic anomaly 45

paleomagnetism 41
positive gravity anomaly 39
positive magnetic anomaly 45
P-wave shadow zone 33
seismic reflection 26
seismic refraction 26
S-wave shadow zone 33

Testing Your Knowledge

Use the questions below to prepare for exams based on this chapter.

1. Describe how seismic reflection and seismic refraction show the presence of layers within Earth.
2. Sketch a cross section of the entire Earth showing the main subdivisions of Earth's interior and giving the name, thickness, and probable composition of each.
3. What facts make it probable that Earth's core is composed of mostly iron?
4. Describe the differences between continental crust and oceanic crust.
5. What is a gravity anomaly, and what does it generally indicate about the rocks in the region where it is found?
6. Discuss seismic-wave shadow zones and what they indicate about Earth's interior.
7. Describe Earth's magnetic field. Where is it generated?
8. What is the temperature distribution with depth into Earth?
9. Heat flow has been found to be about equal through continents and the sea floor. Why was this unexpected? What might cause this equality?
10. What is the Mohorovičić discontinuity?
11. What is the asthenosphere? Why is it important?
12. How does the lithosphere differ from the asthenosphere?
13. What is a magnetic reversal? What is the evidence for magnetic reversals?
14. What is a magnetic anomaly? How are magnetic anomalies measured at sea?
15. *Felsic* and *mafic* are terms used by some geologists to describe
    a. composition of continental and oceanic crust
    b. behavior of earthquake waves
    c. regions in the mantle
16. The boundary that separates the crust from the mantle is called the
    a. lithosphere
    b. asthenosphere
    c. Mohorovičić discontinuity
    d. none of the above
17. The core is probably composed mainly of
    a. silicon
    b. sulfur
    c. oxygen
    d. iron
18. The principle of continents being in a buoyant equilibrium is called
    a. subsidence
    b. isostasy
    c. convection
    d. rebound
19. A positive gravity anomaly indicates that
    a. tectonic forces are holding a region up out of isostatic equilibrium
    b. the land is sinking
    c. local mass deficiencies exist in the crust
    d. all of the above
20. A positive magnetic anomaly could indicate
    a. a body of magnetic ore
    b. the magnetic field strength is higher than the regional average
    c. an intrusion of gabbro
    d. the presence of a granitic basement high
    e. all of the above
21. Which of the following is not an example of the effects of isostasy?
    a. deep mountain roots
    b. magnetic reversals
    c. the postglacial rise of northeastern North America
    d. mountain ranges at subduction zones
22. The S-wave shadow zone is evidence that
    a. the core is made of iron and nickel
    b. the inner core is solid
    c. the outer core is fluid
    d. the mantle is plastic

Expanding Your Knowledge

1. Why does the heat flow from the continents equal that from the oceans?
2. Subsidence of Earth's surface sometimes occurs as reservoirs fill behind newly built dams. Why?
3. What geologic processes might cause the forces that can hold a region out of isostatic equilibrium?
4. Does the correlation of a species extinction with a magnetic reversal prove that the reversal caused the extinction?
5. If the upper mantle is chemically different from the lower mantle, is mantle convection possible?
6. How could you use geophysical techniques to plan the location of a subdivision in an area containing limestone caverns that could collapse if built on?

Exploring Web Resources

www.mhhe.com/plummer9e

Visit the Online Learning Center for Mantle Xenoliths—A Peek at the Deep. Here you can read what geologists believe the upper mantle looks like. You can also check your answers for the Testing Your Knowledge section and click on the direct links to the websites listed below.

http://rses.anu.edu.au/gfd/Gfd_other_pages/Convection_demo/Demo_page_1.html

The Geophysical Fluid Dynamics Group web page contains images of mantle convection models.

http://ees5-www.lanl.gov/IGPP/Geodynamo.html

Core Convection and Geodynamo web page discusses the recent model for reversals of the earth's magnetic field.

Animations

This chapter includes the following animations available on our Online Learning Center at www.mhhe.com/plummer9e.

2.2 Seismic Refraction
2.12 Isostatic Adjustment of Mountains and Sedimentary Basins
2.13 Isostatic Adjustment by Glacial Rebound

CHAPTER

3
The Sea Floor

The rocks and topography of the sea floor are different from those on land. To understand the evidence for plate tectonics in the next chapter, you need to understand the nature of major sea-floor features such as mid-oceanic ridges, oceanic trenches, and fracture zones, as well as the surprisingly young age of the sea-floor rocks.

The material discussed in this chapter and the next are an excellent example of how the scientific method works. This chapter is concerned with the physical *description* of most sea-floor features—the data-gathering part of the scientific method. The next chapter shows how the theory of plate tectonics explains the *origin* of many of these features. Geologists generally agree on the descriptions of features but often disagree on their interpretations. As you read, keep a clear distinction in your mind between *data* and the *hypotheses* used to explain the data.

Opposite: Hydrothermal vent and deep-sea organisms from the Galápagos Rift in the Pacific Ocean. Photograph taken from the *ALVIN* submersible.

Photo © Peter Ryan/Scripps/SPL/Photo Researchers, Inc.

3.1

ASTROGEOLOGY

ORIGIN OF THE OCEAN

As mentioned in the last chapter, geologists are unsure about the thermal history of Earth. The following scenario for the origin of Earth and its oceans is highly speculative. Many geologists would disagree with some of the statements; they are offered merely as an example of what could have happened in Earth's past.

According to a widely accepted theory, about 4.5 billion years ago Earth began to form by the accretion of small, cold chunks of rock and metal that surrounded the sun. As Earth grew it began to heat up because of the heat of collisional impact, gravitational compaction, and radioactive decay of elements such as uranium. The temperature of Earth rose until its iron melted and "fell" to the center to form its core. Violent volcanic activity occurred at this time, releasing great quantities of water vapor and other gases from Earth's interior and perhaps even covering the surface with a thick, red-hot sea of lava. Earth began to cool as its growth and internal reorganization slowed down and as the amount of radioactive material was reduced by decay. Eventually Earth's surface became solid rock, cool enough to permit the condensation of billowing clouds of volcanic water vapor to form liquid water that fell as rain. Thus the modern oceans were born, perhaps 4 billion years ago. Evidence comes from some of the oldest rocks found on Earth including pillow basalts that suggest eruption under water and rocks that were originally sedimentary and formed in a shallow sea. The oceans grew in size as volcanic *degassing* of Earth continued and became salty as the water picked up chlorine from other volcanic gases, and sodium (and calcium and magnesium) from the chemical weathering of minerals on Earth's surface.

Geologists generally agree that the oceans formed early in Earth's history, and that it is salty for the reasons stated. Most would also agree that the oceans resulted from degassing of Earth's interior. However, geologists were surprised to find that the oceanic crust is geologically very young.

METHODS OF STUDYING THE SEA FLOOR

Oceans cover more than 70% of Earth's surface. Even though the rocks of the sea floor are widespread, they are difficult to study. Geologists have to rely on small samples of rock taken from the sea floor and brought to the surface, or they must study the rocks indirectly by means of instruments on board ships. Although the sea floor is difficult to study, its overall structure is relatively simple (as we discuss later in the chapter), so the small number of samples is not nearly as much a problem as it would be in studying continental regions, where the structure is usually much more complex. The study of sea-floor rocks, sediment, and topography provided most of the information that led to the concept of plate tectonics.

Samples of rock and sediments can be taken from the sea floor in several ways. Rocks can be broken from the sea floor by a *rock dredge,* which is an open steel container dragged over the ocean bottom at the end of a cable. Sediments can be sampled with a *corer,* a weighted steel pipe dropped vertically into the mud and sand of the ocean floor (figure 3.1).

Both rocks and sediments can be sampled by means of *sea-floor drilling.* Offshore oil platforms drill holes in the relatively shallow sea floor near shore. A ship with a drilling derrick on its deck can drill a hole in the deep sea floor far from land (figure 3.2*A*). The drill cuts long, rodlike rock cores from the ocean floor. Thousands of such holes have been drilled in the sea floor, and the rock and sediment cores recovered from these holes have revolutionized the field of marine geology. In the 1950s more was known about the moon's surface than about the floor of the sea. Sea-floor drilling has been instrumental in expanding our knowledge of sea-floor features and history. Small research submarines, more correctly called *submersibles,* can take geologists to many parts of the sea floor to observe, photograph, and sample rock and sediment (figure 3.2*B*).

A basic tool for indirectly studying the sea floor is the *echo sounder,* which draws profiles of submarine topography (figure 3.3). A sound sent downward from a ship bounces off the sea floor and returns to the ship. The water depth is determined from the time it takes the sound to make the round trip. *Multibeam sonar,* which sends out and records a variety of sound sources, is able to map the sea floor in even more detail than the single-source echo sounder. A *seismic profiler* works on essentially the same principles as the echo sounders but uses a louder noise at lower frequency. This sound penetrates the bottom of the sea and reflects from layers within the rock and sediment. The seismic profiler gives more information than the echo sounders. It records water depth and reveals the internal structure of the rocks and sediments of the sea floor, such as bedding planes, folds and faults, and unconformities (figure 3.4). *Magnetic, gravity,* and *seismic refraction* surveys (see chapter 2) also can be made at sea. *Deep-sea cameras* can be lowered to the bottom to photograph the rock and sediment (figure 3.5).

A

B

C

FIGURE 3.1

Bottom sampling devices. (*A*) Rock dredge for sampling hard rock. An unusually large rock is caught in the mouth of the dredge. Additional rocks are visible in the chain bag. (*B*) Corer for sampling sea floor sediment. Sediment is caught inside the pipe when the corer is dropped to the sea floor. The bomb-shaped part of the corer adds weight to the core pipe (mostly under water). (*C*) A core of sea floor sediment. The sediment has been forced out of the core barrel and sawed in half lengthwise, so that the sediment layers are visible.

Photo *A* from Scripps Institution of Oceanography, University of California, San Diego; photo *C* by D. Hopkins, U.S. Geological Survey

A

B

FIGURE 3.2

(*A*) The *Joides Resolution* is a ship specially built for sampling both sediment and rock from the deep ocean floor. (*B*) The small research submersible ALVIN of Woods Hole Oceanographic Institution in Massachusetts; it is capable of taking three oceanographers to a depth of about 4,000 meters.

Photo *A* by U.S. Geological Survey, photo *B* by Woods Hole Oceanographic Institution

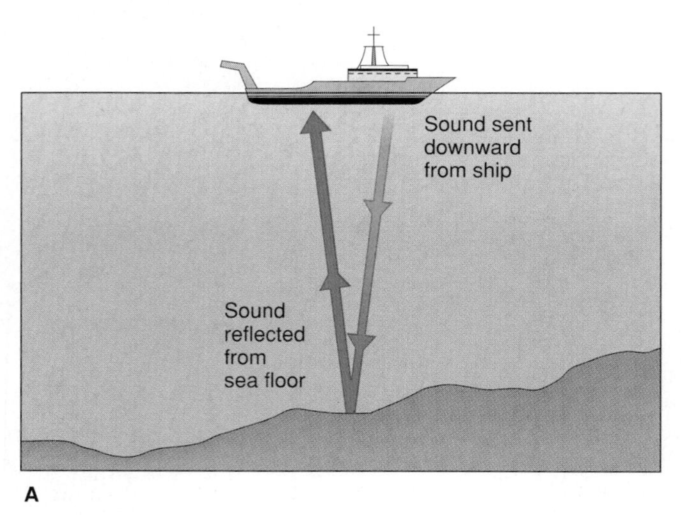

A

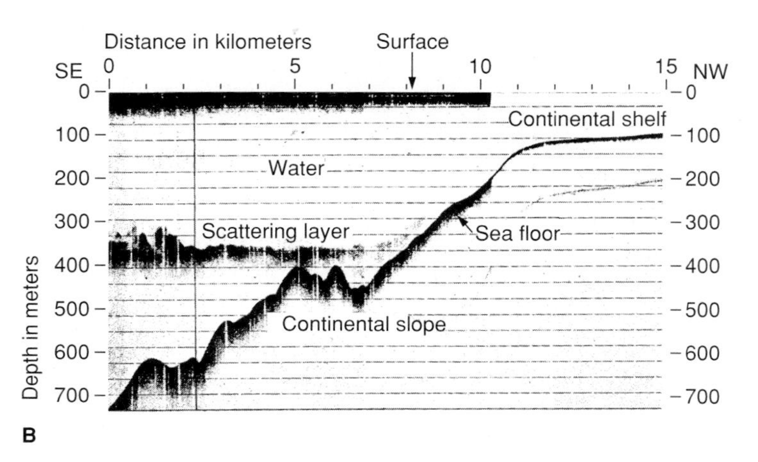

B

FIGURE 3.3

Echo-sounding. (*A*) A sound bounces off the sea floor and returns to a ship. (*B*) An echo-sounder record of the continental shelf and continental slope off Virginia. The scattering layer is caused by a dense concentration of small marine organisms.

Echo-sounder record courtesy of Peter Rona

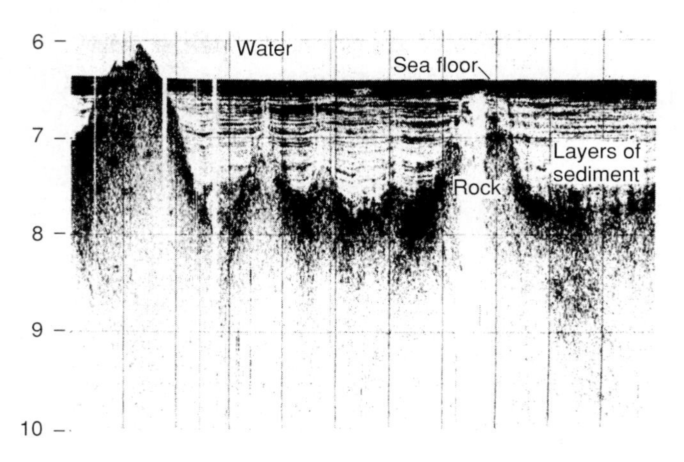

FIGURE 3.4

Seismic profiler record of an abyssal plain, showing sediment layers that have buried an irregular rock surface in the Atlantic Ocean.

From Vogt et al. in Hart, *The Earth's Crust and Upper Mantle,* p. 574, 1969, copyrighted by the American Geophysical Union

FIGURE 3.5

Photograph of the sea floor on the side of a South Pacific island at a depth of 1,000 meters. The rocks are basalt. Rippled sediment between the rocks indicates that strong currents move sea floor sediment here. Field of view is about 1.5 by 2 meters.

From Lamont-Doherty Geological Observatory

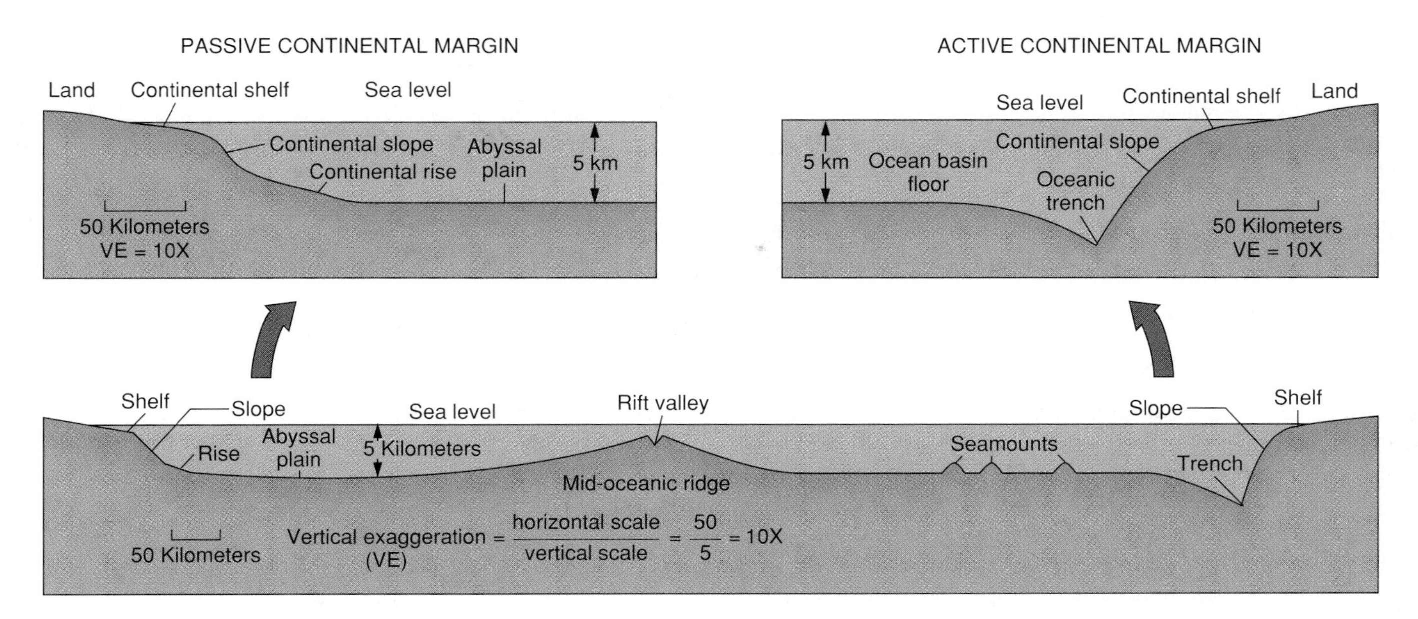

FIGURE 3.6

Profiles of sea floor topography. The vertical scales differ from the horizontal scales, causing vertical exaggeration, which makes slopes appear steeper than they really are. The bars for the horizontal scale are 50 kilometers long, while the same distance vertically represents only 5 kilometers, so the drawings have a vertical exaggeration of 10.

FEATURES OF THE SEA FLOOR

Figure 3.6, a simplified profile of the sea floor, shows that continents have two types of margins. A *passive continental margin,* as found on the east coast of North America, includes a continental shelf, continental slope, and continental rise. An *abyssal plain* usually forms a remarkably flat ocean floor beyond the continental rise. An *active continental margin,* mainly found around the Pacific rim, is associated with earthquakes and volcanoes and has a continental shelf and slope, but the slope extends much deeper to form one wall of an *oceanic trench.* Abyssal plains are seldom found off active margins. The deep ocean floor seaward of trenches is hilly and irregular, lacking the extreme flatness of abyssal plains. Encircling the globe is a *mid-oceanic ridge,* usually (but not always) near the center of an ocean. Conical *seamounts* stick up from the sea floor in some regions. Some important submarine features do not show in this profile view—in particular, fracture zones, submarine canyons, and aseismic ridges.

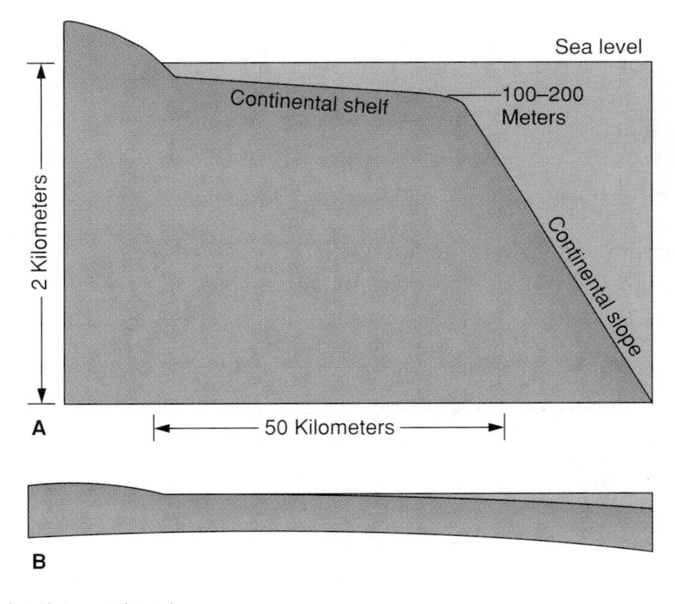

FIGURE 3.7

Continental shelf and continental slope. (*A*) Vertical exaggeration 25×. The continental slope has an actual slope of only 4 or 5 degrees, but the great vertical exaggeration of this drawing makes it appear to be sloping about 60°. (*B*) Same profile with no vertical exaggeration.

CONTINENTAL SHELVES AND CONTINENTAL SLOPES

Almost all continental edges are marked by a gently sloping continental shelf near shore, and a steeper continental slope that leads down to the deep ocean floor (figure 3.7). Figures such as 3.6 and 3.7 are usually drawn with great *vertical exaggeration,* which makes submarine slopes appear much steeper than they really are.

A **continental shelf,** a shallow submarine platform at the edge of a continent, inclines very gently seaward, generally at an angle of 0.1°. Continental shelves vary in width. On the Pacific coast of North America the shelf is only a few kilometers wide, but off Newfoundland in the Atlantic Ocean it is about 500 kilometers wide. Portions of the shelves in the Arctic Ocean off Siberia and northern Europe are even wider. Water depth over a continental shelf tends to increase regularly away from land, with the outer edge of the shelf being about 100 to 200 meters below sea level.

Continental shelves are *topographic features,* defined by their depth, flatness, and gentle seaward tilt. Their *geologic origin* varies from place to place and is related to plate tectonics, so it will be discussed in the next chapter. Some generalities about shelves, however, are worth noting here.

The continental shelves of the world are usually covered with relatively young sediment, in most cases derived from land. The sediment is usually sand near shore, where the bottom is shallow and influenced by wave action. Fine-grained mud is commonly deposited farther offshore in deeper, quieter water.

The outer part of a wide shelf is often covered with coarse sediment that was deposited near shore during a time of lower sea level. The advance and retreat of continental glaciers during the Pleistocene Epoch caused sea level to rise and fall many times by 100 to 200 meters. This resulted in a complex history of sedimentation for continental shelves as they were alternately covered with seawater and exposed as dry land.

Marine seismic surveys and drilling at sea have shown that the young sediments on many continental shelves are underlain by thick sequences of sandstone, shale, and less commonly limestone. These sedimentary rocks are of late Mesozoic and Cenozoic age and appear to have been deposited in much the same way as the modern shelf sediments. Beneath these rocks is continental crust (figure 3.8); the continental shelves are truly part of the continents, even though today the shelves are covered by seawater.

A **continental slope** is a relatively steep slope that extends from a depth of 100 to 200 meters at the edge of the continental shelf down to oceanic depths. The average angle of slope for a continental slope is 4° to 5°, although locally some parts are much steeper.

Because the continental slopes are more difficult to study than the continental shelves, less is known about them. The greater depth of water and the locally steep inclines on the continental slopes hinder rock dredging and drilling and make the results of seismic refraction and reflection harder to interpret. This is unfortunate, for the rocks that underlie the slopes are of particular interest to marine geologists, who believe that in this area the thick continental crust (beneath the land and the continental shelves) grades into thin oceanic crust (underneath the deep ocean floor), as shown in figure 3.8.

Although relatively little is known about continental slopes, it is clear that their character and origin vary greatly from place to place. Because these variations are associated with the differences between divergent plate boundaries and convergent plate boundaries, we will discuss them in the next chapter.

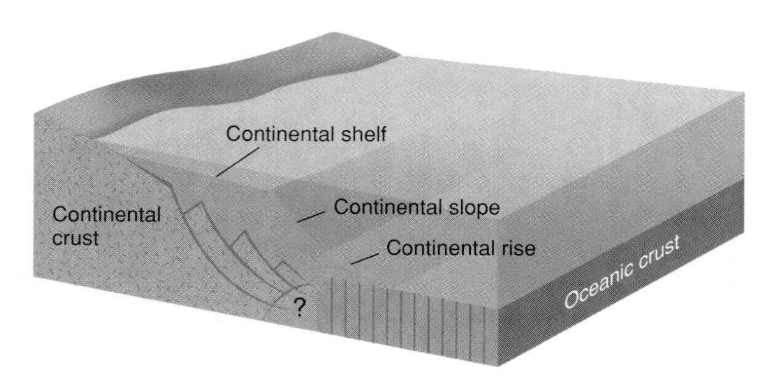

FIGURE 3.8
The continental shelf lies upon continental crust, and the continental rise lies upon oceanic crust. The complex transition from continental crust to oceanic crust lies under the continental slope.

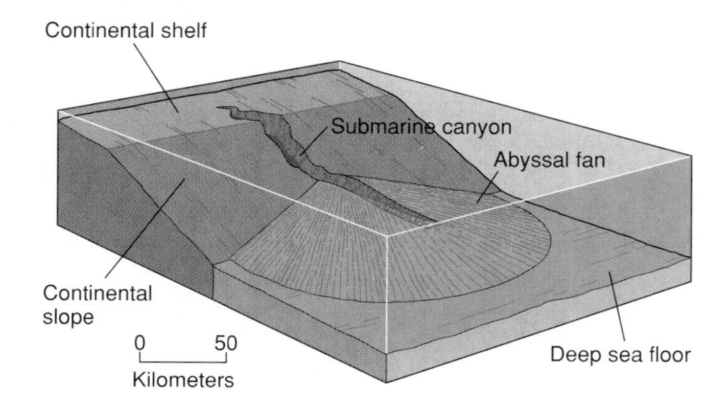

FIGURE 3.9
Submarine canyon and abyssal fan.

SUBMARINE CANYONS

Submarine canyons are V-shaped valleys that run across continental shelves and down continental slopes (figure 3.9). On narrow continental shelves, such as those off the Pacific coast of the United States, the heads of submarine canyons may be so close to shore that they lie within the surf zone. On wide shelves, such as those off the Atlantic coast of the United States, canyon heads usually begin near the outer edge of the continental shelf tens of kilometers from shore. Great fanshaped deposits of sediment called **abyssal fans** are found at the base of many submarine canyons (figure 3.9). Abyssal fans are made up of land-derived sediment that has moved down the submarine canyons. Along continental margins that are cut by submarine canyons, many coalescing abyssal fans may build up at the base of the continental slope.

Submarine canyons are erosional features, but how rock and sediment are removed from the steep-walled canyons is controversial. Erosional agents probably vary in relative importance from canyon to canyon. Divers have filmed *down-canyon movement of sand* in slow, glacierlike flow and in more rapid sand falls (figure 3.10). This sand movement, which has been

FIGURE 3.10
A 10-meter-high sand fall in a submarine canyon near the southern tip of Baja California, Mexico. The sand is beach sand, fed into the nearshore canyon head by longshore currents.
From Scripps Institution of Oceanography, University of California, San Diego

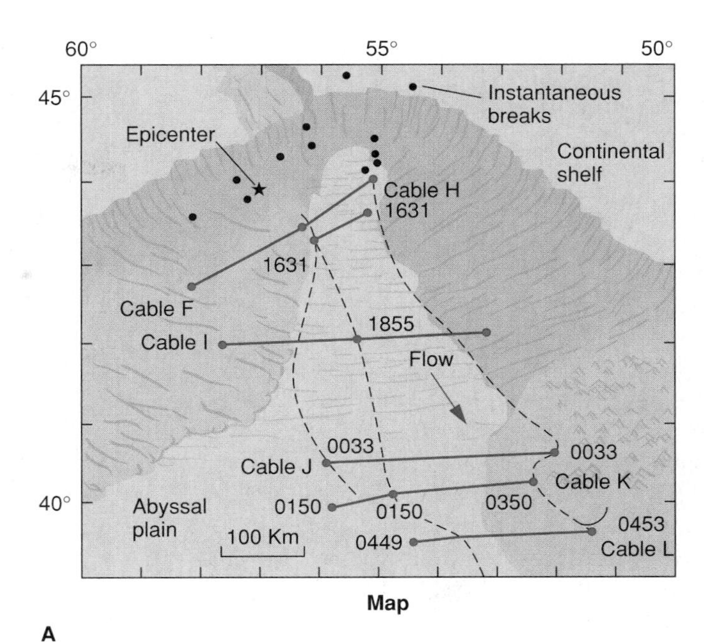

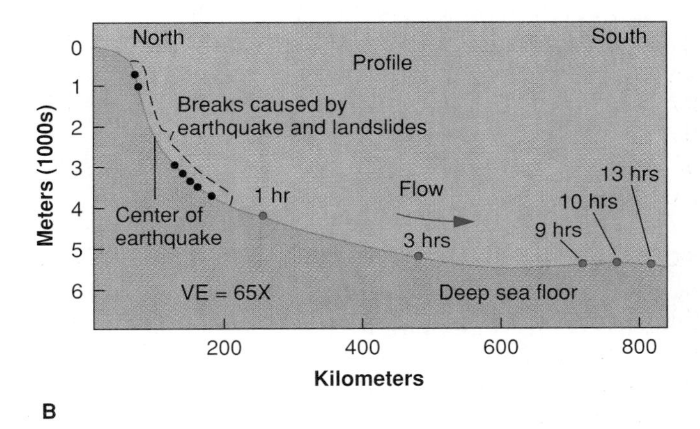

FIGURE 3.11
Submarine cable breaks following the Grand Banks earthquake of 1929. (*A*) Map view of the cable breaks. Black dots near the epicenter show locations of cable breaks that were simultaneous with the earthquake (cables not shown for these breaks). Colored dots show cable breaks that followed the earthquake, with the time of each break shown (on the 24-hour clock). Segments of cable more than 100 kilometers long were broken simultaneously at both ends and then carried away. Dashes show sea floor channels that probably concentrated the flow of a turbidity current, increasing its velocity. (*B*) Profile showing the time elapsed between the quake and each cable break.
A From H. W. Menard, 1964, *Marine Geology of the Pacific,* copyright McGraw-Hill, Inc. *B* from B. C. Heezen and M. Ewing, 1952, *American Journal of Science*

observed to cause erosion of rock, is particularly common in Pacific coast canyons, which collect great quantities of sand from longshore drift. *Bottom currents* have been measured moving up and down the canyons in a pattern of regularly alternating flow, in some cases apparently caused by ocean tides. The origin of these currents is not well understood, but they often move fast enough to erode and transport sediment. *River erosion* may have helped to cut the upper part of canyons when the drop in sea level during the Pleistocene glaciations left canyon heads above the water, such as the extension of the Hudson River into the Atlantic. Many (but not all) submarine canyons are found off land canyons or rivers, which tends to support the view that river erosion helped shape them. It is unlikely, however, that the deeper parts of submarine canyons were ever exposed as dry land.

Turbidity Currents

In addition to the canyon-cutting processes just described, turbidity currents probably play the major role in canyon erosion. **Turbidity currents** are great masses of sediment-laden water that are pulled downhill by gravity. The sediment-laden water is heavier than clear water, so the turbidity current flows down the continental slope until it comes to rest on the flat abyssal plain at the base of the slope (see figure 14.13). Turbidity currents are thought to be generated by underwater earthquakes and landslides, strong surface storms, and floods of sediment-laden rivers discharging directly into the sea on coasts with a narrow shelf. Although large turbidity currents have not been directly observed in the sea, small turbidity currents can be made and studied in the laboratory.

Indirect evidence also indicates that turbidity currents occur in the sea. The best evidence comes from the breaking of submarine cables that carry telephone and telegraph messages across the ocean floor. Figure 3.11 shows a downhill sequence of cable breaks that followed a 1929 earthquake in the Grand Banks region of the northwest Atlantic. This sequence of cable breaks has been interpreted to be the result of an earthquake-caused turbidity current flowing rapidly down the continental slope.

If cable breaks are caused by turbidity currents, they give good evidence of the currents' dramatic size, speed, and energy. Breaks in Grand Banks cables continued for more than 13 hours after the 1929 earthquake, the last of the series occurring more than 700 kilometers from the epicenter. The velocity of the flow that caused the breaks has been calculated to be from 15 to 60 kilometers per hour. Sections of cable more than 100 kilometers long were broken off and carried away, both ends of a missing section being broken simultaneously. Attempts to find broken cable sections were fruitless, and it is assumed that they were buried by sediment.

The Grand Banks 1929 cable breaks are not unique. Cables crossing submarine canyons are broken frequently, particularly after river floods and earthquakes. In the submarine canyons off the Congo (Zaire) River of Africa and off the Magdalena River in Colombia, for example, cables break every few years.

Additional indirect evidence for the existence of turbidity currents comes from the graded bedding and shallow-water fossils in the sediments that make up the continental rise and the abyssal plains, as described in the next section.

PASSIVE CONTINENTAL MARGINS

A **passive continental margin** (figure 3.6) includes a continental shelf, continental slope, and continental rise and generally extends down to an abyssal plain at a depth of about 5 kilometers. It is called a passive margin because it usually develops on geologically quiet coasts that lack earthquakes, volcanoes, and young mountain belts.

Passive margins are found on the edges of most landmasses bordering the Atlantic Ocean. They also border most parts of the Arctic and Indian oceans and a few parts of the Pacific Ocean.

The Continental Rise

Along the base of many parts of the continental slope lies the **continental rise,** a wedge of sediment that extends from the lower part of the continental slope to the deep sea floor. The continental rise, which slopes at about 0.5°, more gently than the continental slope, typically ends in a flat abyssal plain at a depth of about 5 kilometers. The rise rests upon oceanic crust (figure 3.8).

Types of Deposition

Sediments appear to be deposited on the continental rise in two ways—by turbidity currents flowing *down* the continental slope and by *contour currents* flowing *along* the continental slope.

Cores of sediment recovered from most parts of the continental rise show layers of fine sand or coarse silt interbedded with layers of fine-grained mud. The mineral grains and fossils of the coarser layers indicate that the sand and silt came from the shallow continental shelf. Some transporting agent must have carried these sediments from shallow water to deep water. The coarse layers also exhibit graded bedding, which indicates that they settled out of suspension according to size and weight; therefore, the transporting agent for these sediments was most likely turbidity currents. The continental rise in these locations probably formed as turbidity currents deposited abyssal fans at the base of a continental slope.

Sediments in some other parts of the continental rise, however, are uniformly fine-grained and show no graded bedding. This sediment appears to have been deposited by the regular ocean currents that flow along the sea bottom rather than by the intermittent turbidity currents that occasionally flow downslope.

A **contour current** is a bottom current that flows parallel to the slopes of the continental margin—*along* the contour rather than *down* the slope (figure 3.12). Such a current runs south along the continental margin of North America in the Atlantic Ocean. Flowing at the relatively slow speed of a few centimeters per second, this current carries a small amount of fine sediment from north to south. The current is thickest along the continental slope and gets progressively thinner seaward. The thick landward part of the current carries and deposits the most sediment. The thinner seaward edge of the current deposits less sediment. As a result, the deposit of sediment beneath the current is wedge shaped, becoming thinner away from land. Similar contour currents apparently shape parts of the continental rise off other continents as well.

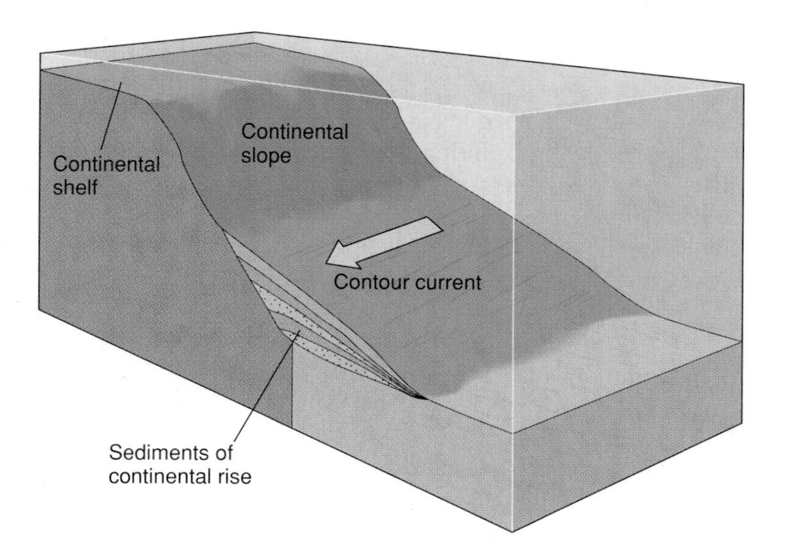

FIGURE 3.12
A contour current flowing along the continental margin shapes the continental rise by depositing fine sediment. Coarse layers within the continental rise were deposited by turbidity currents flowing down the continental slope.

Abyssal Plains

Abyssal plains are very flat regions usually found at the base of the continental rise. Seismic profiling has shown that abyssal plains are formed of horizontal layers of sediment. The gradual deposition of sediment buried an older, more rugged topography that can be seen on seismic profiler records as a rock basement beneath the sediment layers (figure 3.4). Samples of abyssal plain sediment show that it is derived from land. Graded bedding within sediment layers suggests deposition by turbidity currents.

Abyssal plains are the flattest features on Earth. They generally have slopes less than 1:1,000 (less than 1 meter of vertical drop for every 1,000 meters of horizontal distance) and some have slopes of only 1:10,000. Most abyssal plains are 5 kilometers deep.

Not all parts of the deep-ocean basin floor consist of abyssal plains. The deep floor is normally very rugged, broken by faults into hills and depressions and dotted with volcanic seamounts. Abyssal plains form only where turbidity currents can carry in enough sediment to bury and obscure this rugged relief. If the sediment is not available or if the bottom-hugging turbidity currents are stopped by a barrier such as an oceanic trench, then abyssal plains cannot develop.

ACTIVE CONTINENTAL MARGINS

An **active continental margin,** characterized by earthquakes and by a young mountain belt and volcanoes on land, consists of a continental shelf, a continental slope, and an oceanic trench (figure 3.6). An active margin usually lacks a continental rise and an abyssal plain and is associated with convergent plate boundaries.

Active margins are found on the edges of most of the land masses bordering the Pacific Ocean and a few other places in the Atlantic and Indian oceans (figure 3.13). A notable exception in the Pacific Ocean is most of the coast of North America. The active margin's distinctive combination of oceanic trench and land volcanoes is found in three places along the Pacific coast of North America: southern Central America; Washington, Oregon, and northernmost California (the trench here is sediment-filled); and south-central Alaska. But passive margins with abyssal fans and abyssal plains occur along most of the coasts of Mexico, California, British Columbia, and the Alaskan panhandle. These coasts *do* have earthquakes, mostly along strike-slip faults, and probably were full-fledged active margins in the geologic past, but are considered to be passive margins today.

Oceanic Trenches

An **oceanic trench** is a narrow, deep trough parallel to the edge of a continent or an island arc (a curved line of islands like the Aleutians or Japan), as shown in figure 3.13. The continental slope on an active margin forms the landward wall of the trench, its steepness often increasing with depth. The slope is typically 4° to 5° on the upper part, steepening to 10° to 15° or even more near the bottom of the trench. The elongate oceanic trenches, often 8 to 10 kilometers deep, far exceed the average depth of abyssal plains on passive margins. The deepest spots on Earth, more than 11 kilometers below sea level, are in oceanic trenches in the southwest Pacific Ocean.

Associated with oceanic trenches are the *earthquakes of the Benioff seismic zones* (see chapter 7), which begin at a trench and dip landward under continents or island arcs (figure 3.14). *Volcanoes* are found above the upper part of the Benioff zones and typically are arranged in long belts parallel to oceanic trenches. These belts of volcanoes form island arcs or erupt

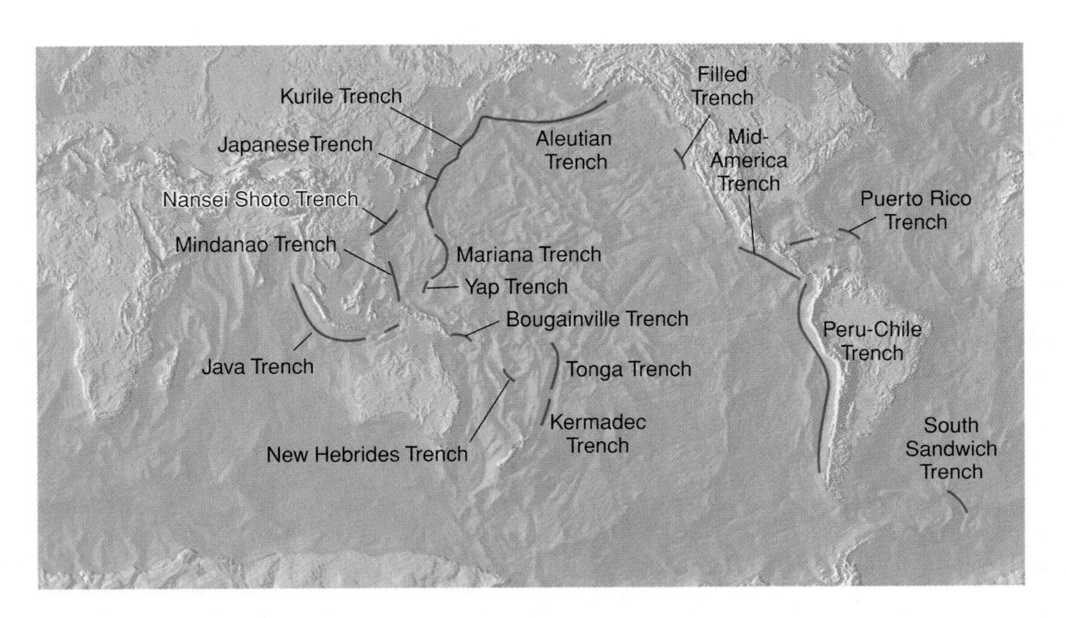

FIGURE 3.13
The distribution of oceanic trenches. Trenches next to continents mark active continental margins and convergent plate boundaries.

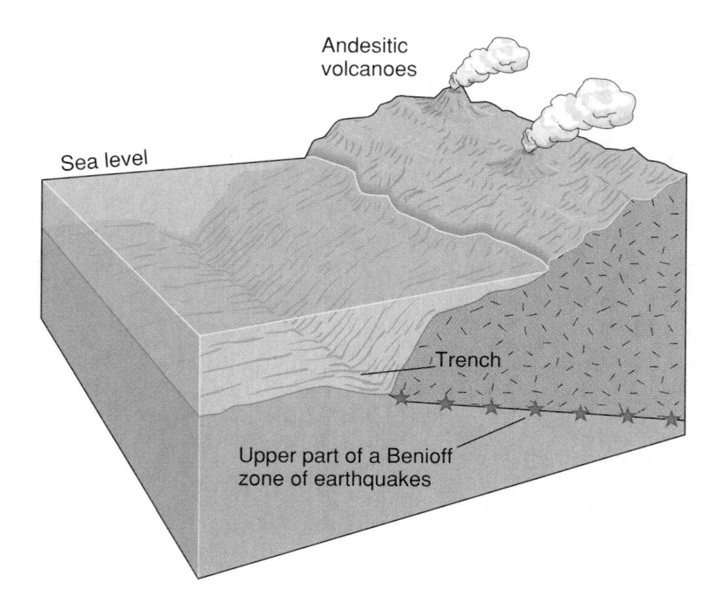

FIGURE 3.14

Active continental margin with an oceanic trench, a Benioff zone of earthquakes (only the upper part is shown), and a chain of andesitic volcanoes on land.

within young mountain belts on the edges of continents. The rock produced by these volcanoes is usually andesite, a type of extrusive rock intermediate in composition between basaltic oceanic crust and "granitic" continental crust.

Oceanic trenches are marked by abnormally *low heat flow* compared to normal ocean crust. This implies that the crust in trenches may be colder than normal crust.

As you learned in the previous chapter, oceanic trenches are also characterized by very large *negative gravity anomalies,* the largest in the world. This implies that trenches are being held down, out of isostatic equilibrium.

THE MID-OCEANIC RIDGE

The **mid-oceanic ridge** is a giant undersea mountain range that extends around the world like the seams on a baseball (figures 3.15 and 3.16). The ridge, which is made up mostly of basalt, is more than 80,000 kilometers long and 1,500 to 2,500 kilometers wide. It rises 2 to 3 kilometers above the ocean floor.

A **rift valley** of tensional origin runs down the crest of the ridge (figures 3.15 and 3.16). The rift valley is 1 to 2 kilometers deep and several kilometers wide—about the dimensions of the Grand Canyon in Arizona. The rift valley on the crest of the mid-oceanic ridge is a unique feature—no mountain range on land has such a valley running along its crest. The rift valley is present in the Atlantic and Indian oceans, but is generally absent in the Pacific Ocean. (This absence is related to the faster rate of plate motion in the Pacific, as we will discuss in chapter 4.)

Geologic Activity on the Ridge

Associated with the rift valley at the crest of the mid-oceanic ridge, and also with the riftless crest of the ridge in the Pacific Ocean, are *shallow-focus earthquakes* from 0 to 20 kilometers below the sea floor.

Careful measurements of the heat loss from Earth's interior through the crust have shown a very *high heat flow* on the crest of the mid-oceanic ridge. The heat loss at the ridge crest is many times the normal value found elsewhere in the ocean; it decreases away from the ridge crest.

Basalt eruptions occur in and near the rift valley on the ridge crest. Sometimes these eruptions build up volcanoes that protrude above sea level as oceanic islands. The large island of Iceland (figure 3.15), which is mostly basaltic, appears to be a section of the mid-oceanic ridge elevated above sea level. Many geologists have studied the active volcanoes, high heat flow, and central rift valley of Iceland to learn about the mid-oceanic

FIGURE 3.15

The mid-oceanic ridge, offset from fracture zones. Darker lines indicate the ridge crest; lighter lines show the fracture zones.

FIGURE 3.16 (OVERLEAF)

The floor of the ocean. Note continental margins, mid-oceanic ridge, fracture zones, and oceanic trenches (compare with figures 3.13 and 3.15). Conical mountains are volcanic seamounts; aligned seamounts are aseismic ridges.

From *World Ocean Floor* by Bruce C. Heezen and Marie Tharp, 1977. Copyright © Marie Tharp 1977. Reproduced by permission of Marie Tharp, 1 Washington Ave., South Nyack, NY 10960

FIGURE 3.17

Underwater photograph of fresh pillow basalt on the floor of the rift valley of the mid-oceanic ridge in the North Atlantic Ocean. A white, tubular sponge grows in the foreground.

Photo by Woods Hole Oceanographic Institution

ridge. As Iceland is above sea level, however, it may not be a typical portion of the ridge.

In the summer of 1974 geologists were able to get a firsthand view of part of the submerged ridge and rift valley. A series of more than forty dives by submersibles, including ALVIN, carried French and American marine geologists directly into the rift valley in the North Atlantic Ocean. The project (called FAMOUS for French-American Mid-Ocean Undersea Study) allowed the ridge rock to be seen, photographed, and sampled directly, rather than indirectly from surface ships (figure 3.17).

The geologists on the FAMOUS project saw clear evidence of extensional faults within the rift valley. These run parallel to the axis of the rift valley and range in width from hairline cracks to gaping fissures that ALVIN dived into. Fresh pillow basalts occur in a narrow band along the bottom of the rift valley, suggesting very recent volcanic activity there, although no active eruptions were observed. It appeared to the geologists that the rift was continuous and that sporadic volcanic activity occurred as a result of the rifting.

Mid-oceanic ridges are often marked by lines of hot springs that carry and precipitate metals. Geologists in submersibles have observed hot springs in several localities along the rift valleys of the mid-oceanic ridges. The hot springs, caused by the high heat flow and shallow basaltic magma beneath the rift valley, range in temperature from about 20°C up to an estimated 350°C (660°F).

As the hot water rises in the rift valley, cold water is drawn in from the sides to take its place. This creates a circulation pattern in which cold seawater is actually drawn *downward* through the cracks in the basaltic crust of the ridge flanks and then moves horizontally toward the rift valley, where it reemerges on the sea floor after being heated (figure 3.18*A*). As the seawater circulates, it dissolves metals and sulfur from the crustal rocks. When the hot, metal-rich solutions contact the cold seawater, metal sulfide particles are discharged into the cold seawater at *black smokers,* which precipitate a chimney-like mound around the hot spring (figure 3.18*B*).

Biologic Activity on the Ridge

The occurrence of black smokers was a surprise to geologists exploring the mid-oceanic ridges, but an even bigger surprise was the presence of exotic, bottom-dwelling organisms surrounding the hot springs. The exotic organisms, including mussels, crabs, starfish, giant white clams, and giant tube worms, are able to survive toxic chemicals, high temperatures, high pressures, and total darkness at depths of more than 2 kilometers (figure 3.18*C*). The organisms live on bacteria that thrive by oxidizing hydrogen sulfide from the hot springs. It is believed that the heat-loving, or *thermophyllic bacteria,* normally reside beneath the sea floor but are blown out of the hot spring when it erupts. Such sulfur-digesting bacteria have also been found in acidic water in mines containing sulfide minerals. Current research in the new field of *geomicrobiology* is examining the role such bacteria may have had in the precipitation of minerals and in the evolution of early lifeforms on Earth and possibly other places in the solar system.

FRACTURE ZONES

Fracture zones are major lines of weakness in Earth's crust that cross the mid-oceanic ridge at approximately right angles (figures 3.15 and 3.16). The rift valley of the mid-oceanic ridge is offset in many places across fracture zones (figure 3.19), and the sea floor on one side of a fracture zone is often at a different elevation than the sea floor on the other side, producing steep cliffs. Shallow-focus earthquakes occur on fracture zones but are confined to those portions of the fracture zones between segments of the rift valley. (The portion of the fracture zone that has earthquakes is known as a *transform fault;* the origin of these faults is discussed in the next chapter.)

Fracture zones extend for thousands of kilometers across the ocean floor, generally heading straight for continental margins. Although fracture zones are difficult to trace where they are buried by the sediments of the abyssal plain and the continental rise, some geologists think that they can trace the extensions of fracture zones onto continents. Some major structural trends on continents appear to lie along the hypothetical extension of fracture zones onto the continents.

SEAMOUNTS, GUYOTS, AND ASEISMIC RIDGES

Conical undersea mountains that rise 1,000 meters or more above the sea floor are called **seamounts** (figures 3.16 and 3.20). Some rise above sea level to form islands. They are scattered on the flanks of the mid-oceanic ridge and on other

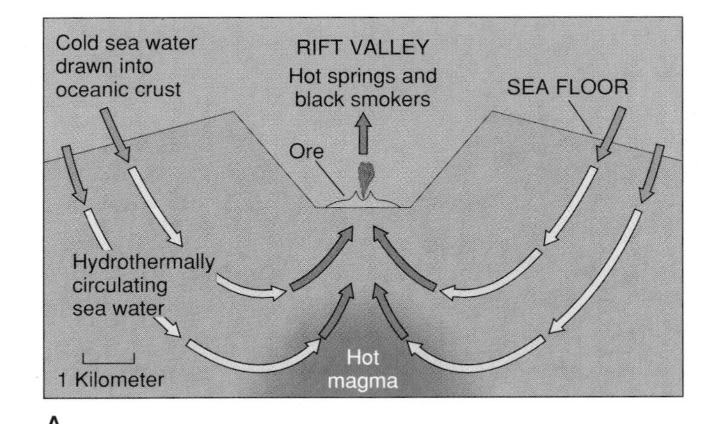

A

B

C

FIGURE 3.18

(*A*) Hydrothermal circulation of sea water at ridge crest creates hot springs and metallic deposits in rift valley. Cold sea water is drawn into fractured crust on ridge flanks. (Size of ore deposit exaggerated.) (*B*) "Black smoker" or submarine hot spring on the crest of the mid-oceanic ridge in the Pacific Ocean near 21° North latitude. The "smoke" is a hot plume of metallic sulfide minerals being discharged into cold sea water from a chimney 0.5 meters high. The large mounds around the chimney are metallic deposits. The instruments in the foreground are attached to the small submarine from which the picture was taken. (*C*) Giant worms, crabs, and clams from Galápagos vent.

Photo *B* by U.S. Geological Survey. Photo *C* © WHOI/D. Foster/Visuals Unlimited

parts of the sea floor, including abyssal plains. One area of the sea floor with a particularly high concentration of seamounts—one estimate is 10,000—is the western Pacific. Rocks dredged from seamounts are nearly always basalt, so it is thought that most seamounts are extinct volcanoes. Of the thousands of seamounts on the sea floor, only a few are active volcanoes. Most of these are on the crests of the mid-oceanic ridges. A few others, such as the two active volcanoes on the island of Hawaii, are found at locations not associated with a ridge.

Guyots are flat-topped seamounts (figure 3.20) found mostly in the western Pacific Ocean. Most geologists think that the flat summits of guyots were cut by wave action. These flat tops are now many hundreds of meters below sea level, well below the level of wave erosion. If the guyot tops were cut by waves, the guyots must have subsided after erosion took place. Evidence of such subsidence comes from the dredging of dead reef corals from guyot tops. Since such corals grow only in shallow water, they must have been carried to their present depths as the guyots sank.

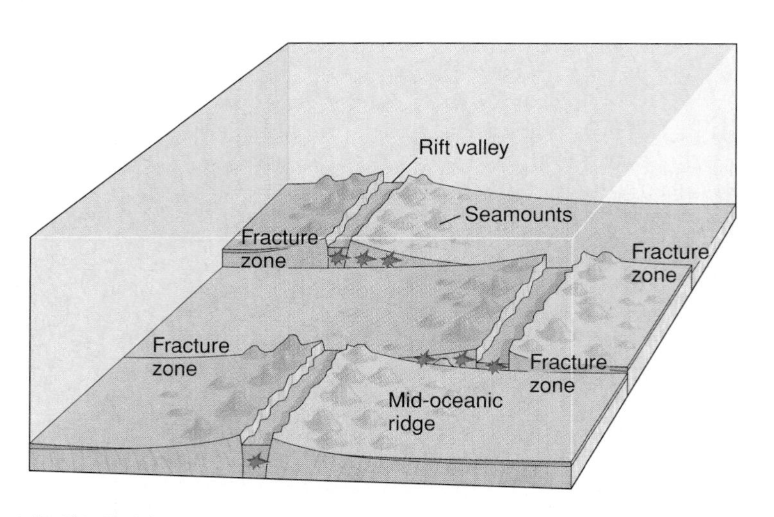

FIGURE 3.19

Fracture zones, which run perpendicular to the ridge crest and connect offset segments of the ridge, are often marked by steep cliffs up to 3 or 4 kilometers high. Shallow focus earthquakes (shown by stars) occur below the rift valley and on the portion of the fracture zone between the mid-oceanic ridge.

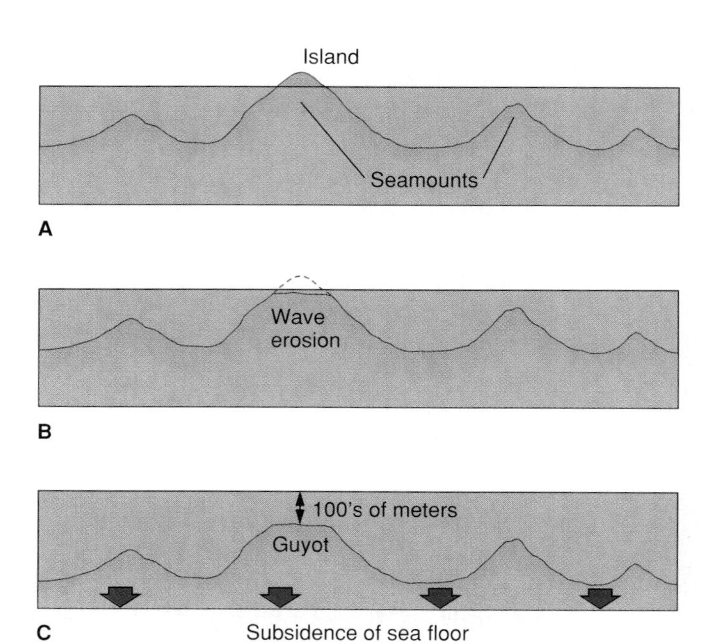

FIGURE 3.20

(*A*) Seamounts are conical mountains on the sea floor, occasionally rising above sea level to form islands. (*B*) The flat summit of a guyot was probably eroded by waves when the top of a seamount was above sea level. (*C*) The present depth of a guyot is due to subsidence.

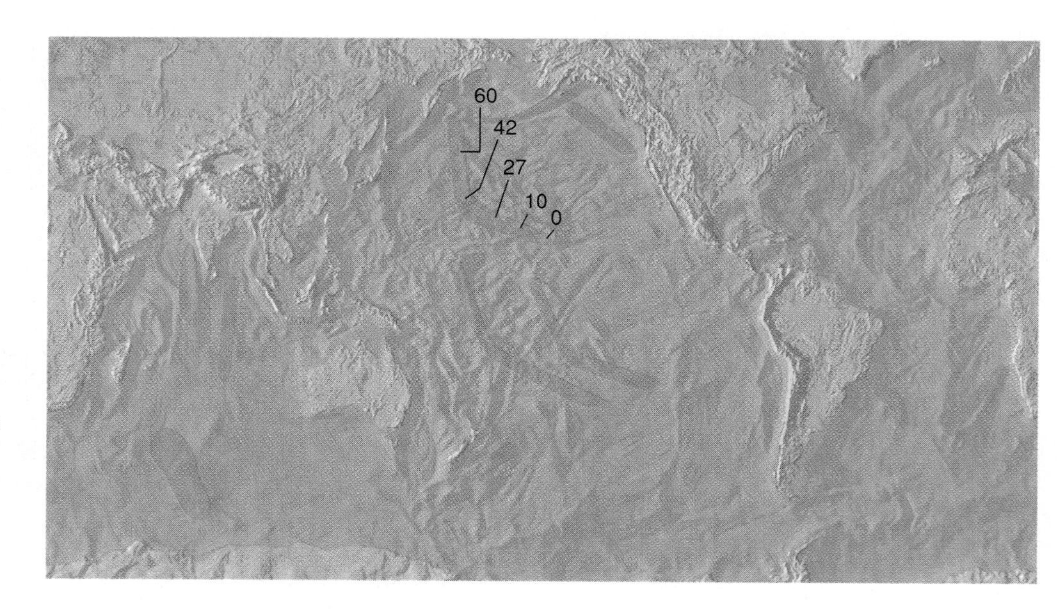

FIGURE 3.21

The distribution of the major aseismic ridges on the sea floor (compare with figure 3.16). The numbers on the Hawaiian-Emperor chain are ages in millions of years (see the end of chapter 4).

From W. Jason Morgan, 1972, *Geological Society of America Memoir 132,* and other sources

Many of the guyots and seamounts on the sea floor are aligned in chains. Such volcanic chains, together with some other ridges on the sea floor, are given the name **aseismic ridges** (figures 3.16 and 3.21); that is, they are submarine ridges that are not associated with earthquakes. The name *aseismic* is used to distinguish these features from the much larger mid-oceanic ridge, where earthquakes occur along the rift valley.

REEFS

Reefs are wave-resistant ridges of coral, algae, and other calcareous organisms. They form in warm, shallow, sunlit water that is low in suspended sediment. Reefs stand above the surrounding sea floor, which is often covered with sediment derived from the reef (see figure 14.18). Three important types are *fringing reefs, barrier reefs,* and *atolls* (figure 3.22).

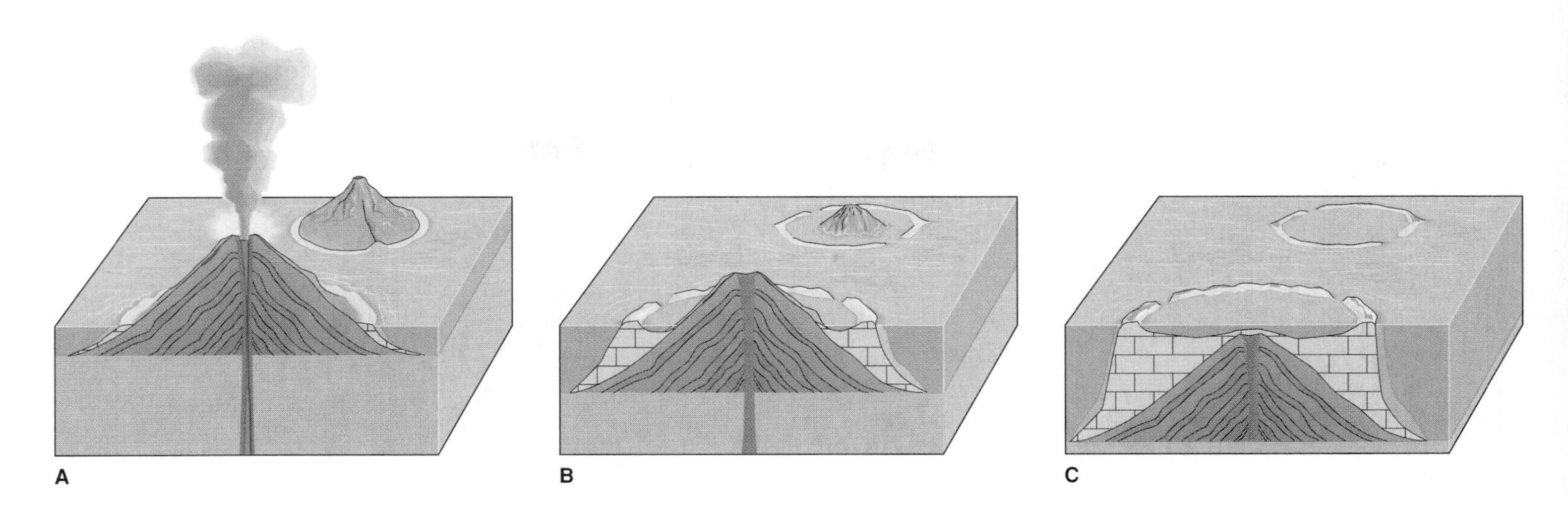

FIGURE 3.22

Types of coral-algal reefs. (*A*) Fringing reefs are attached directly to the island. (*B*) Barrier reefs are separated from the island by a lagoon. (*C*) Atolls are circular reefs with central lagoons. Charles Darwin proposed that the sequence of fringing, barrier, and atoll reefs form by the progressive subsidence of a central volcano, accompanied by the rapid upward growth of corals and algae. (*D*) Reefs on the island of Mooréa in the Society Islands, south-central Pacific (Tahiti in the background). Living corals appear brown. Fringing reef next to shoreline, barrier reef at breaker line; light blue lagoon between reefs is covered with carbonate sand. The island is an extinct volcano, heavily eroded by stream action.

Photo © David Hiser/Stone

Fringing reefs are flat, tablelike reefs attached directly to shore. The seaward edge is marked by a steep slope leading down into deeper water. Many of the reefs bordering the Hawaiian Islands are of this type.

Barrier reefs parallel the shore but are separated from it by wide, deep lagoons. This type of reef is shown in figure 14.18. The lagoon has relatively quiet water because the reef shelters it by absorbing the energy of large, breaking waves. A barrier reef lies about 8 kilometers offshore of the Florida Keys, a string of islands south of Miami. On a much grander scale is the Great Barrier Reef off northeastern Australia. It extends for about 2,000 kilometers along the coast, and its seaward edge lies up to 250 kilometers from shore. Another long barrier reef lies along the eastern coast of the Yucatan Peninsula in Central America, and others surround many islands in the South Pacific.

Atolls are circular reefs that rim lagoons. They are surrounded by deep water. Small islands of calcareous sand may be built by waves at places along the reef ring. The diameter of atolls varies from 1 to more than 100 kilometers. Numerous atolls dot the South Pacific. Bikini and Eniwetok atolls were used through 1958 for the testing of nuclear weapons by the United States.

Following the 4-year cruise of the HMS *Beagle* in the 1830s, Charles Darwin proposed that these three types of reefs are related to one another by subsidence of a central volcanic island, as shown in figure 3.22. A fringing reef initially becomes established near the island's shore. As the volcano slowly subsides because of tectonic lowering of the sea floor, the reef becomes a barrier reef, because the corals and algae grow rapidly upward, maintaining the reef's position near sea level. Less and less of the island remains above sea level, but the reef grows upward into shallow, sunlit water, maintaining its original size and shape. Finally the volcano disappears completely below sea level, and the reef becomes a circular atoll. Drilling through atolls in the 1950s showed that these reefs were built on deeply buried volcanic cores, thus confirming Darwin's hypothesis of 120 years before.

FIGURE 3.23

Photograph (taken through a scanning-electron microscope) of pelagic sediment from the floor of the Pacific Ocean. The sediment is made up of microscopic skeletons of single-celled marine organisms (large objects are foraminifera; smaller, sievelike ones are radiolaria about 0.05 mm in diameter).

SEDIMENTS OF THE SEA FLOOR

The basaltic crust of the sea floor is covered in many places with layers of sediment. This sediment is either *terrigenous,* derived from land, or *pelagic,* settling slowly through seawater.

Terrigenous sediment is land-derived sediment that has found its way to the sea floor. The sediment that makes up the continental rise and the abyssal plains is mostly terrigenous and apparently has been deposited by turbidity currents or similar processes. Once terrigenous sediment has found its way down the continental slope, contour currents may distribute it along the continental rise. On active continental margins, oceanic trenches may act as traps for terrigenous sediment and prevent it from spreading out onto the deep sea floor beyond the trenches.

Pelagic sediment is sediment that settles slowly through the ocean water. It is made up of fine-grained clay and the skeletons of microscopic organisms (figure 3.23). Fine-grained pelagic clay is found almost everywhere on the sea floor, although in some places it is masked by other types of sediments that accumulate rapidly. The clay is mostly derived from land; part of it may be volcanic ash. This sediment is carried out to sea primarily by wind, although rivers and ocean currents also help to distribute it.

Microscopic shells and skeletons of plants and animals also settle slowly to the sea floor when marine organisms of the surface waters die. In some parts of the sea, such as the polar and equatorial regions, great concentrations of these shells have built unusually thick pelagic deposits.

The constant slow rain of pelagic clay and shells occurs in all parts of the sea. Although the rate of accumulation varies from place to place, pelagic sediment should be expected on all parts of the sea floor.

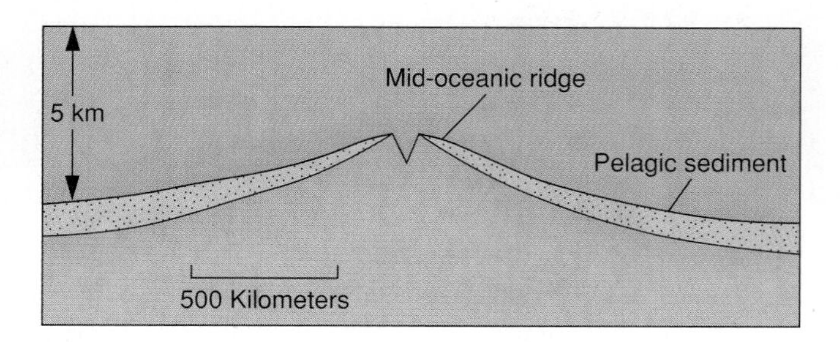

FIGURE 3.24

Pelagic sediment is thin or absent on the crest of the mid-oceanic ridge and becomes progressively thicker away from the ridge crest (distribution of sediment highly simplified).

Surprisingly, however, pelagic sediment is almost completely absent on the crest of the mid-oceanic ridge. Pelagic sediment is found on the flanks of the mid-oceanic ridge, often thickening away from the ridge crest (figure 3.24). But its absence on the ridge crest was an unexpected discovery about sea-floor sediment distribution.

OCEANIC CRUST AND OPHIOLITES

As you have seen in the previous chapter, oceanic crust differs significantly from continental crust; it is both thinner and of a different composition than continental crust. Seismic reflection and seismic refraction surveys at sea have shown the oceanic crust to be about 7 kilometers thick and divided into three layers (figure 3.25).

The top layer (Layer 1), of variable thickness and character, is marine sediment. In an abyssal fan or on the continental rise, Layer 1 may consist of several kilometers of terrigenous sediment. On the upper flanks of the mid-oceanic ridge, there may be less than 100 meters of pelagic sediment. An average thickness for Layer 1 might be 0.5 kilometer.

Beneath the sediment is Layer 2, which is about 1.5 kilometers thick. It has been extensively sampled. This layer consists of pillow basalt overlying dikes of basalt. The basalt pillows, rounded masses that form when hot lava erupts into cold water (figure 3.25), are highly fractured at the top of Layer 2. The dikes in the lower part of Layer 2 are closely spaced, parallel, vertical dikes ("sheeted dikes"). Widely sampled by drilling and dredging, Layer 2 has also been observed directly by geologists in submersibles diving into the rift valley of the mid-oceanic ridge during the FAMOUS expedition.

The lowest layer in the crust is Layer 3. It is about 5 kilometers thick and thought to consist of sill-like gabbro bodies. The evidence for this interpretation is suggestive but not conclusive. Geologists have drilled through 2 kilometers of pillow basalt and basaltic dikes in oceanic crust, but have not yet been able to reach gabbro by drilling through basalt. Gabbro (and other rocks) are exposed on some fault blocks in rift valleys and on some steep submarine cliffs in fracture zones; 0.5 kilometer of gabbro has been drilled on one such cliff in the southern Indian Ocean. Some geologists presume that the gabbro represents Layer 3 exposed by

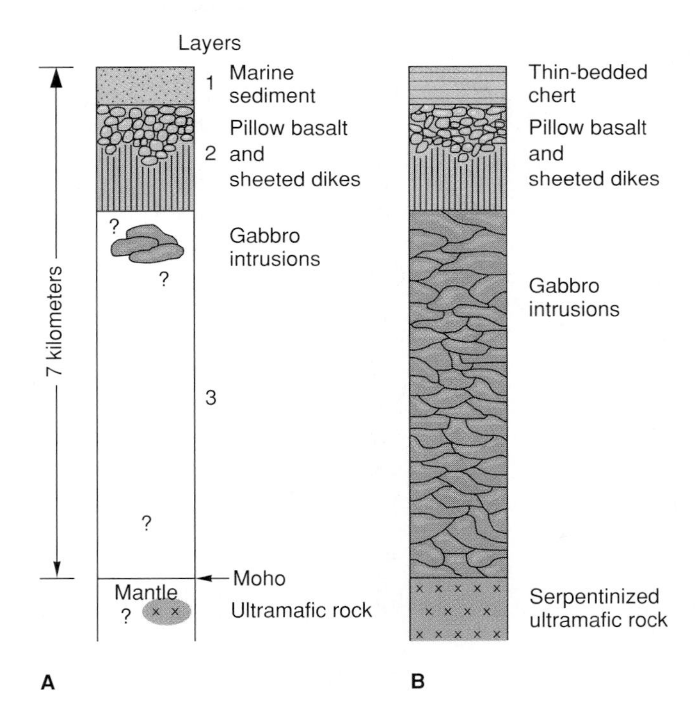

FIGURE 3.25

A comparison of oceanic crust and an ophiolite sequence. (*A*) Structure of oceanic crust, determined from seismic studies and drilling. Gabbro and ultramafic rock drilled and dredged from sea-floor fault blocks may be from Layer 3 and the mantle. (*B*) Typical ophiolite sequence found in mountain ranges on land. Thickness approximate—sequence is usually highly faulted. (*C*) Pillow basalt from the upper part of an ophiolite, northern California. These rocks formed as part of the sea floor, when hot lava cooled quickly in cold seawater.

3.2 ENVIRONMENTAL GEOLOGY

Geologic Riches in the Sea

Many resources are currently being extracted from the sea floor and from seawater, and in some instances there is a great potential for increased extraction.

Offshore oil and gas are the most valuable resources now being taken from the sea. More than one-sixth of the United States' oil production (and more than one-quarter of world production) comes from drilling platforms set up on the continental shelf (box figure 1). Oil and gas have been found within deeper parts of the sea floor, such as the continental slope and continental rise. Producing oil from these deeper regions is much more costly than present production; oil spills from wells in deep water would be especially hard to control.

Other important resources are dredged from the sea floor. *Phosphorite* can be recovered from shallow shelves and banks and used for fertilizers. *Gold, diamonds,* and heavy *black sands* (which are black because they contain metal-bearing minerals) are being separated from the surface sands and gravels of some continental shelves by specially designed ships.

Manganese nodules (box figure 2) cover many parts of the deep sea floor, notably in the central Pacific. These black, potato-sized lumps contain approximately 25% manganese, 15% iron, and up to 2% nickel and 2% copper, along with smaller amounts of cobalt. Although there are international legal problems concerning who owns them, larger industrial countries such as the United States may mine them, particularly for copper, nickel, and cobalt. The United States is also interested in manganese; it imports 95% of its manganese, which is critical to producing some types of steel.

Metallic brines and sediments, first discovered in the Red Sea, are deposited by submarine hot springs active at the rift valley of the mid-oceanic ridge crest. The Red Sea sediments contain more than 1% copper and more than 3% zinc, together with impressive amounts of silver, gold, and lead (worth $25 billion according to a 1983 estimate). Because of their great value, the sediments will probably be mined even though they are at great depth. Deposits similar to those of the Red Sea—although not of such great economic potential—have been found on several other parts of the ridge.

A few substances can be extracted from the salts dissolved in seawater. Approximately two-thirds of the world's production of *magnesium* is obtained from seawater, and in many regions *sodium chloride* (table salt) is obtained by solar evaporation of seawater.

Box 3.2 — FIGURE 1

Offshore oil drilling platform. As many as 50 different wells can be drilled from a single platform.

Photo by U.S. Geological Survey

Box 3.2 — FIGURE 2

Dense concentration of manganese nodules on an abyssal plain in the South Pacific, depth 5,300 meters. Field of view about 1.5 by 2 meters.

From Lamont-Doherty Geological Observatory

faulting, but other geologists are not so sure. There is some evidence that rocks exposed in fracture zones may not be representative of the entire sea floor. Even if the drilled gabbro *does* represent the upper 1/10 of Layer 3, no one has yet sampled the lower 9/10 of the layer.

Seismic velocities of 7 kilometers per second are consistent with the choice of gabbro for Layer 3, but deep drilling on land (see box 2.1) has shown that seismic reflection and refraction records on land are routinely misinterpreted regarding rock type and depths to rock boundaries. These errors extend to oceanic crust as well. A drillhole in the eastern Pacific Ocean has been reoccupied four times in a 12-year span, and has now reached a total depth of 2,000 meters below the sea floor. Seismic evidence suggested that the Layer 2–Layer 3 boundary would be found at a depth of about 1,700 meters, but the drill went well past that depth without finding the contact between the dikes of Layer 2 and the expected gabbro of Layer 3. Either the seismic interpretation or the model of Layer 3's composition must be wrong.

Geologists' ideas about the composition of oceanic crust are greatly influenced by a study of **ophiolites,** distinctive rock sequences found in many mountain chains on land (figure 3.25). The thin top layer of an ophiolite consists of marine sedimentary rock, often including thin-bedded chert. Below the sedimentary rock lies a zone of pillow basalt, which in turn is underlain by a sheeted-dike complex that probably served as feeder dikes for the pillowed lava flows above. The similarities between the upper part of an ophiolite and Layers 1 and 2 of oceanic crust are obvious.

Below the sheeted dikes of an ophiolite is a thick layer of podlike gabbro intrusions, perhaps thick sills. Beneath the gabbro lies ultramafic rock such as peridotite, the top part of which has been converted to serpentine by metamorphism. Geologists have long thought that ophiolites represent slivers of oceanic crust somehow emplaced on land. If this is true, then the gabbro of ophiolites may represent oceanic Layer 3, and the peridotite may represent mantle rock below oceanic crust. The contact between the peridotite and the overlying gabbro would be the Mohorovičić discontinuity.

Recent work has shown that many, if not most, ophiolites are *not* typical sea floor, but a special type of sea floor formed in marginal ocean basins next to continents by the process of backarc spreading (chapter 4). If ophiolites do not represent typical sea floor, then more extensive, deeper sea-floor drilling is needed before a clear picture of oceanic crust can be formed. This is an important goal, for oceanic crust is the most common surface rock, covering 60% of Earth's surface.

THE AGE OF THE SEA FLOOR

As marine geologists began to determine the age of sea-floor rocks (by isotopic dating) and sediments (by fossils), an astonishing fact was discovered. All the rocks and sediments of the sea floor proved to be younger than 200 million years old. This was true only for rocks and sediments from the *deep* sea floor, not those from the continental margins. The rocks and sediments presently found on the deep sea floor formed during the Mesozoic and Cenozoic eras, but not earlier.

By contrast, as will be discussed in chapter 8, Earth is estimated to be 4.5 billion years old. Every continent contains some rocks formed during the Paleozoic Era and the Precambrian. Some of the Precambrian rocks on continents are more than 3 billion years old, and a few are almost 4 billion years old. Continents, therefore, preserve rocks from most of Earth's history. In sharp contrast to the continents is the deep sea floor, which covers more than half of Earth's surface but preserves less than one-twentieth of its history in its rocks and sediment.

THE SEA FLOOR AND PLATE TECTONICS

As we mentioned at the beginning, this chapter is concerned with the *description* of the sea floor. The *origin* of most sea-floor features is related to plate tectonics. The next chapter shows you how the theory of plate tectonics explains the existence and character of continental shelves and slopes, trenches, the mid-oceanic ridge, and fracture zones as well as the very young age of the sea floor itself. In the workings of the scientific method, this chapter largely concerns *data.* The next chapter shows how *hypotheses* and *theories* account for these data.

Summary

The *continental shelf* and the steeper *continental slope* lie under water along the edges of continents. They are separated by a change in slope angle at a depth of 100 to 200 meters.

Submarine canyons are cut into the continental slope and outer continental shelf by a combination of *turbidity currents,* sand flow and fall, bottom currents, and river erosion during times of lower sea level. Graded bedding and cable breaks suggest the existence of turbidity currents in the ocean.

Abyssal fans form as sediment collects at the base of submarine canyons.

A *passive continental margin* occurs off geologically quiet coasts and is marked by a continental rise and abyssal plains at the base of the continental slope.

The *continental rise* and *abyssal plains* may form from sediment deposited by turbidity currents.

The continental rise may also form from sediment deposited by *contour currents* at the base of the continental slope.

An *active continental margin* is marked by an *oceanic trench* at the base of the continental slope; the continental rise and abyssal plains are absent.

Oceanic trenches are twice as deep as abyssal plains, which generally lie at a depth of 5 kilometers. Associated with trenches are *Benioff zones* of earthquakes and *andesitic volcanism,* forming either an island arc or a chain of volcanoes in a young mountain belt near the edge of a continent. Trenches have low heat flow and negative gravity anomalies.

The *mid-oceanic ridge* is a globe-circling mountain range of basalt, located mainly in the middle of ocean basins. The crest of the ridge is marked by a *rift valley,* shallow-focus earthquakes, high heat flow, active *basaltic volcanism,* hydrothermal activity, and exotic organisms.

Fracture zones are lines of weakness that offset the mid-oceanic ridge. Shallow-focus quakes occur on the portion of the fracture zone between the offset ridge segments.

Seamounts are conical, submarine volcanoes, now mostly extinct. *Guyots* are flattopped seamounts, probably leveled by wave erosion before subsiding.

Chains of seamounts and guyots form *aseismic ridges.*

Corals and algae living in warm, shallow water construct *fringing reefs, barrier reefs,* and *atolls.*

Terrigenous sediment is composed of land-derived sediment deposited near land by turbidity currents and other processes. *Pelagic sediment* is made up of wind-blown dust and microscopic skeletons that settle slowly to the sea floor.

The crest of the mid-oceanic ridge lacks pelagic sediment.

Oceanic crust consists of basalt pillows and dikes, probably overlying gabbro.

Ophiolites in continental mountain ranges probably represent slivers of somewhat atypical oceanic crust somehow emplaced on land.

The oldest rocks on the deep sea floor are 200 million years old. The continents, in contrast, contain some rock that is 3 to 4 billion years old.

Terms to Remember

abyssal fan 56
abyssal plain 59
active continental margin 59
aseismic ridge 64
atoll 66
barrier reef 66
continental rise 58
continental shelf 56
continental slope 56
contour current 58
fracture zone 62
fringing reef 66
guyot 63
mid-oceanic ridge 60
oceanic trench 59
ophiolite 69
passive continental margin 58
pelagic sediment 66
reef 64
rift valley 60
seamount 62
submarine canyon 56
terrigenous sediment 66
turbidity current 57

Testing Your Knowledge

Use the questions below to prepare for exams based on this chapter.

1. What is a submarine canyon? How do submarine canyons form?
2. Discuss the appearance, structure, and origin of abyssal plains.
3. Sketch a cross profile of the mid-oceanic ridge, showing the rift valley. Label your horizontal and vertical scales.
4. Sketch an active continental margin and a passive continental margin, labeling all their parts. Show approximate depths.
5. What is a fracture zone? Sketch the relation between fracture zones and the mid-oceanic ridge.
6. In a sketch, show the association between an oceanic trench, a Benioff zone of earthquakes, and volcanoes on the edge of a continent.
7. Describe two different origins for the continental rise.
8. What is a turbidity current? What is the evidence that turbidity currents occur on the sea floor?
9. Describe the appearance and origin of seamounts and guyots.
10. Describe the two main types of sea-floor sediment.
11. How does the age of sea-floor rocks compare with the age of continental rocks? Be specific.
12. Sketch a cross section of a fringing reef, a barrier reef, and an atoll.
13. What is the thickness and composition of oceanic crust?
14. Most ocean water probably came from
    a. degassing of Earth's interior
    b. outer space
15. Which is true of the continental shelf?
    a. it is a shallow submarine platform at the edge of a continent
    b. it inclines very gently seaward
    c. it can vary in width
    d. all of the above
16. The average angle of slope for a continental slope is
    a. 1°–2° b. 3°–4°
    c. 4°–5° d. greater than 10°
17. Great masses of sediment-laden water that are pulled downhill by gravity are called
    a. contour currents b. bottom currents
    c. turbidity currents d. traction currents
18. Oceanic trenches
    a. are narrow deep troughs
    b. run parallel to the edge of a continent or an island arc
    c. are often 8–10 km deep
    d. all of the above

19. Which is characteristic of mid-oceanic ridges?
    a. shallow-focus earthquakes
    b. high heat flow
    c. basalt eruptions
    d. all of the above
20. Reefs parallel to the shore but separated from it by wide, deep lagoons are called
    a. fringing reefs b. barrier reefs
    c. atolls d. lagoonal reefs
21. Pelagic sediment could be composed of
    a. fine-grained clay
    b. skeletons of microscopic organisms
    c. volcanic ash
    d. all of the above
22. What part of the continental margin marks the true edge of the continent?
    a. continental shelf b. continental slope
    c. continental rise d. abyssal plain
23. Distinctive rock sequences of basalt and marine sedimentary rock that may be slices of the ocean floor are
    a. guyots b. ophiolites
    c. seamounts d. fracture zones

Expanding Your Knowledge

1. How many possible origins can you think of for the rift valley on the mid-oceanic ridge?
2. What factors could cause sea level to rise? To fall?
3. Why is the rock of the deep sea floor (60% of Earth's surface) basalt? Where did the basalt come from?
4. How could fracture zones have formed?
5. How many hypotheses can you think of to explain the relatively young age of the sea floor?

Exploring Web Resources

www.mhhe.com/plummer9e
Visit our Online Learning Center for additional readings, media resources, and a great animation to further your understanding of reefs and atolls. Try out the interactive quizzes and crossword puzzles. Click on the direct links to the websites listed below and learn more about the sea floor.

http://seawifs.gsfc.nasa.gov:80/OCEAN_PLANET/HTML/ocean_planet_ocean_science.html
Smithsonian Ocean Planet Exhibit contains many links to sea floor topics such as hydrothermal vents and microbiology of deep-sea vents.

www.esdim.noaa.gov/ocean_page.html
National Oceanographic and Atmospheric Administration (NOAA) has many links to numerous sites relating to the oceans.

www.ngdc.noaa.gov/mgg/mggd.html
NOAA Marine Geology and Geophysics Division website contains world sea floor maps and information on ocean-drilling data and samples.

www-odp.tamu.edu/
Ocean Drilling Program (ODP) website contains information on the nature and history of the ocean crust and provides data from research projects.

www.whoi.edu/VideoGallery/
Woods Hole Oceanographic Institution contains video clips of black smokers, exotic organisms at mid-oceanic ridges, and oceanographic research using vessels and submersibles.

www.pmel.noaa.gov/vents/
NOAA Vents Program website provides photos, video clips, data, and research program activities about the investigation of submarine volcanoes and hydrothermal venting around the world.

Animations

This chapter includes the following animations available on our Online Learning Center at www.mhhe.com/plummer9e.

3.22 Fringing Reefs and Atolls

CHAPTER

4

Plate Tectonics

Once you study volcanoes, igneous, metamorphic and sedimentary rocks, and earthquakes, you will learn how these topics are related to plate tectonics. In this chapter we take a closer look at plates and plate motion. We will pay particular attention to plate boundaries and the possible driving mechanisms for plate motion.

The history of the concept of plate tectonics is a good example of how scientists think and work and how a hypothesis can be proposed, discarded, modified, and then reborn. In the first part of this chapter we trace the evolution of an idea—how the earlier hypotheses of moving continents (continental drift) and a moving sea floor (seafloor spreading) were combined to form the theory of plate tectonics.

Opposite: Satellite image of the Red Sea and Arabia. Plate motion has torn the Arabian Peninsula (right center) away from Africa (left) to form the Red Sea (center).
Photo courtesy Eros Data Center/U.S. Geological Survey

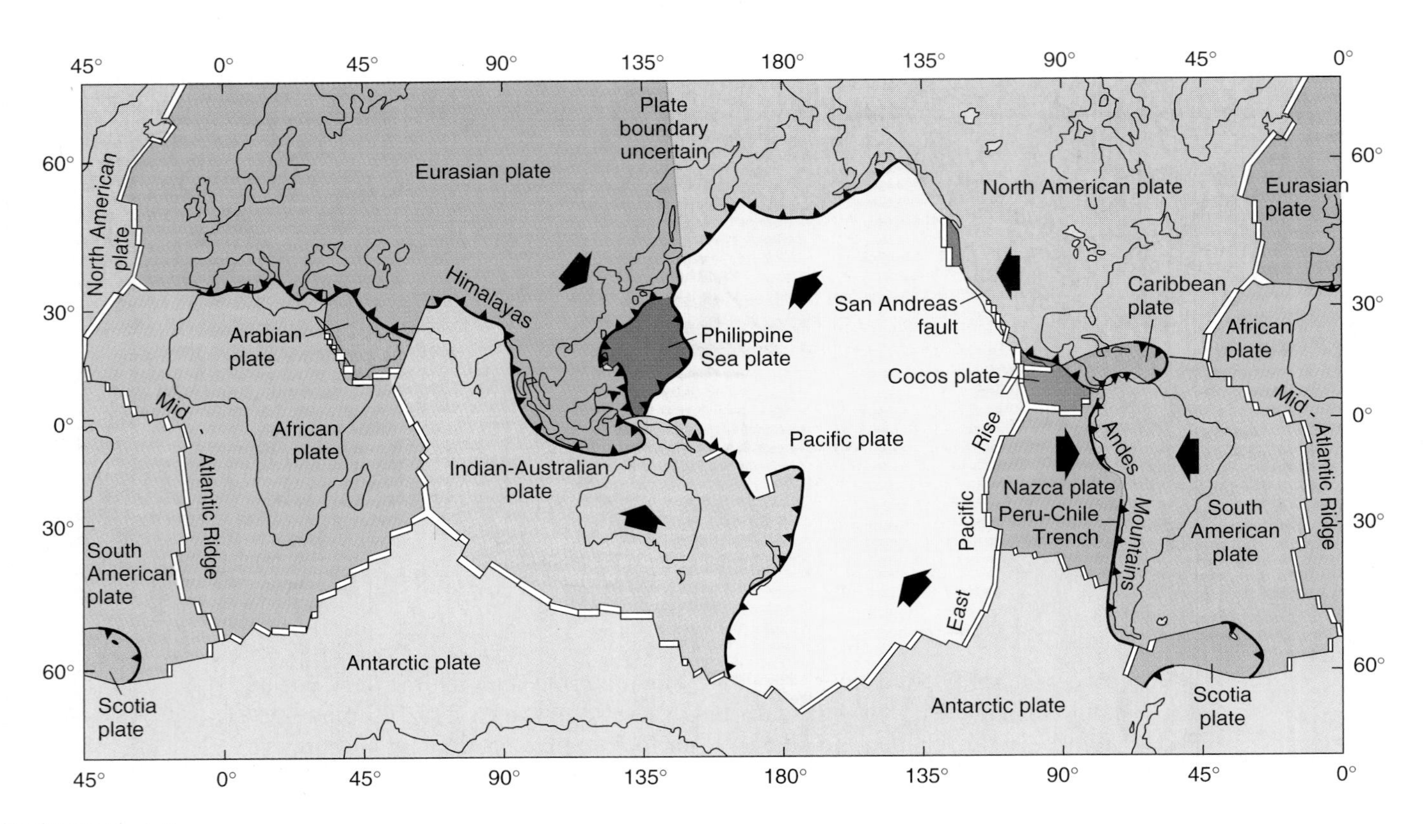

FIGURE 4.1

The major plates of the world. The western edge of the map repeats the eastern edge so that all plates can be shown unbroken. Double lines indicate spreading axes on divergent plate boundaries. Single lines show transform boundaries. Heavy lines with triangles show convergent boundaries, with triangles pointing down subduction zones.

Modified from W. Hamilton, U.S. Geological Survey

Tectonics is the study of the origin and arrangement of the broad structural features of Earth's surface, including not only folds and faults, but also mountain belts, continents, and earthquake belts. Tectonic models such as an expanding Earth or a contracting Earth have been used in the past to explain *some* of the surface features of Earth. Plate tectonics has come to dominate geologic thought today because it can explain so *many* features. The basic idea of **plate tectonics** is that Earth's surface is divided into a few large, thick plates that move slowly and change in size. Intense geologic activity occurs at *plate boundaries* where plates move away from one another, past one another, or toward one another. The eight large plates shown in figure 4.1, plus a few dozen smaller plates, make up the outer shell of Earth (the crust and upper part of the mantle).

The concept of plate tectonics was born in the late 1960s by combining two preexisting ideas—continental drift and sea-floor spreading. **Continental drift** is the idea that continents move freely over Earth's surface, changing their positions relative to one another. **Sea-floor spreading** is a hypothesis that the sea floor forms at the crest of the mid-oceanic ridge, then moves horizontally away from the ridge crest toward an oceanic trench. The two sides of the ridge are moving in opposite directions like slow conveyor belts.

Before we take a close look at plates, we will examine the earlier ideas of moving continents and a moving sea floor because these two ideas embody the theory of plate tectonics.

THE EARLY CASE FOR CONTINENTAL DRIFT

Continents can be made to fit together like pieces of a picture puzzle. The similarity of the Atlantic coastlines of Africa and South America has long been recognized. The idea that continents were once joined together, and have split and moved apart from one another, has been around for more than 130 years (figure 4.2).

In the early 1900s Alfred Wegener, a German meteorologist, made a strong case for continental drift. He noted that South America, Africa, India, Antarctica, and Australia had almost identical late Paleozoic rocks and fossils.

The plant *Glossopteris* is found in Pennsylvanian and Permian-age rock on all five continents and fossil remains of *Mesosaurus,* a freshwater reptile, is found in Permian-age rocks only in Brazil and South Africa (figure 4.3). In addition, fossil remains of land-dwelling reptiles *Lystrosaurus* and *Cynognathus* are found in Triassic-age rocks on all five continents.

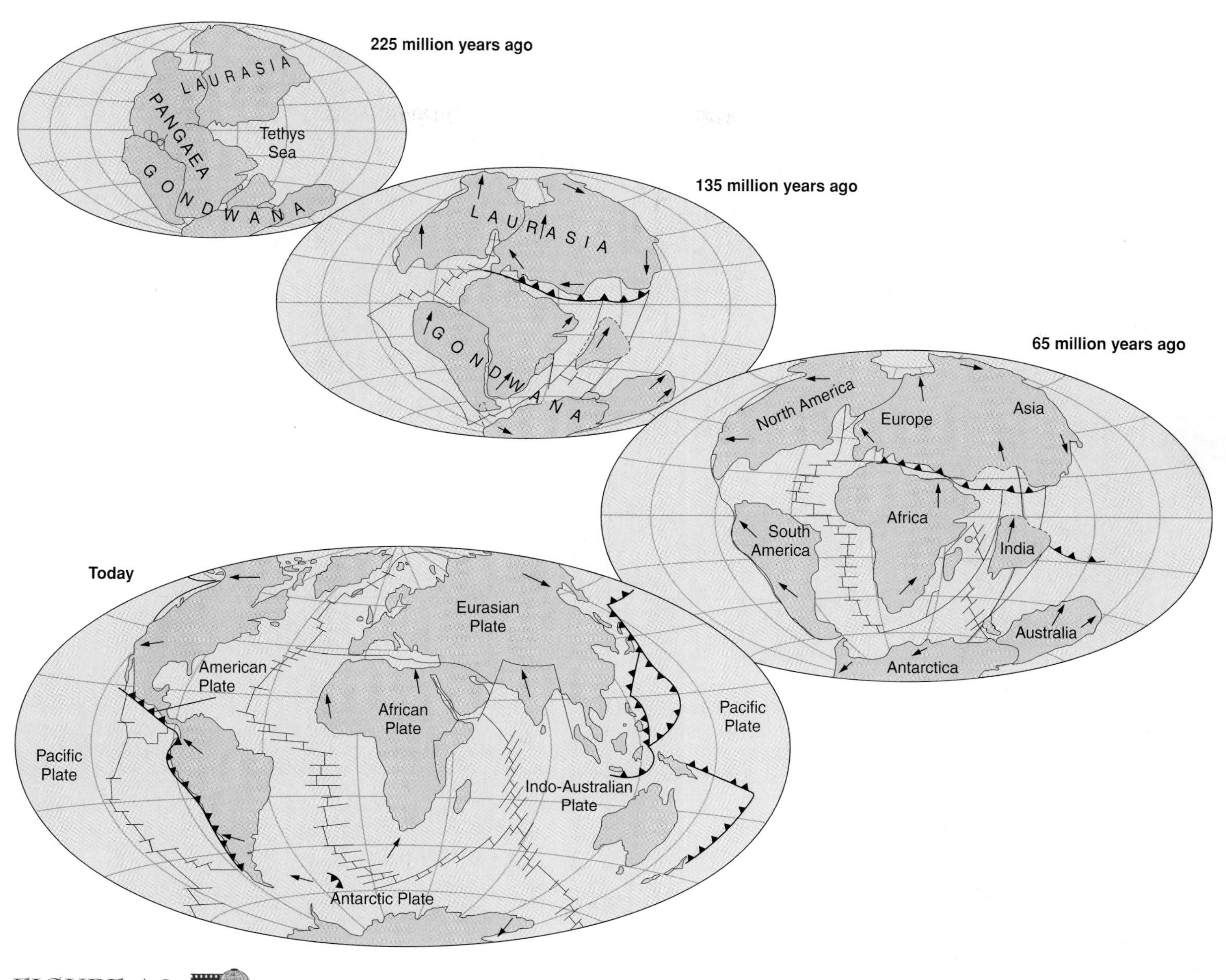

FIGURE 4.2

Pangaea breakup and continental drift.
From American Petroleum Institute

Wegener reassembled the continents to form a giant supercontinent *Pangaea* (also spelled Pangea today). Wegener thought that the similar rocks and fossils were easier to explain if the continents were joined together rather than in their present, widely scattered positions.

Pangaea initially separated into two parts. *Laurasia* was the northern supercontinent, containing what is now North America and Eurasia (excluding India). *Gondwanaland* was the southern supercontinent, composed of all the present-day southern-hemisphere continents and India (which has drifted north).

The distribution of Late Paleozoic glaciation strongly supports the idea of Pangaea (figure 4.4). The Gondwanaland continents (the southern-hemisphere continents and India) all have glacial deposits of Late Paleozoic age. If these continents were spread over Earth in Paleozoic time as they are today, a climate cold enough to produce extensive glaciation would have had to prevail over almost the whole world. Yet no evidence has been found of widespread Paleozoic glaciation in the northern hemisphere. In fact, the late Paleozoic coal beds of North America and Europe were being laid down at that time in swampy, probably warm environments. If the continents are arranged according to Wegener's Pangaea reconstruction, then glaciation in the southern hemisphere is confined to a much smaller area (figure 4.4), and the absence of widespread glaciation in the northern hemisphere becomes easier to explain. Also, the present arrangement of the continents would require that late Paleozoic ice sheets flowed from the oceans toward the continents, which is impossible.

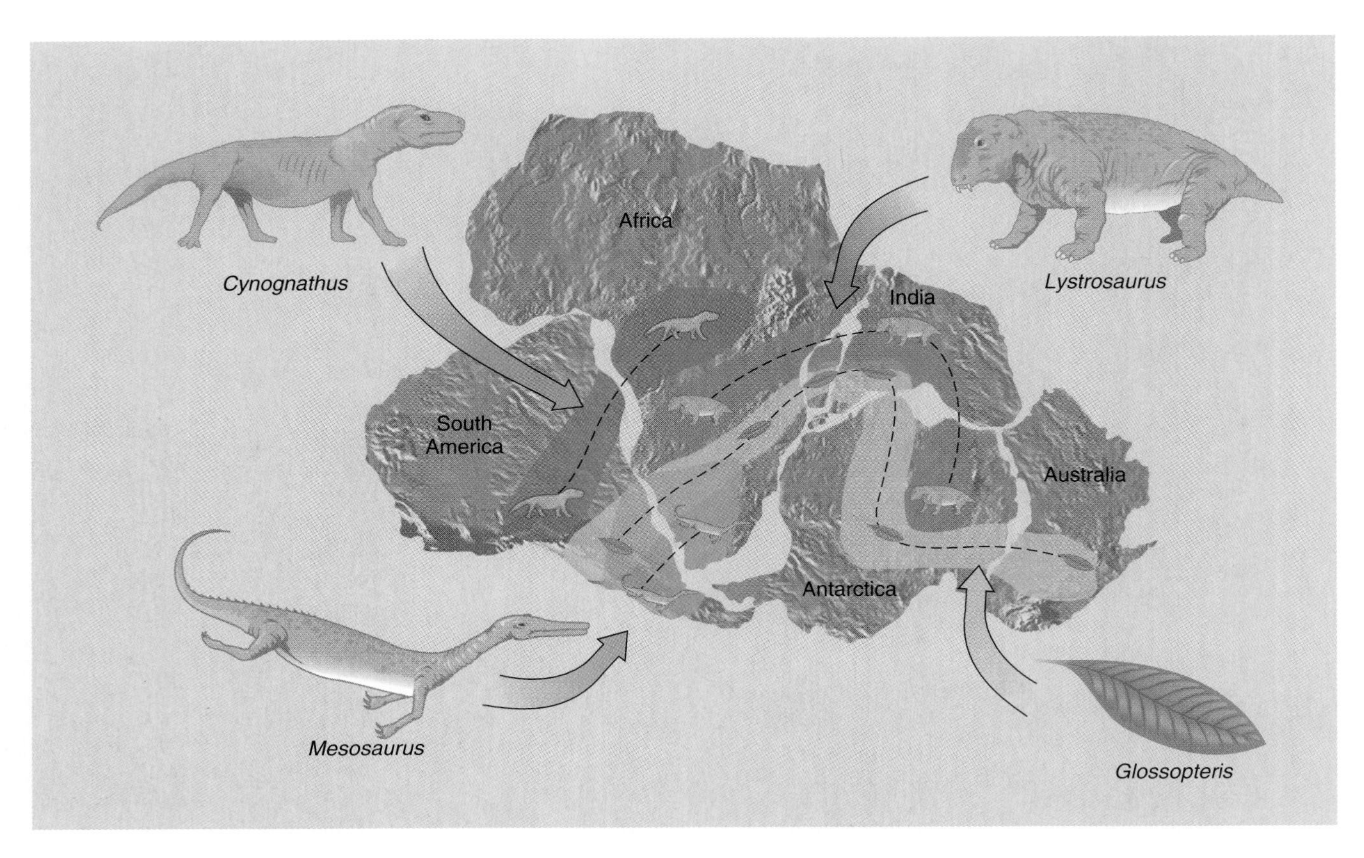

FIGURE 4.3

Distribution of plant and animal fossils that are found on the continents of South America, Africa, Antarctica, India, and Australia give evidence for the southern supercontinent of Gondwana. *Glossopteris,* and other fern-like plants, are found in Permian- and Pennsylvanian-age rocks on all five continents. *Cynognathus* and *Lystrosaurus* were sheep-sized land reptiles that lived during the Early Triassic Period. Fossils of the freshwater reptile *Mesosaurus* are found in Permian-age rocks on the southern tip of Africa and South America.

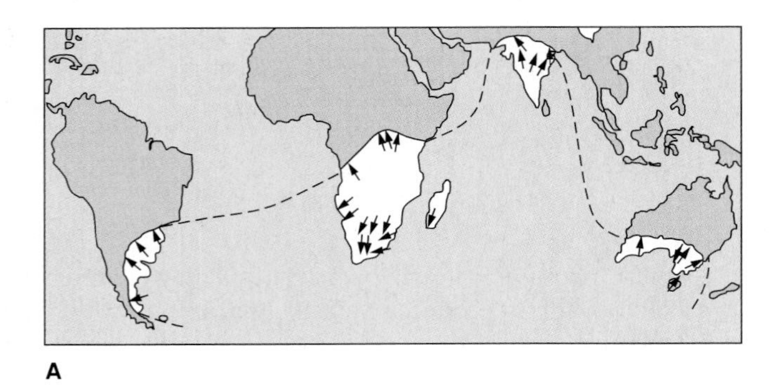

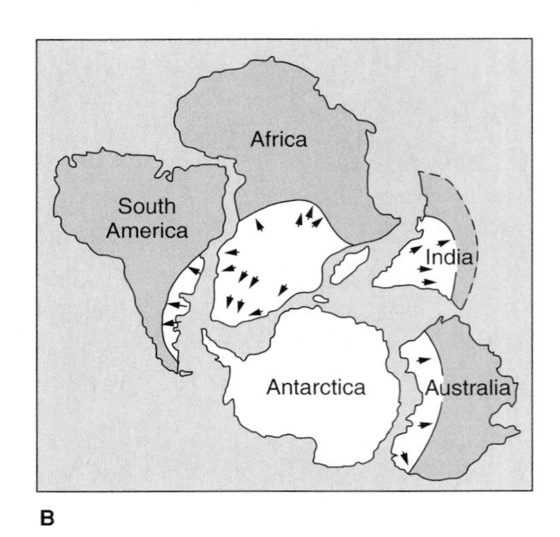

FIGURE 4.4

Distribution of late Paleozoic glaciations; arrows show direction of ice flow. (*A*) Continents in present positions show wide distribution of glaciation (white land areas with flow arrows). (*B*) Continents reassembled into Pangaea. Glaciated region becomes much smaller.

From Arthur Holmes, 1965, *Principles of Physical Geology,* 2d ed., Ronald Press

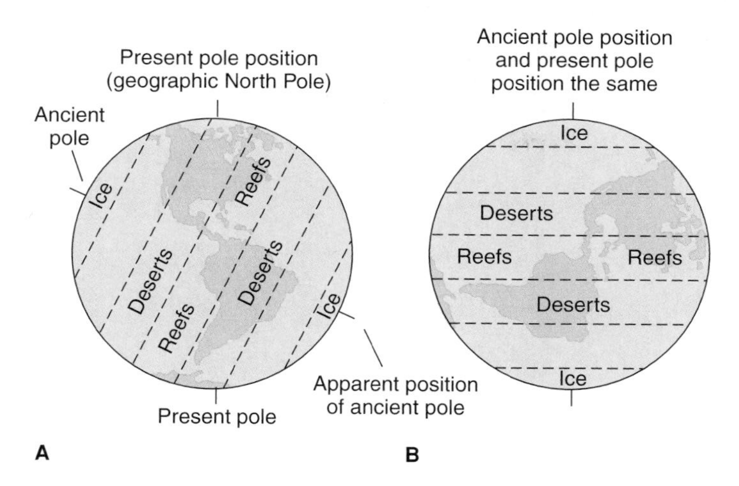

FIGURE 4.5

Two ways of interpreting the distribution of ancient climate belts. (*A*) Continents fixed, poles wander. (*B*) Poles fixed, continents drift. For simplicity, the continents in (*B*) are shown as having moved as a unit, without changing positions relative to one another. If continents move, they should change relative positions, complicating the pattern shown.

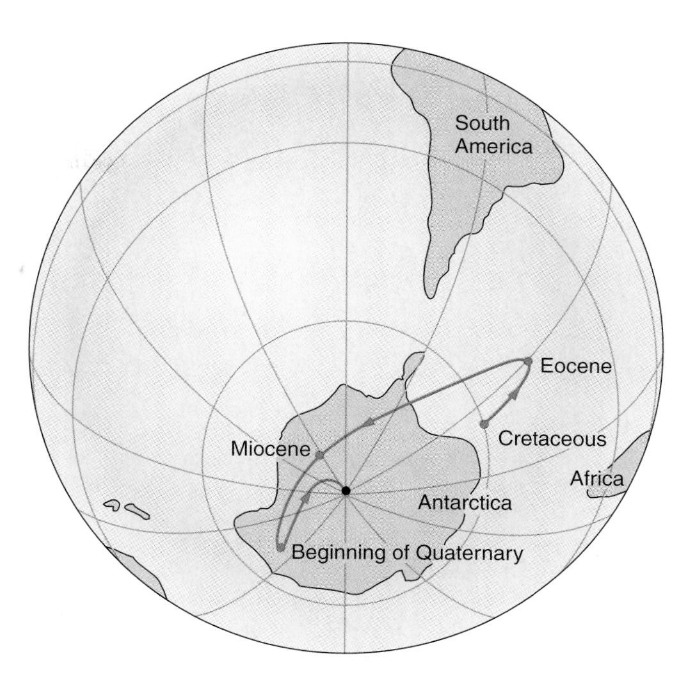

FIGURE 4.6

Apparent wandering of the South Pole since the Cretaceous Period as determined by Wegener from paleoclimate evidence. Wegener, of course, believed that *continents* rather than poles moved.

From A. Wegener, 1928, *The Origins of Continents and Oceans,* reprinted and copyrighted. 1968, Dover Publications

Wegener also reconstructed old climate zones (the study of ancient climates is called *paleoclimatology*). Glacial till and striations indicate a cold climate near the North or South Pole. Coral reefs indicate warm water near the Equator. Crossbedded sandstones can indicate ancient deserts near 30° North and 30° South latitude. If ancient climates had the same distribution on Earth that modern climates have, then sedimentary rocks can show where the ancient poles and Equator were located.

Wegener determined the positions of the North and South Poles for each geologic period. He found that ancient poles were in different positions than the present poles (figure 4.5*A*). This apparent movement of the poles he called **polar wandering.** Polar wandering, however, is a deceptive term. The evidence can actually be explained in the following ways:

1. The continents remained motionless and the poles actually *did* move—polar wandering (figure 4.5*A*).
2. The poles stood still and the continents moved—continental drift (figure 4.5*B*).
3. Both occur—this is the most likely scenario.

Wegener plotted curves of apparent polar wandering (figure 4.6). Since one interpretation of polar wandering data was that the continents moved, Wegener believed that this supported his concept of continental drift. (Notice that in only one interpretation of polar wandering do the poles actually move. You should keep in mind that when geologists use the term *polar wandering* they are referring to an *apparent* motion of the poles, which may or may not have actually occurred.)

Skepticism about Continental Drift

Although Wegener presented the best case possible in the early 1900s for continental drift, much of his evidence was not clear-cut. The presence of land-dwelling reptiles throughout the scattered continents was explained by land bridges, which were postulated to somehow rise up from the sea floor and then subside again. The existence or nonexistence of land bridges was difficult to prove without data on the topography of the sea floor. Also, fossil plants could have been spread from one continent to another by winds or ocean currents. Their distribution over more than one continent does not *require* that the continents were all joined in the supercontinent, Pangaea. In addition, polar wandering might have been caused by moving poles rather than by moving continents. Because his evidence was not conclusive, Wegener's ideas were not widely accepted. This was particularly true in the United States, largely because of the mechanism Wegener proposed for continental drift.

Wegener proposed that continents plowed through the oceanic crust (figure 4.7), perhaps crumpling up mountain ranges on the leading edges of the continents where they pushed against the sea floor. Most geologists in the United States thought that this idea violated what was known about the strength of rocks at the time. The driving mechanism proposed by Wegener for continental drift was a combination of centrifugal force from Earth's rotation and the gravitational forces that cause tides. Careful calculations of these forces showed them to

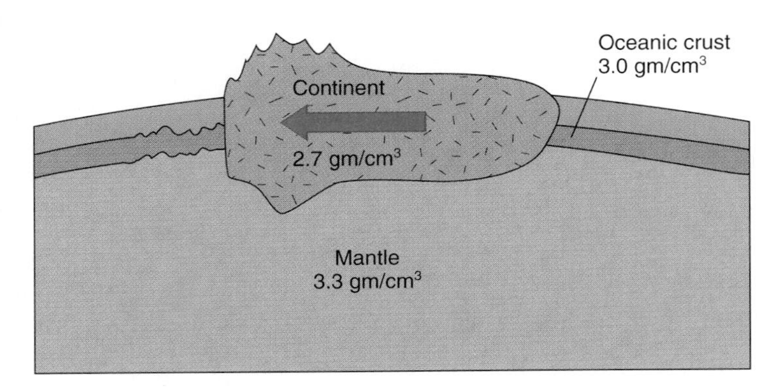

FIGURE 4.7

Wegener's concept of continental drift implied that the less dense continents drifted *through* oceanic crust, crumpling up mountain ranges on their leading edges as they pushed against oceanic crust.

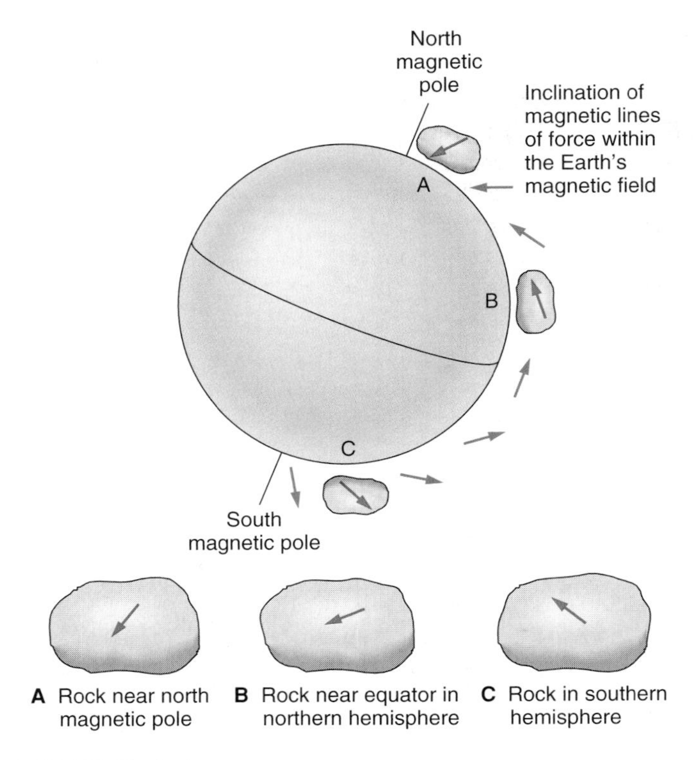

FIGURE 4.8

Magnetic dip (inclination) increases toward the north magnetic pole. Rocks in bottom part of figure are small samples viewed horizontally at locations A, B, and C on the globe. The magnetic dip can therefore be used to determine the distance from a rock to the north magnetic pole.

be too small to move continents. Because of these objections, Wegener's ideas received little support in the first half of the twentieth century in the United States or in much of the northern hemisphere (where the great majority of geologists live). The few geologists in the southern hemisphere, however, where Wegener's matches of fossils and rocks between continents were more evident, were more impressed with the concept of continental drift.

PALEOMAGNETISM AND THE REVIVAL OF CONTINENTAL DRIFT

Much work in the 1940s and 1950s set the stage for the revival of the idea of continental drift and its later incorporation, along with sea-floor spreading, into the new concept of plate tectonics. The new investigations were in two areas: (1) study of the sea floor and (2) geophysical research, especially in relation to rock magnetism.

Convincing new evidence about polar wandering came from the study of rock magnetism. Wegener's work dealt with the wandering of Earth's *geographic* poles of rotation. The *magnetic* poles are located close to the geographic poles, as you saw in the chapter on Earth's interior. Historical measurements show that the position of the magnetic poles moves from year to year, but that the magnetic poles stay close to the geographic poles as they move. As we discuss magnetic evidence for polar wandering, we are referring to an apparent motion of the magnetic poles. Because the magnetic and geographic poles are close together, our discussion will refer to apparent motion of the geographic poles as well.

As we discussed in chapter 2, many rocks record the strength and direction of Earth's magnetic field at the time the rocks formed. Magnetite in a cooling basaltic lava flow acts like a tiny compass needle, preserving a record of Earth's magnetic field when the lava cools below the *Curie point.* Iron-stained sedimentary rocks such as red shale can also record Earth magnetism. The magnetism of old rocks can be measured to determine the direction and strength of the magnetic field in the past. The study of ancient magnetic fields is called *paleomagnetism.*

Because magnetic lines of force dip more steeply as the north magnetic pole is approached, the inclination (dip) of the magnetic alignment preserved in the magnetite minerals in the lava flows can be used to determine the distance from a flow to the pole at the time that the flow formed (figure 4.8).

Old pole positions can be determined from the magnetism of old rocks. The magnetic alignment preserved in magnetite minerals points to the pole, and the dip of the alignment tells how far away the pole was. Figure 4.9 shows how Permian lava flows in North America indicate a Permian pole position in eastern Asia.

For each geologic period, North American rocks reveal a different magnetic pole position; this path of the *apparent* motion of the north magnetic pole through time is shown in figure 4.10. Paleomagnetic evidence thus verifies Wegener's idea of polar wandering (which he based on paleoclimatic evidence).

Like Wegener's paleoclimatic evidence, the paleomagnetic evidence from a *single* continent can be interpreted in two ways: either the continent stood still and the magnetic pole moved or the pole stood still and the continent moved. At first glance, paleomagnetic evidence does not seem to be a significant advance over paleoclimatic evidence. But when paleomagnetic evidence from *different* continents was compared, an important discovery was made.

Although Permian rocks in North America point to a pole position in eastern Asia, Permian rocks in *Europe* point to a different position (closer to Japan), as shown in figure 4.10. Does this mean there were *two* north magnetic poles in the Permian Period? In fact, every continent shows a different position for the Permian pole. A different magnetic pole for each continent seems highly unlikely. A better explanation is that a single pole stood still while continents split apart and rotated as they diverged.

Note the polar wandering paths for North America and Europe in figure 4.10. The paths are of similar shape, but the path for European poles is to the east of the North American path. If we mentally push North America back toward Europe, closing the Atlantic Ocean, and then consider the paths of polar wandering, we find that the path for North America lies exactly on the path for Europe. This strongly suggests that there was one north magnetic pole and that the continents were joined together. There appear to be two north magnetic poles because the rocks of North America moved west; their magnetic minerals now point to a different polar position than they did when the minerals first formed.

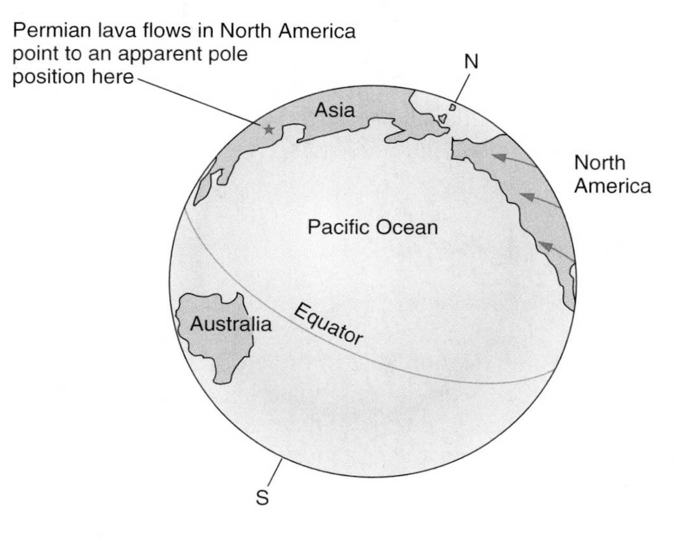

FIGURE 4.9

Paleomagnetic studies of Permian lava flows on North America indicate an apparent position for the north magnetic pole in eastern Asia.

Recent Evidence for Continental Drift

As paleomagnetic evidence revived interest in continental drift, new work was done on fitting continents together. By defining the edge of a continent as the middle of the continental slope, rather than the present (constantly changing) shoreline, a much more precise fit has been found between continents (figure 4.11).

The most convincing evidence for continental drift came from greatly refined rock matches between now-separated continents. If continents are fitted together like pieces of a jigsaw puzzle, the "picture" should match from piece to piece.

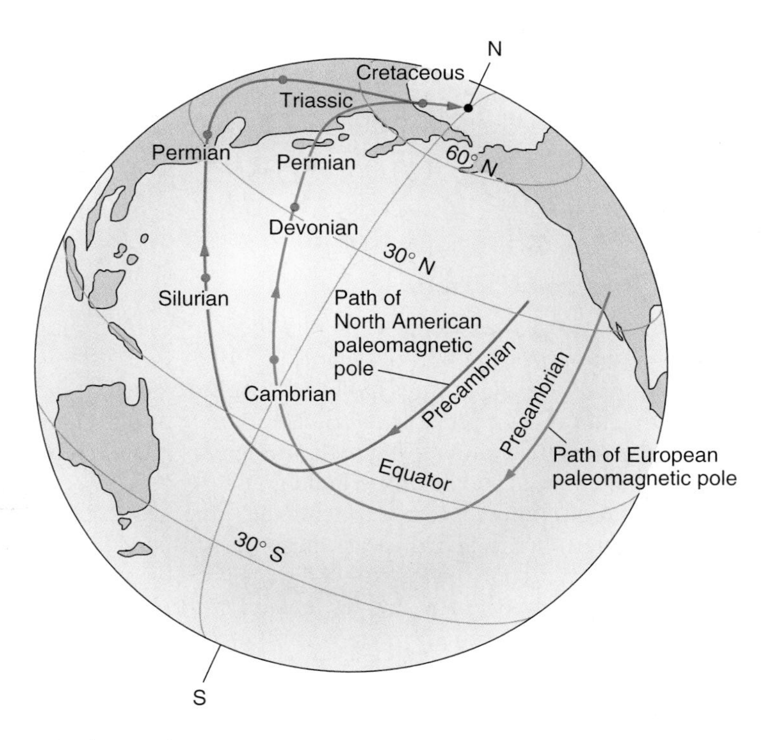

FIGURE 4.10

Polar wandering of the north magnetic pole as determined from measurements of rocks from North America and Europe.

From A. Cox and R. R. Doell, 1960, *Geological Society of America Bulletin*

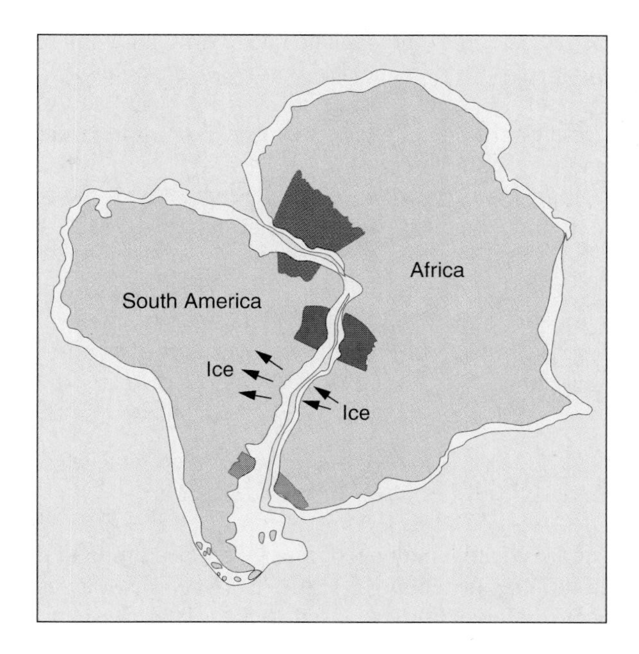

FIGURE 4.11

Jigsaw puzzle fit and matching rock types between South America and Africa. Light-blue areas around continents are continental shelves (part of continents). Colored areas within continents are broad belts of rock that correlate in type and age from one continent to another. Arrows show direction of glacier movement as determined from striations.

The matches between South America and Africa are particularly striking. Some distinctive rock contacts extend out to sea along the shore of Africa. If the two continents are fitted together, the identical contacts are found in precisely the right position on the shore of South America (figure 4.11). Isotopic ages of rocks also match between these continents.

Glacial striations show that during the late Paleozoic Era continental glaciers moved from Africa toward the present Atlantic Ocean, while similar glaciers seemingly moved *from* the Atlantic Ocean *onto* South America (figure 4.11). Continental glaciers, however, cannot move from sea onto land. If the two continents had been joined together, the ice that moved off Africa could have been the ice that moved onto South America. This hypothesis has now been confirmed; from their lithology, many of the boulders in South American tills have been traced to a source that is now in Africa.

Some of the most detailed matches have been made between rocks in Brazil and rocks in the African country of Gabon. These rocks are similar in type, structure, sequence, fossils, ages, and degree of metamorphism. Such detailed matches are convincing evidence that continental drift did, in fact, take place.

There is also an abundance of satellite geodetic data from the Global Positioning Satellite system (GPS) so we can now watch the continents move—it is about as eventful as watching your fingernails grow!

History of Continental Positions

Rock matches show when continents were together; once the continents split, the new rocks formed are dissimilar. Paleomagnetic evidence indicates the direction and rate of drift, allowing maps of old continental positions, such as figure 4.2, to be drawn.

Although Pangaea split up 200 million years ago to form our present continents, the continents were moving much earlier. Pangaea was formed by the collision of many small continents long before it split up. Recent work shows that continents have been in motion for at least the past 2 billion years (some geologists say 4 billion years), well back into Precambrian time. For more than half of Earth's history, the continents appear to have collided, welded together, then split and drifted apart, only to collide again, over and over, in an endless, slow dance.

SEA-FLOOR SPREADING

At the same time that many geologists were becoming interested again in the idea of moving *continents,* Harry Hess, a geologist at Princeton University, proposed that the *sea floor* might be moving, too. This proposal contrasted sharply with the earlier ideas of Wegener, who thought that the ocean floor remained stationary as the continents plowed through it (figure 4.7). Hess's 1962 proposal was quickly named sea-floor spreading, for it suggests that the sea floor moves away from the mid-oceanic ridge as a result of mantle convection (figure 4.12).

According to the initial concept of sea-floor spreading, the sea floor is moving like a conveyor belt away from the crest of the mid-oceanic ridge, down the flanks of the ridge, and across the deep-ocean basin, to disappear finally by plunging beneath a continent or island arc (figure 4.12). The ridge crest, with sea floor moving away from it on either side, has been called a *spreading axis* (or *spreading center*). The sliding of the sea floor beneath a continent or island arc is termed **subduction.** The sea floor moves at a rate of 1 to 10 centimeters per year (your fingernail grows at about 1 cm/year). Although this may seem to be quite slow, it is rapid compared to most geologic processes.

Hess's Driving Force

Why does the sea floor move? Hess's original hypothesis was that sea-floor spreading is driven by deep mantle convection. **Convection** is a circulation pattern driven by the rising of hot material and/or the sinking of cold material. Hot material has a lower density, so it rises; cold material has a higher density and sinks. The circulation of water heating in a pan on a stove is an example of convection. Convection in the mantle was a controversial idea in 1962; for although convection can be easily demonstrated in a pan of water, it was hard to visualize the solid rock of the mantle behaving as a liquid. Over very long periods of time, however, it is possible for the hot mantle rock to flow plastically. A slow convective circulation is set up by temperature differences in the rock, and convection can explain many sea-floor features as well as the young age of the sea-floor rocks. (The heat that flows outward through Earth to drive convection is both original heat from the planet's formation and heat from the decay of radioactive isotopes, as discussed in chapter 9.)

Explanations

The Mid-Oceanic Ridge

If convection drives sea-floor spreading, then hot mantle rock must be rising under the mid-oceanic ridge. Hess showed how the *existence of the ridge* and its *high heat flow* are caused by the rise of this hot mantle rock. The *basalt eruptions* on the ridge crest are also related to this rising rock, for here the mantle rock is hotter than normal and begins to undergo partial melting.

As hot rock continues to rise beneath the ridge crest, the circulation pattern splits and diverges near the surface. Mantle rock moves horizontally away from the ridge crest on each side of the ridge. This movement creates tension at the ridge crest, cracking open the oceanic crust to form the *rift valley* and its associated *shallow-focus earthquakes.*

Oceanic Trenches

As the mantle rock moves horizontally away from the ridge crest, it carries the sea floor (the basaltic oceanic crust) piggyback along with it. As the hot rock moves sideways, it cools and

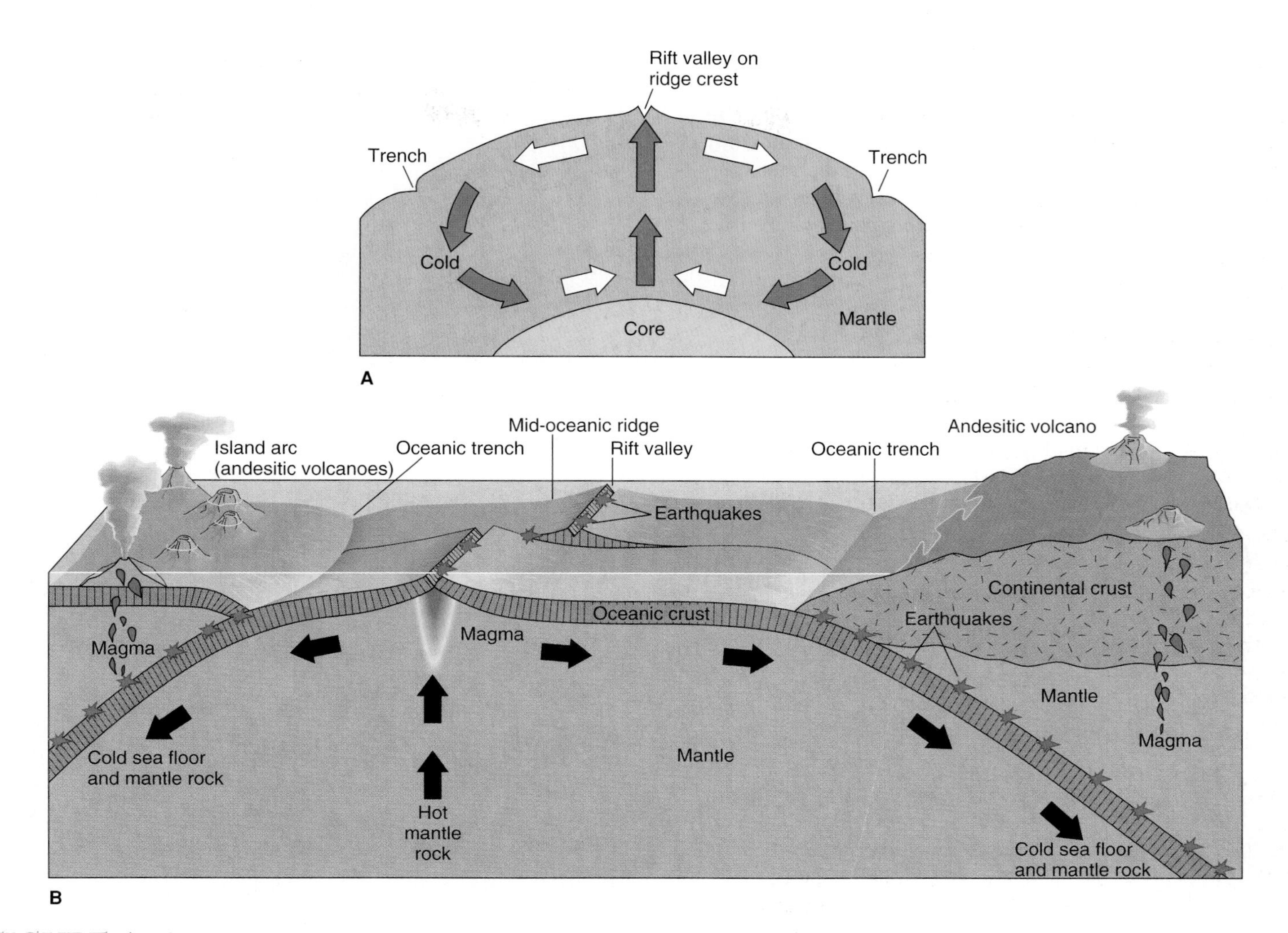

FIGURE 4.12

Sea-floor spreading. (*A*) Hess proposed that convection extended throughout the mantle. (Scale of ridge and trenches is exaggerated.) (*B*) Hot mantle rock rising beneath the mid-oceanic ridge (a spreading axis) causes basaltic volcanism and high heat flow. Divergence of sea floor splits open the rift valley and causes shallow-focus earthquakes (stars on ridge). Sinking of cold rock causes subduction of older sea floor at trenches, producing Benioff zones of earthquakes and andesitic magma.

becomes denser, sinking deeper beneath the ocean surface. Hess thought it would become cold and dense enough to sink back into the mantle. This downward plunge of cold rock accounts for the *existence of the oceanic trenches* as well as their *low heat flow* values. It also explains the large *negative gravity anomalies* associated with trenches, for the sinking of the cold rock provides a force that holds trenches out of isostatic equilibrium (see chapter 2).

As the sea floor moves downward into the mantle along a subduction zone, it interacts with the stationary rock above it. This interaction between the moving sea-floor rock and the stationary rock can cause the *Benioff zones of earthquakes* associated with trenches. It can also produce *andesitic volcanism,* which forms volcanoes either on the edge of a continent or in an island arc (figure 4.12).

Hess's ideas have stood up remarkably well over more than 30 years. We now think of plates moving instead of sea floor riding piggyback on convecting mantle, and we think that several mechanisms cause plate motion, but Hess's explanation of sea-floor topography, earthquakes, and age remain valid today.

Age of the Sea Floor

The *young age of sea-floor rocks* (see the previous chapter) is neatly explained by Hess's sea-floor spreading. New, young sea floor is continually being formed by basalt eruptions at the ridge crest. This basalt is then carried sideways by convection and is subducted into the mantle at an oceanic trench. Thus old sea floor is continually being destroyed at trenches, while new sea floor is being formed at the ridge crest. (This is also the reason for the puzzling lack of pelagic sediment at the ridge crest. Young sea floor at the ridge crest has little sediment because the basalt is newly formed. Older sea floor farther from the ridge crest has been moving under a constant rain of pelagic sediment, building up a progressively thicker layer as it goes.)

Note that sea-floor spreading implies that the youngest sea floor should be at the ridge crest, with the age of the sea floor becoming progressively older toward a trench. This increase in age away from the ridge crest was not known to exist at the time of Hess's proposal but was an important prediction of his hypothesis. This prediction has been successfully tested, as you shall see later in this chapter when we discuss marine magnetic anomalies.

PLATES AND PLATE MOTION

By the mid-1960s the twin ideas of moving continents and a moving sea floor were causing great excitement and emotional debate among geologists. By the late 1960s, these ideas had been combined into a single theory that revolutionized geology by providing a unifying framework for Earth science—the theory of plate tectonics.

As described earlier, a **plate** is a large, mobile slab of rock that is part of Earth's surface (figure 4.1). The surface of a plate may be made up entirely of sea floor (as is the Nazca plate), or it may be made up of both continental and oceanic rock (as is the North American plate). Some of the smaller plates are entirely continental, but all the large plates contain some sea floor.

Plate tectonics has added some new terms, based on rock behavior, to the zones of Earth's interior, as we have discussed in some previous chapters. The plates are part of a relatively rigid outer shell of Earth called the **lithosphere.** The lithosphere includes the rocks of the crust and uppermost mantle (figure 4.13).

The lithosphere beneath oceans increases in both age and thickness with distance from the crest of the mid-oceanic ridge. Young lithosphere near the ridge crest may be only 10 kilometers thick, while very old lithosphere far from the ridge crest may be as much as 100 kilometers thick. An average thickness for oceanic lithosphere might be 70 kilometers, as shown in figure 4.13.

Continental lithosphere is thicker, varying from perhaps 125 kilometers thick to as much as 200 to 250 kilometers thick beneath the oldest, coldest, and most inactive parts of the continents.

Below the rigid lithosphere is the **asthenosphere,** a zone of low seismic-wave velocity that behaves plastically because of increased temperature and pressure. Some geologists think that the asthenosphere is partially molten; the melting of just a few percent of the asthenosphere's volume could account for its properties and behavior. The plastic asthenosphere acts as a lubricating layer under the lithosphere, allowing the plates to move. The asthenosphere, made up of upper mantle rock, is the low-velocity zone described in chapter 2. It may extend from a depth of 70 to 200 kilometers beneath oceans; its thickness, depth, and even existence under continents is vigorously debated. Below the asthenosphere is more rigid mantle rock.

The idea that plates move is widely accepted by geologists, although the reasons for this movement are debated. Plates move away from the mid-oceanic ridge crest or other spreading axes. Some plates move toward oceanic trenches. If the plate is made up mostly of sea floor (as are the Nazca and Pacific plates), the plate can be subducted down into the mantle, forming an oceanic trench and its associated features. If the leading edge of the plate is made up of continental rock (as is the South American plate), that plate will not subduct. Continental rock, being less dense (specific gravity 2.7) than oceanic rock (specific gravity 3.0), is too light to be subducted.

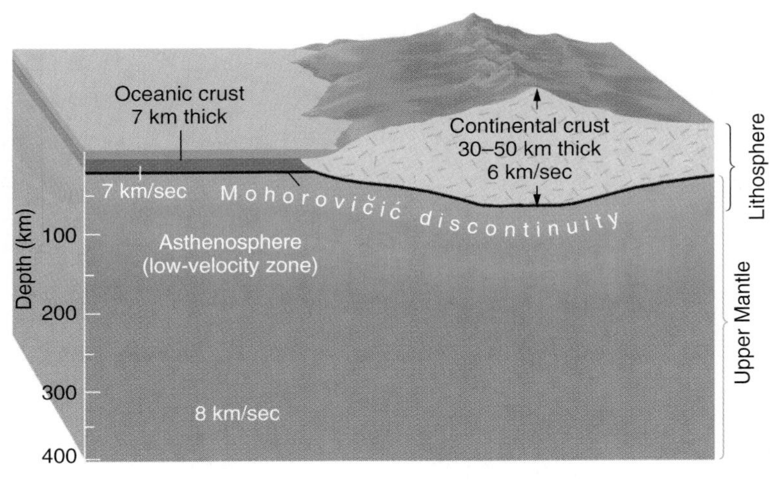

FIGURE 4.13

The rigid lithosphere includes the crust and uppermost mantle; it forms the plates. The plastic asthenosphere acts as a lubricating layer beneath the lithosphere. Oceanic lithosphere averages 70 kilometers thick; continental lithosphere varies from 125 to 250 kilometers thick. Asthenosphere may not be present under continents.

A plate is a rigid slab of rock that moves as a unit. As a result, the interior of a plate is relatively inactive tectonically. Plate interiors generally lack earthquakes, volcanoes, young mountain belts, and other signs of geologic activity. According to plate tectonic theory, these features are caused by plate interactions at plate boundaries.

Earthquakes, volcanoes, and young mountains are generally distributed in narrow belts separated by broad regions of inactivity, as you will see in subsequent chapters. This distribution puzzled geologists for a long time, and many hypotheses were advanced to explain it. The plate tectonic concept is the latest explanation. Plate boundaries are defined and located by mapping narrow belts of earthquakes, volcanoes, and young mountains. The plates themselves are the broad, relatively inactive regions outlined by these belts of geologic activity.

Plate tectonics has become a unifying theory of geology because it can explain so many diverse features of Earth. Earthquake distribution, the origin of mountain belts, the origin of sea-floor topography, the distribution and composition of volcanoes, and many other features can all be related to plate tectonics. It is a convenient framework that unifies geologic thought, associating features that were once studied separately and relating them to a single cause: plate interactions at plate boundaries.

Plate boundaries are of three general types, based on whether the plates move away from each other, move toward each other, or move past each other. A **divergent plate boundary** is a boundary between plates that are moving apart. A **con-**

vergent plate boundary lies between plates that are moving toward each other. A **transform plate boundary** is one at which two plates move horizontally past each other.

HOW DO WE KNOW THAT PLATES MOVE?

The proposal that Earth's surface is divided into moving plates was an exciting, revolutionary hypothesis, but it required testing to win acceptance among geologists. You have seen how the study of paleomagnetism supports the idea of moving continents. In the 1960s two critical tests were made of the idea of a moving sea floor. These tests involved marine magnetic anomalies and the seismicity of fracture zones. These two, successful tests convinced most geologists that plates do indeed move.

Marine Magnetic Anomalies

In the mid-1960s, magnetometer surveys at sea disclosed some intriguing characteristics of marine magnetic anomalies. Most magnetic anomalies at sea are arranged in bands that lie parallel to the rift valley of the mid-oceanic ridge. Alternating positive and negative anomalies (chapter 2) form a stripelike pattern parallel to the ridge crest (figure 4.14).

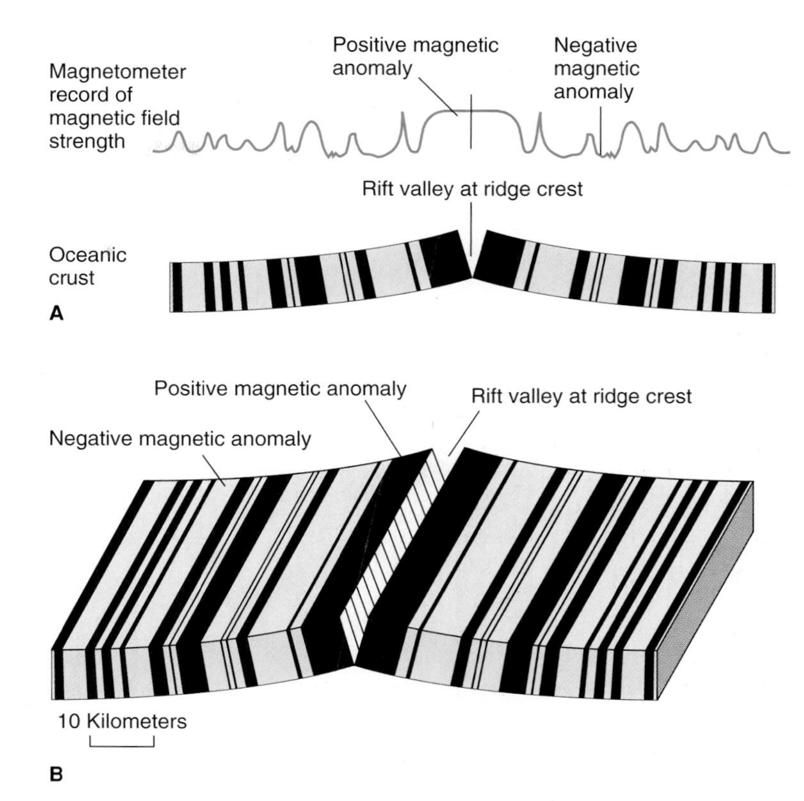

FIGURE 4.14

Marine magnetic anomalies. (*A*) The red line shows positive and negative magnetic anomalies as recorded by a magnetometer towed behind a ship. In the cross section of oceanic crust, positive anomalies are drawn as black bars and negative anomalies are drawn as blue bars. (*B*) Perspective view of magnetic anomalies shows that they are parallel to the rift valley and symmetric about the ridge crest.

The Vine-Matthews Hypothesis

Two British geologists, Fred Vine and Drummond Matthews, made several important observations about these anomalies. They recognized that the pattern of magnetic anomalies was symmetrical about the ridge crest. That is, the pattern of magnetic anomalies on one side of the mid-oceanic ridge was a mirror image of the pattern on the other side (figure 4.14). Vine and Matthews also noticed that the same pattern of magnetic anomalies exists over different parts of the mid-oceanic ridge. The pattern of anomalies over the ridge in the northern Atlantic Ocean is the same as the pattern over the ridge in the southern Pacific Ocean.

The most important observation that Vine and Matthews made was that the pattern of magnetic *anomalies* at sea matches the pattern of magnetic *reversals* already known from studies of lava flows on the continents (figure 4.15 and chapter 2). This correlation can be seen by comparing the pattern of colored bands in figure 4.15 (reversals) with the pattern in figure 4.14 (anomalies).

Putting these observations together with Hess's concept of sea-floor spreading, which had just been published, Vine and Matthews proposed an explanation for magnetic anomalies. They suggested that there is continual opening of tensional cracks within the rift valley on the mid-oceanic ridge crest. These cracks on the ridge crest are filled by basaltic magma from below, which cools to form dikes. Cooling magma in the dikes records Earth's magnetism at the time the magnetic minerals crystallize. The process is shown in figure 4.16.

When Earth's magnetic field has a *normal polarity* (the present orientation), cooling dikes are normally magnetized. Dikes that cool when the field is reversed (figure 4.16) are reversely magnetized. So each dike preserves a record of the polarity that prevailed during the time the magma cooled. Extension produced by the moving sea floor then cracks a dike in two, and the two halves are carried away in opposite directions down the flanks of the ridge. New magma eventually intrudes the newly opened fracture. It cools, is magnetized, and forms a new dike, which in turn is split by continued extension. In this way a system of reversely magnetized and normally magnetized dikes forms parallel to the rift valley. These dikes, in the Vine-Matthews hypothesis, are the cause of the anomalies.

The magnetism of normally magnetized dikes adds to Earth's magnetism, and so a magnetometer carried over such dikes registers a stronger magnetism than average—a *positive* magnetic anomaly. Dikes that are reversely magnetized subtract from the present magnetic field, and so a magnetometer towed over such dikes measures a weaker magnetic field—a *negative* magnetic anomaly. Since sea-floor motion separates these dikes into halves, the patterns on either side of the ridge are mirror images.

Measuring the Rate of Plate Motion

There are two important points about the Vine-Matthews hypothesis of magnetic anomaly origin. The first is that it allows us to

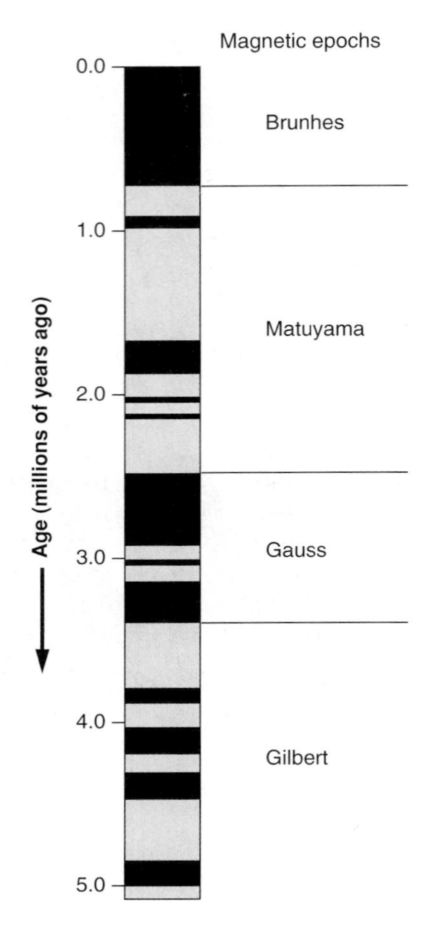

FIGURE 4.15

Magnetic reversals during the past 5 million years. Black represents normal magnetism; blue represents reverse magnetism.

From Robert Butler, 1992, *Paleomagnetism,* Blackwell Scientific Publications, p. 212

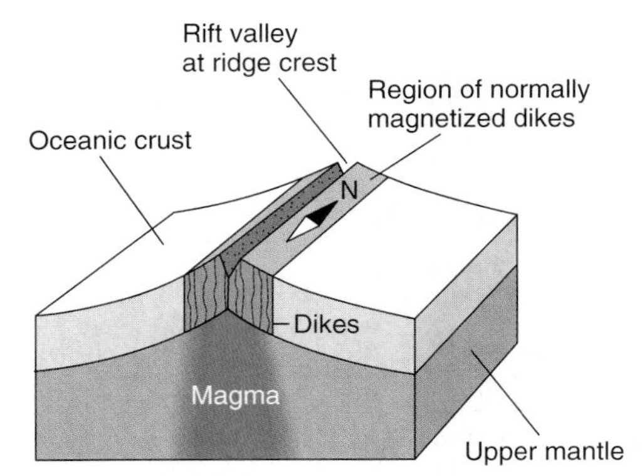

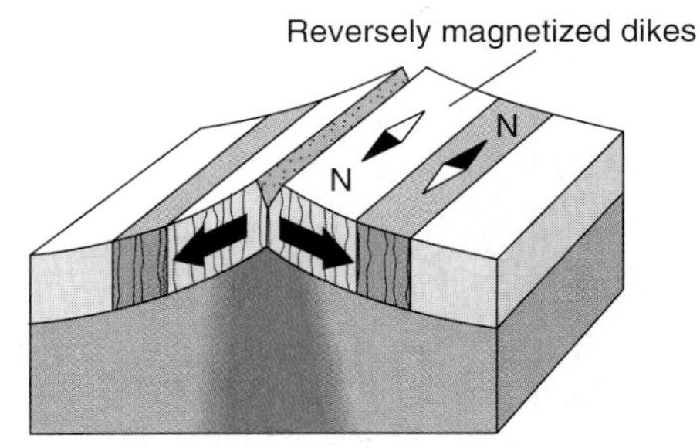

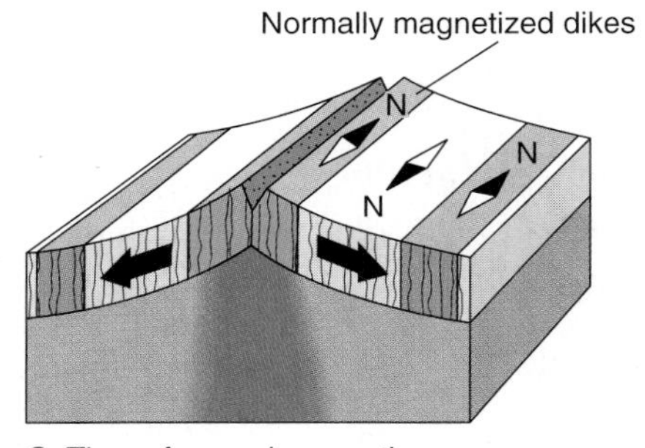

C Time of normal magnetism

FIGURE 4.16

The origin of magnetic anomalies. (*A*) During a time of normal magnetism, a series of basaltic dikes intrudes the ridge crest, becoming normally magnetized. (*B*) The dike zone is torn in half and moved sideways, as a new group of reversely magnetized dikes forms at the ridge crest. (*C*) A new series of normally magnetized dikes forms at the ridge crest. The dike pattern becomes symmetric about the ridge crest.

measure the *rate of sea-floor motion* (which is the same as plate motion, since continents and the sea floor move together as plates).

Because magnetic reversals have already been dated from lava flows on land (figure 4.15), the anomalies caused by these reversals are also dated and can be used to discover how fast the sea floor has moved (figure 4.17). For instance, a piece of the sea floor representing the reversal that occurred 4.5 million years ago may be found 45 kilometers away from the rift valley of the ridge crest. The piece of sea floor, then, has traveled 45 kilometers since it formed 4.5 million years ago. Dividing the distance the sea floor has moved by its age gives 10 km/million years, or 1 cm/year for the rate of sea-floor motion here. In other words, on each side of the ridge, the sea floor is moving away from the ridge crest at a rate of 1 centimeter per year. Such measured rates generally range from 1 to 10 centimeters per year.

Predicting Sea-Floor Age

The other important point of the Vine-Matthews hypothesis is that it *predicts the age of the sea floor* (figure 4.17). Magnetic reversals are now known to have occurred back into Precambrian time. Sea floor of *all* ages is therefore characterized by parallel bands of magnetic anomalies. Figure 2.24 shows the pattern of marine magnetic anomalies (and the reversals that caused them) during the past 160 million years. The distinctive pattern of these anomalies through time allows them to be identified by age, a process similar to dating by tree rings.

Now, even before they sample the sea floor, marine geologists can predict the age of the igneous rock of the sea floor by measuring the magnetic anomalies at the sea surface. Most sections of the sea floor have magnetic anomalies. By matching the measured anomaly pattern with the known pattern that is shown in figure 2.24, the age of the sea floor in the region can be predicted, as shown on the map in figure 4.18.

This is a very powerful test of the hypothesis that the sea floor moves. Suppose, for example, that the sea floor in a par-

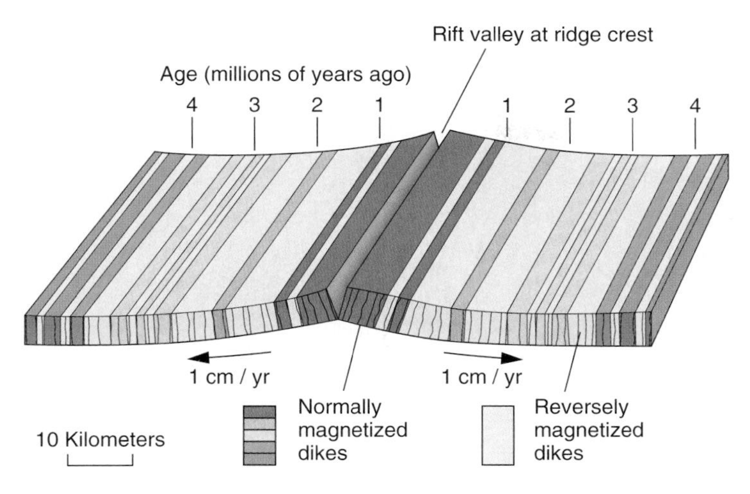

FIGURE 4.17

Correlation of magnetic anomalies with magnetic reversals allows anomalies to be dated. Magnetic anomalies can therefore be used to predict the age of the sea floor and to measure the rate of sea-floor spreading (plate motion).

ticular spot is predicted to be 70 million years old from a study of its magnetic anomalies. If the hypothesis of sea-floor motion and the Vine-Matthews hypothesis of magnetic anomaly origin are correct, a sample of igneous rock from that spot *must* be 70 million years old. If the rock proves to be 10 million years old or 200 million years old or 1.2 billion years old, or any other age except 70 million years, then both these hypotheses are wrong. But if the rock proves to be 70 million years old, as predicted, then both hypotheses have been successfully tested.

Hundreds of rock and sediment cores recovered from holes drilled in the sea floor were used to test these hypotheses. Close correspondence has generally been found between the predicted age and the measured age of the sea floor. (The sea-floor age is usually measured by fossil dating of sediment in the cores rather than by isotopic dating of igneous rock.) This evidence from deep-sea drilling has been widely accepted by geologists as verification of the hypotheses of plate motion and magnetic anomaly origin. Most geologists now think that these concepts are no longer hypotheses but can now be called theories. (A *theory,* as discussed in box 1.4 in connection with the scientific method, is a concept with a much higher degree of certainty than a hypothesis.)

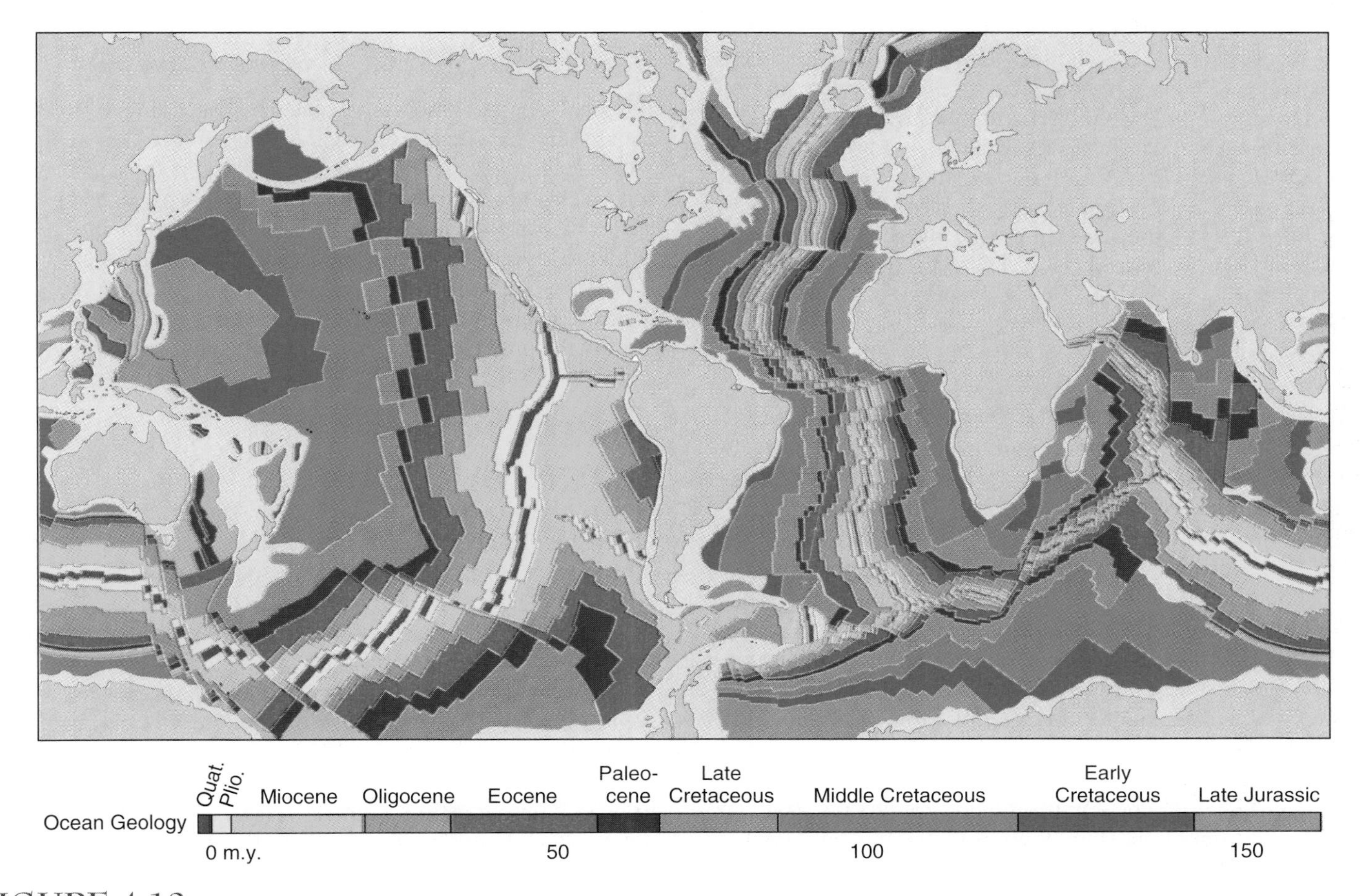

FIGURE 4.18

The age of the sea floor as determined from magnetic anomalies.
After *The Bedrock Geology of the World* by R. L. Larson, W. C. Pitman, III, et al., W. H. Freeman

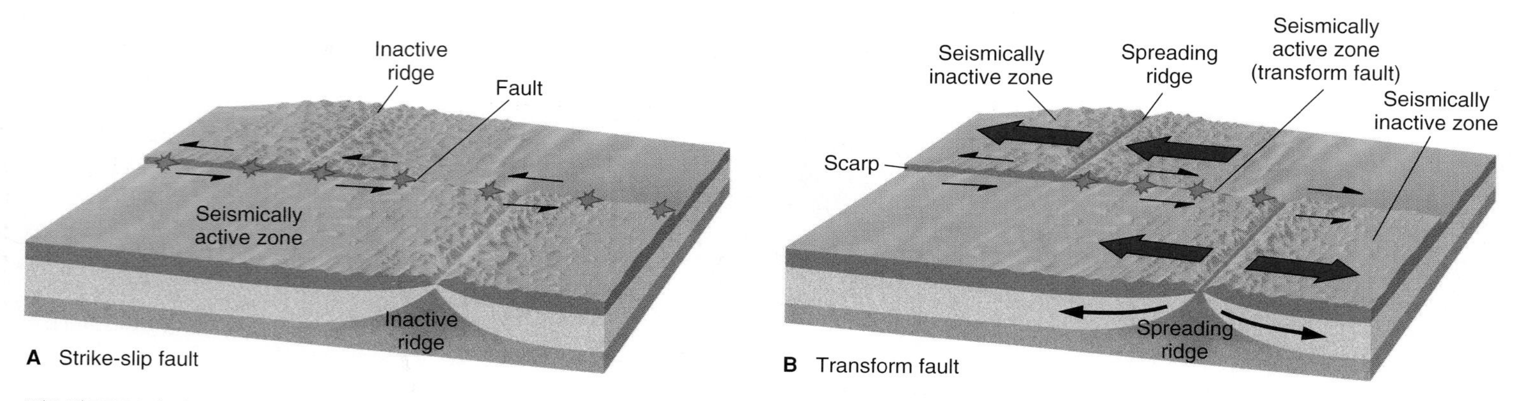

FIGURE 4.19

Two possible explanations for the relationship between fracture zones and the mid-oceanic ridge. (*A*) The expected rock motions and earthquake distribution assuming that the ridge was once continuous across the fracture zone. (*B*) The expected rock motions and earthquake distribution assuming that the two ridge segments were never joined together and that the sea floor moves away from the rift valley segments. Only explanation (*B*) fits the data. The portion of the fracture zone between the ridge segments is a transform fault.

Another Test: Fracture Zones and Transform Faults

Cores from deep-sea drilling tested plate motion by allowing us to compare the actual age of the sea floor with the age predicted from magnetic anomalies. Another rigorous test of plate motion has been made by studying the seismicity of fracture zones.

The mid-oceanic ridge is offset along fracture zones (see figure 3.19). Conceivably, the mid-oceanic ridge was once continuous across a fracture zone but has been offset by strike-slip motion along the fracture zone (figure 4.19*A*). If such motion is occurring along a fracture zone, we would expect to find two things: (1) earthquakes should be distributed along the entire length of the fracture zone, and (2) the motion of the rocks on either side of the fracture zone should be in the direction shown by the arrows in figure 4.19*A*.

In fact, these things are not true about fracture zones. Earthquakes do occur along fracture zones, but only in those segments between offset sections of ridge crest. In addition, first-motion studies of earthquakes (see chapter 7) along fracture zones show that the motion of the rocks on either side of the fracture zone during an earthquake is exactly opposite to the motion shown in figure 4.19*A*. The actual motion of the rocks as determined from first-motion studies is shown in figure 4.19*B*. The portion of a fracture zone between two offset portions of ridge crest is called a **transform fault.**

The motion of rocks on either side of a transform fault was predicted by the hypothesis of a moving sea floor. Note that sea floor moves away from the two segments of ridge crest (figure 4.19*B*). Looking along the length of the fracture zone, you can see that blocks of rock move in opposite directions only on that section of the fracture zone between the two segments of ridge crest. Earthquakes, therefore, occur only on this section of the fracture zone, the transform fault. The direction of motion of rock on either side of the transform fault is exactly predicted by the assumption that rock is moving away from the ridge crests. Verification by first-motion studies of this predicted motion along fracture zones was another successful test of plate motion.

Measuring Plate Motion Directly

In recent years the motion of plates has been directly measured using satellites, radar, lasers, and the Global Positioning System. These techniques can measure the distance between two widely separated points to within 1 centimeter. If two plates move toward each other at individual rates of 2 cm/year and 6 cm/year, the combined rate of convergence is 8 cm/year. The measurement techniques are sensitive enough to easily measure such a rate if measurements are repeated each year. Such measured rates match closely the predicted rates from magnetic anomalies.

DIVERGENT PLATE BOUNDARIES

Divergent plate boundaries, where plates move away from each other, can occur in the middle of the ocean or in the middle of a continent. The result of divergent plate boundaries is to create, or open, new ocean basins. This dynamic process has occurred throughout the geologic past.

When a supercontinent such as Pangaea breaks up, a divergent boundary can be found in the middle of a continent. The divergent boundary is marked by rifting, basaltic volcanism, and uplift. During rifting, the continental crust is stretched and thinned. This extension produces shallow-focus earthquakes on normal faults, and a rift valley forms as a central *graben* (a down-dropped fault block). The faults act as pathways for basaltic magma, which rises from the mantle to erupt on the surface as cinder cones and basalt flows. Uplift at a divergent boundary is usually caused by the upwelling of hot mantle beneath the crust; the surface is elevated by the thermal expansion of the hot, rising rock and of the surface rock as it is warmed from below.

There is current debate on the sequence of events as continents split. Some geologists think that rifting comes first (figure 4.20). The thinning of cold crust would reduce pressure on the underlying rock and allow hot mantle rock to rise passively, elevating the land by thermal expansion; in this case rifting *causes* uplift. Other geologists believe that thermal uplift comes first as hot mantle rock rises actively under uniformly thick crust. As the uplifted crust is stretched, rifting occurs; in this case rifting is the *result* of uplift.

Figure 4.21 shows how a continent might rift to form an ocean. The figure shows rifting before uplift, because recent

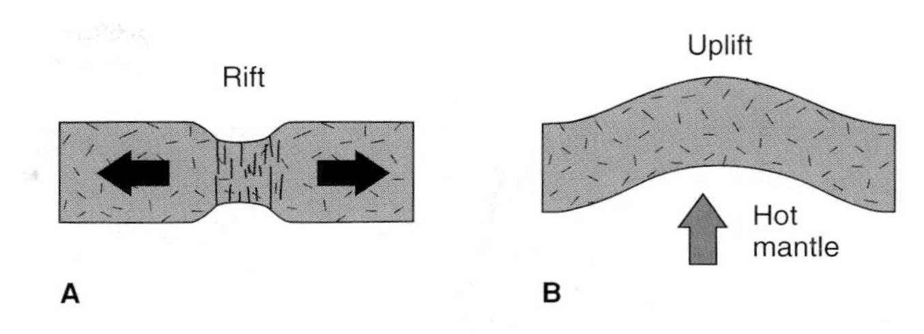

FIGURE 4.20

Two models of the beginning of continental divergence. (*A*) Rifting comes first; uplift may follow. (*B*) Uplift comes first; rifting may follow.

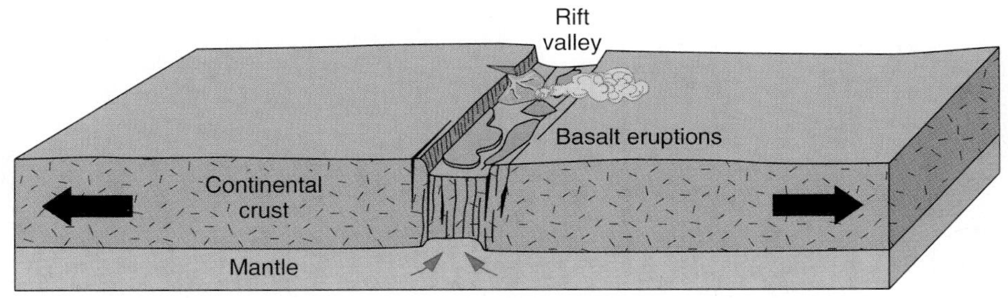

A Continent undergoes extension. The crust is thinned and a rift valley forms (East African Rift Valleys).

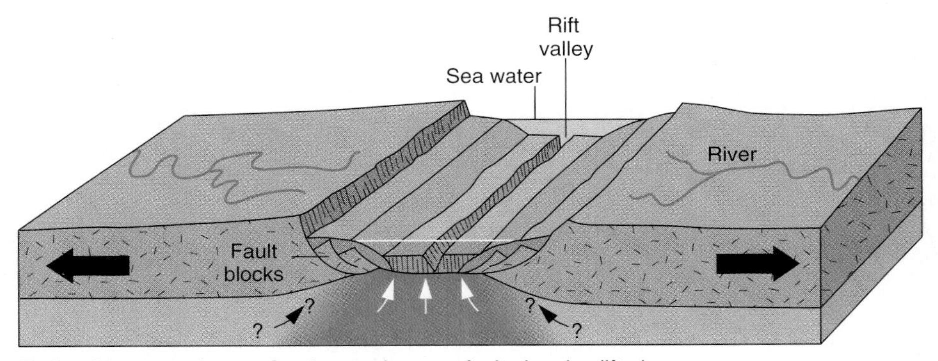

B Continent tears in two. Continent edges are faulted and uplifted. Basalt eruptions form oceanic crust (Red Sea).

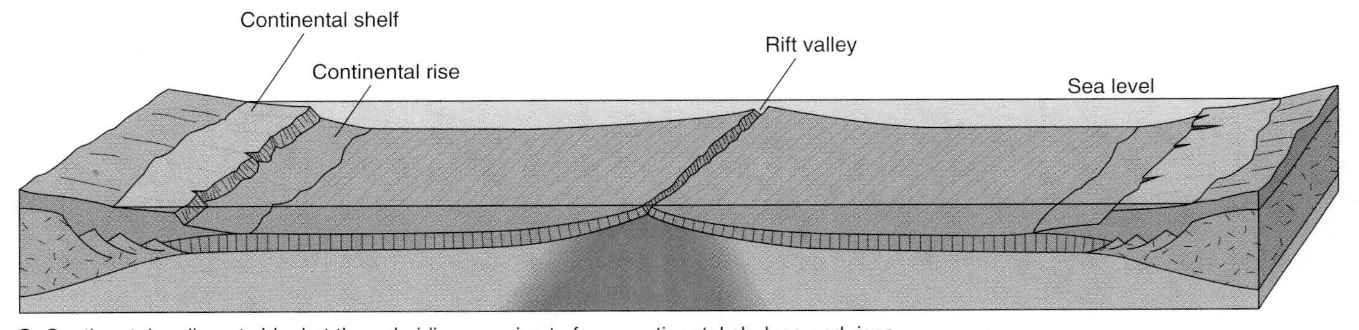

C Continental sediments blanket the subsiding margins to form continental shelves and rises. The ocean widens and a mid-oceanic ridge develops (Atlantic Ocean).

FIGURE 4.21

A divergent plate boundary forming in the middle of a continent will eventually create a new ocean.

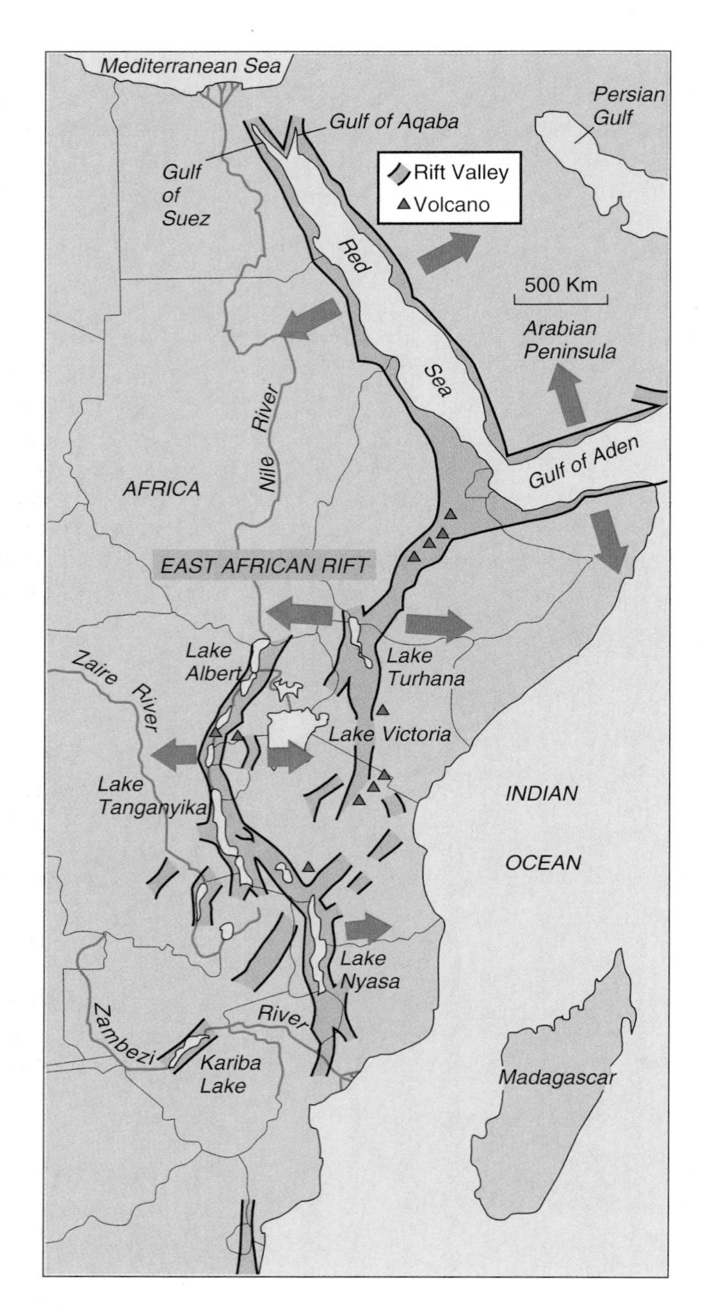

FIGURE 4.22

The East African Rift Valleys and the Red Sea

FIGURE 4.23

Spacecraft photograph looking south along the Red Sea. Gulf of Suez at bottom, Gulf of Aqaba at lower left. Note similarities in the shorelines of the Arabian peninsula (left) and Africa (right), suggesting that the Red Sea was formed by splitting of the continent.
Photo by NASA

work indicates that this was the sequence for the opening of the Red Sea. The crust is initially stretched and thinned. Numerous normal faults break the crust, and the surface subsides into a central graben (figure 4.21*A*). Shallow earthquakes and basalt eruptions occur in this rift valley, which also has high heat flow. An example of a boundary at this stage is the African Rift Valleys in eastern Africa (figure 4.22). The valleys are grabens that may mark the site of the future breakup of Africa.

As divergence continues, the continental crust on the upper part of the plate clearly separates, and seawater floods into the linear basin between the two divergent continents (figure 4.21*B*). A series of fault blocks have rotated along curved fault planes at the edges of the continents, thinning the continental crust. The rise of hot mantle rock beneath the thinned crust causes continued basalt eruptions that create true oceanic crust between the two continents. The center of the narrow ocean is marked by a rift valley with its typical high heat flow and shallow earthquakes. The Red Sea is an example of a divergent margin at this stage (figures 4.22 and 4.23).

The upward rise of basaltic magma from the mantle to form oceanic crust between two diverging continents is analogous to the rise of water between two floating blocks of wood that are moved apart (figure 4.24).

After modest widening of the new ocean, uplift of the continental edges may occur. As continental crust thins by stretching and faulting, the surface initially subsides. At the same time, hot mantle rock wells up beneath the stretched crust (figure 4.21*B*). The rising diapir of hot mantle rock would cause uplift by thermal expansion.

The new ocean is narrow, and the tilt of the adjacent land is away from the new sea, so rivers flow away from the sea (figure 4.21*B*). At this stage the seawater that has flooded into the rift may evaporate, leaving behind a thick layer of rock salt overlying the continental sediments. The likelihood of salt precipitation increases if the continent is in one of the desert belts or if one or both ends of the new ocean should become temporarily blocked, perhaps by volcanism. Not all divergent boundaries contain rock salt, however.

The plates continue to diverge, widening the sea. Thermal uplift creates a mid-oceanic ridge in the center of the sea (figure 4.21*C*). The flanks of the ridge subside as the sea-floor rock cools as it moves.

The trailing edges of the continents also subside as they are lowered by erosion and as the hot rock beneath them cools. Subsidence continues until the edges of the continents are under water. A thick sequence of marine sediment blankets the thinned continental rock, forming a *passive continental margin* (figures 4.21*C* and 4.25; see also previous chapter). The sediment forms a shallow continental shelf, which may contain a deeply buried salt layer. The deep continental rise is formed as sediment is carried down the continental slope by turbidity currents and other mechanisms. The Atlantic Ocean is currently at this stage of divergence (figure 3.16).

A divergent boundary on the sea floor is located on the crest of the mid-oceanic ridge. If the spreading rate is slow, as it is in the Atlantic Ocean (1 cm/year), the crest has a rift valley. Fast spreading, as in the Pacific Ocean (10 cm/year), prevents a rift from forming. A divergent boundary at sea is marked by the same features as a divergent boundary on land—tensional cracks, normal faults, shallow earthquakes, high heat flow, and basaltic eruptions. The basalt forms dikes within the cracks and pillow lavas on the sea floor, creating new oceanic crust on the trailing edge of plates.

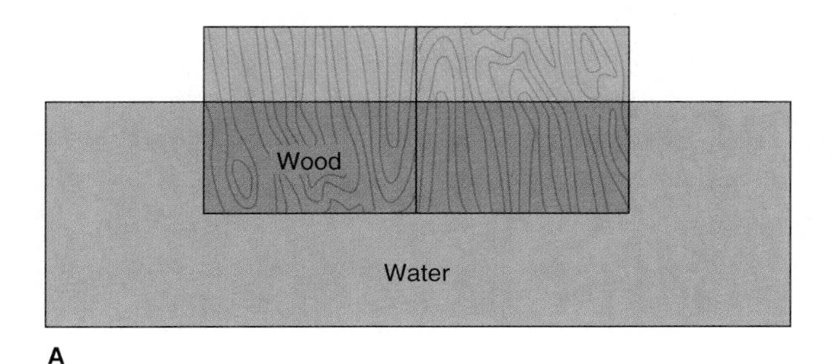

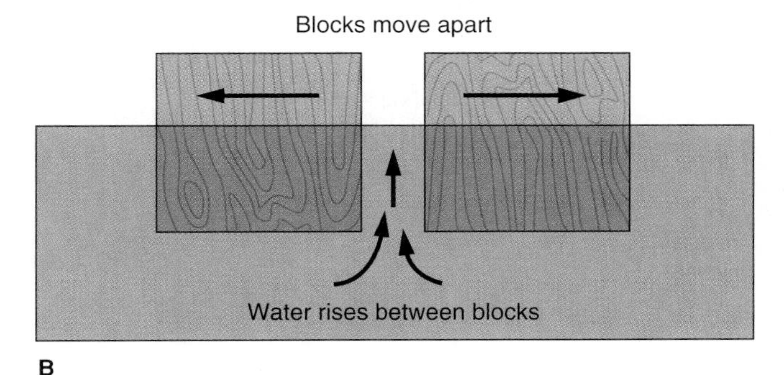

FIGURE 4.24
Water rises upward to fill the gap when (*A*) two floating blocks are (*B*) moved apart. The water moves as a result of the motion of the blocks.

TRANSFORM BOUNDARIES

At transform boundaries, where one plate slides horizontally past another plate, the plate motion can occur on a single fault or on a group of parallel faults. Transform boundaries are marked by shallow-focus earthquakes in a narrow zone for a single fault, or in a broad zone for a group of parallel faults (see figure 7.27). First-motion studies of the quakes indicate strike-slip movement parallel to the faults.

The name *transform fault* comes from the fact that the displacement along the fault abruptly ends or transforms into another kind of displacement. The most common type of

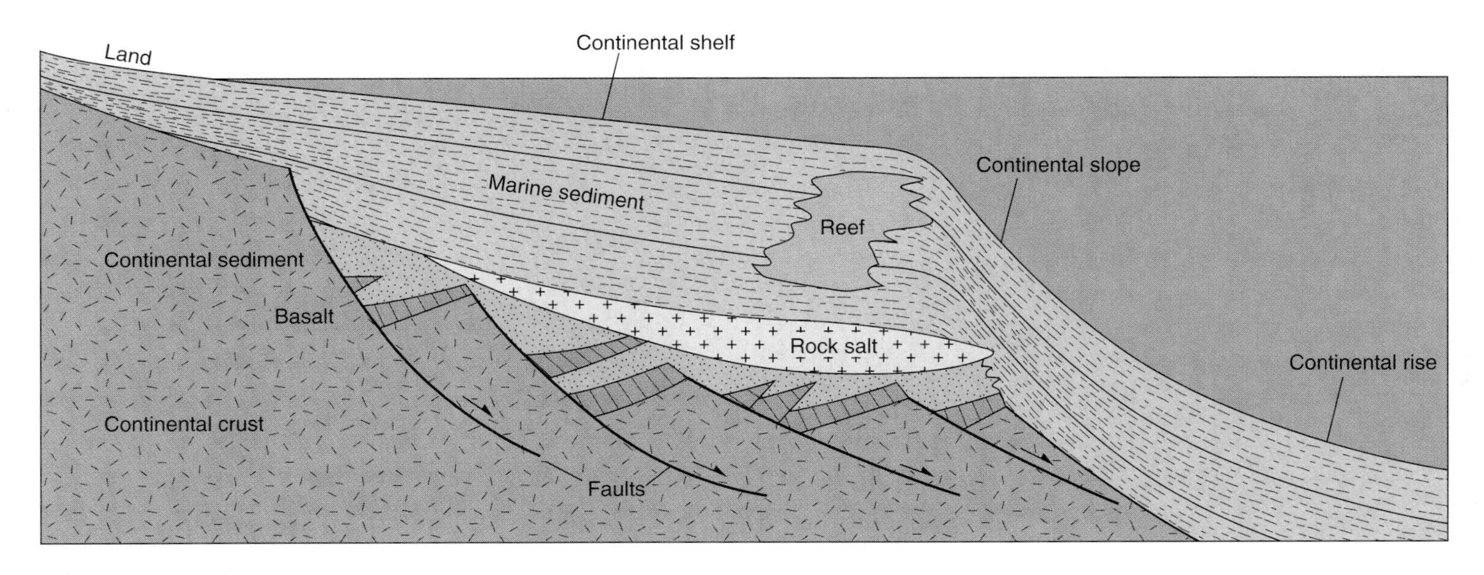

FIGURE 4.25
A passive continental margin formed by continental breakup and divergence. Downfaulted continental crust forms basins, which fill with basalt and sediment. A layer of rock salt may form if a narrow ocean evaporates. A thick sequence of marine sediments covers these rocks and forms the continental shelf, slope, and rise. A reef may form at the shelf edge if the water is warm; buried reefs occur on many parts of the Atlantic shelf of North America.

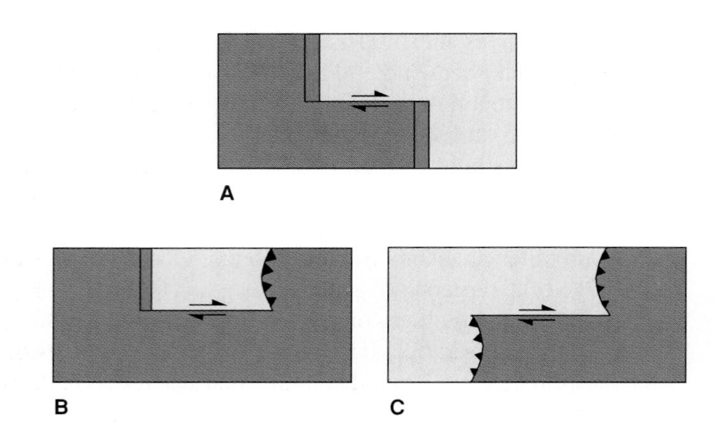

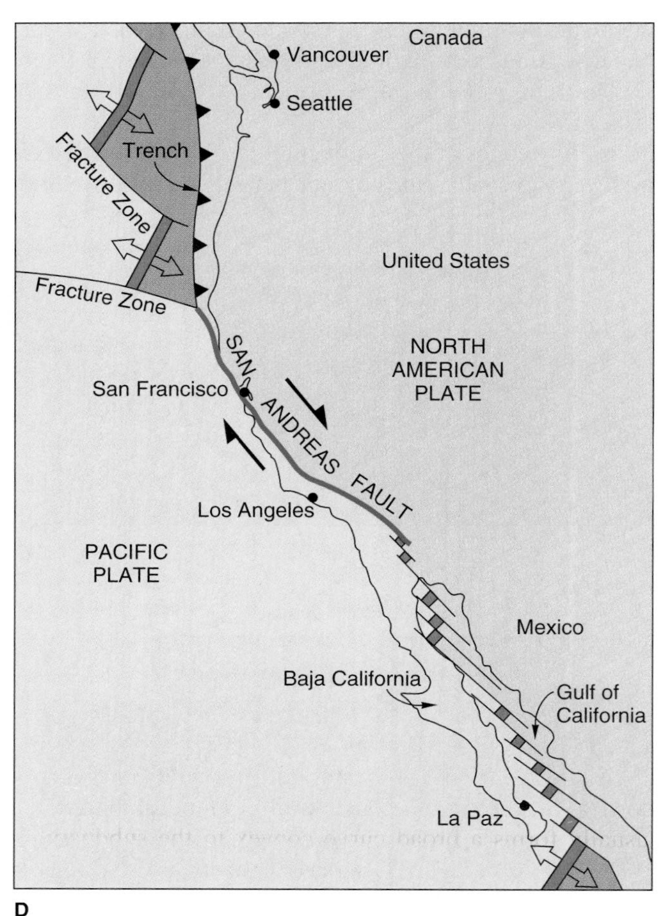

FIGURE 4.26

Transform boundaries (*A*) between two ridges; (*B*) between a ridge and a trench; and (*C*) between two trenches. Triangles on trenches point down subduction zones. Color tones show two plates in each case. (*D*) The San Andreas fault is a transform plate boundary between the North American plate and the Pacific plate. The south end of the San Andreas fault is a ridge segment (shown in red) near the U.S.-Mexico border. The north end of the fault is a "triple junction" where three plates meet at a point. The relative motion along the San Andreas fault is shown by the large black arrows, as the Pacific plate slides horizontally past the North American plate.

Modified from U.S. Geological Survey

transform fault occurs along fracture zones and connects two divergent plate boundaries at the crest of the mid-oceanic ridge (figures 4.26 and 4.19*B*). The spreading motion at one ridge segment is transformed into the spreading motion at the other ridge segment by strike-slip movement along the transform fault.

Not all transform faults connect two ridge segments. As you can see in figure 4.26, a transform fault can connect a ridge to a trench (a divergent boundary to a convergent boundary), or can connect two trenches (two convergent boundaries). The San Andreas fault in California is a transform fault with a complex history (figure 4.26*D*).

What is the origin of the offset in a ridge-ridge transform fault? The offsets appear to be the result of irregularly shaped divergent boundaries (figure 4.27). When two oceanic plates begin to diverge, the boundary may be curved on a sphere. Mechanical constraints prevent divergence along a curved boundary, so the original curves readjust into a series of right-angle bends. The ridge crests align perpendicular to the spreading direction, and the transform faults align parallel to the spreading direction. An old line of weakness in a continent may cause the initial divergent boundary to be oblique to the

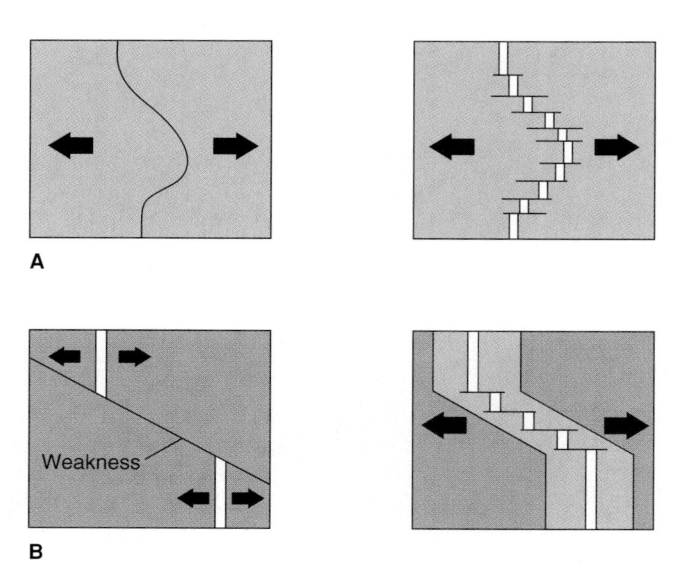

FIGURE 4.27

Divergent boundaries form ridge crests perpendicular to the spreading direction and transform faults parallel to the spreading direction. (*A*) Oceanic plates. (*B*) Continental plates.

spreading direction when the continent splits. The boundary will then readjust into a series of transform faults parallel to the spreading direction.

CONVERGENT PLATE BOUNDARIES

At convergent plate boundaries two plates move toward each other (often obliquely). The character of the boundary depends partly on the type of plates that converge. A plate capped by oceanic crust can move toward another plate capped by oceanic crust, in which case one plate dives (subducts) under the other. If an oceanic plate converges with a plate capped by a continent, the dense oceanic plate subducts under the continental plate. If the two approaching plates are both carrying continents, the continents collide and crumple but neither is subducted.

Ocean-Ocean Convergence

Where two plates capped by sea floor converge, one plate subducts under the other (the Pacific plate sliding under the western Aleutian Islands is an example). The subducting plate bends downward, forming the outer wall of an oceanic trench, which usually forms a broad curve convex to the subducting plate (figures 4.28 and 4.29).

As one plate subducts under another, a Benioff zone of shallow-, intermediate-, and deep-focus earthquakes is created within the upper portion of the downgoing lithosphere (figure 4.28). The reasons for these quakes are discussed in chapter 7. The existence of deep-focus earthquakes to a depth of 670 kilometers tells us that brittle plates continue to (at least) that depth. The pattern of quakes shows that the angle of subduction changes with depth, usually becoming steeper (figure 4.28). Some plates crumple or break into segments as they descend.

As the descending plate reaches depths of at least 100 kilometers, magma is generated in the overlying asthenosphere (figure 4.28). The magma probably forms by partial melting of the asthenosphere, perhaps triggered by dewatering of the downgoing oceanic crust as it is subducted, as described in chapter 11. Differentiation and assimilation may also play an important role in the generation of the magma, which is typically andesitic to basaltic in composition.

The magma works its way upward to erupt as an **island arc,** a curved line of volcanoes that form a string of islands parallel to the oceanic trench (figure 4.28). Beneath the volcanoes are large plutons in the thickened arc crust.

The distance between the island arc and the trench can vary, depending upon where the subducting plate reaches the 100-kilometer depth. If the subduction angle is steep, the plate reaches this magma-generating depth at a location close to the trench, so the horizontal distance between the arc and trench is short (figure 4.30). If the subduction angle is gentle, the arc-trench distance is greater. A thick, buoyant plate (such as a subducting aseismic ridge) may subduct at such a gentle angle that it merely slides horizontally along under another plate. Because the top of the subducting plate never reaches the 100-kilometer depth, such very shallow subduction zones lack volcanism.

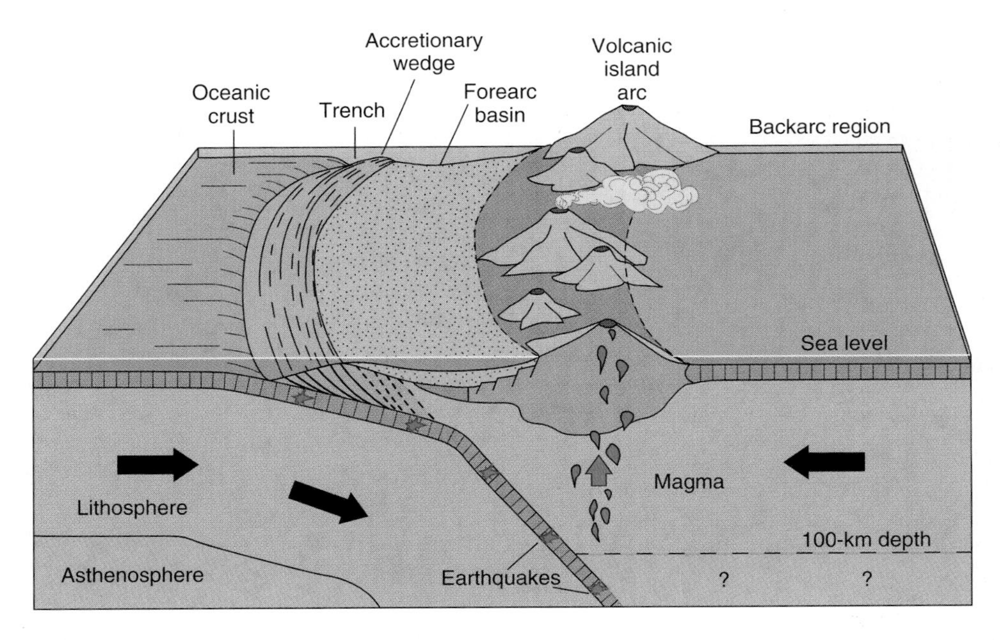

FIGURE 4.28

Ocean-ocean convergence forms a trench, a volcanic island arc, and a Benioff zone of earthquakes.

Modified from W. R. Dickinson, 1977, in *Island Arcs, Deep Sea Trenches and Backarc Basins* (pp. 33–40), copyrighted by American Geophysical Union

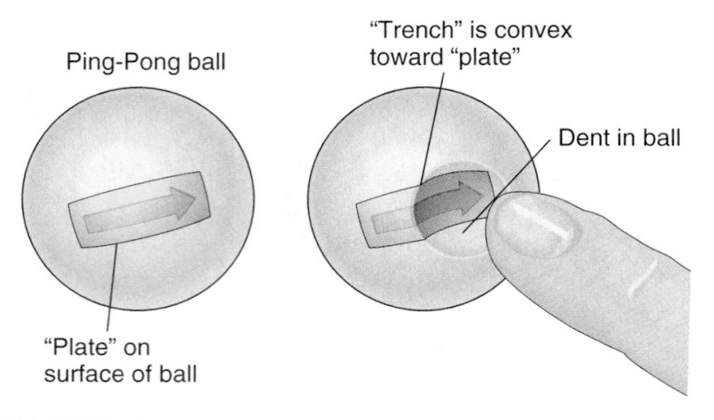

FIGURE 4.29

A dented Ping-Pong ball can show why trenches are curved on a sphere.

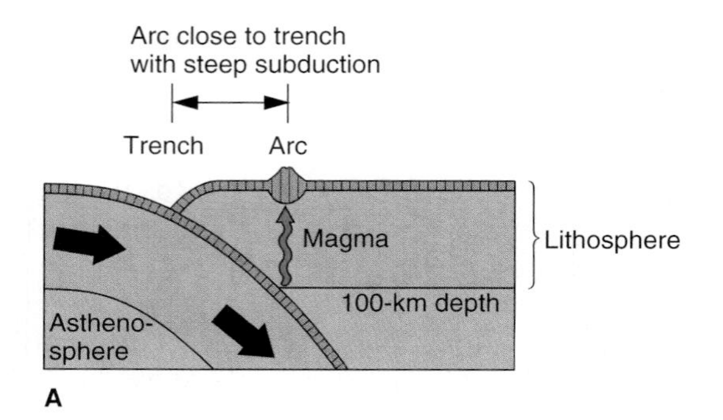

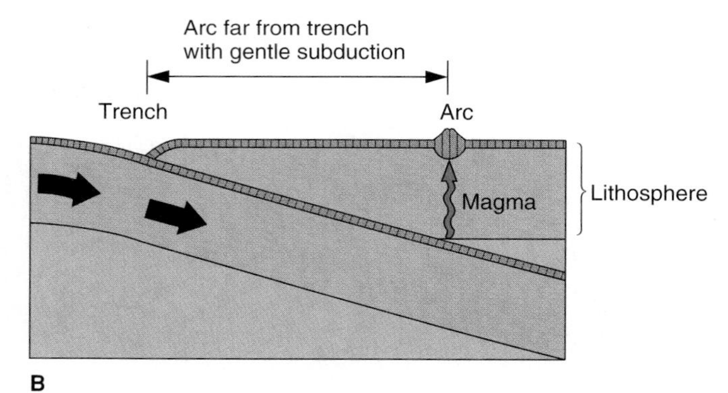

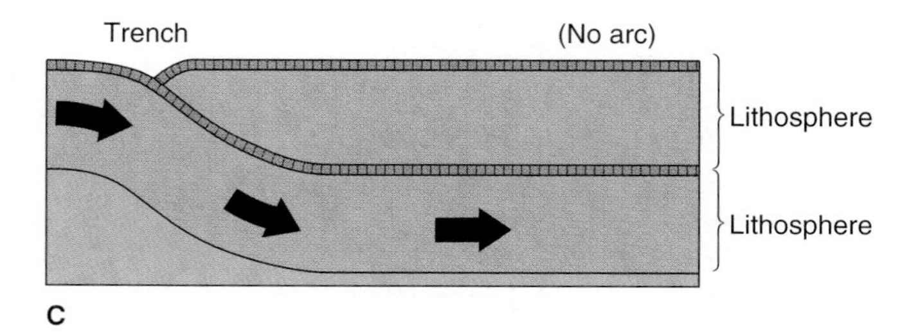

FIGURE 4.30

Andesitic magma is generated where the top of the subducting lithosphere reaches a depth of 100 kilometers, so the subduction angle determines the arc-trench spacing.

When a plate subducts far from a mid-oceanic ridge, the plate is cold, with a low heat flow. Oceanic plates form at ridge crests, then cool and sink as they spread toward trenches. Eventually they become cold and dense enough to sink back into the mantle. Oceanic trenches are marked by strong negative gravity anomalies. These show that trenches are not currently in isostatic equilibrium, but are being actively pulled down. Hess thought that this pulling was caused by a down-turning convection current in the mantle. Today most geologists think that the pulling is caused by the sinking of cold, dense lithosphere.

The inner wall of a trench (toward the arc) consists of an *accretionary wedge* (or *subduction complex*) of thrust-faulted and folded marine sediment and pieces of oceanic crust (figure 4.28). The sediment is snowplowed off the subducting plate by the overlying plate. New slices of sediment are continually added to the bottom of the accretionary wedge, pushing it upward to form a ridge on the sea floor. A relatively undeformed *forearc basin* lies between the accretionary wedge and the volcanic arc. (The trench side of an arc is the forearc; the other side of the arc is the backarc.)

Trench positions change with time (figure 4.31). As one plate subducts, the overlying plate may be moving toward it. The motion of the leading edge of the overlying plate will force the trench to migrate horizontally over the subducting plate (figure 4.31*A*). The Peru-Chile trench is moving over the Nazca plate in this manner as South America moves westward (figure 4.1). There is another reason that trenches move. It is now widely believed that a subducting plate does not sink in a direction parallel to the length of the plate, but falls through the mantle at an angle that is *steeper* than the dip of the downgoing plate (figure 4.31*B*). This steep sinking pulls the subducting plate progressively away from the overlying plate, and causes the hinge line of bending and the oceanic trench to migrate seaward onto the subducting plate. The location at which the subducting plate contacts the 100-kilometer depth to generate andesite also migrates seaward toward the subducting plate, and may cause the position of the island arc to migrate toward the subducting plate as well.

Ocean-Continent Convergence

When a plate capped by oceanic crust is subducted under the *continental* lithosphere, an accretionary wedge and forearc basin form an *active continental margin* between the trench and the continent (figure 4.32). A Benioff zone of earthquakes dips under the edge of the continent, which is marked by andesitic volcanism and a young mountain belt. An example of this type of boundary is the subduction of the Nazca plate under western South America.

The magma that is created by ocean-continent convergence forms a **magmatic arc,** a broad term used both for island arcs at sea and for belts of igneous activity on the edges of continents. The surface expression of a magmatic arc is either a line of andesitic islands (such as the Aleutian Islands) or a line of andesitic continental volcanoes (such as the Cascade volcanoes of the Pacific Northwest). Beneath the volcanoes are large plutons in thickened crust. We see these plutons as batholiths on land when they are exposed by deep erosion. The igneous

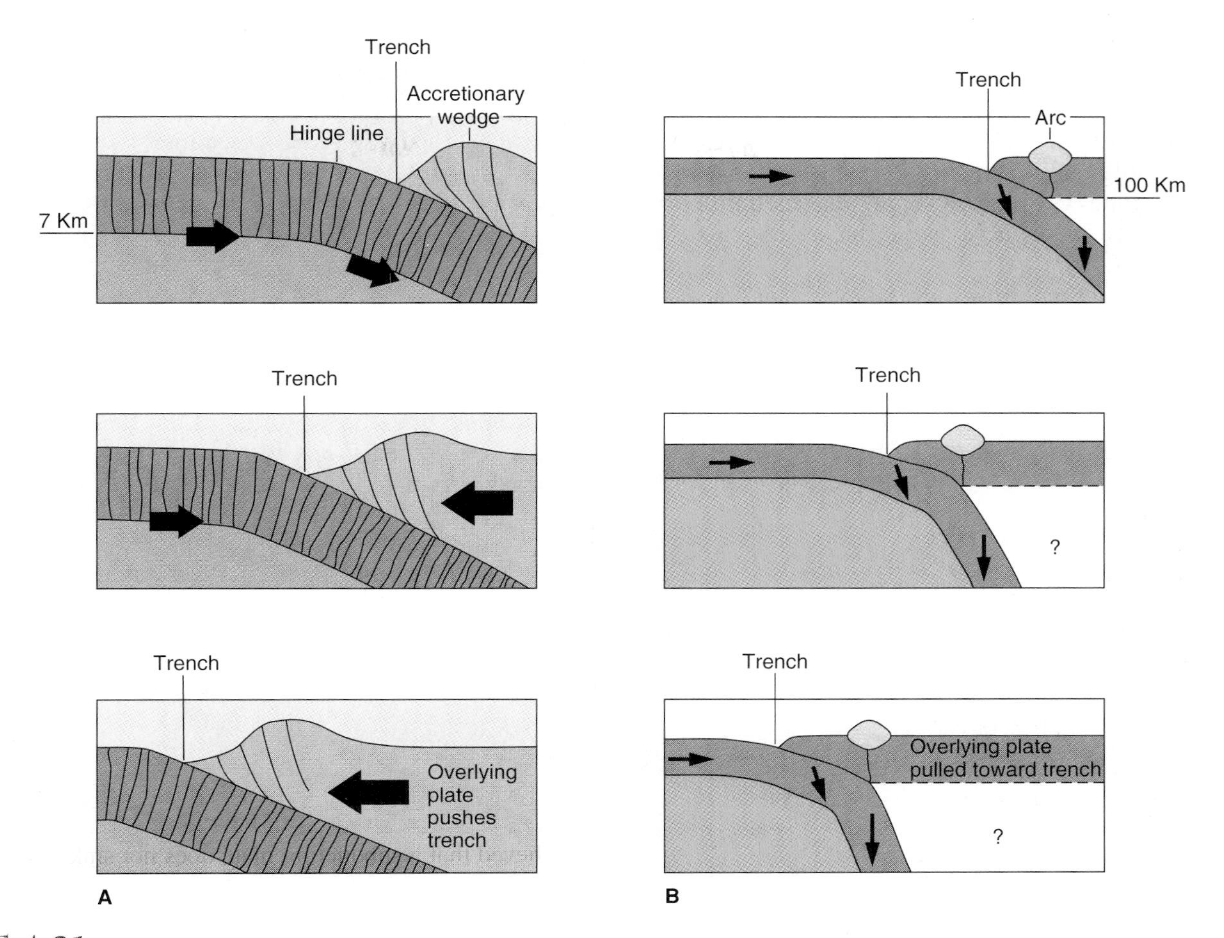

FIGURE 4.31

Migration of trench, hinge line, accretionary wedge, and volcanic arc. (*A*) The motion of the overlying plate can force this migration. (*B*) The cooling, subducting plate sinks at a steeper angle than its dip, pulling the overlying plate toward the subducting plate.

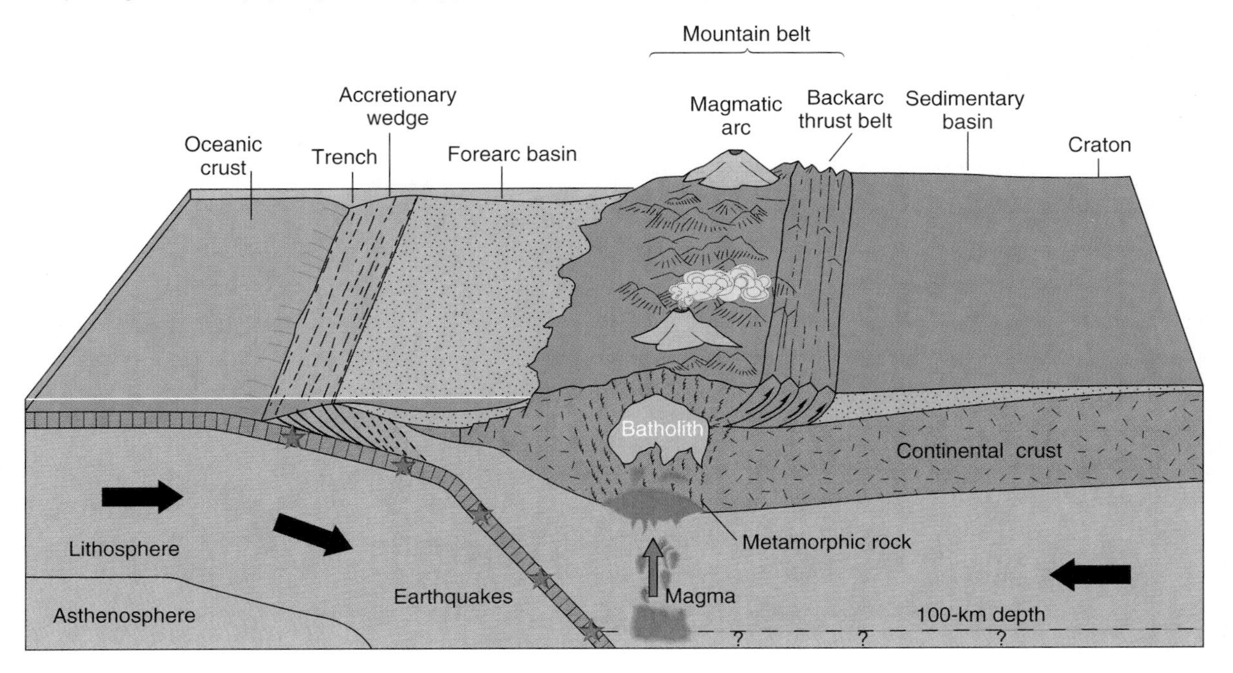

FIGURE 4.32

Ocean-continent convergence forms an active continental margin with a trench, a Benioff zone, a magmatic arc, and a young mountain belt on the edge of the continent.

processes that form the granitic and intermediate magmas of batholiths are described in chapter 11.

The hot magma rising from the subduction zone thickens the continental crust and makes it weaker and more mobile than cold crust. Regional metamorphism takes place within this hot, mobile zone. Crustal thickening causes uplift, so a young mountain belt forms here as the thickened crust rises isostatically.

Another reason for the growth of the mountain belt is the stacking up of thrust sheets on the continental (backarc) side of the magmatic arc (figure 4.32). The thrust faults, associated with folds, move slivers of mountain-belt rocks landward over the continental interior (the *craton*). Underthrusting of the rigid craton beneath the hot, mobile core of the mountain belt may help form the fold-thrust belt.

Inland of the backarc fold-thrust belt, the craton subsides to form a sedimentary basin (sometimes called a *foreland basin*). The weight of the stacked thrust sheets depresses the craton isostatically. The basin receives sediment, some of which may be marine if the craton is forced below sea level. This basin extends the effect of subduction far inland. Subduction of the sea floor off California during the Mesozoic Era produced basin sedimentation as far east as the central Great Plains.

Continent-Continent Convergence

Two continents may approach each other and collide. They must be separated by an ocean floor that is being subducted under one continent and that lacks a spreading axis to create new oceanic crust (figure 4.33). The edge of one continent will

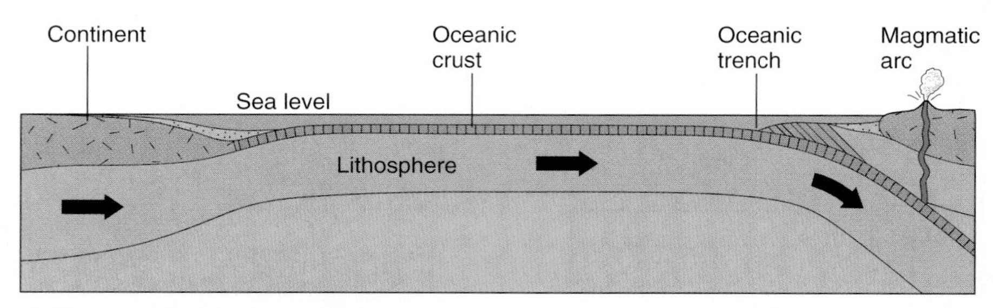

A Ocean-continent convergence

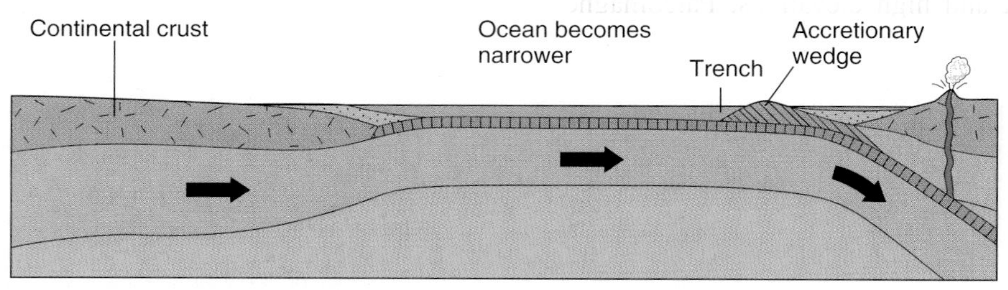

B Ocean-continent convergence

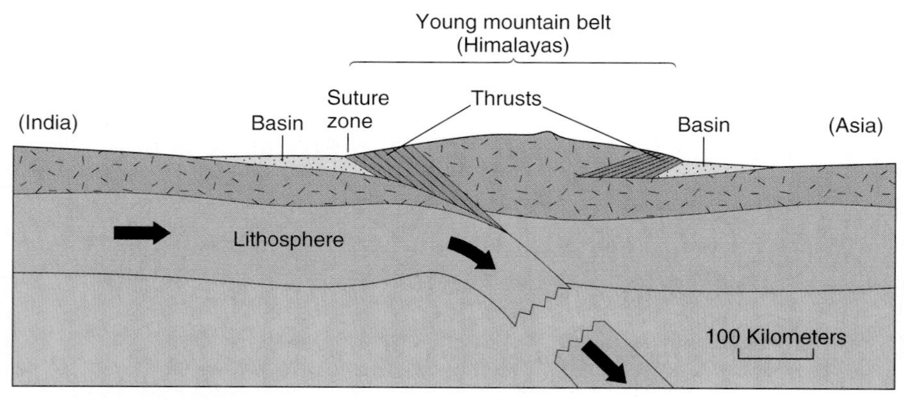

C Continent-continent collision

FIGURE 4.33 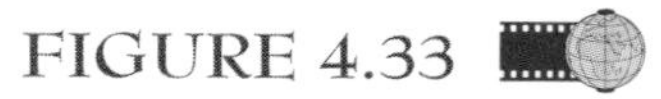

Continent-continent collision forms a young mountain belt in the interior of a new, larger continent.

Modified from W. R. Dickinson, 1977, in *Island Arcs, Deep Sea Trenches and Backarc Basins* (pp. 33–40), copyrighted by American Geophysical Union

initially have a magmatic arc and all the other features of ocean-continent convergence.

As the sea floor is subducted, the ocean becomes narrower and narrower until the continents eventually collide and destroy or close the ocean basin. Oceanic lithosphere is heavy and can sink into the mantle, but continental lithosphere is less dense and cannot sink. One continent may slide a short distance under another, but it will not go down a subduction zone. After collision the heavy oceanic lithosphere breaks off the continental lithosphere and continues to sink, leaving the continent behind (figure 4.33*C*).

The two continents are welded together along a dipping *suture zone* that marks the old site of subduction. Thrust belts and subsiding basins occur on both sides of the original magmatic arc, which is now inactive. The presence of the original arc thickens the crust in the region of impact. The crust is thickened further by the shallow underthrusting of one continent beneath the other and also by the stacking of thrust sheets in the two thrust belts. The result is a mountain belt in the interior of a continent (a new large continent formed by the collision of the two smaller continents). The entire region of impact is marked by a broad belt of shallow-focus earthquakes along the numerous faults, as shown in figure 7.28*A*. A few deeper quakes may occur within the sinking oceanic lithosphere beneath the mountain range.

The Himalaya in central Asia are thought to have formed in this way, as India collided with and underthrust Asia to produce exceptionally thick crust and high elevations. Paleomagnetic studies show that India was once in the southern hemisphere and drifted north to its present position. The collision with Asia occurred after an intervening ocean was destroyed by subduction (figure 4.2).

BACKARC SPREADING

Regional extension occurs within or behind many arcs. This extension can tear an arc in two, moving the two halves in opposite directions (figure 4.34). If it occurs behind an arc, it can move the arc away from a continent. It can split the edge of a continent, moving a narrow strip of the continent seaward (this is apparently how Japan formed). In each case the spreading creates new oceanic crust that is similar, but not identical, to the oceanic crust formed at the crest of mid-oceanic ridges. This backarc oceanic crust is apparently the type of crust found in most ophiolites (chapter 3).

The reason for backarc extension is energetically debated. One suggestion is that extension is caused by a rising and spreading mantle diapir of hot rock or magma somehow generated by the downgoing plate (figure 4.35). The spreading diapir tears open the backarc basin, and the rising magma forms new oceanic crust. Another suggestion is that the subducting plate drags on the overlying asthenosphere, causing it to move in secondary convection cells that stretch and fracture the overlying oceanic crust. A third suggestion, which seems

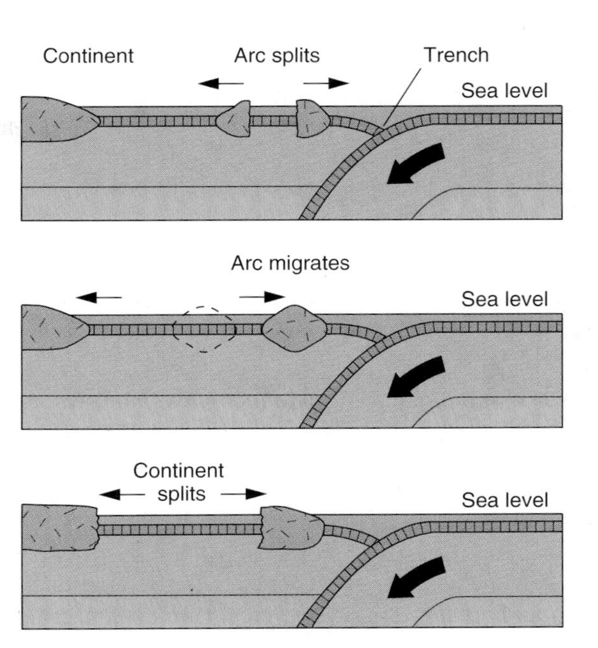

FIGURE 4.34

Backarc spreading. Regional extension in the overlying plate of a subduction zone can split an arc, move an arc offshore, or split a continent.

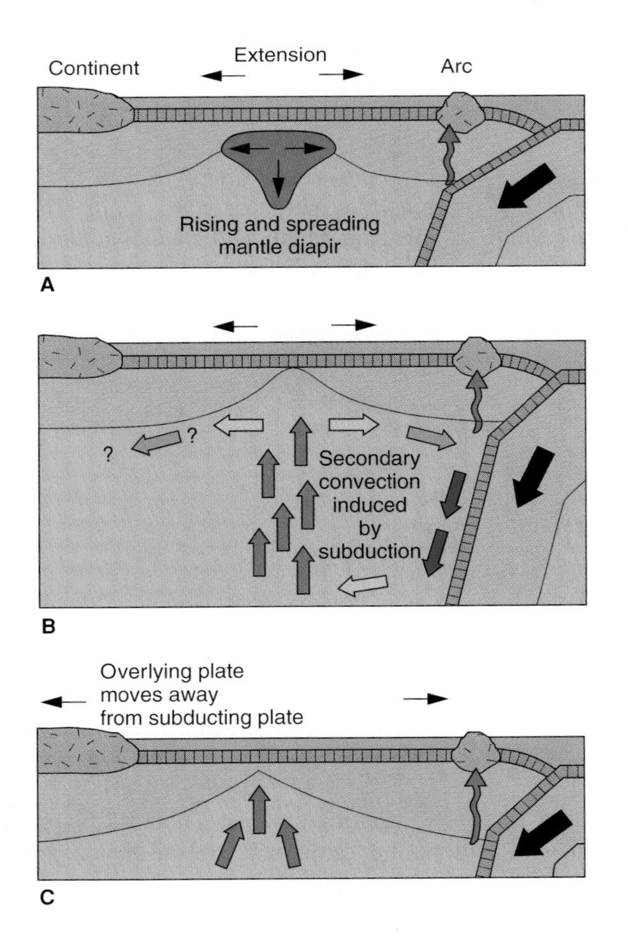

FIGURE 4.35

Causes of backarc spreading. Extension may be caused by a rising mantle diapir, by secondary convection, or by relative plate motions.

the best explanation for the most rapidly spreading backarc basins in the Pacific, is that the overlying plate is retreating away from the subducting plate. If the arc on the overlying plate stays fixed near the subducting plate, the retreat of the overlying plate will tear open the backarc basin.

THE MOTION OF PLATE BOUNDARIES

Almost nothing is fixed in plate tectonics. Not only do plates move, but plate boundaries move as well. Plates may move away from each other at a divergent boundary on a ridge crest for tens of millions of years, but the ridge crest can be migrating across Earth's surface as this occurs. Ridge crests can also jump to new positions. The original ridge crest may suddenly become inactive; the divergence will jump quickly to a new position and create a new ridge crest (the evidence lies in the sea-floor magnetic anomaly pattern).

Convergent boundaries migrate, too, as shown in figure 4.31. As they migrate, trenches and magmatic arcs migrate along with the boundaries. Convergent boundaries can also jump; subduction can stop in one place and begin suddenly in a new place.

Transform boundaries change position, too. California's San Andreas fault has been in its present position about 5 million years. Prior to that, the plate motion was taken up on seafloor faults parallel to the San Andreas (figure 4.36). In the future, the San Andreas may shift eastward again. The 1992 Landers earthquake, on a new fault in the Mojave Desert, and its pattern of aftershocks extending an astonishing 500 miles northward, suggest that the San Andreas may be trying to jump inland again. If it eventually does, most of California will be newly attached to the Pacific plate instead of the North American plate, and California will slide northwestward relative to the rest of North America.

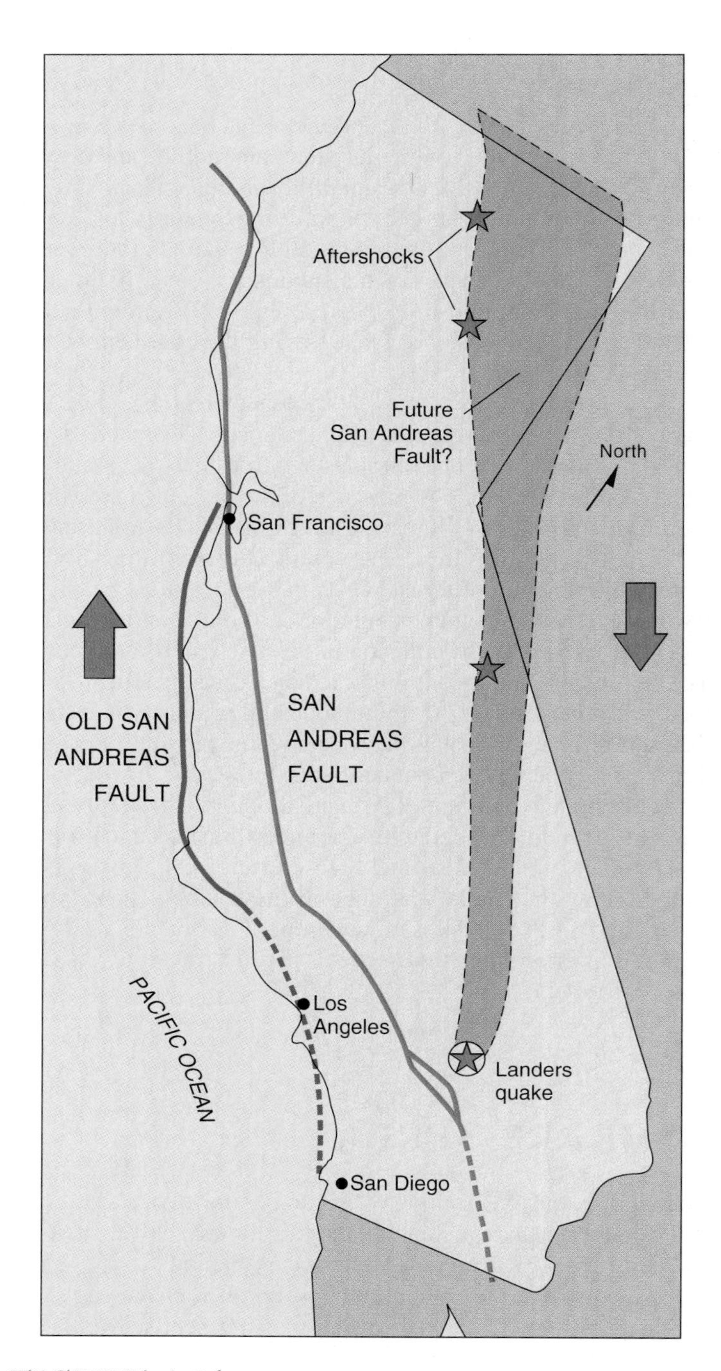

FIGURE 4.36

The San Andreas fault (a transform boundary) has changed position through time. Prior to 5 million years ago, the fault was offshore (blue line). In the future it may jump inland again (green zone).

PLATE SIZE

Plates can change in size. For example, new sea floor is being added on the trailing edge of the North American plate at the spreading axis in the central Atlantic Ocean. Most of the North American plate is not being subducted along its leading edge because this edge is made up of lightweight continental rock. Thus the North American plate is growing in size as it moves slowly westward.

The Nazca plate is getting smaller. The spreading axis is adding new rock along the trailing edge of the Nazca plate, but the leading edge is being subducted down the Peru-Chile Trench. If South America were stationary, the Nazca plate might remain the same size, because the rate of subduction and the rate of spreading are equal. But South America is slowly moving westward because of spreading on the Atlantic Ridge, pushing the Peru-Chile Trench in front of it. This means that the site of subduction of the Nazca plate is gradually coming closer to its spreading axis to the west, and so the Nazca plate is getting smaller. The same thing is probably happening to the Pacific plate as the Eurasian plate moves eastward into the Pacific Ocean.

IN GREATER DEPTH 4.1

Plate Tectonics and Sea Level

Geologists have long known that at certain times in the geologic past the sea covered vast areas of the continents that are now dry land. Much of the interior of the United States, for example, is underlain by marine limestones deposited during parts of the Paleozoic Era. Were the continents lower at these times or was sea level higher?

As you have seen, the subsidence of the craton during subduction can allow vast regions of the continental interior to be flooded with seawater. Some marine deposits on the craton, however, are so extensive that they probably were caused by a rise in sea level.

Although several mechanisms, such as glaciation, can change sea level, the development of plate tectonics has led to a hypothesis that may explain some of the ancient sea-level fluctuations.

During an episode of rapid plate motion, an active spreading axis will be marked by a mid-oceanic ridge caused by the thermal expansion of rock on the rising limb of a convection current. When plate motion stops, convection also stops, and the rock at an old spreading axis cools off and contracts. This means that the mid-oceanic ridge subsides and eventually becomes level sea floor. That is, when plates move, a ridge is present; and when plates stop, the ridge is absent.

When a ridge is present, it displaces seawater, raising sea level and causing the sea to flood land areas. When the ridge is absent, the water returns to the ocean basin and the continents are dry once again.

The plates need not stop completely. A rapid spreading rate would cause a large ridge and a sea-level rise, and a slower spreading rate would cause a smaller ridge and a lower sea level. There is good evidence that some sea-level fluctuations can be correlated to changes in the rate of the sea-floor motion. Not all changes in sea level can be explained by this mechanism. Glaciation and other factors clearly affect sea level too.

THE ATTRACTIVENESS OF PLATE TECTONICS

Most geologists accept the general concept of plate tectonics because it can explain in a general way the distribution and origin of many Earth features. These features are discussed throughout this book, and we summarize them here.

The distribution and composition of the world's *volcanoes* can be explained by plate tectonics. *Basaltic* volcanoes and lava flows form at divergent plate boundaries when hot mantle rock rises at a spreading axis. *Andesitic* volcanoes, particularly those in the circum-Pacific belt, result from subduction of an oceanic plate beneath either a continental plate or another oceanic plate. Although most of the world's volcanoes occur at plate margins, some do not (Hawaii being an example). We will discuss some of these isolated volcanoes later in the chapter when we describe mantle plumes.

Earthquake distribution and first motion can largely be explained by plate tectonics. Shallow-focus earthquakes along normal faults are caused by extension at divergent plate boundaries. Shallow-focus earthquakes also occur on transform faults when plates slide past one another. Broad zones of shallow-focus earthquakes are located where two continents collide. Dipping Benioff zones of shallow-, intermediate-, and deep-focus quakes are found along the giant thrust faults formed when an oceanic plate is subducted beneath another plate. Most of the world's earthquakes (like most volcanoes) occur along plate boundaries, although a few take place within plates and are difficult to explain in terms of plate tectonics.

Young mountain belts—with their associated igneous intrusions, metamorphism, and fold-thrust belts—form at convergent boundaries. "Subduction mountains" form at the edges of continents where sea floor is sliding under continents. "Continental-collision" mountains form in continental interiors when two continents collide to form a larger continent. Old mountain belts mark the position of old, now inactive, plate boundaries.

The major features of the sea floor can also be explained by plate tectonics. The *mid-oceanic ridge* with its rift valley forms at divergent boundaries. *Oceanic trenches* are found where oceanic plates are subducted at convergent boundaries. *Fracture zones* are created at transform boundaries.

Other hypotheses can explain some of these features, but not all of them. Belts of folded mountains have been attributed to compression caused by a contracting Earth. Rift valleys, on the other hand, have been explained by extension caused by an *expanding* Earth. The hypotheses are incompatible with each other and do not give a unifying view of Earth. Plate tectonics explains more features than any other hypothesis or theory, and it provides a unifying framework for the study of Earth. That is why so many geologists support the concept, at least as a working model of how Earth works.

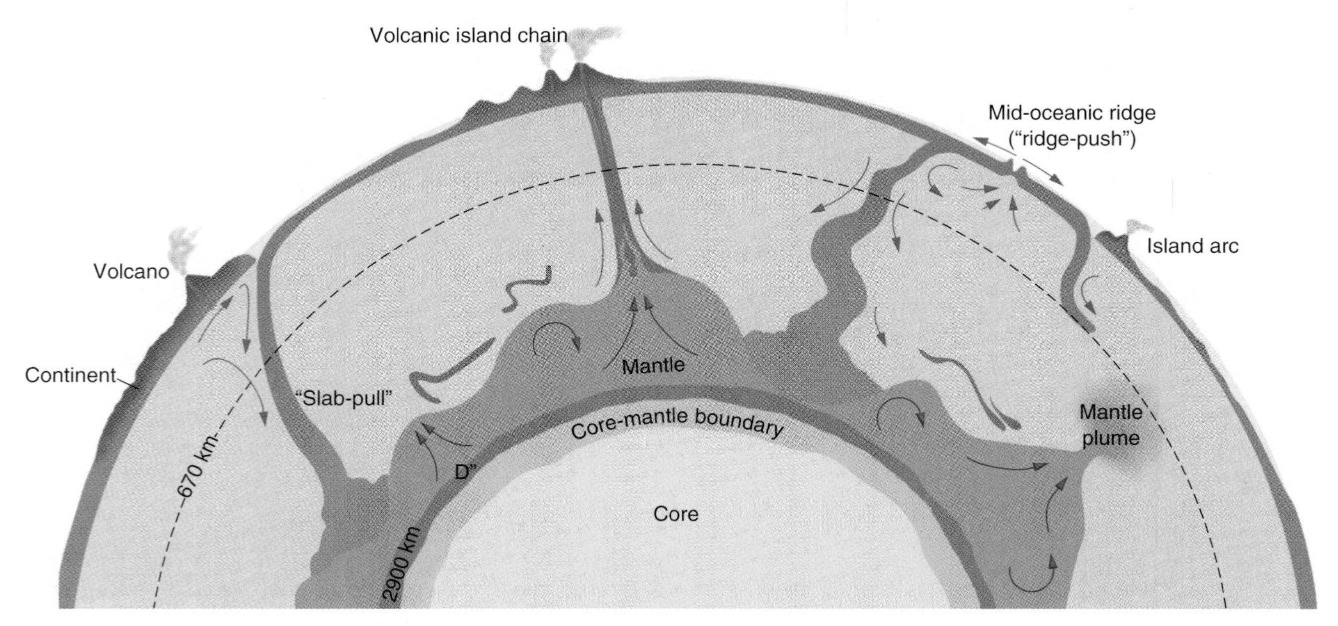

FIGURE 4.37

Model of mantle convection.

Modified from Kellogg, L. H., Hager, B. H., and van der Hilst, R. D., 1999, *Science,* 283:1881–1884

WHAT CAUSES PLATE MOTIONS?

There is currently a great deal of speculation about why plates move. There may be several reasons for plate motion. Any mechanism for plate motion has to explain why:

1. mid-oceanic ridge crests are hot and elevated, while trenches are cold and deep;
2. ridge crests have tensional cracks;
3. the leading edges of some plates are subducting sea floor, while the leading edges of other plates are continents (which cannot subduct).

Convection in the mantle, proposed as a mechanism for sea-floor spreading (figure 4.12), can account for these facts, as we have shown earlier in this chapter. Mantle convection is quite likely because heat loss from Earth's core should heat the overlying mantle, causing it to overturn. The old sea-floor spreading model assumed mantle-deep convection. Recent studies using seismic tomography and computer modeling of seismic waves suggest that the dynamics of convection in the mantle is not simple (figure 4.37). Cold lithospheric plates may subduct down to the core-mantle boundary whereas other less dense (younger) plates may only reach the 670-kilometer boundary. One of the most recent models suggests that the lowermost part of the mantle does not mix with the upper and middle mantle, but acts as a "lava lamp" turned on low fueled by internal heating and heat flow across the core-mantle boundary. Variation in the thickness of this dense layer may control where mantle plumes rise and subducted plates ultimately rest.

Some geologists think that mantle convection is a *result* of plate motion rather than a cause of it. The sinking of a cold, subducting plate can create mantle convection (convection can be driven by either hot, rising material or by cold, sinking material). Hot mantle rock rises at divergent boundaries to take the place of the diverging plates (figure 4.24); however, such plate-caused convection would be shallow rather than mantle-deep.

The basic question in plate motion is why do plates diverge and sink? Two or three different mechanisms may be at work here.

One proposal is called "*ridge-push.*" As a plate moves away from a divergent boundary, it cools and thickens. Cooling sea floor subsides as it moves, and this subsidence forms the broad side slopes of the mid-oceanic ridge. An even more important slope forms on the base of the lithosphere mantle. The mantle thickens as cooling converts asthenospheric mantle to lithospheric mantle. Therefore, the boundary between them is a slope down which the lithosphere slides (figure 4.38). The oceanic plate is thought to slide down this slope at the base of the lithosphere, which may have a relief of 80 to 100 kilometers.

Another mechanism is called "*slab-pull*" (figure 4.39). Cold lithosphere sinking at a steep angle through hot mantle should pull the surface part of the plate away from the ridge crest and then down into mantle as it cools. A subducting plate sinks because it is denser than the surrounding mantle. This density contrast is partly due to the fact that the sinking lithosphere is cold. The subducting plate may also increase its density while it sinks, as low-density materials such as water are lost and as plate minerals collapse into denser forms during subduction. Slab-pull is thought to be at least twice as important as ridge-push in moving an oceanic plate away from a ridge crest. Slab-pull causes rapid plate motion.

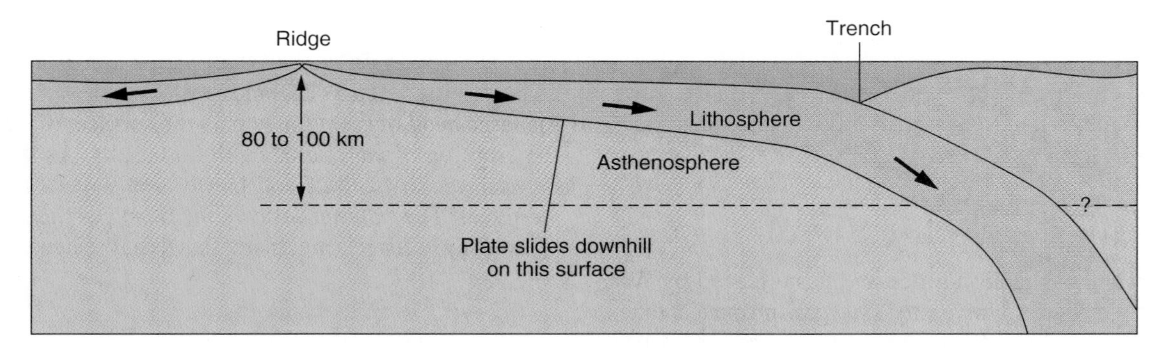

FIGURE 4.38

Ridge-push. A plate may slide downhill on the sloping boundary between the lithosphere and the asthenosphere at the base of the plate.

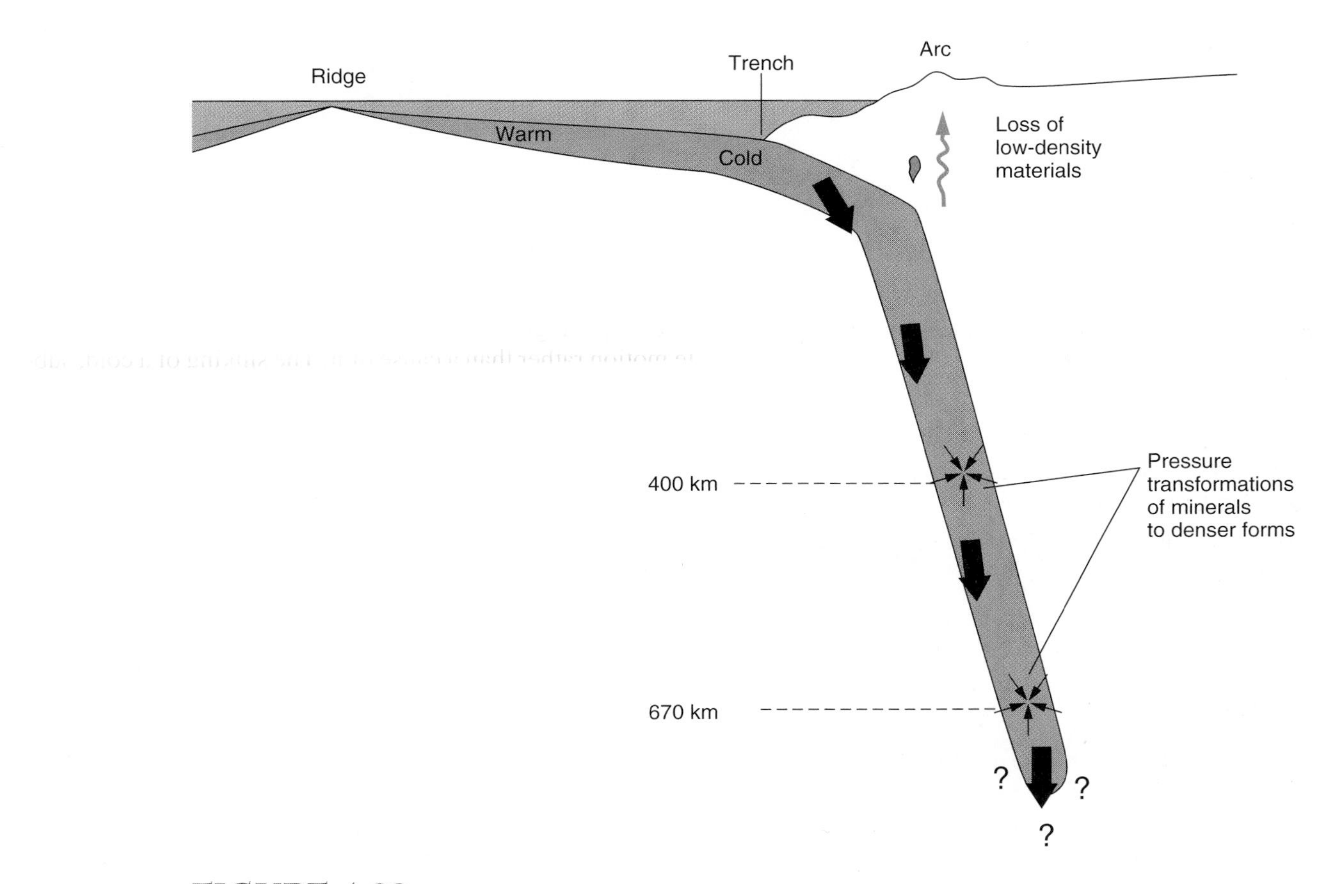

FIGURE 4.39

Slab-pull. The dense, leading edge of a subducting plate pulls the rest of the plate along. Plate density increases due to cooling, loss of low-density material, and pressure transformation of minerals to denser forms.

If subducting plates fall into the mantle at angles steeper than their dip (figure 4.31), then trenches and the overlying plates are pulled horizontally seaward toward the subducting plates. This mechanism has been termed "*trench-suction.*" It is probably a minor force, but may be important in moving continents apart. Divergent continents at the leading edges of plates cannot be moved by slab-pull, because they are not on subducting plates. They might be moved by ridge-push from the rear, and/or trench-suction from the front (figure 4.40). They move much more slowly than subducting plates.

All three of these mechanisms (ridge-push, slab-pull, and trench-suction), particularly in combination, are compatible with high, hot ridges; cold, deep trenches; and tensional cracks at the ridge crest. They can account for the motion of both oceanic and continental plates. In this scheme, plate motions are controlled by variations in lithosphere density and thickness, which, in turn, are controlled largely by cooling. In other words, the reasons for plate motions are the properties of the plates themselves and the pull of gravity. This idea is in sharp contrast to most convection models, which assume that plates are dragged along by the movement of mantle rock beneath the plates.

Mantle Plumes and Hot Spots

A modification of the convection process was suggested by W. Jason Morgan of Princeton University. Morgan proposed that convection occurs in the form of **mantle plumes,** narrow columns of hot mantle rock that rise through the mantle, much like smoke rising from a chimney (figure 4.41). Mantle plumes are now thought to have large spherical or mushroom-shaped heads above a narrow rising tail. They are essentially stationary with respect to moving plates and to each other.

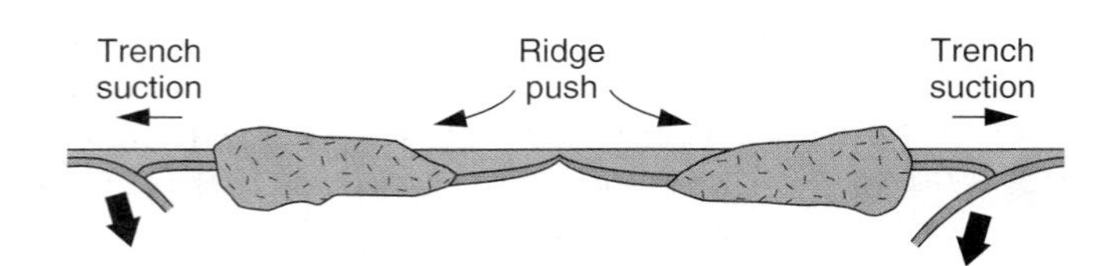

FIGURE 4.40

Divergent continents may be moved by ridge-push from the rear, or trench-suction (caused by steeply sinking plates) from the front.

Plumes may form "hot spots" of active volcanism at Earth's surface. Note in figure 4.42 that many plumes are located in volcanic regions such as Iceland, Yellowstone, and Hawaii. When the large head of the plume nears the surface, it causes uplift and the eruption of vast fields of flood basalts. As the head widens beneath the crust the flood-basalt area widens and the crust is stretched. The tail that follows the head produces a narrow spot of volcanic activity, much smaller than the head.

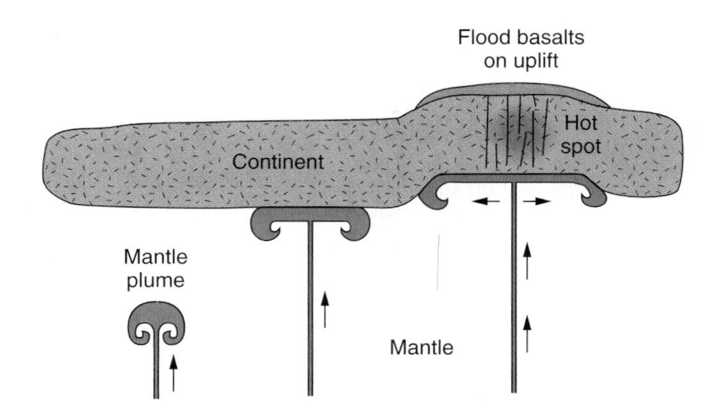

FIGURE 4.41

Mantle plumes rise upward through the mantle. When the large head contacts a continent, it causes uplift and the eruption of flood basalts.

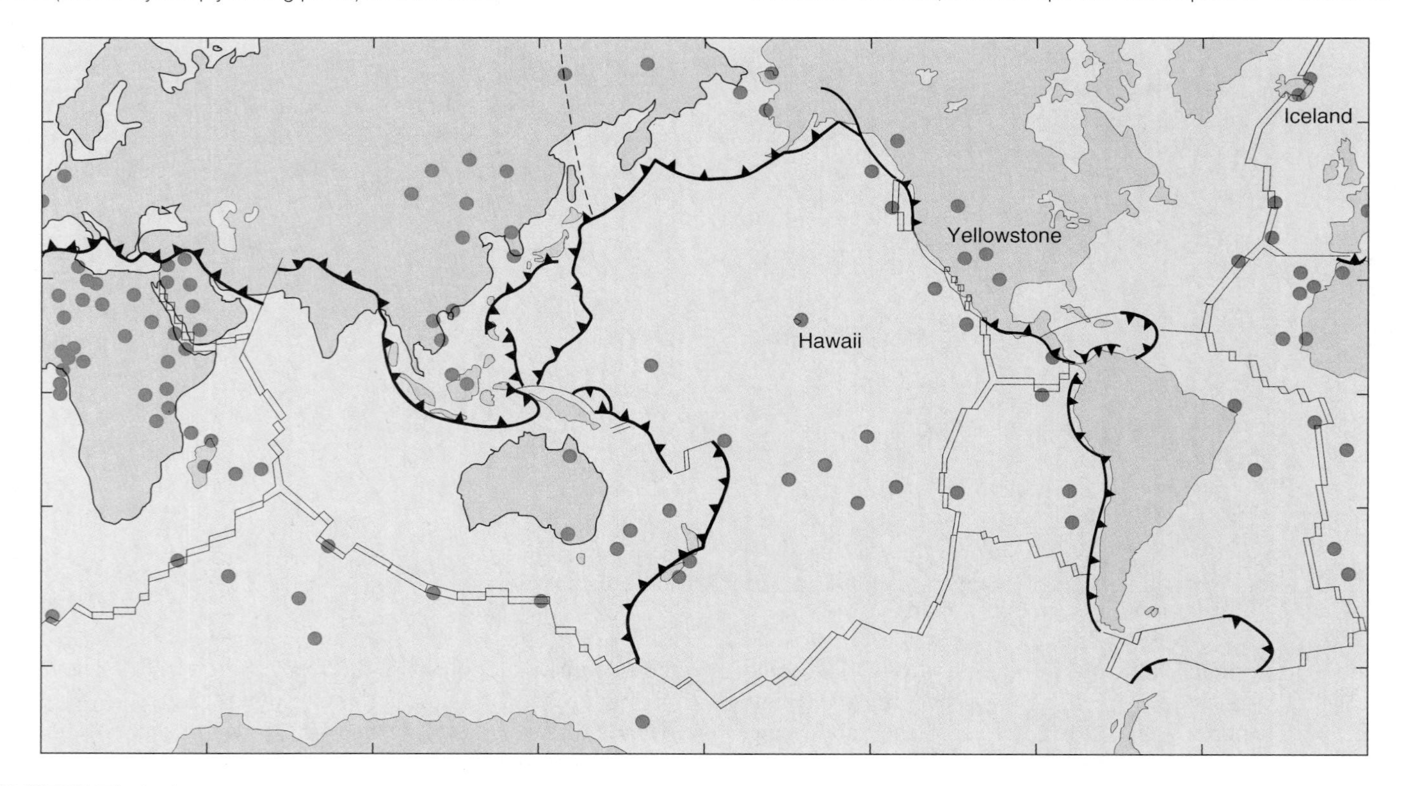

FIGURE 4.42

Distribution of hypothesized mantle plumes, identified by volcanic activity and structural uplift within the past few million years. The hot spots near the poles are not shown.

Compiled by J. T. Wilson and K. Burke, *Scientific American*

The outward, radial flow of the expanding head may be strong enough to break the lithosphere and start plates moving. In Morgan's view a few plumes, such as those on the mid-oceanic ridge in the Atlantic Ocean in figure 4.42, are enough to drive plates apart (in this case, to push the American plates westward).

A mantle plume rising beneath a continent should heat the land and bulge it upward to form a dome marked by volcanic eruptions. As the dome forms, the stretched crust typically fractures in a three-pronged pattern (figure 4.43). Continued radial flow outward from the rising plume eventually separates the crust along two of the three fractures but leaves the third fracture inactive. In this model of continental breakup, the two active fractures become continental edges as new sea floor forms between the divergent continents. The third fracture is a *failed rift* (or *aulacogen*), an inactive rift that becomes filled with sediment.

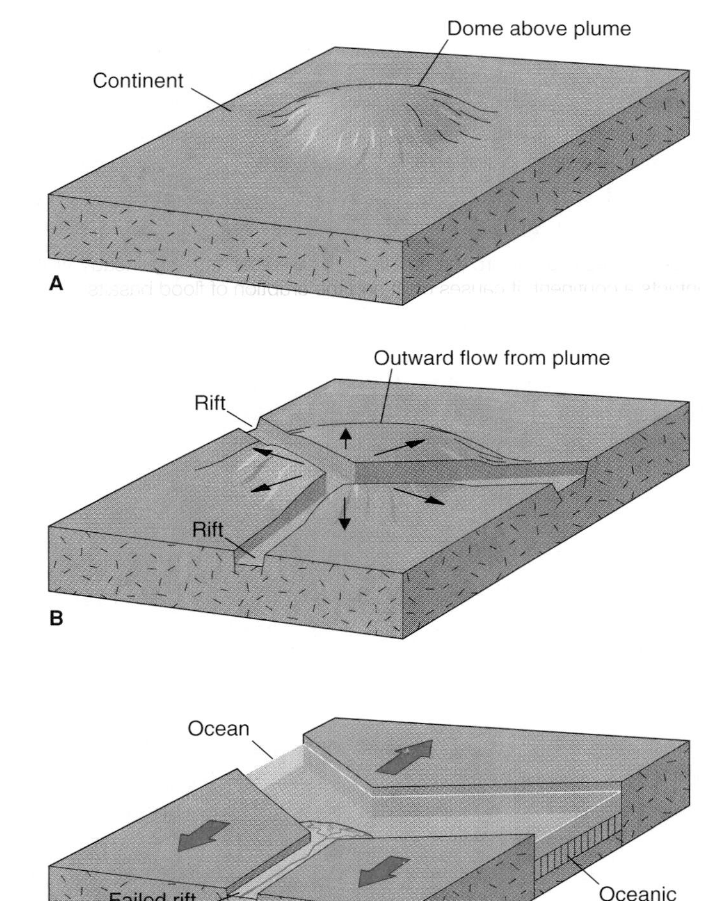

FIGURE 4.43 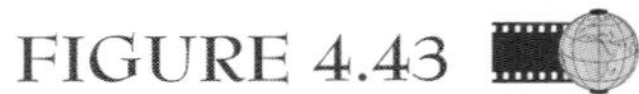

Continental breakup caused by a mantle plume. (*A*) A dome forms over a mantle plume rising beneath a continent. (*B*) Three radial rifts develop due to outward radial flow from the top of the mantle plume. (*C*) Continent separates into two pieces along two of the three rifts, with new ocean floor forming between the diverging continents. The third rift becomes an inactive "failed rift" (or aulacogen) filled with continental sediment.

An example of this type of fracturing may exist in the vicinity of the Red Sea (figure 4.44). The Red Sea and the Gulf of Aden are active diverging boundaries along which the Arabian Peninsula is being separated from northeastern Africa. The third, inactive, rift is the northernmost African Rift Valley, lying at an angle of about 120° to each of the narrow seaways.

Figure 4.45 shows how two plumes might split a continent and begin plate divergence. Local uplift *causes* rifting over each plume. The rifts lengthen with time until the land is torn in two. The two halves begin to diverge from being dragged along from below by the outward radial flow of the plume. Along the long rift segments between plumes, rifting occurs *before* uplift.

A place where a mantle plume might now be rising beneath a continent is in Yellowstone National Park in northwestern Wyoming. The area's volcanism, high elevation, high heat flow, and hot spring and geyser activity all may be due to this plume. Radial flow of mantle rock beneath the western United States may be tearing the continent apart and causing the earthquakes in the region, including the 1959 earthquake near Madison Canyon, Montana. Eventually an ocean may form here as North America is split apart by the plume.

Some plumes rise beneath the centers of oceanic plates. A plume under Hawaii rises in the center of the Pacific plate. As the plate moves over the plume, a line of volcanoes forms, creating an aseismic ridge (figure 4.46 and chapter 3). The volcanoes are gradually carried away from the eruptive center, sinking as they go because of cooling. The result is a line of extinct volcanoes (seamounts and guyots) increasing in age away from an active volcano directly above the plume.

In the Hawaiian island group, the only two active volcanoes are in the extreme southeastern corner (figure 4.47). The

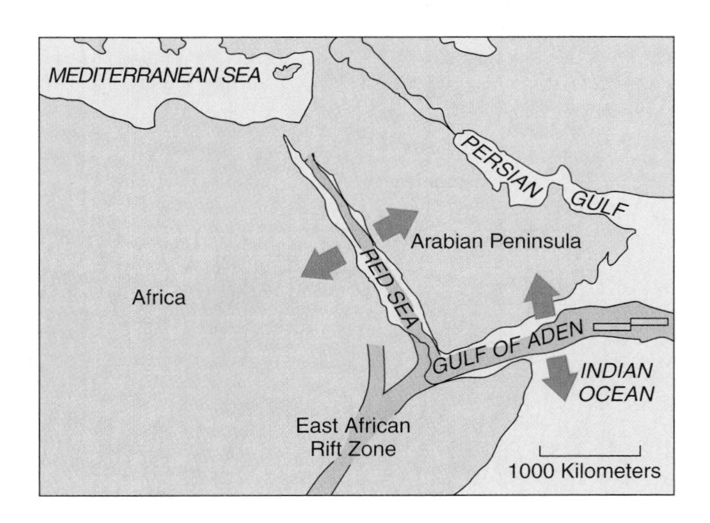

FIGURE 4.44

An example of radial rifts. The Red Sea and the Gulf of Aden are the active rifts, as the Arabian peninsula drifts away from Africa. The Gulf of Aden contains a mid-oceanic ridge and central rift valley. The inactive failed rift is the rift valley shown in Africa.

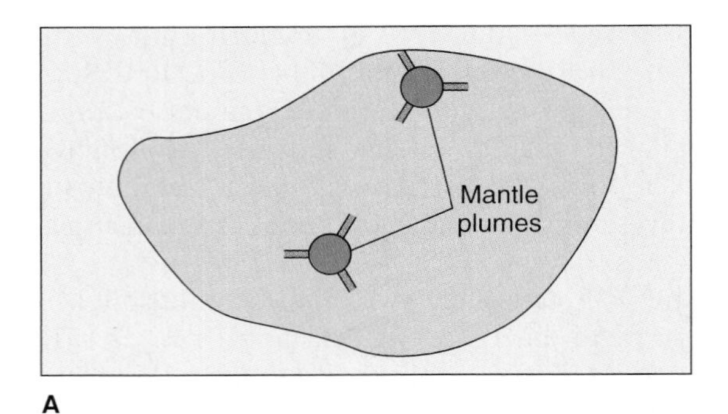

A

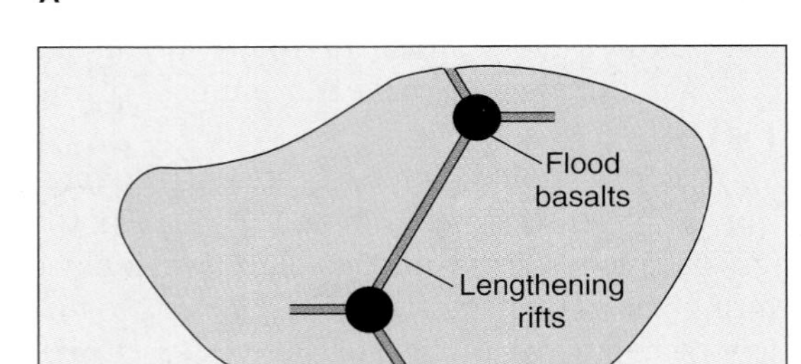

B

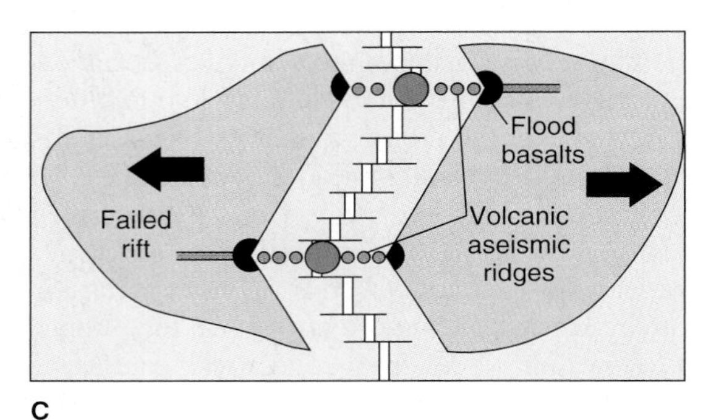

C

FIGURE 4.45

(*A*) Two mantle plumes beneath a continent. (*B*) The rifts lengthen and flood basalts erupt over the plumes. (*C*) The continent splits and failed rifts form. The new ocean is marked by ridge crests, fracture zones, and aseismic ridges (chains of volcanoes).

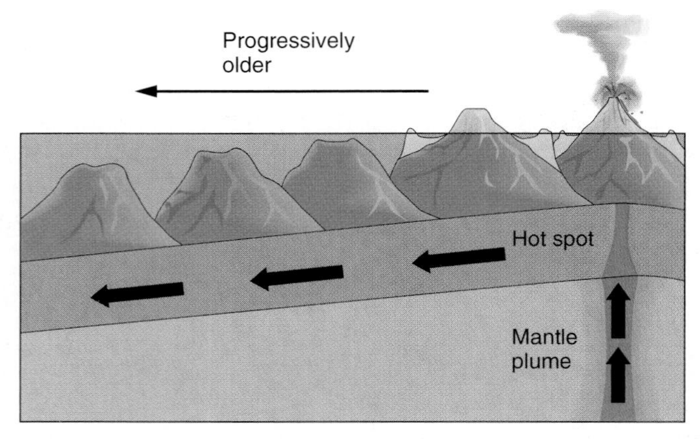

FIGURE 4.46

Sea floor moving over a hot spot forms an aseismic ridge as a chain of volcanoes and guyots.

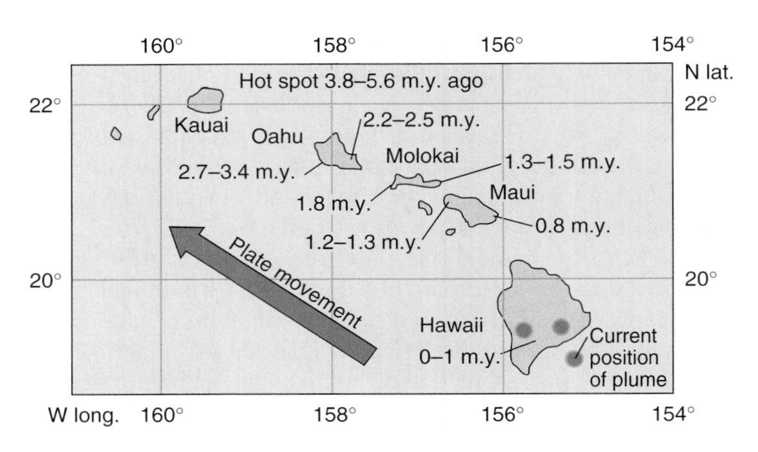

FIGURE 4.47

Ages of volcanic rock of the Hawaiian island group. Ages increase to northwest. Two active volcanoes on Hawaii shown by red dots. See also figure 3.21.

isotopic ages of the Hawaiian basalts increase regularly to the northwest, and a long line of submerged volcanoes forms an aseismic ridge to the northwest of Kauai (figure 3.21). Most aseismic ridges on the sea floor appear to have active volcanoes at one end, with ages increasing away from the eruptive centers. Deep-sea drilling has shown, however, that not all aseismic ridges increase in age along their lengths. This evidence has led to alternate hypotheses for the origin of aseismic ridges. It may pose difficulties for the plume hypothesis itself.

Note in figure 3.21 that the three large aseismic ridges in the Pacific Ocean change direction abruptly. If these ridges formed from three separate stationary plumes, they suggest that the Pacific plate has moved in two directions in the past. Early movement approximately northward, followed by more westerly movement, could have produced the ridge patterns.

THE RELATIONSHIP BETWEEN PLATE TECTONICS AND ORE DEPOSITS

The plate tectonic theory provides an overall model for the origin of metallic ore deposits that has been used to explain the occurrence of known deposits and in exploration for new deposits. Because many ore deposits are associated with igneous activity, there is a close relationship between plate boundaries and metallic ore deposits.

As discussed in chapter 3, *divergent plate boundaries* are often marked by lines of active hot springs in rift valleys that carry and precipitate metallic minerals in mounds around the hot springs. The metals in rift-valley hot springs are predominantly iron, copper, and zinc, with smaller amounts of manganese, gold, and silver. Although the mounds are nearly solid metal sulfide, they are small and widely scattered on the sea floor, so commercial mining of them may not be practical. Occasionally, the ore minerals may be concentrated in richer

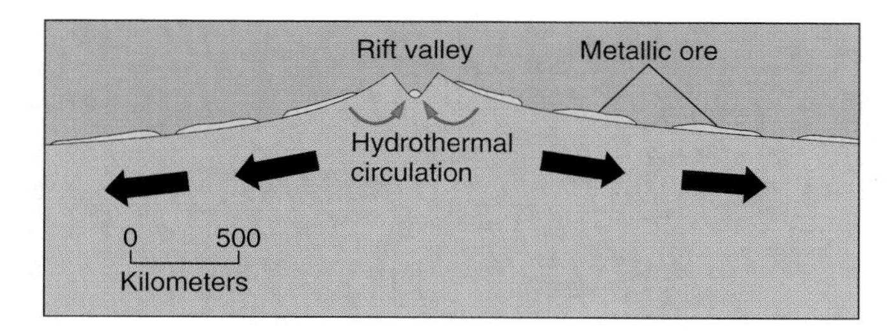

FIGURE 4.48

Divergent oceanic plates carry metallic ores away from rift valley. (Size of ore deposits exaggerated.)

deposits. On the floor of the Red Sea metallic sediments have precipitated in basins filled with hot-spring solutions. Although the solutions are hot (up to 60°C, or 140°F), they are very dense because of their high salt content (they are seven times saltier than sea water), so they collect in sea-floor depressions instead of mixing with the overlying sea water. Although not currently mined, the metallic sediments were estimated in 1983 to be worth $25 billion.

Hot metallic solutions are also found along some divergent continental boundaries. Near the Salton Sea in southern California, which lies along the extension of the mid-oceanic ridge inland, hot water very similar to the Red Sea brines has been discovered underground. The hot water is currently being used to run a geothermal power plant. The high salt and metal content is corrosive to equipment, but metals such as copper and silver may one day be recovered as valuable by-products.

Sea-floor spreading carries the metallic ores away from the ridge crest (figure 4.48), perhaps to be subducted beneath island arcs or continents at *convergent plate boundaries.* Slivers of *ophiolite* on land may contain these rich ore minerals in relatively intact form. A notable example of such ores occurs on the island of Cyprus in the Mediterranean Sea. Banded chromite ores may also be contained in the serpentinized ultramafic rock at the bottom of ophiolites.

Volcanism at *island arcs* can also produce hot-spring deposits on the flanks of the andesitic volcanoes. Pods of very rich ore collect above local bodies of magma, and the ore is sometimes distributed as sedimentary layers in shallow basins (figure 4.49). The circulation pattern and the ore-forming processes are quite similar to those of spreading centers, but the island arc ores usually contain more lead and gold. Rich *massive sulfide deposits* overlying fractured volcanic rock in the Precambrian shield area of Canada may have formed in this way on ancient island arcs.

Subduction of the sea floor beneath a *continent* produces broad belts of metallic ore deposits near the edge of the continent. Figure 4.50 shows how the distribution of some metals in the western United States might be related to depth along a subduction zone (the figure shows only one of several competing models relating continental ore deposits to plate tectonics). The pattern of ore belts in the United States has probably been disturbed by changing subduction angles, strike-slip faulting, and

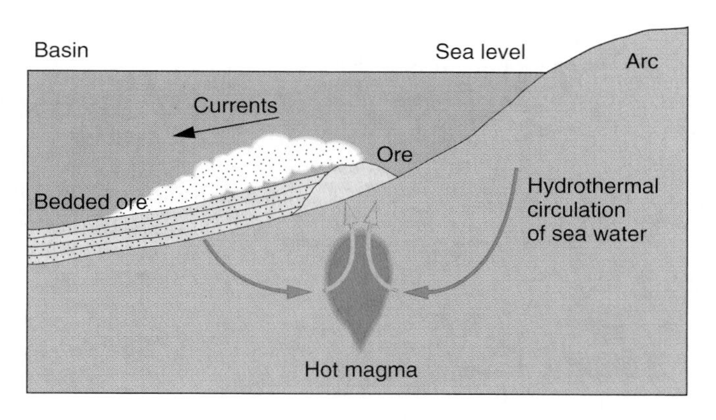

FIGURE 4.49

On island arcs metallic ores can form over hot springs and be redistributed into layers by currents in shallow basins.

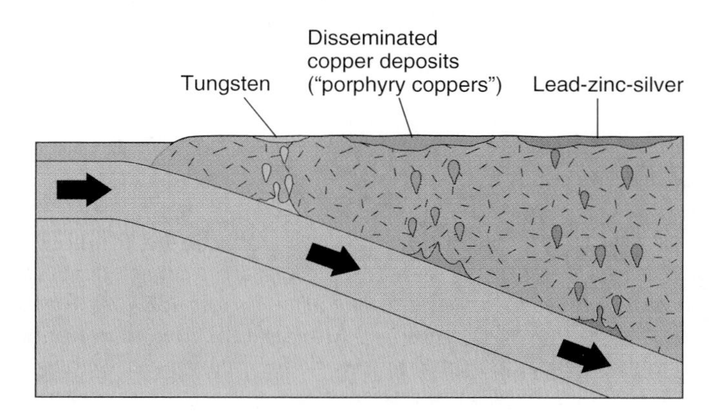

FIGURE 4.50

Possible relation of ore belts in the western United States to depth along the subduction zone. Different metallic ores (and different igneous rocks) are generated at different depths along a subducted plate.

backarc spreading. Similar patterns of ore belts occur in other subduction mountain ranges, notably the Andes.

The origin of the continental ores above a subduction zone is not clear. The hot-spring deposits from the ridge crest are subducted with oceanic crust and could become remobilized to rise into the continent above. The ores may also "distill" off other parts of the descending oceanic crust or upper mantle. The metals may also derive from the continental crust itself or the mantle below it. The metals may be concentrated somehow by the heat of a rising blob of magma or the hydrothermal circulation associated with it.

The connection between some hydrothermal ore deposits and plate tectonics is tenuous at best. The "Mississippi Valley-type" lead-zinc deposits of the continental interior are very puzzling features. Over broad areas metal ore has been emplaced in limestone and dolomite, both by cavity filling and replacement (both are usually considered to be hydrothermal processes). There is no obvious connection between the ores and any igneous rocks, which may be absent in the ore regions. The presence of the ores in the thin sedimentary cover of a supposedly inactive interior of a plate is difficult to explain. The ores do group roughly around the New Madrid earthquakes of southeastern Missouri (figure 7.13); deep faults, perhaps on a failed rift, may have provided pathways for ore-bearing fluids.

It is tempting to think that *mantle plumes* might cause ore deposition, for plumes provide a source of both magma and hydrothermal solutions. The locations of supposed plumes, however, such as Yellowstone and Hawaii, are notable for their *lack* of ore deposits.

A FINAL NOTE

Geologists, like other people, are susceptible to fads. Most geologists believe plate tectonics is an exciting theory and accept it as a working model of Earth. There will undoubtedly continue to be modifications and refinements to the theory. Most geologists today believe that plates exist and move. But widespread belief in a theory does not make it true. Two hundred years ago geologists "knew" that basalt crystallized out of seawater. In the 1800s glacial deposits were thought to be deposited by Noah's flood. Both of these incorrect ideas were finally disproved by decades of exacting field work and often bitter debate.

Forty years ago continental drift rated only a footnote in most introductory textbooks. Now there are many believers in continental motion, and textbooks use it as a framework for the entire field of physical geology. Although the idea of continental stability provided the framework for many past textbooks, today the idea that continents are fixed in position rates only a footnote as an outmoded concept.

Objections were raised to the concept of plate tectonics after it was proposed in the late 1960s. Some sea-floor features did not seem compatible with a moving sea floor. The geology of many continental regions did not seem to fit into the theory of plate tectonics, in some cases not even slightly. But a revolutionary, new idea in science is always controversial. As it progresses from an "outrageous hypothesis" to a more widely accepted theory, after much discussion and testing, a new idea evolves and changes. The newness of the idea wears off, and successful tests and predictions convert skeptics to supporters (sometimes grudgingly). Perhaps equally importantly, dissenters die off.

As refinements were made to plate tectonics, and as more was learned about the puzzling sea-floor features and continental regions, they began to seem more compatible with plate tectonics. Objections died out, and plate tectonics became widely accepted.

It is wise to remember that at the time of Wegener most geologists vehemently disagreed with continental drift. Because Wegener proposed that continents plow through seafloor rock, and because his proposed forces for moving continents proved inadequate, most geologists thought that continental drift was wrong. Although these geologists had sound reasons for their dissent, we now think that due to the mounting evidence, continental drift is more acceptable and that the early *geologists* were wrong.

The evidence for plate tectonics is very convincing. The theory has been rightly called a revolution in Earth science, comparable to the development of the theory of evolution in the biological sciences. It is an exciting time to be a geologist. Our whole concept of Earth dynamics has changed in the last forty years.

Summary

Plate tectonics is the idea that Earth's surface is divided into several large plates that change position and size. Intense geologic activity occurs at plate boundaries.

Plate tectonics combines the concepts of *sea-floor spreading* and *continental drift*.

Alfred Wegener proposed continental drift in the early 1900s. His evidence included coastline fit, similar fossils and rocks in now-separated continents, and paleoclimatic evidence for *apparent polar wandering*. Wegener proposed that all continents were once joined together in the supercontinent *Pangaea*.

Wegener's ideas were not widely accepted until the 1950s, when work in paleomagnetism revived interest in polar wandering.

Evidence for continental drift includes careful fits of continental edges and detailed rock matches between now-separated continents. The positions of continents during the past 200 million years have been mapped.

Hess's hypothesis of *sea-floor spreading* suggests that the sea floor moves away from the ridge crest and toward trenches as a result of mantle convection.

According to the concept of sea-floor spreading, the high heat flow and volcanism of the ridge crest are caused by hot mantle rock rising beneath the ridge. Divergent *convection* currents in the mantle cause the rift valley and earthquakes on the ridge crest, which is a *spreading axis* (or *center*). New sea floor near the rift valley has not yet accumulated pelagic sediment.

Sea-floor spreading explains trenches as sites of sea-floor *subduction*, which causes low heat flow and negative gravity anomalies. Benioff zones and andesitic volcanism are caused by interaction between the subducting sea floor and the rocks above.

Sea-floor spreading also explains the young age of the rock of the sea floor as caused by the loss of old sea floor through subduction into the mantle.

Plates are composed of blocks of *lithosphere* riding on a plastic *asthenosphere*. Plates move away from spreading axes, which add new sea floor to the trailing edges of the plates.

An apparent confirmation of plate motion came in the 1960s with the correlation of marine *magnetic anomalies* to *magnetic reversals* by Vine and Matthews. The origin of magnetic anomalies at sea apparently is due to the recording of normal and reverse magnetization by dikes that intrude the crest of the mid-oceanic ridge, then split and move sideways to give anomaly patterns a mirror symmetry.

The Vine-Matthews hypothesis gives the rate of plate motion (generally 1 to 6 cm/year) and can predict the age of the sea floor before it is sampled.

Deep-sea drilling has apparently verified plate motions and the age predictions made from magnetic anomalies.

Earthquake distribution and first-motion studies on *transform faults* on fracture zones also verify plate motions.

Divergent plate boundaries are marked by rift valleys, shallow-focus earthquakes, high heat flow, and basaltic volcanism.

Transform boundaries between plates sliding past one another are marked by strike-slip (transform) faults and shallow-focus earth quakes.

Convergent plate boundaries can cause *subduction* or *continental collision.* Subducting plate boundaries are marked by trenches, low heat flow, Benioff zones, andesitic volcanism, and young mountain belts or island arcs. Continental-collision boundaries have shallow-focus earthquakes and form young mountain belts in continental interiors.

The distribution and origin of most volcanoes, earthquakes, young mountain belts, and major sea-floor features can be explained by plate tectonics.

Plate motion was once thought to be caused by *mantle convection,* but is now attributed to the cold, dense, leading edge of a subducting plate pulling the rest of the plate along with it (*slab-pull*). Plates near mid-oceanic ridges also slide down the sloping lithosphere-asthenosphere boundary at the ridge (*ridge-push*). *Trench-suction* may help continents diverge.

Mantle plumes are narrow columns of hot, rising mantle rock. They cause flood basalts and may split continents, causing plate divergence.

An aseismic ridge may form as an oceanic plate moves over a mantle plume acting as an eruptive center (hot spot).

Terms to Remember

asthenosphere 82
continental drift 74
convection 80
convergent plate boundary 82
divergent plate boundary 82
island arc 91
lithosphere 82
magmatic arc 92
mantle plume 100
plate 82
plate tectonics 74
polar wandering 77
sea-floor spreading 74
subduction 80
transform fault 86
transform plate boundary 83

Testing Your Knowledge

Use the questions below to prepare for exams based on this chapter.

1. What was Wegener's evidence for continental drift?
2. What is polar wandering? What is the paleoclimatic evidence for polar wandering? What is the magnetic evidence for polar wandering? Does polar wandering require the poles to move?
3. What is the evidence that South America and Africa were once joined?
4. In a series of sketches show how the South Atlantic Ocean might have formed by the movement of South America and Africa.
5. What is Pangaea?
6. In a single cross-sectional sketch, show the concept of sea-floor spreading and how it relates to the mid-oceanic ridge and oceanic trenches.
7. How does sea-floor spreading account for the age of the sea floor?
8. What is a plate in the concept of plate tectonics?
9. Define *lithosphere* and *asthenosphere.*
10. What is the origin of marine magnetic anomalies according to Vine and Matthews?
11. Why does the pattern of magnetic anomalies at sea match the pattern of magnetic reversals (recorded in lava flows on land)?
12. How has deep-sea drilling tested the concept of plate motion?
13. How has the study of fracture zones tested the concept of plate motion?
14. Explain how plate tectonics can account for the existence of the mid-oceanic ridge and its associated rift valley, earthquakes, high heat flow, and basaltic volcanism.
15. Explain how plate tectonics can account for the existence of oceanic trenches as well as their low heat flow, their negative gravity anomalies, the associated Benioff zones of earthquakes, and andesitic volcanism.
16. What is a transform fault?
17. Discuss possible driving mechanisms for plate tectonics.
18. Describe the various types of plate boundaries and the geologic features associated with them.
19. What is a mantle plume? What is the geologic significance of mantle plumes?
20. The southern supercontinent is called
    a. Gondwanaland b. Laurasia
    c. Pangaea d. Glossopteris
21. The sliding of the sea floor beneath a continent or island arc is called
    a. rotation b. subduction
    c. tension d. polar wandering
22. In cross section, the plates are part of a rigid outer shell of the earth called the
    a. lithosphere b. crust
    c. asthenosphere d. mantle

23. The Vine-Matthews hypothesis explains the origin of
    a. polar wandering
    b. sea floor magnetic anomalies
    c. continental drift
    d. mid-ocean ridges

24. The San Andreas fault in California is a
    a. normal fault b. transform fault
    c. reverse fault d. thrust fault

25. What would you most expect to find at ocean-ocean convergence?
    a. suture zone b. mid-ocean ridge
    c. island arc

26. What would you most expect to find at ocean-continent convergence?
    a. magmatic arc b. island arc
    c. suture zone d. mid-ocean ridge

27. What would you most expect to find at continent-continent convergence?
    a. magmatic arc b. island arc
    c. suture zone d. mid-ocean ridge

28. Passive continental margins are created at
    a. divergent plate boundaries
    b. transform faults
    c. convergent plate boundaries

29. The Hawaiian islands are thought to be the result of
    a. subduction
    b. mid-ocean ridge volcanics
    c. mantle plumes
    d. ocean-ocean convergence

30. Metallic ores are created at diverging plate boundaries
    a. through hydrothermal processes
    b. in lava flows
    c. in sedimentary deposits
    d. through metamorphism

Expanding Your Knowledge

1. Plate tectonics helps cool Earth as hot mantle rock rises near the surface at ridge crests and mantle plumes. What can we assume about the internal temperature of other planets that do not seem to have plate tectonics? What would happen to Earth's internal temperature if the plates stopped moving?
2. Are ridge offsets along fracture zones easier to explain with mantle-deep convection *causing* plate motion or with shallow convection occurring as a *result* of plate motion?
3. Why are mantle plumes narrow? What conditions at the core-mantle boundary could cause the formation and rise of a mushroom-shaped plume?
4. The slab-pull and ridge-push mechanisms of plate motion may operate only after a plate starts to move. What starts plate motion?
5. If subducting plates can penetrate the 670-kilometer mantle boundary, and sink all the way to the base of the mantle, why are there no earthquakes deeper than 670 kilometers?

Exploring Web Resources

www.mhhe.com/plummer9e

Visit the Online Learning Center for additional readings, media resources, and some great animations. Check out the videos on plate tectonics, continental drift and plate dynamics. This site also features interactive quizzes, flashcards and the answers to the Testing Your Knowledge section. And, from the Online Learning Center, you can link directly to the sites listed below to further your understanding of plate tectonics.

http://pubs.usgs.gov/publications/text/dynamic.html

This Dynamic Earth: The Story of Plate Tectonics U.S. Geological Survey on-line book by W. J. Kious and R. Tilling provides general information about plate tectonics.

http://cddisa.gsfc.nasa.gov/926/slrtecto.html

Tectonic Plate Motion (explains how plate motion is calculated).

http://vishnu.glg.nau.edu/rcb/globaltext.html

View images of plate tectonic reconstructions by R. Blakely at Northern Arizona University.

Animations

This chapter includes the following animations available on our Online Learning Center at www.mhhe.com/plummer9e.

4.2 Breakup of Pangaea
4.16 Origin of Magnetic Anomalies
4.21 Divergent Plate Boundary
4.28 Ocean-Ocean Convergence
4.32 Ocean-Continent Convergence
4.33 Continent-Continent Convergence
4.37 Mantle Convection
4.43 Continental Breakup by Mantle Plumes
4.46 Hot Spots and Aseismic Ridges

CHAPTER

5

Mountain Belts and the Continental Crust

Mountain belts can evolve from marine-deposited rocks to towering peaks during periods of tens of millions of years. Ultimately the peaks are eroded to plains and become part of the stable interior of a continent. A further appreciation of the long and complex process of mountain building will come from a study of material covered in subsequent chapters. For instance, you must understand structural geology to appreciate what a particular pattern of folds and faults can reveal about the history of mountain building in a particular region. To understand how the rocks formed during the various stages of a mountain belt's history, you must know about volcanism, plutonism, sedimentation, and metamorphism. A study of weathering and erosion will help you understand how mountains are worn away. Plate tectonic theory has been strikingly effective in helping geologists make sense of all the complex aspects of mountain belts and the continental crust. For this reason, you need to thoroughly understand the material in chapter 4 before you can appreciate how continents evolve.

In this chapter, we first point out what geologists have observed of mountain belts. Next, we describe how these observations are interpreted, particularly in light of plate tectonic theory. Finally, we discuss current perceptions of how continents change and grow.

Opposite: Climber in the Karakoram Himalaya, Pakistan.
Photo © Galen Rowell/Mountain Light Photography

FIGURE 5.1
View of glaciated peaks in one of the mountain ranges in the Andes mountain belt. A parallel but much lower range is visible at the extreme right skyline of the picture.
Photo by C. C. Plummer

A mountain, as you know, is a large terrain feature that rises more or less abruptly from surrounding levels. Volcanoes are mountains; so are erosional remnants of plateaus (mesas). In this chapter, we will not focus on individual mountains; rather, we are concerned here with Earth's **major mountain belts,** chains thousands of kilometers long composed of numerous mountain ranges. A **mountain range** is a group of closely spaced mountains or parallel ridges (figure 5.1). A mountain range is likely to be composed of tectonically deformed sedimentary, volcanic, or metamorphic rocks. It may also show a history of intrusive igneous activity.

The map in figure 5.2 shows that most of the world's mountains are in long chains that extend for thousands of kilometers. The Himalaya, the Andes, the Alps, and the Appalachians are examples of major mountain belts, each comprising numerous mountain ranges.

Geologists find working in mountain ranges to be physically and intellectually challenging. High mountains have steep faces and broad exposures of bedrock. This is good because they allow a geologist to decipher complex interrelationships between rock units. But the geologist may have to become a proficient mountain climber to access the good exposures. (Conversely, mountain climbers who develop an interest in the rocks they climb sometimes become geologists.) On the other hand, exposures of bedrock critical to interpreting the local geology may be buried beneath glaciers or talus from rockfall. Furthermore, even in the highest and best exposed mountains, we never see bedrock representative of all of a mountain range. Significant amounts of rock (usually thousands of meters) once overlying the rocks we now see have been eroded away. Moreover, the present exposures are like the proverbial tips of icebergs—there is much more rock below the exposed mountain range that we cannot observe. For instance, the Himalaya, Earth's highest mountain range, rise to 8,000 meters above sea level; yet their roots (Earth's crust beneath the mountains) extend downward 65,000 meters. In other words, at best, we have exposed to us less than 1/8 of the thickness of a mountain range.

Our models of how major mountain belts evolve use data from over a century of studying the geologic structures and rocks exposed in the world's many mountain ranges. Often, a particular study aims to piece together the geologic history of a single mountain range or part of a range. In other field studies, a geologist focuses on a particular type of rock exposed

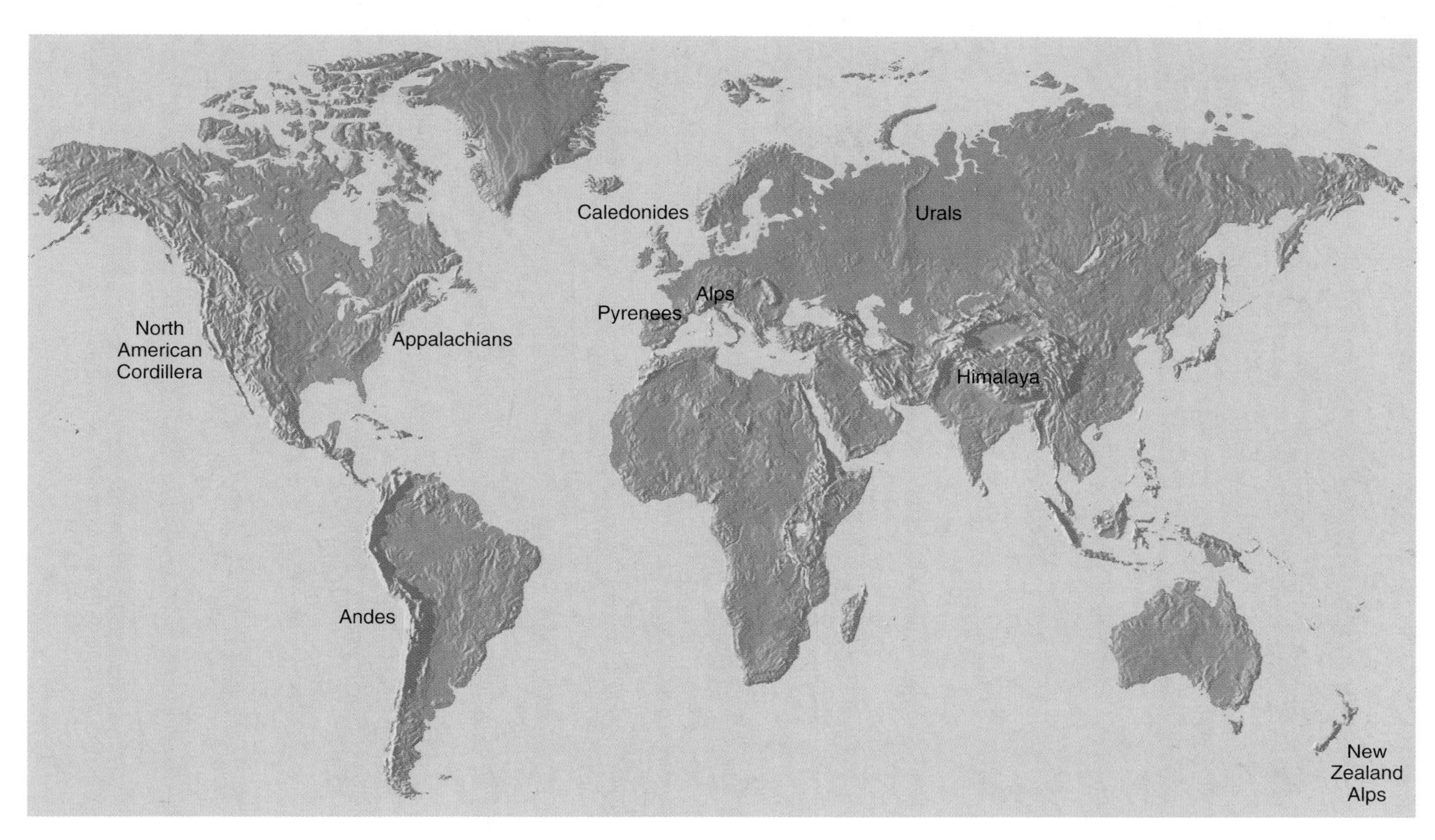

FIGURE 5.2
Map of the world showing major mountain belts.

in a mountain range. For instance, a geologist may study the variations in metamorphic rocks in a mountainous area with the intent of gaining more insight into metamorphic processes. Geologists working on the "big picture," developing hypotheses of how major mountain belts evolve, might use the published results of hundreds or thousands of local studies, using them as pieces of a puzzle. (Science works largely because scientists build on the work of others.) Models that currently are widely accepted regarding the evolution of mountain belts are cast within the broader framework of plate tectonic theory and will be described later in this chapter. But even before the advent of plate tectonic theory, geologists recognized the importance of *isostasy* (described in chapters 1 and 2) in explaining vertical movement of mountain belts. Isostatic adjustment means that thicker continental crust tends to "float" higher on the mantle than thinner crust.

Despite the complexities of individual ranges, there are some characteristics that are shared by most or all major mountain belts. As you read through each of the characteristics, think of what you have learned from previous chapters (especially the chapter on plate tectonics) that might help explain an observed characteristic of mountain belts.

CHARACTERISTICS OF MAJOR MOUNTAIN BELTS

Size and Alignment

Major mountain belts are very long compared to their width. For instance, the mountain belt that forms the western part of North America (the *North American Cordillera*) starts as the Aleutian Island arc in southwestern Alaska. The trend of the Aleutians is continuous with the trend of the mountain ranges that make up much of the Alaskan mainland. (See figure 5.3. You will also want to refer to the geologic map on the inside cover as you read this chapter.) Through Alaska the ranges trend more or less east-west, but in northern Canada they curve into a more north-south trend. The north-south ranges include the many individual ranges of western Canada and western United States. The belt is widest in the United States where it extends from the Coast Ranges of California and Oregon eastward to the Rocky Mountains of Montana, Wyoming, and Colorado. The belt narrows as the north-south trend continues through Mexico.

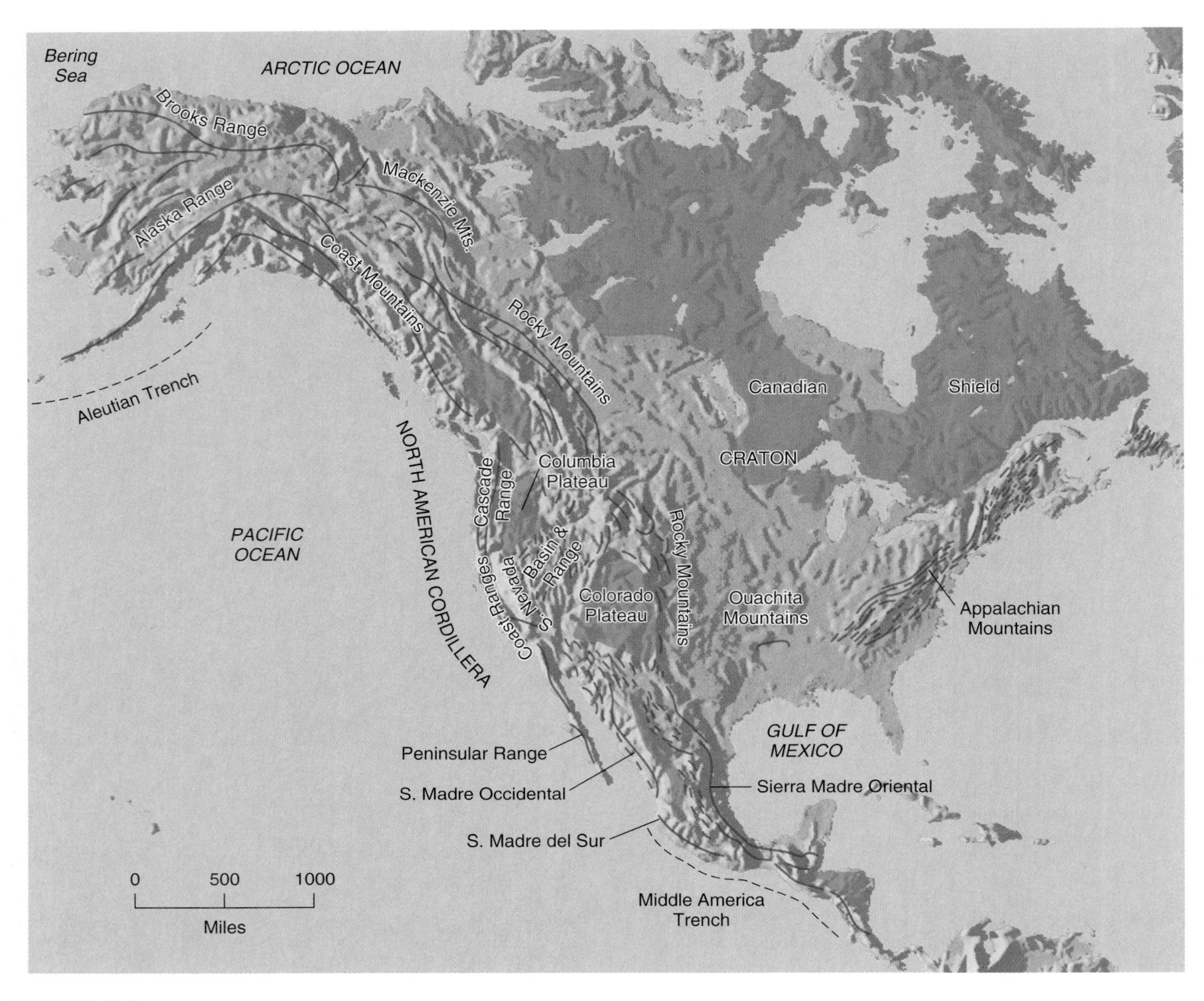

FIGURE 5.3
The mountain belts of North America. Some of the major ranges in the Cordillera are labeled.

Ages of Mountain Belts and Continents

Major mountain belts with higher mountain ranges tend to be geologically younger than those where mountains are lower. Mountain building for the Himalaya, our highest mountain belt, began only about 45 million years ago and is still taking place. Mountain building, other than isostatic adjustment, in the much lower Appalachians ceased around 250 million years ago. Individual ranges within a mountain belt, however, may vary considerably in height even though they are about the same age.

Mountain regions commonly show evidence that they were once high above sea level, were eroded to hills or low plains, and then rose again in a later episode of uplift. Such episodes of uplift and erosion may occur a number of times during the long history of a mountain range. Ultimately, mountain ranges stabilize and are eroded to plains.

On the North American continent, the Appalachian Mountains extend from eastern Canada southward through the eastern United States into Alabama (figure 5.3). In the Appalachians, fossils and isotopic ages of rocks indicate that these mountains began to evolve earlier than the mountain belt along the western coast of North America. The interior plains between the Appalachians and the Rockies are considered to have evolved from mountain belts in the very distant geologic past (during the Precambrian). The once deep-seated roots of the former Precambrian mountain belts are the *basement* rock for the now stable central part of the continent. Layers of Paleozoic and

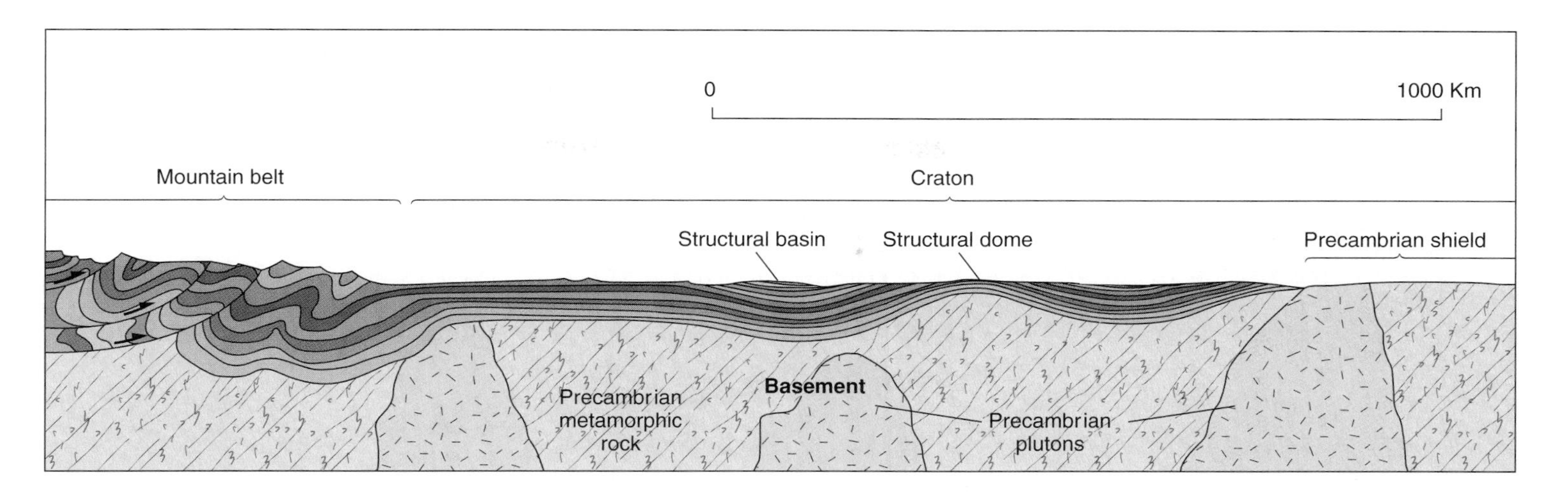

FIGURE 5.4

Schematic cross section through part of a mountain belt (left) and part of a continental interior (craton). Vertical scale is exaggerated.

younger sedimentary rock cover most of that basement. The great age of the mountain-building episodes that preceded the Paleozoic sedimentation is confirmed by isotopically determined dates of over 1 billion years obtained from rocks in the few scattered locations where the basement is exposed. (The most noteworthy are the Grand Canyon in Arizona, the Ozark dome in Missouri, the Black Hills of South Dakota, some ranges in the Rocky Mountains, and the Adirondacks in New York.) The region of a continent that has been structurally stable for a prolonged period of time is called a **craton** (figures 5.3, 5.4 and 5.5). The central part of the United States and Canada is all part of a craton. Other continents are similarly cored by cratons.

Most of the craton in the central United States has a very thin blanket—only 1,000 to 2,000 meters—of sedimentary rock layers overlying its basement. Sediment was mostly deposited in shallow inland seas during Paleozoic time; however, for the craton in much of eastern and northern Canada, as well as Greenland, no sedimentary rocks cover the eroded remnants of old mountain ranges. This region is a **Precambrian shield**—that is, a complex of Precambrian metamorphic and plutonic rocks exposed over a large area. Such shields and basement complexes of cratons represent the roots of mountain ranges that completed the deformation process more than a billion years ago.

Thickness and Characteristics of Rock Layers

The relatively thin cover of sedimentary rock overlying the basement in the craton contrasts sharply with the thick sedimentary sequence typical of a mountain belt. In mountain belts, layered sedimentary rock commonly is more than 10 kilometers thick. The sedimentary rock in cratons may show no deformation, or it may have been gently warped into basins and domes above the basement (figure 5.4). By contrast, mountain belts are

FIGURE 5.5

Satellite image of part of a craton in Western Australia. Metamorphic rock (dark gray) that is 3.5 to 3 billion years old surrounds oval-shaped domes of granite and gneiss (white) that are 2.8 to 3.3 billion years old. Gently dipping sedimentary and volcanic rocks (tan and reddish) unconformably overlie the granite-metamorphic basement complex. The area is 400 km across.

Landsat mosaic produced by the Remote Sensing Applications Centre Department of Land Administration, Western Australia

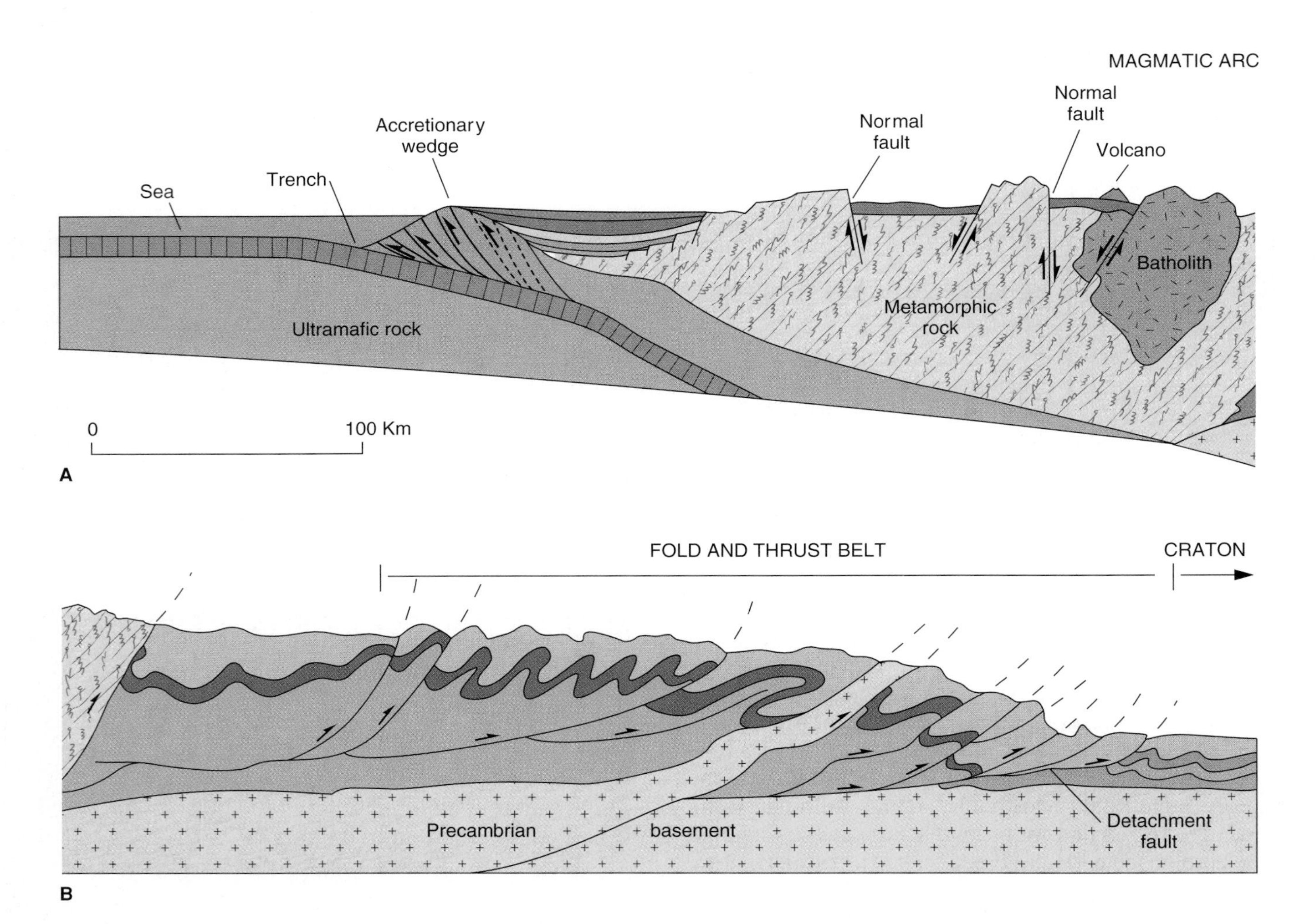

FIGURE 5.6

Cross section of a "typical" mountain belt. For simplicity, only a few of the many layers of sedimentary rock are shown. (*B*) is the continuation of the mountain belt to the right of (*A*).

characterized by a variety of folds and faults that indicate moderate to very intense deformation.

Most of the sedimentary rock in mountains is of marine origin, indicating that today's highlands were once below sea level. In part of a mountain belt, the rocks are similar to the sedimentary rocks found on the craton—limestones, shales, and sandstones. By contrast, another wide zone in a mountain belt contains great thicknesses of volcanically derived rock—volcanic ash, sediments from eroded volcanic rock, as well as lava flows.

Patterns of Folding and Faulting

Reconstructing the original position and determining the thickness of layers of sedimentary and volcanic rock in mountain belts is complicated because in most instances the layered rocks have been folded and faulted at some time after they were deposited. (Refer to figure 5.6 as you read through the following paragraphs.) Folds will be open in those parts of a mountain belt where deformation is not very intense. Tighter folds (figure 5.7) indicate greater deformation. Large overturned and recumbent folds (figure 5.8) may be exposed in more intensely deformed portions of mountain belts. Reverse faults are common, particularly in the intensely folded regions. Especially noteworthy are the **fold and thrust belts** found in many mountainous regions. These are characterized by large thrust faults (reverse faults at a low angle to horizontal), stacked one upon another; the intervening rock usually was folded while it was being transported during faulting. Beneath the lowermost fault, called a *detachment fault,* the rock has remained in place.

Overall, the folds and thrust faults in a mountain belt suggest tremendous squeezing or *crustal shortening* and *crustal thickening.* The sedimentary rocks of the Alps, for instance, are estimated to have covered an area of ocean floor about 500 kilometers wide when they were deposited. They were later compressed into the present width of the Alps, which is less than 200 kilometers.

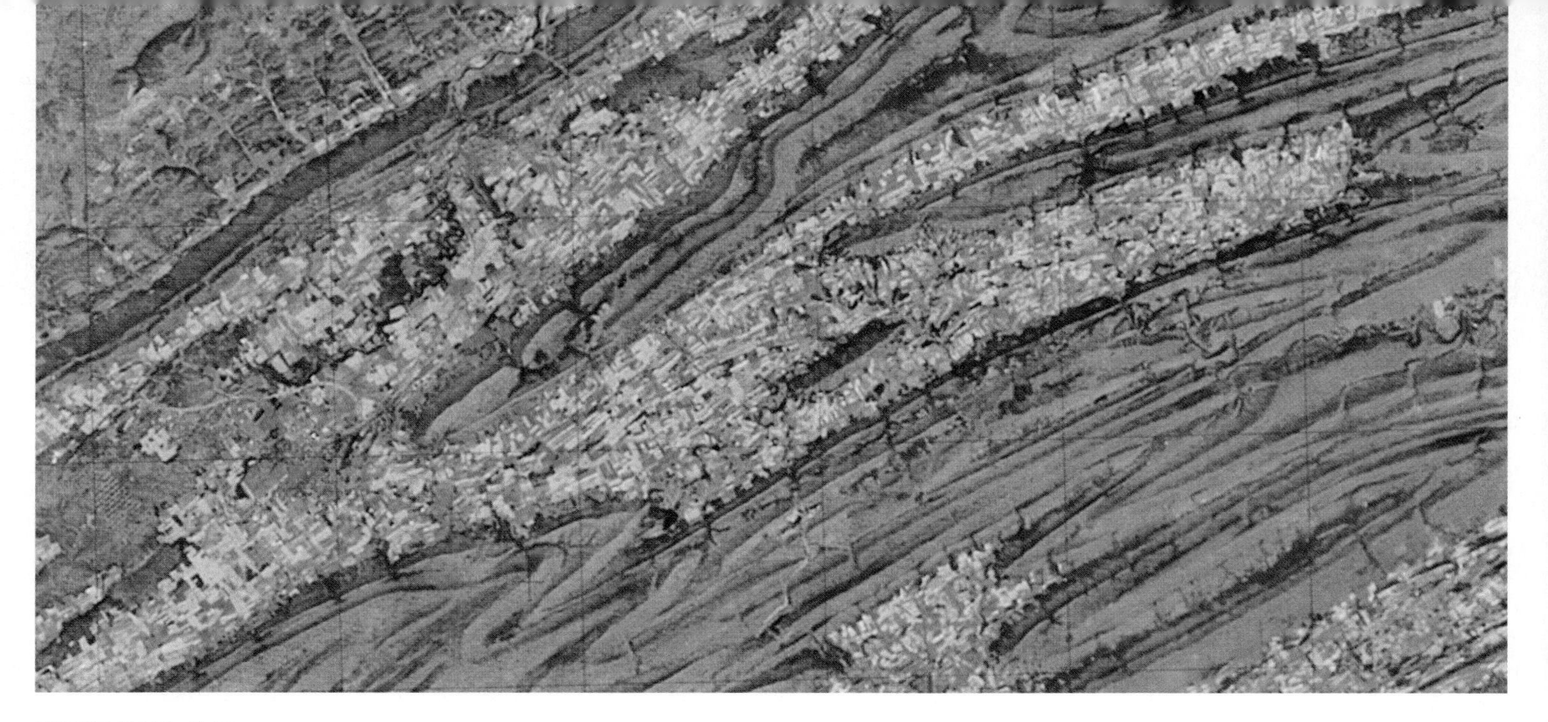

FIGURE 5.7

False color satellite image of part of the Valley and Ridge province of the Appalachian mountain belt, near Harrisburg, Pennsylvania. The ridges are sedimentary beds resistant to erosion. The pattern indicates tight and open folding occurred prior to erosion.

U.S. Geological Survey

FIGURE 5.8

Recumbent folds exposed on a mountainside in the Andes.

Photo by C. C. Plummer

5.1

ENVIRONMENTAL GEOLOGY

Ultramafic Rocks in Mountain Belts—From the Mantle to Talcum Powder

Ultramafic rocks (described in chapter 11 on intrusive rocks) occur commonly in the portions of mountain belts occupied by metamorphic and plutonic rocks. Ultramafic rocks tend to crop out in long, narrow zones that parallel the trend of a mountain belt. Most geologists believe the bodies of ultramafic rocks represent mantle material that was faulted into the crust during the mountain-building process. Some of the ultramafic bodies are found associated with marine-deposited volcanic and sedimentary rocks in an *ophiolite sequence* (described in chapter 3 about the sea floor). This may represent a segment of a former oceanic crust together with its underlying mantle.

Ultramafic rocks in mountain belts commonly show the effects of the metamorphism that has altered adjacent rock units. Two of the foliated metamorphic products of ultramafic rocks are of special interest. One is *serpentinite,* a rock composed of the mineral serpentine. Another is a rock composed mostly of the mineral talc, commonly known as *soapstone.*

Serpentinite is a shiny, mottled, dark green and black rock that looks rather like a snake's skin. It splits apart easily along irregular, slippery, foliation surfaces. Hillsides or slopes with serpentinite as bedrock are sparsely vegetated because constant sliding prevents soil and vegetation from building up. Houses built on serpentinite hillsides (by people without a knowledge of geology) also slide downslope. Serpentinite is the official state rock of California—a state in which a large number of homes have been destroyed because they were built on sliding hillsides. (Serpentinite, however, is seldom to blame.)

Soapstone, which is less common than serpentinite, is valuable mainly because of talc's softness (number 1 on Mohs scale). Many sculptures (most notably Eskimo carvings) are made from soapstone because of the ease with which it can be cut. The best-known product of talc, however, is talcum powder.

Talc's perfect cleavage has led to its unique application in reflecting highway signs and center road stripes. Finely crushed talc is mixed with paint. The many fine, parallel cleavage flakes reflect light in a specific direction depending on how the mixture is brushed on the sign. For the signs showing different daytime and nighttime speed limits, the daytime speed limit is painted with ordinary black paint, and the nighttime limit is painted with talc mixed into the white paint. At night the light from a car's headlights reflects back to the car from the cleavage surfaces of many tiny flakes of talc.

Metamorphism and Plutonism

A complex of regional metamorphic and plutonic rock is generally found in the mountain ranges of the most intensely deformed portions of major mountain belts. Most of the metamorphic rocks were originally sedimentary and volcanic rocks that had been deeply buried and subjected to intense stress and high temperature. Where found, *migmatites* (interlayered granitic and metamorphic rocks) may represent those parts of the mountain belts that were once at even deeper levels in the crust, where higher temperatures caused partial melting of the rocks (as described in chapter 11). Where found, rocks such as these that formed in the lower crust must have been transported into much higher levels of the crust during a mountain-building episode. Batholiths, mostly of granite, attest to magma generated from partial melting of the lower crust (or upper mantle). Magma diapirs traveled upward to collect and solidify at a higher level in the crust.

Normal Faulting

Older portions of some major mountain belts have undergone normal faulting (figure 5.9). Cross-cutting relationships show that the normal faulting occurs after the intense deformation that resulted in tight folding, thrust faulting, and metamorphism and after most batholiths had formed. As shown in figure 5.10, this late stage of normal faulting is a result of *vertical uplift* or *horizontal extension.* Either of these contrasts with the overall shortening that prevailed during folding, thrust faulting, and metamorphism of the original marine rocks.

In addition to the clearly post-shortening normal faulting, normal faulting takes place in the high, central part of a major mountain belt *simultaneous* with the folding and thrust faulting of the outer part of the belt (figure 5.6). This implies that *extension* occurs in the central portion at the same time that *shortening* takes place in the outer portion of an active major mountain belt. This is comparable to what happens to a mound of cold molasses poured on a table. As the molasses flows out from the central thick part, it experiences tensional stress in the central portion.

Thickness and Density of Rocks

Geophysical investigations yield additional information about mountains and the continental crust. As discussed in chapter 2, about Earth's interior, gravity measurements indicate that the rocks of the continental crust (including mountain belts) are lighter (less dense) than those of the oceanic crust. Seismic velocities indicate a composition approximating that of granite for continental crust. Furthermore, evidence from seismic

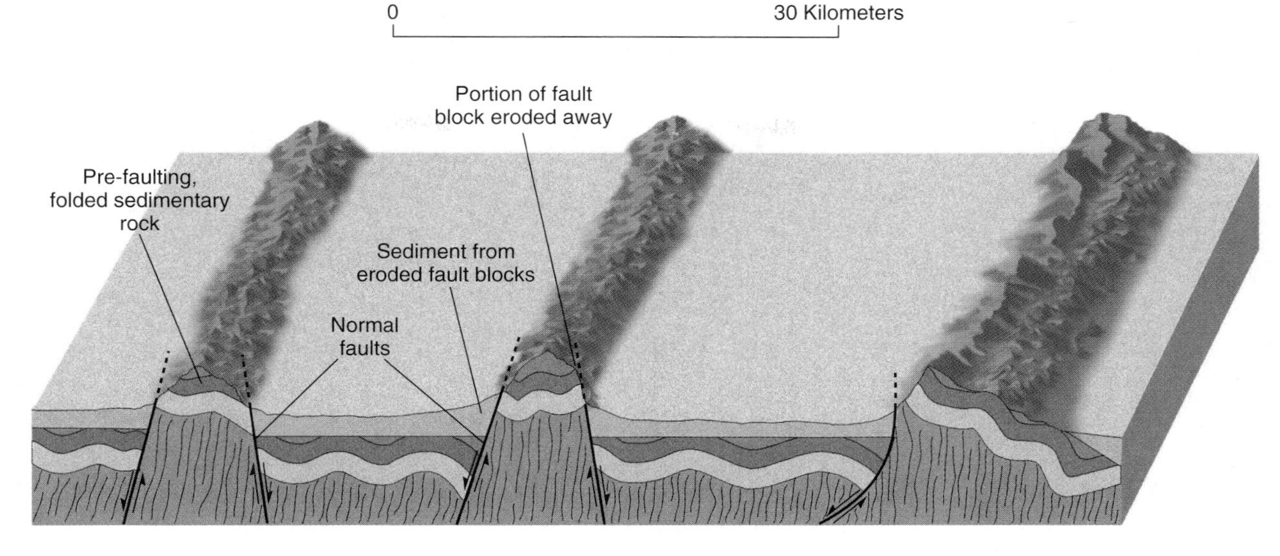

FIGURE 5.9
Fault-block mountains with movement along normal faults.

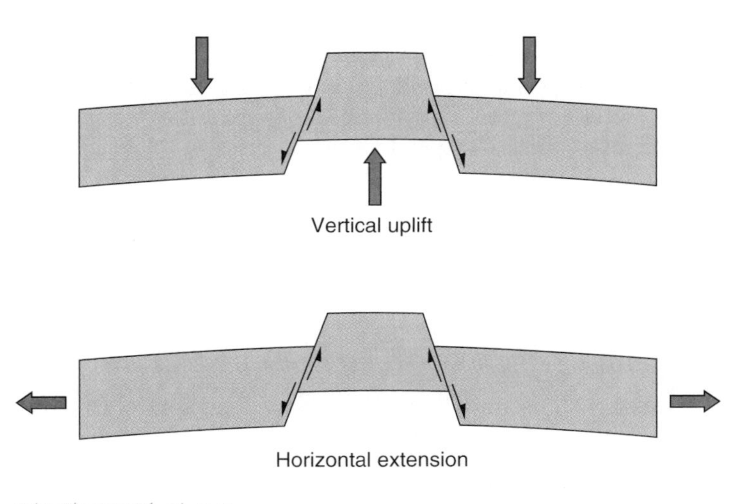

FIGURE 5.10
Normal faults bounding fault-block mountain ranges. Two ways in which forces might be distributed.

studies supports the view that this lighter crust is much thicker beneath mountain belts than under the craton and that the crust is thicker under younger mountain belts than under older ones.

Features of Active Mountain Ranges

Frequent earthquakes are characteristic of portions of mountain belts that are geologically young and considered still active. Also, deep ocean trenches are found parallel to many young mountain belts (the Andes, for example). Trenches lie off the coasts of island arcs, which can be regarded as very young mountain ranges. Isolated active volcanoes perched on top of older rock in a mountain range suggest that melting is still taking place at depth.

THE EVOLUTION OF A MOUNTAIN BELT

Although each mountain belt differs in details, geologists have developed (and continue to modify) models that provide logical explanations of how mountain belts evolve. Plate tectonic theory provides the framework for most of our models that explain various aspects of mountain belts.

The evolution of a mountain belt typically begins when sediment is deposited; it ends hundreds of millions of years later when the former mountain belt has become part of a craton. For instructional purposes we describe the history of a mountain belt in three stages: (1) the accumulation, (2) the orogenic, and (3) the uplift and block-faulting stages. There are, however, no sharp time boundaries between the stages. Processes associated with more than one stage may be occurring at the same time in different parts of a mountain belt.

The Accumulation Stage

As we noted earlier, mountain belts typically contain thick sequences of sedimentary and volcanic rocks. The accumulation of these great thicknesses (several kilometers) of sedimentary or volcanic rocks takes place in the **accumulation stage** of mountain building. Most of the sedimentary rocks and much of the volcanic material accumulate in a marine environment. The source for the sediment deposited in the water must be a nearby landmass, belonging to either an adjoining continent or part of a volcanic island arc.

Accumulation in an Opening Ocean Basin

The east coast of North America is a *passive continental margin* (described in the previous chapter; see figures 4.21 and 4.25). Europe and North America have been moving away from

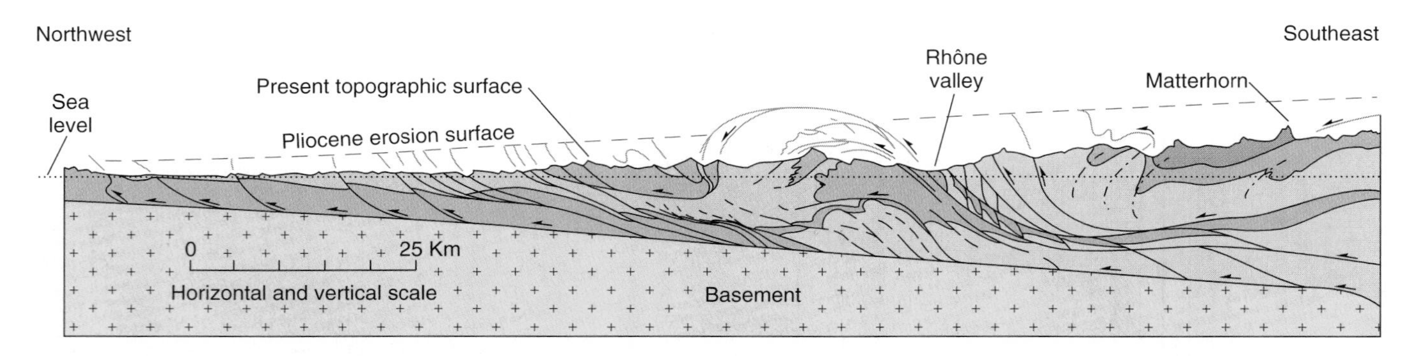

FIGURE 5.11

Cross section through part of the Alps. Thicker lines are thrust faults. Lesser folds are not shown. Movement is from the right to the left of the diagram (southeast to northwest). Only a few arrows are shown to indicate movement of the overriding thrust block.

From S. E. Boyer and D. Elliot, 1982. *AAPG Bulletin.* Reprinted by permission of American Association of Petroleum Geologists

each other for well over 100 million years. During this time sediment eroded from the continents has washed into the adjoining seas. The resulting very thick sequences of sedimentary rock make up continental shelves and slopes. The sedimentary rocks that form at passive margins are predominantly shales, limestones, and sandstones (mostly quartz sandstone). Volcanic rocks are rare or absent. As long as the Atlantic Ocean basin continues to open, sedimentation will continue and more sedimentary layers will accumulate on both sides of the Atlantic.

Where we find a thick sequence of shale, sandstone, and limestone exposed in a mountain range, it is reasonable to assume these rocks were originally deposited at a similar passive continental margin.

Accumulation along a Convergent Boundary

A large variety of rock types are created near a convergent plate boundary. Volcanic rocks—most characteristically andesites—accumulate near the boundary either as pyroclastic layers or as flows. Sedimentary rocks, moreover, may accumulate in equal or greater amounts than volcanic rocks. Limestone, however, is usually absent or present only in minor proportions. Shales and sandstones are the predominant sedimentary rock types. The sandstones are mostly *graywackes* ("dirty" looking sandstones with sand-sized grains of volcanic or other rocks in a fine-grained matrix) rather than quartz sandstone.

The source of this sedimentary as well as volcanic material is a *magmatic arc,* which, as described in chapter 4 (see figure 4.32), is a chain of volcanoes or plutons along a line (usually curved as seen from above).

Sediment eroded from the magmatic arc, as well as newly erupted pyroclastic debris, is transported to and deposited in the basins on either side of the arc. Usually the basin to the seaward side will be underwater, at least initially, and in time will fill with sediment. Once the basin has filled, sediment may be transported beyond it and accumulate outward onto the deep ocean floor.

The Orogenic Stage

Intense deformation follows or is contemporaneous with the accumulation stage. An **orogeny** is an episode of intense deformation of the rocks in a region; the deformation is usually accompanied by metamorphism and igneous activity. Layered rocks are compressed into folds. Reverse faulting (especially thrust faulting) is widespread during an orogeny. Normal faulting may also occur but is not as widespread.

The more deeply buried rocks, subjected to regional metamorphism, are converted to schists and gneisses. Magma generated in the deep crust or the upper mantle may work its way upward to erupt in volcanoes or form batholiths.

Orogenies and Ocean-Continent Convergence

The relationships among igneous activity, metamorphism, and subduction are described in chapters 11 and 15. Plate convergence also accounts for the folded and reverse-faulted layered rocks found in mountain belts. An *accretionary wedge* develops where newly formed layers of marine sediment are folded and faulted as they are snowplowed off the subducting oceanic plate (see figure 4.31 and explanation in the previous chapter).

Rock caught in and pulled down the subduction zone is subjected to intense shearing. If rock is carried further down the subduction zone it becomes metamorphosed (as described in chapter 15).

Fold and thrust belts may develop on the craton (backarc) side of the mountain belt (figure 5.11). Thrusting is away from the magmatic arc toward the craton. The magmatic arc is at a high elevation, because the crust is thicker and composed largely of hot igneous and metamorphic rocks. The large thrust sheets move toward and sometimes over the craton. (In the Rocky Mountains, thrust faulting of the craton itself has taken place.) The thrusting probably is largely due to the crustal shortening caused by convergence. There is, however, some controversy among geologists over additional processes that may take place. Some geologists regard gravity flow (from the high and mobile magmatic arc outward over the low and rigid craton) to

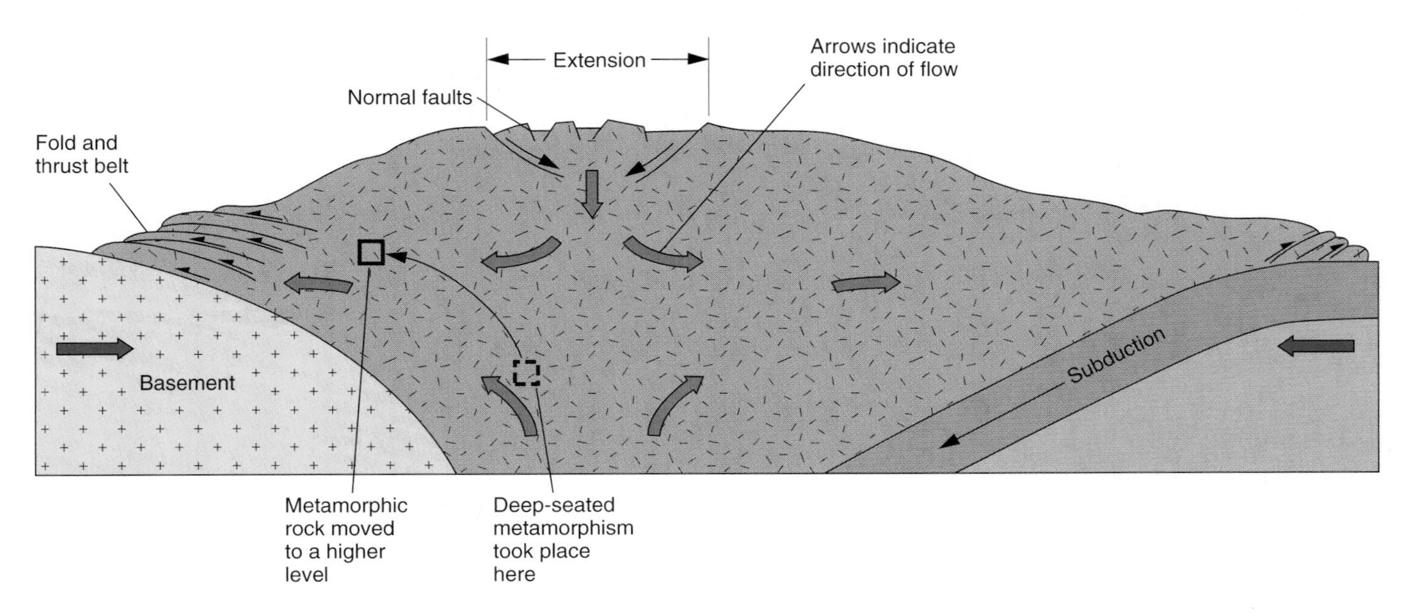

FIGURE 5.12

Schematic cross section of a mountain belt in which gravitational collapse and spreading are taking place during plate convergence. Red arrows indicate flowage of rock. Faulting occurs in brittle rock near the surface. Rock that was metamorphosed at depth flows to a higher level in the mountain belt.

contribute significantly to the process. Others think that the expanding magmatic arc pushes the sedimentary (and sometimes metamorphic and igneous) rocks outward to become the fold and thrust belt. (The magmatic arc is likened to a bulldozer pushing a wedge of loose material outward.)

Normal faulting may take place due to extension in the high, central part.

In the late 1980s and early 1990s geologists developed a model that explains (1) fold and thrust belts, (2) simultaneous normal faulting, and (3) how once deep-seated metamorphic rocks rise to an upper level in a mountain belt. What is believed to occur is that the central part of the mountain belt becomes too high and gravitationally unstable, resulting in **gravitational collapse and spreading.** The mobile central portion becomes increasingly elevated during plate convergence. This is due to compression of sedimentary and metamorphic rocks as well as to volcanic eruptions and emplacement of plutons. After some time, the welt in the mountain belt becomes too high to be supported by the underlying rocks and collapse begins. (Oxford University geologist John Dewey suggests that collapse begins when the welt exceeds 3 kilometers above sea level.) As shown in figure 5.12 the gravitational collapse forces rock outward as well as downward. At deeper levels in the mountain belt, the rock is *ductile* (or *plastic*) and flows; nearer the surface, rock fractures, so movement is through faulting. The rock is pushed outward and helps create, along with crustal shortening, the fold and thrust belt.

In the high, central part, the outward flowing rock results in extension (figure 5.12); therefore the brittle, near-surface rocks fracture, and normal faulting takes place.

The flowage pattern (as shown in figure 5.12) can also explain how once deep-seated metamorphic rocks (migmatites, for example) are found in upper levels of a mountain belt. Lower crustal rocks are squeezed, forcing them to flow upward and outward, bringing them closer to the surface.

It is important to note that where there are convergent boundaries, accumulation and deformation are occurring simultaneously. In other words, the accumulation stage and the orogenic stage are taking place at the same time. This was not evident to early geologists, who did not have the perspective of plate tectonic theory for understanding mountain building. They generally assumed that the accumulation stage ended before the orogenic stage began.

Arc-Continent Convergence

Sometimes an island arc collides with a continent (figure 5.13). If an intervening ocean is destroyed by subduction (the subduction also causes the arc), the arc will approach the continent. When collision occurs, the arc, like a continent, is too buoyant to be subducted. Continued convergence of the two plates may cause the remaining sea floor to break away from the arc and create a new site of subduction and a new trench seaward of the arc (figure 5.13*C*). Note that the direction of the new subduction is opposite to the direction of the original subduction (these are sometimes called flipping subduction zones), but it still may supply the arc with magma. The arc has now become welded to the continent, increasing the size of the continent.

This type of collision apparently occurred in northern New Guinea (north of Australia). A similar collision may have added an island arc to the Sierra Nevada complex in California during Mesozoic time, when a subduction zone may have existed in what now is central California. Many geologists think that much of westernmost North America has formed from a series of arcs colliding with North America (discussed later in this chapter under "terranes").

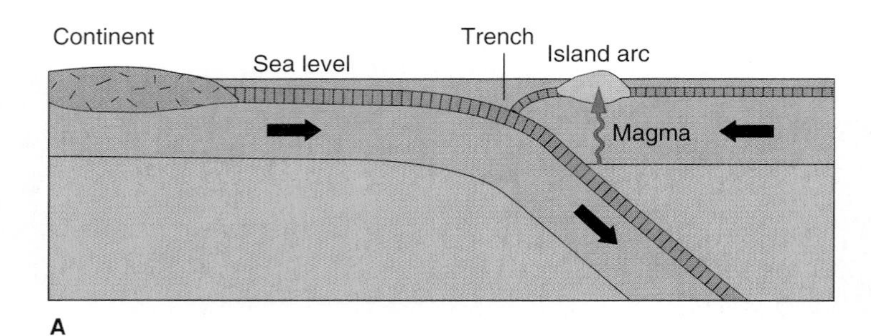

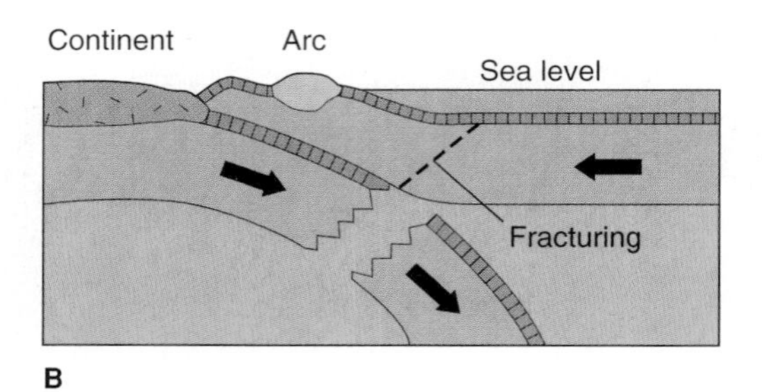

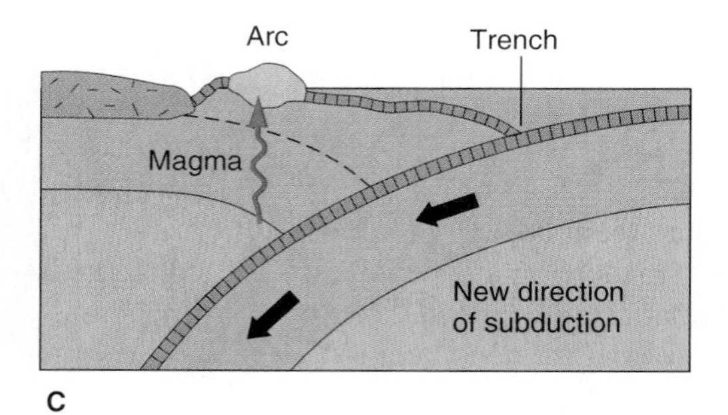

FIGURE 5.13

Arc-continent convergence can weld an island arc onto a continent. The direction of subduction can change after impact.

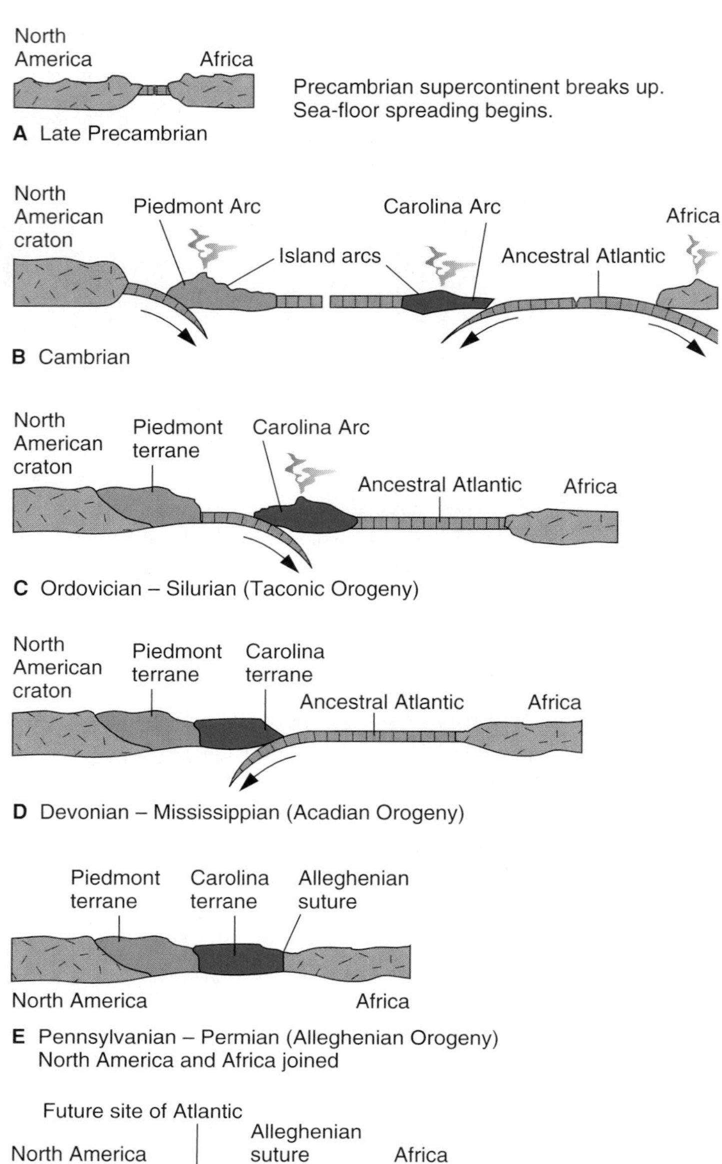

FIGURE 5.14

The geologic evolution of the southern Appalachians.

Modified "Tectonic synthesis of the U.S. Appalachians," figure 9, volume F-2, from *Geology of North America: An Overview* by R. D. Hatcher, Jr. © 1989 by Geological Society of America. Reproduced with permission of the Geological Society of America

Orogenies and Continent-Continent Convergence

As described in the previous chapter (see figure 4.33), some mountain belts form when an ocean basin closes and continents collide. Mountain belts that we find within continents (with cratons on either side) are believed to be products of continent-continent convergence. The Ural Mountains resulted from the collision of Asia and Europe. Convergence of the African and European plates created the Alps. Our highest mountains are in the Himalayan belt. The Himalayan orogeny started around 45 million years ago as India began colliding with Asia (India was originally in the southern hemisphere). The thick sequences of sedimentary rocks that had built up on both continental margins were intensely faulted and folded. Fold and thrust belts developed and were carved by erosion into the mountain ranges that make up the Himalaya. The mountains are still rising and frequent earthquakes attest to continuing tectonic activity. North of the Himalaya, Tibet rose to become what is now the highest plateau in the world. Normal faults in the Tibetan plateau indicate that gravitational collapse is taking place.

The Appalachian Mountains are an example of continent-continent convergence, but with a more complicated history. Arc-continent convergence was also involved and the mountain belt was later split apart by plate divergence.

A condensed version of orogeny in the Appalachians is as follows (figure 5.14): During late Precambrian and earliest Paleozoic time the ancestral Atlantic Ocean developed as sea-floor spreading forced the passive margins of North America, Europe, and Africa away from one another. During the Paleozoic, plate motion shifted, subduction began, and the ocean basin began closing. Island arcs developed between the continents. These became plastered onto the North American craton as the ancestral Atlantic basin closed. A couple hundred million years after subduction began, the ocean basin closed completely, first with Europe and later with Africa crashing into North America. By the end of the Paleozoic, the three continents were sutured together. The Appalachians and what is now the Caledonide mountain belt of Great Britain and Norway were part of a single mountain belt within the supercontinent, *Pangaea.* The mountain belt was comparable to the present-day Himalaya.

Early in the Mesozoic Era, the supercontinent split, roughly parallel to the old suture zone. The present continents moved (and continue to move) farther and farther away from their present divergent boundary, the mid-Atlantic ridge.

What happened to the Appalachians would seem too implausible even for a science fiction plot. Yet, if one accepts the principles of plate tectonic theory and examines the rocks and structures in the Appalachians (and their counterparts in Europe and Africa), the argument for this sequence of events is not only plausible but convincing.

The cycle of splitting of a supercontinent, opening of an ocean basin, followed by closing of the basin and collision of continents is known as the *Wilson Cycle.* Canadian geologist J. Tuzo Wilson proposed the cycle in the 1960s for the tectonic history of the Appalachians.

The Wilson Cycle apparently has occurred before. A question raised is why would a continent split apart more or less along a suture zone where one would expect the crust to be thickest? One recently proposed hypothesis is that this is the zone that is weakened and thinned somewhat by outward flow of rock during gravitational collapse and spreading. (Another hypothesis involves *delamination,* detachment and sinking downward of the underlying lithospheric mantle, as described on p. 123.)

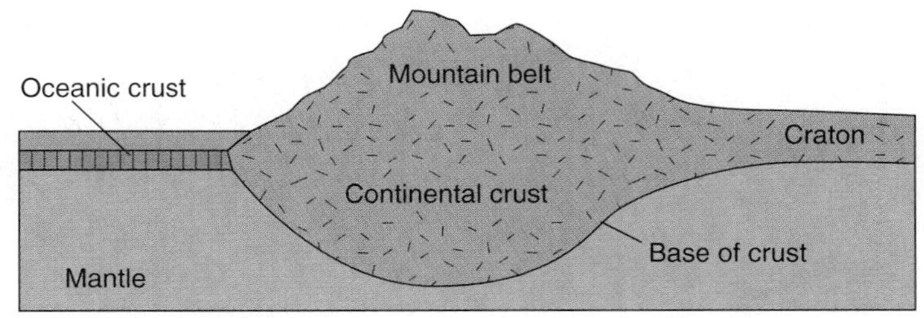

A At end of orogenic stage

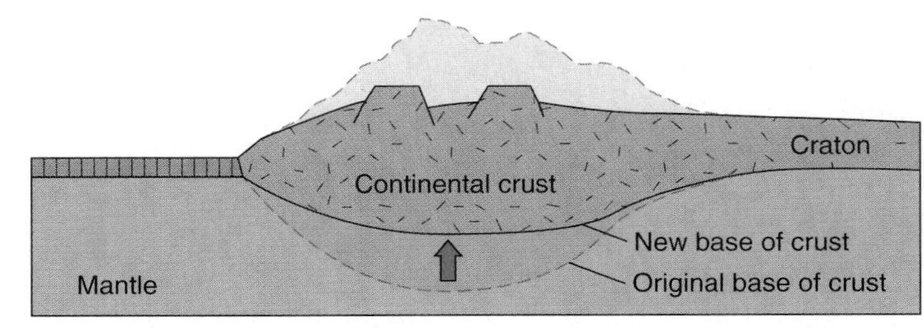

B Mountain belt moves upward

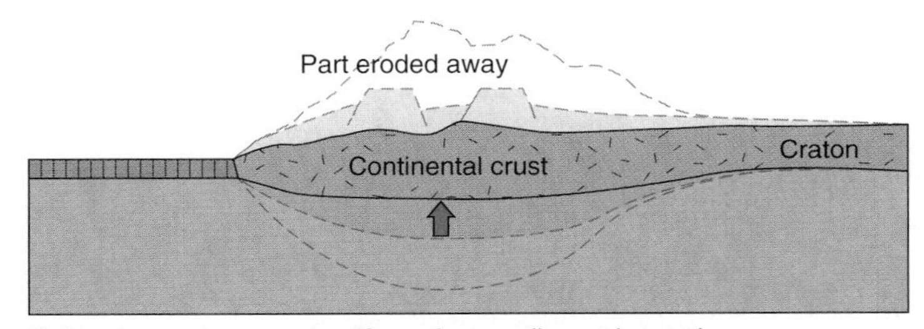

C Erosion and renewed uplift continue until crust beneath mountain belt is the same thickness as that of the craton

FIGURE 5.15

Isostasy in a mountain belt. The thickness of the continental crust is exaggerated.

The Uplift and Block-Faulting Stage

After plate convergence stops and the compressive force of the orogeny is relaxed, there is a long period of uplift accompanied by erosion. During this stage, which lasts many millions of years, large regions in the mountain belt move vertically upward. Erosion may keep pace with uplift and the area remain low. Alternatively, uplift may temporarily outpace erosion, resulting in plateaus or mountain ranges. The present Appalachian Mountains are the result of uplift and erosion that have taken place long after the orogenic stage ended more than 250 million years ago. It is likely that the Appalachian Mountains eroded down to a plain after the last, Paleozoic orogeny. The coastal plain east of the Appalachians is made of young sedimentary rock unconformably overlying metamorphic and igneous rocks that were part of the original mountain belt. This region has remained a plain. The present Appalachians represent rejuvenation following relatively recent uplift in late Tertiary time. The uplift may have been due to reactivation of ancient thrust faults caused by compressive stress within the westward-moving North American plate. The coastal plains have not moved upward, probably due to a lack of thrust faults. So the topography of the Appalachians is geologically quite young while the original structures due to orogenic deformation are quite old. The Adirondack Mountains of northern New York also participated in the uplift and rejuvenation, but the orogeny that they went through is Precambrian—much older than the Appalachian orogenies. Eventually, the entire Appalachian mountain belt will be eroded to a plain and become part of the North American craton.

Isostasy

For most other older mountain belts, we can attribute the geologically most recent uplift to *isostatic adjustment* of the crust that was thickened during the orogenic stage (as explained in chapter 2 and as shown in figure 5.15). According to the concept of isostasy, lighter, less dense continental crust "floats" higher on the mantle than the denser oceanic crust. The craton has achieved an equilibrium and is floating at the proper level for its thickness. Mountains, being thicker continental crust, "float" higher than the stable continent. As material is removed

5.2 ENVIRONMENTAL GEOLOGY

A System Approach to Understanding Mountains

During recent years, geologists have used a system approach to gain insight into the growth and wearing away of mountains. This approach regards mountains as products of three closely interdependent components. The components of the system are (1) tectonics (plate tectonics and isostasy), (2) climate, and (3) erosion.

The tendency in the past has been to concentrate on tectonics to explain the growth of a mountain belt and to relegate climate and erosion to relatively minor roles. Through mountain system analyses, we gain insight into the extent to which each of the three components interacts with and changes the other two components. Climate influences erosion in obvious ways. For instance, if there is a wet climate there will be erosion due to abundant running water at lower elevation and heavy glaciation at higher elevation. If the climate is arid, erosion will be much slower. Tectonics affects climate because if a region is uplifted to a high elevation, the climate there will be cold and glaciers can develop. With less uplift and lower mountains, erosion will be mainly due to running water. A moist climate can also result in heavy vegetation at lower elevations, which would tend to retard erosion.

Erosion and climate can influence tectonics as well. For example, the extent and type of erosion can help determine whether a highland grows higher or lower with time. If a high plateau, dissected by only a few valleys, undergoes erosion, the plateau is eroded downward uniformly (box figure 1). Following erosion, isostasy results in the plateau floating upward, but not up to its original level. Its average surface, which essentially is its actual surface, is at a lower elevation than before erosion took place. If erosion carves many deep valleys and leaves relatively few mountains between the valleys, the entire regional block will

Region that erodes into mountains and deep valleys

Region that erodes almost uniformly downward

Isostatic adjusted highlands before extensive erosion

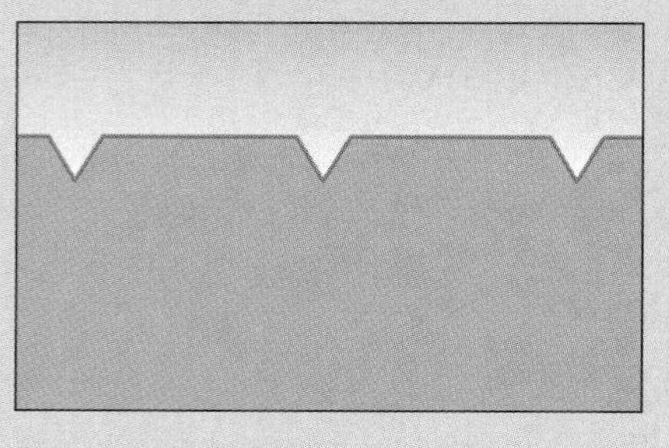

After extensive erosion, before isostatic readjustment

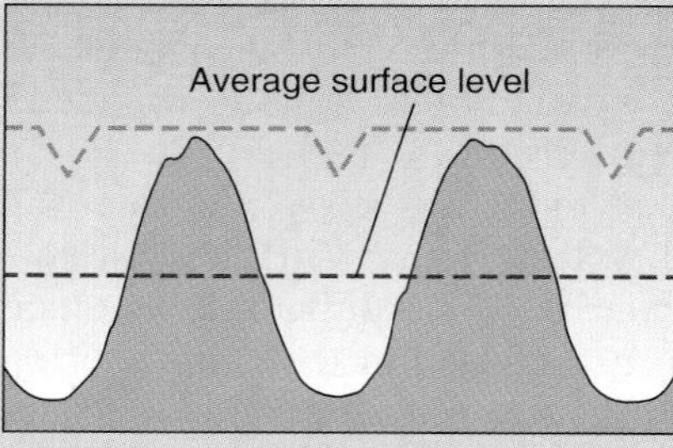

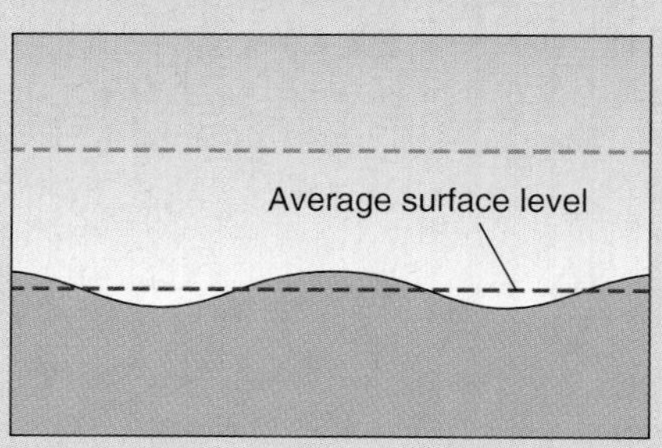

After isostatic readjustment

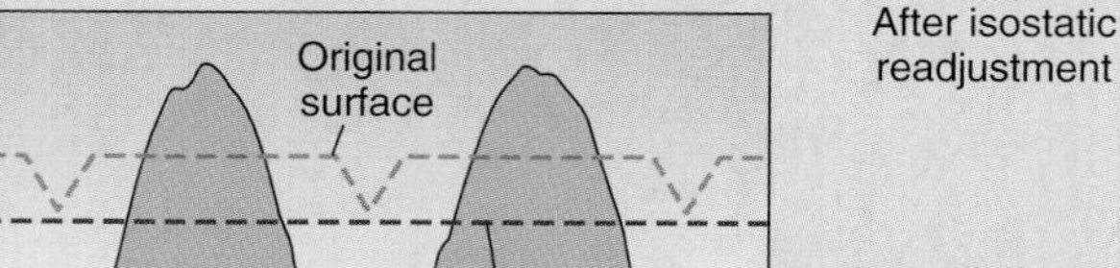

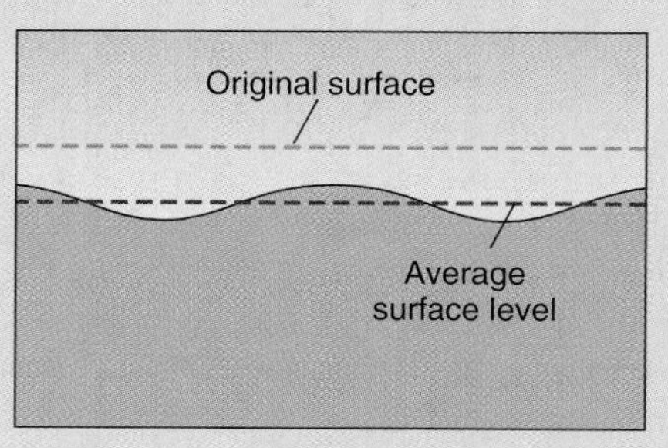

Box 5.2 — FIGURE 1

Comparison between two regions, before extensive erosion, after extensive erosion, and after isostatic readjustment. Region on the left erodes into mountains and deep valleys. Region on the right remains a plateau after approximately uniform erosion.

have less mass and will float isostatically upward. As in the case of the plateau, its average surface would rise to a level lower than before erosion; however, its average surface is somewhere between the peaks and bottoms of valleys. Although the average height of the block rises to a level below its previous average height, the mountains rise to heights greater than before.

Climate enters the picture because, interacting with tectonics, it helps control the type and extent of weathering that takes place. For instance, heavy precipitation takes place in the Himalaya because of the flow of very humid air from the south during the summer monsoon. At the higher elevations, the precipitation in the form of heavy snowfall contributes to extensive and very active glaciation. As described in the chapter on glaciation, glaciers are extremely effective agents of erosion. Deep, glacially carved valleys separate high mountains. The mountains will grow higher during isostatic adjustment at the same time the region as a whole is lowered by erosion. The Tibetan Plateau is north of the Himalaya. It is the highest, largest plateau in the world, with an average elevation of around 5 kilometers—higher than any mountain in the United States except for Alaska. The plateau has not been carved into mountains, because the climate is quite different from that of the Himalaya. The moist air from the south is blocked by the Himalaya, and the Tibetan Plateau is in its rain shadow (see chapter 18). Without water, there are no glaciers or large rivers to carve the plateau into mountains and valleys. So this region is slowly being eroded downward, getting progressively lower as erosion and isostasy balance each other out.

Additional Reading

Pinter, N., and M. T. Brandon. How erosion builds mountains. *Scientific American* (April, 1997): pp. 74–79.

from mountains by erosion, the range floats upward to regain its isostatic balance. This process can be thought of as "the pull of erosion." Isostatic adjustment does not take place instantaneously. Usually there will be a considerable time lag between erosion and isostatic adjustment. As the mountains wear down to a low plain, erosion becomes virtually ineffective and the now thin crust achieves isostatic balance; the former mountain belt becomes part of the craton. The reason the craton consists of plutonic and metamorphic rock is that these were the rocks that formed the deep roots of the former mountain belt.

At most places on continents, the altitude above sea level is related to local crustal thickness. Beneath the 5-kilometer-high Tibetan Plateau, the crust is 75 kilometers thick. Under Kansas, the crust is 44 kilometers thick and beneath Denver, the "mile high city," the crust is 50 kilometers thick. (If the United States ever joins the rest of the world and goes metric, Denver will be known as the "1.6 kilometer high city.") Just west of Denver, the altitude of the Rocky Mountains jumps to 2 kilometers higher than that at Denver. Scientists expected to find a corresponding thickening of the crust beneath these mountains. They were surprised by 1995 seismic studies that indicated that the crust is no thicker beneath the Rockies than at Denver. (Similar discrepancies between crustal thickness and mountain elevations have been reported for the southern Sierra Nevada.) To explain the higher elevations, geologists regard the mantle as hotter and therefore less dense beneath that part of the Rockies. The crust plus less dense mantle are floating on deeper, denser mantle. Seismic wave studies verify that the mantle here is hot and appears to be asthenosphere that is at a shallower level in Earth than usual.

It is important to emphasize that isostatic adjustment takes place during the orogenic stage as well, but the isostatic forces during an orogeny are overshadowed by forces due to plate convergence.

Normal Faulting

Normal faults are characteristic of this stage. The crust breaks into fault-bounded blocks. If an upthrown block is large enough, it becomes a **fault-block mountain range.** The normal faulting implies *horizontal extension,* the regional pulling apart of the crust. Isostatic vertical adjustment of a fault block probably occurs at the same time.

Although most fault-block mountain ranges are bounded by normal faults on either side of the range, some are tilted fault blocks in which the uplift has been great along one side of the range while the other side of the range has pivoted as if hinged (figure 5.16). The Sierra Nevada (California) and Teton (Wyoming) Range are tilted fault-block mountains (figure 5.17).

Isolated volcanic activity may be associated with this stage of a mountain belt's evolution. Eruptions occur along faults extending deep into the crust or the upper mantle.

Uplift is neither rapid nor continuous. Part of a mountain range may suddenly move upward a few centimeters (or, more rarely, a few meters) and then not move again for hundreds of years. Erosion works relentlessly on newly uplifted mountains, carving the block into peaks during the long, spasmodic rise. Over the long time period, the later episodes of renewed faulting and uplift involve successively less and less vertical movement.

Block-faulting is taking place in much of the western United States—the Basin and Range province (also called the Great Basin) of Nevada and parts of Utah, Arizona, New Mexico, Idaho, and California (figure 5.18). Hundreds of small, block-faulted mountain ranges are in evidence. They are separated by valleys that are filling with debris eroded from the mountains. Extension in the Basin and Range is probably due to hotter mantle beneath the crust, as shown in figure 5.19.

Delamination

Researchers in the 1990s were paying considerable attention to the hypothesis of lithospheric delamination and the role it may play in the evolution of mountain belts. For instance, delamination is used to explain the block-faulting, thin crust, and geologically young volcanic activity of the Basin and Range. **Lithospheric delamination** (or simply **delamination**) is the detachment of part of the mantle portion of the lithosphere beneath a mountain belt (figure 5.20). As you know, the lithosphere consists of the crust and the underlying, rigid mantle.

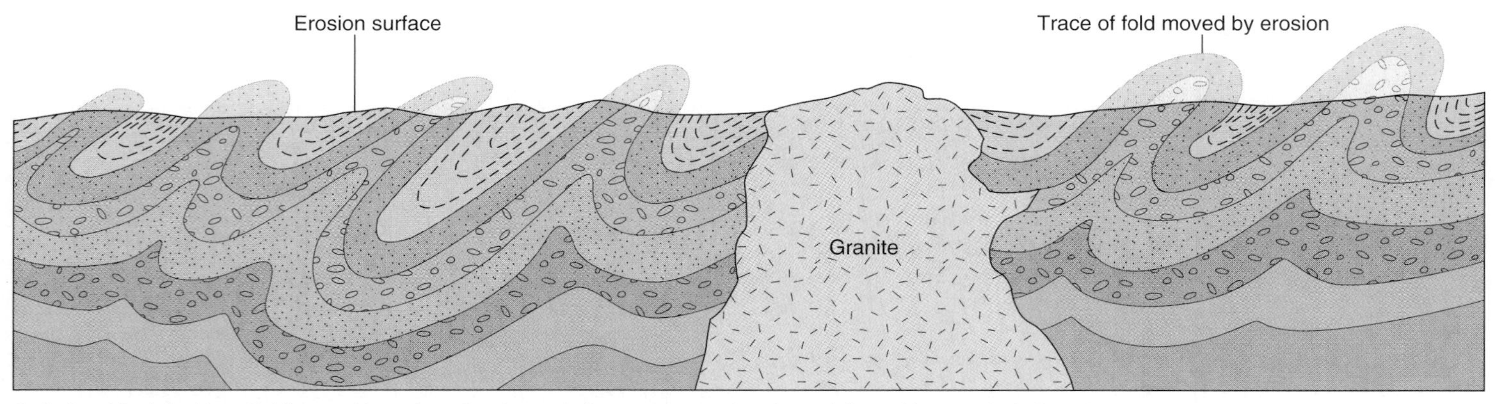

A Before block-faulting. Folding and intrusion of a pluton during an orogeny has been followed by a period of erosion.

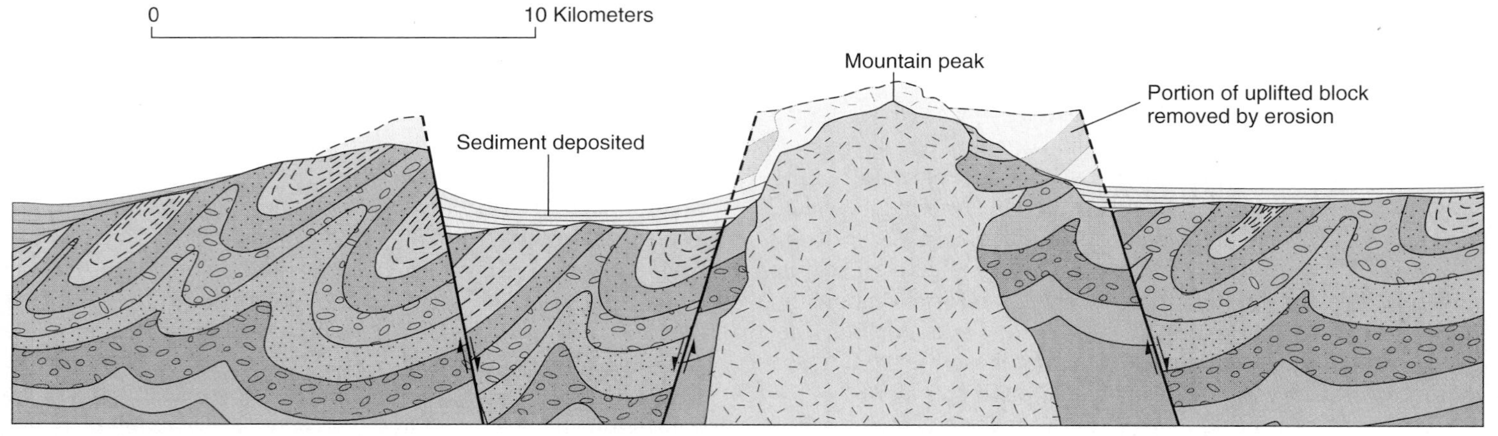

B The same area after block-faulting. Tilted fault-block mountain range on left. Range to right is bounded by normal faults.

FIGURE 5.16

Development of fault-block mountain ranges.

FIGURE 5.17

The Teton Range, Wyoming, a fault-block range. The rocks exposed are Precambrian metamorphic and igneous rocks that were faulted upward. Extensive past glaciation is largely responsible for their rugged nature. For more information go to www.winona.msus.edu/geology/travels/tetons/travel.html.

Photo by C. C. Plummer

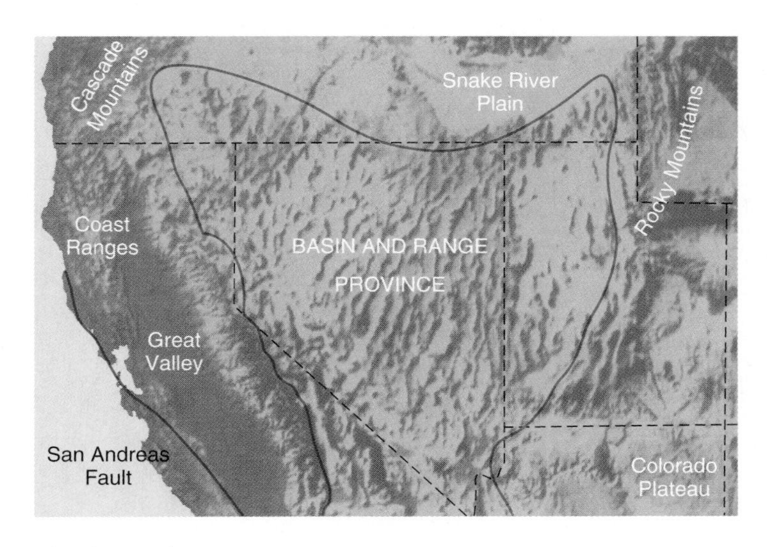

FIGURE 5.18

The Basin and Range and adjoining geological provinces.

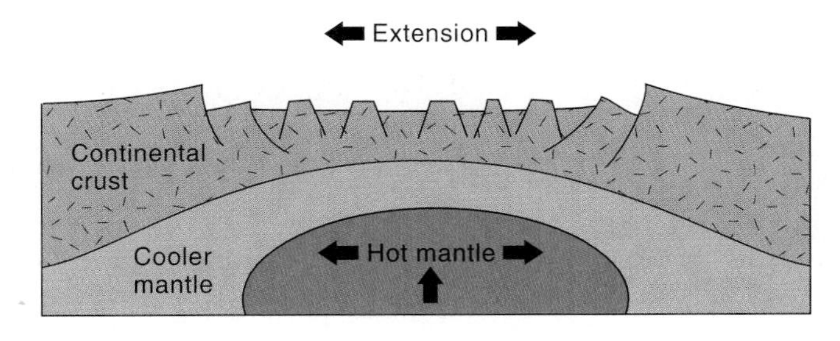

FIGURE 5.19

Hot, buoyant mantle causes extension, thinning, and block-faulting of the overlying crust.

Beneath the lithosphere is the hotter, plastic mantle of the asthenosphere. During an orogeny the crust as well as the underlying lithosphere mantle thickens. The lithosphere mantle is cooler and denser than the asthenosphere mantle. As indicated in figure 5.20, the thickened portion of the lithosphere mantle is gravitationally unstable so it breaks off and sinks through the asthenosphere to a lower level in the mantle. Hot asthenosphere mantle flows in to replace the foundered, colder mantle. Heating of the crust follows, allowing the lower crust to flow. The once thick crust becomes thinner than that of adjoining regions of the mountain belt. Extension results in block-faulting in the upper part of the crust (as in figure 5.19).

Delamination beneath the Basin and Range helps explain the extensive rhyolitic and basaltic eruptions that occurred tens of million years after the end of the orogenic stage. Heating in the lower part of the crust to 700°C would have generated silicic magma that erupted as rhyolite. Basaltic magma would have formed from partial melting of the asthenosphere when it moved upward (replacing the foundered lithosphere mantle) and pressure was reduced (as explained in chapter 11). That the crust was once thicker in the Basin and Range is supported by

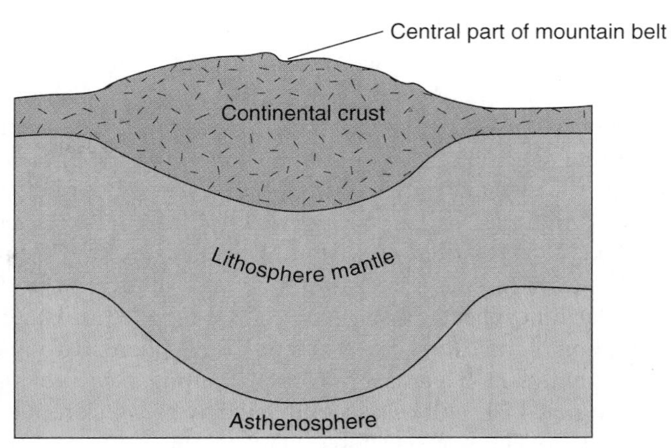

A Thick continental crust of a mountain belt produced during orogeny.

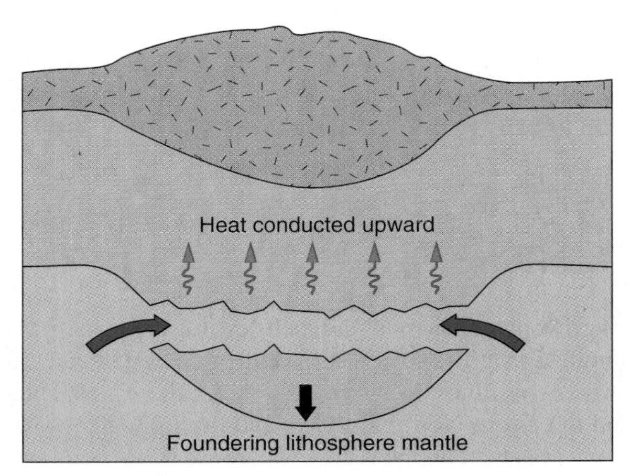

B Delamination of gravitationally unstable lithosphere mantle. Hot asthenosphere flows into place and heats overlying lithosphere.

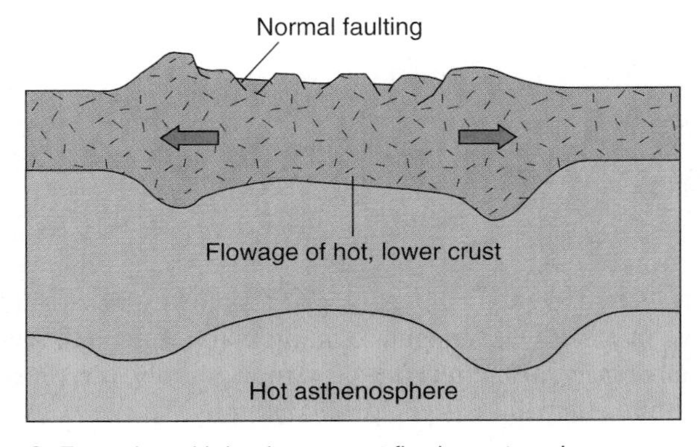

C Extension with hot lower crust flowing outward.

FIGURE 5.20

Delamination and thinning of continental crust following orogeny. Not drawn to scale.

Based on J. F. Dewey, 1988, Extensional collapse of orogens. *Tectonics,* v. 7, pp. 1123–1139, and K. D. Nelson, 1992, Are crustal thickness variations in old mountain belts like the Appalachians a consequence of lithospheric delamination? *Geology,* v. 20, pp. 498–502

recent studies of fossil plants indicating that the Basin and Range was 3 kilometers higher than at present.

Delamination is also being invoked to help explain why, when Pangaea broke up, North America split from Europe and Africa more or less along the old suture zone. *Gravitational collapse* could have contributed to the weakening and thinning of this once thick part of the mountain belt during the orogenic stage. The breakup of the supercontinent began around 30 million years after the orogenic stage ended. Delamination of the underlying lithosphere mantle would have resulted in heating and thinning of the overlying, remaining lithosphere. Rifting of the supercontinent began with normal faulting (see previous chapter, figure 4.21) and was accompanied by basaltic eruptions and intrusions. The Appalachians split from the European Caledonides. Europe, Africa, and North America went their separate ways as the Atlantic opened and widened.

Delamination (like gravity collapse) is an example of an hypothesis that builds on plate tectonic theory. It was proposed because it explains data better than other concepts do. It still needs further testing to become widely accepted as a theory.

THE GROWTH OF CONTINENTS

Continents grow bigger as mountain belts evolve along their margins. Accumulation and igneous activity add new continental crust beyond former coastlines. In the Paleozoic Era the Appalachians were added to eastern North America, and during the Mesozoic and Cenozoic eras the continent grew westward because of accumulation and orogenic processes in many parts of what is now the Cordillera. Therefore, if we isotopically dated rocks that had been through an orogeny, starting in the Canadian Shield and working toward the east and west coasts, we should find the rocks to be progressively younger. In a very general way, this seems to be the case; however, there are some rather glaring exceptions.

Displaced Terranes

In many parts of mountain belts are regions where the age and characteristics of the bedrock appear unrelated to that of adjacent regions. To help understand the geology of mountain belts, geologists have in recent years begun dividing major mountain belts into **tectonostratigraphic terranes** (or, more simply, **terranes**), regions within which there is geologic continuity. The geology in one terrane is markedly different from a neighboring terrane. Terrane boundaries are usually faults. Typically, a terrane covers thousands of square kilometers, but some terranes are considerably smaller. Alaska and western Canada have been subdivided by some geologists into over fifty terranes (figure 5.21). Terranes are named after major geographic features; for instance, Wrangellia, parts of which are now in Alaska and Canada (and with fragments in Washington and Idaho, according to some geologists) was named after the Wrangell Mountains of Alaska.

Many terranes appear to have formed essentially in place as a result of accumulation and orogeny along the continent's margin. Other terranes have rock types and ages that do not seem related to the rest of the geology of the mountain belt and have been called **suspect terranes,** that is, terranes that may not have formed at their present site. If evidence indicates that a terrane did not form at its present site on a continent, it is regarded as an **accreted terrane.** Accreted terranes that can be shown to have traveled great distances are known as *exotic terranes.*

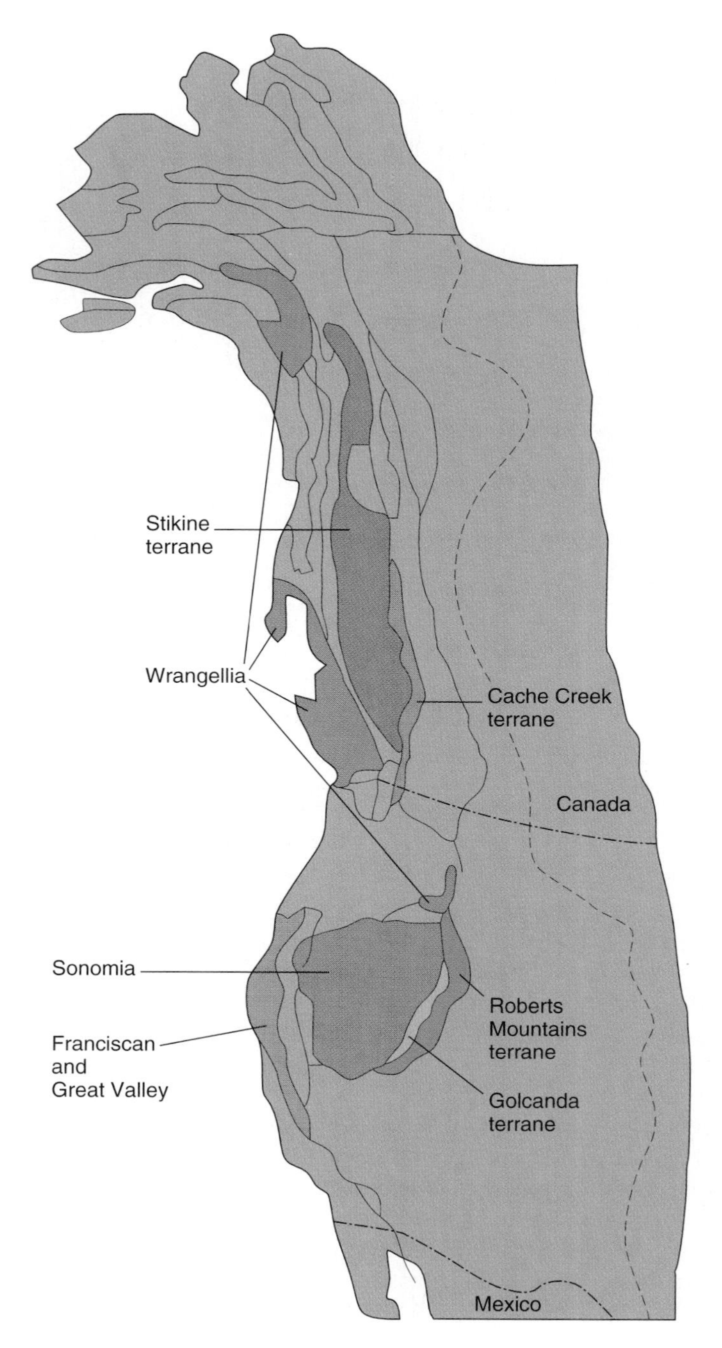

FIGURE 5.21
Some terranes in western North America. Note Wrangellia is in Alaska, British Columbia, and Idaho.
After U.S. Geological Survey Open File Map 83–716

A suspect terrane will have rock types and ages different from adjoining terranes, but to prove that it came from elsewhere in the world (and therefore is an accreted terrane), geologists may compare fossil assemblages or determine the paleomagnetic poles (see chapter 4) of the terrane's rocks. If the terrane is exotic, its fossil assemblage should indicate a very different climatic or environmental setting compared to that of the adjoining terrane. For an exotic terrane, the paleomagnetic poles for the rocks in the terrane will plot at some part of the world very distant from poles of adjoining terranes that formed in place. This indicates that a particular terrane formed in a different part of Earth and drifted into the continent of which it is now a part. Some accreted terranes were island arcs and some might have been *microcontinents* (such as present-day New Zealand) that moved considerable distances before crashing into other landmasses. Others may have been fragments of distant continents that split off and moved a long distance because of transform faulting. Imagine what might happen if the San Andreas fault remains active for another 100 million years or so. Not only would Los Angeles continue northward toward San Francisco and bypass it in about 25 million years (see box 6.3), but the block of coastal California west of the fault would continue moving out to sea, becoming a large island with continental crust that drifts northward across the Pacific. Ultimately it would crash into and suture onto Alaska.

Figure 5.22 shows a tentative reconstruction of how parts of Alaska might have migrated in time. This is based on paleomagnetic data that indicate that parts of Alaska originally formed south of the equator and moved many thousands of kilometers to become part of the Cordillera. Note from the diagram that the path of migration was not simple. Plates split, plates joined, and the direction of movement changed from time to time.

The Appalachians as well as mountain belts in other continents have also been divided into terranes. Even the Canadian Shield has been subdivided into terranes. Some geologists think they can determine, despite the great age and complexity of the shield's rocks, the extent to which some terranes traveled before crashing together.

We should caution the reader that geologists do not always agree on the nature and boundaries of terranes. While most would

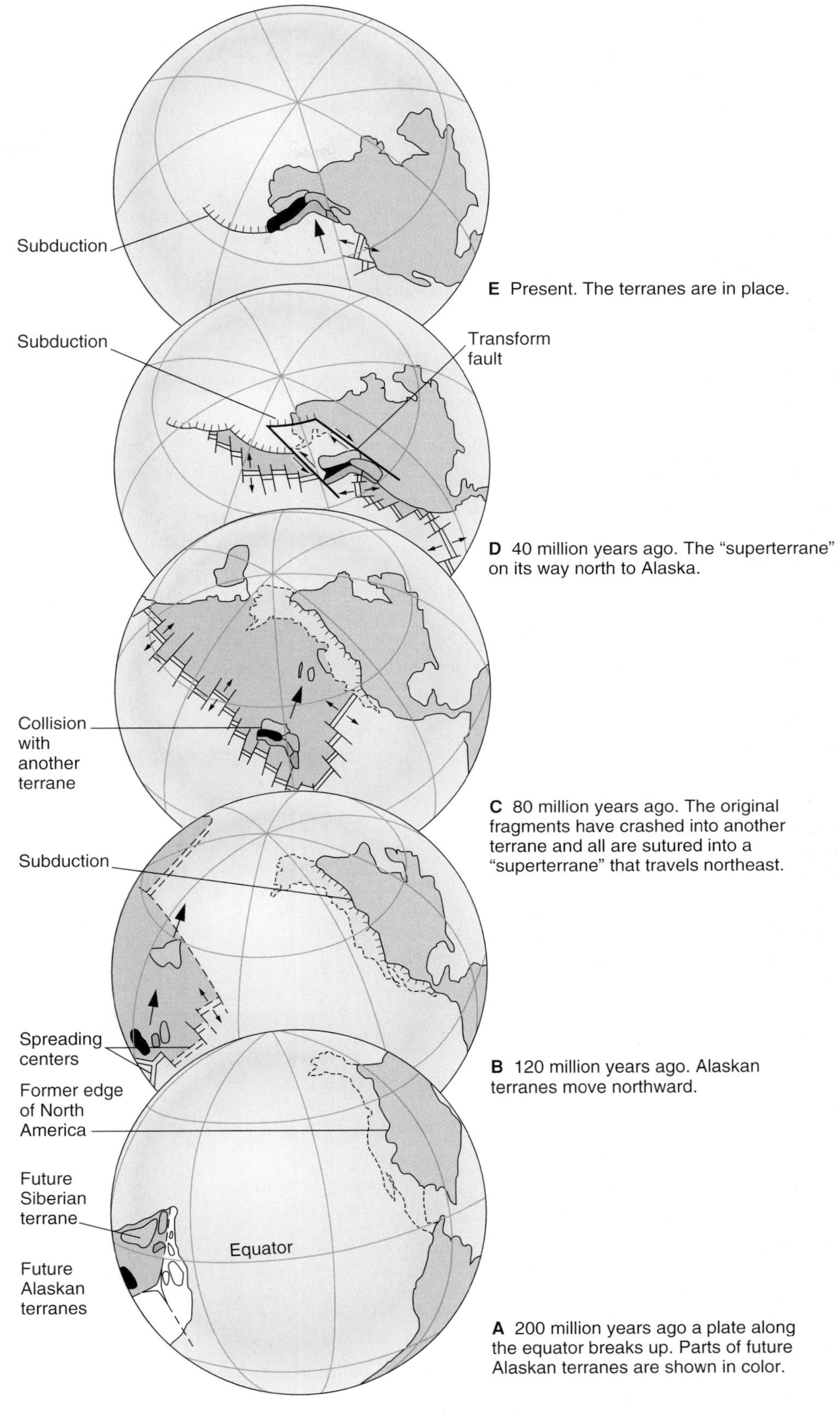

FIGURE 5.22

How fragments of the Southern Pacific Crust may have become part of Alaska.

Modified from D. B. Stone, B. C. Panuska, and D. R. Packer, 1982, "Paleolatitudes versus time for southern Alaska," *Journal of Geophysical Research,* vol. 87 (pp. 3697–3707), copyrighted by American Geophysical Union

probably agree that some terranes are exotic, many geologists think the subdividing of Alaska and western Canada into fifty terranes is overdoing it and not supported by sufficient evidence. Only time and more painstaking gathering of evidence will allow geologists to determine the history of each alleged terrane.

Concluding Comment

Only a couple decades ago, many geologists thought that through the application of plate tectonic theory, we could easily determine the processes at work in each mountain belt and work back in time to understand the history of each of the continents. Some suggested that there would hardly be major problems for Earth scientists to solve in the future. Plate tectonics was a breakthrough, and a great many problems were solved; but with this great leap forward in the science we have identified new problems. New generations of geologists will have no shortage of challenges and no less excitement from solving newly discovered problems than did their predecessors who saw the dawn of the plate tectonics breakthrough. Science present builds on science past.

Summary

Major mountain belts are made up of a number of *mountain ranges.* Mountain belts are generally several thousand kilometers long but only a few hundred kilometers wide.

Mountain belts generally evolve as follows. First, a thick sequence of sedimentary and volcanic rock accumulates (the *accumulation stage*). Second, the accumulation stage is either accompanied or followed by an *orogenic stage,* which involves intense deformation of the layered rocks into folds and reverse (including thrust) faults, along with metamorphism and igneous activity. Third, the area is then subjected to a long period of uplift, often with block-faulting, and erosion. Eventually, the mountain belt is eroded down to a plain and incorporated into the *craton,* or stable interior of the continent.

According to the theory of plate tectonics, mountains on the edge of continents are formed by continent-oceanic convergence, and mountains in the interior of continents are formed by continent-continent collisions.

The uplift of a region following termination of an orogeny is generally attributed to isostatic adjustment of continental crust.

Continents grow larger when new mountain belts evolve along continental margins. They may also grow by the addition of terranes that may have traveled great distances before colliding with a continent.

Terms to Remember

accreted terrane 126
accumulation stage 117
craton 113
fault-block mountain range 123
fold and thrust belts 114
gravitational collapse and spreading 119
lithospheric delamination (or delamination) 123
major mountain belt 110
mountain range 110
orogeny 126
Precambrian shield 113
suspect terrane 126
terrane (tectonostratigraphic terrane) 126

Testing Your Knowledge

Use the questions below to prepare for exams based on this chapter.

1. What does a fold and thrust belt tell us about what occurred during an orogeny?
2. What is the difference between the forces that could explain fault-block mountains and the forces that could account for an orogenic stage?
3. Explain how erosion and isostasy eventually produce stable, relatively thin, continental crust.
4. How do the sequences of sedimentary rocks in cratons differ from those in mountain belts?
5. What sequence of events accounts for a mountain belt that is bounded on either side by cratons?
6. The mountain belt that forms the western part of North America is called the
   a. Appalachians
   b. North American Cordillera
   c. Andes
   d. Rockies
   e. Himalaya
7. The craton
   a. covers the central part of the United States and Canada
   b. has only 1,000–2,000 m of sedimentary rock overlying basement rock
   c. has rock beneath any sedimentary rock that is old plutonic and metamorphic rock
   d. all of the above
8. The Precambrian shield
   a. contains geologically young rocks
   b. occurs only in mountainous regions
   c. is a complex of Precambrian metamorphic and plutonic rocks exposed over a large area
   d. all of the above
9. Folds and reverse faults in a mountain belt suggest
   a. crustal shortening
   b. deep water deposition of the sediment
   c. tensional stress
   d. all of the above

10. Which is not a stage in the history of a mountain belt?
    a. subsidence b. orogenic
    c. accumulation d. uplift and block-faulting

11. To explain fold and thrust belts, simultaneous normal faulting, and how once deep-seated metamorphic rocks rise to an upper level in a mountain belt, geologists use a model called
    a. tectonism b. orogeny
    c. gravitational collapse and spreading d. faulting

12. The Wilson Cycle describes
    a. the cycle of uplift and erosion of mountains
    b. the movement of asthenosphere
    c. the block-faulting that occurs at mountains
    d. the cycle of splitting of a supercontinent, opening of an ocean basin, followed by closing of the basin and collision of continents

13. The detachment of part of the mantle portion of the lithosphere beneath a mountain belt is called
    a. gravitational collapse b. lithospheric delamination
    c rifting d. none of the above

14. Which is not a type of terrane?
    a. accumulated b. tectonostratigraphic
    c. exotic d. accreted
    e. suspect

15. Which is a source for terranes?
    a. microcontinents b. ocean crustal fragments
    c. fragments of distant continents d. all of the above

16. Block-faulting may be due to (choose all that apply)
    a. isostatic adjustment b. gravitational collapse
    c. subduction d. lithospheric delamination

17. A mountain belt formed through ocean-continent convergence may contain (choose all that apply)
    a. fold and thrust belts
    b. thick accumulations of marine sediment
    c. normal faults
    d. metamorphism

18. Place these stages of development of a mountain belt in order:
    a. uplift and block faulting b. orogeny
    c. accumulation

Expanding Your Knowledge

1. How are unconformities used to determine when orogenies occurred?
2. How has seismic tomography contributed to our understanding of mountain belts?
3. How do basalt and ultramafic rocks from the oceanic lithosphere become part of mountain belts?
4. Why is a craton locally warped into basins and domes?
5. How could fossils in a terrane's rocks be used to indicate that it is an exotic terrane?

Exploring Web Resources

www.mhhe.com/plummer9e

Dance of the Continents (with SWEAT). Go to the Online Learning Center and read how building followed by breakup of supercontinents has taken place throughout geologic time. The process is compared to a dance. Each dance cycle, which takes about a billion years, is set to a symphony in four movements in which the continental fragments come together and later "dance" away. The creation, in the Paleozoic, and breakup, in the Mesozoic, of Pangaea is only the latest of the several dance cycles.

Related to this is a hypothesis called SWEAT, which is an acronym for southwest United States-East Antarctic connection. According to this hypothesis, in one of the Precambrian supercontinents, the craton in the southwestern United States adjoined what is now Antarctica. This site also includes answers to the Testing Your Knowledge section, additional quizzing, readings, and a great animation. Click on the links to go directly to the websites listed below.

www.hartwick.edu/geology/work/VFT-so-far/VFT.html

Hartwick College Virtual Field Trip. A field trip through part of the Appalachians in central New York state. The site includes a geologic history for this part of the Appalachians.

http://vishnu.glg.nau.edu/people/jhw/Tibet/Tibet.html

Tibet: A Virtual Field Trip. You can take a trip through the Tibetan Plateau into the Himalaya. Good summaries of the geology of the Tibetan Plateau and the Himalayan Mountains are accessible through this site.

www.winona.msus.edu/geology/travels/tetons/travel.html

Geology of Grand Teton National Park, Wyoming. A photo, map, and text description of the spectacular Grand Teton Range and its geologic history.

Animations

This chapter includes the following animation available on our Online Learning Center at www.mhhe.com/plummer9e.

5.15 Isostasy in a Mountain Belt

CHAPTER

6

Geologic Structures

We now shift our focus to changes in bedrock caused by powerful forces originating deep within Earth. In this chapter we explain how rocks respond to these tectonic forces.

The main purpose of this chapter is to help you recognize certain geologic structures, understand the forces that caused them, and thus determine the geologic history of an area.

Some principles that will be discussed in chapter 8 should help you interpret the way structures develop in an area and the sequence of that development. Recognition of unconformities as well as the principles of original horizontality, superposition, and cross-cutting relationships are as important to structural geology as they are to determining relative time.

Previous and subsequent chapters require an understanding and knowledge of structural geology as presented in this chapter. To understand earthquakes (chapter 7), for instance, you must know about faults. Appreciating how major mountain belts and the continents have evolved (chapter 5) calls for a comprehension of faulting and folding. Understanding plate tectonic theory as a whole (chapter 4) also requires a knowledge of structural geology. (Plate tectonic theory developed primarily to explain certain structural features.) In areas of active tectonics, the location of geologic structures is important in the selection of safe sites for schools, hospitals, dams, bridges, and nuclear power facilities.

Also, understanding structural geology can help us more fully appreciate the problem of finding more of Earth's dwindling natural resources. Chapter 21 discusses the association of certain geologic structures with petroleum deposits and other valuable resources.

Opposite: Rocks that were once horizontal have been contorted into folds during mountain building. Damaraland, Namibia, Africa.
Photo © Michael Fogden/Animals, Animals/Earth Scenes

FIGURE 6.1
Folded and faulted sedimentary beds exposed in a roadcut near Palmdale, California.
Photo by C. C. Plummer

In a broad sense structural geology can be thought of as the study of the architecture of Earth's crust, its deformational features, and their mutual relations and origins. For our purposes, **structural geology** can be defined as the branch of geology concerned with the shapes, arrangement, and interrelationships of bedrock units and the forces that cause them.

TECTONIC FORCES AT WORK

Stress and Strain in the Earth's Crust

Tectonic forces move and deform parts of the crust, particularly along plate margins. When studying deformed rocks, structural geologists typically refer to **stress,** a force per unit area. Where stress can be measured, it is expressed as the force per unit area at a particular point; however, it is difficult to measure stress in rocks that are currently buried. We can observe the effects of past stress (caused by tectonic forces and confining pressure from burial) when bedrock is exposed after uplift and erosion. From our observations we may be able to infer the principal directions of stress that prevailed. We also can observe in exposed bedrock the effect of forces on a rock that was stressed. **Strain** is the change in size (volume) or shape, or both, while an object is undergoing stress. In figure 6.1, originally horizontal rock layers have changed in shape and are strained into wave-like folds that are broken by faults. The layers have been strained, probably by a horizontal stress that pushed or compressed the layers together until they were shortened by buckling.

The relationship between stress and strain can be illustrated by deforming a piece of Silly Putty (figure 6.2). If the

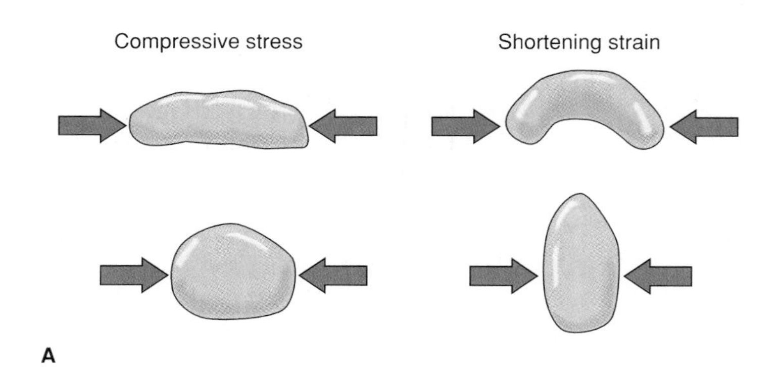

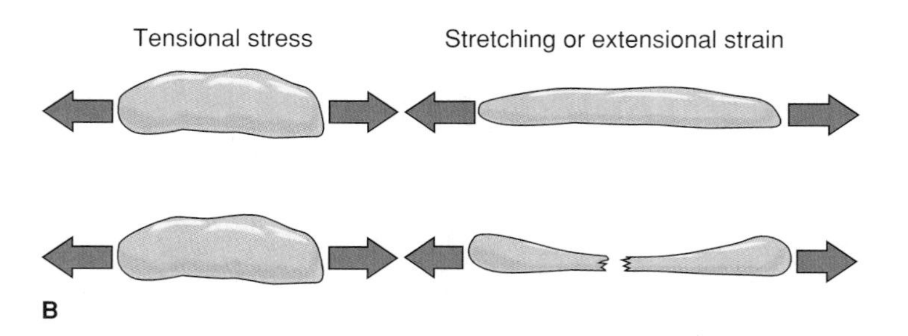

FIGURE 6.2
The effects of compressional and tensional stresses on Silly Putty. (*A*) Compressing Silly Putty results in shortening either by folding or flattening. (*B*) Pulling (tensional stress) Silly Putty causes stretching or extension; if pulled (strained) too fast, or chilled, the Silly Putty will break after first stretching.

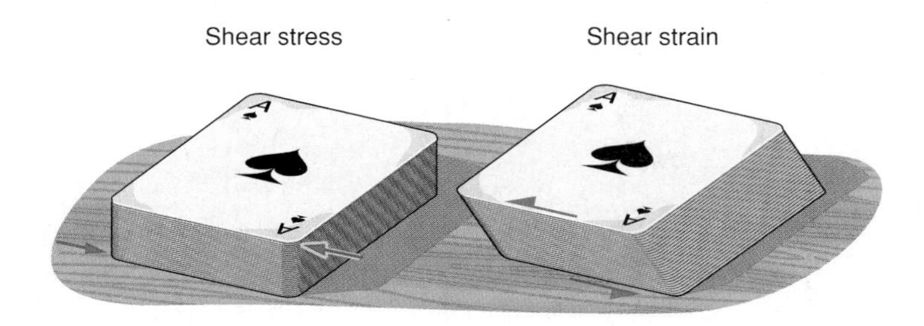

FIGURE 6.3
Shear stress can be modeled by shearing a deck of cards.

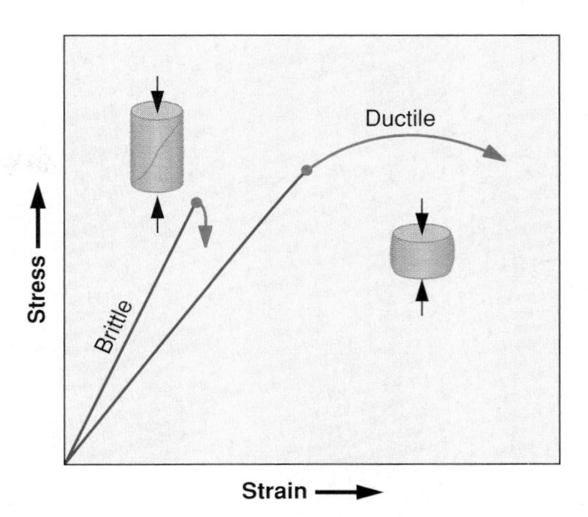

FIGURE 6.4
Graph shows the behavior of rocks with increasing stress and strain. Elastic behavior occurs along the straight line portions (shown in blue) of the graph. At stresses greater than the elastic limit (red points), the rock will either deform as a ductile material or break, as shown in the deformed rock cylinders.

Silly Putty is pushed together or squeezed from opposite directions, we say the stress is **compressive.** Compressive stress is common along convergent plate boundaries and typically results in rocks being deformed by a *shortening strain.* In figure 6.2*A,* an elongate piece of Silly Putty may shorten by bending or folding, whereas a ball of Silly Putty will shorten in the direction parallel to the compressive stress and elongate or stretch in the direction perpendicular to it. Rocks subjected to compressive stresses, particularly along convergent plate boundaries, behave in the same way and are typically shortened in the horizontal direction and elongated in the vertical direction.

A **tensional stress** is caused by forces pulling away from one another in opposite directions (figure 6.2*B*) and results in a *stretching* or *extensional strain.* If we apply a tensional stress on a ball of Silly Putty, it will elongate or stretch parallel to the applied stress. If the tensional stress is applied rapidly, the Silly Putty will first stretch and then break apart (figure 6.2*B*). Tensional stresses are quite rare in the crust; in fact, most stresses directly measured in the Earth are compressive.

The third type of stress is a **shear stress** that is due to forces parallel to one another, but in opposite directions along a discrete surface such as a fault. It is much like putting a deck of cards in your hands and shearing the deck by moving your hands in opposite directions (figure 6.3). A shear stress results in a *shear strain* parallel to the direction of the stresses. Shear stresses are notable along transform plate boundaries and along other actively moving faults.

Behavior of Rocks to Stress and Strain

Rocks behave as elastic, ductile, or brittle materials depending on the amount and rate of stress applied, the type of rock, and the temperature and pressure under which the rock is strained.

If a deformed body recovers its original shape after the stress is reduced or removed, the behavior is **elastic.** For example, if a tensional stress is applied to a rubber band it will stretch as long as the stress is applied, but once the stress is removed the rubber band returns (or recovers) to its original shape and its behavior is elastic. Silly Putty will behave elastically if molded into a ball and bounced. Most rocks can behave in an elastic way at very low stresses (a few kilobars); however, once the stress applied exceeds the **elastic limit** (figure 6.4) the rock will deform in a permanent way.

A rock that behaves in a **ductile** or plastic manner will bend while under stress and does not return to its original shape after relaxation of the stress. Silly Putty behaves as a ductile material unless the rate of strain is rapid. Rocks exposed to elevated pressure and temperature during regional metamorphism also behave in a ductile manner. As shown in figure 6.4, material behaving in a ductile manner does not require much of an increase in stress to continue to strain (relatively flat curve). Ductile behavior results in rocks that are permanently deformed mainly by folding or bending of rock layers (figure 6.1).

A rock exhibiting **brittle** behavior will fracture at stresses higher than its elastic limit, much like a rubber band will break if stretched too far. Rocks typically exhibit brittle behavior at or near Earth's surface where pressure and temperatures are low. Under these conditions, rocks favor breaking rather than bending. Faults and joints are examples of structures that form by brittle behavior of the crust.

A sedimentary rock exposed at Earth's surface is brittle; it will fracture if you hit it with a hammer. How then do sedimentary rocks, such as those shown in figure 6.1, become bent (or deformed in a ductile way)? The answer is that either stress increased very slowly or that the rock was deformed under considerable confining pressure (buried under more rock).

Note, however, that there are some fractures (faults) disrupting the bent layers in figure 6.1. This tells us that although the rock was plastic initially, the stress became too intense or the rate of strain increased and the rock fractured.

A

B

FIGURE 6.5
Ditch at a winery near Hollister, California, displaced by the San Andreas fault. Photo (*A*) was taken in 1975, (*B*) in 1992.
Photos by C. C. Plummer

Present Deformation of the Crust

Geologists often say the crust of Earth is "mobile" or "restless" because bedrock is moving and being deformed in many parts of the world. Bedrock may be displaced suddenly during earthquakes. By contrast, much of the movement of rock within the crust is continuous and very slow, at rates of less than a millimeter per year.

Compared with most geologic processes, the present-day movement of the crust in much of California is very rapid. A large slice of the coastal portion of the state is moving relentlessly northward relative to the rest of North America and has apparently been doing so for millions of years. Some of the movement is jerky and associated with earthquakes (discussed in the following chapter), but elsewhere it is essentially continuous and smooth.

Around Hollister, California, ditches, curbs, homes, and buildings are being slowly torn apart because they straddle an active fault (figure 6.5). A **fault** is a fracture in bedrock along which movement has taken place. The fault that goes through Hollister (part of the San Andreas fault system) is one of several major cracks extending deep into the lithosphere, separating the northward-moving coastal portion of California from the rest of the state. Hollister residents whose homes "ride" the fault report hearing almost constant creaking, evidently due to motion along the fault. Yet the movement—about 1 centimeter per year (roughly the same rate at which a fingernail grows)—is slow enough that walls can be patched with plaster as they crack.

Geologically rapid movement of the crust can be observed in young, developing mountain regions such as the California Coast Ranges, which have been forming throughout the Cenozoic Era (the last 65 million years). In other parts of the world, however, the continents and sea floors are shifting very slowly up and down as well as moving laterally. Some of these motions can be detected only by precise, repeated surveying.

STRUCTURES AS A RECORD OF THE GEOLOGIC PAST

Some geologic structures that give us clues to the past will be described in subsequent chapters. Batholiths, stocks, dikes, and sills, for example, are keys to past igneous activity (see chapter on intrusive activity). In this chapter we are mainly concerned with types of structures that can provide a record of crustal deformation that is no longer active. Very old structures that are now visible at Earth's surface were once buried and are exposed through erosion.

The study of geologic structures is of more than academic interest. The petroleum and mining industries, for example, employ geologists to look for and map geologic structures associated with oil and metallic ore deposits. Understanding and mapping geologic structures is also important for evaluating problems related to engineering decisions and environmental planning, such as determining the most appropriate sites for dams, large bridges, or nuclear reactors, and even the building of houses, schools, and hospitals.

Geologic Maps and Field Methods

In an ideal situation, a geologist studying structures would be able to fly over an area and see the local and regional patterns of bedrock from above. Sometimes this is possible, but very often soil and vegetation conceal the bedrock. Therefore, geologists ordinarily use observations from a number of individual *outcrops* (exposures of bedrock at the surface) in determining the patterns of geologic structures. The characteristics of rock at each outcrop in an area are plotted on a map by means of appropriate symbols. With the data that can be collected, a geologist can make inferences about those parts of the area he or she cannot observe. A **geologic map,** which uses standardized symbols and patterns to represent rock types and geologic structures, is typically produced from the field map for a given area (for example, see the geologic map of North America inside the front cover). On such a map are plotted the type and distribution of rock units, the occurrence of structural features (folds, faults, joints, etc.), ore deposits, and so forth. Sometimes surficial features, such as deposits by former glaciers, are included, but these may be shown separately on a different type of geologic map.

Anyone trained in the use of geologic maps can gain considerable information about local geologic structures because standard symbols and terms are used on the maps and the accompanying reports. For example, the symbol ⊕ on a geologic map denotes horizontal bedding in an outcrop. Different colors and patterns on a geologic map represent distinct rock units.

Strike and Dip

According to the principle of *original horizontality,* sedimentary rocks and some lava flows and ash falls are deposited as horizontal beds or strata. Where these originally horizontal rocks are found tilted, it indicates that tilting must have occurred after deposition and lithification (figure 6.6). Someone studying a geologic map of the area would want to know the extent and direction of tilting. By convention, this is determined by plotting the relationship between a surface of an inclined bed and an imaginary horizontal plane. You can understand the relationship by looking carefully at figure 6.7, which represents sedimentary beds cropping out alongside a lake (the lake surface provides a convenient horizontal plane for this discussion).

FIGURE 6.6

Tilted sedimentary beds along the coast of northern California near Point Arena. Here, the strike is the line formed by the intersection of the tilted sedimentary beds and the horizontal layer of sand in the foreground. The direction of dip is toward the left.

Photo by Diane Carlson

Strike is the compass direction of a line formed by the intersection of an inclined plane with a horizontal plane. In this example, the inclined plane is a bedding plane. You can see from figure 6.7 that the beds are striking from north to south. Customarily only the northerly direction (of the strike line) is given, so we simply say that beds strike north a certain number of degrees east or west (such as N50°E).

Observe that the **angle of dip** is measured downward from the horizontal plane to the bedding plane (an inclined plane). Note that the angle of dip (30° in the figure) is measured within a vertical plane that is perpendicular to both the bedding and the horizontal planes.

The **direction of dip** is the compass direction in which the angle of dip is measured. If you could roll a ball down a bedding surface, the compass direction in which the ball rolled would be the direction of dip.

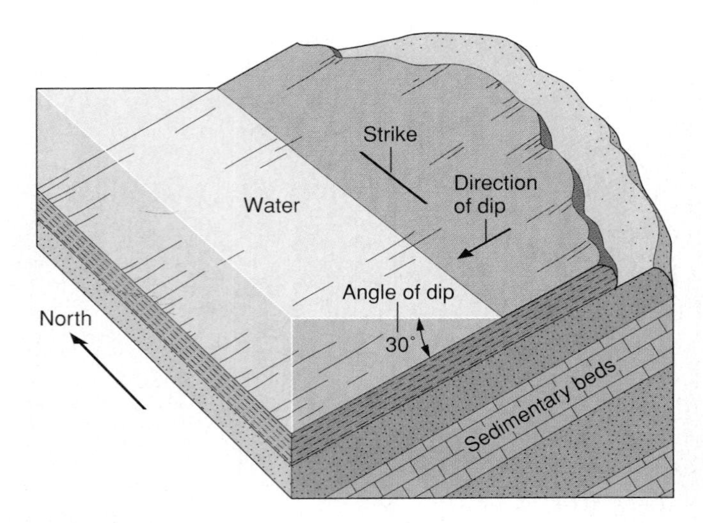

FIGURE 6.7

Strike, angle of dip, and direction of dip.

The dip angle is always measured at a right angle to the strike, that is, perpendicular to the strike line as shown in figure 6.7. Because the beds could dip away from the strike line in either of two possible directions, the general direction of dip is also specified—in this example, west.

A specially designed instrument called a Brunton pocket transit (after the inventor) is used by geologists for measuring the strike and dip (figure 6.8). The Brunton pocket transit contains a compass, a level, and a device for measuring angles of inclination. Besides recording strike and dip measurements in a field notebook, a geologist who is mapping an area draws strike and dip symbols on the field map, such as Y or ⋏ for each outcrop with dipping or tilted beds. On the map, the intersection of the two lines at the center of each strike and dip symbol represents the location of the outcrop where the strike and dip of the bedrock were measured. The long line of the symbol is aligned with the compass direction of the strike. The small tick, which is always drawn perpendicular to the strike line, is put on one side or the other, depending on which of the two directions the beds actually dip. The angle of dip is given as a number next to the appropriate symbol on the map. Thus, ^{25}Y indicates that the bed is dipping 25° from the horizontal toward the northwest and the strike is northeast (assuming that the top of the page is north). The orientation of the bed would be written N45°E, 25°NW. figure 6.9 is a geologic map that shows all the sedimentary layers striking north and dipping 30° to the west (N0°, 30°W).

Beds with vertical dip require a unique symbol because they dip neither to the left nor the right of the direction of strike. The symbol used is ⨯, which indicates that the beds are striking northeast and that they are vertical (N45°E, 90°).

Geologic Cross Sections

A **geologic cross section** represents a vertical slice through a portion of Earth. It is much like a roadcut (see figure 6.1) or the wall of a quarry in that it shows the orientation of rock units and

FIGURE 6.8

Geologist determining the strike and dip of inclined beds in Nevada.
Photo by C. C. Plummer

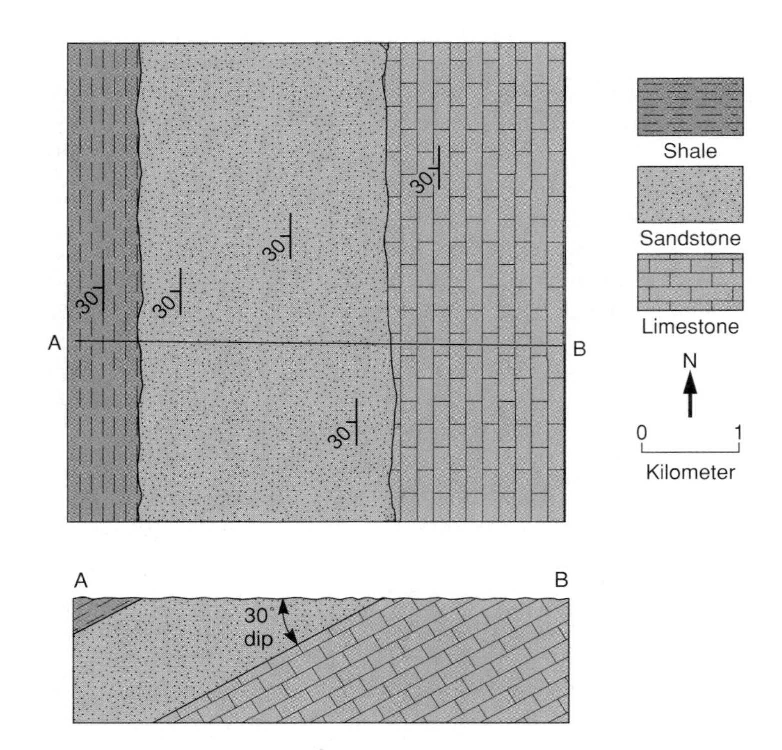

FIGURE 6.9

A geologic map and cross section of an area with three sedimentary formations. (Each formation may contain many individual sedimentary layers, as explained in chapter 14.) Beds strike north and dip 30° to the west. The geologic cross section (vertical cut) is constructed between points A and B on the map.

structures in the vertical dimension. Geologic cross sections are constructed from geologic maps by projecting the dip of rock units into the subsurface (figure 6.9), and are quite useful in helping visualize geology in three dimensions. They are used extensively throughout this book as well as in professional publications.

FOLDS

Folds are bends or wavelike features in layered rock. Folded rock can be compared to several layers of rugs or blankets that have been pushed into a series of arches and troughs. Oftentimes folds in rock can be seen in roadcuts or other exposures (figure 6.10). When the arches and troughs of folds are concealed (or when they exist on a grand scale), geologists can still determine the presence of folds by noticing repeated reversals in the direction of dip taken on outcrops in the field or shown on a geologic map.

The fact that the rock is folded shows that it was strained in a ductile way rather than by elastic or brittle strain. Yet the rock exposed in outcrops is generally brittle and shatters when struck with a hammer. The rock is not metamorphosed (most metamorphic rock is intensely folded because it is plastic under the high pressure and temperature environment of deep burial and tectonic stresses). Perhaps folding took place when the rock was buried at a moderate depth where high confining pressure favors plastic behavior. Alternatively, folding could have taken place close to the surface under a very low rate of strain. (When we strike a rock with a hammer, the strain rate is very high at the point of impact.)

FIGURE 6.10
Folded sedimentary rocks, Lulworth Cove, Dorset, England.
Photo © Tom Bean

Geometry of Folds

Determining the geometry or shape of folds may have important economic implications because many oil and gas deposits and also some metallic mineral deposits are localized in folded rocks (see chapter 21). The geometry of folds is also important in unraveling how a rock was strained and how it might be related to the movement of tectonic plates. Folds are usually associated with compressive stresses along convergent plate boundaries, but are also commonly formed where rock has been sheared along a fault.

Because folds are wave-like forms that usually form by the shortening of rock layers, two basic fold geometries are common—anticlines and synclines (figure 6.11).

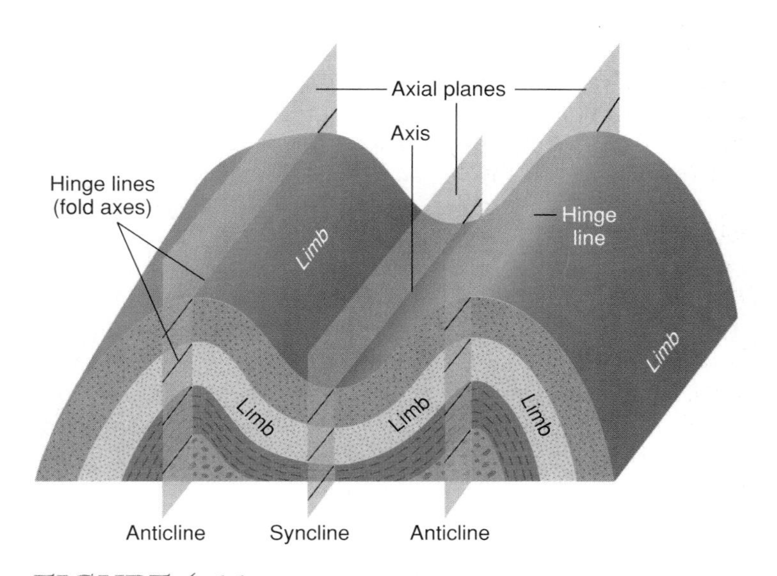

FIGURE 6.11
Diagrammatic sketch of two anticlines and a syncline illustrating the axial planes, hinge lines, and fold limbs.

An **anticline** is an upward arching fold. Usually the rock layers dip away from the **hinge line** (or *axis*) of the fold. The downward-arching counterpart of an anticline is a **syncline,** a troughlike fold. The layered rock usually dips toward the syncline's hinge line. In the series of folds shown in figure 6.11, two anticlines are separated by a syncline. Each anticline and adjacent syncline share a **limb.** Note the hinge lines on the crests of the two anticlines and bottom of the syncline. Similar hinge lines could be located in the hinge areas at the contacts between any two adjacent folded layers. For each anticline and the syncline, the hinge lines are contained within the shaded vertical planes. Each of these planes is an **axial plane,** an imaginary plane containing all of the hinge lines of a fold. The axial plane divides the fold into its two limbs.

It is important to remember that anticlines are not necessarily related to ridges nor synclines to valleys, because valleys and ridges are nearly always erosional features. In an area that

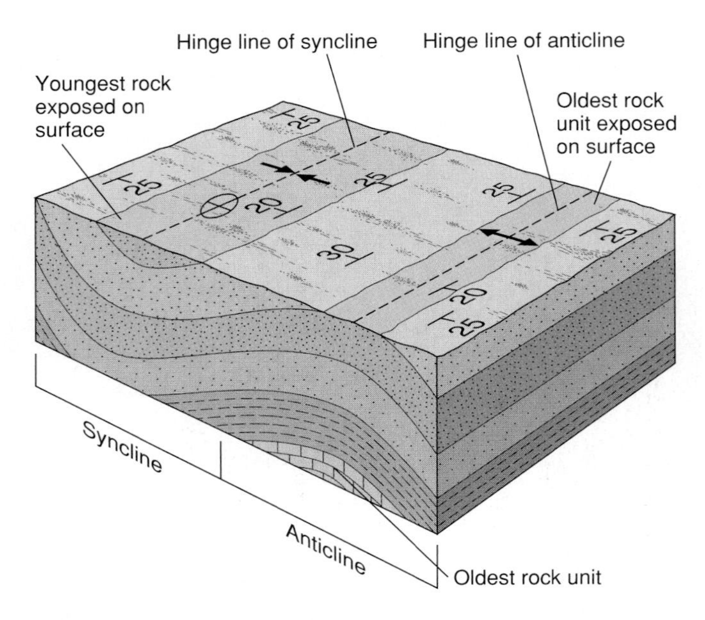

FIGURE 6.12

Folded rock. This view is a *block diagram.* Its top represents the land surface (geologic map) and its two visible sides are vertical cross sections. The surface has been eroded to a nearly horizontal plain. Side views are interpretations based on what the geologist maps on the surface.

has been eroded to a plain, the presence of underlying anticlines and synclines is determined by the direction of dipping beds in exposed bedrock, as shown in figure 6.12. (In the field, of course, the cross sections are not exposed to view as they are in the diagram.)

Figure 6.12 also illustrates how determining the relative ages of the rock layers, or beds, can tell us whether a structure is an anticline or a syncline. Observe that the oldest exposed rocks are along the hinge line of the anticline. This is because lower layers in the originally flat-lying sedimentary or volcanic rock were moved upward and are now in the core of the anticline. The youngest rocks, on the other hand, which were originally in the upper layers, were folded downward and are now exposed along the synclinal hinge line.

Plunging Fold

The examples shown so far have been of folds with horizontal hinge lines. These are the easiest to visualize. In nature, however, anticlines and synclines are apt to be **plunging folds**—that is, folds in which the hinge lines are not horizontal. On a surface leveled by erosion, the patterns of exposed strata (beds) resemble Vs or horseshoes (figures 6.13 and 6.14) rather than the striped patterns of nonplunging folds. However, plunging anticlines and synclines are distinguished from one another in the same way as are nonplunging folds—by directions of dip or by relative ages of beds.

A plunging syncline contains the youngest rocks in its center or core, and the V or horseshoe points in the direction opposite of the plunge. Conversely, a plunging anticline contains the oldest rocks in its core and the V points in the same direction as the plunge of the fold.

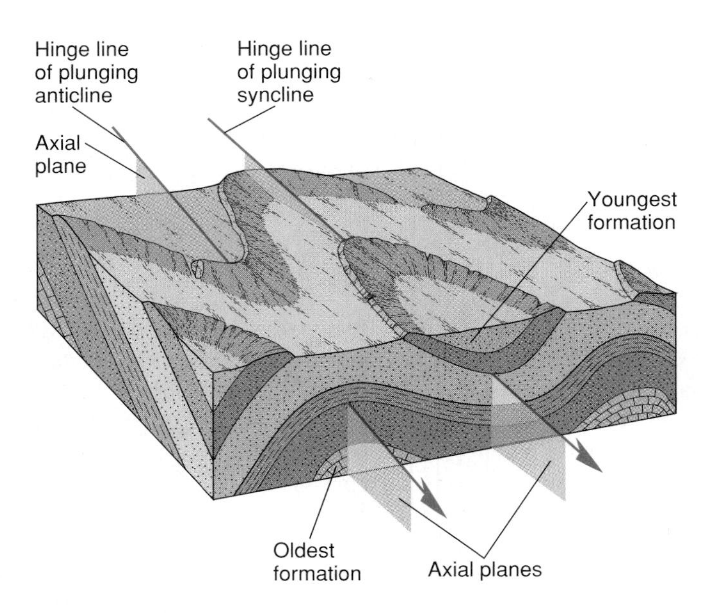

FIGURE 6.13

Plunging folds: anticline on left and right, syncline in center. The hinge lines are at an angle to the block diagram, penetrating the surface and emerging from the front cross section.

Structural Domes and Structural Basins

A **structural dome** is a structure in which the beds dip away from a central point. In cross section, a dome resembles an anticline and is sometimes called a doubly plunging anticline. In a **structural basin,** the beds dip toward a central point (figure 6.15); in cross section, it is comparable to a syncline (doubly plunging syncline). A structural basin is like a set of nested bowls. If the set of bowls is turned upside down, it is analogous to a structural dome.

Domes and basins tend to be features on a grand scale (some are more than a hundred kilometers across), formed by uplift somewhat greater (for domes) or less (for basins) than that of the rest of a region. Michigan's lower peninsula and parts of adjoining states and Ontario are on a large structural basin (see map on the inside front cover). Domes of similar size are found in other parts of the Middle West. Smaller domes are found in the Rocky Mountains (figure 6.16).

Domes and anticlines (as well as some other structures) are important to the world's petroleum resources, as described in chapter 21.

Interpreting Folds

Folds occur in many varieties and sizes. Some are studied under the microscope, while others can have adjacent hinge lines tens of kilometers apart. Some folds are a kilometer or more in height. figure 6.17 shows several of the more common types of folds. **Open folds** (figure 6.17*B*) have limbs that dip gently. All

FIGURE 6.14

Plunging folds. Anticline in Utah plunging in the direction of the upper part of the photo.

Photo by Frank M. Hanna

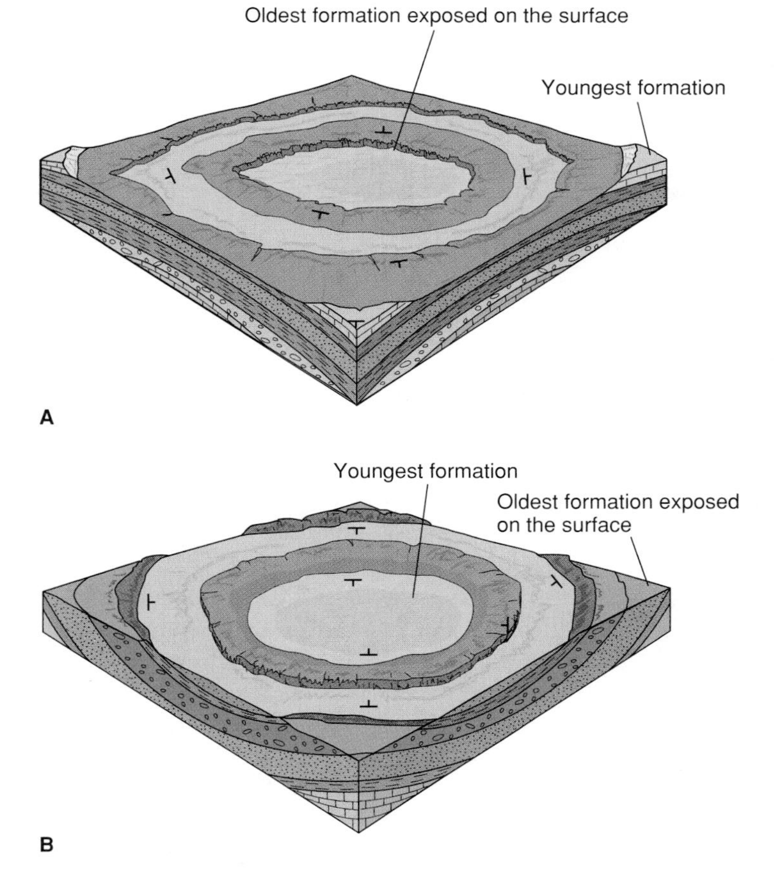

FIGURE 6.15

(*A*) Structural dome. (*B*) Structural basin.

FIGURE 6.16

Dome near Casper, Wyoming. The ridges are sedimentary layers that are resistant to erosion. Beds dip away from the center of the dome.

Photo by D. A. Rahm, courtesy of Rahm Memorial Collection, Western Washington University

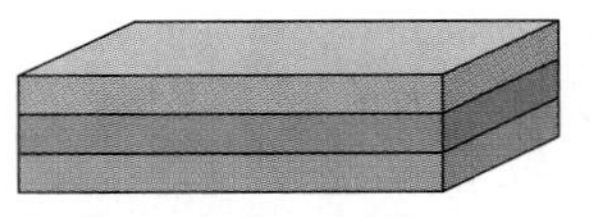

FIGURE 6.17

Various types of folds. The length of the arrows in A through E is proportional to the amount and direction of stresses that caused folding. (*A*) Strata before folding. (*B*) Open folds in Spain (they are plunging away from the people). (*C*) Isoclinal folds from the northern Sierra Nevada, California.

Photo *B* by C. C. Plummer, photo *C* by Diane Carlson

other factors being equal, the more open the fold, the less intense the stress involved. By contrast, an **isoclinal fold,** one in which limbs are parallel to one another, implies intense compressive or shear stress (figure 6.17*C*).

Folds that have vertical axial planes are referred to as upright folds. However, where the axial plane of a fold is not vertical but is inclined, the fold may be classified as *asymmetric*. If the axial plane is inclined to such a degree that the fold limbs dip in the same direction, the fold is classified as an **overturned fold.** Overturned folds imply that unequal compressive stresses or even a shearing stress caused the upper limb of the fold to override the lower limb (figure 6.17*D*). Looking at an outcrop where only the overturned limb of a fold is exposed, you would probably conclude that the youngest bed is at the top. The principles of *superposition* (see chapter 8), however, cannot be applied to determine top and bottom for overturned beds. You must either see the rest of the fold or find primary sedimentary structures within the beds such as mudcracks that indicate the original top or upward direction.

Recumbent folds (figure 6.17*E*) are overturned to such an extent that the limbs are essentially horizontal. Recumbent

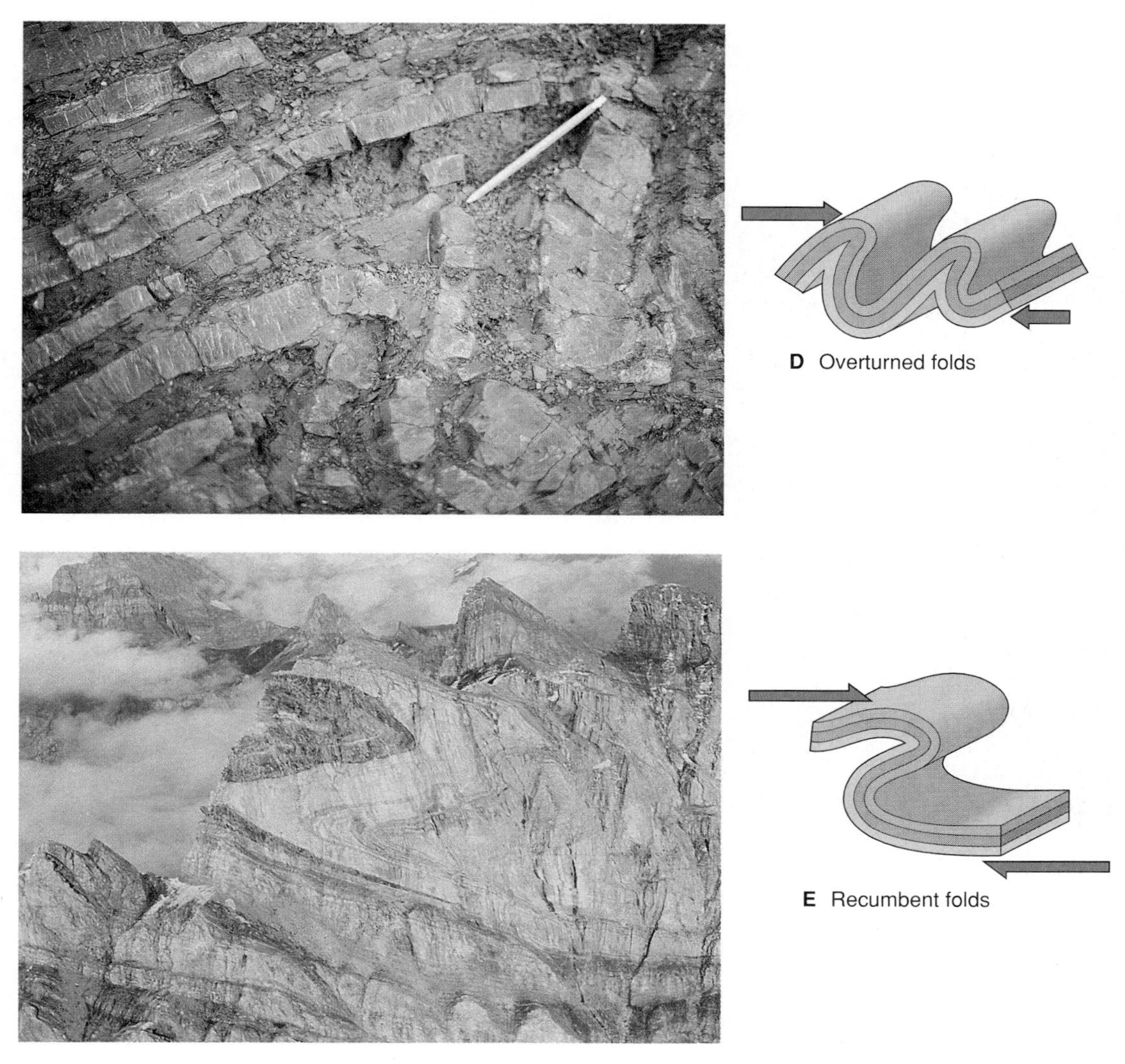

FIGURE 6.17 (CONTINUED) 

(*D*) Overturned anticline from northern California. (*E*) Recumbent folds in the Alps.

Photo *D* by Diane Carlson; photo *E* courtesy of Professor John Ramsay, from *The Techniques of Modern Structural Geology, vol. 2,* J. G. Ramsay & M. J. Huber, © 1987 Academic Press

folds are found in the cores of mountain ranges such as the Canadian Rockies, Alps, and Himalaya and indicate that compressive and/or shear stresses were more intense in one direction, and they probably record shortening of the crust associated with plate convergence.

FRACTURES IN ROCK

If a rock is brittle, or if the strain rate is too rapid for deformation to be accommodated by plastic behavior, the rock fractures. Commonly there is some movement or displacement. If essentially no displacement occurs, a fracture or crack in bedrock is called a **joint.** If the rock on either side of a fracture moves, the fracture is a *fault* (as defined earlier). Most rock at or near the surface is brittle, so nearly all exposed bedrock is jointed to some extent.

Joints

In discussing volcanoes, we describe *columnar jointing,* in which hexagonal columns form as the result of tension and contraction of a cooling, solidified lava flow. *Sheet jointing,* a type of jointing due to expansion (discussed along with weathering in chapter 12) is caused by tension. The pressure release due to removal of overlying rock has the effect of creating tensional stress perpendicular to the land surface.

Columnar and sheet joints are examples of fractures that form from nontectonic stresses and are therefore referred to as primary joints. In this chapter we are concerned with joints that form not from cooling or unloading but from tectonic stresses.

Joints are one of the most commonly observed structures in rocks (figure 6.18). Where joints are oriented approximately parallel to one another, a **joint set** can be defined. Joints are important because they often indicate the direction of compressive stress

FIGURE 6.18

Vertical joints in sedimentary rock of the Colorado Plateau formed in response to tectonic uplift of the region.

Photo by Frank M. Hanna

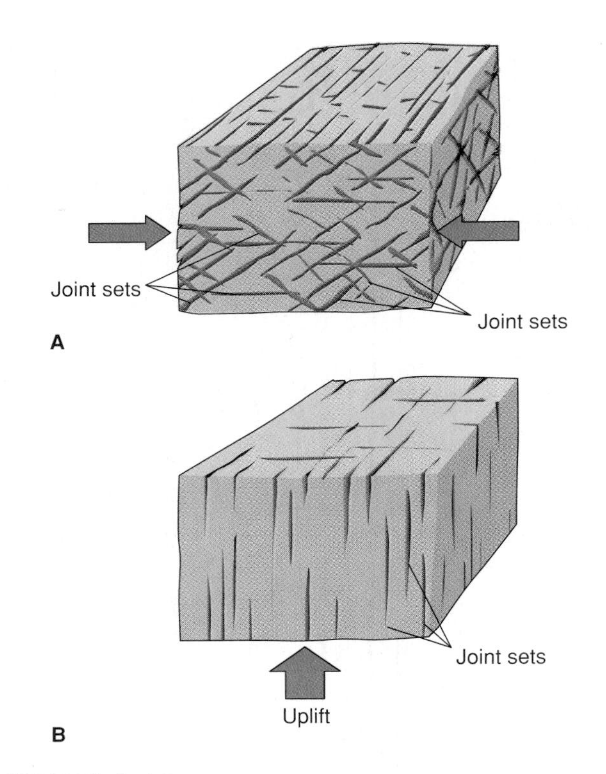

FIGURE 6.19

Joint sets. Arrows indicate directions of principal stress. (*A*) Three joint sets associated with a horizontal compressional stress. (*B*) Two joint sets associated with uplift and a vertical compressional stress.

operative during their formation. For example, in figure 6.19*A* three sets of joints have formed in response to a horizontal compressive stress. Vertical joint sets (figure 6.19*B*) are often associated with tectonic uplift of a region.

Geologists sometimes find valuable ore deposits by studying a joint system. For example, gold-bearing hydrothermal solutions may migrate upward through a set of joints and deposit quartz and gold in the cracks. Accurate information about joints also is important in the planning and construction of large engineering projects, particularly dams and reservoirs. If the bedrock at a proposed location is intensely jointed, the possibility of dam failure or reservoir leakage may make that site too hazardous.

Faults

Faults were defined earlier as fractures in bedrock along which movement has taken place. The displacement may be only several centimeters or may involve hundreds of kilometers. For many geologists, an active fault is regarded as one along which movement has taken place during the last 11,000 years. Most faults, however, are no longer active.

The nature of past movement ordinarily can be discerned where a fault is exposed in an outcrop (figure 6.20). The

FIGURE 6.20

Fault in Big Horn Mountains, Wyoming, is marked by broken, red-stained rocks and by displaced rock layers.

Photo by Diane Carlson

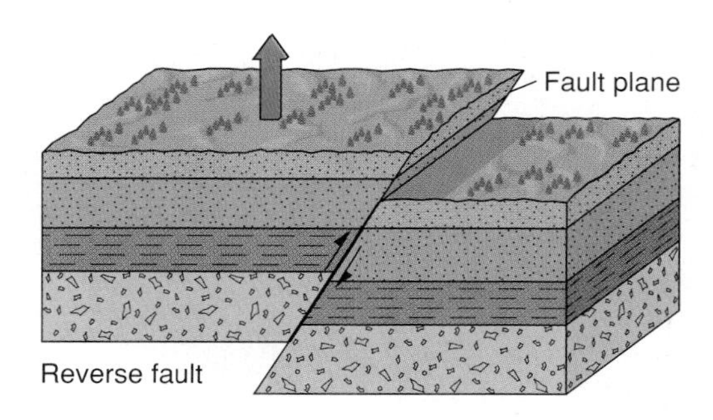

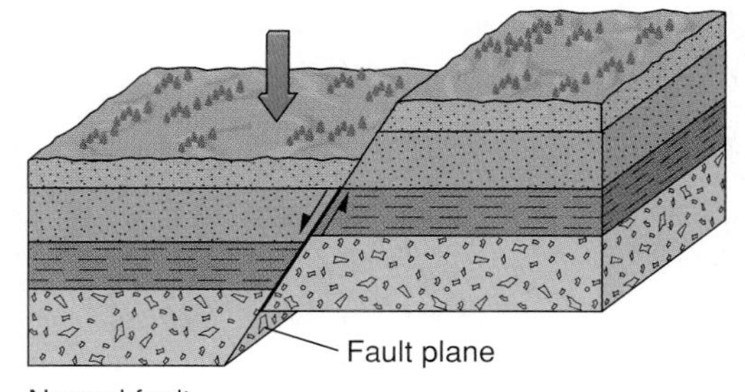

A Dip-slip faults

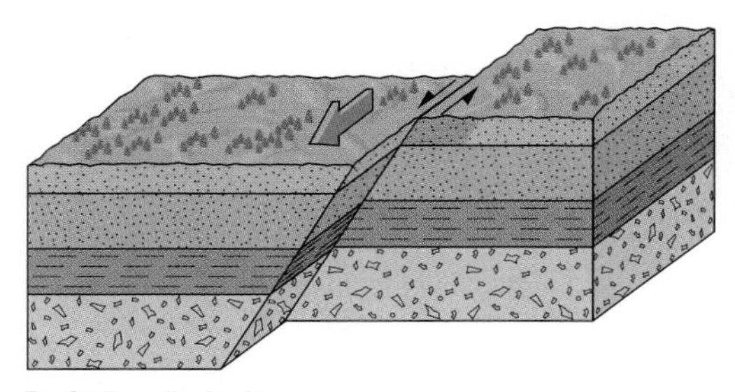

B Strike-slip faults

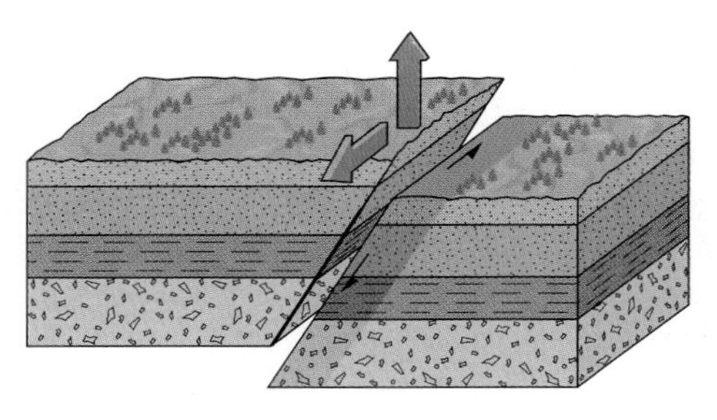

C Oblique-slip fault

FIGURE 6.21

Three types of faults illustrated by displaced blocks. Heavier arrows show direction in which block to the left moved. (*A*) Dip-slip movement. (*B*) Strike-slip movement. (*C*) Oblique-slip movement. Black arrows show dip-slip and strike-slip components of movement.

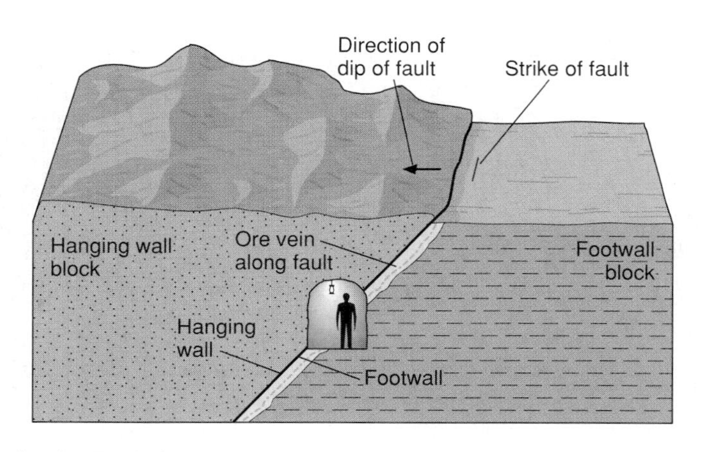

FIGURE 6.22

Relationship between the hanging wall block and footwall block of a fault.

geologist looks for dislocated beds or other features of the rock that might show how much displacement has occurred and the relative direction of movement. In some faults the contact between the two displaced sides is very narrow. In others the rock has been broken or ground to a fractured or pulverized mass sandwiched between the displaced sides.

Geologists describe fault movement in terms of direction of slippage: dip-slip, strike-slip, or oblique-slip (figure 6.21). In a **dip-slip fault,** movement is parallel to the dip of the fault surface. A **strike-slip fault** indicates *horizontal* motion parallel to the strike of the fault surface. An **oblique-slip fault** has both strike-slip and dip-slip components.

Dip-Slip Faults

Normal and reverse faults, the most common types of dip-slip faults, are distinguished from each other on the basis of the relative movement of the *footwall block* and the *hanging-wall block.* The **footwall** is the underlying surface of an inclined fault plane, whereas the overlying surface is the **hanging wall.** These old mining terms brought into geology are illustrated in figure 6.22. If a miner is tunneling along the strike of a fault, his feet are on the footwall and his lantern is hung on the hanging wall.

In a **normal fault** (figures 6.23 and 6.24), the hanging-wall block has moved downward relative to the footwall block. The relative movement is represented on a geological cross section by a pair of arrows, because we cannot generally tell which block actually moved. As shown in figure 6.23, a normal fault results in extension or lengthening of the crust. When there is extension of the crust, the hanging-wall block moves downward along the fault to compensate for the pulling apart of the rocks. Sometimes a block bounded by normal faults will drop down, creating a *graben,* as shown in figure 6.23*C.* (*Graben* is the German word for "ditch.") *Rifts* are grabens associated with diverging plate boundaries, either along mid-oceanic ridges or on continents (see chapters 3 and 4).

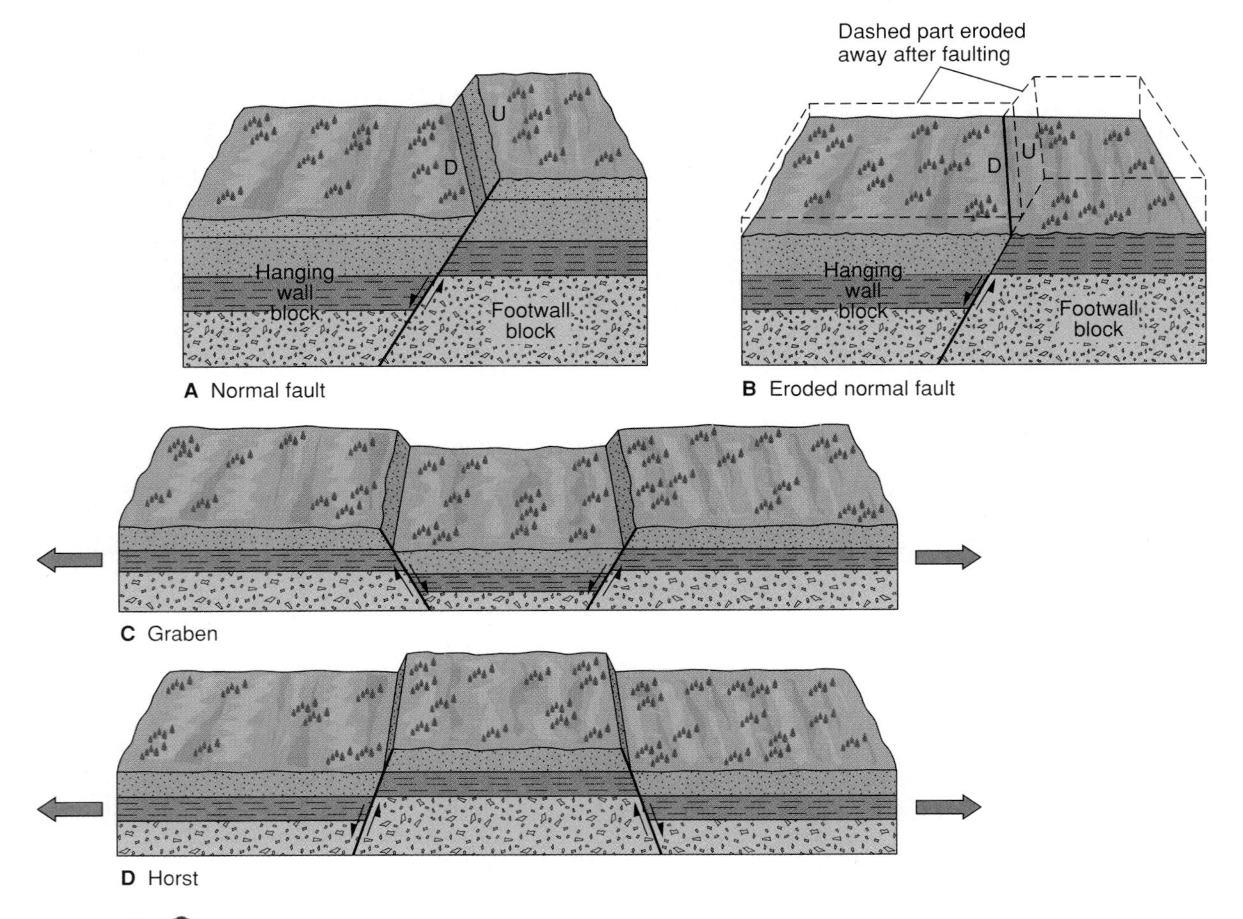

FIGURE 6.23

Normal faults. (*A*) Diagram shows the fault before erosion and the geometric relationships of the fault. (*B*) The same area after erosion. (*C*) A graben. (*D*) A horst. Arrows in *C* and *D* indicate horizontal extension of the crust.

FIGURE 6.24

Normal faults with prominent horst block offsets volcanic ash layers in southern Oregon.

Photo by Diane Carlson

If a block bounded by normal faults is uplifted sufficiently, it becomes a fault-block mountain range. (This is also called a *horst,* the opposite of a graben.) The Basin and Range province of Nevada and portions of adjoining states is characterized by numerous mountain ranges separated from adjoining valleys by normal faults.

In a **reverse fault,** the hanging-wall block has moved upward relative to the footwall block (figures 6.25 and 6.26). As shown in figure 6.25, horizontal compressive stresses cause reverse faults. Reverse faults tend to shorten the crust.

A **thrust fault** is a reverse fault in which the dip of the fault plane is at a low angle to horizontal (figures 6.25*C* and 6.27). In some mountain regions it is not uncommon for the upper plate (or hanging-wall block) of a thrust fault to have overridden the lower plate (footwall block) for several tens of kilometers. Thrust faults typically move or thrust older rocks on top of younger rocks (figure 6.27), and result in an extreme shortening of the crust. Thrust faults commonly form at convergent plate boundaries to accommodate shortening during collision.

IN GREATER DEPTH 6.1

Structures Associated with Salt Domes

Most folding is the product of horizontal compression; what we would expect if rock layers were placed in a giant vise. However, we get vertical compressive stresses when bodies of salt rise through the crust. Domes and normal faults are the characteristic structures that form above the rising bodies of salt.

Blobs (more correctly *diapirs*) of rock salt rise through layers of sedimentary rock. Sometimes these bodies, which are a kilometer (or more) wide, travel upward as much as 10 kilometers. They originate from buried layers of evaporites. Because rock salt is easily deformed plastically, it squeezes out of the original beds. It works its way upward because salt is less dense than the overlying rocks (box figure 1). Where it breaks through rock layers, those layers are upturned adjacent to the rock salt (box figure 2). The rising salt pushes the overlying rock into a dome or, less commonly, an anticline. The compressive stresses caused by the rising salt diapirs may also create normal faults that trap oil (box figure 2).

Many important oil fields along the Gulf coast of Texas and Louisiana are located at salt domes. A few salt-created anticlines are also found in the Colorado plateau.

If the rock salt breaks through the Earth's surface in a dry climate, it will ooze very slowly over the land surface and continue to move, like a glacier, but much more slowly. Layering within the mass is highly contorted from the squeezing that takes place (box figure 3).

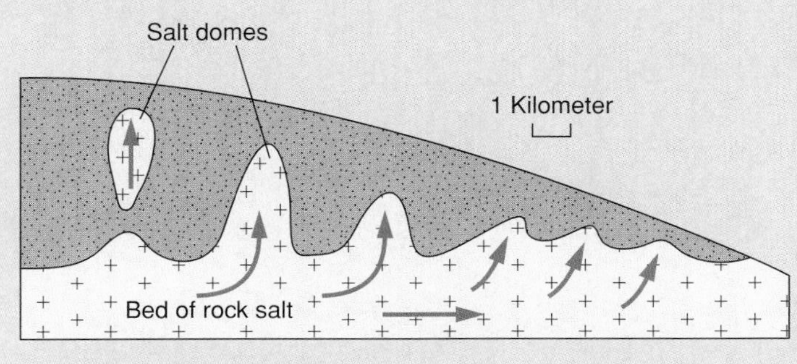

Box 6.1 — Figure 1

Salt domes may form as a bed of rock salt is loaded unevenly by a thick wedge of sediment. The salt flows toward the thin part of the sediment wedge and also flows upward.

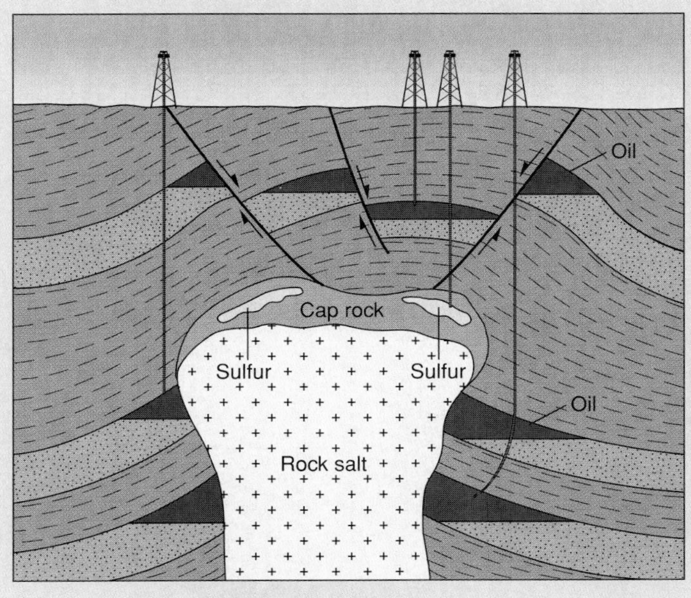

Box 6.1 — Figure 2

A salt dome. Oil and gas are trapped in folds and along faults above the dome and within upturned sandstones along the flanks of the dome. Insoluble cap rock may contain recoverable sulfur.

Related Web Resource

Visit the Applied Geodynamics Laboratory website at the University of Texas–Austin for interesting three-dimensional images and animation of salt domes:

www.utexas.edu/research/beg/giovanni/

Box 6.1 — Figure 3

Rock salt body in Spain.

Photo by C. C. Plummer

6.2

IN GREATER DEPTH

Is There Oil Beneath My Property—First Check the Geologic Structure

An "oil pool" can exist only under certain conditions. Crude oil does not fill caves underground as the term *pool* may suggest; rather, it simply occupies the pore spaces of certain sedimentary rocks, such as poorly cemented sandstone, in which void space exists between grains. Natural gas (less dense) often occupies the pore spaces above the crude oil, while water (more dense) is generally found saturating the rock below the oil pool (box figure 1).

A **source rock,** which is always a sedimentary rock, must be present for oil to form. The sediment of the source rock has to include remains of organisms buried during sedimentation. This organic matter partially decomposes into petroleum and natural gas. Once formed, the droplets of petroleum tend to migrate, following fractures and interconnecting pore spaces. Being less dense than the rock, the petroleum usually migrates upward, although horizontal migration does occur.

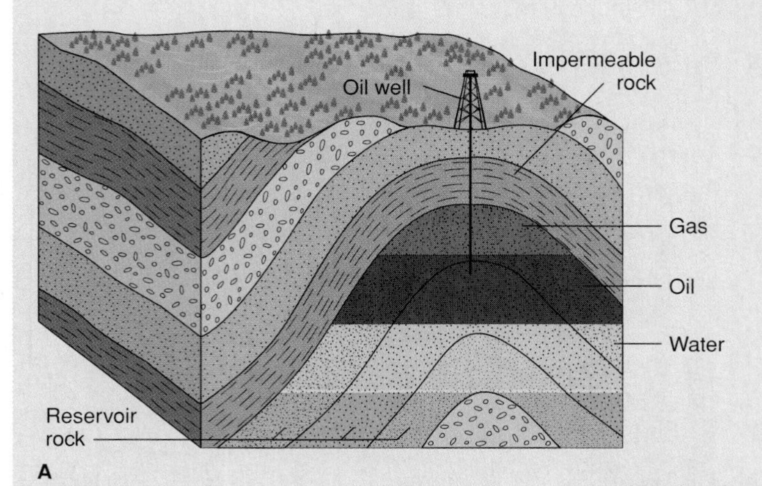

Box 6.2 — FIGURE 1

(*A*) Oil and gas are concentrated or trapped in hinge of anticline. Gas and oil float on water in porous and permeable reservoir rock (sandstone). (*B*) Eroded anticline forms trap in Lander oil field, Wyoming.

Photo *B* by Diane Carlson

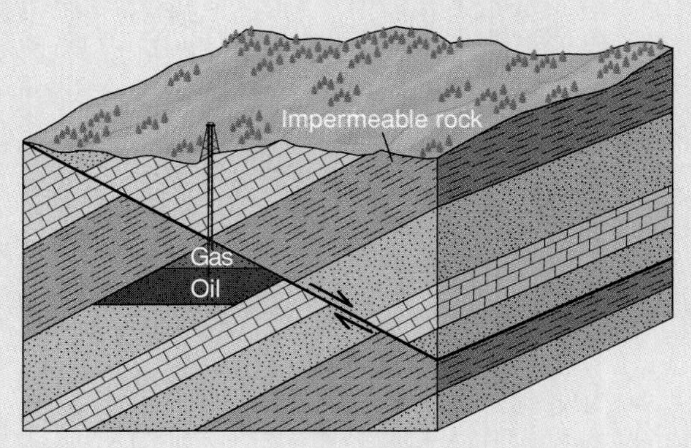

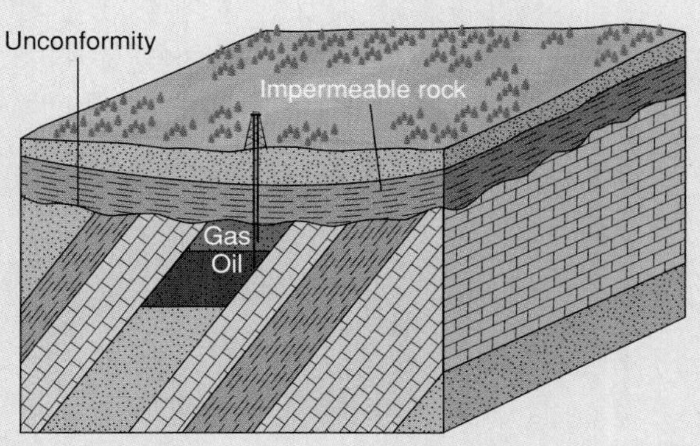

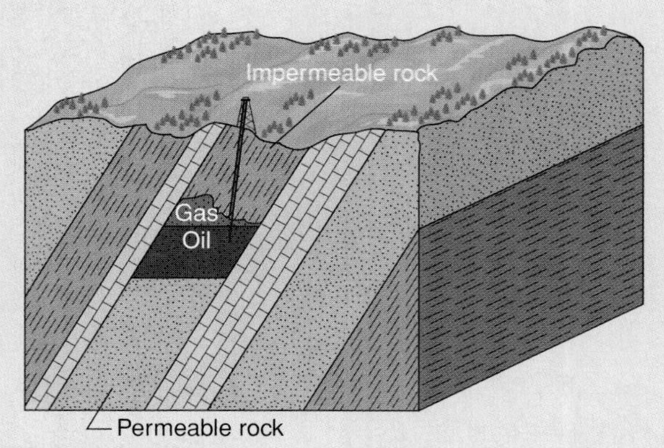

Box 6.2 — FIGURE 2

Structures other than anticlines that trap oil.

If it is not blocked by impermeable rock, the oil may migrate all the way to the surface, where it is dissipated and permanently lost. Natural oil seeps, where leakage of petroleum is taking place, exist both on land and offshore. Where impermeable rock blocks the oil droplets' path of migration, an oil pool may accumulate below the rock, much like helium-filled balloons might collect under a domed ceiling. For any significant amount of oil to collect, the rock below the impermeable rock must be porous as well as permeable. Such a rock, when it contains oil, is called a **reservoir rock.**

Another necessary condition is that the geologic structure must be one that favors the accumulation and retention of petroleum. An "anticlinal trap" is one of the best structures for holding oil. As oil became a major energy source and the demand for it increased, most of the newly discovered wells penetrated anticlinal traps. Geologists discovered these by looking for indication of anticlines exposed at the surface. As time went by, other types of structures were also found to be oil traps. Many of these were difficult to find because of the lack of telltale surface patterns indicating favorable underground structures. Box figure 2 illustrates some structures other than anticlinal traps that might have a potential for oil production.

At present, oil companies rely on detailed and sophisticated geologic studies of an area they hope may have the potential for an "oil strike." The petroleum industry also depends heavily on geophysical techniques (see chapter 2) for determining, by indirect means, the subsurface structural geology.

Even when everything indicates that conditions are excellent for oil to be present underground, there is no guarantee that oil will be found. Eventually an oil company must commit a million dollars or more to drill a deep test well, or "wildcat" well. Statistics indicate that the chance of a test well yielding commercial quantities of oil is much less than 1 in 10. As more and more of the world's supply of petroleum is used up, what is left becomes increasingly harder—and costlier—to find.

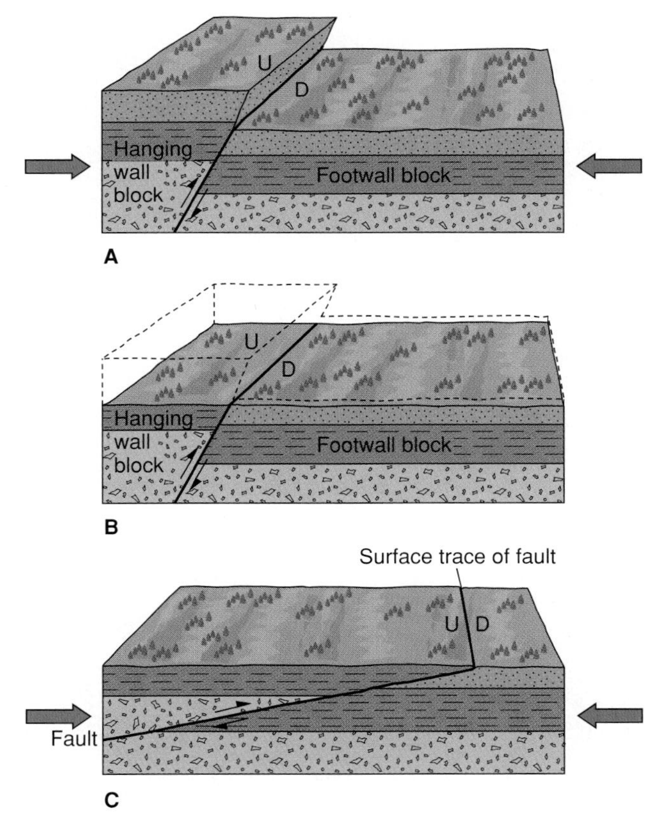

FIGURE 6.25 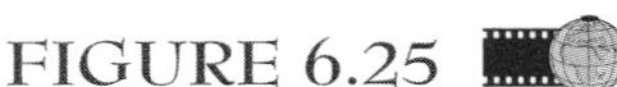

(*A*) A reverse fault. The fault is unaffected by erosion. Arrows indicate compressive stress. (*B*) Diagram shows area after erosion; dashed lines indicate portion eroded away. (*C*) Thrust fault due to horizontal compression.

FIGURE 6.26

Reverse fault in volcanic ash beds, southern Oregon.

Photo by Diane Carlson

A

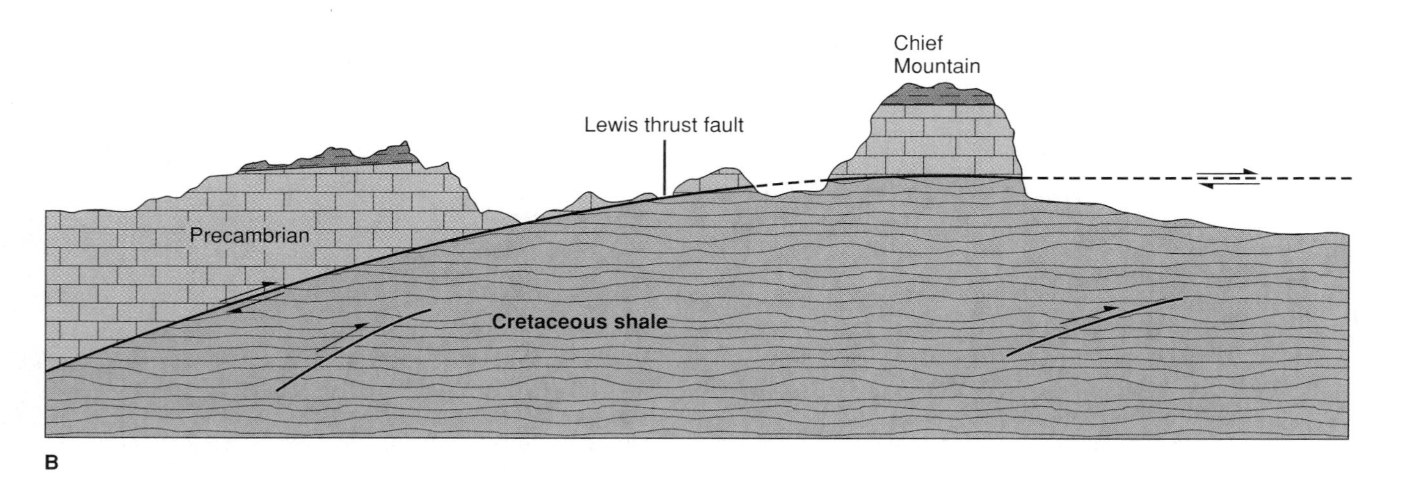

B

FIGURE 6.27
(*A*) Chief Mountain in Glacier National Park, Montana, is an erosional remnant of a major thrust fault. (*B*) Cross section of the area. Older (Precambrian) rocks have been thrust over younger (Cretaceous) rocks. Dashed lines show where the Lewis thrust fault has been eroded away.
Photo by Frank M. Hanna

Strike-Slip Faults

A fault where the movement (or *slip*) is predominantly horizontal and parallel to the strike of the fault, is called a **strike-slip fault.** The displacement along a strike-slip fault is either left-lateral or right-lateral and can be determined by looking across the fault. For instance, if a recent fault displaced a stream (figure 6.28), a person walking along the stream would stop where it is truncated by the fault. If the person looks across the fault and sees the displaced stream to the right, it is a **right-lateral fault.** In a **left-lateral fault,** a stream or other displaced feature would appear to the left across the fault. Again, we cannot tell which side actually moved, so pairs of arrows are used to indicate relative movement.

Large strike-slip faults, like the San Andreas fault in California, typically define a zone of faulting that may be several kilometers wide and hundreds of kilometers long (see box 6.3). The surface trace of an active strike-slip fault is usually defined by a prominent linear valley that has been more easily eroded

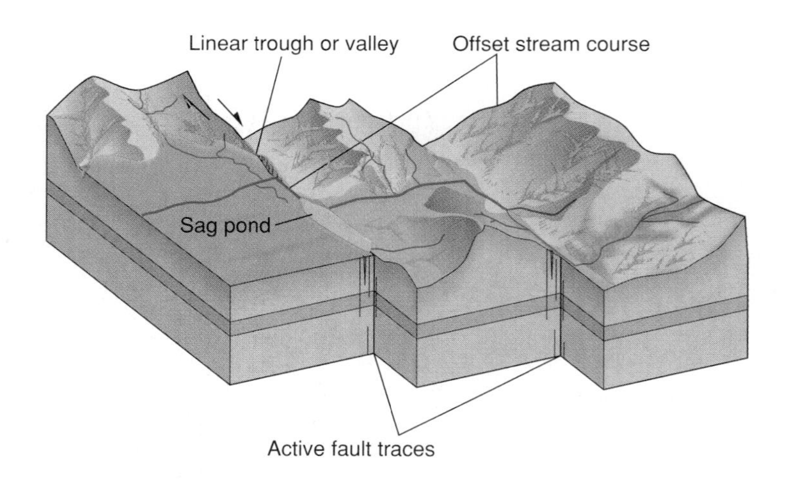

FIGURE 6.28
Right-lateral strike-slip fault showing offset stream channels, sag ponds, and linear valleys and ridges.

IN GREATER DEPTH 6.3

California's Greatest Fault—The San Andreas

The San Andreas fault in California is the best known geologic structure in the United States, but the geologists and seismologists who study it admit that our knowledge of its activity and its history is far from complete. Actually, the San Andreas is the longest of several, subparallel faults that transect western California (box figure 1). Collectively, these right-lateral faults are known as the San Andreas fault system. The system is in a belt approximately 100 kilometers wide and 1,300 kilometers long that extends into Mexico, ending at the Gulf of California.

Los Angeles is slowly moving toward San Francisco because of San Andreas fault motion. At an average rate of movement of about 2 centimeters per year, Los Angeles could be a western suburb of San Francisco (or San Francisco an eastern suburb of Los Angeles) in some 25 million years. Earthquakes are produced by sudden movement within the fault system as explained in chapter 7. Bedrock along the San Andreas fault shifted as much as 4.5 meters in association with the 1906 quake that destroyed much of San Francisco.

The San Andreas fault is not a simple crack, but a belt of broken and ground-up rock, usually a hundred meters or more wide. Its presence is easy to determine throughout most of its length. Along the fault trace are long, straight valleys (formed by erosion and subsidence) that show quite different terrain on either side. Stream channels follow much of the fault zone because the weak, ground-up material along the fault is easily eroded. Locally, elongate lakes (called sag ponds) are found where the ground-up material has settled more than the surface of adjacent parts of the fault zone. The fault was named after one of these ponds, San Andreas Lake, just south of San Francisco (box figure 2).

One can visually follow the fault northward from San Andreas Lake into the southwestern suburbs of San Francisco. There the fault zone is hidden by recently built housing tracts. Apparently the builders and residents have chosen to ignore the hazards of living on the nation's most famous fault.

Geologists have been unable to agree on the total displacement of the fault or on how long it has been active. Some believe movement began in the Mesozoic Era (over 65 million years ago); most think that it began later. The difficulty in establishing an age for the inception of the faulting lies in finding clear evidence of displaced bedrock. What geologists would like to find, if it exists, is a rock unit that can be isotopically dated and that was formed across the fault zone just before the faulting began. Now displaced rock on both sides of the fault zone would have to be clearly identifiable as having been the same unit.

Geologically young features that cross the fault, such as displaced stream channels (box figure 3), are common. Similarly, ancient rocks that undoubtedly were there before faulting began are recognized as having been displaced. Many California geologists believe that the belt of granitic rock just west of the fault was once the southern continuation of the granitic batholiths of the Sierra Nevada (box figure 4*A*), which are more than 80 million

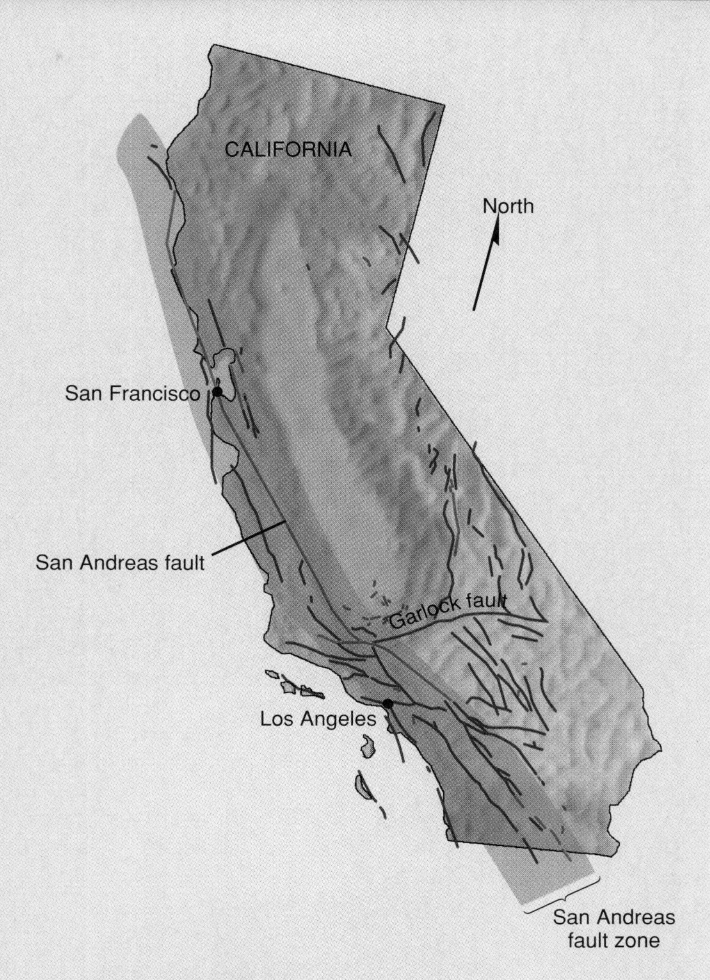

BOX 6.3 — FIGURE 1

California has its faults. Red lines indicate faults that have been active within the last 200 years and blue lines indicate faults that have been active over the last 2 million years.

From California Division of Mines and Geology

years old. But these extremes tell us only that the age of the San Andreas is somewhere between approximately 80 million years and a few thousand years, when the stream channel in box figure 3 carved its course across the fault.

The strongest evidence for long-term faulting comes from almost identical volcanic sequences now 315 kilometers apart. The volcanic activity took place 23.5 million years ago. Using these figures, we can calculate the average rate of motion as 1.3 centimeters per year for the San Andreas fault. (But movement along other faults means the rate of motion is higher for the fault system.) Older rocks that appear to have been offset 560 kilometers have been correlated with less certainty, suggesting that the total offset for the San Andreas fault is at least 560 kilometers.

BOX 6.3 — FIGURE 2

Part of the San Andreas fault. View northward toward San Francisco. Lakes occupy the fault zone. Hills to the left of the fault are moving northward.
Photo by B. Amundson

BOX 6.3 — FIGURE 3

Stream channel displaced by the San Andreas fault. The arrows on either side of the fault trace indicate relative motion.
Photo by C. C. Plummer

How long ago faulting began remains controversial. According to plate tectonic theory, the San Andreas fault is a transform boundary that separates the North American plate from the Pacific plate. One hypothesis places the beginning of strike-slip movement at about 30 million years ago. According to this hypothesis, the Baja California peninsula split from mainland Mexico as sea-floor spreading began (box figure 4*B*). As the Gulf of California widens, the block of crust west of the San Andreas is pushed northward.

Additional Resources

Wallace, R. E., ed. 1990. *The San Andreas fault system,* California: U.S. Geological Survey Professional Paper 1515.

Related Web Resource

Internet version of the San Andreas fault system:
http://pubs.usgs.gov/gip/earthq3/
http://quake.usgs.gov/info/1906/index.html
"What have we learned about the San Andreas fault since 1906" web page shows research sites along the San Andreas Fault.

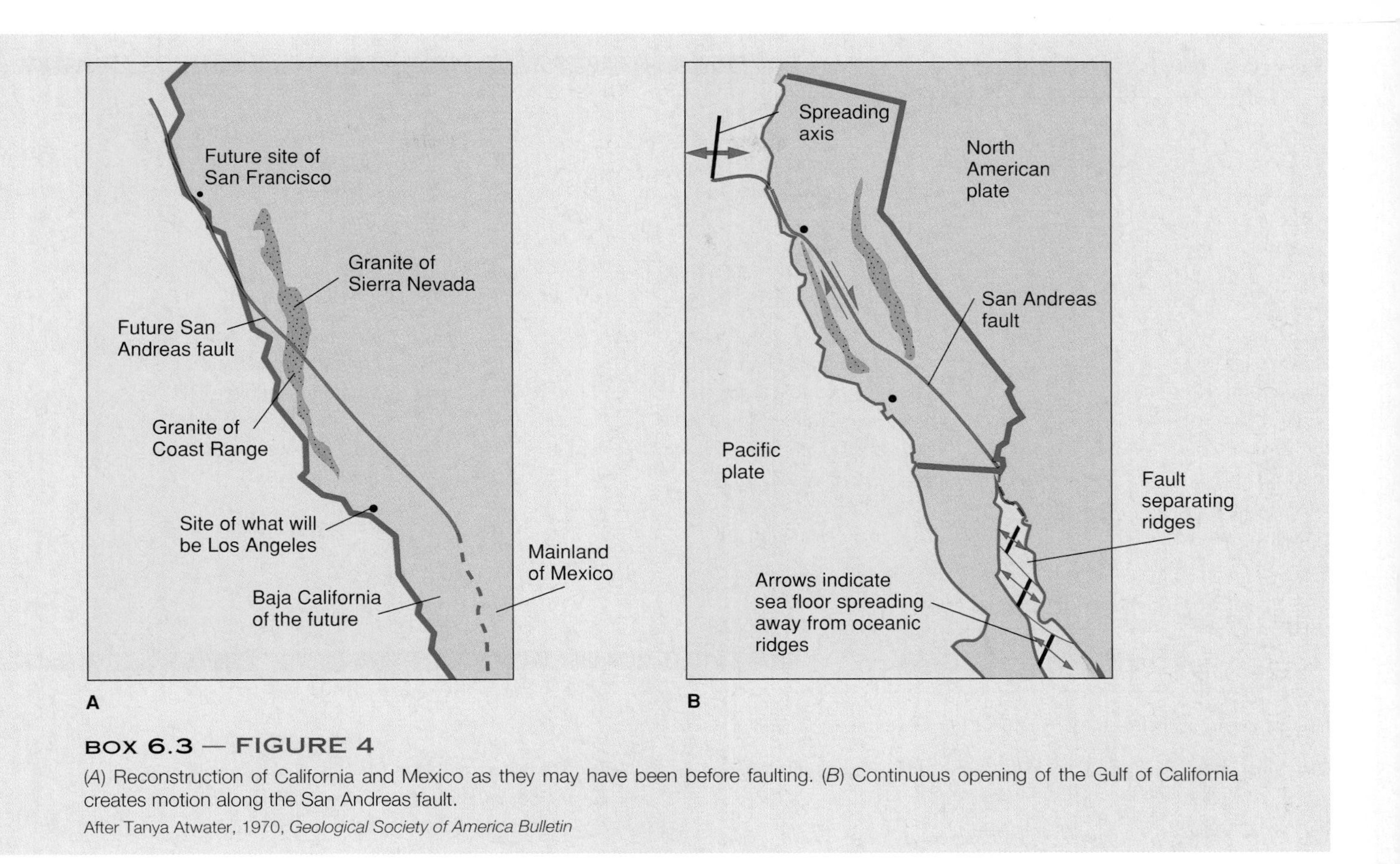

BOX 6.3 — FIGURE 4

(*A*) Reconstruction of California and Mexico as they may have been before faulting. (*B*) Continuous opening of the Gulf of California creates motion along the San Andreas fault.

After Tanya Atwater, 1970, *Geological Society of America Bulletin*

where the rock has been ground up along the fault during movement. The linear valley may contain lakes or sag ponds (figure 6.28) where the highly permeable fault rock allows ground water to freely flow to the surface. The trace of the fault may also be marked by offset surface features such as streams, fences, and roads or by distinctive rock units.

Strike-slip faults accommodate shearing stress along transform plate boundaries where plates slide past one another. One of the most famous examples of a transform fault is the San Andreas Fault. The San Andreas fault is a right-lateral strike-slip fault that forms the boundary between the North American and Pacific plates (see box 6.3 figure 4).

Summary

Tectonic forces result in deformation of the Earth's crust. *Stress* (force per unit area) is a measure of the tectonic force and confining pressure acting on bedrock. Stress can be *compressive, tensional,* or *shearing. Strained* (changed in size or shape) rock records past stresses, usually as joints, faults, or folds.

A geologic map shows the structural characteristics of a region. *Strike* and *dip* symbols on geologic maps indicate the attitudes of inclined surfaces such as bedding planes. The strike and dip of a bedding surface indicate the relationship between the inclined plane and a horizontal plane.

If rock layers bend (plastic behavior) rather than break, they become folded. Rock layers are folded into *anticlines* and *synclines* and recumbent folds. If the hinge line of a fold is not horizontal, the fold is *plunging.* Older beds exposed in the core of a fold indicate an anticline, whereas younger beds in the center of the structure indicate a syncline. In places where folded rock has been eroded to a plain, an anticline can usually be distinguished from a syncline by whether the beds dip toward the center (syncline) or away from the center (anticline).

Fractures in rock are either *joints* or *faults.* A joint indicates that movement has not occurred on either side of the fracture; displaced rock along a fracture indicates a fault. *Dip-slip* faults are either *normal* or *reverse,* depending on the motion of the hanging wall block relative to the footwall block. A reverse fault with a low angle of dip for the fault plane is a *thrust fault.* Reverse faults accommodate horizontal shortening of the crust whereas normal faults accommodate horizontal stretching or extension.

In a *strike-slip* fault, which can be either left-lateral or right-lateral, horizontal movement has occurred.

Terms to Remember

angle of dip 135
anticline 137
axial plane 137
brittle 133
compressive stress 133
dip-slip fault 143
direction of dip 135
ductile 133
elastic 133
elastic limit 133
fault 134
fold 137
footwall 143
geologic cross section 136
geologic map 135
hanging wall 143
hinge line 137
isoclinal fold 140
joint 141
joint set 141
left-lateral fault 148
limb 137
normal fault 143
oblique-slip fault 143
open fold 138
overturned fold 140
plunging fold 138
recumbent fold 140
reservoir rock 147
reverse fault 144
right-lateral fault 148
shear stress 133
source rock 146
strain 132
stress 132
strike 135
strike-slip fault 148
structural basin 138
structural dome 138
structural geology 132
syncline 131
tensional stress 133
thrust fault 144

Testing Your Knowledge

Use the questions below to prepare for exams based on this chapter.

1. Most anticlines have both limbs dipping away from their hinge lines. For which kind of fold is this not the case?
2. What are the four main types of contacts and how would you distinguish between them if you were a geologist doing field work?
3. On a geologic map, if no cross sections were available, how could you distinguish an anticline from a syncline?
4. If you locate a dip-slip fault while doing field work, what kind of evidence would you look for to determine whether the fault is normal or reverse?
5. Name several geologic structures described in earlier chapters.
6. What is the difference between strike, direction of dip, and angle of dip?
7. Draw a simple geologic map, using strike and dip symbols for a syncline plunging to the west.
8. How does a structural dome differ from a plunging anticline?
9. Which of the statements is true?
   a. when forces are applied to an object, the object is under stress
   b. strain is the change in size (volume) or shape, or both, while an object is undergoing stress
   c. stresses can be compressive, tensional, or shear
   d. all of the above
10. The compass direction of a line formed by the intersection of an inclined plane with a horizontal plane is called
    a. strike b. direction of dip
    c. angle of dip
11. Folds in a rock show that the rock behaved in
    a. a ductile way b. an elastic way
    c. a brittle way d. all of the above
12. An anticline is
    a. any fold
    b. an overturned fold
    c. an upward-arched fold
    d. a downward-arched fold
13. A syncline is
    a. an upward-arched fold
    b. an overturned fold
    c. a downward-arched fold
    d. horizontal beds
14. A structure in which the beds dip away from a central point is called a
    a. basin b. anticline
    c. structural dome d. syncline
15. Which is not a type of fold?
    a. open b. isoclinal
    c. overturned d. recumbent
    e. thrust
16. Fractures in bedrock along which movement has taken place are called
    a. joints b. faults
    c. cracks d. crevasses
17. In a normal fault, the hanging-wall block has moved ____ relative to the footwall block.
    a. upward b. downward
    c. sideways

18. Normal faults accommodate what kind of strain?
    a. shortening
    b. extensional
    c. ductile

19. Faults that typically move older rock on top of younger rock are
    a. normal faults
    b. thrust faults
    c. strike-slip faults

Expanding Your Knowledge

1. In what parts of North America would you expect to find the most intensely folded rock?
2. A subduction zone can be regarded as a very large example of what type of fault?
3. Why do some horizontal compressive stresses cause thrust faults while others cause strike-slip faults?
4. Can you identify and name the various geologic structures shown in the figures in chapter 8?

Exploring Web Resources

www.mhhe.com/plummer9e

Visit our Online Learning Center for additional readings and media resources. Check your Testing Your Knowledge answers, and click on the links to go directly to the sites listed below.

http://geology.ou.edu/~ksmart/structure_webpage/

Structural geology site maintained by Kevin J. Smart at University of Oklahoma contains many links to online courses, computer software, bibliographies, and research projects dealing with structural geology.

www.geo.cornell.edu/geology/classes/geol326/326.html

Website for structural geology course taught by the Department of Geological Sciences at Cornell University contains images showing structural features and models of thrust-fault movement.

http://craton.geol.brocku.ca/ctg.html

Canadian Tectonics Group website contains structural geology images, computer software, and a newsletter outlining research projects.

Animations

This chapter includes the following animations available on our Online Learning Center at www.mhhe.com/plummer9e.

6.17 Types of Folds
6.21 Types of Fault Movement
6.23 Normal Faults
6.25 Reverse and Thrust Faults

CHAPTER

7

Earthquakes

This chapter will help you understand the nature and origin of earthquakes. We discuss the seismic waves created by earthquakes and how the quakes are measured and located by studying these waves. We also describe some effects of earthquakes, such as ground motion and displacement, damage to buildings, and quake-caused fires, landslides, and tsunamis.

Earthquakes are largely confined to a few narrow belts on Earth. This distribution was once puzzling to geologists, but here we show how the concept of plate tectonics neatly explains it.

As geologists learn more about earthquake behavior, there is the possibility that we will be able to forecast earthquakes. We conclude the chapter with a look at this developing branch of Earth study.

Opposite: The bottom floors of this high-rise building collapsed, causing it to fall over during a strong, horizontal jolt during the September 1999 earthquake in Taiwan. The magnitude 7.6 earthquake occurred along an east-dipping thrust fault and was the largest to hit central Taiwan since 1600. Damage exceeded $14 billion, and left 2,131 people dead, 8,137 injured, and 600,000 homeless.

Photo © Chan Cheung-On/Liaison Agency

On April 18, 1906, at 5:12 in the morning, part of California slid abruptly past the rest of the state during a great earthquake. A visible scar 450 kilometers long was left where the Earth was torn along coastal northern California. Rock was displaced horizontally as much as 4.5 meters; soil above the rock was displaced up to 6.5 meters. The quake, located on a segment of the San Andreas fault near San Francisco, shook the ground for one full minute. Buildings toppled in San Francisco, and broken gas mains fed fires that raged for three days (figure 7.1*A*). Broken water mains hampered fire fighting. The fires were finally extinguished when buildings were dynamited to create a firebreak. Terrified and homeless people moved to refugee camps set up in city parks. Looters were shot on sight. As the city gradually recovered from the shock of the devastation, it was found that at least 3,000 people had died and $400 million (in 1906 dollars) of damage had been done. Perhaps 90% of the destruction was caused by the fires.

At 5:04 P.M. on October 17, 1989, San Francisco was again severely shaken for 15 seconds by the Loma Prieta earthquake located to the south on the San Andreas fault near Santa Cruz. Although the quake did not tear the ground surface, it collapsed some buildings and freeway overpasses built upon the soft "bay fill" sediment in San Francisco and Oakland (figure 7.1*B*). A section of the Bay Bridge collapsed. Just as in 1906, raging fires were fed by broken gas mains in the Marina district of San Francisco and were difficult to fight because of broken water mains; fireboats helped extinguish them. Very severe damage occurred in small towns near the center of the quake. The death toll was 63, and damage was $6 billion.

At 5:30 P.M. on March 27, 1964, southern Alaska was rocked by an earthquake that lasted for 3 minutes. Although the force of this earthquake was twice as strong as the 1906 San Francisco earthquake, loss of life and property was relatively low because of Alaska's small population—15 people died as a direct result of the shaking, and damage amounted to slightly over $300 million (in 1964 dollars). The tremor was felt over an area of more than 1 million square kilometers (350,000 square miles). A section of the Earth's surface 50 by 200 kilometers was raised as much as 13 meters, and a similar block of land sank 1 to 2 meters. Horizontal movement was slight. In Anchorage, 150 kilometers from the center of the earthquake, landslides wrecked parts of the city. The greatest loss of life was caused by large sea waves (or tsunami) generated by land movement associated with the earthquake—almost 100 people drowned in Alaska (figure 7.1*C*), and a few people drowned as far away as Oregon and northern California, as the waves spread over the Pacific Ocean.

On January 17, 1994, at 4:31 A.M., the Northridge earthquake rocked the San Fernando Valley just north of Los Angeles, California, for 40 seconds. The quake, about 3 kilometers from California State University, Northridge, damaged or destroyed all 53 CSUN buildings and seriously damaged 300 other schools. Numerous freeway overpasses collapsed (including some that had previously collapsed in a nearby 1971 quake), closing four Interstates and seven other highways for months. The two upper stories of the Northridge Meadows Apartments collapsed onto the lower story, killing 16 people (figure 7.1*D*). A 3-story concrete parking garage at a shopping center pancaked down, trapping a worker for hours under 3 meters of concrete. Gas and water mains were broken, triggering several hard-to-fight fires; about 100 homes burned at a Sylmar mobile home park. Fifteen thousand newly homeless people had to live in tents, and tens of thousands of people had no water, electricity, or gas for several days. The death toll of 72 was very low because the quake occurred early in the morning on the Martin Luther King holiday, so very few commuters were on the collapsed freeways. Damage exceeded $25 billion.

At 10:54 A.M. on February 28, 2001, Seattle was jolted by the second largest earthquake to strike the state of Washington in recent history. The Nisqually earthquake was centered 65 kilometers southwest of Seattle near the state capital in Olympia, Washington. Because the quake occurred deep within the Earth (49 kilometers) on the downgoing Juan de Fuca plate, shaking was felt over a broad area. Although classified as a strong earthquake, the depth of the quake minimized the intensity of shaking, and damage was restricted to older buildings constructed of unreinforced brick and concrete (figure 7.1*E*) and bridges that had not been seismically retrofitted. No one died as a direct result of the earthquake and injuries numbered around 250. This was not the "big one." There is still the probability that a great earthquake with much larger ground motions could strike the Pacific Northwest.

CAUSES OF EARTHQUAKES

What causes earthquakes? An **earthquake** is a trembling or shaking of the ground caused by the sudden release of energy stored in the rocks beneath Earth's surface. As described in chapter 6, great forces acting deep in the Earth may put a *stress* on the rock, which may bend or change in shape (*strain*). If you bend a stick of wood, your hands put a stress (the force per unit area) on the stick; its bending (a change in shape) is the strain.

Like a bending stick, rock can deform only so far before it breaks. When a rock breaks, waves of energy are released and sent out through the Earth. These are **seismic waves,** the waves of energy produced by an earthquake. It is the seismic waves that cause the ground to tremble and shake during an earthquake.

The sudden release of energy when rock breaks may cause one huge mass of rock to slide past another mass of rock into a different relative position. As you know from chapter 6 the break between the two rock masses is a *fault.* The classic explanation of why earthquakes take place is called the **elastic rebound theory** (figure 7.2). It involves the sudden release of progressively stored strain in rocks, causing movement along a fault. Deep-seated internal forces (*tectonic forces*) act on a mass of rock over many decades. Initially the rock bends but does not break. More and more energy is stored in the rock as the bending becomes more severe. Eventually the energy stored in the rock exceeds the breaking strength of the rock, and the rock

A

D

B

C

E

FIGURE 7.1

Damage from earthquakes in the United States, (*A*) Damaged buildings and fires in San Francisco after 1906 earthquake. (*B*) Collapsed double-decked Cypress freeway in Oakland after the 1989 Loma Prieta earthquake. (*C*) Tsunami damage from the 1964 Alaska earthquake carried a fishing boat inland in Resurrection Bay at Seward. (*D*) The collapse of the lower story of the Northridge Meadows Apartments killed 16 people in the 1994 earthquake. San Fernando Valley, southern California. (*E*) Damage from falling bricks in downtown Seattle after the February 28, 2001, Nisqually earthquake.

Photo *A* © Arnold Genthe CORBIS. Photo *B* © Lloyd Cluff/CORBIS. Photo *C* © National Geophysical Data Center. Photo *D* © Joel Lugavere/Los Angeles Times. Photo *E* © Associated Press/Stevan Morgain

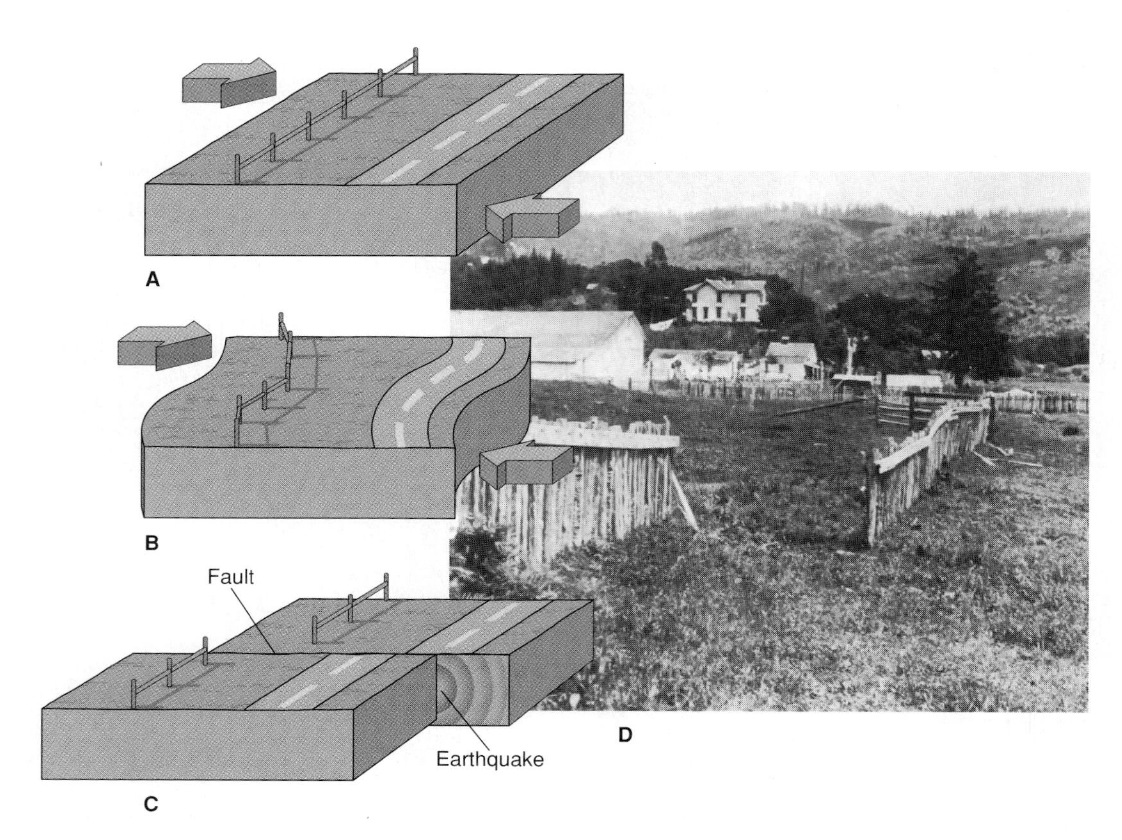

FIGURE 7.2

The elastic rebound theory of the cause of earthquakes. (*A*) Rock with stress acting on it. (*B*) Stress has caused strain in the rock. Strain builds up over a long period of time. (*C*) Rock breaks suddenly, releasing energy, with rock movement along a fault. Horizontal motion is shown; rocks can also move vertically or diagonally. (*D*) Fence offset nearly 3 meters after 1906 San Francisco earthquake.

Photo by G. K. Gilbert, U.S. Geological Survey

FIGURE 7.3

Horizontal offset of rows in a lettuce field, 1979, El Centro, California.

Photo by University of Colorado; courtesy National Geophysical Data Center, Boulder, Colorado

breaks suddenly, causing an earthquake. Two masses of rock move past one another along a fault. The movement may be vertical, horizontal, or both (figure 7.3). The strain on the rock is released; the energy is expended by moving the rock into new positions and by creating seismic waves.

Recently some modifications have been suggested for the sequence of events shown in figure 7.2. The classic model implies that existing faults are strong; a very large stress must act to break rocks along a fault. The new idea is that faults are weak and only need a small stress to cause rupture and an earthquake. The evidence for the new idea is suggestive but not yet conclusive, so we currently have two models for fault behavior. The weak-fault model poses serious problems for earthquake prediction, as you will see later in the chapter.

The brittle behavior of breaking rock is characteristic only of rocks near Earth's surface. Rocks at depth are subject to increased temperature and pressure, which tend to reduce brittleness. Deep rocks deform plastically (*ductile* behavior) instead of breaking (*brittle* behavior); hence, there is a limit to the depth where faults can occur.

Most earthquakes are associated with movement on faults, but in some quakes the connection with faulting may be difficult to establish. Four recent California quakes, including the 1994 Northridge quake, occurred on buried thrust faults, some of which were unknown and none of which involved surface displacement. Most earthquakes in the eastern United States are also not associated with surface displacement. Earthquakes also occur during explosive volcanic eruptions and as magma forcibly fills underground magma chambers prior to many eruptions; these quakes may not be associated with fault movement at all.

Another cause has been recently postulated for deep earthquakes (100 to 670 kilometers below the surface), essentially all of which are found on cold, subducting plates sliding down into the mantle. Although the downgoing plates are colder than the surrounding rock, the high temperature and pressure at depth suggests to some geologists that the rock in the plates should behave plastically rather than breaking in the brittle manner of near-surface rocks. The suggested cause of deep quakes is mineral transformations within the downgoing rock, as pressure collapses one mineral into a denser form. Lab experiments have shown bodies of the new, denser minerals along fractures. Whether the process occurs on a large scale to produce large quakes is unknown. Similar suggestions for the cause of deep quakes include the dehydration of water-containing serpentine and the conversion of serpentine into glass. Both these processes occur suddenly on small fractures in lab experiments.

SEISMIC WAVES

The point within the Earth where seismic waves first originate is called the **focus** (or *hypocenter*) of the earthquakes (figure 7.4). This is the center of the earthquake, the point of initial breakage and movement on a fault. Rupture begins at the focus and then spreads rapidly along the fault plane. The point on the Earth's surface directly above the focus is the **epicenter.**

Two types of seismic waves are generated during earthquakes. **Body waves** are seismic waves that travel through the Earth's interior, spreading outward from the focus in all directions. **Surface waves** are seismic waves that travel on Earth's surface away from the epicenter, like water waves spreading out from a pebble thrown into a pond. Rock movement associated with seismic surface waves dies out with depth into the Earth, just as water movement in ocean waves dies out with depth.

Body Waves

There are two types of body waves, both shown in figure 7.5. A **P wave** is a compressional (or longitudinal) wave in which rock vibrates back and forth *parallel* to the direction of wave propagation. Because it is a very fast wave, traveling through near surface rocks at speeds of 4 to 7 kilometers per second (9,000 to more than 15,000 miles per hour), a P wave is the first (or *primary*) wave to arrive at a recording station following an earthquake.

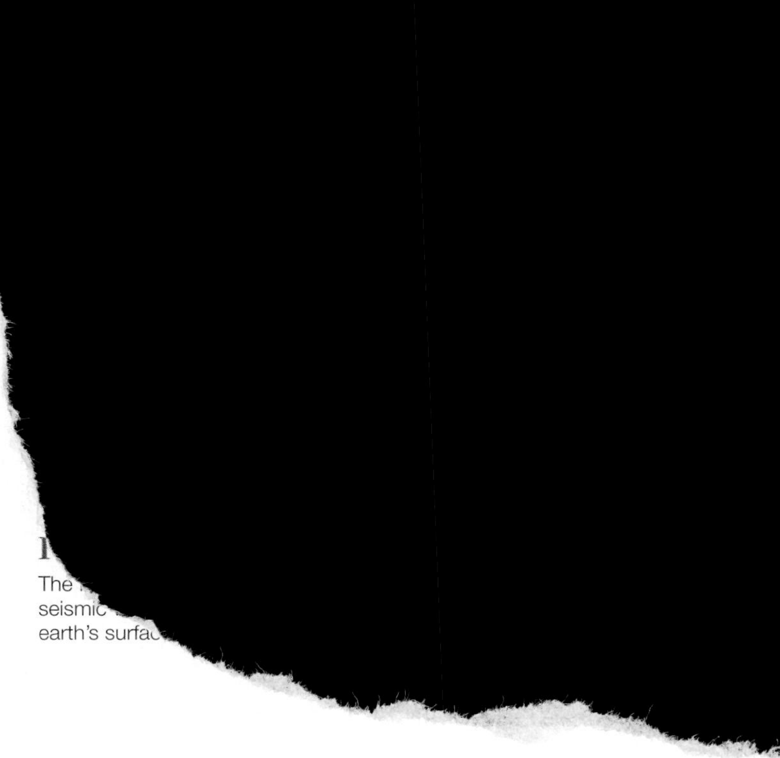

The
seismic
earth's surfac

The second type of body wave is called an **S wave** (*secondary*) and is a slower, transverse wave that travels through near surface rocks at 2 to 5 kilometers per second. An S wave is propagated by a shearing motion much like that in a stretched, shaken rope. The rock vibrates *perpendicular* to the direction of wave propagation, that is, crosswise to the direction the waves are moving.

Both P waves and S waves pass easily through solid rock. A P wave can also pass through a fluid (gas or liquid), but an S wave cannot. We discussed the importance of this fact in chapter 2.

Surface Waves

Surface waves are the slowest waves set off by earthquakes. In general, surface waves cause more property damage than body waves because surface waves produce more ground movement and travel more slowly, so they take longer to pass. The two most important types of surface waves are Love waves and Rayleigh waves, named after the geophysicists who discovered them.

Love waves are most like S waves that have no vertical displacement. The ground moves side to side in a horizontal plane that is perpendicular to the direction the wave is traveling or propagating (figure 7.5*C*). Like S waves, Love waves do not travel through liquids and would not be felt on a body of water. Because of the horizontal movement, Love waves tend to knock buildings off their foundations and destroy highway bridge supports.

Rayleigh waves behave like rolling ocean waves. Unlike ocean waves, Rayleigh waves cause the ground to move in an elliptical path opposite to the direction the wave passes (figure 7.5*D*). Rayleigh waves tend to be incredibly destructive to buildings because they produce more ground movement and take longer to pass.

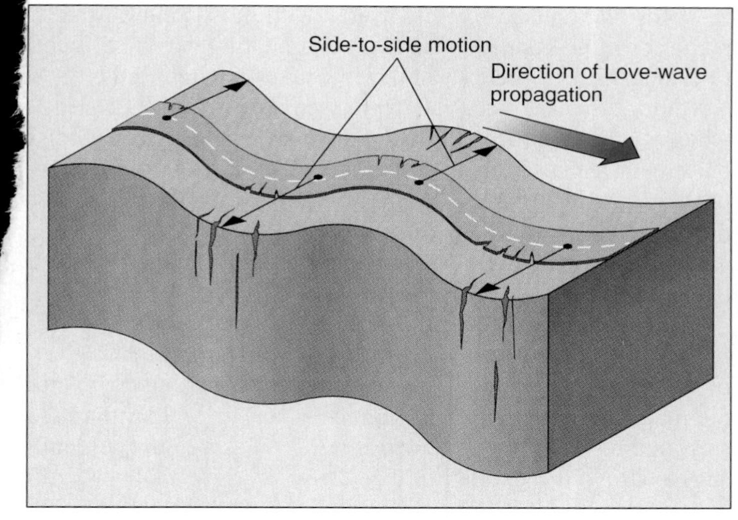

C Love wave

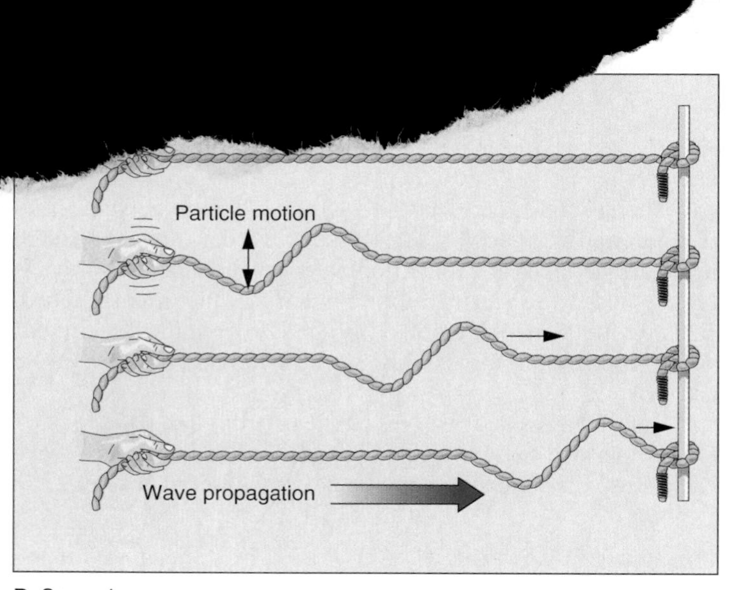

B Secondary wave

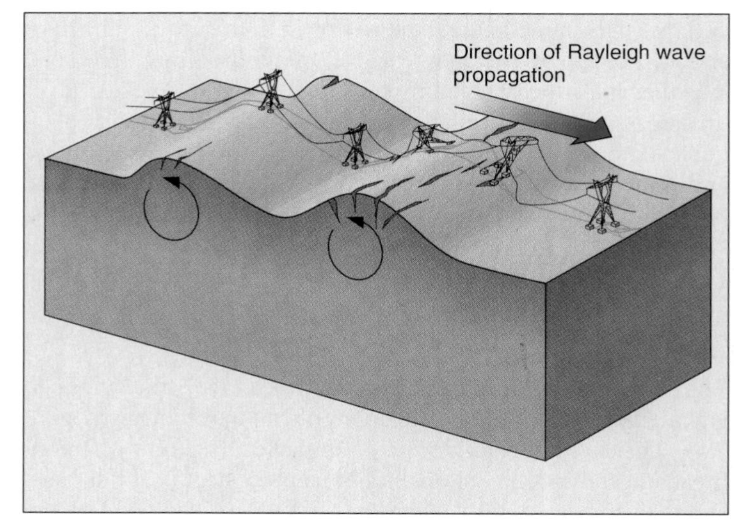

D Rayleigh wave

FIGURE 7.5

Particle motion in seismic waves. (*A*) A P wave is illustrated by a sudden push on the end of a stretched spring or Slinky. The particles vibrate *parallel* to the direction of wave propagation. (*B*) An S wave is illustrated by shaking a loop along a stretched rope. The particles vibrate *perpendicular* to the direction of wave propagation. (*C*) Love waves behave like S waves in that the particle motion is perpendicular to the direction of wave travel along the earth's surface. (*D*) Rayleigh waves are like ocean waves and cause a rolling motion on the earth's surface. The particle motion is elliptical and opposite (counterclockwise) to the direction of wave propagation.

LOCATING AND MEASURING EARTHQUAKES

The invention of instruments that could accurately record seismic waves was an important scientific advance. They measure the amount of ground motion and can be used to find the location, depth, and size of an earthquake.

The instrument used to measure seismic waves is a *seismometer*. The principle of the seismometer is to keep a heavy suspended mass as motionless as possible—suspending it by springs or hanging it as a pendulum from the frame of the instrument (figure 7.6). When the ground moves, the frame of the instrument moves with it; however, the inertia of the heavy mass suspended inside keeps the mass motionless to act as a point of reference in determining the amount of ground motion. Seismometers are usually placed in clusters of three to record the motion along the x, y, and z axes of three-dimensional space.

A seismometer by itself cannot record the motion that it measures. A **seismograph** is a recording device that produces a permanent record of Earth motion detected by a seismometer, usually in the form of a wiggly line drawn on a moving strip of paper (figure 7.7). The paper record of Earth vibration is called

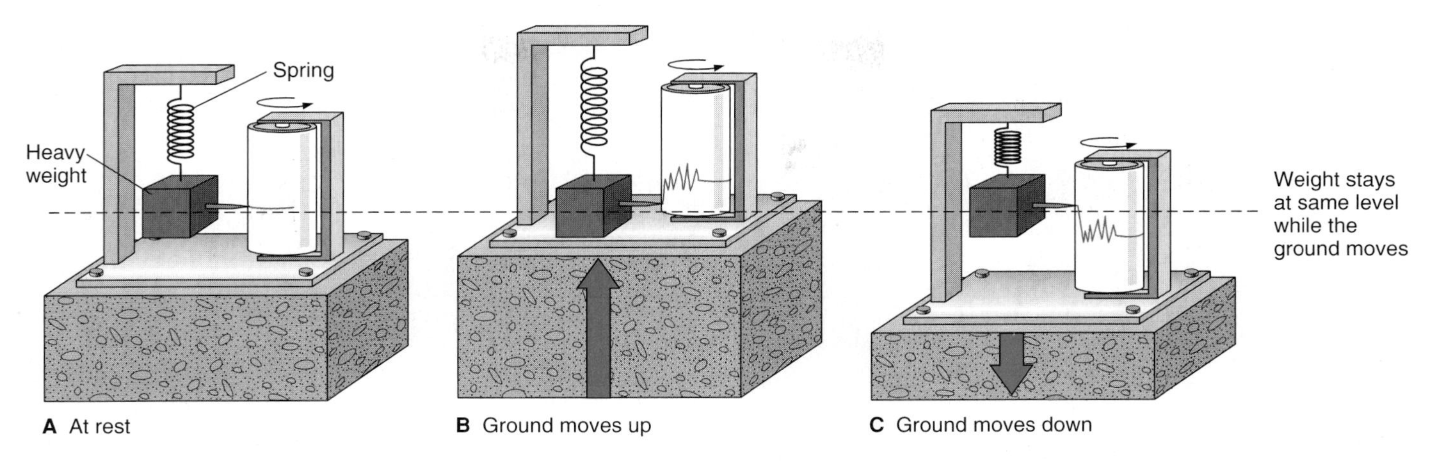

FIGURE 7.6

A simple seismograph for detecting vertical rock motion. The pen records the ground motion on the seismogram as the spring stretches and compresses with up and down movement of the spring. Frame and recording drum move with the ground. Inertia of the weight keeps it and the needle relatively motionless.

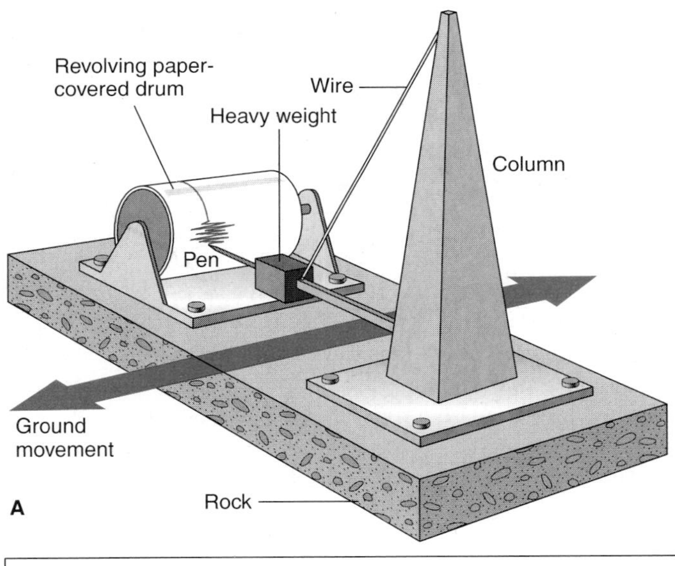

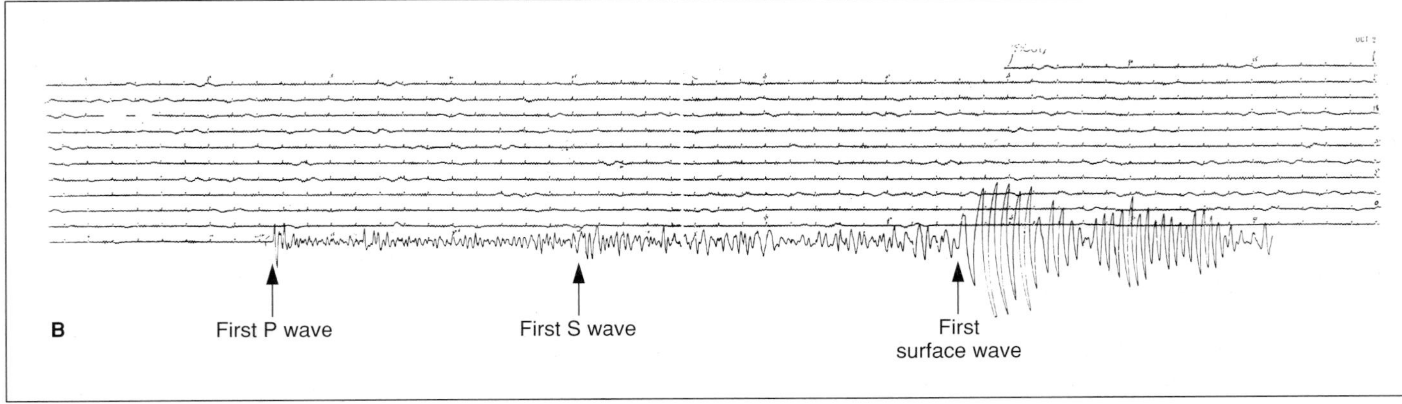

FIGURE 7.7

(*A*) A seismograph for horizontal motion. Modern seismographs record earth motion on moving strips of paper. The mass is suspended by a wire from the column and swings like a pendulum when the ground moves horizontally. A pen attached to the mass records the motion on a moving strip of paper. (*B*) A seismogram of a 1967 earthquake in Taiwan, magnitude 6.2, recorded in Berkeley, California, 6,300 miles away. First arrivals of P, S, and surface waves are shown.

Courtesy University of California, Berkeley

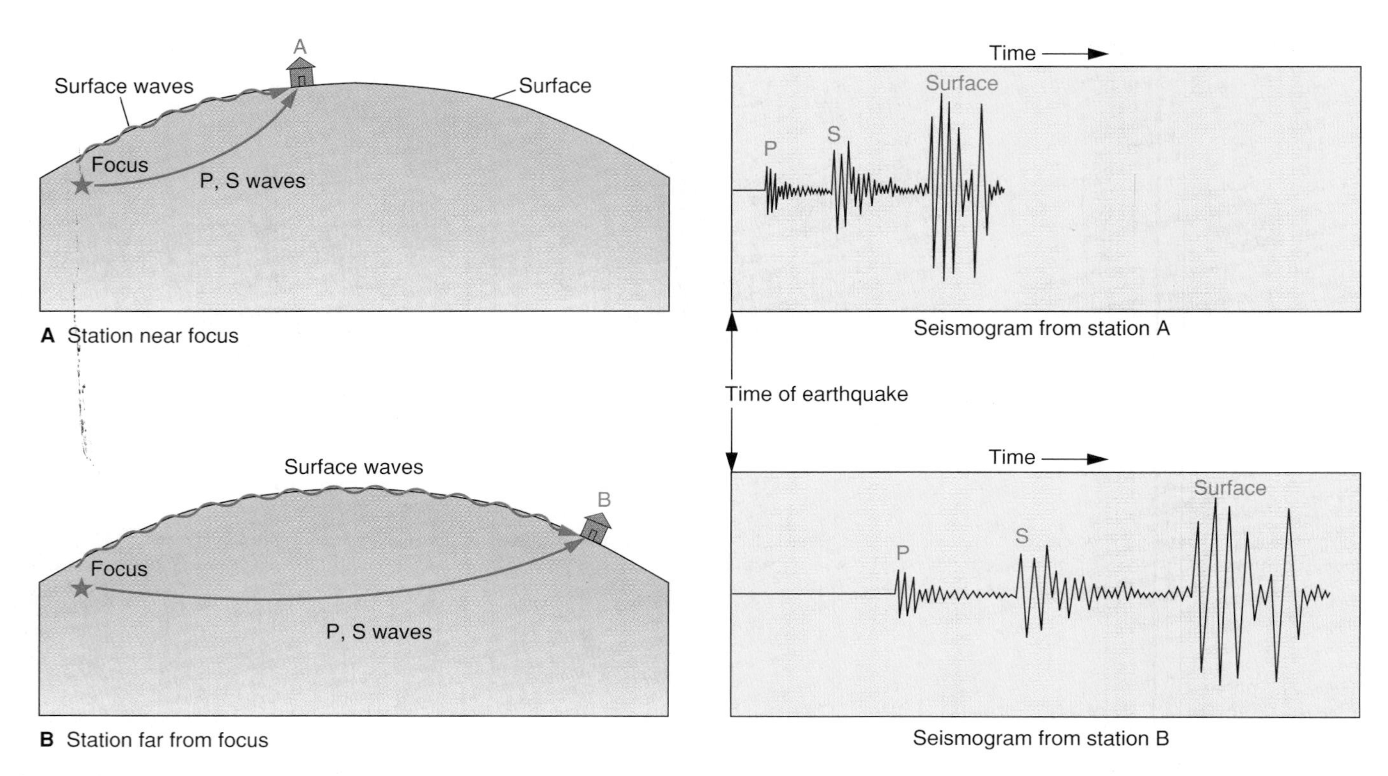

FIGURE 7.8
Intervals between P waves, S waves, and surface waves increase with distance from the focus.

a **seismogram.** The seismogram can be used to measure the strength of the earthquake.

A network of seismograph stations is maintained all over the world to record and study earthquakes (and nuclear bomb explosions). Within minutes after an earthquake occurs, distant seismographs begin to pick up seismic waves. A large earthquake can be detected by seismographs all over the world.

Because the different types of seismic waves travel at different speeds, they arrive at seismograph stations in a definite order, first the P waves, then the S waves, and finally the surface waves. These three different waves can be distinguished on the seismograms. By analyzing these seismograms, geologists can learn a great deal about an earthquake, including its location and size.

Determining the Location of an Earthquake

P and S waves start out from the focus of an earthquake at essentially the same time. As they travel away from the quake, the two kinds of body waves gradually separate because they are traveling at different speeds. On a seismogram from a station close to the earthquake, the first arrival of the P wave is separated from the first arrival of the S wave by a short distance on the paper record (figure 7.8). At a recording station far from the earthquake, however, the first arrivals of these waves will be recorded much farther apart on the seismogram. The farther the seismic waves travel, the longer the time intervals between the arrivals of P and S waves and the more they are separated on the seismograms.

Because the time interval between the first arrivals of P and S waves increases with distance from the focus of an earthquake, this interval can be used to determine the distance from the seismograph station to a quake. The increase in the P-S interval is regular with increasing distance for several thousand kilometers and so can be graphed in a **travel-time curve,** which plots seismic-wave arrival time against distance (figure 7.9).

In practice, a station records the P and S waves from a quake, then a seismologist matches the interval between the waves to a standard travel-time curve. By reading directly from the graph, one can determine, for example, that an earthquake has occurred 5,300 kilometers (3,300 miles) away. This determination can often be made very rapidly, even while the ground is still trembling from the quake.

A single station can determine only the distance to a quake, not the direction. A circle is drawn on a globe with the center of the circle being the station and its radius the distance to the quake (figure 7.10). The scientists at the station know that the quake occurred somewhere on that circle, but from the information recorded they are not able to tell where. With information from other stations, however, they can pinpoint the location of the quake. If three or more stations have determined the distance to a single quake, a circle is drawn for each station. If this is done on a map, the intersection of the circles locates the epicenter.

Analyses of seismograms can also indicate at what depth beneath the surface the quake occurred. Most earthquakes occur

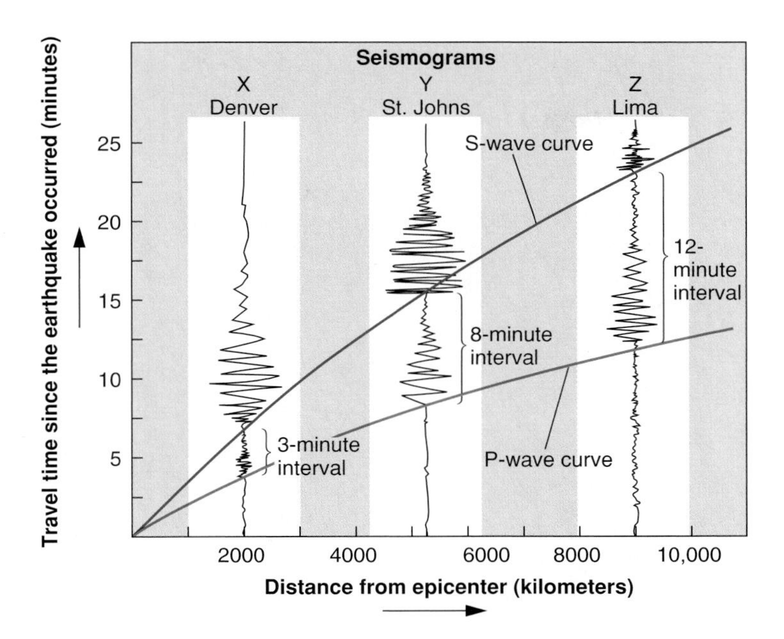

FIGURE 7.9

A travel-time curve is used to determine the distance to an earthquake. Note that the time interval between the first arrival of P and S waves increases with distance from the epicenter. Seismogram X has a 3-minute interval between P and S waves corresponding to a distance of 2,000 km from the epicenter, Y has an interval of 8 minutes, so the earthquake occurred 5,300 km away, and Z an interval of 12 minutes, and is a distance of 9,000 km from the epicenter.

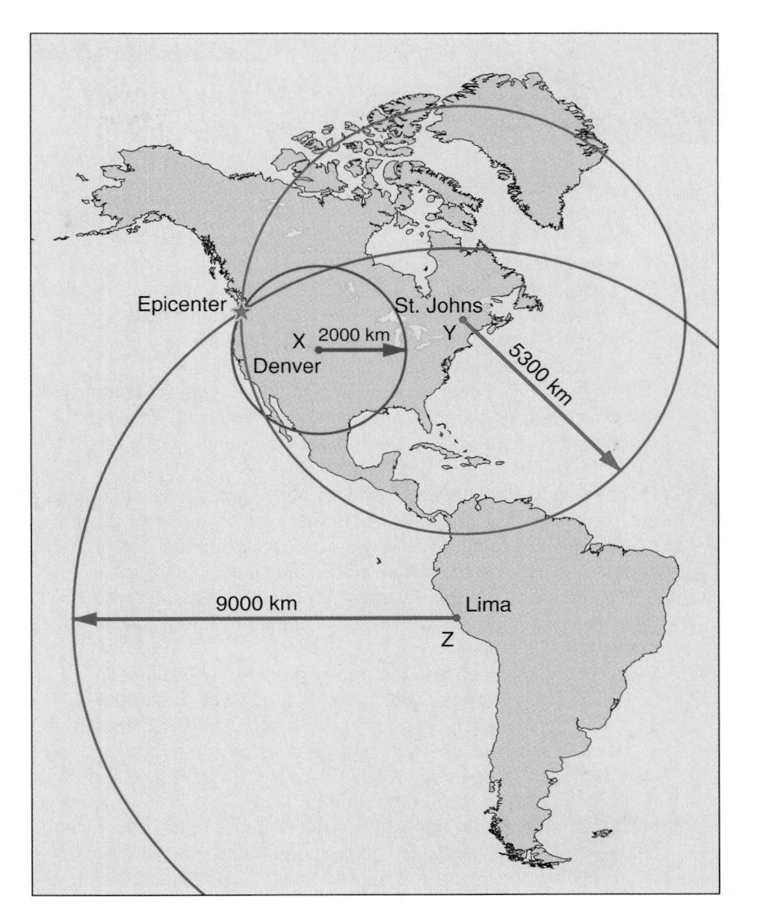

FIGURE 7.10

Locating an earthquake. The distance from each of three stations (Denver, St. Johns, and Lima) is determined from seismograms and the travel-time curves shown in figure 7.9. Each distance is used for the radius of a circle about the station. The location of the earthquake is just offshore of Vancouver, British Columbia, where the three circles intersect.

relatively close to Earth's surface, although a few occur much deeper. The maximum **depth of focus**—the distance between focus and epicenter—for earthquakes is about 670 kilometers. Quakes are classified into three groups according to their depth of focus:

Shallow focus	0–70 km deep
Intermediate focus	70–350 km deep
Deep focus	350–670 km deep

Shallow-focus earthquakes are most common; they account for 85% of total quake energy released. Intermediate- (12%) and deep-focus (3%) quakes are rarer because most deep rocks flow plastically when stressed or deformed; they are unable to store and suddenly release energy as brittle surface rocks do.

Measuring the Size of an Earthquake

The size of earthquakes is measured in two ways. One method is to find out how much and what kind of damage the quake has caused. This determines the **intensity,** which is a measure of an earthquake's effect on people and buildings. Intensities are expressed as Roman numerals ranging from I to XII on the **modified Mercalli scale** (table 7.1); higher numbers indicate greater damage.

Although intensities are widely reported at earthquake locations throughout the world, using intensity as a measure of earthquake strength has a number of drawbacks. Because damage generally lessens with distance from a quake's epicenter, different locations report different intensities for the same earthquake (figure 7.11). Moreover, damage to buildings and other structures depends greatly on the type of geologic material on which a structure was built as well as the type of construction. Houses built on solid rock normally are damaged far less than houses built upon loose sediment, such as delta mud or bay fill. Brick and stone houses usually suffer much greater damage than wooden houses, which are somewhat flexible. Damage estimates are also subjective: people may exaggerate damage reports consciously or unconsciously. Intensity maps can be drawn for a single earthquake to show the approximate damage over a wide region (figure 7.11). Intensity maps are useful for assessing how different areas respond to seismic waves and provide valuable information for earthquake planning. But such maps cannot be drawn for uninhabited areas (the open ocean, for instance), so not all quakes can be assigned intensities. The one big advantage of intensity ratings is that no instruments are required, which

table 7.1 Modified Mercalli Intensity Scale of 1931 (Abridged)

I.	Not felt except by a very few under especially favorable circumstances.
II.	Felt only by a few persons at rest, especially on upper floors of buildings. Delicately suspended objects may swing.
III.	Felt quite noticeably indoors, especially on upper floors of buildings, but many people do not recognize it as an earthquake. Standing motor cars may rock slightly. Vibration like passing of truck. Duration estimated.
IV.	During the day felt indoors by many, outdoors by few. At night some awakened. Dishes, windows, doors disturbed; walls made cracking sound. Sensation like heavy truck striking building. Standing motor cars rocked noticeably.
V.	Felt by nearly everyone; many awakened. Some dishes, windows, etc., broken; a few instances of cracked plaster; unstable objects overturned. Disturbance of trees, poles, and other tall objects sometimes noticed. Pendulum clocks may stop.
VI.	Felt by all; many frightened and run outdoors. Some heavy furniture moved; a few instances of fallen plaster or damaged chimneys. Damage slight.
VII.	Everybody runs outdoors. Damage *negligible* in buildings of good design and construction; *slight* to moderate in well-built ordinary structures; *considerable* in poorly built or badly designed structures; some chimneys broken. Noticed by persons driving motor cars.
VIII.	Damage *slight* in specially designed structures; *considerable* in ordinary substantial buildings with partial collapse; *great* in poorly built structures. Panel walls thrown out of frame structures. Fall of chimneys, factory stacks, columns, monuments, walls. Heavy furniture overturned. Sand and mud ejected in small amounts. Changes in well water. Persons driving motor cars disturbed.
IX.	Damage *considerable* in specially designed structures; well-designed frame structures thrown out of plumb; *great* in substantial buildings, with partial collapse. Buildings shifted off foundations. Ground cracked conspicuously. Underground pipes broken.
X.	Some well-built wooden structures destroyed; most masonry and frame structures destroyed with foundations; ground badly cracked. Rails bent. Considerable landslides from river banks and steep slopes. Shifted sand and mud. Water splashed (slopped) over banks.
XI.	Few, if any, (masonry) structures remain standing. Bridges destroyed. Broad fissures in ground. Underground pipelines completely out of service. Earth slumps and land slips in soft ground. Rails bent greatly.
XII.	Damage total. Waves seen on ground surface. Lines of sight and level distorted. Objects thrown upward into the air.

From Wood and Neumann, 1931, Bulletin Seismological Society of America

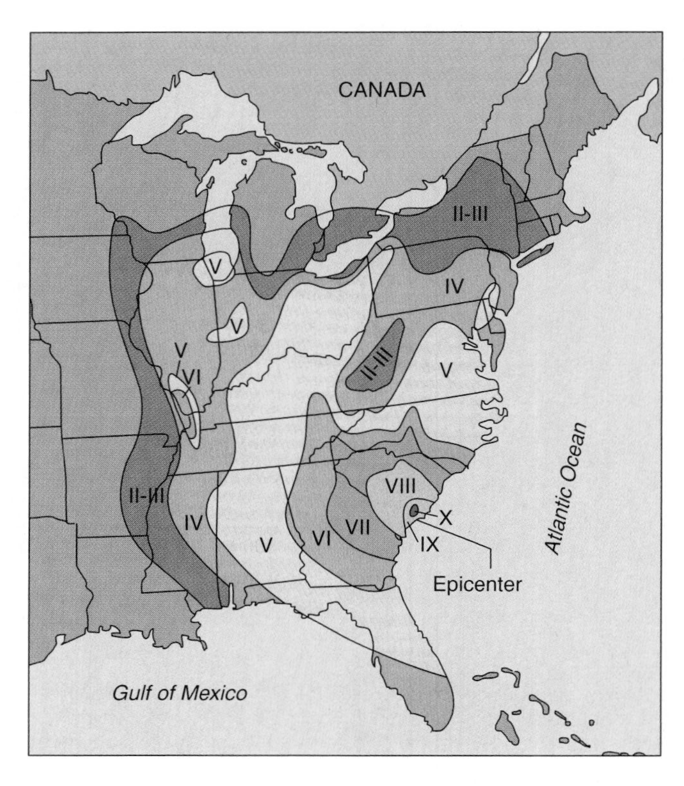

FIGURE 7.11

Zones of different intensity from the 1886 Charleston, South Carolina, earthquake. The map illustrates the general decrease in intensity with increasing distance from the epicenter, and also the effect of different types of earth materials.

U.S. Geological Survey

allows seismologists to estimate the size of earthquakes that occurred before seismographs were available.

The second method of measuring the size of a quake is to calculate the amount of energy released by the quake. This method is usually done by measuring the height (amplitude) of one of the wiggles on a seismogram. The larger the quake, the more the ground vibrates and the larger the wiggle. After measuring a specific wave on a seismogram, and correcting for the type of seismograph and for the distance from the quake, scientists can assign a number called the **magnitude.** It is a measure of the energy released during the earthquake.

For the past several decades magnitude has been reported on the **Richter scale,** a numerical scale of magnitudes. The Richter scale is open ended, meaning there are no earthquakes too large or too small to fit on the scale. The higher numbers indicate larger earthquakes. Very small earthquakes can have negative magnitudes, but these are seldom reported. The largest Richter magnitude measured so far is 8.6. Smaller earthquakes are much more common than large ones (table 7.2).

There are several methods of measuring magnitude, however. The original Richter scale applied only to shallow earthquakes in southern California. Different seismic waves (body or surface) can be measured to make the scale more useful over larger areas, so several different magnitudes are sometimes reported for a single quake. A further complication is that magnitudes calculated from seismograms tend to be inaccurate (usually too low) above magnitude 7.

A new method of calculating magnitude involves the use of the *seismic moment* of a quake, which is determined from the strength of the rock, surface area of the rupture, and the amount of rock displacement along the fault. The **moment magnitude**

table 7.2 **Comparison of Earthquake Magnitude, Description, Intensity, and Expected Annual World Occurrence**

Richter Magnitude	Description	Maximum Expected Mercalli Intensity at Epicenter	Annual Expected Number
2.0	Very Minor	I Usually detected only by instruments	600,000
2.0–2.9	Very Minor	I–II Felt by some indoors, especially on upper floors	300,000
3.0–3.9	Minor	III Felt indoors	49,000
4.0–4.9	Light	IV–V Felt by most; slight damage	6,200
5.0–5.9	Moderate	VI–VII Felt by all; damage minor to moderate	800
6.0–6.9	Strong	VII–VIII Everyone runs outdoors; moderate to major damage	266
7.0–7.9	Major	IX–X Major damage	18
8.0 or higher	Great	X–XII Major and total damage	1 or 2

Source: U.S. Geological Survey

is the most objective way of measuring the energy released by a large earthquake. The 1964 Alaska quake is estimated to have a moment magnitude of 9.2, and the 1960 Chile quake 9.5. Unfortunately the media rarely indicate which type of magnitude they are reporting, and scientists typically revise magnitudes for several weeks after a quake as they receive more information, so trying to find out the "real" magnitude of a recent quake can be confusing. Table 7.3 lists Richter magnitudes and moment magnitudes for many earthquakes of interest.

Because the Richter scale is logarithmic, the difference between two consecutive whole numbers on the scale means an increase of 10 times in the amplitude of Earth's vibrations, particularly below magnitude 5. This means that if the measured amplitude of vibration for certain rocks is 1 centimeter during a magnitude-4 quake, these rocks will move 10 centimeters during a magnitude-5 quake occurring at the same location.

It has been estimated that a tenfold increase in the size of Earth vibrations is caused by an increase of roughly 32 times in terms of energy released. A quake of magnitude 5, for example, releases approximately 32 times more energy than one of magnitude 4. A magnitude-6 quake is about 1,000 times (32×32) more powerful in terms of energy released than a magnitude-4 quake. The actual energy released in earthquakes of varying magnitudes is shown in figure 7.12.

Although a seismograph is usually required to measure magnitude, this measure has many advantages over intensity as an indicator of earthquake strength. A worldwide network of standard seismograph stations now makes determining magnitude a routine matter; and the press reports magnitudes for all earthquakes of interest to the United States. Eventually, a single magnitude number can be assigned to a single earthquake, whereas intensity varies for a single earthquake, depending on the amount and kind of local damage. Magnitudes can be reported for all quakes, even those in distant uninhabited areas where there is no property to affect.

Location and Size of Earthquakes in the United States

Figure 7.13 shows the locations of all damaging earthquakes that have occurred in the United States from 1977 to 1997. Note that only a few localities are relatively free of earthquakes.

Most of the large earthquakes occur in the western states. Quakes in California, Nevada, Utah, Idaho, Montana, Washington, and other western states are related to known faults and usually (but not always) involve surface rupture of the ground. Earthquakes in Alaska occur mainly below the Aleutian Islands where the Pacific plate is converging with and being subducted beneath the North American plate.

Earthquakes east of the Rocky Mountains are more rare and are generally smaller and deeper than earthquakes in the western United States. They usually are not associated with surface rupture. The quakes may be occurring on the deeply buried, relatively inactive faults of old *divergent plate boundaries* and *failed rifts* (*aulacogens*), both of which are described in chapter 4.

Although large quakes are extremely rare in the central and eastern United States, when they do occur they can be very destructive and widely felt, because Earth's crust is older, cooler, and more brittle in the east than in the west and seismic waves travel more efficiently. The Saint Lawrence River Valley along the Canadian border has had several intensity IX and X earthquakes, most recently in 1944. Plymouth, Massachusetts, had an intensity IX quake in 1638, and a quake of intensity VIII occurred in 1775 near Cambridge, Massachusetts. In 1929 in Attica, New York, an earthquake of intensity IX knocked over 250 chimneys. A series of quakes (intensity XI) that occurred near New Madrid, Missouri, in the winter of 1811–1812 were the most widely felt earthquakes to occur in North America in recorded history. The quakes knocked over chimneys as far away as Richmond, Virginia, and rang church bells in Boston, 700 kilometers away.

table 7.3 Earthquake Magnitudes

		Richter Magnitude	Moment Magnitude
1811–12	New Madrid, Missouri area	7.5, 7.3, 7.8	7.7, 7.6, 7.9
1857	Fort Tejon, S. Calif.	7.6	7.9
1872	Lone Pine, Calif.	7.3	7.8
1886	Charleston, South Carolina	6.7	7.0
1906	San Francisco, N. Calif.	8.25	7.7
1915	Pleasant Valley, Nevada	7.7	7.1
1933	Long Beach, S. Calif.	6.3	6.2
1952	Kern County, S. Calif.	7.2	7.5
1954	Dixie Valley, Nevada	7.2, 7.1	7.3, 6.9
1957	Aleutian Islands, Alaska	8.1	8.8
1958	Southeastern Alaska	7.9	8.3
1959	Hebgen Lake, Montana	7.7	7.3
1960	Chile	8.5	9.5
1964	near Anchorage, Alaska	8.6	9.2
1965	Aleutian Islands, Alaska	8.2	8.7
1970	Peru	7.75	7.9
1971	San Fernando Valley, S. Calif.	6.4	6.6
1975	Hawaii	7.2	7.5
1976	China	7.6	7.5
1980	Humboldt County, N. Calif.	6.9	7.2
1983	Coalinga, Calif.	6.7	6.2
1983	Challis, Idaho	7.2	7.0
1983	Adirondack Mountains, New York	5.1	4.9
1983	Hawaii	6.6	
1985	Ixtapa, Mexico		8.1, 7.5
1987	Lawrenceville, Illinois	5.0	5.0
1987	Whittier, S. Calif.	5.9	5.9
1988	Quebec	6.0	
1989	Loma Prieta, N. Calif.	7.0	7.2
1989	Hawaii	6.1	6.4
1992	Humboldt County, N. Calif.	7.1, 6.6, 6.7	
1992	Landers, S. Calif.	7.6, 6.7	7.25, 6.4
1994	Northridge, S. Calif.	6.4	6.7
1995	Kobe, Japan (7.2 on Japanese scale)	6.8	6.9
1998	Papua, New Guinea		7.1
1999	Izmit, Turkey		7.4
1999	Taipei, Taiwan		7.6
2001	El Salvador		7.7
2001	Gujarat, India		7.7
2001	Puget Sound (Nisqually), Washington		6.8

U.S. Geological Survey and other sources

The 1886 quake in Charleston, South Carolina (intensity X), was felt throughout almost half the United States (figure 7.11) and killed sixty people; it was sharply felt in New York City. Moderate quakes hit Arkansas and New Hampshire in 1982, and in 1983 a quake of 5.1 magnitude rocked New York's Adirondack Mountains. A 5.0-magnitude quake near Lawrenceville, Illinois, was felt from Kansas to South Carolina to Ontario in 1987. In 1988 a 6.0-magnitude quake north of Quebec City was felt as far away as Indiana and Washington, D.C.

Geologists have mapped regions of seismic risk in the United States (figure 7.14), and elsewhere throughout the world, primarily on the assumption that large earthquakes will occur in the future in places where they have occurred in the past.

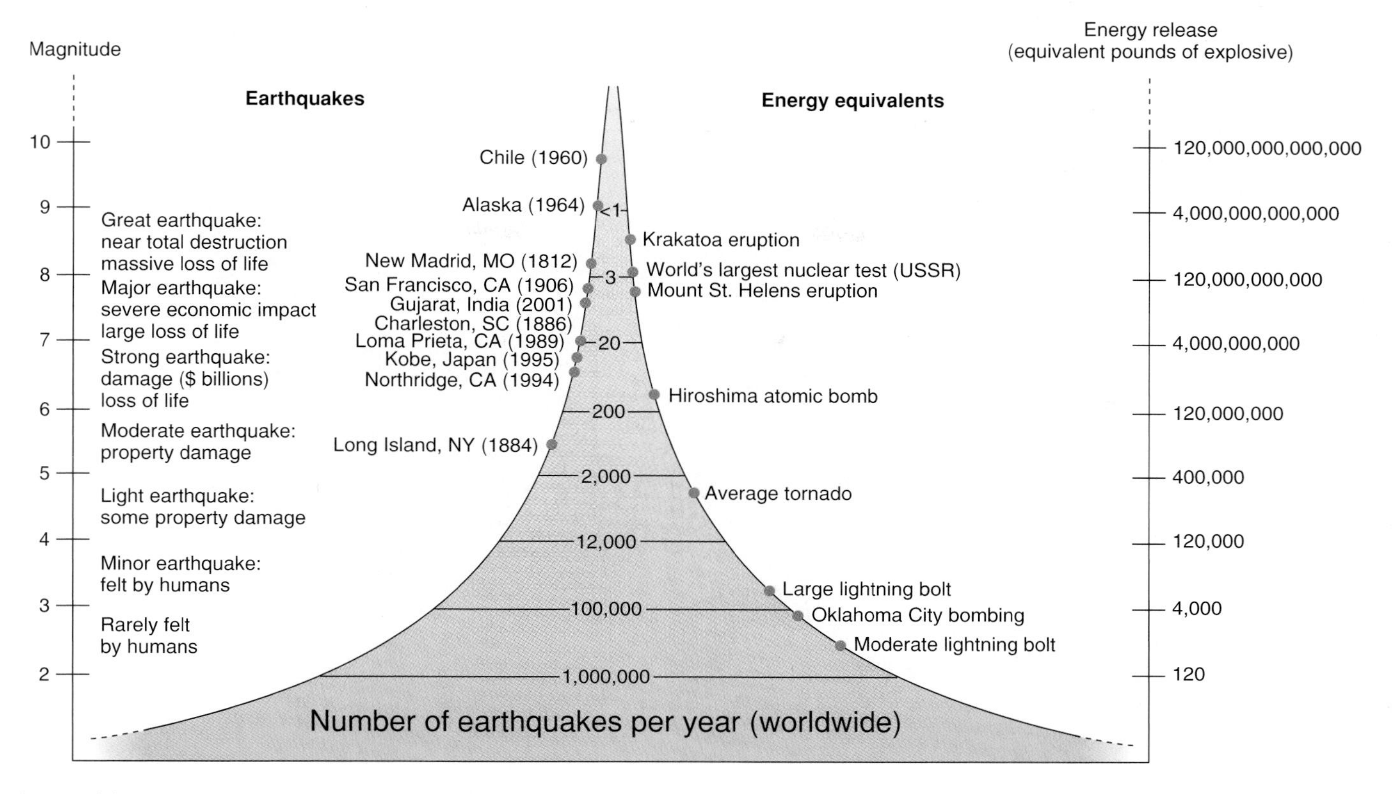

FIGURE 7.12

Diagram shows the relationship between the moment magnitude of an earthquake, the number of earthquakes per year throughout the world, and the energy released during an earthquake.

After IRIS Consortium (www.iris.edu)

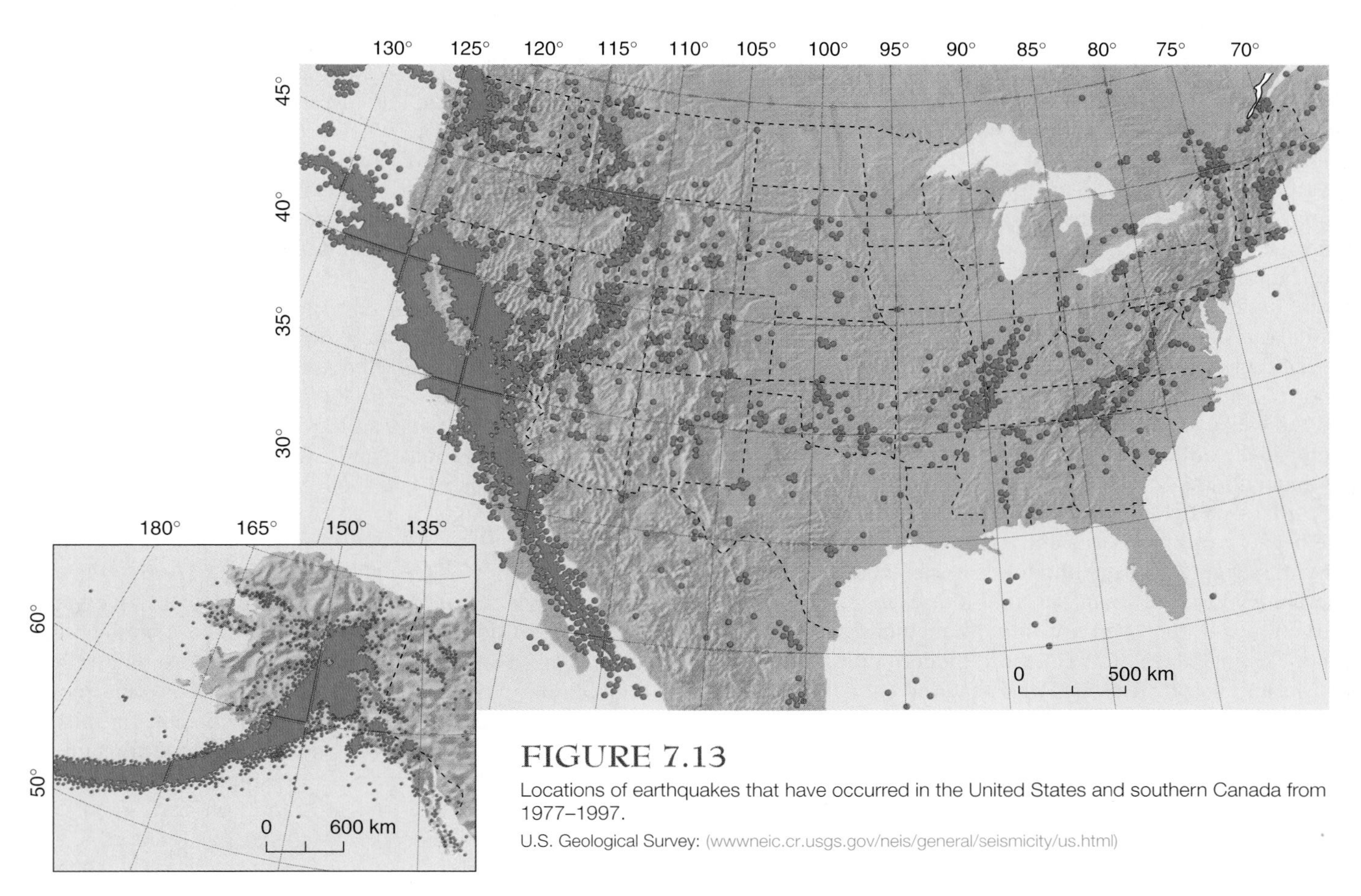

FIGURE 7.13

Locations of earthquakes that have occurred in the United States and southern Canada from 1977–1997.

U.S. Geological Survey: (wwwneic.cr.usgs.gov/neis/general/seismicity/us.html)

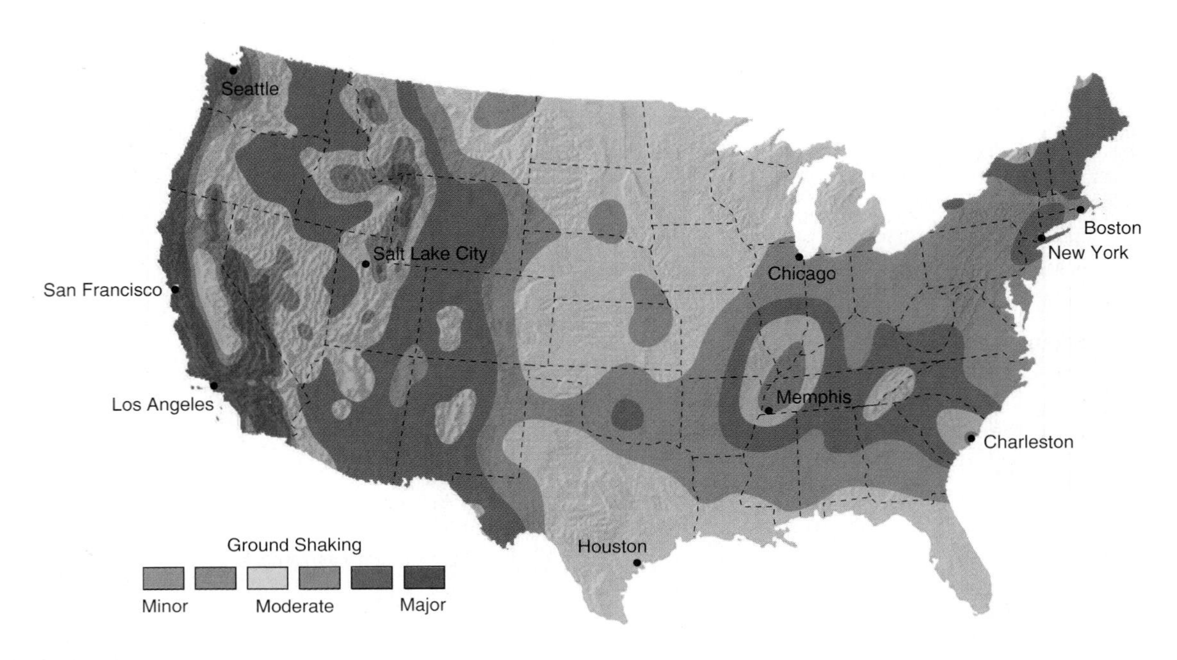

FIGURE 7.14
Map of seismic hazard in the United States based on the expected amount of ground shaking and damage.
USGS-National Seismic Hazard Mapping Project

EFFECTS OF EARTHQUAKES

Ground motion is the trembling and shaking of the land that can cause buildings to vibrate. During small quakes, windows and walls may crack from such vibration. In a very large quake the ground motion may be visible. It can be strong enough to topple large structures such as bridges and office and apartment buildings (figure 7.15). Most people injured or killed in an earthquake are hit by falling debris from buildings. Because proper building construction can greatly reduce the dangers, building codes need to be both strict and strictly enforced in earthquake-prone areas. Much of the damage and loss of life in the recent Turkey, El Salvador, and India earthquakes were due to poorly constructed buildings that did not meet building codes. As we have seen, the location of buildings also needs to be controlled; buildings built on soft sediment are damaged more than buildings on hard rock.

Fire is a particularly serious problem just after an earthquake because of broken gas and water mains and fallen electrical wires (figure 7.16). Although fire was the cause of most of the damage to San Francisco in 1906, changes in building construction and improved fire-fighting methods have reduced (but not eliminated) the fire danger to modern cities. The stubborn Marina district fires in San Francisco in 1989 attest to modern dangers of broken gas and water mains.

Landslides can be triggered by the shaking of the ground (figure 7.17*A*). The 1959 Madison Canyon landslide in Montana was triggered by a nearby quake of magnitude 7.7. Landslides and subsidence caused extensive damage in downtown and suburban Anchorage during the 1964 Alaskan quake (magnitude 8.6). The 1970 Peruvian earthquake (magnitude 7.75) set off thousands of landslides in the steep Andes Mountains, burying more than 17,000 people (see box 13.1). In 1920 in China over 100,000 people living in hollowed-out caves in cliffs of loess (described in chapter 18) were killed when a quake collapsed the cliffs. The 2001 El Salvador quake resulted in nearly 500 landslides, the largest of which occurred in Santa Tecla where 1,200 people were missing after tons of soil and rock fell on a neighborhood.

A special type of ground failure caused by earthquakes is *liquefaction*. This occurs when a water-saturated soil or sediment turns from a solid to a liquid as a result of earthquake shaking. Liquefaction may occur several minutes after an earthquake, causing buildings to sink and underground tanks to float as once-solid sediment flows like water (figure 7.17*B*). Liquefaction was responsible for much of the damage in the 1989 Loma Prieta quake and contributed to the damage in the 1906 San Francisco, the 1964 Alaska, the 1995 Kobe, Japan, and the 2001 Puget Sound, Washington, and Gujarat, India, quakes.

Permanent displacement of the land surface may be the result of movement along a fault. Rocks can move vertically, those on one side of a fault rising while those on the other side drop. Rocks can also move horizontally, those on one side of a fault sliding past those on the other side. Diagonal movement with both vertical and horizontal components can also occur during a single quake. Such movement can affer huge areas, although the displacement in a single earthquake seldom exceeds 8 meters. The trace of a fault on Earth's surface may appear as a low cliff, called a *scarp,* or as a closed tear in the ground (figure 7.18). In rare instances small cracks open during

A

C

B

D

FIGURE 7.15

Earthquake damage to structures from recent major earthquakes throughout the world. (*A*) Elevated highway knocked over by a strong horizontal jolt during the 1995 Kobe, Japan, earthquake. Damage exceeded $400 billion and destroyed or severely damaged more than 88,000 buildings. (*B*) Poorly constructed buildings crumbled during the 1999 Izmit, Turkey, earthquake while structures built to seismic code and old mosques are left standing. (*C*) Many high-rise buildings collapsed during the 1999 Taiwan earthquake. The M 7.6 quake was the largest to hit central Taiwan in the past 400 years and damage exceeded $14 billion. (*D*) One of the many buildings damaged during the January 2001 Gujarat, India, earthquake that caused over $1.3 billion in damage.

Photo *A* by Reuter/Sankei/Shimbun. Photo *B* by AP/Wide World Photo. Photo *C* by Smith Glenn/Corbis SYGMA. Photo *D* by Jaswant Arelekar/IITK, Kanpur, India

7.1

IN GREATER DEPTH

Earthquake Engineering

Damage and loss of life can be substantially reduced by siting structures on solid bedrock or dense soils and by building structures that adhere to strict seismic building codes. In the 7.2 magnitude earthquake that struck Armenia in 1988, 50,000 people lost their lives when poorly constructed buildings crumbled. More recently, on January 26, 2001, the magnitude 7.7 Gujarat, India, quake killed over 18,000, left 600,000 homeless, and destroyed 332,000 houses. In contrast, earthquake-resistant structures and enforcement of seismic building codes in the San Francisco Bay area resulted in only 63 people dying in the 7.2 magnitude Loma Prieta earthquake.

Buildings that are constructed of strong, flexible, and light materials such as steel, wood, and reinforced concrete (strengthened by steel rebar) are the most resistant to damage by seismic shaking. Houses built with unreinforced concrete block or brick, which are only as strong as the mortar holding the blocks and bricks together, tend to lack flexibility and crumble in large earthquakes. During moderate-sized earthquakes, many houses lose their chimneys or brick facades. Buildings with heavy roofs made of tile or slate also tend to collapse. During the Gujarat, India, earthquake, many reinforced concrete buildings failed because the walls, floors, ceilings, and elevator shafts were not well connected to allow the entire building to flex as one (box figure 1*A*).

The 1985 Mexico City earthquake, which killed 5,000 people and caused over $5 billion in damage, is a classic example of the effect soft soils have on the amplification of earthquake waves. The ground shaking was relatively mild in most parts of Mexico, but it was amplified in some buildings and by the lake-bed sediments beneath Mexico City. Most buildings are not rigid but slightly flexible; they sway gently like large pendulums when struck by wind or seismic waves. The time necessary for a single back-and-forth oscillation is called a period, and it varies with a building's height and mass. Natural bodies of rock and sediment vibrate the same way, with periods that vary with the body's size and density.

When earthquake waves struck Mexico City, many were vibrating with a two-second period. The body of lake sediment beneath the city has a natural period of two seconds also, so the wave motion was amplified by the sediments. This type of amplification, or resonance, occurs when you push a child on a swing—if you push gently in time with the swing's natural period, the swing goes higher and higher. The water-saturated sediment began to move like a sloshing waterbed. The ground moved back and forth by 40 centimeters every two seconds, and it did this fifteen to twenty times.

This shaking was devastating to buildings with a natural two-second period (generally those five to twenty stories high), and several hundred buildings collapsed as they further resonated with the shaking (box figure 1*B*). Many structures weakened by the main shock collapsed during the large aftershock. Most shorter and taller buildings had other periods of vibration and rode out the shaking with little damage. Although 800 buildings collapsed or were seriously damaged, most of the city's 600,000 structures survived with little or no damage. The Mexico City quake vividly illustrates the fact that proper building design can greatly lessen earthquake damage.

A

B

Box 7.1 — FIGURE 1

(*A*) Insufficient connection between the reinforced concrete elevator shaft and rest of the building led to separation and partial collapse during the 2001 Gujarat, India, earthquake. (*B*) This 15-story building collapsed completely during the 1985 Mexico City earthquake, crushing all its occupants as its reinforced-concrete floors "pancaked" together.

Photo *A* by C. V .R. Marty/IITK, Kanpur, India. Photo *B* by M. Celebi, U.S. Geological Survey

FIGURE 7.16

Almost 100 homes burned at a Sylmar mobile home park following the Northridge earthquake, southern California, 1994.

Photo © Ken Lubas/Los Angeles Times

a quake (but not to the extent that Hollywood films often portray). Ground displacement during quakes can tear apart buildings, roads, and pipelines that cross faults. Sudden subsidence of land near the sea can cause flooding and drownings.

Aftershocks are small earthquakes that follow the main shock. Although aftershocks are smaller than the main quake, they can cause considerable damage, particularly to structures previously weakened by the powerful main shock. A long period of aftershocks can be extremely unsettling to people who have lived through the main shock. *Foreshocks* are small quakes that precede a main shock. They are usually less common and less damaging than aftershocks but can sometimes be used to help predict large quakes (although not all large quakes have foreshocks).

Tsunami

The sudden movement of the sea floor upward or downward during a submarine earthquake can generate very large sea waves, popularly called "tidal waves." Because the ocean tides have nothing to do with generating these huge waves, the Japanese term *tsunami* is preferred by geologists. Tsunami are also called **seismic sea waves.** They usually are caused by great earthquakes (magnitude 8^+) that disturb the sea floor, but they also result from submarine landslides or volcanic explosions. When a large section of sea floor suddenly rises or falls during a quake, all the water over the moving area is lifted or dropped for an instant. As the water returns to sea level, it sets up long, low waves that spread very rapidly over the ocean (figure 7.19). Because vertical motion of the sea floor is most conducive to the formation of a tsunami, most are associated with subduction zone earthquakes, which tend to be some of the strongest quakes.

A tsunami is unlike an ordinary water wave on the sea surface. A large wind-generated wave may have a wavelength of 400 meters and be moving in deep water at a speed of 90 kilometers per hour (55 miles per hour). The wave height when it breaks on shore may be only 0.6 to 3 meters, although in the middle of hurricanes the waves can be more than 15 meters high. A tsunami, however, may have a wavelength of 160 kilometers, and may be moving at 725 kilometers per hour (450 miles per hour). In deep water the wave height may be only 0.6 to 2 meters, but near shore the tsunami may peak up to heights of 15 to 30 meters. This great increase in wave height near shore is caused by bottom topography; only a few localities have the combination of gently sloping offshore shelf and funnel-shaped bay that force tsunamis to awesome heights (the record height was 85 meters in 1971 in the Ryukyu islands south of Japan). Along most coastlines tsunami height is very small.

A

B

FIGURE 7.17

(*A*) Landslide in Pacific Palisades triggered by the Northridge earthquake, 1994. (*B*) Liquefaction of soil by a 1964 quake in Niigata, Japan, caused earthquake-resistant apartment buildings to topple over intact.

Photo *A* © Al Seib/Los Angeles Times; photo *B* by National Geophysical Data Center

A

B

C

D

FIGURE 7.18

Varieties of ground displacement caused by earthquakes. (*A*) Sixteen-foot scarp (cliff) formed by vertical ground motion, Alaska, 1964. (*B*) Tearing of the ground near Olema, California, 1906. (*C*) Fence compressed by ground movement, Gallatin County, Montana, 1959. (*D*) Compression of concrete freeway, San Fernando Valley, California, 1971.

Photo *A* by U.S. Geological Survey; photo *B* by G. K. Gilbert, U.S. Geological Survey; photo *C* by I. J. Witkind, U.S. Geological Survey

Although the speed of the wave slows drastically as it moves through shallow water, a tsunami can still hit some shores as a very large, very fast wave. Because of its extremely long wavelength, a tsunami does not withdraw quickly as normal waves do. The water keeps on rising for five to ten minutes, causing great flooding before the wave withdraws. The long duration and great height of a tsunami can bring widespread destruction to the entire shore zone.

A tsunami formed by the 1960 Chilean quake crossed the Pacific Ocean and did extensive damage in Japan. One of the most destructive tsunamis of modern times was generated April 1, 1946, by a 7.3-magnitude earthquake offshore from Alaska. It devastated the city of Hilo, Hawaii, causing 159 deaths (figure 7.20*A*). The tsunami following the 1964 Alaskan quake drowned twelve people in Crescent City, California, a small coastal town near the Oregon border. Wave damage near an epicenter can be awesome. The 1946 Alaska tsunami destroyed the Scotch Cap lighthouse on nearby Unimak Island, sweeping it off its concrete base, which was 10 meters above sea level, and killing its five occupants. The wave also swept away a radio tower, whose base was 31 meters above sea level.

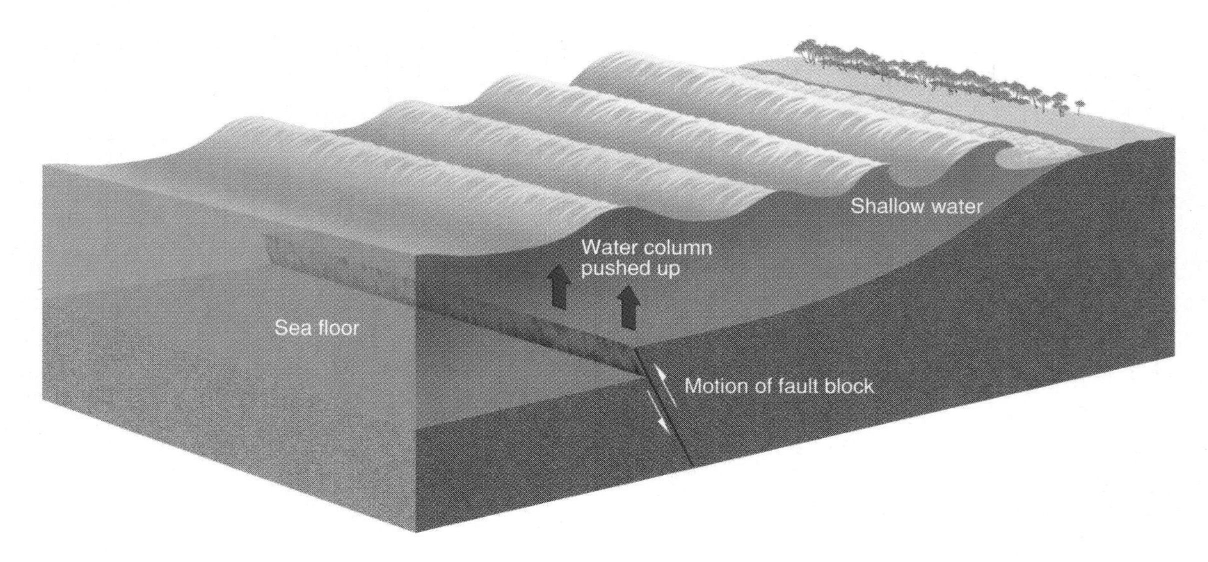

FIGURE 7.19

Tsunami waves are generated by a submarine earthquake that displaces the sea floor and water column above. Long, low waves are formed above the displaced sea floor to compensate for the momentary rise in sea level and spread very rapidly (at the speed of a jet liner) in the deep ocean. In shallower water, the tsunami slows to highway speeds and builds in height until it breaks and crashes onto the shore with incredible force, causing destructive flooding along low-lying coastal areas.

A

B

FIGURE 7.20

(*A*) A tsunami can be seen in the background as it crashes onto shore in Hilo, Hawaii, catching residents by surprise as they run for their lives. The tsunami devastated the city of Hilo and killed 159 people. It was generated by an earthquake in the Aleutian Islands. (*B*) A man salvages possessions from the ruins of this house in Sissano, July 20, in the remote northwest of Papua, New Guinea. Rescue workers believe that up to 3,000 people were swept to their deaths when three giant tsunami waves struck July 17, 1998, and devastated seven coastal towns.

Photo *A* courtesy of Bishop Museum. Photo *B* by Agence France-Presse

Probably the most devastating tsunami of the 20th century, in terms of loss of life, occurred July 17, 1998, in Papua, New Guinea, after a 7.1-magnitude earthquake struck 20 kilometers offshore. Three waves were generated twenty minutes after the earthquake, with the highest wave measuring 15 meters, which completely destroyed three coastal villages and killed more than 2,200 people (figure 7.20*B*). The tremendous loss of life was caused by the lack of warning as the waves hit at dusk on the three villages located on a low-lying sandbar enclosing a bay. The wave was higher than expected for the size of the earthquake, possibly due to the steep bottom topography off the north coast of New Guinea that brought deep water into the gently sloping near shore.

After the 1946 Hilo, Hawaii, tsunami, the U.S. Coast and Geodetic Survey established a Tsunami Early Warning System in an attempt to minimize loss of life in Pacific coastal communities. A network of seismic stations reports large earthquakes that are capable of generating tsunami to the Tsunami Warning Center in Honolulu. An array of tidal gauges and deep-ocean tsunami detectors are then read to determine if a tsunami has been generated. Even though tsunami travel at high speeds,

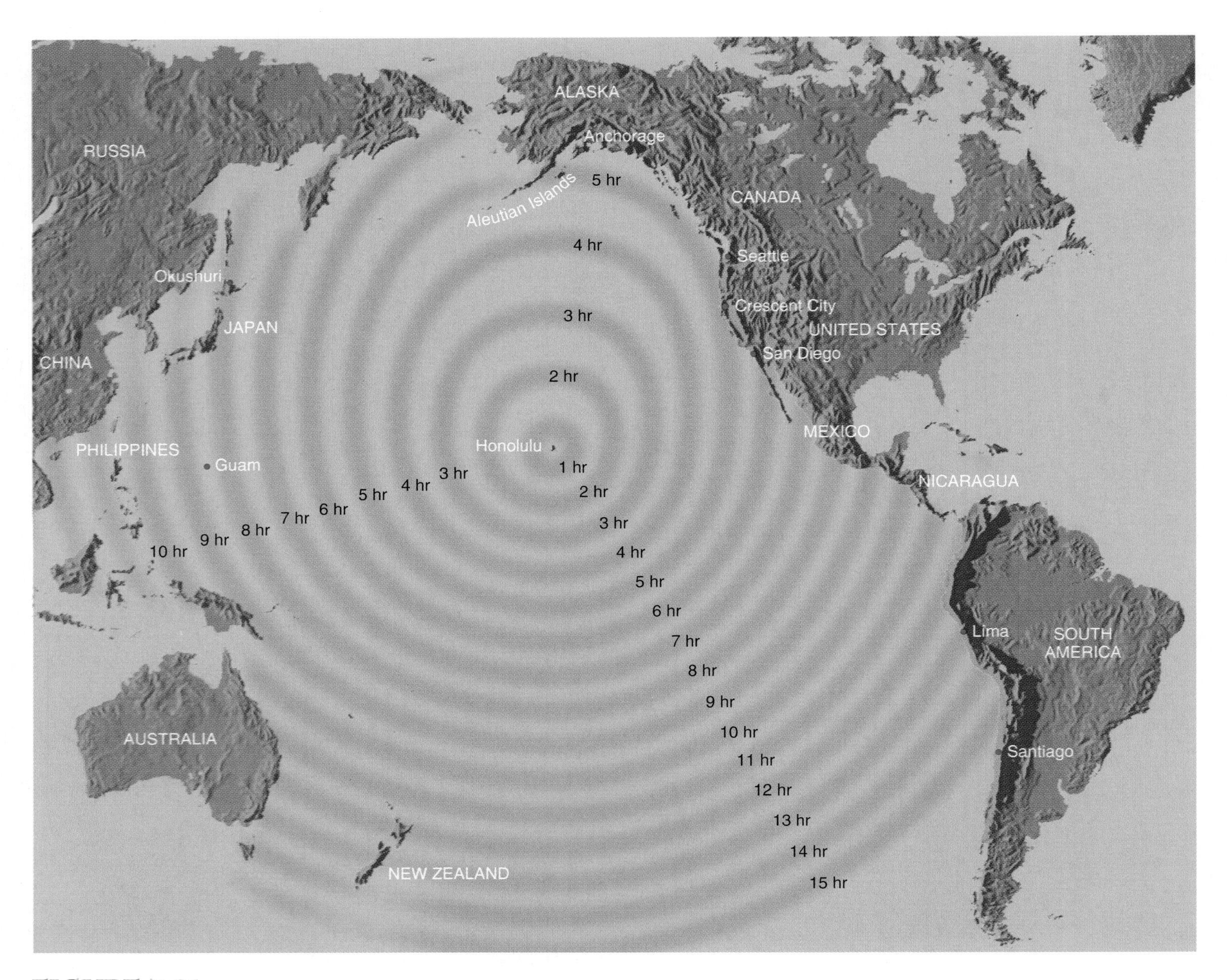

FIGURE 7.21
Tsunami travel times from locations within the Pacific with respect to Honolulu, Hawaii.
From NOAA

there is usually sufficient time to warn low-lying coastal communities of the impending wave.

Travel times of tsunami generated within the Pacific Ocean have been determined with respect to Hawaii (figure 7.21). For example, a tsunami generated in central Japan would reach Hawaii 8 hours later and one generated off the north coast of Peru would strike Hawaii in 12 hours. But a low-lying coastal community located near the epicenter of the tsunami-generating earthquake could have very little warning. Such was the case in 1993 when a magnitude 7.8 earthquake occurred in the Sea of Japan just 15 to 30 kilometers off the coast of Okushiri, Japan. The first waves, 5 to 10 meters high, struck the small fishing village of Aonae in less than five minutes and were followed by waves with a maximum height of 31 meters. The 239 deaths caused by the tsunami could have been much greater if residents had not run to higher ground immediately after feeling the main shock of the earthquake.

WORLD DISTRIBUTION OF EARTHQUAKES

Most earthquakes are concentrated in narrow geographic belts (figure 7.22*A*), although *some* earthquakes have occurred in most regions on Earth. The boundaries of plates in the plate tectonic theory are defined by these earthquake belts (figure 7.22*B*). The most important concentration of earthquakes by far

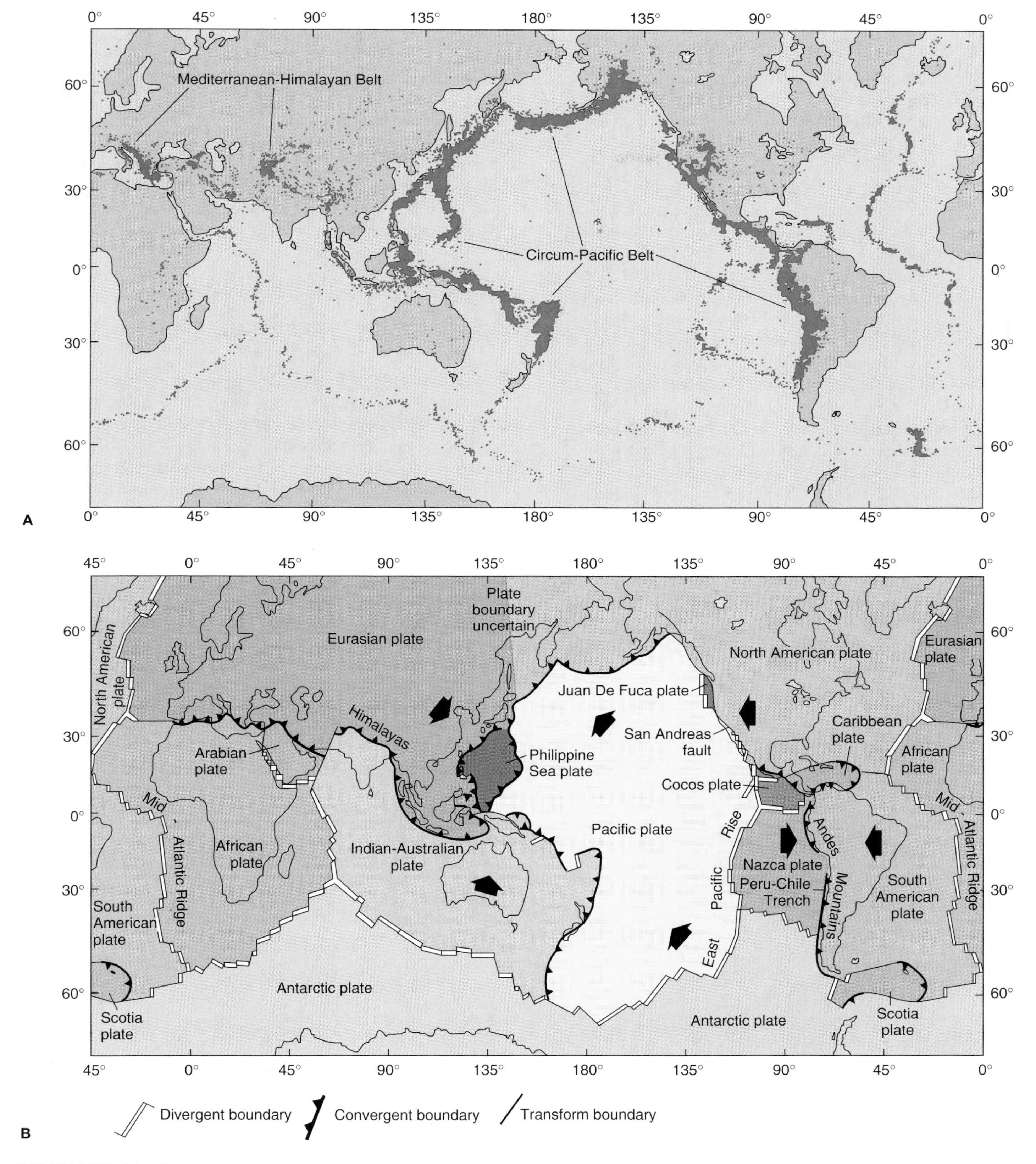

FIGURE 7.22

(*A*) World distribution of earthquakes recorded over a six-year period with focal depths between 0 and 700 kilometers. (*B*) The major plates of the world in the theory of plate tectonics. Compare the locations of plate boundaries with earthquake locations shown in figure 7.22*A*. Double lines show diverging plate boundaries, single lines show transform boundaries. Heavy lines with triangles show converging boundaries; triangles point down subduction zone.

After W. Hamilton, U.S. Geological Survey

is in the **circum-Pacific belt,** which encircles the rim of the Pacific Ocean. Within this belt occur approximately 80% of the world's shallow-focus quakes, 90% of the intermediate-focus quakes, and nearly 100% of the deep-focus quakes.

Another major concentration of earthquakes is in the **Mediterranean-Himalayan belt,** which runs through the Mediterranean Sea, crosses the Mideast and the Himalaya, and passes through the East Indies to meet the circum-Pacific belt north of Australia.

A number of shallow-focus earthquakes occur in two other significant locations on Earth. One is along the summit or crest of the *mid-oceanic ridge,* a huge underwater mountain range that runs through all the world's oceans (see figure 1.10 and chapter 3). A few earthquakes have also been recorded in isolated spots usually associated with basaltic volcanoes, such as those of Hawaii.

In most parts of the circum-Pacific belt, earthquakes, andesitic volcanoes, and *oceanic trenches* (see chapter 3) appear to be closely associated. Careful determination of the locations and depths of focus of earthquakes has revealed the existence of distinct *earthquake zones* that begin at oceanic trenches and slope landward and downward into Earth at an angle of about 30° to 60° (figure 7.23). Such zones of inclined seismic activity are called **Benioff zones** after the man who first recognized them.

Benioff zones slope under a continent or a curved line of islands called an **island arc.** Andesitic volcanoes may form the islands of the island arc, or they may be found near the edge of a continent that overlies a Benioff zone.

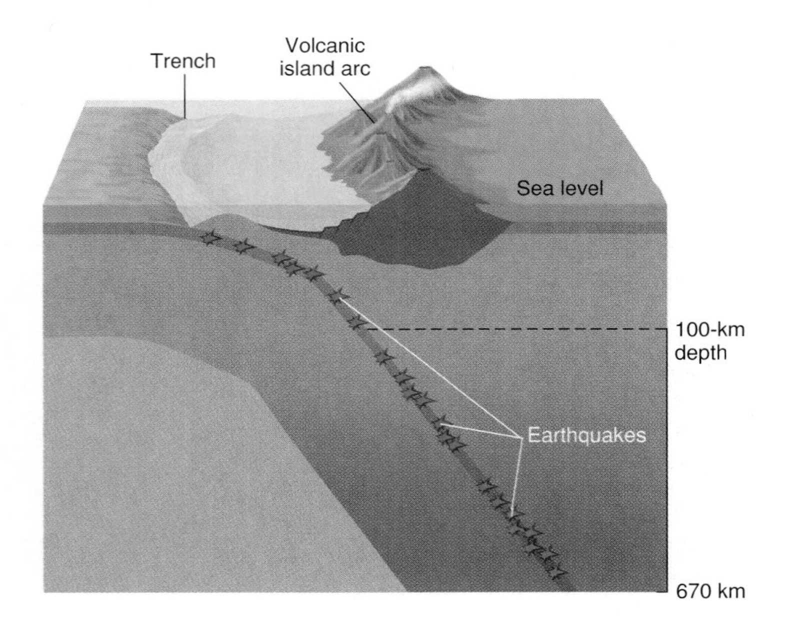

FIGURE 7.23

A Benioff zone of earthquakes begins at an oceanic trench and dips under a continent (such as South America) or a volcanic island arc. Upper part of Benioff zone may extend to a depth of 670 kilometers.

Most of the circum-Pacific belt is made up of Benioff zones associated in this manner with oceanic trenches and andesitic volcanoes. Parts of the Mediterranean-Himalayan belt represent Benioff zones, too, notably in the eastern Mediterranean Sea and in the East Indies. Essentially all the world's intermediate- and deep-focus earthquakes occur in Benioff zones.

FIRST-MOTION STUDIES OF EARTHQUAKES

By studying seismograms of an earthquake on a distant fault, geologists can tell which way rocks moved along that fault. First-motion studies play an important role in determining the overall sense of movement along plate boundaries. Rock motion is determined by examining seismograms from many locations surrounding a quake. Each seismograph station can tell whether the first rock motion recorded there was a push or a pull (figure 7.24). If the rock moved toward the station (a push), then the pen drawing the seismogram is deflected up. If the first motion is away from the station (a pull), then the pen is deflected down.

If an earthquake occurs on a fault as shown in figure 7.25*A,* large areas around the fault will receive a push as first motion, while different areas will receive a pull. Any station within the green area marked *A* will receive a push, for the rock is moving from the epicenter toward those stations, as shown by the arrows in the figure. All stations in area *C* will also receive a push, but areas *B* and *D* will record a pull as the first motion.

In figure 7.25*B* you can see the same pattern of pushes and pulls, but in this case the pattern is caused by a fault with a different orientation. Either fault can cause the same pattern, if the rock moves in the direction shown by the arrows. In other words, there are two possible solutions to any pattern of first motions.

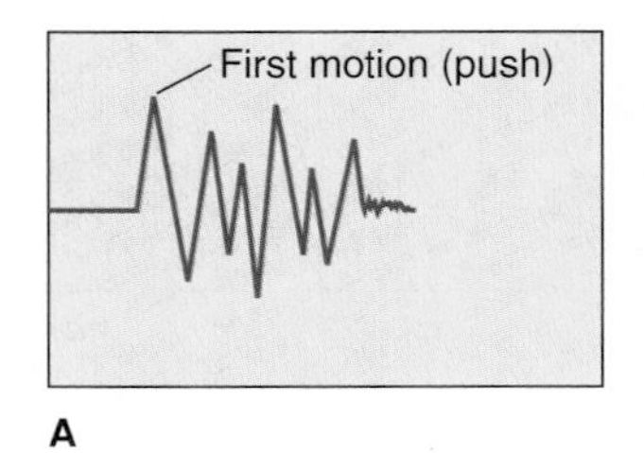

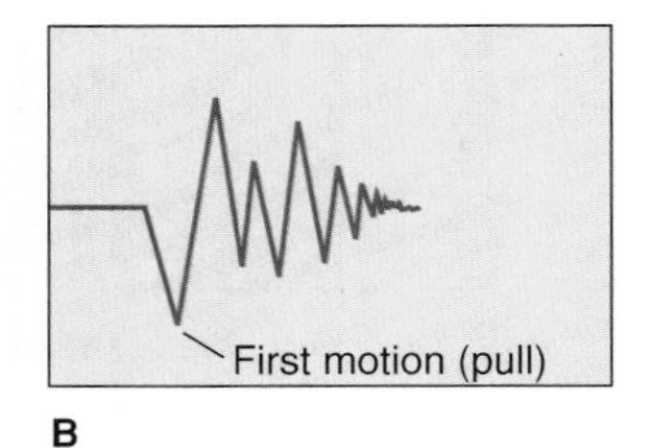

FIGURE 7.24

Seismograms showing how first horizontal motions of rocks along a fault are determined. (*A*) If the first motion is a push (from the epicenter to the seismograph station), the seismogram trace is deflected upward. (*B*) If it is a pull (away from the station), the deflection is downward.

EARTHQUAKE

WHAT TO DO BEFORE, DURING, AND AFTER AN EARTHQUAKE

Being prepared for an earthquake can reduce the damage to your property and chance of serious injury or loss of life. There is a saying, "earthquakes do not kill people, buildings do." If you live in or visit an earthquake-prone area, you should do the following:

BEFORE AN EARTHQUAKE

1. Make sure your house is firmly attached to the foundation with anchor bolts; repair any deep cracks in foundations or ceilings. Brick chimneys should be braced and anchored to the roof joists.
2. Check for hazards inside the home. Tall bookshelves should be bolted to the wall, place heavy objects on the bottom shelves, flammable items and household chemicals should also be on the bottom shelves in locked cabinets, glass and china should be in lower cabinets secured with strong latches, heavy pictures and mirrors should not be hung where people sit or sleep, the water heater should be strapped to wall studs and bolted to the floor.
3. Learn how to turn off all the utilities at your house; flexible gas lines should be used to avoid breaking. Keep an adjustable wrench near the gas main to shut off the gas immediately after an earthquake to avoid fires.
4. Have disaster supplies on hand (keep in a safe place in large, lockable plastic trash container): flashlight and extra batteries, portable radio, first-aid kit and manual, essential medicines, emergency food and water (one gallon per person, per day), nonelectric can opener, sleeping bags and tent, fire extinguisher, matches, portable stove and propane, sturdy shoes, cash and credit cards. (*Check the condition of batteries, water, and food every 6 months.*)
5. Have an emergency communication plan to reunite family members who may be separated from one another during the earthquake. Because it is often easier to call long distance after an earthquake, establish an out-of-state relative or friend to act as the contact person.

DURING AN EARTHQUAKE

1. If you are indoors, DROP, COVER, and HOLD under a heavy piece of furniture positioned against an inside wall, or crouch in a room corner or interior hall or doorway. Stay away from windows or anything that could fall on you. In a high-rise building, do not run to exits or stairways that may be damaged or jammed with people; never use the elevator.
2. If in an unreinforced building or otherwise unsafe building, it may be better to leave the building. Because most injuries result from people leaving buildings and being hit by falling debris or downed utility lines, be alert to possible dangers.
3. If you are outdoors, move to an open area away from buildings, street lights, and utility lines until the shaking stops.
4. If in a moving vehicle, stop quickly and stay in the car. Move away from buildings, trees, bridges, ramps, overpasses, and utility lines.

AFTER AN EARTHQUAKE

1. Be prepared for aftershocks.
2. Help anyone who is injured or trapped; do not move seriously injured persons unless they are in immediate danger of further injury.
3. Check for damage to utilities. If you smell gas, turn off gas valves, open the windows, and leave immediately. If electricity is shorting out, turn off the main power switch at the meter box. If water pipes are broken, turn off the supply at the main valve. In an emergency, water from hot water tanks, toilet bowls, and melted ice cubes can be used. Do not flush the toilet until sewage lines are checked.
4. Carefully inspect your chimney for damage to prevent fire and carbon monoxide poisoning.
5. Listen to the radio for the latest emergency information; use your telephone only for emergency calls.
6. Do not travel unnecessarily; avoid low-lying coastal areas until the threat of a tsunami has passed; also avoid landslide areas and severely damaged structures.

Related Web Resources

For additional safety information visit the American Red Cross website:
www.redcross.org/services/disaster/keepsafe/readyearth.html
or the U. S. Federal Emergency Management Agency site:
www.fema.gov/library/quakef.htm
An online version of the Southern California Earthquake Center's handbook (SCEC) *Putting Down Roots in Earthquake Country* gives detailed information on preparing for earthquakes and specific information for southern California seismicity:
www.scecdc.scec.org/eqcountry.html

From U.S. Federal Emergency Management Agency (FEMA) and the Red Cross

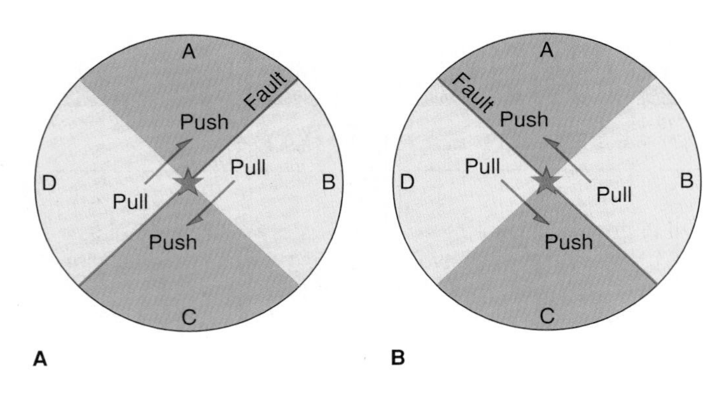

FIGURE 7.25

Map view of two possible solutions for the same pattern of first motion. Each solution has a different fault orientation. If the fault orientation is known, the correct solution can be chosen. The star marks the epicenter, and rock motion is shown by the arrows.

If the orientation of the fault trace on Earth's surface is known, as it is for most faults on land (and for a few on the ocean floor), the correct choice of the two solutions can be made. But if the orientation of the fault is not known—as is the usual case for earthquakes at sea or at great depth—the choice between the two possible solutions can be difficult. One solution may be more *likely,* based on the study of topography, other faults, or similar factors. But such a solution is by no means *proven* if the fault orientation is not known. As you read about first-motion studies, keep in mind that there are always two solutions to any pattern.

EARTHQUAKES AND PLATE TECTONICS

One of the great attractions of the concept of plate tectonics is its ability to explain the distribution of earthquakes and the rock motion associated with them.

As described in chapter 4, the concept of plate tectonics is that Earth's surface is divided into a few giant *plates.* Plates are rigid slabs of rock, thousands of kilometers wide and 70 to 125 (or more) kilometers thick, that move across Earth's surface. Because the plates include continents and sea floors on their upper surfaces, the plate tectonics concept means that the continents and sea floors are moving. The plates change not only position but size and shape.

Earthquakes occur commonly at the edges of plates but only occasionally in the middle of a plate. The close correspondence between plate edges and earthquake belts can be seen by comparing the map of earthquake distribution in figure 7.22*A* with the plate map in figure 7.22*B*.

This correspondence is hardly surprising—plate boundaries are identified and *defined* by earthquakes. According to plate tectonics, earthquakes are caused by the interactions of plates along plate boundaries. Therefore, narrow bands of earthquakes are used to outline plates on plate maps. This can be clearly seen in the east Pacific Ocean off South America, where the Nazca plate (figure 7.22*B*) is almost completely outlined by earthquake epicenters (figure 7.22*A*).

The earthquakes on the western border of the Nazca plate are shallow-focus quakes, and they occur in a narrow belt along the crest of the mid-oceanic ridge here, locally called the East Pacific Rise. The quakes along the eastern boundary occupy a broader belt that lies mostly within South America. This belt includes shallow-, intermediate-, and deep-focus earthquakes in a Benioff zone that begins at the Peru-Chile Trench just offshore and slopes steeply down under South America to the east. The Nazca plate moves eastward, away from the crest of the mid-oceanic ridge and toward the subduction zone at the trench, where the plate plunges down into the mantle. The plate's western boundary is located at the crest of the East Pacific Rise, and its eastern boundary is at the bottom of the Peru-Chile Trench.

Earthquakes at Plate Boundaries

As you have learned, there are three types of plate boundaries, *divergent boundaries* where plates move away from each other, *transform boundaries* where plates move horizontally past each other, and *convergent boundaries* where plates move toward each other. Each type of boundary has a characteristic pattern of earthquake distribution and rock motion.

Divergent Boundaries

At a divergent boundary, where plates move away from each other, earthquakes are shallow, restricted to a narrow band, and much lower magnitude than those that occur at convergent or transform boundaries. A divergent boundary on the sea floor is marked by the crest of the mid-oceanic ridge and the *rift valley* that is often (but not always) found on the ridge crest (figure 7.26*A*). The earthquakes are located along the sides of the rift valley and beneath its floor. The rock motion that is deduced from first-motion studies shows that the faults here are normal faults, parallel to the rift valley. (This is the most likely solution of the first-motion studies. Keep in mind as you read this section that there are always two solutions for first motion; we give only one—the more likely solution.) If this interpretation of the first motions is correct, it implies that the ridge crest is undergoing horizontal extension, which is apparently tearing the sea floor open here, creating the rift valley and causing the earthquakes.

A divergent boundary within a *continent* is usually also marked by a rift valley, shallow-focus quakes, and normal faults (figure 7.26*B*). The African Rift Valleys in eastern Africa (figure 7.22*B*) seem to be such a boundary. Horizontal extension here may be tearing eastern Africa slowly apart, creating the rift valleys, some of which contain lakes.

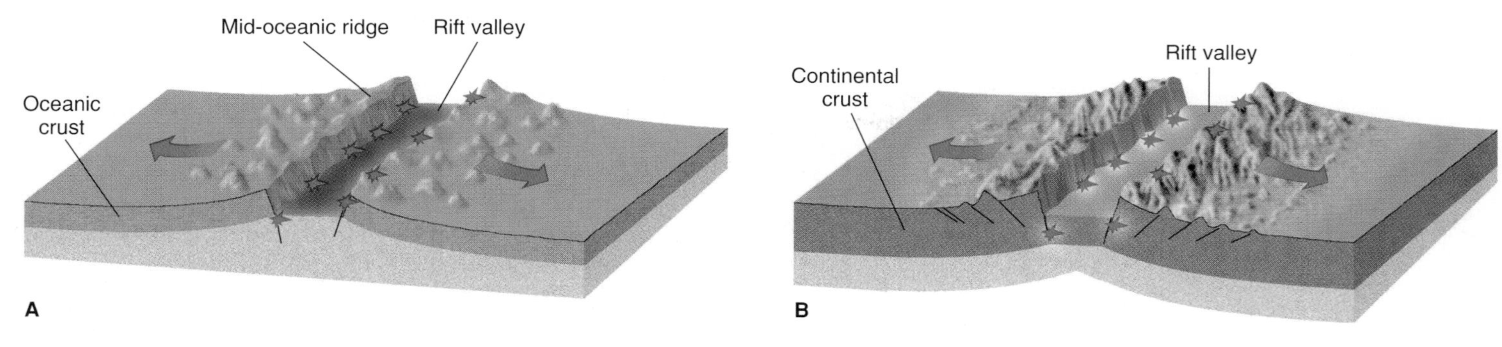

FIGURE 7.26

Divergent plate boundaries. (*A*) On the ocean floor. (*B*) On a continent. Each is marked by a rift valley and shallow-focus earthquakes (shown as stars). Depth of rift valleys is exaggerated.

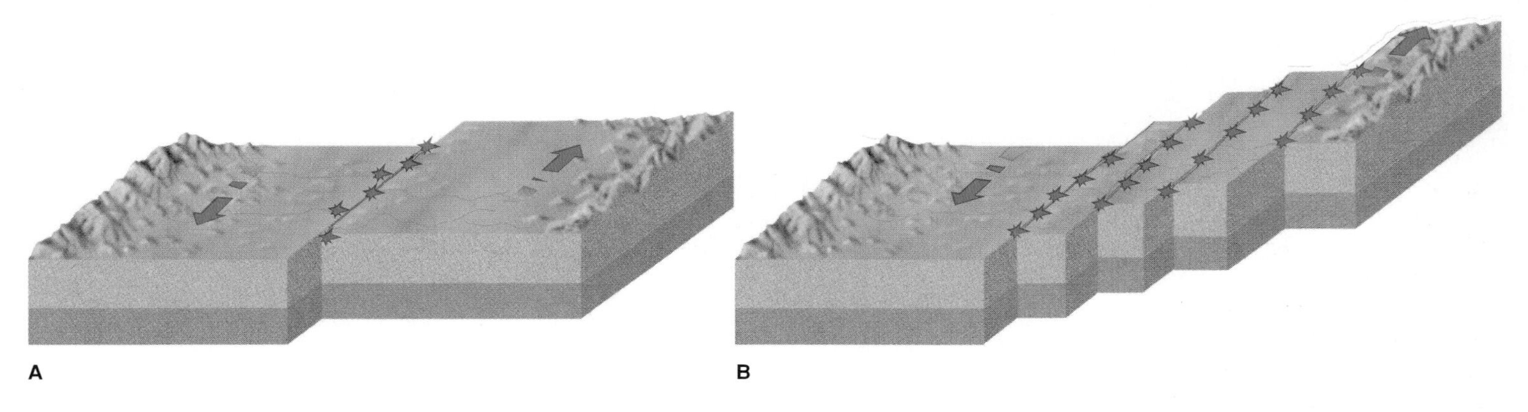

FIGURE 7.27

Transform boundaries. (*A*) Narrow band of shallow-focus earthquakes shown as stars along single fault. (*B*) Broad band of earthquakes along a system of parallel faults.

Transform Boundaries

Where two plates move past each other along a transform boundary, the earthquakes are shallow. First-motion studies indicate strike-slip motion on faults parallel to the boundary. The earthquakes may be aligned in a narrow band along one fault (figure 7.27*A*), or they may form a broader zone if plate motion is taken up by movement along a system of parallel faults (figure 7.27*B*). The San Andreas fault in California (figure 7.22*B*) may be an example of a single fault forming a plate boundary, but many geologists believe that the broad zone of seismicity along the Basin and Range faults in the western United States (figure 7.13) shows that a system of faults both parallel and at an angle to the San Andreas forms the plate boundary here.

Convergent Boundaries

Convergent boundaries are of two general types, one marked by the *collision* of two continents (figure 7.28*A*), the other marked by *subduction* of the ocean floor under a continent (figure 7.28*B*) or another piece of sea floor. Each type has a characteristic pattern of earthquakes.

Collision boundaries are characterized by broad zones of shallow earthquakes on a complex system of faults (figure 7.28*A*). Some of the faults are parallel to the dip of the suture zone that marks the line of collision; some are not. One continent usually overrides the other slightly (continents are not dense enough to be subducted), creating thick crust and a mountain range. The Himalaya are thought to represent such a boundary (figure 7.21*B*). The seismic zone is so broad and complex at such boundaries that other criteria, such as detailed geologic maps, must be used to identify the position of the suture zone at the plate boundary.

During *subduction,* earthquakes occur for several different reasons (figure 7.28*B*). As a dense oceanic plate bends to go down at a trench, it stretches slightly at the top of the bend, and normal faults occur as the rocks are subjected to *tension.* This gives a block-faulted character to the outer (seaward) wall of a trench. For some distance below the trench, the subducting plate is in contact with the overlying plate. First-motion studies of

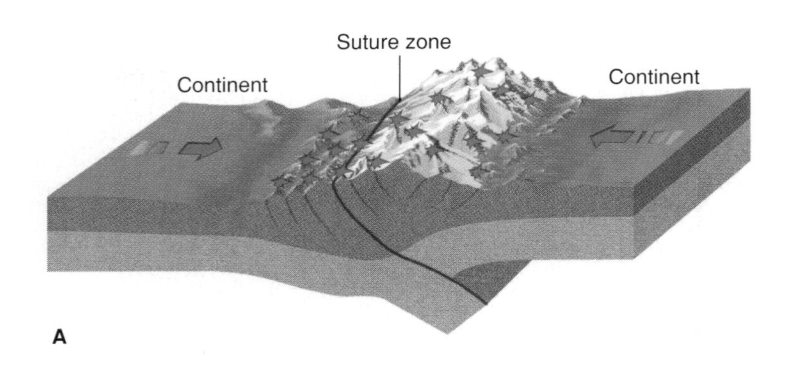

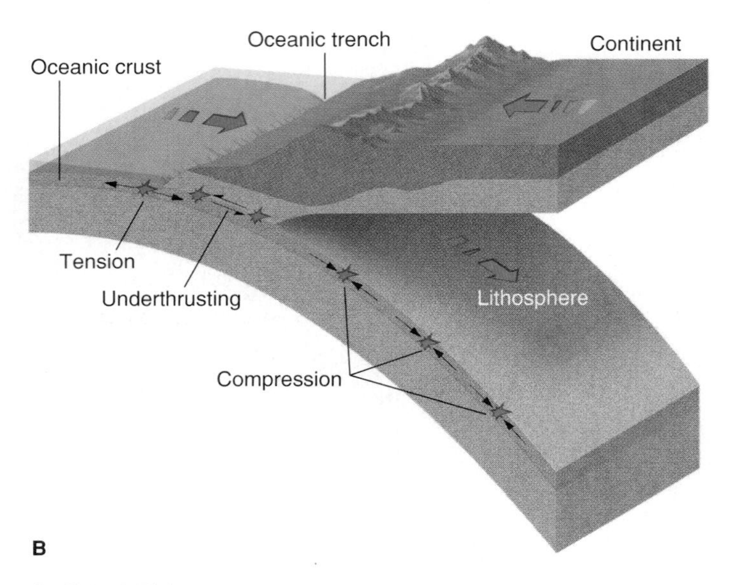

FIGURE 7.28

(*A*) A convergent boundary marked by the collision of two continents. A very broad zone of shallow-focus earthquakes occurs along a complex system of faults. (*B*) A convergent boundary with ocean floor subducting under a continent. Earthquakes occur near the top of the subducting plate due to tension, underthrusting, and compression.

earthquakes at these shallow depths show that the quakes are caused by shallow-angle thrust-faulting. This is the motion expected as one plate slides beneath another, a process commonly called *underthrusting*.

At greater depths, where the descending plate is not in direct contact with the overlying plate, earthquakes are common, but the reasons for them are not obvious. The quakes are confined to a thin zone, only 20 to 30 kilometers thick, within the lithosphere of the descending plate, which is about 100 kilometers thick (figure 7.28*B*). This zone is thought to be near the top of the lithosphere, where the rock is colder and more brittle.

First motions of these quakes imply that along many subduction zones the quakes are caused by *compression* parallel to the dip of the plate (figure 7.28*B*). This could mean that the plates are running into some obstacle at depth (figure 7.29*A*).

Earthquake first motion on other subduction zones, however, implies *tension* parallel to the dip of the descending plates (figure 7.29*B*). This might mean that the plates are being pulled along by their dense leading edges. Cold, descending lithosphere is slightly denser than the surrounding hot mantle. The density of a subducting plate may also be increased by the loss of volatile materials or the pressure collapse of low-density minerals into denser forms.

The distribution of earthquakes with depth on a subducting plate suggests that mineral transformations may cause or control quakes (figure 7.30). Quakes are common near the surface and decrease exponentially to a depth of about 300 kilometers. There are few quakes from 300 to 450 kilometers, but many quakes from 450 to 670 kilometers. Serpentine on a subducting plate should lose its water at depths of 100 to 300 kilometers, perhaps causing quakes if the dehydration is sudden. Olivine collapses to the denser spinel structure at 400 kilometers. Serpentine forms glass at 450 to 670 kilometers, and spinel collapses to perovskite at 670 kilometers. These transformations occur suddenly, and sometimes noisily, in the laboratory. There is debate about whether they can create quakes in real plates. The olivine-spinel and spinel-perovskite transformations seem to mark the lower boundaries of quake depth zones.

Subduction Angle

The horizontal and vertical distribution of earthquakes can be used to determine the angle of subduction of a downgoing plate. Subduction angles vary considerably from trench to trench. Many plates start subducting at a gentle angle, which becomes much steeper with depth (figure 7.30). At a few trenches in the open Pacific, subduction begins (and continues) at almost a vertical angle. Subduction angle is probably controlled by plate density and by the rate of plate convergence.

Some plates crumple into folds as they descend, usually at depths below 200 kilometers. Other plates tear into segments during subduction (figure 7.31), perhaps as a result of density variations within a descending plate. The thick oceanic crust of a submarine volcanic ridge or plateau is buoyant, and subducts at a gentle angle. Thin, normal oceanic crust results in a denser plate segment, which subducts at a steep angle. Folding and tearing of a descending plate may also be required geometrically to allow subduction of a curved rigid plate to successively greater depths. The plate forms with a radius of curvature equal to that of Earth's surface, and during subduction it sinks to regions of progressively smaller radius.

In summary, earthquakes are very closely related to plate tectonics. Most plate boundaries are defined by the distribution of earthquakes, and plate motion can be deduced by the first motions of the quakes. Analysis of first motions can also help determine the type and orientation of stresses that act on plates, such as tension and compression. Quake distribution with depth indicates the angle of subduction and has shown that some plates change subduction angle and even break up as they descend. A few quakes, such as those that occur in the center of plates, cannot easily be related to plate motion.

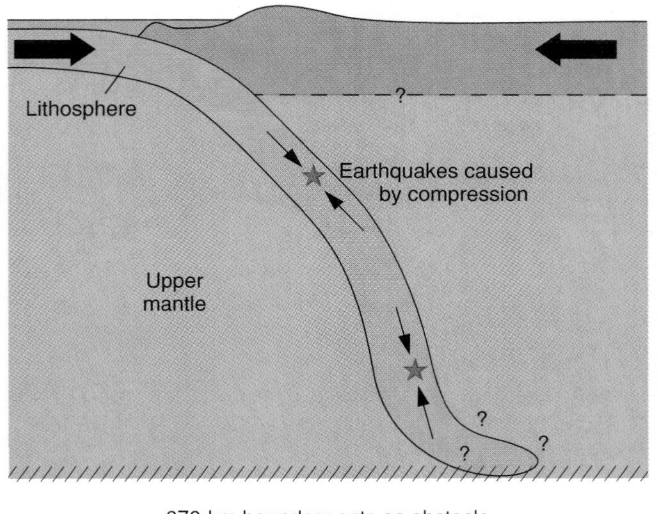

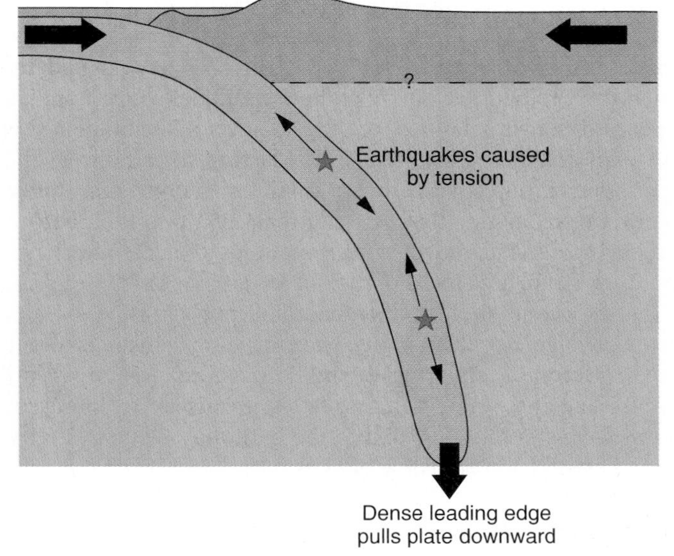

FIGURE 7.29

(*A*) Many subducting plates have earthquakes caused by compression, perhaps from hitting an obstacle. (*B*) Other plates have tensional earthquakes, probably from the plate being pulled downward.

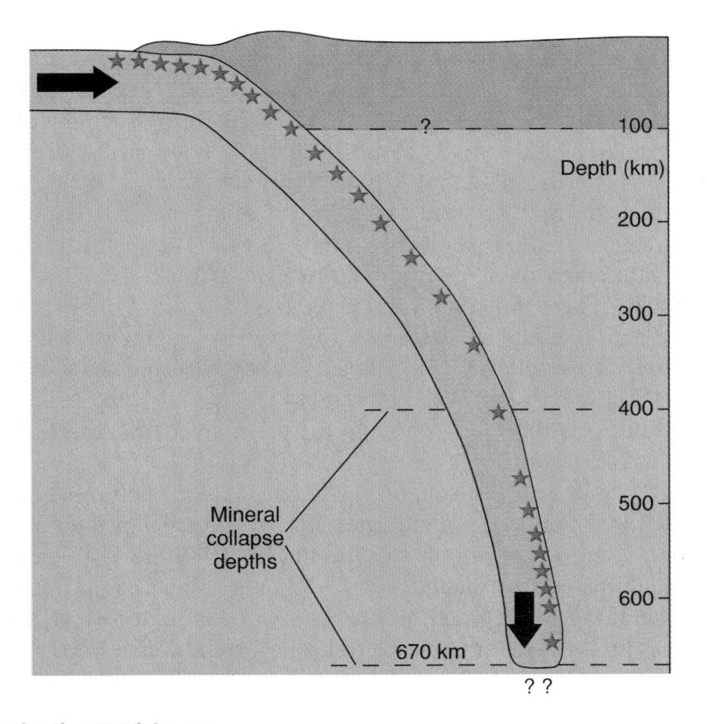

FIGURE 7.30

The distribution of earthquakes with depth.

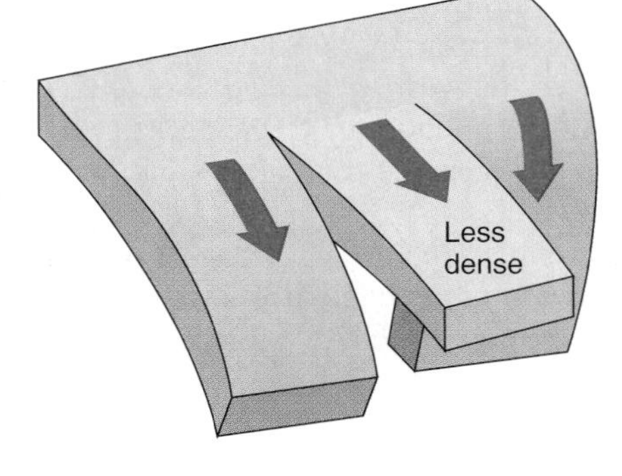

FIGURE 7.31

A descending plate may tear into segments, some with steeper subduction angles than others.

EARTHQUAKE PREDICTION

People who live in earthquake-prone regions are plagued by unscientific predictions of impending earthquakes by popular writers and self-proclaimed prophets. Several techniques are being explored for *scientifically* forecasting a coming earthquake. One group of methods involves monitoring slight changes, or precursors, that occur in rock next to a fault before the rock breaks and moves; these methods that assume large amounts of strain are stored in rock before it breaks (figure 7.2).

Just as a bent stick may crackle and pop before it breaks with a loud snap, a rock may give warning signals that it is about to break. Before a large quake, small cracks may open within the rock, causing small tremors, or *microseisms,* to increase. The *properties of the rock* next to the fault may be changed by the opening of such cracks. Changes in the rock's magnetism, electrical resistivity, or seismic velocity may give some warning of an impending quake.

The opening of tiny cracks changes the rock's porosity, so *water levels in wells* often rise or fall before quakes. The cracks provide pathways for the release of radioactive radon gas from rocks (radon is a product of radioactive decay of uranium and other elements). An *increase in radon emission from wells* may be a prelude to an earthquake. A very local method of predicting quakes is to time the *interval between eruptions of Old Faithful geyser* in Yellowstone National Park. Long-term records of the time between eruptions have shown that this interval changes in a regular way before a large local earthquake, probably because of porosity changes within the surrounding rock.

In some areas the *surface of Earth tilts and changes elevation* slightly before an earthquake. Scientists use highly sensitive instruments to measure this increasing strain, in hopes of predicting quakes.

Chinese scientists claim successful short-range predictions by watching *animal behavior*—horses become skittish and snakes leave their holes shortly before a quake. United States scientists are conducting a few pilot programs along these lines, but many are skeptical of the Chinese claims.

Japanese and Russian geologists were the first to predict earthquakes successfully, and Chinese geologists have made some very accurate predictions. In 1975 a 7.3-magnitude earthquake near Haicheng in northeastern China was predicted five hours before it happened. Alerted by a series of *foreshocks,* authorities evacuated about a million people from their homes; many watched outdoor movies in the open town square. Half the buildings in Haicheng were destroyed, along with many entire villages, but only a few hundred lives were lost. In grim contrast, however, the Chinese program failed to predict the 1976 Tangshan earthquake (magnitude 7.6), which struck with no warning and killed an estimated 250,000 people.

But all of the above methods assume that great strain is being stored in rock before it breaks and that that strain (or its effects) can be measured. As described in an earlier section, the new hypothesis that faults are weak (and break under small stress) contradicts this assumption. Most of these methods were considered very promising in the 1970s but have proved to be of little real help in predicting quakes. A typical quake predictor, such as tilt of the land surface, may precede one quake and then be absent for the next 10 quakes. In addition, each precursor can be caused by forces unrelated to earthquakes (land tilt is also caused by mountain building, mass wasting, and wetting and drying of the land). As a result, enthusiasm and funding for quake prediction are currently declining.

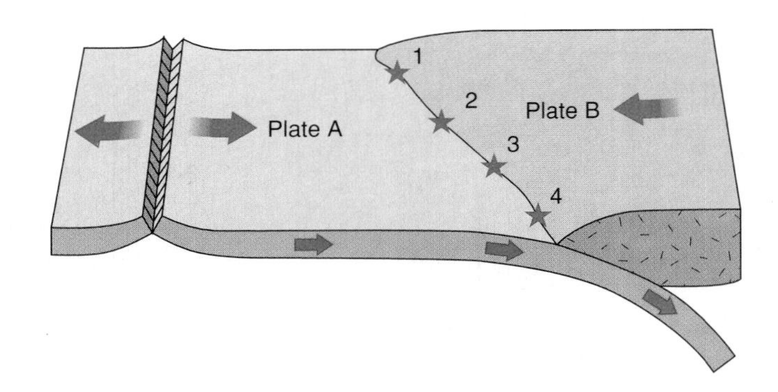

FIGURE 7.32
As plate A slides diagonally under plate B, earthquakes occur in a regular sequence from one to four. The entire sequence may repeat itself.

A fundamentally different method of earthquake risk analysis has been developed that does not depend upon whether faults are weak or strong. It involves determining the history of earthquakes along a fault, and using the history to calculate the probability of a quake at a given location. *Patterns of earthquakes in space and time* are being recognized, particularly on the faults along some plate boundaries. If one plate is sliding diagonally under another plate, a series of earthquakes may occur in a regular way along the plate boundary (figure 7.32). If such a pattern is known to repeat itself in the same area over several centuries, it becomes possible to forecast where the next large quake might occur. Along some long-active faults, there are short inactive segments called *seismic gaps* where earthquakes have not occurred for a long time. These gaps are widely thought to be the most likely sites for future earthquakes, so many of them are being carefully studied (see box 7.3).

By studying the seismic history of faults, geologists in the United States are sometimes able to forecast earthquakes along some segments of some faults. In 1988 the U.S. Geological Survey estimated a 50% chance of a magnitude-7 quake along the segment of the San Andreas fault near Santa Cruz. In 1989 the magnitude-7 Loma Prieta quake occurred on this very section. Since the techniques are new and in some cases only partly understood, some errors will undoubtedly be made. Many faults are not monitored or studied historically because of lack of money and personnel, so we will never have a warning of impending quakes in some regions. For large urban areas near active faults such as the San Andreas, however, earthquake risk analysis may reduce damage and loss of life.

ENVIRONMENTAL GEOLOGY IN CALIFORNIA 7.3

WAITING FOR THE BIG ONE IN CALIFORNIA

The San Andreas fault, running north–south for 1,300 kilometers through California, is a right-lateral fault capable of generating great earthquakes of magnitude 8 or more. The 1906 earthquake (estimated Richter magnitude 8.25) near San Francisco caused a 450-kilometer scar in northern California (box figure 1). The portion of the fault nearest Los Angeles last broke in 1857 in a quake that was probably of comparable size. The ground has not broken in either of these regions since these quakes. Each old break is now a seismic gap, where rock strain is being stored prior to the next giant quake.

The recent California quakes were considerably smaller than the "Big One" long predicted by geologists to be in the magnitude-8 range. The 1906 quake in the north had an estimated Richter magnitude of 8.25, and a moment magnitude of 7.7. The Fort Tejon quake in the south was 7.6 Richter, 7.9 moment (see table 7.3). In contrast, the 1989 Loma Prieta quake on the San Andreas fault near San Francisco was 7.0 Richter, 7.2 moment. The 1994 Northridge quake (not on the San Andreas fault) was 6.4 Richter, 6.7 moment. So recent California quakes have been about magnitude 7 or less, and the Big One should be 8. A magnitude-8 quake has 10 times the ground shaking and 32 times the energy of a magnitude-7 quake. In other words it would take about 32 Loma Prieta quakes to equal the Big One. Comparing moment magnitudes for 1994 and 1857 in southern California, it would take nearly 64 Northridge quakes to equal the Fort Tejon quake.

A great earthquake of magnitude 8 could strike either the northern section or the southern section of the San Andreas fault tomorrow. Which section will break first? Because the southern section has been inactive longer, it may be the likelier candidate. A magnitude-8 quake here could cause hundreds of billions of dollars in damage and kill thousands of people if it struck during weekday business hours when Los Angeles area buildings and streets are crowded with people.

Spotty evidence suggests that great earthquakes have a recurrence interval of about 140 years on the southern portion of the San Andreas fault. Historic records in California do not go very far back in time, and much of the evidence involves isotopic dating of broken beds of carbon-rich lake muds. This number is only an average; 2 or 3 quakes spaced about 60 years apart occur in clusters, with the interval between clusters being 200 to 300 years. Prior to 1857 this portion last broke in about 1745. Adding this 112-year difference to 1857 gives 1969, and adding the average of 140 years to 1857 gives 1997, so the present danger to Los Angeles is clear. If the 1857 quake ended a cluster, however, the next big quake may not occur till about 2050 or later.

The southern section of the fault is heavily instrumented to detect changes in ground tilt, water-well levels, and other earthquake precursors. In the mid-1970s most geologists thought that the Big One was imminent when it was discovered that most of southern California had risen as much as 0.4 meters from 1959 to 1974, with the greatest uplift occurring very close to the fault (the "Palmdale bulge"). Rapid subsidence from 1974 to 1976, however, eliminated at least half of this uplift, and the episode of rising and falling land passed without a large quake.

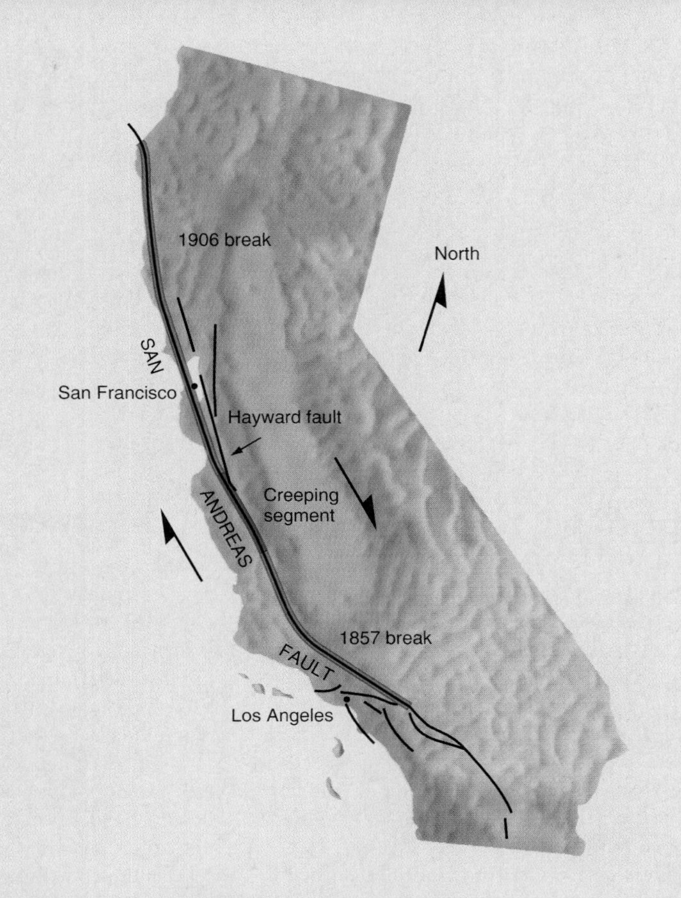

BOX 7.3 — FIGURE 1

The two major breaks on the San Andreas fault in California. Each break occurred during a giant earthquake. Each old break is now a seismic gap (shown in red) where the fault is locked and may be the future site for another major earthquake.
From U.S. Geological Survey

But the northern portion of the San Andreas fault is dangerous, too. Prior to 1906 this section of the fault broke in another giant quake in 1838. These quakes were only 68 years apart, and 1906 plus 68 equals 1974, so the northern section may actually be overdue for a big quake.

Another way of estimating the recurrence interval is by rock displacement. In 1906 rocks were displaced about 4.5 meters at the epicenter, and we know that plate motion across the San Andreas fault is about 5 centimeters per year. It should therefore take about 90 years to store enough strain to move rocks 5 meters, so the next quake should have occurred in 1996.

The U.S. Geological Survey recently estimated that the southern portion of the fault has about a 60% chance of an earthquake of magnitude 7.5 to 8.3 within the next 30 years. The 1992 Landers and the 1994 Northridge quakes have occurred since that prediction, and there is disagreement among geologists as to whether those quakes increase or decrease the likelihood of the Big One in southern California. Even more likely (85% chance) is a 7^+-magnitude quake on any one of several faults that parallel the San Andreas fault and lie closer to (or even under) Los Angeles.

The probability of a repeat of the 1906 quake (8^+) on the northern section of the San Andreas fault may be very low, less than 21% for the next 30 years. Following the 1989 Loma Prieta quake, however, the chance of a 6.7 or greater quake in the San Francisco Bay area in the next 30 years has been raised to 70%. A likely candidate for the quake is not the San Andreas but the Hayward fault across the Bay from San Francisco. Such a quake near or under Bay area cities such as Oakland and Berkeley would cause far greater death and destruction than the 1989 quake.

Although the chance of an 8^+-magnitude Big One may be higher in southern California, the chance of a 7^+-quake killing thousands of people is about equal in the north and the south.

Further Reading

U.S. Geological Survey, 1990. *The San Andreas fault system.* Professional Paper 1515.

Related Web Resources

For more information about the San Andreas Fault and the likelihood of it creating a large earthquake, visit U.S. Geological Survey websites:

http://pubs.usgs.gov/gip/earthq3/safaultgip.html and
http://quake.wr.usgs.gov/hazprep/BayAreaInsert/
http://geopubs.wr.usgs.gov/fact-sheet/fs152-99/

Summary

Earthquakes usually occur when rocks break and move along a fault to release strain that has gradually built up in the rock. Volcanic activity can also cause earthquakes. Deep quakes may be caused by mineral transformations.

Seismic waves move out from the earthquake's *focus. Body waves* (P waves and S waves) move through Earth's interior, and *surface waves* (*Love waves and Rayleigh waves*) move on Earth's surface.

Seismographs record seismic waves on *seismograms,* which can be used to determine an earthquake's strength, location, and depth of focus. Most earthquakes are shallow-focus quakes, but some occur as deep as 670 kilometers below Earth's surface.

The time interval between first arrivals of P and S waves is used to determine the distance between the seismograph and the *epicenter.* Three or more stations are needed to determine the location of earthquakes.

Earthquake *intensity* is determined by assessing damage and is measured on the *modified Mercalli scale.*

Earthquake *magnitude,* determined by the amplitude of seismic waves on a seismogram, is measured on the *Richter scale. Moment magnitudes,* determined by field work, are widely used today and often are larger than Richter magnitudes.

The most noticeable effects of earthquakes are ground motion and displacement (which destroy buildings and thereby injure or kill people), fire, landslides, and *tsunami. Aftershocks* can continue to cause damage months after the main shock.

Earthquakes are generally distributed in belts. The *circum-Pacific belt* contains most of the world's earthquakes. Earthquakes also occur on the Mediterranean-Himalayan belt, the crest of the mid-oceanic ridge, and in association with basaltic volcanoes.

Benioff zones of shallow-, intermediate-, and deep-focus earthquakes are associated with andesitic volcanoes, oceanic trenches, and the edges of continents or island arcs.

The concept of plate tectonics explains most earthquakes as being caused by interactions between two plates at their boundaries. Plate boundaries are generally defined by bands of earthquakes.

Divergent plate boundaries are marked by a narrow zone of shallow earthquakes along normal faults, usually in a rift valley. Transform boundaries are marked by shallow quakes caused by strike-slip motion along one or more faults.

Convergent boundaries where continents collide are marked by a very broad zone of shallow quakes. Convergent boundaries involving deep subduction are marked by Benioff zones of quakes caused by tension, underthrusting, and compression.

The distribution of quakes indicates subduction angles and shows that some plates break up as they descend.

Although not yet too successful, earthquake prediction has been attempted through the measurement of rock properties near faults and the recognition of patterns of quakes at plate boundaries, including seismic gaps.

Terms to Remember

aftershock 171
Benioff zone 176
body wave 159
circum-Pacific belt 176
depth of focus 163
earthquake 156
elastic rebound theory 156
epicenter 159
focus 159
intensity 163
island arc 176
Love wave 159

magnitude 164
Mediterranean-Himalayan belt 176
modified Mercalli scale 163
moment magnitude 164
P wave 159
Rayleigh wave 159
Richter scale 164
seismic sea wave 171
seismic wave 156
seismogram 162
seismograph 160
surface wave 159
S wave 159
travel-time curve 162
tsunami (seismic sea wave) 171

Testing Your Knowledge

Use the questions below to prepare for exams based on this chapter.

1. Describe in detail how earthquake epicenters are located by seismograph stations.
2. What causes earthquakes?
3. Compare and contrast the concepts of intensity and magnitude of earthquakes.
4. Name and describe the various types of seismic waves.
5. Discuss the distribution of earthquakes with regard to location and depth of focus.
6. Show with a sketch how the concept of plate tectonics can explain the distribution of earthquakes in a Benioff zone and on the crest of the mid-oceanic ridge.
7. Describe several techniques that may help scientists predict earthquakes.
8. How may the timing of earthquakes someday be controlled?
9. Describe several ways that earthquakes cause damage.
10. How do earthquakes cause tsunami?
11. What are aftershocks?
12. The elastic rebound theory
    a. explains folding of rocks
    b. explains the behavior of seismic waves
    c. involves the sudden release of progressively stored strain in rocks, causing movement along a fault
    d. none of the above
13. The point within Earth where seismic waves originate is called the
    a. focus b. epicenter
    c. fault scarp d. fold
14. P waves are
    a. compressional b. transverse
    c. tensional
15. What is the minimum number of seismic stations needed to determine the location of the epicenter of an earthquake?
    a. 1 b. 2
    c. 3 d. 5
    e. 10
16. The Richter scale measures
    a. intensity
    b. magnitude
    c. damage and destruction caused by the earthquake
    d. the number of people killed by the earthquake
17. Benioff zones are found near
    a. mid-ocean ridges
    b. ancient mountain chains
    c. interiors of continents
    d. oceanic trenches
18. Most earthquakes at divergent plate boundaries are
    a. shallow focus
    b. intermediate focus
    c. deep focus
    d. all of the above
19. Most earthquakes at convergent plate boundaries are
    a. shallow focus
    b. intermediate focus
    c. deep focus
    d. all of the above
20. A zone of shallow earthquakes along normal faults is typical of
    a. diverging boundaries
    b. transform boundaries
    c. subduction zones
    d. collision boundaries
21. A seismic gap is
    a. the time between large earthquakes
    b. a segment of an active fault where earthquakes have not occurred for a long time
    c. the center of a plate where earthquakes rarely happen
22. Which of the following is not true of tsunami?
    a. very long wavelength
    b. high wave height in deep water
    c. very fast moving
    d. continued flooding after wave crest hits shore

Expanding Your Knowledge

1. What are some arguments in favor of and against predicting earthquakes? What would happen in your community if a prediction were made today that within a month a large earthquake would occur nearby?
2. Most earthquakes occur at plate boundaries where plates interact with each other. How might earthquakes be caused in the interior of a rigid plate?
3. How can you prepare for an earthquake in your own home?
4. Suppose you want to check for earthquake danger before buying a new home. How can you check the regional geology for earthquake dangers? The actual building site? The home itself?

Exploring Web Resources

www.mhhe.com/plummer9e
Visit our Online Learning Center for additional readings and media resources. Check your answers for Testing Your Knowledge, and click on the links to the other great websites listed below.

http://quake.wr.usgs.gov/hazprep/BayAreaInsert/
U.S. Geological Survey, 1990. *The next big earthquake.* This magazine-like pamphlet is available free from Earthquakes, U.S.G.S., 345 Middlefield Road, Menlo Park, CA 94025.

http://pubs.usgs.gov/gip/earthq3/safaultgip.html
U.S. Geological Survey, 1990. *The San Andreas fault system.* Professional Paper 1515.

www.geophys.washington.edu/seismosurfing.html
Exhaustive list of worldwide Internet sites for information about earthquakes.

http://quake.wr.usgs.gov/
U.S. Geological Survey Earthquake Information. Gives information on reducing earthquake hazards, earthquake preparedness, latest quake information, historical earthquakes, and how earthquakes are studied. Also a good starting place for links to other earthquake sites.

http://quake.wr.usgs.gov/recenteqs/faq.html
Frequently Asked Questions about recent earthquakes maintained by the U.S. Geological Survey.

www.seismo.unr.edu/
University of Nevada, Reno Seismological Laboratory site contains information about recent earthquakes, earthquake preparedness, and links to other earthquake sites.

www.seismo.berkeley.edu/seismo/Homepage.html
Seismographic information page maintained by U. C.–Berkeley that has many links to other earthquake sites (particularly in California), 3-D earthquake movie, Northridge earthquake rupture movies, and information on earthquake preparedness.

http://vquake.calstatela.edu/
California State University, Los Angeles *Virtual Earthquake.* Create and analyze an earthquake.

http://pubs.usgs.gov/gip/earthq4/severitygip.html
General information about the size of an earthquake. Discussion of Richter and Mercalli scales.

http://pubs.usgs.gov/publications/text/dynamic.html
General information about plate tectonics.

http://geopubs.wr.usgs.gov/circular/c1187/
U.S. Geological online version of Tsunami Circular.

http://walrus.wr.usgs.gov/tsunami/PNGhome.html
U.S. Geological Survey web page gives information about the devastating July 17, 1998, tsunami at Papua, New Guinea, and links to other sites.

Animations

This chapter includes the following animations available on our Online Learning Center at www.mhhe.com/plummer9e.

- 7.2 Elastic Rebound Theory
- 7.4 Focus of an Earthquake
- 7.5 Particle Motion in Seismic Waves
- 7.6 Seismograph for Detecting Vertical Rock Motion
- 7.7 Seismograph for Detecting Horizontal Rock Motion
- 7.19 Tsunami

CHAPTER

8

Time and Geology

The immensity of geologic time is hard for humans to perceive. It is unusual for someone to live a hundred years, but a person would have to live 10,000 times that long to observe a geologic process that takes a million years. In this chapter we try to help you develop a sense of the vast amounts of time over which geologic processes have been at work.

Geologists working in the field or with maps or illustrations in a laboratory are concerned with relative time—unraveling the sequence in which geologic events occurred. For instance, a geologist looking at the photo of Arizona's Grand Canyon on the facing page can determine that the tilted sedimentary rocks are older than the horizontal sedimentary rocks and that the lower layers of the horizontal sedimentary rocks are older than the layers above them. But this tells us nothing about how long ago any of the rocks formed. To determine how many years ago rocks formed, we need the specialized techniques of radioactive isotope dating. Through isotopic dating we have been able to determine that the rocks in the lowermost part of the Grand Canyon are well over a billion years old.

This chapter explains how to apply several basic principles to decipher a sequence of events responsible for geologic features. These principles can be applied to many aspects of geology—as, for example, in understanding geologic structures (chapter 6). Understanding the complex history of mountain belts (chapter 5) also requires knowing the techniques for determining relative ages of rocks.

Determining age relationships between geographically widely separated rock units is necessary for understanding the geologic history of a region, a continent, or the whole Earth. Substantiation of the plate tectonics theory depends on intercontinental correlation of rock units and geologic events, piecing together evidence that the continents were once one great body.

Widespread use of fossils led to the development of the standard geologic time scale. Originally based on relative age relationships, the subdivisions of the standard geologic time scale have now been assigned numerical ages in thousands, millions, and billions of years through isotopic dating. Think of the geologic time scale as a sort of calendar to which events and rock units can be referred. Its major subdivisions are referred to elsewhere in this book.

Opposite: Grand Canyon, Arizona. Horizontal Paleozoic beds (top of photo) overlie tilted Precambrian beds (bottom of photo) and older, Precambrian metamorphic rock (at far left of photo).

THE KEY TO THE PAST

We who are part of present-day Western civilization are not so resistant to the concept of immense geologic time as were those who lived two or three centuries ago. Then it was commonly believed in the Western world that Earth was only a few thousand years old (although Chinese and Hindu cultures had a better concept of the vast age of the universe). In early Christendom, geologic events were placed within a biblical chronology, and catastrophic happenings were blamed for features of the landscape. When rocks several thousand meters above sea level seemed to have been formed of sediment deposited in water (even containing fossils somewhat like modern marine organisms), the explanation was that a worldwide inundation had simply drowned all Earth's mountains in a matter of days. Because no known physical laws could account for such events, they were attributed to divine intervention. In the eighteenth century, however, James Hutton, a Scotsman who is regarded as the father of modern geology, realized that geologic features could be explained through present-day processes. He recognized that our mountains are not permanent, but have been carved into their present shapes and will be worn down by the slow agents of erosion now working on them. He realized that the great thicknesses of sedimentary rock we find on the continents are products of sediment removed from land and deposited as mud and sand in seas. The time required for these processes to take place had to be incredibly long. Hutton upset conventional thinking (the world was believed to be less than 6,000 years old) by writing in 1788, "We find no sign of a beginning—no prospect for an end." Hutton's writings received scant notice until his ideas were given widespread attention by Charles Lyell in a landmark book, *Principles of Geology.* Lyell referred to Hutton's concept that geologic processes operating at present are the same processes that operated in the past as the principle of **uniformitarianism.** The principle is stated more succinctly as "The present is the key to the past."

The term *uniformitarianism* is a bit unfortunate, because it suggests that changes take place at a uniform *rate.* Hutton recognized that sudden, violent events, such as a major, short-lived volcanic eruption, also influence Earth's history. In some countries *actualism* is used in place of uniformitarianism. The term actualism comes closer to conveying Hutton's principle that the same processes and natural laws that operated in the past are those we can actually observe or infer from observation as operating at present. It is based on the assumption, central to the sciences, that physical laws are independent of time and location. Under present usage, uniformitarianism has the same meaning as actualism for most geologists.

We now realize that geology involves time periods much greater than a few thousand years. But how long? For instance, were rocks near the bottom of the Grand Canyon (chapter opening photo) formed closer to 10,000 or 100,000 or 1,000,000 or 1,000,000,000 years ago? What geologists needed was some "clock" that began working when rocks formed. Such a "clock" was found when radioactivity was discovered. Dating based on radioactivity (discussed later in this chapter) allows us to determine a rock's **numerical age** (also known as *absolute age*)—age given in years or some other unit of time. Geologists working in the field or in a laboratory with maps, cross-sections, and photographs are concerned with **relative time,** the *sequence* in which events took place, than with the number of years involved.

These statements show the difference between absolute age and relative time:

"The American Revolutionary War took place after the signing of the Magna Carta but before World War II." This statement gives the time of an event (the Revolutionary War) relative to other events.

But in terms of numerical age we could say: "The Revolutionary War took place about two and a half centuries ago." Note that a numerical age does not have to be an *exact* age, merely age given in units of time. Because most geologic problems are concerned with the sequence of events, we discuss relative time first.

RELATIVE TIME

The geology of an area may seem, at first glance, to be hopelessly complex. A nongeologist might think it impossible to decipher the sequence of events that created such a geologic pattern; however, a geologist has learned to approach seemingly formidable problems by breaking them down to a number of simple problems. As an example, the geology of the Grand Canyon, shown diagrammatically in figure 8.18 and in the chapter opening photo, can be analyzed in four parts: (1) horizontal layers of rock; (2) inclined layers; (3) rock underlying the inclined layers (plutonic and metamorphic rock); and (4) the canyon itself, carved into these rocks.

After you have studied the following section, return to the photo of the Grand Canyon and see if you can determine the sequence of geologic events that took place.

Principles Used to Determine Relative Age

Most of the individual parts of the larger problem are solved by applying several simple principles while studying the exposed rock. In this way the sequence of events or the relative time involved can be determined. Contacts are particularly useful for deciphering the geologic history of an area. (**Contacts,** as will be described in subsequent chapters, are the surfaces separating two different rock types or rocks of different ages.) To explain various principles, we will use a fictitious place that bears some resemblance to the Grand Canyon. We will call this place, represented by the block diagram of figure 8.1, Minor Canyon. The formation names are also fictitious. (**Formations,** as described in chapter 14, are bodies of rock of

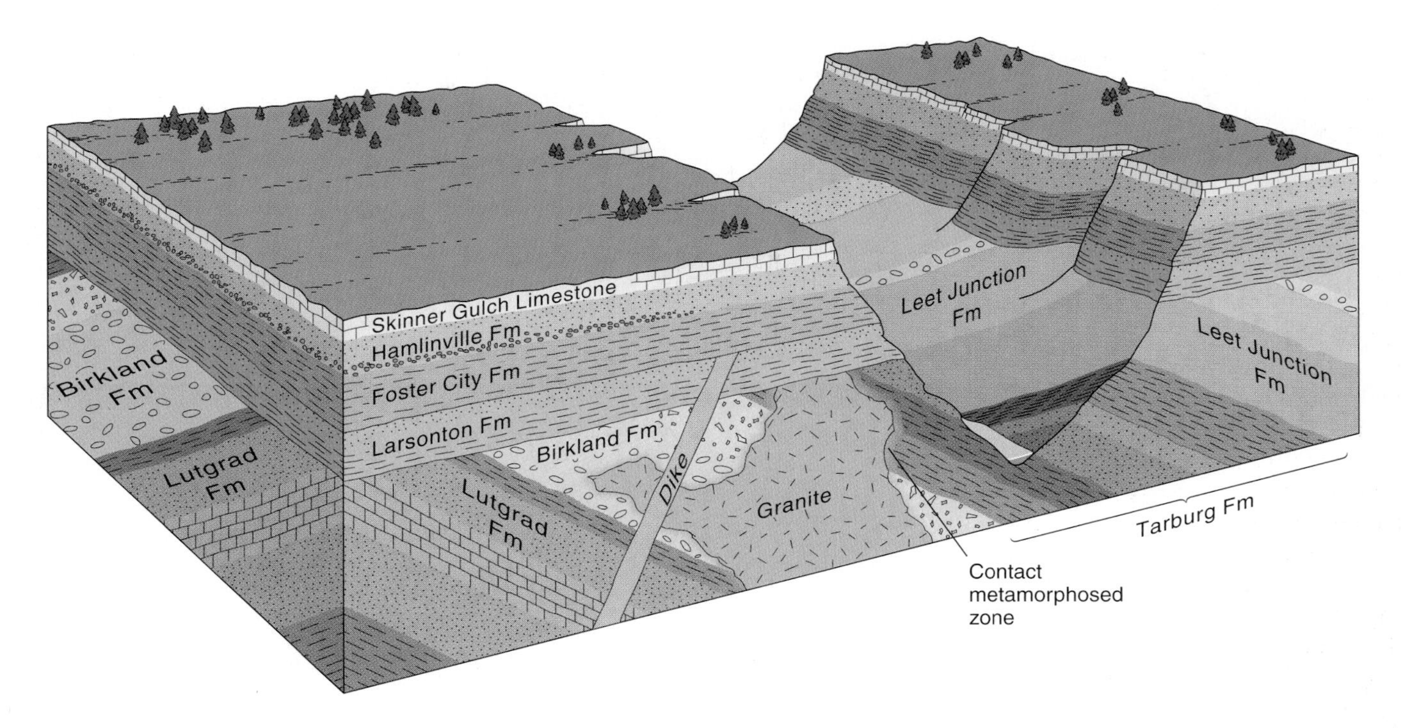

FIGURE 8.1

Block diagram representing the Minor Canyon area.

considerable thickness with recognizable characteristics that make each distinguishable from adjacent rock units. They are named after local geographic features, such as towns.) Note the contacts between the tilted formations, the horizontal formations, the granite, and the dike. What sequence of events might be responsible for the geology of Minor Canyon? (You might briefly study the block diagram and see how much of the geologic history of the area you can decipher before reading further.)

To determine the relationship of geologic events to one another, four basic principles are applied. These are the principles of (1) original horizontality, (2) superposition, (3) lateral continuity, and (4) cross-cutting relationships. These principles will be used in interpreting figure 8.1.

Original Horizontality

The principle of **original horizontality** states that beds of sediment deposited in water formed as horizontal or nearly horizontal layers (as described in chapter 14). (All the layers in figure 8.1 represent sedimentary rock originally deposited in a marine environment.)

Note in figure 8.1 that the Larsonton Formation and overlying rock units (Foster City Formation, Hamlinville Formation, and Skinner Gulch Limestone) are horizontal. Evidently their original horizontal attitude has not changed since they were deposited; however, the Lutgrad, Birkland, Tarburg, and Leet Junction Formations must have been tilted after they were deposited as horizontal layers. By applying the principle of original horizontality, we have determined that a geologic event—tilting of bedrock—occurred after the Leet Junction, Tarburg, Birkland, and Lutgrad Formations were deposited on a sea floor. We can also see that the tilting event did not affect the Larsonton and overlying formations. (A reasonable conclusion is that tilting was accompanied by uplift and erosion, all before renewed deposition of younger sediment.)

Superposition

The principle of **superposition** states that within a sequence of undisturbed sedimentary or volcanic rocks, the layers get younger going from bottom to top.

Obviously, if sedimentary rock is formed by sediment settling onto the sea floor, then the first (or bottom) layer must be there before the next layer can be deposited on top of it. The principle of superposition also applies to layers formed by multiple lava flows, where one lava flow is superposed on a previously solidified flow.

Applying the principle of superposition, we can determine that the Skinner Gulch Limestone is the youngest layer of sedimentary rock in the Minor Canyon area. The Hamlinville Formation is the next oldest formation, and the Larsonton Formation is the oldest of the still horizontal sedimentary rock units. Similarly, we assume that the inclined layers were originally horizontal (by the first principle). By mentally restoring them to their horizontal position (or "untilting" them), we can see that the youngest formation of the sequence is the Leet Junction Formation and that the Tarburg, Birkland, and Lutgrad Formations are progressively older.

table 8.1 Relative Ages of Features in Figure 8.1 Determinable by Cross-cutting Relationships

Feature	Is Younger Than	But Older Than
Valley (canyon)	Skinner Gulch Ls.	
Foster City Fm.	Dike	Hamlinville Fm.
Dike	Larsonton Fm.	Foster City Fm.
Larsonton Fm.	Leet Junction Fm. and granite	Dike
Granite	Tarburg Fm.	Larsonton Fm.

Lateral Continuity

The principle of **lateral continuity** states that an original sedimentary layer extends laterally until it tapers or thins at its edges. This is what we expect at the edges of a depositional environment, or where one type of sediment interfingers laterally with another type of sediment as environments change. In figure 8.1 the bottom bed of the Hamlinville Formation tapers as we would expect from this principle. We are not seeing any other layers taper, either because we are not seeing their full extent within the diagram or because they have been truncated (cut off abruptly) due to later events.

Cross-Cutting Relationships

The fourth principle can be applied to determine the remaining age relationships at Minor Canyon. The principle of **cross-cutting relationships** states that a disrupted pattern is older than the cause of disruption. A layer cake (the pattern) has to be baked (established) before it can be sliced (the disruption).

To apply this principle, look for disruptions in patterns of rock. Note that the valley in figure 8.1 is carved into the horizontal rocks as well as into the underlying tilted rocks. The sedimentary beds on either side of the valley appear to have been sliced off, or *truncated,* by the valley. The principle of lateral continuity tells us that sedimentary beds normally become thinner toward the edges rather than stop abruptly. So the event that caused the valley must have come after the sedimentation responsible for deposition of the Skinner Gulch Limestone and underlying formations. That is, the valley is younger than these layers. We can apply the principle of cross-cutting relationships to contacts elsewhere in figure 8.1 with the results shown in table 8.1.

We can now describe the geological history of the Minor Canyon area represented in figure 8.1 on the basis of what we have learned through applying the principles. Figures 8.2 through 8.11 show how the area changed over time, progressing from oldest to youngest events.

By *superposition,* we know that the Lutgrad Formation, the lowermost rock unit in the tilted sequence, must be the oldest of the sedimentary rocks as well as the oldest rock unit in the diagram. From the principle of *original horizontality,* we infer that these layers must have been tilted after they formed. Figure 8.2 shows initial sedimentation of the Lutgrad Formation taking place. If the entire depositional basin were shown, the layer would be tapered at its edges, according to the principle of *lateral continuity.*

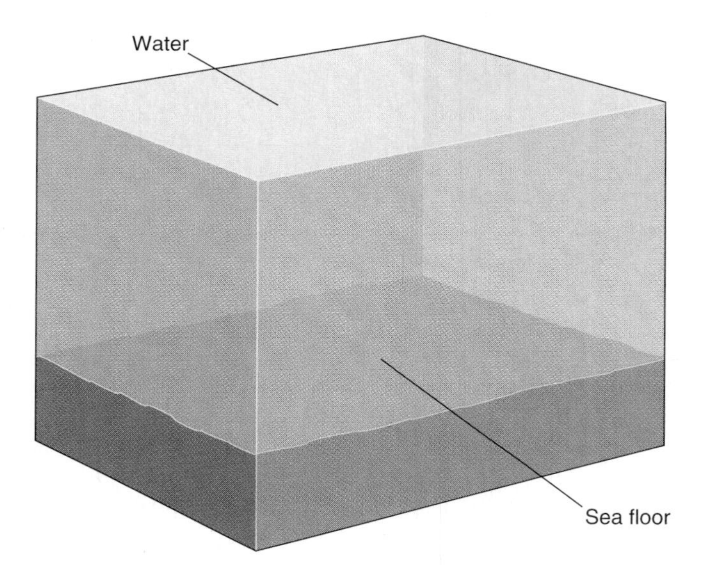

FIGURE 8.2
The area during deposition of the initial sedimentary layer of the Lutgrad Formation.

Superposition indicates that the Birkland Formation was deposited on top of the Lutgrad Formation. Deposition of the Tarburg and Leet Junction Formations followed in turn (figure 8.3).

The truncation of bedding in the Lutgrad, Birkland, and Tarburg Formations by the granite tells us that the granite intruded sometime after the Tarburg Formation was formed (this is an *intrusive contact*). Although figure 8.4 shows that the granite was emplaced before tilting of the layered rock, we cannot determine from looking at figure 8.1 whether the

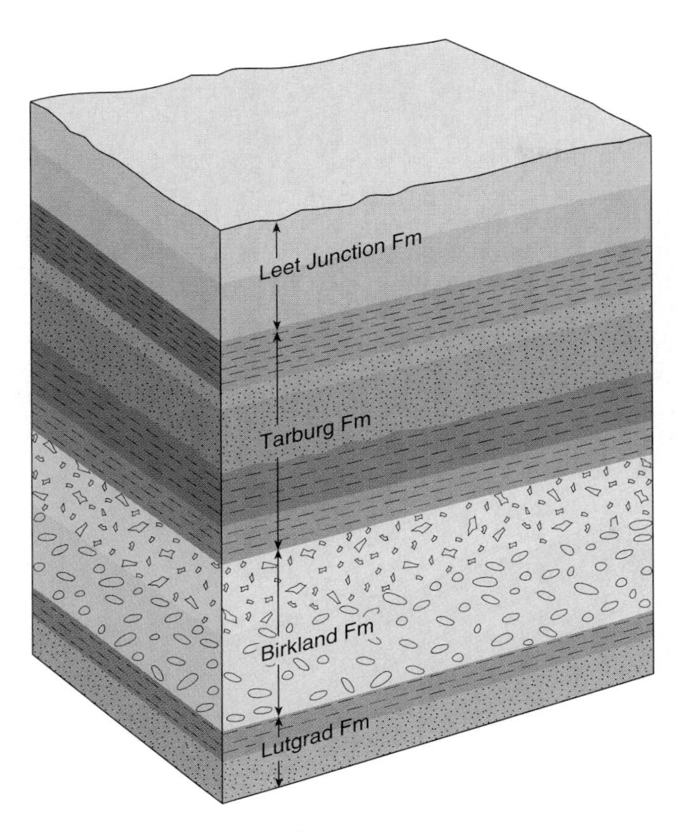

FIGURE 8.3
The area before intrusion of the granite.

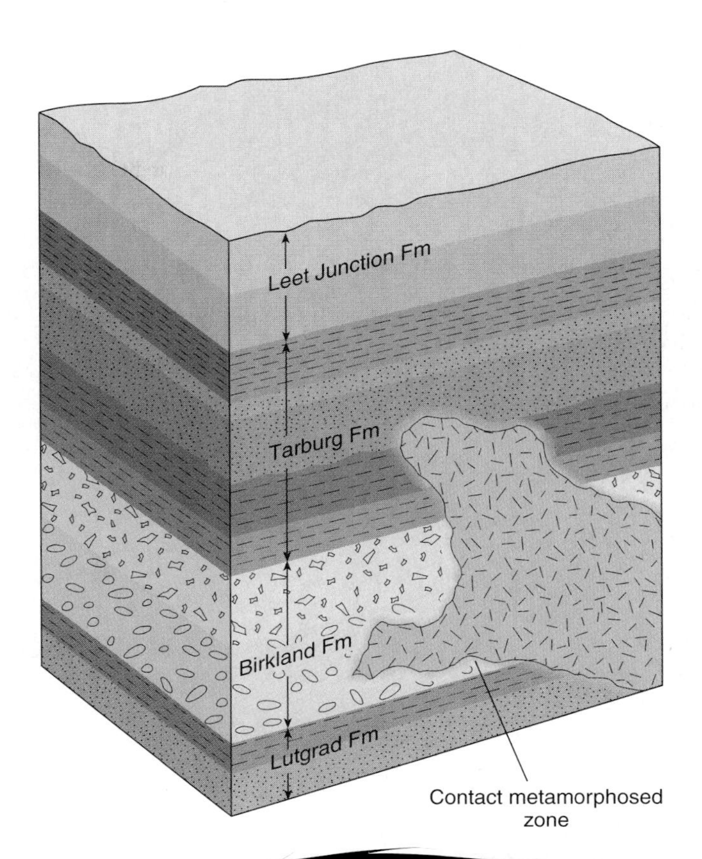

FIGURE 8.4
The area befor

granite intruded the sedimentary rocks before or after tilting. We can, however, determine through *cross-cutting relationships* that tilting and intrusion of the granite occurred before deposition of the Larsonton Formation. Figure 8.5 shows the rocks in the area have been tilted and erosion has taken place. Sometime later, sedimentation was renewed and the lowermost layer of the Larsonton Formation was deposited on the erosion surface, as shown in figure 8.6. Contacts representing buried erosion surfaces such as these are called *unconformities* and are discussed in more detail later in this chapter.

After the Larsonton Formation was deposited, an unknown additional thickness of sedimentary layers was deposited, as shown in figure 8.7. This can be determined through application of cross-cutting relationships. The dike is truncated by the Foster City Formation; therefore, it must have extended into some rocks that are no longer present, such as shown in figure 8.8. Figure 8.9 shows the area after the erosion that truncated the dike took place.

Once again, sedimentation took place as the lowermost layer of the Foster City Formation blanketed the erosion surface (figure 8.10). Sedimentation continued until the uppermost layer (top of the Skinner Gulch Limestone) was deposited. At some later time, the area was raised above sea level and the stream

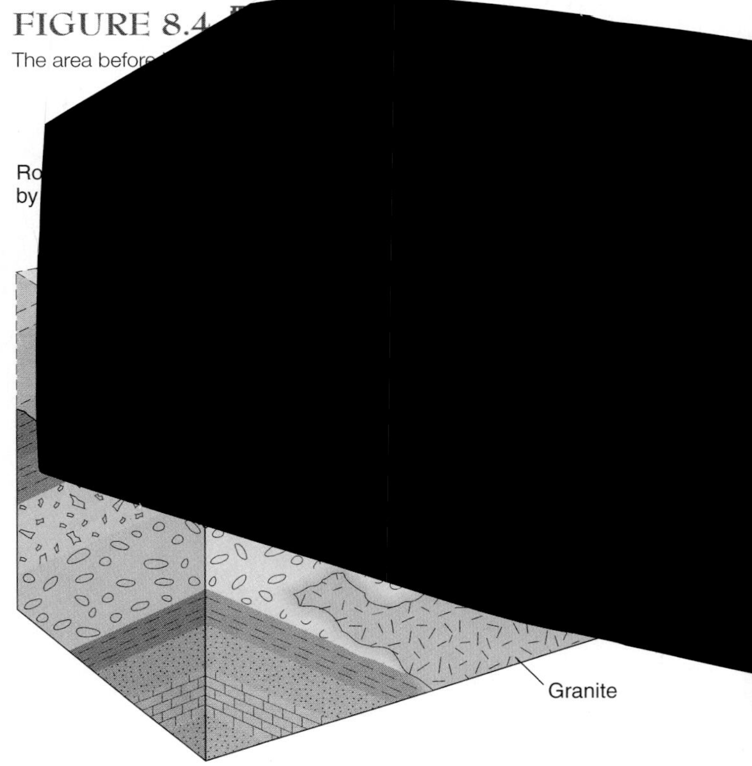

FIGURE 8.5
The area before deposition of the Larsonton Formation. Dashed lines show rock probably lost through erosion.

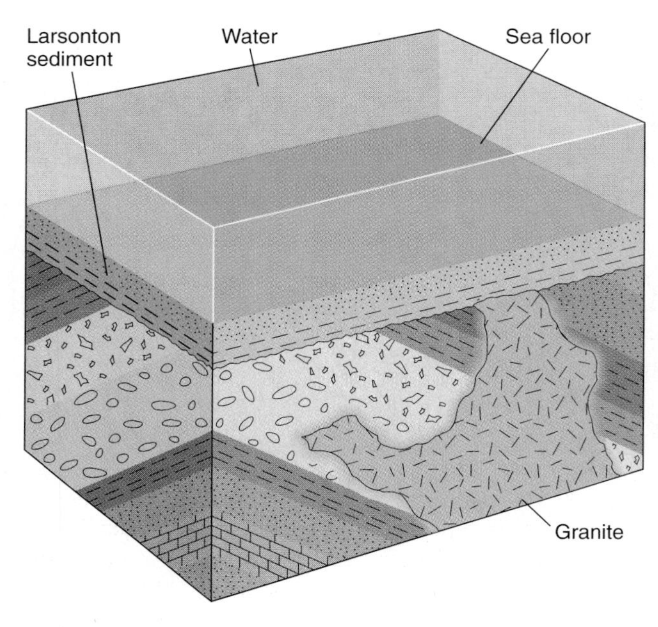

FIGURE 8.6
The area at the time the Larsonton Formation was being deposited.

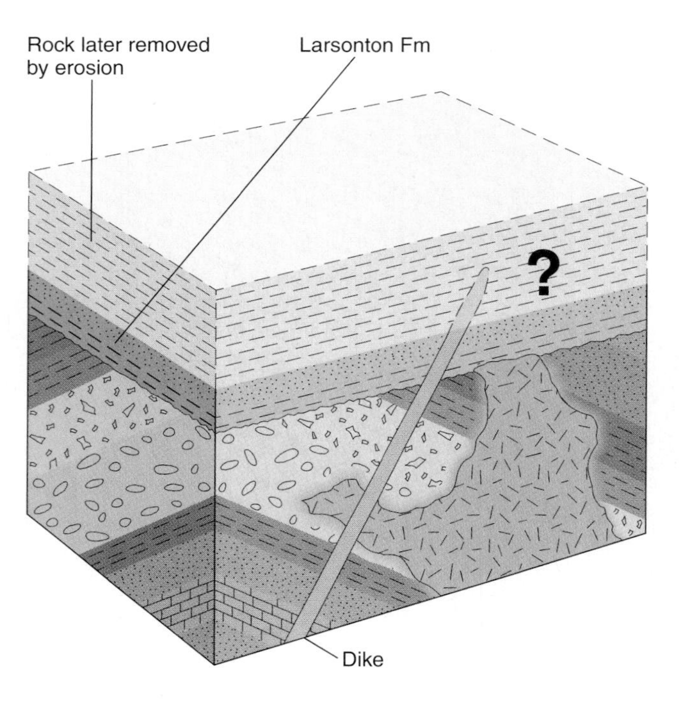

FIGURE 8.8
Dike intruded into the Larsonton Formation and preexisting overlying layers of indeterminable thickness.

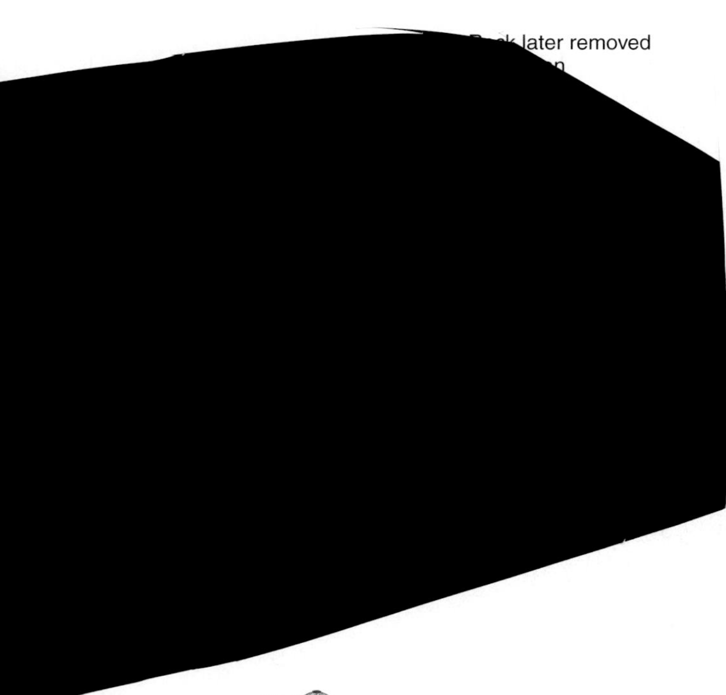

FIGURE 8.7
Area before intrusion of dike. Thickness of layers above the Larsonton Formation is indeterminable.

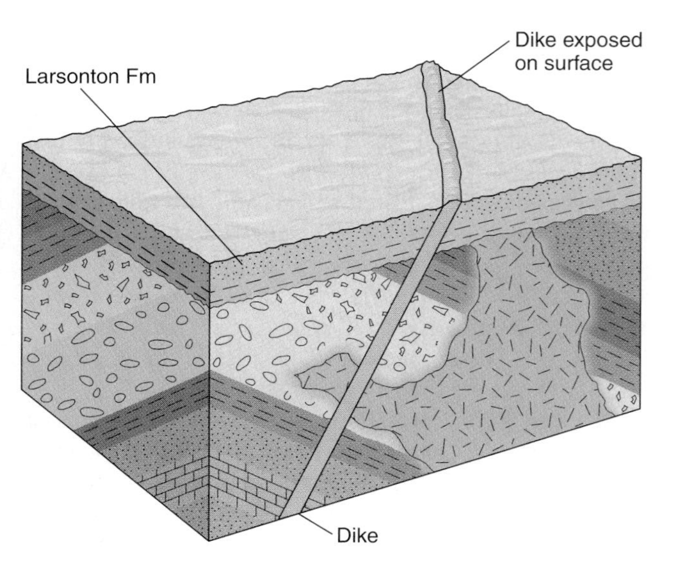

FIGURE 8.9
The area after rock overlying the Larsonton Formation, along with part of the dike, was removed by erosion.

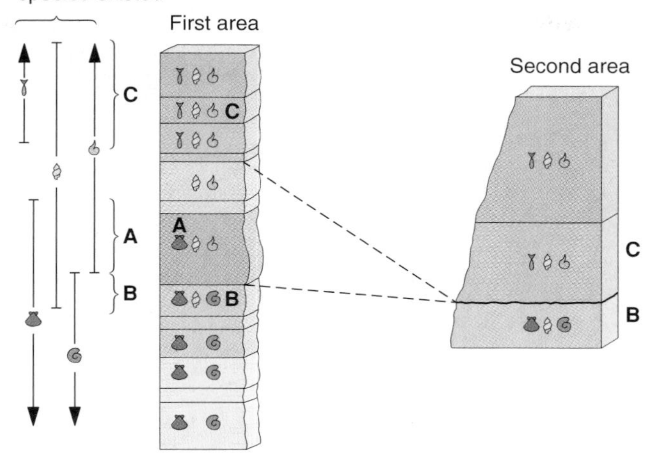

FIGURE 8.20

The use of fossil assemblages for determining relative ages. Rock *A* contains . Therefore, it must have formed during time interval *A*. Rock *B* contains . Therefore, it must have formed during time interval *B*. Rock *C* contains . Therefore, it must have formed sometime during time interval *C*. In the second area, sequence *A* is missing. Therefore, there is an unconformity between sequence *B* and sequence *C*.

discovering a shark's tooth in a rock is not very helpful in determining the rock's relative age.

A geologist is likely to find a **fossil assemblage,** several different fossil species in a rock layer. A fossil assemblage is generally more useful for dating rocks than a single fossil is, because the sediment must have been deposited at a time when all the species represented existed (figure 8.20).

Some fossils are restricted in geographic occurrence, representing organisms adapted to special environments. But many former organisms apparently lived over most of the earth, and fossil assemblages from these may be used for worldwide correlation. Fossils in the lowermost horizontal layers of the Grand Canyon are comparable to ones collected in Wales, Great Britain, and many other places in the world (the trilobites in figure 8.21 are an example). We can therefore correlate these rock units and say they formed during the same general span of geologic time.

The Standard Geologic Time Scale

Geologists can use fossils in rock to refer the age of the rock to the **standard geologic time scale,** a worldwide relative time scale. Based on fossil assemblages, the geologic time scale subdivides geologic time. On the basis of fossils found, a geologist can say, for instance, that the rocks of the lower portion of horizontal layers in the Grand Canyon formed during the *Cambrian Period.* This implicitly correlates these rocks with certain rocks in Wales (in fact, the period takes its name from Cambria, the Latin name for Wales) and elsewhere in the world where similar fossils occur.

The geologic time scale, shown in a somewhat abbreviated form in table 8.2, has had tremendous significance as a unifying concept in the physical and biological sciences. The working out of the evolutionary chronology by successive generations of geologists and other scientists has been a remarkable human achievement. The geologic time scale, representing an extensive fossil record, consists of three **eras,** which are subdivided into **periods,** which are, in turn, divided into **epochs.** (Remember that this is a relative time scale.)

FIGURE 8.21

Trilobite fossils from the Cambrian Period.

Courtesy Ward's Natural Science Est. Inc., Rochester, N.Y.

table 8.2 **Geologic Time Scale**

Era	Period	Epoch
Cenozoic	*Quaternary*	Recent (Holocene) Pleistocene
	Tertiary	Pliocene Miocene Oligocene Eocene Paleocene
Mesozoic	*Cretaceous* *Jurassic* *Triassic*	
Paleozoic	*Permian* *Pennsylvanian* } Carboniferous* *Mississippian* } *Devonian* *Silurian* *Ordovician* *Cambrian*	
Precambrian Time		

*Outside of North America, Carboniferous Period is used rather than Pennsylvanian and Mississippian.

8.2 IN GREATER DEPTH

Demise of the Dinosaurs—Was It Extraterrestrial?

Dinosaurs dominated the continents during the Mesozoic Era. Now they prey on the imaginations of children of all ages. It's hard to accept that beings as powerful and varied as dinosaurs existed and were wiped out. But the fossil record is clear—when the Mesozoic came to a close, dinosaurs became extinct. Not a single of the numerous dinosaur species survived into the Cenozoic Era. Not only did the dinosaurs go, but about 75% of all plant and animal species, marine as well as terrestrial, were extinguished. This was one of Earth's "great dyings"—an even "greater dying" was when the Paleozoic Era ended with the extinction of over 95% of all species. Most major extinctions have been gradual, and scientists usually have attributed them to climate changes.

A decade ago, geologist Walter Alvarez, his father, physicist Luis Alvarez, and two other scientists proposed a hypothesis that the dinosaur extinction was caused by the impact of an asteroid. This was based on the chemical analysis of a thin layer of clay marking the boundary between the Mesozoic and Cenozoic eras (usually referred to as the K-T boundary—it separates the Cretaceous [K] and Tertiary [T] periods). The K-T boundary clay was found to have about 30 times the amount of the rare element iridium as is normal for crustal rocks. Iridium is relatively abundant in meteorites and other extraterrestrial objects such as comets, and the scientists suggested that the iridium was brought in by an extraterrestrial body.

The scientists proposed a scenario in which an asteroid 10 kilometers in diameter struck Earth. This generated a gigantic dust cloud, which darkened the earth long enough to disrupt plant life growth and reproduction as well as drop the temperature of Earth worldwide. Creatures perished because of the disruption in their food supply or because of their inability to withstand the sudden climate change.

Other scientists proposed an alternative hypothesis blaming exceptional volcanic activity for the extinction as well as for the high iridium in the K-T boundary layer. (Some Hawaiian eruptions have a high iridium content.) Debate between proponents of the two alternate hypotheses became heated and further evidence to support each was sought. K-T layers throughout the world were found to have grains of quartz that have been subjected to shock metamorphism (see box 15.2), supporting the asteroid hypothesis. Concentrations of some elements common in volcanic rocks but not in meteorites were found in the K-T boundary layers, supporting the volcanic hypothesis.

The asteroid hypothesis advocates predicted that a large meteorite crater should be found someplace on Earth that could be dated as having formed around 65 million years ago when the Mesozoic ended.

In 1990, evidence was found suggesting a large (120 km in diameter) crater centered along the coast of Mexico's Yucatan peninsula at a place called Chicxulub. The alleged crater at Chicxulub, now buried beneath sediment, would be the right size to have been formed by a 10-kilometer asteroid. Geologists were quick to search for and present evidence in support of the Chicxulub crater. Among evidence cited was sediment that appeared to have been deposited in various locations by giant sea waves. Also cited were widespread microscopic spheres of glass that formed when rock melted from the impact and droplets were thrown high into the air.

Geologists checking Mexican records compiled during drilling for oil in Yucatan find there are breccias in the Chicxulub area. Breccias, due to meteorite impact, are common at known meteorite craters. Also, the age of the impact seems to be correct for the K-T boundary. The evidence for an asteroid impact is overwhelming. However, the impact may not be entirely to blame for dinosaur extinction.

We do know that climates were changing toward the end of the Cretaceous and dinosaur species were decreasing. Perhaps exceptional volcanic activity led to climate changes that were not favorable to dinosaur survival and the asteroid dealt the final unfortunate blow to dinosaurs.

"Unfortunate" is from the perspective of dinosaurs, not humans. The only mammals in the Cretaceous were inconsequential, rat-sized creatures. They survived the K-T extinction and, with dinosaurs no longer dominating the land, evolved into the many mammal species that populate Earth today, including humans.

Related Web Resource
Walking with Dinosaurs
www.bbc.co.uk/dinosaurs/

Precambrian denotes the vast amount of time that preceded the Paleozoic Era (which begins with the Cambrian Period). The **Paleozoic Era** (meaning "old life") began with the appearance of complex life (trilobites, for example), as indicated by fossils. Rocks older than Paleozoic contain few fossils. This is because creatures with shells or other hard parts, which are easily preserved as fossils, did not evolve until the beginning of the Paleozoic.

The **Mesozoic Era** (meaning "middle life") followed the Paleozoic. On land, dinosaurs became the dominant animals of the Mesozoic. We live in the **Recent** (or **Holocene**) **Epoch** of the **Quaternary Period** of the **Cenozoic Era** (meaning "new life"). The Quaternary also includes the most recent ice ages, which were part of the **Pleistocene Epoch.**

It is noteworthy that the fossil record indicates that mass extinctions, in which a large number of species have become extinct, have occurred a number of times in the geologic past. The two greatest mass extinctions define the boundaries between the three eras (see boxes 8.1 and 8.2).

Fossils have been used to determine ages of the horizontal rocks in Grand Canyon. All are Paleozoic. The lowermost horizontal formations (figure 8.18) are Cambrian, above which are Mississippian, Pennsylvanian, and Permian rock units. By referring to the geologic time scale (table 8.2), we can see that Ordovician, Silurian, and Devonian rocks are not represented. Thus an unconformity (buried erosion surface) is present within the horizontally layered rocks of Grand Canyon.

NUMERICAL AGE

Counting annual growth rings in a tree trunk will tell you how old a tree is. Similarly, layers of sediment deposited annually in glacial lakes can be counted to determine how long those lakes existed (*varves,* as these deposits are called, are explained in chapter 19). But only within the few decades following the discovery of radioactivity have scientists been able to determine numerical ages of rock units. We have subsequently been able to assign numerical values to the geologic time scale and determine how many years ago the various eras, periods, and epochs began and ended. We can now state that the Cenozoic Era began some 65 million years ago, the Mesozoic Era started about 250 million years ago, and the Precambrian ended (or the Paleozoic began) about 545 million years ago (until recently, the beginning of the Paleozoic was regarded as 570 million years old). The Precambrian includes most of geologic time, because the age of Earth is commonly regarded as about 4.5 to 4.6 billion years.

The oldest rocks found on Earth are from northwestern Canada and have been dated at 4.03 billion years old. In 2001, the oldest known mineral was dated at 4.4 billion years old, which is much older than the oldest rock dated so far. The mineral, a zircon crystal from Australia, was likely originally in a granite. Scientists who have studied this mineral think that its chemical makeup indicates that the granite formed from a magma that had a component of melted sedimentary rock. This would indicate that seas existed much earlier than geologists had previously thought possible.

Isotopic Dating

Radioactivity provides a "clock" that begins working when radioactive elements are sealed into newly crystallized minerals. The rates at which radioactive elements decay can be measured and duplicated in many different laboratories. Therefore, if we can determine the ratio of a particular radioactive element and its decay products in a mineral, we can calculate how long ago that mineral crystallized.

Determining the age of a rock through its radioactive elements is known as **isotopic dating** (previously, and somewhat inaccurately, called *radiometric dating*). Geologists who specialize in this important field are known as *geochronologists.*

Isotopes and Radioactive Decay

As will be discussed in chapter 9, every atom of a given element possesses the same number of protons in its nucleus. The number of neutrons, however, need not be the same in all atoms of the same element. The **isotopes** of a given element have different numbers of neutrons, but the same number of protons.

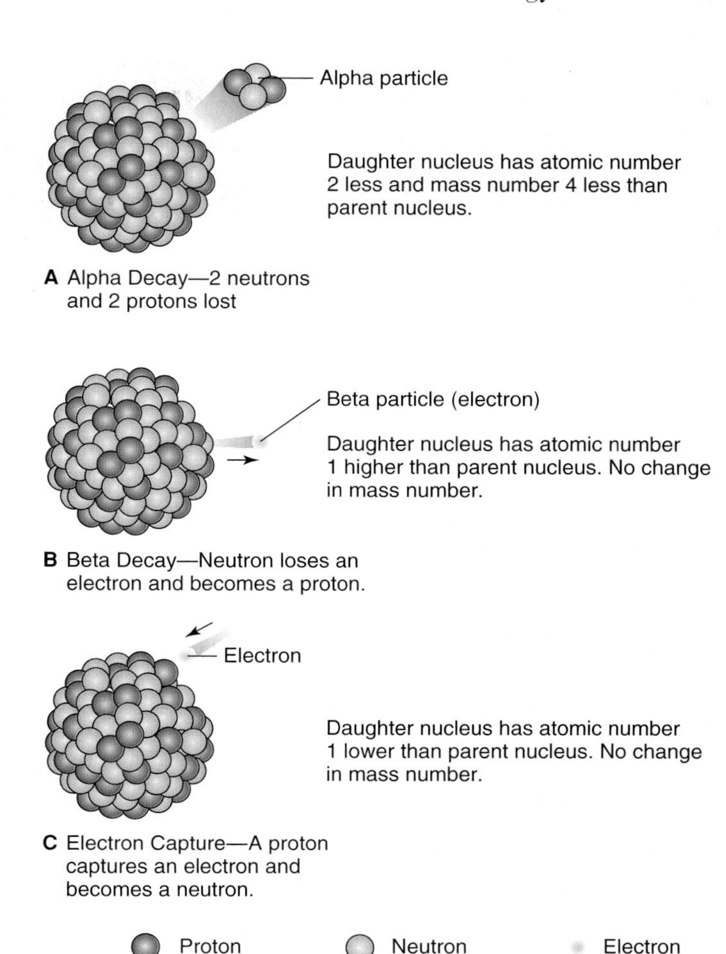

FIGURE 8.22
Three modes of radioactive decay.

Uranium, for example, commonly occurs as two isotopes, uranium -238 (^{238}U) and uranium -235 (^{235}U). The former has 238 protons and neutrons in its nucleus, whereas the latter has 235. Of these particles, 92 (the atomic number of uranium) must be protons and the rest neutrons. ^{238}U weighs slightly more than ^{235}U. Because of the difference in mass, the two isotopes may be separated with an apparatus called a mass spectrometer.

Radioactive decay is the spontaneous nuclear change of isotopes with unstable nuclei. Energy is produced with radioactive decay. Emissions from radioactive elements can be detected by a Geiger counter or similar device and, in high concentrations, can damage or kill humans (see box 8.3).

Nuclei of radioactive isotopes change primarily in three ways (figure 8.22). An *alpha* (α) *emission* is the ejection of two protons and two neutrons from a nucleus. When an alpha emission takes place the atomic number of the atom is reduced by two and its atomic mass number is reduced by four. (You can refer to the periodic table of elements, appendix D, to determine what

8.3

ENVIRONMENTAL GEOLOGY

Radon, a Radioactive Health Hazard

Radon is an odorless, colorless gas. Every time you breathe outdoors you inhale a harmless, minute amount of radon. If the concentration of radon that you breathe in a building is too high, however, you could, over time, develop lung cancer. It is one of the intermediate daughter products in the radioactive disintegration of ^{238}U to ^{206}Pb. It has a half-life of only 3.8 days.

Concentrations of radon are highest in areas where the bedrock is granite, gneiss, limestone, black shale, or phosphate-rich rock—rocks in which uranium is relatively abundant. Concentrations are also high where glacial deposits are made of fragments of these rocks. Even in these areas, radon levels are harmless in open, freely circulating air. Radon may dissolve in ground water or build up to high concentrations in confined air spaces.

The United States Environmental Protection Agency (EPA) regards 5 million American homes to have unacceptable radon levels in the air. Scientists outside of EPA have concluded that the standards the EPA is using are too stringent. They think that a more reasonably defined danger level means that only 50,000 homes have radon concentrations that pose a danger to their occupants.

Radon was first recognized in the 1950s as a health hazard in uranium mines, where the gas would collect in poorly ventilated air spaces. Radon lodges in the respiratory system of an individual, and as it deteriorates into daughter products, the subatomic particles given off cause damage to lung tissue. Three-quarters of the uranium miners studied were smokers. Thus, it is difficult to determine the extent to which smoking or radon induced lung cancer. (All studies show, however, that smoking and exposure to high radon levels are more likely to cause lung cancer than either alone.)

Interpolating the high rates of cancer incidence from the uranium miners to the population exposed to the very much lower radium levels in homes, as the EPA has done, is scientifically questionable.

What should you do if you are living in a high radon area? First have your house checked to see what the radon level is. Then read up on what acceptable standards should be. In most buildings with a high radon level, the gas seeps in from the underlying soil through the building's foundation. If a building's windows are kept open and fresh air circulates freely, radon concentrations cannot build up. But, houses are often kept sealed for air conditioning during the summer and heating during the winter. Air circulation patterns are such that a slight vacuum sucks the gases from the underlying soil into the house. Thus, radon concentrations might build up to dangerous levels.

The problem may be solved in several ways (aside from leaving windows open winter and summer). Basements can be made airtight so that gases cannot be sucked into the house from the soil. Air circulation patterns can be altered so that gases are not sucked in from underlying soil, or they are mixed with sufficient fresh, outside air.

If you are purchasing a new house, it would be a good idea to have it tested for radon before buying, particularly if the house is in an area of high-uranium bedrock or soil.

Related Web Resource

Radon in Earth, Air, and Water

http://sedwww.cr.usgs.gov/radon/radonhome.html

new element the isotope becomes.) After an alpha emission, ^{238}U becomes ^{234}Th (thorium), which has an atomic number of 90. The original isotope (^{238}U) is referred to as the *parent isotope.* The new isotope (^{234}Th) is the *daughter product.*

Beta (β) *emissions* involve the release of an electron from a nucleus. To understand this, we need to explain that electrons, which have virtually no mass and are usually in orbit around the nucleus, are also in the nucleus as part of a neutron. A neutron is a proton with an electron inside it; thus it is electrically neutral. If an electron is emitted from a neutron during radioactive decay, the neutron becomes a proton and the atom's atomic number is increased by one. For example, when ^{234}Th undergoes a beta emission, it becomes ^{234}Pa, an element with an atomic number of 91. Note that the atomic mass number has not changed. This is because the weight of an electron is negligible.

The third mode of change is *electron capture,* whereby a proton in the nucleus captures an orbiting electron. The proton becomes a neutron. The atom becomes a different element, one that has an atomic number one less than its parent isotope. An example of this is the potassium-argon system, discussed below, in which ^{40}K become ^{40}Ar. The parent isotope, potassium, has an atomic number of 19 and the atomic number of argon, the daughter product, is 18.

Figure 8.23 shows how ^{238}U decays to ^{206}Pb (lead-206) in a series of alpha and beta emissions. The important point is not the intermediate steps but the starting and ending isotopes. In the process, ^{238}U loses 10 protons, so that the daughter product has an atomic number of 82 (which is lead) and a total of 32 protons and neutrons so the new atomic mass number is 206. ^{206}Pb can only be produced by the decay of ^{238}U, and can be distinguished from other lead isotopes with a mass spectrometer.

Radioactive elements are used for determining age because in a large number of atoms of a radioactive element, the *proportion* of atoms that will radioactively decay is constant. Even

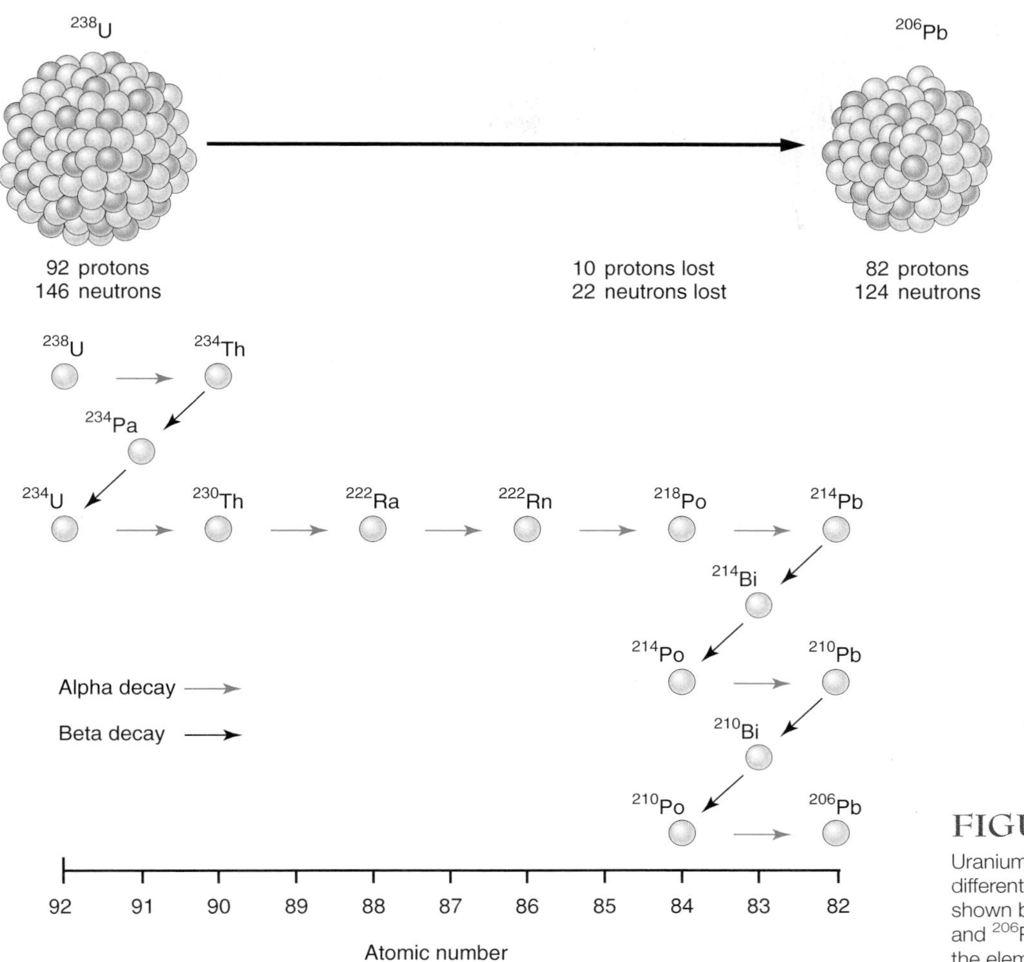

FIGURE 8.23

Uranium 238 decays to Lead 206. The different intermediate steps in the process are shown below the models of the nuclei of ^{238}U and ^{206}Pb. Refer to appendix C for names of the elements.

a rock possessing a small fraction of a percent of uranium contains millions of uranium atoms. The proportional amount of atoms that periodically decay is apparently unaffected by chemical reactions or by the high pressures and high temperatures of Earth's interior.

The rate of proportional decay for isotopes is expressed as **half-life,** the time it takes for a given amount of a radioactive isotope to be reduced by one-half. (The other half disintegrates into daughter products and energy.) The half-lives of some isotopes created in nuclear reactors are in fractions of a second. Naturally occurring isotopes that we find in rocks have very long half-lives (table 8.3). ^{40}K has a half-life of 1.3 billion years. If you began with one milligram of ^{40}K, 1.3 billion years later one-half milligram of ^{40}K would remain. After another 1.3 billion years, there would be one-fourth of a milligram, and after another half-life only one-eighth of a milligram. (Note that two half-lives do not equal a whole life.)

To determine the age of a rock by using ^{40}K, the amount of ^{40}K in that rock must first be determined by chemical analysis. The amount of ^{40}Ar (the daughter product) must also be determined and then used to calculate how much ^{40}K was present when the rock formed. By knowing how much ^{40}K was originally present in the rock and how much is still there, we can calculate the age of the rock on the basis of its half-life mathematically (see box 8.4). The graph in figure 8.24*A* demonstrates the mathematical relationship between a radioactively decaying isotope and time.

Radiocarbon Dating

Because of its short half-life of 5,730 years, radiocarbon dating is useful only in dating things and events accurately back to about 40,000 years—about seven half-lives. (However, new techniques allow some scientists to push the limit to nearly double that time.) The technique is most useful in archaeological dating and for very young geologic events (Recent, or Holocene, volcanic and glacial features for instance). It is also used to date historical artifacts. For instance, the Dead Sea Scrolls, the oldest of the surviving biblical manuscripts, were radiocarbon dated and their ages ranged from the third century B.C. to 68 A.D. These ages are consistent with estimates previously made by archaeologists and other scholars.

8.4

IN GREATER DEPTH

CALCULATING THE AGE OF A ROCK

The relationship between time and radioactive decay of an isotope is expressed by the following equation (which is used to plot curves such as shown in figure 8.24).

$$N = N_0 e^{-\lambda t}$$

N is the number of atoms of the isotope at time t, the time elapsed. N_0 is the number of atoms of that isotope present when the "clock" was set. The mathematical constant e has a value of 2.718. λ is a decay constant—that portion of the isotope that decays per unit time.

The relationship between λ and the half-life (t_{hl}) is

$$\lambda = \frac{\ln 2}{t_{hl}} = \frac{0.693}{t_{hl}}$$

Replacing λ in the first equation and converting that equation to natural logarithmic (to the base e) form, we get

$$t = \frac{t_{hl}}{.693} \ln \frac{N}{N_0}$$

N/N_0 is the ratio of parent atoms at present to the original number of parent atoms.

As an example, we will calculate the age of a mineral using ^{235}U decaying to ^{207}Pb. Table 8.3 indicates that the half-life is 713 million years. A laboratory determines that, at present, there are 440,000 atoms of ^{235}U and that the amount of ^{207}Pb indicates that when the mineral crystallized there were 1,200,000 atoms of ^{235}U. (We assume that there was no ^{207}Pb in the mineral at the time the mineral crystallized.) Plugging these values into the formula, we get

$$t = \frac{713{,}000{,}000}{.693} \ln \frac{440{,}000}{1{,}200{,}000}$$

Solving this gives us 1,032,038,250 years. The technique is not that precise, but rounded off, we can say the mineral formed 1.032 billion years ago.

table 8.3 Radioactive Isotopes Commonly Used for Determining Ages of Earth's Materials

Parent Isotope	Half-Life	Daughter Product	Effective Dating Range (years)
K-40 ^{40}K	1.25 billion years	^{40}Ar	100,000–4.6 billion
U-238 ^{238}U	4.5 billion years	^{206}Pb	10 million–4.6 billion
U-235 ^{235}U	713 million years	^{207}Pb	10 million–4.6 billion
Th-232 ^{232}Th	14.1 billion years	^{208}Pb	10 million–4.6 billion
Rb-87 ^{87}Rb	49 billion years	^{87}Sr	10 million–4.6 billion
C-14 ^{14}C	5,730 years	^{14}N	100–40,000

Radiocarbon dating is fundamentally different from the parent-daughter systems described previously in that ^{14}C is being created continuously in the atmosphere. Carbon (atomic number 6) is in the air as part of CO_2. It is mostly the stable isotope ^{12}C. However, ^{14}C is created in the atmosphere when cosmic radiation bombards nitrogen (N), atomic number 7. A neutron strikes and is captured by an ^{14}N atom. A proton is expelled from the nucleus and the atom becomes ^{14}C. The nucleus of the newly created carbon atom is unstable and will, sooner or later, through a beta emission, revert to ^{14}N. The electron is emitted from the atom as radiation. The rate of production of ^{14}C approximately balances the rate at which ^{14}C reverts to ^{14}N so that the level of ^{14}C remains essentially constant in the atmosphere.

Living matter incorporates ^{12}C and ^{14}C into its tissues; the ratios of ^{12}C and ^{14}C in the new tissues are usually the same as in the atmosphere. On dying, the plant or animal ceases to build new tissue. The ^{14}C disintegrates radioactively at the fixed rate of its half-life (5,730 years). The radioactive emissions per gram of carbon from the plant or animal remains are determined. The greater the radioactivity, the younger the sample. By comparing the radioactivity per gram

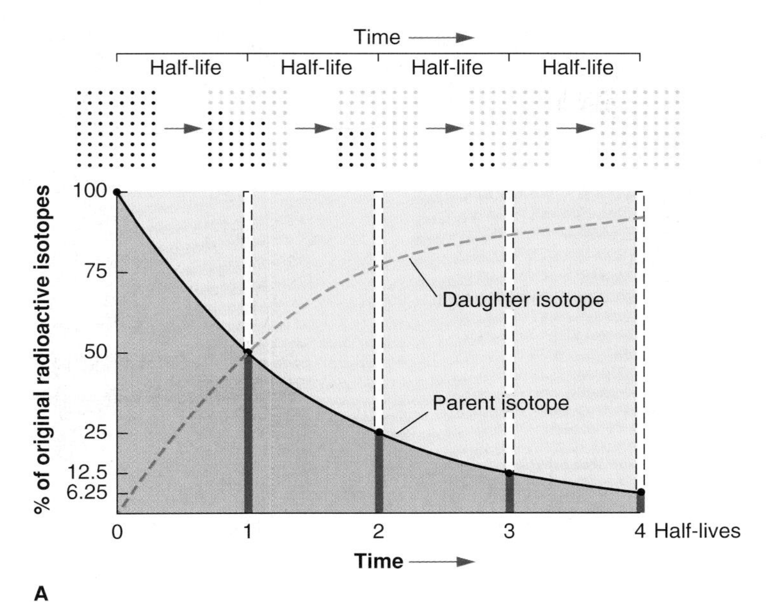

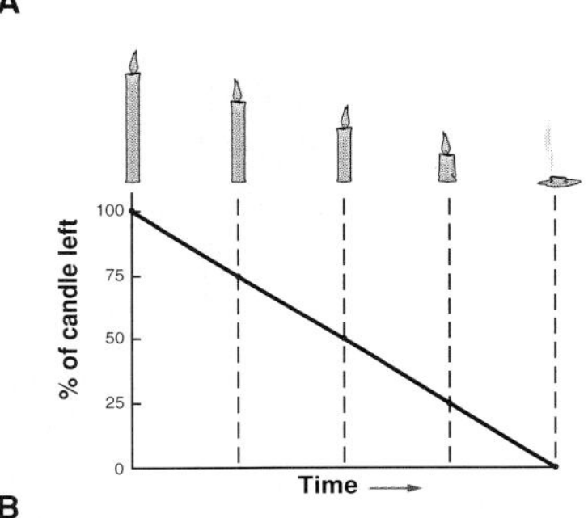

FIGURE 8.24

(*A*) The curve used to determine the age of a rock by comparing the percentage of radioactive isotope remaining in time to the original amount. Darker bars show the amount left after each half-life. Dashed bars show the amount disintegrated into daughter product and lost nuclear particles. The numbers of dots in the squares above the graph are proportional to the numbers of atoms. (*B*) For comparison, a candle burns at a linear rate.

to previously determined standards, we can indirectly determine the ratio of ^{12}C to radioactive ^{14}C in organic remains, and we can determine the time elapsed since the death of the organism.

Uses of Isotopic Dating

Some techniques determine isotopic ratios for a whole rock; others use single minerals within a rock. Exactly what is being dated depends on the isotope analyzed. Usually, an isotopic date determines how long ago the rock or mineral became a closed system. That is how long ago it was sealed off so that neither parent nor daughter isotopes could enter or leave the mineral or rock. Different isotopic pairs have different closure temperatures; when a rock cools below that temperature the system is closed and the "clock" starts. For instance, the $^{40}K-^{40}Ar$ isotopic pair has closure temperatures ranging from 150°C to 550°C, depending on the mineral. (Ar is a gas and gets trapped in different crystal structures at different temperatures.)

The best dates are obtained from igneous rocks. For a lava flow, which cools and solidifies rapidly, the age determined gives us the time of eruption and solidification. On the other hand, plutonic rocks, which may take over a million years to solidify, will not necessarily yield the time of intrusion, but the time at which a mineral cooled below the closure temperature. Dating metamorphic rocks usually means determining the time of cooling following the end of a metamorphic episode. Sedimentary rocks are difficult to date reliably.

In order for an isotopic age determination to be accurate, several conditions must be met. To ensure that the isotopic system has remained closed, the rock collected must show no signs of weathering or hydrothermal alteration. Second, one should be able to infer there were no daughter isotopes in the system at the time of closure or to make corrections for probable amounts of daughter isotopes present before the "clock" was set. Third, there must be sufficient parent and daughter atoms to be measurable by the mass spectrometer being used. And, of course, technicians and geochronologists must be highly skilled at working sophisticated equipment and collecting and processing rock specimens.

Whenever possible, geochronologists will use more than one isotope pair for a rock. The two U-Pb systems (table 8.3) can usually be used together and provide an internal cross-check on the age determinate. Using the K-Ar or another system would provide more confidence in our age determination.

How Reliable Is Isotopic Dating?

Half-lives of radioactive isotopes, whether short-lived, such as used in medicine, or long-lived, such as used in isotopic dating, have been found not to vary beyond statistical expectations. The half-life of each of the isotopes we use for dating rocks has not changed with physical conditions or chemical activity nor could the rates have been different in the distant past. It would violate laws of physics for decay rates (half-lives) to have been different in the past. Moreover, when several isotopic dating systems are painstakingly done on a single ancient igneous rock and when the same age is obtained or we understand the reason for differences in ages, it confirms that the decay constants for each system are indeed constant.

Comparing isotopic ages with relative age relationships confirms the reliability of isotopic dating. For instance, a dike that crosscuts rocks containing Cenozoic fossils gives us a relatively young isotopic age (less than 65 million years old) whereas a pluton truncated by overlying sedimentary rocks with Paleozoic fossils yields a relatively old age (greater than 250 million years). Many thousands of similar determinations have confirmed the reliability of radiometric dating.

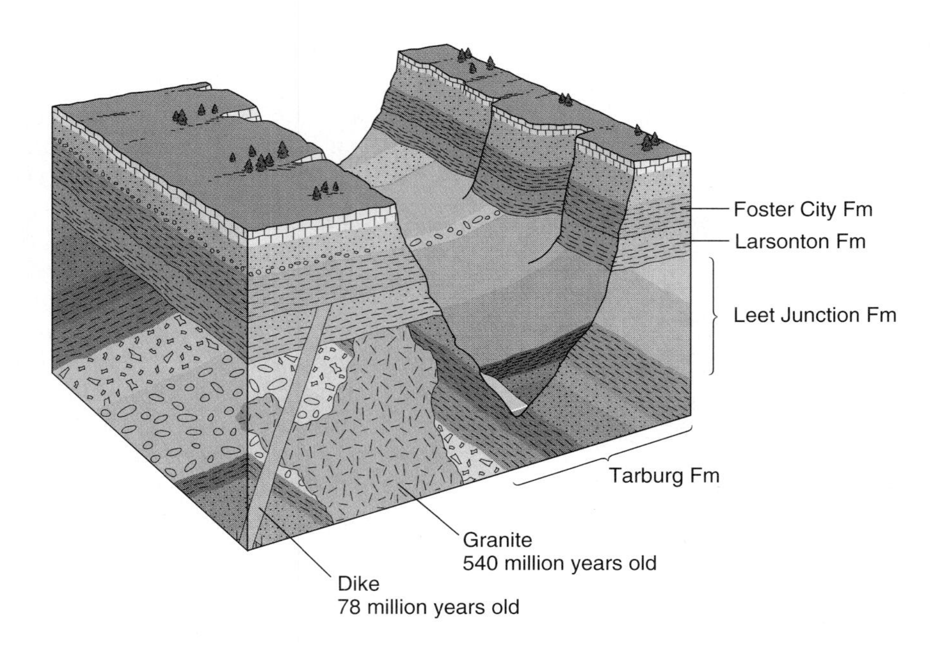

FIGURE 8.25
The Minor Canyon area as shown in figure 8.1 but with isotopic dates for igneous rocks indicated.

COMBINING RELATIVE AND NUMERICAL AGES

Radiometric dating can provide numerical time brackets for events whose relative ages are known. Figure 8.25 adds radiometric dates for each of the two igneous bodies in the fictitious Minor Canyon area of figure 8.1. The date obtained for the granite is 540 million years B.P. (before present), while the dike formed 78 million years ago. We can now state that the Tarburg Formation and older tilted layers formed before 540 million years ago (though we cannot say how much older they are). We still do not know whether the Leet Junction Formation is older or younger than the granite because of the lack of cross-cutting relationships. The Larsonton Formation's age is bracketed by the age of the granite and the age of the dike. That is, it is between 540 and 78 million years old. The Foster City and overlying formations are younger than 78 million years old; how much younger we cannot say.

Isotopic dates from volcanic ash layers or lava flows interlayered between fossiliferous sedimentary rocks have been used to assign numerical ages to the geologic time scale (figure 8.26). Isotopic dating has also allowed us to extend the time scale back into the Precambrian. There is, of course, a margin of uncertainty in each of the given dates. The beginning of the Paleozoic, for instance, was regarded until recently to be 570 million years ago, but with an uncertainty of ±30 million years. Recent work has fixed the age as closer to 545 million years. There are inherent limitations on the dating techniques as well as problems in finding the ideal rock for dating. For instance, if you wanted to obtain the date for the end of the Paleozoic Era and the beginning of the Mesozoic Era, the ideal rock would be found where there is no break in deposition of sediments between the two eras, as indicated by fossils in the rocks. But the difficulties in dating sedimentary rock mean you would be unlikely to date such rocks. Therefore, you would need to date volcanic rocks interlayered with sedimentary rocks found as close as possible to the transitional sedimentary strata. Alternatively, isotopically dated intrusions, such as dikes, whose cross-cutting relationships indicate that the age of intrusion is close to that of the transitional sedimentary layers, could be used to approximate the numerical age of the transition.

Isotopic dating has shown that the Precambrian took up most of geologic time (87%). Obviously, the Precambrian needed to be subdivided. The three major subdivisions of the Precambrian are the **Prearchean** (or **Hadean**), the **Archean,** and the **Proterozoic** (Greek for beginning life). Each is regarded as an **eon,** the largest unit of geological time. A fourth, and youngest, eon is the **Phanerozoic** (Greek for visible life). The Phanerozoic eon is all of geologic time with an abundant fossil record; in other words, it is made up of the three eras that followed the Precambrian.

AGE OF THE EARTH

In 1625 Archbishop James Ussher determined that Earth was created in the year 4004 B.C. His age determination was made by counting back generations in the Bible. This would make Earth 6,000 years old at present. That very young age of Earth was

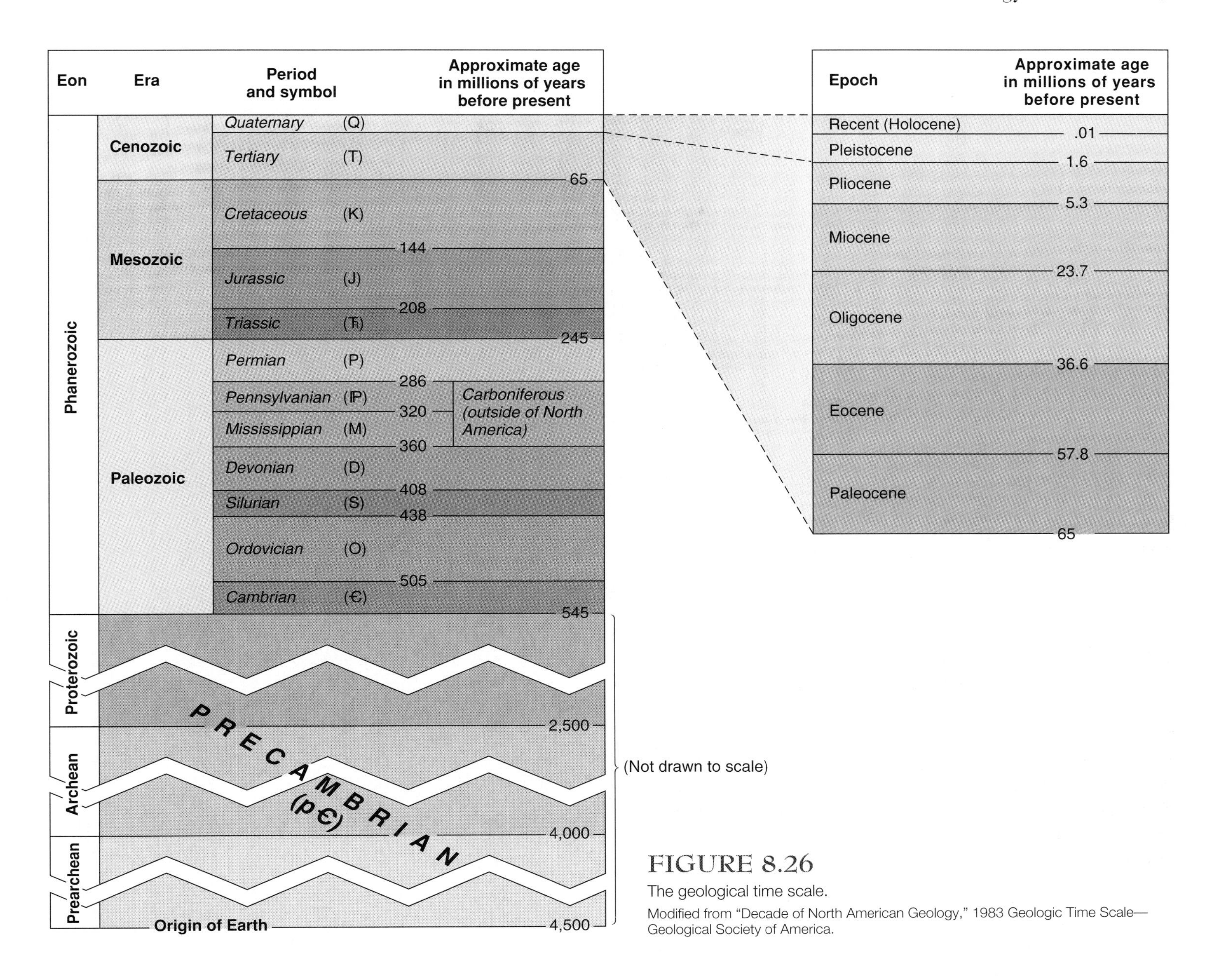

FIGURE 8.26

The geological time scale.

Modified from "Decade of North American Geology," 1983 Geologic Time Scale—Geological Society of America.

largely taken for granted by Western countries. By contrast, Hindus at the time regarded Earth as very old. According to an ancient Hindu calendar, the year A.D. 2000 would be year 1,972,949,101.

With the popularization of uniformitarianism in the early 1800s, Earth scientists began to realize that Earth must be very old—at least in the hundreds of millions of years. They were dealt a setback by the famous English physicist, Lord Kelvin. Kelvin, in 1866, calculated from the rate at which Earth loses heat that Earth must have been entirely molten between 20 and 100 million years ago. He later refined his estimate to between 20 and 40 million years. He was rather arrogant in scoffing at Earth scientists who believed that uniformitarianism indicated a much older age for Earth. The discovery of radioactivity in 1896 invalidated Kelvin's claim because it provided a heat source that he had not known about. When radioactive elements decay, heat is given off and that heat is added to the heat already in Earth. The amount of radioactive heat given off at present approximates the heat Earth is losing. So, for all practical purposes, Earth is not getting cooler.

The discovery of radioactivity also provided the means to determine how old Earth is. In 1905, the first crude isotopic dates were done and indicated an age of 2 billion years. Isotopic dating of meteorites has shown that they are between 4.5 and 4.6 billion years old. Some of the meteorites dated are planetary fragments. It is probable that all of the planets formed at the same time.

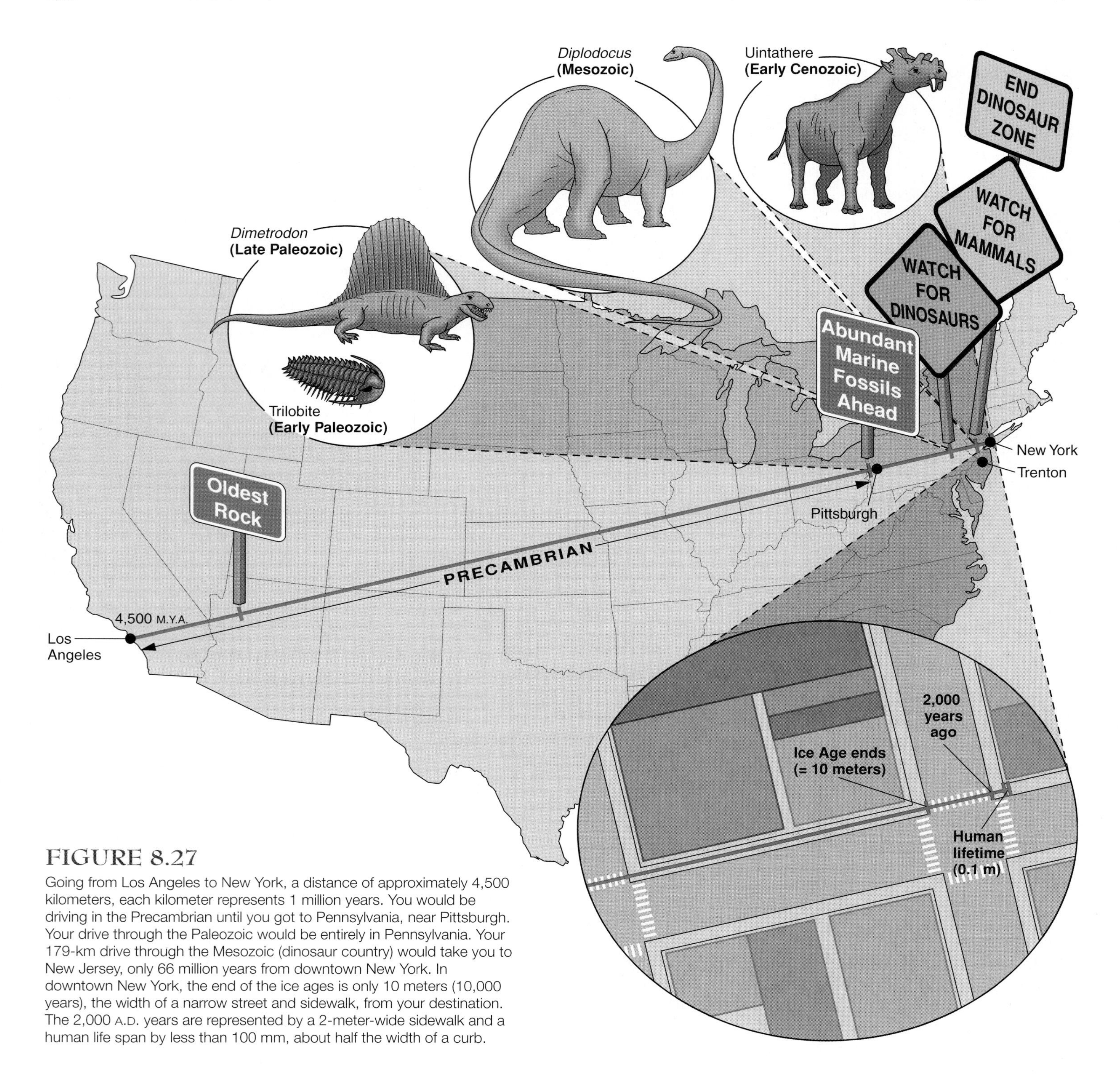

FIGURE 8.27

Going from Los Angeles to New York, a distance of approximately 4,500 kilometers, each kilometer represents 1 million years. You would be driving in the Precambrian until you got to Pennsylvania, near Pittsburgh. Your drive through the Paleozoic would be entirely in Pennsylvania. Your 179-km drive through the Mesozoic (dinosaur country) would take you to New Jersey, only 66 million years from downtown New York. In downtown New York, the end of the ice ages is only 10 meters (10,000 years), the width of a narrow street and sidewalk, from your destination. The 2,000 A.D. years are represented by a 2-meter-wide sidewalk and a human life span by less than 100 mm, about half the width of a curb.

Comprehending Geologic Time

The vastness of geologic time (sometimes called deep time) is difficult for us to comprehend. One way of visualizing deep time is to visualize driving from Los Angeles to New York, a distance of approximately 4,500 kilometers. Each kilometer represents 1 million years—this is a very, very slow trip. The highlights of the trip corresponding to Earth's history are shown in figure 8.27. Note that if you live to be 100, your life is represented by less than the width of a curb at the edge of a sidewalk.

Another way to get a sense of geologic time is to compare it to a motion picture. A movie is projected at a rate of 32 frames per second; that is, each image is flashed on the screen for only 1/32 of a second, giving the illusion of continuous motion. But suppose that each frame represented 100 years. If you lived 100 years, one frame would represent your whole lifetime.

If we were able to show the movie on a standard projector, each 100 years would flash by in 1/32 of a second. It would take only one-sixteenth of a second to go back to the signing of the Declaration of Independence. The 2,000-year-old Christian era would be on screen for 3/4 of a second. A section showing all time back to the last major ice age would only be less than seven seconds long. However, you would have to sit through almost six hours of film to view a scene at the close of the Mesozoic Era (perhaps you would see the last dinosaur die). And to give a complete record from the beginning of the Paleozoic Era, this epic film would have to run continuously for two days. You would have to spend over two weeks (sixteen days) in the theater, without even a popcorn break between reels, to see a movie entitled "The Complete Story of Earth, from Its Birth to Modern Civilization."

Thinking of our lives as taking less than a frame of such a movie can be very humbling. From the perspective of being stuck in that one last frame, geologists would like to know what the whole movie is like or, at least, get a synopsis of the most dramatic parts of the film.

Summary

The principle of *uniformitarianism*, a fundamental concept of geology, states that the present is the key to the past.

Relative time, or the sequence in which geologic events occur in an area, can be determined by applying the principles of *original horizontality, superposition, lateral continuity,* and *cross-cutting relationships.*

Unconformities are buried erosion surfaces that help geologists determine the relative sequence of events in the geologic past. Beds above and below a *disconformity* are parallel, generally indicating less intense activity in Earth's crust. An *angular unconformity* implies that folding or tilting of rocks took place before or around the time of erosion. A *nonconformity* implies deep erosion because metamorphic or plutonic rocks have been exposed and subsequently buried by younger rock.

Rocks can be correlated by determining the physical continuity of rocks between the two areas (generally this works only for a short distance). A less useful means of correlation is similarity of rock types (which must be used cautiously).

Fossils are used for worldwide correlation of rocks. Sedimentary rocks are assigned to the various subdivisions of the *geologic time scale* on the basis of fossils they contain, which are arranged according to the principle of *faunal succession.*

Numerical age—how many years ago a geologic event took place—is generally obtained by using *isotopic dating* techniques. Isotopic dating is accomplished by determining the ratio of the amount of a radioactive isotope presently in a rock or mineral being dated to the amount originally present. The time it takes for a given amount of an isotope to decay to half that amount is the *half-life* for that isotope. Numerical ages have been determined for the subdivisions of the geologic time scale.

Terms to Remember

angular unconformity 196
Archean Eon 208
Cenozoic Era 202
contacts 190
correlation 197
cross-cutting relationship 192
disconformity 196
eon 208
epoch 201
era 201
faunal succession 200
formation 190
fossil assemblage 201
half-life 205
inclusion 196
index fossil 200
isotope 203
isotopic dating 203
lateral continuity 192
Mesozoic Era 202
nonconformity 197
numerical (or absolute) age 190
original horizontality 191
Paleozoic Era 202
period 201
Phanerozoic Eon 208
physical continuity 197
Pleistocene Epoch 202
Prearchean (Hadean) Eon 208
Precambrian 202
Proterozoic Eon 208
Quaternary Period 202
radioactive decay 203
Recent (Holocene) Epoch 202
relative time 190
standard geologic time scale 201
superposition 191
unconformity 196
uniformitarianism 190

Testing Your Knowledge

Use the questions below to prepare for exams based on this chapter.

1. Why is it desirable to find an index fossil in a rock layer? In the absence of index fossils, why is it desirable to find several fossils in a rock unit to determine relative age?

2. Suppose you had a radioactive isotope *X* whose half-life in disintegrating to daughter product *Y* is 120,000 years. By calculating how much it took to make the present amount of *Y*, you determine that, originally, the rock contained 8 grams of isotope *X*. At present only 1/4 gram of *X* is in the rock. How many half-lives have gone by? How old is the rock?

3. By applying the various principles, draw a cross section of an area in which the following sequence of events occurred:
   a. Several layers of sedimentary rocks were deposited in the Cambrian through Devonian periods on a much older Precambrian basement of metamorphic rock.
   b. No record exists of Mississippian through Triassic rocks.
   c. A stock intruded in the Jurassic Period.
   d. Tilting and erosion to a flat plain preceded deposition of a sedimentary layer in late Cretaceous.
   e. In the Tertiary Period, more and steeper tilting affected the entire area.
   f. Erosion during the Quaternary Period created slightly hilly terrain.
   g. Following erosion, a volcano with a feeder dike erupted in recent time.

4. Name as many types of contacts (e.g., intrusive contact) as you can.

5. Using information from box 8.4, calculate the age of a feldspar. At present, there are 1.2 million atoms of ^{40}K. The amount of ^{40}Ar in the mineral indicates that originally there were 1.9 million ^{40}K atoms in the rock. Use a half-life of 1.3 billion years. (Hint: The answer is 862 million years.)

6. "Geological processes operating at present are the same processes that have operated in the past" is the principle of
   a. correlation
   b. catastrophism
   c. uniformitarianism
   d. none of the above

7. "Within a sequence of undisturbed sedimentary rocks, the layers get younger going from bottom to top" is the principle of
   a. original horizontality
   b. superposition
   c. cross-cutting
   d. none of the above

8. If rock A cuts across rock B, then rock A is ____ rock B.
   a. younger than
   b. the same age as
   c. older than

9. Which is a method of correlation?
   a. physical continuity
   b. similarity of rock types
   c. fossils
   d. all of the above

10. Eras are subdivided into
    a. periods
    b. eons
    c. ages

11. Periods are subdivided into
    a. eras
    b. epochs
    c. ages

12. Which division of geologic time was the longest?
    a. Precambrian
    b. Paleozoic
    c. Mesozoic
    d. Cenozoic

13. Which is a useful radioactive decay scheme?
    a. ^{238}U/^{206}Pb
    b. ^{235}U/^{207}Pb
    c. ^{40}K/^{40}Ar
    d. ^{87}Rb/^{87}Sr
    e. all of the above

14. C-14 dating can be used on all of the following except
    a. wood
    b. shell
    c. the Dead Sea Scrolls
    d. granite
    e. bone

15. Concentrations of radon are highest in areas where the bedrock is
    a. granite
    b. gneiss
    c. limestone
    d. black shale
    e. phosphate-rich rock
    f. all of the above

16. Which is not a type of unconformity?
    a. disconformity
    b. angular unconformity
    c. nonconformity
    d. triconformity

17. A geologist could use the principle of inclusion to determine the relative age of
    a. fossils
    b. metamorphism
    c. shale layers
    d. xenoliths

18. The oldest abundant fossils of complex multicellular life date from the
    a. Precambrian
    b. Paleozoic
    c. Mesozoic
    d. Cenozoic

19. A contact between parallel sedimentary rock that records missing geologic time is
    a. a disconformity
    b. an angular unconformity
    c. a nonconformity

Expanding Your Knowledge

1. How much of the ^{238}U originally part of Earth is still present?
2. As indicated by fossil records, why have some ancient organisms survived through very long periods of time whereas others have been very short-lived?
3. Suppose a sequence of sedimentary rock layers was tilted into a vertical position by tectonic forces. How might you determine (a) which end was originally up and (b) the relative ages of the layers?
4. Note that in table 8.3 the epochs are given only for the Cenozoic Era (as is commonly done in geology textbooks). Why are the epochs for the Mesozoic and Paleozoic considered less important and not given?
5. Moon rocks have been dated that are considerably older than any dated on Earth. Give a hypothesis to explain this.

Exploring Web Resources

www.mhhe.com/plummer9e

Visit the Online Learning Center to learn more about Geological Time. Review your answers to Testing Your Knowledge and try some of the other quizzes. Explore the additional readings, media resources, and direct links to the other sites listed below.

www.ucmp.berkeley.edu/exhibit/exhibits.html

Paleontology without Walls. University of California Museum of Paleontology virtual exhibit. Click on Geologic Time.

http://vearthquake.calstatela.edu/VirtualDating/

Virtual dating. This site provides an excellent, interactive way of learning how isotopic dating works. You can change data presented and watch graphs and other illustrations change accordingly. Quizzes help you understand the material.

www.talkorigins.org/origins/faqs-youngearth.html

The age of the earth. Topics include isotopic dating, the geologic time scale, changing views of the age of Earth. The site also explores the creation/evolution controversy.

http://asa.calvin.edu/ASA/resources/Wiens.html

Radiometric dating: a Christian perspective. At this website you can get a very thorough knowledge of isotopic dating, how it works, and how it has been used to determine the age of Earth and other events. The author addresses concerns of people who feel that an old Earth is incompatible with their religious beliefs.

Animations

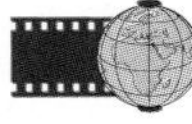

This chapter includes the following animations available on our Online Learning Center at www.mhhe.com/plummer9e.

8.1–8.11 The Geologic History of "Minor Canyon"

8.13 Disconformity

CHAPTER

9

Atoms, Elements, and Minerals

This chapter is the first of six on the material of which Earth is made. The previous chapters have been mostly about rocks. Nearly all rocks are made of minerals. Therefore, to understand rocks, you must learn what minerals are as well as the characteristics of some of the most common minerals.

In this chapter, you are introduced to some basic principles of chemistry (for those of you who have not had a chemistry course). This will help you understand material covered in the chapters on rocks, weathering, and the composition of Earth's crust and its interior. You will discover that each mineral is composed of specific chemical elements, the atoms of which are in a remarkably orderly arrangement. A mineral's chemistry and the architecture of its internal structure determine the physical properties used to distinguish it from other minerals. You should learn how to readily determine physical properties and use them to identify common minerals. (Appendix A is a further guide to identifying minerals.)

Opposite: Crystals of calcite (white), galena (dark gray) and sphalerite (orange) from the tri-state mining district of Kansas, Missouri, and Oklahoma. Note the cubic cleavage in galena and the rhombohedral cleavage cracks in calcite.

Photo ©Raymond Coveney

Figure 9.1 shows a specimen of the common rock *granite.* It forms from magma solidifying within the earth's crust. Granite is made up mostly of the minerals *feldspar* and *quartz.* But what is a mineral? A **mineral** is defined as a naturally occurring solid that is crystalline (which is to say that it has a periodically repeating arrangement of atoms) and has a specific chemical composition. We will return to the crystalline part of the definition later after discussing the chemical composition of minerals.

Quartz is made up exclusively of oxygen and silicon atoms. More precisely, quartz contains twice as many oxygen (O) atoms as silicon (Si) atoms. Therefore, the chemical formula for quartz is SiO_2, its specific composition.

Rock salt is another rock, quite different from granite. It forms when salt water is evaporated. Unlike granite, it is made of grains of only one mineral. Rock salt is a consolidated aggregate of grains of the mineral *halite*—you know it as table salt. Halite's definite chemical composition is indicated by its formula, NaCl. This means that it is composed of equal numbers of sodium (Na) and chlorine (Cl) atoms. These atoms are arranged in an orderly 3-dimensional pattern (meaning halite is crystalline). Each sodium atom is surrounded by six chlorine atoms and each chlorine atom is surrounded by six sodium atoms. Billions of each type of atom are necessary to form a salt crystal the size of a pinhead. Halite tends to form as cubic crystals because of the particular orderly way in which chlorine and sodium atoms are packed together (figure 9.2). This and other physical and chemical properties of halite are caused by (1) the pattern of repeating atoms, (2) the way the atoms are bonded to neighboring atoms, and (3) the chemical properties of the elements chlorine and sodium.

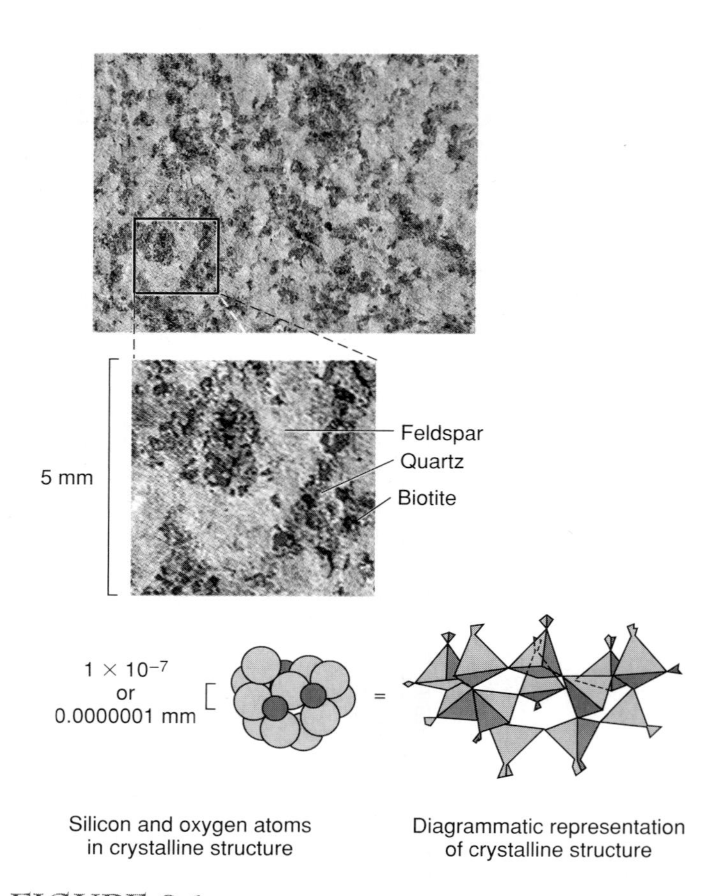

FIGURE 9.1
Specimen of granite showing the relationship among rock, minerals, and crystal structures of atoms. The diagrammatic representation as tetrahedrons is explained later in the chapter.

ATOMS AND ELEMENTS

Halite (or table salt) crystals can be separated chemically into sodium and chlorine, both of which are elements. An **element** is a substance that cannot be broken down into other substances by ordinary chemical methods. (It is interesting to note that salt, Na and Cl combined, is edible, whereas, chlorine alone is a deadly gas and pure sodium is explosive if exposed to air.)

An **atom** is the smallest possible particle of an element that retains the properties of that element. All atoms of the element chlorine are essentially identical to all other chlorine atoms; the same is true for sodium. Water (H_2O) can be chemically broken down to the elements oxygen and hydrogen (two hydrogen atoms for every oxygen atom). When two hydrogen atoms and one oxygen atom are bonded, as in liquid water (see box 9.1), they are part of a *molecule,* the smallest possible unit of a compound that has the properties of that substance.

Atoms are far too small to be seen with even the most powerful optical microscope. Our concepts of atoms are really models. A *model* in science is an image—graphic, mathematical, or verbal—that is consistent with the known data.

Models of atoms constructed by chemists and physicists show three types of subatomic particles—protons, neutrons,

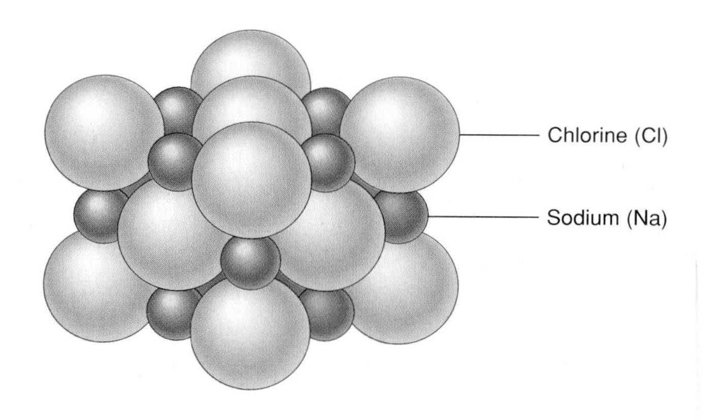

FIGURE 9.2
Model of the crystal structure of halite (or table salt).

and electrons. A **proton** is a subatomic particle that contributes mass and a single positive electrical charge to an atom. A **neutron** is a subatomic particle that contributes mass to an atom but is electrically neutral. An **electron** is a single negative electric charge that contributes a tiny percentage of mass to an atom. Electrons are regarded as moving very rapidly within specific energy levels, which are depicted as shells (figure 9.3*A*).

IN GREATER DEPTH 9.1

Water and Ice—Molecules and Crystals

Earth is often called the blue planet because oceans cover most of its surface. Ice dominates our planet's polar regions. It is fortunate that water is so abundant because life would be impossible without it. In fact, we humans are made up mostly of water. The nature and behavior of water molecules helps explain why water is vital to life on Earth.

In a water molecule the two hydrogen atoms are tightly bonded to the oxygen atom. However, the shape of the molecule is asymmetrical, with the two hydrogen atoms on the same side of the atom (box figure 1). This means the molecule is polarized, with a slight excessive positive charge at the hydrogen side of the molecule and a slight excess negative charge at the opposite side. Because of the slight electrical attraction of water molecules, other substances are readily attracted to the molecules and are dissolved or carried away by water. Water has been called the universal solvent. Dirt washes out of clothing; water, in blood, carries nutrients to our muscles and transports waste to our kidneys and out of our bodies.

When water is in its liquid state, the molecules are moving about; however, because of the polarity, molecules are slightly attracted to one another. For this reason, water molecules are closer together than they are in most other liquids. However, in ice the water molecules are not as tightly packed together as in liquid water.

When water freezes, the bonding hydrogen atoms are shared between adjacent water molecules, resulting in an orderly 3-dimensional pattern that is hexagonal, like in a honeycomb. (This explains the hexagonal shape of snowflakes.) The openness of the honeycomb-like crystalline structure of ice contrasts with the more closely packed molecules in liquid water. For this reason, ice is less dense than liquid water. This is an unusual solid-liquid relationship. For most substances, the solid is denser than its liquid phase.

The fact that ice is less dense than liquid water has profound implications. Ice floats rather than sinks in liquid water. Icebergs float in the ocean. Lakes freeze from the top down. Ice on a lake surface acts as an insulating layer that retards the freezing of underlying water. If ice sank, lakes would freeze much more readily and thaw much more slowly. Our climate would be very different if ice sank. The Arctic Ocean surface freezes during the winter but only at its surface. If the ice were to sink, more ocean water would be exposed to the cold atmosphere and would freeze and sink. Eventually, the entire Arctic Ocean would freeze and would not thaw during the summer. If this were the case, life as we know it probably would not exist.

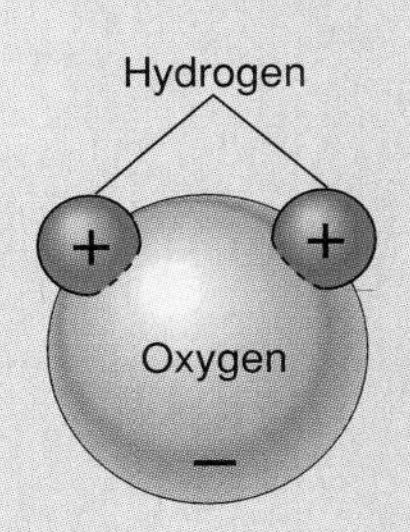

Box 9.1 — FIGURE 1
Water molecule.

When water freezes, it expands. A bottled beverage placed in a freezer breaks its container upon freezing. When water trapped in cracks in rock freezes, it will expand and will help break up the rock (as explained in the chapter on weathering).

Related Web Resources

Snow Crystal Research Nice images taken with an electron microscope.
www.lpsi.barc.usda.gov/emusnow/

Snow Crystals Caltech's site. More about ice and nice pictures of snow crystals.
www.its.caltech.edu/~atomic/snowcrystals/

Protons and neutrons form the **nucleus** of an atom. Although the nucleus occupies an extremely tiny fraction of the volume of the entire atom, practically all the mass of the atom is concentrated in the nucleus. The **atomic mass number** is the total number of neutrons and protons in an atom. The atomic mass number of the oxygen atom in figure 9.3*B* is 16 (8 protons plus 8 neutrons). Heavier elements have more neutrons and protons than do lighter ones. For example, the heavy element gold has an atomic mass number of 197, whereas helium has an atomic mass number of 4.

The number of protons controls the "character" of an element more than does the number of other subatomic particles. The **atomic number** of an element is the number of protons in each atom. We can refine our earlier definition of an element by adding that each atom of an element has the *same number of protons.* Gold has an atomic number of 79, or 79 protons per atom; oxygen always has 8 protons; hydrogen always has 1 proton; chlorine has 17; and sodium has 11. (Other atomic numbers are listed in appendix C as well as in appendix D, the periodic table of elements.)

The number of neutrons (and therefore the mass of an element) can vary within limits. **Isotopes** of an element are atoms *containing different numbers of neutrons* but the *same number of protons.* For example, the most common isotope of oxygen has 8 neutrons; far less abundant is the oxygen isotope with 10 neutrons. In geology, radioactive isotopes are important for determining the age of rocks, as described in the chapter on geologic time.

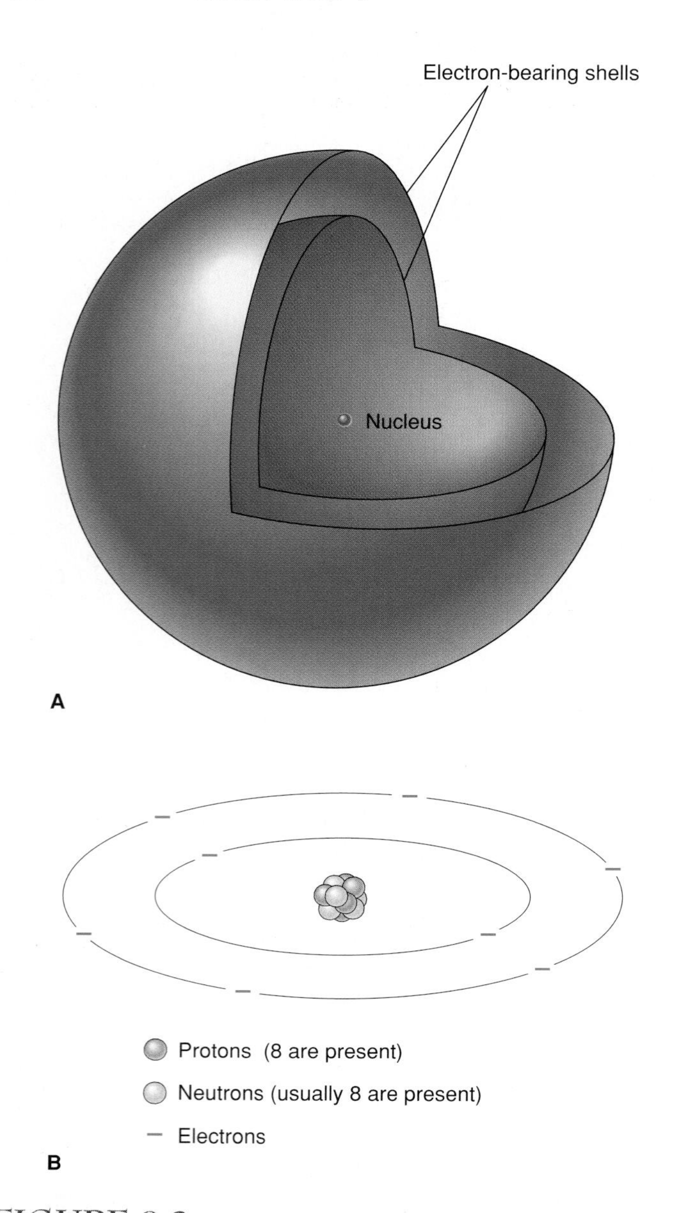

FIGURE 9.3

(*A*) Model of an oxygen atom. The nucleus, composed of neutrons and protons, is actually much smaller than indicated relative to the volume of the atom. The hollow spheres represent the two electron-bearing shells. (*B*) Schematic representation of the oxygen atom. The two circles containing electrons represent the electron-bearing shells.

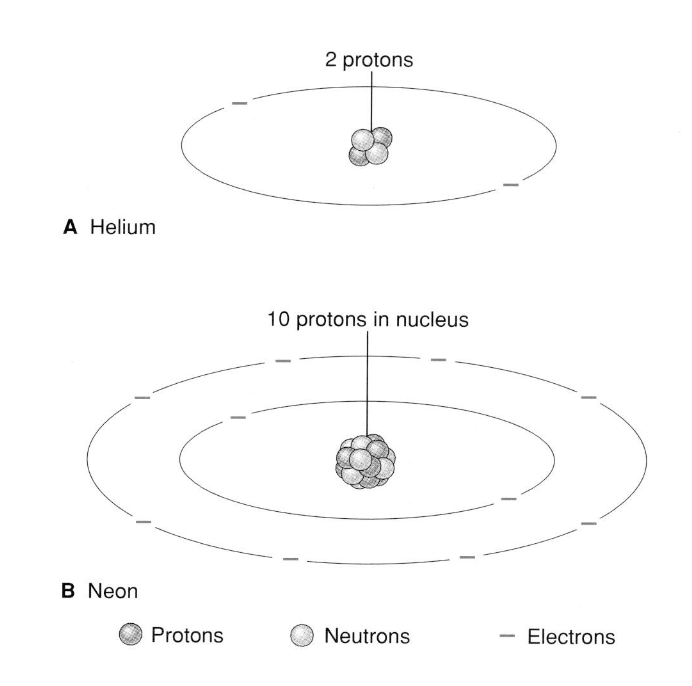

FIGURE 9.4

(*A*) Helium atom and (*B*) neon atom.

An element's atomic weight is closely related to the mass number. **Atomic weight** is the weight of an *average* atom of an element, given in atomic mass units. Because sodium has only one naturally occurring isotope, its atomic mass number and its atomic weight are the same—23. On the other hand, chlorine has two common isotopes, with mass numbers of 35 and 37. The atomic weight of chlorine, which takes into account the abundance of each isotope, is 35.5 (the lighter isotope is more common than the heavier one).

Electrons, each having 1/1836 the mass of a proton, are visualized as occupying shells around the nucleus. Virtually the entire volume of an atom is taken up by the space in which these tiny, electrically negative charges move. The number of electrons in an atom is related to the number of protons in the nucleus. Each electron is opposite in charge but equal in strength to a proton.

If the number of electrons in orbit around a nucleus is equal to the number of protons in the nucleus, the positive and negative charges balance each other and the atom is electrically neutral. Helium has 2 electrons, exactly balancing its 2 protons. Most elements, however, are not able to maintain an electrical balance between protons and electrons within a single atom.

Chemical Activity

Many geological processes can be explained as chemical reactions. Some rocks form as a result of chemical reactions between substances. Chemical reactions also play a part in weathering (see chapter 12). Understanding a few basic concepts of chemistry will clarify why and under what conditions chemical reactions occur.

Atoms that are not electrically neutral tend to react (or combine) with other atoms to neutralize the electrical imbalance. In addition to electrical neutrality, an atom is most stable when each of its shells is filled with electrons. The innermost shell is full when it possesses 2 electrons. Outer shells each require 8 electrons for an atom to be nonreactive (this is true for the second and third shells; elements having additional shells are more complicated).

Helium, for example, is a *stable* element because 2 protons are balanced by 2 electrons, and the 2 electrons exactly fill one shell. Neon is also stable; its 10 protons are balanced by 2 electrons in the inner shell and 8 electrons in the next shell (see figure 9.4). Normally, neither of these elements reacts with other elements.

Ions

Chlorine and sodium are more typical elements because if an electron shell is complete, the atom is electrically out of balance. Note that sodium in figure 9.5 has a complete inner shell with 2 electrons and a second shell, also filled, with 8 electrons. One more electron would neutralize all 11 protons in the nucleus, but an eleventh electron alone in a shell is extremely unstable, so the sodium atom normally does without it. In each sodium atom, then, the 11 protons (11+) and 10 electrons (10−) add up to a single excess positive charge (+1). Such an atom is an **ion,** an electrically charged atom or group of atoms. The sodium ion can be abbreviated as Na^+.

Chlorine, with an atomic number of 17, has a complete inner shell with 2 electrons and a complete second shell of 8 electrons around this. A neutral chlorine atom would have only 7 electrons in the third shell, but this shell requires 8 electrons so an extra electron is captured and incorporated in it. The chlorine ion then contains 18 electrons and 17 protons, and so has a single excess negative charge (Cl^-).

Positive and negative ions are attracted to each other. In a crystalline structure the mutual attraction is one way atoms are held in place or bonded to one another. **Bonding** is the attachment of an atom to one or more adjacent atoms (see box 9.2).

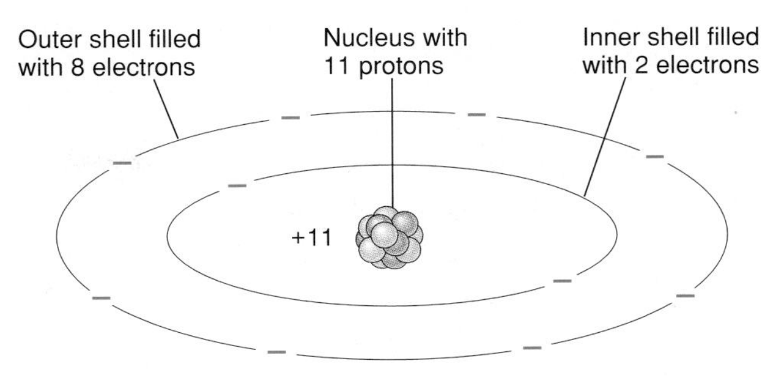

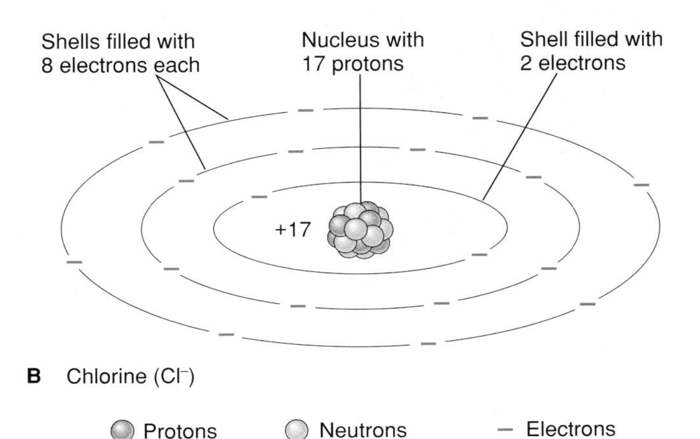

FIGURE 9.5

(*A*) Sodium (Na^+) ion. Ten electrons fill two shells. The nucleus contains eleven protons. (*B*) Chlorine (Cl^-) ion. Electron shown in red completes the outer shell of the chlorine atom, making it an ion.

CHEMICAL COMPOSITION OF THE EARTH'S CRUST

Estimates of the chemical composition of Earth's crust are based on many chemical analyses of the rocks exposed on Earth's surface. (Models for the composition of the interior of the Earth—the core and the mantle—are based on more indirect evidence.) Table 9.1 lists the generally accepted estimates of the

table 9.1 **Crustal Abundance of Elements**

Element	Symbol	Percentage by Weight	Percentage by Volume	Percentage of Atoms
Oxygen	O	46.6	93.8	60.5
Silicon	Si	27.7	0.9	20.5
Aluminum	Al	8.1	0.8	6.2
Iron	Fe	5.0	0.5	1.9
Calcium	Ca	3.6	1.0	1.9
Sodium	Na	2.8	1.2	2.5
Potassium	K	2.6	1.5	1.8
Magnesium	Mg	2.1	0.3	1.4
All other elements		1.5	—	3.3

9.2

IN GREATER DEPTH

Bonding

Ions may be regarded as tiny spheres that behave much like magnets. Positively charged ions attract negatively charged ions so that their electrical charges can be neutralized. In salt water, equal numbers of sodium ions (Na^+) and chlorine ions (Cl^-) move about freely. The electrical neutrality of the water is maintained because positive sodium ions exactly balance negative chlorine ions. If the water evaporates, the sodium and chlorine are electrically attracted to each other and crystallize into halite. The crystalline structure is the most orderly way for chlorine and sodium ions to pack themselves together and neutralize their collective charges.

A chlorine ion and a sodium ion are fixed in place by their electrical attraction to each other. This is called **ionic bonding** because it is brought about by an attraction between positively and negatively charged ions (box figure 1).

Ionic bonding is the most common type of bonding in minerals; however, in most minerals the bonds between atoms are not purely ionic. Atoms are also commonly bonded together by **covalent bonding,** or bonding in which adjacent atoms *share* electrons. Diamond is composed exclusively of covalently bonded carbon atoms (box figure 2). Carbon has an atomic number of 6, which means that the innermost shell is full with 2 electrons. Four more electrons are required to maintain electrical neutrality. In a diamond, each carbon atom has 4 electrons in the outer shell to maintain neutrality, while the need for 8 electrons in that shell is satisfied by electrons that are shared with adjacent carbon atoms. Neighboring carbon atoms are so close together that each of the outer-shell electrons spends half its time orbiting one atom and half orbiting an adjacent atom. Electrical neutrality is maintained, and each atom, in a sense, has 8 electrons in the outer shell (even though they are not all there at the same time). Covalent bonds in the diamond are extremely strong, and diamond is the hardest natural substance on Earth. However, covalent bonds are not necessarily stronger than ionic bonds.

A third type of bonding, *metallic bonding,* is not as important to geology. In metals, such as iron or gold, the atoms are closely packed and the electrons move freely throughout the crystal so as to hold the atoms together. The ease with which electrons move accounts for the high electrical conductivity of metals.

Finally, after all atoms have bonded together, there may be weak, attractive forces remaining. This is the very weak force that holds adjacent, tightly bonded sheets of mica or graphite together. (An analogy is when you rub a balloon over your hair, the electrical charges allow the balloon to stick to a wall.)

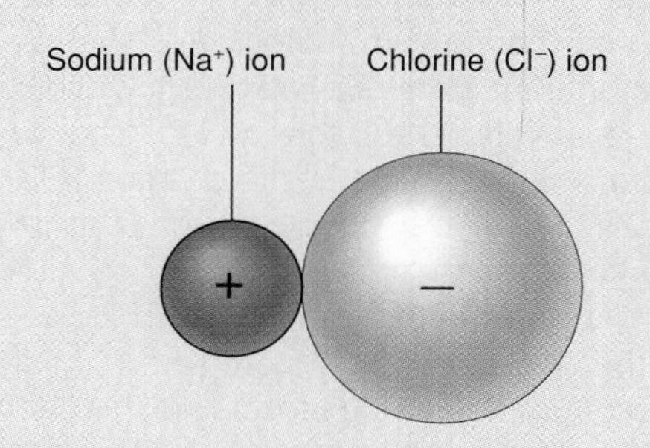

Box 9.2 — Figure 1
Ionic bonding between sodium (Na^+) and chlorine (Cl^-).

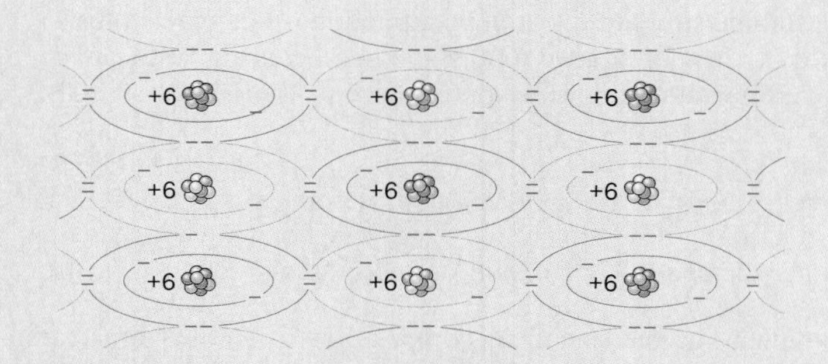

Box 9.2 — Figure 2
Carbon atoms covalently bonded, as in diamond.

Graphite, like diamond, is pure carbon. Graphite is used in pencils and as a lubricant. Amazingly, the hardest mineral and one of the softest have the same composition. The distinction is in the bonding.

Related Web Resource

To see how diamond and graphite differ, go to Mineral Web www.man.ac.uk/Geology/MineralWeb/Mineral_Web.html to see rotating, crystal structures in 3-D. (First you must download and install Chime software, which is easy to do from the website.) From the pull-down menu, select *graphite.* Note the rotating crystal structure. The rods connecting carbon atoms represent strong bonds. You can use your mouse to stop the rotation and view the structure from any perspective. Then go to *diamond.* How do the two crystal structures differ? Can you see why plates of tightly bonded graphite slide easily past one another?

abundance of elements in the Earth's crust. At first glance, the chemical composition of the crust (and, therefore, the average rock) seems quite surprising.

We think of oxygen as the O_2 molecules in the air we breathe. Yet most rocks are composed largely of oxygen, as it is the most abundant element in the Earth's crust. Unlike the oxygen gas in air, oxygen in minerals is strongly bonded to other elements. By weight, oxygen accounts for almost half the crust, but it takes up 93% of the volume of an average rock. This is because oxygen's electron shells take up a large amount of space relative to its weight. (Note how much bigger oxygen atoms are relative to other atoms in figure 9.7 and others.) It is not an exaggeration to regard the crust as a mass of oxygen with other elements occupying positions in crystalline structures between oxygen atoms.

Silicon is the element used to make computer chips. **Silica** is a term for *oxygen combined with silicon.* Because silicon is the second most abundant element in the crust, most minerals contain silica. The common mineral quartz (SiO_2) is pure silica that has crystallized. Quartz is one of many minerals that are **silicates,** substances that contain silica (as indicated by their chemical formulas). Most silicate minerals also contain one or more other elements.

Note that the third most abundant element is aluminum, which is more common in rocks than iron is. Knowing this, you might assume that aluminum would be less expensive than iron, but of course this is not the case. Common rocks are not mined for aluminum because it is so strongly bonded to oxygen and other elements. The amount of energy required to break these bonds and separate the aluminum makes the process too costly for commercial production. Aluminum is mined from the uncommon deposits where aluminum-bearing rocks have been weathered, producing compounds in which the crystalline bonds are not so strong.

Collectively, the eight elements listed in table 9.1 account for more than 98% of the weight of the crust. All the other elements total only about 1.5%. Absent from the top eight elements are such vital elements as hydrogen (tenth by weight) and carbon (seventeenth by weight).

The element copper is only twenty-seventh in abundance, but our industrialized society is highly dependent on this metal. Most of the wiring in electronic equipment is copper, as are many of the telephone and power cables that crisscross the continent. However, the Earth's crust is not homogeneous, and geological processes have created concentrations of elements such as copper in a few places. Exploration geologists are employed by mining companies to discover where (as well as why) ore deposits of copper and other metals occur (see chapter 21).

CRYSTALLINITY

By definition, a mineral must be crystalline. In fact, most solids are crystalline. A **crystalline** substance is one in which the atoms are arranged in a 3-dimensional, regularly repeating, orderly pattern. The print by M. C. Escher (figure 9.6) vividly expresses the principle of crystallinity. You can visualize what crystallinity is in nature by mentally substituting identical clusters of atoms for each fish and imagining the clusters packed together.

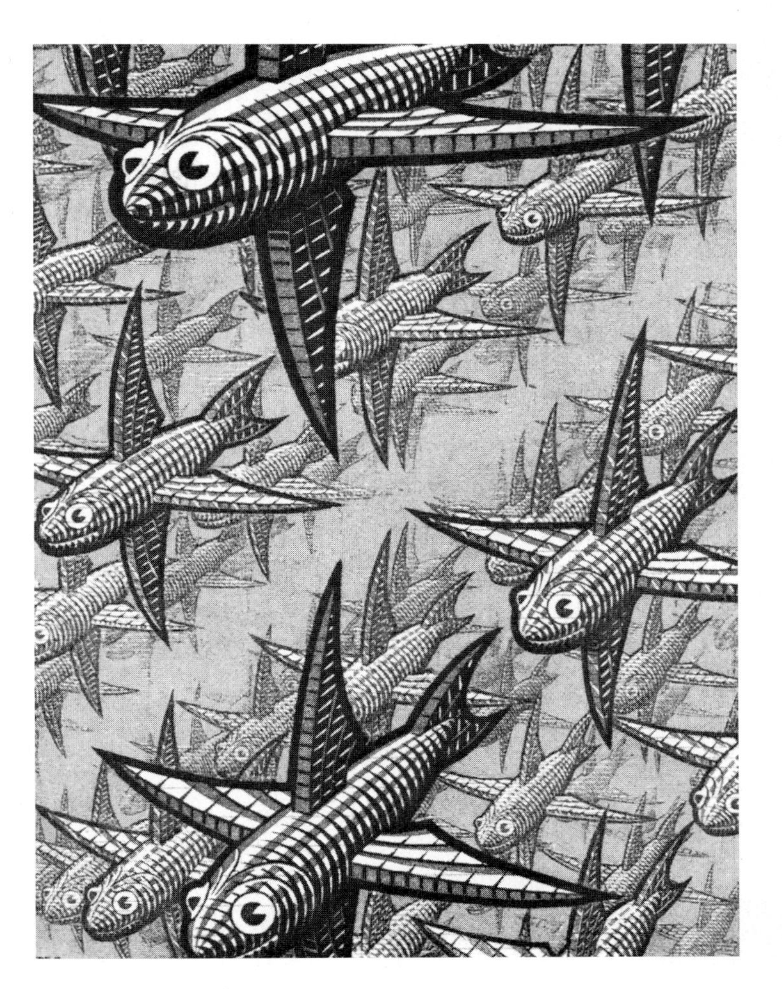

FIGURE 9.6

Depth, printed by M. C. Escher.

As you can tell from figure 9.2, halite (or table salt) is crystalline. What controls the type (or "architecture") of crystal structure of a substance is principally the relative size of adjacent atoms. Thus, if all the sodium atoms in figure 9.2 were about the same size as the chlorine atoms, the particular crystal structure would be different (bearing in mind that neighboring atoms must "touch" one another). Liquids are not crystalline because their atoms or molecules are free to move about; nor is glass, because the atoms in glass are as randomly arranged as those in a liquid, only "frozen" into place. The structure of glass is comparable to what would happen if fish like those in the Escher drawing (figure 9.6) were swimming freely and randomly distributed when the water suddenly froze.

Of particular importance are the crystal structures derived from the two most common elements in the Earth's crust—oxygen and silicon.

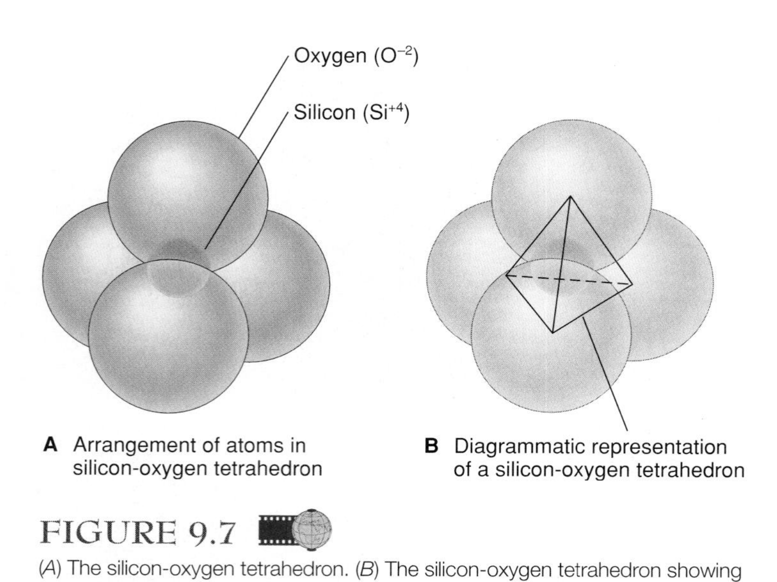

FIGURE 9.7

(*A*) The silicon-oxygen tetrahedron. (*B*) The silicon-oxygen tetrahedron showing the corners of the tetrahedron coinciding with the centers of oxygen ions.

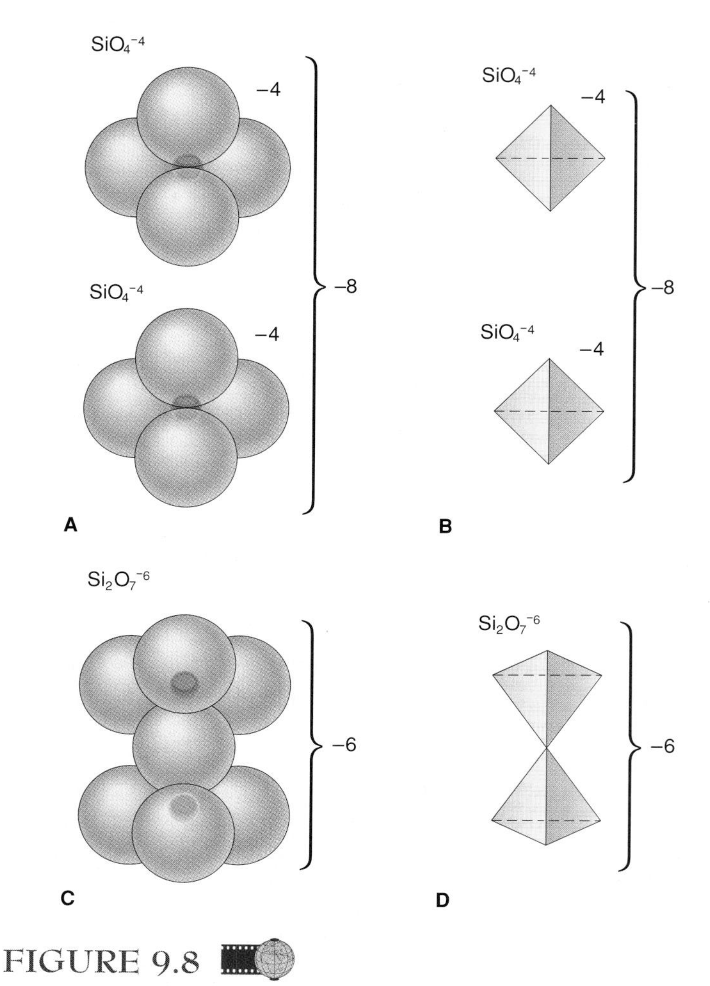

FIGURE 9.8

Two single tetrahedrons (*A* and *B*) require more positively charged ions to maintain electrical neutrality than do two tetrahedrons sharing an oxygen atom (*C* and *D*). *B* and *D* are the schematic representations of *A* and *C* respectively.

The Silicon-Oxygen Tetrahedron

The two most abundant elements of the crust, silicon and oxygen, combine to form the basic building block for most common minerals. In each "building block," four oxygen atoms are packed together around a single, much smaller, silicon atom, as shown in figure 9.7*A*. The four-sided, pyramidal, geometric shape called a *tetrahedron* is used to represent the four oxygen atoms surrounding a silicon atom. Each *corner* of the tetrahedron represents the *center* of an oxygen atom (figure 9.7*B*). This basic building block of a crystal is called a **silicon-oxygen tetrahedron** (also known as the silica tetrahedron).

The atoms of the tetrahedron are strongly bonded together. Within a silicon-oxygen tetrahedron the negative charges exceed the positive charges (see figures 9.7*A* and 9.8*A*). A single silicon-oxygen tetrahedron is a complex ion with a formula of SiO_4^{-4} because silicon has a charge of +4 and the four oxygen ions have 8 negative charges (−2 for each oxygen atom).

A silicon-oxygen tetrahedron can bond either with positively charged ions, such as iron or aluminum, or with other silicon-oxygen tetrahedrons. In other words, for the silicon-oxygen tetrahedron to be stable within a crystal structure, it must either (1) be balanced by enough positively charged ions or (2) share oxygen atoms with adjacent tetrahedrons (as shown in figure 9.8*C* and *D*) and therefore reduce the need for extra, positively charged ions. The structures of silicate minerals range from an *isolated silicate structure,* which depends entirely on positively charged ions to hold the tetrahedrons together, to *framework silicates* (quartz, for example), in which all oxygen atoms are shared by adjacent tetrahedrons. The various types of silicate structures are shown diagrammatically in figure 9.9 and are discussed next.

Isolated Silicate Structure

Silicate minerals that are structured so that none of the oxygen atoms are shared by tetrahedrons have an ***isolated silicate structure.*** The individual silicon-oxygen tetrahedrons are bonded together by positively charged ions (figure 9.10). The common mineral **olivine,** for example, contains two ions of either magnesium (Mg^{+2}) or iron (Fe^{+2}) for each silicon-oxygen tetrahedron. The formula for olivine is $(Fe, Mg)_2SiO_4$.

Chain Silicates

A **chain silicate structure** forms when two of a tetrahedron's oxygen atoms are shared with adjacent tetrahedrons to form a chain (figures 9.9 and 9.11). Each chain, which extends indefinitely, has a net excess of negative charges. Minerals may have a single- or double-chain structure. For single-chain silicate structures, the ratio of silicon to oxygen (as figure 9.11 shows) is 1:3; therefore, each mineral in this group (the *pyroxene* group) incorporates SiO_3^{-2} in its formula, and it must be electrically balanced by the positive ions that hold the parallel chains together.

One pyroxene mineral, for example, has a formula of $MgSiO_3$. This pyroxene may form in a cooling magma when earlier formed crystals of olivine, Mg_2SiO_4, react with silica (SiO_2) in the remaining melt (as discussed in the chapter on the origin of igneous rocks). To accommodate the additional silica, olivine's

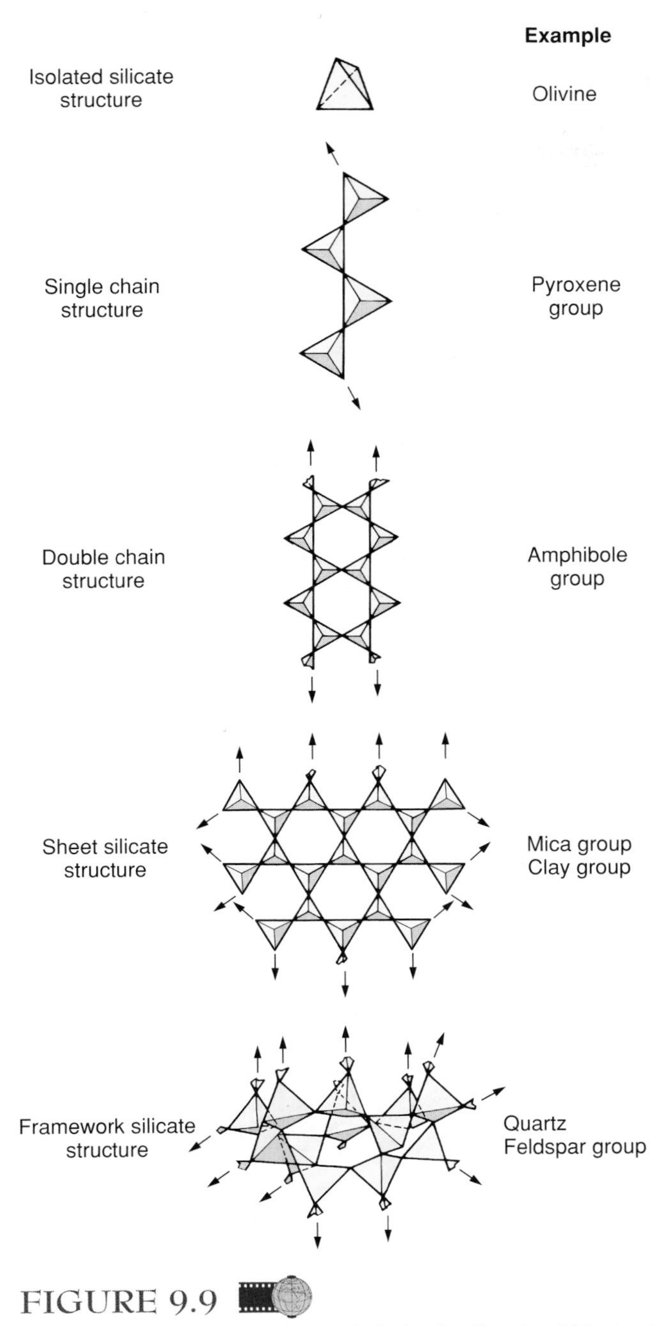

FIGURE 9.9

Common silicate structures. Arrows indicate directions in which structure repeats indefinitely.

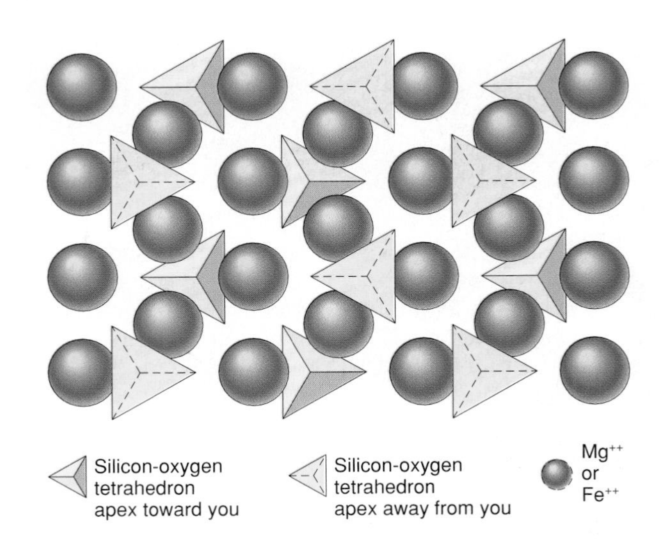

FIGURE 9.10

Diagram of the crystal structure of olivine, as seen from one side of the crystal.

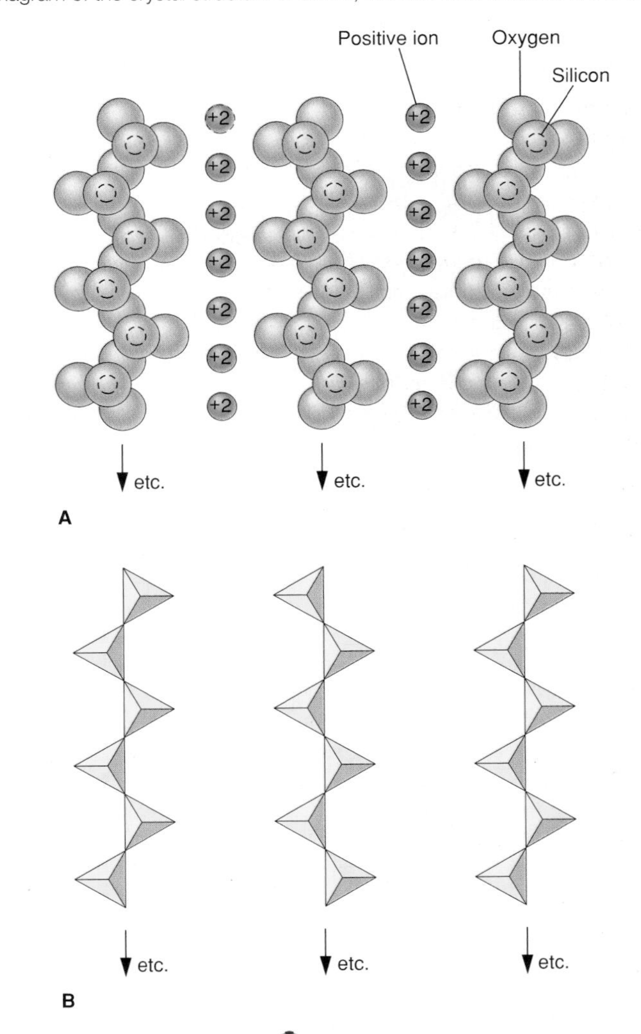

FIGURE 9.11

Single-chain silicate structure. (*A*) Model of a single-chain silicate mineral. (*B*) The same chain silicate shown diagrammatically as linked tetrahedrons; positive ions between the chains are not shown.

isolated silicate structure is rearranged into the single-chain silicate structure of pyroxene. In this case, Mg^{+2} ions occupy the positive ion positions between chains as shown in figure 9.11*A*.

The *amphibole* group is characterized by *two* parallel chains (double-chain silicate structure) in which every other tetrahedron along a chain shares an oxygen atom with the adjacent chain (see figure 9.9). In even a small amphibole crystal, millions of parallel double chains are bonded together by positively charged ions.

Chain silicates tend to be shaped like columns, needles, or even fibers. The long structure of the external form corresponds to the linear dimension of the chain structure. Fibrous aggregates of minerals are called *asbestos* (see box 9.3).

9.3

ENVIRONMENTAL GEOLOGY

Asbestos—How Hazardous?

Asbestos is a generic name for fibrous aggregates of minerals (box figure 1). Because it does not ignite or melt in fire, asbestos has a number of valuable industrial applications. Woven into cloth, it may be used to make suits for firefighters. It can also be used as a fireproof insulation for homes and other buildings and has commonly been used in plaster for ceilings. Five of the six commercial varieties of asbestos are amphiboles, known commercially as "brown" and "blue" asbestos. The sixth variety is *chrysotile*—which is not a chain silicate and which belongs to the *serpentine* family of minerals—and is more commonly known as "white asbestos." White asbestos is, by far, the most commonly used in North America (about 95% of that used in the United States).

Public fear of asbestos in the United States has resulted in its being virtually outlawed by the federal government. Tens of billions of dollars have been spent (probably unnecessarily) to remove or seal off asbestos from schools and other public buildings.

Asbestos's bad reputation comes from the high death rate among asbestos workers exposed, without protective attire, to extremely high levels of asbestos dust. Some of these workers, whose bodies were covered with fibers, were called "snowmen." In Manville, New Jersey, children would catch the "snow" (white asbestos dust released from the asbestos factory there) in their mouths. The high death rates among asbestos workers are attributed to *asbestosis* and lung cancer. Asbestosis is similar to silicosis contracted by miners; essentially, the lungs become clogged with asbestos dust after prolonged heavy exposure. The incidence of cancer has been especially high among asbestos workers who were also smokers. It's not clear that heavy exposure to white asbestos caused cancer among nonsmoking asbestos workers. However, brown and blue (amphibole) varieties, which are not mined for asbestos in North America, have been linked to cancer for heavy exposure (even if for a short term).

What are the hazards of asbestos to an individual in a building where walls or ceilings contain asbestos? Recent studies from a wide range of scientific disciplines indicate that the risks are minimal to nonexistent, at least for exposure to white asbestos. The largest asbestos mines in the world are at Thetford Mines, Quebec. A study of longtime Thetford Mines residents, whose houses border the waste piles from the asbestos mines, indicated that their incidence of cancer was no higher than that of Canadians overall. Nor have studies in the United States been able to link nonoccupational exposure to asbestos and cancer. One estimate of the risk of death from cancer due to exposure to asbestos dust is one per 100,000 lifetimes. (Compare this to the risk of death from lightning of 4 per 100,000 lifetimes or automobile travel—1,600 deaths per 100,000 lifetimes.)

Box 9.3 — Figure 1
Chrysotile asbestos.
Photo by C. C. Plummer

In California, a closed-down white asbestos mining site designated for Environmental Protection Agency Superfund cleanup is a short distance from where asbestos is being mined cleanly and efficiently. The asbestos is packaged and shipped to Japan. It cannot be used in the United States, because the United States is the only industrialized nation whose laws do not distinguish between asbestos types and permit the use of chrysotile.

A reason why chrysotile is less hazardous than amphibole asbestos is that chrysotile fibers will dissolve in lungs and amphibole will not. Recent experiments by scientists at Virginia Polytechnic Institute indicate that it takes about a year for chrysotile fibers to dissolve in lung fluids, whereas glass fibers of the same size will dissolve only after several hundred years. Yet fiberglass is being used increasingly as a substitute for asbestos.

Related Web Resource

Welcome to the Asbestos Institute

www.asbestos-institute.ca/main.html

Sheet Silicates

In a **sheet silicate structure** each tetrahedron shares three oxygen atoms to form a sheet (figure 9.9). The *mica* group and the *clay* group of minerals are sheet silicates. The positive ions that hold the sheets together are "sandwiched" between the silicate sheets.

Framework Silicates

When all four oxygen ions are shared by adjacent tetrahedrons, a **framework silicate structure** is formed. Quartz is a framework silicate mineral. A feldspar is a framework silicate as well. However, its structure is slightly more complex because aluminum substitutes for some of the silicon atoms in some of the tetrahedrons. The same kind of substitution also takes place in amphiboles and micas, which helps account for the wide variety of silicate minerals.

MINERALS

According to our earlier definition, three conditions must be satisfied for a substance to be a mineral:

1. It must be a *crystalline solid.*
2. It must *occur naturally.*
3. It must have a *specific chemical composition.*

There are other meanings for the word *mineral* that conflict with the geologist's definition. The "minerals" listed on cereal boxes, for instance, have nothing to do with what geologists or chemists mean when they talk about minerals. Nor, for that matter, does the geologist's definition agree with the miner's definition of a mineral. To a miner, a "mineral" is anything of commercial value that is extracted from the ground.

In this book, the term *mineral* is used strictly in the geologic sense.

Crystalline Solid

We have already discussed crystallinity. In particular, we have described the various orderly arrangements of atoms for the silicate minerals. Nature is not always accommodating to definitions, however, and some substances exist that are not crystalline but otherwise meet the criteria for a mineral. These not-quite-minerals are called *mineraloids.* An example of a mineraloid is *opal,* which does not have a very orderly arrangement of its atoms.

Natural Occurrence

This criterion needs little explanation. Human-made crystalline compounds are not regarded as minerals.

Specific Chemical Composition

A specific chemical composition is to be expected because of the inherent orderliness of crystalline substances. Essentially, this means that chemical analysis of any sample of a given mineral will always produce the same ratios of elements (in quartz, for example, two atoms of oxygen for every atom of silicon). In other words, the composition of any mineral can be expressed as a chemical formula. Quartz and halite have simple formulas, SiO_2 and NaCl, respectively. Some formulas are more complex because atoms of one element may substitute for those of another element in a particular mineral. This is known as *solid solution.* For instance, we gave the formula for the mineral olivine as $(Fe, Mg)_2SiO_4$. Because magnesium and iron ions are about the same size, they can substitute freely for each other without distorting the crystal structure and significantly altering the properties of the mineral. Some chemical formulas appear much more complex than the crystal structures they represent because several such substitutions can occur.

Orderly variations in chemical compositions can exist within single crystals. This property is known as *zoning.* Zoning is very common in plagioclase feldspar (the crust's most abundant mineral, described later in this chapter). In plagioclase, calcium (Ca) and sodium (Na) can substitute for one another within the crystal structure (figure 9.12). Commonly, plagioclase in igneous rocks is zoned with the center of the crystal being high in Ca; toward the edges, the crystal is progressively richer in Na (and poorer in Ca). This is interpreted as indicating that the crystal started growing when crystallization of magma began at a high temperature. (High temperatures favor Ca and low temperatures favor Na in plagioclase.) As the magma cooled, the plagioclase crystal continued to grow, but

FIGURE 9.12
Zoning in plagioclase feldspar, as seen under a polarizing microscope. The concentric color bands each indicate different amounts of Ca and Na in the crystal structure.

the new layers of plagioclase incorporated progressively more Na and less Ca.

Chemical analysis can aid the identification of minerals. But, for a variety of reasons, including the cost and difficulty of chemical analysis, physical properties (which reflect chemical composition as well as bonding and crystalline structure) are more generally used in identifying minerals. These properties are described later in this chapter.

The Important Minerals

It is useful to be able to associate the names of important minerals with the physical properties that identify them. Of course, what constitutes an "important" mineral depends on your perspective. To a miner or prospector, an important mineral is one that is commercially valuable (and is, by implication, relatively uncommon). A "rock hound" is interested in collecting any mineral that is pretty or unusual. A gemologist specializes in those varieties of minerals that are of gem quality (diamonds, emeralds, etc.). A mineralogist is a scientist who studies the chemistry and crystallographic structure of minerals. In this book, the minerals we regard as important are those that help us understand the nature of the Earth. We are particularly interested in the *rock-forming minerals* because they make up most of the rocks of the Earth's crust.

Of the several thousand identifiable minerals on Earth, most are rare and not important to geology (many occur at only a single site on the globe). Only a few hundred are classified as rock-forming minerals. Even most of these are relatively uncommon in comparison with the few minerals that make up the vast bulk of Earth's crust. The five mineral groups listed in the upper third of table 9.2 account for well over 90% of Earth's crust. These are the minerals whose names recur most often in this book. All are silicates. (The formulas and characteristic properties of minerals listed in table 9.2 are in appendix A.)

Quartz may be the only familiar name among the most common minerals, unless you have already had some exposure to geology. Like people, however, each mineral has its own character or physical properties. As you become more familiar with them, they will become more than just strange names.

As shown in table 9.2, minerals with similar crystal structures and compositions are grouped under a common name. The most abundant group of minerals in the crust is the **feldspar group.** Like quartz, the feldspars are framework silicates; however, aluminum has substituted for some of the silicon in the linked tetrahedrons (aluminum ions and silicon ions are close to the same size). In addition to silicon, aluminum, and oxygen, feldspars contain sodium, calcium, or potassium. Those that contain potassium are called **potassium feldspar.** (Two potassium feldspars that have subtle crystallographic differences are *microcline* and *orthoclase.*) If sodium or calcium (or both) are incorporated into the feldspar crystal structure, then the mineral is a **plagioclase feldspar.**

table 9.2 Minerals of the Earth's Crust

Name	Chemical Composition	Type of Silicate Structure or Chemical Group
The most common rock-forming minerals. (These make up more than 90% of the Earth's crust.)		
Feldspar group		
Plagioclase	Ca and Na Al silicate	Framework silicate
Potassium feldspar (orthoclase, microcline)	K Al silicate	Framework silicate
Pyroxene group (augite most common)	Fe, Mg silicate (some with Al, Na, Ca)	Single-chain silicate
Amphibole group (hornblende most common)	Complex Fe, Mg, Al silicate hydroxide	Double-chain silicate
Quartz	Silica	Framework silicate
Mica group		
Muscovite	K Al silicate hydroxide	Sheet silicate
Biotite	K Fe, Mg Al silicate hydroxide	Sheet silicate
Other common rock-forming minerals.		
Silicates		
Olivine	Mg, Fe silicate	Isolated silicate
Garnet group	Complex silicates	Isolated silicate
Clay minerals group	Complex Al silicate hydroxides	Sheet silicate
Nonsilicates		
Calcite	$CaCO_3$	Carbonate
Dolomite	$CaMg(CO_3)_2$	Carbonate
Gypsum	$CaSO_4 \cdot 2H_2O$	Sulfate
Much less common minerals of commercial value.		
Halite	NaCl	Chloride
Diamond	C	Native element
Gold	Au (gold)	Native element
Hematite	Iron oxide (Fe_2O_3)	Oxide
Magnetite	Iron oxide (Fe_3O_4)	Oxide
Chalcopyrite	Cu, Fe sulfide	Sulfide
Sphalerite	Zn sulfide	Sulfide
Galena	Pb sulfide	Sulfide

ENVIRONMENTAL GEOLOGY 9.4

CLAY MINERALS THAT SWELL

Clay minerals are very common at the earth's surface; they are a major component of soil. There are a great number of different clay minerals. What they all have in common is that they are sheet silicates. They differ by which ions hold sheets together and by the number of sheets "sandwiched" together.

Ceramic products and bricks are made from clay. Surprisingly, some clay minerals are edible; some are used in the manufacturing of pills. *Kaolinite,* a clay mineral, is the main ingredient in Kaopectate, a remedy for upset stomachs. Popular fast-food chains use clay minerals as a thickener for shakes (you can tell which ones, because the chains do not call them "milk shakes"—they do not use milk).

Montmorillonite is one of the more interesting clay minerals. It is better known as *expansive clay* or *swelling clay.* If water is added to the montmorillonite, the water molecules are absorbed into the spaces between silicate layers (box figure 1). This results in a large increase in volume, sometimes up to several hundred percent. The pressure generated can be up to 50,000 kilograms per square meter. This is sufficient to lift a good-sized building.

If a building is erected on expansive clay that subsequently gets wet, a portion of the building will be shoved upward. In all likelihood the building will break. Some people think that expansive soils have caused more damage than earthquakes and landslides combined.

On the other hand, swelling clays can be put to use. Montmorillonite, mixed with water, can be pumped into fractured rock or concrete. When the water is absorbed, swelling clay expands to fill and seal the crack. The technique is particularly useful where dams have been built against fractured bedrock. Sealing the cracks with expansive clays ensures that water will stay in the reservoir behind the dam.

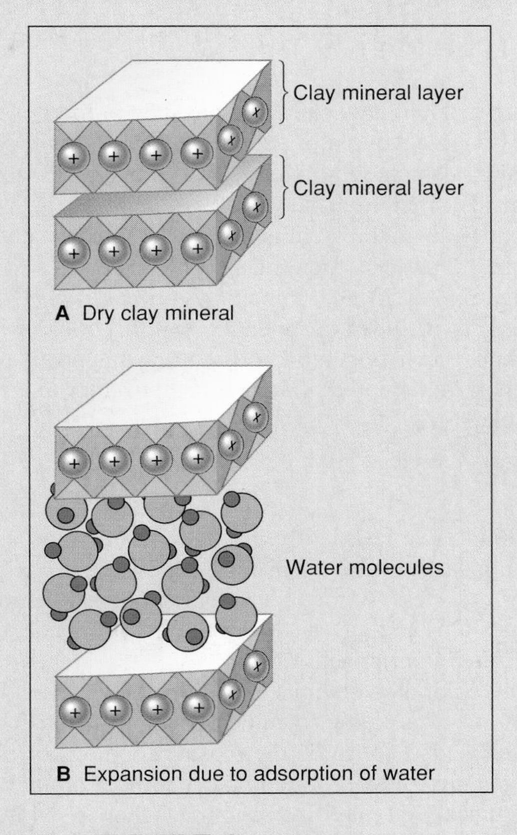

Box 9.4 — FIGURE 1

Expansive clays. (The red ion represents aluminum in the clay layers and is not drawn to scale.)

The **pyroxene group** and the **amphibole group,** which are single- and double-chain silicates, respectively, each contains a number of minerals. *Augite* is the most common pyroxene, and *hornblende* is the most common amphibole.

The **mica group** is characterized by minerals with a sheet silicate structure. The two most common micas are biotite and muscovite. **Biotite** is a dark-colored, iron/magnesium-bearing mica. **Muscovite** mica lacks iron and magnesium and is transparent or white.

The **clay mineral group** is another group of sheet silicates (see box 9.4). Clays are abundant on the Earth's surface and in sedimentary rocks but make up only a minor percentage of the crust as a whole.

Olivine is not among the most common minerals in the crust. It is, however, the predominant mineral in the upper mantle; therefore, it is vastly more abundant in Earth as a whole than the minerals that form most of the crust.

Nonsilicate minerals include *native elements,* which are minerals composed of only one element. Gold is a native element, as are diamond and graphite, both of which are composed solely of carbon. Other nonsilicates are classified according to the predominant negatively charged ions in their crystal structures. For instance, halite is a chloride because the negatively charged ions in the crystal are Cl^{-}. If the mineral contains CO_3^{-2} ions, it is a *carbonate. Sulfides* have S^{-2} ions, *sulfates* SO_4^{-2}, and *oxides* O^{-2} (but without Si, S, or C bonded to the oxygen atoms).

Nonsilicate minerals are also more abundant on the Earth's surface than in the crust as a whole. **Calcite** (calcium carbonate, or $CaCO_3$) is the most common nonsilicate mineral and is usually found at or near the Earth's surface. Limestone and marble are rocks composed mainly of calcite.

Ore minerals, or economic minerals, are minerals of commercial value; most are not silicates. Among the ore minerals are iron oxides (the minerals magnetite and hematite) mined for iron and a copper-iron sulfide (the mineral chalcopyrite) that is the main source of copper. Lead comes from galena (lead sulfide) and zinc is derived from sphalerite (a zinc sulfide).

THE PHYSICAL PROPERTIES OF MINERALS

The best approach to understanding physical properties of minerals is to obtain a sample of each of the most common rock-forming minerals named in table 9.2. The properties described can then be identified in these samples.

To identify an unknown mineral, you should first determine its physical properties, then match the properties with the appropriate mineral, using a mineral identification key or chart such as the ones included in appendix A of this book. With a bit of experience, you may get to know the few diagnostic tests for each common mineral and no longer need to refer to an identification table.

Color

The first thing most people notice about a mineral is its color. For some minerals, color is a useful property. Muscovite mica is white or colorless. Most of the **ferromagnesian minerals** (iron/magnesium-bearing), such as augite, hornblende, olivine, and biotite, are either green or black.

Because color is so obvious, beginning students tend to rely too heavily on it as a key to mineral identification. Unfortunately, color is also apt to be the most ambiguous of physical properties (figure 9.13). If you look at a number of quartz crystals, for instance, you may find specimens that are white, pink, black, yellow, or purple. Color is extremely variable in quartz and many other minerals because even minute chemical impurities can strongly influence it. Obviously, it is poor procedure to attempt to identify quartz strictly on the basis of color.

Streak

A pulverized mineral gives a color, called a **streak,** that usually is more reliable than the color of the specimen itself. Scraping the edge of a mineral sample across an unglazed porcelain plate leaves a streak that may be diagnostic of the mineral. For instance, hematite always leaves a reddish brown streak though the sample may be brown or red or silver.

Unfortunately, few of the silicate minerals—the most common minerals—leave an identifying streak because most are harder than the porcelain streak plate.

Luster

The quality and intensity of *light* that is reflected from the surface of a mineral is termed **luster.** (A photograph cannot show this quality.) The luster of a mineral is described by comparing it to familiar substances.

FIGURE 9.13

Why color may be a poor way of identifying minerals. These are all corundum gems including ruby and sapphire.

Photo by Dane A. Pentland © 1992 Smithsonian Institution

Luster is either *metallic* or *nonmetallic.* A **metallic luster** gives a substance the appearance of being made of metal. Metallic luster may be very shiny, like a chrome car part, or less shiny, like the surface of a broken piece of iron.

Nonmetallic luster is more common. The most important type is **glassy** (also called **vitreous**) **luster,** which gives a substance a glazed appearance, like glass or porcelain. Most silicate minerals have this characteristic. The feldspars, quartz, the micas, and the pyroxenes and amphiboles all have a glassy luster.

Less common is an **earthy luster.** This resembles the surface of unglazed pottery and is characteristic of the various clay minerals. Some uncommon lusters include *resinous* luster (appearance of resin), *silky* luster, and *pearly* luster.

Hardness

The property of "scratchability," or **hardness,** can be tested fairly reliably. For a true test of hardness, the harder mineral or substance must be able to make a groove or scratch on a smooth, fresh surface of the softer mineral. For example, quartz can always scratch calcite or feldspar. Substances can be compared to **Mohs' hardness scale** (table 9.3), on which ten minerals are designated as standards of hardness. The softest mineral, talc (used for talcum powder because of its softness), is designated as 1. Diamond, the hardest natural substance on earth, is 10 on the scale.

Rather than carry samples of the ten standard minerals, a geologist doing fieldwork usually relies on common objects to test for hardness (table 9.3). A fingernail usually has a hardness of about 2 1/2. If you can scratch the smooth surface of a mineral with your fingernail, the hardness of the mineral must be less than 2 1/2 (figure 9.14). A copper coin or a penny has a hardness between 3 and 4; however, the brown oxidized surface

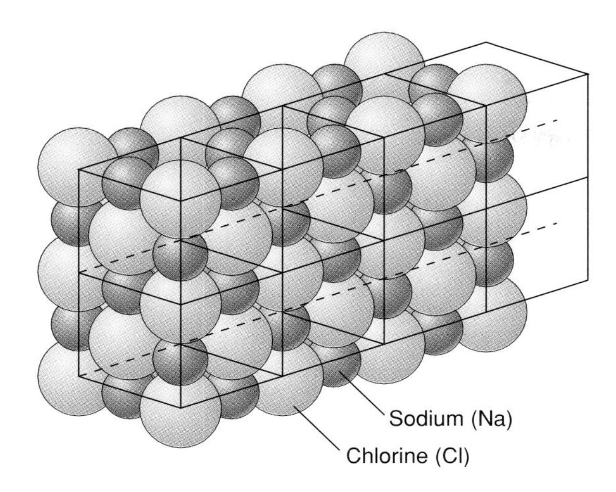

FIGURE 9.18
Relationship of cubes to part of a halite crystal. The crystal structure of halite can be represented as cubes stacked together in three dimensions.

age. The micas, however, are easily split apart into sheets (figure 9.19). If we could look at the arrangement of atoms in the crystalline structure of micas, we would see that the individual silicon-oxygen tetrahedrons are strongly bonded to one another within each of the silicate sheets. The bonding *between* adjacent sheets, however, is very weak; therefore it is easy to pull the mineral apart parallel to the plane of the sheets.

Cleavage is one of the most useful diagnostic tools because it is identical for a given mineral from one sample to another. Cleavage is especially useful for identifying minerals when they are small grains in rocks.

The wide variety of combinations of cleavage and *quality* of cleavage also increases the diagnostic value of this property. Mica has a single direction of cleavage, and its quality is perfect (figure 9.20*A*). Other minerals are characterized by one, two, or more cleavage directions; the quality can range from perfect to poor (poor cleavage is very hard for anyone but a well-trained mineralogist to detect).

Three of the most common mineral groups—the feldspars, the amphiboles, and the pyroxenes—have two directions of cleavage (figure 9.20*B* and *C*). In feldspars, the two directions are at angles of about 90° to each other, and both directions are of very good quality. In pyroxenes, the two directions are also at about right angles, but the quality is only fair. In amphiboles (figure 9.21), the quality of the cleavage is very good and the two directions are at an angle of 56° (or 124° for the obtuse angle).

Halite is an example of a mineral with three excellent cleavage directions, all at 90° to each other. This is called *cubic cleavage* (figure 9.20*D*). Halite's cleavage tells us that the bonds are weak in the planes parallel to the cube faces shown in figure 9.18.

Calcite also has three cleavage directions, each excellent. But the angles between them are clearly not right angles. Calcite's cleavage is known as *rhombohedral* cleavage (figures 9.20*E* and 9.22).

Some minerals have more than three directions of cleavage (figures 9.20*F* and *G*). Diamond has very good cleavage in four

A

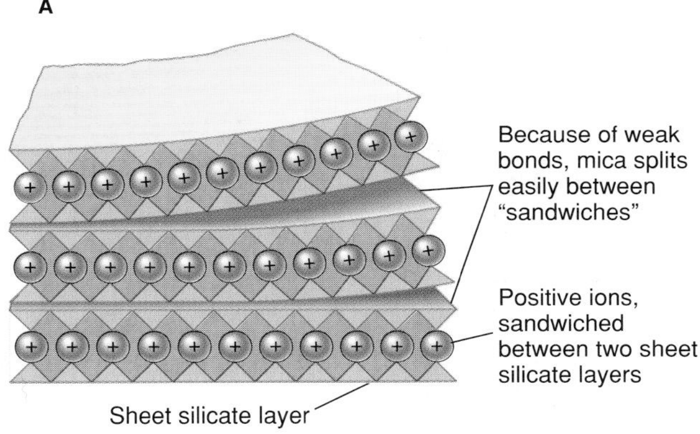

B

FIGURE 9.19
(*A*) Mica pulled apart along cleavage planes. (*B*) Relationship of mica to cleavage. Mica crystal structure is simplified in this diagram.

directions (ironically, the hardest natural substance on Earth can be easily shattered into small cleavage fragments). Sphalerite, the principal ore of zinc, has six directions.

Recognizing cleavage and determining angular relationships between cleavage directions takes some practice. Students new to mineral identification tend to ignore cleavage because it is not as immediately apparent to the eye as color. But determining cleavage is frequently the key to identifying a mineral, so the small amount of practice needed to develop this skill is worthwhile.

Fracture

Fracture is the way a substance breaks where not controlled by cleavage. Minerals that have no cleavage commonly have an *irregular fracture.*

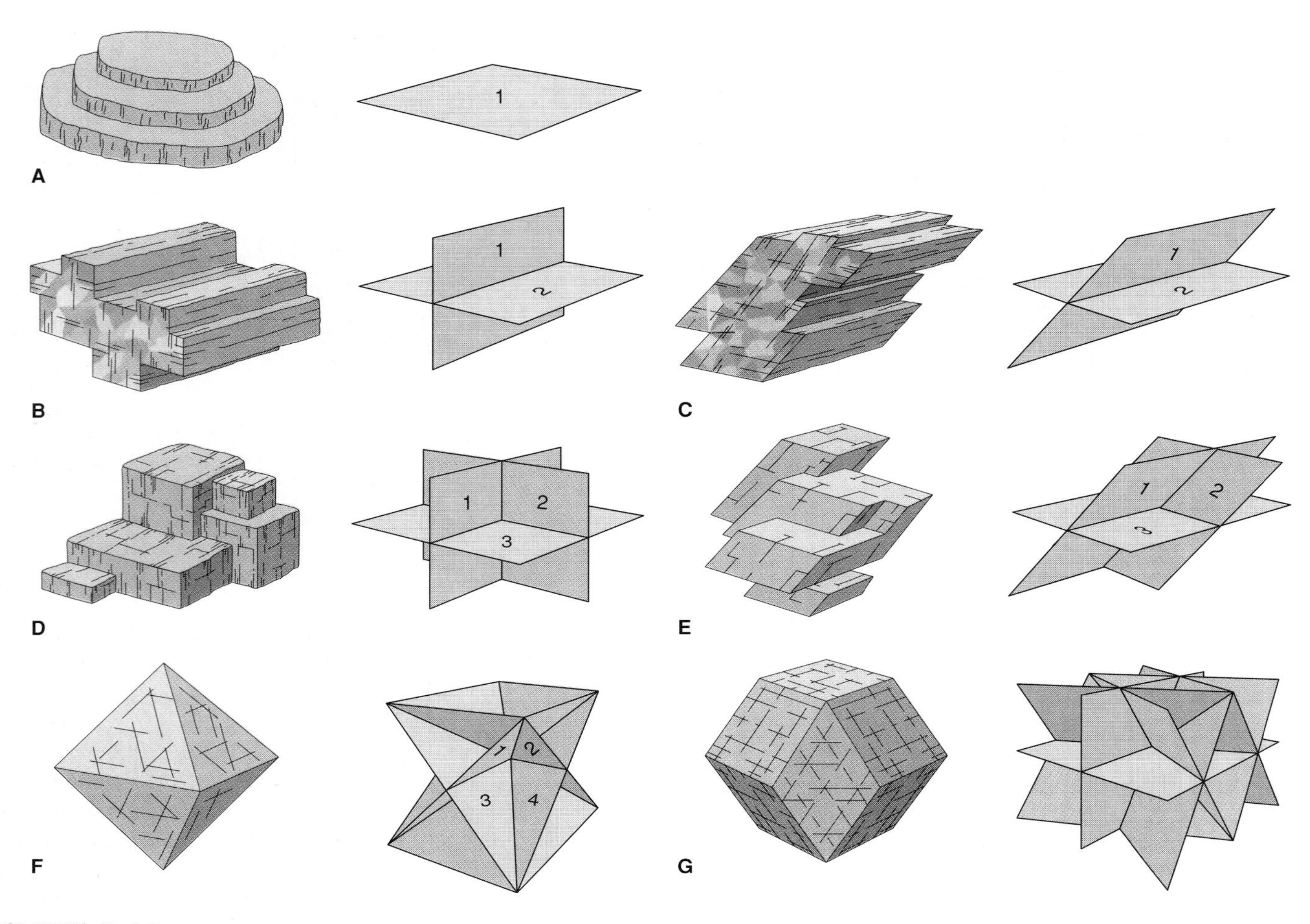

FIGURE 9.20

Possible types of mineral cleavage. (*A*) One direction of cleavage. (*B*) Two directions of cleavage that intersect at 90° angles. Feldspar is an example. (*C*) Two directions of cleavage that do not intersect at 90° angles. Amphibole is an example. (*D*) Three directions of cleavage that intersect at 90° angles. Halite is an example. (*E*) Three directions of cleavage that do not intersect at 90° angles. Calcite is an example. (*F*) Four directions of cleavage. Diamond is an example. (*G*) Six directions of cleavage. Sphalerite is an example.

Reprinted by permission from R. D. Dallmeyer, *Physical Geology Laboratory Manual,* Dubuque, Iowa: Kendall-Hunt Publishing Company, 1978

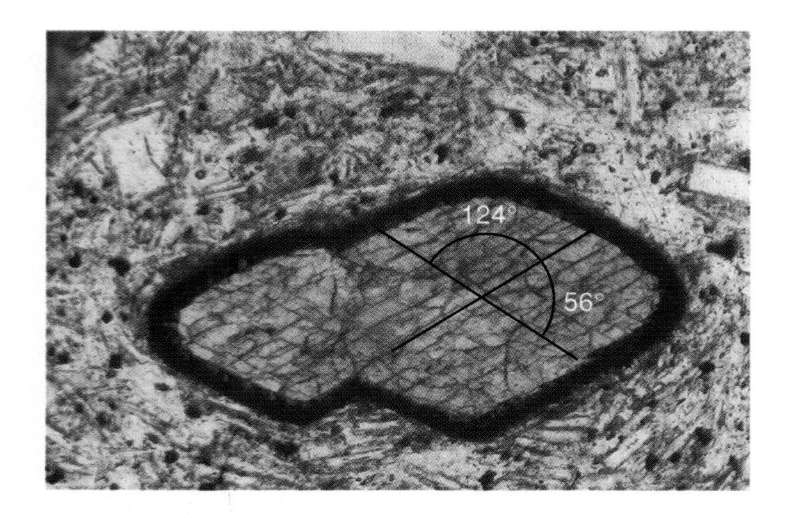

FIGURE 9.21

Amphibole cleavage as seen in a polarizing microscope.

FIGURE 9.22

Cleavage fragments of calcite.

FIGURE 9.23
Conchoidal fracture in glass.

FIGURE 9.24
Plagioclase striations.

Some minerals break along curved fracture surfaces known as *conchoidal fractures* (figure 9.23). These look like the inside of a clam or conch shell. This type of fracture is commonly observed in quartz and garnet (but these minerals also show irregular fractures). Conchoidal fracture is particularly common in glass, including obsidian (volcanic glass).

Specific Gravity

It is easy to tell that a brick is heavier than a loaf of bread just by hefting each of them. The brick has a higher **density,** weight per given volume, than the bread. Density is commonly expressed as **specific gravity,** the ratio of a mass of a substance to the mass of an equal volume of water.

Liquid water has a specific gravity of 1. (Ice, being lighter, has a specific gravity of about 0.9.) Most of the common silicate minerals weigh about two and a half times as much as equal volumes of water: quartz has a specific gravity of 2.65; the feldspars range from 2.56 to 2.76. Special scales are needed to determine specific gravity precisely. However, a person can easily distinguish by hand very dense minerals such as galena (a lead sulfide with a specific gravity of 7.5) from the much less dense silicate minerals.

Gold, with a specific gravity of 19.3, is much heavier than galena. Because of its high density, gold can be collected by "panning." While the lighter clay and silt particles in the pan are sloshed out with the water, the gold dust lags behind in the bottom of the pan.

Special Properties

Some properties only apply to one or a few minerals. Smell is one. Some clay minerals have a characteristic "earthy" smell when they are moistened. A few have a distinctive taste. If you lick halite, it tastes salty, because it is, of course, table salt.

Plagioclase feldspar commonly exhibits **striations**—straight, parallel lines on the flat surfaces of one of the two cleavage directions (figure 9.24). The lines appear to be etched by a delicate scriber. In plagioclase, they are caused by a systematic change within the pattern of crystalline structure.

The mineral **magnetite** (an iron oxide) owes its name to its characteristic physical property of being attracted to a magnet. Where large bodies of magnetite are found in the Earth's crust, compass needles point toward the magnetite body rather than to magnetic north. Airplanes navigating by compass have become lost because of the influence of large magnetite bodies. Some other minerals are weakly magnetic; their magnetism can only be detected by specialized magnetometers, similar to metal detectors in airports. Magnetism is important to modern civilization. We use magnetic tape (coated with magnetite or other magnetic material) for sound and video recordings as well as for magnetic memory disks in our computers. In previous chapters, you have seen how magnetite in igneous rocks has preserved a record of Earth's magnetic field through geologic time; this has been an important part of the verification of plate tectonic theory. Some bacteria create magnetite, and this fact has been used to support the hypothesis that life has existed on Mars (see box 9.5).

Other Properties

A clear crystal of calcite exhibits an unusual property. If you place transparent calcite over an image on paper, you will see two images (figure 9.25). This phenomenon is known as *double refraction* and is caused by light splitting into two components when it enters some crystalline materials. Each of the components is traveling through the mineral at different velocities. Most minerals possess double refraction, but it is usually slight and can be observed using polarizing filters, notably in polarizing microscopes. Polarizing microscopes are very useful to professional geologists and advanced students for identifying minerals and interpreting how rocks formed. Photomicrographs elsewhere in this book were taken through polarizing

9.5 ASTROGEOLOGY

Magnetite on Mars—Does a Meteorite Found on Ice Contain Evidence of Martian Life?

A meteorite collected on an Antarctic glacier in 1986 is one of the most extensively studied rocks on Earth (see also box 19.5). Carbonate globules found in the rock indicate to some scientists that they are fossils, former living organisms; however, the evidence is ambiguous and other scientists favor alternate hypotheses for the origin of the globules.

Recently, the hypothesis that Mars had living organisms received a strong boost when submicroscopic magnetite crystals were discovered in the rock through the use of an electron microscope. The size and shape of the magnetite crystals are identical to ones found on Earth that are only formed by bacteria.

The bacteria on Earth grow magnetite crystals in order to find an optimum environment in a body of water. They do this by using Earth's magnetic field, much as we do when we use a compass to navigate. If the hypothesis that the magnetite in the Martian meteorite also formed by bacteria is correct, that indicates several things about what Mars was like 4 billion years ago when the rock formed. First, Mars would have had life earlier than any well-documented life on Earth. Second, there would have been bodies of water on Mars at the time. Third, Mars would have had to have a strong magnetic field (as Earth does at present, but Mars does not) when the rock formed.

Related Web Resource

For excellent photos and more information visit NASA's **Martian micro-magnets** site

http://science.nasa.gov/headlines/y2000/ast20dec_1.htm

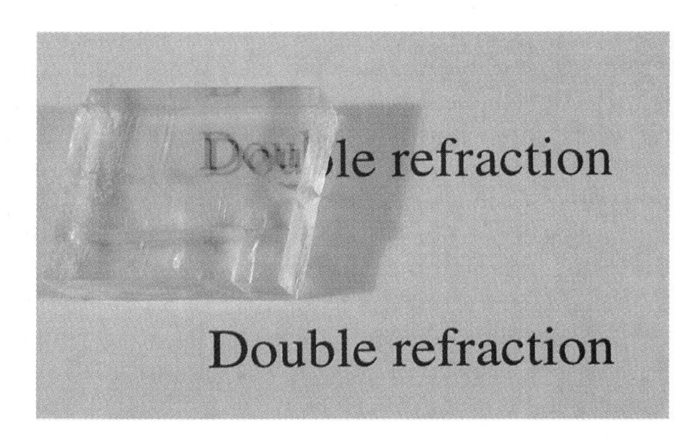

FIGURE 9.25

Double refraction in calcite. Two images of the letters are seen through the transparent calcite crystal.

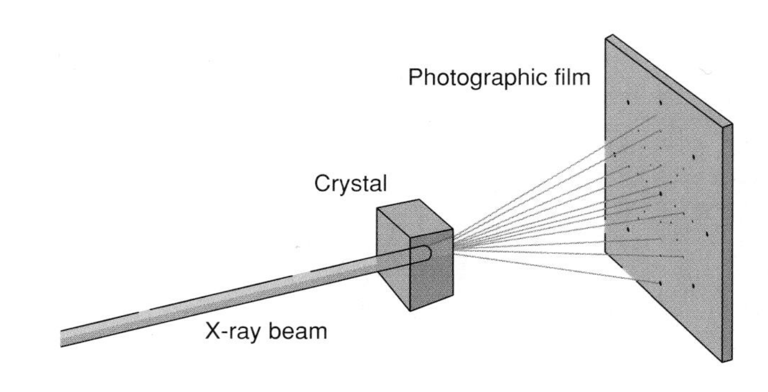

FIGURE 9.26

An X-ray beam passes through a crystal and is deflected by the atoms into a pattern of beams. The dots exposed on the film are an orderly pattern used to identify the particular mineral.

microscopes (for example, figure 11.5*B*). Explaining optical phenomena such as this is beyond the scope of this book, but if interested, you can go to the **Molecular Expressions Microscopy Primer** site www.micro.magnet.fsu.edu/primer/virtual/virtualpolarized.html.

Specialized equipment is needed to determine some properties. Perhaps most important are the characteristic effects of minerals on X rays, which we can explain only briefly here. X rays entering a crystalline substance are deflected by planes of atoms within the crystal. The X rays leave the crystal at precise and measurable angles controlled by the orientation of the planes of atoms that make up the internal crystalline structure (figure 9.26). The pattern of X rays exiting can be recorded on photographic film or by various recording instruments. Each mineral has its own pattern of reflected X rays, which serves as an identifying "fingerprint."

Chemical Tests

One chemical reaction is routinely used for identifying minerals. The mineral calcite, as well as some other carbonate minerals (those containing CO_3^{-2}), reacts with a weak acid to produce carbon dioxide gas. In this test, a drop of dilute hydrochloric acid applied to the sample of calcite bubbles vigorously, indicating that CO_2 gas is being formed. Normally this is the only chemical test that geologists do during field research.

Accurate chemical analyses of minerals and rocks are done in labs using a wide range of techniques. A chemical analysis can accurately tell us the amount of each element present in a mineral. However, chemical analysis alone cannot be used to conclusively identify a mineral. We also need to know about the mineral's crystalline structure. Diamond and graphite have an identical composition but very different crystalline structures.

Summary

Atoms are composed of *protons* (+), *neutrons*, and *electrons* (−). A given element always has the same number of protons. An atom in which the positive and negative electric charges do not balance is an *ion*.

Ions or atoms bond together in very orderly, 3-dimensional structures that are *crystalline*.

A crystalline substance is considered a *mineral* (in geologic terms) if it is naturally occurring and has a specific chemical composition.

The three most abundant elements in the Earth's crust are oxygen, silicon, and aluminum. Most minerals are silicates, having the silicon-oxygen tetrahedron as their basic building block.

Feldspars are the most common minerals in the Earth's crust. The next most abundant minerals are quartz, the pyroxenes, the amphiboles, and the micas. All are silicates.

Minerals are usually identified by their physical properties. Cleavage is perhaps the most useful physical property for identification purposes. Other important physical properties are external crystal form, fracture, hardness, luster, color, streak, and specific gravity.

Terms to Remember

amphibole group 227
atom 216
atomic mass number 217
atomic number 217
atomic weight 218
biotite 227
bonding 219
calcite 227
chain silicate structure 222
clay mineral group 227
cleavage 230
covalent bonding 220
crystal form 229
crystalline 221
density 233
earthy luster 228
electron 216
element 216
feldspar group 226
ferromagnesian mineral 228
fracture 231
framework silicate structure 225
glassy (vitreous) luster 228
hardness 228
ion 220
ionic bonding 220
isolated silicate structure 222
isotope 217
luster 228
magnetite 233
metallic luster 228
mica group 227
mineral 216
Mohs' hardness scale 228
muscovite 227
neutron 216
nonmetallic luster 228
nucleus 217
olivine 222
ore mineral 227
plagioclase feldspar 226
potassium (orthoclase) feldspar 226
proton 216
pyroxene group 227
quartz 226
sheet silicate structure 225
silica 221
silicates 221
silicon-oxygen tetrahedron 222
specific gravity 233
streak 228
striations 233

Testing Your Knowledge

Use the questions below to prepare for exams based on this chapter.

1. Compare feldspar and quartz.
   How do they differ chemically?
   What type of silicate structure does each have?
   How would you distinguish between them on the basis of cleavage?
2. How do the crystal structures of pyroxenes and amphiboles differ from one another?
3. How do the various feldspars differ from one another chemically?
4. Distinguish the following terms:
   silica — silicate
   silicon — silicon-oxygen tetrahedron
5. What is the distinction between cleavage and external crystal form?
6. How would you distinguish the following on the basis of physical properties? (You might refer to appendix A.)
   feldspar/quartz
   muscovite/feldspar
   calcite/feldspar
   pyroxene/feldspar
7. Using triangles to represent tetrahedrons, start with a single triangle (to represent isolated silicate structure) and, by drawing more triangles, build on the triangle to show a single-chain silicate structure. By adding more triangles, convert that to a double-chain structure. Turn your double-chain structure into a sheet silicate structure.
8. What major factor controls chemical activity between atoms?
9. What are the three most common elements (by number and approximate percentage) in the Earth's crust?

10. What are the next five most common elements?

11. A substance that cannot be broken down into other substances by ordinary chemical methods is a(n)
    a. atom
    b. element
    c. molecule
    d. compound

12. The subatomic particle that contributes mass and a single positive electrical charge is the
    a. proton
    b. neutron
    c. electron

13. Atoms containing different numbers of neutrons but the same number of protons are called
    a. compounds
    b. ions
    c. elements
    d. isotopes

14. Atoms with either a positive or negative charge are called
    a. compounds
    b. ions
    c. elements
    d. isotopes

15. The bonding between Cl and Na in halite is
    a. ionic
    b. covalent
    c. metallic
    d. male

16. Which is not true of a single silicon-oxygen tetrahedron?
    a. the atoms of the tetrahedron are strongly bonded together
    b. it has a net negative charge
    c. the formula is SiO_4
    d. it has four silicon atoms

17. Which is not a type of silicate structure?
    a. isolated
    b. single chain
    c. double chain
    d. sheet
    e. framework
    f. pentagonal

18. The most common mineral in the Earth's crust is
    a. quartz
    b. feldspar
    c. pyroxene
    d. amphibole
    e. biotite

19. On Mohs' hardness scale ordinary window glass has a hardness of about
    a. 2–3
    b. 3–4
    c. 5–6
    d. 7–8

20. The ability of a mineral to break along preferred directions is called
    a. fracture
    b. crystal form
    c. hardness
    d. cleavage

21. Striations are associated with
    a. quartz
    b. mica
    c. potassium feldspar
    d. plagioclase

22. Glass is
    a. atoms randomly arranged
    b. crystalline
    c. ionically bonded
    d. covalently bonded

23. Crystalline substances are always
    a. ionically bonded
    b. minerals
    c. made of repeating patterns of atoms
    d. made of glass

Expanding Your Knowledge

1. Why are nonsilicate minerals more common on the surface of the Earth than within the crust?
2. How does oxygen in the atmosphere differ from oxygen in rocks and minerals?
3. What happens to the atoms in water when it freezes? Is ice a mineral? Is a glacier a rock?
4. How would you expect the appearance of a rock high in iron and magnesium to differ from a rock with very little iron and magnesium?

Exploring Web Resources

www.mhhe.com/plummer9e
Ever wonder why your watch has "quartz" on it? A small slice of quartz in it permits it to keep incredibly accurate time. This is because a small electric current to the quartz causes it to vibrate at a very precise rate (close to 100,000 vibrations per second). Visit the boxed readings on the website for the full story as well as other articles and media resources. Check your answers for the Testing Your Knowledge section, and click on the direct links to explore the sites listed below.

www.rockhounds.com/
Bob's Rock Shop. Contains a great amount of information for mineral collectors. Click on "crystallography and mineral crystal systems" for a more in-depth study of crystallography than presented in this book.

www.man.ac.uk/Geology/MineralWeb/Mineral_Web.html
Mineral Web. Crystal structures are displayed in 3-D. The structures rotate and you can manipulate the rotation using a mouse. You must install Chime, a program for viewing the structures. Chime can be downloaded easily from this site. Once installed, take a look at the crystal structures of diamond, olivine, muscovite, and other minerals. You can also observe the various silicon-oxygen tetrahedron structures.

http://web.wt.net/~daba/Mineral/
Mineral Database. There are descriptions of close to 4,000 mineral species. The descriptions include mineral properties beyond the scope of an introductory geology course; however, there are links to other sites that include pictures of minerals.

www.theimage.com/
The Image. Photos of minerals and gems. Click on Mineral Gallery and choose a mineral to view photos and properties of that mineral. The Gemstone Gallery has photos of gem minerals.

Animations

This chapter includes the following animations available on our Online Learning Center at www.mhhe.com/plummer9e.

9.7–9.9, 9.11 Silicon-Oxygen Tetrahedrons

CHAPTER

10

Volcanism and Extrusive Rocks

Chapters 10 and 11 cover igneous activity. Either may be read before the other. Chapter 11 emphasizes intrusive activity, but it also covers igneous rock classification and the origin of magmas, which are applicable both to volcanic and intrusive phenomena. Chapter 10 concentrates on volcanoes and related extrusive activity.

Volcanic eruptions, while awesome natural spectacles, also provide important information on the workings of the earth's interior. Volcanic eruptions vary in nature and in degree of explosive violence. A strong correlation exists between the chemical composition of magma (or lava) and the violence of an eruption. The size and shape of volcanoes and lava flows and their pattern of distribution on the Earth's surface also correspond to the composition of their lavas.

Understanding volcanism provides a background for theories relating to mountain building, the development and evolution of continental and oceanic crust, and how the crust is deformed. Our observations of volcanic activity fit nicely into plate tectonic theory as described in chapter 11.

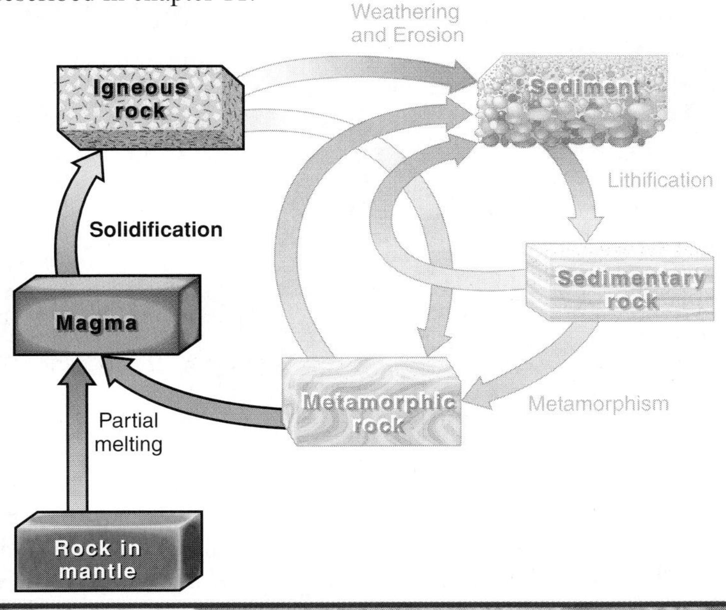

Opposite: Pele dancing. Braided lava flow of the 1984 eruption of Mauna Loa, Hawaii. The pattern suggests a woman with arms raised. Pele is the Hawaiian goddess of volcanoes. Photo by Katia Krafft who, along with her husband, volcanologist Maurice Krafft, was killed by a pyroclastic flow during an eruption of Mt. Unzen in Japan.

A

B

FIGURE 10.1

Contrasting styles of volcanic eruptions. (*A*) May 18, 1980. Exploding magma blasts out of the side of Mount St. Helens. (*B*) Lava flow in Hawaii, 1969. A lava fountain is at the source of lava cascading over a cliff.

A © 1980 Keith Ronnholm.
B Photo by D. A. Swanson, U.S. Geological Survey

The May 18, 1980, eruption of Mount St. Helens (figure 10.1*A* and box 10.1) was a spectacular release of energy from the Earth's interior. The plate tectonic explanation is that North America is overriding a portion of the Pacific Ocean floor. Melting of previously solid rock takes place at depth, just above the subducting plate. (This is described briefly in chapter 1 and more thoroughly in chapter 11.) At least some of the **magma** (molten rock or liquid that is mostly silica) works its way upward to the Earth's surface to erupt. Magma does not always reach the Earth's surface before solidifying, but when it does it is called **lava.**

At Mount St. Helens the lava solidified quickly as it was blasted explosively by gases into the air, producing rock fragments known as **pyroclasts** (from the Greek *pyro,* "fire," and *clast,* "broken"). Pyroclastic debris is also known as *tephra.* In Hawaii, lava extrudes out of fissures in the ground as **lava flows** (figure 10.1*B*), Pyroclastic debris and rock formed by solidification of lava are collectively regarded as **extrusive rock,** surface rock resulting from volcanic activity.

The most obvious landform created by **volcanism** is a **volcano,** a hill or mountain formed by the extrusion of lava or ejection of rock fragments from a vent; however, volcanoes are not the only volcanic landforms. Very fluid lava may flow out of the Earth and flood an area, solidifying into a nearly horizontal layer of extrusive rock. Successive layers of lava flows may accumulate, building a lava plateau.

Volcanic activity is important to the science of geology for several reasons. Landforms are created and portions of the earth's surface built up. Less commonly, as at Mount St. Helens, landforms are destroyed by violent eruptions. Volcanoes are important because they provide clues about the nature of the Earth's inaccessible interior and help us understand how the Earth's internal processes work. By studying the magma, gases, and rocks from eruptions, we can infer the chemical conditions as well as the temperatures and pressures within the Earth's crust or underlying mantle.

EFFECTS ON HUMANS

Not surprisingly, myths and religions relating gods to volcanoes flourish in cultures that live with volcanoes. In Iceland, Loki, of Norse mythology, is regarded as imprisoned underground, blowing steam and lava up through fissures. Pacific Northwest Indians regarded the Cascade volcanoes as warrior gods who would sometimes throw red-hot boulders at each other. They also had a romantic side. Mount Hood and Mount Adams fought over Mount St. Helens, the youngest and prettiest of the volcano gods. In Hawaii, Madame Pele is regarded as a goddess who controls eruptions. (This chapter's opening photo is titled "Pele dancing.") According to legend, Pele and her sister tore up the ocean floor to produce the Hawaiian island chain. Today, many fervently believe that Pele dictates when and where an eruption will take place. In the 1970s, when Kilauea began erupting near a village, residents chartered an airplane and dropped flowers and a bottle of gin into the lava vent to appease Pele.

Volcanism is also relevant to human affairs in very tangible ways. Its effects can be catastrophic or, surprisingly, beneficial.

The Growth of Hawaii

Although occasionally a field or village is overrun by outpourings of lava, the overall effects of volcanism have been favorable to humans in Hawaii. Kilauea volcano has been active since 1983, and over 1.5 billion cubic meters of lava erupted—enough to build a highway that circles the world four times. There were 181 houses destroyed in the 1980s and 1990s, but no one was killed or injured. Nevertheless, the weathered volcanic ash and lava produce excellent fertile soil. Moreover, Hawaii's periodically erupting volcanoes (which are relatively safe to watch) are great spectacles that attract both tourists and scientists, benefiting the island's economy (figure 10.1*B*).

Were it not for volcanic activity, Hawaii would not exist. The islands are the crests of a series of volcanoes that have been built up from the bottom of the Pacific Ocean over millions of years (the vertical distance from the summit of Mauna Loa volcano to the ocean floor greatly exceeds the height above sea level of Mount Everest). When lava flows into the sea and solidifies, more land is added to the islands. Hawaii is, quite literally, growing.

Geothermal Energy

In other areas of recent volcanic activity, underground heat generated by igneous activity is harnessed for human needs. In Italy, Mexico, New Zealand, Argentina, Japan, and California, geothermal installations produce electric power. Steam or superheated water trapped in layers of hot volcanic rock is tapped by drilling and then piped out of the ground to power turbines that generate electricity. Naturally heated geothermal fluids can also be tapped for space or domestic water heating or industrial use, as in paper manufacturing.

Effect on Climate

Occasionally, a volcano will spew large amounts of fine, volcanic dust and gas into the high atmosphere. Winds can keep fine particles suspended over the earth for years. The 1991 eruption of Mount Pinatubo in the Philippines produced noticeably more colorful sunsets worldwide (see description in chapter 1). More significantly, it reduced solar radiation that penetrates the atmosphere. Measurements indicated that the worldwide average temperature dropped approximately half a degree Celsius for a couple years. While this drop may not seem like much, it was enough to temporarily offset the global warming trend of the past 100 years.

The 1815 eruption of Tambora in Indonesia was the largest, single eruption in a millenium—40 cubic kilometers of material was blasted out of a volcanic island, leaving a six-kilometer-wide depression. The following year, 1816, became known as "the year without summer." In New England, snow in June was widespread and frosts throughout the summer ruined crops. Parts of Europe suffered famine because of the cold weather effects on agriculture.

10.1

ENVIRONMENTAL GEOLOGY

MOUNT ST. HELENS BLOWS UP

Before 1980, Mount St. Helens, in southern Washington, had not erupted since 1857. On March 27, 1980, ash and steam eruptions began and continued for the next six weeks. These were minor eruptions in which magma was not erupted. Rather, they were due to exploding gas blasting out the volcano's previously formed rock; however, the steam and the pattern of earthquakes indicated magma was working its way upward beneath the volcano.

After several weeks, the peak began swelling—like a balloon being inflated—indicating magma was now inside the volcano. The northern flank of the volcano bulged outward at a rate of 1.5 meters per day. Bulging continued until the surface of the northern slope was displaced outward over a hundred meters from its original position. The bulge was too steep to be stable and the U.S. Geological Survey warned of another hazard—a mammoth landslide.

On May 18 a monumental blast destroyed the summit and north flank of Mount St. Helens (see figure 10.1). Seconds after the eruption began, an area extending northward 10 kilometers was stripped of all vegetation and soil.

Although the sequence of events was exceedingly rapid, it is now clear what happened (box figure 1). A fairly strong earthquake loosened the bulging north slope, triggering a landslide. (Aside from eruption, the landslide, known as a debris avalanche, would have been among the largest ever to take place.) The landslide stripped away the rock that had sealed in the magma, and with the protective lid removed, gases that had been dissolved in the magma were suddenly released. The magma exploded into a violent froth while blasting out of the volcano's north flank. The huge lateral blast of hot gas and volcanic rock debris destroyed all organic matter close to the volcano and knocked down the forest beyond the scorched zone.

For the next 30 hours, exploding gases propelled frothing magma and volcanic ash vertically into the high atmosphere. The mushroom-shaped cloud of ash was blown northeastward by winds. A rain of ash went on for days, causing damage as far away as Montana. Volcanic mudflows also caused damage during and after the eruption. The mudflows resulted from water from melted snow and glacier ice mixing with volcanic debris to form a slurry having the consistency of wet cement. Mudflows flowed down river valleys, carrying away steel bridges and other structures.

Damage was in the hundreds of millions of dollars and 63 people were killed. The death toll might have been much worse had not scientists warned public officials about the potential hazards, causing them to evacuate the danger zone before the eruption. For comparison, 29,000 people were killed during an eruption of Mount Pelé (described later in this chapter), and 23,000 lives were lost in a 1985 volcanic mudflow in Colombia.

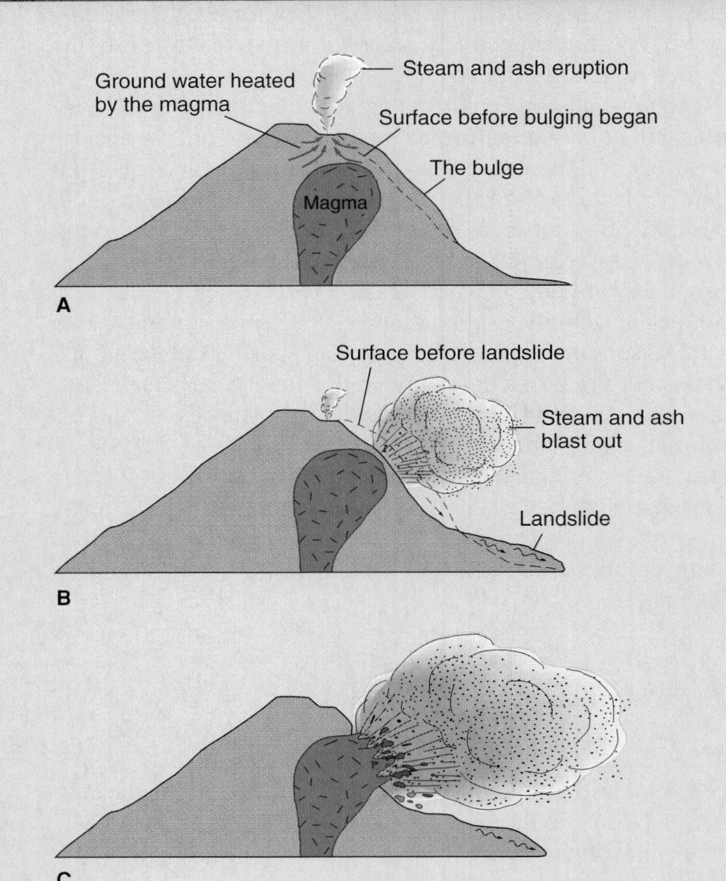

BOX 10.1 — FIGURE 1

Sequence of events at Mount St. Helens, May 18, 1980. (*A*) Just before the May 18, 1980 eruption. (*B*) The landslide relieves the pressure on the underlying magma. (*C*) Magma blasts outward

Perhaps Mount St. Helens will remain quiescent for decades or a century; however, other volcanoes in the Pacific Northwest could erupt and be disastrous to nearby cities. Seattle and Tacoma are close to Mount Rainier. Mount Hood is practically in Portland, Oregon's suburbs. Vancouver, British Columbia, could be in danger if either Mount Garibaldi to the north or Mount Baker in Washington to the south erupt.

Related Web Resource

For a more in-depth account of the Mount St. Helens eruption go to the extended box on the book's website at www.mhhe.com/plummer9e.

Volcanic Catastrophes

While the eruption of Mount St. Helens in 1980 was indeed awesome, its effects were not nearly as disastrous as a number of historical eruptions elsewhere in the world. For instance, the Roman city of Pompeii and at least four other towns near Naples in Italy were destroyed in A.D. 79 when Mount Vesuvius erupted (figure 10.2). Before the eruption vineyards on the flanks of the apparently "dead" volcano extended to the summit. After it erupted without warning, Pompeii was buried under 5 to 8 meters of hot ash. Seventeen centuries later the town was rediscovered.

Excavation revealed molds of people suffocated by the ashfall, many with facial expressions of terror. This eruption was not the end of Vesuvius's activity. The volcano was active almost continually from 1631 to 1944, with major twentieth-century eruptions in 1906, 1929, and 1944.

The island of Krakatoa in the western Pacific, composed of three apparently inactive volcanoes, erupted in 1883 with the force of several hydrogen bombs. This Indonesian island, which formerly rose 800 meters above sea level, was blown apart. Only one-third of the island remained after the eruption. An estimated 13 cubic kilometers of rock collapsed into the subsurface magma chamber that had been emptied by the eruption, leaving an underwater depression 300 meters deep where the major part of the island had been. The explosion was heard 5,000 kilometers away. Over 34,000 people died as a result of the giant sea waves (tsunamis) generated by the explosion.

A similar series of explosions in prehistoric time (about 6,600 years ago) was at least partially responsible for creating Crater Lake in Oregon. Volcanic debris covering more than a million square kilometers in Oregon and neighboring states has been traced to those eruptions. The original volcano, named Mount Mazama by geologists, is estimated to have been about 2,000 meters higher than the present rim of Crater Lake. Collapse of the volcano, as well as explosions, accounts for the depression the present-day lake occupies (figures 10.3 and 10.4).

The southern Cascade Mountains, where Crater Lake is located, have been built up by eruptions over the past 30 to 40 million years (figure 10.5; see also the geologic map, inside cover). Only the youngest peaks (those built within the past 2 million years), such as Mount St. Helens, Mount Rainier, Mount Shasta, and Mount Hood, still stand out as

FIGURE 10.2
Pompeii with Mount Vesuvius in the background.
Photo by R. W. Decker

FIGURE 10.3
Crater Lake, Oregon. Figure 10.4 shows its geologic history.
Photo by © Greg Vaughn/Tom Stack & Associates

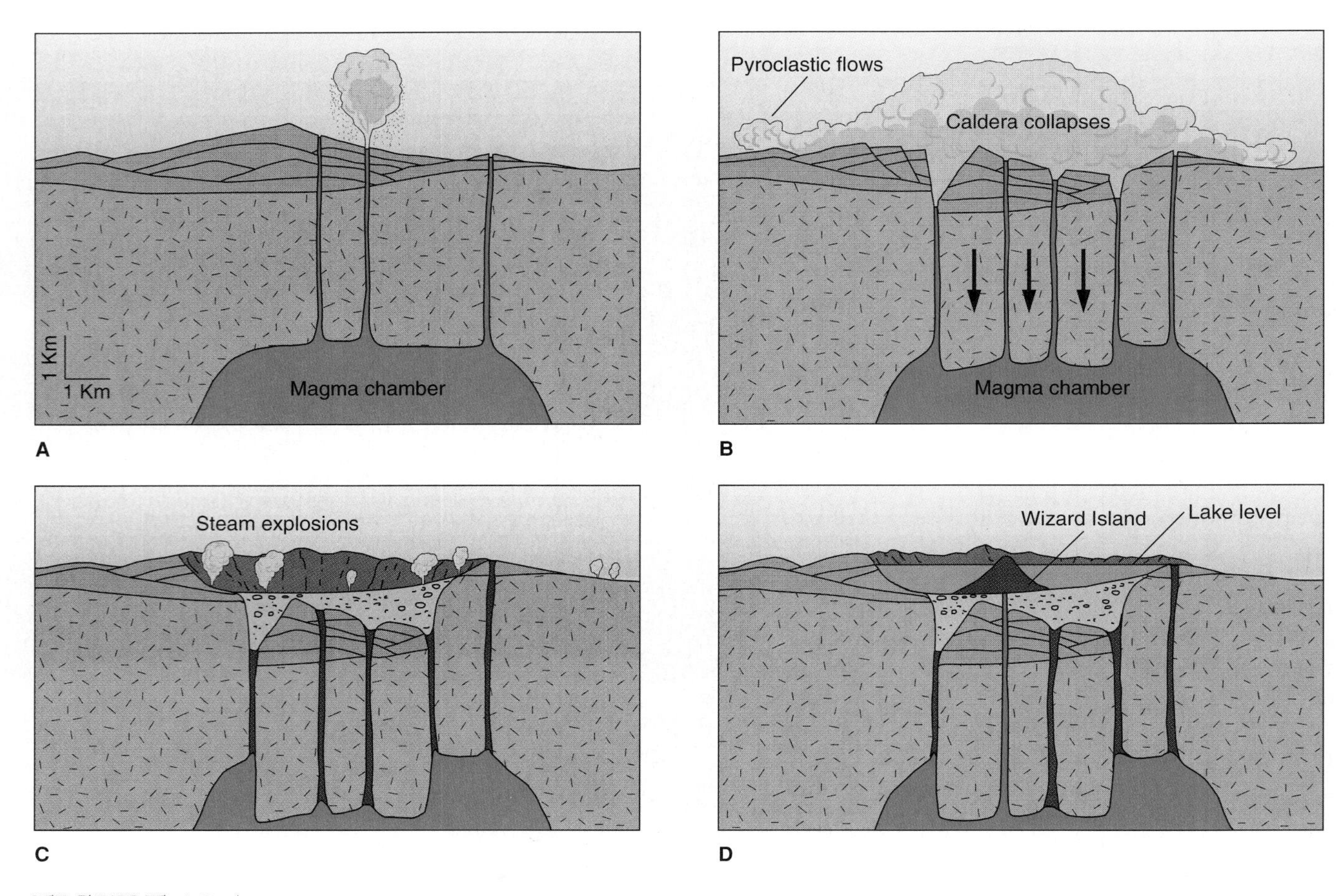

FIGURE 10.4

The development of Crater Lake. (*A*) Cluster of overlapping volcanoes form. (*B*) Collapse into the partially emptied magma chamber is accompanied by violent eruptions. (*C*) Volcanic activity ceases, but steam explosions take place in the caldera. (*D*) Water fills the caldera to become Crater Lake, and minor renewed volcanism builds a cinder cone (Wizard Island).

After C. Bacon, U.S. Geological Survey

cones. As Mount St. Helens has demonstrated, any of these could again become active.

The Record of Fatalities

Figure 10.6 shows the results of research at the Smithsonian Institute and Macquarie University, Australia. Note the dramatic increase in fatalities during the recent centuries (figure 10.6*A*). This is not due to increasing volcanic activity but to increasing population and more people living near volcanoes. Figure 10.6*B*, which shows the cumulative number of deaths during the last seven centuries, also shows that most of the fatalities have been caused by seven major eruptions.

Volcanoes can kill in a number of ways. Figure 10.6*C* indicates that pyroclastic flows account for the most fatalities. A *pyroclastic flow,* described later in this chapter, is a mixture of hot gas and pyroclastic debris that rapidly flows down a volcano's flanks. Famine and other indirect causes account for the next greatest number of fatalities. Widespread destruction of crops and farm animals can cause regional famine (as occurred with the eruption of Tambora in 1815). Note that relatively few events (specific eruptions) have caused the large number of deaths attributable to famine.

Pyroclastic material accounts for the largest number of deadly events; however, few people die in each event, so the total number of deaths is not great. Most of the deaths due to pyroclastic material are caused by collapse of ash-covered roofs or by being hit by falling, pyroclastic fragments.

Eruptive Violence and Physical Characteristics of Lava

What determines the degree of violence associated with volcanic activity? Why can we state confidently that active volcanism in Hawaii poses only slight danger to humans, but expect violent explosions to occur in the Cascade Mountains? Whether eruptions are very explosive or relatively "quiet" is largely determined by two factors: (1) the amount of gas in the lava or magma and (2) the ease or difficulty with which the gas can escape to the

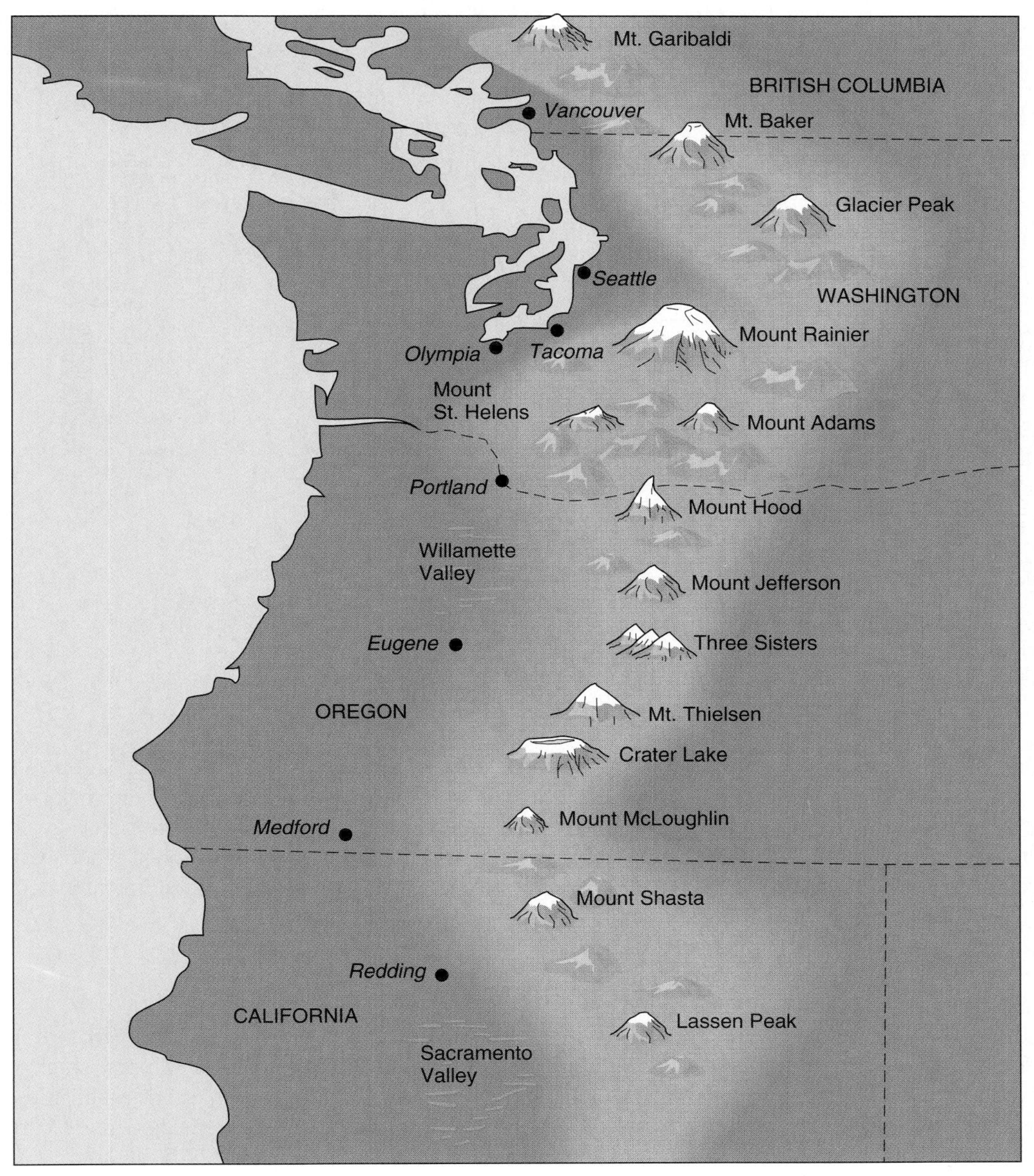

FIGURE 10.5
The Cascade volcanoes.

atmosphere. The **viscosity,** or resistance to flow, of a lava determines how easily the gas escapes. The more viscous the lava and the greater the volume of gas trying to escape, the more violent the eruption. Later we will show how these factors not only determine the degree of violence of an eruption but also influence the shape and height of a volcano.

The two most important factors that influence viscosity are (1) the silica (SiO_2) content of the lava and (2) the temperature of the lava relative to the cooler temperature at which it solidifies. (A third factor is gas dissolved in magma—the greater the dissolved gas content, the more fluid the lava.) If the lava being extruded is considerably hotter than its solidification temperature, the lava is less viscous (more fluid) than when its temperature is near its solidification point. Temperatures at which lavas solidify range from about 700°C for silicic rocks to 1,200°C for mafic rocks.

Volcanic rocks, and the magma from which they formed, have a silica content that ranges from 45% to 75% by weight. **Silicic** (or felsic) rocks are silica-rich (65% or more SiO_2) rocks. *Rhyolite* is the most abundant silicic volcanic rock. **Mafic rocks** are *silica-deficient* rocks. Their silica content is close to 50%. *Basalt* is the most common mafic rock. **Intermediate rocks** have a chemical content between that of felsic and mafic rocks. The most common intermediate rock is *andesite.* Chapter 11 contains

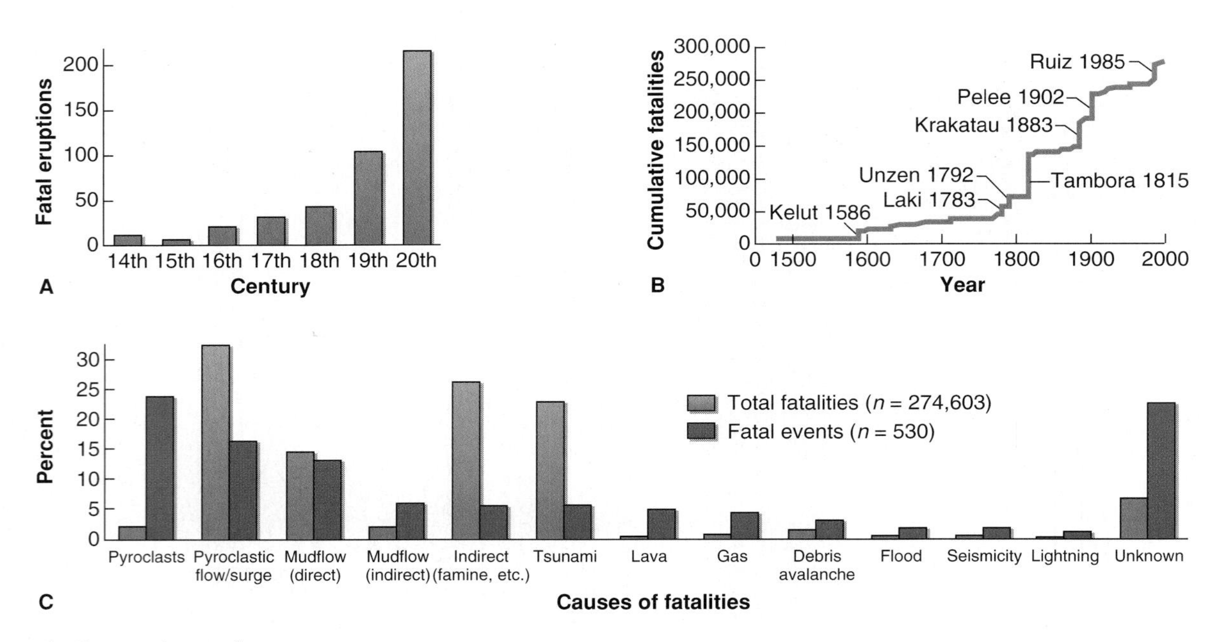

FIGURE 10.6

Volcano fatalities. (*A*) Fatal volcano eruptions per century. (*B*) Cumulative volcano fatalities. Note the big jumps with the seven most deadly eruptions. These were eruptions that killed over 10,000 people and account for two-thirds of the total. (*C*) The causes of volcano fatalities.

Reprinted with permission from "Volcano Fatalities" by T. Simkin, L. Siebert, and R. Blong, *Science,* v. 291: p. 255. Copyright © 2001 American Association for the Advancement of Science

a more complete description of the chemistry of igneous rocks and their relationship to the mineral content of rocks.

Mafic lavas, which are relatively low in SiO_2, tend to flow easily. Conversely, felsic lavas are much more viscous and flow sluggishly. Mafic lava is around 10,000 times as viscous as water, whereas silicic magma is around 100 million times the viscosity of water. Lavas rich in silica are more viscous because even before they have cooled enough to allow crystallization of minerals, silicon-oxygen tetrahedrons have begun to form small frameworks in the lava. Although too few atoms are involved for the structures to be considered crystals, the total effect of these silicate structures is to make the liquid lava more viscous, much the way that flour or cornstarch thickens gravy.

Because silicic magmas are the most viscous, they are associated with the most violent eruptions. Mafic magmas are the least viscous and commonly erupt as lava flows (such as in Hawaii). Eruption associated with intermediate magma can be violent or can produce lava flows. The Cascade volcanoes are predominantly composed of intermediate rock.

EXTRUSIVE ROCKS AND GASES

Scientific Investigation of Volcanism

Volcanoes and lava flows, unlike many other geologic phenomena, can be observed directly, and samples can be collected without great difficulty (at least for the quiet, Hawaiian-type eruption). We can measure the temperature of lava flows, collect samples of gases being given off, observe the lava solidifying into rock, and take newly formed rock samples into the laboratory for analysis and study. By comparing rocks observed solidifying from lava with similar ones from other areas of the world (and even with samples from the moon) where volcanism is no longer active, we can infer the nature of volcanic activity that took place in the geologic past.

Gases

From active volcanoes we have learned that most of the gas released during eruptions is water vapor, which condenses as steam. Other gases, such as carbon dioxide, sulfur dioxide, hydrogen sulfide (which smells like rotten eggs), and hydrochloric acid, are given off in lesser amounts with the steam.

Surface water introduced into a volcanic system can greatly increase the explosivity of an eruption, as exemplified by the devastation of the island of Krakatoa (described earlier).

Gases and Pyroclastics

During an eruption, expanding, hot gases may propel pyroclastics high into the atmosphere as a column rising from a volcano. At high altitudes, the pyroclastics often spread out into a dark, mushroom cloud. The fine particles are transported by high atmosphere winds. Eventually debris settles back to earth under gravity's influence as *ashfall* (or sometimes *pumice fall*) deposits.

A **pyroclastic flow** is a mixture of gas and pyroclastic debris that is so dense that it hugs the ground as it flows rapidly into low areas (figure 10.7). As figure 10.7 shows, there are two ways in which pyroclastic flows develop. An exploding froth of gas and

IN GREATER DEPTH 10.2

Volcanoes and Flying

There have been several occasions in which jumbo jets have flown into volcanic ash clouds with nearly disastrous results. In 1989, a KLM Boeing 747 unknowingly entered an ash plume over Mount Redoubt, a volcano in Alaska, at an altitude of 8,000 meters (26,000 feet). The pilot applied full power hoping to climb out of the plume. After climbing a thousand meters all four engines stopped. The plane dropped to an altitude of 4,000 meters (13,000 feet) in eight tension-filled minutes before the flight crew was able to restart the engines. Although the plane landed safely in Anchorage, the cost to repair it was $80 million. Its engines, which had to be replaced, contained glassy coatings that turned out to be melted and resolidified ash. When full power was applied, the engines became very hot—hotter than the melting temperature of the ash. After this discovery, the standard procedure now is to reduce power to keep the engine temperature well below the melting point of volcanic ash and lessen the chances of engine failure. It is, of course, preferable to fly around pyroclastics.

Another, less serious problem is what appears to be extensive scratching of airplane windows. The enormous amount of sulfuric acid aerosol that was belched into the atmosphere by Mount Pinatubo in 1991 caused scratching. Acid attacks the windows, made of acrylic, and etches fine lines in them. Although Pinatubo is in the Philippines, airplanes flying above 10 kilometers throughout the northern hemisphere have had their windows damaged; an annoyance to passengers and costly to airlines.

FIGURE 10.7
Pyroclastic flow descending Mayon volcano, Philippines, in 1984. Insert shows two ways that pyroclastic flows can form.
Photo by Chris Newhall, U.S. Geological Survey

FIGURE 10.8
The ruins of St. Pierre in 1902. Mount Pelée is in the clouds.
Photo by Underwood & Underwood, courtesy Library of Congress

magma can blast out from under a solid or very viscous plug capping a volcano. Or it may be caused by gravitational collapse of a column of gas and pyroclastic debris that was initially blasted vertically into the air. These turbulent masses can travel over 100 kilometers per hour and are extremely dangerous. In 1991, a pyroclastic flow at Japan's Mount Unzen killed 31 people, including 3 geologists. Far worse was the destruction of St. Pierre (figure 10.8) on the Caribbean island of Martinique where about 28,000 people were killed by a pyroclastic flow in 1902 (see box 10.5).

EXTRUSIVE ROCKS

Most extrusive rocks are named and identified on the basis of their composition and texture. But some names are based solely on texture (e.g., pumice).

Composition

The amount of silica in a lava largely controls not only the viscosity of lava and the violence of eruptions but also which particular rock is formed. Chapter 11 describes how igneous rocks are identified based on the minerals present and their relative abundance in the rock. Because extrusive igneous rocks are generally fine-grained, a specialized microscope is usually needed for precise identification of the component minerals. In most cases, however, we can guess the probable mineral content by noting how dark or light in color an extrusive rock is. Most silicic rocks are light-colored because they contain abundant feldspar and quartz (both of which are silica-rich) and few dark minerals (which contain iron and magnesium and are silica-deficient). Mafic rocks, on the other hand, tend to be dark because of the abundance of ferromagnesian minerals.

table 10.1 Summary of Textures in Volcanic Rocks

Name	Description
Fine-grained (adjective)	Mosaic of interlocking minerals that are smaller than 1 mm.
Porphyritic (adjective)	Some crystals, phenocrysts, are larger than 1 mm (usually considerably larger). Most grains are smaller than 1 mm. Or phenocrysts are enclosed in glass.
Obsidian	Glass. Atoms are disordered.
Vesicular (adjective)	Holes in rock due to trapped gas.
Pumice	Frothy glass.
Tuff	Consolidated fine pyroclastic material.
Volcanic breccia	Consolidated pyroclastic debris that includes blocks or bombs.

Rhyolite, a silicic rock, is usually cream-colored, tan, or pink; it is made up mostly of feldspar but always includes some quartz. Note that the rhyolite (and granite) portion of figure 11.6 is larger than the areas shown for andesite and basalt. Geologists commonly subdivide this portion of the classification system. For example, *dacite,* the rock associated with the 1980 Mount St. Helens eruptions, contains more ferromagnesian minerals and plagioclase but less potassium feldspar and quartz than the average rhyolite. In our classification system, dacite corresponds to the right portion of the area in figure 11.6 assigned to rhyolite.

A **basalt** has a relatively low amount (about 50% by weight) of SiO_2. Much of that silica is bonded to iron and magnesium to form ferromagnesian minerals, such as *olivine* or *pyroxene,* which are green or black. The remaining silica plus aluminum is bonded predominantly with calcium to form calcium-rich *plagioclase feldspar* (which tends to be darker gray than the white or pink potassium or sodium feldspars associated with felsic rocks). Basalt does not contain quartz because no silica is left over after the other minerals have formed. Because of the preponderance of dark minerals in basalt, this rock is usually dark gray to black.

Andesite, which crystallizes from an intermediate lava, can be recognized by its moderately gray or green color. It is this color because a little over half the rock is light- to medium-gray plagioclase feldspar, while the rest is ferromagnesian minerals (usually *pyroxene* or *amphibole*).

Textures

Texture refers to a rock's appearance with respect to the size, shape, and arrangement of its grains or other constituents. Table 10.1 is a summary of the textures described on pages 249–250.

FIGURE 10.9
Obsidian.
Photo by C. C. Plummer

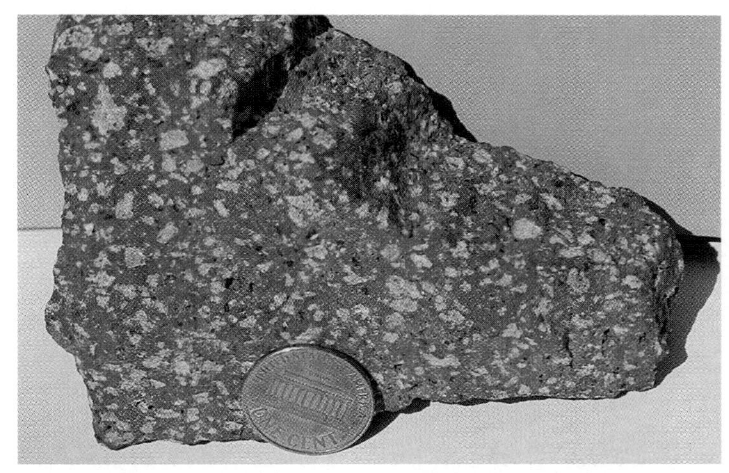

A

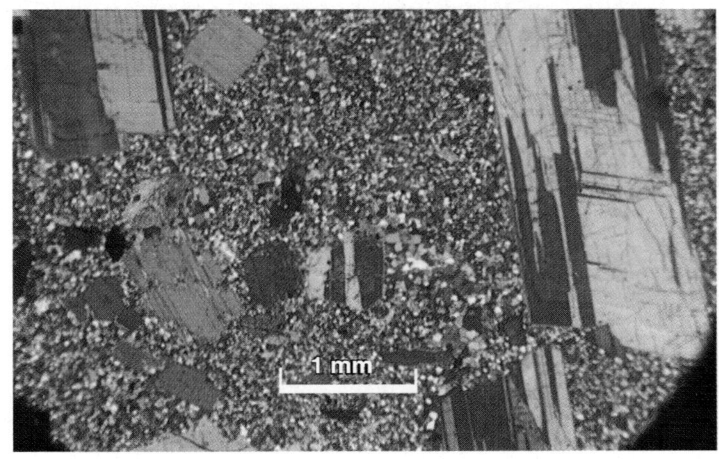

B

FIGURE 10.10
Porphyritic andesite. A few large crystals (phenocrysts) are surrounded by a great number of fine grains. (*A*) Hand specimen. (*B*) Photomicrograph (using polarized light) of the same rock. The black and white striped phenocrysts are plagioclase and the green ones are ferromagnesian minerals.
Photos by C. C. Plummer

Some extrusive rocks (such as obsidian and pumice) are classified solely on the basis of their textures, but most are classified by composition *and* texture. *Grain size* is a rock's most important textural characteristic. For the most part, extrusive rocks are fine-grained or else made of glass.

A **fine-grained rock** is one in which most of the mineral grains are smaller than 1 millimeter. In some, the individual minerals are distinguishable only with a microscope. **Obsidian** (figure 10.9), which is volcanic glass, is one of the few rocks that is not composed of minerals. A fine-grained or glassy texture distinguishes extrusive rocks from most intrusive rocks.

Two critical factors determine grain size during the solidification of igneous rocks: rate of cooling and viscosity. If lava cools rapidly, the atoms have time to move only a short distance; they bond with nearby atoms, forming only small crystals. With extremely rapid or almost instantaneous cooling, individual atoms in the lava are "frozen" in place, forming glass rather than crystals.

Grain size is controlled to a lesser extent by the viscosity of the lava. Atoms in a highly viscous lava cannot move as freely as those in a more fluid lava. Hence, a rock formed from viscous lava is more likely to be obsidian or of finer grains than one formed from more fluid lava. Most obsidian, when chemically analyzed, has a very high silica content and is silicic, the chemical equivalent of rhyolite.

Porphyritic Textures

Extrusive rock that does not have a uniformly fine-grained texture throughout is described as porphyritic. A **porphyritic rock** is one in which larger crystals are enclosed in a *matrix* (or *groundmass*) of much finer-grained minerals or obsidian. The larger crystals are termed **phenocrysts.** A porphyritic rock looks rather like raisin bread; the matrix or groundmass is the bread, the phenocrysts are the raisins. In the porphyritic andesite shown in figure 10.10, phenocrysts of feldspar and ferromagnesian minerals are enclosed in a matrix of crystals too fine-grained to distinguish with the naked eye but visible under a microscope.

Porphyritic texture usually indicates two stages of solidification. Slow cooling takes place while the magma is underground. Minerals that form at higher temperatures crystallize and grow to form phenocrysts in the still partly fluid magma. If the entire mass is then erupted, the remaining liquid portion cools rapidly and forms the fine-grained matrix.

FIGURE 10.11
Vesicular basalt.
Photo by C. C. Plummer

FIGURE 10.12
Pumice.
Photo by C. C. Plummer

FIGURE 10.13
Volcanic bombs.
Photo by C. C. Plummer

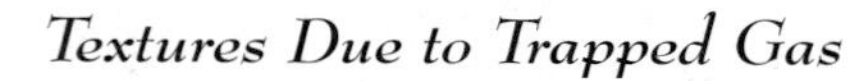

Textures Due to Trapped Gas

A magma deep underground is under high pressure, generally high enough to keep all its gases in a dissolved state. On eruption, the pressure is suddenly released and the gases come out of solution. This is analogous to what happens when a bottle of beer or soda is opened. Because the drink was bottled under pressure, the gas (carbon dioxide) is in solution. Uncapping the drink relieves the pressure, and the carbon dioxide separates from the liquid as gas bubbles. If you freeze the newly opened drink very quickly, you have a piece of ice with small, bubble-shaped holes. Similarly, when a lava solidifies while gas is bubbling through it, holes are trapped in the rock, creating a distinctive vesicular texture. **Vesicles** are cavities in extrusive rock resulting from gas bubbles that were in lava, and the texture is called *vesicular.* A vesicular rock has the appearance of Swiss cheese (whose texture is caused by trapped carbon dioxide gas). *Vesicular basalt* is quite common (figure 10.11). *Scoria,* a highly vesicular basalt, actually contains more gas space than rock.

In more viscous lavas, where the gas cannot escape as easily, the lava is churned into a froth (like the head in a glass of beer). When cooled quickly, it forms **pumice** (figure 10.12), a frothy glass with so much void space that it floats in water. Powdered pumice is used as an abrasive because it can scratch metal or glass.

Fragmental Textures

Pyroclasts, the fragments formed by volcanic explosion, can be almost any size. *Dust* and *ash* are the finest particles; *cinders* are about the size of sand grains; *bombs* and *blocks* are large pyroclasts. When solid rock has been blasted apart by a volcanic explosion, the pyroclastic fragments are *angular,* with no rounded edges or corners and are called **blocks.** If lava is ejected into the air, a molten blob becomes streamlined during flight, solidifies, and falls to the ground as a **bomb,** a spindle- or lens-shaped pyroclast (figure 10.13).

When pyroclastic material (ash, bombs, etc.) accumulates and is cemented or otherwise consolidated, the new rock is

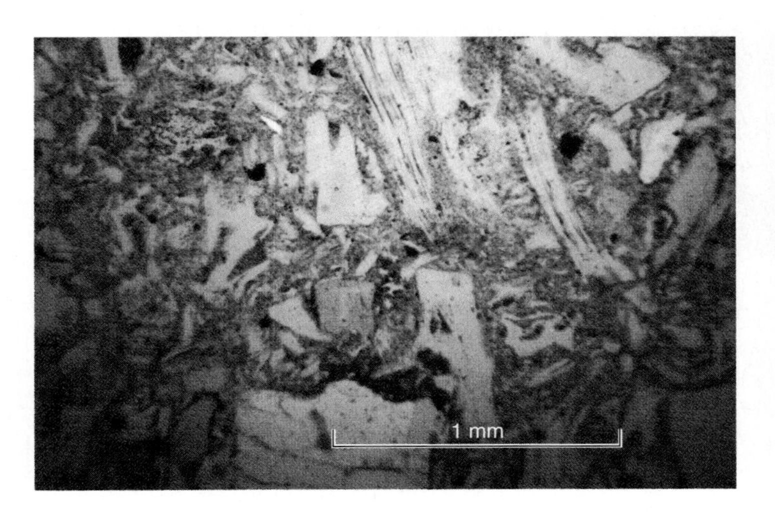

FIGURE 10.14

Photomicrograph of a tuff. Fragments of different rocks and minerals are angular and variously colored.

Photo by C. C. Plummer

FIGURE 10.15

Crater and caldera in Kamchatka, Russia. In the foreground is the crater on Karymsky volcano. In the background is a lake-filled caldera.

Photo by C. Dan Miller, U.S. Geological Survey

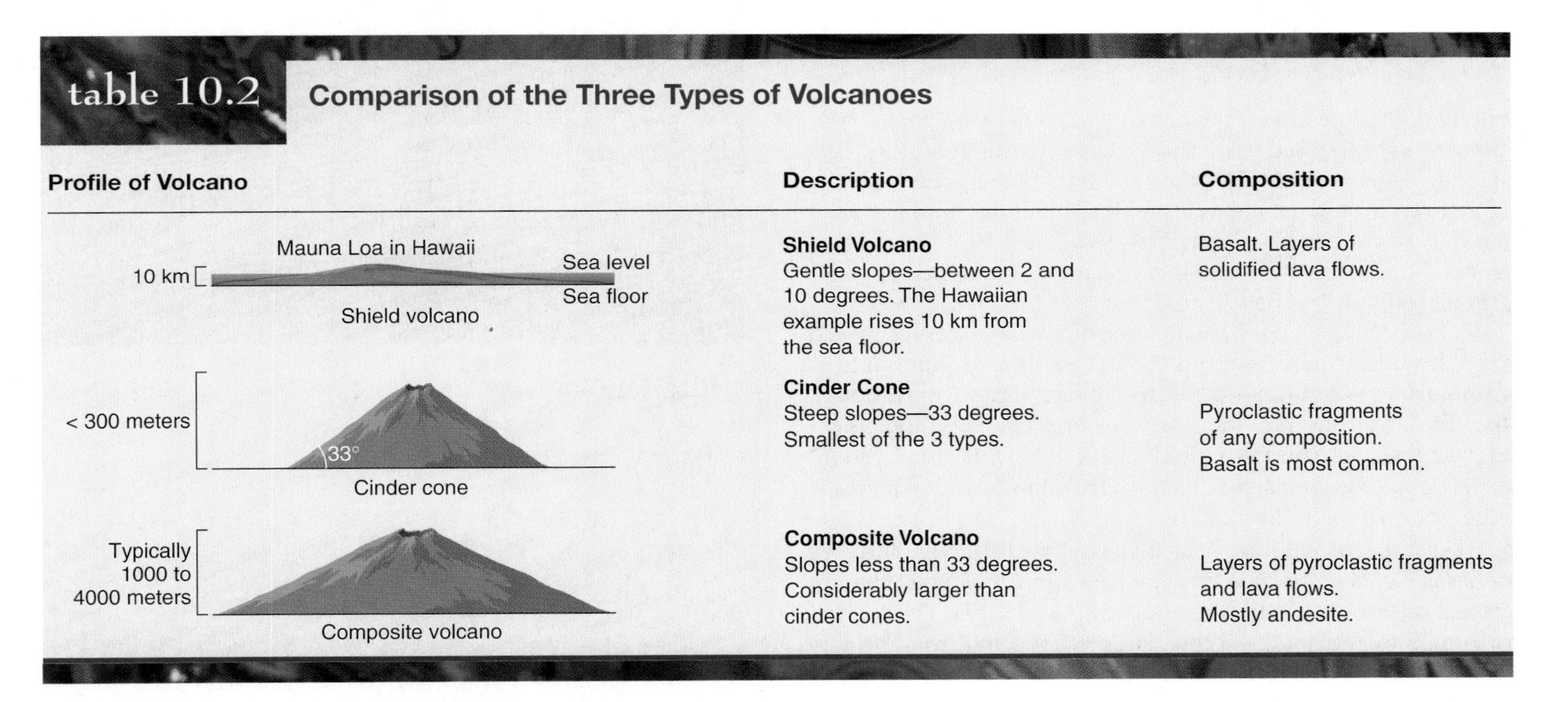

table 10.2 Comparison of the Three Types of Volcanoes

Profile of Volcano	Description	Composition
Mauna Loa in Hawaii; 10 km; Sea level; Sea floor; Shield volcano	**Shield Volcano** Gentle slopes—between 2 and 10 degrees. The Hawaiian example rises 10 km from the sea floor.	Basalt. Layers of solidified lava flows.
< 300 meters; 33°; Cinder cone	**Cinder Cone** Steep slopes—33 degrees. Smallest of the 3 types.	Pyroclastic fragments of any composition. Basalt is most common.
Typically 1000 to 4000 meters; Composite volcano	**Composite Volcano** Slopes less than 33 degrees. Considerably larger than cinder cones.	Layers of pyroclastic fragments and lava flows. Mostly andesite.

called *tuff* or *volcanic breccia,* depending on the size of the fragments. A **tuff** (figure 10.14) is a rock composed of fine-grained pyroclastic particles. A **volcanic breccia** is a rock that includes larger pieces of volcanic rock (blocks, bombs).

TYPES OF VOLCANOES

Volcanic material that is ejected from and deposited around a central vent produces the conical shape typical of volcanoes. The **vent** is the opening through which an eruption takes place. The **crater** of a volcano is a basinlike depression over a vent at the summit of the cone (figure 10.15). Material is not always ejected from the central vent. In a **flank eruption,** lava pours from a vent on the side of a volcano.

A **caldera** is a volcanic depression much larger than the original crater, having a diameter of at least one kilometer. (The most famous caldera in the United States is misnamed "Crater Lake.") A caldera can be created when a volcano's summit is blown off by exploding gases, as occurred at Mount St. Helens in May 1980, or, as in the case of Crater Lake, when a volcano (or several volcanoes) collapses into a vacated magma chamber (see figure 10.4).

The three major types of volcanoes (shield, cinder cone, and composite) that are discussed on pages 252–253 and that are compared in table 10.2 are markedly distinct from one another

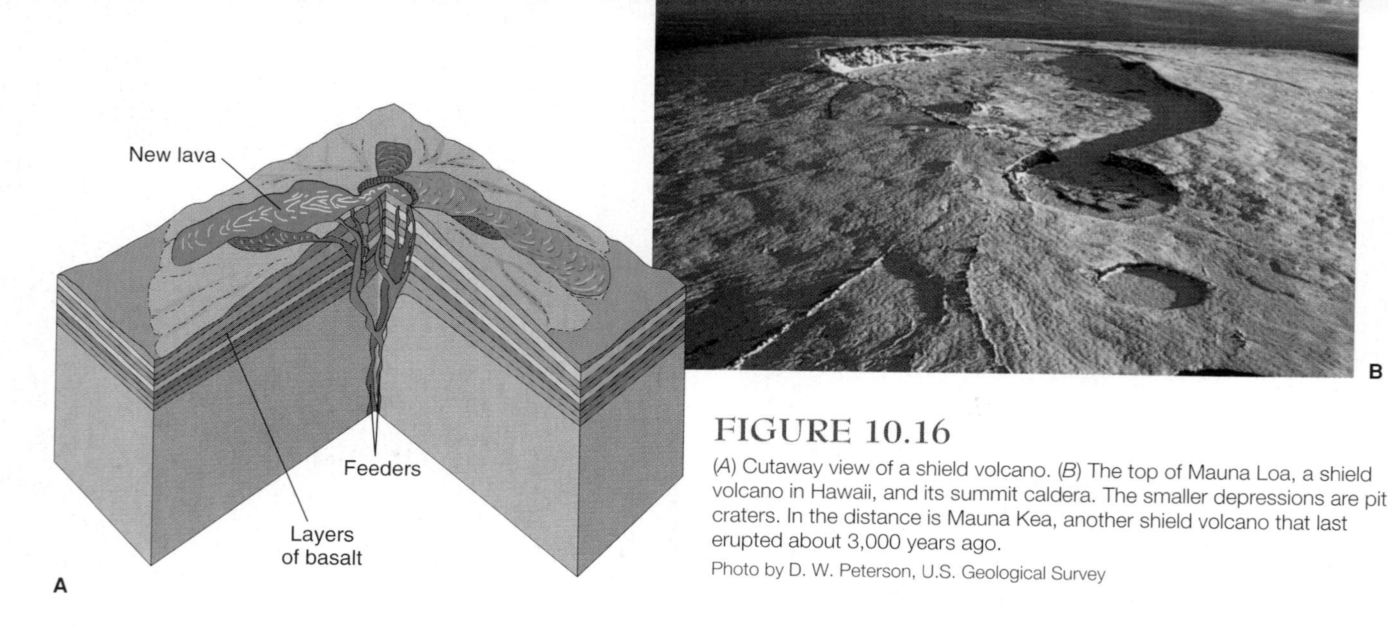

FIGURE 10.16

(*A*) Cutaway view of a shield volcano. (*B*) The top of Mauna Loa, a shield volcano in Hawaii, and its summit caldera. The smaller depressions are pit craters. In the distance is Mauna Kea, another shield volcano that last erupted about 3,000 years ago.

Photo by D. W. Peterson, U.S. Geological Survey

in size, shape, and, usually, composition. Although volcanic domes are not cones, they are associated with volcanoes and are also examined in this section.

Shield Volcanoes

Shield volcanoes are broad, gently sloping volcanoes constructed of solidified lava flows. During eruptions, the lava spreads widely and thinly due to its low viscosity. Because the lava flows from a central vent, without building up much near the vent, the slopes are usually between 2° and 10° from the horizontal, producing a volcano in the shape of a flattened dome or "shield" (figure 10.16).

The islands of Hawaii are essentially a series of shield volcanoes built upward from the ocean floor by intermittent eruptions over millions of years (figure 10.16*B*). Although spectacular to observe, the eruptions are relatively nonviolent because the lavas are fairly fluid (less viscous). By implication, then, the shield volcanoes of the Hawaiian Islands are composed of a series of layers of basalt.

Hawaiian names have been given to two distinctive surfaces of basalt flows. *Pahoehoe* (pronounced *pah-hoy-hoy*) is characterized by a ropy or billowy surface (figure 10.17). The surface is formed by the quick cooling and solidification from the surface downward of a lava flow or pool of lava that was fully liquid. By contrast, basalt that is cool enough to have partially solidified moves as a slow, pasty mass. Its largely solidified front is shoved forward as a pile of rubble. A flow such as this is called *aa* (pronounced *ah-ah*) and has a jagged, rubbly surface (figure 10.18).

A minor feature called a *spatter cone,* a small, steep-sided cone built from lava sputtering out of a vent (figure 10.19), will occasionally develop on a solidifying lava flow. When a small concentration of gas is trapped in a cooling lava flow, lava is belched out of a vent through the solidified surface of the flow. Falling lava plasters itself onto the developing cone and solidifies. The sides of a spatter cone can be very steep, but they are rarely over 10 meters high.

FIGURE 10.17

Pahoehoe from a 1972 eruption in Hawaii.

Photo by D. W. Peterson, U.S. Geological Survey

FIGURE 10.18

An *aa* flow in Hawaii, 1983.

Photo by J. D. Griggs, U.S. Geological Survey

FIGURE 10.19
A spatter cone (approximately 1 meter high) erupting in Hawaii.
Photo by J. B. Judd, U.S. Geological Survey

Cinder Cones

A **cinder cone** (less commonly called a *pyroclastic cone*) is a volcano constructed of pyroclastic fragments ejected from a central vent (figure 10.20). In contrast to the gentle slopes of shield volcanoes, cinder cones commonly have slopes of about 30°. Most of the ejected material lands near the vent during an eruption, building up the cone to a peak. The steepness of slopes of accumulating loose material is limited by gravity to about 33°. Cinder cones tend to be very much smaller than shield volcanoes. In fact, cinder cones are commonly found on the flanks and in the calderas of Hawaii's shield volcanoes. Few cinder cones exceed a height of 500 meters.

Cinder cones form by pyroclastic material accumulating around a vent. They form because of a buildup of gases and are independent of composition. Most cinder cones are associated with mafic or intermediate lava. Silicic cinder cones, which are made of fragments of pumice, are also known as pumice cones.

The life span of an active cinder cone tends to be short. The local concentration of gas is depleted rather quickly during the eruptive periods. Moreover, as landforms, cinder cones are temporary features in terms of geologic time. The unconsolidated pyroclasts are eroded relatively easily.

Composite Volcanoes

A **composite volcano** (also called **stratovolcano**) is one constructed of alternating layers of pyroclastic fragments and solidified lava flows (figure 10.21*A*). The slopes are intermediate in steepness compared with cinder cones and shield volcanoes. Pyroclastic layers build steep slopes as debris collects near the vent, just as in cinder cones; however, subsequent lava flows

B

FIGURE 10.20
Cerro Negro, a cinder cone in Nicaragua. (*A*) View from the air; (*B*) nighttime eruption of pyroclastics at the summit.
Photo *A* by Mark Hurd Aerial Surveys Corp. courtesy California Division of Mines and Geology; Photo *B* by R. W. Decker

10.3 ASTROGEOLOGY

Extraterrestrial Volcanic Activity

Volcanic activity has been a common geologic process operating on the Moon and on several other bodies in the solar system. Approximately one-third of the Moon's surface consists of nearly circular, dark-colored, smooth, relatively flat lava plains. The lava plains, found mostly on the near side of the Moon, are called *maria* (singular, *mare;* literally, "seas"). This kind of terrain represents a significant period in the Moon's early history when large lava flows flooded parts of the Moon's surface, filling in low places. There are also a few extinct shield volcanoes on the Moon.

Elongate trenches or cracklike valleys called *rilles* are found mainly in the smoother portions of the lunar maria. They range in length from a few kilometers to hundreds of kilometers. Some are arc-shaped or crooked and may be collapsed lava tubes, channels eroded by pyroclastic flows, or fractures along which gas has escaped.

Mercury, the innermost planet, also has extensive areas covered by maria.

Radar images of Venus show a surface that is young and volcanically active. More than three-fourths of that surface is covered by continuous plains formed by enormous floods of lava. Close examination of these plains reveals extensive networks of lava channels and individual lava flows as much as 300 kilometers long.

Large shield volcanoes, some in chains along a great fault, have been identified on Venus, and molten lava lakes may exist. In other places, thick lavas have oozed out to form kilometer-high pancake-shaped domes. Radar studies have shown that some of these domes are composed of a glassy substance mixed with bubbles of trapped gas. Fan-shaped deposits adjacent to some volcanoes may be pyroclastic debris.

Several of Venus's volcanoes are active, emitting large amounts of sulfur gases and causing the almost continuous lightning that has been observed by spacecraft.

Nearly half of the planet Mars may be covered with volcanic material. There are areas of extensive lava flows similar to the lunar maria and a number of volcanoes, some with associated lava flows.

Mars has at least nineteen large shield volcanoes, probably composed of basalt. The largest one, Olympus Mons (box figure 1), is three times the height of Mount Everest and wider than Texas. Its caldera is more than 90 kilometers across.

Ten volcanoes have been observed by spacecraft on Jupiter's moon Io (box figure 2) and seven of those have erupted for periods of at least four months. Material rich in sulfur compounds is thrown at least 500 kilometers into space at speeds of up to 3,200 kilometers per hour. This material often forms umbrella-shaped clouds as it spreads out and falls back to the surface. Lakes of molten sulfur and huge multicolored (black, yellow, red, orange, and brown) lava flows of sulfur or a sulfur-silicate mixture are common. More than 100 calderas larger than 25 kilometers across have been observed, including one that vents sulfur gases. Clouds of sulfur and sulfur dioxide often form and precipitate reddish sulfur dioxide "snow." The energy source for Io's volcanoes may be the gravitational pulls of Jupiter and two of its other larger satellites, causing Io to heat up much as a piece of wire will do if it is flexed continuously.

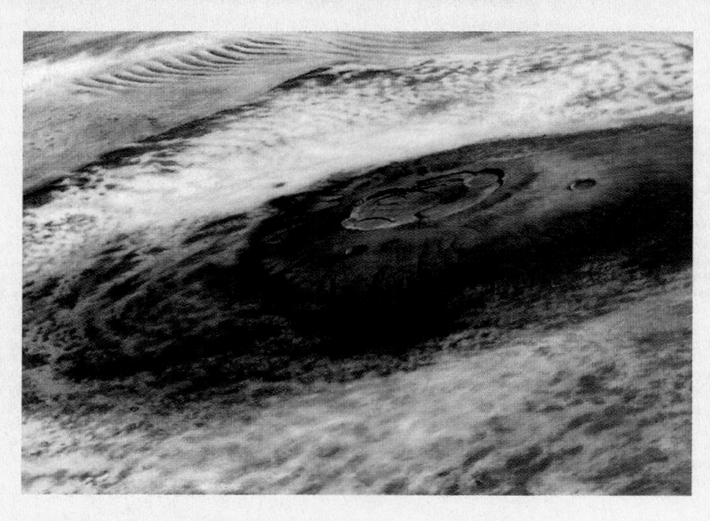

Box 10.3 — FIGURE 1
Olympus Mons on Mars.
Photo by NASA

Box 10.3 — FIGURE 2
A volcanic eruption on Jupiter's moon Io.
Photo by NASA

Neptune's moon Triton is the fourth object in the solar system that has active volcanoes. There, "ice volcanoes" erupt what is probably liquid nitrogen, dust, or methane compounds.

Related Web Resource
Introduction to the Nine Planets
http://seds.lpl.arizona.edu/nineplanets/nineplanets/intro.html

partially flatten the profile of the cone as the downward flow builds up the height of the flanks more than the summit area. The solidified lava acts as a protective cover over the loose pyroclastic layers, making composite volcanoes less vulnerable to erosion than cinder cones.

Composite volcanoes are built over long spans of time. Eruption is intermittent, with hundreds or thousands of years of inactivity separating a few years of intense activity. During the quiet intervals between eruptions, composite volcanoes may be eroded by running water, landslides, or glaciers. These surficial processes tend to alter the surface, shape, and form of the cone. But because of their long lives and relative resistance to erosion, composite cones can become very large. Aconcagua, a composite volcano in the Andes, is 6,960 meters (22,835 feet) above sea level and the highest peak in the western hemisphere.

The extrusive material that builds composite cones is predominantly of intermediate composition, although there may be local minor silicic and mafic eruptions. Therefore, *andesite* is the rock most associated with composite volcanoes. If the lava is especially hot, the relatively low viscosity fluid flows easily from the crater down the slopes. On the other hand, if enough gas pressure exists, an explosion may litter the slopes with pyroclastic andesite, particularly if the lava has fully or partially solidified and clogged the volcano's vent.

The composition as well as eruptive history of individual volcanoes can vary considerably. For instance, Mount Rainier is composed of 90% lava flows and only 10% pyroclastic layers. Conversely, Mount St. Helens was built mostly from pyroclastic eruptions—reflecting a more violent history. As would be expected, the composition of the rocks formed during the 1980 eruptions of Mount St. Helens is somewhat higher in silica than average for Cascade volcanoes.

Distribution of Composite Volcanoes

Nearly all the larger and better known volcanoes of the world are composite volcanoes. They tend to align along two major belts on the Earth (figure 10.22). The **circum-Pacific belt,** or "Ring of Fire," is the larger. The Cascade Range volcanoes described earlier make up a small segment of the circum-Pacific belt.

Several composite volcanoes in Mexico rise higher than 5,000 meters, including Orizaba (third highest peak in North America) and Popocatépetl (see box 10.4).

The circum-Pacific belt includes many volcanoes in Central America, western South America (including Nevado del Ruiz in Colombia), and Antarctica. Mount Erebus, in Antarctica, is the southernmost active volcano in the world (figure 10.23).

The western portion of the Pacific belt includes volcanoes in New Zealand, Indonesia, the Philippines (with Pinatubo, Mayon, whose 1993 eruptions killed over 30 people, and many

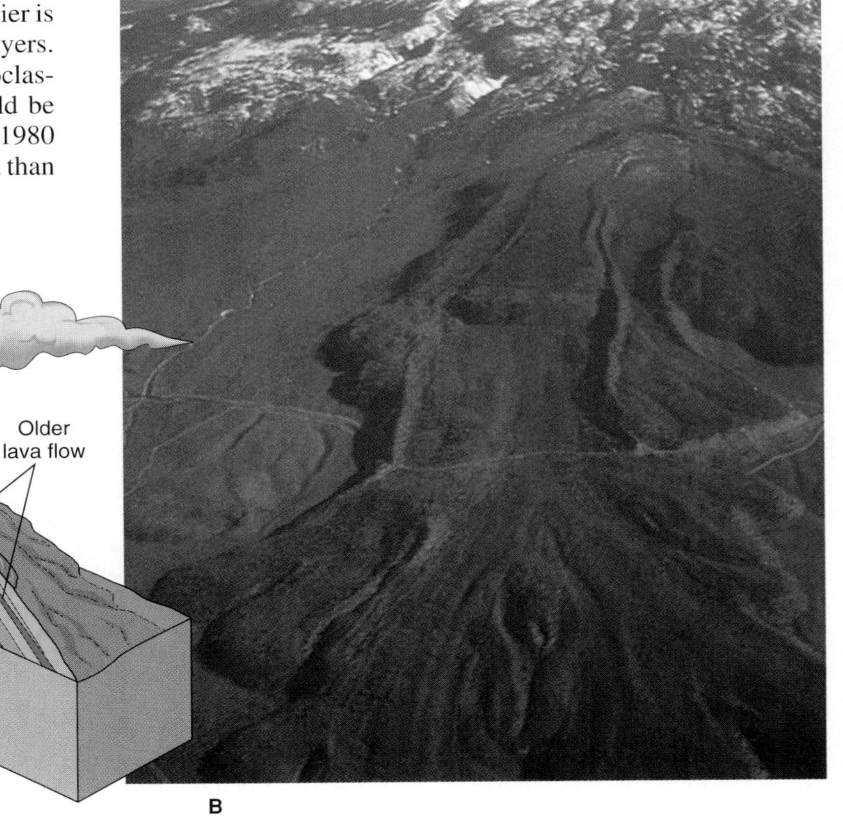

FIGURE 10.21

(*A*) Cutaway view of a composite volcano. Light-colored layers are pyroclastics. (*B*) Mount Shasta, a composite volcano in California. Shastina on Mount Shasta's flanks is a subsidiary cone, largely made of pyroclastics. Note the lava flow that originated on Shasta and extends beyond the volcano's base.

Photo by B. Amundson

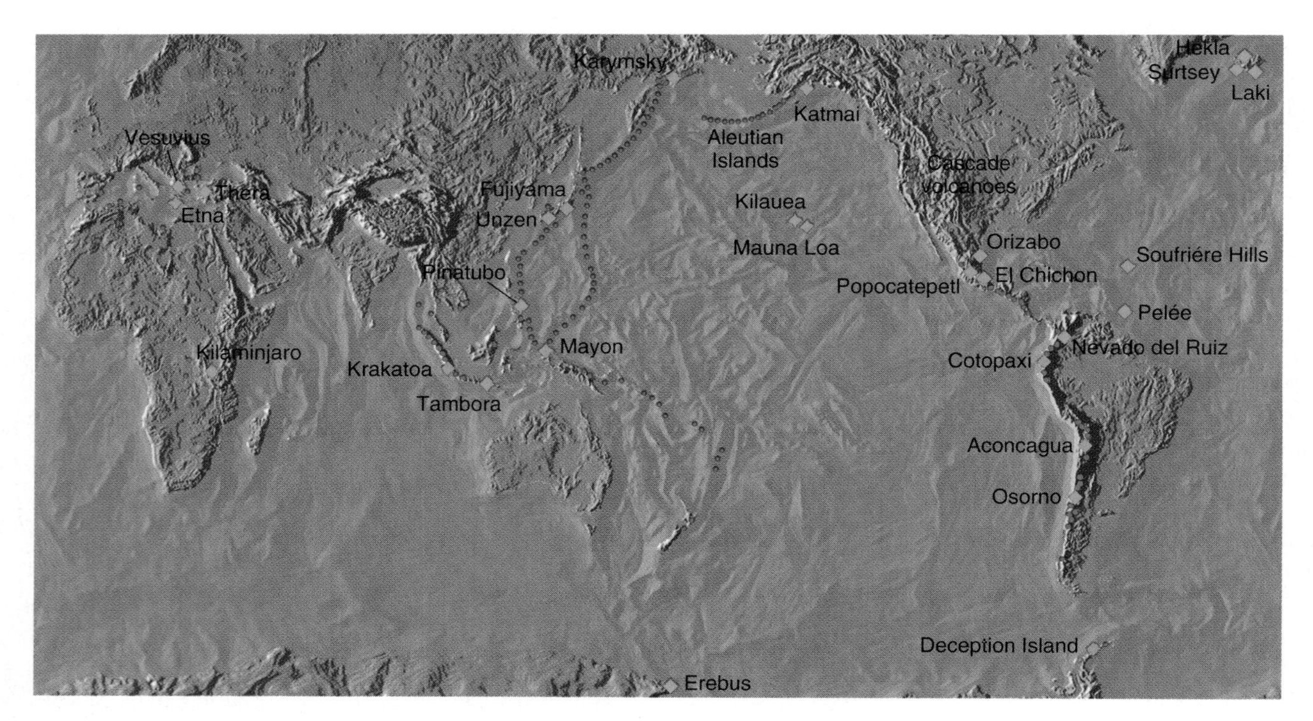

FIGURE 10.22
Map of the world showing recently active major volcanoes.

FIGURE 10.23
Mount Erebus, Antarctica, the southernmost active volcano in the world. The photo is taken on sea ice. The summit is 3,794 meters (12,444 feet) above sea level. A summit crater contains a convecting lava lake.
Photo by Philip R. Kyle

FIGURE 10.24
Mount Fuji, woodblock print by Japanese artist Hiroshige (1797–1858).

other volcanoes), and Japan. The beautifully symmetrical Fujiyama, in Japan, is probably the most frequently painted volcano in the world (figure 10.24). The northernmost part of the circum-Pacific belt includes active volcanoes in Russia (see figure 10.15) and on Alaska's Aleutian Islands.

The second major volcanic belt is the **Mediterranean belt,** which includes Mount Vesuvius. An exceptionally violent eruption of Mount Thera, an island in the Mediterranean, may have destroyed an important site of early Greek civilization. (Some archaeologists consider Thera the original "lost continent" of Atlantis.) Mount Etna, on the island of Sicily, has been called the "lighthouse of the Mediterranean" because of its frequent eruptions throughout the centuries. The largest eruption in 300 years began in 1991 and lasted for 473 days. Some 250 million cubic meters of lava covered 7 km^2 of land. A town was saved from the lava by heroic efforts that included building a dam to retain the lava (the lava quickly overtopped it), plugging some natural channels, and diverting the lava into other, newly constructed channels.

ENVIRONMENTAL GEOLOGY

10.4

Popocatépetl—Will It Erupt Big Time?

Popocatépetl, located 55 kilometers east of Mexico City, one of the world's largest cities, and 45 miles west of the city of Puebla, began erupting in 1994. Some 30 million people live within view of Popocatépetl (Aztec for "smoking mountain"). A major eruption could endanger hundreds of thousands of those people.

Popocatépetl, affectionately called "Popo," at 5,484 meters (17,991 feet) above sea level, is one of North America's highest mountains. Not only does Popocatépetl provide a majestic scenic presence (box figure 1*A*), but it figures prominently in Mexico's history, art, and culture. According to Aztec legend, Popocatépetl is a warrior eternally guarding his sleeping lover, the neighboring mountain Ixtaccihuatl (Aztec for "white lady") whose outline resembles that of a supine woman. Cortez sent his men to climb Popocatépetl during the Spanish conquest of Mexico in 1521. They were lowered into the smoking crater and returned with sulfur used to make gunpowder. (This was the first recorded ascent in the world of a major mountain.)

The volcano began awakening from a long period of dormancy in December 1994 with a minor dusting of ash on Puebla. Some 75,000 people living on the eastern flank of the volcano were temporarily evacuated. An extensive monitoring network of instruments was deployed and teams of Mexican scientists assisted by members of the U.S. Geological Survey Volcanic Disaster Assistance Program began assessing the potential hazards.

The threat of a disaster is taken very seriously because in 1982 an apparently insignificant, jungle-covered, 1,000-meter-high volcano in southern Mexico called *El Chichon* erupted with a series of violent explosions. Towns near the previously inactive volcano were buried by heavy ashfall or blasted by searing, gas-charged pyroclastic flows. The death toll could only be roughly estimated to be in the thousands.

By determining the size and extent of ancient pyroclastic deposits and dating them, geologists have determined that Popocatépetl produced major explosive eruptions every 1,000 to 3,000 years for the last 10,000 years. Each has produced widespread pumice falls, pyroclastic flows, and mudflows. Preconquest population centers were repeatedly destroyed by these catastrophic eruptions. Since the year 1345, records indicate that there have been some 30 small eruptions before the present activity. Volcanologists consider it one of the world's most dangerous volcanoes. Will the current activity culminate in a colossal event that takes place every thousand years or so?

If such an event were to occur, pyroclastic flows and mudflows would destroy villages and could kill thousands of people if they are not evacuated in time. Heavy ashfall would cause further damage. Mexico City is not likely to be affected by pyroclastic flows or mudflows, but if ash is blown over and deposited in the city there could be serious consequences. Air traffic to and from Mexico City International Airport would be threatened. Water supplies, electrical power grids, and sewer systems could be damaged or destroyed.

A

B

Box 10.4 — FIGURE 1

Popocatépetl in 1960 (*A*) and during the December 19, 2000, eruption (*B*). Snow-covered glaciers that stand out in the 1960 photo are now covered with ash.

Photo *A* by C. C. Plummer, *B* Photo © AP/Wide World Photos

Minor steam and ash eruptions from Popocatépetl continued through 1995. On March 29th, 1996, a new lava dome was identified in the summit crater of the volcano. A month later, an explosion of the new dome killed five climbers, who were at the summit of the volcano. Small ash-producing explosions took place throughout 1996 and into 1997. In April 1997, several slightly larger explosions took place. These generated ash columns that rose as much as 5 kilometers above the mountain. Volcanic bombs showered the flanks of the cone; some started grass fires, scaring rural residents.

In late 2000, activity increased and on December 18, Popo's largest eruption in over 1,000 years took place with spectacular night-time displays of incandescent lava expelled from the mountain. By this time 14,000 people had been evacuated to shelters. Evacuation of high-risk towns had begun days earlier, and by December 21 some 50,000 people had been evacuated. Concern developed over the potential for a large mudflow because of melting of a glacier. On January 31, 2001, a pyroclastic flow descended the volcano to within 8 kilometers of a town.

One author's (Plummer) reflections on the mountain:

One of the indelible memories of growing up in Mexico City is that of the huge, magnificent volcanoes on the eastern skyline—Popocatépetl and Ixtaccihuatl. They were always visible when the sky was clear, which was frequent in the days before Mexico became one of the world's smoggiest cities. At age 15, I was fortunate enough to join some experienced mountaineers and climb the snow-covered mountain. I was stunned by the debilitating effect high altitude has on climbers. I would count out ten steps upward, then collapse over my ice axe panting for several minutes before taking another ten steps. When we reached the summit, everyone with me felt nauseated and had splitting headaches—altitude sickness. I didn't feel too bad. After the climb I felt an enormous sense of accomplishment. My life changed. The beauty of mountains and the challenge of climbing them became the focal point of my existence. At college, I became interested in geology (after taking an introductory physical geology course) as a natural extension of my love of mountains. Ultimately, my interest in geology overtook my interest in climbing.

Related Web Resource

CENAPRED Volcano Site

www.cenapred.unam.mx/mvolcan.html

Although a summary of reports can be accessed in English, you can get a daily report in Spanish. You can also see the volcano live by clicking on *Tamano B* under *Imagen del volcan.*

FIGURE 10.25

The dome in Mount St. Helens crater as seen from a biplane flying inside the crater in 1993.

Photo by C. C. Plummer

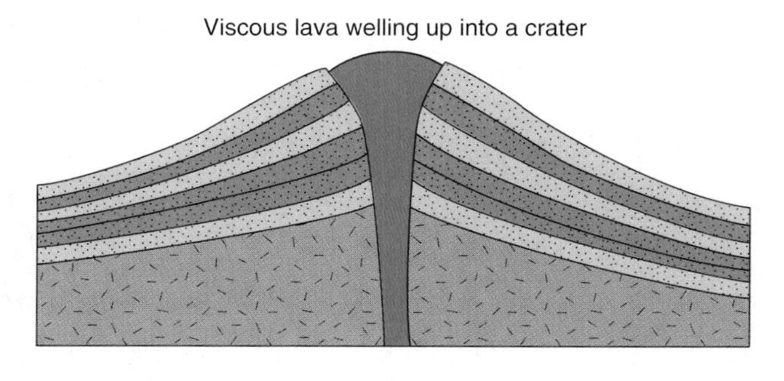

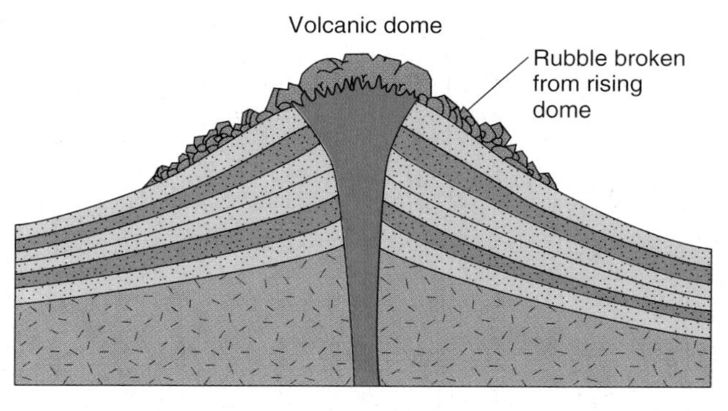

FIGURE 10.26

A volcanic dome forming in the crater of a cinder cone.

Volcanic Domes

Volcanic domes are steep-sided, dome- or spine-shaped masses of volcanic rock formed from viscous lava that solidifies in or immediately above a volcanic vent. A volcanic dome grew within the caldera of Mount St. Helens after the climactic eruption of May 1980 (figure 10.25). This was expected because of the high viscosity of the lava from the St. Helens eruptions. In 1983 alone, the dome increased its elevation by 200 meters. Most of the viscous lavas that form volcanic domes are high in silica. They solidify as rhyolite or, less commonly, andesite if minerals crystallize, or as obsidian if no minerals crystallize.

Because the thick, pasty lava that squeezes from a vent is too viscous to flow, it builds up a steep-sided dome or spine (figure 10.26). Some volcanic domes act like champagne corks, keeping gases from escaping. If the plug is removed or broken, the gas escapes suddenly and violently. Some of the most destructive volcanic explosions known have been associated with volcanic domes (see box 10.5).

ENVIRONMENTAL GEOLOGY 10.5

A Tale of Two Volcanoes—Lives Lost and Lives Saved in the Caribbean

Montserrat and Martinique are two of the tropical islands that are part of a volcanic island arc (box figure 1). Both islands had major eruptions during the twentieth century that destroyed towns. Violent and deadly pyroclastic flows associated with growth of volcanic domes caused most of the destruction. For one island the death toll was huge and for the other it was minimal.

In 1902, the port city of St. Pierre on the island of Martinique was destroyed after a period of dome growth and pyroclastic flows on Mount Pelée (no relationship to Pele, Hawaii's goddess of volcanoes). A series of pyroclastic flows broke out of a volcanic dome and flowed down the sides of the volcano. Searingly hot pyroclastic flows can travel at over 100 km per hour and will destroy any living things in their paths. After the pyroclastic flows began, the residents of St. Pierre became fearful and many wanted to leave the island. The authorities claimed there was no danger and prevented evacuation. The climax came on the morning of May 8, when great exploding clouds descended like an avalanche down the mountainside, raced down a stream valley, through the port city and onto the harbor. St. Pierre and the ships anchored in the harbor were incinerated (see figure 10.8). Temperatures within the pyroclastic flow were estimated at 700°C. About 28,000 people were burned to death or suffocated (of the four survivors, one was a condemned prisoner in a poorly ventilated dungeon).

Ninety-three years later, in July 1995, small steam-ash eruptions began at Soufriére Hills volcano on the neighboring island of Montserrat. As a major eruption looked increasingly likely, teams of volcanologists from France, the U.K., the U.S. (including members of the U.S. Geological Survey's Volcano Disaster Assistance Team that had successfully predicted the eruption of Mt. Pinatubo in the Philippines, as described in

Box 10.5 — FIGURE 1

Eruption of Soufriére Hills volcano on Montserrat, August 4, 1997. An ash cloud billows upward above a ground-hugging pyroclastic flow. Map of the West Indies showing location of Montserrat, Martinique, and Soufriére Hills volcano. Photo by AP/Kevin West

chapter 1), and elsewhere flew in to study the volcano and help assess the hazards. An unprecedented array of modern instruments (including seismographs, tiltmeters, gas analyzers) were deployed around the volcano. In November 1995 viscous, andesitic lava built a dome over the vent. Pyroclastic flows began when the dome collapsed in March 1996. Pyroclastic flows continued with more dome building and collapsing. By 1997, nearly all the people in the southern part of the island were evacuated, following advice from the scientific teams. In June 1997 large eruptions took place and pyroclastic flows destroyed the capital city of Plymouth. In contrast to the tragedy of St. Pierre, only 19 people were killed.

In August 1997, major eruptions resumed. This time the northern part of the island, previously considered safe, was faced with pyroclastic flows (box figure 1), and more people were evacuated from the island.

Related Web Resource

Montserrat Volcano Observatory home page

www.geo.mtu.edu/volcanoes/west.indies/soufriere/govt

FIGURE 10.27

Basalt layers in the Columbia plateau, Washington.

Photo by P. Weis, U.S. Geological Survey

LAVA FLOODS

Not all extrusive rocks are associated with volcanoes. Lava that is very nonviscous and flows almost as easily as water does not build a cone around its vents. Such lava is, of course, mafic (low in silica).

Plateau basalts were produced during the geologic past by vast outpourings of lava. The Columbia plateau area of Washington, Idaho, and Oregon (see inside cover), for example, is constructed of layer upon layer of basalt (figure 10.27), in places as thick as 3,000 meters. Each individual flood of lava added a layer usually between 15 and 100 meters thick and sometimes thousands of square kilometers in extent.

Basalt layers give the landscape a striking appearance in most places where they are exposed. Instead of stacked-up slabs or tablets of solid, unbroken rock, the individual layers may appear to be formed of parallel, vertical columns, mostly six-sided. This characteristic of basalt is called **columnar structure** or **columnar**

A

B

FIGURE 10.28
Columnar jointing at Devil's Postpile, California. (*A*) The columns as seen from above. (Scratches were caused by glacial erosion as described in chapter 12.) (*B*) Side view.
Photos by C. C. Plummer

A

B

FIGURE 10.29
(*A*) A collapsed roof of a lava tube in Hawaii shows a stream of lava during the 1969–71 eruptions. (*B*) Lava tube at Lava Beds National Monument, California.
Photo *A* © Krafft/HOA-QUI/Photo Researchers, Inc. *B* by C. C. Plummer

jointing (figure 10.28). The columns can be explained by the way in which basalt contracts as it cools *after* solidifying. Basalt solidifies completely at temperatures below about 1,200°C. The hot layer of rock then continues to cool to temperatures normal for the Earth's surface. Like most solids, basalt contracts as it cools. The layer of basalt is easily able to accommodate the shrinkage in the narrow vertical dimension; but the cooling rock cannot "pull in" its edges, which may be many kilometers away. The tension fractures the rock into an orderly hexagonal pattern.

Lava tubes, tunnel-like caves within lava flows, may develop during the late stages of solidification of a lava flow either on a lava plateau or a shield volcano (figure 10.29). After most of the lava flow has solidified, the remaining lava flows in a channel roofed by basalt that solidified at the surface of the flow.

SUBMARINE ERUPTIONS

Submarine eruptions, notably those occurring along mid-oceanic ridges, almost always consist of mafic lavas that create basalt. As described in chapter 11, basaltic rock, thought to have been formed from lava erupting along mid-oceanic ridges or solidifying underground beneath the ridges, makes up virtually the entire crust underlying the oceans. In a few places—Iceland,

10.6 ENVIRONMENTAL GEOLOGY

Fighting a Volcano in Iceland—and Winning

In 1973 a volcano began erupting on a small island in Iceland. Go to the book's website www.mhhe.com/plummer9e to learn about:

- how a town was almost buried by ash;
- what volunteers did to keep roofs from collapsing from heavy ash deposits;
- a lava flow that threatened to seal off the harbor and end the town's thriving fishing industry;
- an unprecedented effort to halt the lava flow;
- the cleanup and rebuilding of the town;
- how the residents get heat and hot water from the lava flow.

Box 10.6 — FIGURE 1
Lava fountaining behind the town on Heimaey. The glow behind the town in the left part of the photo is the lava flow advancing to the harbor.
Photo © Solarfilma

FIGURE 10.30
Pillow basalt in Iceland.
Photo by R. W. Decker

FIGURE 10.31
Pillow basalt on a mid-oceanic ridge. Photo taken from a submersible vessel.
Courtesy of Woods Hole Oceanographic Institution

for example—volcanic islands rise above the otherwise submerged system (see box 10.6).

Pillow Basalts

Figure 10.30 shows **pillow structure**—rocks, generally basalt, occurring as pillow-shaped rounded masses closely fitted together. From observations of submarine eruptions by divers, we know how the pillow structure is produced. Elongate blobs of lava break out of a thin skin of solid basalt over the top of a flow that is submerged in water. Each blob is squeezed out like toothpaste, and its surface is chilled to rock within seconds. A new blob forms as more lava inside breaks out. Each new pillow settles down on the pile, with little space left in between. Some pillow basalt forms in lakes and rivers; however, most forms at mid-oceanic ridge crests (figure 10.31). According to plate tectonic theory, basalt magma flows up the fracture that develops at a divergent boundary (explained in chapter 11). The magma that reaches the sea floor solidifies as pillow basalt. The rest solidifies in the fracture as a dike. Pillow basalt that is overlying a series of dikes is sometimes found in mountain ranges. These probably formed during sea-floor spreading in the distant past followed, much later, by uplift.

Summary

Lava is molten rock that reaches the earth's surface, having been formed as *magma* from rock within the Earth's crust or from the uppermost part of the mantle.

More people have been killed by pyroclastic flows and, indirectly, by famine than by other volcanic hazards.

Lava contains 45% to 75% *silica* (SiO_2). The more silica, the more viscous the lava. Viscosity is also determined by the temperature of the lava. Viscous lavas are associated with more violent eruptions than are fluid lavas. *Volcanic domes* form from the extrusion of very viscous lavas.

A *mafic* lava, relatively low in silica, crystallizes into *basalt,* the most abundant extrusive igneous rock. Basalt, which is dark in color, is composed of minerals that are relatively high in iron, magnesium, and calcium.

Rhyolite, a light-colored rock, forms from *silicic* lavas that are high in silica but contain little iron, magnesium, or calcium. Because potassium and sodium are important elements in rhyolite, its constituent minerals are mostly potassium- and sodium-rich feldspars and quartz.

A lava with a composition between mafic and silicic crystallizes to *andesite,* a moderately dark rock. Andesite contains about equal amounts of ferromagnesian minerals and sodium- and calcium-rich feldspars.

Extrusive rocks are characteristically fine-grained. *Porphyritic* rock contains some larger crystals in an otherwise fine-grained rock. Rocks that solidified too rapidly for crystals to develop form a natural glass called *obsidian.* Gas trapped in rock forms *vesicles.*

Pyroclasts are the result of volcanic explosions. *Tuff* is volcanic ash that has consolidated into a rock. If large pyroclastic fragments have reconsolidated, the rock is a *volcanic breccia.*

A *cinder cone* is composed of loose pyroclastic material that forms steep slopes as it falls back around the crater. Cinder cones are not as large as the other two major types of cones.

A *shield volcano* is built up by successive eruptions of mafic lava. Its slopes are gentle but its volume generally large.

Composite cones are made of alternating layers of pyroclastic material and solidified lava flows. They are not as steep as cinder cones but steeper than shield volcanoes. Young composite volcanoes, predominantly composed of andesite, are aligned along the circum-Pacific belt and, less extensively, in the Mediterranean belt.

Plateau basalts are thick sequences of lava floods. *Columnar jointing* develops in solidified basalt flows. Basalt that erupts underwater forms a *pillow structure.* Pillow basalts are common along the crests of mid-oceanic ridges.

Terms to Remember

andesite 248
basalt 248
block 250
bomb 250
caldera 251
cinder cone 253
circum-Pacific belt 255
columnar structure (columnar jointing) 260
composite volcano (stratovolcano) 253
crater 251
extrusive rock 241
fine-grained rock 249
flank eruption 251
intermediate rock 245
lava 240
lava flows 241
mafic rock 245
magma 240
Mediterranean belt 256
obsidian 249
phenocryst 249
pillow structure (pillow basalts) 262
plateau basalts 260
porphyritic rock 249
pumice 250
pyroclastic flow 246
pyroclasts 241
rhyolite 248
shield volcano 252
silicic (felsic) 245
texture 248
tuff 251
vent 251
vesicle 250
viscosity 245
volcanic breccia 251
volcanic dome 258
volcanism 241
volcano 241

Testing Your Knowledge

Use the questions below to prepare for exams based on this chapter.

1. Compare the hazards of lava flows to pyroclastic flows.
2. What roles do gases play in volcanism?
3. What do pillow structures indicate about the environment of volcanism?
4. Name the minerals, and the approximate percentage of each, that you would expect to be present in each of the following rocks: andesite, rhyolite, basalt.
5. What property (or characteristic) of obsidian makes it an exception to the usual geologic definition of *rock?*
6. What determines the viscosity of a lava?
7. What determines whether a series of volcanic eruptions builds a shield volcano, a composite volcano, or a cinder cone? Describe each type of volcanic cone.
8. Explain how a vesicular porphyritic andesite might have formed.
9. Why are extrusive igneous rocks fine-grained?
10. Why don't flood basalts build volcanic cones?
11. Mount St. Helens
    a. last erupted violently in 1980
    b. is part of the Cascade Range
    c. is located in southern Washington
    d. all of the above
12. Volcanic eruptions can affect the climate because
    a. they heat the atmosphere
    b. volcanic dust and gas can reduce the amount of solar radiation that penetrates the atmosphere
    c. they change the elevation of the land
    d. all of the above
13. Whether volcanic eruptions are very explosive or relatively quiet is largely determined by
    a. the amount of gas in the lava or magma
    b. the ease or difficulty with which the gas escapes to the atmosphere
    c. the viscosity of a lava
    d. all of the above
14. Temperatures at which lavas solidify range from about ____°C for silicic rocks to ____°C for mafic rocks.
    a. 100; 200 b. 300; 1,000
    c. 700; 1,200 d. 1,000; 2,000
15. One gas typically not released during a volcanic eruption is
    a. water vapor b. carbon dioxide
    c. sulfur dioxide d. hydrogen sulfide
    e. oxygen
16. Mafic rocks contain about ____% silica.
    a. 10 b. 25
    c. 50 d. 65
    e. 80
17. Silicic rocks contain about ____% silica.
    a. 10 b. 25
    c. 50 d. 70
    e. 80
18. Which is not an extrusive igneous rock?
    a. granite b. rhyolite
    c. basalt d. andesite
19. Which is not a major type of volcano?
    a. shield b. cinder cone
    c. composite d. stratovolcano
    e. spatter cone
20. A typical example of a shield volcano is
    a. Mount St. Helens b. Kilauea in Hawaii
    c. El Chichón d. Mount Vesuvius
21. An example of a composite volcano is
    a. Mount St. Helens b. El Chichón
    c. Mount Vesuvius d. all of the above
22. Which volcano is not usually made of basalt?
    a. shield b. composite cone
    c. spatter cone d. cinder cone
23. An igneous rock made of pyroclasts has a texture called
    a. fragmental b. vesicular
    c. porphyritic d. fine-grained

Expanding Your Knowledge

1. What might explain the remarkable alignment of the Cascade volcanoes?
2. What would the present-day environmental effects be for an eruption such as that which created Crater Lake?
3. Why are there no active volcanoes in the eastern parts of the United States and Canada?
4. Why are continental igneous rocks richer in silica than oceanic igneous rocks?

Exploring Web Resources

www.mhhe.com/plummer9e

Visit the Online Learning Center for additional readings and article updates and for direct links to the sites listed below and other media resources.

http://volcano.und.nodak.edu/

Volcano World. This is an excellent site to learn about volcanoes. At the home page you may click on "Volcano of the Week" or "Volcano World Starting Points." "Starting Points" presents a menu that includes currently active volcanoes, volcano video clips, interviews with volcanologists, and more. You can also subscribe to being alerted to new eruptions by e-mail.

www.geo.mtu.edu/volcanoes/

Michigan Tech Volcanoes Page. The focus for this site is on scientific and educational information relative to volcanic hazard mitigation. Clicking on "volcanic humor" will show the lighter side of volcanology.

Animations

This chapter includes the following animations available on our Online Learning Center at www.mhhe.com/plummer9e.

Box 10.1, figure 1 Mount St. Helens Eruption

CHAPTER

11

Igneous Rocks, Intrusive Activity, and the Origin of Igneous Rocks

Chapters 10 and 11 are about igneous rocks and igneous processes. (Either chapter may be read first.) Chapter 10 focuses on volcanoes and igneous activity that takes place at the earth's surface. Chapter 11 describes igneous processes that take place underground. However, you will learn early in this chapter how volcanic as well as intrusive rocks are classified based on their grain size and mineral content.

We begin the chapter by introducing the rock cycle. This is a conceptual device that shows the interrelationship between igneous, sedimentary, and metamorphic rocks. We then begin focusing on igneous rocks. After the section on igneous rock classification, we describe structural relationships between bodies of intrusive rock and other rocks in the earth's crust. This is followed by a discussion of how magmas form and are altered. We conclude by discussing various hypotheses that relate igneous activity to plate tectonic theory.

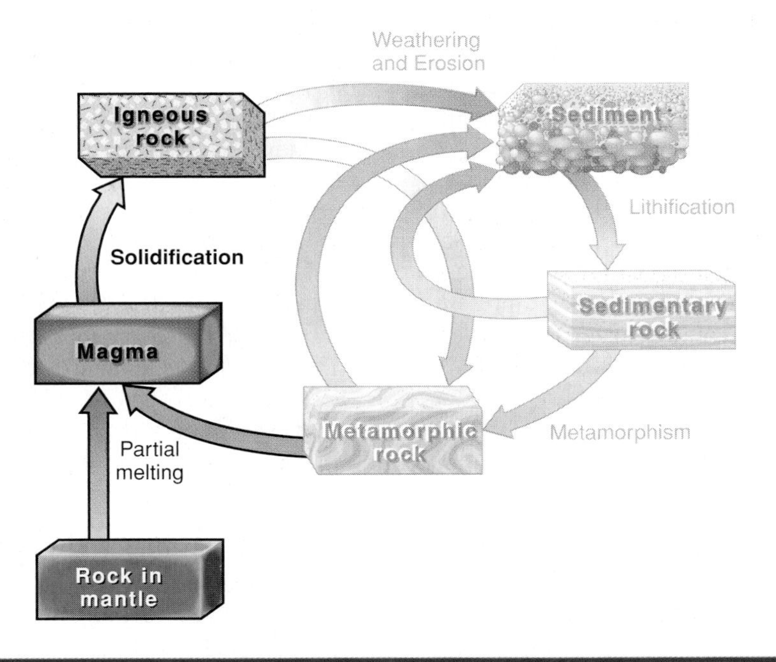

Opposite: A geologist investigating intrusive rocks in northern Victoria Land, Antarctica.
Photo by C. C. Plummer

THE ROCK CYCLE

A **rock** is naturally formed, consolidated material usually composed of grains of one or more minerals. You will see how some minerals break down chemically and form new minerals when a rock finds itself in a new physical setting. For instance, feldspars that may have formed at high temperatures deep within the Earth can react with surface waters to become clay minerals at the earth's surface.

As mentioned in chapter 1, the Earth changes because of its internal and external heat engines. If the Earth's internal engine had died (and tectonic forces had therefore stopped operating), the external engine plus gravity would long ago have leveled the continents, and the resulting sediment would have been deposited on the sea floor. Everything would be at rest. Nothing would be changing. That is to say, everything would be in *equilibrium* (and geology would be a dull subject). But this is not the case. The internal and external forces continue to interact, forcing substances out of equilibrium. Therefore, the Earth has a highly varied and ever-changing surface. Minerals and rocks change as well.

A useful aid in visualizing these relationships is the **rock cycle** shown in figure 11.1. The three major rock types—igneous, metamorphic, and sedimentary—are shown. As you see, each may form at the expense of another if it is forced out of equilibrium with its physical or climatic environment by either internal or surficial forces.

As described in chapter 1, *magma* is molten rock. *Igneous rocks* form when magma solidifies. If the magma is brought to the surface by a volcanic eruption, it may solidify into an *extrusive* igneous rock. Magma may also solidify very slowly beneath the surface. The resulting *intrusive* igneous rock may be exposed later after uplift and erosion remove the overlying rock (as shown in figure 1.2). The igneous rock, being out of equilibrium, may then undergo *weathering* and *erosion,* and the debris produced is transported and ultimately deposited (usually on a sea floor) as *sediment.* If the unconsolidated sediment becomes *lithified* (cemented or otherwise consolidated into a rock), it becomes a *sedimentary rock.* As the rock is buried by additional layers of sediment and sedimentary rock, heat and pressure increase. Tectonic forces may also increase the temperature and pressure. If the temperature and pressure become high enough, usually at depths greater than several kilometers below the surface, the original sedimentary rock is no longer in equilibrium and recrystallizes. The new rock that forms is called a *metamorphic rock.* If the temperature gets very high, the rock partially melts producing magma, completing the cycle.

The cycle can be repeated, as implied by the arrows in figure 11.1. However, there is no reason to expect all rocks to go through each step in the cycle. For instance, sedimentary rocks might be uplifted and exposed to weathering, creating new sediment.

We should emphasize that the rock cycle is a conceptual device to help students place the common rocks and how they form in perspective. As such it is a simplification and does not encompass all geologic processes. For instance, most magma comes from partial melting of the mantle. Note from the diagram that this magma does not come from recycled rocks so, strictly speaking, is not part of a "cycle."

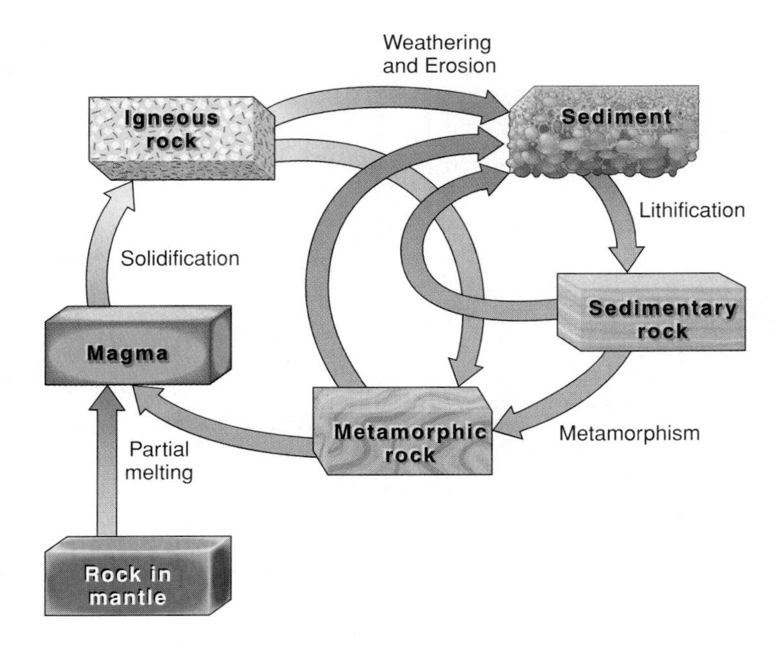

FIGURE 11.1
The rock cycle.

A Plate Tectonic Example

One way of relating the rock cycle to plate tectonics is illustrated by an example from what happens at a convergent plate boundary (figure 11.2). *Magma* is created in the zone of melting above the subduction zone. The magma, being less dense than adjacent rock, works its way upward. A volcanic eruption takes place if magma reaches the surface. The magma solidifies into *igneous rock.* The igneous rock is exposed to the atmosphere and subjected to *weathering* and *erosion.* The resulting *sediment* is transported and then deposited in low-lying areas. In time, the buried layers of sediment solidify into *sedimentary rock.* The sedimentary rock becomes increasingly more deeply buried as more sediment accumulates. After the sedimentary rock is buried to depths of several kilometers, the heat and pressure become too great and the rock recrystallizes into a *metamorphic rock.* As the depth of burial becomes even greater (several tens of kilometers), the metamorphic rock may find itself in a zone of melting. Here, temperatures are high enough to partially melt the metamorphic rock and magma is created, thus completing the cycle.

The rock cycle diagram reappears on the opening pages of chapters 10 through 12 as well as 14 and 15. The highlighted portion of the diagram will indicate where the material covered in each chapter fits into the rock cycle.

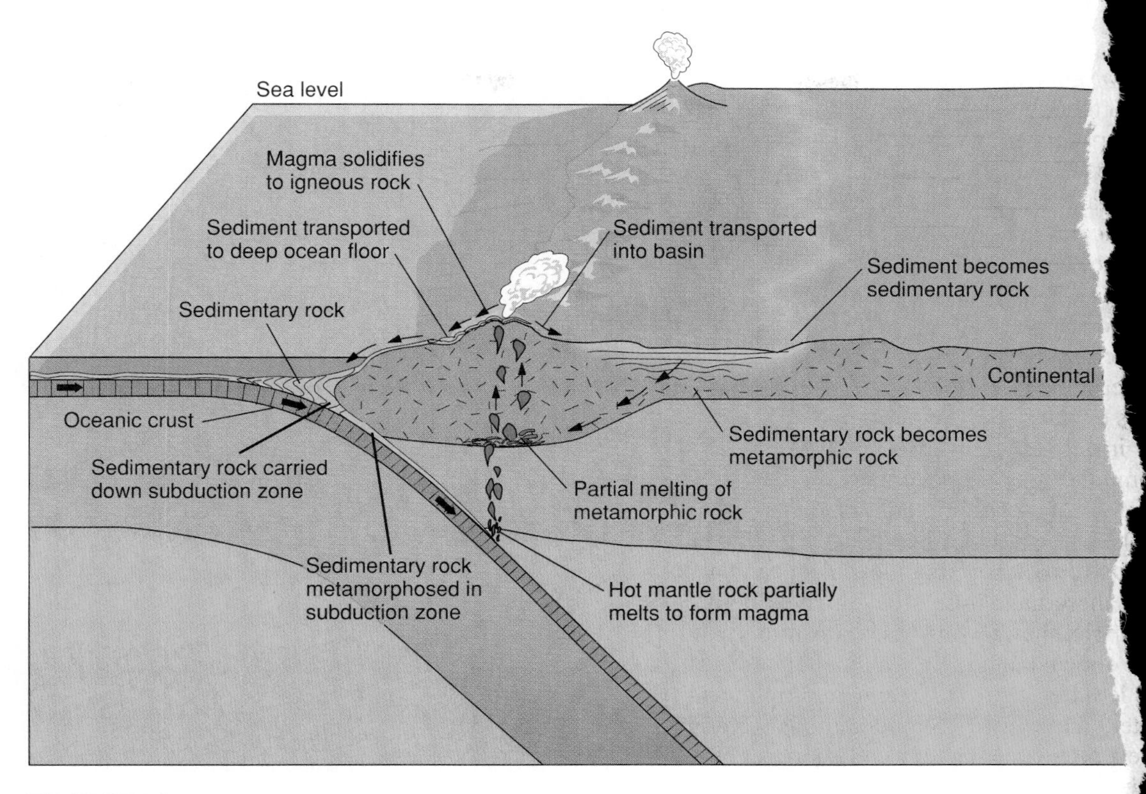

FIGURE 11.2

The rock cycle with respect to a convergent plate boundary. Magma solidifies as igneous rock at the volcano. Sediment fro volcano collects in the basin to the right of the diagram. Sediment converts to sedimentary rock as it is buried by more sedime buried sedimentary rocks are metamorphosed. The most deeply buried metamorphic rocks partially melt and the magma moves u An alternate way the rock cycle works is shown on the left of the diagram. Sediment from the continent (and volcano) becomes sedimentary rock, some of which is carried down the subduction zone. It is metamorphosed as it descends. It may contribute to the magma that forms in the mantle above the subduction zone.

IGNEOUS ROCKS

If you go to the island of Hawaii, you might observe red hot lava flowing over the land and, as it cools, solidifying into the fine-grained (the grains are less than 1 mm across), black rock we call basalt. Basalt is an **igneous rock,** rock that has solidified from magma. **Magma** is molten rock, usually rich in silica and containing dissolved gases. (**Lava** is magma on the Earth's surface.) Igneous rocks may be either **extrusive** if they form at the Earth's surface (e.g., basalt) or **intrusive** if magma solidifies underground. **Granite,** a coarse-grained (the grains are larger than 1 mm) rock composed predominantly of feldspar and quartz, is an intrusive rock. In fact, granite is the most abundant intrusive rock found in the continents.

Unlike the volcanic rock in Hawaii, nobody has ever seen magma solidify into intrusive rock. So what evidence suggests that bodies of granite (and other intrusive rocks) solidified underground from magma? (1) Mineralogically and chemically, intrusive rocks are essentially identical to volcanic rocks. (2) Volcanic rocks are fine-grained (or glass) due to their rapid solidification; intrusive rocks are generally coarse-grained, which is inferred to mean that the magma crystallized slowly underground. (3) Experiments have confirmed that most of the minerals in these rocks can form only at high temperatures. Other experiments indicate that some of the minerals could have formed only under high pressures, implying they were deeply buried. More evidence comes from examining *intrusive contacts,* such as shown in figures 11.3 and 11.4. (A **contact** is a surface separating different rock types. Other types of contacts are described elsewhere in this book.) (4) Preexisting solid rock, *country rock,* appears to have been forcibly broken by an intruding liquid, with the magma flowing into the fractures that developed. **Country rock,** incidentally, is an accepted term for any older rock into which an igneous body intruded. (5) Close examination of the country rock immediately adjacent to the intrusive rock usually indicates that it appears "baked" (*metamorphosed* is the correct term) close to the contact with the intrusive rock. (6) Rock types of the country rock often match **xenoliths,** fragments of rock that are distinct from the body of igneous rocks in which they are enclosed. (7) In the intrusive rock adjacent to contacts with country rock are **chill zones,** finer-grained rocks that indicate magma solidified more quickly here because of the rapid loss of heat to cooler rock.

, experiments have greatly increased
w igneous rocks form, geologists have
n the laboratory an artificial rock iden-
ry fine-grained rocks containing the
been made from artificial magmas, or
and pressure at which granite appar-
cated in the laboratory—but not the
o calculations, a large body of magma
ears to solidify completely. This very
e coarse-grained texture of most intru-
cesses involving silicates are known to
et another problem in trying to apply
s to real rocks is determining the role of
in the crystallization of rocks such as
mount of gases are retained in rock crys-
om a magma, but large amounts of gas
) are released during volcanic eruptions.
usive rock forming; hence we can only
these gases might have played before
mple shows why gases are important.
shown that granite can melt at temper-
if water is present and under high pres-
ter, the melting temperature is several
er. Not knowing how much water was
ization makes accurate determination of
and speculative.

ock Textures

ture refers to a rock's appearance with respect to the size, shape, and arrangement of its grains or other constituents. The most significant aspect of texture in igneous rocks is grain size. Extrusive rocks typically are **fine-grained rocks,** in which most of the grains are smaller than 1 millimeter. The grains are small because magma cools rapidly at the Earth's surface and larger crystals don't have time to form. Some intrusive rocks are also fine-grained; these occur as smaller bodies that apparently solidified near the surface upon intrusion into relatively cold country rock (probably within a couple kilometers of the Earth's surface). *Basalt, andesite,* and *rhyolite* are the common fine-grained igneous rocks. Igneous rocks that formed at considerable depth—usually more than several kilometers—are called **plutonic rocks** (after Pluto, the Roman god of the underworld). Characteristically, these rocks are coarse-grained, reflecting the slow cooling and solidification of magma. For our purposes, **coarse-grained rocks** are defined as those in which most of the grains are larger than 1 millimeter. The crystalline grains of plutonic rocks are commonly interlocked in a mosaic pattern (figure 11.5).

Some rocks are *porphyritic* and have larger crystals enclosed in a much finer-grained matrix (as described more fully in chapter 10). An analogy would be sugar cubes in a bowl of salt and pepper. If the matrix is fine-grained, extrusive rock names are used. For instance, figure 11.7*E* shows a *porphyritic andesite.* Porphyritic extrusive rocks are usually interpreted as having begun crystallizing slowly underground followed by eruption and rapid solidification of the remaining magma at the Earth's surface.

FIGURE 11.3

Granite (light-colored rock) solidified from magma that intruded dark-colored country rock. Torres del Paine, Chile. The dark-colored country rock is shale deposited in a marine environment. The spires are erosional remnants of rock that was once deep underground.

Photo by Kay Kepler

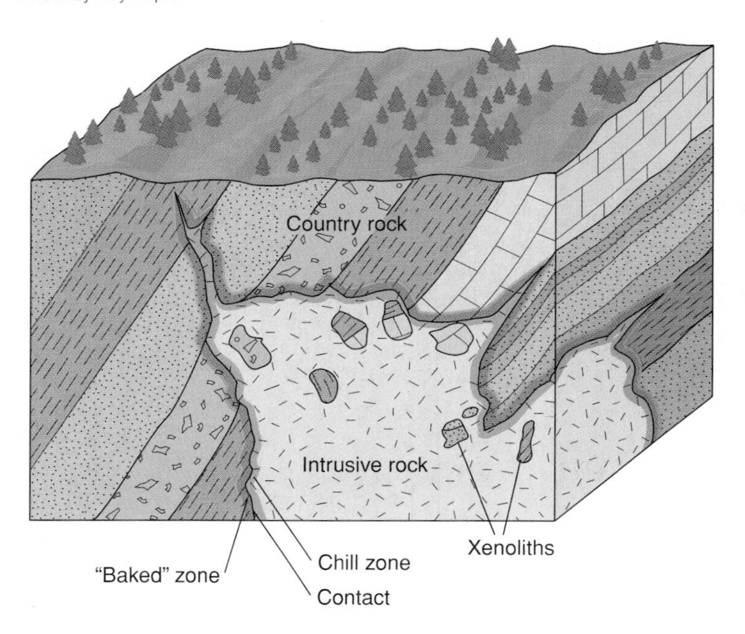

FIGURE 11.4

Igneous rock apparently intruded preexisting rock (country rock) as a liquid. (Xenoliths are usually much smaller than indicated.)

Identification of Igneous Rocks

Igneous rock names are based on texture (notably grain size) and mineralogical composition (which reflects chemical composition). Mineralogically (and chemically) equivalent rocks are *granite-rhyolite, diorite-andesite,* and *gabbro-basalt.* The relationships between igneous rocks are shown in figure 11.6.

Because of their larger mineral grains, plutonic rocks are easier to identify than extrusive rocks. The physical properties of each mineral in a plutonic rock can be determined more readily. And, of course, knowing what minerals are present makes rock identification a simpler task. For instance, **gabbro** is formed of coarse-grained ferromagnesian minerals and gray, plagioclase

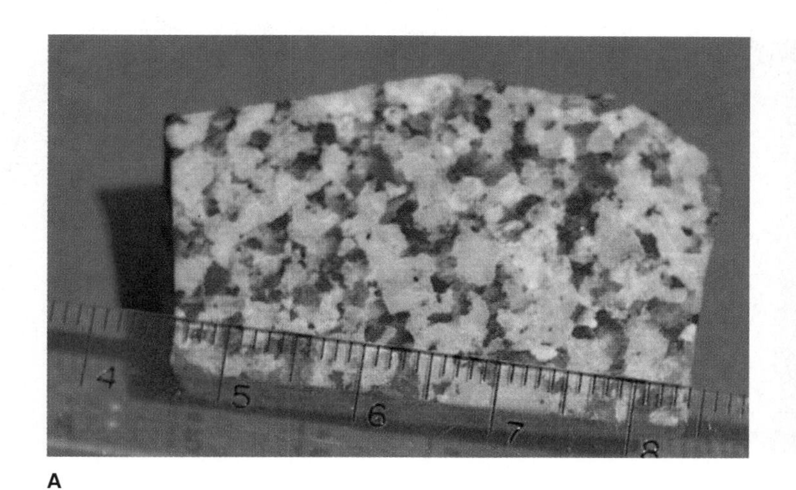

A

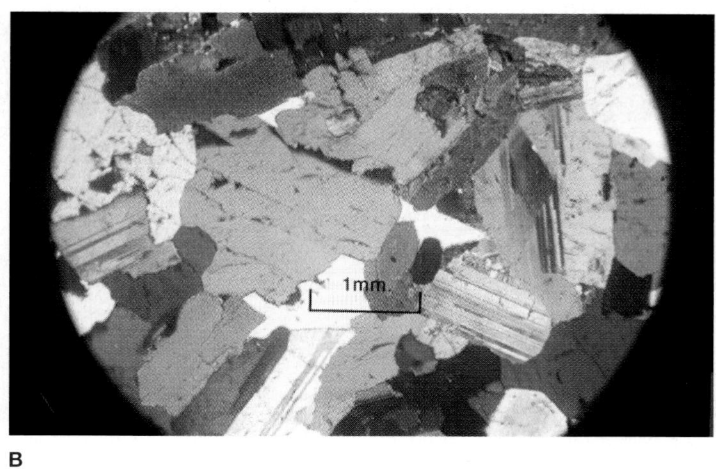

B

FIGURE 11.5

(*A*) Coarse-grained texture characteristic of plutonic rock. Plagioclase is white and quartz is gray; potassium feldspar has been stained yellow. Small gradations on ruler are millimeters. (*B*) A similar rock seen through a polarizing microscope. Note the interlocking crystal grains of individual minerals.

Photos by C. C. Plummer

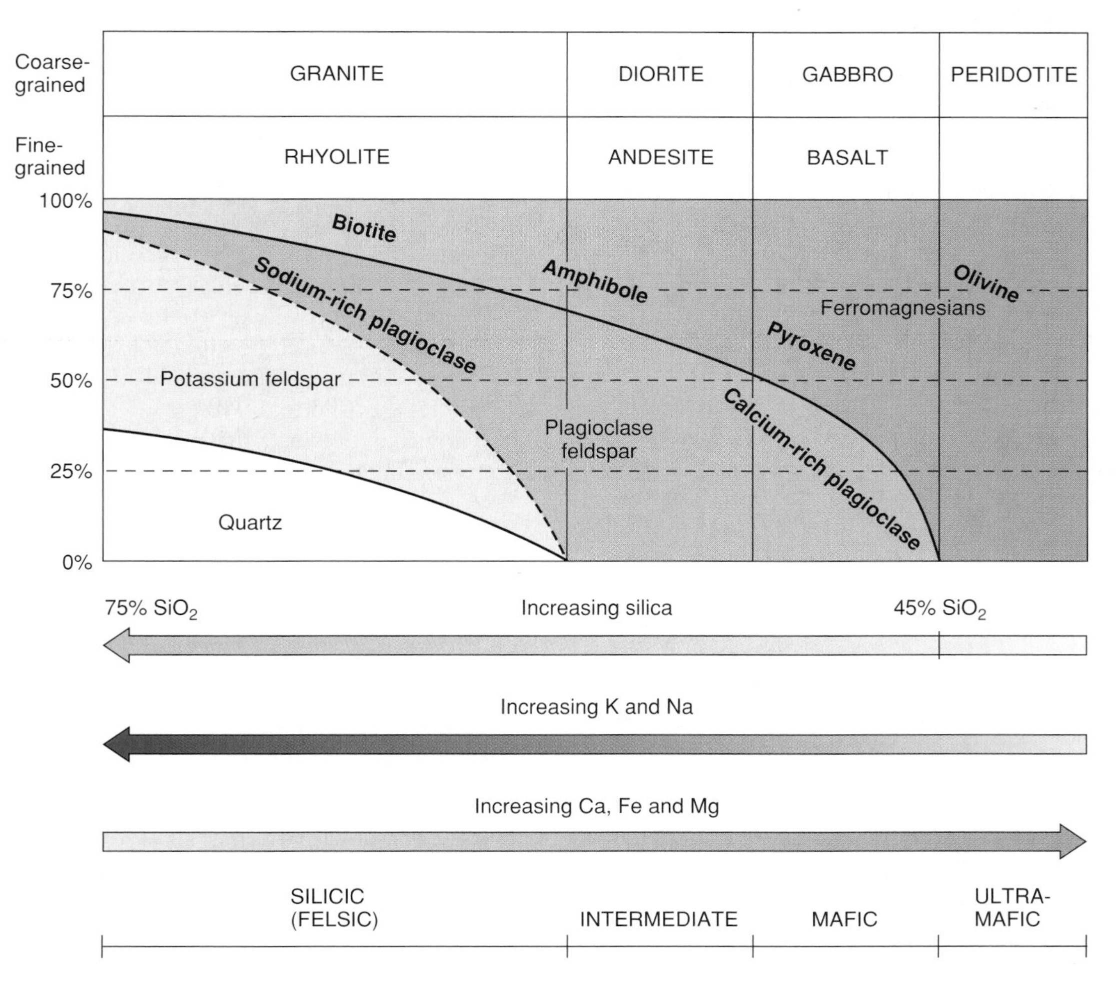

FIGURE 11.6

Classification chart for the most common igneous rocks. Rock names based on special textures are not shown. Sodium-rich plagioclase is associated with silicic rocks whereas calcium-rich plagioclase is associated with mafic rocks. The names of the particular ferromagnesian minerals indicate the approximate composition of the rocks in which they are most likely to be found.

table 11.1 Identification of Igneous Rocks

Coarse-Grained	Granite	Diorite	Gabbro	Peridotite
Fine-Grained	Rhyolite	Andesite	Basalt	—
Mineral Content	Quartz, feldspars (white, light gray, or pink). Minor ferromagnesian minerals.	Feldspars (white or gray) and about 35–50% ferromagnesian minerals. No quartz.	Predominance of ferromagnesian minerals. Rest of rock is plagioclase feldspar (medium to dark gray).	Entirely ferromagnesian minerals (olivine and pyroxene).
Color of Rock (most commonly)	Light-colored	Medium-gray or medium-green	Dark gray to black	Green to black

feldspar. (Recall from the mineral chapter that ferromagnesian minerals are silicates that contain iron and magnesium—amphibole, pyroxene, olivine, and biotite.) One can positively identify the feldspar on the basis of cleavage and, with practice, verify that no quartz is present. Gabbro's fine-grained counterpart is **basalt,** which is also composed of ferromagnesian minerals and plagioclase. The individual minerals cannot be identified by the naked eye, however, and one must use the less reliable attribute of color—basalt is usually dark gray to black.

As you can see from figure 11.6, granite and **rhyolite** are composed predominantly of feldspars (usually white or pink) and quartz. Granite, being coarse-grained, can be positively identified by verifying that quartz is present. Rhyolite is usually cream-colored, tan, or pink. Its light color indicates that ferromagnesian minerals are not abundant. **Diorite** and **andesite** are composed of feldspars and significant amounts of ferromagnesian minerals (30–50%). The minerals can be identified and their percentages estimated to indicate diorite. Andesite, being fine-grained, can usually be identified by its medium-gray or medium-green color. Its appearance is intermediate between light-colored rhyolite and dark basalt.

Use the chart in figure 11.6 along with table 11.1 to identify common igneous rocks. You may also find it helpful to turn to appendix B, which includes a key for identifying common igneous rocks. (Photos of typical igneous rocks are shown in figure 11.7.)

Varieties of Granite

Granite and rhyolite occupy a larger area in the classification chart than do the other rocks. This reflects their greater variation in composition. For instance, a granite whose composition corresponds to the right side of the field in figure 11.6 (that is, nearer to diorite/andesite) contains much more plagioclase than potassium feldspar and somewhat more ferromagnesian minerals than does a rock whose composition plots in the left side of the granite field. Geologists have arbitrarily subdivided the field of granite and named each of the varieties; a rock in the right portion of the field, for example, is called granodiorite.

Any classification system is, of course, a human device, and for this reason, classification systems differ somewhat among groups of geologists. We define the boundary between granite and diorite by the presence or absence of quartz; but we could just as easily have placed the boundary slightly to the left, so that a rock with 10% or less quartz would be diorite.

Chemistry of Igneous Rocks

The chemical composition of the magma determines which minerals and how much of each will crystallize when an igneous rock forms. For instance, the presence of quartz in a rock indicates that the magma was enriched in silica (SiO_2). The lower part of figure 11.6 shows the relationship of chemical composition to rock type. Chemical analyses of rocks are reported as weight percentages of oxides (e.g., SiO_2, MgO, Na_2O, etc.) rather than as separate elements (e.g., Si, O, Mg, Na). For virtually all igneous rocks, SiO_2 (silica) is the most abundant component. The amount of SiO_2 varies from about 45% to 75% of the total weight of common volcanic rocks. The variations between these extremes account for striking differences in the appearance and mineral content of the rocks.

Mafic Rocks

Rocks with a silica content close to 50% (by weight) are considered *silica-deficient,* even though SiO_2 is, by far, the most abundant constituent. Chemical analyses show that the remainder is composed mostly of the oxides of aluminum (Al_2O_3), calcium (CaO), magnesium (MgO), and iron (FeO and Fe_2O_3). (These oxides generally combine with SiO_2 to form the silicate minerals as described in chapter 9.) Rocks in this group are called **mafic**—silica-deficient igneous rocks with a relatively high content of magnesium, iron, and calcium. (The term *mafic* comes from magnesium and ferric.) Basalt and gabbro are, of course, mafic rocks.

Silicic (Felsic) Rocks

At the other extreme, the *silica-rich* (65% or more of SiO_2) rocks tend to have only very small amounts of the oxides of calcium, magnesium, and iron. The remaining 25% to 35% of these rocks is mostly aluminum oxide (Al_2O_3) and oxides of sodium (Na_2O) and potassium (K_2O). These are called **silicic**

Granite

Diorite

Gabbro

FIGURE 11.7

Common igneous rocks.

Photo *diorite/gabbro* by Larry Davis

Rhyolite

Andesite (porphyritic)

Basalt

11.1

IN GREATER DEPTH

PEGMATITE—A ROCK MADE OF GIANT CRYSTALS

Pegmatites are extremely coarse-grained igneous rocks. In some pegmatites, crystals are as large as 10 meters across. Strictly speaking, a pegmatite can be of diorite, gabbro, or granite; however, the vast majority of pegmatites are silicic, with very large crystals of potassium feldspar, sodium-rich plagioclase feldspars, and quartz. Hence, the term *pegmatite* generally refers to a rock of granitic composition (if otherwise, a term such as *gabbroic pegmatite* is used). Pegmatites are interesting as geological phenomena and important as minable resources.

The extremely coarse texture of pegmatites is attributed to both slow cooling and the low viscosity of the fluid from which they form. Lava solidifying to rhyolite is very viscous. Magma solidifying to granite, being chemically similar, should be viscous.

Pegmatites, however, probably crystallize from a fluid composed largely of water under high pressure. Water molecules and ions from the parent, granitic magma make up a residual magma. Geologists believe the following sequence of events accounts for most pegmatites.

As a granite pluton cools, increasingly more of the magma solidifies into the minerals of a granite. By the time the pluton is well over 90% solid, the residual magma contains a very high amount of silica and ions of elements that will crystallize into potassium and sodium feldspars. Also present are elements that could not be accommodated into the crystal structures of the common minerals that formed during the normal solidification phase of the pluton. Fluids, notably water, that were in the original magma are left over as well. If no fracture above the pluton permits the fluids to escape, they are sealed in, as in a pressure cooker. The watery residual magma has a low viscosity, which allows appropriate atoms to migrate easily toward growing crystals. The crystals add more and more atoms and grow very large.

Pegmatite bodies are generally quite small. Many are podlike structures, located either within the upper portion of a granite pluton or within the overlying country rock near the contact with granite, the fluid body evidently having squeezed into the country rock before solidifying. Pegmatite dikes are fairly common, especially within granite plutons, where they apparently filled cracks that developed in the already solid granite. Some pegmatites form small dikes along contacts between granite and country rock, filling cracks that developed as the cooling granite pluton contracted.

Most pegmatites contain only quartz, feldspar, and perhaps mica. Minerals of considerable commercial value are found in a few pegmatites. Large crystals of muscovite mica are mined from pegmatites. These crystals are called "books" because the cleavage flakes (tens of centimeters across) look like pages. Because muscovite is an excellent insulator, the cleavage sheets are used in electrical devices, such as toasters, to separate uninsulated electrical wires. Even the large feldspar crystals in pegmatites are mined for various industrial uses, notably the manufacture of ceramics.

BOX 11.1 — FIGURE 1

Pegmatite in northern Victoria Land, Antarctica. The knife is 8 centimeters long. The black crystals are tourmaline. Quartz and feldspar are light colored. Photo by C. C. Plummer

Many rare elements are mined from pegmatites. These elements were not absorbed by the minerals of the main pluton and so were concentrated in the residual pegmatitic magma, where they crystallized as constituents of unusual minerals. Minerals containing the element lithium are mined from pegmatites. Lithium becomes part of a sheet silicate structure to form a pink or purple variety of mica (called lepidolite). Uranium ores, similarly concentrated in the residual melt of magmas, are also extracted from pegmatites.

Some pegmatites are mined for gemstones. Emerald and aquamarine, varieties of the mineral beryl, occur in pegmatites that crystallized from a solution containing the element beryllium. A large number of the world's very rare minerals are found only in pegmatites, many of these in only one known pegmatite body. These rare minerals are mainly of interest to collectors and museums.

Hydrothermal veins (described in chapter 15) are closely related to pegmatites. Veins of quartz are common in country rock near granite. Many of these are believed to be caused by water that escapes from the magma. Silica dissolved in the very hot water cakes on the walls of cracks as the water cools while traveling surfaceward. Sometimes valuable metals such as gold, silver, lead, zinc, and copper are deposited with the quartz in veins.

or **felsic rocks**—silica-rich igneous rocks with a relatively high content of potassium and sodium (the *fel* part of the name comes from *feldspar,* which crystallizes from the potassium, sodium, aluminum, and silicon oxides; *si* in *felsic* is for silica). The silicic rocks rhyolite and granite are light-colored because of the low amount of ferromagnesian minerals.

Intermediate Rocks

Rocks with a chemical content between that of felsic and mafic are classified as **intermediate rocks.** *Andesite,* which is usually green or medium gray, is the most common intermediate volcanic rock.

Ultramafic Rocks

An **ultramafic rock** is composed entirely or almost entirely of ferromagnesian minerals. No feldspars are present and, of course, no quartz. **Peridotite,** a coarse-grained rock composed of pyroxene and olivine, is the most abundant ultramafic rock. Chemically, these rocks contain less than 45% silica.

Note that the chart (figure 11.6) does not include a fine-grained counterpart. This is because ultramafic extrusive rocks are mostly restricted to the very early history of the Earth and are quite rare. For our purposes they can be ignored.

Some ultramafic rocks form from differentiation (explained later in this chapter) of a basaltic magma at very high temperatures. Most ultramafic rocks come from the mantle, rather than from the Earth's crust (see box 2.2).

Where we find large bodies of ultramafic rocks, the usual interpretation is that a part of the mantle has traveled upward as solid rock.

INTRUSIVE BODIES

Intrusions, or **intrusive structures,** are bodies of intrusive rock whose names are based on their size and shape as well as their relationship to surrounding rocks. They are important aspects of the architecture, or *structure,* of the Earth's crust. The various intrusions are named and classified on the basis of the following considerations: (1) Is the body large or small? (2) Does it have a particular geometric shape? (3) Did the rock form at a considerable depth or was it a shallow intrusion? (4) Does it follow layering in the country rock or not?

Shallow Intrusive Structures

Some igneous bodies apparently solidified near the surface of the Earth (probably at depths of less than 2 kilometers). These bodies appear to have solidified in the subsurface "plumbing systems" of volcanoes or lava flows. Shallow intrusive structures tend to be relatively small compared with those that formed at considerable depth. Because the country rock near the Earth's surface generally is cool, intruded magma tends to chill and solidify relatively rapidly. Also, smaller magma bodies will cool faster than larger bodies, regardless of depth. For both these reasons, shallow intrusive bodies are likely to be fine-grained.

A **volcanic neck** is an intrusive structure apparently formed from magma that solidified within the throat of a volcano. One of the best examples is Ship Rock in New Mexico (figure 11.8). Here is how geologists interpret the history of this feature. A

A

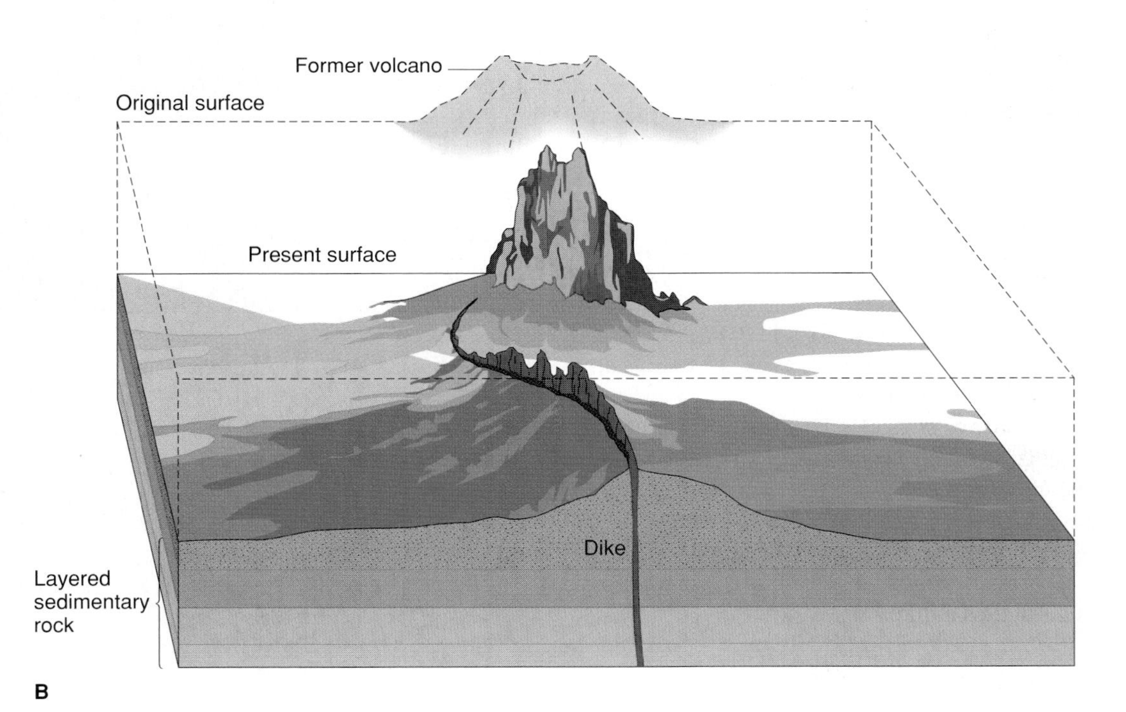

B

FIGURE 11.8

(*A*) Ship Rock in New Mexico, which rises 420 meters (1,400 feet) above the desert floor. (*B*) Relationship to the former volcano.
Photo by Frank M. Hanna

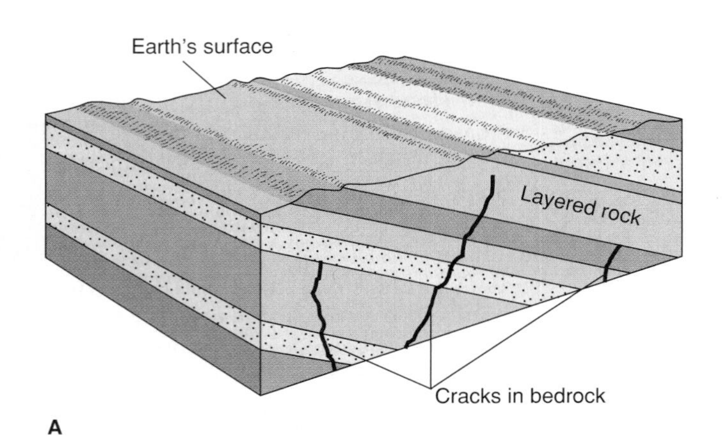

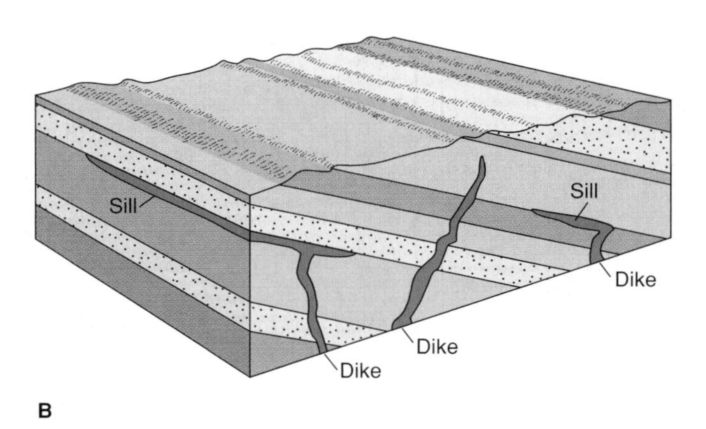

FIGURE 11.9
(*A*) Cracks or planes of weakness before intrusion of magma. (*B*) The concordant intrusion where magma has intruded between sedimentary layers is a sill; the discordant intrusions are dikes.

volcano formed above what is now Ship Rock. The magma for the volcano moved upward through a more or less cylindrical "plumbing system." Eruptions ceased and the magma underground solidified into what is now Ship Rock. In time, the volcano and its underlying rock—the country rock around Ship Rock—eroded away. The more resistant igneous body eroded more slowly into its present shape. Weathering and erosion are continuing (falling rock has been a serious hazard to rock climbers).

Dikes and Sills

Another, and far more common, intrusive structure can also be seen at Ship Rock. The low, wall-like ridge extending outward from Ship Rock is an eroded dike. A **dike** is a tabular (shaped like a tabletop), discordant, intrusive structure (figure 11.9). *Discordant* means that the body is not parallel to any layering in the country rock. (Think of a dike as cutting across layers of country rock.) Dikes may form at shallow depths and be fine-grained, such as those at Ship Rock, or form at greater depths and be coarser-grained. Dikes need not appear as walls protruding from the ground (figure 11.10). The ones at Ship Rock do so only because they are more resistant to weathering and erosion than the country rock.

A **sill** is also a tabular intrusive structure, but it is *concordant.* That is, sills, unlike dikes, are parallel to any planes or layering in the country rock (figures 11.9 and 11.11). Typically, the country rock bounding a sill is layered sedimentary rock. As magma squeezes into a crack between two layers, it solidifies into a sill.

If the country rock is not layered, a tabular intrusion is regarded as a dike.

FIGURE 11.10
Dikes (light-colored rocks) in northern Victoria Land, Antarctica.
Photo by C. C. Plummer

Intrusives That Crystallize at Depth

A **pluton** is a body of magma or igneous rock that crystallized at considerable depth within the crust. Most plutons have no particular shape, unlike dikes and sills. Where plutons are exposed at the Earth's surface, they are arbitrarily distinguished by size. A **stock** is a small discordant pluton with an outcrop area (i.e., the area over which it is exposed to the atmosphere) of less than 100 square kilometers. If the outcrop area is greater than 100 square kilometers, the body is called a **batholith** (figure 11.12), a large discordant pluton.

Most batholiths crop out over areas vastly greater than the minimum 100 square kilometers.

FIGURE 11.11
A sill (dark layer) in sedimentary layered rock, Glacier National Park, Montana.
Photo © William E. Ferguson

Although batholiths often contain mafic and intermediate rocks, they almost always are predominantly composed of granite. Detailed studies of batholiths indicate that they are formed of numerous, coalesced plutons. Apparently, large blobs of magma worked their way upward through the lower crust and collected 5 to 30 kilometers below the surface where they solidified (figure 11.13). These blobs of magma, known as **diapirs,** are less dense than the surrounding rock that is shouldered aside as the magma rises. Batholiths occupy large portions of North America, particularly in the west. Over half of California's Sierra Nevada mountains (figure 11.14) is a batholith whose individual plutons were emplaced during a period of over 100 million years. An even larger batholith extends almost the entire length of the mountain ranges of Canada's west coast and southeastern Alaska—a distance of 1,800 kilometers. Smaller batholiths are also found in the Appalachian Mountains in eastern North America. (The extent and location of North American batholiths are shown on the geologic map on the inside cover.)

Granite is considerably more common than rhyolite, its volcanic counterpart. Why is this? Silicic magma is much more *viscous* (that is, more resistant to flow) than mafic magma. Therefore, a silicic magma body will travel upward through the crust more slowly and with more difficulty than mafic or intermediate magma. Unless it is exceptionally hot, a silicic magma will not be able to work its way through the relatively cool and rigid rocks of the upper few kilometers of crust. Instead, it is much more likely to solidify slowly into a pluton.

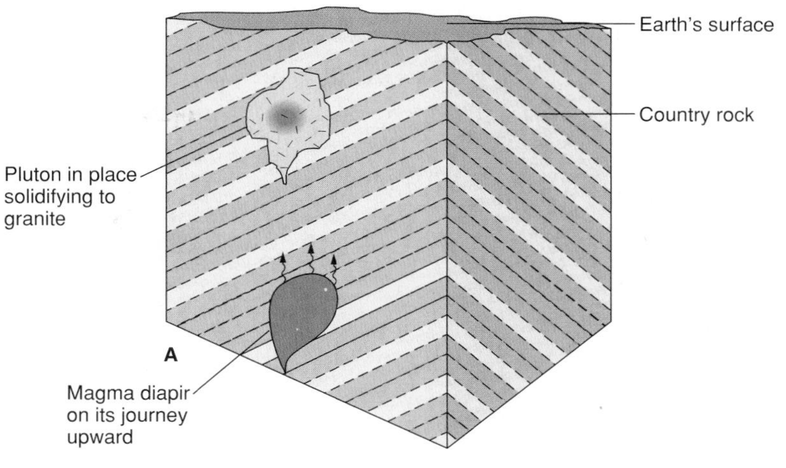

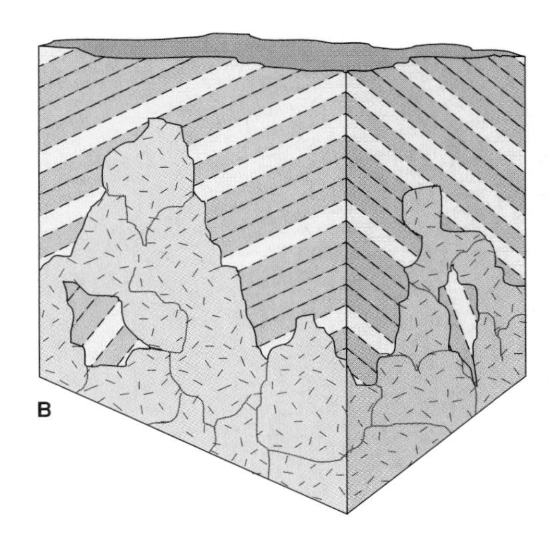

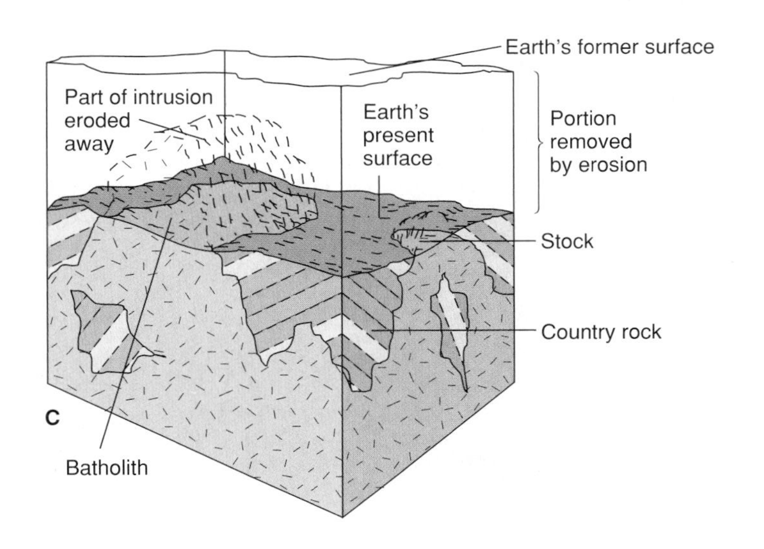

FIGURE 11.12
(*A*) The first of numerous magma diapirs has worked its way upward and is emplaced in the country rock. (*B*) Other magma diapirs have intruded, coalesced, and solidified into a solid mass of plutonic rock. (*C*) After erosion, surface exposures of plutonic rock are a batholith and a stock.

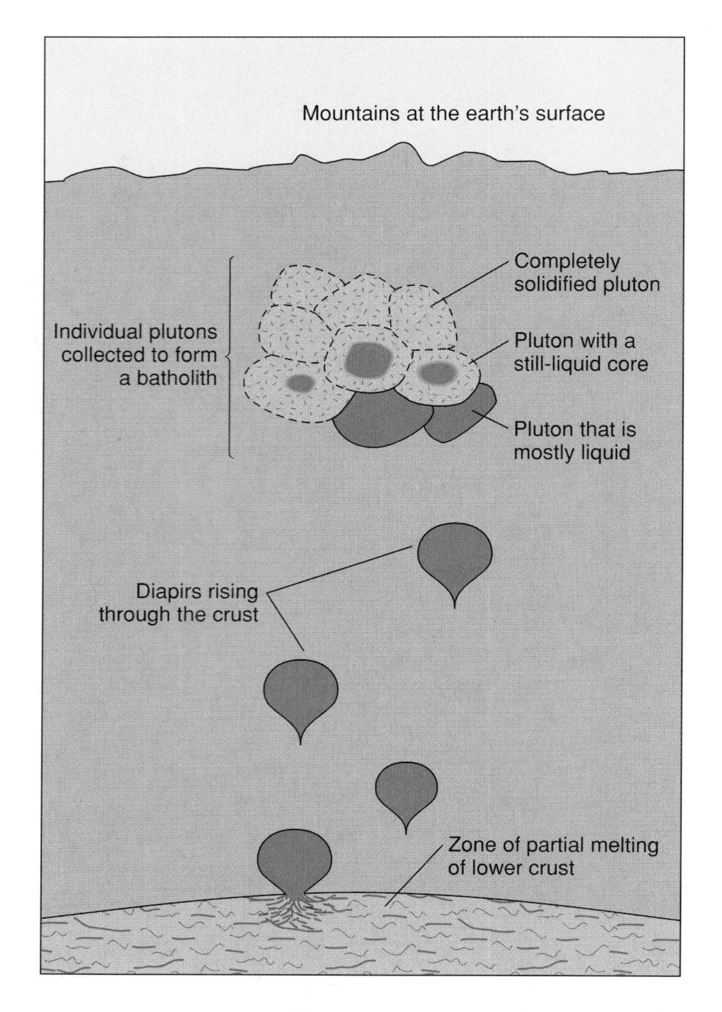

FIGURE 11.13

Diapirs of magma travel upward from the lower crust and solidify in the upper crust.

ABUNDANCE AND DISTRIBUTION OF PLUTONIC ROCKS

Granite is the most abundant igneous rock in mountain ranges. It is also the most commonly found igneous rock in the interior lowlands of continents. Throughout the lowlands of much of Canada, very old plutons have intruded even older metamorphic rock. As explained in the chapter on mountains and the continental crust, very old mountain ranges have, over time, eroded and become the stable interior of a continent. Metamorphic and plutonic rocks similar in age and complexity to those in Canada are found in the Great Plains of the United States. Here, however, they are mostly covered by a veneer (a kilometer or so) of younger, sedimentary rock. These "basement" rocks are exposed to us in only a few places. In Grand Canyon, Arizona, the Colorado River has eroded through the layers of sedimentary rock to expose the ancient plutonic and metamorphic basement. In the Black Hills of South Dakota, local uplift and subsequent erosion has exposed similar rocks.

Granite, then, is the predominant igneous rock of the continents. As described in chapter 10, basalt and gabbro are the predominant rocks underlying the oceans. Andesite (usually along continental margins) is the building material of most young volcanic mountains. Underneath the crust, ultramafic rocks make up the upper mantle.

FIGURE 11.14

Part of the Sierra Nevada batholith. All light-colored rock shown here (including that under the distant snow-covered mountains) is granite. The extent of the Sierra Nevada batholith is shown in the inset.

Inset from geology.wr.usgs.gov.

HOW MAGMA FORMS

If a rock is heated sufficiently, it begins melting to form magma. Under ideal conditions, rock can melt and yield a granitic magma at temperatures as low as 625°C. Temperatures over 1,000°C are required to create basaltic magma; however, there are several factors that control the melting temperature of rock. Pressure, amount of gas (particularly water) present, and the particular mix of minerals all influence when melting takes place. These factors are discussed later in this chapter.

Heat for Melting Rock

Most of the heat that contributes to the generation of magma comes from the very hot Earth's core (where temperatures are estimated to be greater than 5,000°C). Heat is conducted toward the Earth's surface through the mantle and crust. This is comparable to the way heat is conducted through the wall from a hot room into a cooler room or through the metal of a frying pan. Heat is also brought from the lower mantle when part of the mantle flows upward, either through convection (described in chapters 1 and 4) or by hot mantle plumes. The geothermal gradient, described below, is a manifestation of heat transfer in the mantle.

Geothermal Gradient

A miner descending a mine shaft notices a rise in temperature. This is due to the **geothermal gradient,** the rate at which temperature increases with increasing depth beneath the surface. Data show the geothermal gradient, on the average, to be about 3°C for each 100 meters (30°C/km) of depth in the upper part of the crust. The geothermal gradient is not the same everywhere. Figure 11.15 shows geothermal gradients for two regions. The curve for the volcanic region indicates a higher geothermal gradient than that for the continental interior. Temperatures high enough to melt rock would be expected at a relatively shallow depth beneath the volcanic region. You would have to go deeper in the continental interior to reach the same temperature; however, the rock there does not melt because of the increased pressure at that depth.

One reason for a higher geothermal gradient is that deeper, and therefore hotter, mantle rock has worked its way upward closer to the Earth's surface due either to mantle convection or mantle plumes. "Hot spots" in the crust (where the geothermal gradient is locally very high) are believed to be caused by hot **mantle plumes,** which are narrow upwellings of hot material within the mantle. Hot mantle plumes help account for some igneous activity, such as the oceanic eruptions that built up the Hawaiian Islands. Volcanism in the middle of continents may also be attributable to mantle plumes. Yellowstone National Park, in Wyoming, is a product of silicic eruptions. The eruptions were much larger and more violent than any of historical time. Geologists attribute these eruptions to a hot mantle plume that caused melting of the crust beneath this area.

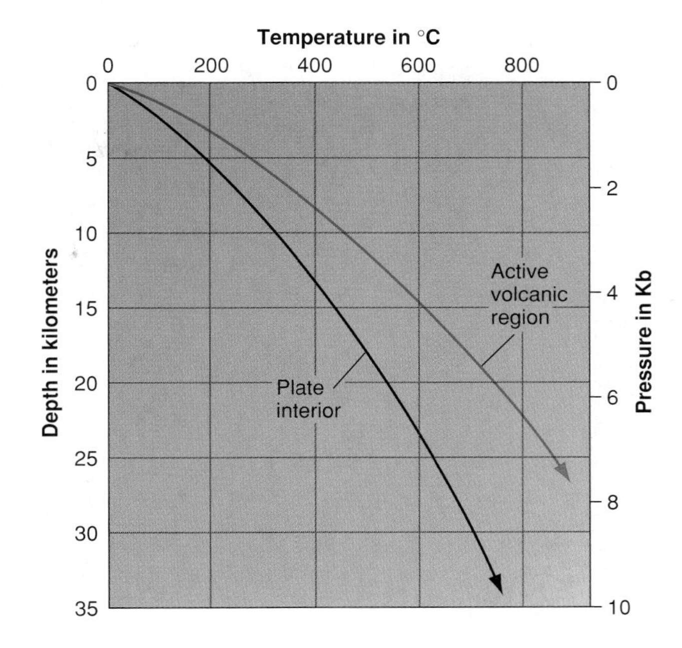

FIGURE 11.15

Geothermal gradients at two parts of the earth's crust.

Factors That Control Melting Temperatures

Pressure

The melting point of a mineral generally *increases* with increasing pressure. Pressure increases with depth in the Earth's crust, just as temperature does. So a rock that melts at a given temperature at the surface of the Earth requires a higher temperature to melt deep underground. Rock will not melt where the geothermal gradient for the plate interior is applicable, because the melting temperature is always going to be higher than the temperature of rock at any given depth.

Hawaiian volcanic activity attributable to the underlying mantle plume illustrates how *reduced* pressure contributes to the creation of magma. Solid rock that was once very deep in the mantle (and, therefore, very hot) has worked its way upward. Most of its heat has been retained during the upward journey. However, the pressure decreases as the rock body travels upward. As it approaches 50 kilometers or so from the earth's surface, pressure is sufficiently reduced so that melting takes place.

Water under Pressure

If enough gas, especially water vapor, is present and under high pressure, a dramatic change occurs in the melting process. Water sealed in under high pressure helps break the silicon-oxygen bonds causing the crystal to liquefy. A mineral's melting temperature is significantly lowered by water under high pressure (figure 11.16).

Experiments have shown that, under moderately high pressure, water mixed with granite lowers the melting point of

11.2 ENVIRONMENTAL GEOLOGY

HARNESSING MAGMATIC ENERGY

Buried magma chambers indirectly contribute the heat for today's geothermal electric generating plants. As explained in the chapter on ground water, water becomes heated in hot rocks. The heat source is usually presumed to be an underlying magma chamber. The rocks containing the hot water are penetrated by drilling. Steam exiting the hole is used to generate electricity.

Why not drill into and tap magma itself for energy? The amount of energy stored in a body of magma is enormous. The U.S. Geological Survey estimates that magma chambers in the United States within 10 kilometers of Earth's surface contain about 5,000 times as much energy as the country consumes each year. Our energy problems could largely be solved if significant amounts of this energy were harnessed.

There are some formidable technical difficulties in drilling into a magma chamber and converting the heat into useful energy. Despite these difficulties, the United States has considered developing magmatic energy. Experimental drilling has been carried out in Hawaii through the basalt crust of a lava lake that formed in 1960.

As drill bits approach a magma chamber, they must penetrate increasingly hotter rock. The drill bit must be made of special alloys to prevent it from becoming too soft to cut rock. The rock immediately adjacent to a basaltic magma chamber is around 1,000°C, even though that rock is solid. Drilling into the magma would require a special technique. One device currently being experimented with is a jet-augmented drill. As the drill enters the magma chamber, it simultaneously cools and solidifies the magma in front of the drill bit. Thus the drill bit creates a column of rock that extends downward into the magma chamber and simultaneously bores a hole down the center of this column. Once the hollow column is deep enough within the magma chamber, a boiler is placed in the hole. The boiler is protected from the magma by the jacket of the column of rock. Water would be pumped down the hole and turned to water vapor in the boiler by heat from the magma. Steam emerging from the hole would be used to generate electricity.

In principle, the idea is fairly simple, but there are serious technical problems. For one thing, high pressures would have to be maintained on the drill bit during drilling and while the boiler system was being installed; otherwise, gases within the magma might blast the magma out of the drill hole and create a manmade volcano. (The closest thing to a human-made volcano occurred in Iceland when a small amount of magma broke into a geothermal steam well and erupted briefly at the well head, showering the area with a few tons of volcanic debris.)

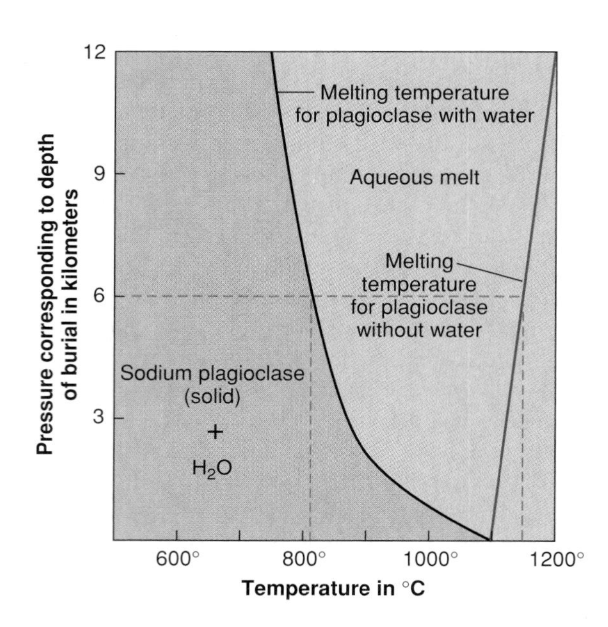

FIGURE 11.16

Melting temperature of a mineral with and without water present. The curve on the left is for melting of plagioclase saturated with water under the pressure corresponding to depth of burial. The line on the right corresponds to melting of dry plagioclase. The dashed, red line indicates that at pressures corresponding to 6 kilometers plagioclase with water melts at just over 800°C, whereas dry plagioclase melts at around 1150°C.
After Tuttle and Brown, Geologic Society of America, 1958

granite from over 900°C (when dry) to as low as 650°C when saturated with water under pressure equivalent to that of 10,000 atmospheres or *bars.* (Pressure at depth is usually expressed in *kilobars;* one kilobar is equal to 1,000 bars.) Ten kilobars corresponds to a depth of approximately 35 kilometers.

Effect of Mixed Minerals

Two metals—as in solder—can be mixed in a ratio that lowers their melting temperature far below that of the melting points of the pure metals. Minerals behave similarly. Experiments have shown that in some cases mixed fragments of two minerals melt at a lower temperature than either mineral alone. Figure 11.17 shows the melting temperatures for quartz and potassium feldspar mixed in various proportions. If the mixture is 42% quartz (58% potassium feldspar), melting takes place at just above 1,000°C. On the other hand, 1,500°C is needed to liquefy a mixture of 75% quartz and 25% potassium feldspar (corresponding to the right edge of the diagram). Pure quartz requires even higher temperatures to melt.

HOW MAGMAS OF DIFFERENT COMPOSITIONS EVOLVE

A major topic of investigation for geologists is why igneous rocks are so varied in composition. On a global scale magma composition is clearly controlled by geologic setting. But why?

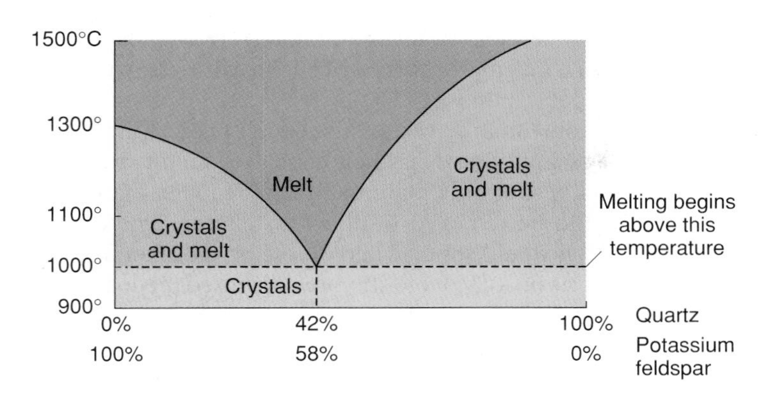

FIGURE 11.17
Melting temperatures for mixtures of quartz and potassium feldspar at atmospheric pressure.
Modified from Schairer and Bowen, 1956. V 254, p. 16. *American Journal of Science*

Why are basaltic magmas associated with oceanic crust whereas granitic magmas are common in the continental crust? On a local scale igneous bodies often show considerable variation in rock type. For instance, individual plutons typically display a considerable range of compositions, mostly varieties of granite, but many also will contain minor amounts of gabbro or diorite. In this section, we describe processes that result in differences in composition of magmas. The following section relates these processes to plate tectonics for the larger view of igneous activity.

Sequence of Crystallization and Melting

Early in the 20th century, N. L. Bowen conducted a series of experiments that determined the sequence in which minerals crystallize in a cooling magma. The sequence became known as **Bowen's reaction series** and is shown in figure 11.18. A simplified explanation of the series and its importance to igneous rocks is presented below. For a more in-depth presentation, go to our website at www.mhhe.com/plummer9e.

Bowen's experiments showed that in a cooling magma, certain minerals are stable at higher melting temperatures and crystallize before those stable at lower temperatures. Looking at the *discontinuous branch,* which contains only ferromagnesian minerals, we can see that olivine crystallizes before pyroxene and pyroxene crystallizes before amphibole. A complication is that early formed crystals *react* with the remaining melt and recrystallize as cooling proceeds. For instance, early formed olivine crystals react with the melt and recrystallize to pyroxene when pyroxene's temperature of crystallization is reached. Upon further cooling, pyroxene continues to crystallize until all of the melt is used up or the melting temperature of amphibole is reached. At this point, pyroxene reacts with the remaining melt and amphibole forms at its expense. If all of the melt is used up before all of the pyroxene recrystallizes to amphibole, the solid rock would contain pyroxene and amphibole.

Crystallization in the discontinuous and the *continuous branch* takes place at the same time. The continuous branch contains only plagioclase feldspar. Plagioclase is a *solid-solution* mineral (discussed in the chapter on minerals) in which either sodium or calcium atoms can be accommodated in its crystal structure, along with aluminum, silicon, and oxygen. The composition of plagioclase changes as magma is cooled and earlier formed crystals react with the melt. The first plagioclase crystal to form as a hot melt cools contain calcium but little or no sodium. As cooling continues, the early formed crystals grow and incorporate progressively more sodium into their crystal structures.

Any magma left after the crystallization is completed along the two branches is richer in silicon than the original magma and also contains abundant potassium and aluminum. The potassium and aluminum combine with silicon to form *potassium feldspar.* (If the water pressure is high, *muscovite* may also form at this stage.) Excess SiO_2 crystallizes as *quartz.*

From Bowen's reaction series we can derive several important concepts that are necessary to understand igneous rocks and processes:

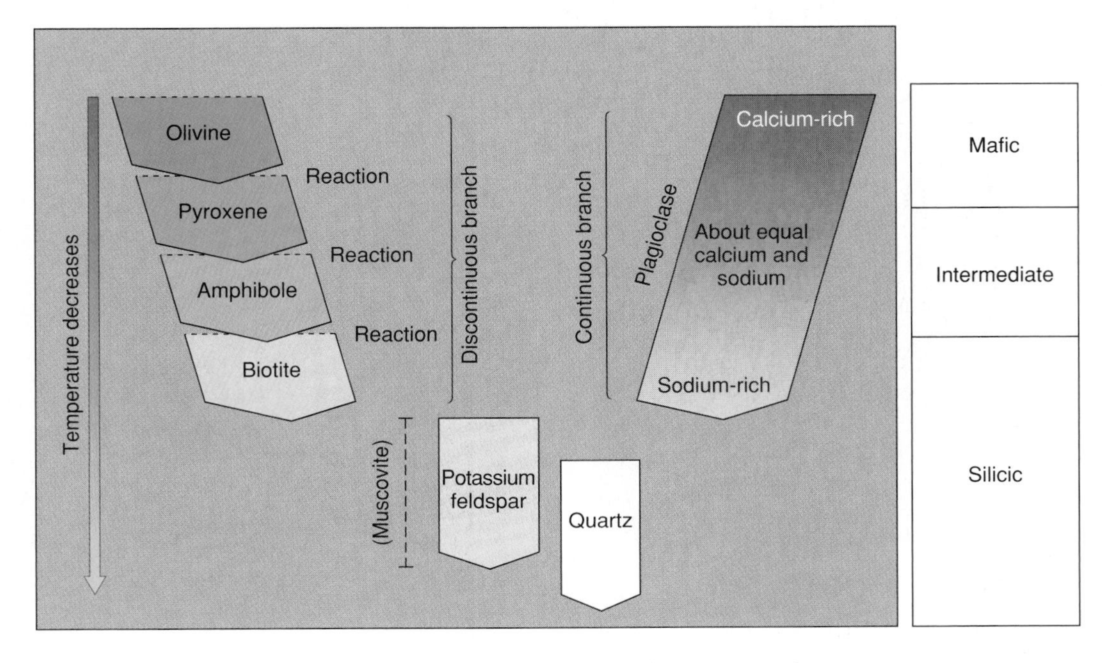

FIGURE 11.18
Bowen's reaction series.

- A mafic magma will crystallize into pyroxene (with or without olivine) and calcium-rich plagioclase, that is, basalt or gabbro, if the early formed crystals are not removed from the remaining magma. Similarly, an intermediate magma will crystallize into diorite or andesite, if early formed minerals are not removed.
- If minerals are separated from a magma, the remaining magma is more silicic than the original magma. For example, if olivine and calcium-rich plagioclase are removed, the residual melt would be richer in silicon and sodium and poorer in iron and magnesium.
- If you heat a rock, the minerals will melt in reverse order. In other words you would be going up the series as diagrammed in figure 11.18. Quartz and potassium feldspar would melt first. If the temperature is raised further, biotite and sodium-rich plagioclase would contribute to the melt. Any minerals higher in the series would remain solid unless the temperature is raised further.
- Bowen's reaction series can be used to show how two important processes that create and modify magma composition work. These are *differentiation* and *partial melting.*

Differentiation

The process by which different ingredients separate from an originally homogenous mixture is **differentiation.** An example is the separation of whole milk into cream and nonfat milk. Differentiation in magmas takes place through **crystal settling,** the downward movement of minerals that are denser (heavier) than the magma from which they crystallized.

If crystal settling takes place in a mafic magma chamber, olivine and, perhaps, pyroxene crystallize and settle to the bottom of the magma chamber (figure 11.19). This makes the remaining magma more silicic. Calcium-rich plagioclase also separates as it forms. The remaining magma is, therefore, depleted of calcium, iron, and magnesium. Because these minerals were economical in using the relatively abundant silica, the remaining magma becomes richer in silica as well as in sodium and potassium.

It is possible that by removing enough mafic components, the residual magma would be silicic enough to solidify into granite (or rhyolite). But, it is more likely only enough mafic components would be removed to allow an intermediate residual magma, which would solidify into diorite or andesite. The lowermost portions of some large sills are composed predominantly of olivine and pyroxene, whereas upper levels are considerably less mafic. Even in large sills, however, differentiation has rarely progressed far enough to produce granite within the sill.

Ore Deposits Due to Crystal Settling

Crystal settling accounts for important ore deposits that are mined for chromium and platinum. Most of the world's chromium and platinum come from a huge sill in South Africa. The sill, the famous Bushveldt Complex, is 8 kilometers thick and 500 kilometers long. Layers of chromite (a chromium-bearing mineral) up to 2 meters thick are found, and mined, at the base of the sill. Layers containing platinum overlie the chromite-rich layers.

Partial Melting

As mentioned earlier, progressing upward through Bowen's reaction series (going from cool to hot), gives us the sequence in which minerals in a rock melt. As might be expected, the first portion of a rock to melt as temperatures rise forms a liquid with the chemical composition of quartz and potassium feldspar. The oxides of silicon plus potassium and aluminum "sweated out" of the solid rock could accumulate into a pocket of silicic magma. If higher temperatures prevailed, more mafic magmas would be created. Small pockets of magma could merge and form a large enough mass to rise as a diapir. In nature, temperatures rarely rise high enough to entirely melt a rock.

Partial melting of the lower continental crust likely produces silicic magma that eventually solidifies into granite or

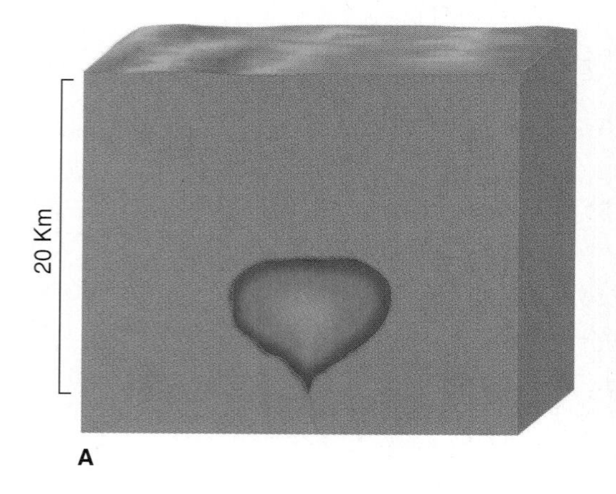

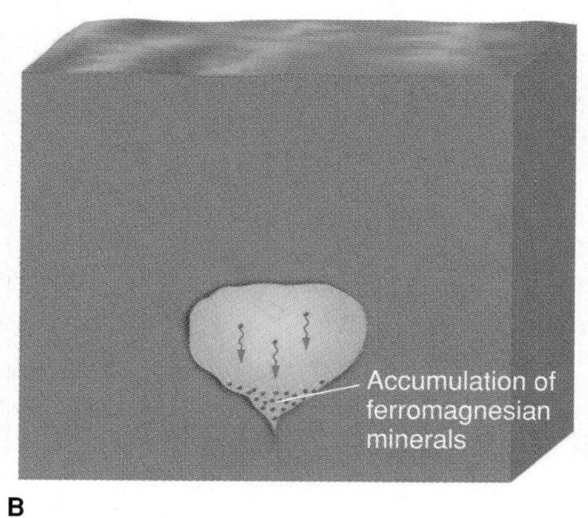

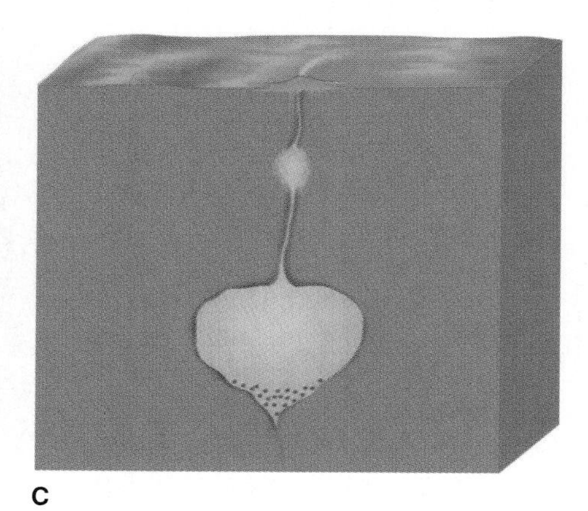

FIGURE 11.19

Differentiation of a magma body. (*A*). Recently intruded mafic magma is completely liquid. (*B*) Upon slow cooling, ferromagnesian minerals such as olivine crystallize and sink to the bottom of the magma chamber. The remaining liquid is now an intermediate magma. (*C*) Some of the intermediate magma moves upward to form a smaller magma chamber at a higher level and a volcano.

rhyolite. But geologists generally regard basaltic magma (Hawaiian lava, for example) as the product of partial melting of ultramafic rock in the mantle at temperatures hotter than those in the crust. The solid residue left behind in the mantle when the basaltic magma is removed is an even more silica-deficient ultramafic rock.

Assimilation

A very hot magma may melt some of the country rock and *assimilate* the newly molten material into the magma (figure 11.20). This is like putting a few ice cubes into a cup of hot coffee. The ice melts and the coffee cools as it becomes diluted. Similarly, if a hot basaltic magma, perhaps generated from the mantle, melts portions of the continental crust, the magma simultaneously becomes richer in silica and cooler. Possibly intermediate magmas such as are associated with circum-Pacific andesite volcanoes may derive from assimilation of some crustal rocks by a basaltic magma.

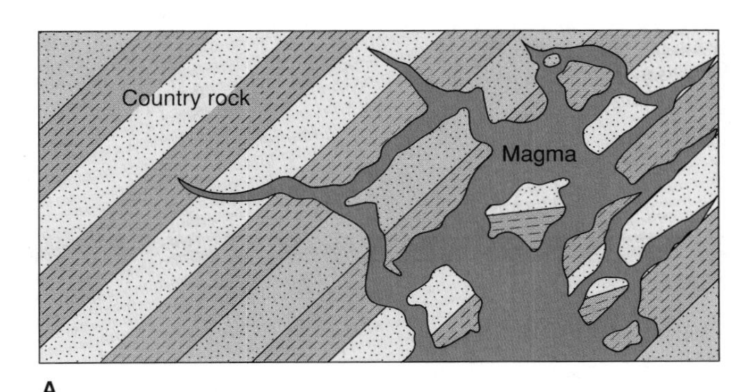

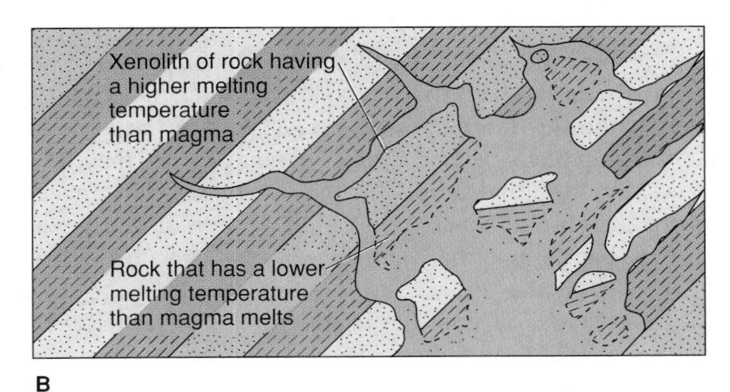

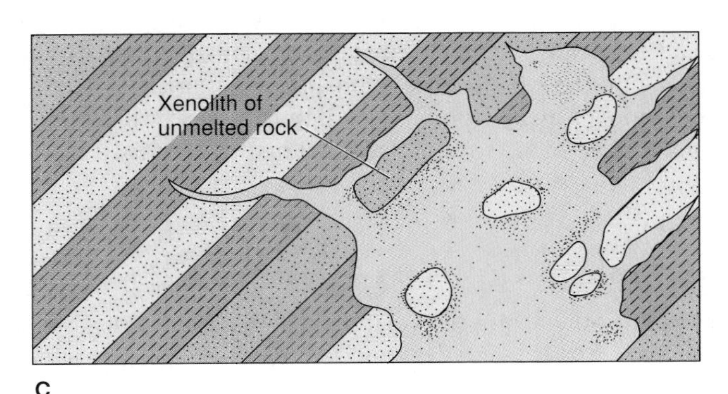

FIGURE 11.20

Assimilation. Magma formed is intermediate in composition between the original magma and the absorbed country rock. (*A*) Ascending magma breaks off blocks of country rock (the process is called *stoping*). (*B*) Xenoliths of country rock with melting temperatures lower than the magma melt. (*C*) The molten country rock blends with the original magma, leaving unmelted portions as inclusions.

Mixing of Magmas

Some of our igneous rocks may be "cocktails" of different magmas. The concept is quite simple. If two magmas meet and merge within the crust, the combined magma will be compositionally intermediate (figure 11.21). If you had approximately equal amounts of a granitic magma mixing with a basaltic magma, the resulting magma should crystallize underground as diorite or erupt on the surface to solidify as andesite.

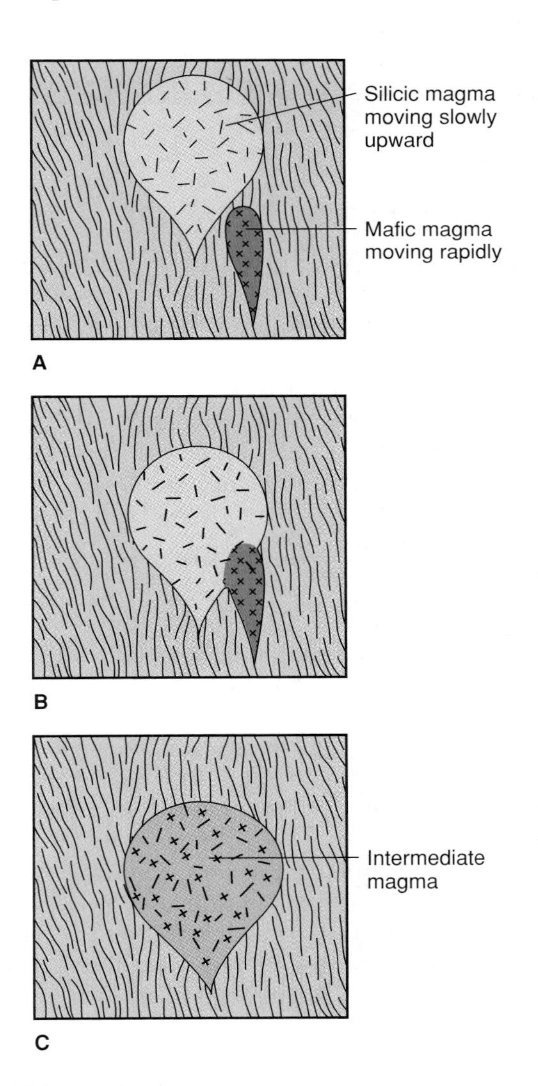

FIGURE 11.21

Mixing of magmas. (*A*) Two bodies of magma moving surfaceward. (*B*) The mafic magma catches up with the silicic magma. (*C*) The two magmas combine and become an intermediate magma.

table 11.2 Relationships between Rock Types and Their Usual Plate Tectonic Setting

Rock	Original Magma	Final Magma	Processes	Plate Tectonic Setting
basalt & gabbro	mafic	mafic	partial melting of mantle (asthenosphere)	1. divergent boundary—oceanic crust created 2. Intraplate • plateau basalt • volcanic island chains (e.g., Hawaii)
andesite & diorite	mafic (usually)	intermediate	partial melting of mantle (asthenosphere) followed by: • differentiation or • assimilation or • magma mixing	convergent boundary
granite & rhyolite	silicic	silicic	partial melting of lower crust	1. convergent boundary 2. intraplate • over mantle plume

EXPLAINING IGNEOUS ACTIVITY BY PLATE TECTONICS

One of the appealing aspects of the theory of plate tectonics is that it accounts reasonably well for the variety of igneous rocks and their distribution patterns. Divergent boundaries are associated with creation of basalt and gabbro of the oceanic crust. Andesite and granite are associated with convergent boundaries. Table 11.2 summarizes the relationships.

Igneous Processes at Divergent Boundaries

Geologists agree that basaltic magma produced at divergent boundaries is due to partial melting of the asthenosphere. The *asthenosphere,* as described in chapter 1, is the plastic zone of the mantle beneath the rigid *lithosphere* (the upper mantle and crust that make up a plate). Along divergent boundaries, the asthenosphere is relatively close (5 to 10 kilometers) to the surface (figure 11.22).

The probable reason the asthenosphere is plastic or "soft" is that temperatures there are only slightly lower than the temperatures required for partial melting of mantle rock.

If extra heat is added, or pressure is reduced, partial melting should take place. The asthenosphere beneath divergent boundaries probably represents mantle material that has welled upward from deeper levels of the mantle. As the hot asthenosphere gets close to the surface, decrease in pressure results in partial melting. The magma that forms is mafic and will solidify as basalt or gabbro. The portion that did not melt remains behind as a silica-depleted, iron and magnesium–enriched ultramafic rock.

Some of the basaltic magma erupts along a submarine ridge to form pillow basalts (described in chapter 10), while some fills near-surface fissures to create dikes. Deeper down, magma solidifies more slowly into gabbro. The newly solidified rock is pulled apart by spreading plates; more magma fills the new fracture and erupts on the sea floor. The process is repeated, resulting in a continuous production of mafic crust.

The basalt magma that builds the oceanic crust is removed from the underlying mantle, depleting the mantle beneath the ridge of much of its calcium, aluminum, and silicon oxides. The unmelted residue (olivine and pyroxene) becomes depleted mantle, but it is still a variety of ultramafic rock. The rigid ultramafic rock, the overlying gabbro and basalt, and any sediment that may have deposited on the basalt collectively are the lithosphere of an oceanic plate, which moves away from a spreading center over the asthenosphere. (The nature of the oceanic crust is described in more detail in chapter 3.)

Intraplate Igneous Activity

Hawaiian volcanism is unusual because Hawaii is not at a plate boundary. Most geologists think that basaltic magma there is created because the Pacific plate is overriding a hot mantle plume in a process described and illustrated in chapter 4.

The huge volume of mafic magma that erupted to form the Columbia plateau basalts of Washington and Oregon (described in chapter 10) is attributed to a past hot mantle plume, according to a recent hypothesis (figure 11.23). In this case, the large

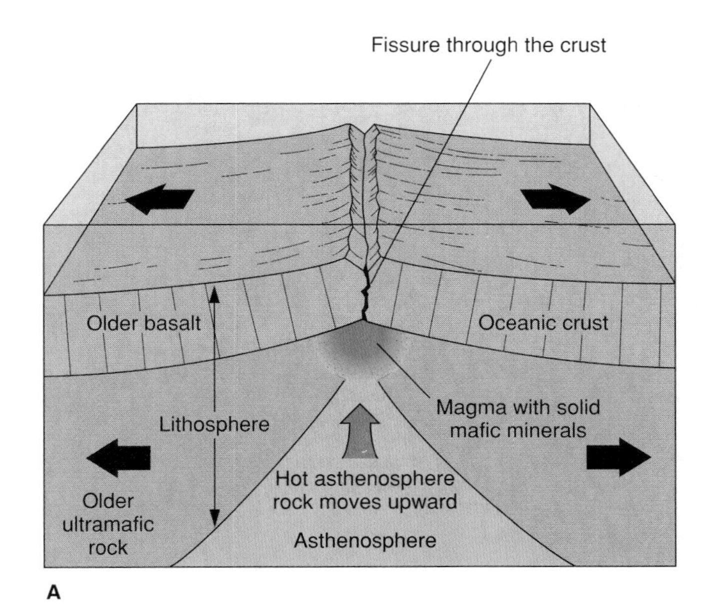

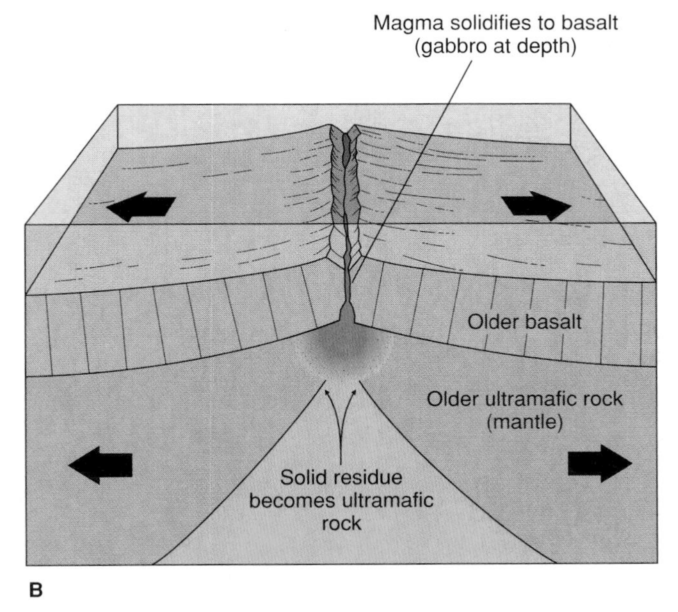

FIGURE 11.22

Schematic representation of how basaltic oceanic crust and the underlying ultramafic mantle rock form at a divergent boundary. The process is more continuous than the two-step diagram implies. (*A*) Partial melting of asthenosphere takes place beneath a mid-oceanic ridge. (*B*) The magma squeezes into the fissure system. Solid mafic minerals are left behind as ultramafic rock.

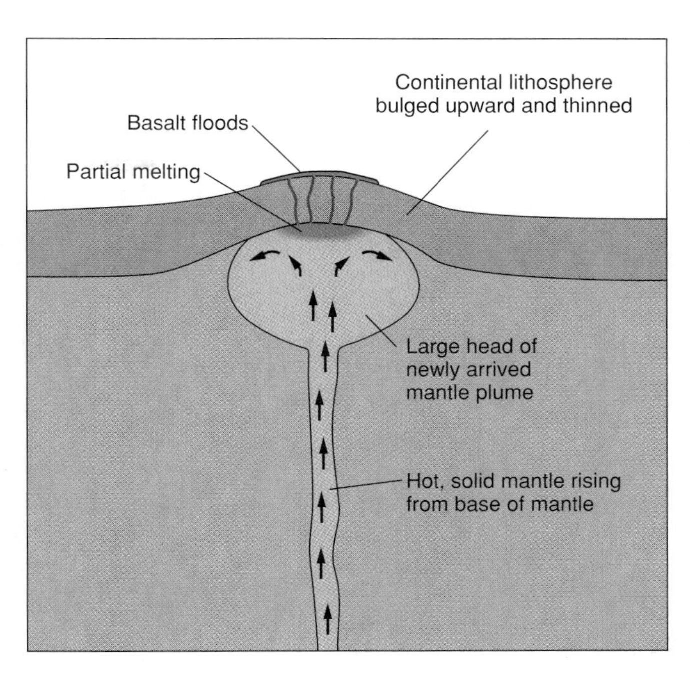

FIGURE 11.23

A hot mantle plume with a large head rises from the lower mantle. When it reaches the base of the lithosphere it uplifts and stretches the overlying lithosphere. The reduced pressure results in partial melting, producing basaltic magma. Large volumes of magma travel through fissures and flood the Earth's surface.

volume of basalt is due to the arrival beneath the lithosphere of a mantle plume with a large head on it (sort of a mega-diapir).

Very explosive rhyolitic eruptions at Yellowstone National Park that occurred several hundred thousand years ago are also credited to a hot mantle plume that caused extensive partial melting of the continental crust. The resulting silicic magma bodies reached the surface (unlike most granitic magmas that solidify as plutons) as huge eruptions.

Igneous Processes at Convergent Boundaries

Intermediate and silicic magmas are clearly related to the convergence of two plates and subduction; however, exactly what takes place is debated by geologists. Compared to divergent boundaries, there is less agreement about how magmas are generated at converging boundaries. The scenarios that follow are currently believed by geologists to be the best explanations of the data.

The Origin of Andesite

Magma for most of our andesitic composite volcanoes (such as are found along the west coast of the Americas) seems to originate from a depth of about 100 kilometers. This coincides with the depth at which the subducted oceanic plate is sliding under the asthenosphere (figure 11.24). Partial melting of the asthenosphere takes place, resulting in a mafic magma. In most cases, melting occurs because the subducted oceanic crust releases water into the asthenosphere. The water collected in the oceanic crust when it was beneath the ocean and is driven out as the descending plate is heated. The water lowers the melting temperature of the ultramafic rocks in this part of the mantle. Partial melting produces a mafic magma.

But how can we keep producing magma from ultramafic rock after those rocks have been depleted of the constituents of the mafic magma? The answer is that hot asthenospheric

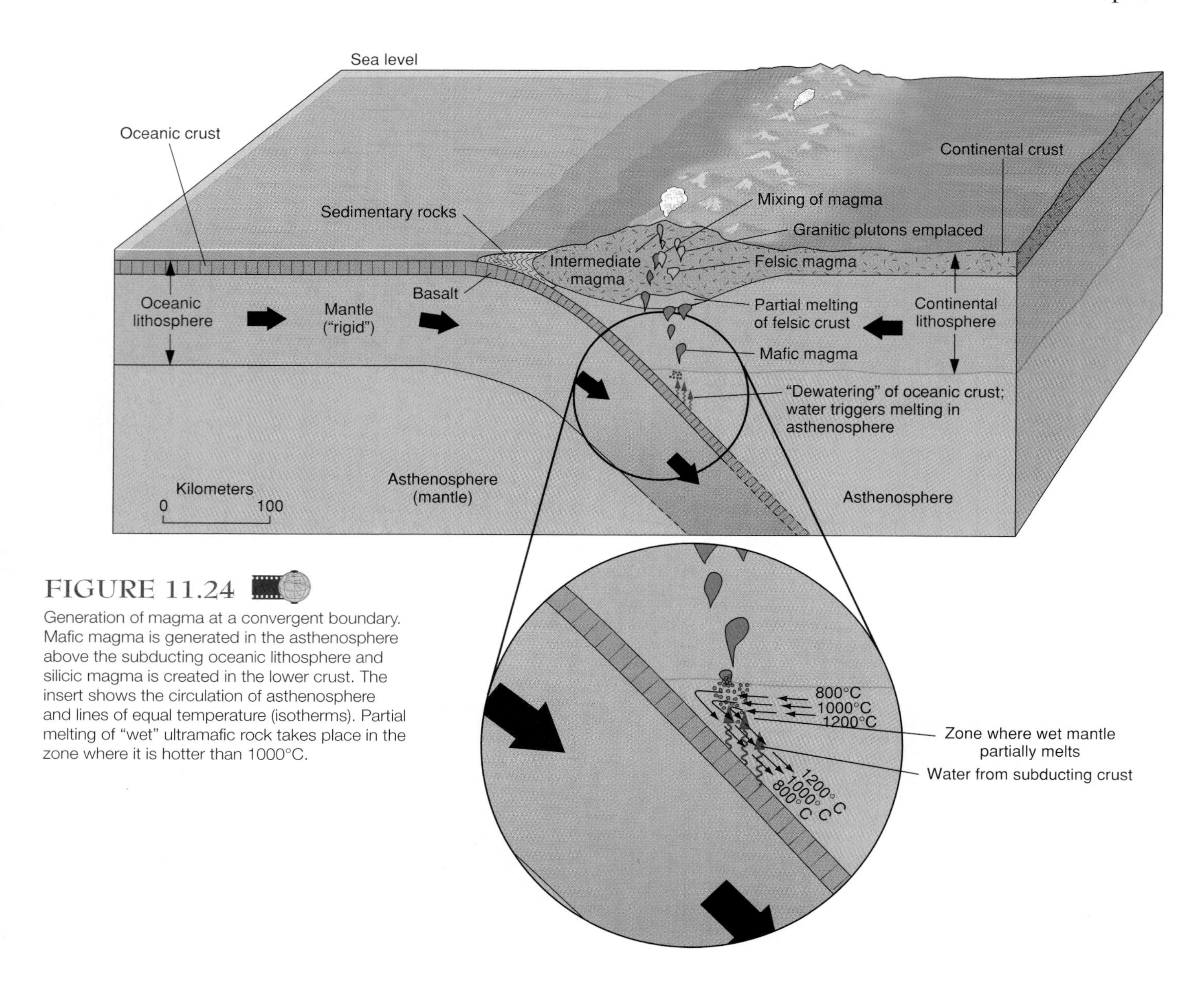

FIGURE 11.24

Generation of magma at a convergent boundary. Mafic magma is generated in the asthenosphere above the subducting oceanic lithosphere and silicic magma is created in the lower crust. The insert shows the circulation of asthenosphere and lines of equal temperature (isotherms). Partial melting of "wet" ultramafic rock takes place in the zone where it is hotter than 1000°C.

rock continues to flow into the zone of partial melting. As shown in figure 11.24, asthenospheric ultramafic rock is dragged downward by the descending lithospheric slab. More ultramafic rock flows laterally to replace the descending material. A continuous flow of hot, "fertile" (containing the constituents of basalt) ultramafic rock is brought into the zone where water, moving upward from the descending slab, lowers the melting temperature. After being depleted of basaltic magma, the solid, residual, ultramafic rock continues to sink deeper into the mantle.

On its slow journey through the crust, the mafic magma evolves into an intermediate magma by differentiation, assimilation of silicic crustal rocks, and by magma mixing.

Under special circumstances basalt of the descending oceanic crust can partially melt to yield an intermediate magma. In most subduction zones, the basalt remains too cool to melt, even at a depth of over 100 km. But geologists believe that partial melting of the subducted crust produces the magma for andesitic volcanoes in South America. Here, the oceanic crust is much younger and considerably hotter than normal. The spreading axis where it was created is not far from the trench. Because the lithosphere has not traveled far before being subducted it is still relatively hot. As can be seen from figure 11.25, subduction is at a shallower angle, because this hotter crust is more buoyant than the usual case (as in figure 11.24).

The reason that partial melting of subducted basalt is unusual is that this kind of subduction and magma generation is, geologically speaking, short-lived. Subduction will end when the overriding plate crashes into the mid-oceanic ridge. Most subduction zones are a long distance from the divergent boundaries of their plates, so steep subduction and magma production from the asthenosphere is the norm.

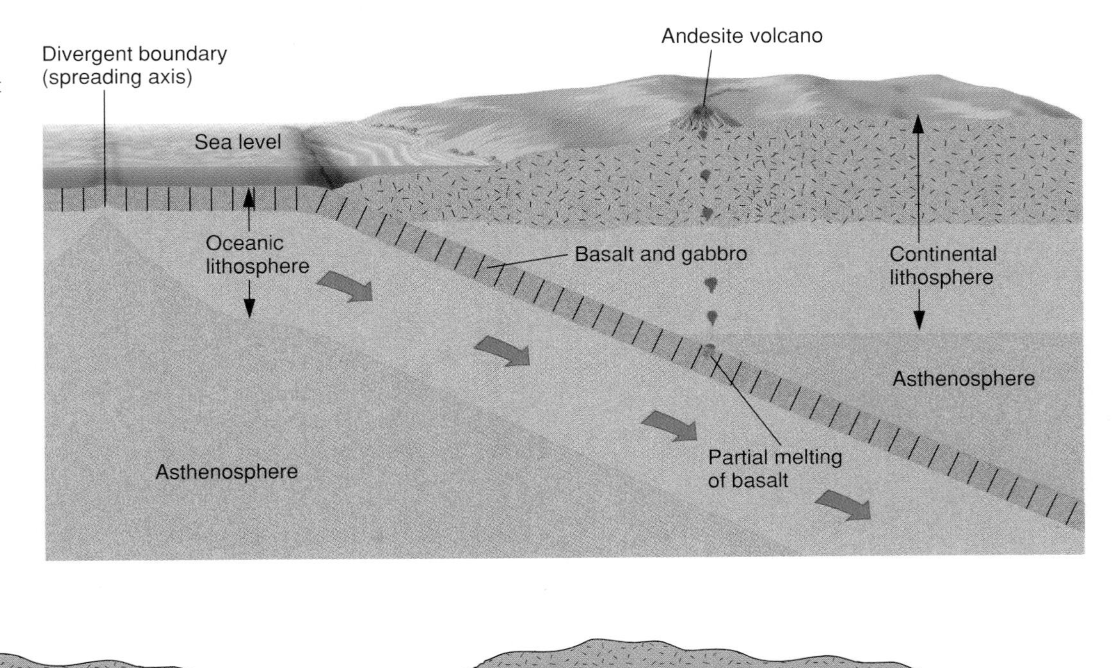

FIGURE 11.25

Young, hot, oceanic lithosphere is buoyant and subducts at a shallow angle. Basalt partially melts when it is heated further by the overlying asthenosphere.

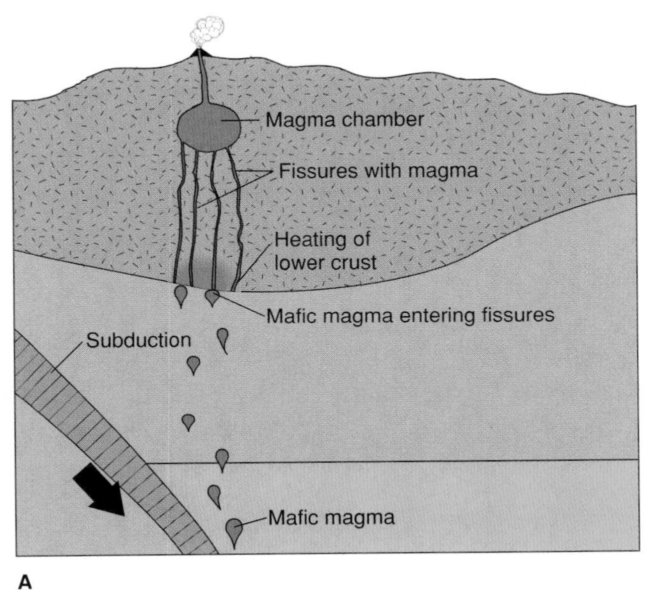

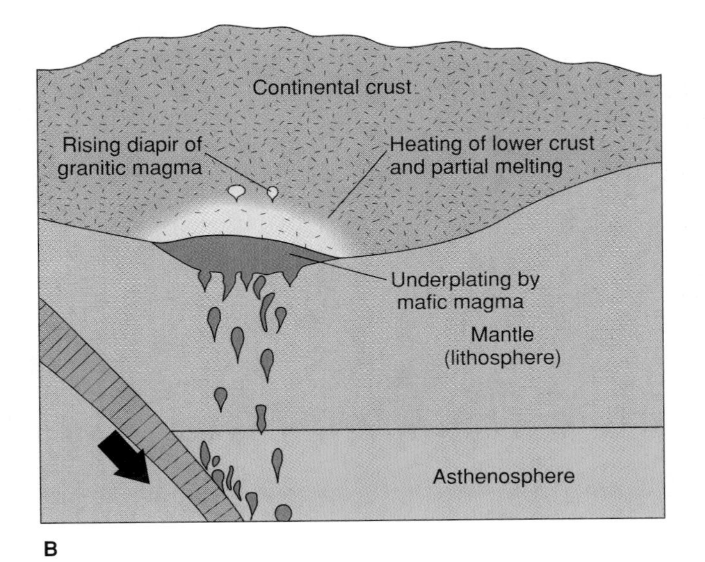

FIGURE 11.26

How mafic magma could add heat to the lower crust and result in partial melting to form a granitic magma. (*A*) Mafic magma from the asthenosphere rises through closely spaced fissures in the lower crust (widths are highly exaggerated in diagram). (*B*) Magmatic underplating of the continental crust.

The Origin of Granite

To explain the great volumes of granitic plutonic rocks, most geologists think that partial melting of the lower continental crust must take place. The continental crust contains the high amount of silica needed for a silicic magma. As the silicic rocks of the continental crust have relatively low melting temperatures (especially if water is present), partial melting of the lower continental crust is likely. Currently, geologists think that *magmatic underplating* by andesitic or basaltic magma plays an important role by providing the extra heat source needed to generate granitic magmas in the lower continental crust. Initially, some of the mafic magma coming from the asthenosphere works its way through fissures systems in the lower crust (figure 11.26*A*). As the lower crust gets hotter, the rock becomes more plastic and melting begins. Fissures are sealed. The denser, mafic magma then pools under (underplates) the lighter, partially molten, lowest crust (figure 11.26*B*). Heat from the cooling and crystallizing mafic magma is conducted upward to create larger volumes of silicic magma by partially melting more of the continental crust. The silicic magma, in turn, separates from its solid residue and works its way upward in diapirs to a higher level of the crust where it slowly solidifies to a pluton, usually as part of a batholith.

We should emphasize that the picture we have presented is not an observation but a reasoned interpretation of available data.

Summary

The interaction between the internal and external forces of the Earth is illustrated by the rock cycle, a conceptual device relating igneous, sedimentary, and metamorphic rocks to each other, to surficial processes such as weathering and erosion, and to internal processes such as tectonic forces. Changes take place when one or more processes force Earth's material out of equilibrium.

Igneous rocks form from solidification of magma. If the rock forms at the Earth's surface it is *extrusive. Intrusive rocks* are igneous rocks that formed underground. Some intrusive rocks have solidified near the surface as a direct result of volcanic activity. Volcanic *necks* solidified within volcanoes. Fine-grained *dikes* and *sills* may also have formed in cracks during local extrusive activity. A sill is *concordant*—parallel to the planes within the country rock. A dike is *discordant.* Both are tabular bodies. Coarser grains in either a dike or a sill indicate that it probably formed at considerable depth.

Most intrusive rock is *plutonic*—that is, coarse-grained rock that solidified slowly at considerable depth. Most plutonic rock exposed at the Earth's surface is in *batholiths*—large plutonic bodies with no particular shape. A smaller, irregular body is called a *stock.*

Silicic (or felsic) rocks are rich in silica, whereas mafic rocks are silica deficient. Most igneous rocks are named on the basis of their mineral content, which in turn reflects the chemical composition of the magmas from which they formed, and on grain sizes. *Granite, diorite,* and *gabbro* are the coarse-grained equivalents of *rhyolite, andesite,* and *basalt,* respectively. *Peridotite* is an *ultramafic* rock, made entirely of ferromagnesian minerals and is mostly associated with the mantle.

Basalt and gabbro are strongly predominant in the oceanic crust. Granite strongly predominates in the continental crust. Younger granite batholiths occur mostly within younger mountain belts. Andesite is largely restricted to narrow zones along convergent plate boundaries.

The *geothermal gradient* is the increase in temperature with increase in depth. Hot *mantle plumes* and magma at shallow depths in volcanic regions locally raise the geothermal gradient.

No single process can satisfactorily account for all igneous rocks. In the process of *differentiation,* based on *Bowen's reaction series,* a residual magma more silicic than the original mafic magma is created when the early-forming minerals separate out of the magma. In *assimilation,* a hot, original magma is contaminated by picking up and absorbing rock of a different composition. *Magma mixing* produces a magma whose composition is intermediate, between that of the two types of magma that were mixed.

Partial melting of the mantle usually produces basaltic magma whereas granitic magma is most likely produced by partial melting of the lower crust.

The theory of *plate tectonics* incorporates the above concepts. Basalt is generated where hot mantle rock partially melts, most notably along divergent boundaries. The fluid magma rises easily through fissures, if present. The ferromagnesian portion that stays solid remains in the mantle as ultramafic rock. Granite and andesite are associated with subduction. Differentiation, assimilation, partial melting, and mixing of magmas may each play a part in creating the appropriate rocks.

Terms to Remember

Testing Your Knowledge

Use the questions below to prepare for exams based on this chapter.

1. Why do mafic magmas tend to reach the surface much more often than silicic magmas?
2. What role does the asthenosphere play in generating magma at (a) a convergent boundary; (b) a divergent boundary?
3. How do batholiths form?
4. How would you distinguish, on the basis of minerals present, among granite, gabbro, and diorite?
5. How would you distinguish andesite from a diorite?
6. What rock would probably form if magma that was feeding volcanoes above subduction zones solidified at considerable depth?
7. Why is a higher temperature required to form magma at the oceanic ridges than in the continental crust?
8. What is the difference between feldspar found in gabbro and feldspar found in granite?
9. What is the difference between a dike and a sill?
10. Describe the differences between the continuous and the discontinuous branches of Bowen's reaction series.
11. A surface separating different rock types is called a
    a. xenolith b. contact
    c. chill zone d. none of the above
12. The major difference between intrusive igneous rocks and extrusive igneous rocks is
    a. where they solidify b. chemical composition
    c. type of minerals d. all of the above
13. Which is not an intrusive igneous rock?
    a. gabbro b. diorite
    c. granite d. andesite
14. By definition, stocks differ from batholiths in
    a. size b. shape
    c. chemical composition d. all of the above
15. Which is not a source of heat for melting rock?
    a. geothermal gradient b. the hotter mantle
    c. mantle plumes d. water under pressure
16. The geothermal gradient is, on the average, about
    a. 1°C/km b. 10°C/km
    c. 30°C/km d. 50°C/km
17. The continuous branch of Bowen's reaction series contains the mineral
    a. pyroxene b. plagioclase
    c. amphibole d. biotite
18. The discontinuous branch of Bowen's reaction series contains the mineral
    a. pyroxene b. amphibole
    c. biotite d. all of the above
19. The most common igneous rock of the continents is
    a. basalt b. granite
    c. rhyolite d. ultramafic
20. Granite and rhyolite are different in
    a. texture b. chemistry
    c. mineralogy d. the kind of magma that each crystallized from
21. The difference in texture between intrusive and extrusive rocks is primarily due to
    a. different mineralogy
    b. different rates of cooling and crystallization
    c. different amounts of water in the magma
22. How can geologists recognize shallow intrusives from deeper intrusives? (choose all that apply)
    a. mineralogy b. size
    c. rock texture d. chemistry
23. A change in magma composition due to melting of surrounding country rock is called
    a. magma mixing b. assimilation
    c. crystal setting d. partial melting

Expanding Your Knowledge

1. In parts of major mountain belts there are sequences of rocks that geologists interpret as slices of ancient oceanic lithosphere. Assuming that such a sequence formed at a divergent boundary and was moved toward a convergent boundary by plate motion, what rock types would you expect to make up this sequence, going from the top downward?
2. Explain what would happen, according to Bowen's reaction series, under the following circumstances: olivine crystals form and only the surface of each crystal reacts with the melt to form a coating of pyroxene that prevents the interior of olivine from reacting with the melt.

Exploring Web Resources

www.mhhe.com/plummer9e
Visit the Online Learning Center for article updates and direct links to the sites listed below as well as additional readings and media resources.

http://uts.cc.utexas.edu/~rmr/
Rob's Granite Page. This site has a lot of information on granite and related igneous activity. The site is useful for people new to geology as well as for professionals. There are numerous images of granite. Click on "Did you know that granite is like ice cream?" for an interesting comparison. The page also has photos of various granites and links to other sites that have more images.

www.geolab.unc.edu/Petunia/IgMetAtlas/mainmenu.html
Atlas of rocks, minerals, and textures (from University of North Carolina). This site contains some photomicrographs of plutonic and volcanic rocks. The images are thin sections (slices of rock so thin that most minerals are transparent) seen in a polarizing microscope. Most images are taken from cross-polarized light, which causes many minerals to appear in distinctive, bright colors. For some of the rocks (gabbro, for instance), you can also see what it looks like under plain polarized light by clicking the circle with the horizontal gray lines.

http://seis.natsci.csulb.edu/basicgeo/IGNEOUS_TOUR.html
Igneous Rocks Tour. This site has some hand specimen images of common igneous rocks and should provide a useful review for rock identification.

Animations

This chapter includes the following animations available on our Online Learning Center at www.mhhe.com/plummer9e.

11.22 Divergent Boundary

11.24 Generation of Magma

C H A P T E R

12

Weathering and Soil

In this chapter, you will study several visible signs of weathering in the world around you, including the cliffs and slopes of the Grand Canyon and the rounded edges of boulders. As you study these features, keep in mind that weathering processes made the planet suitable for human habitation. From the weathering of rock eventually came the development of soil, upon which the world's food supply depends.

How does rock weather? You learned in chapters 10 and 11 that the minerals making up igneous rocks crystallize at relatively high temperatures and sometimes at high pressures as magma and lava cool. Although these minerals are stable when they form, most of them are not stable during prolonged exposure at Earth's surface. In this chapter you see how minerals and rocks change when they are subjected to the physical and chemical conditions existing at Earth's surface. Rocks undergo mechanical weathering (physical disintegration) and chemical weathering (decomposition) as they are attacked by air and water. Your knowledge of the chemical composition and atomic structure of minerals will help you understand the reactions that occur during chemical weathering.

Weathering processes create sediments (primarily mud and sand) and soil. Sedimentary rocks, which form from sediments, are discussed in chapter 14. In a general sense, weathering prepares rocks for erosion and is a fundamental part of the rock cycle, transforming rocks into the raw material that eventually becomes sedimentary rocks.

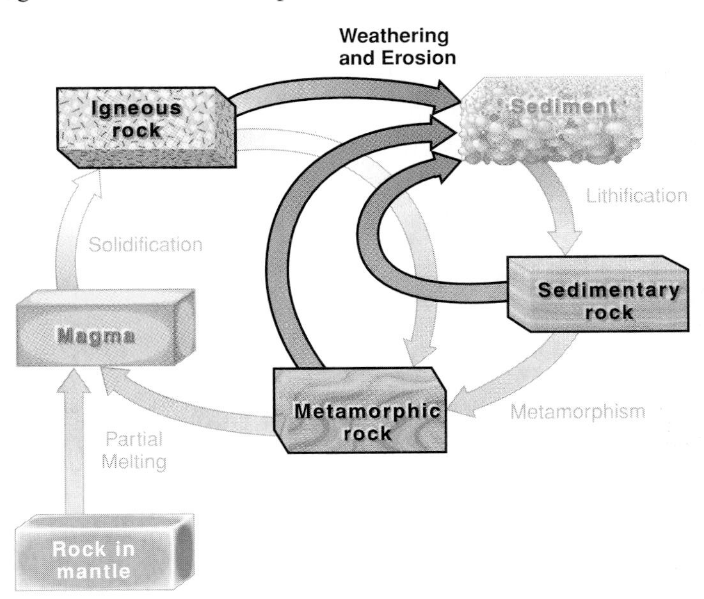

Opposite: Differential weathering at Bryce Canyon National Park in Utah has produced spires in the sandstone beds.
Dennis Flaherty/Photo Researchers, Inc.

WEATHERING, EROSION, AND TRANSPORTATION

Rocks exposed at Earth's surface are constantly being altered by water, air, changing temperature, and other environmental factors. The term **weathering** refers to the group of destructive processes that change the physical and chemical character of rock at or near Earth's surface. The tightly bound crystals of any rock can be loosened and altered to new minerals by weathering.

It is important to distinguish between weathering and *erosion,* and between erosion and *transportation.* Weathering breaks down rocks that are either stationary or moving. **Erosion** is the picking up or *physical removal* of rock particles by an agent such as streams or glaciers. Weathering helps break down a solid rock into loose particles that are easily eroded. Most eroded rock particles are at least partially weathered, but rock can be eroded before it has weathered at all. A stream can erode weathered or unweathered rock fragments.

After a rock fragment is picked up (eroded), it is transported. **Transportation** is the movement of eroded particles by agents such as rivers, waves, glaciers, or wind. Weathering processes continue during transportation. A boulder being transported by a stream can be physically worn down and chemically altered as it is carried along by the water.

HOW WEATHERING ALTERS ROCKS

Rocks undergo both mechanical weathering and chemical weathering. **Mechanical weathering** (or physical disintegration) includes several processes that break rock into smaller pieces. The change in the rock is physical; there is little or no chemical change. For example, water freezing and expanding in cracks can cause rocks to disintegrate physically. **Chemical weathering** is the decomposition of rock from exposure to water and atmospheric gases (principally carbon dioxide, oxygen, and water vapor). As rock is decomposed by these agents, new chemical compounds form.

Mechanical weathering breaks up rock but does not change the composition. A large mass of granite may be broken into smaller pieces by frost action, but its original crystals of quartz, feldspar, and ferromagnesian minerals are unchanged. On the other hand, if the granite is being chemically weathered, some of the original minerals are chemically changed into different minerals. Feldspar, for example, will change into a clay mineral (with a crystal structure similar to mica). In nature, mechanical and chemical weathering usually occur together, and the effects are interrelated.

Weathering is a relatively long, slow process. Typically, cracks in rock are enlarged gradually by frost action or plant growth (as roots pry into rock crevices), and as a result, more surfaces are exposed to attack by chemical agents. Chemical weathering initially works along contacts between mineral grains. Tightly bound crystals are loosened as weathering products form at their contacts. Mechanical and chemical weathering then proceed together, until a once tough rock slowly crumbles into individual grains.

Solid minerals are not the only products of chemical weathering. Some minerals—calcite, for example—dissolve when chemically weathered. We can expect limestone and marble, rocks consisting mainly of calcite, to weather chemically in quite a different way than granite.

EFFECTS OF WEATHERING

The results of chemical weathering are easy to find. Look along the edges or corners of old stone structures for evidence. The inscriptions on statues and gravestones that have stood for several decades may no longer be sharp (figure 12.1). Building blocks of limestone or marble exposed to rain and atmospheric gases may show solution effects of chemical weathering in a surprisingly short time. Granite and slate gravestones and building materials are much more resistant to weathering due to the strong silicon-oxygen bonds in the silicate minerals. However, after centuries the mineral grains in granite may be loosened, cracks enlarged, and the surface discolored and dulled by the products of weathering. Surface discoloration is also common on rock *outcrops,* where rock is exposed to view, with no plant or soil cover. That is why field geologists carry rock hammers—to break rocks to examine unweathered surfaces.

We tend to think of weathering as destructive because it mars statues and building fronts. As rock is destroyed, however, valuable products can be created. Soil is produced by rock weathering, so most plants depend on weathering for the soil they need in order to grow. In a sense, then, all agriculture depends upon weathering. Weathering products transported to the sea by rivers as dissolved solids make seawater salty and serve as nutrients for many marine organisms. Some metallic ores, such as those of copper and aluminum, are concentrated into economic deposits by chemical weathering.

Many weathered rocks display interesting shapes. **Spheroidal weathering** occurs where rock has been rounded by weathering from an initial blocky shape. It is rounded because chemical weathering acts more rapidly or intensely on the corners and edges of a rock than on the smooth rock faces (figure 12.2).

Differential weathering is the term for varying rates of weathering in an area where some rocks are more resistant to weathering than others. Resistant rocks weather slowly, and may protrude above softer rocks that weather rapidly. Figures 12.3 and 12.4 show some striking landforms produced by erosion of rocks that weather at different rates.

A

B

FIGURE 12.1

(*A*) The effects of chemical weathering are obvious in the marble gravestone on the right but not in the slate gravestone on the left which still retains its detail. Both gravestones date back to the 1780s. (*B*) This marble statue has lost most of the fine detail on the face by chemical weathering.

Photo *A* by C. C. Plummer

A

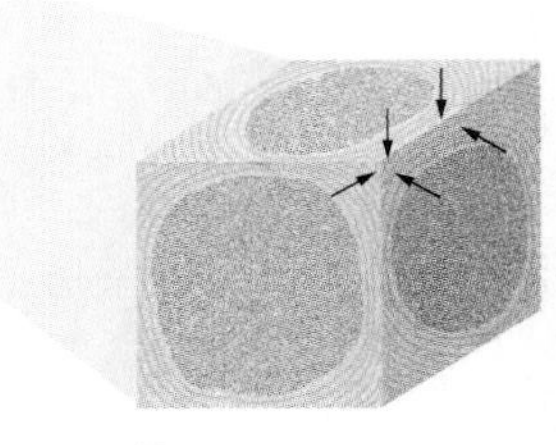

B

C

D

FIGURE 12.2

(*A*) Water penetrating along cracks at right angles to one another in an igneous rock produces spheroidal weathering of once-angular blocks. (*B*) Because of the increased surface area, chemical weathering attacks the corners and edges more rapidly than the flat faces, creating the spheroidal shapes shown in (*C*). (*D*) Spheroidally weathered granite exposed along the Salt River Canyon, Arizona

FIGURE 12.3
Pedestal rock near Lees Ferry, Arizona. Resistant sandstone cap protects weak shale pedestal from weathering and erosion. Hammer for scale is barely visible at base of pedestal.

FIGURE 12.4
Sedimentary rocks in the Grand Canyon, Arizona. In the foreground, layers of sandstone resist weathering and form steep cliffs. Less resistant layers of shale weather to form gentler slopes between cliffs.

MECHANICAL WEATHERING

Of the many processes that cause rocks to disintegrate, the most effective are frost action and pressure release.

Frost Action

Frost action—the mechanical effect of freezing water on rocks—commonly occurs as frost wedging or frost heaving. In **frost wedging** the expansion of freezing water pries rock apart. Most rock contains a system of cracks called *joints,* caused by the slow flexing of brittle rock by deep-seated Earth forces (see chapter 6). Water that has trickled into a joint in a rock can freeze and expand by as much as 9% when the temperature drops below 0°C (32°F). The expanding ice wedges the rock apart, extending the joint or even breaking the rock into pieces (figures 12.5 and 12.6). Frost wedging is most effective in regions with many days of freezing and thawing (mountaintops and midlatitude regions with pronounced seasons). Partial thawing during the day adds new water to the ice in the crack; refreezing at night adds new ice to the old ice.

Frost heaving lifts rock and soil vertically. Solid rock conducts heat better than soil, so on a cold winter day the bottom of a partially buried rock will be much colder than soil at the same depth. As the ground freezes in winter, ice forms first under large rock fragments in the soil. The expanding ice layers push boulders out of the ground, a process well known to New England farmers and other residents of rocky soils. Frost heaving bulges the ground surface upward in winter, breaking up roads and leaving lawns spongy and misshapen after the spring thaw.

Pressure Release

The reduction of pressure on a body of rock can cause it to crack as it expands; **pressure release** is a significant type of mechanical weathering. A large mass of rock, such as a batholith, may originally form under great pressure from the weight of several kilometers of rock above it. This batholith is gradually exposed by tectonic uplift of the region followed by erosion of the overlying rock (figure 12.7). The removal of the great weight of rock above the batholith, usually termed *unloading,* allows the granite to expand upward. Cracks called **sheet joints** develop parallel to the outer surface of the rock as the outer part of the rock expands more than the inner part (figures 12.7 and 12.8). On slopes, gravity may cause the rock between such joints to break loose in concentric slabs from the underlying granite mass. This process of spalling off of rock layers is called **exfoliation;** it is somewhat similar to peeling layers from an onion. **Exfoliation domes** (figure 12.9) are large, rounded landforms developed in massive rock, such as granite, by exfoliation. Some famous examples of exfoliation domes include Stone Mountain in Georgia and Half Dome in Yosemite.

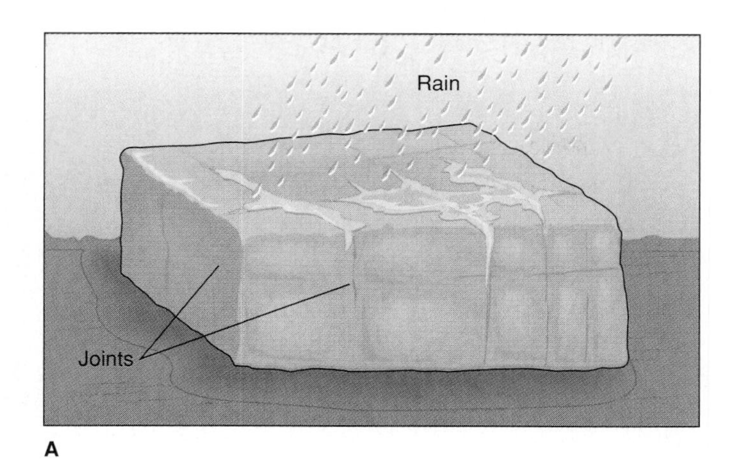

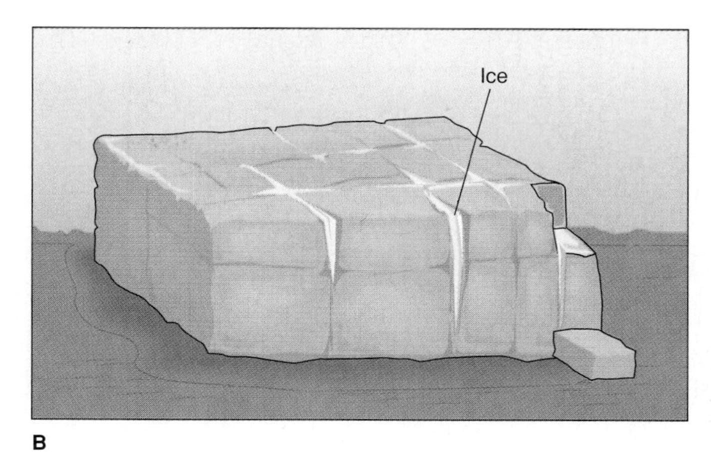

FIGURE 12.5
Frost wedging occurs when (*A*) water fills joints (cracks) in a rock and then freezes. (*B*) Expanding ice wedges the rock apart.

FIGURE 12.6
Frost wedging has broken this granite into large, angular fragments as ice expanded in rock joints. Sierra Nevada, California.
Photo by B. Amundson

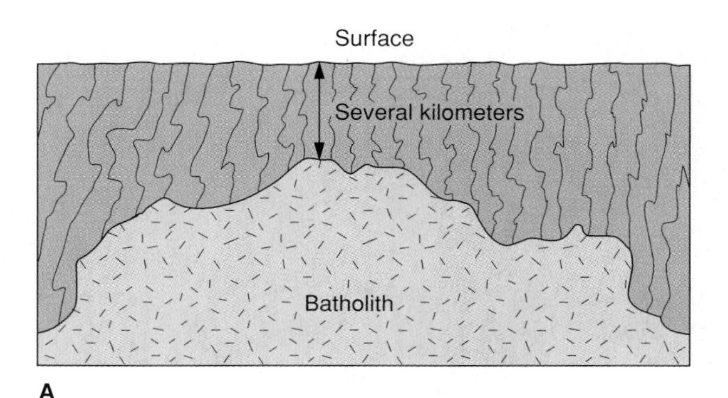

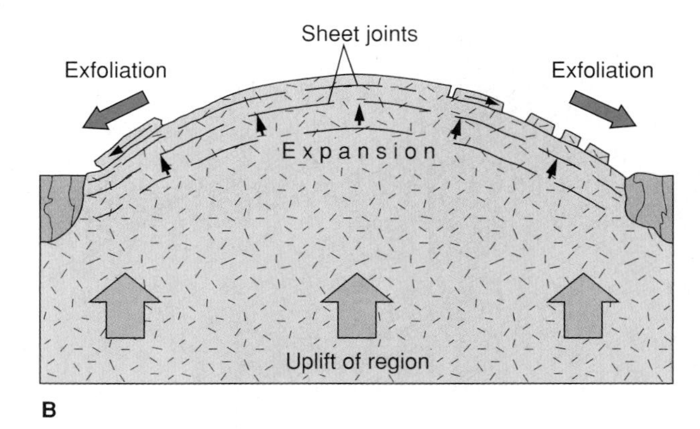

FIGURE 12.7
Sheet joints caused by pressure release. A granite batholith (*A*) is exposed by regional uplift followed by the erosion of the overlying rock (*B*). Unloading reduces pressure on the granite and causes outward expansion. Sheet joints are closely spaced at the surface where expansion is greatest. Exfoliation of rock layers produces rounded exfoliation domes.

Other Processes

Several other processes mechanically weather rock but in most environments are less effective than frost action and pressure release. *Plant growth,* particularly roots growing in cracks (figure 12.10*A*), can break up rocks, as can *burrowing animals.* Such activities help to speed up chemical weathering by enlarging passageways for water and air. The *pressure of salt crystals* formed as water evaporates inside small spaces in rock also helps to disintegrate desert rocks (figure 12.10*B*). *Extreme changes in temperature,* as in a forest fire, can cause a rock to expand or contract until it cracks. Whatever

FIGURE 12.8

Sheet joints in a granite outcrop near the top of the Sierra Nevada, California. The granite formed several kilometers below the surface, and expanded outward when it was exposed by uplift and erosion. Note that the sheet joints are closer together near the top of the outcrop, where the pressure release is the greatest.

FIGURE 12.9

Exfoliation dome, Yosemite National Park, California. Onionlike layers of rock are peeling off the dome.

A

B

FIGURE 12.10

(*A*) Tree roots will pry this rock apart as they grow within the rock joints, Tuolemne Meadows, California. (*B*) This rock is being broken by the growth of salt crystals, which precipitate as water evaporates within the cracks in the rock. Death Valley, California.

Photo *A* by Diane Carlson. Photo *B* by © by Frank M. Hanna

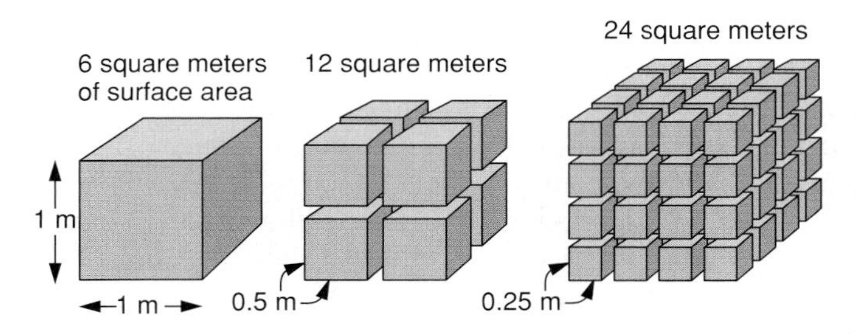

FIGURE 12.11
Mechanical weathering can increase the surface area of a rock, increasing the rate of chemical weathering. As a cube breaks up into smaller pieces, its volume remains the same but its surface area increases.

processes of mechanical weathering are at work, as rocks disintegrate into smaller fragments the total surface area increases (figure 12.11), allowing more extensive chemical weathering by water and air.

CHEMICAL WEATHERING

The processes of chemical weathering, or *rock decomposition,* transform rocks and minerals exposed to water and air into new chemical products. A mineral that crystallized deep underground from a water-deficient magma may eventually be exposed at the surface, where it can react with the abundant water there to form a new, different mineral. A mineral containing very little oxygen may react with oxygen in the air, extracting oxygen atoms from the atmosphere and incorporating them into its own crystal structure, thus forming a different mineral. These new minerals are weathering products. They have adjusted to physical and chemical conditions at (or near) Earth's surface. Minerals change gradually at the surface until they come into *equilibrium,* or balance, with the surrounding conditions.

Role of Oxygen

Oxygen is abundant in the atmosphere and quite active chemically, so it often combines with minerals or with elements within minerals that are exposed at Earth's surface.

The rusting of an iron nail exposed to air is a simple example of chemical weathering. Oxygen from the atmosphere combines with the iron to form iron oxide, the reaction being expressed as follows:

$$\underset{\text{iron}}{4Fe} + \underset{\text{oxygen}}{3O_2} \rightarrow \underset{\text{iron oxide}}{2Fe_2O_3}$$

Iron oxide formed in this way is a weathering product of numerous minerals containing iron, such as the ferromagnesian group (pyroxenes, amphiboles, biotite, and olivine). The iron in the ferromagnesian silicate minerals must first be separated from the silica in the crystal structure before it can oxidize. The iron oxide (Fe_2O_3) formed is the mineral **hematite,** which has a brick-red color when powdered. If water is present, as it usually is at Earth's surface, the iron oxide combines with water to form **limonite,** which is the name for a group of mostly amorphous, hydrated iron oxides (often including the mineral *goethite*), which are yellowish-brown when powdered. The general formula for this group is $Fe_2O_3 \cdot nH_2O$ (the *n* represents a small, whole number such as 1, 2, or 3 to show a variable amount of water). The brown, yellow, or red color of soil and many kinds of sedimentary rock is commonly the result of small amounts of hematite and limonite released by the weathering of iron-containing minerals (figure 12.12).

FIGURE 12.12
Sandstone has been colored red by hematite and released by the chemical weathering of ferromagnesian minerals. Southern Utah.
Photo by Diane Carlson

Role of Acid

The most effective agent of chemical weathering is acid. Acids are chemical compounds that give off hydrogen ions (H^+) when they dissociate, or break down, in water. Strong acids produce a great number of hydrogen ions when they dissociate, and weak acids produce relatively few such ions.

The hydrogen ions given off by natural acids disrupt the orderly arrangement of atoms within most minerals. Because a hydrogen ion has a positive electrical charge and a very small size, it can substitute for other positive ions (such as Ca^{++}, Na^+, or K^+) within minerals. This substitution changes the chemical composition of the mineral and disrupts its atomic structure. The mineral decomposes, often into a different mineral, when it is exposed to acid.

Some strong acids occur naturally on Earth's surface, but they are relatively rare. Sulfuric acid and hydrofluoric acid are strong acids emitted during many volcanic eruptions. They can kill trees and cause intense chemical weathering of rocks near volcanic vents. The bubbling mud of Yellowstone National Park's mudpots (figure 12.13) is produced by rapid weathering

FIGURE 12.13

A mudpot of boiling mud is created by intense chemical weathering of the surrounding rock by the acid gases dissolved in a hot spring. Yellowstone National Park, Wyoming.

caused by acidic sulfur gases that are given off by some hot springs. Strong acids also drain from some mines when sulfur-containing minerals such as pyrite oxidize and form acids at the surface (figure 12.14). Uncontrolled mine drainage can kill fish and plants downstream and accelerate rock weathering.

The most important natural source of acid for rock weathering at Earth's surface is dissolved carbon dioxide (CO_2) in water. Water and carbon dioxide form *carbonic acid* (H_2CO_3), a weak acid that dissociates into the hydrogen ion and the bicarbonate ion (see equation *A* in table 12.1). Even though carbonic acid is a weak acid, it is so abundant at Earth's surface that it is the single most effective agent of chemical weathering.

Earth's atmosphere (mostly oxygen and nitrogen) contains 0.03% carbon dioxide. Some of this carbon dioxide dissolves in rain as it falls, so most rain is slightly acidic when it hits the ground. Large amounts of carbon dioxide also dissolve in water that percolates through soil. The openings in soil are filled with a gas mixture that differs from air. Soil gas has a much higher content of carbon dioxide (up to 10%) than does air, because carbon dioxide is produced by the decay of organic matter and the respiration of soil organisms, such as worms. Rainwater that has trick-

FIGURE 12.14

Spring Creek debris dam collects acid mine drainage from the Iron Mountain Mines Superfund site in northern California.
Photo by Charles Alpers, U.S. Geological Survey

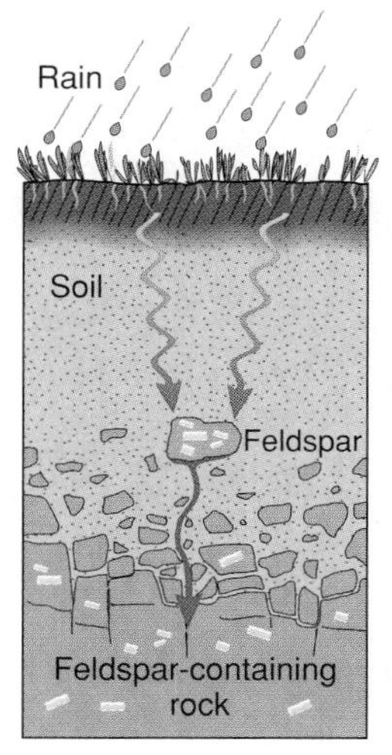

FIGURE 12.15
Chemical weathering of a feldspar. Water percolating through the soil alters the feldspar to a clay mineral and carries away soluble ions and silica.

led through soil is therefore usually acidic and readily attacks minerals in the unweathered rock below the soil (figure 12.15).

Solution Weathering

Some minerals are completely dissolved by chemical weathering. *Calcite,* for instance, goes into solution when exposed to carbon dioxide and water, as shown in equation *B* in table 12.1. The carbon dioxide and water combine to form carbonic acid, which dissociates into the hydrogen ion and the bicarbonate ion, as you have seen, so the equation for the solution of calcite can also be written as equation *C* in table 12.1.

There are no solid products in the last part of the equation, indicating that complete solution of the calcite has occurred. Caves can form underground when flowing ground water dissolves the sedimentary rock limestone, which is mostly calcite. Rain can discolor and dissolve statues and tombstones carved from the metamorphic rock marble, which is also mostly calcite (see figure 12.1).

Chemical Weathering of Feldspar

The weathering of feldspar is an example of the alteration of an original mineral to an entirely different type of mineral as the weathered product. When feldspar is attacked by the hydrogen ion of carbonic acid (from carbon dioxide and water), it forms clay minerals. In general, a **clay mineral** is a hydrous aluminum silicate with a sheet-silicate structure like that of mica. Therefore, the entire silicate structure of the feldspar crystal is altered by weathering: feldspar is a framework silicate, but the clay mineral product is a sheet silicate, differing both chemically and physically from feldspar. Partly because of the complexity of the reaction, the chemical weathering of feldspar proceeds at a much slower rate than the solution weathering illustrated by calcite.

Let us look in more detail at the weathering of feldspar (equation *D* in table 12.1). Rainwater percolates down through soil, picking up carbon dioxide from the atmosphere and the upper part of the soil. The water, now slightly acidic, comes in contact with feldspar in the lower part of the soil (figure 12.15), as shown in the first part of the equation. The acidic water reacts with the feldspar and alters it to a clay mineral.

The hydrogen ion (H^+) attacks the feldspar structure, becoming incorporated into the clay mineral product. When the hydrogen moves into the crystal structure, it releases potassium (K) from the feldspar. The potassium is carried away in solution as a dissolved ion (K^+). The bicarbonate ion from the original

table 12.1 **Chemical Equations Important to Weathering**

A. Solution of Carbon Dioxide in Water to Form Acid

CO_2	+	H_2O	←	H_2CO_3	←	H^+	+	HCO_3^-
carbon dioxide		water		carbonic acid		hydrogen ion		bicarbonate ion

B. Solution of Calcite

$CaCO_3$	+	CO_2	+	H_2O	⇄	Ca^{++}	+	$2HCO_3^-$
calcite		carbon dioxide		water		calcium ion		bicarbonate ion

C. Solution of Calcite

$CaCO_3$	+	H^+	+	HCO_3^-	⇄	Ca^{++}	+	$2HCO_3^-$

D. Chemical Weathering of Feldspar to Form a Clay Mineral

$2KAlSi_3O_8$	+	$2H^+ + 2HCO_3^-$	+	H_2O	→	$Al_2Si_2O_5(OH)_4$	+	$2K^+ + 2HCO_3^-$	+	$4\ SiO_2$
potassium feldspar		(from CO_2 and H_2O)				clay mineral		(soluble ions)		silica in solution or as fine solid particles

12.1

ENVIRONMENTAL GEOLOGY

ACID RAIN

The burning of coal, oil, and natural gas (the *fossil fuels*) adds a great deal of carbon dioxide to the atmosphere (box figure 1). As you have seen in table 12.1, this carbon dioxide combines with water to form carbonic acid in rain. Coal and oil can also contain nitrogen and sulfur, which are given off as gases (NO_2 and SO_2) when these fuels are burned, forming nitric acid and sulfuric acid in rain. These two acids are much stronger than carbonic acid.

The strength of an acidic solution is measured on the pH scale from 0 to 14 (box figure 2). A solution of pH 7 is chemically neutral, neither acidic nor alkaline. Values below 7 are acidic; the lower the number the more acidic the solution. Values above 7 are alkaline or basic. The pH scale is logarithmic, so a change of 1 on the scale means a change of 10 in the concentration of H^+ ions that make a solution acidic.

Ordinary rain has a pH of about 5.5 to 6 from the small amount of carbon dioxide given off during respiration (every time we exhale, we add a little CO_2 to the atmosphere), and from natural sources of acidic sulfur gases such as volcanoes and coastal marshes. Ordinary rain is about as acidic as milk—hardly a strong acid.

In cities and downwind of industrial smokestacks (often for hundreds of miles) the increased amount of acid gases can reduce the pH of rain to 4, 3, or even 2 (the pH of lemon juice or vinegar). This is the environmental problem termed *acid rain* (although all rain is really acid). This rain in turn can lower the pH of streams, lakes, and soils. Such low pH values are hard on organisms; fish may die in streams and lakes polluted by acid rain, and forests suffer under acid rain. This is particularly a problem where rocks and soil do not buffer the acid. For example, areas with exposed limestone are least affected.

Chemical weathering is accelerated by acid rain. Statues and stone buildings in cities weather many times faster than stone structures in rural areas free of acid rain. See figure 12.1.

Related Web Sources

http://minerals.er.usgs.gov/acid1.html
http://bqs/usgs.gov/acidrain/index.htm

BOX 12.1 — FIGURE 1

The burning of fossil fuels releases carbon dioxide and other acid gases, such as sulfur dioxide and nitrous oxide, that combine with water to produce acid rain.

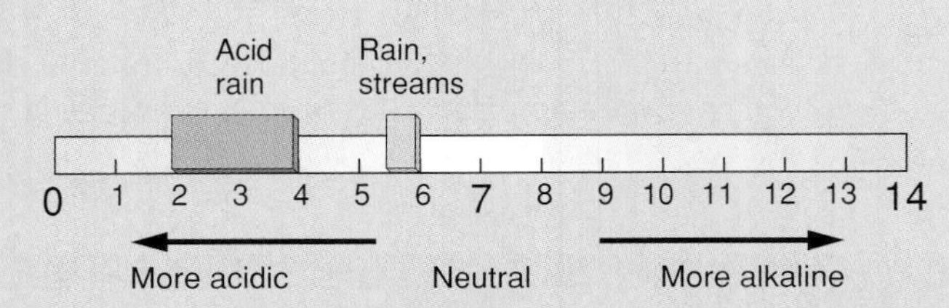

BOX 12.1 — FIGURE 2

The pH scale.

carbonic acid does not enter into the reaction; it reappears on the right side of the equation. The soluble potassium and bicarbonate ions are carried away by water (ground water or streams).

All the silicon from the feldspar cannot fit into the clay mineral, so some is left over and is carried away as silica (SiO_2) by the moving water. This excess silica may be carried in solution or as extremely small solid particles.

The weathering process is the same regardless of the type of feldspar: K-feldspar forms potassium ions; Na-feldspar and Ca-feldspar (plagioclase) form sodium ions and calcium ions, respectively. The ions that result from the weathering of Ca-feldspar are calcium ions (Ca^{++}) and bicarbonate ions (HCO_3^-), both of which are very common in rivers and in underground water, particularly in humid regions.

Chemical Weathering of Other Minerals

The weathering of ferromagnesian or dark minerals is much the same as that of feldspars. Two additional products are found on the right side of the equation—magnesium ions and iron oxides (hematite, limonite, and goethite).

The susceptibility of the rock-forming minerals to chemical weathering is dependent upon the strength of the mineral's chemical bonding within the crystal framework. Because of the strength of the silicon-oxygen bond, quartz is quite resistant to chemical weathering. Thus, quartz (SiO_2) is the rock-forming mineral least susceptible to chemical attack at Earth's surface. Ferromagnesian minerals such as olivine, pyroxene, and amphi-

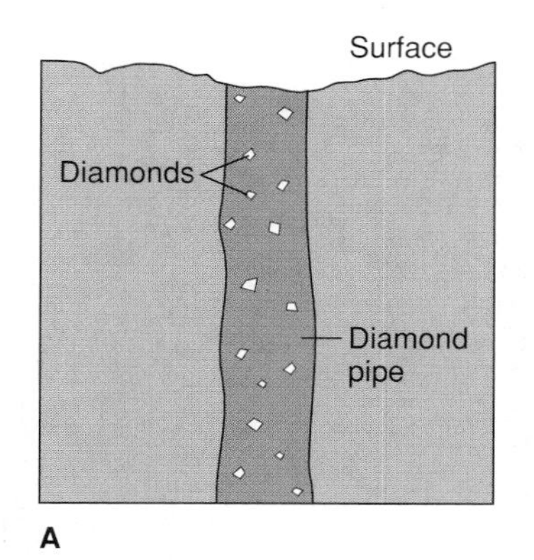

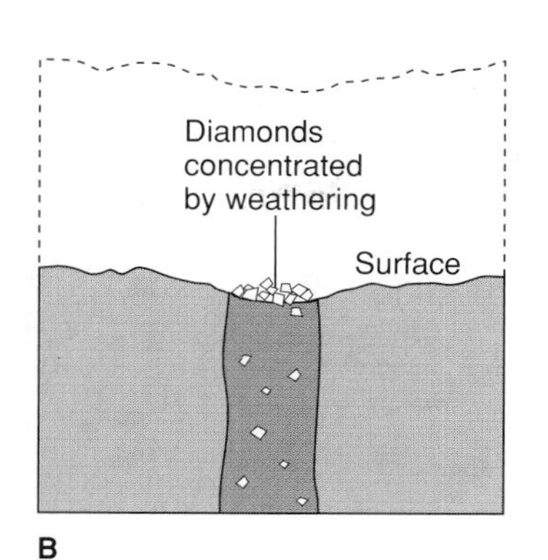

FIGURE 12.16

Residual concentration by weathering. (*A*) Cross-sectional view of diamonds widely scattered within diamond pipe. (*B*) Diamonds concentrated on surface by removal of rock by weathering and erosion.

bole include other positively charged ions such as Al, Fe, Mg, and Ca. The presence of these positively charged ions in the crystal framework makes these minerals vulnerable to chemical attack due to the weaker chemical bonding between these ions and oxygen, as compared to the much stronger silicon-oxygen bonds. For example olivine $(Fe, Mg)_2SiO_4$ weathers rapidly because its isolated silicon-oxygen tetrahedra are held together by relatively weak ionic bonds between oxygen and iron and magnesium. These ions are replaced by H^+ ions during chemical weathering similar to that described for the feldspars.

Weathering and Diamond Concentration

Diamond is the hardest mineral known and is also extremely resistant to weathering. This is due to the very strong covalent bonding of carbon, as described in chapter 9. But, diamonds are often concentrated by weathering as illustrated in figure 12.16. Diamonds are brought to the surface of Earth in *diamond pipes,* columns of brecciated or broken ultramafic rock that have risen from the upper mantle. Diamonds are widely scattered in diamond pipes when they form. At Earth's surface the ultramafic rock in the pipe is preferentially weathered and eroded away. The diamonds, being more resistant to weathering, are left behind, concentrated in rich deposits on top of the pipes. Rivers may redistribute and reconcentrate the diamonds, as in South Africa and India. In Canada, diamond pipes have been eroded by glaciers, and diamonds may be found widely scattered in glacial deposits.

Weathering and Climate

Chemical weathering is most intense where there is abundant liquid water. If water is scarce, as in deserts, or is frozen into solid ice, chemical weathering is slow or even absent. As a result, the resistance of different rock types to chemical weathering varies from climate to climate. In the wet climate of central Pennsylvania, for example, limestone dissolves readily to form valleys; shale forms low hills; and the ridges of the Appalachian Mountains are mostly formed of resistant, quartz-rich sandstone. In the dry climate of Arizona, however, limestone is as resistant as quartz sandstone, so both limestone and sandstone form the resistant cliffs within the Grand Canyon, while less resistant shale forms the gentle slopes between cliffs (see figure 12.4).

Weathering Products

Table 12.2 summarizes weathering products for the common minerals. Note that quartz and clay minerals commonly are left after complete chemical weathering of a rock. Sometimes other solid products, such as iron oxides, also are left after weathering.

The solution of calcite supplies substantial amounts of calcium ions (Ca^{++}) and bicarbonate ions (HCO_3^-) to underground water. The weathering of Ca-feldspars (plagioclase) into clay minerals can also supply Ca^{++} and HCO_3^- ions, as well as silica (SiO_2), to water. Under ordinary chemical circumstances, the dissolved Ca^{++} and HCO_3^- can combine to form solid $CaCO_3$ (calcium carbonate), the mineral calcite. Dissolved silica can also precipitate as a solid from underground

table 12.2 Weathering Products of Common Rock-Forming Minerals

Original Mineral	Under Influence of CO_2 and H_2O	Main Solid Product		Other Products (Mostly Soluble)
Feldspar	→	Clay mineral	+	Ions (Na^+, Ca^{++}, K^+), SiO_2
Ferromagnesian minerals (including biotite mica)	→	Clay mineral	+	Ions (Na^+, Ca^{++}, K^+, Mg^{++}), SiO_2, Fe oxides
Muscovite mica	→	Clay mineral	+	Ions (K^+), SiO_2
Quartz	→	Quartz grains (sand)		
Calcite	→	—		Ions (Ca^{++}, HCO_3^-)

water. This is significant because calcite and silica are the most common materials precipitated as *cement,* which binds loose particles of sand, silt, and clay into solid sedimentary rock (see chapter 14). The weathering of calcite, feldspars, and other minerals is a likely source for such cement.

If the soluble ions and silica are not precipitated as solids, they remain in solution and may eventually find their way into a stream and then into the ocean. Enormous quantities of dissolved material are carried by rivers into the sea (one estimate is 4 billion tons per year). This is the main reason seawater is salty.

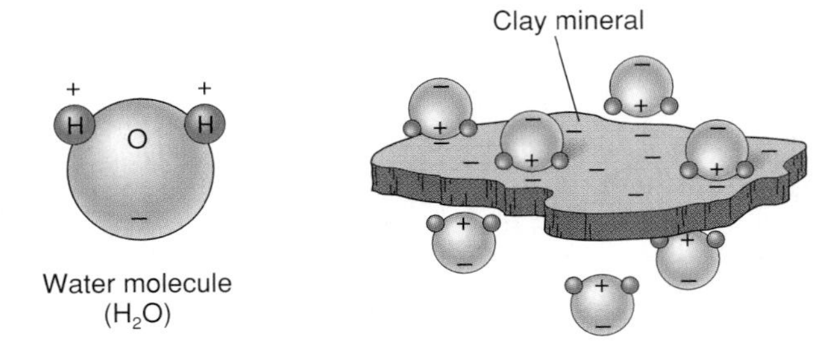

FIGURE 12.17

Negative charges on a clay mineral attract positive ends of water molecules.

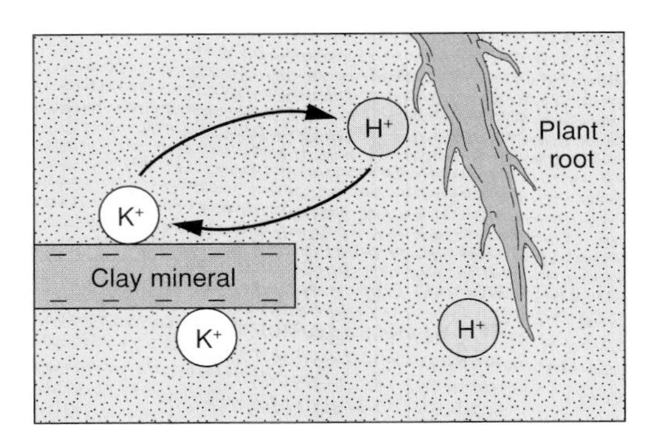

FIGURE 12.18

Ion exchange between plant root and clay mineral.

SOIL

In civil engineering and construction, soil is the usual name for any kind of loose, unconsolidated earth material; but most geologists commonly use the term **soil** for a layer of *weathered,* unconsolidated material on top of bedrock (a general term for rock beneath soil). Soil scientists further restrict the term *soil* to horizons of weathered, unconsolidated material that contains organic matter and is capable of supporting plant growth. (If this definition is used, then the term *regolith* can be applied to any loose surface sediment; soil would be the upper part of the regolith.) A mature, fertile soil is the product of centuries of mechanical and chemical weathering of rock, combined with the addition and decay of plant and other organic matter.

The term **loam** refers to a soil of approximately equal amounts of sand, silt, and clay. (*Clay-sized* particles usually consist of *clay minerals*). Loamy soils are often well-drained, may contain organic matter, and are often very fertile. *Topsoil* is the upper part of the soil and is more fertile than the underlying *subsoil,* which is often stony and lacks organic matter.

Clay minerals and quartz, the two minerals usually remaining after complete weathering of rock (table 12.2), have important roles in soil development and plant growth. Quartz crystals form sand grains that help keep soil loose and aerated, allowing good water drainage. (Partially weathered crystals of feldspar and other minerals can also form sand-sized grains.)

Clay minerals help to hold water and plant nutrients in a soil. Clay minerals occur as microscopic plates. Because of ion substitution within their sheet silicate structure, most clay minerals have a negative electrical charge on the flat faces of the plates. This negative charge attracts water and nutrient ions to the clay mineral.

The water molecule, made up of two hydrogen atoms and one oxygen atom, is neutral in charge but has a positive end and a negative end. The negative charge on the flat faces of the clay mineral attracts the positive ends of the water molecules to the clay flake (figure 12.17). The clay holds the water loosely enough that most of it is available for uptake by plant roots.

Plant nutrients, such as Ca^{++} and K^{+}, commonly supplied by the weathering of minerals such as feldspar, are also held loosely on the surface of clay minerals. A plant root is able to release H^{+} from organic acids and exchange it for the Ca^{++} and K^{+} that the plant needs for healthy growth (figure 12.18).

Soil Horizons and Classification

Most soils take a long time to form. The rate of soil formation is controlled by rainfall, temperature, slope, and to some extent the type of bedrock that weathers to form soil. High temperature and abundant rainfall speed up soil formation, but in most places a fully developed soil that can support plant growth takes hundreds or thousands of years to form.

As soils mature, distinct layers appear in them (figure 12.19). Soil layers are called **soil horizons** and can be distinguished from one another by appearance and chemical composition. Boundaries between soil horizons are usually transitional rather than sharp. By observing a vertical cross section, or *soil profile,* various horizons can be identified.

The **O horizon** is the uppermost layer that consists entirely of non-decomposed and highly decomposed organic material. For example, fallen leaves and needles along with ground vegetation would constitute the O horizon in a forested area.

The **A horizon** is the dark-colored soil layer that is rich in organic material and forms just below surface vegetation. This horizon contains decomposed plant material, or *humus,* and contributes to the formation of organic acids that accelerate leaching in the lower part of the A horizon (called *E horizon*). The lower part of the **A horizon,** or **zone of leaching,** is characterized by the downward movement of water. Part of the rain falling on the ground percolates downward through the soil. This tends to leach, or carry dissolved chemicals downward to

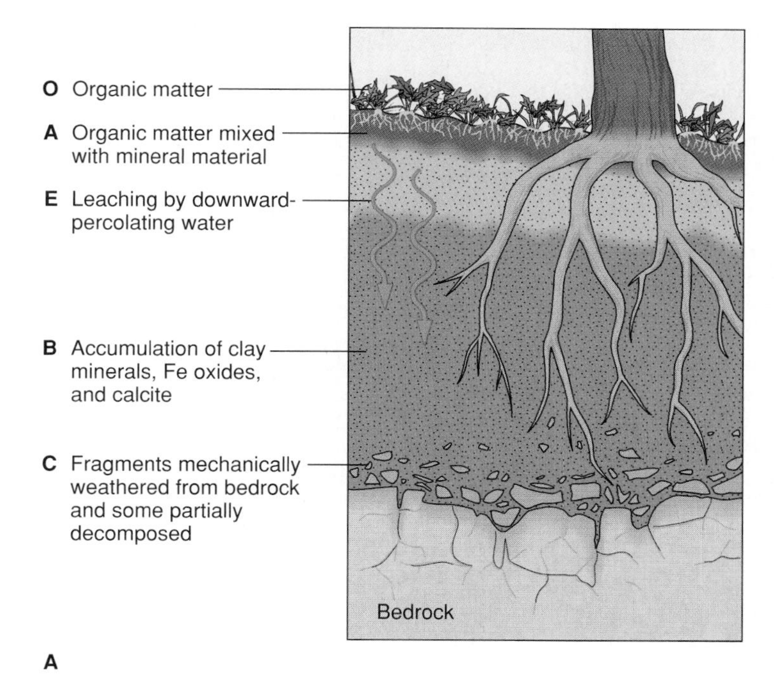

FIGURE 12.19

(*A*) Horizons (*O, A, B,* and *C*) in a soil profile that form in a humid climate. (*B*) Soil profile exposed in Illinois that shows the *A* horizon stained dark by humus. The lower part of Horizon *A* is lighter in color, sandy, and crumbly. The hammer rests on the clayey horizon *B,* which is stained red by hematite, leached downward from bottom of horizon *A.*

lower levels in the soil profile. In a humid (wet) climate, iron oxides and dissolved calcite are most typically leached downward; clays are also transported downward. Leaching may make the lower part of the A horizon pale and sandy, but the uppermost part is often darkened by humus.

The **B horizon,** or **zone of accumulation,** is a soil layer characterized by the accumulation of material leached downward from the A horizon above. This layer is often quite clayey and stained red or brown by hematite and limonite. Calcite may also build up in B horizons.

The **C horizon** is incompletely weathered parent material lying below the B horizon. The parent material is commonly the underlying bedrock, which is subjected to mechanical and chemical weathering from frost action, roots, plant acids, and other agents. In such a case, the C horizon is transitional between unweathered bedrock below and developing soil above.

The Soil Conservation Service of the U.S. Department of Agriculture has developed a soil classification system to group soils with similar properties so that soils can be mapped in a systematic way. There are 12 large groups called *orders* that are distinguished by the characteristics of the horizons present in soil profiles. Brief descriptions of the orders are given in table 12.3, and their distribution in the continental United States is shown on the map in figure 12.20. Each order can be further subdivided into many subdivisions, or suborders, which are defined by even more specific diagnostic physical and chemical properties observed in the soil.

Residual and Transported Soils

A **residual soil** is one that develops from weathering of the rock directly beneath it. Figure 12.19*A* is a diagram of a residual soil developing in a humid climate from a bedrock source. Although this is a typical situation, a number of important agricultural regions in the United States and elsewhere have developed on **transported soils,** which did not form from the local rock but from regolith brought in from some other region. Transported soils usually form on sediment deposited by running water, wind, or glacial ice. For example, mud deposited by a river during times of flooding can form an excellent agricultural soil next to the river after floodwaters recede. The soil-forming mud was not weathered from the rock beneath its present location but was carried downstream from regions perhaps hundreds of miles away. Transported wind deposits called *loess* (see chapter 18) are the parent material for some of the most valuable food-producing soils in the Midwest and the Pacific Northwest.

Soils, Parent Material, Time, and Slope

The character of a soil depends partly on the parent material from which it develops. A soil developing on weathering granite will be sandy, as sand-sized particles of quartz and partially weathered feldspar are released from the granite. As time

table 12.3 Soil Orders

Soil Orders (meaning of name)	Description
Gelisols (frozen soils)	Soils with permafrost within 2 meters of the surface.
Histosols (organic soils)	Wet, organic soils such as peat in swamps and marshes.
Spodosols (ashy soils)	Acid soils low in plant nutrient ions with subsurface accumulation of organic matter and compounds of aluminum and iron. Cool, humid forests.
Andisols (volcanic ash)	Soils formed in volcanic ash.
Oxisols (oxide soils)	Heavily weathered soils low in plant nutrient ions and rich in aluminum and iron oxides. Tropical, usually moist.
Vertisols (inverted soils)	Clayey soils that swell when wet and shrink when dry, forming wide, deep cracks.
Aridisols (arid soils)	Dry, desert soils low in organic matter and with carbonate horizons.
Ultisols (ultimate soils)	Strongly weathered soils low in plant nutrient ions with clay accumulation in the subsurface. Usually moist.
Mollisols (soft soils)	Nearly black surface horizon rich in organic matter and plant nutrient ions. Subhumid to subarid grasslands.
Alfisols (pedalfers)	Gray to brown surface horizon, subsurface horizon of clay accumulation. Medium to high in plant nutrient ions. Usually moist, as in humid forests.
Inceptisols (beginning soils)	Very young soils that have weak horizons. Usually moist.
Entisols (recent soils)	Soils that have no horizons.

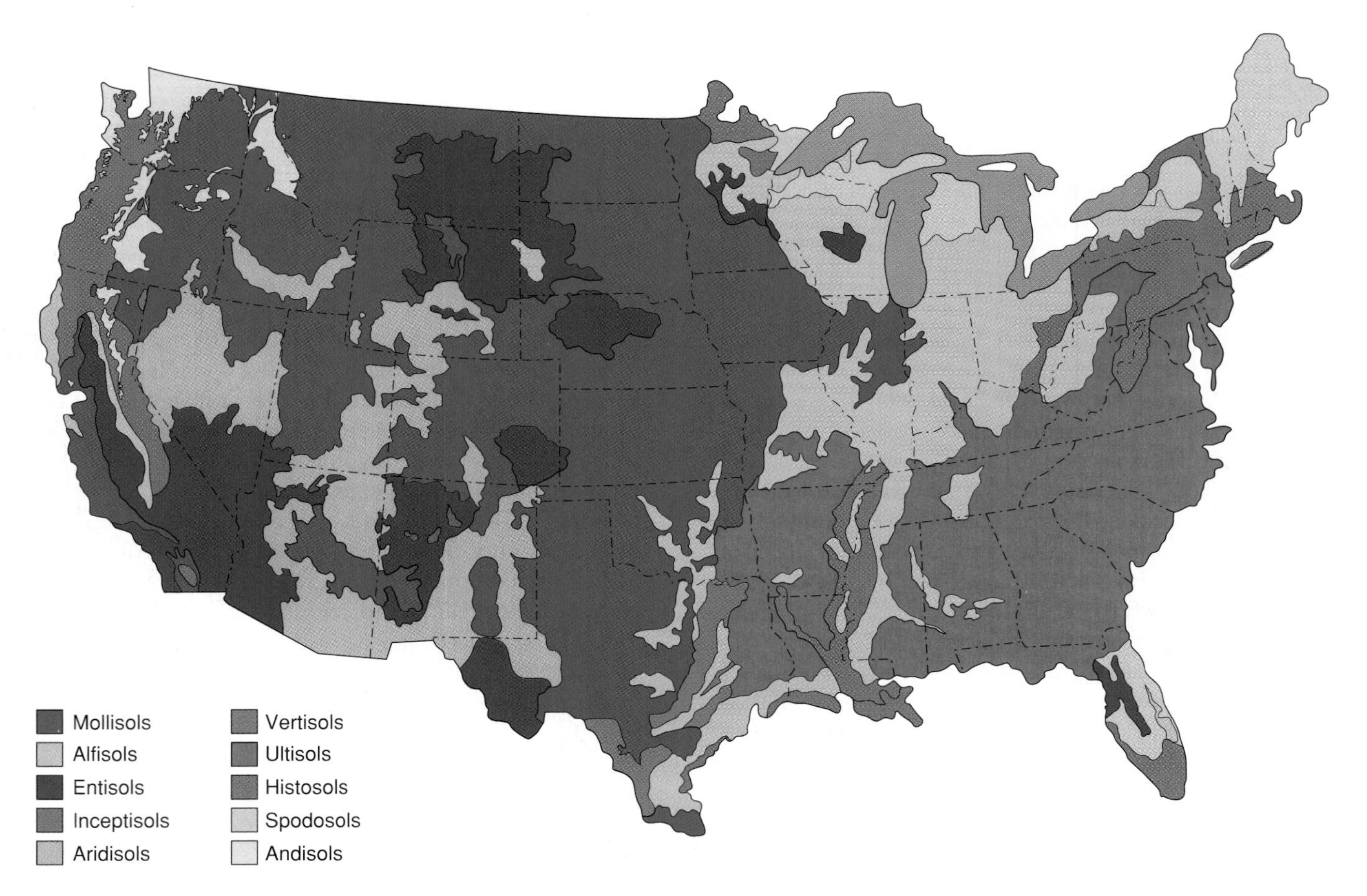

FIGURE 12.20
Distribution of soil orders in the United States.
U.S.D.A. Soil Conservation Map.

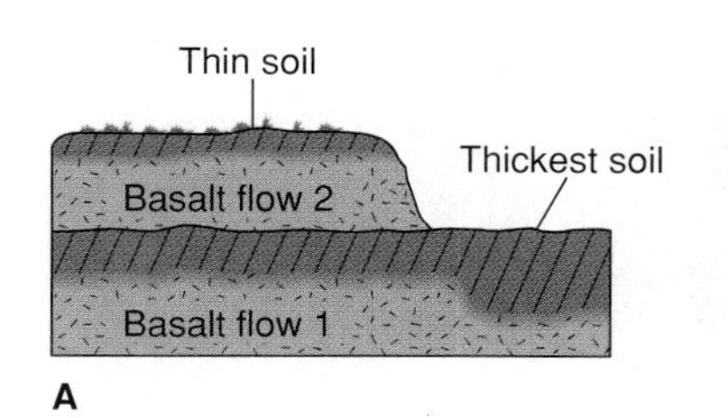

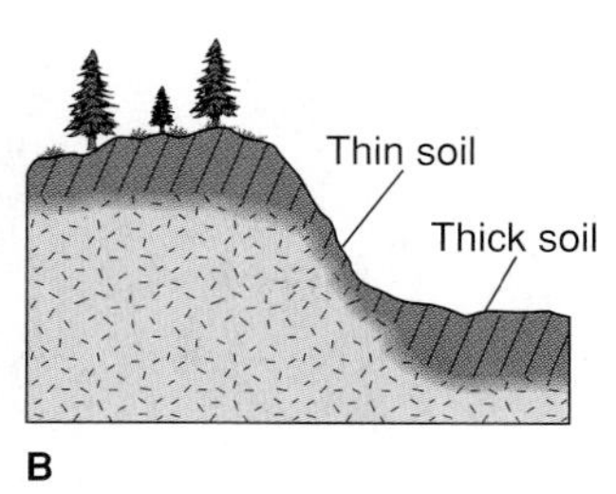

FIGURE 12.21

Soil thickness. (*A*) Soil thickens with time. Basalt flow 1 was exposed to soil-forming processes for a longer time than flow 2. The soil developed below flow 2 is an example of a buried soil. (*B*) Steep slopes have thin soil.

passes, the partially weathered feldspar grains weather completely, forming fine-grained clay minerals. The quartz does not weather, so the resulting soil has both sand and clay (and perhaps silt) in it.

A soil forming on basalt may never be sandy, even in its early stages of development (this depends on the relative rates of chemical weathering versus mechanical weathering). The fine-grained feldspars in the basalt weather to fine-grained clay minerals. Since the parent rock had no coarse-grained minerals and no quartz to start with, the resulting soil may lack sand. Such a soil may not drain well, although it can be quite fertile.

Note that the character of a soil changes with time. A soil developing from granite begins as a sandy soil and becomes more clayey with time. Over very long periods, the type of parent rock becomes less and less important. Given enough time, soils forming from many different kinds of igneous, metamorphic, and sedimentary rocks can become quite similar (in the same climate). The presence or absence of coarse grains of quartz in the parent rock becomes the only characteristic of the parent rock to have long-term significance.

With time, soils tend to become thicker (figure 12.21*A*); most modern soils have taken centuries to form. A new deposit of volcanic ash, which is very fine-grained and rich in plant nutrients, may be covered with grass and other low plants in just a few years, but a new lava flow, which weathers much more slowly than ash, may not have enough soil to support grass for many decades. It may take centuries for the lava-flow soil to thicken enough to support shrubs, and thousands of years to support trees. The fertile agricultural soils of the Canadian plains and the northern United States took more than 10,000 years to develop on glacial deposits after the thick continental ice sheets melted (see figure 19.34).

Another factor controlling soil thickness is the slope of the land surface (figure 12.21*B*). Soils tend to be thick on flat land where erosion is slow and water can collect, and thin on steep slopes where gravity pulls water and soil particles downhill.

Organic Activity

Organisms contribute to soil development. Plant roots break up rocks and burrowing organisms such as ants, worms, and rodents bring soil particles to the surface and create passageways for water and air to get underground, thus speeding up chemical weathering. Respiration of soil organisms and decay of plant and animal material adds carbon dioxide gas to soil, creating carbonic acid. Plants and humus also release organic acids that increase chemical weathering. Once soil begins to develop on a newly exposed rock, it attracts plants and soil organisms that increase chemical weathering, accelerating the rate of soil development. Partially decayed organic matter provides plant nutrients, increasing soil fertility.

Soils and Climate

Climate affects soil thickness and character. Soils in wet climates, as in Europe, most of Canada, and the eastern United States, tend to be thick and are generally characterized by downward movement of water through Earth materials (figure 12.19 shows such a soil). In general these soils tend to have a high content of aluminum and iron oxides and are marked by effective downward leaching due to high rainfall and to the acids produced by the decay of abundant humus.

In arid (dry) climates, as in many parts of the western United States, soils tend to be thin and are characterized by little leaching, scant humus, and the *upward* movement of soil water beneath the land surface. The water is drawn up by subsurface evaporation and capillary action.

The evaporation of water beneath the land surface can cause the precipitation of salts within the soil. These salts are usually calcium salts such as calcite (figure 12.22). An extreme example of salt buildup can be found in desert *alkali soils,* in which heavy concentrations of toxic sodium salts may prevent plant growth.

Hardpan

Hardpan is a general term for a hard layer of Earth material that is difficult to dig or drill. Geologists usually restrict the term to a hard, often clayey, layer of cemented soil particles. Such a layer may be too hard for even backhoes to dig through; planting a tree in a lawn with a hardpan layer may require a jackhammer. Hardpan layers in wet climates are usually formed of clay minerals, silica, and iron compounds that have accumulated in the B horizon. In arid climates a different type of hardpan forms from the cementing of soil by calcium carbonate and other salts that precipitate in the soil as water evaporates. Both types of hardpan are really layers of rock within loose soil. A hardpan layer can break plows, prevent water drainage through the soil, and act as a barrier to plant roots. Tree roots may grow laterally along rather than down through hardpan; such shallow-rooted trees are easily uprooted by wind.

Laterites

In tropical regions where temperatures are high and rainfall is abundant, highly leached soils called **laterites** (*oxisols*) form. Under such conditions weathering is deep and intense. Laterites are usually red and are composed almost entirely of iron and aluminum oxides, generally the least soluble products of rock

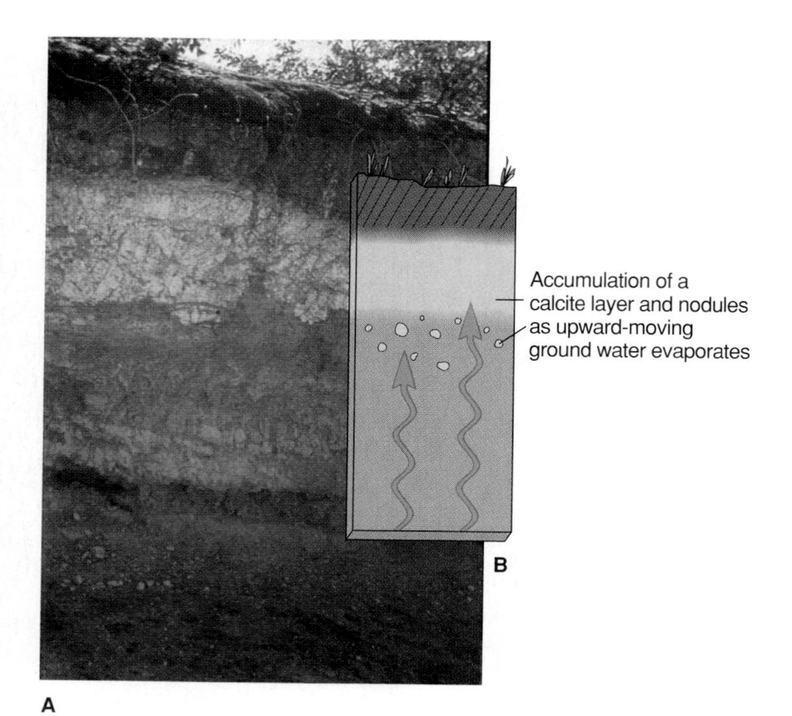

FIGURE 12.22

Soil profile marked by upward-moving ground water that evaporates underground in a drier climate, precipitating calcium carbonate within the soil, sometimes forming a light-colored layer.

Photo by D. Yost, USDA–Soil Conservation Service

FIGURE 12.23

Laterite soil (oxisol) develops in very wet climates, where intense downward leaching carries away all but iron and aluminum oxides. Many laterites are deep red.

Photo by D. Yost, USDA–Soil Conservation Service

weathering in tropical climates (figure 12.23). If the soil is rich in hematite, it can be mined as iron ore, but tropical rainfall usually hydrates the hematite to limonite, which is seldom rich enough to mine. However, aluminum is sometimes found in nearly pure layers of *bauxite* ($Al_2O_3 \cdot nH_2O$, the principal ore of aluminum), particularly in laterites formed by the weathering of aluminum-rich volcanic tuffs.

Under tropical conditions of high rainfall and high temperature, most weathering products are soluble—even silica. The least soluble product is aluminum oxide, which remains on top of the weathering rocks, forming bauxite in a soil very rich in aluminum. Like the diamonds, the aluminum has been concentrated residually by the removal of everything else. The aluminum ores may be redistributed slightly by running water (figure 12.24).

Because bauxite forms under conditions of tropical weathering, the United States has very little aluminum ore and depends almost entirely on recycling and tropical countries for its aluminum supply. A small percentage of the U.S. aluminum supply has come from bauxite deposits near Little Rock, Arkansas, that formed on igneous rock with a high aluminum content approximately 50 million years ago when the region had a tropical climate.

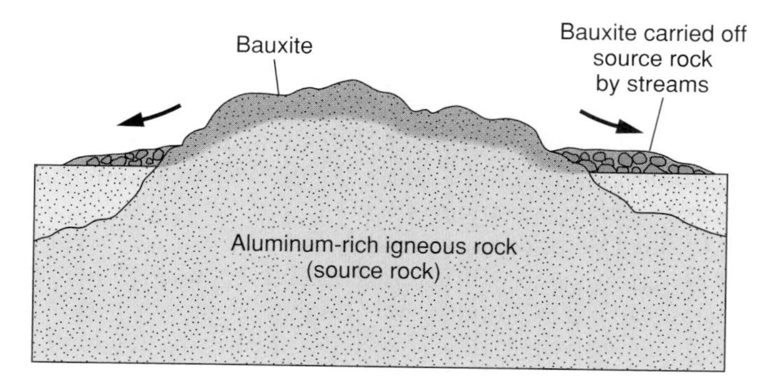

FIGURE 12.24

Bauxite forms by intense tropical weathering of an aluminum-rich source rock such as a volcanic tuff.

Laterites are relatively nonproductive soils. This may seem strange when you think of the lush jungle growth that often exists on tropical lateritic soils. Jungle vegetation, though, is nourished largely by a layer of humus on top of the soil. If the jungle and the humus layer are cleared away or burned—an increasingly common practice in tropical regions—the laterite quickly becomes incapable of sustaining plant growth, making tropical agriculture very difficult (figure 12.25). Laterite exposed to the sun is apt to bake into a permanent, bricklike layer that makes digging nearly impossible. This hard layer can be quarried, however, and makes a durable building material.

FIGURE 12.25
Soil erosion caused by clear cutting of rain forest north of Kuantan, Malaysia.
Photo © George Loun/Visuals Unlimited

Buried Soils

A soil may become buried by volcanic ash, windblown dust, glacial deposits, other sediment, or lava (see figure 12.21). A buried soil is called a *paleosol* (paleo = ancient). Such soils may be distinctive and traceable over wide regions, making them useful for dating rocks and sediments and for interpreting past climates and topography.

Summary

When rocks that formed deep in Earth become exposed at the Earth's surface, they are altered by *mechanical* and *chemical weathering.*

Weathering processes produce *spheroidal weathering, differentially weathered* landforms, *sheet joints,* and *exfoliation domes.*

Mechanical weathering, largely caused by *frost action* and *pressure release* after unloading, disintegrates (breaks) rocks into smaller pieces.

By increasing the exposed surface area of rocks, mechanical weathering helps speed chemical weathering.

Chemical weathering results when a mineral is unstable in the presence of water and atmospheric gases. As chemical weathering proceeds, the mineral's components recombine into new minerals that are more in equilibrium.

Weak acid, primarily from the solution of carbon dioxide in water, is the most effective agent of chemical weathering.

Calcite dissolves when it is chemically weathered. Most of the silicate minerals form *clay minerals* when they chemically weather. Quartz is very resistant to chemical weathering.

Soil develops by chemical and mechanical weathering of a parent material. Some definitions of soil require that it contain organic matter and be able to support plant growth.

Soils, which can be *residual* or *transported,* usually have distinguishable layers, or *horizons,* caused in part by water movement within the soil.

Climate is the most important factor determining soil type. Other factors in soil development are parent material, time, slope, and organic activity.

Laterites form under conditions of intense tropical weathering; they are usually red from concentrated iron oxides. Bauxite, the ore of aluminum, may be found in laterites.

Terms to Remember

A horizon (zone of leaching) 304
B horizon (zone of accumulation) 305
chemical weathering 294
C horizon 305
clay mineral 301
differential weathering 294
erosion 294
exfoliation 296
exfoliation dome 296
frost action 296
frost heaving 296
frost wedging 296
hematite 299
laterite 307
limonite 299
loam 304
mechanical weathering 294
O horizon 304
pressure release 296
residual soil 305
sheet joints 296
soil 304
soil horizon 304
spheroidal weathering 294
transportation 294
transported soil 305
weathering 294

Testing Your Knowledge

Use the questions below to prepare for exams based on this chapter.

1. Why are some minerals stable several kilometers underground but unstable at Earth's surface?
2. Describe what happens to each mineral within granite during the complete chemical weathering of granite in a humid climate. List the final products for each mineral.
3. Explain what happens chemically when calcite dissolves. Show the reaction in a chemical equation.
4. Why do stone buildings tend to weather more rapidly in cities than in rural areas?
5. Describe at least three processes that mechanically weather rock.
6. How can mechanical weathering speed up chemical weathering?
7. Name at least three natural sources of acid in solution. Which one is most important for chemical weathering?
8. What is the difference between a residual soil and a transported soil?
9. What is a laterite and how does it form?
10. What are soil horizons?
11. Name the soil horizons. How do they form?
12. Physical disintegration of rock into smaller pieces is called
    a. chemical weathering b. transportation
    c. deposition d. mechanical weathering
13. The decomposition of rock from exposure to water and atmospheric gases is called
    a. chemical weathering b. transportation
    c. deposition d. mechanical weathering
14. Which is not a type of mechanical weathering?
    a. frost wedging b. frost heaving
    c. pressure release d. oxidation
15. The single most effective agent of chemical weathering at Earth's surface is
    a. carbonic acid H_2CO_3 b. water H_2O
    c. carbon dioxide CO_2 d. hydrochloric acid HCl
16. The most common end product of the chemical weathering of feldspar is
    a. clay minerals b. pyroxene
    c. amphibole d. calcite
17. The most common end product of the chemical weathering of quartz is
    a. clay minerals b. pyroxene
    c. amphibole d. calcite
    e. quartz does not usually weather chemically
18. Soil with approximately equal amounts of sand, silt, and clay along with a generous amount of organic matter is called
    a. loam b. inorganic
    c. humus d. caliche
19. Which is characteristic of soil horizons?
    a. they can be distinguished from one another by appearance and chemical composition
    b. boundaries between soil horizons are usually transitional rather than sharp
    c. they are classified by letters
    d. all of the above
20. The soil horizon containing only organic material is the
    a. A horizon b. B horizon
    c. C horizon d. O horizon
21. Hardpan forms in the
    a. A horizon b. B horizon
    c. C horizon d. O horizon
22. Tropical soils are typically
    a. rich in organic material b. very fertile
    c. deeply leached d. easily replenished

Expanding Your Knowledge

1. Which mineral weathers faster—hornblende or quartz? Why?
2. Compare and contrast the weathering rate and weathering products for Ca-rich plagioclase in the following localities:
    a. central Pennsylvania with 40 inches of rain per year;
    b. Death Valley with 2 inches of rain per year;
    c. an Alaskan mountaintop where water is frozen year-round.
3. The amount of carbon dioxide gas has been increasing in the atmosphere for the past 40 years as a result of the burning of fossil fuels. What effect will the increase in CO_2 have on the rate of chemical weathering? The increase in CO_2 may cause global warming in the future. What effect would a warmer climate have on the rate of chemical weathering? Give the reasons for your answers.
4. In a humid climate, is a soil formed from granite the same as one formed from gabbro? Discuss the similarities and possible differences with particular regard to mineral content and soil color.

Exploring Web Resources

www.mhhe.com/plummer9e
Visit the Online Learning Center to review your Testing Your Knowledge answers. This website also has additional quizzing and direct links to the sites listed below as well as reading articles and other media resources.

http://soils.ag.uidaho.edu/soilorders/
University of Idaho Soil Science Division. Web page contains photos, descriptions, and surveys of the 12 major soil orders.

http://res.agr.ca/cansis/_overview.html
Canadian Soil Information System provides links to detailed soil surveys and land inventories.

CHAPTER

13

Mass Wasting

When material on a hillside has weathered (the process described in chapter 12), it is likely to move downslope because of the pull of gravity. Soil or rock moving in bulk at Earth's surface is called mass wasting. Mass wasting is one of several surficial processes. Other processes of erosion, transportation, and deposition—involving streams, glaciers, wind, and ocean waves—are discussed in the following chapters.

Landsliding is the best known type of mass wasting. Landslides destroy towns and kill people. While these disasters involve relatively rapid movement of debris and rock, mass wasting can also be very slow. Creep is a type of mass wasting too slow to be called a landslide.

In this chapter we describe how different types of mass wasting shape the land and alter the environment and what factors control the rapidity or slowness of the process. Understanding mass wasting and its possible hazards is particularly important in hilly or mountainous regions.

Opposite: January 13, 2001 landslide at Santa Tecla, El Salvador that was triggered by an earthquake.
Photo © AP/Wide World Photos

You may recall from previous chapters that mountains are products of tectonic forces. Most mountains are associated with present or past converging plate boundaries. If tectonism were not at work, the surfaces of the continents would long ago have been reduced to featureless plains due to weathering and erosion. We consider the material on mountain slopes or hillsides to be out of equilibrium with respect to gravity. Because of the force of gravity, the various agents of erosion (moving water, ice, and wind) work to make slopes gentler and therefore increasingly more stable. The process of erosion discussed in this chapter is mass wasting.

Mass wasting (also called mass movement) is movement in which bedrock, rock debris, or soil moves downslope in bulk, or as a mass, because of the pull of gravity. Mass wasting includes movement so slow that it is almost imperceptible (called *creep*) as well as **landslides,** a general term for the slow to very rapid descent of rock or soil.

Mass wasting affects humans in many ways. Its effects range from the devastation of a killer landslide (see box 13.1) to the nuisance of having a fence slowly pulled apart by soil creep. The cost in lives and property from landslides is surprisingly high. According to the U.S. Geological Survey, more people in the United States died from landslides during the last three months of 1985 than were killed during the last twenty years by all other geologic hazards, such as earthquakes and volcanic eruptions. Over time, landslides have cost Americans triple the combined costs of earthquakes, hurricanes, floods, and tornadoes. On average, the annual cost of landslides in the United States has been 1.5 billion dollars and 25 lost lives. In many cases of mass wasting, a little knowledge of geology, along with appropriate preventive action, could have averted destruction.

CLASSIFICATION OF MASS WASTING

A number of systems are used by geologists, engineers, and others for classifying mass wasting, but none has been universally accepted. Some are very complex and useful only to the specialist.

The classification system used here and summarized in table 13.1 is based on (1) rate of movement, (2) type of material, and (3) nature of the movement.

Rate of Movement

A landslide like the one in Peru (box 13.1) clearly involves rapid movement. Just as clearly, movement of soil at a rate of less than a centimeter a year is slow movement. Between these extremes is a wide range of velocities.

Type of Material

Mass wasting processes are usually distinguished on the basis of whether the descending mass started as bedrock (as in a rockslide) or as debris. The term **debris,** as applied to mass wasting processes, means any unconsolidated material at Earth's surface, such as soil and rock fragments (weathered or unweathered) of any size.

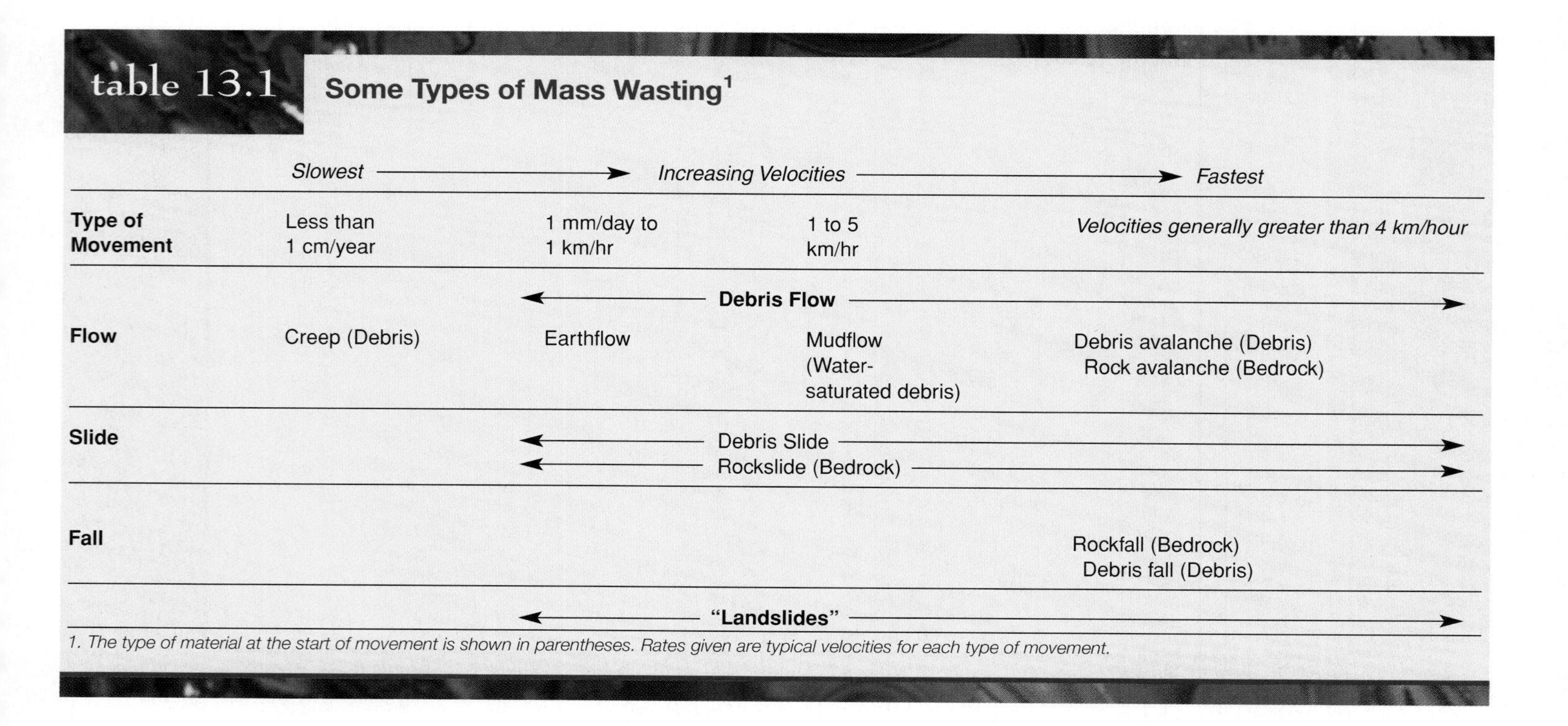

table 13.1 Some Types of Mass Wasting[1]

	Slowest ⟶		*Increasing Velocities* ⟶	*Fastest*
Type of Movement	Less than 1 cm/year	1 mm/day to 1 km/hr	1 to 5 km/hr	*Velocities generally greater than 4 km/hour*
		⟵ **Debris Flow** ⟶		
Flow	Creep (Debris)	Earthflow	Mudflow (Water-saturated debris)	Debris avalanche (Debris) Rock avalanche (Bedrock)
Slide		⟵ Debris Slide ⟶ ⟵ Rockslide (Bedrock) ⟶		
Fall				Rockfall (Bedrock) Debris fall (Debris)
		⟵ **"Landslides"** ⟶		

1. The type of material at the start of movement is shown in parentheses. Rates given are typical velocities for each type of movement.

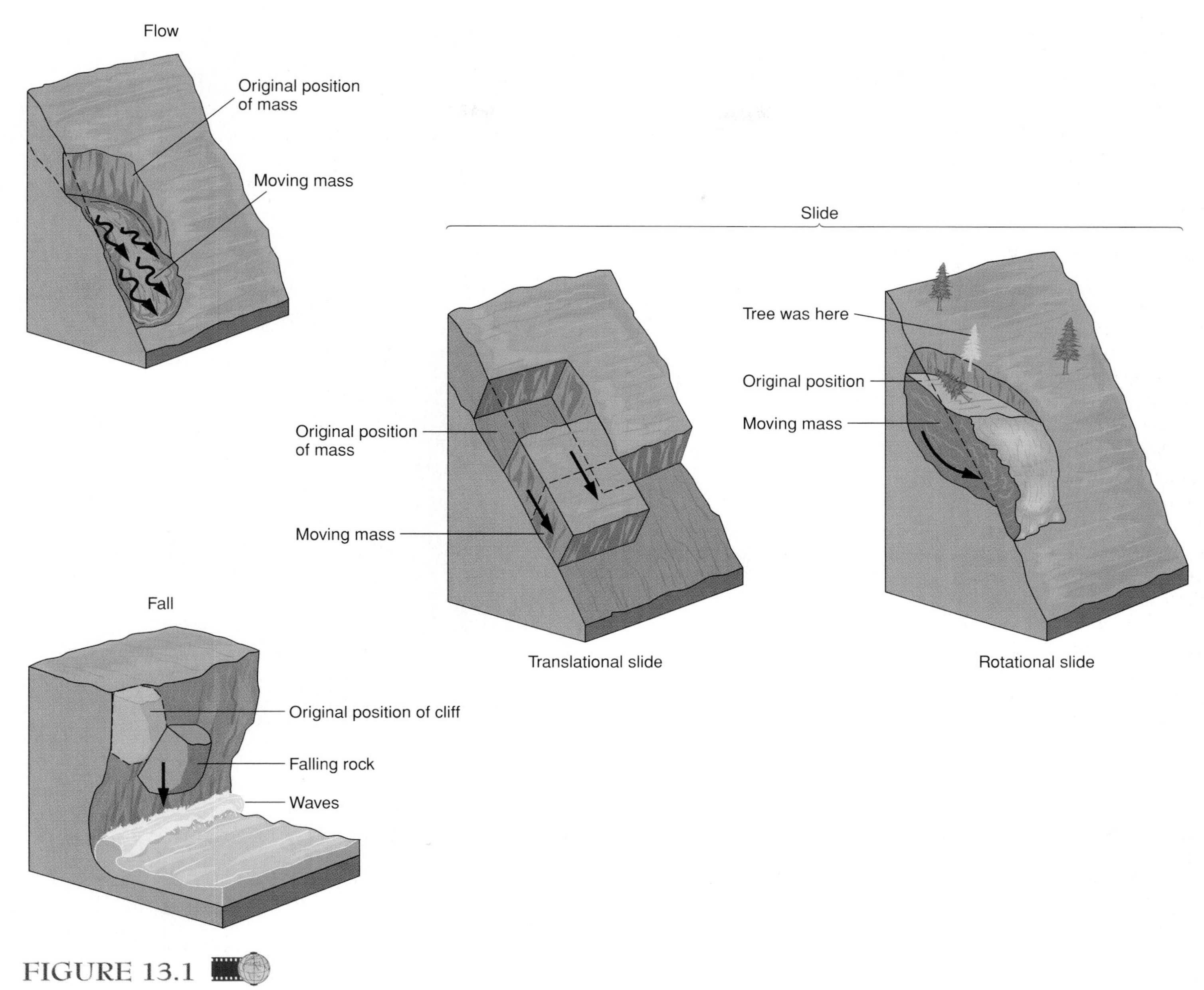

FIGURE 13.1
Flow, slide, and fall.

The amount of water (or ice and snow) in a descending mass strongly influences the rate and type of movement.

Type of Movement

In general, the type of movement in mass wasting can be classified as mainly flow, slide, or fall (figure 13.1). A **flow** implies that the descending mass is moving downslope as a viscous fluid. **Slide** means the descending mass remains relatively intact, moving along one or more well-defined surfaces. A **fall** occurs when material free-falls or bounces down a cliff.

Two kinds of slip are shown in figure 13.1. In a **translational slide,** the descending mass moves along a plane approximately parallel to the slope of the surface. A **rotational slide** (also called a *slump*) involves movement along a curved surface, the upper part moving downward while the lower part moves outward.

CONTROLLING FACTORS IN MASS WASTING

Table 13.2 summarizes the factors that influence the likelihood and the rate of movement of mass wasting. The table makes apparent some of the reasons why the landslide (a

13.1

ENVIRONMENTAL GEOLOGY

DISASTER IN THE ANDES

As a result of a tragic combination of geological conditions, one of the most devastating landslides in history destroyed the town of Yungay in Peru in 1970. Yungay was one of the most picturesque towns in the Santa River Valley, which runs along the base of the highest peaks of the Peruvian Andes. Heavily glaciated Nevado Huascarán, 6,663 meters (21,860 feet) above sea level, rises steeply above the populated narrow plains along the Santa River.

In May 1970 a sharp earthquake occurred. The earthquake was centered offshore from Peru about 100 kilometers from Yungay. Although the tremors in this part of the Andes were no stronger than those that have done only light damage to cities in the United States, many poorly constructed homes collapsed. Because of the steepness of the slopes, thousands of small rockfalls and rockslides were triggered.

The greatest tragedy began when a slab of glacier ice about 800 meters wide, perched near the top of Huascarán, was dislodged by the shaking. (A few years earlier American climbers returning from the peak had warned that the ice looked highly unstable. The Peruvian press briefly noted the danger to the towns below, but the warning was soon forgotten.)

The mass of ice rapidly avalanched down the extremely steep slopes, breaking off large masses of rock debris, scooping out small lakes and loose rock that lay in its path. Eyewitnesses described the mass as a rapidly moving wall the size of a ten-story building. The sound was deafening. More than 50 million cubic meters of muddy debris traveled 3.7 kilometers (12,000 feet) vertically and 14.5 kilometers (9 miles) horizontally in less than four minutes, attaining speeds between 200 and 435 kilometers per hour (125 to 270 miles per hour). The main mass of material traveled down a steep valley until it came to rest blocking the Santa River and burying about 1,800 people in the small village of Ranrahirca (box figure 1). A relatively small part of the mass of mud and debris that was moving especially rapidly shot up the valley sidewall at a curve and overtopped a ridge. The mass was momentarily airborne before it fell on the town of Yungay, completely burying it under several meters of mud and loose rock. Only the top of the church and tops of palm trees were visible, marking where the town center was buried (box figure 2). Ironically, the cemetery was not buried because it occupied the high ground. The few survivors were people who managed to run to the cemetery.

The estimated death toll at Yungay was 17,000. This was considerably more than the town's normal population, because it was Sunday, a market day, and many families had come in from the country.

For several days after the slide the debris was too muddy for people to walk on, but within three years grass had grown over the site. Except for the church steeple and the tops of palm trees that still protrude above the ground, and the crosses erected by families of those buried in the landslide, the former site of Yungay appears to be a scenic meadow overlooking the Santa River. The U.S. Geological Survey and Peruvian geologists found evidence that Yungay itself had been built on top of debris left by an even bigger slide in the recent geologic past. More slides will almost surely occur here in the future.

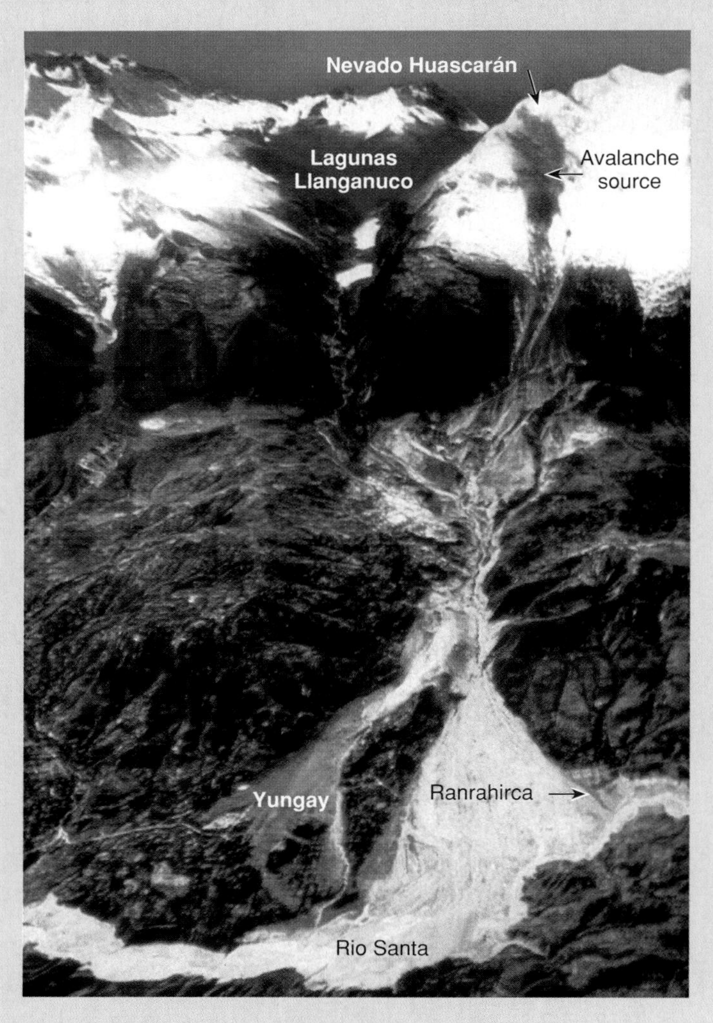

Box 13.1 — FIGURE 1

Air photo showing the 1970 debris avalanche in Peru, which buried Yungay. The main mass of debris destroyed the small village of Ranrahirca.

Photo by Servicio Aerofotografico de Peru, courtesy of U.S. Geological Survey

Further Reading

Ericksen, G. E., G. Plafker, and J. Fernandez Concha, 1970. *Preliminary report on the geologic events associated with the May 31, 1970, Peru, earthquake.* U.S. Geological Survey Circular 639.

Box 13.1 — Figure 2

(*A*) Yungay is completely buried, except for the cemetery and a few houses on the small hill in the lower right of the photograph. (*B*) Behind the palm trees is the top of a church buried under 5 meters of debris at Yungay's central plaza. (*C*) Three years later.

Photos *A* and *B* by George Plafker, U.S. Geological Survey; photo *C* by C. C. Plummer

debris avalanche) in Peru (box 13.1) occurred and why it moved so rapidly. (1) The slopes were exceptionally steep, and (2) the **relief** (the vertical distance between valley floor and mountain summit) was great, allowing the mass to pick up speed and momentum. (3) Water and ice not only added weight to the mass of debris but also acted as lubricants. (4) Abundant loose rock and debris were available in the course of the slide. (5) Where the slide began, there were no plants with roots to anchor loose material on the slope. Finally, (6) the area is earthquake prone. Although the slide would have occurred eventually even without an earthquake, it was triggered by an earthquake.

Other factors influence susceptibility to mass wasting as well as its rate of movement. The orientation of planes of weakness in bedrock (bedding planes, foliation planes, etc.) is important if the movement involves bedrock rather than debris. Fractures or bedding planes oriented so that slabs of rock can slide easily along these surfaces greatly increase the likelihood of mass wasting.

Climatic controls inhibit some types of mass wasting and aid others (table 13.2). Climate influences how much and what kinds of vegetation grow in an area and what type of weathering occurs. Infrequent but heavy rainfall aids mass wasting because it quickly saturates debris that lacks the protective vegetation found in wetter climates. By contrast, rain that drizzles intermittently much of the year results in vegetation that tends to inhibit mass wasting. In cold climates, freezing and thawing contribute to downslope movement.

table 13.2 Summary of Controls of Mass Wasting

Driving Force: Gravity

Contributing Factors	Most Stable Situation	Most Unstable Situation
Slope angle	Gentle slopes or horizontal surface	Steep or vertical
Local relief	Low	High
Thickness of debris over bedrock	Slight thickness (usually)	Great thickness
Orientation of planes of weakness in bedrock	Planes at right angles to hillside slopes	Planes parallel to hillside slopes
Climatic factors:		
Ice	Temperature stays above freezing	Freezing and thawing for much of the year
Water in soil or debris	Film of water around fine particles	Saturation of debris with water
Precipitation	Frequent but light rainfall or snow	Long periods of drought with rare episodes of heavy precipitation
Vegetation	Heavily vegetated	Sparsely vegetated

Triggering Mechanisms: (1) earthquakes; (2) weight added to upper part of a slope; (3) undercutting of bottom of slope.

Gravity

Gravity is the driving force for mass wasting. Figure 13.2*A* and *B* show gravity acting on a block on a slope. The length of the vertical arrow is proportional to the force—the heavier the material, the longer the arrow. The effect of gravity is resolvable into two component forces, indicated by the black arrows. One, the *normal force,* is perpendicular to the slope and its value indicates the block's ability to stay in place because of frictional considerations. The other, called the **shear force,** is parallel to the slope, and indicates the block's ability to move. The length of the arrows is proportional to the strength of each force. The steeper the slope (and the heavier the block), the greater the shear force and the greater the tendency of the block to slide. Friction counteracts the shear force. If friction is greater than the shear force, the block will not move. If the force of friction is reduced (for instance, with water) so that it is less than the shear force, the block will slide. Similar forces act on debris on a hillside (figure 13.2*C*). The resistance to movement or deformation of that debris is its **shear strength.** Shear strength is controlled by factors such as the cohesiveness of the material, friction between particles, and the anchoring effect of plant roots. Shear strength is also related to the normal force. The larger the normal force, the greater the shear strength. If the shear strength is greater than the shear force, the debris will not move or be deformed. On the other hand, if shear strength is less than shear force, the debris will flow or slide.

Building a heavy structure high on a slope demands special precautions. To prevent movement of both the slope and the building, pilings may have to be sunk through the debris, perhaps even into bedrock. Developers may have to settle for fewer buildings than planned if the weight of too many structures will make the slope unsafe.

Water

Water is a critical factor in mass wasting. When debris is saturated with water (as from heavy rain or melting snow), it becomes heavier and is more likely to flow downslope. The added gravitational shear force from the increased weight, however, is probably less important than the reduction in shear strength. This is due to increased *pore pressure* in which water forces grains of debris apart.

Paradoxically, a small amount of water in soil can actually prevent downslope movement. When water does not completely fill the pore spaces between the grains of soil, it forms a thin film around each grain (as shown in figure 13.3). Loose grains adhere to one another because of the *surface tension* created by the film of water, and shear strength increases. Surface tension of water between sand grains is what allows you to build a sand castle. The sides of the castle can be steep or even vertical because surface tension holds the moist sand grains in place. Dry sand cannot be shaped into a sand castle because the sand grains slide back into a pile that generally slopes at an angle of about 30° to 35° from the horizontal. On the other hand, an experienced sand castle builder also knows that it is impossible to build anything with sand that is too wet. In this case the water completely occupies the pore space between sand grains, forcing them apart and allowing them to slide easily past one another. When the tide comes in, or someone pours a pail of water on your sand castle, all you have is a puddle of wet sand.

Similarly, as the amount of water in debris increases, rate of movement tends to increase. Damp debris may not move at all, whereas moderately wet debris moves slowly downslope.

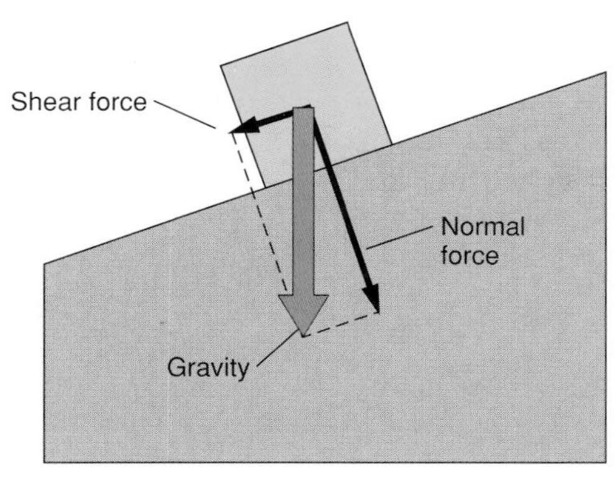

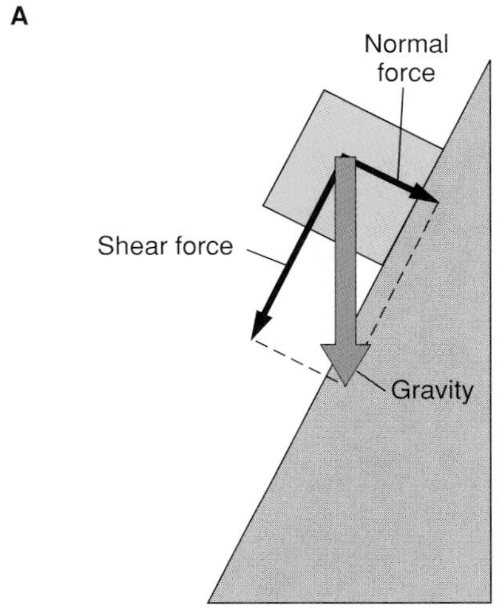

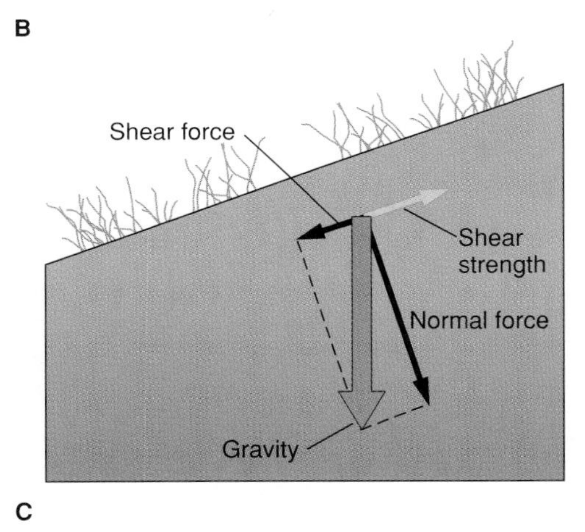

FIGURE 13.2

Relationship of shear force and normal force to gravity. (*A*) For a block on a gently inclined surface. (*B*) For a block on a steep surface. (*C*) Forces acting at a point in debris. Shear strength is represented by a yellow arrow. If that arrow is longer than the one represented by shear force, debris at that point will not slide or be deformed.

Slow types of mass wasting, such as creep, are generally characterized by a relatively low ratio of water to debris. Mudflows always have high ratios of water to debris. A mudflow that continues to gain water eventually becomes a muddy stream.

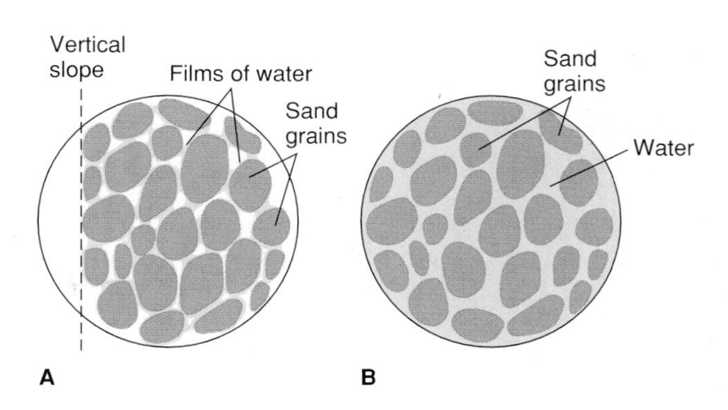

FIGURE 13.3

The effect of water in sand. (*A*) Unsaturated sand held together by surface tension of water. (*B*) Saturated sand grains forced apart by water; mixture flows easily.

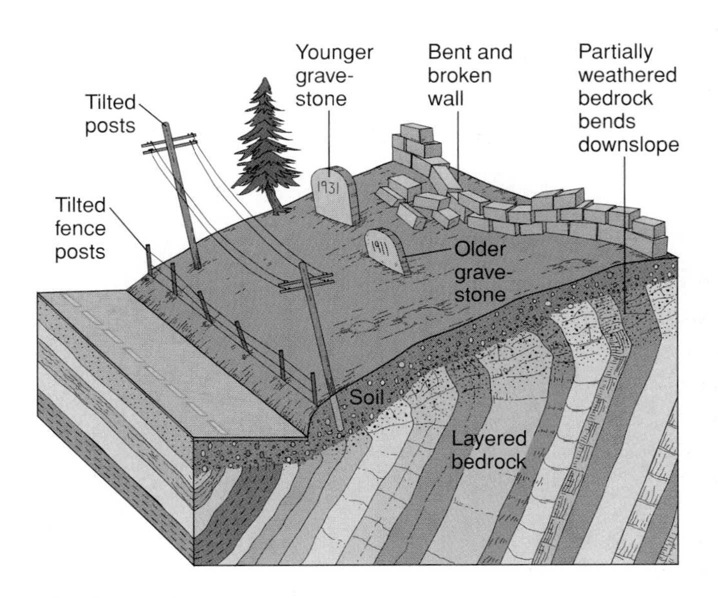

FIGURE 13.4

Indicators of creep.

After C. F. S. Sharpe

COMMON TYPES OF MASS WASTING

The common types of mass wasting are shown in table 13.1. Here we will describe each type in detail.

Creep

Creep is very slow, downslope movement of soil or unconsolidated debris. Shear forces, over time, are only slightly greater than shear strengths. The rate of movement is usually less than a centimeter per year and can be detected only by observations taken over months or years. When conditions are right, creep can take place along nearly horizontal slopes. Some indicators of creep are illustrated in figures 13.4 and 13.5.

A

B

FIGURE 13.5

(*A*) Tilted gravestones in a churchyard at Lyme Regis, England (someone probably straightened the one upright gravestone). Grassy slope is inclined gently to the left. (*B*) Soil and partially weathered, nearly vertical sedimentary strata have crept downslope.

Photo *A* by C. C. Plummer; photo *B* by Frank M. Hanna

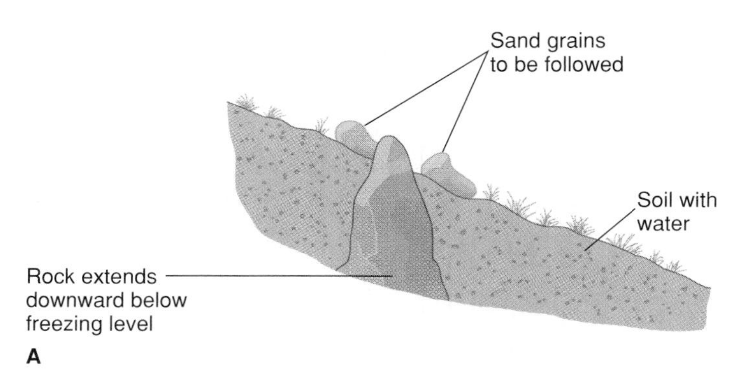

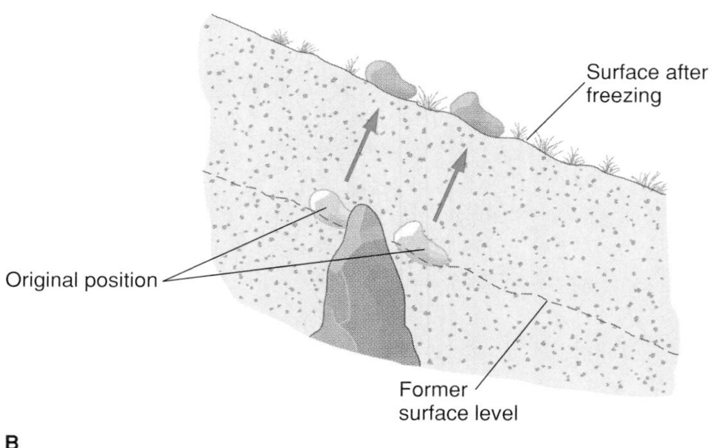

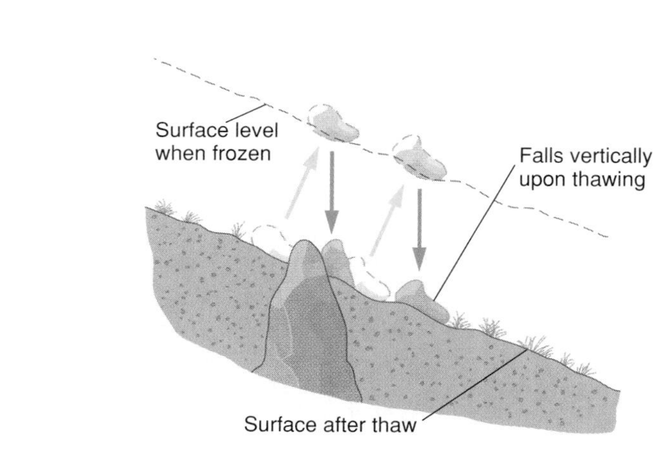

FIGURE 13.6

Downslope movement of soil, illustrated by following two sand grains (each less than a millimeter in size) during a freeze-thaw cycle. Movement downward might not be precisely vertical if adjacent grains interfere with each other.

Two factors that contribute significantly to creep are water in the soil and daily cycles of freezing and thawing. As we have said, water-saturated ground facilitates movement of soil downhill. What keeps downslope movement from becoming more rapid in most areas is the presence of abundant grass or other plants that anchor the soil. (Understandably, overgrazing can severely damage sloping pastures.)

Although creep does take place in year-round warm climates, the process is more active where the soil freezes and thaws during part of the year. During the winter in regions such as the northeastern United States, the temperature may rise above and fall below freezing once a day. When there is moisture in the soil, each freeze-thaw cycle moves soil particles a minute amount downhill, as shown in figure 13.6.

Debris Flow

The general term **debris flow** is used for mass wasting in which motion is taking place throughout the moving mass (flow). The common varieties earthflow, mudflow, and debris avalanche are described in this section.

Earthflow

In an **earthflow,** debris moves downslope as a viscous fluid; the process can be slow or rapid. Earthflows usually occur on

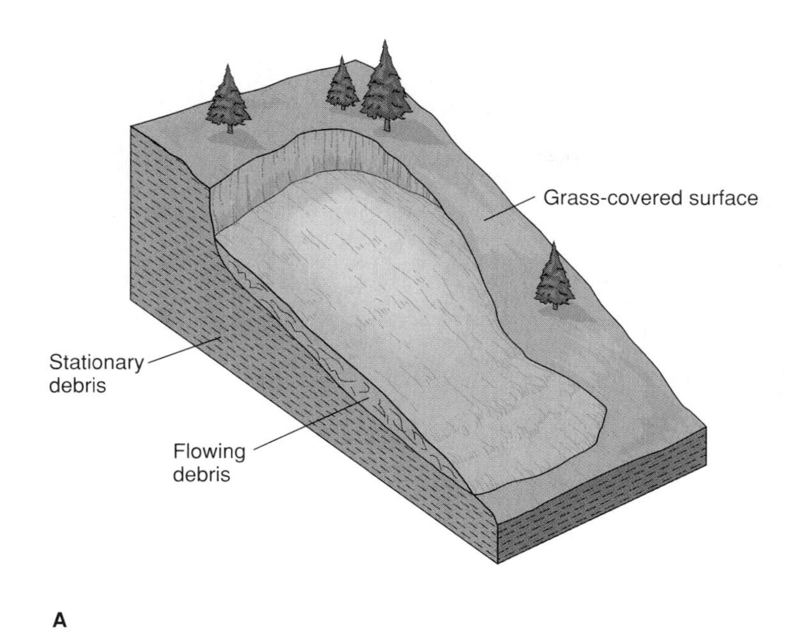

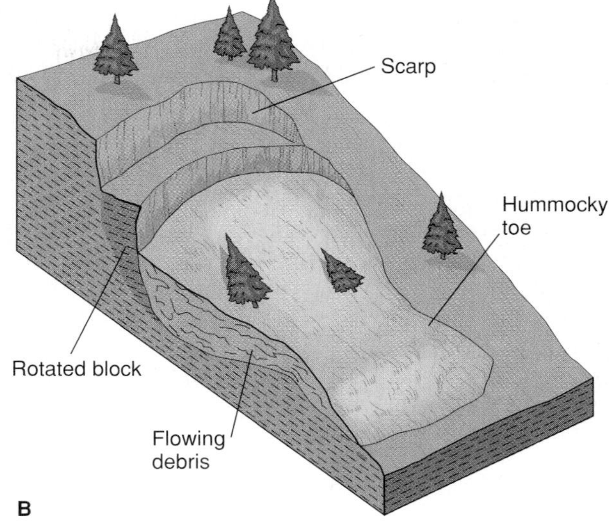

FIGURE 13.7
Earthflow. (*A*) Earthflow involving only flow. (*B*) Earthflow and rotational sliding.

hillsides that have a thick cover of debris, often after heavy rains have saturated the soil. Typically, the flowing mass remains covered by a blanket of vegetation, with a *scarp* (steep cut) developing where the moving debris has pulled away from the stationary upper slope.

A landslide may be entirely an earthflow, as in figure 13.7*A*, with debris particles moving past one another roughly parallel to the slope. Commonly, however, rotational sliding takes place above the earthflow as in figures 13.7*B* and 13.8. This example is a *debris slide* (upper part) and an *earthflow* (lower part). In such cases, debris remains in a relatively coherent block or blocks that rotate downward and outward, forcing the debris below to flow.

A hummocky lobe usually forms at the toe or front of the earthflow where debris has accumulated. An earthflow can be active over a period of hours, days, or months; in some earthflows intermittent slow movement continues for years.

Humans can trigger earthflows by adding too much water to soil from septic tank systems or by overwatering lawns. In one case, in Los Angeles, a man departing on a long trip forgot to turn off the sprinkler system for his hillside lawn. The soil became saturated, and both house and lawn were carried downward on an earthflow whose lobe spread out over the highway below.

Earthflows, like other kinds of landslides, can be triggered by undercutting at the base of a slope. The undercutting can be caused by waves breaking along shorelines or streams eroding and steepening the base of a slope. Along coastlines, mass wasting commonly destroys buildings (figure 13.9). Entire housing developments and expensive homes built for a view of the ocean are lost. A home buyer who knows nothing of geology may not realize that the sea cliff is there because of the

FIGURE 13.8
Earthflow that destroyed several houses in March 1995 at La Conchita, California.
Photo by Robert L. Schuster, U.S. Geological Survey

13.2 ENVIRONMENTAL GEOLOGY

Los Angeles, A Mobile Society

The following satirical newspaper column was written by humorist Art Buchwald in 1978, a year, like the "El Niño" year of 1998, in which southern California had many landslides because of unusually wet weather.

Los Angeles—I came to Los Angeles last week for rest and recreation, only to discover that it had become a rain forest.

I didn't realize how bad it was until I went to dinner at a friend's house. I had the right address, but when I arrived there was nothing there. I went to a neighboring house where I found a man bailing out his swimming pool.

I beg your pardon, I said. Could you tell me where the Cables live?

"They used to live above us on the hill. Then, about two years ago, their house slid down in the mud, and they lived next door to us. I think it was last Monday, during the storm, that their house slid again, and now they live two streets below us, down there. We were sorry to see them go—they were really nice neighbors."

I thanked him and slid straight down the hill to the new location of the Cables' house. Cable was clearing out the mud from his car. He apologized for not giving me the new address and explained, "Frankly, I didn't know until this morning whether the house would stay here or continue sliding down a few more blocks."

Cable, I said, you and your wife are intelligent people, why do you build your house on the top of a canyon, when you know that during a rainstorm it has a good chance of sliding away?

"We did it for the view. It really was fantastic on a clear night up there. We could sit in our Jacuzzi and see all of Los Angeles, except of course when there were brush fires.

"Even when our house slid down two years ago, we still had a great sight of the airport. Now I'm not too sure what kind of view we'll have because of the house in front of us, which slid down with ours at the same time."

But why don't you move to safe ground so that you don't have to worry about rainstorms?

"We've thought about it. But once you live high in a canyon, it's hard to move to the plains. Besides, this house is built solid and has about three more good mudslides in it."

Still, it must be kind of hairy to sit in your home during a deluge and wonder where you'll wind up next. Don't you ever have the desire to just settle down in one place?

"It's hard for people who don't live in California to understand how we people out here think. Sure we have floods, and fire and drought, but that's the price you have to pay for living the good life. When Esther and I saw this house, we knew it was a dream come true. It was located right on the tippy top of the hill, way up there. We would wake up in the morning and listen to the birds, and eat breakfast out on the patio and look down on all the smog.

"Then, after the first mudslide, we found ourselves living next to people. It was an entirely different experience. But by that time we were ready for a change. Now we've slid again and we're in a whole new neighborhood. You can't do that if you live on solid ground. Once you move into a house below Sunset Boulevard, you're stuck there for the rest of your life.

"When you live on the side of a hill in Los Angeles, you at least know it's not going to last forever."

Then, in spite of what's happened, you don't plan to move out?

"Are you crazy? You couldn't replace a house like this in L.A. for $500,000."

What happens if it keeps raining and you slide down the hill again?

"It's no problem. Esther and I figure if we slide down too far, we'll just pick up and go back to the top of the hill, and start all over again; that is, if the hill is still there after the earthquake."

Further Reading

John McPhee's *The control of nature* contains a factual, and highly readable, account of 1978 landslides in southern California.

relentless erosion of waves along the shoreline. Nor is the person likely to be aware that a steepened slope creates the potential for landslides.

Bulldozers can undercut the base of a slope more rapidly than wave erosion, and such oversteepening of slopes by human activity has caused many landslides. Unless careful engineering measures are taken at the time a cut is made, road-cuts or platforms carved into hillsides for houses may bring about disaster (figure 13.18).

Solifluction and Permafrost

One variety of earthflow is usually associated with colder climates. **Solifluction** is the flow of water-saturated debris over impermeable material. Because the impermeable material beneath the debris prevents water from draining freely, the debris between the vegetation cover and the impermeable material becomes saturated (figure 13.10). Even a gentle slope is susceptible to movement under these conditions (figure 13.11).

The impermeable material beneath the saturated soil can be either impenetrable bedrock or, as is more common, **permafrost,** ground that remains frozen for many years. Most solifluction takes place in areas of permanently frozen ground, such as in Alaska and northern Canada. Permafrost occurs at depths ranging from a few centimeters to a few meters beneath the surface. The ice in permafrost is a cementing agent for the debris. Permafrost is as solid as concrete.

FIGURE 13.9
Landslide with houses and roads moved on slide blocks. Point Fermin, Los Angeles, California.
Photo by Frank M. Hanna

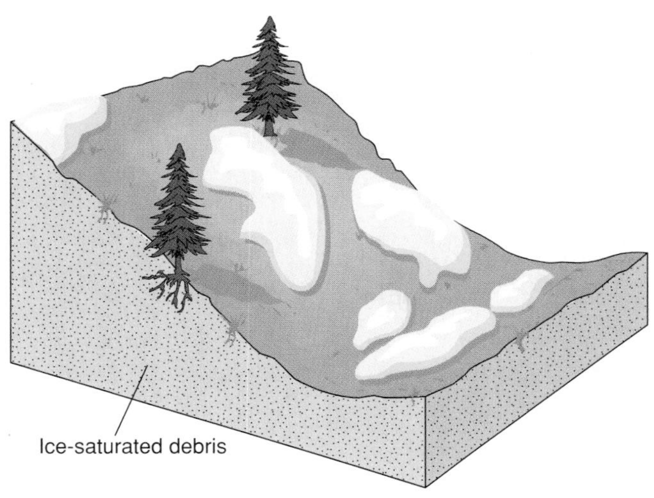

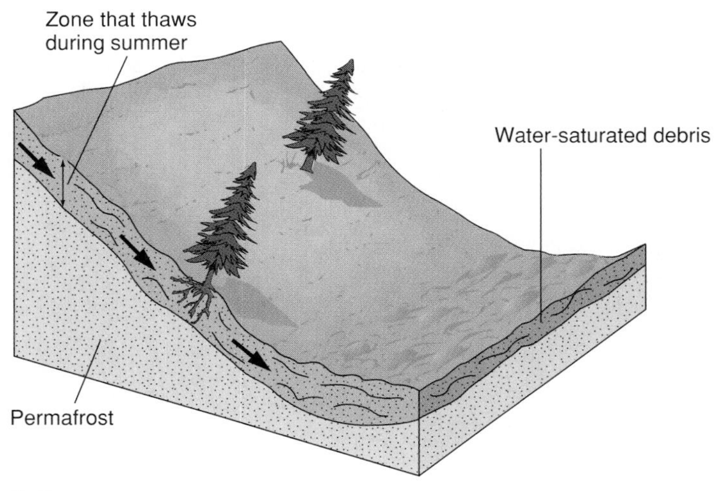

FIGURE 13.10
Solifluction due to thawing of ice-saturated debris.

FIGURE 13.11
A railroad built on permafrost terrain in Alaska.
Photo by Lynn A. Yehle, U.S. Geological Survey

Above the permafrost is a zone that, if the debris is saturated, is frozen during the winter and indistinguishable from the underlying permafrost. When this zone thaws during the summer, the water, along with water from rain and runoff, cannot percolate downward through the permafrost, and so the slopes become susceptible to solifluction.

Because solifluction movement is not rapid enough to break up the overlying blanket of vegetation into blocks, the water-saturated debris flows downslope, pulling vegetation along with it and forming a wrinkled surface. Gradually the debris collects at the base of the slope, where the vegetated surface bulges into a hummocky lobe.

Solifluction is not the only hazard associated with permafrost. Great expanses of flat terrain in arctic and subarctic climates become swampy during the summer because of permafrost, making overland travel very difficult. Building and maintaining roads is an engineering headache. In the preliminary stages of planning the Alaska pipeline, a road was bulldozed across permafrost terrain during the winter, removing the vegetation from the rock-hard ground. It was an excellent truck route during the winter, but when summer came, the road became a quagmire several hundred kilometers long. The strip can never be used by vehicles as planned, nor will the vegetation return for many decades. Building structures on permafrost terrain presents serious problems. For instance, heat from a building can melt underlying permafrost; the building then sinks into the mud.

Mudflow

A **mudflow** is a flowing mixture of debris and water, usually moving down a channel (figure 13.12). It can be visualized as a stream with the consistency of a thick milkshake. Most of the solid particles in the slurry are clay and silt (hence the muddy appearance), but coarser sediment commonly is part of the mixture. Usually after a heavy rainfall a slurry of debris and water forms and begins moving down a slope. Most mudflows quickly become channeled into valleys. They then move downvalley like a stream except that, because of the heavy load of debris, they are

13.3 ASTROGEOLOGY

Permafrost Ice on Mars

On Earth most underground water is in the liquid form, with the only abundant underground ice occurring as permafrost in the polar regions. In these areas underground ice is responsible for producing several distinctive landforms; most notable are surfaces with polygonal cracks called patterned ground (box figure 1*A*).

On Mars, because of its cooler temperatures, ice is the normal state of underground water and is apparently quite abundant. Many features of the Martian landscape have been attributed to this underground ice, including polygonally patterned ground (box figure 1*B*), rampart craters, and chaotic terrain.

Rampart craters are Martian meteorite craters that are surrounded by material that appears to have flowed from the point of impact. It has been suggested that the underground ice in the surface layers of Mars melted as a result of the impact and, along with rock fragments, sloshed outward from the impact area.

Patches of jumbled and broken angular slabs and blocks called *chaotic terrain* occur in some places on Mars (box figure 2). Some channels originate in these areas, and it is believed that this terrain may be caused by melting of permafrost and consequent collapse of the ground. Subsidence due to withdrawal of magma has also been suggested as the cause of chaotic terrain.

Some astrogeologists think that there is a good chance that Mars is the only other planet besides Earth that has life. According to their hypothesis, primitive organisms evolved during an earlier time when Mars had liquid water. The microorganisms might be dormant in the permafrost and perhaps could be revived by melting the ice. This hypothesis received a boost in 1996 when scientists studying a meteorite from Mars reported that it revealed possible signs of former life (see boxes 9.5 and 19.5). Hardly the alien creatures pictured in science fiction movies or books but organic life nevertheless.

A

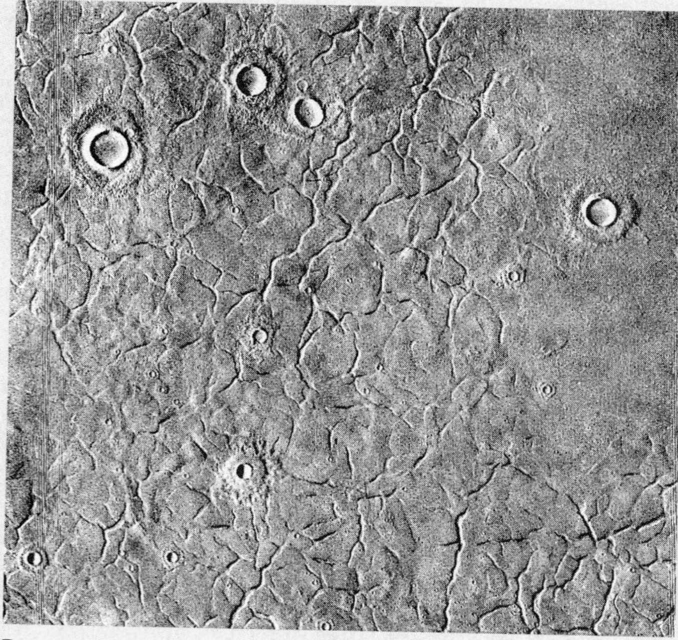

B

Box 13.3 — FIGURE 1

(*A*) Patterned ground in Alaska. Coarse rock forms the edges of polygons; fine-grained material in center supports vegetation. (*B*) Patterned ground on Mars. The polygons in photo *B* are over 100 times larger than those in photo *A*. Note the impact craters.

Photo *A* by C. C. Plummer; Photo *B* by NASA

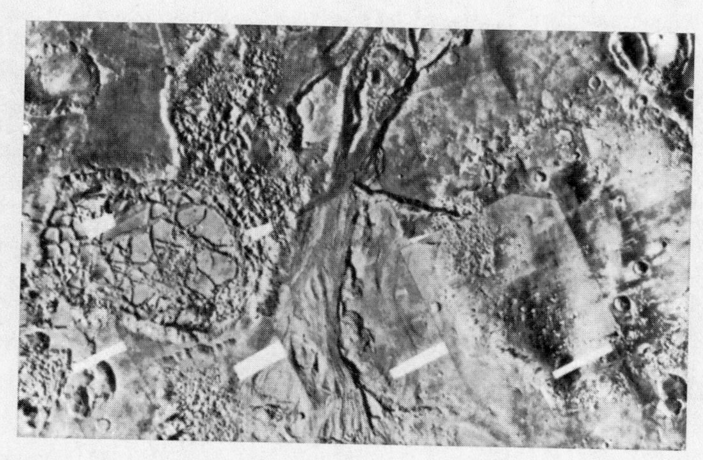

Box 13.3 — FIGURE 2

Chaotic terrain with associated channels on Mars. Photo is a composite of several images. White rectangles are where images were not taken.

Photo by NASA

FIGURE 13.12
A dried mudflow in the Peruvian Andes.
Photo by C. C. Plummer

FIGURE 13.13
Man examining a 75-meter-long bridge on Washington state highway 504, across the North Fork of the Toutle River. The bridge was washed out by mudflow during the May 18, 1980, eruption of Mount St. Helens. The steel structure was carried about 0.5 kilometers downstream and partially buried by the mudflow.
Photo by Robert L. Schuster, U.S. Geological Survey

more viscous. Mud moves more slowly than a stream but, because of its high viscosity, can transport boulders, automobiles, and even locomotives. Houses in the path of a mudflow will be filled with mud, if not broken apart and carried away.

Mudflows are most likely to occur in places where debris is not protected by a vegetative cover. For this reason, mudflows are more likely to occur in arid regions than in wet climates. A hillside in a desert environment, where it may not have rained for many years, may be covered with a blanket of loose material. With sparse desert vegetation offering little protection, a sudden thunderstorm with drenching rain can rapidly saturate the loose debris and create a mudflow in minutes.

Mudflows frequently occur on young volcanoes that are littered with ash. Water from heavy rains mixes with pyroclastic debris as at Mount Pinatubo in 1991. (For over a decade after the big eruption, mudflows near Mount Pinatubo continue to cost lives and destroy property.) Or the water can come from glaciers that are melted by lava or hot pyroclastic debris, as occurred at Mount St. Helens in 1980 (figure 13.13) and at Colombia's Nevado del Ruiz in 1985, which cost 23,000 lives. Mudflows also occur after forest fires destroy slope vegetation that normally anchors soil in place. Burned-over slopes are extremely vulnerable to mudflow if heavy rains fall before the vegetation is restored.

The year 1978 was particularly bad for debris flows in southern California. One mudflow roared through a Los Angeles suburb carrying almost as many cars as large boulders. A sturdily built house withstood the onslaught but began filling with muddy debris. Two of its occupants were pinned to the wall of a bedroom and could do nothing as the room filled slowly with mud. The mud stopped rising just as it was reaching their heads. Hours later they were rescued. (John McPhee's *The Control of Nature,* listed in *Exploring Web Resources* on this book's website [www.mhhe.com/plummer9e], is a highly readable account of this and other debris flows in southern California.)

Debris Avalanche

The fastest variety of debris flow is a **debris avalanche,** a very rapidly moving, turbulent mass of debris, air, and water. The best modern example is the one that buried Yungay (see box 13.1, pgs. 316–317).

Some geologists have suggested that in very rapidly moving rock avalanches, air trapped under the rock mass creates an air cushion that reduces friction. This could explain why some landslides reach speeds of several hundred kilometers per hour. But other geologists have contended that the rock mass is too turbulent to permit such an air cushion to form.

Rockfalls and Rockslides

Rockfall

When a block of bedrock breaks off and falls freely or bounces down a cliff, it is a **rockfall** (figure 13.14). Cliffs may form naturally by the undercutting action of a river, wave action, or glacial erosion. Highway or other construction projects may also oversteepen slopes. Bedrock commonly has cracks (joints) or other planes of weakness such as foliation (in metamorphic rocks) or sedimentary bedding planes. Blocks of rock will break off along these planes. In colder climates rock is effectively broken apart by frost wedging (as explained in chapter 12).

Commonly, an apron of fallen rock fragments, called **talus,** accumulates at the base of a cliff (figure 13.15).

A spectacular rockfall took place in Yosemite National Park in the summer of 1996, killing one man and injuring several other people. The rockfall originated from near Glacier Point

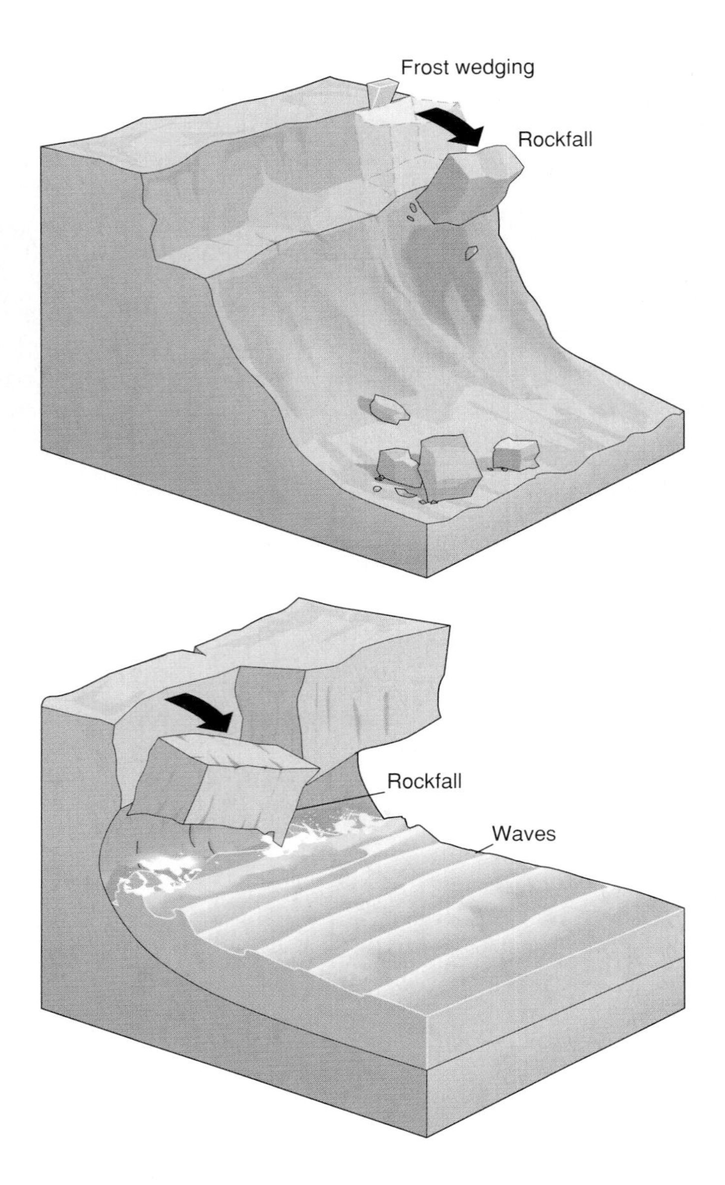

FIGURE 13.14

Two examples of rockfall.

FIGURE 13.15

Talus.

Photo by C. C. Plummer

FIGURE 13.16

Small dust clouds linger high above Yosemite Valley where rock slabs broke loose and fell to the valley floor, which, upon impact, created the debris-laden blast of air climbing up the other side of the valley. The photo was taken by a rock climber on a nearby cliff.

Photo by Ed Youmans

(the place where the photo for figure 19.1 was taken). Two huge slabs (weighing approximately 80,000 tons) of an overhanging arch broke loose just seconds apart. (The arch was a product of exfoliation, and broke loose along a sheet joint—see chapter 12.) The slabs slid a short distance over steep rock from which they were launched outward, as if from a ski jump, away from the vertical cliffs. The slabs fell free for around 500 meters (1,700 feet) and hit the valley floor 30 meters out from the base of the cliff (you would not have been hit if you were standing at the base of the cliff). They shattered upon impact and created a dust cloud (figure 13.16) that obscured visibility for hours. A powerful air blast was created as air between the rapidly falling rock and the ground was compressed. The debris-laden wind felled a swath of trees between the newly deposited talus and a nature center building. In 1999, another rockfall in the same area killed one rock climber and injured three others.

Rockslide

A **rockslide** is, as the term suggests, the rapid sliding of a mass of bedrock along an inclined surface of weakness, such as a bedding plane (figure 13.17), a major fracture in the rock, or a foliation plane (box 13.4). Once sliding begins, a rock slab usually breaks up into rubble. Like rockfalls, rockslides can be caused by undercutting at the base of the slope from erosion or construction.

Some rockslides travel only a few meters before halting at the base of a slope. In country with high relief, however, a rockslide may travel hundreds or thousands of meters before reaching a valley floor. If movement becomes very rapid, the rockslide may break up and become a rock avalanche. A **rock avalanche** is a very rapidly moving, turbulent mass of broken-up bedrock. Movement in a rock avalanche is flowage on a grand scale. The only difference between a rock avalanche and a debris avalanche is that a rock avalanche begins its journey as bedrock.

Ultimately, a rockslide or rock avalanche comes to rest as the terrain becomes less steep. Sometimes the mass of rock fills the bottom of a valley and creates a natural dam. If the

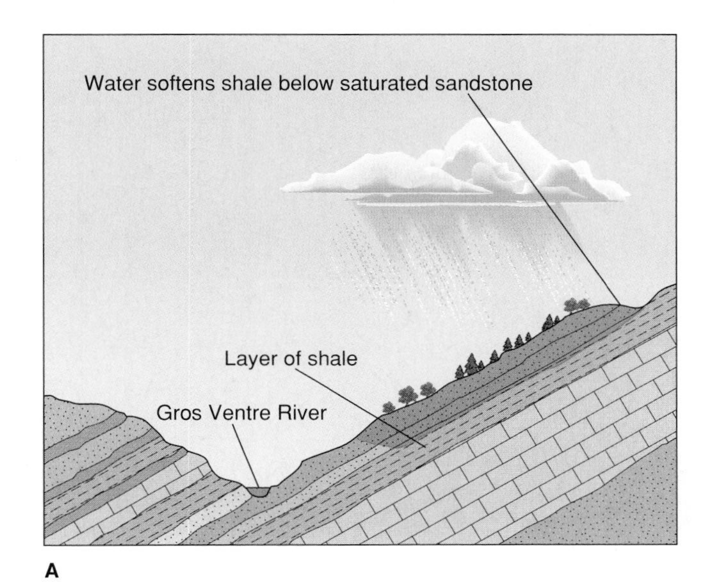

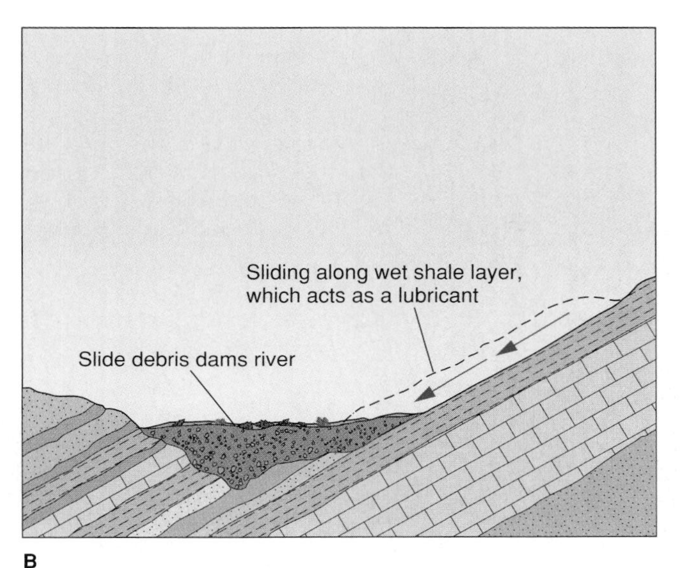

FIGURE 13.17

(*A*) and (*B*) Diagram of the Gros Ventre, Wyoming, slide. (*C*) Photo of Gros Ventre slide.

A and *B* after W. C. Alden, U.S. Geological Survey. Photo *C* by D. A. Rahm, courtesy of Rahm Memorial Collection, Western Washington University

13.4 ENVIRONMENTAL GEOLOGY

FAILURE OF THE ST. FRANCIS DAM—A TRAGIC CONSEQUENCE OF GEOLOGY IGNORED

In 1928 the St. Francis Dam near Los Angeles, California, broke, only a year after it had been completed (box figure 1). The concrete dam was about 60 meters (200 feet) high, and the wall of water that roared down the valley killed about 400 people in two counties.

The eastern edge of the dam had been built against a metamorphic rock with foliation planes parallel to the sides of the valley. Landslide scars in the valley should have been ample warning to the builders that the metamorphic rock moved even under only the force of gravity. A competent engineer worries as much about the stability of the rock against which a dam is built as about the strength of the dam itself. Water pressure at the base of the dam exerted a force of 5.7 tons per square foot against the dam. With pressure such as this, the dam and part of the bordering foliated rock could easily slide. Movement would be parallel to the weak foliation planes, just as if the dam had been anchored against a giant deck of cards.

Ironically, investigators never found out for sure whether this was what caused the failure of the dam. Many other blunders had been made in construction, and any one of them could have caused the dam to break. The base of the dam was on a fault with ground-up rock; and, incredibly, the other side of the dam was built against rock that disintegrates in water. This is but one of many instances in which ignorance of geology cost lives and money. Had professional geological advice been sought, the dam probably would not have been built in that spot.

Related Web Resource

http://seis.natsci.csulb.edu/VIRTUAL_FIELD/Francesquito_Dam/franmain.htm

St. Francis Dam virtual field trip

BOX 13.4 — FIGURE 1

The St. Francis Dam, California, after its failure. The former water level in the reservoir is visible on the hills in the background.

Photo by H. T. Stearns, U.S. Geological Survey

rock mass suddenly enters a lake or bay, it can create a huge wave that destroys lives and property far beyond the area of the original landslide.

An example is a disastrous landslide that took place in northern Italy in 1963. A huge layer of limestone broke loose parallel to its bedding planes. The translational slide involved around 250 million cubic meters that slid into the Vaiont Reservoir, creating a giant wave. The 175-meter-high wave (almost two football fields high) overtopped the Vaiont Dam (it was the world's highest dam, rising 265 meters above the valley floor). Three thousand people were killed in the villages that it flooded in the valleys below. The dam was not destroyed, a tribute to excellent engineering, but the men in charge of the building project were convicted of criminal negligence for ignoring the landslide hazards.

As in slower mass movements, water can play an important role in causing a rockslide. In 1925, exceptionally heavy rains in the Gros Ventre Mountains of Wyoming caused water to seep into a layer of sandstone, wetting the underlying layer of shale, greatly reducing its shear strength (figure 13.17). The layers of sedimentary rock were inclined roughly parallel to the hillside. With the wet shale acting as a lubricant, the overlying sedimentary rock and its soil cover slid into the valley, blocking the river. The slide itself merely created a lake, but the natural dam broke two years later and the resulting flood destroyed the small town of Kelly several kilometers downstream. Several residents who were standing on a bridge watching the floodwaters come down the valley were killed.

Debris Slides and Debris Falls

As the names suggest, debris slides and debris falls behave similarly to rockslides and rockfalls, except that they involve debris that moves as a coherent mass (at least initially). A **debris fall** is a free-falling mass of debris.

A **debris slide** is a coherent mass of debris moving along a well-defined surface (or surfaces). If the movement is along a curved surface the landslide is a *rotational debris slide.* Debris slides were mentioned earlier with earthflows, with which they are commonly associated (figure 13.7). Debris may slide, however, without an earthflow taking place.

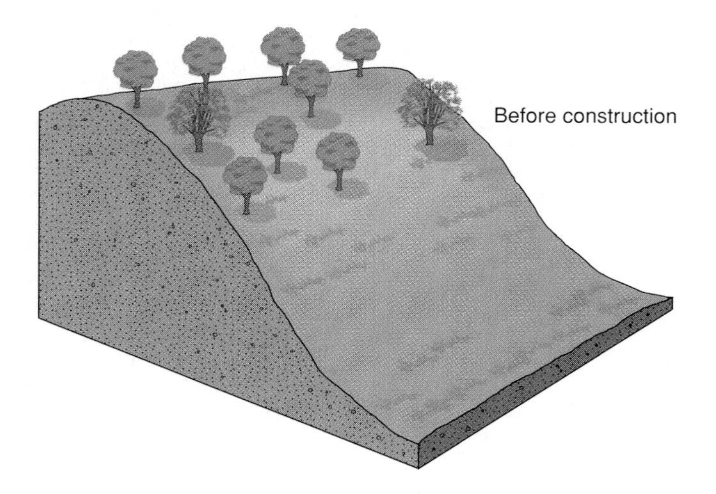

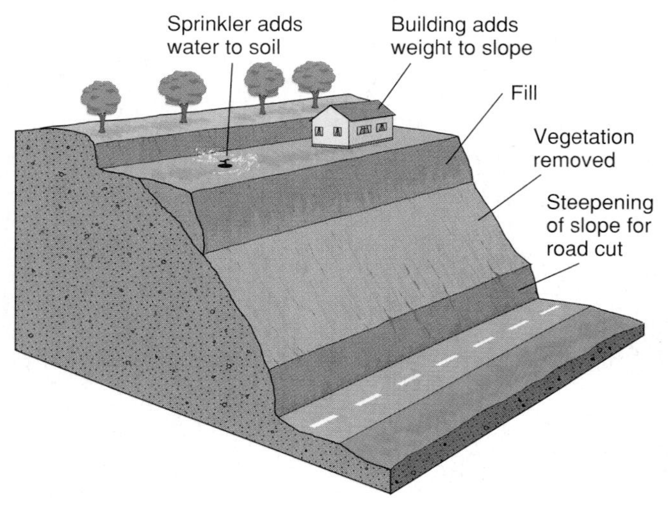

FIGURE 13.18
A hillside becomes vulnerable to mass wasting due to construction activities.

PREVENTING LANDSLIDES

Preventing Mass Wasting of Debris

Usually mass movements of debris can be prevented. Proper engineering is essential when the natural environment of a hillside is altered by construction. As shown in figure 13.18, construction generally makes a slope more susceptible to mass wasting of debris in several ways: (1) the base of the slope is undercut, removing the natural support for the upper part of the slope; (2) vegetation is removed during construction; (3) buildings constructed on the upper part of a slope add weight to the potential slide; and (4) extra water may be allowed to seep into the debris.

Some preventive measures can be taken during construction. A retaining wall is usually built where a cut has been made in the slope, but this alone is seldom as effective a deterrent to downslope movement as people hope. If, in addition, drain pipes are put through the retaining wall and into the hillside, water can percolate through and drain away rather than collecting in the debris behind the wall (figure 13.19). Without drains, excess water results in decreased shear strength and the whole soggy mass can easily burst through the wall.

Another practical preventive measure is to avoid oversteepening the slope. The hillside can be cut back in a series of terraces rather than in a single steep cut. This not only reduces the slope angle but also reduces the shear force by removing much of the overlying material. It also prevents loose material (such as boulders dislodged from the top of the cut) from rolling to the base. Road cuts constructed in this way are usually reseeded with rapidly growing grass or plants whose roots help anchor the slope. A vegetation cover also minimizes erosion from running water.

Preventing Rockfalls and Rockslides on Highways

Rockslides and rockfalls are a major problem on highways built through mountainous country. Steep slopes and cliffs are created when road-cuts are blasted and bulldozed into

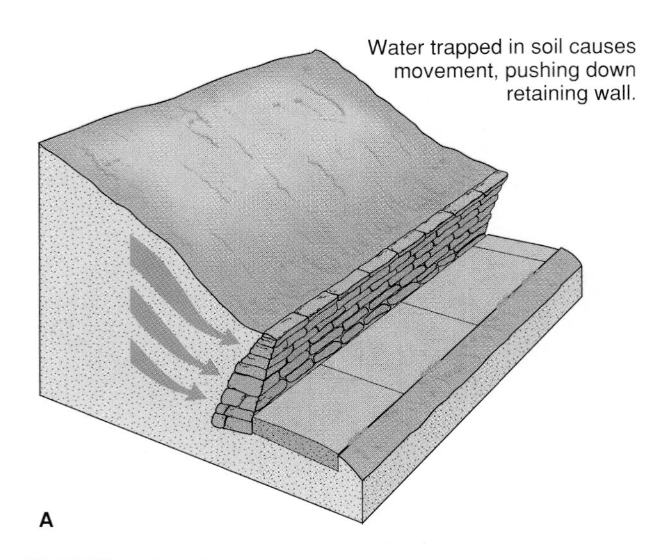

Water drains through pipe, allowing wall to keep slope from moving.

B

FIGURE 13.19
Use of drains to help prevent mass wasting.

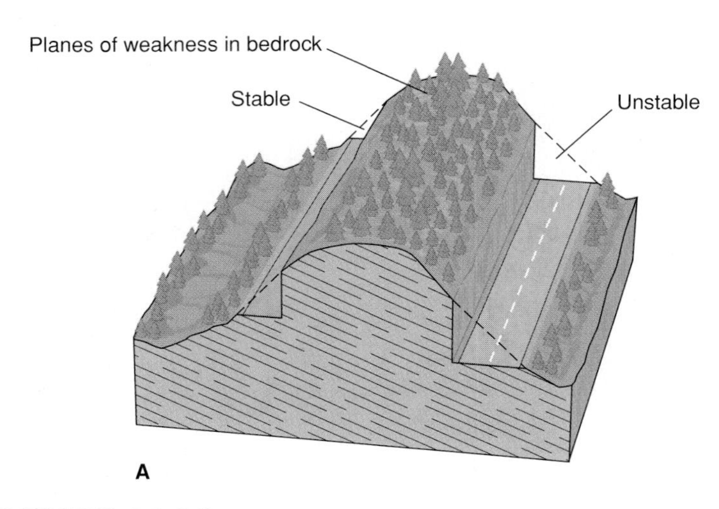

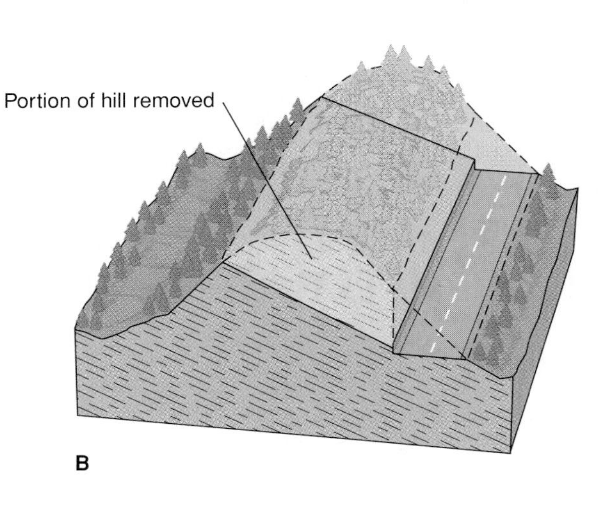

FIGURE 13.20
(*A*) Cross section of a hill showing a relatively safe road cut on the left and a hazardous road cut on the right. (*B*) The same hazardous road cut after removal of rock that might slide.

mountain sides. If the bedrock has planes of weakness (such as joints, bedding planes, or foliation planes), the orientation of these planes relative to the road-cut determines whether there is a rockslide hazard (as in figure 13.20*A*). If the planes of weakness are inclined into the hill, there is no chance of a rockslide. On the other hand, where the planes of weakness are approximately parallel to the slope of the hillside, a rockslide may occur.

Various techniques are used to prevent rockslides. By doing a detailed geologic study of an area before a road is built, builders might avoid a hazard by choosing the least dangerous route for the road. If a road-cut must be made through bedrock that appears prone to sliding, all of the rock that might slide could be removed (sometimes at great expense), as shown in figure 13.20*B*.

In some instances, slopes prone to rock sliding have been "stitched" in place by the technique shown in figure 13.21.

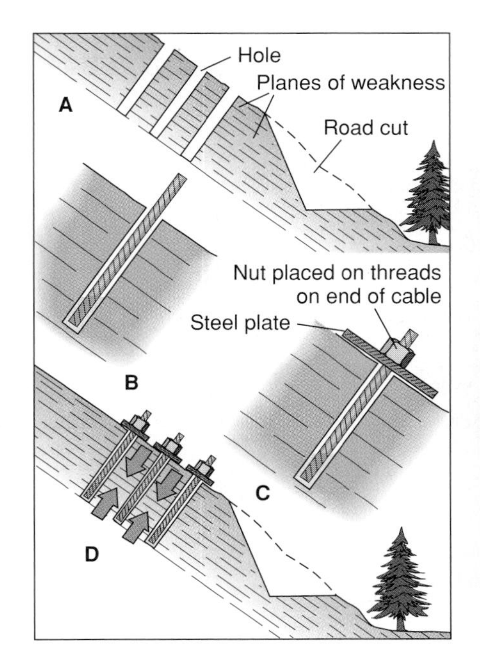

FIGURE 13.21

"Stitching" a slope to keep bedrock from sliding along planes of weakness. (*A*) Holes are drilled through unstable layers into stable rock. (*B*) Expanded view of one hole. A cable is fed into the hole and cement is pumped into the bottom of the hole and allowed to harden. (*C*) A steel plate is placed over the cable and a nut tightened. (*D*) Tightening all the nuts pulls unstable layers together and anchors them in stable bedrock. (*E*) Stabilized road-cut along Richardson Highway near Valdez, Alaska.
Photo by Paul G. Bauer

Summary

Mass wasting is the movement of a mass of debris (soil and loose rock fragments) or bedrock toward the base of a slope. Movement can take place as a flow, slide, or fall. Gravity is the driving force. The component of gravitational force that propels mass wasting is the *shear force,* which occurs parallel to the slopes. The resistance to that force is the *shear strength* of rock or debris. If shear force exceeds shear strength, mass wasting takes place. Water is an important factor in mass wasting.

A number of other factors determine whether movement will occur and, if it does, the rate of movement.

The slowest type of movement, *creep,* occurs mostly on relatively gentle slopes, usually aided by water in the soil. In colder climates, repeated freezing and thawing of water within the soil contributes to creep. *Landsliding* is a general term for more rapid mass wasting of rock, debris, or both. *Debris flows* include earthflows, mudflows, and debris avalanches. *Earthflows* vary greatly in velocity although they are not as rapid as *debris avalanches,* which are turbulent masses of debris, water, and air. *Solifluction,* a special variety of earthflow, usually takes place in arctic or subarctic climates where ground is permanently frozen (*permafrost*). A *mudflow* is a slurry of debris and water. Most mudflows flow in channels much as streams do.

Rockfall is the fall of broken rock down a vertical or near-vertical slope. A *rockslide* is a slab of rock sliding down a less-than-vertical surface. *Debris falls* and *debris slides* involve unconsolidated material rather than bedrock.

Terms to Remember

creep 319
debris 314
debris avalanche 325
debris fall 329
debris flow 320
debris slide 329
earthflow 320
fall 315
flow 315
landslide 314
mass wasting 314
mudflow 323
permafrost 322
relief 317
rock avalanche 327
rockfall 325
rockslide 327
rotational slide (slump) 315
shear force 318
shear strength 318
slide 315
solifluction 322
talus 325
translational slide 315

Testing Your Knowledge

Use the questions below to prepare for exams based on this chapter.

1. Describe the effect on shear strength of the following:
   a. thickness of debris;
   b. orientation of planes of weakness;
   c. water in debris; and
   d. vegetation.
2. Compare the shear force to the force of gravity (drawing diagrams similar to figure 13.2) for the following situations:
   a. a vertical cliff;
   b. a flat horizontal plane; and
   c. a 45° slope.
3. How does a rotational slide differ from a translational slide?
4. What role does water play in each of the types of mass wasting?
5. Why is solifluction more common in colder climates than in temperate climates?
6. List and explain the key factors that control mass wasting.
7. What is the slowest type of mass wasting process?
   a. debris flow b. rockslide
   c. creep d. rockfall
   e. avalanche
8. Any unconsolidated material at Earth's surface of any size is called
   a. debris b. sediment
   c. soil d. talus
9. A descending mass moving downslope as a viscous fluid is referred to as a
   a. fall b. landslide
   c. flow d. slide
10. The driving force behind all mass wasting processes is
    a. gravity b. slope angle
    c. type of bedrock material d. presence of water
    e. vegetation
11. The resistance to movement or deformation of debris is its
    a. mass b. shear strength
    c. shear force d. density
12. Flow of water-saturated debris over impermeable material is called
    a. solifluction b. flow
    c. slide d. fall
13. A flowing mixture of debris and water, usually moving down a channel, is called a
    a. mudflow b. slide
    c. fall d. debris flow
14. An apron of fallen rock fragments that accumulates at the base of a cliff is called
    a. debris b. sediment
    c. soil d. talus
15. How does construction destabilize a slope?
    a. adds weight to the top of the slope
    b. decreases water content of the slope
    c. adds weight to the bottom of the slope
    d. increases the shear strength of the slope
16. How can landslides be prevented during construction? (choose all that apply)
    a. retaining walls
    b. cut steeper slopes
    c. install water drainage systems
    d. add vegetation

Expanding Your Knowledge

1. Why do people fear earthquakes, hurricanes, and tornadoes more than they fear landslides?
2. If you were building a house on a cliff, what would you look for to ensure that your house would not be destroyed through mass wasting?
3. Why isn't the land surface of Earth flat after millions of years of erosion by mass wasting as well as by other erosional agents?
4. Can any of the indicators of creep be explained by processes other than mass wasting?

Exploring Web Resources

www.mhhe.com/plummer9e

Go to the Online Learning Center to access the answers for the Testing Your Knowledge section. This site also has additional quizzes and readings and media resources to further your understanding of mass wasting, as well as direct links to all the sites listed below.

http://landslides.usgs.gov/

Geologic hazards, landslides, U.S. Geological Survey. You can get to several useful sites from here. Reports on recent landslides can be accessed by clicking on the ones listed. Click on "National Landslide Information Center" for photos of landslides, including some described in this chapter. Watch animation of a landslide. You can access sources of information on landslides and other geologic features for any state, usually from a state's geologic survey.

http://sts.gsc.nrcan.gc.ca/page1/geoh/slide.htm

Landslides and snow avalanches in Canada. Geological Survey of Canada's site has generalized descriptions and some photos of significant Canadian landslides.

Animations

This chapter includes the following animation available on our Online Learning Center at www.mhhe.com/plummer9e.

13.1 Flow, Slide, and Fall

CHAPTER

14

Sediments and Sedimentary Rocks

The rock cycle is a theoretical model of the constant recycling of rocks as they form, are destroyed, and then reform. We began our discussion of the rock cycle with igneous rock (chapters 10 and 11), and we now discuss sedimentary rocks. Metamorphic rocks, the third major rock type, are the subject of the next chapter.

You saw in chapter 12 how weathering produces sediment. In this chapter we explain more about sediment origin as well as the erosion, transportation, sorting, deposition, and eventual lithification of sediments to form sedimentary rock. Because they have such diverse origins, sedimentary rocks are difficult to classify. We divide them into clastic, chemical, and organic sedimentary rocks, but this classification is not entirely satisfactory. Furthermore, despite their great variety, only three sedimentary rocks are very common—shale, sandstone, and limestone.

Sedimentary rocks contain numerous clues to their origin and the environment in which they were deposited. Geologists determine this information from the shape and sequence of rock layers and from the sediment grains and the sedimentary structures such as fossils, cross-beds, ripple marks, and mud cracks that are contained in the rock.

Sedimentary rocks are important because they are widespread and because many of them, such as coal and limestone, are economically important. About three-fourths of the surface of continents is blanketed with a relatively thin skin of sedimentary rocks. Concentrated in sedimentary rocks are important resources such as crude oil, natural gas, ground water, salt, gypsum, uranium, and iron ore.

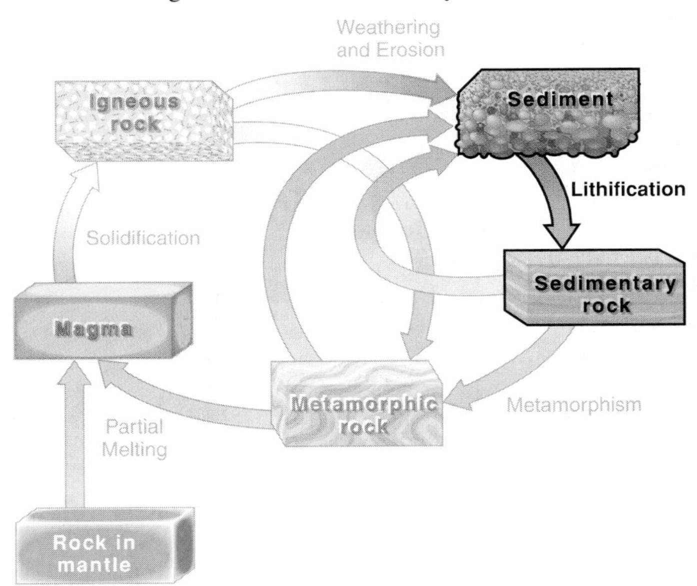

Opposite: Eroded sandstone formations, Colorado Plateau, Arizona.
Carr Clifton/Minden Pictures

table 14.1 Sediment Particles and Clastic Sedimentary Rocks

Diameter (mm)	Sediment		Sedimentary Rock
256	Boulder	Gravel	**Breccia** (angular particles) or **Conglomerate** (rounded particles)
64	Cobble		
2	Pebble		
1/16	Sand		**Sandstone**
1/256	Silt	"Mud"	Siltstone (mostly silt)
	Clay		**Shale** or mudstone (mostly clay)

Sandstone and shale are quite common; the others are relatively rare.

FIGURE 14.1

These boulders have been rounded by abrasion as wave action rolled them against one another on this beach.

SEDIMENT

Most sedimentary rocks form from loose grains or chemical precipitation of sediment. Sediment includes such particles as sand on beaches, mud on a lake bottom, boulders frozen into glaciers, pebbles in streams, and dust particles settling out of the air. An accumulation of clam shells on the sea bottom offshore is sediment, as are coral fragments broken from a reef by large storm waves.

Sediment is the collective name for loose, solid particles that originate from:

1. Weathering and erosion of preexisting rocks.
2. Chemical precipitation from solution, including secretion by organisms in water.

These particles usually collect in layers on Earth's surface. An important part of the definition is that the particles are loose. Sediments are said to be *unconsolidated,* which means that the grains are separate, or unattached to one another.

Sediment particles are classified and defined according to the size of individual fragments. Table 14.1 shows the precise definitions of particles by size.

Gravel includes all rounded particles coarser than 2 mm in diameter, the thickness of a U.S. nickel. (Angular fragments of this size are called *rubble.*) *Pebbles* range from 2 to 64 mm (about the size of a tennis ball). *Cobbles* range from 64 to 256 mm (about the size of a basketball), and *boulders* are coarser than 256 mm.

Sand grains are from 1/16 mm (about the thickness of a human hair) to 2 mm in diameter. Grains of this size are visible and feel gritty between the fingers. **Silt** grains are from 1/256 to 1/16 mm. They are too small to see without a magnifying device, such as a geologist's hand lens. Silt does not feel gritty between the fingers, but it does feel gritty between the teeth (geologists often bite sediments to test their grain size). **Clay** is the finest sediment, at less than 1/256 mm, too fine to feel gritty to fingers or teeth. *Mud* is a term loosely used for wet silt and clay.

Note that we have two different uses of the word *clay*—a *clay-sized particle* (table 14.1) and a *clay mineral.* A clay-sized particle can be composed of any mineral at all provided its diameter is less than 1/256 mm. A clay mineral, on the other hand, is one of a small group of silicate minerals with a sheet-silicate structure. Clay minerals usually fall into the clay-size range.

Quite often the composition of sediment in the clay-size range turns out to be mostly clay minerals, but this is not always the case. Because of its resistance to chemical weathering, quartz may show up in this fine-size grade. (Most silt is quartz.) Intense mechanical weathering can break down a wide variety of minerals to clay size, and these extremely fine particles may retain their mineral identity for a long time if chemical weathering is slow. The great weight of glaciers is particularly effective at grinding minerals down to the clay-size range, producing "rock flour," which gives a milky appearance to glacial meltwater streams (see chapter 19).

Weathering, erosion, and transportation are some of the processes that affect the character of sediment. Both mechanically weathered and chemically weathered rock and sediment can be eroded, and weathering continues as erosion takes place. Sand being transported by a river also can be actively weathering, as can mud on a lake bottom. The character of sediment can also be altered by *rounding* and *sorting* during transportation, and by eventual *deposition.*

Transportation

Rounding is the grinding away of sharp edges and corners of rock fragments during transportation. Rounding occurs in sand and gravel as rivers, glaciers, or waves cause particles to hit and scrape against one another (figure 14.1) or against a rock surface, such as a rocky streambed. Boulders in a stream may show substantial rounding in less than one mile of travel. Because rounding during transportation is so rapid, it is a much more important process than spheroidal weathering (see chapter 12), which also tends to round off sharp edges.

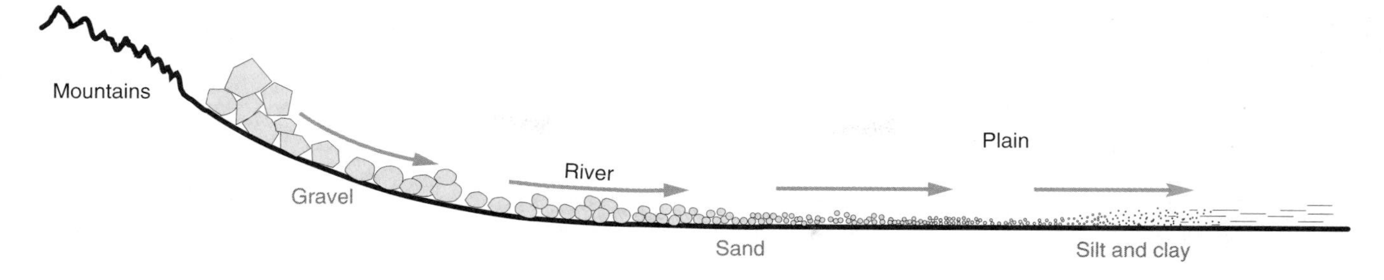

FIGURE 14.2

Sorting of sediment by a river. The coarse sediment is deposited first, and the finest sediment is carried in farthest.

Sorting is the process by which sediment grains are selected and separated according to grain size (or grain shape or specific gravity) by the agents of transportation, especially by running water. Because of their high viscosity and manner of flow, glaciers are poor sorting agents. Glaciers deposit all sediment sizes in the same place, so glacial sediment usually consists of a mixture of clay, silt, sand, and gravel. Such glacial sediment is considered *poorly sorted.* Sediment is considered *well-sorted* when the grains are nearly all the same size. A river, for example, is a good sorting agent, separating sand from gravel, and silt and clay from sand. Sorting takes place because of the greater weight of larger particles. Boulders weigh more than pebbles and are more difficult for the river to transport, so a river must flow more rapidly to move boulders than to move pebbles. Similarly, pebbles are harder to move than sand, and sand is harder to move than silt and clay.

Figure 14.2 shows the sorting of sediment by a river as it flows out of steep mountains onto a gentle plain, where the water loses energy and slows down. As the river loses energy, the heaviest particles of sediment are deposited. The boulders come to rest first (figure 14.3). As the river continues to slow and becomes less turbulent, cobbles and then pebbles are deposited. Sand comes to rest as the river loses still more energy (figure 14.4). Finally, the river is carrying only the finest sediment—silt and clay (figure 14.5). The river has sorted the original sediment mix by grain size.

FIGURE 14.3

Coarse gravel (boulder size) is deposited first along a river's course as the river sorts out the various sediment sizes. River gravel is usually deposited in or near steep mountains.

Deposition

When transported material settles or comes to rest, **deposition** occurs. Sediment is deposited when running water, glacial ice, waves, or wind loses energy and can no longer transport its load.

Deposition also refers to the accumulation of chemical or organic sediment, such as clam shells on the sea floor or plant material on the floor of a swamp. Such sediments may form as organisms die and their remains accumulate, perhaps with no transportation at all. Deposition of salt crystals can take place as seawater evaporates. A change in the temperature, pressure, or chemistry of a solution may also cause precipitation—hot springs may deposit calcite or silica as the warm water cools.

The **environment of deposition** is the location in which deposition occurs. A few examples of environments of deposition are the deep sea floor, a desert valley, a river channel, a

FIGURE 14.4

Deposition of sand occurs when a river loses energy as it flows across a gentle plain.

FIGURE 14.5

The river on the right is carrying only silt and clay as it enters the clear river on the left. This fine sediment may come to rest at the mouth of a river where it enters a lake or the sea.

Photo by C. W. Montgomery

coral reef, a lake bottom, a beach, and a sand dune. Each environment is marked by characteristic physical, chemical, and biological conditions. You might expect mud on the sea floor to differ from mud on a lake bottom. Sand on a beach may differ from sand in a river channel. Some differences are due to varying sediment sources and transporting agents, but some are the result of conditions in the environments of deposition themselves.

One of the most important jobs of geologists studying sedimentary rocks is to try to determine the ancient environment of deposition of the sediment in which the rock formed. Factors that can help in determining this are a detailed knowledge of modern environments, the shape and vertical sequences of rock layers in the field, the features (including fossils) found within the rock, the mineral composition of the rock, and the size, shape, and surface texture of the individual sediment grains. Later in the chapter we give a few examples of interpreting sedimentary rocks.

Preservation

Not all sediments are preserved as sedimentary rock. Gravel in a river may be deposited when a river is low, but then may be eroded and transported by the next flood on the river. Many sediments on land, particularly those well above sea level, are easily eroded and carried away, so they are not commonly preserved. Sediments on the sea floor are easier to preserve. In general, continental and marine sediments are most likely to be preserved if they are deposited in a *subsiding* (sinking) *basin* and if they are covered or *buried by later sediments.*

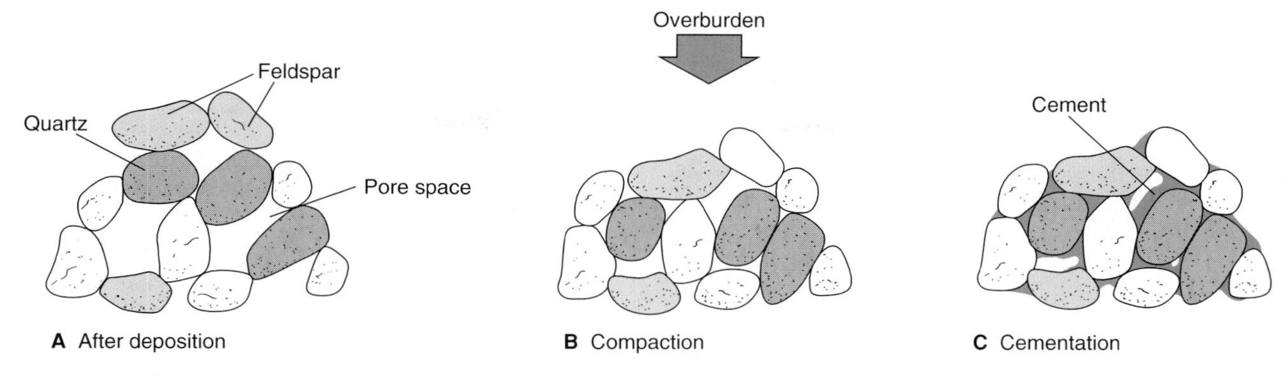

FIGURE 14.6

Lithification of sand grains to become sandstone. (*A*) Loose sand grains are deposited with open pore space between the grains. (*B*) The weight of overburden compacts the sand into a tighter arrangement, reducing pore space. (*C*) Precipitation of cement in the pores by ground water binds the sand into the rock sandstone, which has a clastic texture.

Lithification

Lithification is the general term for the processes that convert loose sediment into sedimentary rock. Most sedimentary rocks are lithified by a combination of *compaction,* which packs loose sediment grains tightly together, and *cementation,* in which the precipitation of cement around sediment grains binds them into a firm, coherent rock. *Crystallization* of minerals from solution, without passing through the loose-sediment stage, is another way that rocks may be lithified.

As sediment grains settle slowly in a quiet environment such as a lake bottom, they form an arrangement with a great deal of open space between the grains (figure 14.6*A*). The open spaces between grains are called *pores,* and in a quiet environment, a deposit of sand may have 40% to 50% of its volume as open **pore space.** (If the grains were traveling rapidly and impacting one another just before deposition, the percentage of pore space will be less.) As more and more sediment grains are deposited on top of the original grains, the increasing weight of this *overburden* packs the original grains together, reducing the amount of pore space. This shift to a tighter packing, with a resulting decrease in pore space, is called **compaction** (figure 14.6*B*). As pore space decreases, some of the interstitial water that usually fills sediment pores is driven out of the sediment.

As underground water moves through the remaining pore space, solid material called **cement** can precipitate in the pore space and bind the loose sediment grains together to form a solid rock. The cement attaches very tightly to the grains, holding them in a rigid framework. As cement partially or completely fills the pores, the total amount of pore space is further reduced (figure 14.6*C*), and the loose sand forms a hard, coherent sandstone by **cementation.**

Sedimentary rock cement is often composed of the mineral calcite or of other carbonate minerals. Dissolved calcium and bicarbonate ions are common in surface and underground waters. If the chemical conditions are right, these ions may recombine to form solid calcite, as shown in the following reaction.

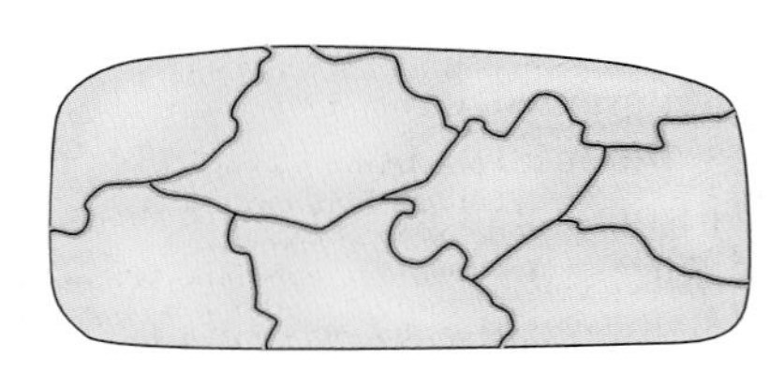

FIGURE 14.7

Crystalline texture. The rock is held together by interlocking crystals, which grew as they precipitated from solution. Such a rock has no cement or pore space.

$$\underbrace{Ca^{++} + 2HCO_3^{-}}_{\text{dissolved ions}} \rightarrow \underbrace{CaCO_3}_{\text{calcite}} + H_2O + CO_2$$

Silica is another common cement. Iron oxides and clay minerals can also act as cement but are less common than calcite and silica. The dissolved ions that precipitate as cement originate from the chemical weathering of minerals such as feldspar and calcite. This weathering may occur within the sediments being cemented, or at a very distant site, with the ions being transported tens or even hundreds of miles by water before precipitating as solid cement.

A sedimentary rock that consists of sediment grains bound by cement into a rigid framework is said to have a **clastic texture.** Usually such a rock still has some pore space because cement rarely fills the pores completely (figure 14.6*C*).

Some sedimentary rocks form by **crystallization,** the development and growth of crystals by precipitation from solution at or near Earth's surface (the term is also used for igneous rocks that crystallize as magma cools). These rocks have a **crystalline texture,** an arrangement of interlocking crystals that develops as crystals grow and interfere with each other (figure 14.7). Crystalline rocks lack cement. They are held together by the interlocking of crystals. Such rocks have no pore space because the crystals have grown until they fill all available space. Some sedimentary rocks with a crystalline texture are the result of *recrystallization,* the growth of new crystals that form from and then destroy the original clastic grains of a rock that has been buried.

TYPES OF SEDIMENTARY ROCKS

Sedimentary rocks are formed from (1) lithification of sediment, (2) precipitation from solution, or (3) consolidation of the remains of plants or animals. These different types of sedimentary rocks are called, respectively, *clastic, chemical,* and *organic* rocks.

Most sedimentary rocks are **clastic sedimentary rocks,** formed from cemented sediment grains that are fragments of preexisting rocks. The rock fragments can be either identifiable pieces of rock, such as pebbles of granite or shale, or individual mineral grains, such as sand-sized quartz and feldspar crystals loosened from rocks by weathering and erosion. Clay minerals formed by chemical weathering are also considered fragments of preexisting rocks. During transportation the grains may have been rounded and sorted. Table 14.1 shows the clastic rocks, such as conglomerate, sandstone, and shale, and shows how these rocks vary in grain size.

Chemical sedimentary rocks are deposited by precipitation of minerals from solution. An example of inorganic precipitation is the formation of *rock salt* as seawater evaporates. Chemical precipitation can also be caused by organisms. The sedimentary rock *limestone,* for instance, can form by the precipitation of calcite within a coral reef by corals and algae. Such a rock is classified as a *biochemical* limestone.

Chemical rocks may or may not have once been sediment. Rock salt may form from sediment where individual salt crystals form in evaporating water until they grow large enough to interlock into a solid rock.

Organic sedimentary rocks are rocks that accumulate from the remains of organisms. *Coal* is an organic rock that forms from the compression of plant remains, such as moss, leaves, twigs, roots, and tree trunks.

Appendix B describes and helps you identify the common sedimentary rocks. The standard geologic symbols for these rocks (such as dots for sandstone, and a "brick wall" symbol for limestone) are shown in appendix F and are used throughout the book.

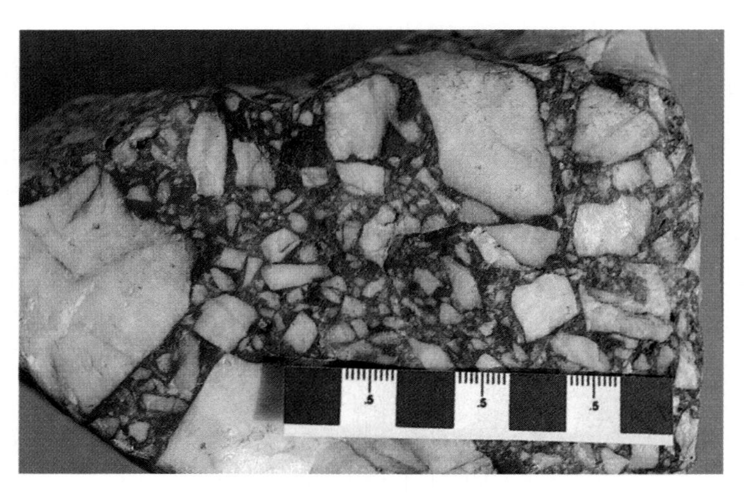

FIGURE 14.8
Breccia is characterized by coarse, angular fragments. The cement in this rock is colored by hematite. The wide black and white bars on the scale are 1 centimeter long, the small divisions are 1 millimeter. Note that most grains exceed 2 mm (table 14.1).

FIGURE 14.9
An outcrop of conglomerate. Note the rounding of cobbles, which vary in composition. Long scale bar 10 cm, short bars 1 cm.

CLASTIC ROCKS

Breccia and Conglomerate

Sedimentary breccia is a coarse-grained sedimentary rock formed by the cementation of coarse, angular fragments of rubble (figure 14.8). Because grains are rounded so rapidly during transport, it is unlikely that the angular fragments within breccia have moved very far from their source. Sedimentary breccia might form from fragments that have accumulated at the base of a steep slope of rock that is being mechanically weathered. Landslide deposits also might lithify into sedimentary breccia. This type of rock is not particularly common.

Conglomerate is a coarse-grained sedimentary rock formed by the cementation of rounded gravel. It can be distinguished from breccia by the definite roundness of its particles (figure 14.9). Because conglomerates are coarse-grained, the particles may not have traveled far; but some transport was necessary to round the particles. Angular fragments that fall from a cliff and then are carried a few miles by a river or pounded by waves crashing in the surf along a beach are quickly rounded. Gravel that is transported down steep submarine canyons, or carried by glacial ice, however, can be transported tens or even hundreds of kilometers before deposition.

Sandstone

Sandstone is formed by the cementation of sand grains (figure 14.10). Any deposit of sand can lithify to sandstone. Rivers

FIGURE 14.10

Types of sandstone. (*A*) Quartz sandstone; more than 90% of the grains are quartz. (*B*) Arkose; the grains are mostly feldspar and quartz. (*C*) Graywacke; the grains are surrounded by dark, fine-grained matrix. (Small scale divisions are 1 millimeter; most of the sand grains are about 1 mm in diameter.)

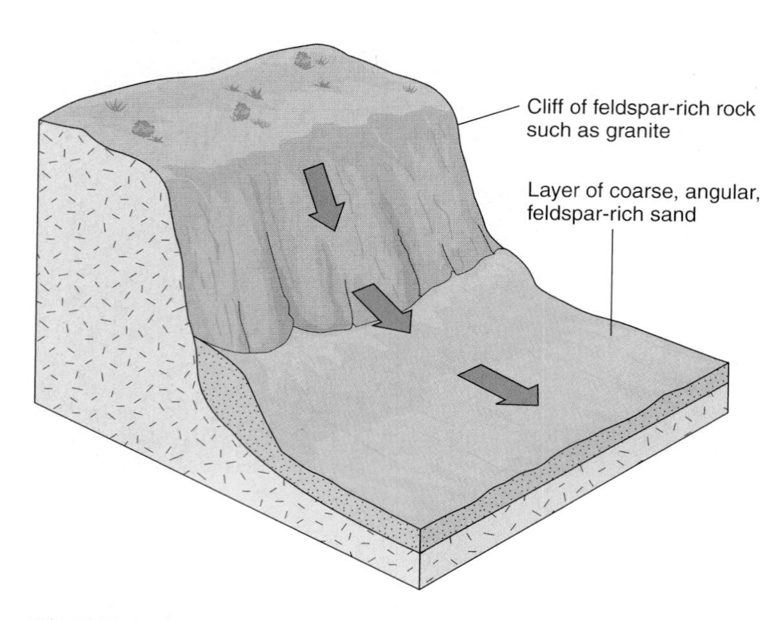

FIGURE 14.11

Feldspar-rich sand (arkose) may accumulate from the rapid erosion of feldspar-containing rock such as granite. Steep terrain accelerates erosion rates so that feldspar may be eroded before it is completely chemically weathered into clay minerals.

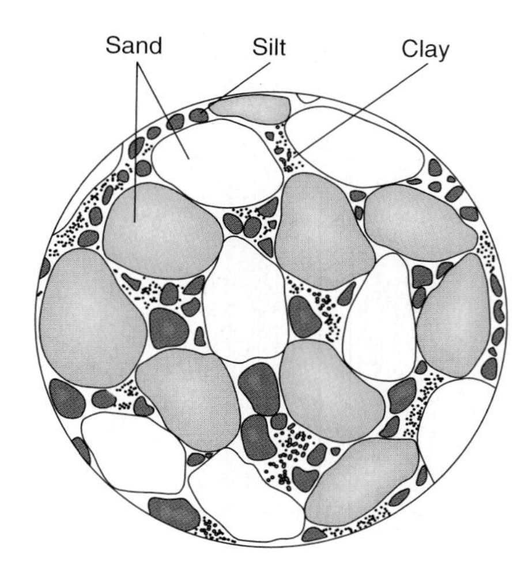

FIGURE 14.12

A poorly sorted sediment of sand grains surrounded by a matrix of silt and clay grains. Lithification of such a sediment would produce a "dirty sandstone."

deposit sand in their channels, and wind piles up sand into dunes. Waves deposit sand on beaches and in shallow water. Deep-sea currents spread sand over the sea floor. As you might imagine, sandstones show a great deal of variation in mineral composition, degree of sorting, and degree of rounding.

Quartz sandstone is a sandstone in which more than 90% of the grains are quartz (figure 14.10*A*). Because quartz is resistant to chemical weathering, it tends to concentrate in sand deposits as the less resistant minerals such as feldspar are weathered away. The quartz grains in a quartz sandstone are usually well-sorted and well-rounded because they have been transported for great distances. Most quartz sandstone was deposited as beach sand or dune sand.

A sandstone with more than 25% of the grains consisting of feldspar is called *arkose* (figure 14.10*B*). Because feldspar grains are preserved in the rock, the original sediment obviously did not undergo severe chemical weathering, or the feldspar would have been destroyed. Cliffs of granite in a desert could be a source for such a sediment, for the rapid erosion associated with rugged terrain would allow feldspar to be mechanically weathered and eroded before it is chemically weathered (a dry climate slows chemical weathering). Most arkoses contain coarse, angular grains, so transportation distances were probably short. An arkose may have been deposited at or near the base of a cliff, as shown in figure 14.11.

Sandstones may contain a substantial amount of **matrix** in the form of fine-grained silt and clay in the space between larger sand grains (figure 14.12). A matrix-rich sandstone is poorly sorted and often dark in color. It is sometimes called a "dirty sandstone."

Graywacke (pronounced "gray-wacky") is a type of sandstone in which more than 15% of the rock's volume consists of

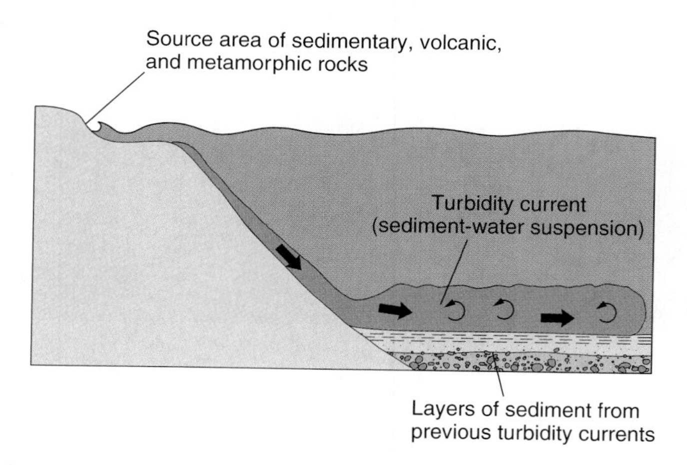

FIGURE 14.13

A turbidity current flow downslope along the sea floor. Dense sediment-laden water is heavier than the clear water that it flows beneath.

fine-grained matrix (figure 14.10*C*). Graywackes are often tough and dense and are generally dark gray or green. The sand grains may be so coated with matrix that they are hard to see, but they typically consist of quartz, feldspar, and sand-sized fragments of other fine-grained sedimentary, volcanic, and metamorphic rocks.

Most graywackes probably formed from sediments transported by **turbidity currents,** dense masses of sediment-laden water that flow downslope along the sea floor. The sediment-water mixture is heavier than clear water, so it is pulled downslope by gravity until it comes to rest on the sea floor at the base of the slope (figure 14.13). Turbidity currents may be generated by underwater landslides, perhaps triggered by earthquakes, or by violent surface storms such as hurricanes that stir up bottom sediment. Sediment-laden rivers discharging directly into the sea may also cause turbidity currents.

A

B

FIGURE 14.14

(*A*) An outcrop of shale near Pittsburgh, Pennsylvania. Note how this fine-grained rock tends to split into very thin layers. (*B*) Shale pieces; note the very fine grain (scale in centimeters), very thin layers (laminations) on the edge of the large piece, and tendency to break into small, flat pieces (fissility).

Photo *A* by C. C. Plummer

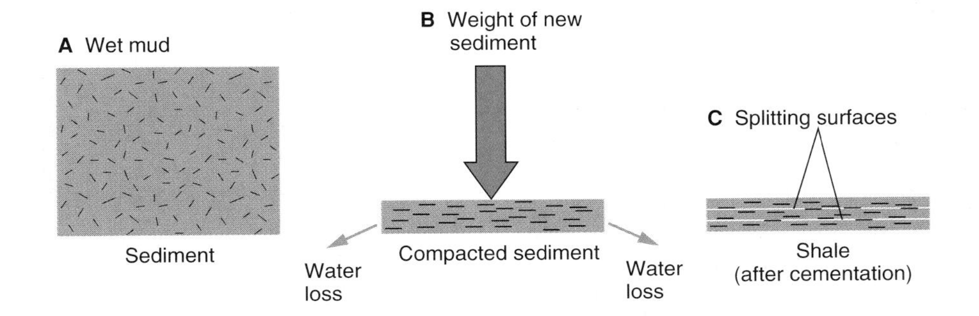

FIGURE 14.15

Lithification of shale from the compaction and cementation of wet mud. (*A*) Randomly oriented silt and clay particles in wet mud. (*B*) Particles reorient, water is lost, and pore space decreases during compaction caused by the weight of new sediment deposited on top of the wet mud. (*C*) Splitting surfaces in cemented shale form parallel to the oriented mineral grains.

The Fine-Grained Rocks

Rocks consisting of fine-grained silt and clay are called *shale, siltstone, claystone,* and *mudstone.*

Shale is a fine-grained sedimentary rock notable for its ability to split into layers (called *fissility*). Splitting takes place along the surfaces of very thin layers (called *laminations*) within the shale (figure 14.14). Most shales contain both silt and clay (averaging about $^2/_3$ clay-sized clay minerals, $^1/_3$ silt-sized quartz) and are so fine-grained that the surface of the rock feels very smooth. The silt and clay deposits that lithify as shale accumulate on lake bottoms, at the ends of rivers in deltas, on river floodplains, and on quiet parts of the deep ocean floor.

Fine-grained rocks such as shale typically undergo pronounced compaction as they lithify. Figure 14.15 shows the role of compaction in the lithification of shale from wet mud. Before compaction, as much as 80% of the volume of the wet mud may have been pore space filled with water. The flakelike clay minerals were randomly arranged within the mud. Pressure from overlying material packs the sediment grains together and reduces the overall volume by squeezing water out of the pores. The clay minerals are reoriented perpendicular to the pressure, becoming parallel to one another like a deck of cards. The fissility of shale is due to weaknesses between these parallel clay flakes.

Compaction by itself does not generally lithify sediment into sedimentary rock. It does help consolidate clayey sediments by pressing the microscopic clay minerals so closely together that attractional forces at the atomic level tend to bind them together. Even in shale, however, the primary method of lithification is cementation.

A rock consisting mostly of silt grains is called *siltstone.* Somewhat coarser-grained than most shales, siltstones lack the fissility and laminations of shale. *Claystone* is a rock composed predominately of clay-sized particles, but lacking the fissility of shale. *Mudstone* contains both silt and clay, having the same grain size and smooth feel of shale but lacking shale's laminations and fissility. Mudstone is massive and blocky, while shale is visibly layered and fissile.

CHEMICAL SEDIMENTARY ROCKS

Chemical sedimentary rocks are precipitated from an aqueous environment. Chemical sedimentary rocks are either precipitated directly by inorganic processes or by the actions of organisms. Chemical rocks include *carbonates, chert,* and *evaporates.*

FIGURE 14.16
Corals precipitate calcium carbonate to form limestone in a reef. Water depth about 25 feet (8 meters). San Salvador Island, Bahamas.

Carbonate Rocks

Carbonate rocks contain CO_3 as part of their chemical composition. The two main types of carbonates are limestone and dolomite.

Limestone

Limestone is a sedimentary rock composed mostly of calcite ($CaCO_3$). Limestones are either precipitated by the actions of organisms or are precipitated directly as the result of inorganic processes. Thus, the two major types of limestone can be classified as either *biochemical* or *inorganic limestone.*

Biochemical limestones are precipitated through the actions of organisms. Most biochemical limestones are formed on continental shelves in warm, shallow water. Biochemical limestone may be precipitated directly as a solid rock in the core of a reef by corals, by encrusting algae, or by other shell-forming organisms (figure 14.16). Such a rock would have a crystalline texture and would contain the fossil remains of organisms preserved in growth position.

The great majority of limestones are biochemical limestones formed of wave-broken fragments of algae, corals, and shells. The fragments may be of any size (gravel, sand, silt, and clay) and are often sorted and rounded as they are transported by waves and currents across the sea floor (figure 14.17). The action of these waves and currents and subsequent cementation of these fragments into rock give these limestones a clastic texture. These *bioclastic* (or *skeletal*) *limestones* take a great variety of appearances. They may be relatively coarse-grained with recognizable fossils (figure 14.18) or uniformly fine-grained and dense from the accumulation of microscopic fragments of coralline algae (figures 14.18 and 14.19). A variety of limestone called *coquina* forms from the cementation of shells and shell fragments that accumulated on the shallow sea floor near shore (figure 14.20).

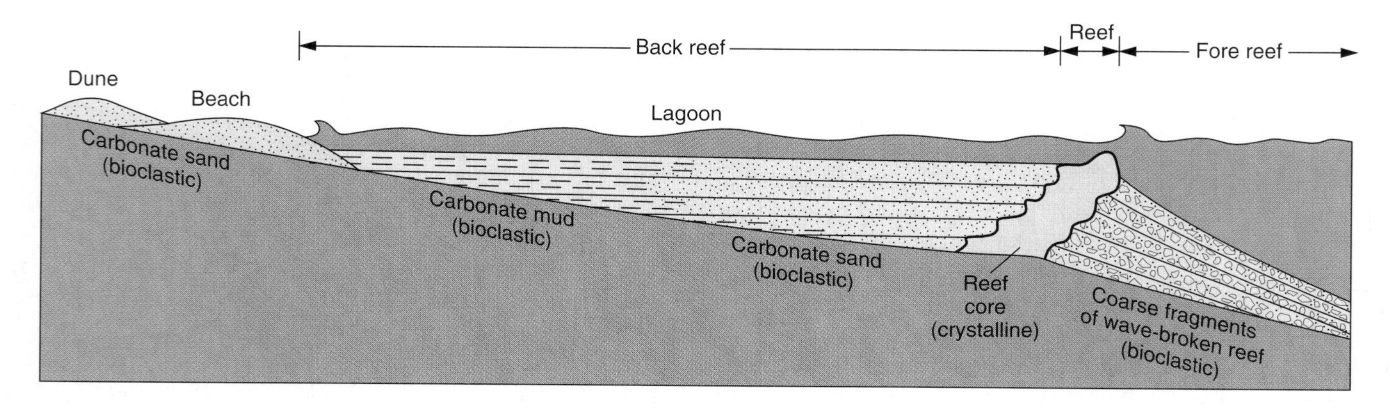

FIGURE 14.17
A living coral-algal reef sheds bioclastic sediment into the fore-reef and back-reef environments. The fore reef consists of coarse, angular fragments of reef. Coralline algae are the major contributors of carbonate sand and mud in the back reef. Beaches and dunes are often bioclastic sand. The sediments in each environment can lithify to form highly varied limestones.

FIGURE 14.18
Bioclastic limestones. The two on the left are coarse-grained and contain visible fossils of corals and shells. The limestone on the right consists of fine-grained carbonate mud formed by coralline algae.

FIGURE 14.19
Coralline algae on the sea floor in 3 meters of water on the Bahama Banks. The "shaving brush" alga is Penicillus, which produces great quantities of fine-grained carbonate mud.

FIGURE 14.20
Coquina, a variety of bioclastic limestone, is formed by the cementation of coarse shells.

FIGURE 14.21

Chalk is a fine-grained variety of bioclastic limestone formed of the remains of microscopic marine organisms that live near the sea surface.

It has a clastic texture and is usually coarse-grained, with easily recognizable shells and shell fragments in it. *Chalk* is a light-colored, porous, very fine-grained variety of bioclastic limestone that forms from the sea-floor accumulation of microscopic marine organisms that drift near the sea surface (figure 14.21).

Inorganic limestones are precipitated directly as the result of inorganic processes. *Oolitic limestone* is a distinctive variety of inorganic limestone formed by the cementation of sand-sized *oolites* (or *ooids*), small spheres of calcite inorganically precipitated in warm, shallow seawater (figure 14.22). Strong tidal currents roll the oolites back and forth, allowing them to maintain a nearly spherical shape as they grow. Wave action may also contribute to their shape.

Oolitic limestone has a clastic texture. *Tufa* and *travertine* are inorganic limestones that form from fresh water. Tufa is precipitated from solution in the water of a continental spring, lake, or from percolating ground water. Travertine may form in caves when carbonate-rich water loses CO_2 to the cave atmosphere. Tufa and travertine both have a crystalline texture; however, tufa is generally more porous, cellular, or open than travertine, which tends to be more dense.

Limestones are particularly susceptible to **recrystallization,** the process by which new crystals, often of the same composition as the original grains, develop in a rock. Calcite grains recrystallize easily, particularly in the presence of water and under the weight of overlying sediment. The new crystals that form are often large and can be easily seen in a rock as light reflects off their broad, flat faces. Because recrystallization often destroys the original clastic texture and fossils of a rock, replacing them with a new crystalline texture, the geologic history of such a rock can be very difficult to determine.

A

B

FIGURE 14.22

(*A*) Aerial photo of underwater dunes of oolites (ooids) chemically precipitated from seawater on the shallow Bahama Banks, south of Bimini. Tidal currents move the dunes. (*B*) An oolitic limestone formed by the cementation of oolites (small spheres). Small divisions on scale are 1 millimeter wide.

Dolomite

The term **dolomite** (table 14.2) is used to refer to both a sedimentary rock and the mineral that composes it, $CaMg(CO_3)_2$. (Some geologists use *dolostone* for the rock.) Dolomite often forms from limestone as the calcium in calcite is partially replaced by magnesium, usually as water solutions move through the limestone.

$$\underset{\text{magnesium in solution}}{Mg^{++}} + \underset{\text{calcite}}{2CaCO_3} \rightarrow \underset{\text{dolomite}}{CaMg(CO_3)_2} + \underset{\text{calcium in solution}}{Ca^{++}}$$

ENVIRONMENTAL GEOLOGY 14.1

Valuable Sedimentary Rocks

Many sedimentary rocks have uses that make them valuable. *Limestone* is widely used as building stone and is also the main rock type quarried for crushed rock for road construction. Pulverized limestone is the main ingredient of cement for mortar and concrete and is also used to neutralize acid soils in the humid regions of the United States. *Coal* is a major fuel, used widely for generating electrical power and for heating. Plaster and plasterboard for home construction are manufactured from *gypsum,* which is also used to stabilize the shrink-swell characteristics of clay-rich soils in some areas. Huge quantities of *rock salt* are consumed by industry, primarily for the manufacture of hydrochloric acid. More familiar uses of rock salt are for table salt and for melting ice on roads.

Some *chalk* is used in the manufacture of blackboard chalk, although most classroom chalk is now made from pulverized limestone. The filtering agent for beer brewing and for swimming pools is likely to be made of *diatomite,* an accumulation of the siliceous remains of microscope diatoms.

Clay from *shale* and other deposits supplies the basic material for ceramics of all sorts, from hand-thrown pottery and fine porcelain to sewer pipe. *Sulfur* is used for matches, fungicides, and sulfuric acid; and *phosphates* and *nitrates* for fertilizers are extracted from natural occurrences of special sedimentary rocks (although other sources also are used). Potassium for soap manufacture comes largely from *evaporites,* as does boron for heat-resistant cookware and fiberglass, and sodium for baking soda, washing soda, and soap. *Quartz sandstone* is used in glass manufacturing and for building stone.

Many *metallic ores,* such as the most common iron ores, have a sedimentary origin. The pore space of sedimentary rocks acts as a reservoir for ground water (chapter 17), crude oil, and natural gas. In chapter 21 we take a closer look at these resources and other useful earth materials.

table 14.2 Chemical Sedimentary Rocks

Rock	Composition	Texture	Origin
Inorganic Sedimentary Rocks			
Limestone	$CaCO_3$	Crystalline	May be precipitated directly from seawater. Cementation of oolites (ooids) precipitated chemically from warm shallow seawater (*oolitic limestone*). Also forms in caves as *travertine* and in springs, lakes, or percolating ground water as *tufa.*
Dolomite	$CaMg(CO_3)_2$	Crystalline	Alteration of limestone by Mg-rich solutions (usually)
Evaporites			Evaporation of seawater or a saline lake.
Rock salt	$NaCl$	Crystalline	
Rock gypsum	$CaSO_4 \cdot 2H_2O$	Crystalline	
Biochemical Sedimentary Rocks			
Limestone	$CaCO_3$ (calcite)	Clastic or crystalline	Cementation of fragments of shells, corals, and coralline algae (*bioclastic limestone* such as coquina and chalk). Also precipitated directly by organisms in reefs.
Chert	SiO_2 (silica)	Crystalline (usually)	Cementation of microscopic marine organisms; rock usually recrystallized.

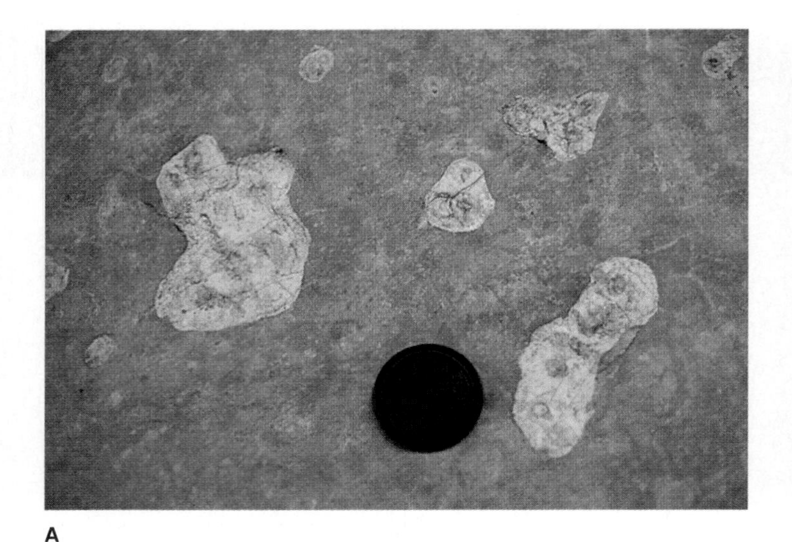
A

B

FIGURE 14.23
(*A*) Chert nodules in Redwall Limestone, Grand Canyon. (*B*) Bedded chert from the Coast Ranges, California. Camera lens cap (5.5 cm) for scale.

Regionally extensive layers of dolomite are thought to form in one of two ways:

1. As magnesium-rich brines created by solar evaporation of seawater trickle through existing layers of limestone.
2. As chemical reactions take place at the boundary between fresh underground water and seawater; the Mg ions could migrate through layers of limestone as sea level rises or falls. This replacement process tends to cause recrystallization of the preexisting limestone, so evidence of the rock's original environment of deposition is often obscured.

Chert

A hard, compact, fine-grained sedimentary rock formed almost entirely of silica, **chert** occurs in two principal forms—as irregular, lumpy nodules within other rocks and as layered deposits like other sedimentary rocks (figure 14.23). The nodules, often found in limestone, probably formed from inorganic precipitation as underground water replaced part of the original rock with silica. The layered deposits typically form from the accumulation of hard, shell-like parts of microscopic marine organisms on the sea floor.

Microscopic fossils composed of silica are abundant in some cherts. But because chert is susceptible to recrystallization, the original fossils are easily destroyed, and the origin of many cherts remains doubtful.

Evaporites

Rocks formed from crystals that precipitate during evaporation of water are called **evaporites.** They form from the evaporation of seawater or a saline lake (figure 14.24), such as Great Salt Lake in Utah. *Rock gypsum,* formed from the mineral gypsum ($CaSO_4 \cdot 2H_2O$), is a common evaporite. *Rock salt,* composed of the mineral halite ($NaCl$), may also form if evaporation continues. Other less common evaporites include the borates, potassium salts, and magnesium salts. All evaporites have a crystalline texture.

FIGURE 14.24
Salt (and mud) deposited on the floor of a dried-up desert lake, Bonneville salt flats, Utah.

FIGURE 14.25
A bed of coal near Trinidad, Colorado. Hammer at bottom for scale.

ORGANIC SEDIMENTARY ROCKS

Coal

Coal is a sedimentary rock that forms from the compaction of plant material that has not completely decayed (figure 14.25). Rapid plant growth and deposition in water with a low oxygen content are needed, so shallow swamps or bogs in a temperate or tropical climate are likely environments of deposition. The plant fossils in coal beds include leaves, stems, tree trunks, and stumps with roots often extending into the underlying shales, so apparently most coal formed right at the place where the plants grew. Coal usually develops from *peat,* a brown, lightweight, unconsolidated or semiconsolidated deposit of moss and other plant remains that accumulate in wet bogs. Peat is transformed into coal largely by compaction after it has been buried by sediments.

Partial decay of the abundant plant material uses up any oxygen in the swamp water, so the decay stops and the remaining organic matter is preserved. Burial by sediment compresses the plant material, gradually driving out any water or other volatile compounds. The coal changes from brown to black as the amount of carbon in it increases. Several varieties of coal are recognized on the basis of the type of original plant material and the degree of compaction (chapter 21).

The Origin of Oil and Gas

Oil and natural gas seem to originate from organic matter in marine sediment. Microscopic organisms, such as diatoms and other single-celled algae, settle to the sea floor and accumulate in marine mud. The most likely environments for this are restricted basins with poor water circulation, particularly on continental shelves. The organic matter may partially decompose, using up the dissolved oxygen in the sediment. As soon as the oxygen is gone, decay stops and the remaining organic matter is preserved.

Continued sedimentation buries the organic matter and subjects it to higher temperatures and pressures, which convert the organic matter to oil and gas. As muddy sediments compact, the gas and small droplets of oil may be squeezed out of the mud and may move into more porous and permeable sandy layers nearby. Over long periods of time large accumulations of gas and oil can collect in the sandy layers. Both oil and gas are less dense than water, so they generally tend to rise upward through water-saturated rock and sediment. Natural gas represents the end point in petroleum maturation.

Details of the origin of coal, oil, and gas are discussed in chapter 21.

SEDIMENTARY STRUCTURES

Sedimentary structures are features found within sedimentary rock. They usually form during or shortly after deposition of the sediment, but before lithification. Structures found in sedimentary rocks are important because they provide clues that help geologists determine the means by which sediment was transported and also its eventual resting place, or environment of deposition. Sedimentary structures may also reveal the orientation, or upward direction of the deposit, which helps geologists unravel the geometry of rocks that have been folded and faulted in tectonically active regions.

One of the most prominent structures, seen in most large bodies of sedimentary rock, is **bedding,** a series of visible layers within rock (figure 14.26). Most bedding is horizontal because the sediments from which the sedimentary rocks formed were originally deposited as horizontal layers. The principle of **original horizontality** states that most water-laid sediment is deposited in horizontal or near-horizontal layers that are essentially parallel to Earth's surface. In many cases this is also true for sediments deposited by ice or wind. If each new layer of sediment buries previous layers, a stack of horizontal layers will develop with the oldest layer on the bottom and the layers becoming younger upward. This is the principle of **superposition.** Sedimentary rocks formed from such sediments preserve the horizontal layering in the form of beds (figure 14.26). A **bedding plane** is a nearly flat surface of deposition separating two layers of rock. A change in the grain size or composition of the particles being deposited, or a pause during deposition, can create bedding planes.

A specialized type of bedding that is not horizontal is **cross-bedding,** a series of thin, inclined layers within a larger bed of rock. The cross-beds form a distinct angle to the horizontal bedding planes of the larger rock unit (figure 14.27).

FIGURE 14.26

Bedding in sandstone and shale, Utah. The horizontal layers formed as one type of sediment buried another in the geologic past. The layers get younger upwards.

FIGURE 14.27

Cross-bedded sandstone in Zion National Park, Utah. Note how the thin layers have formed at an angle to the more extensive bedding planes (also tilted) in the rock. This cross-bedding was formed in sand dunes deposited by the wind.

ASTROGEOLOGY 14.2

Sedimentary Rocks: The Key to Mars' Past

Sedimentary rocks on Mars will provide the historical record to help geologists unravel the planet's past. The latest images from the Mars Orbiter Camera (MOC) aboard the Mars Global Surveyor spacecraft reveal thin (a few to about 200 meters thick) repeated layers of horizontal sediment. The most extensive exposures of the layered deposits on Mars are in the western Candor Chasma (box figure 1), where exposures of horizontal layers have been exposed through erosion. These laterally continuous layers are similar to sedimentary beds that were deposited in shallow seas on Earth and are now exposed in places like the Grand Canyon. The widespread occurrence of the sedimentary layers in craters and low-lying areas on Mars suggest that the sediment may have been deposited by water. Deep seas or oceans usually deposit thick layers of sediment, whereas lakes and shallow seas typically deposit relatively thin horizontal layers, such as those observed in the western Candor Chasma area. There are other processes, however, that could deposit layered sediments, such as explosive volcanic eruptions and dust storms.

Violent volcanic eruptions of pyroclastic material can produce horizontally layered deposits of repeated beds; however, it is unlikely that this type of deposition is responsible for the widespread uniformly thick layers observed on Mars because pyroclastic deposits are usually quite thick near the vent and much thinner away from the volcano. Also, there have been no volcanic vents observed nearby that would have been capable of such a large eruption.

Large dust storms occur on Mars today and undoubtedly occurred in the past. If the layered sediments were deposited by the wind, the layers should contain cross-bedding structures similar to those shown in figure 14.27. The fact that cross-bedding has not been observed, and that wind usually does not deposit laterally continuous, regularly repeated layers of sediment, argues for deposition in water.

Scientists are excited by these latest images of layered sedimentary rocks from Mars that are similar to sediments deposited in lakes or shallow seas on Earth. Was there once water and life on Mars? Because most of the evidence of past life, such as fossils, are found in sedimentary rocks that formed in lakes and shallow seas on Earth, the Candor Chasma area may be the best place to look for evidence that life once existed on Mars.

Additional Information

Malin, M. C., and Edgett, K. S. 2000. Sedimentary rocks of early Mars. *Science* 290 (5498): 1927–1937.

Related Web Resources

Visit the NASA Science News headlines website for the latest information and images from Mars.
http://science.nasa.gov/headlines

Information about the Mars Global Surveyor and the latest images from the Mars Orbiter Camera are available at the Jet Propulsion Laboratory/NASA website.
http://mars.jpl.nasa.gov/mgs/

Visit the Malin Space Science Systems website for an extensive collection of archived Mars Orbiter Camera images.
www.msss.com/

Box 14.2 — FIGURE 1
Layered sedimentary rock exposed in the Candor Chasma. Each layer is approximately 10 meters thick.
Photo by NASA/JPL/Malin Space Systems

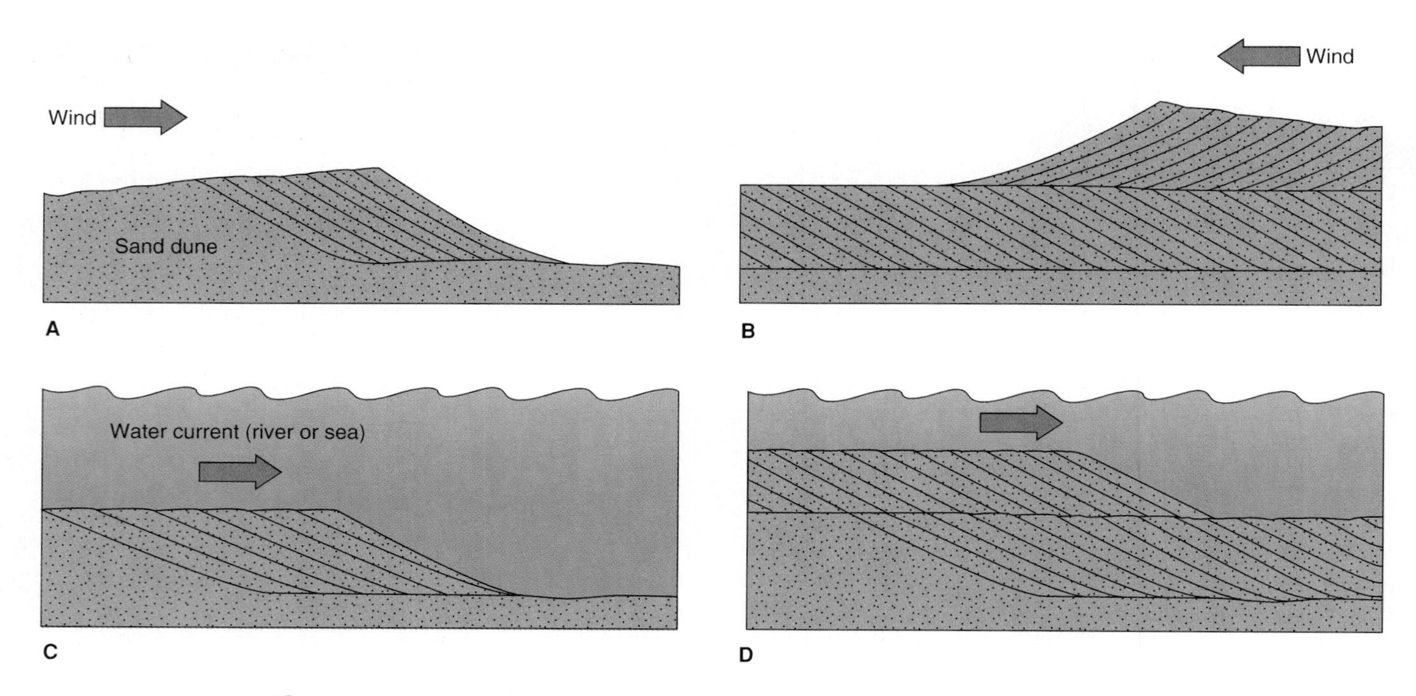

FIGURE 14.28

The development of cross-bedding in wind-blown sand (*A* and *B*) and current-deposited sand (*C* and *D*). (*A*) Sand is deposited in inclined layers on the downwind side of a dune. (*B*) Second dune covers first. Cross-bedding may change orientation if wind direction shifts. (*C*) A water current fills in a depression on river bottom or sea floor with sediment. (*D*) Continued sedimentation may cover first set of cross-beds with another set.

Cross-bedding is found most often in sandstone. It develops as sand is deposited on steep, local slopes (figure 14.28). The distinctive patterns of cross-bedding can form in many places—in sand dunes deposited by wind; in sand ridges deposited by ocean currents on the sea floor; in sediment bars and dunes deposited by rivers in their channels; and in deltas that form at the mouths of rivers.

A **graded bed** is a layer with a vertical change in particle size, usually from coarse grains at the bottom of the bed to progressively finer grains toward the top (figure 14.29). A single bed may have gravel at its base and grade upward through sand and silt to fine clay at the top. A graded bed may build up as sediment is deposited by a gradually slowing current. This seems particularly likely to happen during deposition from a *turbidity current* on the deep sea floor (figure 14.13). Figure 14.30 shows the development of a graded bed by turbidity-current deposition.

Mud cracks are a polygonal pattern of cracks formed in very fine-grained sediment as it dries (figure 14.31). Because drying requires air, mud cracks form only in sediment exposed above water. Mud cracks may form in lake-bottom sediment as the lake dries up; in flood-deposited sediment as a river level drops; or in marine sediment exposed to the air, perhaps temporarily by a falling tide. Cracked mud can lithify to form shale, preserving the cracks. The filling of mud cracks by sand can form casts of the cracks in an overlying sandstone.

FIGURE 14.29

A graded bed has coarse grains at the bottom of the bed and progressively finer grains toward the top. Coin for scale.

Ripple marks are small ridges formed on the surface of a sediment layer by moving wind or water. The ridges form perpendicular to the motion. Ripple marks, found in gravel, sand, or silt, can be caused by either waves or currents of water or

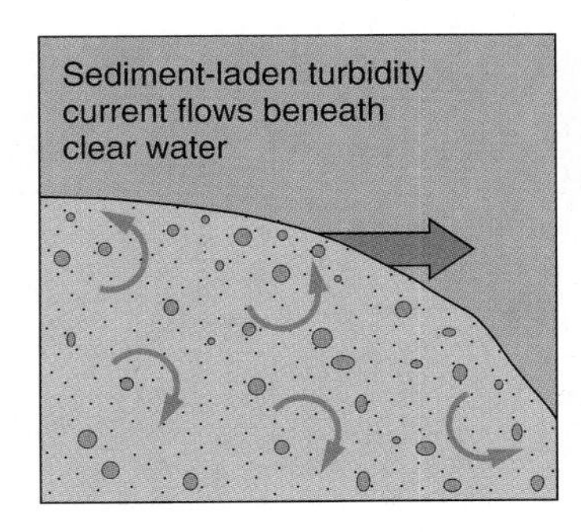

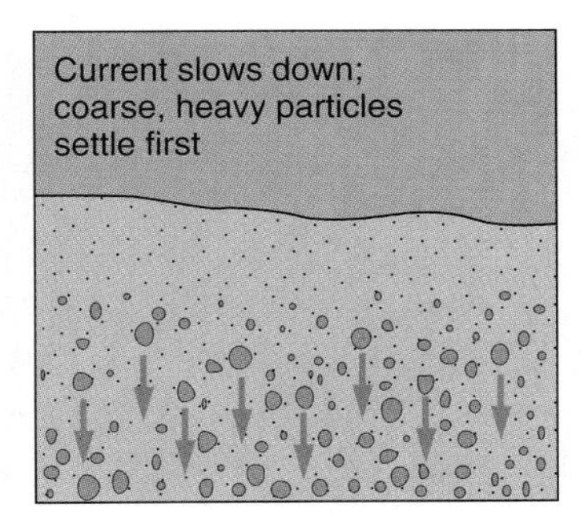

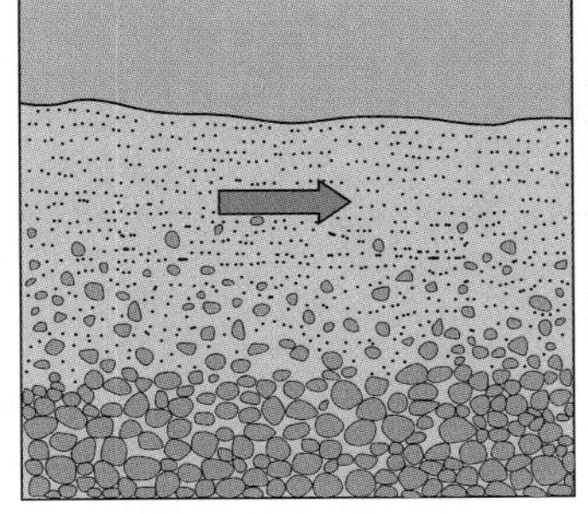

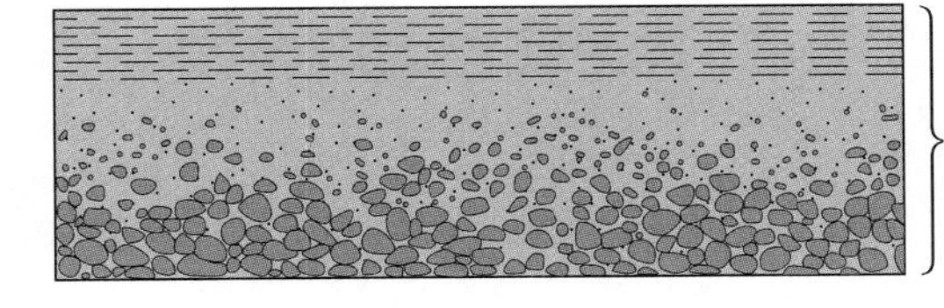

FIGURE 14.30

Development of a graded bed of sediment deposited by a turbidity current.

wind (figure 14.32*A,B*). Wave-caused ripple marks are symmetric ridges; current-caused ripple marks are asymmetric, with steeper sides in the down-current direction. Either type can be preserved in rock (figure 14.32*C,D*).

Fossils, traces of plants or animals preserved in rock, are relatively common sedimentary structures. The hard parts of organisms are most likely to be fossilized. If a bone or shell is covered by sediment before it decomposes, it may be preserved in the rock that forms from the sediment (figure 14.33). The original bone or shell material is seldom preserved unaltered; calcite or silica may fill the pore spaces of the fossil or may completely replace the original material. Sometimes fossils dissolve entirely, leaving an open void or mold in a rock. A mold may be filled in later by silica or calcite from underground water, forming a cast of the original fossil. Sometimes flat fossils of fish or leaves are formed as the weight of overburden presses most of the original material out of the fossil, leaving a thin film of black carbon

A

B

FIGURE 14.31

(*A*) Mud cracks in recently dried mud. (*B*) Mud cracks preserved in shale; they have been partially filled with sediment.

FIGURE 14.32

Development of ripple marks in loose sediment. (*A*) Symmetric ripple marks form beneath waves. (*B*) Asymmetric ripple marks, forming beneath a current, are steeper on their down-current sides. (*C*) Ripple marks on a bedding plane in sandstone, Capitol Reef National Park, Utah. Scale in centimeters. (*D*) Current ripples in wet sediment of a tidal flat, Baja California. Current flowed to right.

Photo *D* by Frank M. Hanna

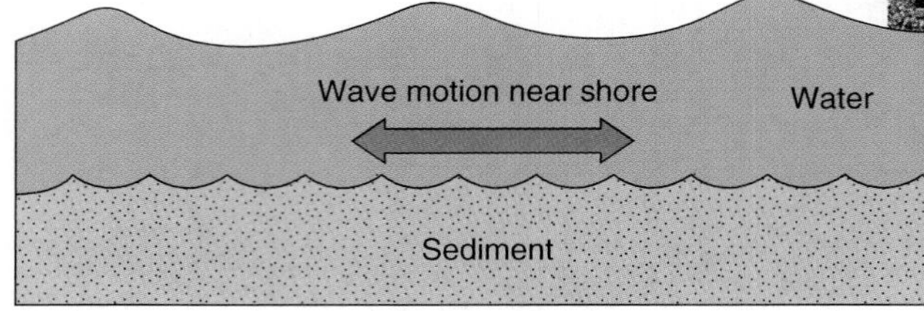

A

C

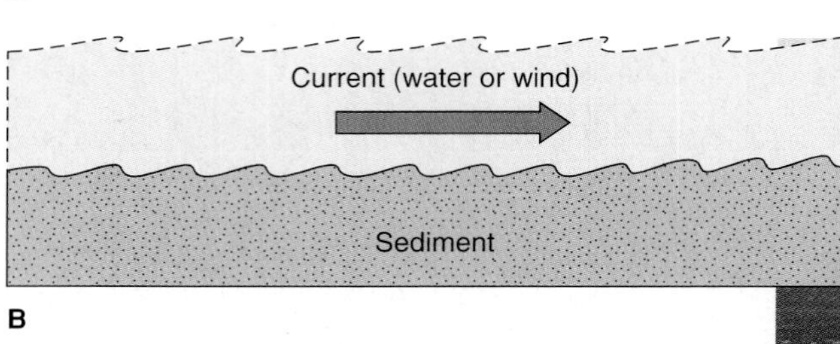

B

D

behind (figure 14.34). Preserved footprints, trails, and burrows are also fossils.

Because many limestones are composed of shell and coral debris, fossils are particularly common in these limestones, but they also occur in shales and more rarely in sandstones. Because coal, crude oil, and natural gas form from ancient organic matter, they are sometimes called *fossil fuels*.

FORMATIONS

A **formation** is a body of rock of considerable thickness with characteristics that distinguish it from adjacent rock units and is large enough to be mappable. Although a formation is usually composed of one or more beds of sedimentary rock, units of metamorphic and igneous rock are also called formations. It is a convenient unit for mapping, describing, or interpreting the geology of a region.

Formations are often based on rock type. A formation may be a single thick bed of rock such as sandstone. A sequence of several thin sandstone beds could also be called a formation, as could a sequence of alternating limestone and shale beds.

The main criterion for distinguishing and naming a formation is some visible characteristic that makes it a recognizable unit. This characteristic may be rock type or sedimentary structures or both. For example, a thick sequence of shale may be overlain by basalt flows and underlain by sandstone. The shale, the basalt, and the sandstone are each a different formation. Or a sequence of thin limestone beds, with a total thickness of hundreds of feet, may have recognizable fossils in the lower half and distinctly different fossils in the upper half. The limestone sequence is divided into two formations on the basis of its fossil content.

FIGURE 14.33

Fossil shells of clams in sandstone, California. Some of the fossils retain the original white shells; in others the shells have dissolved to form open molds.

Formations are given proper names: the first name is often a geographic location where the rock is well exposed, and the second the name of a rock type, such as Navajo Sandstone, Austin Chalk, Baltimore Gneiss, Onondaga Limestone, or Chattanooga Shale. If the formation has a mixture of rock types, so that one rock name does not accurately describe it, it is called simply "formation," as in the Morrison Formation or the Martinsburg Formation.

A **contact** is the boundary surface between two different rock types or ages of rocks. In sedimentary rock formations, the contacts are usually bedding planes.

Figure 14.35 shows the three formations that make up the upper part of the canyon walls in Grand Canyon National Park in Arizona. The contacts between formations are also shown.

INTERPRETATION OF SEDIMENTARY ROCKS

Sedimentary rocks are important in interpreting the geologic history of an area. Geologists examine sedimentary formations to look for clues such as fossils; sedimentary structures; grain shape, size, and composition; and the overall shape and extent of the formation. These clues are useful in determining the source area of the sediment, environment of deposition, and the possible plate tectonic setting at the time of deposition.

Source Area

The **source area** of a sediment is the locality that eroded and provided the sediment. The most important things to determine about a source area are the type of rocks that were

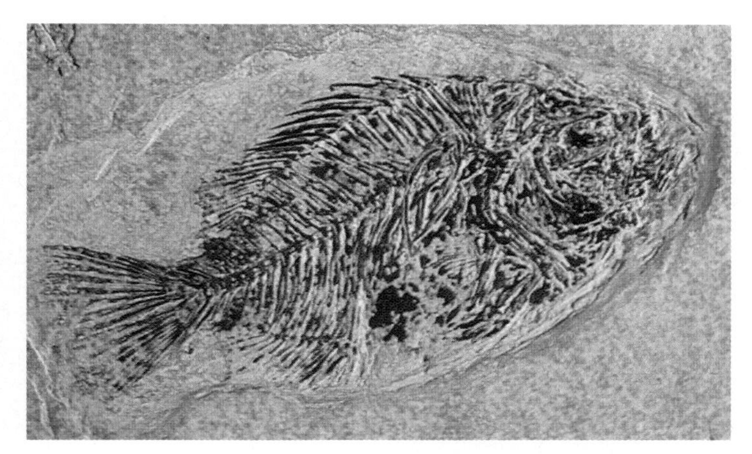

A

B

FIGURE 14.34

(*A*) Fossil fish in a rock from western Wyoming. (*B*) Dinosaur footprint in shale, Tuba City, Arizona. Scale in centimeters.

Photo *A* by U.S. Geological Survey

exposed in it and its location and distance from the site of eventual deposition.

The *rock type* exposed in the source area determines the character of the resulting sediment. The composition of a sediment can indicate the source area rock type, even if the source area has been completely eroded away. A conglomerate may contain cobbles of basalt, granite, and chert; these rock types were obviously in its source area. An arkose containing coarse feldspar, quartz, and biotite may have come from a granitic source area. Furthermore, the presence of feldspar indicates the source area was not subjected to extensive chemical weathering and that erosion probably took place in an arid environment with high relief. A quartz sandstone containing well-rounded quartz grains, on the other hand, probably represents the erosion and deposition of quartz grains from preexisting sandstone. Quartz is a hard, tough mineral very resistant to rounding by abrasion, so if quartz grains are well-rounded they have undergone many cycles of erosion, transportation, and deposition, probably over tens of millions of years.

FIGURE 14.35

The upper three formations in the cliffs of the Grand Canyon, Arizona. The Kaibab Limestone and the Coconino Sandstone are resistant in the dry climate and form cliffs. The Toroweap Formation contains some shale and is less resistant, forming slopes. The black lines are drawn to show the boundaries between the formations.

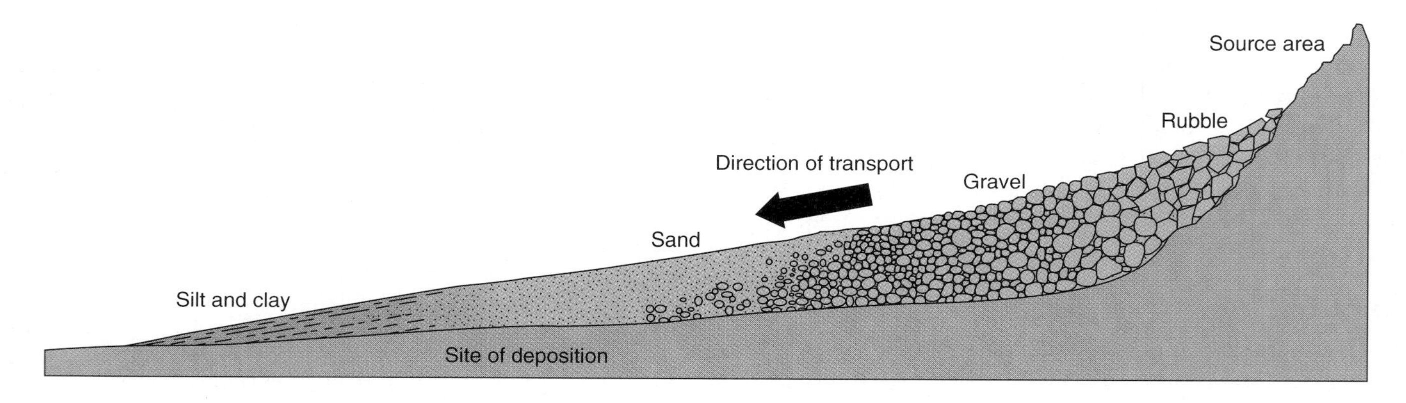

FIGURE 14.36

Sediment deposits often become thinner away from the source area, and sediment grains usually become finer and more rounded. The rocks that form from these sediments would change with distance from the source area from breccia to conglomerate to sandstone to shale. See appendix F for rock symbols.

Sedimentary rocks are also studied to determine the *direction* and *distance* to the source area. Figure 14.36 shows how several characteristics of sediment may vary with distance from a source area. Many sediment deposits get thinner away from the source, and the sediment grains themselves usually become finer and more rounded.

Sedimentary structures often give clues about the directions of old currents (*paleocurrents*) that deposited sediments. Refer back to figure 14.27 and notice how cross-beds slope downward in the direction of current flow. Old current direction can also be determined from asymmetric ripple marks (figure 14.32).

Figure 14.37 shows how three of these characteristics were used to determine the location of the source area for a particular rock unit in the southwestern United States. The unit is the Salt Wash Member of the Morrison Formation (a *member* is a

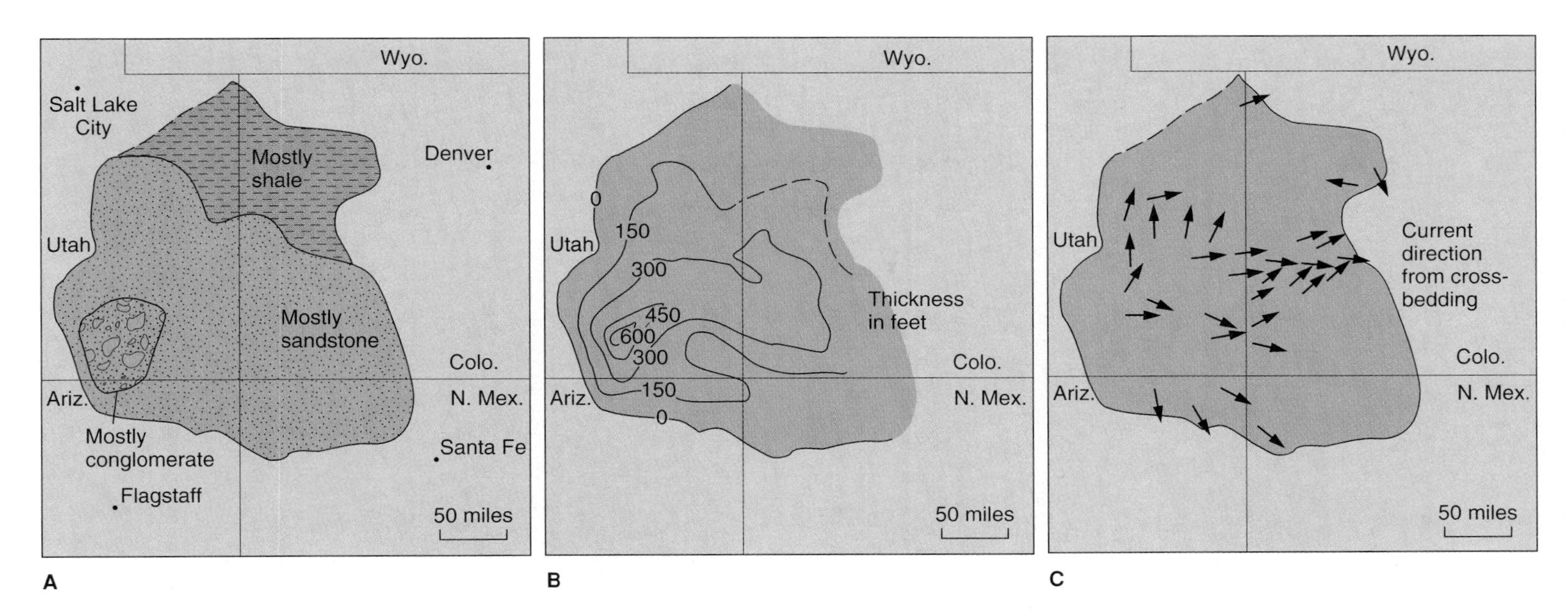

FIGURE 14.37

Characteristics of the Salt Wash Member of the Morrison Formation that help locate its source area. (*A*) The sediment grains become coarser to the southwest. (*B*) The deposit becomes thicker to the southwest. The contour lines show the thickness of the Salt Wash Member in feet. (*C*) Cross-bedding shows that the depositing currents came mostly from the southwest (arrows point down current).

Redrawn and simplified from Craig and others, 1955, U.S. Geological Survey Bulletin

subdivision of a formation). It is an important rock unit, for it contains a great deal of uranium, deposited within the rock by ground water long after the rock formed. The unit thickens and coarsens to the southwest, and cross-beds show that the old currents that deposited the sediment came from rivers that flowed from the southwest. These three facts strongly suggest that the source area was to the southwest. This information helps exploration geologists search for uranium within the Salt Wash Member. The Morrison Formation is also world famous for its dinosaur fossils.

Environment of Deposition

Figure 14.38 shows the common environments in which sediments are deposited. Geologists study modern environments in great detail so that they can interpret ancient rocks. Clues to the ancient environment of deposition come from a rock's composition, the size and sorting and rounding of the grains, the sedimentary structures and fossils present, and the shape and vertical sequence of the sedimentary layers.

Continental environments include alluvial fans, river channels, floodplains, lakes, and dunes. Sediments deposited on land are subject to erosion, so they often are destroyed. The great bulk of sedimentary rocks comes from the more easily preserved shallow marine environments, such as deltas, beaches, lagoons, shelves, and reefs. The characteristics of major environments are covered in detail in chapters 3, 16, and 18 through 20. In this section we describe the main sediment types and sedimentary structures found in each environment.

Glacial Environments

Glacial ice often deposits narrow ridges and layers of sediment in valleys and widespread sheets of sediment on plains. Glacial sediment (*till*) is an unsorted mix of unweathered boulders, cobbles, pebbles, sand, silt, and clay. The boulders and cobbles may be scratched from grinding over one another under the great weight of the ice.

Alluvial Fan

As rivers emerge from mountains onto flatter plains, they deposit broad, fan-shaped piles of sediment. The sediment often consists of coarse, arkosic sandstones and conglomerates, marked by coarse cross-bedding and lens-like channel deposits (figure 14.39).

River Channel and Floodplain

Rivers deposit elongate lenses of conglomerate or sandstone in their channels (figure 14.40). The sandstones may be arkoses or may consist of sand-sized fragments of fine-grained rocks. River channel deposits typically contain cross-beds and current ripple marks. Broad, flat floodplains are covered by periodic floodwaters, which deposit thin-bedded shales characterized by mud cracks and fossil footprints of animals. Hematite may color floodplain deposits red.

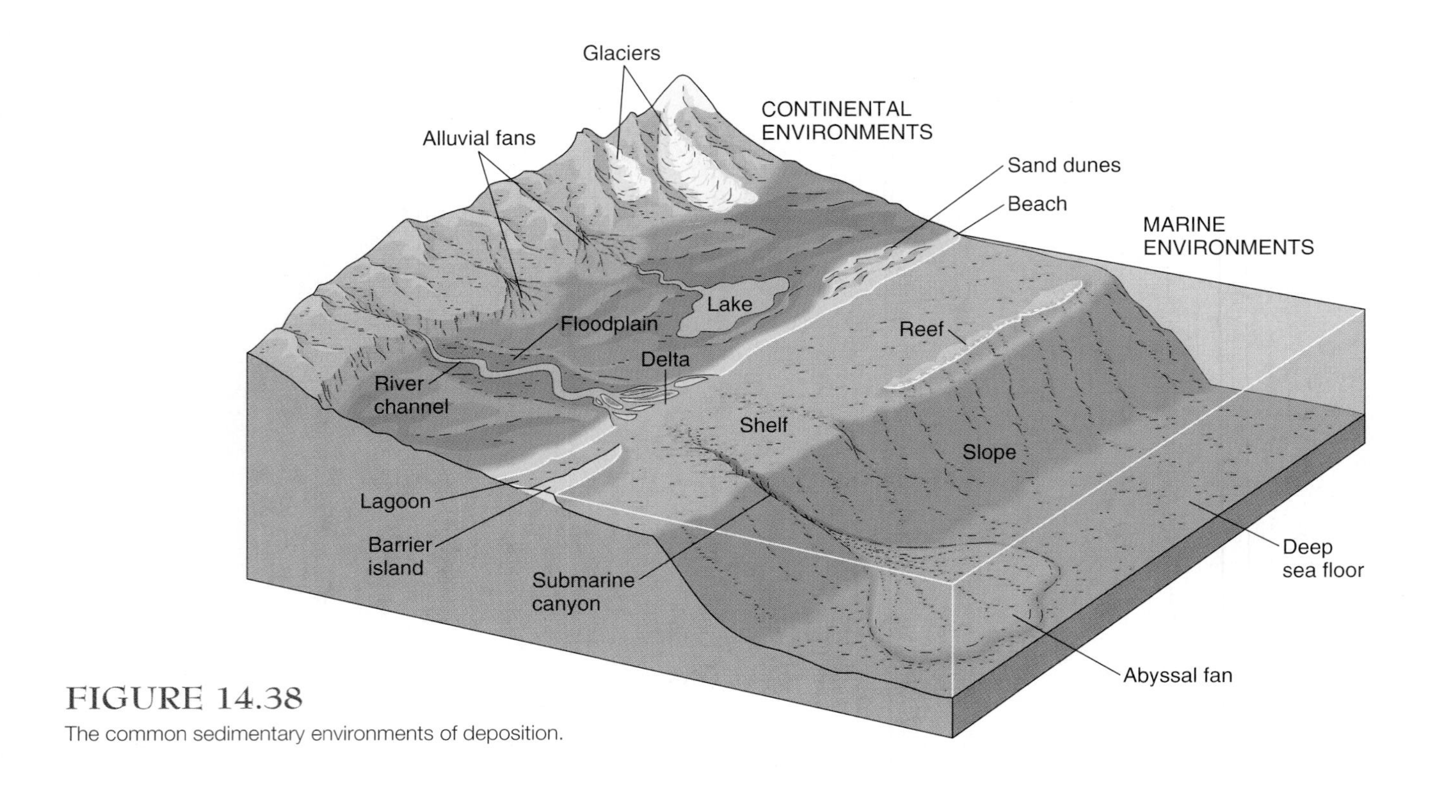

FIGURE 14.38
The common sedimentary environments of deposition.

FIGURE 14.39
Alluvial fan deposits, Baja California. A channel deposit of conglomerate occurs within the coarse-grained sequence.

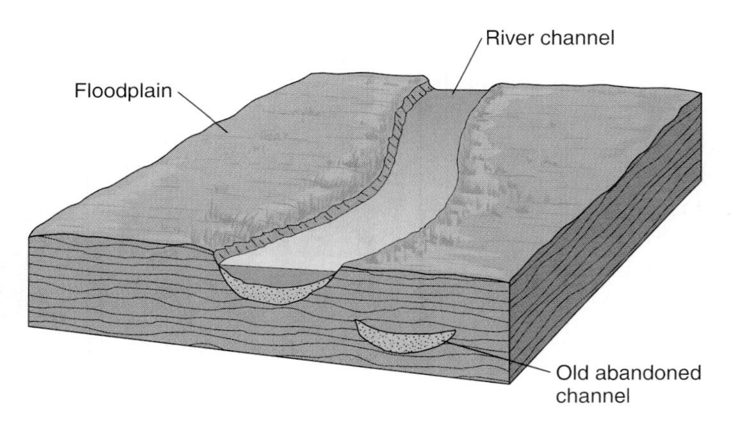

FIGURE 14.40
A river deposits an elongate lens of sand and gravel in its channel. Fine-grained silt and clay are deposited beside the channel on the river's floodplain.

Lake

Thin-bedded shale, perhaps containing fish fossils, is deposited on lake bottoms. If the lake periodically dries up, the shales will be mud-cracked and perhaps interbedded with evaporites such as gypsum or rock salt.

Delta

A delta is a body of sediment deposited when a river flows into standing water, such as the sea or a lake. Most deltas contain a great variety of subenvironments but are generally made up of thick sequences of siltstone and shale, marked by low-angle cross-bedding and cut by coarser channel deposits. Delta sequences may contain beds of peat or coal as well as marine fossils such as clam shells.

Beach, Barrier Island, Dune

A barrier island is an elongate bar of sand built by wave action. Well-sorted quartz sandstone with well-rounded grains is deposited on beaches, barrier islands, and dunes. Beaches and

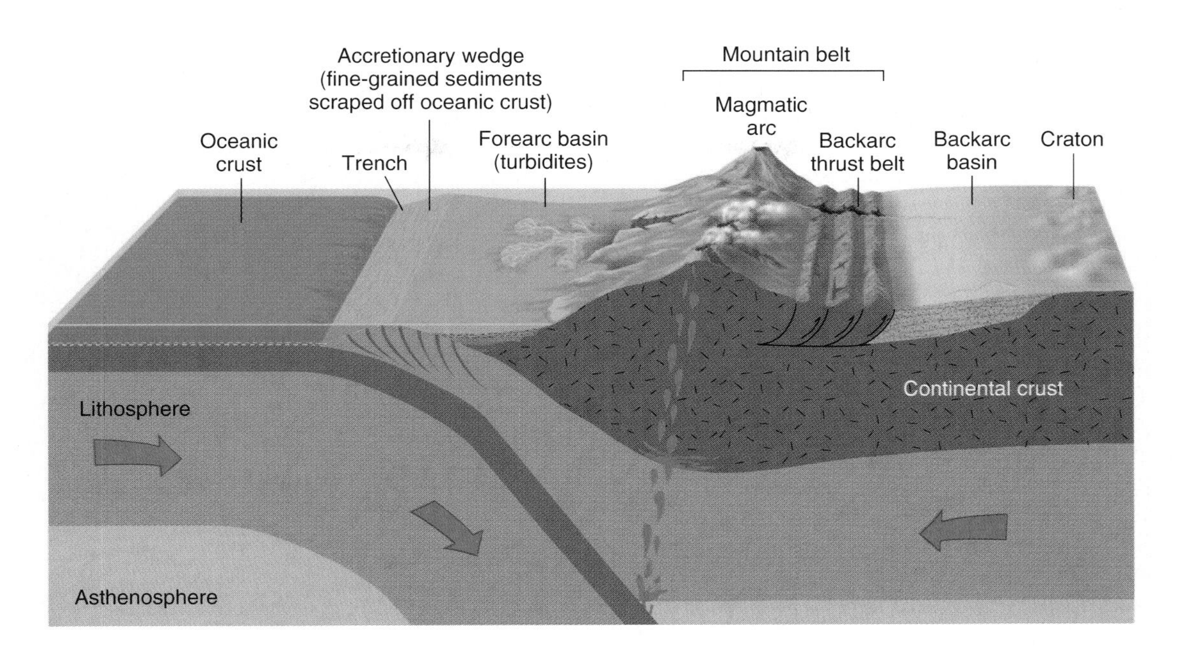

FIGURE 14.41
Sedimentary basins associated with convergent plate boundary include a forearc basin on the oceanward side that contains mainly clastic sediments deposited by streams and turbidity currents from an eroding magmatic arc. Toward the craton (continent) a backarc basin also collects clastic sediment derived from the uplifted mountain belt and craton.

barrier islands are characterized by cross-bedding (often low-angle) and marine fossils. Dunes have both high-angle and low-angle cross-bedding and occasionally contain fossil footprints of land animals such as lizards. All three environments can also contain carbonate sand in tropical regions, thus yielding cross-bedded clastic limestones.

Lagoon

A semi-enclosed, quiet body of water between a barrier island and the mainland is a lagoon. Fine-grained dark shale, cut by tidal channels of coarse sand and containing fossil oysters and other marine organisms, is formed in lagoons. Limestones may also form in lagoons adjacent to reefs (figure 14.17).

Shallow Marine Shelves

On the broad, shallow shelves adjacent to most shorelines, sediment grain size decreases offshore. Widespread deposits of sandstone, siltstone, and shale can be deposited on such shelves. The sandstone and siltstone contain symmetrical ripple marks, low-angle cross-beds, and marine fossils such as clams and snails. If fine-grained *tidal flats* near shore are alternately covered and exposed by the rise and fall of tides, mud-cracked marine shale will result.

Reefs

Massive limestone is deposited in reef cores, with steep beds of limestone breccia forming seaward of the reef, and horizontal beds of sand-sized and finer-grained limestones forming landward (figure 14.17). All these limestones are full of fossil fragments of corals, coralline algae, and numerous other marine organisms.

Deep Marine Environments

On the deep sea floor are deposited shale and graywacke sandstones. The graywackes are deposited by turbidity currents (figure 14.13) and typically contain graded bedding and current ripple marks.

Plate Tectonics and Sedimentary Rocks

The dynamic forces that move plates on Earth are also responsible for the distribution of many sedimentary rocks. As such, the distribution of sedimentary rocks often provides information that helps geologists reconstruct past plate tectonic settings.

In tectonically active areas, particularly along *convergent plate boundaries,* erosion is typically quite rapid, resulting in thick accumulations of clastic sediments that record the uplift and erosion of mountain ranges. Coarse-grained clastic sediments are transported mainly by streams and turbidity currents and are usually deposited in low-lying basins adjacent to mountains (figure 14.41). For example, uplift of the ancestral Sierra Nevada and Klamath mountain ranges in California is recorded by the thick accumulation of turbidite deposits preserved in basins to the west of the mountains. There, graywacke sandstone deposited by turbidity currents contains mainly volcanic clasts in the lower part of the sedimentary sequence and abundant feldspar clasts in the upper part of the sequence. This indicates that a cover of volcanic rocks was first eroded from the ancestral mountains and then, as uplift and erosion continued, the underlying plutonic rocks were exposed and eroded. Other eroded mountains, such as the Appalachians, have left similar records of uplift and erosion in the sedimentary record.

It is not uncommon for rugged mountain ranges such as the Canadian Rockies, European Alps, and Himalaya that stand several thousand meters above sea level to contain sedimentary rocks of marine origin that were originally deposited below sea level. The presence of marine sedimentary rocks such as limestone, chert, and shale containing marine fossils at high elevations attests to the tremendous uplift associated with mountain building at convergent plate boundaries (see chapter 5).

Transform plate boundaries are also characterized by rapid rates of erosion and deposition of sediments as fault-bounded basins open and subside rapidly with continued plate motion. Because of the rapid rate of deposition and the rapid burial of organic material, fault-bounded basins are good places to explore for petroleum. Many of the petroleum occurrences in California are related to basins that formed as the San Andreas transform fault developed.

A *divergent plate boundary* may result in the splitting apart of a continent and formation of a new ocean basin. In the initial stages of continental divergence, a rift valley forms and fills with thick wedges of gravel and coarse sand along its fault-bounded margins; lake bed deposits and associated evaporite rocks may form in the bottom of the rift valley (figure 14.42). In the early stages, continental rifts will have extensive volcanics that contribute to the sediments in the rift. The Red Sea is located along the East African Rift Zone and is a good example of the features and sedimentary rocks formed during the initial stages of continental rifting.

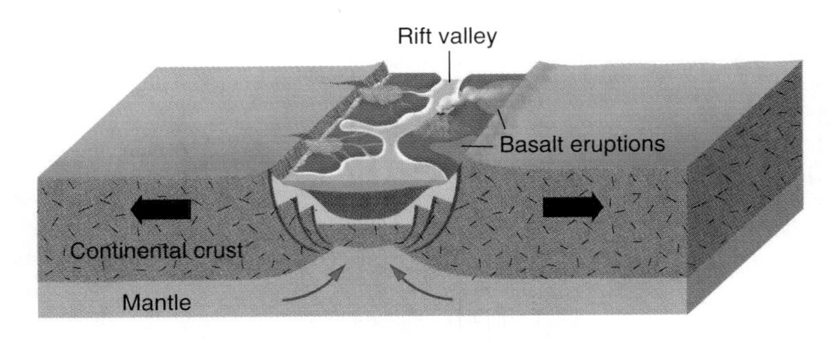

FIGURE 14.42
Divergent plate boundary showing thick wedges of gravel and coarse sand along fault-bounded margins of developing rift valley. Lake bed deposits and evaporite rocks are located on the floor of the rift valley. Refer to figure 4.25 for more detail of faulted margin and sediments deposited along a rifted continental margin.

Summary

Sediment forms by the weathering and erosion of preexisting rocks and by chemical precipitation, sometimes by organisms.

Gravel, sand, silt, and *clay* are sediment particles defined by grain size.

The composition of sediment is governed by the rates of chemical weathering, mechanical weathering, and erosion. During transportation, grains can become rounded and sorted.

Sedimentary rocks form by *lithification* of sediment, by *crystallization* from solution, or by consolidation of remains of organisms. Sedimentary rocks may be *clastic, chemical,* or *organic.*

Clastic sedimentary rocks form mostly by *compaction* and *cementation* of grains. *Matrix* can partially fill the *pore space* of clastic rocks.

Conglomerate forms from coarse, rounded sediment grains that often have been transported only a short distance by a river or waves. *Sandstone* forms from sand deposited by rivers, wind, waves, or turbidity currents. *Shale* forms from river, lake, or ocean mud.

Limestone consists of calcite, formed either as a chemical precipitate in a reef or, more commonly, by the cementation of shell and coral fragments or of oolites. *Dolomite* usually forms from the alteration of limestone by magnesium-rich solutions.

Chert consists of silica and usually forms from the accumulation of microscopic marine organisms. *Recrystallization* often destroys the original texture of chert (and some limestones).

Evaporites, such as rock salt and gypsum, form as water evaporates. *Coal,* a major fuel, is consolidated plant material.

Sedimentary rocks are usually found in *beds* separated by *bedding planes* because the original sediments are deposited in horizontal layers.

Cross-bedding forms where sediment is deposited on a sloping surface in a sand dune, delta, or river bar.

A *graded bed* forms as coarse particles fall from suspension before fine particles, perhaps in a turbidity current.

Mud cracks form in drying mud. *Ripple marks* form beneath waves or currents.

Fossils are the traces of an organism's hard parts or tracks preserved in rock.

A *formation* is a convenient rock unit for mapping and describing rock. Formations are lithologically distinguishable from adjacent rocks; their boundaries are *contacts.*

Geologists try to determine the *source area* of a sedimentary rock by studying its grain size, composition, and sedimentary structures. The source area's rock type and location are important to determine.

The *environment of deposition* of a sedimentary rock is determined by studying bed shape and sequence, grain composition and rounding, and sedimentary structures. Typical environments include alluvial fans, river channels, floodplains, lakes, dunes, deltas, beaches, shallow marine shelves, reefs, and the deep sea floor.

Plate tectonics plays an important role in the distribution of sedimentary rocks; the occurrence of certain types of sedimentary rocks is used by geologists to construct past plate tectonic settings.

Terms to Remember

bedding 349
bedding plane 349
cement 339
cementation 339
chemical sedimentary rock 340
chert 348
clastic sedimentary rock 340
clastic texture 339
clay 336
coal 349
compaction 339
conglomerate 340
contact 355
cross-bedding 349
crystalline texture 339
crystallization 339
deposition 337
dolomite 346
environment of deposition 337
evaporite 348
formation 354
fossil 353
graded bed 352
gravel 336
limestone 344
lithification 339
matrix 342
mud crack 352
organic sedimentary rock 340
original horizontality 349
pore space 339
recrystallization 346
ripple mark 352
rounding 336
sand 336
sandstone 340
sediment 336
sedimentary breccia 340
sedimentary rock 340
sedimentary structure 349
shale 343
silt 336
sorting 337
source area 355
superposition 349
turbidity current 342

Testing Your Knowledge

Use the questions below to prepare for exams based on this chapter.

1. Quartz is a common mineral in sandstone. Under certain circumstances, feldspar is common in sandstone, even though it normally weathers rapidly to clay. What conditions of climate, weathering rate, and erosion rate could lead to a feldspar-rich sandstone? Explain your answer.
2. Describe with sketches how wet mud compacts before it becomes shale.
3. What do mud cracks tell about the environment of deposition of a sedimentary rock?
4. How does a graded bed form?
5. List the clastic sediment particles in order of decreasing grain size.
6. How does a sedimentary breccia differ in appearance and origin from a conglomerate?
7. Describe three different origins for limestone.
8. How does dolomite usually form?
9. What is the origin of coal?
10. Sketch the cementation of sand to form sandstone.
11. How do evaporites form? Name two evaporites.
12. Name the three most common sedimentary rocks.
13. What is a formation?
14. Explain two ways that cross-bedding can form.
15. Particles of sediment from 1/16 mm to 2 mm in diameter are of what size?
    a. gravel
    b. sand
    c. silt
    d. clay
16. Rounding is
    a. the rounding of a grain to a spherical shape
    b. the grinding away of sharp edges and corners of rock fragments during transportation
    c. a type of mineral
    d. none of the above
17. Compaction and cementation are two common processes of
    a. erosion
    b. transportation
    c. deposition
    d. lithification
18. Which is not a chemical or organic sedimentary rock?
    a. rock salt
    b. shale
    c. limestone
    d. gypsum
19. The major difference between breccia and conglomerate is
    a. size of grains
    b. rounding of the grains
    c. composition of grains
    d. all of the above
20. Which is not a type of sandstone?
    a. quartz sandstone
    b. arkose
    c. graywacke
    d. coal
21. Shale differs from mudstone in that
    a. shale has larger grains
    b. shale is visibly layered and fissile; mudstone is massive and blocky
    c. shale has smaller grains
    d. there is no difference between shale and mudstone

22. The chemical element found in dolomite not found in limestone is
    a. Ca
    b. Mg
    c. C
    d. O
    e. Al

23. In a graded bed the particle size decreases
    a. upward
    b. downward
    c. in the direction of the current
    d. particle size stays the same

24. A body or rock of considerable thickness with characteristics that distinguish it from adjacent rock units is called a/an
    a. formation
    b. contact
    c. bedding plane
    d. outcrop

25. If sea level drops or the land rises, what is likely to occur?
    a. a flood
    b. a regression
    c. a transgression
    d. no geologic change will take place

26. Thick accumulations of graywacke and volcanic sediments can indicate an ancient
    a. divergent plate boundary
    b. convergent boundary
    c. transform boundary

27. A sedimentary rock made of fragments of preexisting rocks is
    a. organic
    b. chemical
    c. clastic

28. Clues to the nature of the source area of sediment can be found in
    a. the composition of the sediment
    b. sedimentary structures
    c. rounding of sediment
    d. all of the above

Expanding Your Knowledge

1. How might graded bedding be used to determine the tops and bottoms of sedimentary rock layers in an area where sedimentary rock is no longer horizontal? What other sedimentary structures can be used to determine the tops and bottoms of tilted beds?
2. Which would weather faster in a humid climate, a quartz sandstone or an arkose? Explain your answer.
3. A cross-bedded quartz sandstone may have been deposited as a beach sand or as a dune sand. What features could you look for within the rock to tell if it had been deposited on a beach? On a dune?
4. Why is burial usually necessary to turn a sediment into a sedimentary rock?
5. Why are most beds of sedimentary rock formed horizontally?
6. Discuss the role of sedimentary rocks in the rock cycle, diagramming the rock cycle as part of your answer. What do sedimentary rocks form from? What can they turn into?

Exploring Web Resources

www.mhhe.com/plummer9e

Visit the Online Learning Center website for an animation and discussion on how sediments are deposited along a broad, shallow marine shelf with the rising and falling sea level. The Online Learning Center also has Testing Your Knowledge answers, additional quizzes, and reading and media resources relating to sedimentary rocks.

http://darkwing.uoregon.edu/~dogsci/dorsey/SedResources.html

Web Resources for Sedimentary Geology site contains a comprehensive listing of resources available on the worldwide web.

www.lib.utexas.edu/Libs/GEO/FolkReady/TitlePage.html

Online version of *Petrology of Sedimentary Rocks* by Professor Robert Folk at the University of Texas at Austin.

http://walrus.wr.usgs.gov/seds/

Visit the *U.S. Geological Survey Bedform and Sedimentology* site for computer and photographic images and movies of sedimentary structures.

http://zircon.geology.union.edu/Gildner/stack.html

For a virtual field trip of the Ordovician-age rocks in the Mohawk Valley of New York state, visit Union College's geology website.

www.palaeo.de/edu/JRP/index.html

A virtual field trip of the Jurassic Park Reef in Germany. Learn how fossil reefs preserved in sedimentary rocks are used to interpret past ecosystems.

Animations

This chapter includes the following animations available on our Online Learning Center at www.mhhe.com/plummer9e.

14.28 Development of Cross-Bedding

14.30 Development of Graded Beds

CHAPTER

15

Metamorphism, Metamorphic Rocks, and Hydrothermal Rocks

This chapter on metamorphic rocks, the third major category of rocks in the rock cycle, completes our description of earth materials (rocks and minerals). The information on igneous and sedimentary processes in previous chapters should help you understand metamorphic rocks, which form from *preexisting* rocks.

After reading the chapter on weathering, you know how rocks are altered when exposed at Earth's surface. *Metamorphism* (a word from Latin and Greek that means literally "changing of form") also involves alterations, but the changes are due to deep burial, tectonic forces, and/or high temperature rather than surface conditions.

As you study this chapter, try to keep clearly in mind how the chemical composition of a rock and the temperature, pressure, and water present each contributes to the metamorphic process and the resultant metamorphic rock.

We also discuss hydrothermally deposited rocks and minerals, which are usually found in association with both igneous and metamorphic rocks. Hydrothermal ore deposits, while not volumetrically significant, are of great importance to the world's supply of metals.

Because nearly all metamorphic rocks form deep within the earth's crust, they provide geologists with many clues about conditions at depth. Therefore, an understanding of metamorphism contributes to an understanding of the geologic processes involving Earth's internal forces. Metamorphic rocks are a feature of the oldest exposed rocks of the continents and of major mountain belts. They are especially important in providing evidence of what happens during subduction and plate convergence.

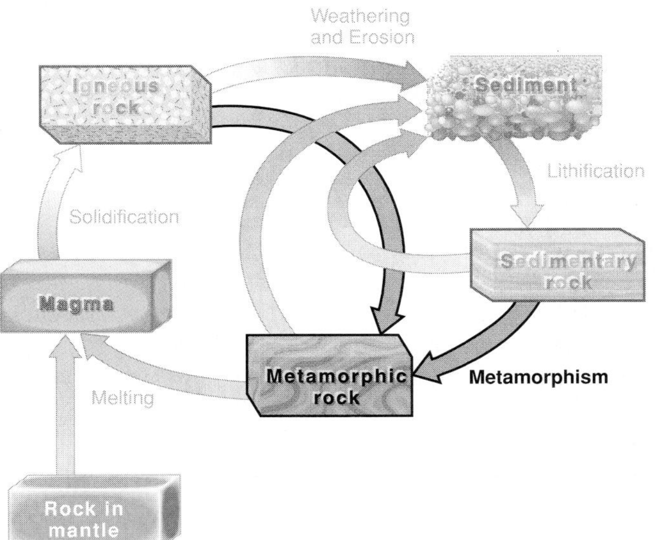

Opposite: Photo taken through a polarizing microscope of a metamorphic rock that was once shale. Micas (brightly-colored crystals) grew while the rock was being folded during regional metamorphism. The area shown is approximately 2 cm wide.
Photo by C.C. Plummer

From your study so far of Earth materials and the rock cycle, you know that rocks change, given enough time, when their physical environment changes radically. In chapter 11, you saw how deeply buried rocks melt (or partially melt) to form magma when temperatures are high enough. What happens to rocks that are deeply buried but are not hot enough to melt? They become metamorphosed. **Metamorphism** refers to changes to rocks that take place in Earth's interior. The changes may be new textures, new mineral assemblages, or both. Transformations occur in the solid state (meaning the rock does not melt). The new rock is a **metamorphic rock.**

As most metamorphism takes place in moderate to great depths in Earth's crust, metamorphic rocks provide us with a window to processes that take place deep underground, beyond our direct observation. Erosion of mountain belts along with uplift due to isostatic adjustment expose metamorphic rocks over large regions. In fact the cores of the continents are largely metamorphic rocks and granitic plutons. As described in the chapter on mountain belts and the continental crust, these form the stable interior of North America, the central lowlands between the Appalachians and the Rocky Mountains and other ranges of western North America. Very ancient (Precambrian) complexes of metamorphic and intrusive igneous rocks are exposed over much of Canada (known as the *Canadian Shield*). The inside front cover shows the Canadian Shield as the region underlain by Precambrian rocks. In the Great Plains of the United States similar rocks form the *basement* underlying a veneer of younger sedimentary rocks (see the brown area on the inside front cover map that the legend indicates is "Platform deposits on Precambrian basement").

In nearly all cases, a metamorphic rock has a texture clearly different from that of the *original* rock, or **parent rock.** When limestone is metamorphosed to marble, for example, the fine grains of calcite coalesce and recrystallize into larger calcite crystals. The calcite crystals are interlocked in a mosaic pattern that gives marble a texture distinctly different from that of the parent limestone. If the limestone is composed entirely of calcite, then metamorphism into marble involves no new minerals, only a change in texture.

More commonly, the various elements of a parent rock react chemically and crystallize into new minerals, thus making the metamorphic rock distinct both mineralogically and texturally from the parent rock. This is because the parent rock is unstable in its new environment. The old minerals recrystallize into new ones that are at *equilibrium* in the new environment. For example, clay minerals form at Earth's surface (see chapter 12). Therefore, they are stable at the low temperature and pressure conditions both at and just below Earth's surface. When subjected to the temperatures and pressures deep within Earth's crust, the clay minerals of a shale can recrystallize into coarse-grained mica. Another example is that under appropriate temperature and pressure conditions, a quartz sandstone with a calcite cement metamorphoses as follows:

FIGURE 15.1

Metamorphic rock from Greenland. Metamorphism took place 3,700 million years ago—it is one of the oldest rocks on Earth.
Photo by C. C. Plummer

$$\underset{\text{calcite}}{CaCO_3} + \underset{\text{quartz}}{SiO_2} \rightarrow \underset{\substack{\text{wollastonite}\\\text{(a mineral)}}}{CaSiO_3} + \underset{\substack{\text{carbon}\\\text{dioxide}}}{CO_2}$$

No one has observed metamorphism taking place, just as no one has ever seen a granite pluton form. What, then, leads us to believe that metamorphic rocks form in a solid state (i.e., without melting) at high pressure and temperature? Many metamorphic rocks found on Earth's surface exhibit contorted banding (figure 15.1). Commonly, banding in metamorphic rocks can be demonstrated to have originally been flat-lying sedimentary layering (even though the rock has since recrystallized). These rocks, now hard and brittle, would shatter if smashed with a hammer. But they must have been **ductile** (or *plastic*), capable of being bent and molded under stress, to have been folded into such contorted patterns. Because high temperature and pressure are necessary to make rocks ductile, a reasonable conclusion is that these rocks formed at considerable depth, where such conditions exist. Moreover, crystallization of a magma would not produce contorted layering.

ENVIRONMENTAL GEOLOGY 15.1

Metamorphic Exhalations May Have Changed the Climate

The world's highest mountain chain, the Himalaya, may have been responsible for a period of global warming some 60 million years ago. As explained in chapter 5, collision between India and Asia created the Himalaya. Before the collision, a great thickness of sedimentary rocks built up on the ocean floor that separated the two landmasses. Limestones were especially abundant. Upon collision, the layers of sedimentary rocks were crumpled, as if caught in a giant vise. Rocks that were deeply buried by the process were metamorphosed.

Derrill Kerrick and his colleagues at Pennsylvania State University have proposed a hypothesis that links the global warming of the time to the release of CO_2 during metamorphism. If limestone contains significant amounts of quartz or clay in addition to the prevailing calcite, chemical reactions during metamorphism produce new minerals as well as carbon dioxide gas. The scientists estimated that several hundred million tons of carbon dioxide per year may have been released into the atmosphere over a period of around 10 million years. This amount of CO_2 added to the atmosphere, and its contribution to the greenhouse effect, would account for the warmer climate inferred for that part of Earth's history.

If global warming is taking place at present (as is believed by many) due to increasing CO_2 levels, then our cars, factories, and cutting down of forests can in a few decades do what metamorphism took millions of years to accomplish.

FACTORS CONTROLLING THE CHARACTERISTICS OF METAMORPHIC ROCKS

A metamorphic rock owes its characteristic texture and particular mineral content to several factors, the most important being (1) the composition of the parent rock before metamorphism, (2) temperature and pressure during metamorphism, (3) the effects of tectonic forces, and (4) the effects of fluids, such as water.

Composition of the Parent Rock

Usually no new elements or chemical compounds are added to the rock during metamorphism, except perhaps water. (Metasomatism, discussed later in this chapter, does involve the addition of other elements.) Therefore, the mineral content of the metamorphic rock is controlled by the chemical composition of the parent rock. For example, a basalt always metamorphoses into a rock in which the new minerals can collectively accommodate the approximately 50% silica and relatively high amounts of the oxides of iron, magnesium, calcium, and aluminum in the original rock. On the other hand, a limestone, composed essentially of calcite ($CaCO_3$), cannot metamorphose into a silica-rich rock.

Temperature

Heat, necessary for metamorphic reactions, comes primarily from the outward flow of geothermal energy from Earth's deep interior. The deeper a rock is beneath the surface, the hotter it will be. The particular temperature for rock at a given depth depends on the local *geothermal gradient* (described in chapter 11). Additional heat could be derived from magma, if magma bodies are locally present.

A mineral is said to be *stable* if, given enough time, it does not react with another substance or convert to a new mineral or substance. Any mineral is stable only within a given temperature range. The stability temperature range of a mineral varies with factors such as pressure and the presence or absence of other substances. Some minerals are stable over a wide temperature range. Quartz, if not mixed with other minerals, is stable at atmospheric pressure (i.e., at Earth's surface) up to about 800°C. At higher pressures, quartz remains stable to even higher temperatures. Other minerals are stable over a temperature range of only 100° or 200°C.

By knowing (from results of laboratory experiments) the particular temperature range in which a mineral is stable, a geologist may be able to deduce the temperature of metamorphism for a rock that includes that *index mineral* (see box 15.3).

Minerals stable at higher temperatures tend to be less dense (or have a lower specific gravity) than chemically identical minerals stable at lower temperatures. As temperature increases, the ions vibrate more within their sites in the crystal structure. A more open (less tightly packed) crystal structure, such as high-temperature minerals tend to have, allows greater vibration of ions. (If the heat and resulting vibrations become too great, the bonds between atoms in the crystal break and the substance becomes liquid.)

The upper limit on temperature in metamorphism overlaps the temperature of partial melting of a rock. If partial melting takes place, the component that melts becomes a magma; the solid residue remains a metamorphic rock. Temperatures at which the igneous and metamorphic realms can coexist vary considerably. For an ultramafic rock (containing only ferromagnesian silicate minerals) the temperature will be over 1,200°C; for a granite or rhyolite under high water pressure, it could be as low as 650°C.

Pressure

Usually, when we talk about pressure, we mean **confining pressure;** that is, pressure applied equally on all surfaces of a substance as a result of burial or submergence. A diver

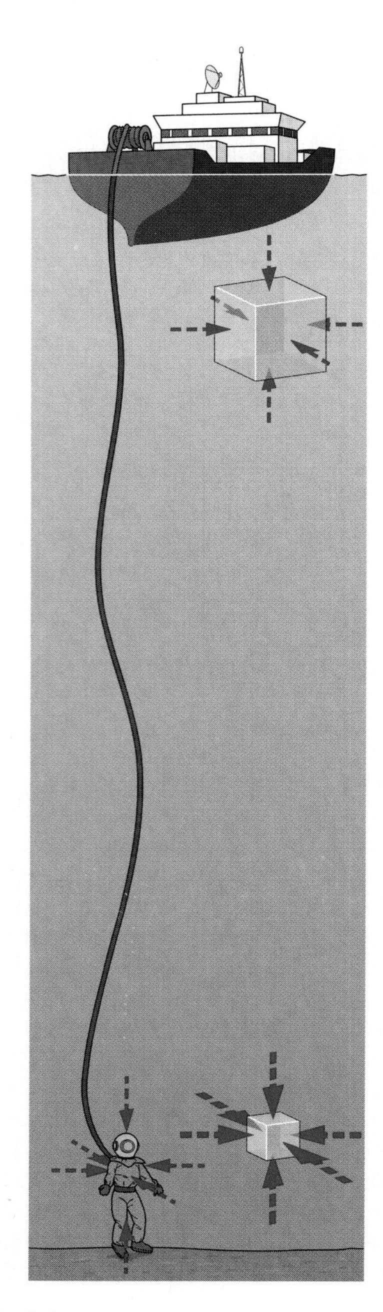

FIGURE 15.2

Confining pressure. The diver's suit is pressurized to counteract hydrostatic pressure. Object (cube) has a greater volume at low pressure than at high pressure.

senses confining pressure proportional to the weight of the overlying water (figure 15.2). The pressure uniformly squeezes the diver's entire body surface. Likewise, an object buried deeply within Earth's crust is compressed by strong confining pressure, called *lithostatic pressure,* which forces grains closer together and eliminates pore space. The *pressure gradient,* the increase in lithostatic pressure with depth, is approximately 1,000 atmospheres per each 3.3 kilometers of burial in crustal rock.

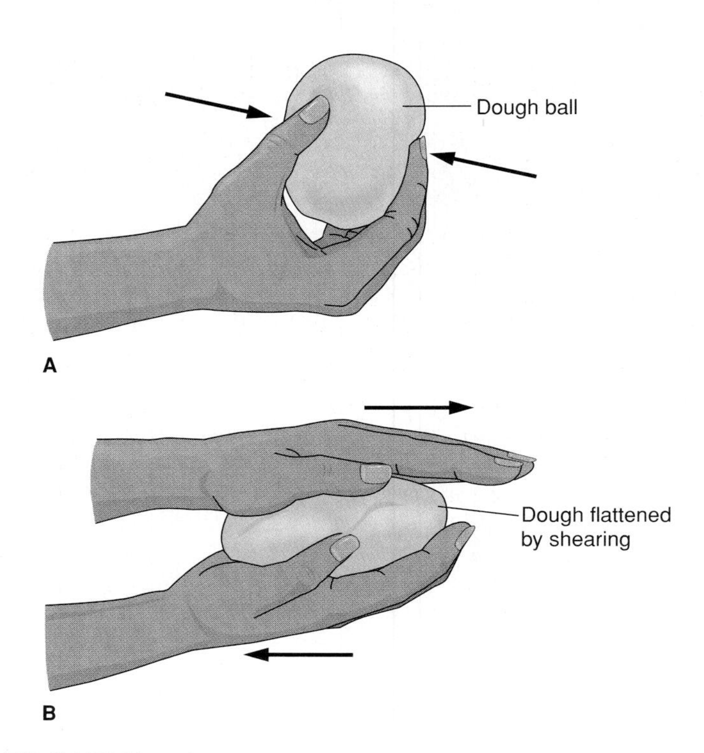

FIGURE 15.3

(*A*) Compressive stress. More force is exerted in the direction of the arrows than elsewhere on the ball. (*B*) Shearing. Parallel arrows indicate the direction of forces.

Any new mineral that has crystallized under high-pressure conditions tends to occupy less space than did the mineral or minerals from which it formed. The new mineral is denser than its low-pressure counterparts because the pressure forces atoms closer together into a more closely packed crystal structure.

Differential Stress

Most metamorphic rocks show the effects of tectonic forces. When forces are applied to an object, the object is under **stress.** If the forces on a body are stronger or weaker in different directions, a body is subjected to **differential stress.** Differential stress tends to deform objects into oblong or flattened forms. If you squeeze a rubber ball between your thumb and forefinger, the ball is under differential stress (figure 15.3*A*). If you squeeze a ball of dough, it will remain flattened after you stop squeezing, because dough is plastic. To illustrate the difference between confining pressure and differential stress, visualize a drum filled with water. If you place a ball of putty underwater in the bottom of the drum, the ball will not change its shape (its volume will decrease slightly due to the weight of the overlying water). Now take the putty ball out of the water and place it under the drum. The putty will be flattened into the shape of a pancake due to the differential stress. In this case, the putty is subjected to *compressive differential stress* or, more simply, **compressive stress** (as is the dough ball shown in figure 15.3*A*).

Differential stress is also caused by **shearing,** which causes parts of a body to move or slide relative to one another across a

FIGURE 15.4

Metamorphosed conglomerate in which the pebbles have been flattened (sometimes called a stretched pebble conglomerate). Compare to figure 14.9.
Photo by C. C. Plummer

plane. An example of shearing is when you spread out a deck of cards on a table with your hand moving parallel to the table. Figure 15.3*B* shows that initially dough is flattened at a low angle to the shear force (the moving hands). As shearing continues, the flattened dough is rotated toward *parallelism* to the shear force. In contrast, compressive stress flattens objects *perpendicular* to the applied force.

Foliation

Differential stress has a very important influence on the texture of a metamorphic rock because it forces the constituents of the rock to become parallel to one another. For instance, the pebbles in the metamorphosed conglomerate shown in figure 15.4 were originally more spherical but have been flattened by differential stress. When a rock has a planar texture, it is said to be **foliated.** Foliation is manifested in various ways. If a platy mineral (such as mica) is crystallizing within a rock that is undergoing differential stress, the mineral grows in such a way that it remains parallel to the direction of shearing or perpendicular to the direction of compressive stress (figure 15.5). Any platy mineral attempting to grow against shearing is either ground up or forced into alignment. Minerals that crystallize in needlelike shapes (for example, hornblende) behave similarly, growing with their long axes parallel to the plane of shearing or perpendicular to compressive stress. The three very different textures described below (from lowest to highest degree of metamorphism) are all variations of foliation and are important in classifying metamorphic rocks:

1. If the rock splits easily along nearly flat and parallel planes, indicating that preexisting, microscopic, platy minerals were pushed into alignment during metamorphism, we say the rock is **slaty,** or that it possesses **slaty cleavage.**
2. If visible platy or needle-shaped minerals have grown essentially parallel to a plane due to differential stress, the rock is **schistose** (figure 15.6).

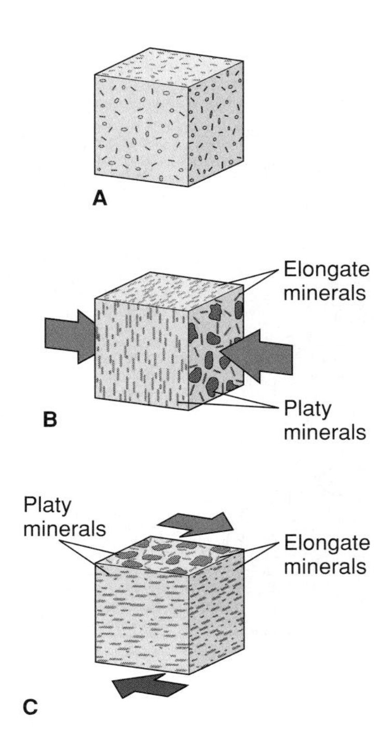

FIGURE 15.5

Orientation of platy and elongate minerals in metamorphic rock. (*A*) Platy minerals randomly oriented (e.g., clay minerals before metamorphism). No differential stress involved. (*B*) Platy minerals (e.g., mica) and elongate minerals (e.g., amphibole) have crystallized under the influence of compressive stress. (*C*) Platy and elongate minerals developed under the influence of shearing.

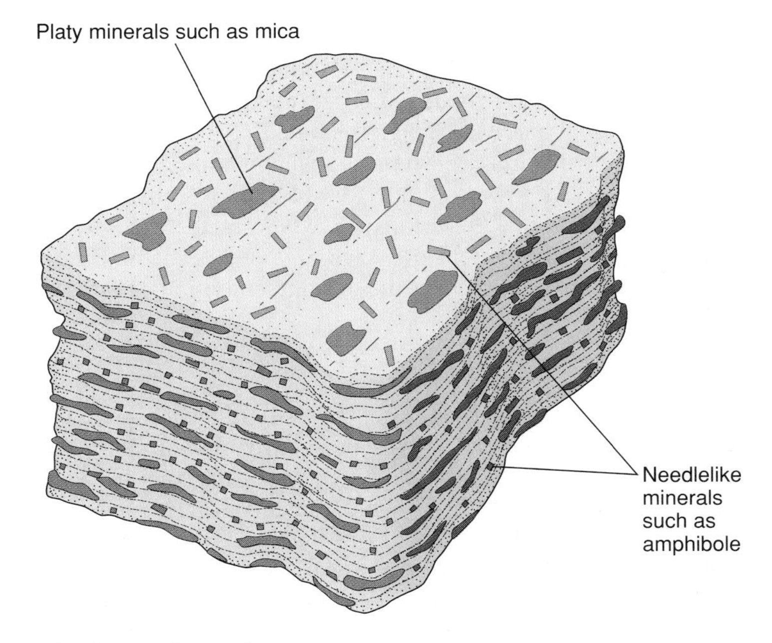

FIGURE 15.6

Schistose texture.

A

B

FIGURE 15.7

Photomicrographs of metamorphic rocks in thin sections of (*A*) nonfoliated rock and (*B*) foliated rock.
Photos by C. C. Plummer

3. If the rock became very ductile and the new minerals separated into distinct (light and dark) layers or lenses, the rock has a layered or **gneissic** texture, such as in figure 15.14.

Fluids

Hot water (as vapor) is the most important fluid involved in metamorphic processes, although other gases, such as carbon dioxide, sometimes play a role. The water may have been trapped in a parent sedimentary rock or given off by a cooling pluton.

Water is thought to help trigger metamorphic chemical reactions. Water, moving through fractures and along grain margins, is a sort of intra-rock rapid transit for ions. Under high pressure, it moves between grains, dissolves ions from one mineral, and then carries these ions elsewhere in the rock where they can react with the ions of a second mineral. The new mineral that forms is stable under the existing conditions.

Time

The effect of time on metamorphism is hard to comprehend. Most metamorphic rocks are composed predominantly of silicate minerals, and silicate compounds are notorious for their sluggish chemical reaction rates. Recently, garnet crystals taken from a metamorphic rock collected in Vermont were analyzed and scientists calculated a growth rate of 1.4 millimeters per million years. The garnets' growth was sustained over a 10.5-million-year period. Many laboratory attempts to duplicate metamorphic reactions believed to occur in nature have been frustrated by the time element. The several million years during which a particular combination of temperature and pressure may have prevailed in nature are impossible to duplicate.

CLASSIFICATION OF METAMORPHIC ROCKS

As we noted before, the kind of metamorphic rock that forms is determined by the metamorphic environment (primarily the particular combination of pressure, stress, and temperature) and by the chemical constituents of the parent rock. Many kinds of metamorphic rocks exist because of the many possible combinations of these factors. These rocks are classified based on broad similarities. (Appendix B contains a systematic procedure for identifying common metamorphic rocks.)

First consider the texture of a metamorphic rock. Is it *foliated* or *nonfoliated* (figure 15.7)? If the rock is nonfoliated, it is named on the basis of its composition. For example, a nonfoliated quartz-rich metamorphic rock is a *quartzite;* one composed almost entirely of calcite is a *marble.*

If the rock is foliated, you must determine the type of foliation. For example, a schistose rock is called a *schist.* But this name tells us nothing about what minerals are in this rock; so we add adjectives to describe the composition. Thus, a *garnet-mica schist* (whose parent was shale) is easily distinguishable from an *amphibole schist* (a metamorphic product of basalt). The relationship of texture to rock name is summarized in table 15.1.

table 15.1 **Classification and Naming of Metamorphic Rocks (Based Primarily on Texture)**

Nonfoliated

Name Based on Mineral Content of Rock

Usual Parent Rock	Rock Name	Predominant Minerals	Identifying Characteristics
Limestone Dolomite	Marble Dolomite marble	Calcite Dolomite	Coarse interlocking grains of calcite (or, less commonly, dolomite). Calcite (or dolomite) has rhombohedral cleavage; hardness intermediate between glass and fingernail. Calcite effervesces in weak acid.
Quartz sandstone	Quartzite	Quartz	Rock composed of interlocking small granules of quartz. Has a sugary appearance and vitreous luster; scratches glass.
Shale Basalt	Hornfels Hornfels	Fine-grained micas Fine-grained ferromagnesian minerals, plagioclase	A fine-grained, dark rock that generally will scratch glass. May have a few coarser minerals present.

Foliated

Name Based Principally on Kind of Foliation Regardless of Parent Rock. Adjectives Describe the Composition (e.g., biotite-garnet schist)

Texture	Rock Name	Typical Characteristic Minerals	Identifying Characteristics
Slaty	Slate	Clay and other sheet silicates	A very fine-grained rock with an earthy luster. Splits easily into thin, flat sheets.
Intermediate between slaty and schistose	Phyllite	Muscovite mica	Fine-grained rock with a silky luster. Generally splits along wavy surfaces.
Schistose	Schist	Biotite and muscovite, amphibole	Composed of visible platy or elongated minerals that show planar alignment. A wide variety of minerals can be found in various types of schist (e.g., garnet-mica schist, hornblende schist, etc.).
Gneissic	Gneiss	Feldspar	Light and dark minerals are found in separate, parallel layers or lenses. Commonly, the dark layers include biotite and hornblende; the light-colored layers are composed of feldspars and quartz. The layers may be folded or appear contorted.

TYPES OF METAMORPHISM

The two most common types of metamorphism are contact metamorphism and regional metamorphism. Hydrothermal processes, in which hot water plays a major role during metamorphism, are discussed later in this chapter.

Contact Metamorphism

Contact metamorphism (also known as *thermal* metamorphism) is metamorphism in which high temperature is the dominant factor. Confining pressure may influence which new minerals crystallize; however, the confining pressure is usually relatively low. This is because contact metamorphism mostly takes place not too far beneath Earth's surface (less than 10 kilometers). Contact metamorphism occurs adjacent to a pluton when a body of magma intrudes relatively cool country rock. The process can be thought of as the "baking" of country rock adjacent to an intrusive contact; hence the term *contact metamorphism.* The zone of contact metamorphism (also called an *aureole*) is usually quite narrow—generally from 1 to 100 meters wide. Differential stress is rarely significant; therefore, these rocks typically are nonfoliated.

During contact metamorphism, shale is changed into the very fine-grained metamorphic rock **hornfels.** Characteristically, only microscopically visible micas develop. Sometimes a few minerals grow large enough to be seen with the naked eye; these are minerals that are especially capable of crystallizing

15.2 ASTROGEOLOGY

IMPACT CRATERS AND SHOCK METAMORPHISM

The spectacular collision of the comet Shoemaker-Levy with Jupiter in 1994 served to remind us that asteroids and comets occasionally collide with a planet. Earth is not exempt from collisions. Large meteorites have produced impact craters when they have collided with Earth's surface. One well-known meteorite crater is Meteor Crater in Arizona, which is a little more than a kilometer in diameter (box figure 1). Many much larger craters are known in Canada, Germany, Australia, and other places.

Impact craters display an unusual type of metamorphism called *shock metamorphism.* The sudden impact of a large extraterrestrial body results in brief but extremely high pressures. Quartz may recrystallize into the rare SiO_2 mineral coesite. Quartz that is not as intensely impacted suffers damage (detectable under a microscope) to its crystal lattice.

The impact of a meteorite also may generate enough heat to locally melt rock. Molten blobs of rock are thrown into the air and become streamlined in the Earth's atmosphere before solidifying into what are called *tektites.* Tektites may be found hundreds of kilometers from the point of meteorite impact.

A large meteorite would also throw large quantities of dust high into the atmosphere. According to one theory, the change in global climate due to a meteorite impact around 65 million years ago caused extinctions of many varieties of creatures (see box 8.2 on the extinction of dinosaurs). Evidence for this impact includes finding tiny fragments of shock metamorphosed quartz and tektites in sedimentary rock that is 65 million years old.

Shock metamorphosed rock fragments are much more common on the Moon than on Earth. There may be as many as 400,000 craters larger than a kilometer in diameter on the Moon. Mercury's surface is remarkably similar to that of the Moon. Our two neighboring planets, Venus and Mars, are not as extensively cratered as is the Moon.

BOX 15.2 — FIGURE 1
Meteor crater in Arizona.
Photo by Frank M. Hanna

Related Web Resources

Meteor Crater

www.meteorcrater.com/

See an animation of the meteorite impact. Go to "reference information" for details about the meteorite impact.

under the particular temperature attained during metamorphism. Hornfels can also form from basalt, in which case amphibole, rather than mica, is the predominant fine-grained mineral produced.

Limestone recrystallizes during metamorphism into **marble,** a coarse-grained rock composed of interlocking calcite crystals (figure 15.8). Dolomite, less commonly found, recrystallizes into a *dolomitic marble.* Marble has long been valued as a building material and as a material for sculpture, partly because it is easily cut and polished and partly because it reflects light in a shimmering pattern, a result of the excellent cleavage of the individual calcite crystals. Marble is, however, highly susceptible to chemical weathering (see chapter 12).

Quartzite is produced when grains of quartz in sandstone are welded together while the rock is subjected to high temperature. This makes it as difficult to break along grain boundaries as through the grains. Therefore quartzite, being as hard as a single quartz crystal, is difficult to crush or break. It is the most durable of common rocks used for construction, both because of its hardness and because quartz is not susceptible to chemical weathering.

Marble and quartzite also form under conditions of regional metamorphism. When grains of calcite or quartz recrystallize, they tend to be equidimensional rather than elongate or platy. For this reason, marble and quartzite do not usually exhibit foliation, even though subjected to differential stress during metamorphism.

Regional Metamorphism

The great majority of the metamorphic rocks found on Earth's surface are products of **regional metamorphism** (also known as *dynamothermal* metamorphism), which is metamorphism that takes place at considerable depth underground (generally greater than 5 kilometers). Regional metamorphic rocks are almost always foliated, indicating differential stress during

FIGURE 15.8
Marble.
Photo by C. C. Plummer

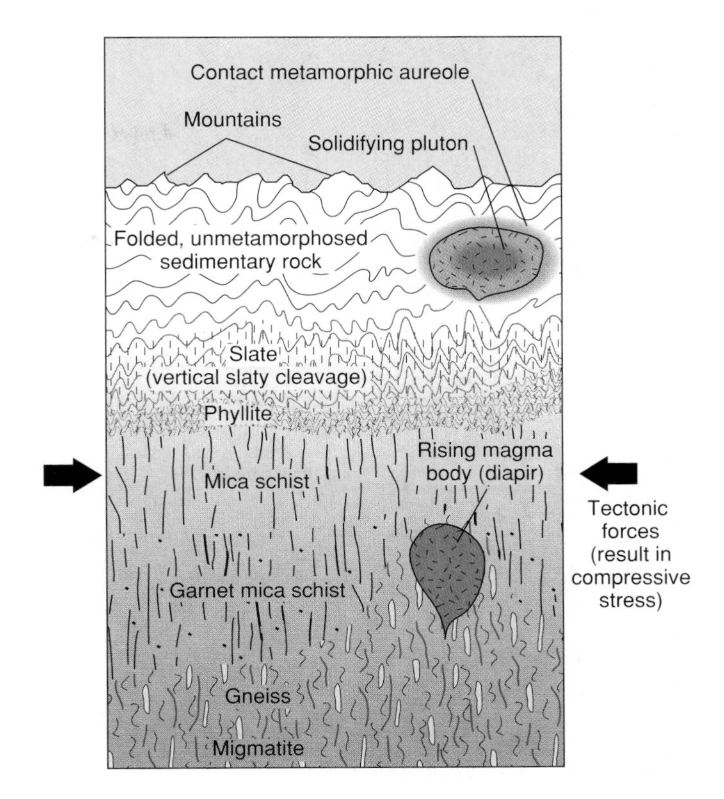

FIGURE 15.9
Schematic cross section representing an approximately 30-kilometer portion of Earth's crust during metamorphism. Rock names given are those produced from shale.

recrystallization. Metamorphic rocks are prevalent in the most intensely deformed portions of mountain ranges. They are visible where once deeply buried cores of mountain ranges are exposed by erosion. Furthermore, large regions of the continents are underlain by metamorphic rocks, thought to be the roots of ancient mountains long since eroded down to plains or rolling hills.

Temperatures during regional metamorphism vary widely. Usually, the temperatures are in the range of 300 to 800°C. Temperature at a particular place depends to a large extent on depth of burial and the geothermal gradient of the region (see box 15.4). Locally, temperature may also increase because of heat given off by nearby magma bodies. The high confining pressure is due to burial under 5 or more kilometers of rock. The differential stress is due to tectonism; that is, the constant movement and squeezing of the crust during mountain-building episodes.

Depending on the pressure and temperature conditions during metamorphism, a particular parent rock may recrystallize into one of several metamorphic rocks. For example, if basalt is metamorphosed at relatively low temperatures and pressures, it will recrystallize into a *greenschist,* a schistose rock containing chlorite (a green sheet-silicate), actinolite (a green amphibole), and sodium-rich plagioclase. At higher temperatures and pressures, the same basalt would recrystallize into an *amphibole schist* (also called *amphibolite*), a rock composed of hornblende, plagioclase feldspar, and perhaps garnet. Metamorphism of other parent rocks under conditions similar to those that produce amphibole schist from basalt should produce the metamorphic rocks shown in table 15.2.

Progressive Metamorphism

To show how rocks are changed by regional metamorphism, we look at what happens to shale during *progressive metamorphism*—that is, as progressively greater pressure and temperature act on a rock type with increasing depth in Earth's crust (figure 15.9).

table 15.2 Regional Metamorphic Rocks That Form under Approximately Similar Pressure and Temperature Conditions

Parent Rock	Rock Name	Predominant Minerals
Basalt	Amphibole schist (amphibolite)	Hornblende, plagioclase, garnet
Shale	Mica schist	Biotite, muscovite, quartz, garnet
Quartz sandstone	Quartzite	Quartz
Limestone or dolomite	Marble	Calcite or dolomite

FIGURE 15.10
Slate outcrop in Antarctica.
Photo by P. D. Rowley, U.S. Geological Survey

FIGURE 15.11
Slate.
Photo by C. C. Plummer

Shale, the parent rock, is formed largely of submicroscopic, platy, clay minerals. The metamorphic rock associated with the lowest pressure and temperature conditions of regional metamorphism is **slate,** a very fine-grained rock that splits easily along flat, parallel planes (figures 15.10 and 15.11). Slate develops under temperatures and pressures only slightly greater than those found in the sedimentary realm. The temperatures are not high enough for the rock to thoroughly recrystallize. The important controlling factor is differential stress. The original clay minerals partially recrystallize into equally fine-grained micas. Under differential stress, the old and new platy minerals are aligned, creating slaty cleavage in the rock. A slate indicates that a relatively cool and brittle rock has been subjected to intense tectonic activity.

Because of the ease with which it can be split into thin, flat sheets, slate is used for making chalkboards, pool tables, and roofs.

Phyllite is a rock in which the newly formed micas are larger than in slate, but still cannot be seen with the naked eye. This requires a further increase in temperature over that needed for slate to form. The very fine-grained mica imparts a silky sheen to the rock, which may otherwise closely resemble slate (figure 15.12). But the slaty cleavage may be crinkled in the process of conversion of slate to phyllite.

A **schist** is characterized by megascopically visible, approximately parallel-oriented minerals. Platy or elongate minerals that crystallize from the parent rock are clearly visible to the naked eye. Shale may recrystallize into several mineralogically distinct varieties of schist. Which minerals form depends on the particular combination of temperature and pressure prevailing during recrystallization. For instance, if the rock is a *mica schist,* metamorphism probably took place at only slightly higher temperatures and pressures than those at which a phyllite forms. A *garnet-mica schist* (figure 15.13) indicates that the temperature and pressure were somewhat greater than necessary for a mica schist to form (see box 15.3).

In a **gneiss,** a rock consisting of light and dark mineral layers or lenses, the highest temperatures and pressures have changed the rock so that minerals have separated into layers. Platy or elongate minerals (such as mica or amphibole) in dark layers alternate with layers of light-colored minerals of no particular shape. Within the light-colored layers coarse feldspars have crystallized. In composition, a gneiss may

IN GREATER DEPTH 15.3

Index Minerals

Certain minerals can only form under a restricted range of pressure and temperature. Stability ranges of these minerals have been determined in laboratories. When found in metamorphic rocks, these minerals can help us infer, within limits, what the pressure and temperature conditions were during metamorphism. For this reason, they are known as *index minerals.* Among the best known are *andalusite, kyanite,* and *sillimanite.* All three have an identical chemical composition (Al_2SiO_5), but different crystal structures. They are found in metamorphosed shales that have an abundance of aluminum. Box figure 1 is a phase diagram showing the pressure/temperature fields in which each is stable.

If andalusite is found in a rock, this indicates that pressures and temperatures were relatively low. Andalusite is often found in contact metamorphosed shales (hornfels). Kyanite, when found in schists, is regarded as an indicator of high pressure; but note that the higher the temperature of the rock, the greater the pressure needed for kyanite to form. Sillimanite is an indicator of high temperature and can be found in some contact metamorphic rocks adjacent to very hot intrusions as well as in regionally metamorphosed schists and gneisses that formed at considerable depths.

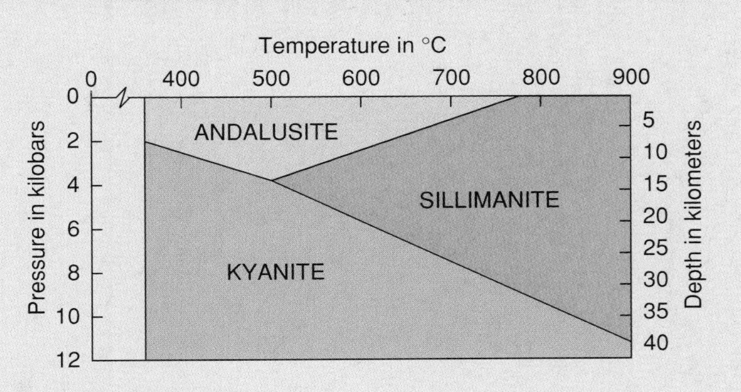

Box 15.3 — FIGURE 1
Phase diagram showing the stability relationships for the Al_2SiO_5 minerals.
M. J. Holdaway, 1971, *American Journal of Science,* v. 271. Reprinted by permission of *American Journal of Science*

Note that if you find all three minerals in the same rock and could determine that they were mutually stable, you could infer that the temperature was close to 500°C and the confining pressure was almost 4 kilobars during metamorphism. (A kilobar is equivalent to the pressure of approximately 1,000 atmospheres.)

FIGURE 15.12
Phyllite, exhibiting a crinkled, silky-looking surface.
Photo by C. C. Plummer

FIGURE 15.13
Garnet-mica schist. Small, subparallel flakes of mica reflect light. Garnet crystals give the rock a "raisin bread" appearance.
Photo by C. C. Plummer

FIGURE 15.14
Gneiss.
Photo by C. C. Plummer

FIGURE 15.15
Migmatite in the Daniels Range, Antarctica.
Photo by C. C. Plummer

resemble granite or diorite, but it is distinguishable from those plutonic rocks by its foliation (figure 15.14).

Temperature conditions under which a gneiss develops approach those at which granite solidifies. It is not surprising, then, that the same minerals are found in gneiss and in granite. In fact, a previously solidified granite can be converted to a gneiss under appropriate pressure and temperature conditions and if the rock is under differential stress.

If the temperature is high enough, partial melting of rock may take place, and a magma collects in layers within the foliation planes of the solid rock. After the magma solidifies, the rock becomes a **migmatite,** a mixed igneous and metamorphic rock (figure 15.15). A migmatite can be thought of as a "twilight zone" rock that is neither fully igneous nor entirely metamorphic.

The metamorphic rocks that we see usually have minerals that formed at or near the highest temperature reached during metamorphism. But why doesn't a rock recrystallize to one stable at lower temperature and pressure conditions during its long journey to the surface where we now find it? The answer is that water, which facilitates chemical reactions, no longer gets into rock. Tectonic forces at work during the peak of metamorphism fracture the rock sufficiently to permit water to get to the mineral grains. After tectonic forces are relaxed, the rocks move upward as a large block as isostatic adjustment takes place. It is unusual to find rocks that indicate *retrogressive metamorphism.* These are rocks that recrystallized under lower temperature and pressure conditions than during the peak of metamorphism. They were fractured during their ascent, permitting water to trigger reactions to new, lower grade minerals.

PLATE TECTONICS AND METAMORPHISM

The model of Earth's crust and upper mantle that we get from plate tectonic theory allows us to explain many of the observed characteristics of metamorphic rocks. To demonstrate the relationship between regional metamorphism and plate tectonics, we will look at what is believed to take place at a convergent boundary in which oceanic lithosphere is subducted beneath continental lithosphere, as shown in figure 15.16.

Differential stress, which is responsible for foliation, occurs wherever rocks are being squeezed between the two plates or wherever rocks are sliding past one another. Shearing is expected in the subduction zone where the oceanic crust slides beneath the continental lithosphere (figure 15.16). In the chapter on mountains (chapter 5), we describe the concept of *gravitational collapse and spreading,* in which the central part of a mountain belt becomes too high after plate convergence and is gravitationally unstable. This forces rock downward and outward as shown by the arrows in figure 15.16 (see also figure 5.20). At deeper levels the plastic rock flows during metamorphism, and foliation should develop parallel to the direction of flowage. Also, vertical foliation may develop throughout much

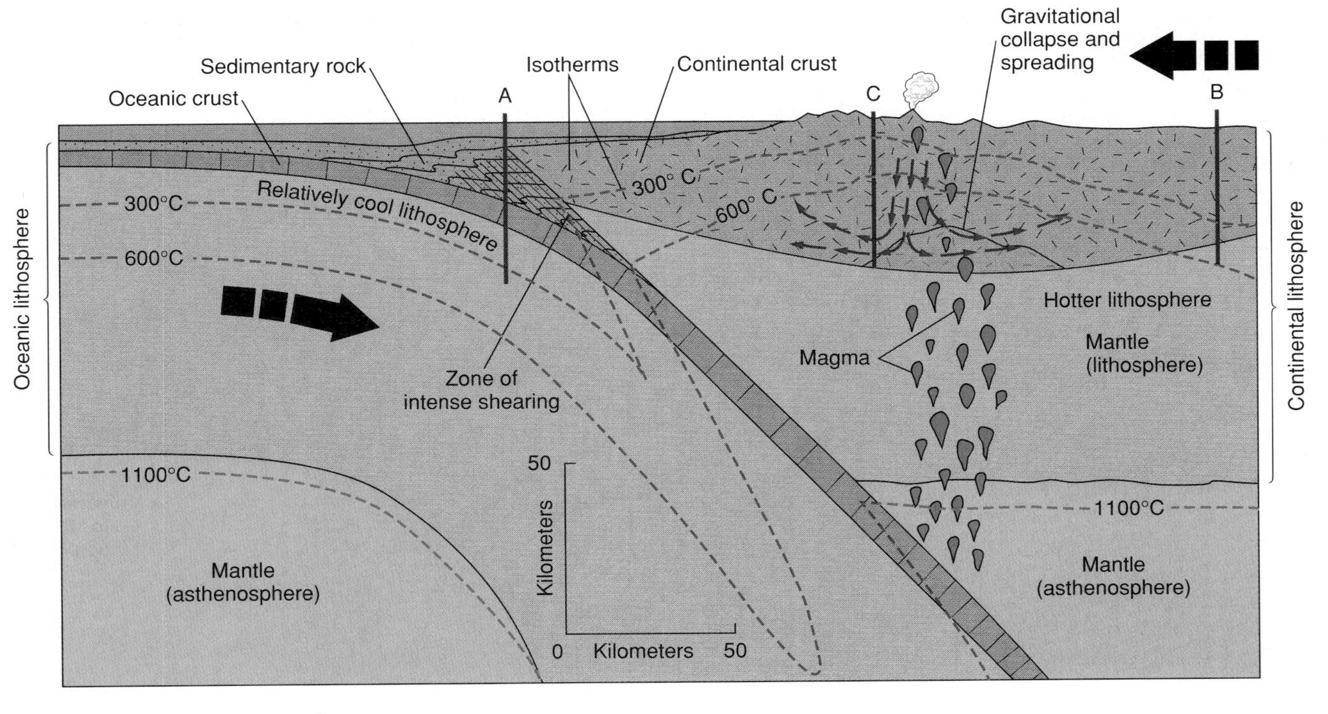

FIGURE 15.16

Metamorphism across a convergent plate boundary.

From W. G. Ernst. *Metamorphism and Plate Tectonic Regimes.* Stroudsberg, Pa.: Dowden, Hutchinson & Ross, 1975; p. 425. Reprinted by permission of the author

of the region due to the compressive stress in which the crust is caught, as if in a vise (see figure 15.9).

Confining pressure is directly related to depth. For this reason, we would expect the same pressure at a depth of 20 kilometers beneath a volcanic area as beneath the relatively cool rocks of a plate's interior.

Metamorphic rocks indicate wide ranges of temperatures for a given pressure. This is understandable if we realize that the geothermal gradient is not the same everywhere in the world. Each of the three places (*A, B,* and *C*) in figure 15.16 would have a different geothermal gradient. If you were somehow able to push a thermometer through the lithosphere, you would find the rock is hotter at shallower depths in areas with higher geothermal gradients than at places where the geothermal gradient is low. As indicated in figure 15.16, the geothermal gradient is higher progressing downward through an active volcanic-plutonic complex (for instance, the Cascade Mountains of Washington and Oregon) than it is in the interior of a plate (beneath the Great Plains of North America, for example).

Heat, then, is the most variable of the controlling factors of metamorphism. Figure 15.16 shows the temperatures calculated for a convergent boundary. A line connecting points that have the same temperature is called an **isotherm.** Note how the 300°C, 600°C, and 1100°C isotherms change depth radically across the subduction zone. This is because rocks that were cool because they were relatively near Earth's surface have been transported rapidly to depth. ("Rapidly" in the geologic sense—movement being around 1 cm/year.) The oceanic lithosphere along a subduction zone has not had time to heat up and come into thermal equilibrium with the relatively hot rocks elsewhere at these depths. On the top of the subducted oceanic slab, the extra heat provided by friction along with the heat that is normal for the overlying continental lithosphere and asthenosphere cause the isotherms to rise sharply toward the surface. The isotherms are bowed upward in the region of the volcanic-plutonic complex because magma created along the lower levels of the subduction zone works its way upward and brings heat from the asthenosphere into the mantle and crust of the continental lithosphere.

To understand why metamorphic rocks can form under such a wide variety of temperatures and pressures, study figure 15.16. You may observe that the bottom of line *A* is at a depth of about 50 kilometers, and if a hypothetical thermometer were here, it would read just over 300° because it would be just below the 300° isotherm. Compare this to vertical line *C* in the volcanic-plutonic complex. The confining pressure at the base of this line would be the same as at the base of line *A,* yet the temperature at the base of line *C* would be well over 600°. The minerals that could form at the base of line *A* would not be the same as those that could form at line *C.* Therefore, we would expect quite different metamorphic rocks in the two places, even if the parent rock had been the same (box 15.4).

15.4 IN GREATER DEPTH

METAMORPHIC FACIES AND THE RELATIONSHIP TO PLATE TECTONICS

During the early part of the twentieth century, geologists in Scandinavia introduced the concept of metamorphic facies. They noted that metamorphic rocks that were once basalts had one set of minerals in some parts of Europe, but in other areas, metabasalts had quite different mineral assemblages. As these rocks were *chemically* similar, the different *mineral assemblages* were regarded as indicating significantly different pressure and temperature conditions during metamorphism. Rocks having the same mineral assemblage are regarded as belonging to the same **metamorphic facies,** implying that they formed under broadly similar pressure and temperature conditions. The name for each facies is based on the assemblage of minerals or the name of a rock common to that facies. For instance, a schistose metabasalt composed mostly of the minerals chlorite, actinolite, and epidote (all of which are green minerals) belongs to the *greenschist facies.* On the other hand, rocks of the same chemical composition (metabasalts) belonging to the *amphibolite facies* are largely made up of hornblende and garnet. (Do not try to remember the names of the facies or their compositions; your aim should be to understand the concept.)

Based on the geologic setting, early workers inferred that the temperature conditions during metamorphism were lower for the greenschist facies than for the amphibolite facies. Laboratory work has since confirmed this as well as determined the pressure and temperature *stability fields* for each of the facies (box figure 1).

The concept of metamorphic facies is analogous to defining climatic zones by the combinations of plants found in each zone. A place where ferns, palm trees, and vines flourish corresponds to a climate with warm temperatures and abundant rainfall. On the other hand, a combination of palm trees, cactus, and sagebrush implies a hot, dry climate.

By identifying the metamorphic facies of rocks presently cropping out on the surface, geologists can infer, within broad limits, the depth at which metamorphism took place. They may also (again, within broad limits) be able to determine the corresponding temperature.

The concept of metamorphic facies preceded plate tectonic theory by several decades. Although earlier geologists were able to relate the individual facies to pressure and temperature combinations, they had no satisfactory explanation for the environments that produced the various combinations. Figure 15.16, which relates the temperatures of regional metamorphism to plate tectonics, may be used to infer the environment for each of the metamorphic facies shown in box figure 1. Box figure 2 shows the likely distribution of metamorphic facies across the same converging boundary as in figure 15.16. To understand the relationship, study box figures 1 and 2 as well as figure 15.16.

If one were to determine the geothermal gradient represented by the three vertical lines marked *A, B,* and *C* on figure 15.16 and box figure 2, the temperatures for particular depths should plot on the corresponding arrows shown in box figure 1. Follow

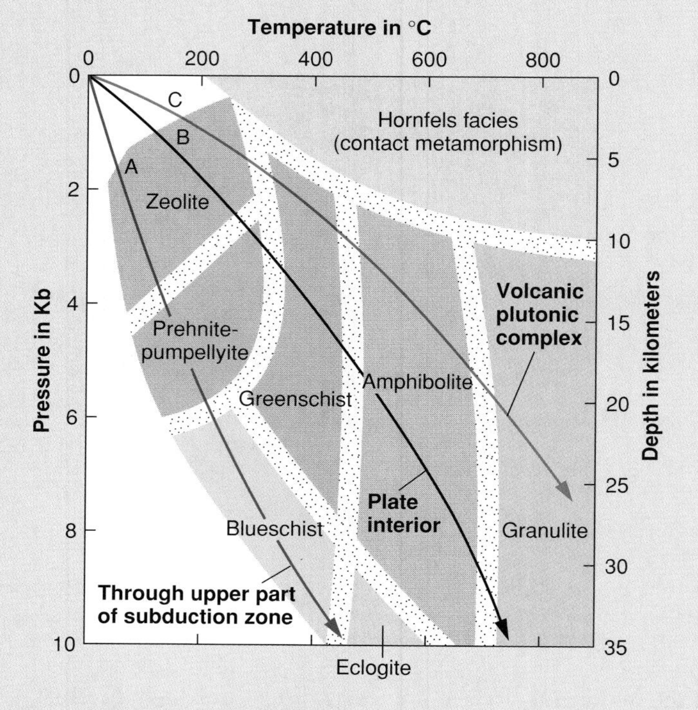

BOX 15.4 — FIGURE 1

The metamorphic facies. Facies are named after minerals (prehnite, zeolite, pumpellyite) or rock types (e.g., blueschist, granulite). Boundaries between facies are approximate. The arrows represent increases in temperature with depth for the three lines labeled *A, B, C* in box figure 2 and in figure 15.16.

From W. G. Ernst, *Metamorphism and Plate Tectonic Regimes.* Stroudsberg, Pa.: Dowden, Hutchinson & Ross, 1975; p. 425. Reprinted by permission of the author

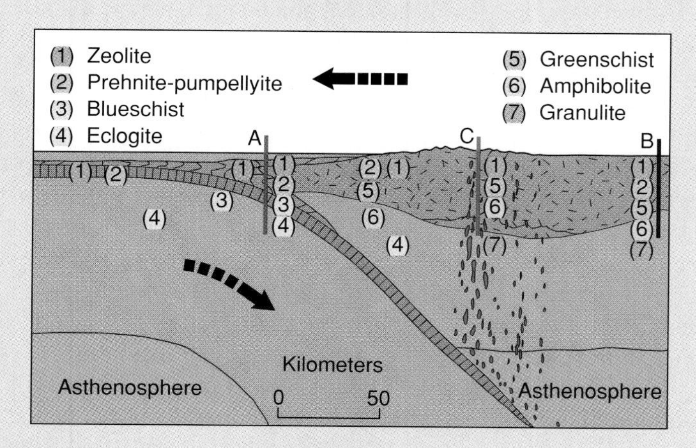

BOX 15.4 — FIGURE 2

Schematic representation of the distribution of facies across a convergent plate boundary.

From W. G. Ernst, *Metamorphism and Plate Tectonic Regimes.* Stroudsberg, Pa.: Dowden, Hutchinson & Ross, 1975; p. 426. Reprinted by permission of the publisher

arrow *A* in box figure 1. This means that if you were able to drill vertically downward along line *A,* you would find, beneath unmetamorphosed rocks, rocks of the zeolite facies. At a greater depth would be the boundary between the zeolite and prehnite-pumpellyite facies. You would reach this boundary at a depth where the pressure is about 4 kilobars and the temperature is approximately 150°C. Your drill would penetrate rocks of the prehnite-pumpellyite facies until you reached the blueschist facies. In hole *B,* in the interior of the plate, the progression would be from zeolite facies to prehnite-pumpellyite facies to greenschist facies to amphibolite facies to granulite facies. Hole *C,* in the volcanic-plutonic complex, would not pass through the prehnite-pumpellyite facies but would go from zeolite facies to greenschist facies to amphibolite facies to granulite facies.

HYDROTHERMAL PROCESSES

Rocks that have precipitated from hot water or have been altered by hot water passing through are hard to classify. As described earlier, hot water is involved to some extent in most metamorphic processes. Beyond metamorphism, hot water also plays an important role creating new rocks and minerals. These form entirely by precipitation of ions derived from hydrothermal solutions. *Hydrothermal minerals* can form in void spaces or between the grains of a host rock. An aggregate of hydrothermal minerals, a **hydrothermal rock,** may crystallize within a preexisting fracture in a rock to form a hydrothermal **vein.**

Hydrothermal processes are summarized in table 15.3. As we have seen, water is important for metamorphic processes not only because water transports ions from one mineral to another, but because many of the minerals (the micas, for instance) that crystallize during metamorphism incorporate water into their crystal structures.

Hydrothermal Activity at Divergent Plate Boundaries

Hydrothermal processes are particularly important at mid-oceanic ridges (which are also divergent plate boundaries). As shown in figure 15.17, cold seawater moves downward through cracks in the basaltic crust and is cycled upward by heat from magma beneath the ridge crest. Very hot water returns to the ocean at submarine hot springs.

Hot water traveling through the basalt and gabbro of the oceanic lithosphere helps metamorphose these rocks while they are close to the divergent boundary. During metamorphism the ferromagnesian igneous minerals, olivine and pyroxene, become converted to *hydrous* (water-bearing) minerals such as amphibole. An important consequence of this is that the hydrous minerals may eventually contribute to magma generation at convergent boundaries. After oceanic crust is subducted the minerals are dehydrated deep in a subduction zone (see figure 15.22). The water produced moves upward into the overlying asthenosphere and contributes to melting and magma generation, as described in chapter 11.

Ore Deposits at Divergent Plate Boundaries

As the seawater moves through the crust, it dissolves metals and sulfur from the crustal rocks and magma. When the hot, metal-rich solutions contact cold seawater, metal sulfides are precipitated in a mound around the hot spring. This process has been filmed in the Pacific, where some springs spew clouds of fine-grained ore minerals that look like black smoke (figure 15.18).

The metals in rift-valley hot springs are predominantly iron, copper, and zinc, with smaller amounts of manganese, gold, and silver. Although the mounds are nearly solid metal sulfide, they are small and widely scattered on the sea floor, so commercial mining of them may not be practical.

table 15.3 Hydrothermal Processes

Role of Water	Name of Process or Product
Water transports ions between grains in a rock. Some water may be incorporated into crystal structures.	Metamorphism
Water brings ions from outside the rock, and they are added to the rock during metamorphism. Other ions may be dissolved and removed.	Metasomatism
Water passes through cracks or pore spaces in rock and precipitates minerals on the walls of cracks and within pore spaces.	Hydrothermal rocks

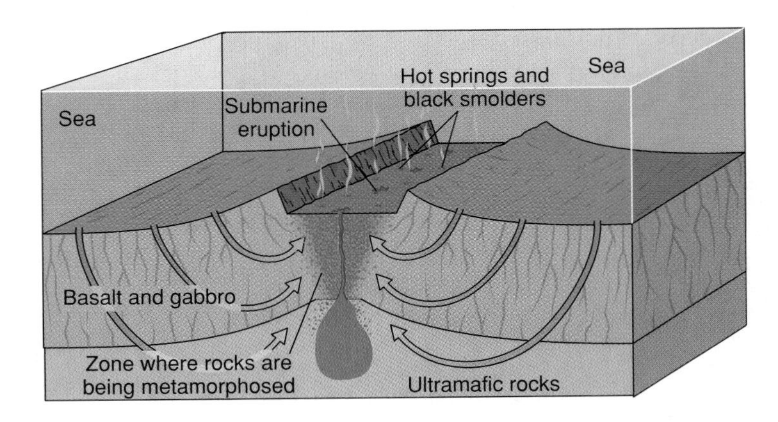

FIGURE 15.17

Cross section of a mid-oceanic ridge (divergent plate boundary). Water descends through fractures in the oceanic lithosphere, is heated by magma and hot igneous rocks, and rises.

FIGURE 15.18

"Black smoker" or submarine hot spring on the crest of the mid-oceanic ridge in the Pacific Ocean near 21° North latitude. The "smoke" is a hot plume of metallic sulfide minerals being discharged into cold seawater from a chimney 0.5 meters high. The large mounds around the chimney are metal deposits. The instruments in the foreground are attached to the small submarine from which the picture was taken.

Photo © WH01-D. Foster/Visuals Unlimited

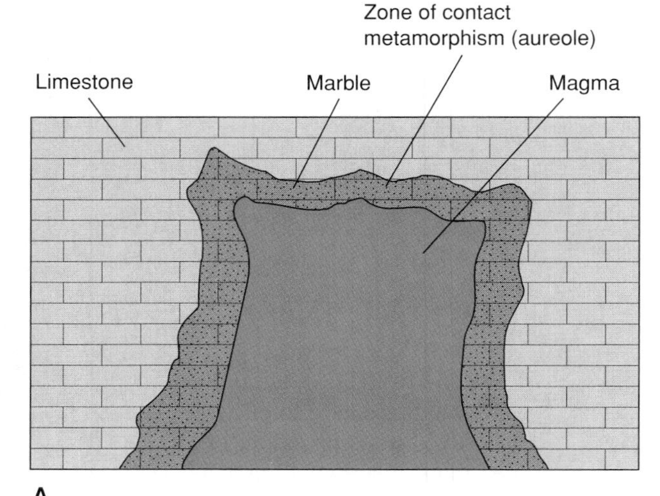

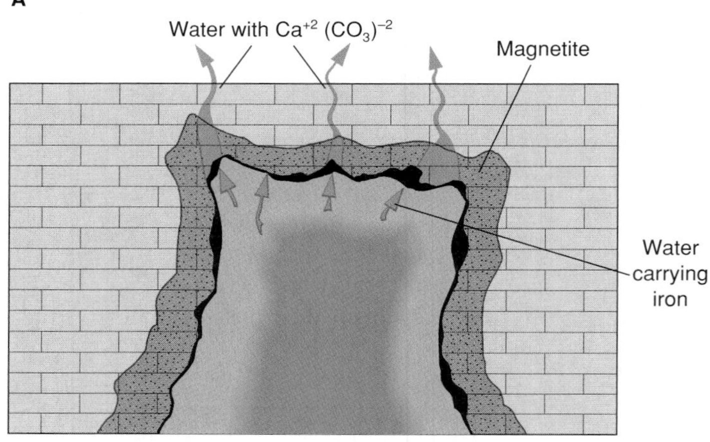

FIGURE 15.19

Development of a contact metasomatic deposit of iron (magnetite). (*A*) Magma intrudes country rock (limestone), and marble forms along contact. (*B*) As magma solidifies, gases bearing ions of iron leave the magma, dissolve some of the marble, and deposit iron as magnetite.

Metasomatism

Metasomatism is metamorphism coupled with the introduction of *ions* from an external source. The ions are brought in by water from outside the immediate environment and are incorporated into the newly crystallizing minerals. Often, metasomatism involves ion exchange. Newly crystallizing minerals replace preexisting ones as water simultaneously dissolves and replaces ions.

When metasomatism takes place during regional metamorphism, very hot water travels through a rock while gneiss or schist is crystallizing. Ions (typically K^+, Na^+, Si^{+4}, and O^{-2}) are carried by the water and participate in metamorphic reactions. Large feldspar crystals may grow in schist due to the addition of potassium or sodium ions.

If metasomatism is associated with contact metamorphism, the ions are introduced from a cooling magma. Some important commercially mined deposits of metals such as iron, tungsten, copper, lead, zinc, and silver are attributed to metasomatism. Figure 15.19 shows how magnetite (iron oxide) ore bodies have formed through metasomatism. Ions of the metal are transported by water and react with minerals in the host rock. Elements within the host rock are simultaneously dissolved out of the host rock and replaced by the metal ions brought in by the fluid. Because of the solubility of calcite, marble commonly serves as a host for metasomatic ore deposits.

Hydrothermal Rocks and Minerals

Quartz veins (figure 15.20) are especially common where igneous activity has occurred. These can form from hot water given off by a cooling magma. They also are produced by ground water heated by a pluton and circulated by convection,

FIGURE 15.20

Ore-bearing veins in a mine.

Photo by C. C. Plummer

ENVIRONMENTAL GEOLOGY 15.5

THE WORLD'S LARGEST HUMAN-MADE HOLE— THE BINGHAM CANYON COPPER MINE

The biggest human-made hole in the world is located at Bingham Canyon near Salt Lake City, Utah. The 800 meter (½ mile) deep open pit mine is 4 km (2½ miles) wide at the top and continues to be enlarged. The reason for this hole is copper.

About 40,000 kilograms of explosives are used per day to blast apart over 60,000 tons of ore (copper-bearing rock) and an equal amount of waste rock. A conveyor belt system moves up to 10,000 tons of crushed rock per hour through a tunnel out of the pit for processing.

Mining began here as a typical underground operation in 1863. The shafts and tunnels of the mine followed a series of veins. Originally ores of silver and lead were mined. Later it was discovered that fine-grained, copper-bearing minerals (chalcopyrite and other copper sulfide minerals) were disseminated in tiny veinlets throughout a granite stock. Although the percentage of copper in the rock was small, the total volume of copper was recognized as huge. With efficient earth-moving techniques, large volumes of ore-bearing rock can be moved and processed. Today mining is still going on, and the company is able to make a profit even though 0.6% of the rock being mined is copper. Since 1904, over 12 million tons of copper have been mined, processed, and sold. The mine has also produced impressive amounts of gold, silver, and other metals.

Such an operation is not without environmental problems. Some people regard the huge hole in the mountains as an eyesore (but it is a popular tourist attraction). Disposing of the waste—over 99% of the rock material mined—creates problems. Wind stirs up dust storms from the piles of finely crushed waste rock unless it is kept wet. The nearby smelter that extracts the pure copper from the sulfide minerals has created a toxic smoke containing sulfuric acid fumes. During most of the twentieth century the toxic smoke was released into the atmosphere; occasionally wind blew polluted air to Salt Lake City. Now, over 99% of the sulfur fumes are removed at the smelter.

Related Web Resource
Mining Technology—Bingham Canyon
www.mining-technology.com/projects/bingham/

BOX 15.5 — FIGURE 1
Bingham Canyon copper mine in Utah.
Courtesy Kennecott Copper Company

as shown in figure 15.21. Where the water is hottest, the most material (notably silica) is dissolved. As water vapor continues upward through increasingly cooler rocks during its journey toward Earth's surface, pressure decreases and heat is lost. Fewer ions can be carried in solution, and so the silicon and oxygen leave the water and cake onto the walls of the crack as silica (SiO_2), forming a quartz vein.

Veins consisting only of quartz are the most widespread, although some quartz veins contain other minerals. Veins with no quartz are not as common and are composed of calcite or some other minerals.

Hydrothermal veins are very important economically. In them we find most of the world's great deposits of zinc, lead, silver, gold, tungsten, tin, mercury, and to some extent, copper.

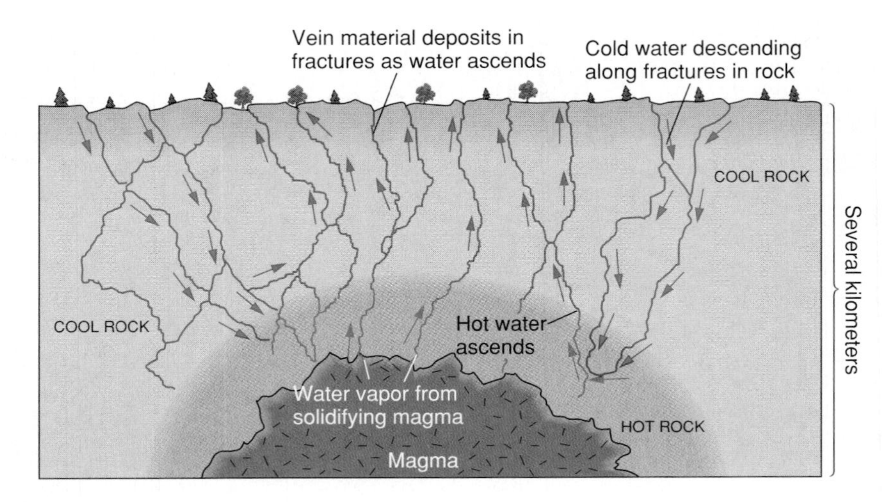

FIGURE 15.21

How veins form. Cold water descends, is heated, dissolves material, ascends, and deposits material as water cools and pressure drops upon ascending.

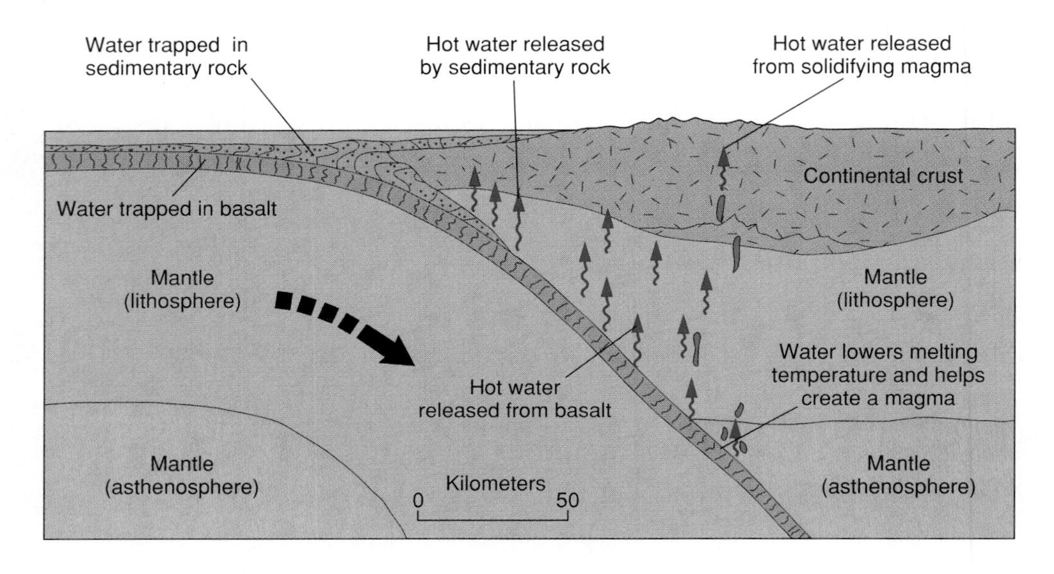

FIGURE 15.22

Water at a convergent boundary. Seawater trapped in the oceanic crust is carried downward and released upon heating at various depths within the subduction zone.

Ore minerals containing these metals are usually found in quartz veins. Veins containing commercially extractable amounts of metals are by no means common (and might even be regarded as freaks of nature).

Some ore-bearing solutions percolate upward between the grains of the rock and deposit very fine grains of ore mineral throughout. These are called *disseminated ore deposits* and account for the largest copper mines in the world (see box 15.5) as well as some very large gold mines. Many disseminated ore deposits are in the upper parts of plutons. This is part of the pluton that solidified earliest. As crystallization continued in the underlying magma, water carried upward ions of metals and other elements that could not be accommodated in feldspars and other igneous minerals. The ore minerals crystallized in tiny fractures and between grains in the overlying, previously solidified igneous rock.

Sources of Water

Where does the water come from? The following is a logical explanation. *Ground water* seeps downward from Earth's surface through pores and fractures in rocks; however, the depth to which surface-derived water can penetrate is quite limited.

Plate tectonics can account for water at deeper levels in the lithosphere as seawater trapped in the oceanic crust can be carried to considerable depths through subduction (figure 15.22).

Water trapped in sediment and in sedimentary rocks lying on basalt may be carried down with the descending crust. However, recent studies indicate that most of the water is carried by hydrous minerals (amphibole, for example) in the basaltic crust. When the rocks get hot enough the hydrous minerals recrystallize, releasing water. The water vapor works its way upward through the overlying continental lithosphere through fissures. In the process of ascending, water assists in the metamorphism of rocks, dissolves minerals, and carries the ions to interact during metasomatism, or it deposits quartz and other minerals in fissures as veins. The water can also lower the melting points of rocks at depth, allowing magma to form (as described in chapter 11 on igneous rocks).

Summary

Metamorphic rocks form from other rocks that are subjected to high temperature generally accompanied by high confining pressure. Recrystallization takes place in the solid state although water, usually present, aids metamorphic reactions. *Foliation* in metamorphic rocks is due to *differential stress* (either *compressive stress* or *shearing*). Slate, phyllite, schist, and gneiss are foliated and distinguished from one another by the type of foliation.

Contact metamorphic rocks are produced during metamorphism usually without significant differential stress but with high temperature. Contact metamorphism occurs in rocks immediately adjacent to intruded magmas.

Regional metamorphism, which involves heat, confining pressure, and differential stress, has created most of the metamorphic rock of Earth's crust. Different parent rocks as well as widely varying combinations of pressure and temperature result in a large variety of metamorphic rocks. Combinations of minerals in a rock can indicate what the pressure and temperature conditions were during metamorphism. Extreme metamorphism, where the rock partially melts, can result in *migmatites.*

Hydrothermal veins form when hot water precipitates material that crystallizes into minerals. During *metasomatism,* hot water introduces ions into a rock being metamorphosed, changing the chemical composition of the metasomatized rock from that of the parent rock.

Plate tectonic theory accounts for the features observed in metamorphic rocks and relates their development to other activities in Earth. In particular, plate tectonics explains (1) the deep burial of rocks originally formed at or near Earth's surface; (2) the intense squeezing necessary for the differential stress, implied by foliated rocks; (3) the presence of water deep within the lithosphere; and (4) the wide variety of pressures and temperatures believed to be present during metamorphism.

Terms to Remember

compressive stress 368
confining pressure 367
contact metamorphism 371
differential stress 368
ductile (plastic) 366
foliation 369
gneiss 374
gneissic 370
hornfels 371
hydrothermal rock 379
isotherm 377
marble 372
metamorphic facies 378
metamorphic rock 366
metamorphism 366
metasomatism 380
migmatite 376
parent rock 366
phyllite 374
quartzite 372
regional metamorphism 372
schist 374
schistose 369
shearing 368
slate 374
slaty 369
slaty cleavage 369
stress 368
vein 379

Testing Your Knowledge

Use the questions below to prepare for exams based on this chapter.

1. What are the effects on metamorphic minerals and textures of temperature, confining pressure, and differential stress?
2. What are the various sources of heat for metamorphism?
3. How do regional metamorphic rocks differ in texture from contact metamorphic rocks?
4. Why is such a variety of combinations of pressure and temperature environments possible during metamorphism?
5. How would you distinguish
   a. schist and gneiss?
   b. slate and phyllite?
   c. quartzite and marble?
   d. granite and gneiss?

6. Why is an edifice built with blocks of quartzite more durable than one built of marble blocks?
7. Metamorphism of limestone may contribute to global warming by the release of
   a. oxygen
   b. sulfuric acid
   c. nitrogen
   d. CO_2
8. Shearing is a type of
   a. compressive stress
   b. confining pressure
   c. lithostatic pressure
   d. differential stress
9. Metamorphic rocks with a planar texture (the constituents of the rock are parallel to one another) are said to be
   a. concordant
   b. foliated
   c. discordant
   d. nonfoliated
10. Metamorphic rocks are classified primarily on
   a. texture—the presence or absence of foliation
   b. mineralogy—the presence or absence of quartz
   c. environment of deposition
   d. chemical composition
11. Which is not a foliated metamorphic rock?
   a. gneiss
   b. schist
   c. quartzite
   d. slate
12. Limestone recrystallizes during metamorphism into
   a. hornfels
   b. marble
   c. quartzite
   d. schist
13. Quartz sandstone is changed during metamorphism into
   a. hornfels
   b. marble
   c. quartzite
   d. schist
14. The correct sequence of rocks that are formed when shale undergoes progressive metamorphism is
   a. slate, gneiss, schist, phyllite
   b. phyllite, slate, schist, gneiss
   c. slate, phyllite, schist, gneiss
   d. schist, phyllite, slate, gneiss
15. The major difference between metamorphism and metasomatism is
   a. temperature at which each takes place
   b. the minerals involved
   c. the area or region involved
   d. metasomatism is metamorphism coupled with the introduction of ions from an external source
16. Ore bodies at divergent plate boundaries can be created through
   a. contact metamorphism
   b. regional metamorphism
   c. hydrothermal processes
17. Metamorphic rocks with the same mineral assemblage belong to the same
   a. metamorphic facies
   b. progressive metamorphism
   c. schistosity
18. A metamorphic rock that has undergone partial melting to produce a mixed igneous-metamorphic rock is a
   a. gneiss
   b. hornfels
   c. schist
   d. migmatite

Expanding Your Knowledge

1. Should ultramafic rocks in the upper mantle be regarded as metamorphic rocks rather than igneous rocks?
2. Where were the metals before they were concentrated in hydrothermal vein ore deposits?
3. What happens to originally horizontal layers of sedimentary rock when they are subjected to the deformation associated with regional metamorphism?
4. Where in Earth's crust would you expect most migmatites to form?

Exploring Web Resources

www.mhhe.com/plummer9e

Visit this website for answers to Testing Your Knowledge as well as additional quizzes, readings, media resources, and direct links to the sites listed below.

www.science.ubc.ca/~geol202/meta/metamorphic.html

University of British Columbia's *Metamorphic Rocks Home Page.* This site is meant for a geology course on the study of rocks. Although it is at a more advanced level, it can be used to reinforce some of the concepts covered in this chapter.

www.geolab.unc.edu/Petunia/IgMetAtlas/mainmenu.html

University of North Carolina's *Atlas of Rocks, Minerals, and Textures.* Click on "Metamorphic microtextures." Click on terms covered in this chapter (e.g., *foliation, gneiss, phyllite, marble, quartzite, slate*) to see excellent photomicrographs taken through a polarizing microscope.

Animations

This chapter includes the following animations available on our Online Learning Center at www.mhhe.com/plummer9e.

15.16 Metamorphism and Plate Convergence
15.17 Mid-Oceanic Ridge
15.21 How Veins Form

CHAPTER

16

Streams and Floods

Running water, aided by mass wasting, is the most important geologic agent in eroding, transporting, and depositing sediment. Almost every landscape on Earth shows the results of stream erosion or deposition. Although other agents—ground water, glaciers, wind, and waves—can be locally important in sculpturing the land, stream action and mass wasting are the dominant processes of landscape development.

The first part of this chapter deals with the various ways that streams erode, transport, and deposit sediment. The second part describes landforms produced by stream action, such as valleys, flood plains, deltas, and alluvial fans, and shows how each of these is related to changes in stream characteristics. The chapter also includes a discussion of the causes and effects of flooding and of various measures used to control flooding.

Opposite: The Gooseberry River flows over a resistant ledge of Precambrian-age lava flows at Gooseberry Falls State Park, Minnesota.

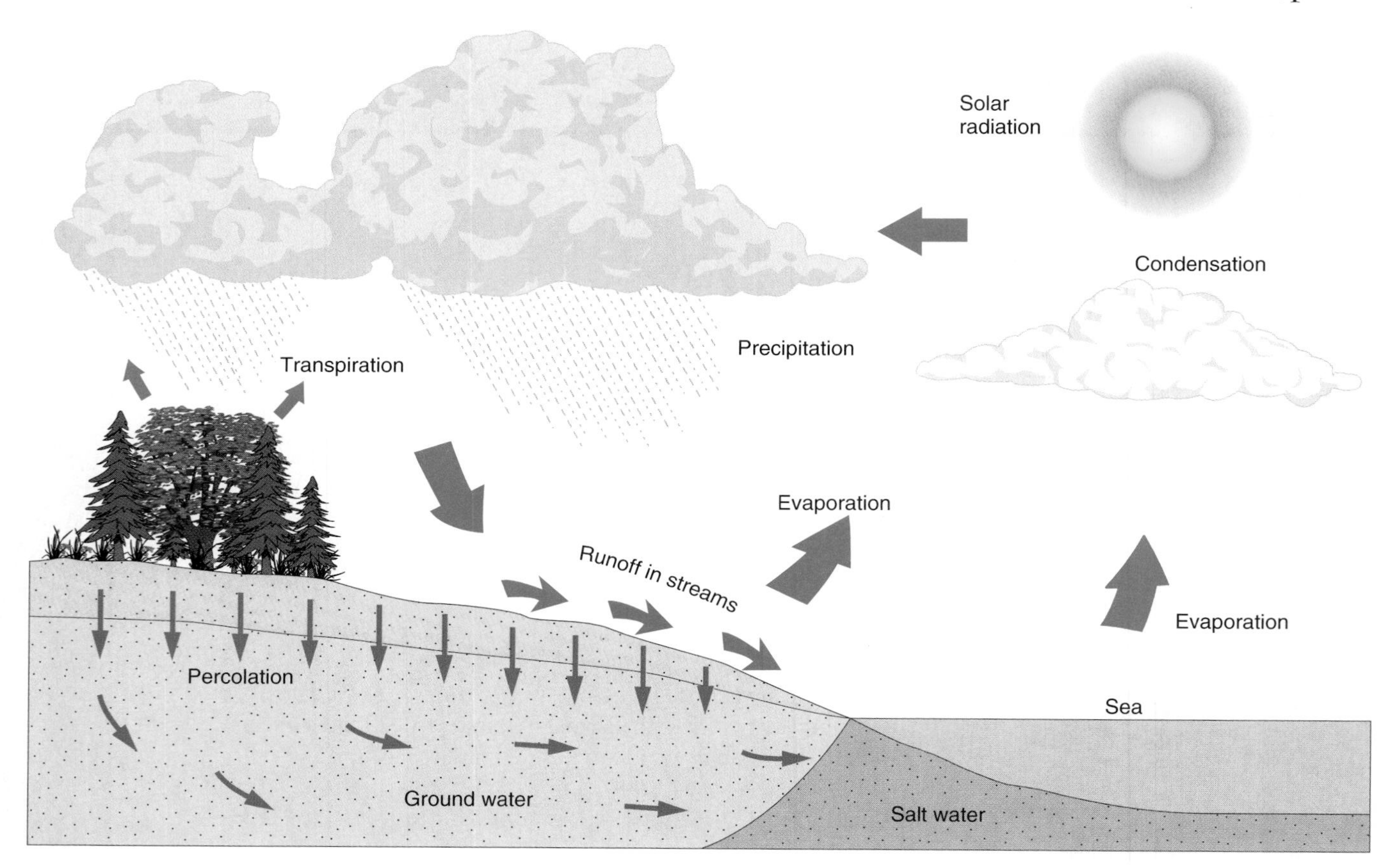

FIGURE 16.1

The hydrologic cycle. Water vapor evaporates from the land and sea, condenses to form clouds, and falls as precipitation (rain and snow). Water falling on land runs off over the surface as streams or percolates into the ground to become ground water. It returns to the atmosphere again by evaporation and transpiration (the loss of water to the air by plants).

THE HYDROLOGIC CYCLE

The movement of water and water vapor from the sea to the atmosphere, to the land, and back to the sea and atmosphere again is called the **hydrologic cycle** (figure 16.1). When rain (or snow) falls on the land surface, more than half the water returns rather rapidly to the atmosphere by evaporation or by transpiration from plants. The remainder either flows over the land surface as runoff in streams or percolates down into the ground to become ground water.

Only about 15% to 20% of rainfall normally ends up as surface runoff in rivers, although the amount of runoff can range from 2% to more than 25% with variations in climate, steepness of slope, soil and rock type, and vegetation. Steady, continuous rains can saturate the ground and the atmosphere, however, and lead to floods as runoff approaches 100% of rainfall.

CHANNEL FLOW AND SHEET FLOW

A **stream** is a body of running water that is confined in a channel and moves downhill under the influence of gravity. In some parts of the country, *stream* implies size: rivers are large, streams somewhat smaller, and brooks or creeks even smaller. Geologists, however, use *stream* for any body of running water, from a small trickle to a huge river.

Figure 16.2*A* shows a *longitudinal profile* of a typical stream viewed from the side. The stream begins in steep mountains and flows out across a gentle plain into the sea. The *headwaters* of a stream are the upper part of the stream near its source in the mountains. The *mouth* is the place where a stream enters the sea, a lake, or a larger stream. A *cross section* of a stream in steep mountains is usually a V-shaped valley cut into solid rock, with the stream channel occupying the narrow bottom of the valley; there is little or no flat land next to the stream on the valley bottom (figure 16.2*B*). Near its mouth a stream usually flows within a broad, flat-floored valley. The stream channel is surrounded by a flat *flood plain* of sediment deposited by the stream (figure 16.2*C*).

A stream normally stays in its **stream channel,** a long narrow depression eroded by the stream into rock or sediment. The stream *banks* are the sides of the channel; the stream *bed* is the bottom of the channel. During a flood the waters of a stream may rise and spill over the banks onto the flat flood plain of the valley floor (figure 16.3).

Not all water that moves over the land surface is confined to channels. Sometimes, particularly during heavy rains, water runs

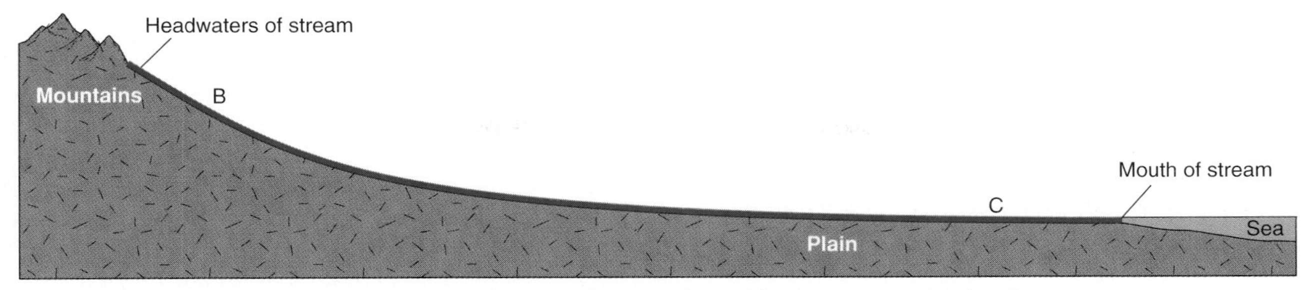

A Longitudinal profile (dark blue line) of a stream beginning in mountains and flowing across a plain into the sea.

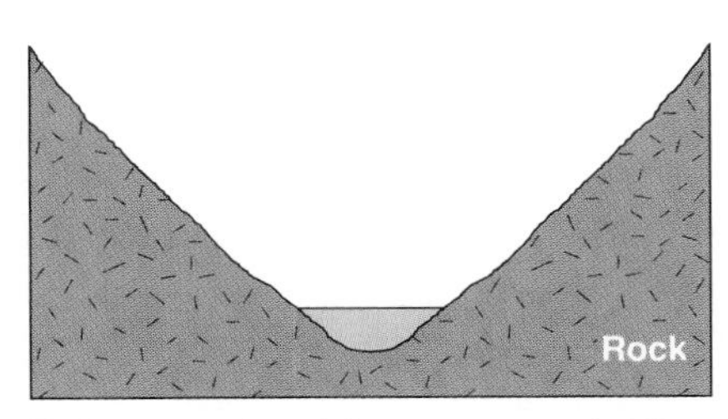

B Cross section of the stream at point B. The channel is at the bottom of a V-shaped valley cut into rock.

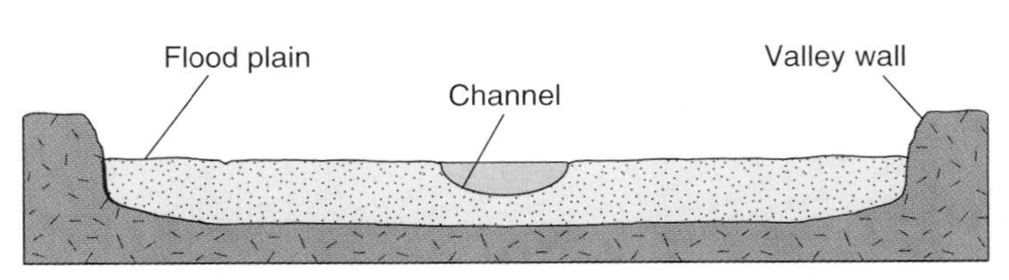

C Cross section at point C. The channel is surrounded by a broad flood plain of sediment.

FIGURE 16.2

Longitudinal and cross sections of a typical stream.

A

B

FIGURE 16.3

A stream normally stays in its channel, but during a flood it can spill over its banks onto the adjacent flatland. Todd Valley in New Zealand during normal discharge (*A*) and also in flood stage when 15 centimeters of rain fell in 3 hours (*B*).

Photos © G. R. 'Dick' Roberts

off as **sheetwash,** a thin layer of unchanneled water flowing downhill. Sheetwash is particularly common in deserts, where the lack of vegetation allows rainwater to spread quickly over the land surface. It also occurs in humid regions during heavy thunderstorms when water falls faster than it can soak into the ground. A series of closely spaced storms can also promote sheetwash; as the ground becomes saturated, more water runs over the surface.

Sheetwash, along with the violent impact of raindrops on the land surface, can produce considerable *sheet erosion,* in which a thin layer of surface material, usually topsoil, is removed by the flowing sheet of water. This gravity-driven movement of sediment is a process intermediate between mass wasting and stream erosion.

Overland sheetwash becomes concentrated in small channels, forming tiny streams called *rills.* Rills merge to form small streams, and small streams join to form larger streams. Most regions are drained by networks of coalescing streams.

DRAINAGE BASINS

Each stream, small or large, has a **drainage basin,** the total area drained by a stream and its tributaries (a **tributary** is a small stream flowing into a larger one). A drainage basin can be outlined on a map by drawing a line around the region drained by all the tributaries to a river (figure 16.4). The Mississippi River's drainage basin, for example, includes all the land area drained by the Mississippi River itself and by all its tributaries, including the Ohio and Missouri Rivers. This great drainage system includes more than one-third the land area of the contiguous 48 states.

A ridge or strip of high ground dividing one drainage basin from another is termed a **divide** (figure 16.4). The best known in the United States is the Continental Divide, a line separating streams that flow to the Pacific Ocean from those that flow to the Atlantic and the Gulf of Mexico. The Continental Divide,

FIGURE 16.4

The drainage basin of the Mississippi River is the land area drained by the river and all its tributaries, including the Ohio and Missouri Rivers; it covers more than 1 million square miles. Heavy rain in any part of the basin can cause flooding on the lower Mississippi River in the states of Mississippi and Louisiana. The Continental Divide separates rivers that flow into the Pacific from rivers that flow into the Atlantic and the Gulf of Mexico.

which extends from the Yukon Territory down into Mexico, crosses Montana, Idaho, Wyoming, Colorado, and New Mexico in the United States. Road signs indicating the crossing of the Continental Divide have been placed at numerous points where major highways intersect the divide.

DRAINAGE PATTERNS

The arrangement, in map view, of a river and its tributaries is a **drainage pattern.** A drainage pattern can, in many cases, reveal the nature and structure of the rocks underneath it.

Most tributaries join the main stream at an acute angle, forming a V (or Y) pointing downstream. If the pattern resembles branches of a tree or nerve dendrites, it is called **dendritic** (figures 16.4 and 16.5*A*). Dendritic drainage patterns develop on uniformly erodible rock or regolith and are the most common type of pattern. A **radial pattern,** in which streams diverge outward like spokes of a wheel, forms on high conical mountains, such as composite volcanoes and domes (figure 16.5*B*). A **rectangular pattern,** in which tributaries have frequent 90° bends and tend to join other streams at right angles, develops on regularly fractured rock (figure 16.5*C*). A network of fractures meeting at right angles forms pathways for streams because fractures are eroded more easily than unbroken rock. A **trellis pattern** consists of parallel main streams with short tributaries meeting them at right angles (figure 16.5*D*). A trellis pattern forms in a region where tiled layers of resistant rock such as sandstone alternate with nonresistant rock such as shale. Erosion of such a region results in a surface topography of parallel ridges and valleys.

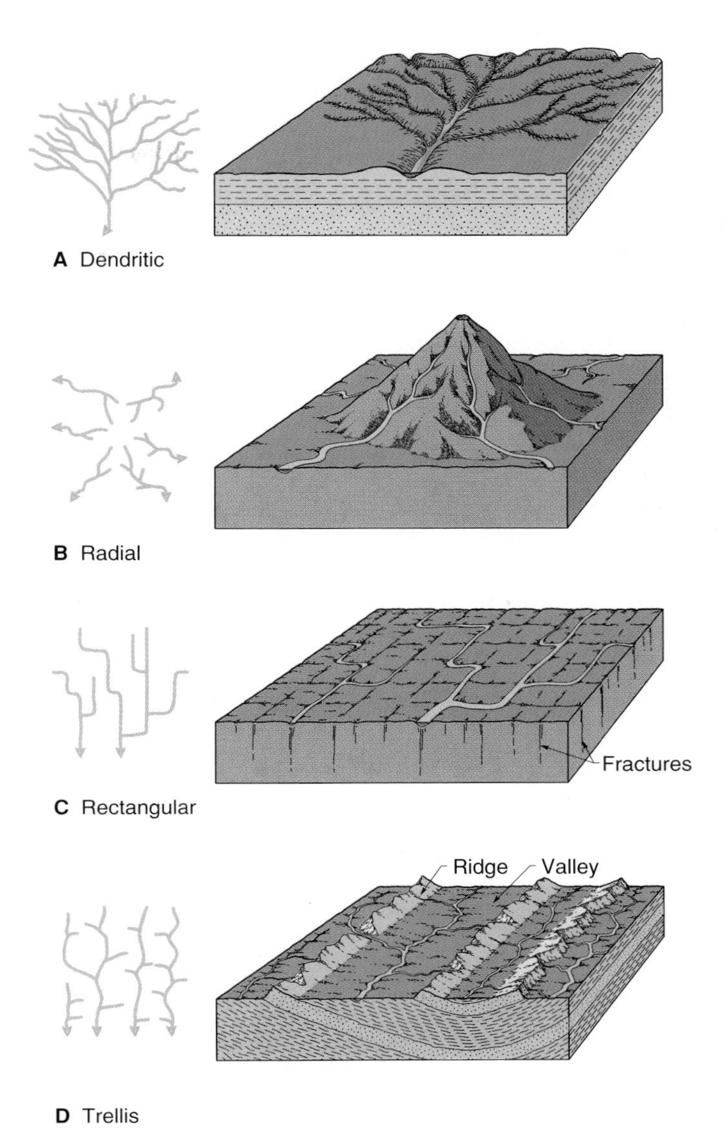

FIGURE 16.5
Drainage patterns can reveal something about the rocks underneath. (*A*) Dendritic pattern develops on uniformly erodible rock. (*B*) A radial pattern develops on a conical mountain or dome. (*C*) A rectangular pattern develops on regularly fractured rock. (*D*) A trellis pattern develops on alternating ridges and valleys caused by the erosion of resistant and nonresistant tilted rock layers.

FACTORS AFFECTING STREAM EROSION AND DEPOSITION

Stream erosion and deposition are controlled primarily by a river's *velocity* and, to a lesser extent, by its *discharge.* Velocity is largely controlled by the stream *gradient,* channel shape, and channel roughness.

Velocity

The distance water travels in a stream per unit time is called the **stream velocity.** A moderately fast river flows at about 5 kilometers per hour (3 miles per hour). Rivers flow much faster during flood, sometimes exceeding 25 kilometers per hour (15 miles per hour).

The cross-sectional views of a stream in figure 16.6 show that a stream reaches its maximum velocity near the middle of the channel. When a stream goes around a curve, the region of maximum velocity is displaced by centrifugal force toward the outside of the curve. Velocity is the key factor in a stream's ability to erode, transport, and deposit. High velocity (meaning greater energy) generally results in erosion and transportation; low velocity causes sediment deposition. Slight changes in velocity can cause great changes in the sediment load carried by the river.

Figure 16.7 shows the stream velocities at which sediments are eroded, transported, and deposited. For each grain size, these velocities are different. The upper curve represents the minimum velocity needed to erode sediment grains. This curve shows the velocity at which previously stationary grains are first picked up by moving water. The lower curve represents the velocity below which deposition occurs, when moving grains come to rest. Between the two curves the water is moving fast enough to transport grains that have already been eroded. Note

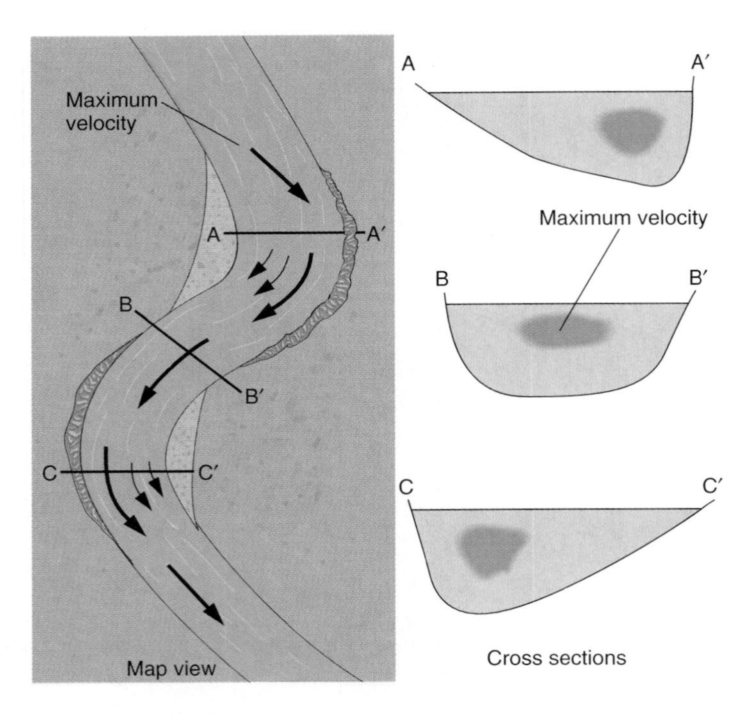

FIGURE 16.6

Regions of maximum velocity in a stream. Arrows on the map show how the maximum velocity shifts to the outside of curves. Sections show maximum velocity on outside of curves and in the center of the channel on a straight stretch of stream.

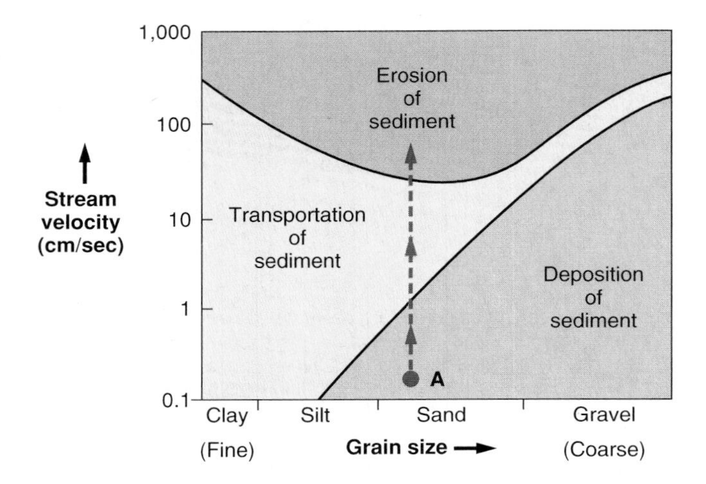

FIGURE 16.7

Logarithmic graph showing the stream velocities at which erosion and deposition of sediment occur. These velocities vary with the grain size of the sediment. See text for a discussion of point *A* and the dashed red line above it.

that it takes a higher stream velocity to erode grains (set them in motion) than to transport grains (keep them in motion).

Point A on figure 16.7 represents fine sand on the bed of a stream that is barely moving. The vertical red arrows represent a flood with gradually increasing stream velocity. No sediment moves until the velocity is high enough to intersect the *upper* curve and move into the area marked "erosion." As the flood recedes, the velocity drops below the upper curve and into the transportation area. Under these conditions the sand that was already eroded continues to be transported, but no new sand is eroded. As the velocity falls below the lower curve, all the sand is deposited again, coming to rest on the stream bed.

The right half of the diagram shows that coarser particles require progressively higher velocities for erosion and transportation, as you might expect (boulders are harder to move than sand grains). The erosion curve also rises toward the left of the diagram, however. This shows that fine-grained silt and clay are actually harder to erode than sand. The reason is that molecular forces tend to bind silt and clay into a smooth, cohesive mass that resists erosion. Once silt and clay are eroded, however, they are easily transported. As you can see from the lower curve, the silt and clay in a river's suspended load are not deposited until the river virtually stops flowing.

Gradient

One factor that controls a stream's velocity is the **stream gradient,** the downhill slope of the bed (or of the water surface, if the stream is very large). A stream gradient is usually measured in feet per mile in the United States, because these units are used on U.S. maps (elsewhere, gradients are expressed in meters per kilometer). A gradient of 5 feet per mile means that the river drops 5 feet vertically for every mile that it travels horizontally. Mountain streams may have gradients as steep as 50 to 200 feet per mile (10 to 40 m/km). The lower Mississippi River has a very gentle gradient, 0.5 foot/mile (0.1 m/km) or less.

A stream's gradient usually decreases downstream. Typically, the gradient is greatest in the headwater region and decreases toward the mouth of the stream (see figure 16.2). Local increases in the gradient of a stream are usually marked by rapids.

Channel Shape and Roughness

The *shape of the channel* also controls stream velocity. Flowing water drags against the stream banks and bed, and the resulting friction slows the water down. In figure 16.8, the streams in *A* and *B* have the same cross-sectional area, but stream *B* flows slower than *A* because the wide, shallow channel in *B* has more surface for the moving water to drag against.

A stream may change its channel width as it flows across different rock types. Hard, resistant rock is difficult to erode, so a stream may have a relatively narrow channel in such rock. As a result it flows rapidly (figure 16.9*A*). If the stream flows onto a softer rock that is easier to erode, the channel may widen, and the river will slow down because of the increased surface area dragging on the flowing water. Sediment may be deposited as the velocity decreases.

The width of a stream may be controlled by factors external to the stream. A landslide may carry debris onto a valley floor, partially blocking a stream's channel (figure 16.9*B*). The constriction causes the stream to speed up as it flows past the

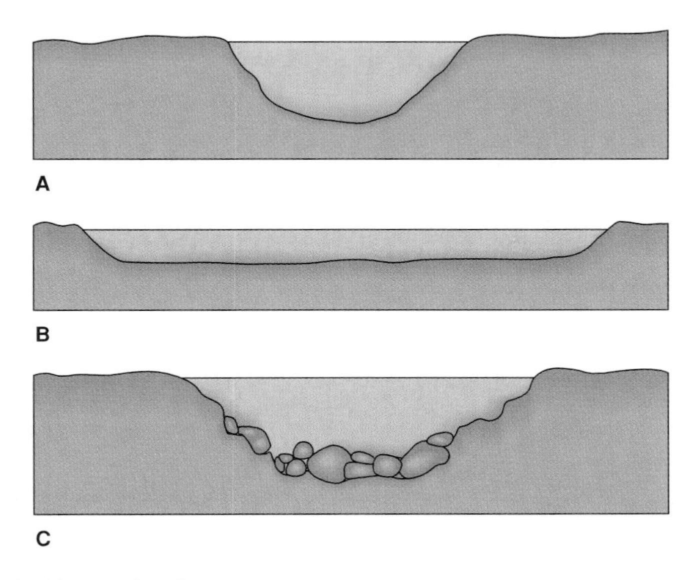

FIGURE 16.8
Channel shape and roughness influence stream velocity. (*A*) Semi-circular channel allows stream to flow rapidly. (*B*) Wide, shallow channel increases friction, slowing river down. (*C*) Rough, boulder-strewn channel slows river.

slide, and the increased velocity may quickly erode the landslide debris, carrying it away downstream. Human interference with a river can promote erosion and deposition. Construction of a culvert or bridge can partially block a channel, increasing the stream's velocity (figure 16.9*C*). If the bridge was poorly designed, it may increase velocity to the point where erosion may cause the bridge to collapse.

The *roughness of the channel* also controls velocity. A stream can flow rapidly over a smooth channel, but a rough, boulder-strewn channel floor creates more friction and slows the flow (see figure 16.8*C*). Coarse particles increase the roughness more than fine particles, and a rippled or wavy sand bottom is rougher than a smooth sand bottom.

Discharge

The **discharge** of a stream is the volume of water that flows past a given point in a unit of time. It is found by multiplying the cross-sectional area of a stream by its velocity (or width × depth × velocity). Discharge can be reported in cubic feet per second (cfs), which is standard in the United States, or in cubic meters per second (m^3/sec).

$$\text{Discharge (cfs)} = \text{average stream width (ft)} \times \text{average depth (ft)} \times \text{average velocity (ft/sec)}$$

A stream 100 feet wide and 15 feet deep flowing at 4 miles per hour (6 ft/sec) has a discharge of 9,000 cubic feet per second (cfs). In streams in humid climates, discharge increases downstream for two reasons: (1) water flows out of the ground into the river through the stream bed; and (2) small tributary streams flow into a larger stream along its length, adding water to the stream as it travels.

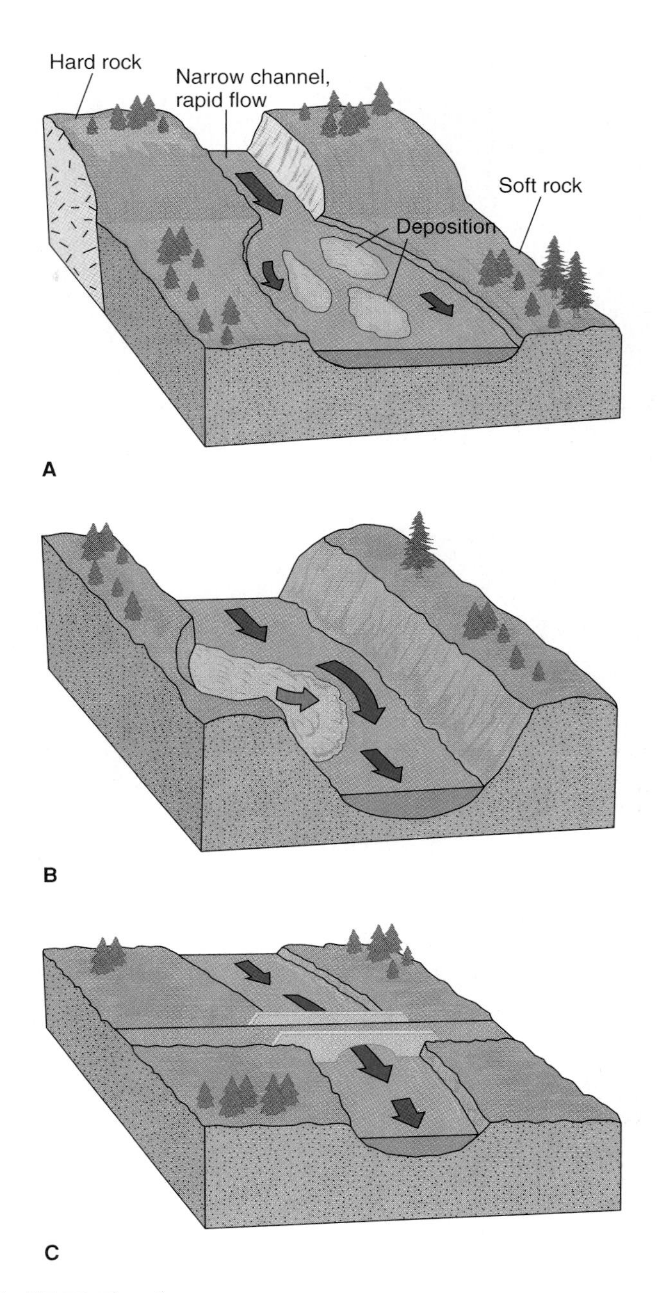

FIGURE 16.9
Channel width variations caused by rock type and obstructions. Length of arrow indicates velocity. (*A*) A channel may widen in soft rock. Deposition may result as stream velocity drops. (*B*) Landslide may narrow a channel, increasing stream velocity. Resulting erosion usually removes landslide debris. (*C*) Bridge piers (or other obstructions) will increase velocity and sometimes erosion next to the piers.

To handle the increased discharge, these streams increase in width and depth downstream. Some streams surprisingly increase slightly in velocity downstream, as a result of the increased discharge (the increase in discharge and channel size, and the typical downstream smoothness of the channel, override the effect of a lessening gradient).

FIGURE 16.10
These large boulders of granite in a mountain stream are moved only during floods. Note the rounding of the boulders and the scoured high-water mark of floods on the valley walls. Note people for scale.

During floods a stream's discharge and velocity increase, usually as a result of heavy rains over the stream's drainage basin. Flood discharge may be 50 to 100 times normal flow. Stream erosion and transportation generally increase enormously as a result of a flood's velocity and discharge. Swift mountain streams in flood can sometimes move boulders the size of automobiles (figure 16.10). Flooded areas may be intensely scoured, with river banks and adjacent lawns and fields washed away. As floodwaters recede, both velocity and discharge decrease, leading to the deposition of a blanket of sediment, usually mud, over the flooded area.

In a dry climate, a river's discharge can decrease in a downstream direction as river water evaporates into the air and soaks into the dry ground (or is used for irrigation). As the discharge decreases, the load of sediment is gradually deposited.

STREAM EROSION

A stream usually erodes the rock and sediment over which it flows. In fact, streams are one of the most effective sculptors of the land. Streams cut their own valleys, deepening and widening them over long periods of time and carrying away the sediment that mass wasting delivers to valley floors. The particles of rock and sediment that a stream picks up are carried along to be deposited farther downstream. Streams erode rock and sediment in three ways—*hydraulic action, solution,* and *abrasion.*

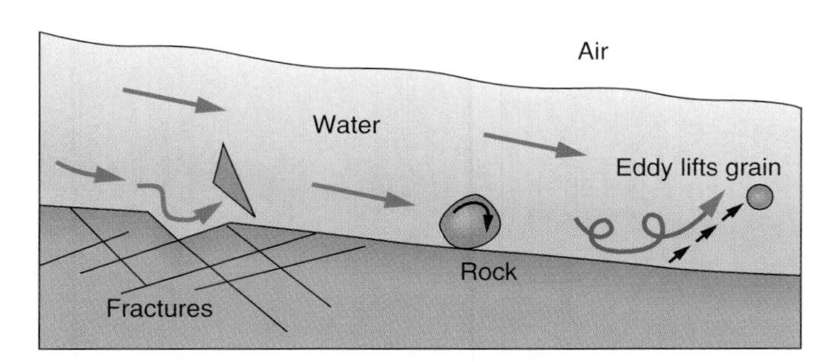

FIGURE 16.11
Hydraulic action can loosen, roll, and lift grains from the stream bed.

Hydraulic action refers to the ability of flowing water to pick up and move rock and sediment (figure 16.11). The force of running water swirling into a crevice in a rock can crack the rock and break loose a fragment to be carried away by the stream. Hydraulic force can also erode loose material from a stream bank on the outside of a curve. The pressure of flowing water can roll or slide grains over a stream bed, and a swirling eddy of water may exert enough force to lift a rock fragment above a stream bed. The great force of falling water makes hydraulic action particularly effective at the base of a waterfall (figure 16.12), where it may erode a deep plunge pool. You may be able to hear the results of hydraulic action by standing beside a swift mountain stream and listening to boulders and cobbles hitting one another as they tumble along downstream.

From what you have learned about weathering, you know that some rocks can be dissolved by water. **Solution,** although ordinarily slow, can be an effective process of weathering and erosion (weathering because it is a response to surface chemical conditions; erosion because it removes material). A stream flowing over limestone, for example, gradually dissolves the rock, deepening the stream channel. A stream flowing over other sedimentary rocks, such as sandstone, can dissolve calcite cement, loosening grains that can then be picked up by hydraulic action.

The erosive process that is usually most effective on a rocky stream bed is **abrasion,** the grinding away of the stream channel by the friction and impact of the sediment load. Sand and gravel tumbling along near the bottom of a stream wear away the stream bed much as moving sandpaper wears away wood. The abrasion of sediment on the stream bed is generally much more effective in wearing away the rock than hydraulic action alone. The more sediment a stream carries, the faster it is likely to wear away its bed.

FIGURE 16.12
Hydraulic action at the base of Niagara Falls in northwestern New York state along the U.S.–Canadian border.
Photo by Larry Davis

The coarsest sediment is the most effective in stream erosion. Sand and gravel strike the stream bed frequently and with great force, while the finer-grained silt and clay particles weigh so little that they are easily suspended throughout the stream and have little impact when they hit the channel.

Potholes are depressions that are eroded into the hard rock of a stream bed by the abrasive action of the sediment load (figure 16.13). During high water when a stream is full, the swirling water can cause sand and pebbles to scour out smooth, bowl-shaped depressions in hard rock. Potholes tend to form in spots where the rock is a little weaker than the surrounding rock. Although potholes are fairly uncommon, you can see them on the beds of some streams at low water level. Potholes may contain sand or an assortment of beautifully rounded pebbles.

FIGURE 16.13
Potholes scoured along bed of McDonald River in Glacier National Park, Montana.
Photo © Joe McDonald/Visuals Unlimited

STREAM TRANSPORTATION OF SEDIMENT

The sediment load transported by a stream can be subdivided into *bed load, suspended load,* and *dissolved load.* Most of a stream's load is carried in suspension and in solution.

The **bed load** is the large or heavy sediment particles that travel on the stream bed (figure 16.14). Sand and gravel, which form the usual bed load of streams, move by either *traction* or *saltation.*

Large, heavy particles of sediment, such as cobbles and boulders, may never lose contact with the stream bed as they move along in the flowing water. They roll or slide along the stream bottom, eroding the stream bed and each other by abrasion. Movement by rolling, sliding, or dragging is called **traction.**

Sand grains move by traction, but they also move downstream by **saltation,** a series of short leaps or bounces off the bottom (see figure 16.14). Saltation begins when sand grains are momentarily lifted off the bottom by *turbulent* water (eddying, swirling flow). The force of the turbulence temporarily counteracts the downward force of gravity, suspending the grains in water above the stream bed. The water soon slows down because the velocity of water in an eddy is not constant; then gravity overcomes the lift of the water, and the sand grain once again falls to the bed of the stream. While it is suspended, the grain moves downstream with the flowing water. After it lands on the bottom, it may be picked up again if turbulence increases, or it may be thrown up into the water by the impact of another falling sand grain. In this way sand grains saltate downstream in leaps and jumps, partly in contact with the bottom and partly suspended in the water.

The **suspended load** is sediment that is light enough to remain lifted indefinitely above the bottom by water turbulence (see figure 16.14). The muddy appearance of a stream during a flood or after a heavy rain is due to a large suspended load. Silt and clay usually are suspended throughout the water, while the

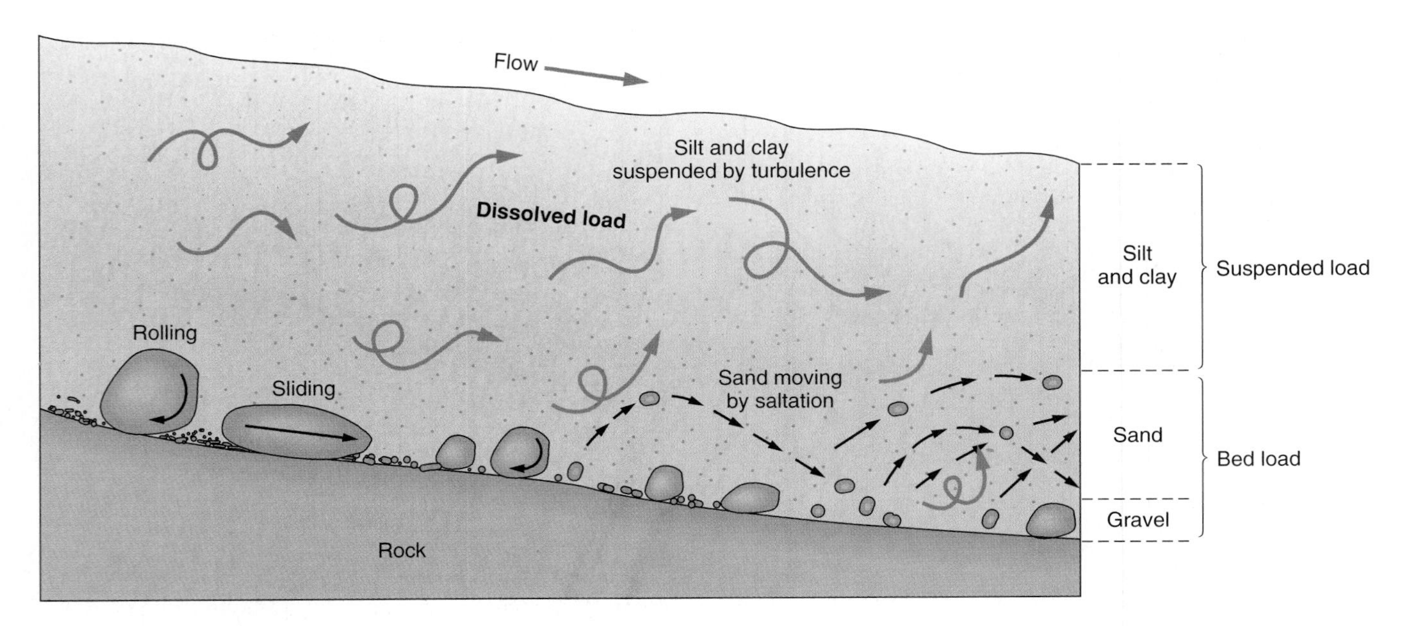

FIGURE 16.14

A stream's bed load consists of sand and gravel moving on or near the stream bed by traction and saltation. Finer silt and clay form the suspended load of the stream. The dissolved load of soluble ions is invisible.

coarser bed load moves on the stream bottom. Suspended load has less effect on erosion than the less visible bed load, which causes most of the abrasion of the stream bed. Vast quantities of sediment, however, are transported in suspension.

Soluble products of chemical weathering processes can make up a substantial **dissolved load** in a stream. Most streams contain numerous ions in solution, such as bicarbonate, calcium, potassium, sodium, chloride, and sulfate. The ions may precipitate out of water as evaporite minerals if the stream dries up, or they may eventually reach the ocean. Very clear water may in fact be carrying a large load of material in solution, for the dissolved load is invisible. Only if the water evaporates does the material become visible as crystals begin to form.

One estimate is that rivers in the United States carry about 250 million tons of solid load and 300 million tons of dissolved load each year. (It would take a freight train eight times as long as the distance from Boston to Los Angeles to carry 250 million tons.)

STREAM DEPOSITION

The sediments transported by a stream are often deposited temporarily along the stream's course (particularly the bed load sediments). Such sediments move sporadically downstream in repeated cycles of erosion and deposition, forming *bars* and *flood-plain deposits.* At or near the end of a stream, sediments may be deposited more permanently in a *delta* or an *alluvial fan.*

Bars

Stream deposits may take the form of a **bar,** a ridge of sediment, usually sand and gravel, deposited in the middle or along the banks of a stream (figure 16.15). Bars are formed by deposition when a stream's discharge or velocity decreases. During a flood, a river can move all sizes of sediment, from silt and clay up to huge boulders, because the greatly increased volume of water is moving very rapidly. As the flood begins to recede, the water level in the stream falls and the velocity drops. With the stream no longer able to carry all its sediment load, the larger boulders drop down on the stream bed, slowing the water locally even more. Finer gravel and sand are deposited between the boulders and downstream from them. In this way, deposition builds up a sand and gravel bar that may become exposed as the water level falls.

The next flood on the river may erode most of the sediment in this bar and move it farther downstream. But as the flood slows, it may deposit new gravel in approximately the same place, forming a new bar (figure 16.16). After each flood, river fishermen and boat operators must relearn the size and position

FIGURE 16.15

Gravel bars along the banks and in the middle of a stream. Note the sand deposited on the boulders at the right.

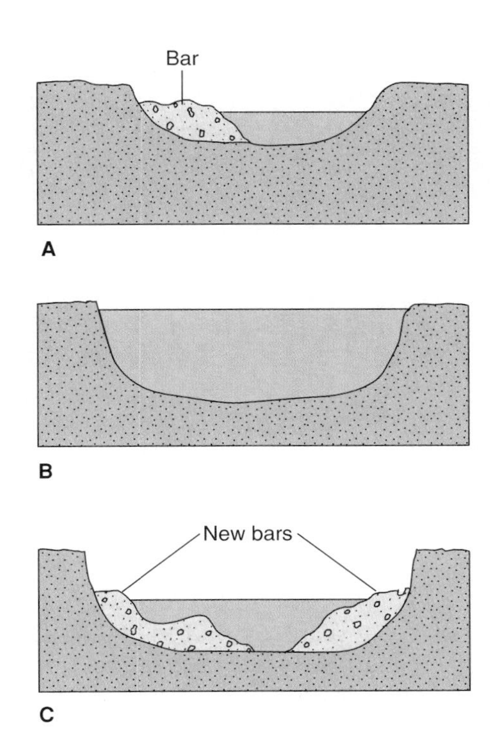

FIGURE 16.16

A flood can wash away bars in a stream, depositing new bars as the water recedes. (*A*) Normal water flow with sand and gravel bar. (*B*) Increased discharge and velocity during flood moves all sediment downstream. Channel deepens and widens if banks erode easily. (*C*) New bars are deposited as water level drops and stream slows down.

of the bars. Sometimes gold panners discover fresh gold in a mined-out river bar after a flood has shifted sediment downstream. A dramatic example of the shifting of sand bars occurred during the planned flood on the Colorado River downstream from the Glen Canyon Dam (box 16.1).

Placer Deposits

Placer deposits are found in streams where the running water has mechanically concentrated heavy sediment. The heavy sediment is concentrated in the stream where the velocity of the water is high enough to carry away lighter material but not the heavy sediment. Such places include river bars on the inside of meanders, plunge pools below waterfalls, and depressions on a stream bed (figure 16.17). Grains concentrated in this manner include gold dust and nuggets, native platinum, diamonds and other gemstones, and worn pebble or sand grains composed of the heavy oxides of titanium and tin.

Braided Streams

Deposition of a bar in the center of a stream (a *midchannel bar*) diverts the water toward the sides where it washes against the stream banks with greater force, eroding the banks and widening the stream (figure 16.18). A stream heavily loaded with sediment may deposit many bars in its channel, causing the stream to widen continually as more bars are deposited.

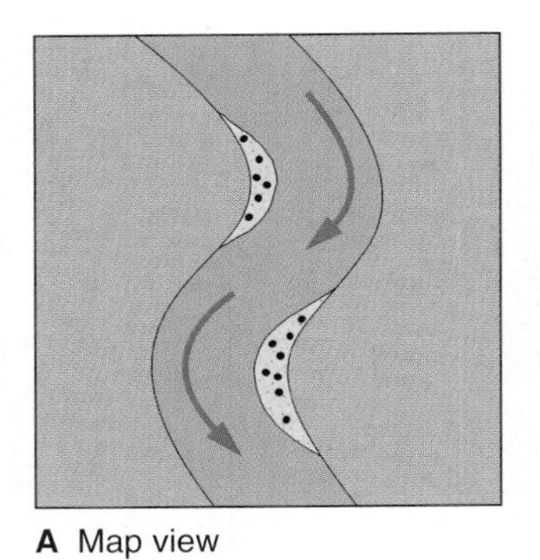

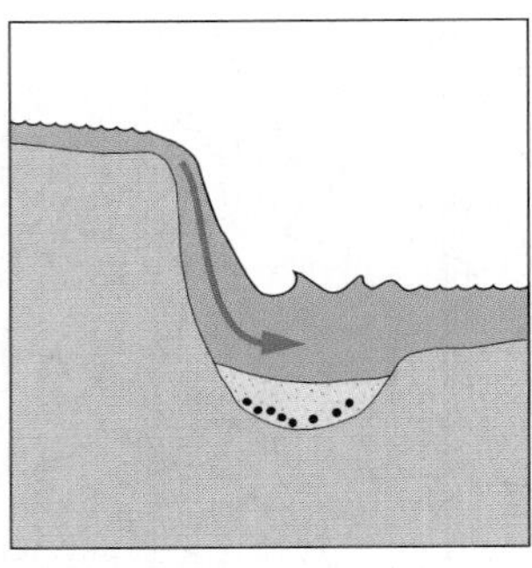

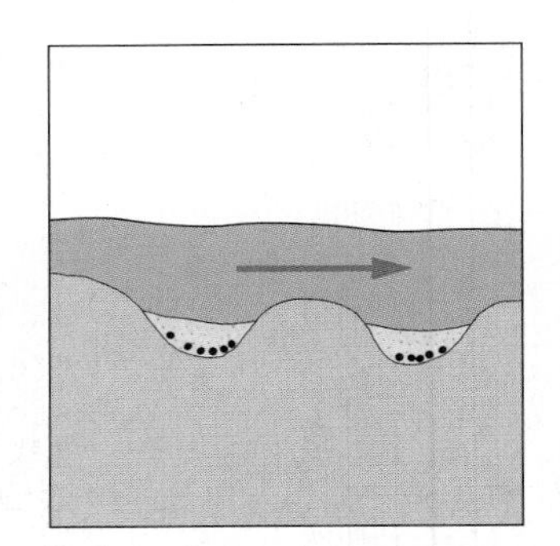

FIGURE 16.17

Types of placer deposits. (*A*) Stream bar. (*B*) Below waterfall. (*C*) Depressions on stream bed. Valuable grains shown in black.

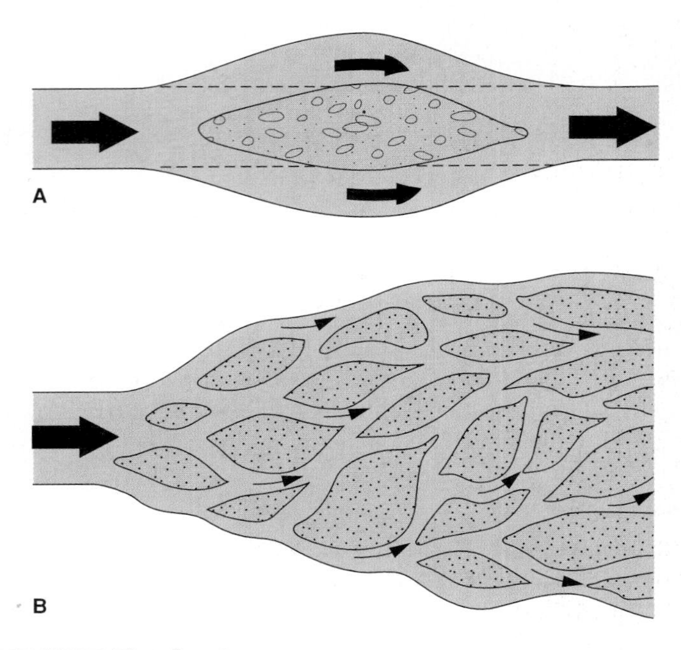

FIGURE 16.18

(*A*) A midchannel bar can divert a stream around it, widening the stream. (*B*) Braided stream occurs where there is an excess of sediment. Bars split main channel into many smaller channels, greatly widening the stream.

FIGURE 16.19

A braided stream carrying a heavy suspended load of sand and gravel from melting glaciers, Denali National Park, Alaska.

Photo © Michael Giannechini/Photo Researchers, Inc.

Such a stream typically goes through many stages of deposition, erosion, deposition, and erosion, especially if its discharge fluctuates. The stream may fill its main channel and become a **braided stream,** flowing in a network of interconnected rivulets around numerous bars (figures 16.18*B* and 16.19). A braided stream characteristically has a wide, shallow channel.

A stream tends to become braided when it is heavily loaded with sediment (particularly bed load) and has banks that are easily eroded. The braided pattern develops in deserts as a sediment-laden stream loses water through evaporation and through percolation into the ground. In meltwater streams flowing off glaciers, braided patterns tend to develop when the discharge from the melting glaciers is low relative to the great amount and ranges of size of sediment the stream has to carry.

ENVIRONMENTAL GEOLOGY 16.1

A Controlled Flood in the Grand Canyon: A Bold Experiment to Restore Sediment Movement in the Colorado River

On March 26, 1996, one of the largest experiments ever conducted on a river took place along the Colorado River below the Glen Canyon Dam (box figure 1). For six days the discharge from the Glen Canyon Dam was increased from 8,000 cfs to 45,000 cfs (a spike flow) to emulate the effects of a flood on the Colorado River (box figure 2). One of the main goals of this controlled flooding experiment was to determine whether the higher flows would result in bed scour and redeposition of sand bars and beaches along the sides of the channel. Another goal was to measure and observe how rocks move along the bed of the river bed with increasing discharge and velocity of floodwaters.

The Colorado River had not experienced its usual summertime floods since the Glen Canyon Dam was completed in 1963. The construction of the dam controlled peak discharges or flows on the Colorado River, which resulted in sand being deposited mainly along the bed or bottom of the river and erosion of beaches along the banks of the river. The Glen Canyon Dam cuts off a significant percentage of the sand supply to the lower Colorado River such that most of the downstream sand is supplied by two tributary streams, the Paria and Little Colorado Rivers. In August 1992 the Paria River flooded and deposited 330,000 tons of sand into the Colorado River, and in January 1993 a flood on the Little Colorado River deposited 10 million tons of sediment below its confluence with the Colorado River. The influx of sediment, coupled with the relatively low discharges from the dam (8–20,000 cfs), resulted in sand being concentrated along the bed of the Colorado River.

One of the main predictions of the experiment proved true. That is, sand caught in deep pools in the bottom of the main channel was scoured and carried in suspension downstream where it was redeposited along the river banks as beaches (box figure 3). Deposition of sand along the banks occurs due to back eddies that upwell and move upstream along the river banks and decrease the velocity of the downstream flow so that deposition can occur. Most of the scouring and deposition occurred in the first three days of the experiment; however, when flows were reduced back to 8,000 cfs beaches began to erode and redeposition occurred once again in the deep pools on the bottom of the river.

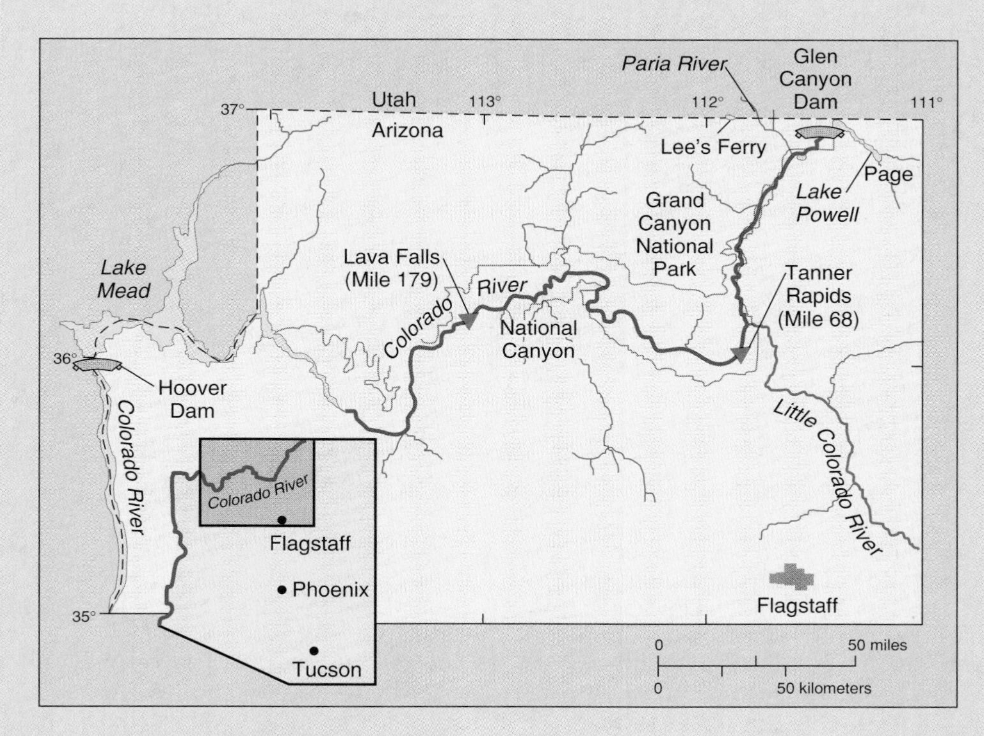

Box 16.1 — FIGURE 1

Location map of the Grand Canyon controlled-flood experiment.

U.S. Geological Survey

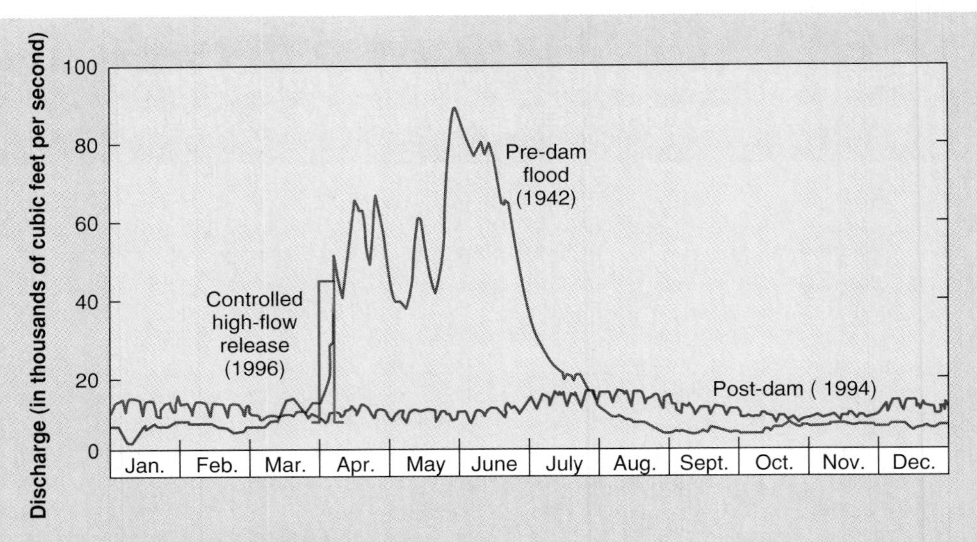

Box 16.1 — Figure 2

Graph of annual peak discharges before (black) and after Glen Canyon Dam (blue) and the 1996 controlled high-flow release.

U.S. Geological Survey

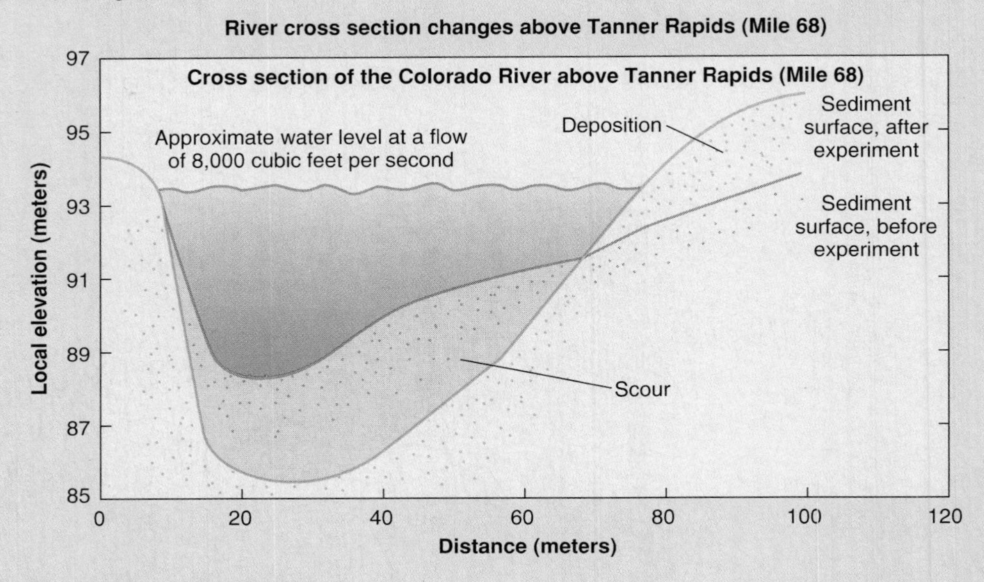

Box 16.1 — Figure 3

Cross section of the channel downstream of the confluence with the Little Colorado River. Increased flows have scoured the bottom sediment and redeposited it as a beach along the river bank.

U.S. Geological Survey

Downstream at Lava Falls, an experiment was set up to determine how and if large boulders deposited in the main channel from a debris flow would move with the increased discharge and velocity of the floodwater. Holes were drilled into 150 basalt boulders and radio tags were inserted (box figure 4) so their movement could be monitored and correlated with the increase in discharge and velocity of the river. Surface velocity measurements were taken by kayaking the river and charting the speed at which floating balls moved. The surface velocities were used to calculate the velocity of the water close to the river bed where the boulders were positioned. Dye was also injected into the river at peak flows to determine the average velocity of the water. The dye indicated that the velocity of the water increased downstream, particularly at the Lava Falls debris flow. This is because the floodwater accelerated as it flowed downstream, pushing the river water in front of it, which increased the downstream velocity. The first crest of water actually arrived behind Hoover Dam at Lake Mead a day ahead of the floodwater marked with a red dye.

The experiment was deemed a success and for the first time a flood was studied as it happened. The experimental flood, even though smaller than the size of a naturally occurring flood (box figure 2), showed that beaches could be restored below a dam and that boulders could be moved out of rapids much like that which occurs on an undammed river during a seasonal flood. It is proposed that other dammed rivers would benefit from periodic floods to help restore their natural conditions and thus minimize the adverse effects of damming a river.

Additional Information

Flooding in Grand Canyon. *Scientific American,* January 1997, pp. 82–89.

Grand Canyon Flood! *NOVA Video,* 1997.

Webb, R. H., J. C. Schmidt, G. R. Marzolf, and R. A. Valdez, eds. 1999. *The Controlled Flood in Grand Canyon.* Geophysical Monograph Series 110.

Related Web Resources

For an overview and details of the specific experiments conducted during the planned flood, visit the following websites:

http://water.usgs.gov/wid/FS_089-96/FS_089-96.html

www.pbs.org/kuat/grandcanyonflood

A

B

Box 16.1 — FIGURE 4

(*A*) Hole being drilled into a basalt boulder and (*B*) radio tag installed to track the movement of boulders as the discharge and velocity of the Colorado River increases at the Lava Falls debris flow locality.

Photos courtesy of KUAT-TV. University of Arizona, photo by Dan Duncan

Meandering Streams and Point Bars

Rivers that carry fine-grained silt and clay in suspension tend to be narrow and deep and to develop pronounced, sinuous curves called **meanders** (figure 16.20). In a long river, sediment tends to become finer downstream, so meandering is common in the lower reaches of a river.

You have seen in figure 16.16 that a river's velocity is higher on the outside of a curve than on the inside. This high velocity can erode the river bank on the outside of a curve, often rapidly (figure 16.21).

The low velocity on the inside of a curve promotes sediment deposition. The sand bars in figure 16.22 have been deposited on the inside of curves because of the lower velocity there. Such a bar is called a **point bar** and usually consists of a series of arcuate ridges of sand or gravel.

The simultaneous erosion on the outside of a curve and deposition on the inside can deepen a gentle curve into a hairpin-like meander (see figure 16.22). Meanders are rarely fixed in position. Continued erosion and deposition cause them to migrate back and forth across a flat valley floor, as well as downstream, leaving scars and arcuate point bars to mark their former positions.

At times, particularly during floods, a river may form a **meander cutoff,** a new, shorter channel across the narrow neck of a meander (figure 16.23). The old meander may be abandoned as sediment separates it from the new, shorter channel. The cutoff meander becomes a crescent-shaped **oxbow lake** (figure 16.24). With time, an oxbow lake may fill with sediment and vegetation.

FIGURE 16.20

Meanders in a stream. These sinuous curves develop because a stream's velocity is highest on the outside of curves, promoting erosion there.

Photo © Glenn M. Oliver/Visuals Unlimited

Flood Plains

A **flood plain** is a broad strip of land built up by sedimentation on either side of a stream channel. During floods, flood plains may be covered with water carrying suspended silt and clay (figure 16.25*A*). When the floodwaters recede, these fine-grained sediments are left behind as a horizontal deposit on the flood plain (figure 16.25*B*).

Some flood plains are constructed almost entirely of horizontal layers of fine-grained sediment, interrupted here and

A

B

FIGURE 16.21

River erosion on the outside of a curve, Newaukum River, Washington. Pictures were taken in (*A*) January and (*B*) March 1965.

Photos by P. A. Glancy, U.S. Geological Survey

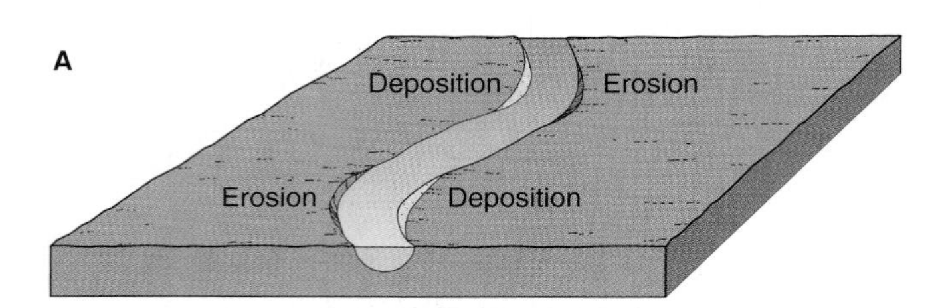

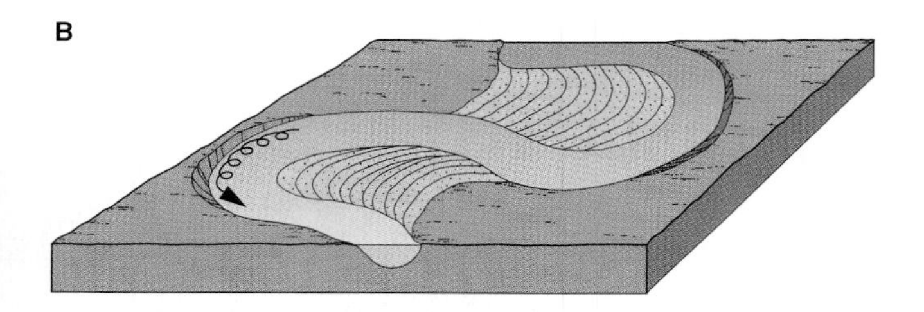

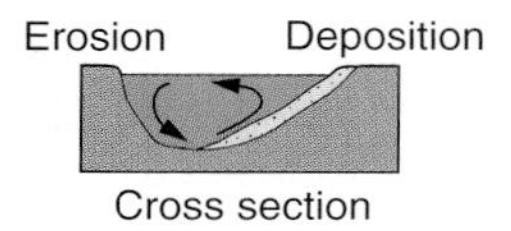

FIGURE 16.22

(*A*) Development of river meanders and point bars by erosion and deposition on curves. (*B*) A point bar of gravel has been deposited on the inside of this curve.

Photo *B* by Frank Hanna

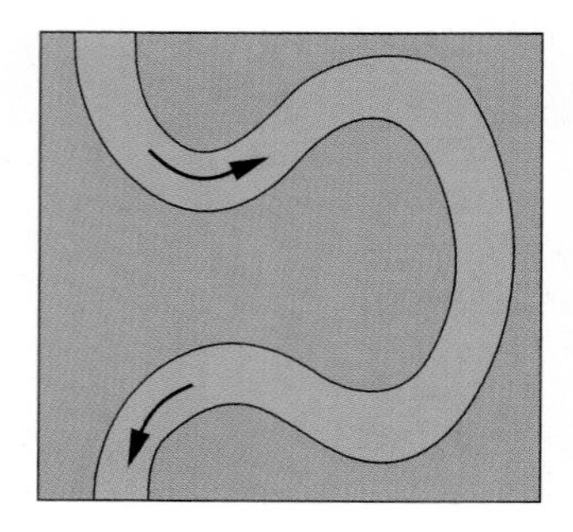

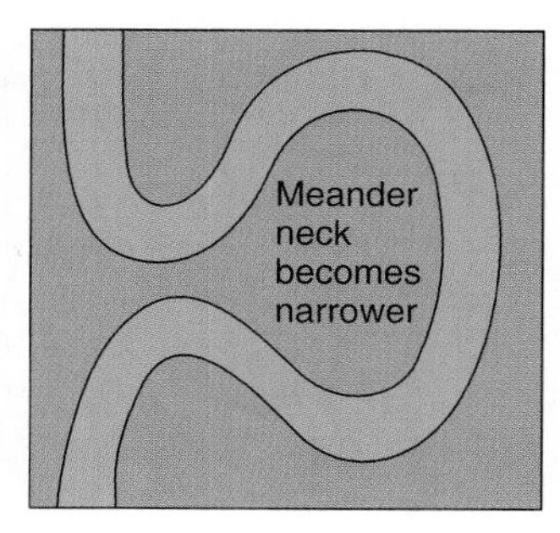

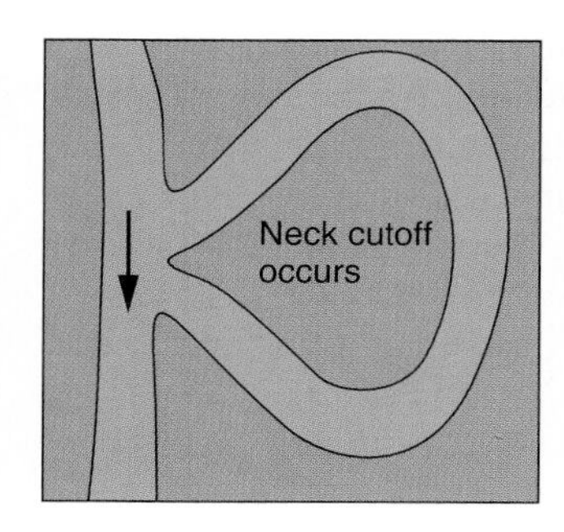

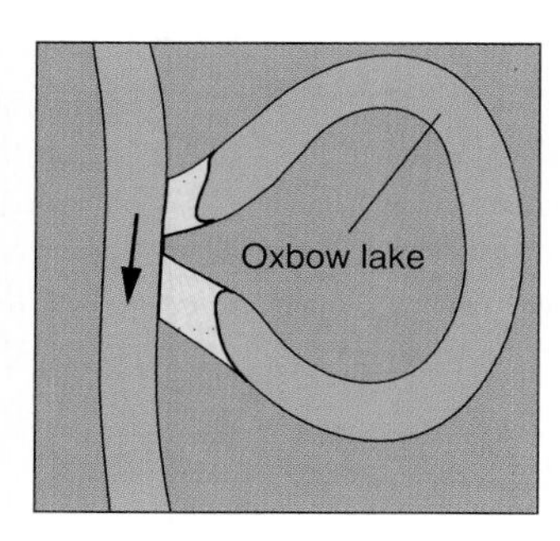

FIGURE 16.23 

Creation of an oxbow lake by a meander neck cutoff. Old channel is separated from river by sediment deposition.

FIGURE 16.24

An oxbow lake marks the old position of a river meander, Sweetwater River, Wyoming.

Photo by W. R. Hansen, U.S. Geological Survey

there by coarse-grained channel deposits (figure 16.26*A*). Other flood plains are dominated by meanders shifting back and forth over the valley floor and leaving sandy point bar deposits on the inside of curves. Such a river will deposit a characteristic fining-upward sequence of sediments: coarse channel deposits are gradually covered by medium-grained point bar deposits, which in turn are overlain by fine-grained flood deposits (figure 16.26*B*).

As a flooding river spreads over a flood plain, it slows down. The velocity of the water is abruptly decreased by friction as the water leaves the deep channel and moves in a thin sheet over the flat valley floor. The sudden decrease in velocity of the water causes the river to deposit most of its sediment near the main channel, with progressively less sediment deposited away from the channel (figure 16.27). A series of floods may build up **natural levees**—low ridges of flood-deposited sediment that form on either side of a stream channel and thin away from the channel. The sediment near the river is coarsest, often sand and silt, while the finer clay is carried farther from the river into the flat, lowland area (the back swamp).

Deltas

Most streams ultimately flow into the sea or into large lakes. A stream flowing into quiet water usually builds a **delta,** a body of sediment deposited at the mouth of a river when the river's velocity decreases (figure 16.28).

The surface of most deltas is marked by **distributaries**—small, shifting channels that carry water away from the main river channel and distribute it over the surface of the delta (figure 16.29). Sediment deposited at the end of a distributary tends to block the water flow, causing distributaries and their sites of sediment deposition to shift periodically.

The shape of a marine delta in map view depends on the balance between sediment supply from the stream and the erosive power of waves and tides (figure 16.30). Some deltas, like that of the Nile river, are broadly triangular; this delta's resemblance to the Greek letter *delta* (Δ) is the origin of the name.

A

B

FIGURE 16.25

River flood plains. (*A*) Flooded flood plain of the Animas River, Colorado. (*B*) Sediment deposited on a flood plain by the Trinity River in California during a 1964 flood.

Photo *A* by D. A. Rahm, courtesy Rahm Memorial Collection, Western Washington University; photo *B* by A. O. Waananen, U.S. Geological Survey

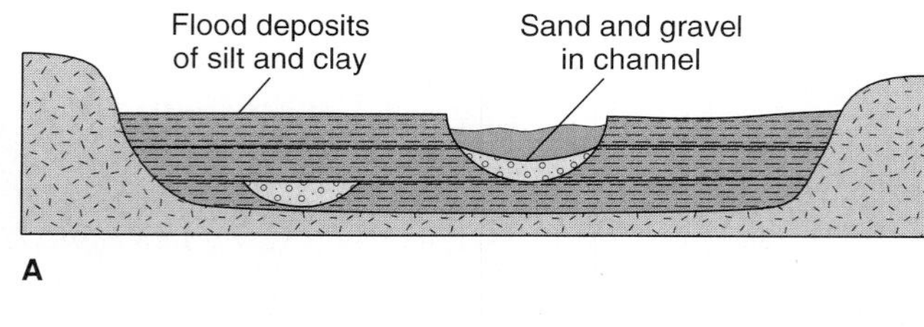

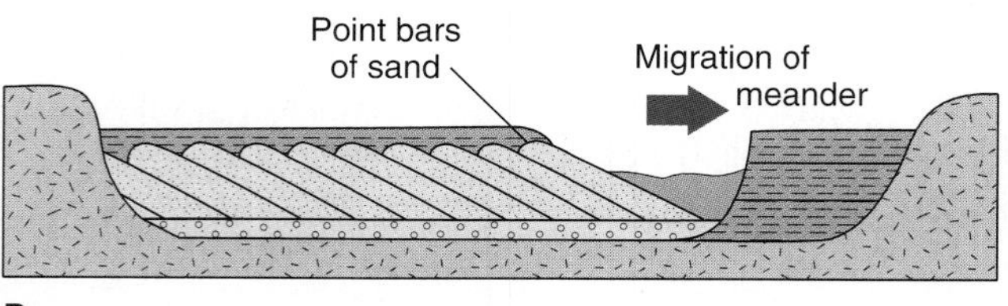

FIGURE 16.26

Flood plains. (*A*) Horizontal layers of fine-grained flood deposits with lenses of coarse-grained channel deposits. (*B*) A fining-upward sequence deposited by a migrating meander. Channel gravel is overlain by sandy point bars, which are overlain by fine-grained flood deposits.

FIGURE 16.28

Delta of Tongariro River, New Zealand. Note the sediment-laden water, and how the land is being built outward by river sedimentation. A river typically divides into several channels (distributaries) on a delta.
Photo © G. R. 'Dick' Roberts

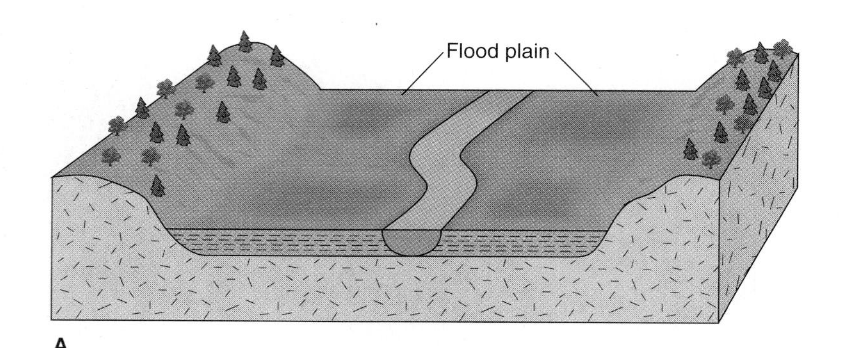

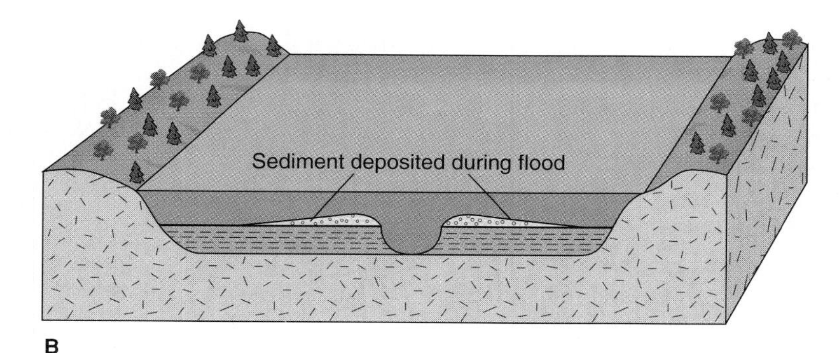

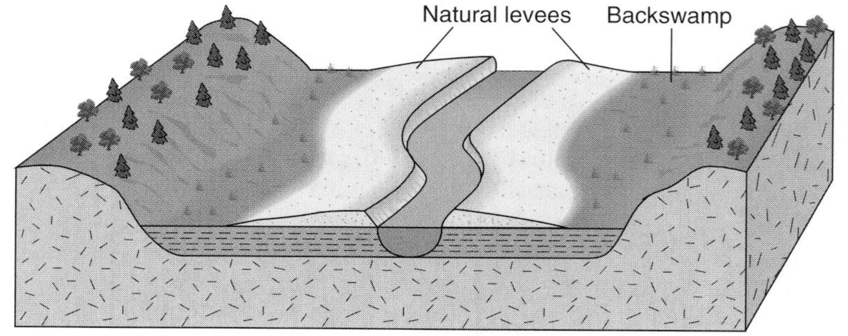

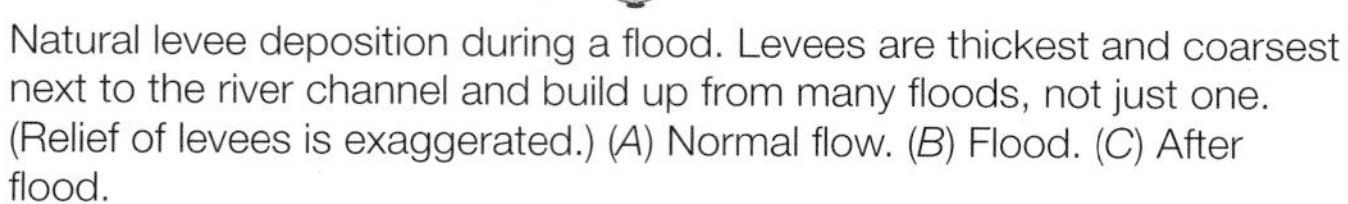

FIGURE 16.27

Natural levee deposition during a flood. Levees are thickest and coarsest next to the river channel and build up from many floods, not just one. (Relief of levees is exaggerated.) (*A*) Normal flow. (*B*) Flood. (*C*) After flood.

an alluvial fan. The loss of velocity is due to the widening or branching of the channel as it leaves the narrow mountain canyon. The gradual loss of water as it infiltrates into the fan also promotes sediment deposition. On large fans, deposits are graded in size within the fan, with the coarsest sediment dropped nearest the mountains and the finer material deposited progressively farther away. Small fans do not usually show such grading.

FLOODING

Many of the world's cities, such as Pittsburgh, St. Louis, and Portland, Oregon, are built beside rivers and therefore can be threatened by floods. Rivers are important transportation routes for ships and barges, and flat flood plains have excellent agricultural soil and offer attractive building sites for houses and industry.

Flooding does not occur every year on every river, but flooding is a natural process on all rivers and must be prepared for by river cities and towns. Heavy rains and the rapid melting of snow in the springtime are the usual causes of floods. The rate and volume of rainfall, and the geographic path of rainstorms, often determine whether flooding will occur.

Floods are described by *recurrence interval,* the average time between floods of a given size. A "100-year flood" is one that can occur, on the *average,* every 100 years (box 16.2). A 100-year flood has a 1-in-100, or one percent, chance of occurring in any given year. It is perfectly possible to have two 100-year floods in successive years—or even in the same year. If a 100-year flood occurs this year on the river you live beside, you should not assume that there will be a 99-year period of safety before the next one.

Flood erosion is caused by the high velocity and large volume of water in a flood. Although relatively harmless on an uninhabited flood plain, flood erosion can be devastating to a city. As a river undercuts its banks, particularly on the outside of curves where water velocity is high, buildings, piers, and bridges may fall into the river. As sections of flood plain are washed away, highways and railroads are cut.

High water covers streets and agricultural fields and invades buildings, shorting out electrical lines and backing up sewers. Water-supply systems may fail or be contaminated. Water in your living room will be drawn upward in your walls by capillary action in wall plasterboard and insulation, creating a soggy mess that has to be torn out and replaced. High water on flat flood plains often drains away very slowly; street travel may be by boat for weeks. If floodwaters are deep enough, houses may float away (figure 16.32).

Flood deposits are usually silt and clay. A new layer of wet mud on a flood plain in an agricultural region can be beneficial in that it renews the fields with topsoil from upstream as used to be the case with the Nile River until the Aswan dam was built. The same mud in a city will destroy lawns, furniture, and machinery. Cleanup is slow; imagine shoveling four inches of worm-filled mud that smells like sewage out of your house.

FIGURE 16.32
January 1997 flood waters were so deep in the Arboga, California area that a house floated off its foundation.
Photo courtesy of Robert A. Eplett, Governor's Office of Emergency Services

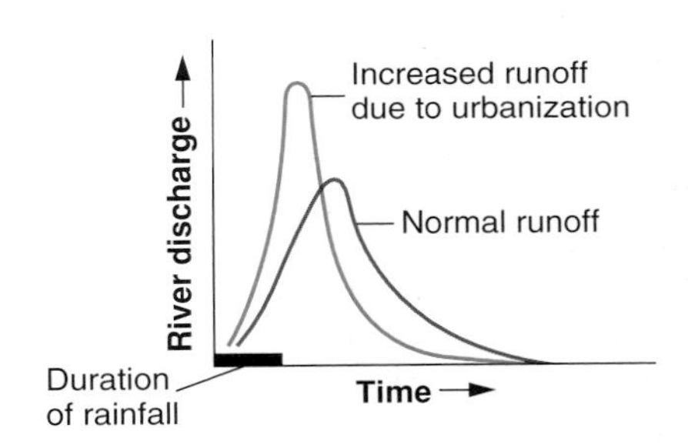

FIGURE 16.33
The presence of a city can increase the chance of floods. The blue curve shows the normal increase in a river's discharge following a rainstorm (black bar). The red curve shows the great increase in runoff rate and amount caused by pavement and storm sewers in a city.

Urban Flooding

Urbanization contributes to severe flooding. Paved areas and storm sewers increase the amount and rate of surface runoff of water, making river levels higher during storms (figure 16.33). Such rapid increases in runoff or discharge to a river is called a "flashy" discharge. Storm sewers are usually designed for a 100-year storm; however, large storms that drop a lot of rain in a short period of time (cloudburst) may overwhelm sewer systems and cause localized flooding. Rising river levels may block storm sewer outlets and also add to localized flooding problems.

Bridges, docks, and buildings built on flood plains can also constrict the flow of floodwaters, increasing the water height and velocity and promoting erosion.

16.2

IN GREATER DEPTH

ESTIMATING THE SIZE AND FREQUENCY OF FLOODS

Because people have encroached on the flood plains of many rivers, flooding is one of the most universally experienced geologic hazards. To minimize flood damage and loss of life, it is useful to know the potential size of large floods and how often they might occur. This is often a difficult task because of the lack of long-term records for most rivers. The U.S. Geological Survey monitors the stage (water elevation) and discharge of rivers and streams throughout the United States in order to collect data that can be used to attempt to predict the size and frequency of flooding and to make estimates of water supply.

Hydrologists designate floods based on their *recurrence interval,* or *return period.* For example, a 100-year flood is the largest flood expected to occur within a period of 100 years. This does not mean that a 100-year flood occurs once every century, but that there is a 1-in-100 chance, or a one percent probability, each year that a flood of this size will occur. Usually flood control systems are built to accommodate a 100-year flood because that is the minimum margin of safety required by the federal government if an individual wants to obtain flood insurance subsidized by the Federal Emergency Management Agency (FEMA).

To calculate the recurrence interval of flooding for a river, the annual peak discharges (largest discharge of the year) are collected and ranked according to size (box figure 1 and table 1). The largest annual peak discharge is assigned a rank (m) of one, the second a two, and so on until all the discharges are assigned a rank number. The *recurrence interval* (**R**) of each annual peak discharge is then calculated by adding one to the *number of years of record* (**n**) and dividing by its *rank* (**m**).

$$R = \frac{n + 1}{m}$$

Box 16.2 Table 1 Annual Peak Discharges and Recurrence Intervals in Rank Order for the Cosumnes River at Michigan Bar, California.

Year	Peak Discharge (cfs)	Magnitude Rank (m)	Recurrence Interval
1997	93,000	1	91.0
1907	71,000	2	45.5
1986	45,100	3	30.33
1956	42,000	4	22.75
1963	39,400	5	18.42
1909	28,400	10	9.20
1943	22,900	20	4.60
1970	16,800	30	3.07
1960	11,200	40	2.30
1971	8,590	50	1.84
1991	6,670	60	1.53

Source: Preliminary data from Richard Hunrichs, hydrologist, U.S. Geological Survey and U.S. Geological Survey Water-Data Report, CA-97-3

For example, the Cosumnes River in California has 90 years of record (**n = 90**), and in 1907 the second largest peak discharge (**m = 2**) of 71,000 cfs occurred. The recurrence interval (**R**), or expected frequency of occurrence, for a discharge this large is 45.5 years:

$$R = \frac{90 + 1}{2} = 45.5$$

That is, there is a 1-in-45.5, or 2 percent, chance each year of a peak discharge of 71,000 cfs or greater occurring on the Cosumnes River.

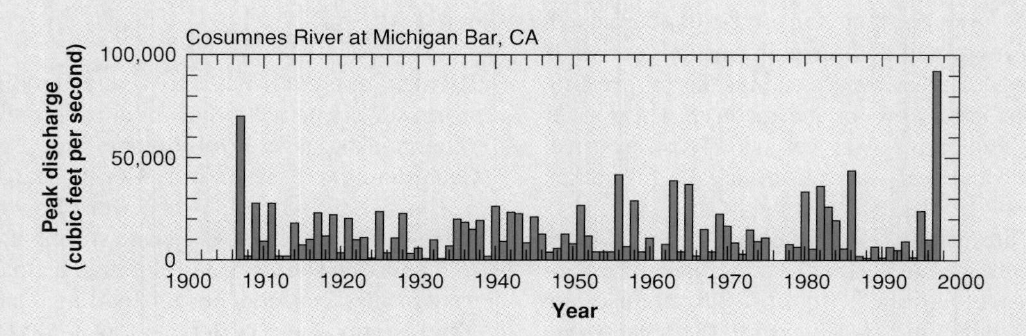

BOX 16.2 — FIGURE 1

Annual peak discharge for the Cosumnes River.

After U.S. Geological Survey Water-Data Report, CA-97-3

Box 16.2 — FIGURE 2

Levee break along the Cosumnes River.

Courtesy of Robert A. Eplett, Governor's Office of Emergency Services

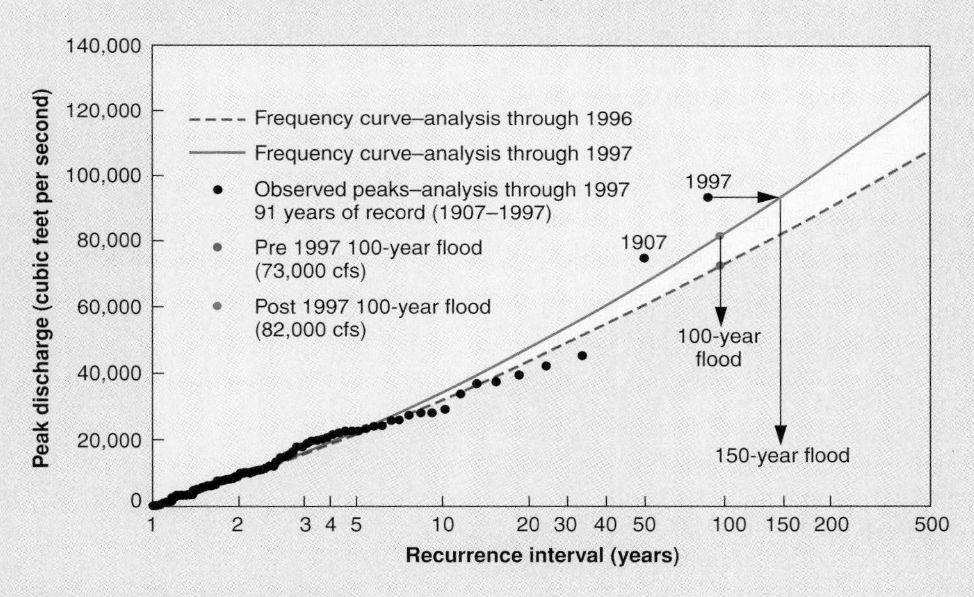

Box 16.2 — FIGURE 3

Flood-frequency curves for the Cosumnes River.

Preliminary data from Richard Hunrichs, hydrologist, U.S. Geological Survey and U.S. Geological Survey Water-Data Report, CA-97-3

The flood of record (largest recorded discharge) occurred on January 2, 1997, when heavy, unseasonably warm rains rapidly melted snow in the Sierra Nevada and caused flooding in much of northern California. A peak discharge of 93,000 cfs in the Cosumnes River resulted in levee breaks and widespread flooding of homes and agricultural areas (box figure 2). The recurrence interval for the 1997 flood (93,000 cfs) is 91 years:

$$\mathbf{R = \frac{90 + 1}{1} = 91\ years}$$

A *flood-frequency curve* can be useful in providing an estimate of the discharge and the frequency of floods. The flood-frequency curve is generated by plotting the annual peak discharges against the calculated recurrence intervals (box figure 3). Because most of the data points defining the curve plot in the lower range of discharge and recurrence interval, there is some uncertainty in projecting larger flood events. Two flood-frequency curves are drawn in box figure 3; the red line represents the best-fit curve for all of the data, whereas the dashed blue line excludes the 1997 flood of record. Notice that the curve has a steeper slope when the 1997 data is included, and that the size of the 100-year flood has increased from 73,000 cfs to 82,000 cfs based on the one additional year of record. Because large floods do not occur as often as small floods, the rare large flood can have a dramatic effect on the shape of the flood-frequency curve and the estimate of a 100-year event. This is particularly true for a river like the Cosumnes that has had only two large events, one in 1907 and the other 90 years later in 1997.

The 100-year flood plain is based on the estimate of the discharge of the 100-year flood and on careful mapping of the flood plain. Changes in the estimated size of the 100-year flood could result in property that no longer has 100-year flood protection. In this case, property owners may be prevented from getting flood insurance or money to rebuild from FEMA unless new flood control structures are built to provide additional protection or houses are raised or even relocated out of the flood plain.

Related Web Resource

To find data sets to calculate the recurrence interval for rivers throughout the United States, access the U.S. Geological Survey Water Data Retrieval Website:

http://water.usgs.gov/usa/nwis/

Further Readings

Water Resources Data for California, Water Year 1997. U.S. Geological Survey Water-Data Report CA-97-3.

Riggs, H. C. 1968. *Frequency Curves.* Techniques of Water Resources Investigations of the U.S. Geological Survey. Book 4, Hydrologic Analysis and Interpretation.

Flash Floods

Some floods occur rapidly and die out just as quickly. *Flash floods* are local, sudden floods of large volume and short duration, often triggered by heavy thunderstorms. A startling example occurred in 1976 in north-central Colorado along the Big Thompson River (figure 16.34). Strong winds from the east pushed moist air up the front of the Colorado Rockies, causing thunderstorms in the steep mountains. The storms were unusually stationary, allowing as much as 30 cm of rain to fall in 2 days. Some areas received 19 cm in just over an hour. Little of this torrential rainfall could soak into the ground. The volume of water in the Big Thompson River swelled to four times the previously recorded maximum, and the river's velocity rose to an impressive 25 kilometers per hour for a few hours on the night of July 31. By the next morning the flood was over, and the appalling toll became apparent—139 people dead, 5 missing, and more than $35 million in damages (figure 16.35*A*).

On July 29, 1997, just 2 days before the 21st anniversary of the Big Thompson River flood, Fort Collins, Colorado, was struck by a flash flood when 20 cm of rain fell in only 5 hours. A 4-meter wall of water rushed down Spring Creek, a tributary to the Cache la Poudre River, nearly devastating two trailer parks. Five people lost their lives and 40 were injured when a 5-meter high railroad embankment that had temporarily dammed Spring Creek broke and sent the wall of water into the trailer park (figure 16.35*B*). Unlike the Big Thompson Canyon flood, the heavy rains fell over the city of Fort Collins rather than upstream in the steep mountain canyons.

Flood area
Denver
COLORADO

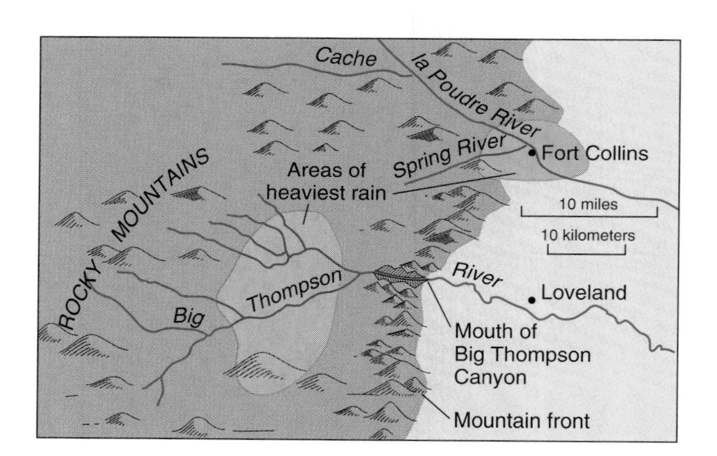

FIGURE 16.34

Location map of the 1976 flash flood on the Big Thompson River in Colorado and the 1997 flash flood in Fort Collins.

Controlling Floods

Flood-control structures can partially reduce the dangers of floodwaters and sedimentation to river cities (figure 16.36). Upstream dams can trap water and release it slowly after the storm. (A dam also catches sediment, which eventually fills its reservoir and ends its life as a flood-control structure.) Artificial levees are embankments built along the banks of a river channel to contain floodwaters within the channel. Protective walls of stone (riprap) or concrete are often constructed along river banks, particularly on the outside of curves, to slow erosion. Floodwalls, walls of concrete, may be used to protect cities from flooding; however, these flood-control structures may constrict the channel and cause water to flow faster with more erosive power downstream. Bypasses are also used along the Mississippi and other rivers to reduce the discharge in the main channel by diverting water into designated basins in the flood plain. The bypasses serve to give part of the natural flood plain back to the river.

A

B

FIGURE 16.35

(*A*) A cabin sits crushed against a bridge following the Big Thompson Canyon flash flood of 1976. (*B*) Devastation along Spring Creek after the 1997 Fort Collins flash flood.

Photos by W. R. Hansen, U.S. Geological Survey

Dams and levees are designed to control certain specified floods. If the flood-control structures on your river were designed for 75-year floods, then a much larger 100-year flood will likely overtop these structures and may destroy your home. The disastrous floods along the Missouri and Mississippi Rivers and their tributaries north of Cairo, Illinois, in 1993 resulted from many such failures in flood control.

Wise land-use planning and zoning for flood plains should go hand in hand with flood control. Wherever possible, buildings should be kept out of areas that might someday be flooded by 100-year floods.

The Great Flood of 1993

In the spring of 1993, heavy rains soaked the ground in the upper Midwestern United States. In June and July a stationary weather pattern created a shifting band of thunderstorms that dumped as much as 4 inches of rain in a single day on localities such as Bismarck, North Dakota; Cedar Rapids, Iowa; and Manhattan, Kansas (figure 16.37). These torrential rains falling on already-saturated ground created the worst flood disaster in U.S. history, as swollen rivers flooded 6.6 million acres in 9 states, killing 38 and causing $12 billion in damage to houses and crops.

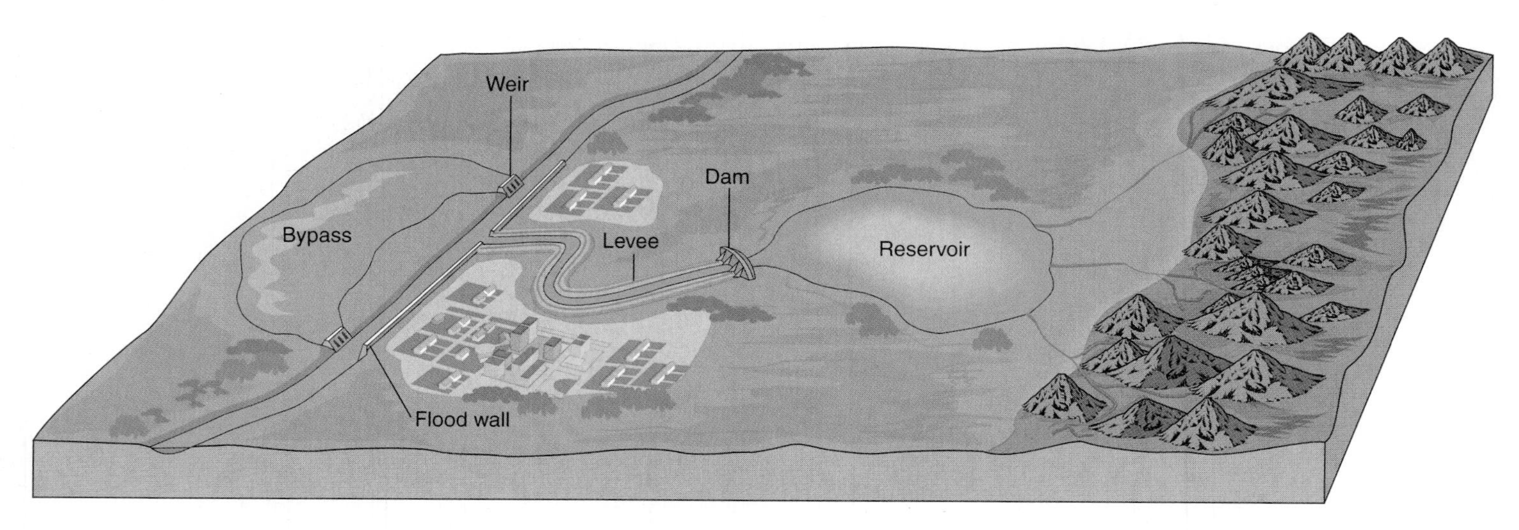

FIGURE 16.36
Examples of flood control structures.

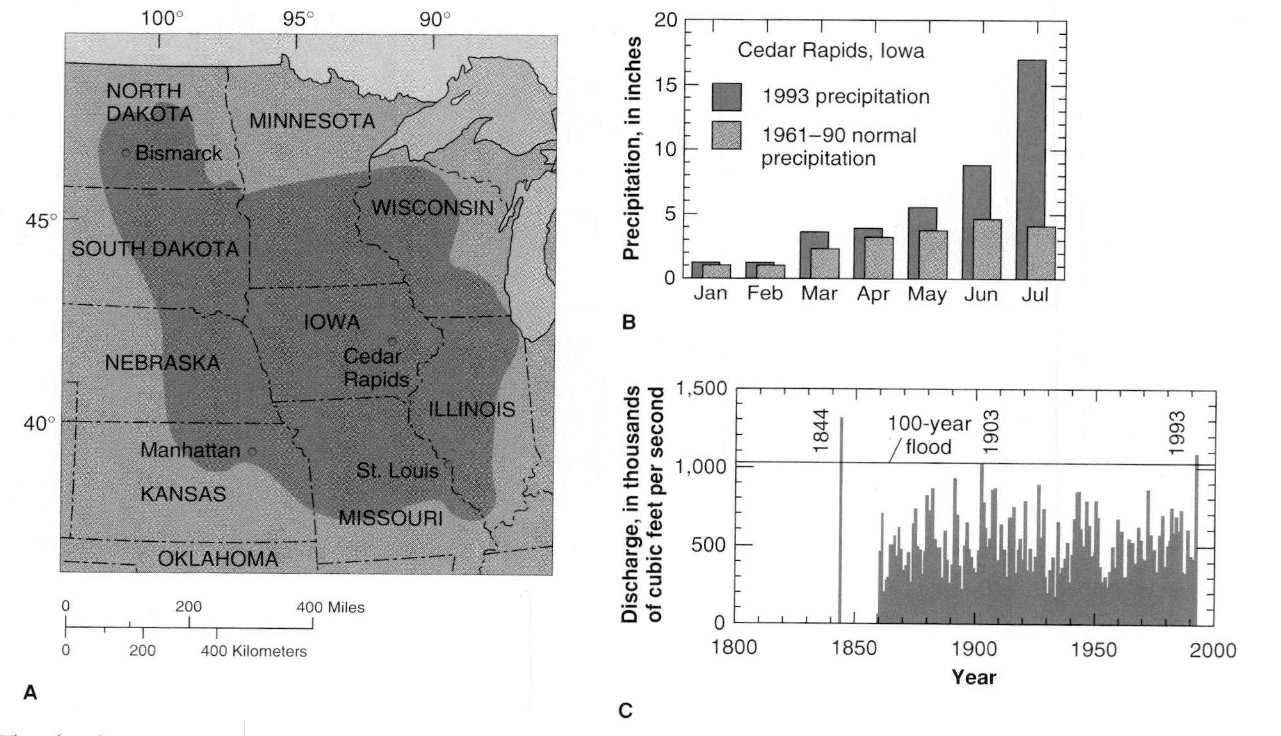

FIGURE 16.37
The Great Flood of 1993. (*A*) Area of flood. (*B*) 1993 rainfall at Cedar Rapids, Iowa, compared to normal rainfall. (*C*) Discharge of the Mississippi River at St. Louis compared to 100-year flood.
Data from U.S. Geological Survey

River discharges exceeded 100-year discharges on many rivers including the Mississippi, the Missouri, the Iowa, the Platte, and the Raccoon. At St. Louis, the Mississippi River discharge was greater than 1,000,000 cubic feet per second on August 1 (six times normal flow), and the river crested 20 feet above flood stage (and 23 feet above flood stage farther downstream at Chester, Illinois). At Hannibal, Missouri, where the 500-year flood height had been determined to be 30 feet, the new flood-control levee 31 feet high was completed in April. The river crested at 32 feet. Some rivers remained above flood stage for months.

The high water in major rivers such as the Mississippi and the Missouri caused many smaller tributary streams to back up, flooding numerous small towns. Many levees broke as water

FIGURE 16.38
Mississippi River water pours through a broken levee onto a farm near Columbia, Illinois, 1993.
Photo © St. Louis Post-Dispatch

overtopped them or just physically pushed the saturated levee sediment aside (figure 16.38). After the rains stopped, in some places the floodwaters took weeks or even months to recede. Some entire towns have been relocated to drier ground.

STREAM VALLEY DEVELOPMENT

Valleys, the most common landforms on the earth's surface, are usually cut by streams. By removing rock and sediment from the stream channel, a stream deepens, widens, and lengthens its own valley.

Downcutting and Base Level

The process of deepening a valley by erosion of the stream bed is called **downcutting.** If a stream removes rock from its bed, it can cut a narrow *slot canyon* down through rock (figures 16.39*A* and *B*). Such narrow canyons do not commonly form because mass wasting and sheet erosion usually remove rock from the valley walls. These processes widen the valley from a narrow, vertical-walled canyon to a broader, open, V-shaped canyon (figures 16.39*C* and *D*). Slot canyons persist, however, in very resistant rock with favorably oriented fractures or in regions where downcutting is rapid.

Downcutting cannot continue indefinitely because the headwaters of a stream cannot cut below the level of the stream bed at the mouth. If a river flows into the ocean, sea level becomes the lower limit of downcutting. The river cannot cut below sea level, or it would have to flow uphill to get to the sea. For most streams, sea level controls the level to which the land can be eroded.

The limit of downcutting is known as **base level;** it is a theoretical limit for erosion of the earth's surface (figure 16.40*A*). Downcutting will proceed until the stream bed reaches base level. If the stream is well above base level, downcutting can be quite rapid; but as the stream approaches base level, the rate of downcutting slows down. For streams that reach the ocean, base level is close to sea level, but since streams need at least a gentle gradient in order to flow, base level slopes gently upward in an inland direction.

During the glacial fluctuations of the Pleistocene Epoch (see chapter 19), sea level rose and fell as water was removed from the sea to form the glaciers on the continents and returned to the sea when the glaciers melted. This means that base level rose and fell for streams flowing into the sea. As a result, the lower reaches of such rivers alternated between erosion (caused by low sea level) and deposition (caused by high sea level). Since the glaciers advanced and retreated several times, the cycle of erosion and deposition was repeated many times, resulting in a complex history of cutting and filling near the mouths of most old rivers.

Base levels for streams that do not flow into the ocean are not related to sea level. In Death Valley in California (figure 16.40*B*), base level for in-flowing streams corresponds to the lowest point in the valley, 86 meters (282 feet) *below* sea level (the valley has been dropped below sea level by tectonic movement along faults). On the other hand, base level for a stream

FIGURE 16.39

Downcutting, mass wasting, and sheet erosion shape canyons and valleys. (*A*) Downcutting can create slot canyons in resistant rock, particularly where downcutting is rapid during flash floods and fractures in the rock are favorably oriented. (*B*) Stream erosion has cut this unusual slot canyon through porous sandstone, Zion National Park, Utah. (*C*) Downslope movement of rock and soil on valley walls widens most canyons into V-shaped valleys. (*D*) The waterfall and rapids on the Yellowstone River in Wyoming indicate that the river is ungraded and actively downcutting. Note the V-shaped cross-profile and lack of flood plain due to the downslope movement of volcanic rock.

Photo *B* by Allen Hagood, Zion Natural History Association

above a high reservoir or a mountain lake can be hundreds of feet above sea level. The surface of the lake or reservoir serves as temporary base level for all the water upstream (figure 16.40*C*). The base level of a tributary stream is governed by the level of its junction with the main stream. A ledge of resistant rock may act as a temporary base level if a stream has difficulty eroding through it.

The Concept of a Graded Stream

As a stream begins downcutting into the land, its longitudinal profile is usually irregular with rapids and waterfalls along its course (see figure 16.39*D*). Such a stream, termed *ungraded,* is using most of its erosional energy in downcutting to smooth out these irregularities in gradient.

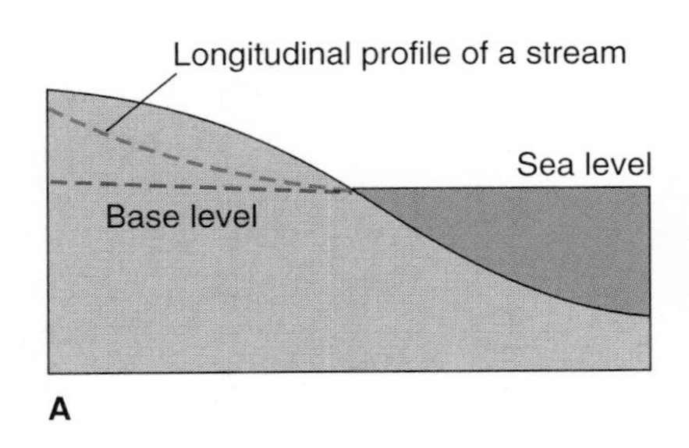

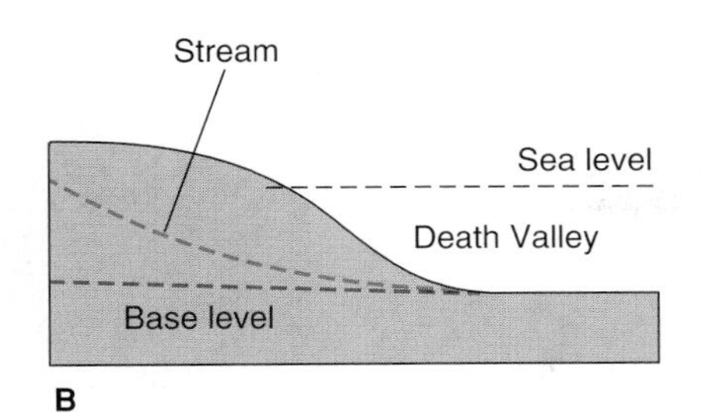

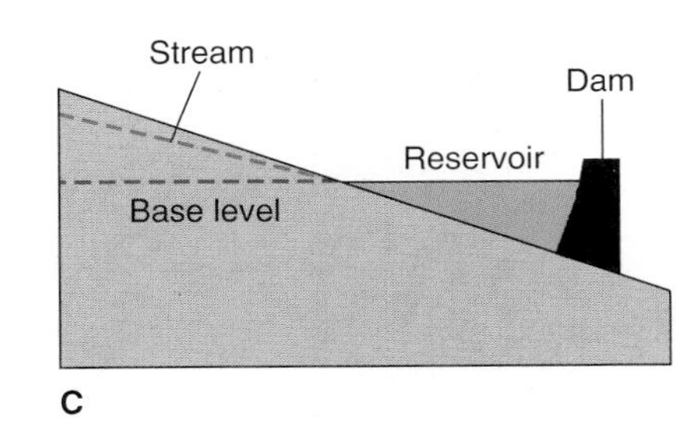

FIGURE 16.40
Base level is the lowest level of downcutting.

As the stream smoothes out its longitudinal profile to a characteristic concave-upward shape (figure 16.41), it becomes graded. A **graded stream** is one that exhibits a delicate balance between its transporting capacity and the sediment load available to it. This balance is maintained by cutting and filling any irregularities in the smooth longitudinal profile of the stream.

Earlier in this chapter you learned how changes in a stream's gradient can cause changes in its sediment load. An increase in gradient causes an increase in a stream's velocity, allowing the stream to erode and carry more sediment. A balance is maintained—the greater load is a result of the greater transporting capacity caused by the steeper gradient.

The relationship also works in reverse—a change in sediment load can cause a change in gradient. For example, a decrease in sediment load may bring about erosion of the stream's channel, thus lowering the gradient. Because dams trap sediment in the calm reservoirs behind them, most streams are almost completely sediment-free just downstream from dams. In some streams this loss of sediment has caused severe channel erosion below a dam as the stream adjusts to its new, reduced load.

A river's energy is used for two things—transporting sediment and overcoming resistance to flow. If the sediment load decreases, the river has more energy for other things. It may use this energy to erode more sediment, deepening its valley. Or it may change its channel shape or length, increasing resistance to flow, so that the excess energy is used to overcome friction. Or the river may increase the roughness of its channel, also increasing friction. The response of a river is not always predictable, and construction of a dam can sometimes have unexpected and perhaps harmful results.

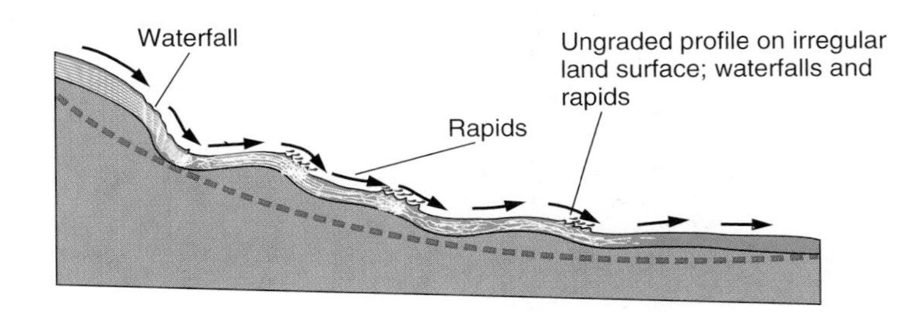

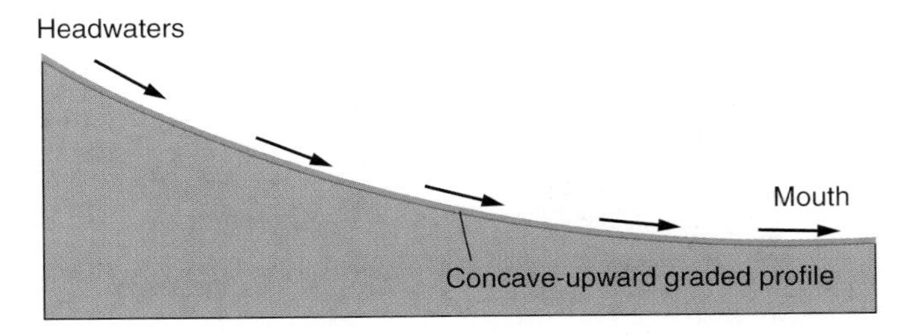

FIGURE 16.41
An ungraded stream has an irregular longitudinal profile with many waterfalls and rapids. A graded stream has smoothed out its longitudinal profile to a smooth, concave-upward curve.

Lateral Erosion

A graded stream can be deepening its channel by downcutting while part of its energy is also widening the valley by **lateral erosion,** the erosion and undercutting of a stream's banks and valley walls as the stream swings from side to side across its valley floor. The stream channel remains the same width as it moves across the flood plain, but the valley widens by erosion, particularly on the outside of curves, and meanders where the stream impinges against the valley walls (figure 16.42). The valley widens as its walls are eroded by the stream and as its walls retreat by mass wasting triggered by stream undercutting. As a valley widens, the stream's flood plain increases in width also.

Headward Erosion

Building a delta or alluvial fan at its mouth is one way a river can extend its length. A stream can also lengthen its valley by **headward erosion,** the slow uphill growth of a valley above its original source through gullying, mass wasting, and sheet erosion (figure 16.43). This type of erosion is particularly difficult to stop. When farmland is being lost to gullies that are eroding headward into fields and pastures, farmers must divert sheet flow and fill the gully heads with brush and other debris to stop, or at least retard, the loss of topsoil.

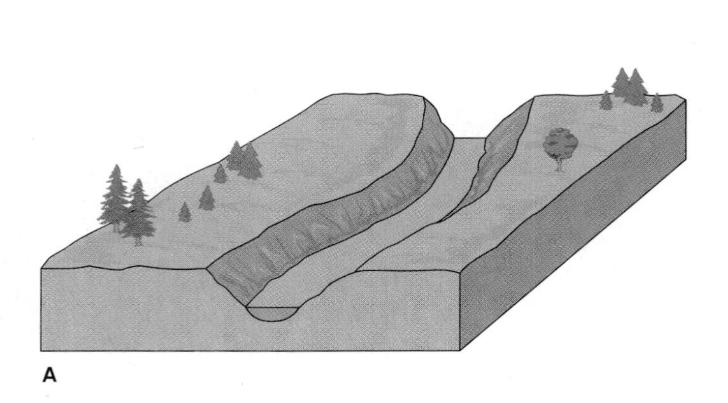

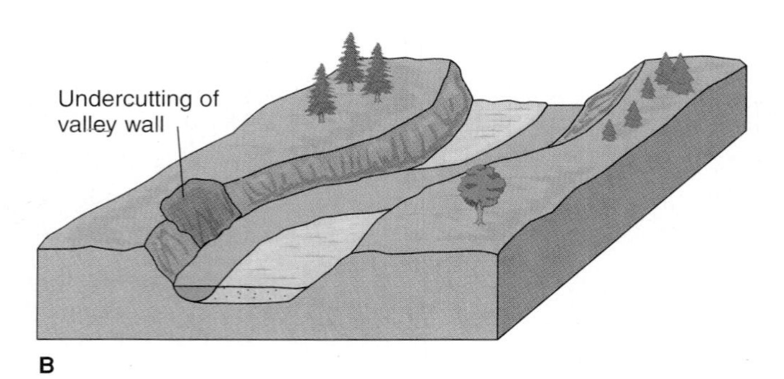

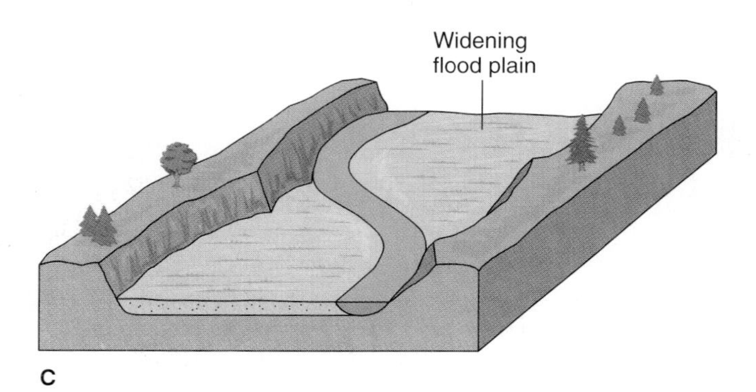

FIGURE 16.42

Lateral erosion can widen a valley by undercutting and eroding valley walls.

FIGURE 16.43

Headward erosion is lengthening this stream channel. Note the dendritic drainage pattern that is developing in the headwaters of the streams, New Plymouth, New Zealand.

Photo © G. R. "Dick" Roberts

FIGURE 16.44

Stream terraces near Jackson Hole, Wyoming. The stream has cut downward into its old flood plain.

Photo by Diane Carlson

STREAM TERRACES

Stream terraces are steplike landforms found above a stream and its flood plain (figure 16.44). Terraces may be benches cut in rock (sometimes sediment-covered), or they may be steps formed in sediment by deposition and subsequent erosion.

Figure 16.45 shows how one type of terrace forms as a river cuts downward into a thick sequence of its own flood-plain deposits. Originally the river deposited a thick section of flood-plain sediments. Then the river changed from deposition to erosion and cut into its old flood plain, parts of which remain as terraces above the river.

Why might a river change from deposition to erosion? One reason might be regional uplift, raising a river that was once meandering near base level to an elevation well above base level. Uplift would steepen a river's gradient, causing the river to speed up and begin erosion. But there are several other reasons why a river might change from deposition to erosion. A change from a dry to a wet climate may increase discharge and cause a river to begin eroding. A drop in base level (such as lowering of sea level) can have the same effect. A situation like that shown in figure 16.45 can develop in a recently glaciated region. Thick valley fill such as glacial outwash (see chapter 19) may be deposited in a stream valley and later, after the glacier stops producing large amounts of sediment, be dissected into terraces by the river.

Terraces can also develop from erosion of a bedrock valley floor. Bedrock benches are usually capped by a thin layer of flood-plain deposits.

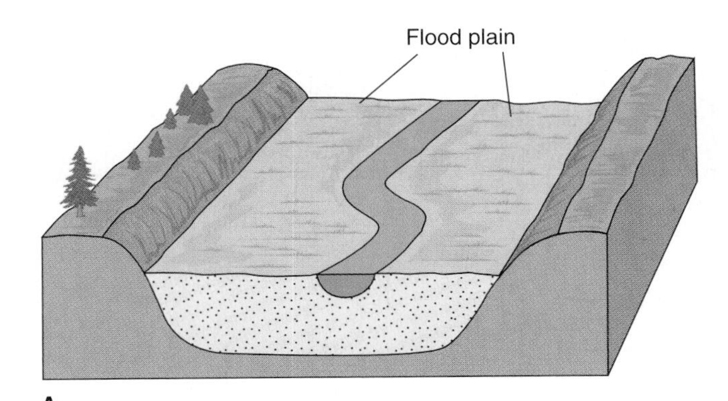

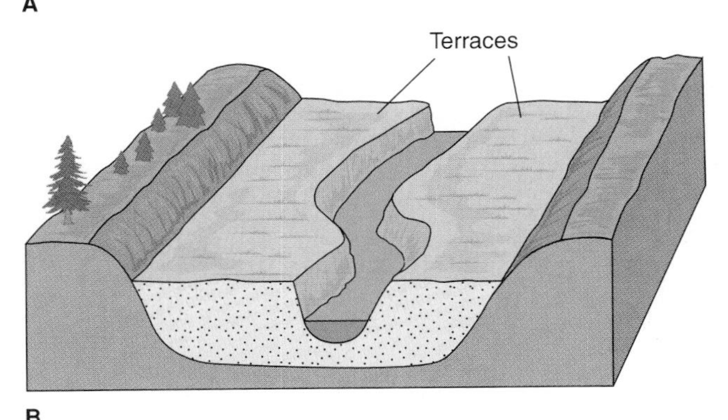

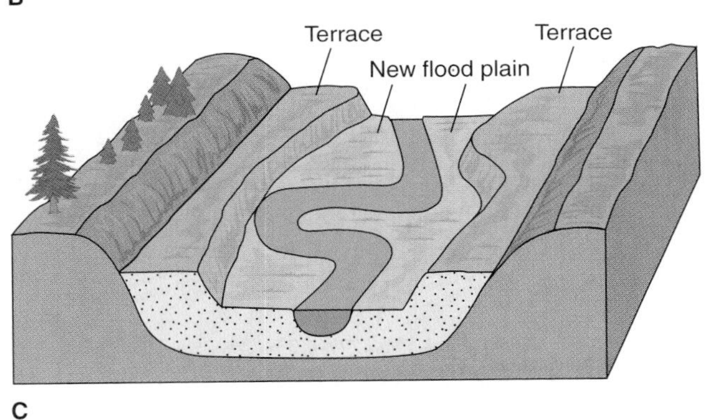

FIGURE 16.45

Terraces formed by a stream cutting downward into its own flood-plain deposits. (*A*) Stream deposits thick, coarse, flood-plain deposits. (*B*) Stream erodes its flood plain by downcutting. Old flood-plain surface forms terraces. (*C*) Lateral erosion forms new flood plain below terraces.

INCISED MEANDERS

Incised meanders are meanders that retain their sinuous pattern as they cut vertically downward below the level at which they originally formed. The result is a meandering *valley* with essentially no flood plain, cut into the land as a steep-sided canyon (figure 16.46).

Some incised meanders may be due to the profound effects of a change in base level. They may originally have been formed as meanders in a laterally eroding river flowing over a flat flood plain, perhaps near base level. If regional uplift elevated the land high above base level, the river would begin downcutting and might be able to maintain its characteristic meander pattern while deepening its valley (figure 16.47). A drop in base level

FIGURE 16.46

Incised meanders of the Colorado River ("The Loop"), Canyonlands National Park, southwestern Utah.

Photo by Frank M. Hanna

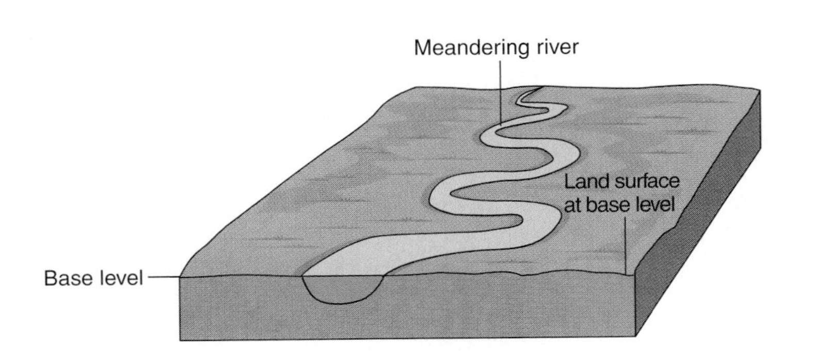

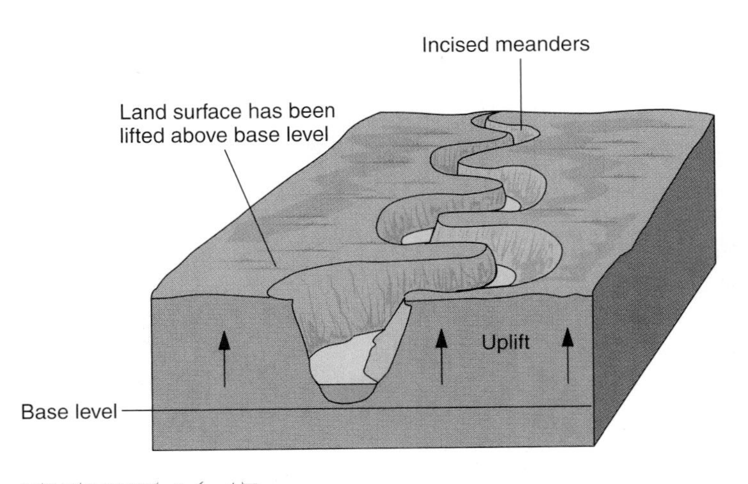

FIGURE 16.47

Incised meanders can form by uplift.

16.3 ASTROGEOLOGY

Stream Features on the Planet Mars

There is probably no liquid water on the surface of Mars today. With the present surface temperatures, atmospheric pressures, and water content in the Martian atmosphere, any liquid water would immediately evaporate. There are some indications, however, that conditions may have been different in the past and that liquid water existed on Mars, at least temporarily. Certain features on Mars, called *channels,* closely resemble certain types of stream channels on Earth. They have tributary systems and meanders and are sometimes braided. The channels trend downslope and tend to get wider toward their mouths. These Martian channels are restricted to certain areas and appear to have been formed by intermittent episodes of erosion.

One type of stream channel on Mars appears to have formed by large flooding events and is similar in appearance to those observed in the Channeled Scablands of Washington state. The Channeled Scablands were formed by extensive flooding during the Pleistocene glacial ages when a naturally formed ice dam broke and released water from a large lake. The mouth of Ares Vallis, an ancient Martian flood channel similar to those observed in the Channeled Scablands, was selected for the July 4, 1997, landing of the Pathfinder spacecraft and Sojourner Rover. It was postulated that a variety of rock types should be present in the mouth of an ancient flood channel. The first photos from the Mars Pathfinder Lander and Sojourner Rover, a "robotic field geologist," revealed a variety of rock types (box figure 1) in what does appear to be an ancient outflow channel.

Box 16.3 — FIGURE 2
Meandering channel and flat terraces within the Nanedi Vallis canyon, which resemble stream features cut by running water on Earth.
Photo courtesy NASA

A second kind of Martian channel (box figure 2), a meandering streamlike feature, occurs on the older surfaces of Mars (more than 3.5 billion years old) and may indicate that early in the history of Mars temperature and atmospheric conditions were such that rainfall could have occurred and long-lived river systems

Box 16.3 — FIGURE 1
View from the Mars Pathfinder Lander showing the Sojourner Rover and a variety of rocks from Ares Vallis.
Photo courtesy of Jet Propulsion Laboratory/NASA

could have existed. On June 8, 1998, the Mars Orbiter Camera (MOC) aboard the Mars Global Surveyor spacecraft captured an image of what appears to be a meandering stream channel and stream terraces inside the Nanedi Vallis canyon. The 2.5 kilometer-wide canyon is located approximately 1,600 kilometers (1,000 miles) south of where the Pathfinder landed. The presence of the narrow stream-like channel and associated terraces within the Nanedi Vallis canyon strongly suggest that a river of water repeatedly flowed down the canyon. But the lack of smaller tributary channels may argue that the channel, like many others on Mars, may have instead formed by surface collapse caused by frozen water underground (see Astrogeology box 13.3 for a discussion of underground ice on Mars).

Related Web Resources

For more information on the possibility of water on Mars, visit the NASA Goddard Institute for Space Studies research site: www.giss.nasa.gov/research/intro/gornitz.03/
Information about the Mars Global Surveyor can be found at the Jet Propulsion Laboratory/NASA site: http://mars.jpl.nasa.gov/mgs

Further Reading

Golombek, M. P. 1998. The Mars Pathfinder Mission. *Scientific American* 7.

without land uplift (possibly because of a lowering of sea level) could bring about the same result.

Although uplift may be a key factor in the formation of many incised meanders, it may not be *required* to produce them. Lateral erosion certainly seems to become more prominent as a river approaches base level, but some meandering can occur as soon as a river develops a graded profile. A river flowing on a flat surface high above base level may develop meanders early in its erosional history, and these meanders may become incised by subsequent downcutting. In such a case uplift is not necessary.

SUPERPOSED STREAMS

Some narrow mountain ranges have steep-sided river valleys slicing directly across them. The origin of these valleys is often puzzling. Why did the river erode *through* the mountain range rather than flow around it?

The usual reason is that the river has been **superposed** on the range from a gently sloping plain that used to exist above it (figure 16.48). If sediment once buried the range, a river could develop on the smooth surface of the sediment layer. As the sediment gradually was eroded away by the river, the river itself would be let down onto the once-buried range and could cut a canyon through the more resistant rocks in the underlying range.

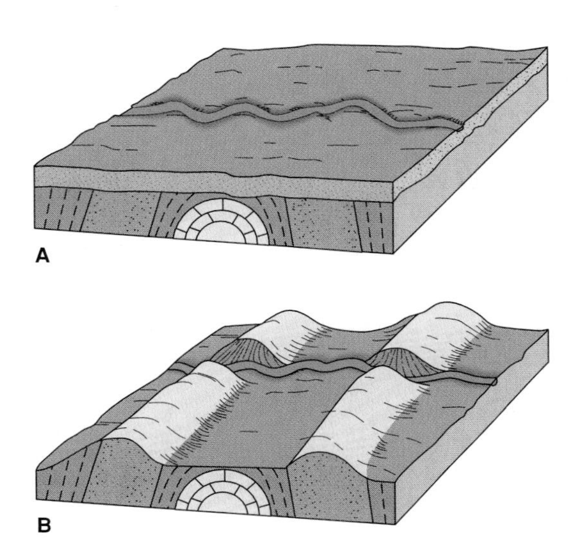

FIGURE 16.48
Development of a superposed stream. (*A*) Stream initially cuts through the horizontal sediment. (*B*) Continued erosion removes horizontal strata and stream cuts through underlying rock forming narrow valleys in resistant rock.

The folded rocks of the Appalachians were worn away by erosion and then partially covered by flat sediments during the Mesozoic. Gradual uplift in the Cenozoic has caused rivers such as the Potomac and Susquehanna to cut down into the once-buried range.

Summary

Normally stream *channels* are eroded and shaped by the streams that flow in them. Unconfined sheet flow can cause significant erosion.

Drainage basins are separated by *divides.*

A river and its tributaries form a *drainage pattern.* A *dendritic* drainage pattern develops on uniform rock, a *rectangular* pattern on regularly jointed rock. A *radial* pattern forms on conical mountains, while a *trellis* pattern usually indicates erosion of folded sedimentary rock.

Stream *velocity* is the key factor controlling sediment erosion, transportation, and deposition. Velocity is in turn controlled by several factors.

An increase in a stream's *gradient* increases the stream's velocity. *Channel shape* and *roughness* affect velocity by increasing or lessening friction. As tributaries join a stream, the stream's *discharge* increases downstream. Floods increase stream discharge and velocity.

Streams erode by *hydraulic action, abrasion,* and *solution.* They carry coarse sediment by *traction* and *saltation* as *bed load.* Finer-grained sediment is carried in *suspension.* A stream can also have a substantial *dissolved load.*

Streams create features by erosion and deposition. *Potholes* form by abrasion of hard rock on a stream bed. *Bars* form in the middle of streams or on stream banks, particularly on the inside of curves where velocity is low (*point bars*). A *braided pattern* can develop in streams with a large amount of bed load.

Meanders are created when a laterally eroding stream shifts across the flood plain, sometimes creating cutoffs and oxbow lakes.

A *flood plain* develops by both lateral and vertical deposition. *Natural levees* are built up beside streams by flood deposition.

A *delta* forms when a stream flows into standing water. The shape and internal structure of deltas are governed by river deposition and wave and current erosion. *Alluvial fans* form, particularly in dry climates, at the base of mountains as a stream's channel widens and its velocity decreases.

Rivers deepen their valleys by *downcutting* until they reach *base level,* which is either sea level or a local base level.

A *graded stream* is one with a delicate balance between its transporting capacity and its available load.

Lateral erosion widens a valley after the stream has become graded.

A valley is lengthened by both *headward erosion* and sediment deposition at the mouth.

Stream terraces can form by erosion of rock benches or by dissection of thick valley deposits during downcutting.

Incised meanders form as (1) river meanders are cut vertically downward following uplift or (2) lateral erosion and downcutting proceed simultaneously.

Superposed streams form as a stream cuts down through a sediment layer into the underlying, more resistant rock of a once-buried mountain range.

Terms to Remember

abrasion 394
alluvial fan 406
bar 396
base level 413
bed load 395
braided stream 398
delta 403
dendritic pattern 391
discharge 393
dissolved load 396
distributary 403
divide 390
downcutting 413
drainage basin 390
drainage pattern 391
flood plain 401
graded stream 415
headward erosion 415
hydraulic action 394
hydrologic cycle 388
incised meander 417
lateral erosion 415
meander 401
meander cutoff 401
natural levee 403
oxbow lake 401
point bar 401
pothole 395
radial pattern 391
rectangular pattern 391
saltation 395
sheetwash 390
solution 394
stream 388
stream channel 388
stream gradient 392
stream terrace 416
stream velocity 391
superposed stream 419
suspended load 395
traction 395
trellis pattern 391
tributary 390

Testing Your Knowledge

Use the questions below to prepare for exams based on this chapter.

1. What factors control a stream's velocity?
2. Describe how bar deposition creates a braided stream.
3. In what part of a large alluvial fan is the sediment the coarsest? Why?
4. What does a trellis drainage pattern tell about the rocks underneath it?
5. Describe one way that incised meanders form.
6. How does a meander neck cutoff form an oxbow lake?
7. How does a natural levee form?
8. Describe how stream terraces form.
9. Describe three ways in which a river erodes its channel.
10. Name and describe the three main ways in which a stream transports sediment.
11. How does a stream widen its valley?
12. What is base level?
13. The total area drained by a stream and its tributaries is called the
    a. hydrologic cycle
    b. tributary area
    c. divide
    d. drainage basin
14. Stream erosion and deposition are controlled primarily by a river's
    a. velocity
    b. discharge
    c. gradient
    d. channel shape
    e. channel roughness

15. What is the gradient of a stream that drops 10 vertical feet over a 2-mile horizontal distance?
    a. 20 feet per mile
    b. 10 feet per mile
    c. 5 feet per mile
    d. 2 feet per mile

16. What are typical units of discharge?
    a. miles per hour
    b. cubic meters
    c. cubic feet per second
    d. meters per second

17. Hydraulic action, solution, and abrasion are all examples of stream
    a. erosion
    b. transportation
    c. deposition

18. Cobbles are more likely to be transported in a stream's
    a. bed load
    b. suspended load
    c. dissolved load
    d. all of the above

19. A river's velocity is ____ on the outside of a meander curve compared to the inside.
    a. higher
    b. equal
    c. lower

20. Sandbars deposited on the inside of meander curves are called
    a. dunes
    b. point bars
    c. cutbanks
    d. none of the above

21. Which is not a drainage pattern?
    a. dendritic
    b. radial
    c. rectangular
    d. trellis
    e. none of the above

22. The broad strip of land built up by sedimentation on either side of a stream channel is
    a. a flood plain
    b. a delta
    c. an alluvial fan
    d. a meander

23. The average time between floods of a given size is
    a. the discharge
    b. the gradient
    c. the recurrence interval
    d. the magnitude

24. A platform of sediment formed where a stream flows into standing water is
    a. an alluvial fan
    b. a delta
    c. a meander
    d. a flood plain

Expanding Your Knowledge

1. Several rivers have been set aside as "wild rivers" on which dams cannot be built. Give at least four arguments against building dams on rivers. Give at least four arguments in favor of building dams.
2. Discuss the similarities between deltas and alluvial fans. Describe the differences between them.
3. How is the recurrence interval for a flood determined? How may new data affect the flood-frequency curve?

Exploring Web Resources

www.mhhe.com/plummer9e

Go to the Online Learning Center to review your answers for the Testing Your Knowledge section. This website also contains additional quizzing as well as reading, media resources, and some really great animations to further your understanding of streams and floods. Click on the links to go directly to the websites listed below.

http://water.usgs.gov/

Contains extensive information on water issues throughout the United States and many links to USGS data and online publications.

http://water.usgs.gov/public/realtime.html

Contains real-time stream-flow data from USGS gauging stations throughout the United States.

http://water.usgs.gov/usa/nwis/

Contains historical stream-flow data from USGS gauging stations throughout the U.S.

www.dartmouth.edu/artsci/geog/floods/

The *Dartmouth Flood Observatory* website contains information on flood detection and satellite images of floods and flood damage from around the world.

http://vcourseware.calstatela.edu/VirtualRiver/

California State University, Los Angeles *Virtual River* exercise. Analyze stream-flow data and observe animations of flowing streams.

Animations

This chapter includes the following animations available on our Online Learning Center at www.mhhe.com/plummer9e.

16.14 Movement of Sediment (Load) in a Stream
16.22 & 16.23 Creation of Meander Cut-Off and Oxbow Lake
16.27, 16.42, 16.45 Development of a Flood Plain and Stream Terraces

CHAPTER

17

Ground Water

Surprisingly, water underground is about 60 times as plentiful as fresh water in lakes and rivers on the land surface (not including water stored as ice in glaciers). Ground water is a tremendously important resource. How it gets underground, where it is stored, how it moves while underground, how we look for it, and, perhaps most important of all, why we need to protect it are the main topics of this chapter.

Also important is how ground water is related to surface rivers and springs. Ground water can form distinctive geologic features, such as caves, sinkholes, and petrified wood. It also can appear as hot springs and geysers. Hot ground water can be used to generate power.

World Distribution of Water (%)

Ocean	97.2
Glaciers and other ice	2.15
Ground water	.61
Lakes	
fresh	.009
saline	.008
Soil moisture	.005
Atmosphere	.001
Rivers	.0001

Opposite: Hot ground water is released from Africa Geyser in the Norris Geyser Basin of Yellowstone National Park as pressure builds underground. Thermophyllic bacteria thrive in the warm pools adjacent to the Geyser.
Photo © Michael Fogden/Bruce Coleman, Inc.

table 17.1 **Porosity and Permeability of Sediments and Rocks**

Sediment	Porosity (%)	Permeability
Gravel	25 to 40	excellent
Sand (clean)	30 to 50	good to excellent
Silt	35 to 50	moderate
Clay	35 to 80	poor
Glacial till	10 to 20	poor to moderate
Rock		
Conglomerate	10 to 30	moderate to excellent
Sandstone		
Well-sorted, little cement	20 to 30	good to very good
Average	10 to 20	moderate to good
Poorly sorted, well-cemented	0 to 10	poor to moderate
Shale	0 to 30	very poor to poor
Limestone, dolomite	0 to 20	poor to good
Cavernous limestone	up to 50	excellent
Crystalline rock		
Unfractured	0 to 5	very poor
Fractured	5 to 10	poor
Volcanic rocks	0 to 50	poor to excellent

Many communities obtain the water they need from rivers, lakes, or reservoirs, sometimes using aqueducts or canals to bring water from distant surface sources. Another source of water lies directly beneath most towns. This resource is **ground water,** the water that lies beneath the ground surface, filling the pore space between grains in bodies of sediment and clastic sedimentary rock and filling cracks and crevices in all types of rock.

Ground water is a major economic resource, particularly in the dry western areas of the United States and Canada where surface water is scarce. Many towns and farms pump great quantities of ground water from drilled wells. Even cities next to large rivers may pump their water from the ground because ground water is commonly less contaminated and more economical to use than surface water.

The source of ground water is rain and snow that falls to the ground. A portion of this precipitation percolates down into the ground to become ground water (see figure 16.1 and the hydrologic cycle discussion in chapter 16). How much precipitation soaks into the ground is influenced by climate, land slope, soil and rock type, and vegetation. In general, approximately 15% of the total precipitation ends up as ground water, but that varies locally and regionally from 1% to 20%.

POROSITY AND PERMEABILITY

Porosity, the percentage of rock or sediment that consists of voids or openings, is a measurement of a rock's ability to hold water. Most rocks can hold some water. Some sedimentary rocks, such as sandstone, conglomerate, and many limestones, tend to have a high porosity and therefore can hold a considerable amount of water. A deposit of loose sand may have a porosity of 30% to 50%, but this may be reduced to 10% to 20% by compaction and cementation as the sand lithifies (table 17.1). A sandstone in which pores are nearly filled with cement and fine-grained matrix material may have a porosity of 5% or less. Crystalline rocks, such as granite, schist, and some limestones, do not have pores but may hold some water in joints and other openings.

Although most rocks can hold some water, they vary a great deal in their ability to allow water to pass through them. **Permeability** refers to the capacity of a rock to transmit a fluid such as water or petroleum through pores and fractures. In other words, permeability measures the relative ease of water flow and indicates the degree to which openings in a rock interconnect. The distinction between porosity and permeability is important. A rock that holds much water is called *porous;* a rock that allows water to flow easily through it is described as *permeable.* Most sandstones and conglomerates are both porous and permeable. An *impermeable* rock is one that does not allow water to flow through it easily. Unjointed granite and schist are impermeable. Shale can have substantial porosity, but it has low permeability because its pores are too small to permit easy passage of water.

THE WATER TABLE

Responding to the pull of gravity, water percolates down into the ground through the soil and through cracks and pores in the rock. Several kilometers down in the crust percolation stops. With increasing depth, sedimentary rock pores tend to be closed

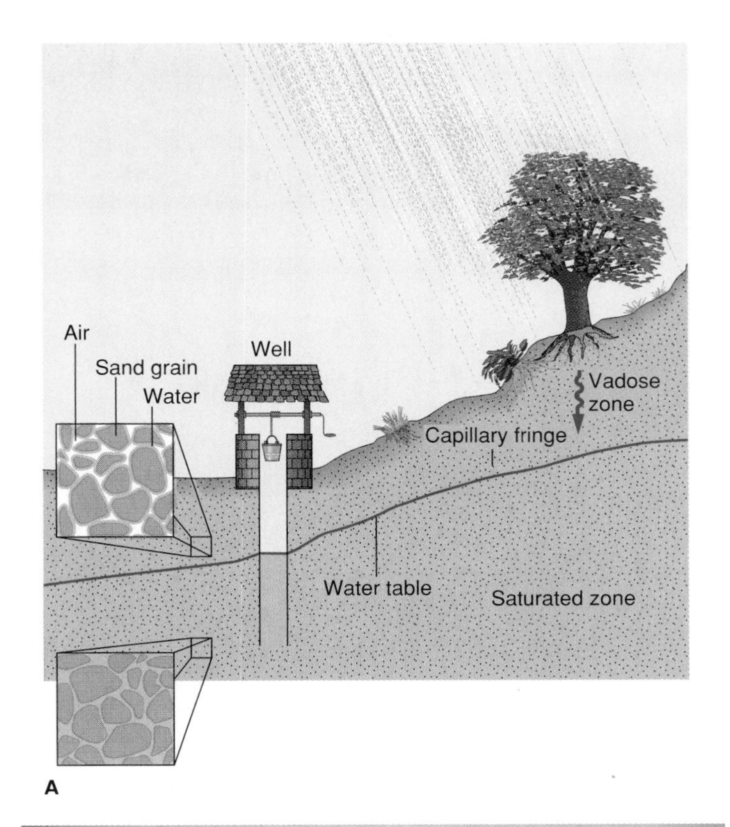

FIGURE 17.1

(*A*) The water table marks the top of the saturated zone, in which water completely fills the rock pore space (inset figure). Above the water table is the vadose zone in which rock openings typically contain both air and water. (*B*) Ground water fills this abandoned Indiana limestone quarry, which extends below the water table.

by increasing amounts of cement and by the weight of the overlying rock. Moreover, sedimentary rock overlies igneous and metamorphic crystalline basement rock, which usually has very low porosity. The depth where downward percolation stops is generally about 5 kilometers below the surface, although in thick wedges of sedimentary rock, such as those found along the Gulf Coast of the United States, percolation may reach a depth of 10 kilometers or more.

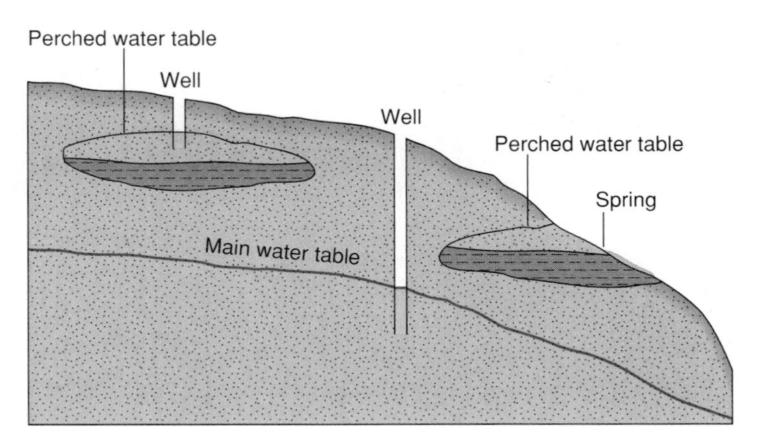

FIGURE 17.2

Perched water tables above lenses of less permeable shale within a large body of sandstone. Downward percolation of water is impeded by the less permeable shale.

The subsurface zone in which all rock openings are filled with water is called the **saturated zone** (figure 17.1*A*). If a well were drilled downward into this zone, ground water would fill the lower part of the well. The water level inside the well marks the upper surface of the saturated zone; this surface is the **water table.**

Most rivers and lakes intersect the saturated zone. Rivers and lakes occupy low places on the land surface, and ground water flows out of the saturated zone into these surface depressions. The water level at the surface of most lakes and rivers coincides with the water table. Ground water also flows into mines and quarries cut below the water table (figure 17.1*B*).

Above the water table there is a zone that is generally unsaturated and is referred to as the **vadose zone** (figure 17.1*A*). Within the vadose zone, surface tension causes water to be held above the water table. The *capillary fringe* is a transition zone with higher moisture content at the base of the vadose zone just above the water table. Some of the water in the capillary fringe has been drawn or wicked upward from the water table (much like water rising up a paper towel if the corner is dipped in water). The capillary fringe is generally less than a meter thick, but may be much thicker in fine-grained sediments and thinner in coarse-grained sediments such as sand and gravel.

Plant roots generally obtain their water from the belt of soil moisture near the top of the vadose zone, where fine-grained clay minerals hold water and make it available for plant growth. Most plants "drown" if their roots are covered by water in the saturated zone; plants need both water and air in soil pores to survive. (The water-loving plants of swamps and marshes are an exception.)

A **perched water table** is the top of a body of ground water separated from the main water table beneath it by a zone that is not saturated (figure 17.2). It may form as ground water collects above a lens of less permeable shale within a more permeable rock, such as sandstone. If the perched water table intersects the land surface, a line of springs can form along the upper contact of the shale lens. The water perched above a shale lens can provide a limited water supply to a well; it is an unreliable long-term supply.

17.1 IN GREATER DEPTH

Darcy's Law and Fluid Potential

In 1856 Henry Darcy, a French engineer, found that the velocity at which water moves depends upon the *hydraulic head* of the water and upon the permeability of the material that the water is moving through.

The hydraulic head of a drop of water is equal to the elevation of the drop plus the water pressure on the drop:

Hydraulic head = elevation + pressure.

In box figure 1*A* the points A and B are both on the water table, so the pressure at both points is zero (there is no water above points A and B to create pressure). Point A is at a higher elevation than B, so A has a higher hydraulic head than B. The difference in elevation is equal to the difference in head, which is labeled **h.** Water will move from point A to point B (as shown by the dark blue arrow), because water moves from a region of high hydraulic head to a region of low head. The distance the water moves from A to B is labeled **L.** The *hydraulic gradient* is the difference in head between two points divided by the distance between the two points:

$$\textbf{Hydraulic gradient} = \frac{\textbf{difference in head}}{\textbf{distance}} = \frac{\textbf{h}}{\textbf{L}}$$

In box figure 1*B* the two points have equal elevation, but the pressure on point C is higher than on point D (there is more water to create pressure above point C than point D). The head is higher at point C than at point D, so the water moves from C to D. In box figure 1*C* point F has a lower elevation than point G, but F also has a higher pressure than G. The difference in pressure is greater than the difference in elevation, so F has a higher head than G, and water moves from F to G. Note that

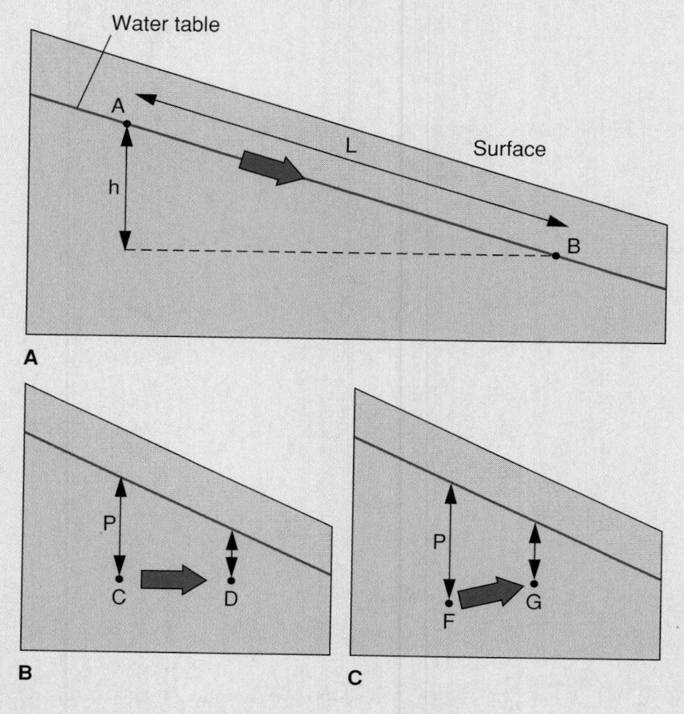

Box 17.1 — Figure 1

Ground water moves in response to hydraulic head (elevation plus pressure). Water movement shown by dark blue arrows. (*A*) Points A and B have the same pressure, but A has a higher elevation; therefore water moves from A to B. (*B*) Point C has a higher pressure (arrow marked P) than D; therefore water moves from C to D at the same elevation. (*C*) Pressure also moves water upward from F to G.

THE MOVEMENT OF GROUND WATER

Compared to the rapid flow of water in surface streams, most ground water moves relatively slowly through rock underground. Because it moves in response to differences in water pressure and elevation, water within the upper part of the saturated zone tends to move downward following the slope of the water table (figure 17.3). See box 17.1 for Darcy's Law.

The circulation of ground water in the saturated zone is not confined to a shallow layer beneath the water table. Ground water may move hundreds of feet vertically downward before rising again to discharge as a spring or to seep into the beds of rivers and lakes at the surface (figure 17.3) due to the combined effects of gravity and the slope of the water table.

The *slope of the water table* strongly influences groundwater velocity. The steeper the slope of the water table, the faster ground water moves. Water-table slope is controlled largely by topography—the water table roughly parallels the land surface (particularly in humid regions), as you can see in figure 17.3. Even in highly permeable rock, ground water will not move if the water table is flat.

How fast ground water flows also depends on the *permeability* of the rock or other materials through which it passes. If rock pores are small and poorly connected, water moves slowly. When openings are large and well connected, the flow of water is more rapid. One way of measuring ground-water velocity is to introduce a tracer, such as a dye, into the water and then watch for the color to appear in a well or spring some distance away. Such experiments have shown that the velocity of ground water varies widely, averaging a few centimeters to many meters a day. Nearly

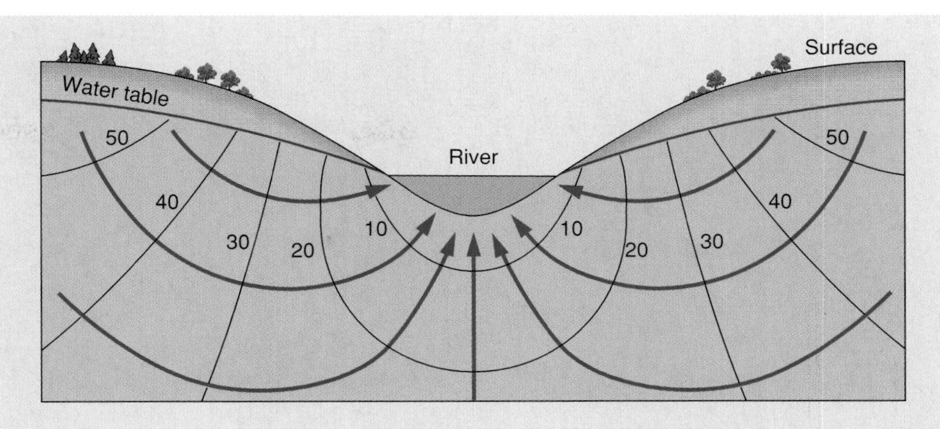

Box 17.1 — FIGURE 2

Dark blue arrows are flow lines, which show direction of ground water flow. Flow is perpendicular to equipotential lines (black lines with numbers), which show regions of equal fluid potential. Ground water generally flows from hilltops toward valleys, emerging from the ground as springs into stream beds and banks, lakes, and swamps.

underground water may move downward, horizontally, or upward in response to differences in head, but that it always moves in the direction of the downward slope of the water table above it. One of the first goals of ground-water geologists, particularly in ground-water contamination investigations, is to find the slope of the local water table in order to determine the direction (and velocity) of ground-water movement.

The velocity of ground-water flow is controlled by both the permeability of the sediment or rock and the hydraulic gradient. *Darcy's Law* states that the velocity equals the permeability multiplied by the hydraulic gradient. This gives the Darcian velocity (or the velocity of water flowing through an open pipe). To determine the actual velocity of ground water, since ground water only flows through the openings in sediment or rock, the Darcian velocity must be divided by the porosity.

ground-water velocity = permeability/porosity × hydraulic gradient

$$V = \frac{K}{n}\frac{h}{L}$$

(Darcy called **K** the hydraulic conductivity; it is a measure of permeability and is specific to a particular aquifer. The porosity is represented by **n** in the equation.)

Most ground-water literature today, including most of the chapter-end supplementary readings on the Online Learning Center at www.mhhe.com/plummer9e, relate ground-water movement to *fluid potential.* Potential is found by multiplying the acceleration of gravity (**g**) times the hydraulic head (**h**):

potential = acceleration of gravity × hydraulic head
= gh

For ground water the acceleration of gravity is constant, so potential is essentially identical to head. Modern figures show ground-water movement in relation to *equipotential lines* (lines of equal potential). Ground water moves from regions of high potential to regions of low potential (from high to low head). Box figure 2 shows how *flow lines,* which show ground-water movement, cross equipotential lines at right angles as water moves from high to low potential.

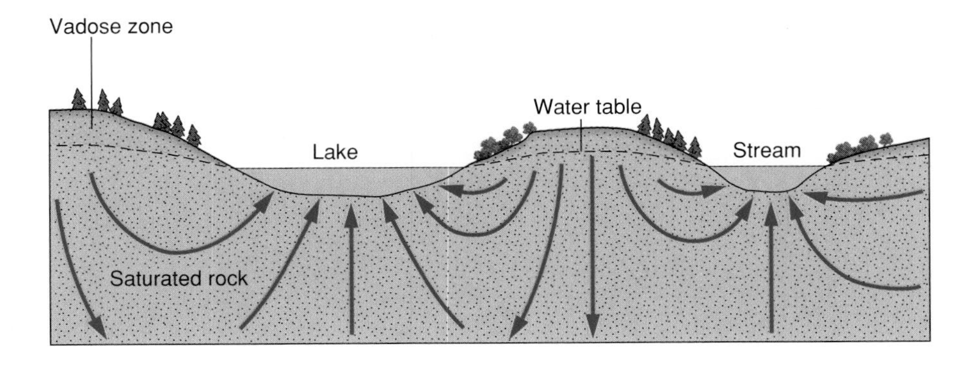

FIGURE 17.3

Movement of ground water beneath a sloping water table in uniformly permeable rock. Near the surface ground water tends to flow parallel to the sloping water table.

impermeable rocks may allow water to move only a few centimeters per *year,* but highly permeable materials, such as unconsolidated gravel or cavernous limestone, may permit flow rates of hundreds or even thousands of meters per day.

AQUIFERS

An **aquifer** is a body of saturated rock or sediment through which water can move easily. Aquifers are both highly permeable and saturated with water. A well must be drilled into an aquifer to reach an adequate supply of water (figure 17.4). Good aquifers include sandstone, conglomerate, well-jointed limestone, bodies of

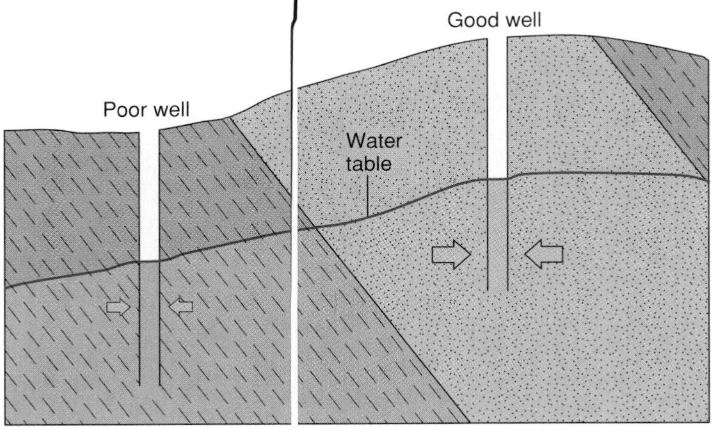

FIGURE 17.4

A well must be installed in an aquifer to obtain water. The saturated part of the highly permeable sandstone is an aquifer, but the less permeable shale is not. Although the shale is saturated, it will not readily transmit water.

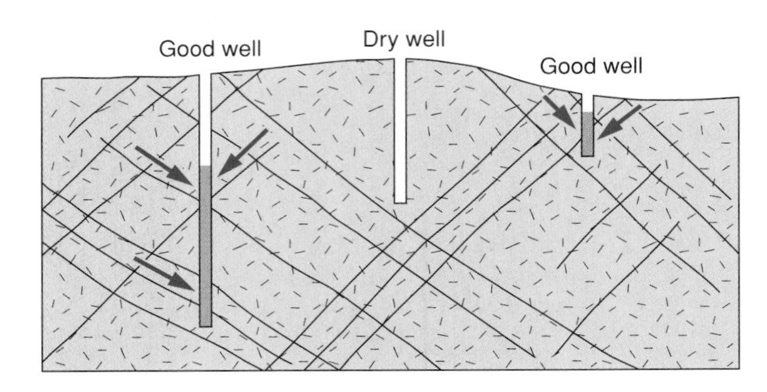

FIGURE 17.5

Wells can obtain some water from fractures in crystalline rock. Wells must intersect fractures to obtain water.

sand and gravel, and some fragmental or fractured volcanic rocks such as columnar basalt (table 17.1). These favorable geologic materials are sought in "prospecting" for ground water or looking for good sites to drill water wells.

Wells drilled in shale beds are not usually very successful because shale, although sometimes quite porous, is relatively impermeable (figure 17.4). Wet mud may have a porosity of 80% to 90% and even when compacted to form shale may still have a high porosity of 30%. Yet the extremely small size of the pores, together with the electrostatic attraction that clay minerals have for water molecules (see chapter 12), prevents water from moving readily through the shale into a well.

Because they are not very porous, crystalline rocks such as granite, gabbro, gneiss, schist, and some types of limestone are not good aquifers. The porosity of such rocks may be 1% or less. (Shale and crystalline rocks are sometimes called *aquitards* because they retard the flow of ground water.) Crystalline rocks that are highly fractured, however, may be porous and permeable enough to provide a fairly dependable water supply to wells (figure 17.5).

Figure 17.6 shows the difference between an **unconfined aquifer,** which has a water table because it is only partly filled with water, and a **confined aquifer,** which is completely filled with water under pressure, and which is usually separated from

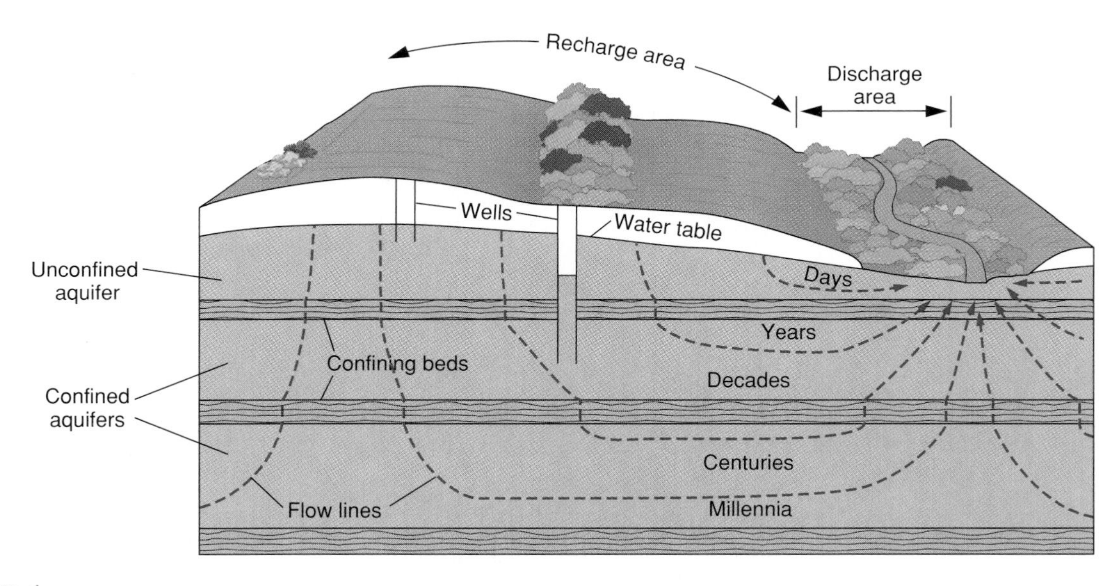

FIGURE 17.6

An unconfined aquifer is exposed to the surface and is only partly filled with water; water in a shallow well will rise to the level of the water table. A confined aquifer is separated from the surface by a confining bed, and is completely filled with water under pressure; water in wells rises above the aquifer. Flow lines show direction of ground water flow. Days, years, decades, centuries, and millennia refer to the time required for ground water to flow from the recharge area to the discharge area. Water enters aquifers in recharge areas and flows out of aquifers in discharge areas.

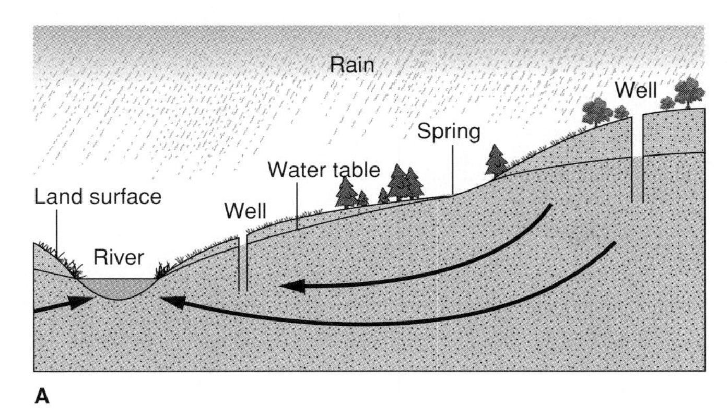

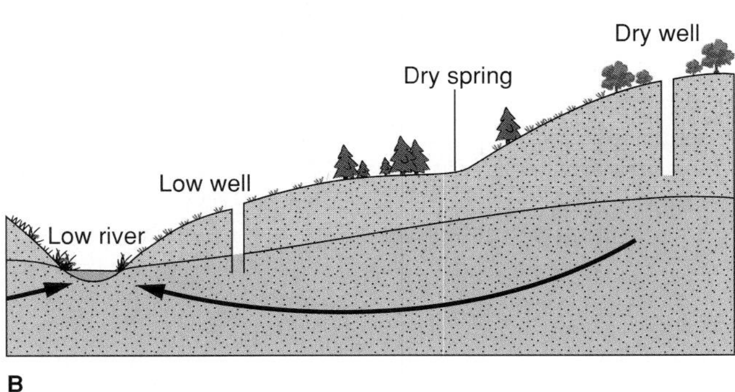

FIGURE 17.7

The water table in an unconfined aquifer rises in wet seasons and falls in dry seasons as water drains out of the saturated zone into rivers. (*A*) Wet season: water table and rivers are high; springs and wells flow readily. (*B*) Dry season: water table and rivers are low; some springs and wells dry up.

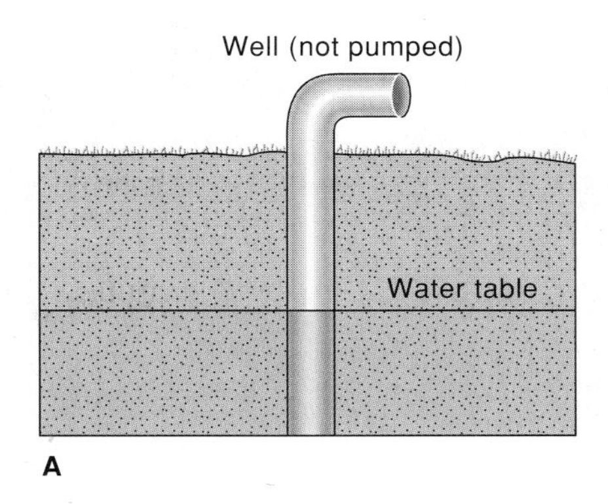

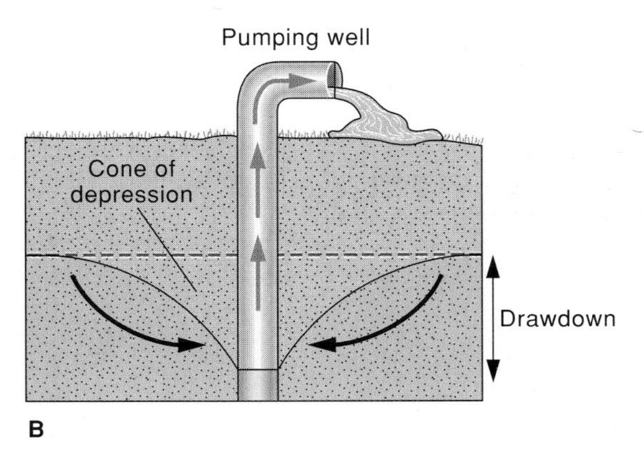

FIGURE 17.8

Pumping well lowers the water table into a cone of depression.

the surface by a relatively impermeable confining bed, or aquitard, such as shale. An unconfined aquifer is recharged rapidly by precipitation, has a rising and falling water table during wet and dry seasons, and has relatively rapid movement of ground water through it (figure 17.7). A confined aquifer is recharged slowly through confining shale beds. With very slow movement of ground water, a confined aquifer may have no response at all to wet and dry seasons.

WELLS

A **well** is a deep hole, generally cylindrical, that is dug or drilled into the ground to penetrate an aquifer within the saturated zone (figure 17.4). Usually water that flows into the well from the saturated rock must be lifted or pumped to the surface. As figure 17.7 shows, a well dug in a valley usually has to go down a shorter distance to reach water than a well dug on a hilltop. During dry seasons the water table falls as water flows out of the saturated zone into springs and rivers. Wells not deep enough to intersect the lowered water table go dry, but the rise of the water table during the next rainy season normally returns water to the dry wells. The addition of new water to the saturated zone is called **recharge.**

When water is pumped from a well, the water table is typically drawn down around the well into a depression shaped like an inverted cone known as a **cone of depression** (figure 17.8). This local lowering of the water table, called **drawdown,** tends to change the direction of ground-water flow by changing the slope of the water table. In lightly used wells that are not pumped, drawdown does not occur and a cone of depression does not form. In a simple, rural well with a bucket lowered on the end of a rope, water cannot be extracted rapidly enough to significantly lower the water table. A well of this type is shown in figure 17.1*A*.

In unconfined aquifers, water rises in shallow wells to the level of the water table. In confined aquifers, the water is under pressure and rises in wells to a level above the top of the aquifer (figure 17.6). Such a well is called an **artesian well** and confined aquifers are also called *artesian aquifers.*

In some artesian wells the water rises above the land surface, producing a flowing well that spouts continuously into the air unless it is capped (figure 17.9). Flowing wells used to occur in South Dakota, when the extensive Dakota Sandstone aquifer was first tapped (figure 17.10), but continued use has lowered the water pressure surface below the ground surface in most parts of the state. Water still rises above the aquifer, but does not reach the land surface.

17.2

ENVIRONMENTAL GEOLOGY

Prospecting for Ground Water

Many wells are drilled or dug without any effort to locate a good aquifer. Many of these wells are successful (especially if only small amounts of water are needed) because most rocks hold some water, which flows into wells that intersect the water table. If a large and dependable supply of water is needed—as for a city water system—specialists in ground-water geology (hydrogeologists) may be called in to locate a promising well site. Hydrogeologists use many methods to locate aquifers. A detailed knowledge of the local rocks is necessary. Therefore, a geologist may map the surface rocks and use electrical, magnetic, and seismic surveys to study sub-surface rocks to determine the presence of possible aquifers. Sometimes a small-diameter test well is drilled before the larger, more expensive supply well is sunk. Hydrogeologists look for potentially high-producing aquifers rather than searching for water directly, which would be much more difficult. In some regions, however, the presence of certain plants may be a useful guide to locating water, particularly the depth of the water table.

Some people search for water by water witching, or *dowsing,* with a divining rod (also sometimes used to search for metals or lost objects). Usually the dowser holds a forked stick horizontally in the hands while walking over an area. The stick is supposed to deflect or twist downward of its own accord when the dowser passes over water. This method has been tried for centuries, the only modification being that a twisted metal rod, often made of a coat hanger, now may be substituted for the stick. Carefully controlled tests conducted by workers in psychic research have shown that water witchers' "success" is equal to or less than pure chance, while geologists' results are superior both to witching and to chance. Records kept on thousands of wells in Australia in the early 1900s show that of wells that were not divined, more than 83% produced flows of 100 gallons per hour and 7.4% were failures, finding no water at all. Of wells that were divined by water witchers, only about 70% produced more than 100 gallons per hour, and 14.7% were dry. In the early part of the twentieth century the U.S. Geological Survey concluded that any future testing of the results of water witching would be a misuse of public funds.

Despite such findings, many people believe strongly in dowsing. Water witchers themselves devoutly believe they can find water, and in some regions of the United States almost no wells are drilled without a witcher's advice. Dowsers are helped in locating water by the fact that most rocks hold some water, and dowsers often have a long-standing knowledge of a particular region and its potential water resources. This is not to say that dowsers are deliberate frauds. Many are convinced that they perform a valuable public service, and some do not charge for their services. Scientists see no reason for dowsing to work and are skeptical about dowsers' "success." Geologists would almost unanimously urge you not to pay for a dowser's service if a fee were charged.

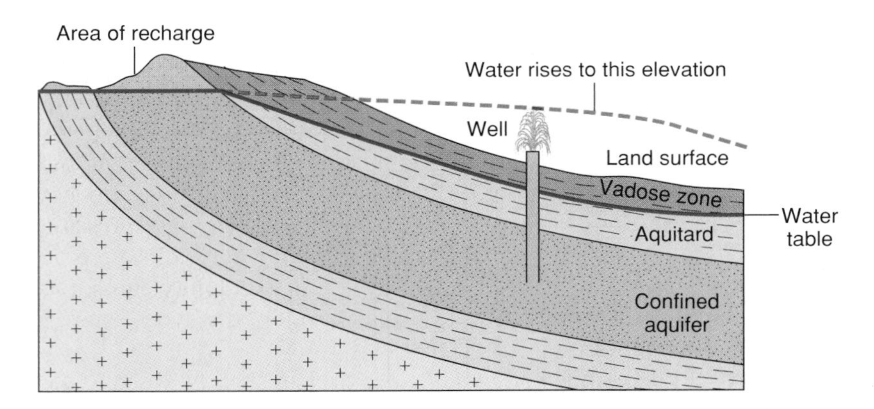

FIGURE 17.9

The Dakota Sandstone in South Dakota is a relatively unusual type of confined aquifer because it is tilted and exposed to the surface by erosion. Water in most wells rose above the land surface when the aquifer was first tapped in the 1800s.

FIGURE 17.10

Artesian well spouts water above land surface in South Dakota, early 1900s. Heavy use of this aquifer has reduced water pressure so much that spouts do not occur today.

Photo by N. H. Darton, U.S. Geological Survey

A

B

FIGURE 17.11

(*A*) A large spring issuing from a cavern in limestone, Jasper National Park, Alberta, Canada. (*B*) A line of springs seeps from the ground at the contact between less permeable shale and the overlying permeable sandstone. Southern Utah.

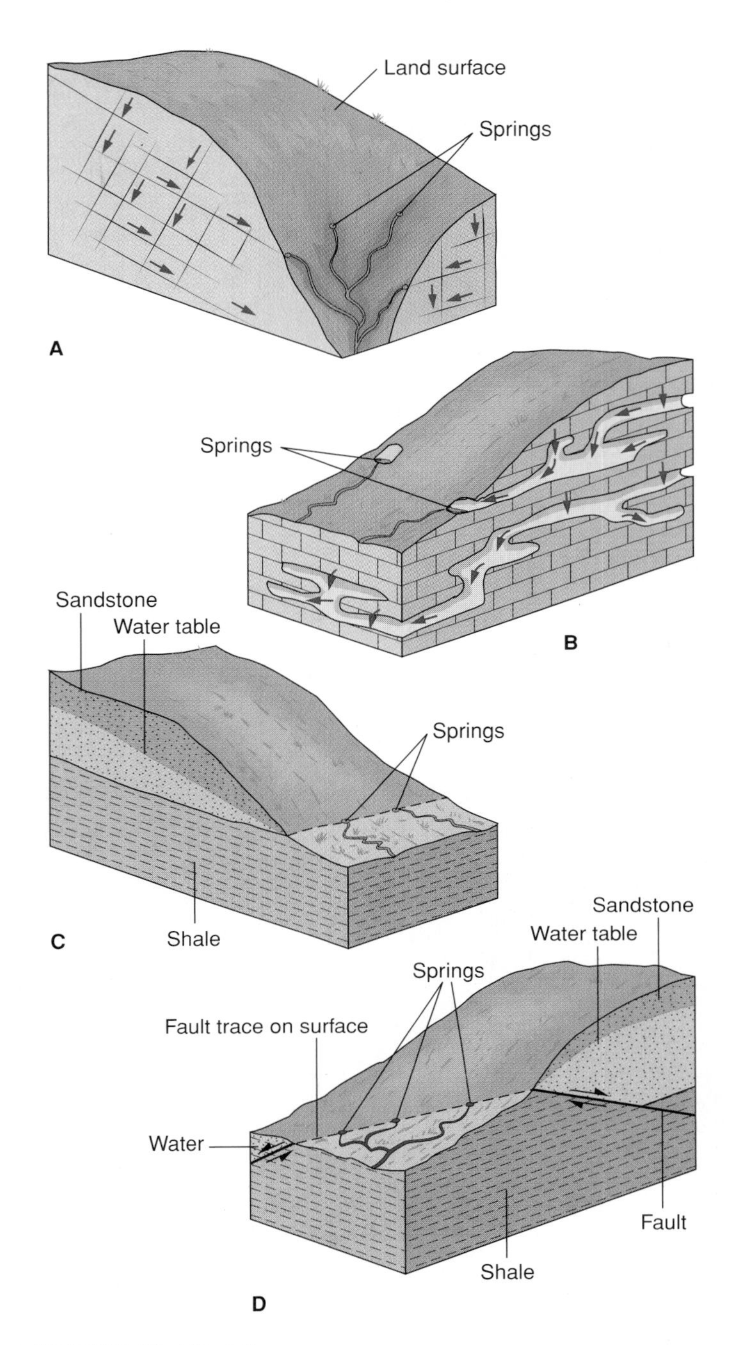

FIGURE 17.12

Springs can form in many ways. (*A*) Water moves along fractures in crystalline rock and forms springs where the fractures intersect the land surface. (*B*) Water enters caves along joints in limestone and exits as springs at the mouths of caves. (*C*) Springs form at the contact between a permeable rock such as sandstone and an underlying less permeable rock such as shale. (*D*) Springs can form along faults when permeable rock has been moved against less permeable rock. Arrows show relative motion along fault.

SPRINGS AND STREAMS

A **spring** is a place where water flows naturally from rock onto the land surface (figure 17.11). Some springs discharge where the water table intersects the land surface, but they also occur where water flows out from caverns or along fractures, faults, or rock contacts that come to the surface (figure 17.12).

Climate determines the relationship between stream flow and the water table. In rainy regions most streams are **gaining streams;** that is, they receive water from the saturated zone (figure 17.13*A*). The surface of these streams coincides with the water table. Water from the saturated zone flows into the stream through the stream bed and banks that lie below the water table. Because of the added ground water, the discharge of these streams increases downstream. Where the water table intersects the land surface over a broad area, ponds, lakes, and swamps are found.

In drier climates rivers tend to be **losing streams;** that is, they are losing water to the saturated zone (figure 17.13*B*). The channels of losing streams lie above the water table. The water

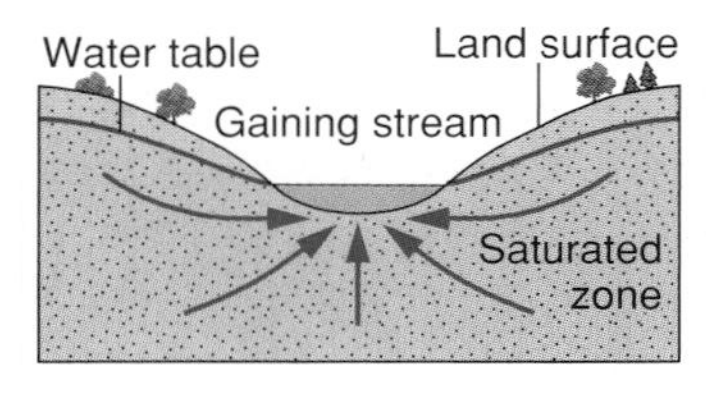

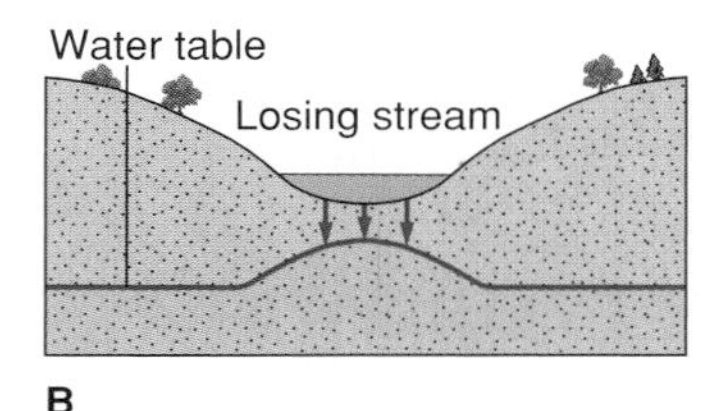

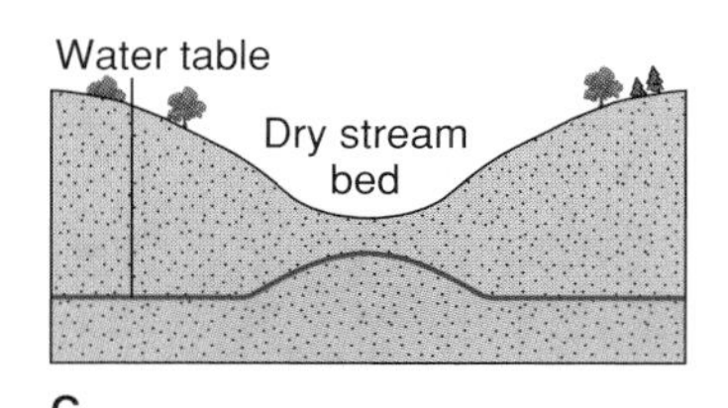

FIGURE 17.13

Gaining and losing streams. (*A*) Stream gaining water from saturated zone. (*B*) Stream losing water through stream bed to saturated zone. (*C*) Water table can be close to the land surface beneath a dry stream bed.

percolating into the ground beneath a losing stream may cause the water table below the stream to rise. This ground-water mound remains beneath the stream even when the stream bed is dry, and in a desert the nearest source of water may lie a short distance under a dry stream bed (figure 17.13*C*).

CONTAMINATION OF GROUND WATER

Ground water in its natural state tends to be relatively free of contaminants in most areas. Because it is a widely used source of drinking water, contamination of ground water can be a very serious problem.

Pesticides and *herbicides* (such as DDT and 2,4-D) applied to agricultural crops (figure 17.14*A*) can find their way into ground water when rain or irrigation water leaches the poisons downward into the soil. *Fertilizers* are also a concern. Nitrate, one of the most widely used fertilizers, is harmful in even small quantities in drinking water.

A

B

C

D

FIGURE 17.14

Some sources of ground-water pollution. (*A*) Pesticides. (*B*) Household garbage. (*C*) Animal waste. (*D*) Industrial toxic waste.

Photo *A* by Michael Stimmann; photo *B* by Frank M. Hanna; photos *C* and *D* from USDA-Soil Conservation Service

Rain can also leach pollutants from city landfills into ground-water supplies (figure 17.14*B*). Consider for a moment some of the things you threw away last year. A partially empty aerosol can of ant poison? The can will rust through in the landfill, releasing the poison into the ground and into the saturated zone below. A broken thermometer? The toxic mercury may eventually find its way to the ground-water supply. A half-used can of oven cleaner? The dried-out remains of a can of lead-base paint? *Heavy metals* such as mercury, lead, chromium, copper, and cadmium, together with household chemicals and poisons, can all be concentrated in ground-water supplies beneath landfills (figure 17.15).

Liquid and solid wastes from septic tanks, sewage plants, and animal feedlots and slaughterhouses may contain *bacteria, viruses,* and *parasites* that can contaminate ground water (figure 17.14*C*). Liquid wastes from industries (figure 17.14*D*) and military bases can be highly toxic, containing high concentrations of heavy metals and compounds such as cyanide and *PCBs* (polychlorinated biphenyls), which are widely used in industry. A degreaser called *TCE* (trichloroethylene) has been increasingly found to pollute both surface and underground water in numerous regions. Toxic liquid wastes are often held in surface ponds or pumped down deep disposal wells. If the ponds leak, ground water can become polluted. Deep wells may be safe for liquid waste disposal if they are deep enough, but in some localities improper design of the disposal wells has resulted in contamination of drinking water supplies and even surface water.

Acid mine drainage from coal and metal mines can contaminate both surface and ground water. It is usually caused by

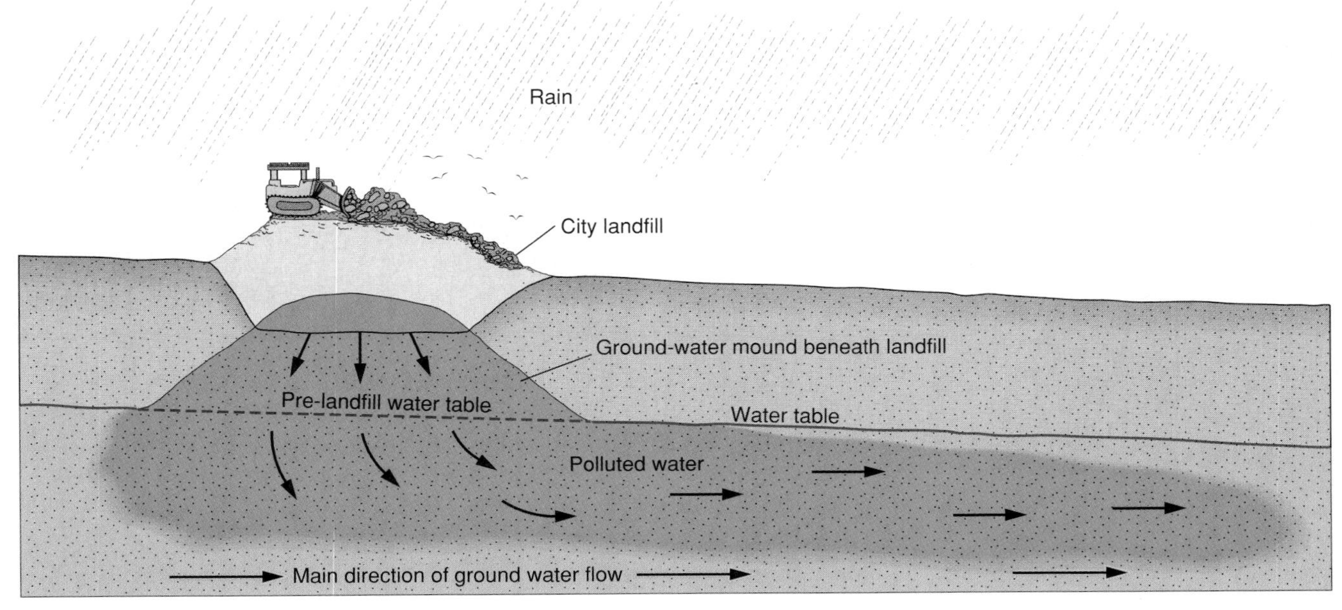

A Cross section

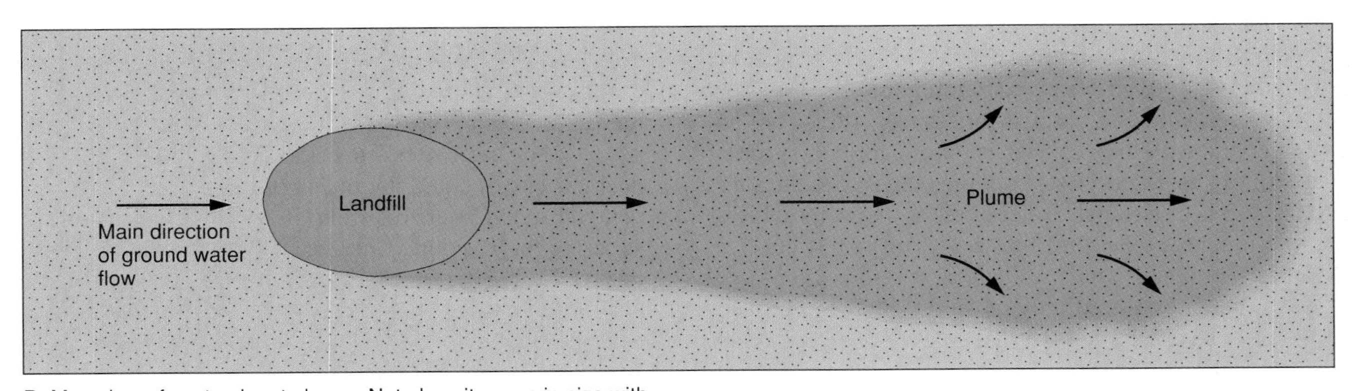

B Map view of contaminant plume. Note how it grows in size with distance from the pollution source.

FIGURE 17.15

Waste piled on the land surface creates a ground-water mound beneath it because the landfill forms a hill and because the waste material is more porous and permeable than the surrounding soil and rock. Rain leaches pollutants into the saturated zone. A plume of contaminated water will spread out in the direction of ground water flow.

17.3 ENVIRONMENTAL GEOLOGY

Hard Water and Soapsuds

"Hard water" is water that contains relatively large amounts of dissolved calcium (often from the chemical weathering of calcite or dolomite) and magnesium (from the ferromagnesian minerals or dolomite). Water taken from the ground-water supply or from a stream for home use may contain enough of these ions to prevent soap from lathering. Calcium and magnesium ions in hard water form gray curds with soap. The curd continues to form until all the calcium and magnesium ions are removed from the water and bound up in the curd. Only then will soap lather and clean laundry. Cleaning laundry in hard water therefore takes an excessively large amount of soap.

Hard water may also precipitate a scaly deposit inside teakettles and hot-water tanks and pipes (box figure 1). The entire hot-water piping system of a home in a hard-water area eventually can become so clogged that the pipes must be replaced.

"Soft water" may carry a substantial amount of ions in solution but not the ions that prevent soap from lathering. Water softeners in homes replace calcium and magnesium ions with sodium ions, which do not affect lathering or cause scale. But water containing a large amount of sodium ions, whether from a softener or from natural sources, may be harmful if used as drinking water by persons on a "salt-free" (low sodium) diet for health reasons.

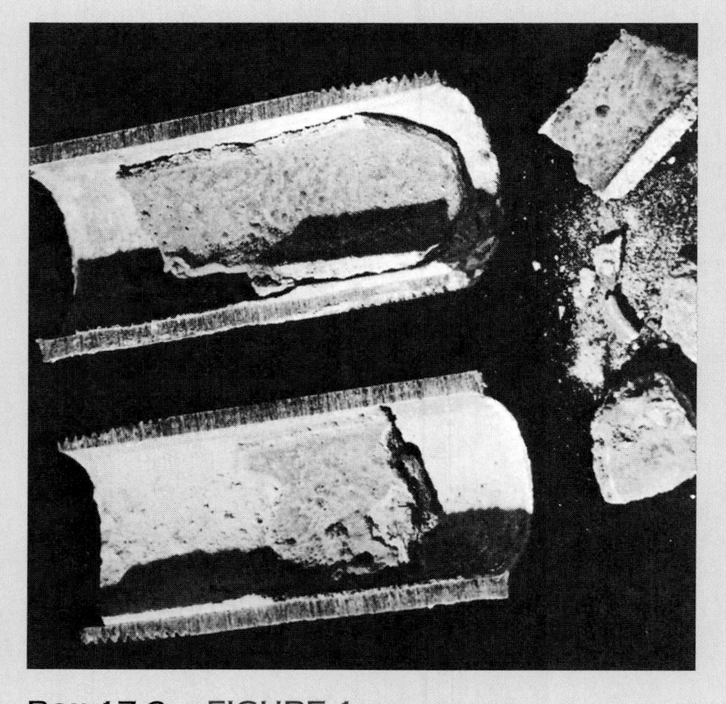

Box 17.3—FIGURE 1
Scale caused by hard water coats the inside of this hot-water pipe. Photo by Hauser Water Conditioning, Inc.

sulfuric acid formed by the oxidation of sulfur in pyrite and other sulfide minerals when they are exposed to air by mining activity. Fish and plants are often killed by the acid waters draining from long-abandoned mines.

Radioactive waste is both an existing and a very serious potential source of ground-water contamination. The shallow burial of *low-level* solid and liquid radioactive wastes from the nuclear power industry has caused contamination of ground water, particularly as liquid waste containers leak into the saturated zone and as the seasonal rise and fall of the water table at some sites periodically covers the waste with ground water. The search for a permanent disposal site for solid, *high-level* radioactive waste (now stored temporarily on the surface) is a major national concern for the United States. The permanent site will be deep underground and must be isolated from ground-water circulation for thousands of years. Salt beds, shale, glassy tuffs, and crystalline rock deep beneath the surface have all been studied, particularly in arid regions where the water table is hundreds of meters below the land surface. The likely site for disposal of high-level waste, primarily spent fuel from nuclear reactors, is Yucca Mountain, Nevada, 180 km northwest of Las Vegas. The site would be deep underground in volcanic tuff well above the current (or predicted future) water table, and in a region of very low rainfall. The U.S. Congress, under intense political pressure from other candidate states who did not want the site, essentially chose Nevada in late 1988 by eliminating the funding for the study of all alternative sites, but the final decision regarding the safety of Yucca Mountain will not be made until after much additional study. Even if the site is deemed safe, it could not open before the year 2010. It could be delayed much later than this: in 1992, a 5.6-magnitude aftershock of the Landers, California, earthquake occurred 19 kilometers from the proposed disposal site. The quake caused $1 million damage to a U.S. Energy Department office building near the site and may indicate that the region is too seismically active for the site to be built here at all.

Not all ground-water contaminants form plumes within the saturated zone as shown in figure 17.15. *Gasoline,* which leaks from gas station storage tanks at tens of thousands of U.S. locations, is less dense than water, and floats upon the water table (figure 17.16). Some liquids such as TCE are heavier than water and sink to the bottom of the saturated zone, perhaps traveling in unpredicted

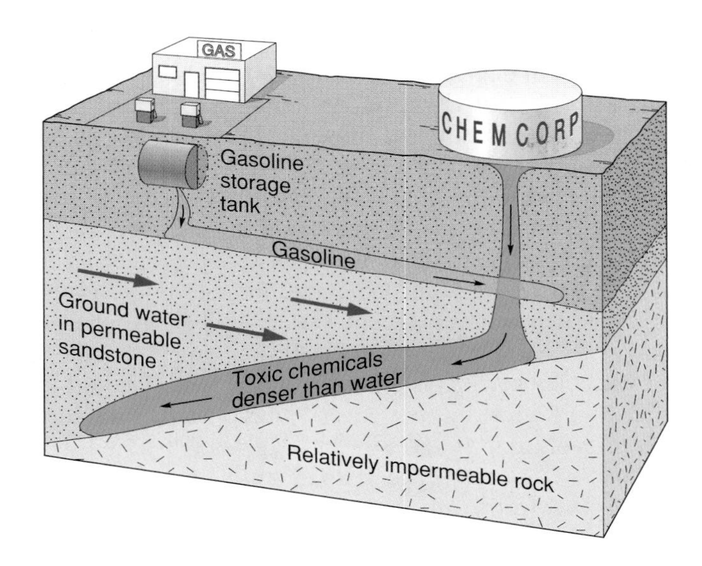

FIGURE 17.16

Not all contaminants move within the saturated zone as shown in figure 17.15. Gasoline floats on water; many dense chemicals move along impermeable rock surfaces below the saturated zone.

directions upon the surface of an impermeable layer (figure 17.16). Determining the extent and flow direction of ground-water pollution is a lengthy process requiring the drilling of tens, or even hundreds, of costly wells for each contaminated site.

Not all sources of ground-water contamination are human-made. Naturally occurring *minerals within rock and soil* may contain elements such as arsenic, selenium, mercury, and other toxic metals. Circulating ground water can leach these elements out of the minerals and raise their concentrations to harmful levels within the water. Not all spring water is safe to drink. Like a "bad waterhole" depicted in a Western movie, some springs contain such high levels of toxic elements that the water can sicken or kill humans and animals that drink it. Many desert springs contain such high concentrations of sodium chloride or other salts that their water is undrinkable.

Soil and rock filter some contaminants out of ground water. This filtering ability depends on the permeability and mineral composition of rock and soil. Under ideal conditions, human sewage can be purified by only 30 to 45 meters of travel through a sandy loam soil (a mixture of clay minerals, sand, and organic humus). The sewage is purified by filtration, ion absorption by clay minerals and humus, and decomposition by soil organisms (figure 17.17). On the other hand, extremely permeable rock, such as highly fractured granite or cavernous limestone, has

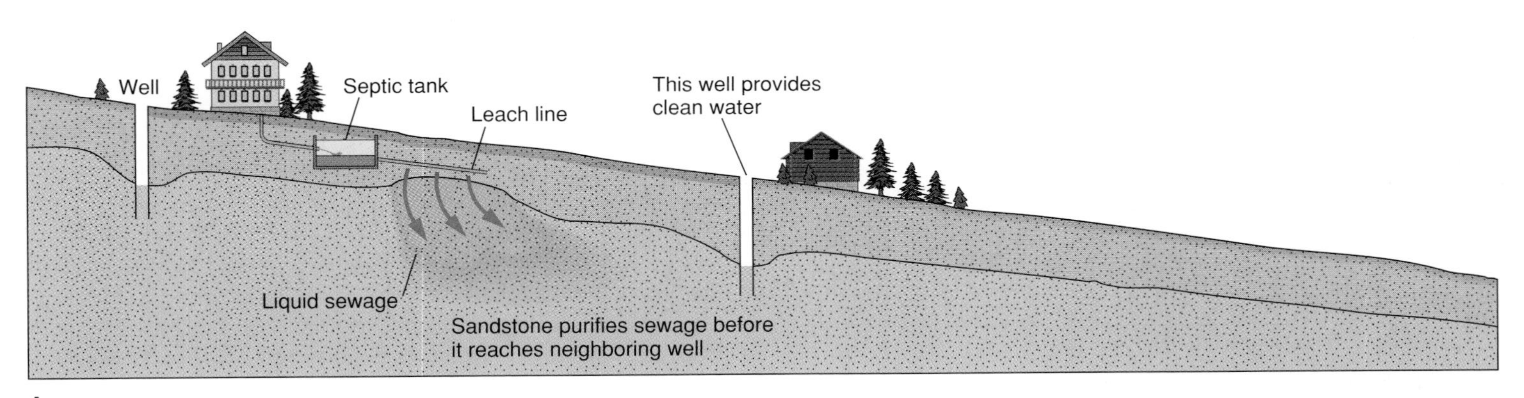

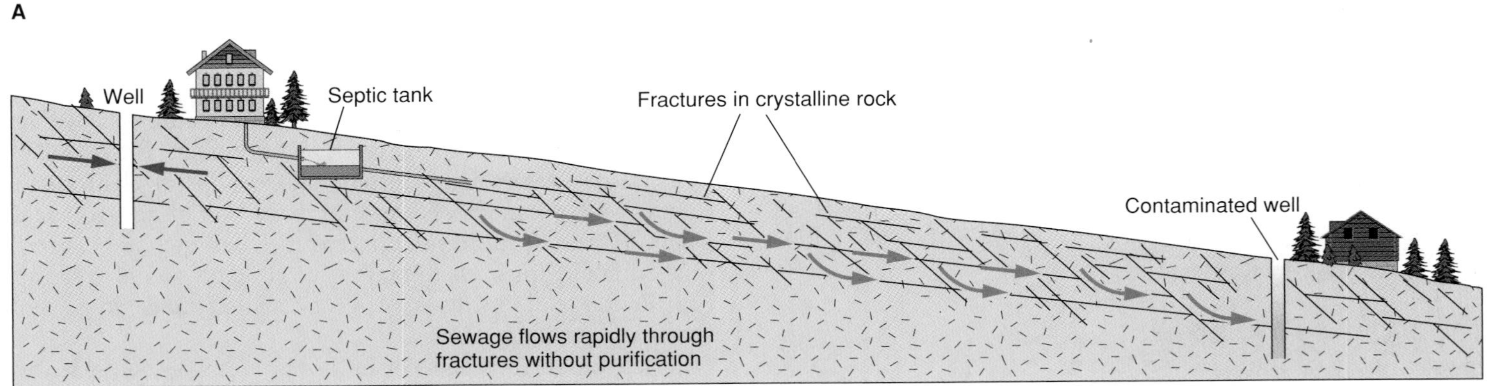

FIGURE 17.17

Rock type and distance control possible sewage contamination of neighboring wells. (*A*) As little as 30 meters of movement can effectively filter human sewage in sandstone and some other rocks and sediments. (*B*) If the rock has large open fractures, contamination can occur many hundreds of meters away.

little purifying effect on sewage. Ground water flows so rapidly through such rocks that it is not purified even after hundreds of meters of travel. Some pesticides and toxic chemicals are not purified by passage through rock and soil at all, even soil rich in humus and clay minerals.

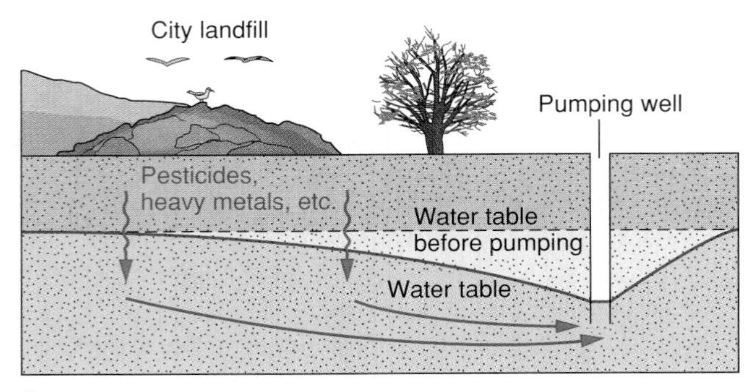

A

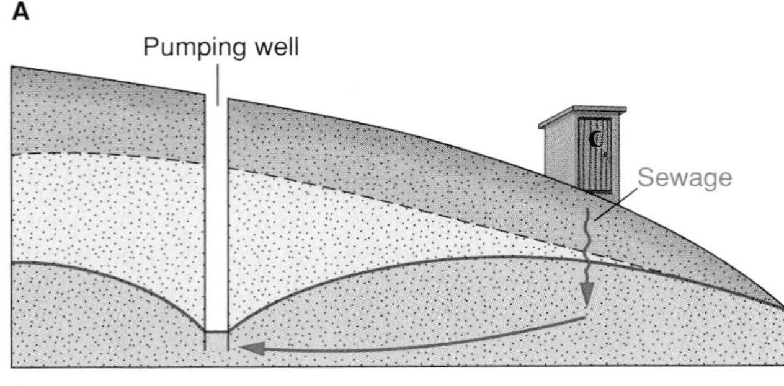

B

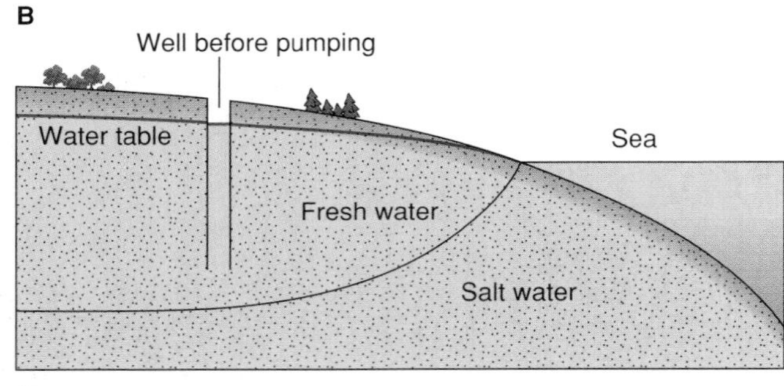

C

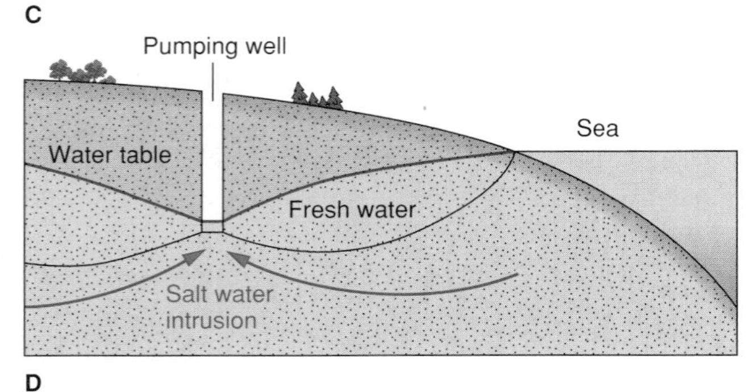

D

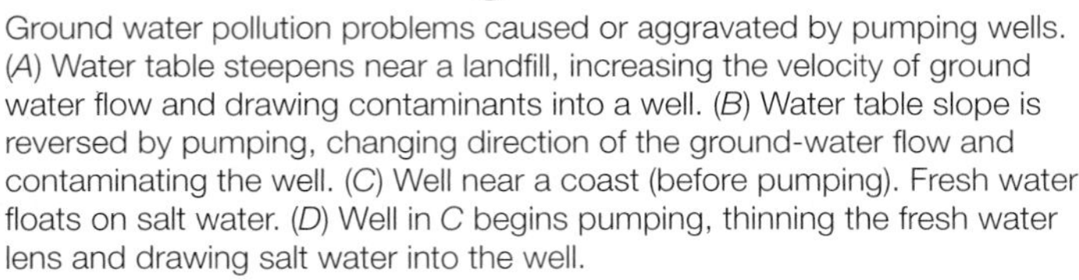

FIGURE 17.18

Ground water pollution problems caused or aggravated by pumping wells. (*A*) Water table steepens near a landfill, increasing the velocity of ground water flow and drawing contaminants into a well. (*B*) Water table slope is reversed by pumping, changing direction of the ground-water flow and contaminating the well. (*C*) Well near a coast (before pumping). Fresh water floats on salt water. (*D*) Well in *C* begins pumping, thinning the fresh water lens and drawing salt water into the well.

Contaminated ground water is extremely difficult to clean up. Networks of expensive wells may be needed to pump contaminated water out of the ground and replace it with clean water. Because of the slow movement and large volume of ground water, the cleanup process for a large region can take decades and tens of millions of dollars to complete.

Ground-water contamination can be largely prevented with careful thought and considerable expense. A city landfill can be sited high above the water table and possible flood levels, or located in a region of ground-water discharge rather than recharge. A site can be sealed below by impermeable (and expensive) clay barriers and plastic liners, and sealed off from rainfall by an impermeable cover. Dikes can prevent surface runoff through or from the site. Although sanitary landfills are expensive, it is much cheaper than ground-water cleanup.

Pumping wells can cause or aggravate ground-water contamination (figure 17.18). Well drawdown can increase the slope of the water table locally, thus increasing the rate of ground-water flow and giving the water less time to be purified underground before it is used (figure 17.18*A*). Drawdown can even reverse the original slope of the water table, perhaps contaminating wells that were pure before pumping began (figure 17.18*B*). Heavily pumped wells near a coast can be contaminated by *saltwater intrusion* (figure 17.18*C* and *D*).

BALANCING WITHDRAWAL AND RECHARGE

A local supply of ground water will last indefinitely if it is withdrawn for use at a rate equal to or less than the rate of recharge to the aquifer. If ground water is withdrawn faster than it is being recharged, however, the supply is being reduced and will one day be gone.

Heavy use of ground water causes a regional water table to drop. In parts of western Texas and eastern New Mexico the pumping of ground water has caused the water table to drop 30 meters over the past few decades. The lowering of the water table means that wells must be deepened and more electricity must be used to pump the water to the surface. Moreover, as water is withdrawn, the ground surface may settle because the water no longer supports the rock and sediment. Mexico City has subsided more than 7 meters and portions of California's Central Valley 9 meters because of extraction of ground water (figure 17.19). Such *subsidence* can crack building foundations, roads, and pipelines. Overpumping of ground water also causes compaction and porosity loss in rock and soil and can permanently ruin good aquifers.

To avoid the problems of falling water tables, subsidence, and compaction, many towns use *artificial recharge* to increase the natural rate of recharge. Natural floodwaters or treated industrial or domestic wastewaters are stored in infiltration ponds on the surface to increase the rate of water percolation into the ground. Reclaimed, clean water from sewage

FIGURE 17.19
Subsidence of the land surface caused by the extraction of ground water, near Mendota, San Joaquin Valley, California. Signs on the pole indicate the positions of the land surface in 1925, 1955, and 1977. The land sank 9 meters in 52 years. Since the late 1970s, subsidence decreased to less than a meter due to reduced ground water pumping and increased use of surface water for irrigation.
Photo by Richard O. Ireland, U.S. Geological Survey

treatment plants is commonly used for this purpose. In some cases, especially in areas where ground water is under confined conditions, water is actively pumped down into the ground to replenish the ground-water supply. This is more expensive than filling surface ponds, but it reduces the amount of water lost through evaporation.

EFFECTS OF GROUND-WATER ACTION

Caves, Sinkholes, and Karst Topography

Caves (or **caverns**) are naturally formed underground chambers. Most caves develop when slightly acidic ground water dissolves limestone along joints and bedding planes, opening up cavern systems as calcite is carried away in solution (figure 17.20). Natural ground water is commonly slightly acidic because of dissolved carbon dioxide (CO_2) from the atmosphere or from soil gases (see chapter 12).

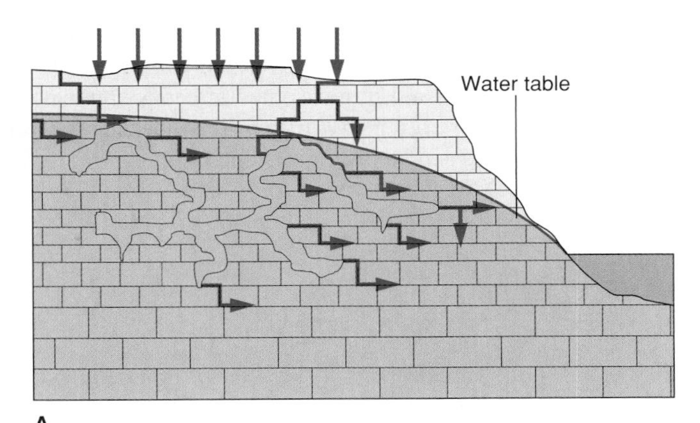

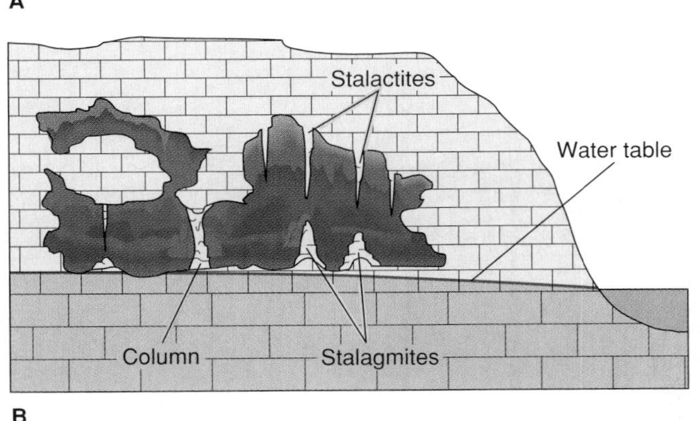

FIGURE 17.20
Solution of limestone to form caves. (*A*) Water moves along fractures and bedding planes in limestone, dissolving it to form caves below the water table. (*B*) Falling water table allows cave system, now greatly enlarged, to fill with air. Calcite precipitation forms stalactites, stalagmites, and columns above the water table.

Geologists disagree whether limestone caves form above, below, or at the water table. Most caves probably are formed by ground water circulating below the water table, as shown in figure 17.20. If the water table drops or the land is elevated above the water table, the cave may begin to fill in again by calcite precipitation. The accompanying equation can be read from left to right for calcite solution, and from right to left for the calcite precipitation reaction (see also table 12.1).

Ground water with a high concentration of calcium (Ca^{++}) and bicarbonate (HCO_3^-) ions may drip slowly from the ceiling of an air-filled cave. As a water drop hangs on the ceiling of the cave, some of the dissolved carbon dioxide (CO_2) may be lost into the cave's atmosphere. The CO_2 loss causes a small amount of calcite to precipitate out of the water onto the cave ceiling. When the water drop falls to the cave floor, the impact may cause more CO_2 loss, and another small amount of calcite may precipitate on the cave floor. A falling water drop, therefore, can precipitate small amounts of calcite on both the cave ceiling and the cave floor and each subsequent drop adds more calcite to the first deposits.

Deposits of calcite (and, rarely, other minerals) built up in caves by dripping water are called *dripstone* or **speleothems. Stalactites** are icicle-like pendants of dripstone hanging from cave ceilings (figure 17.20*B*). They are generally slender and are commonly aligned along cracks in the ceiling, which act as conduits for ground water. **Stalagmites** are cone-shaped masses of dripstone formed on cave floors, generally directly below stalactites. Splashing water precipitates calcite over a large area on the cave floor, so stalagmites are usually thicker than the stalactites above them. As a stalactite grows downward and a stalagmite grows upward, they may eventually join to form a *column* (figure 17.20*B*). Figure 17.21 shows some of the intriguing features formed in caves.

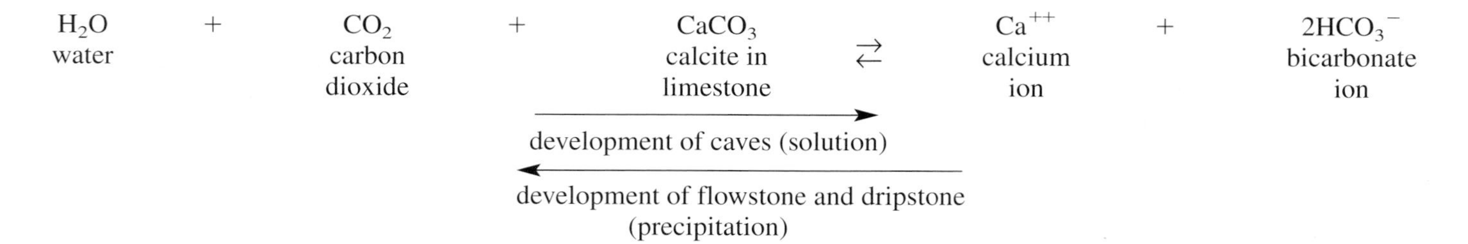

FIGURE 17.21
Stalactites, stalagmites, and flowstone in Great Onyx Cave, Kentucky.
Photo courtesy Stanley Fagerlin

In parts of some caves, water flows in a thin film over the cave surfaces rather than dripping from the ceiling. Sheetlike or ribbonlike *flowstone* deposits develop from calcite that is precipitated by flowing water on cave walls and floors.

The floors of most caves are covered with sediment, some of which is *residual clay,* the fine-grained particles left behind as insoluble residue when a limestone containing clay dissolves. (Some limestone contains only about 50% calcite.) Other sediment, including most of the coarse-grained material found on cave floors, may be carried into the cave by streams, particularly when surface water drains into a cave system from openings on the land surface.

Solution of limestone underground may produce features that are visible on the surface. Extensive cavern systems can undermine a region so that roofs collapse and form depressions in the land surface above. **Sinkholes** are closed depressions found on land surfaces underlain by limestone (figure 17.22). They form either by the *collapse* of a cave roof or by *solution* as descending water enlarges a crack in limestone. Limestone regions in Florida, Missouri, Indiana, and Kentucky are heavily dotted with sinkholes. Sinkholes can also form in regions underlain by gypsum or rock salt, which are also soluble in water.

An area with many sinkholes and with cave systems beneath the land surface is said to have **karst topography** (figure 17.23). Karst areas are characterized by a lack of surface streams, although one major river may flow at a level lower than the karst area.

Streams sometimes disappear down sinkholes to flow through caves beneath the surface. In this specialized instance, a true *underground stream* exists. Such streams are quite rare, however, as most ground water flows very slowly through pores and cracks in sediment or rock. You may hear people with wells describe the "underground stream" that their well penetrates, but this is almost never the case. Wells tap ground water in the rock pores and crevices, not underground streams. If a well did tap a true underground river in a karst region, the water would probably be too polluted to

A

B

FIGURE 17.22
(*A*) Sinkholes formed in limestone near Timaru, New Zealand. (*B*) A collapse sinkhole that formed suddenly in Winter Park, Florida, in 1981.
Photo *A* by G. R. 'Dick' Roberts. Photo *B* by Richard A. Davis, Jr.

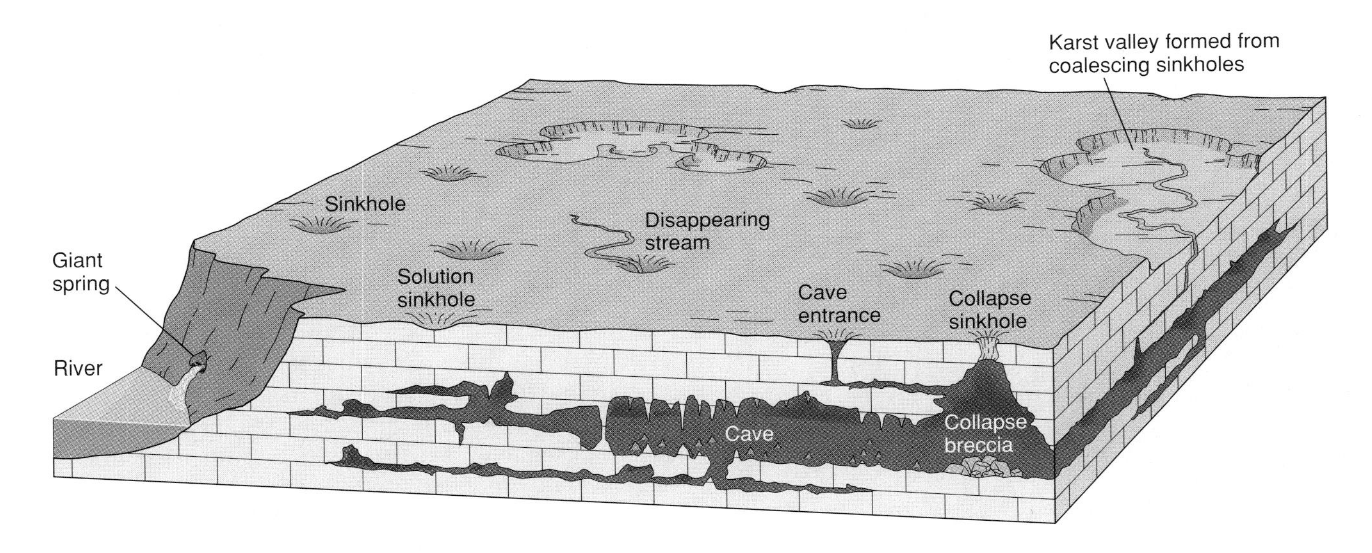

FIGURE 17.23
Karst topography is marked by underground caves and numerous surface sinkholes. A major river may cross the region, but small surface streams generally disappear down sinkholes.

drink, especially if it had washed down from the surface into a cavern without being filtered through soil and rock.

Other Effects

Ground water is important in the preservation of *fossils* such as **petrified wood,** which develops when porous buried wood is either filled in or replaced by inorganic silica carried in by ground water (figure 17.24). The result is a hard, permanent rock, commonly preserving the growth rings and other details of the wood. Calcite or silica carried by ground water can also replace the original material in marine shells and animal bones.

Sedimentary rock *cement,* usually silica or calcite, is carried into place by ground water. When a considerable amount of cementing material precipitates locally in a rock, a hard rounded mass called a **concretion** develops, typically around an organic nucleus such as a leaf, tooth, or other fossil (figure 17.25).

Geodes are partly hollow, globe-shaped bodies found in some limestones and locally in other rocks. The outer shell is amorphous silica, and well-formed crystals of quartz, calcite, or other minerals project inward toward a central cavity

FIGURE 17.24
Petrified log in the Painted Desert, Arizona. The log was replaced by silica carried in solution by ground water. Small amounts of iron and other elements color the silica in the log.
Photo © Eric & David Hosking/Corbis Media

FIGURE 17.25
Concretions that have weathered out of shale. Concretions contain more cement than the surrounding rock and therefore are very resistant to weathering.

(figure 17.26). The origin of geodes is complex but clearly related to ground water. Crystals in geodes may have filled original cavities or have replaced fossils or other crystals.

In arid and semiarid climates, *alkali soil* may develop because of the precipitation of great quantities of sodium salts by evaporating ground water. Such soil is generally unfit for plant growth. Alkali soil generally forms at the ground surface in low-lying areas. (See chapter 12.)

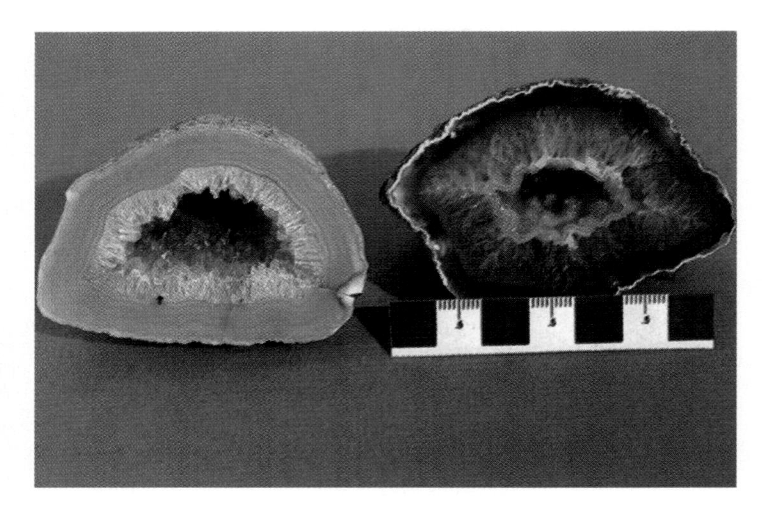

FIGURE 17.26
Geodes. Concentric layers of amorphous silica are lined with well-formed quartz crystals growing inward toward a central cavity. (Scale is in centimeters.)

HOT WATER UNDERGROUND

Hot springs are springs in which the water is warmer than human body temperature. Water can gain heat in two ways while it is underground. First, and more commonly, ground water may circulate near a magma chamber or a body of cooling igneous rock. In the United States most hot springs are found in the western states where they are associated with relatively recent volcanism. The hot springs and pools of Yellowstone National Park in Wyoming are of this type.

Ground water can also gain heat if it circulates unusually deeply in the earth, perhaps along joints or faults. As discussed in chapter 11, the normal geothermal gradient (the increase in temperature with depth) is 25°C/kilometer (about 75°F/mile). Water circulating to a depth of 2 or 3 kilometers is warmed substantially above normal surface water temperature. The famous springs at Warm Springs, Georgia, have been warmed by deep circulation. Warm water, regardless of its origin, is lighter than cold water and readily rises to the surface.

A **geyser** is a type of hot spring that periodically erupts hot water and steam. The water is generally near boiling (100°C). Eruptions may be caused by a constriction in the underground "plumbing" of a geyser, which prevents the water from rising and cooling. The events thought to lead to a geyser eruption are illustrated in figure 17.27. Water gradually seeps into a partially emptied geyser chamber and heat supplied from below slowly warms the water. Bubbles of water vapor and other gases then begin to form as the temperature of the water rises. The bubbles may clog the constricted part of the chamber until the upward pressure of the bubbles pushes out some of the water above in a gentle surge, thus lowering

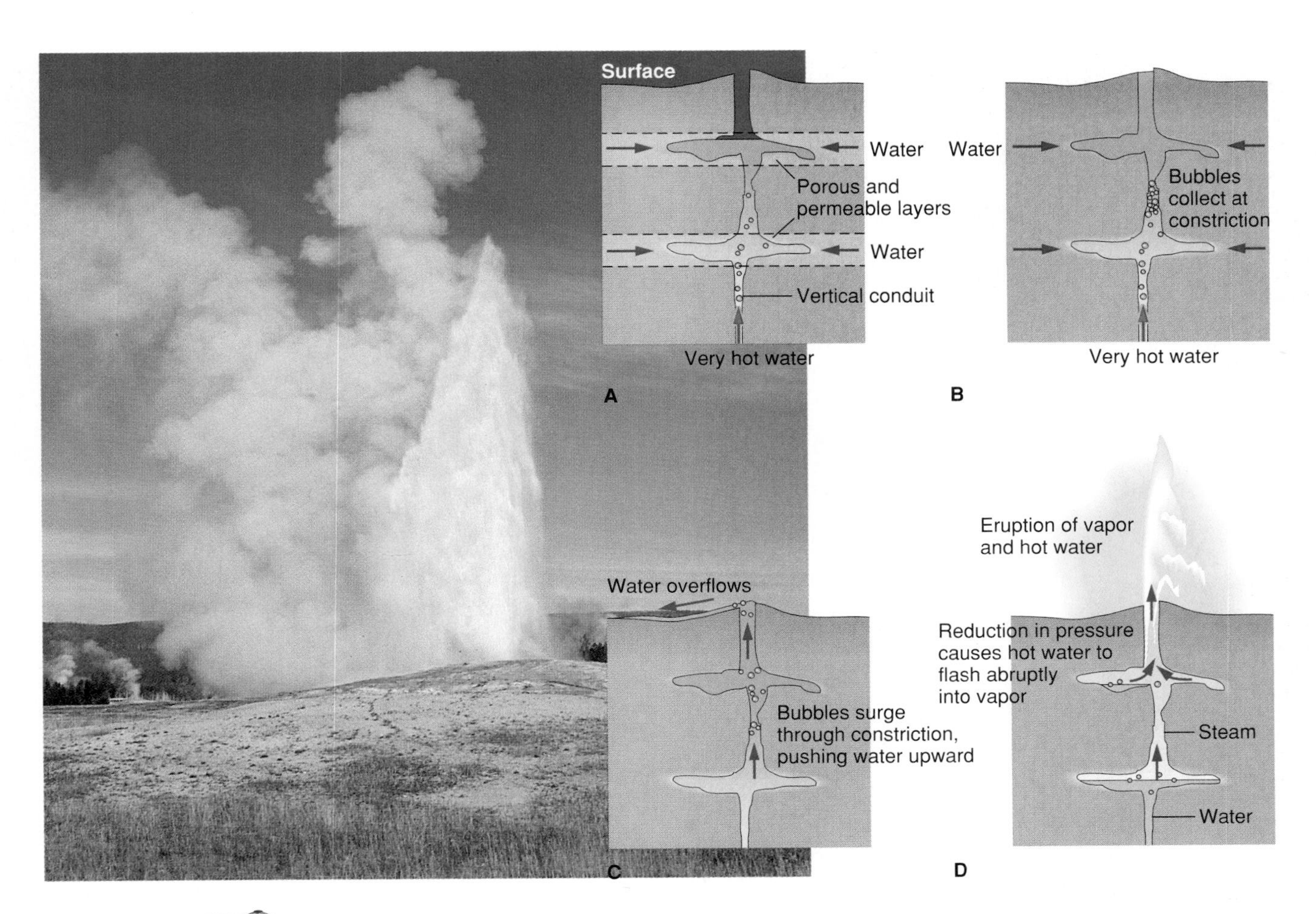

FIGURE 17.27

Eruptive history of a typical geyser in (*A*) through (*D*). Photo shows the eruption of Old Faithful geyser in Yellowstone National Park, Wyoming. See text for explanation.

Photo © Hal Beral/Visuals Unlimited

the pressure on the water in the lower part of the chamber. This drop in pressure causes the chamber water, now very hot, to flash into vapor. The expanding vapor blasts upward out of the chamber, driving hot water with it and condensing into visible steam. The chamber, now nearly empty, begins to fill again and the cycle is repeated. The entire cycle may be quite regular, as it is in Yellowstone's Old Faithful geyser, which averages about 79 minutes between eruptions (though it varies from about 45 to 105 minutes depending on the amount of water left in the chamber after an eruption). Many geysers, however, erupt irregularly, some with weeks or months between eruptions.

As hot ground water comes to the surface and cools, it may precipitate some of its dissolved ions as minerals. *Travertine* is a deposit of *calcite* that often forms around hot springs (figure 17.28), while dissolved *silica* precipitates as *sinter* (called *geyserite* when deposited by a geyser, as shown in figure 17.29). The composition of the subsurface rocks generally determines which type of deposit forms, although sinter can indicate higher subsurface temperatures than

FIGURE 17.28

Precipitation of calcite in the form of travertine terraces around a hot spring (Mammoth Hot Springs, Yellowstone National Park). Thermophyllic bacteria living in the hot water provide the color.

Photo by Diane Carlson

FIGURE 17.29

Geyserite deposits around the vent of Castle Geyser, Yellowstone National Park.

travertine because silica is harder to dissolve than calcite. Both deposits can be stained by the pigments of bacteria that thrive in the hot water. These thermophyllic bacteria are some of the most primitive of living bacteria and suggest that life may have arisen near hot springs.

A *mudpot* is a special type of hot spring that contains thick, boiling mud. Mudpots are usually marked by a small amount of water and strongly sulfurous gases, which combine to form strongly acidic solutions. The mud probably results from intense chemical weathering of the surrounding rocks by these strong acids (see figure 12.16).

FIGURE 17.30

Geothermal power plant at The Geysers, California. Underground steam, piped from wells to the power plant, is discharging from the cooling towers in the background.

Photo by M. Smith, U.S. Geological Survey

Geothermal Energy

Electricity can be generated by harnessing naturally occurring steam and hot water in areas that are exceptionally hot underground. In such a *geothermal area,* wells can tap steam (or superheated water that can be turned into steam) that is then piped to a powerhouse where it turns a turbine that spins a generator, creating electricity.

Geothermal energy production requires no burning of fuel, so the carbon dioxide emissions of power plants that burn coal, oil, or natural gas are not produced. Although geothermal energy is relatively clean, it has some environmental problems. Workers need protection from toxic hydrogen sulfide gas in the steam, and the hot water commonly contains dissolved ions and metals, such as lead and mercury, that can kill fish and plants if discharged on the surface. Geothermal fluids are often highly corrosive to equipment, and their extraction can cause land subsidence. Pumping the cooled wastewater underground can help reduce subsidence problems.

Geothermal fields can be depleted. The largest field in the world is at The Geysers in California (figure 17.30), 120 kilometers north of San Francisco. The Geysers field increased its capacity in recent years to 2,000 megawatts of electricity (enough for 2 million people), but production has declined, and the field may soon run out of steam.

Nonelectric uses of geothermal energy include space heating (in Boise, Idaho; Klamath Falls, Oregon; and Reykjavik, the capital of Iceland) as well as paper manufacturing, ore processing, and food preparation.

Summary

About 15 percent of the water that falls on land percolates underground to become ground water. Ground water fills pores and joints in rock, creating a large reservoir of usable water in most regions.

Porous rocks can hold water. *Permeable* rocks permit water to move through them.

The *water table* is the top surface of the *saturated zone* and is overlain by the *vadose zone.*

Local variations in rock permeability may develop a *perched water table* above the main water table.

Ground-water velocity depends on rock permeability and the slope of the water table.

An *aquifer* is porous and permeable and can supply water to wells. A *confined aquifer* holds water under pressure, which can create *artesian wells.*

Gaining streams, springs, and lakes form where the water table intersects the land surface. *Losing streams* contribute to the ground water in dry regions.

Ground water can be contaminated by city landfills, agriculture, industry, or sewage disposal. Some pollutants can be filtered out by passage of the water through moderately permeable geologic materials.

A pumped well causes a *cone of depression* that in turn can cause or aggravate ground-water pollution. Near a coast, it can cause *saltwater intrusion.*

Artificial *recharge* can help create a balance between withdrawal and recharge of ground-water supplies, and help prevent subsidence.

Solution of limestone by ground water forms *caves, sinkholes,* and *karst topography.* Calcite precipitating out of ground water forms *speleothems* such as *stalactites* and *stalagmites* in caves.

Precipitation of material out of solution by ground water helps form petrified wood, other fossils, sedimentary rock cement, concretions, geodes, and alkali soils.

Geysers and *hot springs* occur in regions of hot ground water. *Geothermal energy* can be tapped to generate electricity.

Terms to Remember

aquifer 427
artesian well 429
cave (cavern) 437
concretion 439
cone of depression 429
confined (artesian) aquifer 428
drawdown 429
gaining stream 431
geode 439
geyser 440
ground water 424
hot spring 440
karst topography 438
losing stream 431
perched water table 425
permeability 424
petrified wood 439
porosity 424
recharge 429
saturated zone 425
sinkhole 438
speleothem 438
spring 431
stalactite 438
stalagmite 438
unconfined aquifer 428
vadose zone 425
water table 425
well 429

Testing Your Knowledge

Use the questions below to prepare for exams based on this chapter.

1. What conditions are necessary for an artesian well?
2. What distinguishes a geyser from a hot spring? Why does a geyser erupt?
3. What is karst topography? How does it form?
4. What chemical conditions are necessary for caves to develop in limestone? For stalactites to develop in a cave?
5. What causes a perched water table?
6. Describe several ways in which ground water can become contaminated.
7. Discuss the difference between porosity and permeability.
8. What is the water table? Is it fixed in position?
9. Sketch four different origins for springs.
10. What controls the velocity of ground-water flow?
11. Name several geologic materials that make good aquifers. Define *aquifer.*
12. How does petrified wood form?
13. What happens to the water table near a pumped well?
14. How does a confined aquifer differ from an unconfined aquifer?
15. Porosity is
    a. the percentage of a rock's volume that is openings
    b. the capacity of a rock to transmit a fluid
    c. the ability of a sediment to retard water
    d. none of the above

16. Permeability is
    a. the percentage of a rock's volume that is openings
    b. the capacity of a rock to transmit a fluid
    c. the ability of a sediment to retard water
    d. none of the above

17. The subsurface zone in which all rock openings are filled with water is called the
    a. saturated zone
    b. water table
    c. vadose zone

18. An aquifer is
    a. a body of saturated rock or sediment through which water can move easily
    b. a body of rock that retards the flow of ground water
    c. a body of rock that is impermeable

19. Which rock type would make the best aquifer?
    a. shale
    b. mudstone
    c. sandstone
    d. all of the above

20. Which of the following determines how quickly ground water flows?
    a. elevation
    b. water pressure
    c. permeability
    d. all of the above

21. Ground water flows
    a. always downhill
    b. from areas of high hydraulic head to low hydraulic head
    c. from high elevation to low elevation
    d. from high permeability to low permeability

22. The drop in the water table around a pumped well is the
    a. drawdown
    b. hydraulic head
    c. porosity
    d. fluid potential

Expanding Your Knowledge

1. Describe any difference between the amounts of water that would percolate downward to the saturated zone beneath a flat meadow in northern New York and beneath a rocky hillside in southern Nevada. Discuss the factors that control the amount of percolation in each case.
2. Where should high-level nuclear waste from power plants be stored? If your state or community uses nuclear power, where is your local waste stored?
3. Should all contaminated ground water be cleaned up? How much money has been set aside by the federal government for cleaning polluted ground water? Who should pay for ground-water cleanup if the company that polluted the water no longer exists? Should some aquifers be deliberately left contaminated if there is no current use of the water, or if future use could be banned?
4. Why are most of North America's hot springs and geysers in the western states and provinces?

Exploring Web Resources

www.mhhe.com/plummer9e

Visit our Online Learning Center for additional readings, interactive quizzes, and answers to Testing your Knowledge. Watch animated presentations on artesian wells and geyser eruptions. Explore the other sites listed below by clicking on their direct link found on this website.

http://toxics.usgs.gov/toxics/

Various sites and information about cleanup of toxics in surface and ground water.

http://water.usgs.gov/public/wid/html/bioremed.html

Information about using bioremediation to clean up toxics in the soil, surface, and ground water.

http://water.wr.usgs.gov/gwatlas/index.html

Ground Water Atlas for the United States. Good general information about aquifers.

http://water.usgs.gov/

Good general website that has a lot of links to water topics in the United States from the USGS.

www.caves.org/

Home page of the *National Speleological Society* contains links to web pages of local interest and access to the NSS bookstore.

Animations

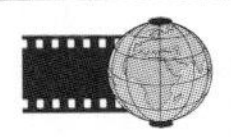

This chapter includes the following animations available on our Online Learning Center at www.mhhe.com/plummer9e.

17.9 Artesian Well and Confined Aquifer
17.18 Pumping Wells
17.27 Geyser Eruption

CHAPTER

18

Deserts and Wind Action

Deserts have a distinctive appearance because a dry climate controls erosional and depositional processes and the rates at which they operate. Although it seldom rains in the desert, running water is the dominant agent of land sculpture. Flash floods cause most desert erosion and deposition, even though they are rare events.

In chapters 13, 16, and 17 you have seen how the land is sculptured by mass wasting, streams, and ground water; in chapter 19, you will learn about glaciers. Here we discuss the fifth agent of erosion and deposition: wind. Deserts and wind action are discussed together because of the wind's particular effectiveness in dry regions. But wind erosion and deposition can be very significant in other climates as well.

Opposite: Totem Pole Buttes, Monument Valley Tribal Park, Navajo Nation, Arizona.
Photo by © Craig Blacklock/Larry Ulrich Stock

FIGURE 18.1

Typical appearance of the southwestern United States desert. Widely spaced plants have adapted to less than 25 centimeters of rain per year.

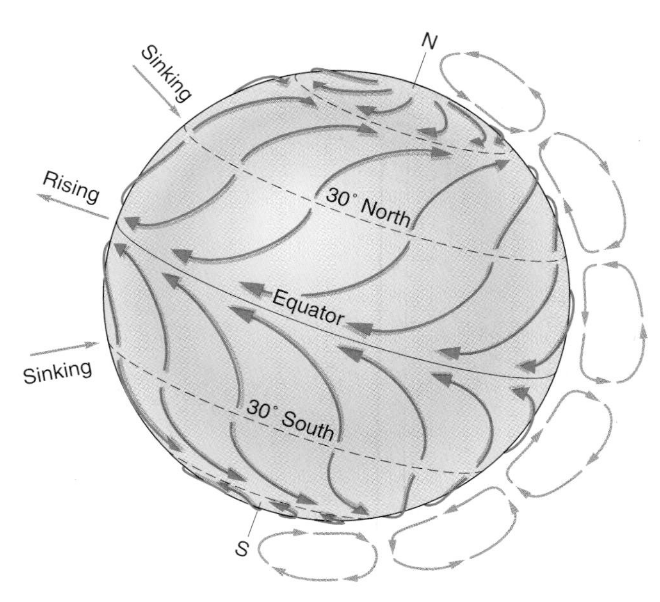

FIGURE 18.2

Global air circulation. Red arrows show surface winds. Blue arrows show vertical circulation of air. Air sinks at 30°N and 30°S latitude (and at the poles).

The word *desert* may suggest shifting sand dunes. Although moviemakers usually film sand dunes to represent deserts, only small portions of most deserts are covered with dunes. Actually a **desert** is any region with low rainfall. A region is usually classified as a desert if it has a dry or *arid climate* with less than 25 centimeters of rain per year. Few plants can tolerate low rainfall, so most deserts have a barren look.

Some specialized types of plants, however, grow well in desert climates despite the dryness. These plants are generally salt-tolerant; and they have extensive root systems to conserve water, so they often are widely spaced (figure 18.1). The leaves are usually very small, minimizing water loss by transpiration; they may even drop off the plants between rainstorms. During much of the year many desert plants look like dead, dry sticks. When rain does fall on the desert, the plants become green, and many will bloom.

DISTRIBUTION OF DESERTS

The location of most deserts is related to descending air. The global pattern of air circulation is shown in simplified form in figure 18.2. The equator receives the sun's heat more directly than the rest of the earth. Air warms and rises at the equator, then moves both northward and southward to sink near 30° North latitude and 30° South latitude. The world's best-known deserts lie in a belt 10°–15° wide centered on 30° North and South latitude (figure 18.3).

Air sinking down through the atmosphere is compressed by the weight of the air above it. As air compresses, it warms up; and as it warms, it is able to hold more water vapor. Evaporation of water from the land surface into the warm, dry air is so great under belts of sinking air that moisture seldom falls back to earth in the form of rain. The two belts at 30° North and South latitude characteristically have clear skies, much sunshine, little rain, and high evaporation.

In contrast to the belts centered on 30°, the equator is marked by rising air masses that expand and cool as they rise. In cooling, the air loses its moisture, causing cloudy skies and heavy precipitation. Thus a belt of high rainfall at the equator separates the two major belts of deserts.

Not all deserts lie near the 30° latitude belts. Some of the world's deserts are the result of the **rain shadow** effect of mountain ranges (figure 18.4). As moist air is forced up to pass over a mountain range, it expands and cools, losing moisture as it rises. The dry air coming down the other side of the mountain compresses and warms, bringing high evaporation with little or no rainfall to the downwind side of the range. This dry region downwind of mountains is the *rain shadow zone.* Parts of the southwestern United States desert in Nevada and northern Arizona are largely the result of the rain shadow effect of the Sierra Nevada range in eastern California.

Great distance from the ocean is another factor that can create deserts, since most rainfall comes from water evaporated from the sea. The dry climate of the large arid regions in China, well north of 30° North latitude, is due to their location in a continental interior and to the rain shadow effect of mountains such as the Himalaya.

Deserts also tend to develop on tropical coasts next to *cold ocean currents.* Cold currents run along the western edges of continents, cooling the air above them. The cold marine air warms up as it moves over land, causing high evaporation and little rain on the coasts. This effect is particularly pronounced on the Pacific coast of South America and the Atlantic coast of Africa and to a lesser extent western Australia.

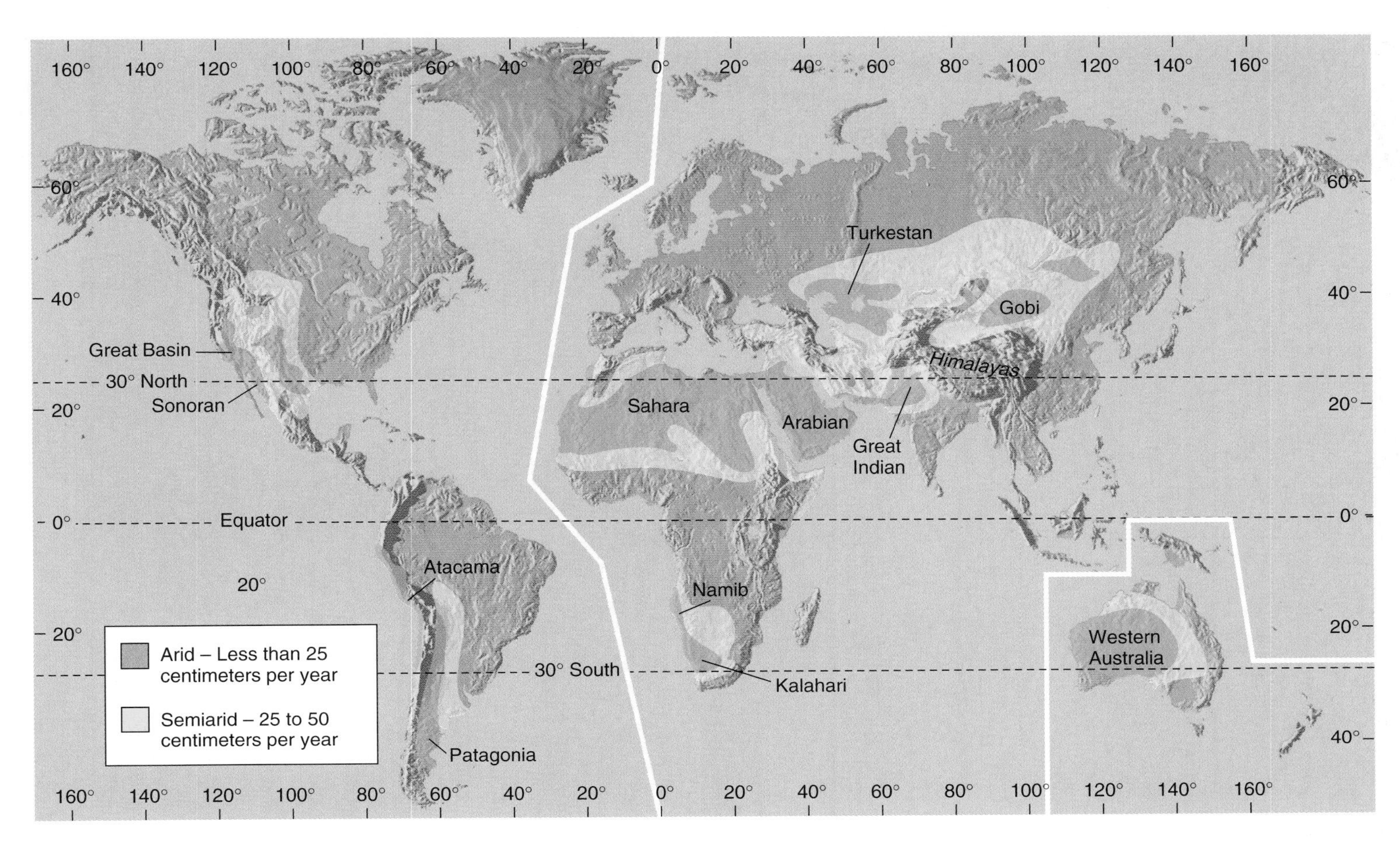

FIGURE 18.3

World distribution of nonpolar deserts. Most deserts lie in two bands near 30°N and 30°S.

From map by U.S. Department of Agriculture

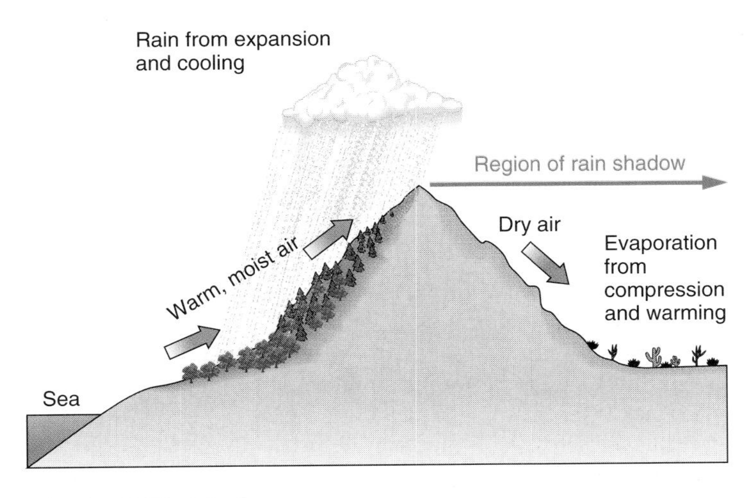

FIGURE 18.4

Rain shadow causes deserts on the downwind side of mountain ranges. Prevailing winds are from left to right.

Not all deserts are hot. The cold, descending air near the North and South Poles (figure 18.2) creates *polar deserts* that have an arid climate along with a snow or ice cover. The entire continent of Antarctica is a desert, as are most of Greenland and the northernmost parts of Alaska, Canada, and Siberia.

SOME CHARACTERISTICS OF DESERTS

Because of their low rainfall, deserts have characteristic drainage and topography that differ from those of humid regions. Desert streams usually flow intermittently. Water runs over the surface after storms, but during most of the year stream beds are dry. As a result, most deserts *lack through-flowing streams.* The Colorado River in the southwestern United States and the Nile River in Egypt are notable exceptions. Both are fed by heavy rainfall in distant mountains. The runoff is great enough to sustain stream flow across dry regions with high evaporation.

18.1 ENVIRONMENTAL GEOLOGY

EXPANDING DESERTS

Many geologists and geographers use a two-part definition of *desert.* A desert must have less than 25 centimeters of rain per year or must be so devoid of vegetation that few people can live there. Many dry regions have supported marginally successful agriculture and moderate human populations in the past but are being degraded into barren deserts today by overgrazing and overpopulation. The expansion of barren deserts into once-populated regions is called *desertification.*

Limited numbers of people can exist in dry regions through careful agricultural practices that protect water sources and limit grazing of sparse vegetation. Overuse of the land by livestock and humans, however, can strip it bare and make it uninhabitable. The large desert in northern Africa, shown in figure 18.3, is the Sahara Desert, and the semiarid region (25 to 50 centimeters of rain per year) to the south of it is the Sahel. In the early 1960s a series of abnormally wet years encouraged farmers in the Sahel to expand their herds and grazing lands.

A severe drought throughout the 1970s and 1980s caused devastation of the plant life of the region as starving livestock searched desperately for food, and humans gathered the last remaining sticks for firewood (box figure 1). Vast areas that were once covered with trees and sparse grass became totally barren, and an acute famine began, killing more than 100,000 people. The desert expanded southward, advancing in some places as much as 50 kilometers per year. The denuded soil in many regions became susceptible to wind erosion, leading to choking dust storms and new, advancing dune fields (some even migrating into cities).

Some of the same problems afflicted the Midwestern United States in the 1930s, as intense land cultivation coupled with a prolonged drought produced the barren Dust Bowl during the time of the Great Depression. Renewed rains and improved soil-conservation practices have reversed the trend in the United States, but the area is still vulnerable to a future drought, and the possibility of future dust storms in prairie states is very real.

Box 18.1 — Figure 1
Desertification in Africa after a period of drought has left inhabitants desperately searching for water.
Photo © Walt Anderson/Visuals Unlimited

Drought accelerates desertification but is not necessary for it to occur. Overloading the land with livestock and humans can strip marginal regions of vegetation even in wet years.

Related Web Resource

http://pubs.usgs.gov/gip/deserts/desertification/

Good overview of desertification around the world. Includes photographs.

Many desert regions have *internal drainage;* the streams drain toward landlocked basins instead of toward the sea. The surface of an enclosed basin acts as a local base level. Because each basin is generally filled to a different level than the neighboring basins, desert erosion may be controlled by many different *local base levels.* As a basin fills with sediment, its surface rises, leading to a *rising base level,* which is a rare situation in humid (wet) regions.

The limited rainfall that does occur in deserts often comes from violent thunderstorms, with a high volume of rain falling in a very short time. Desert thunderstorms may dump more than 13 centimeters of rain in one hour. Such a large amount of rain cannot soak readily into the sun-baked caliche soil, so the water runs rapidly over the land surface, particularly where vegetation is sparse. This high runoff can create sudden local floods of high discharge and short duration called **flash floods.** Flash floods are more common in arid regions than in humid regions. They can turn normally dry stream beds into raging torrents for a short time after a thunderstorm. Because soil particles are not held in place by plant roots, these occasional floods can effectively erode the land surface in a desert region. As a result, desert streams normally are very heavily laden with sediment. Flash floods can easily erode enough sediment to become *mudflows* (see chapter 13).

Desert stream channels are distinctive in appearance because of the great erosive power of flash floods and the intermittent nature of streamflow. Most stream channels are normally dry and covered with sand and gravel that is moved only

A

B

FIGURE 18.5

Desert features. (*A*) Desert stream channel showing dry, gravel-covered floor and steep, vertical sides (cut in sandstone and conglomerate) in Death Valley, California. (*B*) Badland topography (sharp ridges and V-shaped channels) eroded on shale in a dry region where plants are scarce. Badlands National Park, South Dakota.

Photo *B* © Dick Scott, Visuals Unlimited

during occasional flash floods. Rapid downcutting by sediment-laden floodwaters tends to produce narrow canyons with vertical walls and flat, gravel-strewn floors (figure 18.5*A*). Such channels are often called *arroyos* or *dry washes*.

Newcomers to deserts sometimes get into serious trouble in desert canyons in rainy weather. Imagine for a moment that you have camped on the canyon floor in figure 18.5*A* to get out of the strong desert winds. Later that night a towering thunderhead cloud forms, and heavy rain falls on the mountains several miles upstream from you. Although no rain has fallen on you, you are awakened several minutes later by a distant roar. The roar grows louder until a 3-meter "wall" of water rounds a bend in the canyon, heading straight for you at the speed of a galloping horse. Boulders, brush, and tree trunks are being swept along in this raging flash flood. The walls of the canyon are too steep to climb. Several hikers died during the summer of 1997 when such a wall of water roared down a side canyon in Grand Canyon National Park in Arizona. Stay out of desert canyons if there is any sign of rain; sleeping in such canyons is particularly dangerous!

The resistance of some rocks to weathering and erosion is partly controlled by climate. In a humid (wet) climate limestone dissolves easily, forming low places on Earth's surface. In a desert climate the lack of water makes limestone resistant, so it stands up as ridges and cliffs in the desert just as sandstone and conglomerate do. Lava flows and most igneous and metamorphic rock are also resistant. Shale is the least resistant rock in a desert so it usually erodes more deeply than other rock types and forms gentler slopes or badland topography (figure 18.5*B*).

Although intersecting joints form angular blocks of rock in all climates, desert topography characteristically looks more angular than the gently rounded hills and valleys of a humid region. This may be due indirectly to the low rainfall in deserts. Shortage of water slows chemical weathering processes to the point where few minerals break down to form fine-grained clay minerals. Soils are coarse and rocky with few chemically weathered products. Plants, which help bind soil into a cohesive layer in humid climates, are rare in deserts, and so desert soils are easily eroded by wind and rainstorms. Downhill creep of thick, fine-grained soil is partly responsible for softening the appearance of jointed topography in humid climates. With thin, rocky soil and slow rates of creep, desert topography remains steep and angular.

As you learned in chapter 16, climate is only one of many things that determine the shape and appearance of the land. Rock structure is another. As an example, in the next section we will look closely at two different structural regions within the desert of the southwestern United States.

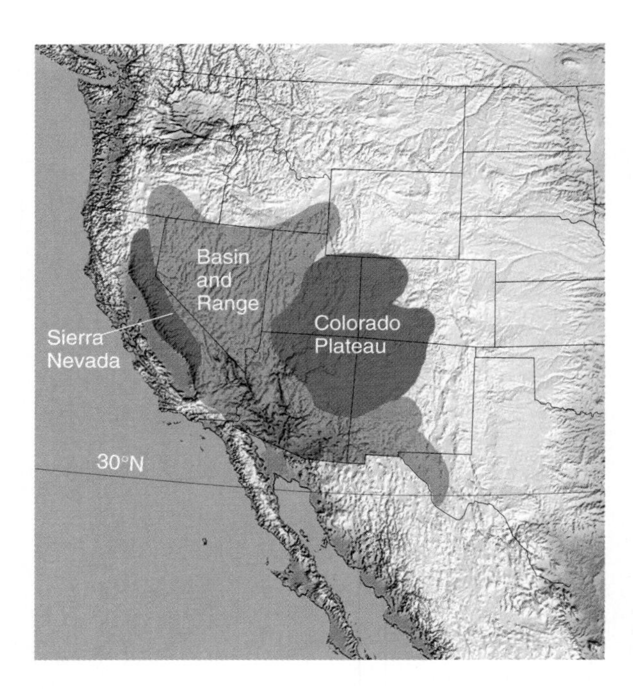

FIGURE 18.6

The Colorado Plateau and the Basin and Range province in the southwestern United States.

After Thelin and Pike, U.S. Geological Survey

DESERT FEATURES IN THE SOUTHWESTERN UNITED STATES

Much of the southwestern United States has an arid (or semiarid) climate, partly because it is close to 30° North latitude and partly because of the rain shadow effect of the Sierra Nevada and other mountain ranges. Within this region of low rainfall are two areas of markedly different geologic structure. One area is the Colorado Plateau and the other is the Basin and Range province, a mountainous region centered on the state of Nevada. The boundaries of these two areas are shown in figure 18.6.

The *Colorado Plateau* centers roughly on the spot known as the Four Corners, where the states of Utah, Colorado, Arizona, and New Mexico meet at a common point. The rocks near the surface of the Colorado Plateau are mostly flat-lying beds of sedimentary rock over 1,500 meters above sea level. These rocks are well exposed at the Grand Canyon in Arizona.

Because the rock layers are well above sea level, they are vulnerable to erosion by the little rain that does fall in the region. Flat-lying layers of resistant rock, such as sandstone, limestone, and lava flows, form **plateaus**—broad, flat-topped areas elevated above the surrounding land and bounded, at least in part, by cliffs. As erosion removes the rock at its base, the cliff is gradually eroded back into the plateau (figure 18.7*A*). Remnants of the resistant rock layer may be left behind, forming flat-topped mesas or narrow buttes (figure 18.7*B*). A **mesa** is a broad, flat-topped hill bounded by cliffs and capped with a resistant rock layer. A **butte** is a narrow hill of resistant rock with a flat top and very steep sides. Most buttes form by continued erosion of mesas. (The term *butte* is also used in other parts of the country for any isolated hill.)

The Colorado Plateau is also marked by peculiar steplike folds (bends in rock layers) called *monoclines.* Erosion of monoclines (and other folds) leaves resistant rock layers protruding above the surface as ridges (figure 18.8). A steeply tilted resistant layer erodes to form a *hogback,* a sharp ridge that has steep slopes. A gently tilted resistant layer forms a *cuesta,* with one steep side and one gently sloping side.

Note that plateaus, monoclines, hogbacks, and cuestas are not unique to deserts. They are surface features found in all climates, but are particularly well exposed in deserts because of thin soil and sparse vegetation.

The *Basin and Range province* is characterized by rugged mountain ranges separated by flat valley floors (figure 18.9). The blocks of rock that form the mountain ranges and the valley floors are bounded by **faults,** cracks in the soil along which some rock movement has taken place. (Chapter 6 discusses faults in more detail.) In the Basin and Range province, movement on the faults has dropped the valleys down relative to the adjacent mountain ranges to accommodate crustal thinning and extension (figure 18.10). Fault-controlled topography is found throughout the Basin and Range province, which covers almost all of Nevada and portions of bordering states as well as New Mexico and a small portion of Texas (figure 18.6). The numerous ranges in this province create multiple rain shadow zones and therefore a very dry climate.

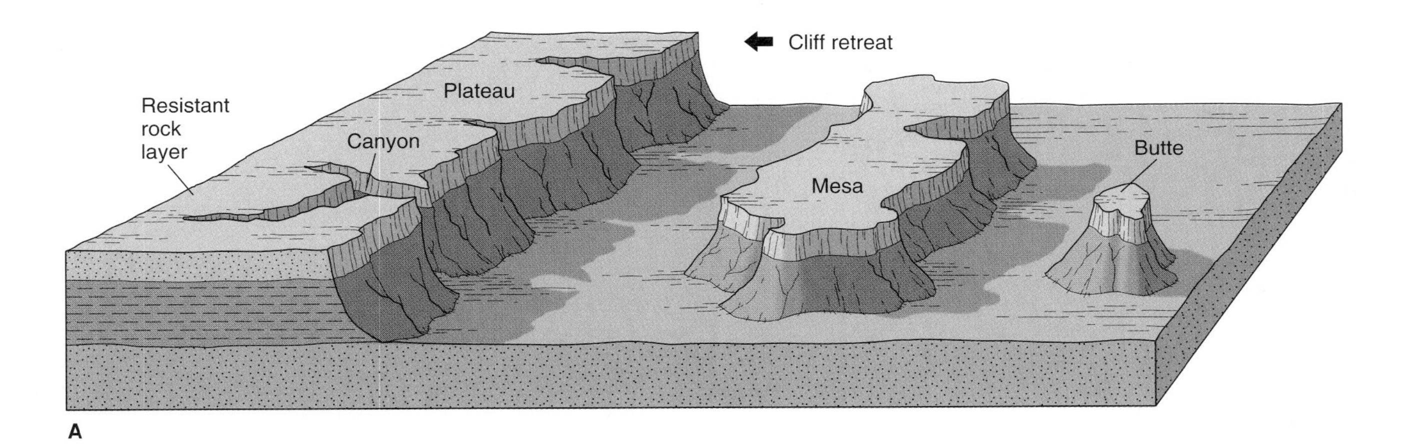

FIGURE 18.7
Characteristic landforms of the Colorado Plateau. (*A*) Erosional retreat of a cliff at the edge of a plateau can leave behind mesas and buttes as erosional remnants of the plateau. (*B*) Mesas and buttes in Monument Valley, Arizona, an area of eroded, horizontal, sedimentary rocks.

Heavy rainfall from occasional thunderstorms in the mountain ranges causes rapid erosion of the steep mountain fronts and resulting deposition on the valley floors (figure 18.11). Rock debris from the mountains, picked up by flash floods and mudflows, is deposited at the base of the mountain ranges in the form of alluvial fans. Alluvial fans (described in chapter 16) build up where stream channels abruptly widen as they flow out of narrow canyons onto the open valley floors, causing a decrease in velocity and rapid deposition of sediment.

Although most of the sediment carried by runoff is deposited in alluvial fans, some fine clay may be carried in suspension onto the flat valley floor. If no outlet drains the valley, runoff water may collect and form a **playa lake** on the valley floor. Playa lakes are usually very shallow and temporary, lasting for only a few days after a rainstorm. After the lake evaporates, a thin layer of fine mud may be left on the valley floor. The mud dries in the sun, forming a **playa,** a very flat surface underlain by hard, mud-cracked clay

FIGURE 18.8

(*A*) Steplike monocline folds often erode so that resistant rock layers form hogbacks and cuestas (these features are not unique to deserts). (*B*) Monocline near the Big Horn Mountains, Wyoming.
Photo by Diane Carlson

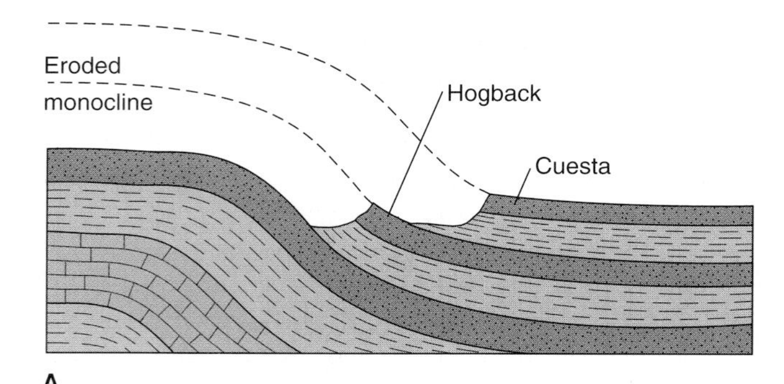

FIGURE 18.9

Basin and Range topography in Death Valley, California. In the distance the fault-bounded Panamint Mountains rise more than 3 kilometers (11,000 feet) above Death Valley. Giant alluvial fans at the base of the mountains show a braided stream pattern. Fine-grained sediments and salt deposits underlie the playa in the foreground.

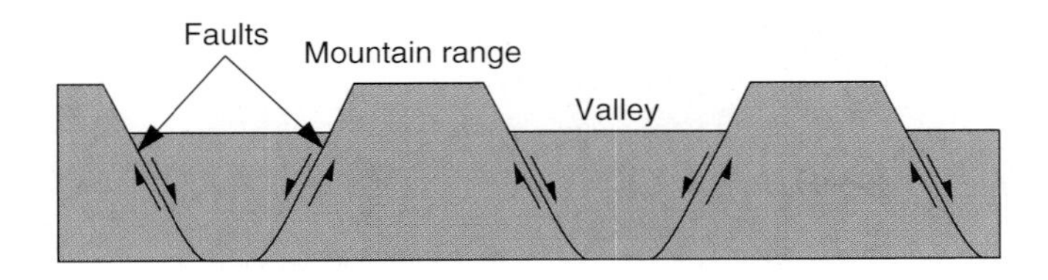

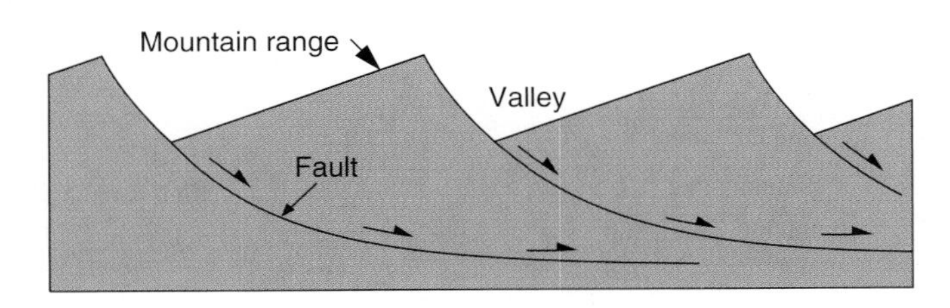

FIGURE 18.10

Two possible geometries of faults that control the mountains and valleys of the Basin and Range province. Movement along the faults is shown by arrows.

(figures 18.12 and 18.13). If the runoff contained a large amount of dissolved salt or if seeping ground water brings salt to the surface, the flat playa surface may be underlain by a bright white layer of dried salt instead of mud, as on the Bonneville Salt Flats in Utah (see figure 14.24).

Continued deposition near the base of the mountains may form a **bajada,** a broad, gently sloping depositional surface formed by the coalescing of individual alluvial fans (figure 18.11). A bajada is much more extensive than a single alluvial fan and may have a gently rolling surface resulting from the merging of the cone-shaped fans.

Erosion of the mountain can eventually form a **pediment,** which is a gently sloping surface, commonly covered with a veneer of gravel, cut into the solid rock of the mountain (figure 18.11). A pediment develops uphill from a bajada as the mountain front retreats. It can be difficult to distinguish a pediment from the surface of the bajada downhill, because both have the same slope and gravel cover. The pediment, however, is an erosional surface, usually underlain by solid rock, while the bajada surface is depositional and may be underlain by hundreds of meters of sediment.

An abrupt change in slope marks the upper limit of the pediment, where it meets the steep mountain front. Many geologists who have studied desert erosion believe that as this steep mountain front erodes, it retreats uphill, maintaining a relatively constant angle of slope.

Notice that rock structure, not climate, largely controls the fact that plateaus and cliffs are found in the Colorado Plateau, while mountain ranges, broad valleys, alluvial fans, and pediments are found in the Basin and Range province. Features such as plateaus, mesas, and alluvial fans can also be found in humid climates wherever the rock structure is favorable to their development; they are not controlled by climate. Features such as steep canyons, playa lakes, thin soil, and sparse vegetation, however, *are* typically controlled by climate.

WIND ACTION

Wind can be an important agent of erosion and deposition in any climate, as long as sediment particles are loose and dry. Wind differs from running water in two important ways. Because air is less dense than water, wind can erode only fine sediment—sand, silt, and clay. But wind is not confined to channels as running water is, so wind can have a widespread effect over vast areas.

In general, the faster the wind blows, the more sediment it can move. Wind velocity is determined by differences in air pressure caused by differences in air temperature. As air warms and cools, it changes density, and these density changes create air pressure differences that cause wind. Wet climates and cloud cover help buffer changes in air temperature, but in dry climates daily temperature changes can be extreme. In a desert, the temperature may range from 10°C (50°F) at night to more than 40°C (100°F) in the daytime. Because of these temperature fluctuations, wind is generally stronger in deserts than in humid regions, typically exceeding 100 kilometers per hour (60 miles per hour). The scarcity of vegetation in deserts to slow wind velocity by friction increases the effectiveness of desert winds.

Although strong winds are also associated with rainstorms and hurricanes, these winds seldom erode sediment because rain wets the surface sediment. Wet sediment is heavy and cohesive and will not be blown away. Strong winds in the desert, however, blow over loose, dry sediment, so wind is an effective erosional agent in dry climates. (As we said earlier in the chapter, running water in the form of flash floods is a far more important erosive agent than wind, even though the wind can be very strong.)

Wind Erosion and Transportation

Thick, choking *dust storms* are one example of wind action (figure 18.14). "Dust Bowl" conditions in the 1930s in the agricultural prairie states lasted for several years due to droughts and poor soil-conservation practices. Loose silt and clay are easily picked up from barren dry soil, such as in a cultivated field. Wind erosion is even greater if the soil is disturbed by animals or vehicles. Silt and clay can remain suspended in turbulent air for a long time, so a strong wind may carry a dust cloud hundreds of meters upward and hundreds of kilometers horizontally. Dust storms of the 1930s frequently blacked out the midday sun, fertile soil was lost over vast regions ruining many farms, and streets and rivers downwind were filled with thick dust deposits.

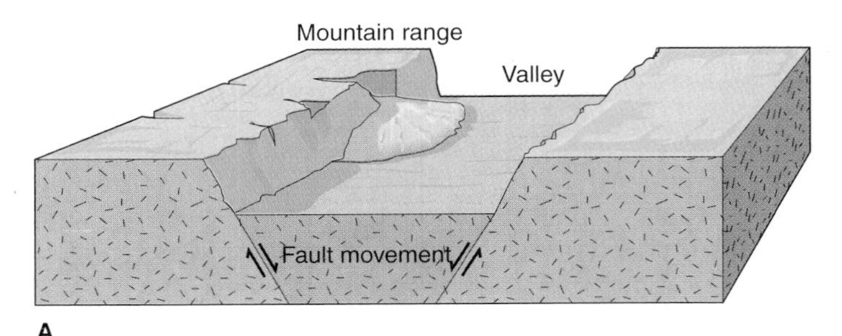

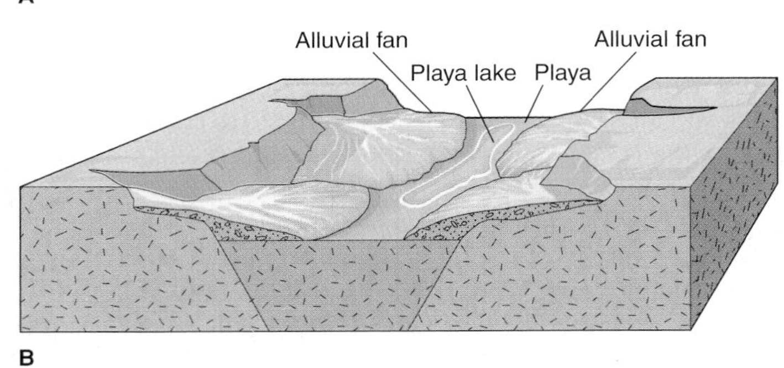

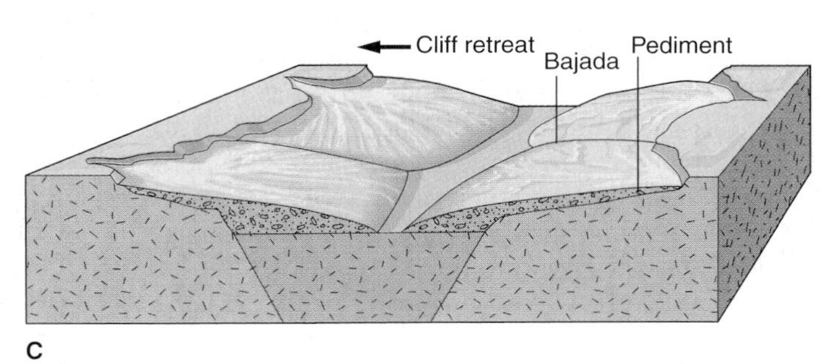

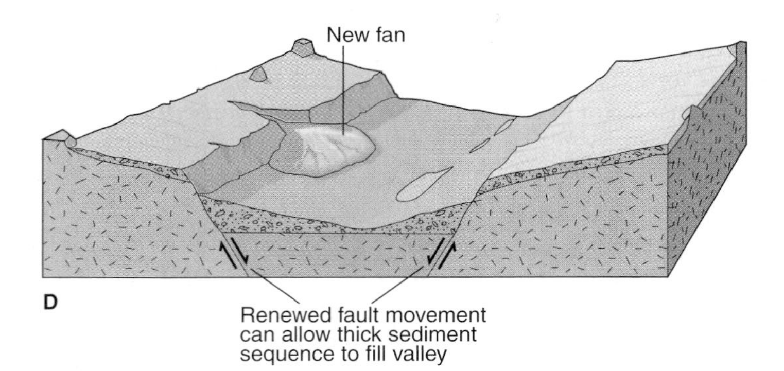

FIGURE 18.11

Development of landforms associated with Basin and Range topography.

FIGURE 18.12

Alluvial fans at the base of mountain canyons, Death Valley, California. The white salt flat in the foreground is part of a playa.

Photo by Frank M. Hanna

FIGURE 18.13

Mud-cracked playa surface.

Photo © Bill Ross/Westlight

A

B

FIGURE 18.14

(*A*) Wind erosion of soil from a cultivated field. (*B*) Approaching dust storm, Prowers County, Colorado, late afternoon (1930s). The storm lasted almost 3 hours, with wind speeds of approximately 50 kilometers per hour.

Photos *A* and *B* by U.S. Department of Agriculture, Soil Conservation Service

Wind-blown sediment is sometimes picked up on land and carried out to sea. Particles from the Sahara Desert in Africa have been collected from the air over the islands of the West Indies after having been carried across the entire Atlantic Ocean. A substantial amount of the fine-grained sediment that settles to the sea floor is land-derived sediment that the wind has deposited on the sea surface. Ships 800 kilometers offshore have reported dustfalls a few millimeters thick covering their decks.

Volcanic ash can be carried by wind for very great distances. An explosive volcanic eruption can blast ash more than 15 kilometers upward into the air. Such ash may be caught in the high-altitude *jet streams,* narrow belts of strong winds with velocities sometimes greater than 300 kilometers per hour. Following the 1980 eruption of Mount St. Helens in western Washington, a visible ash layer blanketed parts of Washington, Idaho, and Montana to the east. At high altitudes, St. Helens ash could be detected blowing over New York and out over the Atlantic Ocean, 5,000 kilometers from the volcano. But the St. Helens eruption was a relatively small one. Ash from the 1883 Krakatoa eruption in Indonesia circled the globe for two years, causing spectacular sunsets and a slight, but measurable, drop in temperature as the ash reflected sunlight back into space. The 1815 eruption of Tambora, also in Indonesia, put so much ash into the air that there were summer blizzards, crop failures, and famine in New England and northern Europe in 1816, "the year without a summer." Lower temperatures (1°F) and brilliant sunsets also marked the 1991 eruption of Pinatubo in the Philippines.

Because sand grains are heavier than silt and clay, sand moves close to the ground in the leaping pattern called *saltation* (as does some sediment in streams). High-speed winds can cause *sandstorms,* clouds of sand moving rapidly near the land surface. The high-speed sand in such a storm can sandblast smooth surfaces on hard rock and scour the windshields and paint of automobiles. Because of the weight of the sand grains, however, sand rarely rises more than 1 meter above a flat land surface, even under extremely strong winds. Therefore, most of the sandblasting action of wind occurs close to the ground (figure 18.15). Telephone poles in regions of wind-driven sand often are severely abraded near the ground. To prevent such abrasion, desert residents pile stones or wrap sheet metal around the base of the poles.

Wind seldom moves particles larger than sand grains, but wind-blown sand may sculpture isolated pebbles, cobbles, or boulders into **ventifacts**—rocks with flat, wind-abraded surfaces (figure 18.16). If the wind direction shifts, or if the stone is turned, more than one flat face may develop on the ventifact.

Deflation

The removal of clay, silt, and sand particles from the land surface by wind is called **deflation.** If the sediment at the land surface is made up only of fine particles, the erosion of these particles by the wind can lower the land surface substantially. A **blowout** is a depression on the land surface caused by wind erosion (figure 18.17). A *pillar,* or erosional remnant of the former land, may be left at the center of a blowout (figures 18.17 and 18.18).

Blowouts are common in the Great Plains states (figure 18.18). One in Wyoming measures 5 by 15 kilometers and is 45 meters deep. The enormous Qattara Depression in northwestern Egypt, more than 250 kilometers long and more than 100 meters *below* sea level, has been attributed to wind deflation. Deflation can continue to deepen a blowout in fine-grained sediment until it reaches wet, cohesive sediment at the water table.

Wind Deposition

Loess

Loess is a deposit of wind-blown silt and clay composed of unweathered, angular grains of quartz, feldspar, and other minerals weakly cemented by calcite. Loess has a high porosity, typically near 60%. Deposits of loess may blanket hills and valleys downwind of a source of fine sediment, such as a desert or a region of glacial outwash.

China has extensive loess deposits (figure 18.19), more than 100 meters thick in places. Wind from the Gobi Desert car-

A

B

FIGURE 18.15

(*A*) Wind erosion near the ground has sandblasted the lower 1 meter of this basalt outcrop, Death Valley, California. Hammer for scale. (*B*) Power pole with its base wrapped in an abrasion-resistant material to minimize wind erosion.

Photo *B* courtesy Paul Bauer

FIGURE 18.16

Ventifacts eroded by blowing sand, Death Valley, California. Most show two flat sides joined by a sharp ridge. Attached sand shows original depth of burial.

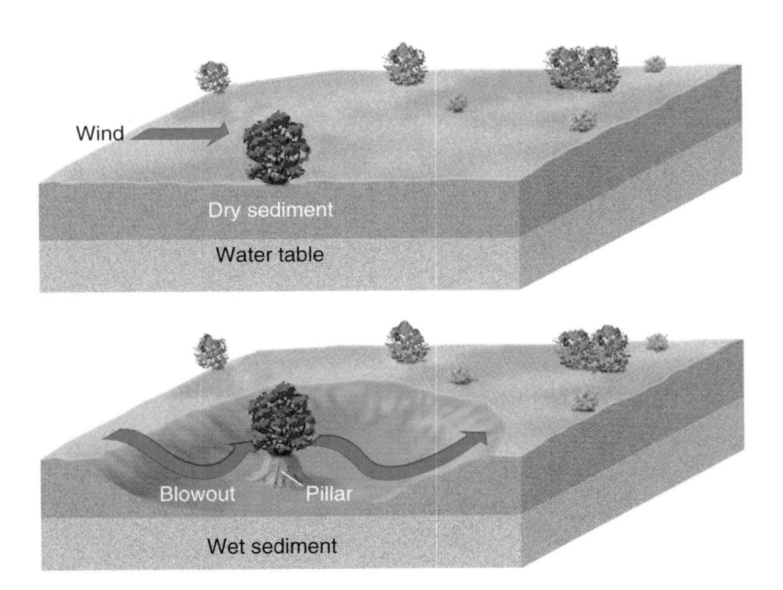

FIGURE 18.17

Deflation by wind erosion can form a blowout in loose, dry sediment. Deflation stops at the water table. A pillar, or erosional remnant, may be found in the center of a blowout.

ried the silt and clay that formed these deposits. Loess is easy to dig into and has the peculiar ability to stand as a vertical cliff without slumping (figure 18.20), perhaps because of its cement or perhaps because the fine, angular, sediment grains interlock with one another. For centuries the Chinese have dug cavelike homes in loess cliffs. When a large earthquake shook China in 1920, however, many of these cliffs collapsed, burying alive about 100,000 people.

FIGURE 18.18

Large blowout near Harrison, Nebraska. Pillar top is the original level of land, before wind erosion lowered the land surface by more than 3 meters. The pillar is the erosional remnant at the center of the blowout.

Photo by N. H. Darton, U.S. Geological Survey

During the glacial ages of the Pleistocene Epoch, the rivers that drained what is now the midwestern United States transported and deposited vast amounts of glacial outwash. Later, winds eroded silt and clay (originally glacial rock flour) from the flood plains of these rivers and blanketed large areas of the Midwest with a cover of loess (figure 18.20). Soils that have developed from the loess are usually fertile and productive. The grain fields of much of the Midwest and in the Palouse area in eastern Washington are planted on these rich soils. Wind erosion of cultivated, loess-covered hills in the Palouse region is a serious problem that has locally removed fertile soil from the hilltops.

Sand Dunes

Sand dunes are mounds of loose sand grains heaped up by the wind. Dunes are most likely to develop in areas with strong winds that generally blow in the same direction. Patches of dunes are scattered throughout the southwestern United States desert. More extensive dune fields occur on some of the other deserts of the world, such as the Sahara Desert of Africa, which contains vast *sand seas.* Dunes are also commonly found just landward of beaches (figure 18.21), where sand is blown inland. Beach dunes are common along the shores of the Great Lakes and along both coasts of the United States. Braided rivers (see chapter 16) can also be sources of sand for dune fields.

The mineral composition of the sand grains in sand dunes depends on both the character of the original sand source and the intensity of chemical weathering in the region. Many dunes, particularly those near beaches in humid regions, are composed

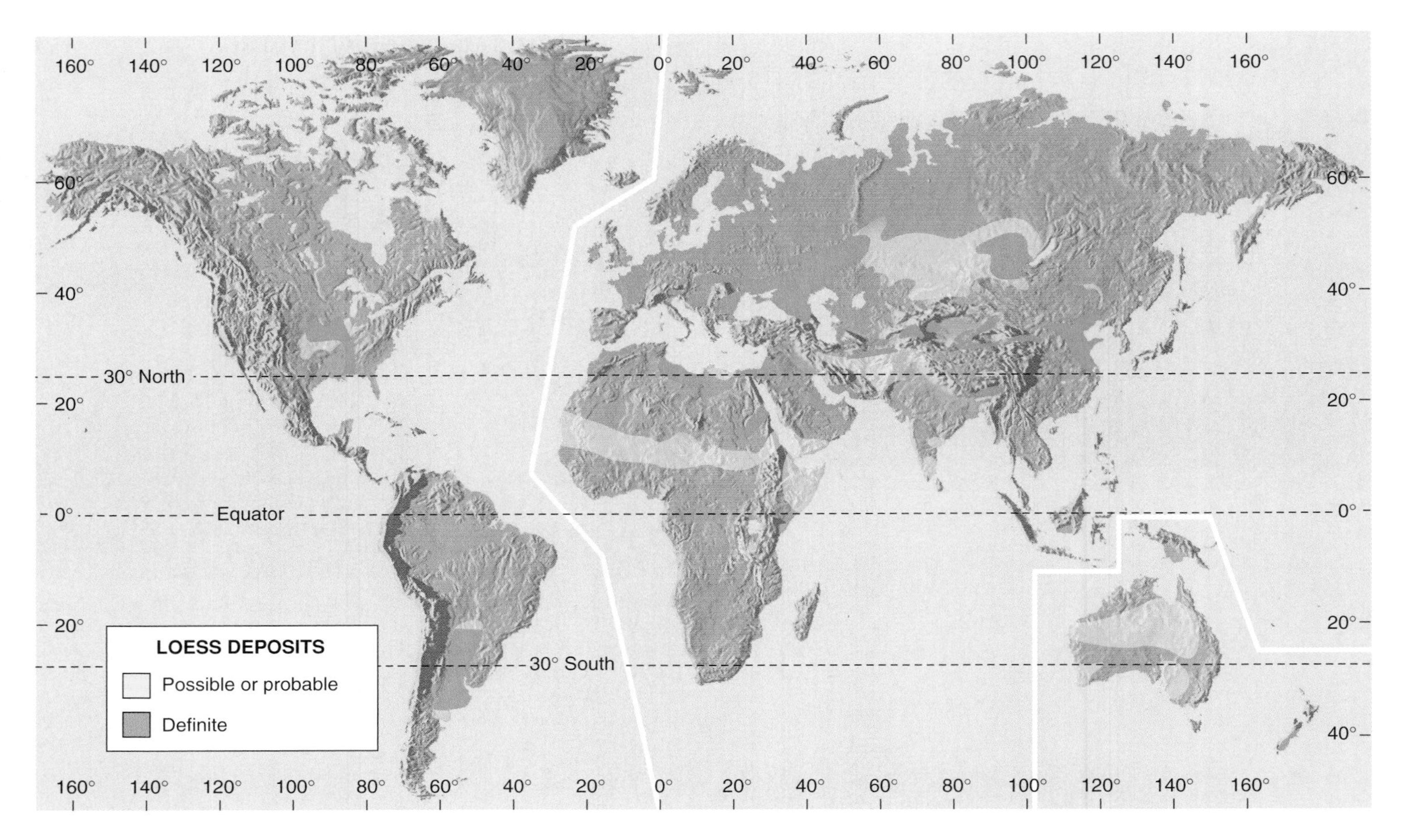

FIGURE 18.19

Major loess-covered areas in the world.

Source: Data from map from U.S. Department of Agriculture; Smalley, I. J., *Loess Lithology and Genesis,* fig. 1, p. 768, Halsted Press, 1975

FIGURE 18.20

Vertical scarps in loess deposits from western Iowa.

Photo © Tom Till

FIGURE 18.21

Coastal dunes formed from beach sand blowing inland, Pismo Beach, California.

Photo by Frank M. Hanna

IN GREATER DEPTH 18.2

Desert Pavement and Desert Varnish

Two intriguing features can be seen in many deserts, particularly on the surface of old alluvial fans no longer receiving new sediment.

Desert pavement is a thin, surface layer of closely packed pebbles (box figure 1). The pebbles were once thought to be residual, left behind as strong winds blew away all the fine grains of a rocky soil. The pebbles are now thought to be brought to the surface by cycles of wetting and drying, which cause the soil to swell and shrink as water is absorbed and lost by soil particles. Swelling soil lifts pebbles slightly; drying soil cracks, and fine grains fall into the cracks. In this way pebbles move up, while fine grains move down. The surface layer of pebbles protects the land from wind erosion and deflation. When the pavement is disturbed (as in the 1991 Gulf War), dust storms and new sand dunes may result.

Many rocks on the surface of deserts are darkened by a chemical coating known as *desert varnish.* Although the interior of the rocks may be light colored, a hard, often shiny, coating of dark iron and manganese oxides and clay minerals can build up on the rock surface over long periods of time (box figure 2). Although no one is quite certain how this coating develops, it seems to be added to the rocks from the outside, for even white quartzite pebbles with no internal source of iron, manganese or clay minerals can develop desert varnish. One current hypothesis is that the clay is windblown, perhaps sticking to rocks dampened by dew. A film of clay on a rock may draw iron and manganese-containing solutions upward from the soil by capillary action, and the presence of the clay minerals may help deposit the dark manganese oxide that cements the clay to the rock. Another hypothesis is that the oxide is deposited biologically by microbes. Regardless of how the varnish forms, the longer a rock is exposed on a desert land surface, the darker it becomes.

Box 18.2 — FIGURE 1

Desert pavement on an old alluvial fan surface. The surface pebbles are closely packed; fine sand underlies the pebbles.

Photo by Frank M. Hanna

Box 18.2 — FIGURE 2

Petroglyphs carved on this rock cut through the dark desert varnish to show the lighter color of the interior of the rock, Valley of Fire, Nevada.

Photo by J. Freeberg, U.S. Geological Survey

largely of quartz grains because quartz is so resistant to chemical weathering. Inland dunes, such as the Great Sand Dunes National Monument in Colorado, often contain unstable feldspar and rock fragments in addition to quartz. Some dunes are formed mostly of carbonate grains, particularly those near tropical beaches. At White Sands, New Mexico, dunes are made of gypsum grains, eroded by wind from playa lake beds.

Sand grains found in dunes are commonly well-sorted and well-rounded because wind is very selective as it moves sediment. Fine-grained silt and clay are carried much farther than sand, and grains coarser than sand are left behind when sand moves. The result is a dune made solely of sand grains, commonly all very nearly the same size. The prevalence of well-rounded grains in many dunes also may be due to selective sorting by the wind. Rounded grains roll more easily than angular grains, and so the wind may remove only the rounded grains from a source to form dunes. Wind will often selectively roll oolitic grains from a carbonate beach of mixed oolitic and skeletal grains.

Most sand dunes are asymmetric in cross section, with a gentle slope facing the wind and a steeper slope on the downwind side. The steep downwind slope of a dune is called the **slip face** (figure 18.22). It forms from loose, cascading sand that generally keeps the slope at the *angle of repose,* which is about 34° for loose, dry sand. Sand is blown up the gentle slope and over the top of the dune. Sand grains fall like snow onto the slip face when they encounter the calm air on the downwind side of the dune. Loose sand settling on the top of the slip face may become oversteepened and slide as a small avalanche down the slip face. These processes form high-angle cross-bedding within the dune. When found in sandstone, such cross-bedding strongly suggests deposition as a dune (see figure 14.27).

In passing over a dune, the wind erodes sand from the gentle upwind slope and deposits it downwind on the slip face. As a result, the entire dune moves slowly in a downwind direction. The rate of dune motion is much slower than the speed of the wind, of course, because only a thin layer of sand on the surface of the dune moves at any one time. The dune may move only 10 to 15 meters per year. Over many years, however, the movement of dunes can be significant, a fact not always appreciated by people who build homes close to moving sand dunes.

If a dune becomes overgrown with grass or other vegetation, movement stops. The Sand Hills of north-central Nebraska are large dunes, formed during the Pleistocene or Holocene Epochs, that have become stabilized by vegetation. The migration of many beach dunes toward beach homes and roads has been stopped by planting a cover of beach grass over the dunes. Dune-buggy tires can uproot and kill the grass, however, and start the dunes moving again.

Sand moving over a dune surface typically forms *wind ripples*—small, low ridges of sand produced by saltation of the grains (figure 18.23). The ripples are similar to those formed in sediment by a water current (see chapter 14). Because sand moves perpendicularly to the long dimension of the ripples, a rippled sand surface indicates the direction of sand movement.

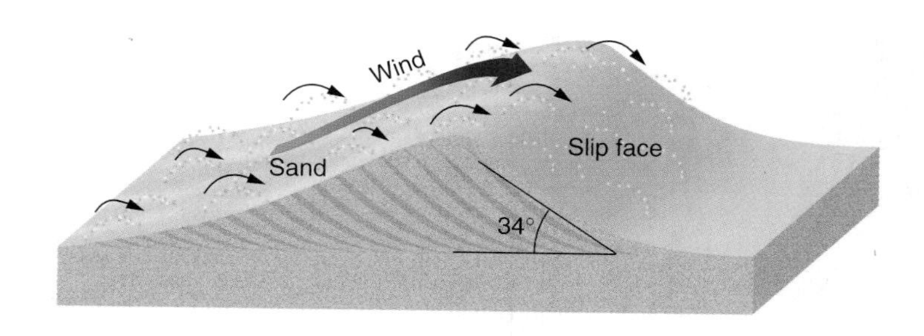

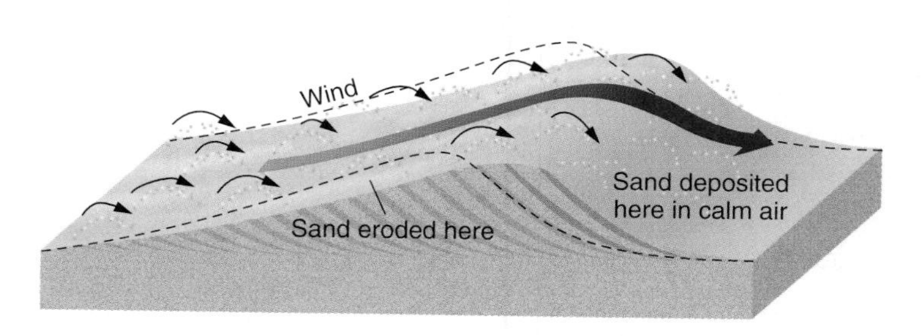

FIGURE 18.22

(*A*) A sand dune forms with a gentle upwind slope and a steeper slip face on the downwind side. Sand eroded from the upwind side of the dune is deposited on the slip face, forming cross-beds. Movement of sand causes the dune to move slowly downwind. (*B*) Strong desert winds (60 miles per hour) blowing to the right remove sand from the gently sloping upwind side of this dune. The sand settles onto the steep slip face on the right.

Types of Dunes

As figure 18.24 shows, dunes tend to develop certain characteristic shapes, depending on (1) the wind's velocity and direction (that is, whether constant or shifting); (2) the sand supply available; and (3) how the vegetation cover, if any, is distributed.

Where the sand supply is limited, a type of dune called a **barchan** generally develops. The barchan is crescent-shaped with a steep slip face on the inward or concave side. The horns on a barchan dune point in the downwind direction (figure 18.24*A*). Barchan dunes are usually separated from one another and move across a barren surface (figure 18.25). If more sand is available, the wind may develop a **transverse dune,** a relatively straight, elongate dune oriented perpendicular to the wind direction (figures 18.24*B* and 18.26).

A **parabolic dune** is somewhat similar in shape to a barchan dune, except that it is deeply curved and is convex in the downwind direction. The horns point upwind and are commonly anchored by vegetation (figure 18.24*C*). The parabolic dune requires abundant sand and commonly forms around a blowout. Because they require abundant sand and strong winds, parabolic dunes are typically found inland from an ocean beach (figure 18.27). All three of these dune shapes develop in areas having steady wind direction, and all three have steep slip faces on the downwind side.

One of the largest types of dunes is the **longitudinal dune** or *seif* (figure 18.24*D*), which is a symmetrical ridge of sand that forms parallel to the prevailing wind direction. Longitudinal dunes occur in long, parallel ridges that are exceptionally straight and regularly spaced. They are typically separated by barren ground or desert pavement. Longitudinal dunes in the Sahara Desert (figure 18.28) are as high as 200 meters and more

FIGURE 18.23
Wind ripples on sand surface, Monument Valley, Utah.
Photo by © Doug Sherman

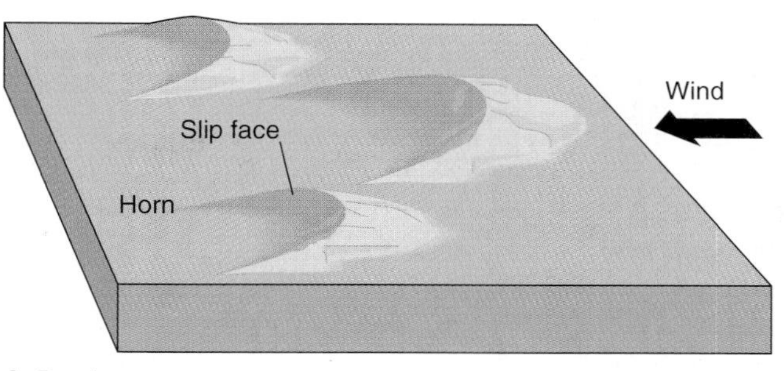

A Barchans

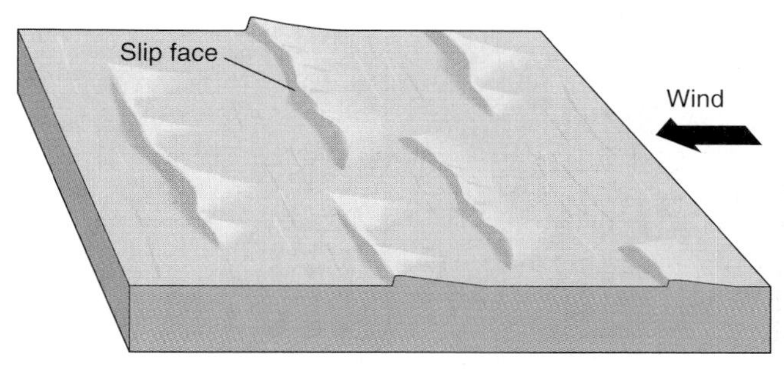

B Transverse dunes

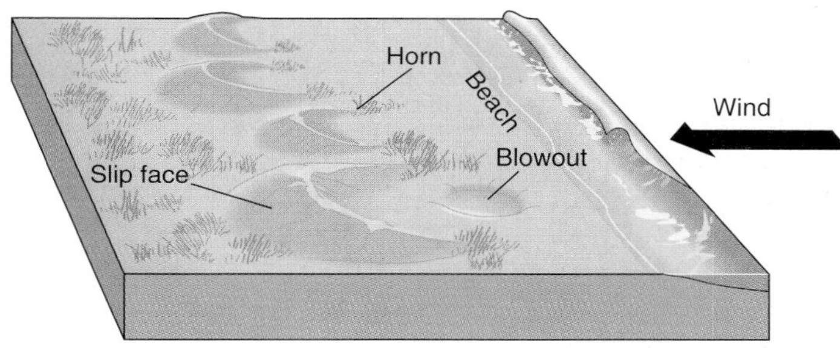

C Parabolic dunes

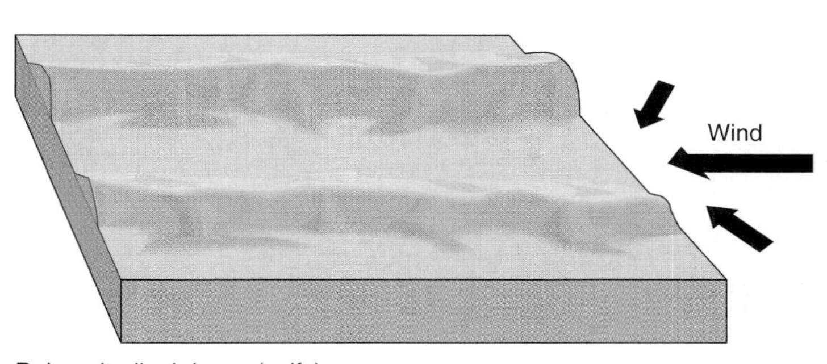

D Longitudinal dunes (seifs)

FIGURE 18.24
Types of sand dunes.

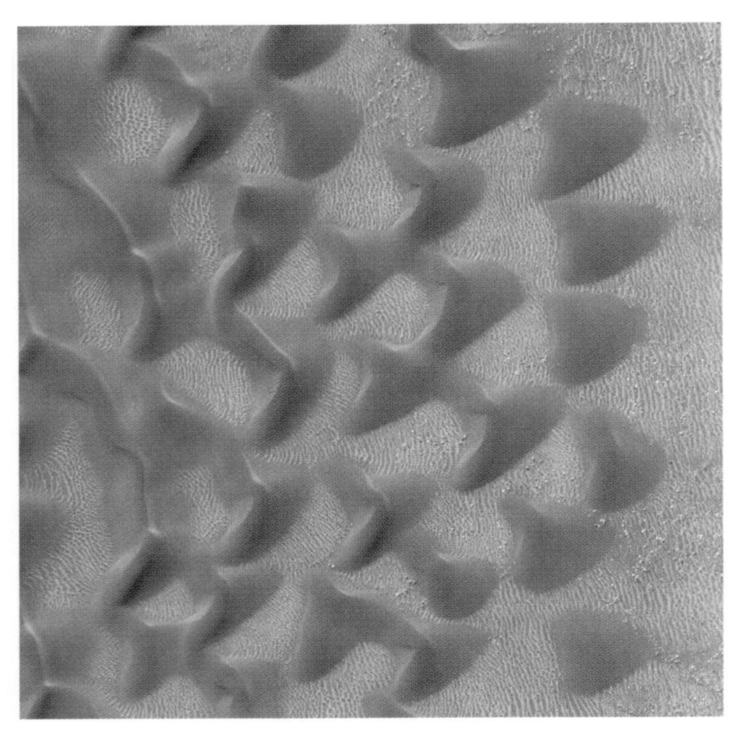

FIGURE 18.25

Barchan dunes on the right resemble "sharks teeth" and become transverse dunes on the left where more sand is available. Mars Orbiter camera captured this image from the edge of a much larger dune field in Proctor Crater.

Photo by *NASA/JPL/Malin Space Science* Systems

FIGURE 18.26

Transverse dunes in the Great Sand Dunes National Monument, Colorado. Wind blows from left to right.

Photo © Adriel Heisey Photography

FIGURE 18.27

Parabolic dunes near Pismo Beach, central California. Wind blows from left to right. The ocean and a sand beach are just to the left of the photo.

Photo by Frank M. Hanna

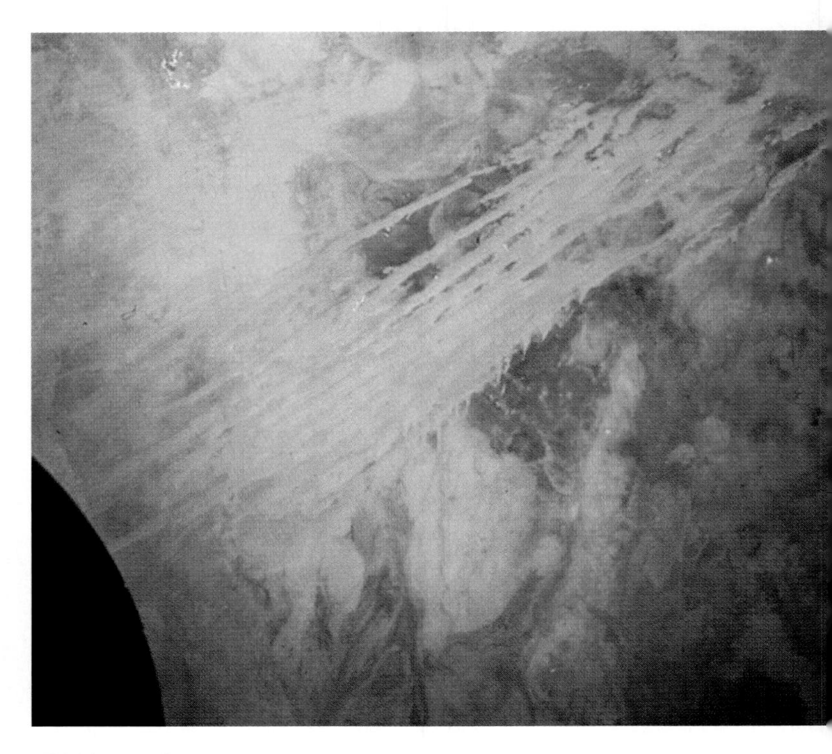

FIGURE 18.28

Longitudinal dunes in the Sahara Desert, Algeria. Photo from Gemini spacecraft at an altitude of about 100 kilometers.

Photo by NASA

than 120 kilometers in length. Numerous hypotheses have been proposed to explain the development of longitudinal dunes, but none can adequately explain their spectacular size and regular spacing. It appears that crosswinds are important in piling up sand, which adds to the height of longitudinal dunes, whereas the more constant prevailing wind direction redistributes the sand down the length of the dunes. Smoke bomb experiments to analyze airflow have shown that the wind spirals down the intervening troughs between longitudinal dunes and may control the regularity of their spacing.

Not all dunes can be classified by an easily recognizable shape. Many of them are quite irregular.

Summary

Deserts are located in regions where less than 25 cm of rain falls in a year. Such regions are found primarily in belts of descending air at 30° North and South latitude. Arid regions also may be due to the *rain shadow* of a mountain range, great distance from the sea, and proximity to a cold ocean current. Descending air forms cold deserts at the poles.

Desert landscapes differ from those of humid regions in that they lack through-flowing streams and they have internal drainage and many local, rising base levels. *Flash floods* caused by desert thunderstorms are effective agents of erosion despite the low rainfall. Limestone is resistant in deserts. Thin soil and slow rates of creep may give desert topography an angular look.

Parts of the southwestern United States are desert, the topography determined primarily by rock structure. Flat-lying sedimentary rocks of the Colorado Plateau are sculptured into cliffs, *plateaus, mesas,* and *buttes*. The fault-controlled topography of the Basin and Range province is marked by *alluvial fans, bajadas, playas,* and *pediments*.

Although wind erosion can be intense in regions of low moisture, streams are usually more effective than wind in sculpturing landscapes, regardless of climate.

Fine-grained sediment can be carried long distances by wind, even across entire continents and oceans.

Sand moves by *saltation* close to the ground, occasionally carving *ventifacts*.

Wind can *deflate* a region, creating a *blowout* in fine sediment.

Sand dunes move slowly downwind as sand is removed from the gentle upwind slope and deposited on the steeper *slip face* downwind.

Dunes are classified as *barchans, transverse dunes, parabolic dunes,* and *longitudinal dunes,* but many dunes do not resemble these types. Dune type depends on wind strength and direction, sand supply, and vegetation.

Terms to Remember

bajada 455
barchan 462
blowout 458
butte 452
deflation 458
desert 448
fault 452
flash flood 450
loess 458
longitudinal dune (seif) 462
mesa 452
parabolic dune 462
pediment 455
plateau 452
playa 453
playa lake 453
rain shadow 448
sand dune 459
slip face 461
transverse dune 462
ventifact 458

Testing Your Knowledge

Use the questions below to prepare for exams based on this chapter.

1. What are two reasons why parts of the southwestern United States have an arid climate?
2. Sketch a cross section of an idealized dune, labeling the slip face and indicating the wind direction. Why does the dune move?
3. Describe the geologic structure and sketch the major landforms of
   a. the Colorado Plateau b. the Basin and Range province.
4. How does a flash flood in a dry region differ from most floods in a humid region?
5. Give two reasons why wind is a more effective agent of erosion in a desert than in a humid region.
6. Name four types of sand dunes and describe the conditions under which each forms.

7. The defining characteristic of a desert is
   a. shifting sand dunes b. high temperatures
   c. low rainfall d. all of the above
   e. none of the above
8. Which is characteristic of deserts?
   a. internal drainage b. limited rainfall
   c. flash floods d. slow chemical weathering
   e. all of the above
9. The major difference between a mesa and a butte is one of
   a. shape b. elevation
   c. rock type d. size
10. The Basin and Range province covers almost all of
    a. Utah b. Nevada
    c. Texas d. Colorado
11. A very flat surface underlain by a dry lake bed of hard, mud-cracked clay is called a
    a. ventifact b. plateau
    c. playa d. none of the above
12. Rocks with flat, wind-abraded surfaces are called
    a. ventifacts b. pediments
    c. bajadas d. none of the above
13. The removal of clay, silt, and sand particles from the land surface by wind is called
    a. deflation b. depletion
    c. deposition d. abrasion
14. Which is not a type of dune?
    a. barchan b. transverse
    c. parabolic d. longitudinal
    e. all of the above are dunes
15. Much of the Southwestern United States is desert because (choose as many as apply)
    a. it is near 30 degrees North
    b. the western mountains create a rain shadow
    c. cold ocean currents in the Pacific cause high evaporation rates in the land
    d. they are a great distance from the ocean
16. A broad ramp of sediment formed at the base of mountains when alluvial fans merge is
    a. a playa b. a bajada
    c. a pediment d. an arroyo
17. A surface layer of closely packed pebbles is called
    a. desert varnish b. deflation
    c. a blowout d. desert pavement

Expanding Your Knowledge

1. Study the photos of sand dunes in this chapter. Which way does the prevailing wind blow in each case?
2. Can deserts be converted into productive agricultural regions? How? Are there any environmental effects from such conversion?
3. At what relative depth is ground water likely to be found in a desert? Why? Is the water likely to be drinkable? Why?

Exploring Web Resources

www.mhhe.com/plummer9e
Visit our Online Learning Center to check your answers for the Testing Your Knowledge section and click on the direct link for another great site on deserts and sand dunes.

http://pubs.usgs.gov/gip/deserts/contents/
Online version of *Deserts: Geology and Resources* by A. S. Walker provides a good overview of deserts, processes, and mineral resources.

Animations

This chapter includes the following animations available on our Online Learning Center at www.mhhe.com/plummer9e.

18.11 Development of Basin and Range Features

18.22 Formation of a Sand Dune

CHAPTER

19

Glaciers and Glaciation

In chapters 13, 16, 17, and 18 you have seen how the surface of the land is shaped by mass wasting, running water, and to some extent, ground water. Running water is regarded as the erosional agent most responsible for shaping Earth's land surface. Where glaciers exist, however, they are far more effective agents of erosion, transportation, and deposition. Geologic features characteristic of glaciation are distinctly different from the features formed by running water. Once recognized, they lead one to appreciate the great extent of glaciation during the recent geologic past (that age popularly known as the Ice Age).

Immense and extensive glaciers, covering as much as a third of Earth's land surface, had a profound effect on the landscape and on our present civilization. Moreover, worldwide climatic changes during the glacial ages distinctively altered landscapes in areas far from the glacial boundaries. For instance, water stored as ice in glaciers came from the oceans, so sea levels were lowered and more land was above sea level.

These episodes of glaciation took place within only the last couple million years, ending about 10,000 years ago. Preserved in the rock record, however, is evidence of extensive older glaciations. The chapter on plate tectonics shows how the record of these ancient glaciations supports the theory of plate tectonics.

To understand how glacial erosion and deposition could have created the features regarded as evidence for past glaciation, you must first appreciate how present-day glaciers erode, transport, and deposit material. In other words, you must apply the principle of uniformitarianism to your study of glaciation.

Opposite: A glacier and rock transported by the glacier in the Peruvian Andes. The mountain climber and camp are on a lateral moraine, which is composed of debris carried along the side of a glacier.
Photo by C. C. Plummer

FIGURE 19.1

Yosemite Valley, as seen from Glacier Point, Yosemite National Park, California. Its U-shaped cross profile is typical of glacially carved valleys.
Photo by C. C. Plummer

A **glacier** is a large, long-lasting mass of ice, formed on land, that moves under its own weight. It develops as snow is compacted and recrystallized. Glaciers can develop anyplace where, over a period of years, more snow accumulates than melts away or is otherwise lost.

There are two types of *glaciated* terrain on the earth's surface. **Alpine glaciation** is found in mountainous regions, while **continental glaciation** exists where a large part of a continent (thousands of square kilometers) is covered by glacial ice. In both cases the moving masses of ice profoundly and distinctively change the landscape.

THE THEORY OF GLACIAL AGES

In the early 1800s the hypothesis of past extensive continental glaciation of Europe was proposed. Among the many people who regarded the hypothesis as outrageous was the Swiss naturalist Louis Agassiz. But after studying the evidence in Switzerland, he changed his mind. In 1837, he published a discourse that eventually led to wide acceptance of the theory. Later, Agassiz traveled widely throughout Europe and North America promoting and extending the theory. Agassiz and his colleagues had observed the characteristic erosional and depositional features of present glaciers in the Alps and had compared these with similar features found in northern Europe and the British Isles, well beyond the farthest extent of the Alpine glaciers. Based on these observations, Agassiz proposed that very large glaciers had covered most of Europe. Agassiz had to overcome skepticism over past climates being quite different from those of today. At the time, the hypothesis seemed to many geologists to be a violation of the principle of uniformitarianism. Agassiz later came to North America and worked with American geologists who had found similar indications of large-scale past glaciation on this continent.

As more evidence accumulated, the hypothesis became accepted as a theory that today is seldom questioned. The **theory of glacial ages** states that at times in the past, colder climates prevailed during which much more of the land surface of Earth was glaciated than at present.

Because the last episode of glaciation was at its peak only about 18,000 years ago, its record has remained largely undestroyed by subsequent erosion and so provides abundant evidence to support the theory. This most recent glacial episode was the last of several glacial ages that alternated with periods of warmer climate (similar to today's climate) around the world.

The glacial ages are not just a scientific curiosity. Our lives and environment today have been profoundly influenced by their effects. For example, much of the fertile soil of the northern Great Plains of the United States developed on the loose debris transported and deposited by glaciers that moved southward from Canada. The thick blankets of sediment left in the Midwest store vast amounts of ground water. The Great Lakes and the thousands of lakes in Minnesota and neighboring states and provinces are the products of past glaciation. The spectacularly scenic areas in many North American national parks owe much of their beauty to glacial action. Yosemite Valley in California might have been another nondescript valley if glaciers had not carved it into its present shape (figure 19.1).

Before we can understand how a continental glacier was responsible for much of the soil in the Midwest, or how a glacier confined to a valley could carve a Yosemite, we must learn something about present-day glaciers.

GLACIERS—WHERE THEY ARE, HOW THEY FORM AND MOVE

Distribution of Glaciers

Glaciers occur both in polar regions, where there is little melting during the summer, and in temperature climates that have heavy snowfall during the winter months. They are found where more snow falls during the cold time of year than can be melted during warm months.

Washington has more glaciers than any other state except Alaska, because of the extensively glaciated mountains of western Washington. Washington's mountains have warmer winters but much more precipitation in the higher elevations than do the Rocky Mountains. There is more snow left after summer melting in Washington than in states to the east of it. Glaciers are common even near the equator in the very high mountains of South America and Africa because of the low temperatures at high altitudes.

ENVIRONMENTAL GEOLOGY 19.1

Glaciers as a Water Resource

Few people think of glaciers as frozen reservoirs supplying water for irrigation, hydroelectric power, recreation, and industrial and domestic use. In the state of Washington, however, streamflow from the approximately 800 glaciers there amounts to about 470 billion gallons of water during a summer, according to the U.S. Geological Survey. More water is stored in glacier ice in Washington than in all of the state's lakes, reservoirs, and rivers.

One important aspect of glacier-derived water is that it is available when needed most. Snow accumulates on glaciers during the wet winter months. During the winter, streams at lower elevations, where rain falls rather than snow, are full and provide plenty of water. During the summer, the climate in the Pacific Northwest is hotter and drier. Demand for water increases, especially for irrigation of crops. Streams that were fed by rainwater may have dried up. Yet in the heat of summer, the period of peak demand, snow and ice on glaciers are melting, and streams draining glaciers are at their highest level.

Paradoxically, the greater the snowfall on a glacier during a winter, the smaller the amount of meltwater during the summer. A larger blanket of white snow reflects the sun's radiation more effectively than the darker, bare glacier ice, which absorbs more of the heat of the sun. Experiments have shown that melting can be greatly increased by darkening the snow surface, for instance, by sprinkling coal dust on it. Similarly, the melting of a glacier can be slowed artificially by covering it with highly reflective material. Such means of controlling glacial meltwater have been proposed to benefit power generating stations or to provide additional irrigation.

These ideas are appealing from a shortsighted point of view. However, the long-term effect of tampering with a glacier's natural regime can adversely affect the overall environment. It is conceivable, for example, that we could melt a glacier out of existence. Because most glaciers in the United States are in national parks or wilderness areas, glaciers have not been tampered with to control the output of meltwater.

Further Reading

U.S. Geological Survey. 1973. *Glaciers, a water resource.* U.S. Geological Survey Information Pamphlet.

Glaciation is most extensive in polar regions, where little melting takes place at any time of year. At present about one-tenth of the land surface on Earth is covered by glaciers (compared with about one-third during the peak of the glacial ages). Approximately 85% of the present-day glacier ice is on the Antarctic continent, covering an area larger than the combined areas of western Europe and the United States; 10% is in Greenland. All the remaining glaciers of the world amount to only about 5% of the world's freshwater ice. This means that Antarctica is in fact storing most of Earth's fresh water in the form of ice. Some have suggested that ice from the Antarctic, towed as icebergs, could be brought to areas of dry climate to alleviate water shortages. It is worth noting that if all of Antarctica's ice were to melt, sea level around the world would rise over 60 meters (200 feet). This would flood the world's coastal cities and significantly decrease the land surface available for human habitation.

FIGURE 19.2

Valley glacier on the flanks of Mount Logan, Canada's highest mountain. Photo by C. C. Plummer

Types of Glaciers

A simple criterion—whether or not a glacier is restricted to a valley—is the basis for classifying glaciers by form. A **valley glacier** is a glacier that is confined to a valley and flows from a higher to a lower elevation. Like streams, small valley glaciers may be tributaries to a larger trunk system. Valley glaciers are prevalent in areas of alpine glaciation. As might be expected, most glaciers in the United States and Canada, being in mountains, are of the valley type (figure 19.2).

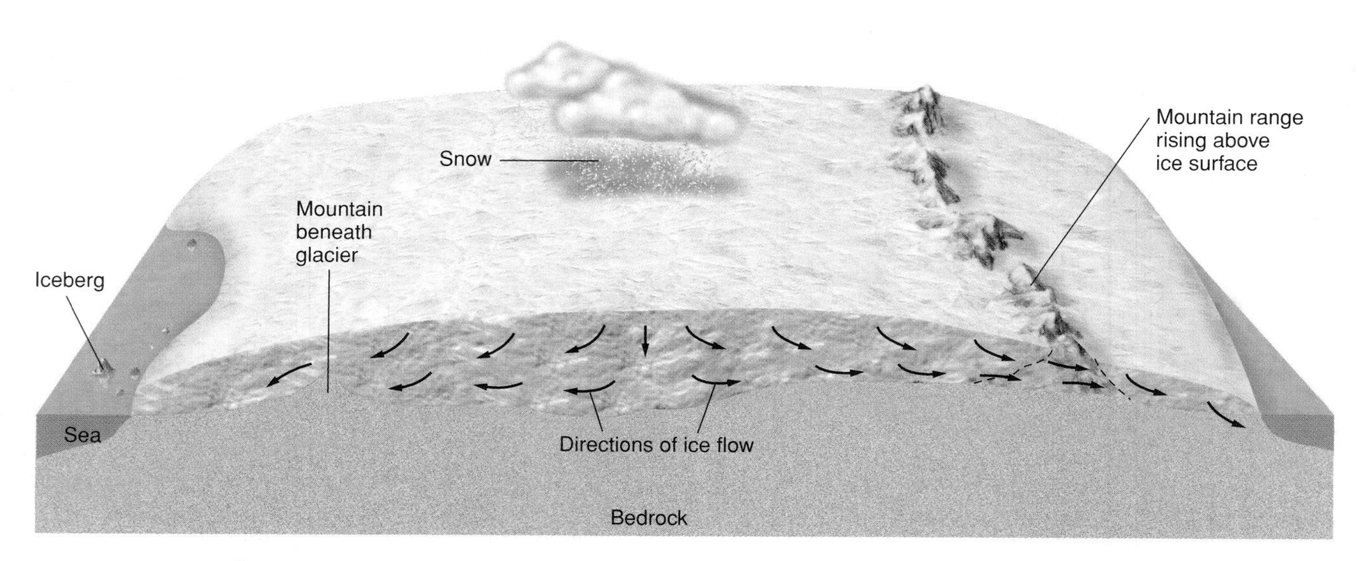

FIGURE 19.3

Diagrammatic cross section of an ice sheet. Vertical scale is highly exaggerated.

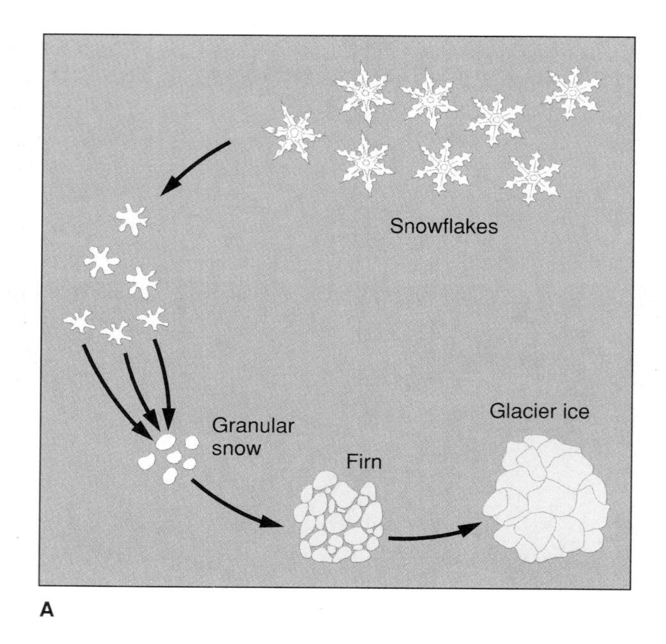

FIGURE 19.4

(*A*) Conversion of snow to glacier ice. (*B*) Thin slice of glacier ice in polarized light. In polarized light, the colors of individual ice grains vary depending on their crystallographic orientation.
Photo by C. C. Plummer

In contrast, an **ice sheet** is a mass of ice that is not restricted to a valley but covers a large area of land (over 50,000 square kilometers). Ice sheets are associated with continental glaciation. Only two places on Earth now have ice sheets: Greenland and Antarctica. A similar but smaller body is called an **ice cap.** Ice caps (and valley glaciers as well) are found in a few mountain highlands and on islands in the Arctic Ocean off Canada, Russia, and Scandinavia. An ice cap or ice sheet flows downward and outward from a central high point, as figure 19.3 shows.

Formation and Growth of Glaciers

Snow converts to glacier ice in somewhat the same way that sediment turns into a sedimentary rock and then into metamorphic rock; figure 19.4 shows the process. A snowfall can be compared to sediment settling out of water. A new snowfall may be in the form of light "powder snow," which consists mostly of air trapped between many six-pointed snowflakes. In a short time the snowflakes settle by compaction under their

FIGURE 19.5
An iceberg in southern Chile.
Photo by C. C. Plummer

own weight and much of the air between them is driven out. Meanwhile, the sharp points of the snowflakes are destroyed as flakes reconsolidate into granules. In warmer climates, partial thawing and refreezing results in coarse granules—the "corn snow" of spring skiing. In colder climates where little or no melting takes place, the snowflakes will recrystallize into fine granules. After the granular snow is buried by a new snowpack, usually during the following winter, the granules are compacted and weakly "cemented" together by ice. The compacted mass of granular snow, transitional between snow and glacier ice, is called *firn.* Firn is analogous to a sedimentary rock such as sandstone.

Through the years, the firn becomes more deeply buried as more snow accumulates. More air is expelled, the remaining pore space is greatly reduced and granules forced together recrystallize into the tight, interlocking mosaic of *glacier ice* (figure 19.4*B*). The recrystallization process involves little or no melting and is comparable to metamorphism. Glacier ice is texturally similar to the metamorphic rock, quartzite.

Under the influence of gravity, glacier ice moves downward and is eventually **ablated,** or lost. For glaciers in all but the coldest parts of the world, ablation is due mostly to melting, although some ice evaporates directly into the atmosphere. If a moving glacier reaches a body of water, blocks of ice break off (or *calve*) and float free as **icebergs** (figure 19.5). In most of the Antarctic, ablation takes place largely through calving of icebergs and direct evaporation. Only along the coast does melting take place, and there for only a few weeks of the year.

Glacial Budgets

If, over a period of time, the amount of snow a glacier gains is greater than the amount of ice and water it loses, the glacier's budget is *positive* and it expands. If the opposite occurs, the glacier decreases in volume and is said to have a *negative budget.* Glaciers with positive budgets push outward and downward at their edges; they are called **advancing glaciers.** Those with negative budgets grow smaller and their edges melt back; they are **receding glaciers.** Bear in mind that the glacial ice moves downvalley, as shown in figure 19.6, whether the glacier is advancing or receding. In a receding glacier, however, the rate of flow of ice is insufficient to replace all of the ice lost in the lower part of the glacier. If the amount of snow retained by the glacier equals the amount of

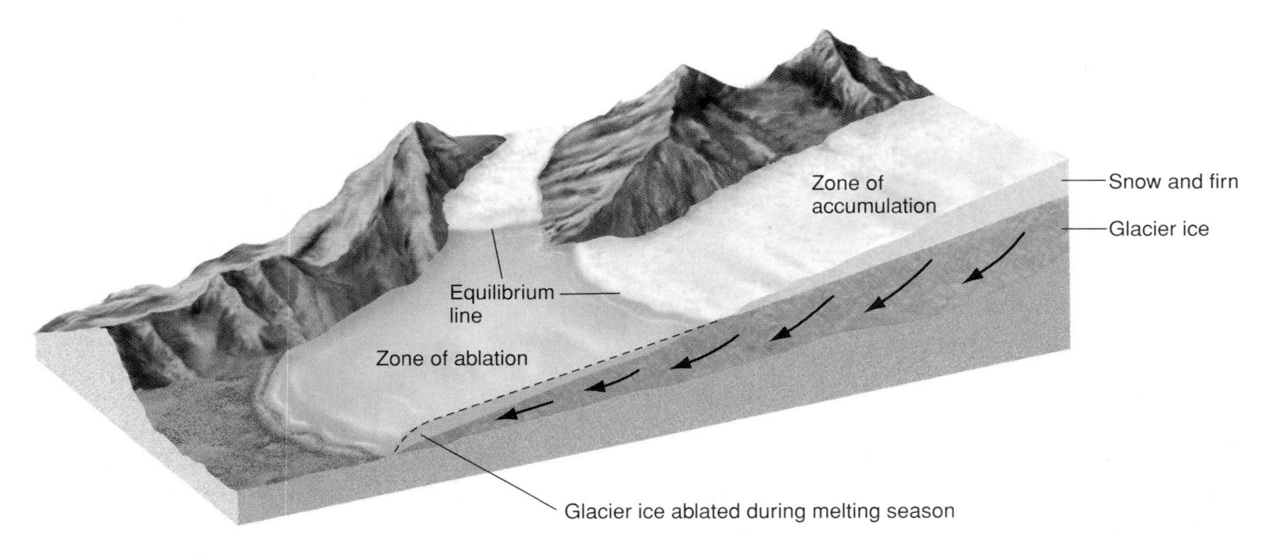

FIGURE 19.6
A valley glacier as it would appear at the end of a melt season. Below the equilibrium line, glacier ice and snow have been lost during the melting season. In the zone of accumulation above that line, firn is added to the glacier from the previous winter snowfall.

A

B

FIGURE 19.7

South Cascade Glacier, Washington. If the photos were taken at the end of the melt season, the snow line would be the boundary between white snow and darker glacier ice. These two photos were taken 23 years apart. Photo (*A*) was taken in 1957; note that the glacier extended into the lake and that small icebergs calved from it. Photo (*B*) was taken in 1980; notice that the glacier has shrunk and receded. During the interval between photos, the glacier lost approximately 7.5 meters of ice averaged over its surface, or the equivalent of 18.7 million cubic meters of water for the entire glacier. For a comparison to the glacier in 1992, go to http://hymet.com/glacier.htm/.

Photos by U.S. Geological Survey

ice and water lost, the glacier has a *balanced budget* and is neither advancing nor receding.

The upper part of a glacier, called the **zone of accumulation,** is the part of the glacier with a perennial snow cover (figure 19.6). The lower part is the **zone of ablation,** for there ice is lost, or ablated, by melting, evaporation, and calving.

The boundary between these two altitudinal zones of a glacier is an irregular line called the **equilibrium line,** which marks the highest point at which the glacier's winter snow cover is lost during a melt season (figure 19.7).

The equilibrium line may shift up or down from year to year, depending on whether there has been more accumulation or more ablation. Its location therefore indicates whether a glacier has a positive or negative budget. An equilibrium line migrating upglacier over a period of years is a sign of a negative budget, whereas an equilibrium line migrating downglacier indicates that the glacier has a positive budget. If an equilibrium line remains essentially in the same place year after year, the glacier has a balanced budget.

The **terminus,** the lower edge of a glacier, moves farther downvalley when a valley glacier has a positive budget. In a receding glacier the terminus melts back upvalley. Because most glaciers move slowly, migration of the terminus tends to lag several years behind a change in the budget.

An ice sheet with a positive budget increases in volume, advancing its outer margins. If the expanding ice sheet extends into the ocean, an increasing number of icebergs break off and float away in the open sea.

Advancing or receding glaciers are significant and sensitive indicators of climatic change; however, an advancing glacier does not necessarily indicate that the climate is getting colder. It may mean that the climate is getting wetter, or that more precipitation is falling during the winter months, or that the summers are cloudier. It is estimated that a worldwide decrease in the mean annual temperature of about 5°C could bring about a new ice age. Conversely, global climate warming of just a few degrees could significantly reduce the world's glaciers.

Movement of Valley Glaciers

Valley glaciers move downslope under the influence of gravity, the rate being variable, generally ranging from less than a few millimeters a day to 15 meters a day. The upper part of a glacier—where the volume of ice is greater and slopes tend to be

ASTROGEOLOGY 19.2

Ice Elsewhere in the Solar System

The polar regions of Mars are covered with ice caps, which are only a few meters thick and composed mostly of frozen carbon dioxide (dry ice). During the summer on each hemisphere, the ice caps shrink markedly as the carbon dioxide vaporizes; however, a small cap remains. This small residual cap (400 kilometers in diameter) is probably composed of water ice.

Two distinctive types of terrain can be observed in the Martian polar regions. *Laminated terrain* is the name given to areas where series of alternating light and dark layers can be seen. The layers are essentially horizontal, and each is about 15 to 35 meters thick. As many as fifty layers have been counted in one location. The layers are thought to represent alternating beds of high dust content and high ice content, and their alternation may be due to some kind of climatic change. That the layers are stratified outwash from glaciers has also been suggested. Near the margins of the polar caps are large troughs and ridges that could be glacial valleys and moraines.

Underlying the laminated terrain is another terrain, which is characterized by small pits. The pits of this *etch-pitted terrain* may be due to wind erosion, and if so, the pits would be deflation basins (blowouts). The pits may also be glacial kettles or collapse pits caused by the loss of underground ice.

There is some evidence suggesting that the moon has ice in polar craters that are not visible from Earth. But a recent experiment using a spacecraft failed to confirm this. If there is ice on the moon it would significantly reduce problems of establishing lunar bases or colonization by providing a source of water for humans.

Comets (such as Hale-Bopp, which graced nighttime skies in 1997) are essentially chunks of water ice, frozen carbon dioxide (dry ice), methane, ammonia, and rock dust that travel through space. Although they are usually only a kilometer or so in diameter, their tails of evaporating gases and dust extend over 100 million kilometers.

Some of the satellites for the large outer planets have abundant ice and some may have liquid water (raising the possibility that life could exist on these satellites). Europa, one of Jupiter's moons (box figure 1) is covered with ice. Images (received in 1997) from the spacecraft Galileo show that the ice has been cracked, bulged, and disrupted, indicating movement from below. Some scientists think that a deep ocean underlies the ice and that motion in the ocean causes disruption of the ice. An alternate hypothesis is that the ice layer is at least 100 kilometers thick and warm enough to move and churn the surface ice.

Box 19.2 — FIGURE 1

Composite image of Jupiter and its moons. From top to bottom: Io, Europa, Ganymede, and Callisto.

JPL/NASA

For Further Information

Nine Planets

www.nineplanets.org/

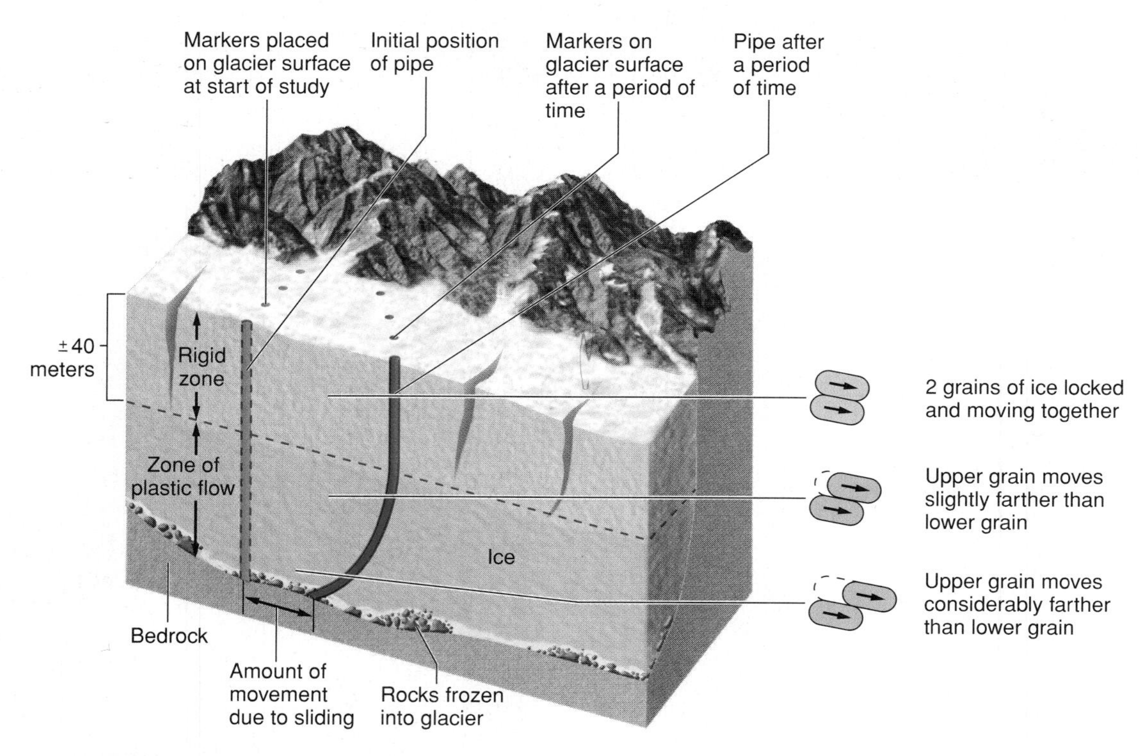

FIGURE 19.8

Movement of a glacier. Markers on the glacier indicate that the center of the glacier moves faster than its side. Cross-sectional view shows movement within the glacier.

steeper—generally moves faster than ice farther down or on gentler slopes. In this way ice from the higher altitudes keeps replenishing ice lost in the zone of ablation.

Glaciers in temperate climates—where the temperature of the glacier is at or near the melting point for ice—tend to move faster than those in colder regions—where the ice temperature stays well below freezing.

Velocity also varies within the glacier itself (figure 19.8). The central portion of a valley glacier moves faster than the sides (as water does in a stream), and the surface moves faster than the base. How ice moves within a valley glacier has been demonstrated by studies in which holes are drilled through the glacier ice and flexible pipes inserted. Changes in the shape and position of the pipes are measured periodically. The results of these studies are shown diagrammatically in figure 19.8.

Note in the diagram that the base of the pipe has moved downglacier. This indicates **basal sliding,** which is the sliding of the glacier as a single body over the underlying rock. A thin film of meltwater that develops along the base from the pressure of the overlying glacier facilitates basal sliding. Think of a large bar of wet soap sliding down an inclined board.

Note that the lower portion of the pipe is bent in a downglacier direction. The bent pipe indicates **plastic flow** of ice, movement that occurs within the glacier due to the plastic or "deformable" nature of the ice itself. Visualize two neighboring grains of ice within the glacier, one over the other. Both are moving—carried along by the ice below them; however, the higher of our two ice grains slides over its underlying neighbor a bit further. The reason the pipe is bent more sharply near the base of the glacier is that pressure from overlying ice results in greater flowage with increasing depth. Deep in the glacier, ice grains are sliding past their underlying neighbors further than similar ice grains higher up where the pipe is less bent. We should point out that a glacier flows not only because ice grains slide past one another but also because ice grains deform and recrystallize.

In the **rigid zone,** or upper part of the glacier, the pipe has been moved downglacier; however, it has remained unbent. The ice nearer the top apparently rides along passively on the plastically moving ice closer to the base. In the rigid zone, grains of ice do not move relative to their neighbors.

Crevasses

Along its length, a valley glacier moves at different rates in response to changes in the steepness of the underlying rock. Typically a valley glacier rides over a series of rock steps. Where the glacier passes over a steep part of the valley floor, it

ENVIRONMENTAL GEOLOGY 19.3

Water Beneath Glaciers: Floods, Giant Lakes, and Galloping Glaciers

A Galloping Glacier

Glacial motion is often used as metaphor for slowness ("The trial proceeded at a glacial pace"). But, some glaciers will *surge,* that is, move very rapidly for short periods following years of barely moving at all. The most extensively documented surge (or "galloping glacier") was that of Alaska's Bering Glacier in 1993–94. The Bering Glacier is the largest glacier in continental North America and it surges on a 20–30 year cycle. After its previous surge in 1967, its terminus retreated 10 kilometers. In August 1993, the latest surge began. Ice traveled at velocities up to 100 meters per day for short periods of time and sustained velocities of 35 meters per day over a period of several months. The terminus advanced 9 kilometers by the time the surge ended in November 1994. When glaciers surge, the previously slow moving, lower part of a glacier breaks into a chaotic mass of blocks (box figure 1). Surges are usually attributed to a buildup of water beneath part of a glacier, floating it above its bed. In July 1994, a large flood of water burst from Bering Glacier's terminus, carrying with it blocks of ice up to 25 meters across.

A Flood

Glacial outburst floods are not always associated with surges. In October 1996, a volcano erupted beneath a glacier in Iceland. The glacier, which is up to 500 meters thick, covers one-tenth of Iceland. Emergency teams prepared for the flood that geologists predicted would follow the eruption. The expected flood took place early in November with a peak flow of 45,000 cubic meters per second (over 1.5 million cubic feet per second)! The flood lasted only a few hours; however, it caused between 10 and 15 million dollars worth of damage. Three major bridges were destroyed or damaged and 10 kilometers of roads were washed away. Because people had been kept away from the expected flood path there were no casualties.

A Giant Lake

One of the world's largest lakes was only recently discovered. But don't expect to take a dip or go windsurfing on it. It lies below the thickest part of the East Antarctic Ice Sheet and is named after the Russian research station, Vostok, which is 4,000 meters above the lake at the coldest and most remote part of Antarctica. Lake Vostok was discovered in the 1970s through ice-penetrating radar; however, its extent was unknown until 1996 when satellite-borne radar revealed how large a lake it is. Studies indicate that the lake is 200 km long and 50 km wide—about the size of Lake Ontario. At its deepest, it is 510 meters, placing it among the ten deepest lakes in the world. Recently, more, but smaller, lakes beneath the East Antarctic Ice Sheet have been discovered.

The lake has been sealed off from the rest of the world for around a million years, and it is likely that it contains organisms, such as microbes, dating back to that time. These organisms (and their genes) would not have been affected by modern pollution or nuclear bomb fallout. By coincidence, the world's deepest ice hole (over 3 kilometers) was being drilled from Vostok Station above the lake when the size of Lake Vostok was being determined. The ice core from this hole should add to the findings from the Greenland drilling projects (box 19.4) and provide an even greater picture of Earth's climate during the ice ages. When the hole was completed in 1997, drilling was halted short of reaching the lake due to fear of contaminating it and harming whatever living organisms might be in the very old water. Study of the lake and its organisms is curtailed until a future generation of scientists can devise means of sampling the waters without altering its ecosystem.

Box 19.3 — FIGURE 1
Part of a glacier after a surge (lower part of photo). The debris-covered ice has been broken up into a chaotic mass of blocks. In the background is a small glacier that has retreated up its valley. Photo taken near the Canada-Alaska border.
Photo by C. C. Plummer

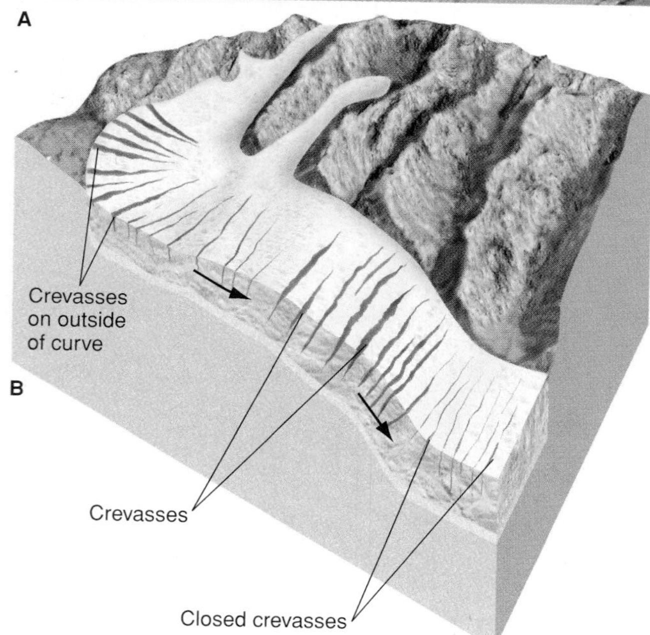

FIGURE 19.9 (A) Crevasses on a glacier, looking down from Mount Logan, Yukon Territory, Canada. (B) Crevasses along the course of a glacier.
Photo by C. C. Plummer

moves faster. The upper rigid zone of ice, however, cannot stretch to move as rapidly as the underlying plastic-flowing ice. Being brittle, the ice of the rigid zone is broken by the tensional forces. Open fissures, or **crevasses,** develop (figure 19.9). Crevasses also form along the margins of glaciers in places where the path is curved, as shown in part of figure 19.9. This is because ice (like water) flows faster toward the outside of the curve. For glaciers in temperate climates, a crevasse should be no deeper than about 40 meters, the usual thickness of the rigid zone. If you are falling down a crevasse, it may be of some consolation that, as you are hurtling to death or injury, you realize on the way down that you will not fall more than 40 meters.

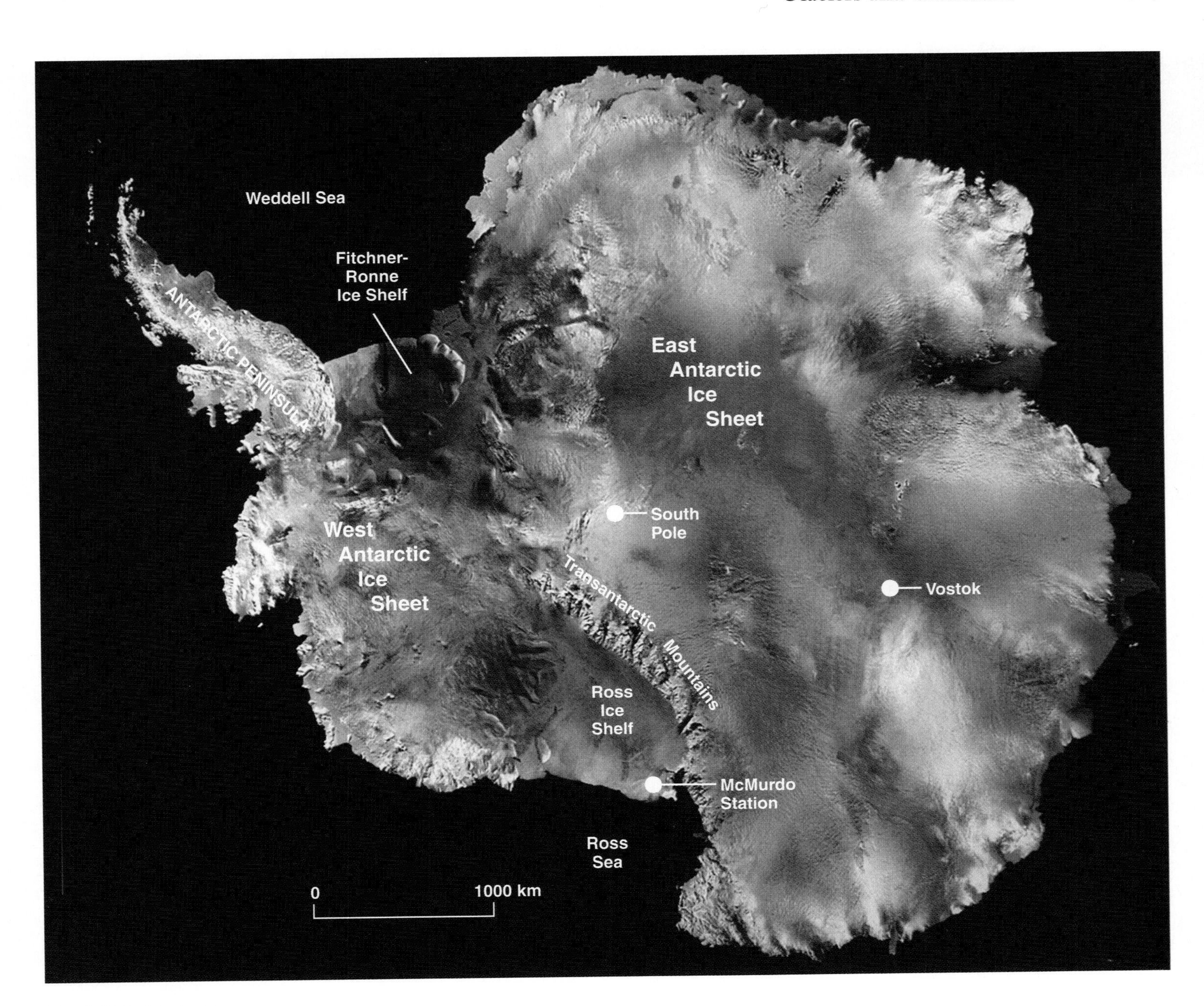

FIGURE 19.10

The Antarctic continent and its ice sheets. Vostok is at the highest part of the East Antarctic Ice Sheet.

U.S. Geological Survey/NASA

After the ice has passed over a steep portion of its course, it slows down, and compressive forces close the crevasses.

Movement of Ice Sheets

An ice sheet or ice cap moves like a valley glacier except that it moves downward and outward from a central high area toward the edges of the glacier (as shown in figure 19.3).

Glaciological research in Antarctica has determined how ice sheets grow and move. Antarctica has two ice sheets: the West Antarctic Ice Sheet is separated by the Transantarctic Mountains from the much larger East Antarctic Ice Sheet (figure 19.10). The two ice sheets join in the low areas between mountain ranges. Both are nearly completely within a zone of accumulation because so little melting takes place (ablation is largely by calving of icebergs) and because occasional snowfalls nourish their high central parts. The ice sheets mostly overlie interior lowlands, but also completely bury some mountain ranges. Much of the base of the West Antarctic Ice Sheet is on bedrock that is below sea level. At least one active

19.4 IN GREATER DEPTH

DRILLING THROUGH ICE SHEETS FOR A RECORD OF THE PAST

Go to the book's website www.mhhe.com/plummer9e to find out about drilling through Greenland and Antarctic ice sheets to collect cylindrical cores of glacier ice from the several-kilometer-deep holes. Analysis of these ice cores has

- provided a record of climate change for the last 250,000 years;
- indicated that ice ages end abruptly—over a few years, rather than centuries as previously thought;
- provided a detailed record of when major volcanic eruptions took place prior to recorded history;
- indicated that cooler temperatures between 8,000 and 9,000 years ago correlate with a period of heavier than usual volcanic activity.

FIGURE 19.11
The South Pole. Actually, the true South Pole is several kilometers from here. The moving ice sheet has carried the striped pole away from the site of the true South Pole, where the pole was erected in 1956.
Photo by C. C. Plummer

volcano underlies the West Antarctic Ice Sheet (resulting in a depression in the ice sheet). Where mountain ranges are higher than the ice sheet, the ice flows through as valley glaciers.

At the South Pole (figures 19.10 and 19.11)—neither the thickest part nor the center of the East Antarctic Ice Sheet—the ice is 2,700 meters thick. The thickest part of the East Antarctic Ice Sheet is 4,776 meters.

Most of the movement of the East Antarctic Ice Sheet is by means of plastic flow. It has been thought that most of the ice sheet is frozen to the underlying rocks and basal sliding takes place only locally. But the recent discovery of a giant lake and other lakes beneath the thickest part of the Antarctic Ice Sheet (box 19.3) indicates that liquid water at its base is more widespread and basal sliding might be more important than previously regarded.

GLACIAL EROSION

Wherever basal sliding takes place, the rock beneath the glacier is abraded and modified. As meltwater works into cracks in bedrock and refreezes, pieces of the rock are broken loose and frozen into the base of the moving glacier, a process known as *plucking*. While being dragged along by the moving ice, the rock within the glacier grinds away at the underlying rock (figure 19.12). The thicker the glacier, the more pressure on the rocks and the more effective the grinding and crushing.

Pebbles and boulders that are dragged along are *faceted,* that is, given a flat surface by abrasion. Bedrock underlying a glacier is *polished* by fine particles and *striated* (scratched) by sharp-edged, larger particles. Striations and grooves on bedrock indicate the direction of ice movement (figure 19.13).

The grinding of rock across rock produces a powder called **rock flour.** Rock flour is composed largely of very fine (silt- and clay-sized) particles of unaltered minerals (pulverized from

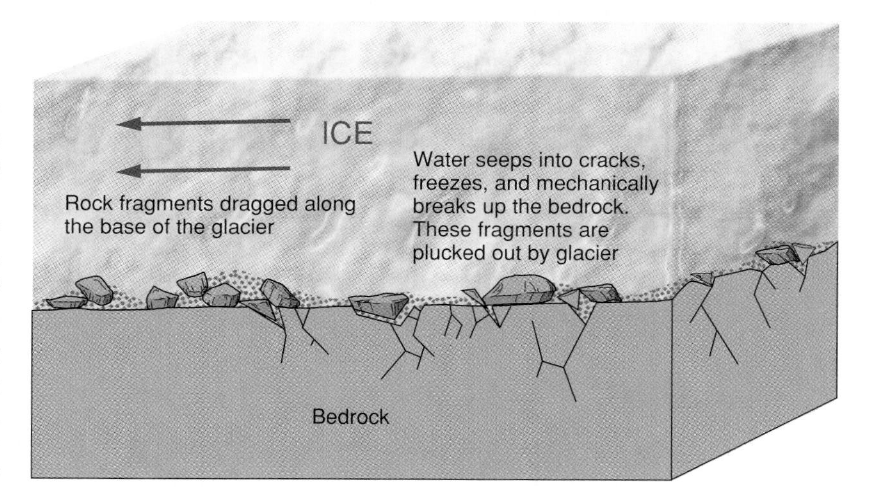

FIGURE 19.12
Plucking and abrasion beneath a glacier.

FIGURE 19.13

Striated and polished bedrock surface in South Australia. Unlike glacial striations commonly found in North America, these were caused by late Paleozoic glaciation.

chemically unweathered bedrock). When *meltwater* washes rock flour from a glacier, the streams draining the glacier appear milky and lakes into which glacial meltwater flows often appear a milky green color.

Not all glacier-associated erosion is caused directly by glaciers. Mass wasting takes place on steep slopes created by downcutting glaciers. Frost wedging breaks up bedrock ridges and cliffs above a glacier, causing frequent rockfalls. Snow avalanches bring down loose rocks onto the glacier surface, where they ride on top of the ice. Debris may also fall into crevasses to be transported within or at the base of a glacier, as shown in figure 19.19.

Erosional Landscapes Associated with Alpine Glaciation

We are in debt to glaciers for the rugged and spectacular scenery of high mountain ranges. Figure 19.14 shows how glaciation has radically changed a previously unglaciated mountainous region. The striking and unique features associated with mountain glaciation are due to the erosional effects of glaciers as well as frost wedging on exposed rock.

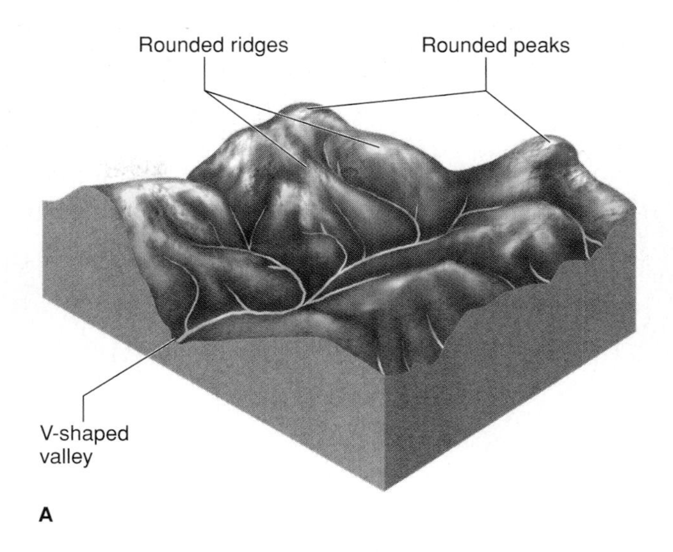

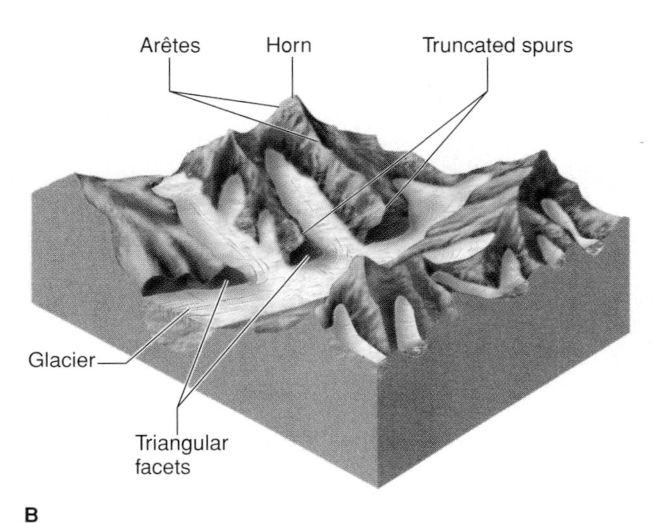

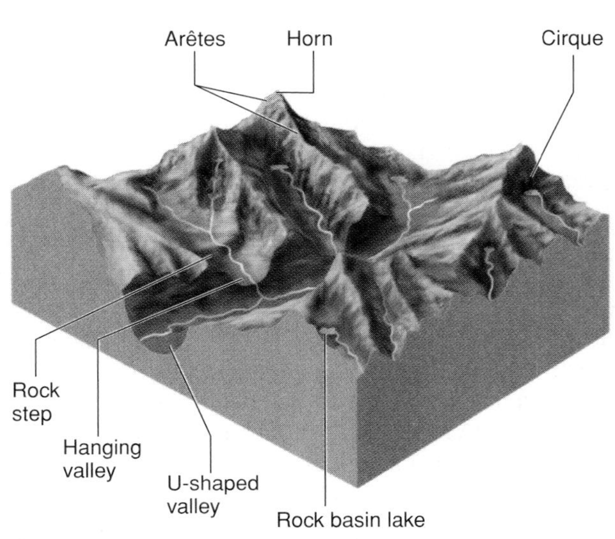

FIGURE 19.14

(*A*) A stream-carved mountain landscape before glaciation. (*B*) The same area during glaciation. Ridges and peaks become sharper due to frost wedging. (*C*) The same area after glaciation.

FIGURE 19.15

Glacially carved valley in Grand Teton National Park, Wyoming. Rounded knobs produced from glacial erosion are visible in the lower slopes, in front of the U-shaped valley.
Photo by C. C. Plummer

Glacial Valleys

Glacially carved valleys are easy to recognize. A **U-shaped valley** (in cross profile) is characteristic of glacial erosion (figure 19.15), just as a **V**-shaped valley is characteristic of stream erosion.

The thicker a glacier is, the more erosive force it exerts on the underlying valley floor, and the more bedrock is ground away. For this reason, a large trunk glacier erodes downward more rapidly and carves a deeper valley than do the smaller tributary glaciers that join it. After the glaciers disappear, these tributaries remain as **hanging valleys** high above the main valley (figure 19.16).

Valley glaciers, which usually occupy valleys formerly carved by streams, tend to straighten the curves formed by running water. This is because the mass of ice of a glacier is too sluggish and inflexible to move easily around the curves. In the process of carving the sides of its valley, a glacier erodes or "truncates" the lower ends of ridges that extended to the valley. **Truncated spurs** are ridges that have *triangular facets* produced by glacial erosion at their lower ends (figure 19.14*B*).

Although a glacier tends to straighten and smooth the side walls of its valley, ice action often leaves the surface of the

FIGURE 19.16

A hanging valley in Yosemite National Park, California.
Photo by C. C. Plummer

underlying bedrock carved into a series of steps. This is due to the variable resistance of bedrock to glacial erosion. Figure 19.17 shows what happens when a glacier abrades a relatively weak rock with closely spaced fractures. Water seeps into cracks in the bedrock, freezes there, and enlarges fractures or makes new ones. Rock frozen into the base of the glacier grinds and loosens more pieces. After the ice has melted back, a chain of **rock-basin lakes** (also known as **tarns**) may occupy the depressions carved out of the weaker rock. A series of such lakes, reminiscent of a string of prayer beads, is sometimes called *paternoster lakes.*

Areas where the bedrock is more resistant to erosion stand out after glaciation as *rounded knobs* (see figures 19.15 and 19.22), usually elongated parallel to the direction of glacier flow. These are also known as *roches moutonnées.* (The term was used to describe an assemblage of rounded knobs in the Alps that resembled grazing sheep. In French, *roche* is rock and *moutonnée* means fleecy or curled.) On stronger, less fractured rock, glacial erosion works in a similar way but is less effective because of the bedrock's resistance to grinding and crushing.

Cirques, Horns, and Arêtes

A **cirque** is a steep-sided, half-bowl-shaped recess carved into a mountain at the head of a valley carved by a glacier (figure 19.18). In this unique, often spectacular, topographic feature, a large percentage of the snow that accumulates eventually converts to glacier ice and spills over the threshold as the valley glacier starts its downward course.

A cirque is not entirely carved by the glacier itself but is also shaped by the weathering and erosion of the rock walls above the surface of the ice. Frost wedging and avalanches break up the

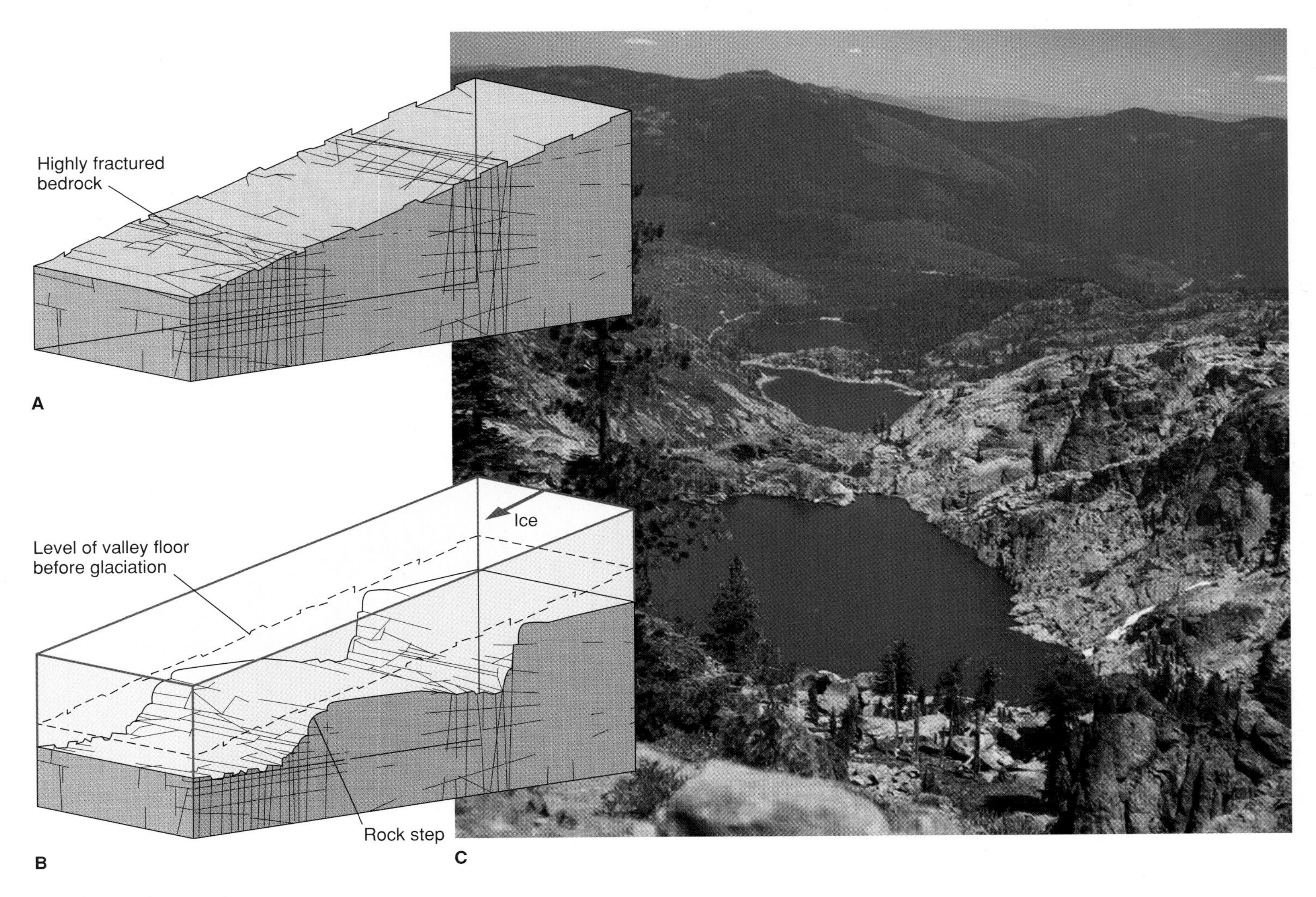

FIGURE 19.17
Development of rock steps. (*A*) Valley floor before glaciation. (*B*) During glaciation. (*C*) Rock steps and rock-basin lakes, Sierra Nevada, California.
A and *B* after F. E. Matthes, 1930, U.S. Geological Survey; Photo by C. C. Plummer

FIGURE 19.18
A cirque occupied by a small glacier in the Canadian Rocky Mountains. The glacier was much larger during the ice ages.
Photo by C. C. Plummer

rock and steepen the slopes above the glacier. Broken rock tumbles onto the valley glacier and becomes part of its load, and some rock may fall into a crevasse that develops where the glacier is pulling away from the cirque wall (figure 19.19).

The headward erosional processes that enlarge a cirque also help create the sharp peaks and ridges characteristic of glaciated mountain ranges. A **horn** is the sharp peak that remains after cirques have cut back into a mountain on several sides (figure 19.20).

Frost wedging works on the rock exposed above the glacier, steepening and cutting back the side walls of the valley. Sharp ridges called **arêtes** separate adjacent glacially carved valleys (figure 19.21).

Erosional Landscapes Associated with Continental Glaciation

In contrast to the rugged and angular nature of glaciated mountains, an ice sheet tends to produce rounded topography. The rock underneath an ice sheet is eroded in much the same way as

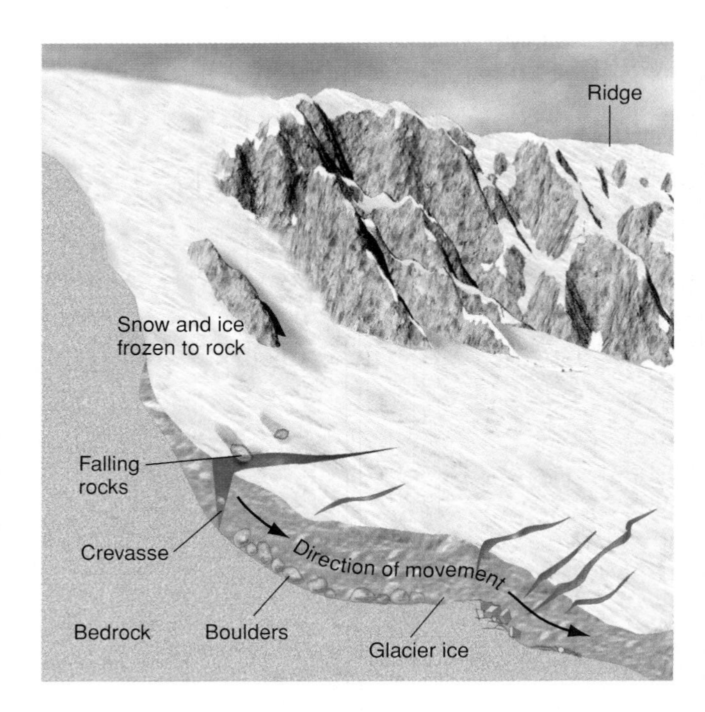

FIGURE 19.19
Cutaway view of a cirque.

FIGURE 19.21
An arête on Mount Logan, Yukon Territory, Canada.
Photo by C. C. Plummer

FIGURE 19.20
The Matterhorn in Switzerland. Engraving by Edward Whymper, who, in 1865, became the first person to climb the Matterhorn.

the rock beneath a valley glacier; however, the weight and thickness of the ice sheet may produce more pronounced effects. Rounded knobs are common (figure 19.22) as are grooved and striated bedrock. Some grooves are actually channels several meters deep and many kilometers long. The orientation of grooves and striations indicates the direction of movement of a former ice sheet.

An ice sheet may be thick enough to bury mountain ranges, rounding off the ridges and summits and perhaps streamlining them in the direction of ice movement. Much of northeastern Canada, with its rounded mountains and grooved and striated bedrock surface, shows the erosional effects of ice sheets that formerly covered that part of North America (figure 19.23).

FIGURE 19.22

Rounded knob (*roche moutonnée*) in southern Ontario, Canada.

Photo by C.C. Plummer

FIGURE 19.23

Glacially scoured terrain in Canadian Arctic. Bedrock is covered by spongy tundra vegetation.

Photo by Paula J. Noble

GLACIAL DEPOSITION

The rock fragments scraped and plucked from the underlying bedrock and carried along at the base of the ice make up most of the load carried by an ice sheet, but only part of a valley glacier's load. Much of a valley glacier's load comes from rocks broken from the valley walls.

Most of the rock fragments carried by glaciers are angular, as the pieces have not been tumbled around enough for the edges and corners to be rounded. The debris is unsorted, and clay-sized to boulder-sized particles are mixed together (figure 19.24). The unsorted and unlayered rock debris carried or deposited by a glacier is called **till.**

Glaciers are capable of carrying virtually any size of rock fragment, even boulders as large as a house. An **erratic** is an ice-transported boulder that has not been derived from underlying bedrock. If its bedrock source can be found, the erratic indicates the direction of movement of the glacier that carried it.

Moraines

When till occurs as a body of unsorted and unlayered debris either on a glacier or left behind by a glacier, the body is regarded as one of several types of **moraines. Lateral moraines** are elongate, low mounds of till that form along the sides of a valley glacier (figures 19.24, 19.25, 19.26, and 19.27).

FIGURE 19.24
Till transported on top of and alongside a glacier in Peru. View is downglacier. The lake is dammed up by an end moraine at its far end.
Photo by C.C. Plummer

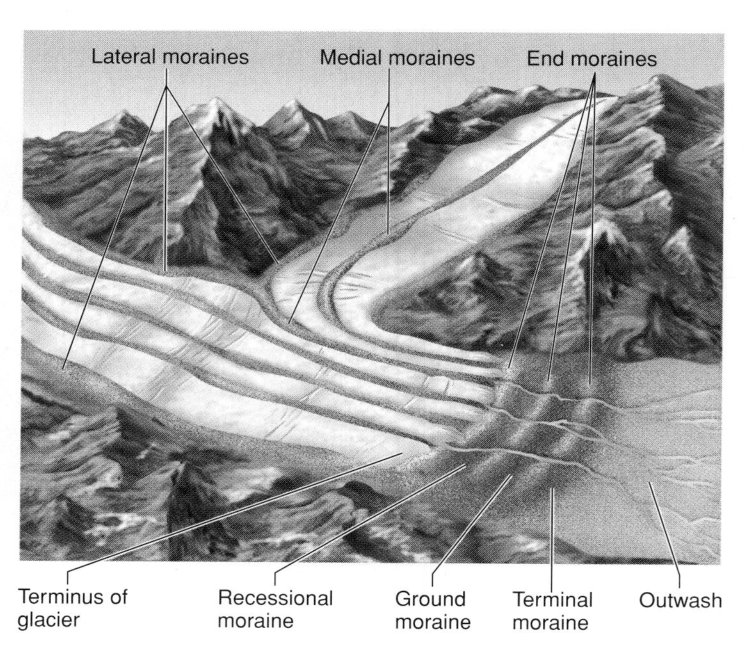

FIGURE 19.25
Moraines associated with valley glaciers.

FIGURE 19.26
Medial and lateral moraines on valley glaciers, Yukon Territory, Canada. Ice is flowing toward viewer and to lower right.
Photo by C. C. Plummer

Rockfall debris from the steep cliffs that border valley glaciers accumulates along the edges of the ice to form lateral moraines.

Where tributary glaciers come together, the adjacent lateral moraines join and are carried downglacier as a single long ridge of till known as a **medial moraine.** In a large trunk glacier that has formed from many tributaries, the numerous medial moraines give the glacier the appearance from the air of a multilane highway (figures 19.25 and 19.26).

An actively flowing glacier brings debris to its terminus. If the terminus remains stationary for a few years or advances, a distinct **end moraine,** a ridge of till, piles up along the front edge of the ice. Valley glaciers build end moraines that are crescent-shaped or sometimes horseshoe-shaped (figures 19.25 and 19.27). The end moraine of an ice sheet takes a similar lobate form, but is much longer and more irregular than that of a valley glacier (figure 19.28).

FIGURE 19.27
In the foreground, end moraines (recessional moraines) curve into two long lateral moraines. The two lateral moraines extend back to a glacially carved valley in the Sierra Nevada, California.
Photo by C. C. Plummer

Geologists distinguish two special kinds of end moraines. A *terminal moraine* is the end moraine marking the farthest advance of a glacier. A *recessional moraine* is an end moraine built while the terminus of a receding glacier remains temporarily stationary. A single receding glacier can build several recessional moraines (as in figures 19.24, 19.25, 19.27, and 19.28).

As ice melts, rock debris that has been carried by a glacier is deposited to form a **ground moraine,** a fairly thin, extensive layer or blanket of till (figures 19.25 and 19.28). Very large areas that were once covered by an ice sheet now have the gently rolling surface characteristic of ground moraine deposits.

In some areas of past continental glaciation there are bodies of till shaped into streamlined hills called **drumlins** (figures 19.28 and 19.29). A drumlin is shaped like an inverted spoon aligned parallel to the direction of ice movement of the former glacier. Its gentler end points in the downglacier direction. Because we cannot observe drumlins forming beneath present ice sheets there is uncertainty how till becomes shaped into these streamlined hills.

Outwash

In the zone of wastage, large quantities of meltwater usually run over, beneath, and away from the ice. The material deposited by the debris-laden meltwater is called **outwash.** Because it has the characteristic layering and sorting of stream-deposited sediment, outwash can be distinguished easily from the unlayered and unsorted deposits of till. Because outwash is fairly well sorted and the particles generally are not chemically weathered, it is an excellent source of aggregate for building roadways and for mixing with cement to make concrete.

An outwash feature of unusual shape associated with former ice sheets and some very large valley glaciers is an **esker,** a long, sinuous ridge of water-deposited sediment (figures 19.28 and 19.30). Eskers can be up to 10 meters high and are formed of cross-bedded and well-sorted sediment. Evidently eskers are deposited in tunnels within or under glaciers, where meltwater loaded with sediment flows under and out of the ice.

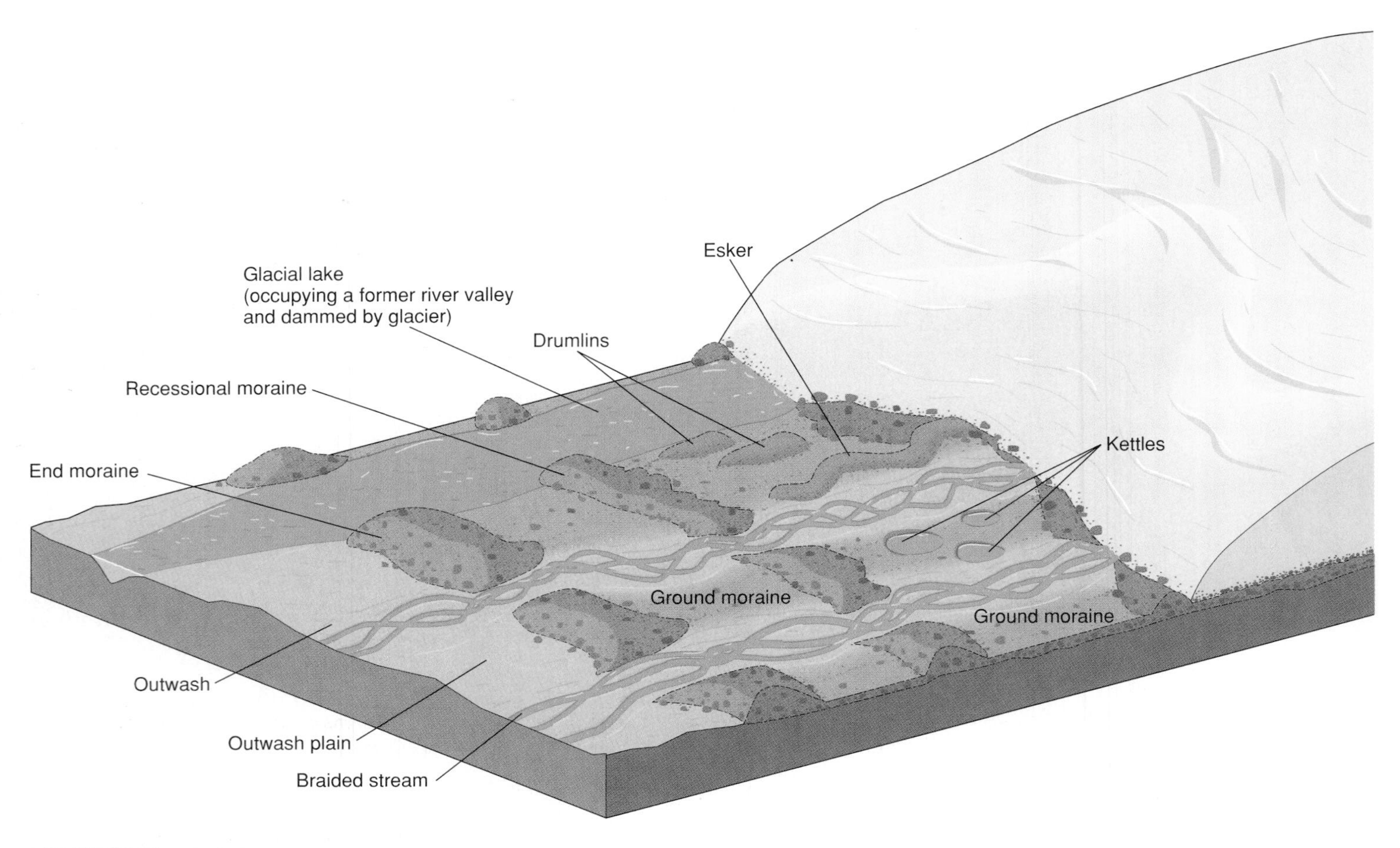

FIGURE 19.28

Depositional features in front of a receding ice sheet.

FIGURE 19.29

Drumlin in New York state.

Photo by Ward's Natural Science Est., Inc., Rochester, N.Y.

FIGURE 19.30

An esker in northeastern Washington.

Photo by D. A. Rahm, courtesy Rahm Memorial Collection, Western Washington University

FIGURE 19.31
A kettle (*foreground*) and outwash (*background* and *left*) from a glacier. Stagnant ice underlies much of the till. Yukon Territory, Canada.
Photo by C. C. Plummer

As meltwater builds thick deposits of outwash alongside and in front of a retreating glacier, blocks of stagnant ice may be surrounded and buried by sediment. When the ice block finally melts (sometimes years later), a depression called a **kettle** forms (figures 19.28 and 19.31). Many of the small scenic lakes in the upper Middle West of the United States are kettle lakes.

The streams that drain glaciers tend to be very heavily loaded with sediment, particularly during the melt season. As they come off the glacial ice and spread out over the outwash deposits, the streams form a braided pattern (see chapter 16 on streams).

The large amount of rock flour that these streams carry in suspension settles out in quieter waters. In dry seasons or drought, the water may dry up and the rock flour deposits may be picked up by the wind and carried long distances. Some of the best agricultural soil in the United States has been formed by rock flour that has been redeposited by wind. Such fine-grained, wind-blown deposits of dust are called *loess* (see chapter 18).

Glacial Lakes and Varves

Lakes often occupy depressions carved by glacial erosion but can also form behind dams built by glacial deposition. Commonly a lake forms between a retreating glacier and an earlier end moraine (see figure 19.24).

In the still water of the lake, clay and silt settle on the bottom in two thin layers—one light-colored, one dark—that are characteristic of glacial lakes. Two layers of sediment representing one year's deposition in a lake are called a **varve** (figure 19.32). The light-colored layer consists of slightly coarser sediment (silt) deposited during the warmer part of the year when the nearby glacier is melting and sediment is transported to the lake. The silt settles within a few weeks or so after reaching the lake. The dark layer is finer sediment (clay)—material that sinks down more slowly during the winter after the lake surface freezes and the supply of fresh, coarser sediment stops due to lack of meltwater. The dark color is attributed to fine organic matter mixed with the clay.

FIGURE 19.32
Varves from a former glacial lake. Each pair of light and dark layers represents a year's deposition. Gradations on ruler are centimeters.
Photo by Brian Atwater, U.S. Geological Survey

Because each varve represents a year's deposit, varves are like tree rings and indicate how long a glacial lake existed.

EFFECTS OF PAST GLACIATION

As the glacial theory gained general acceptance during the latter part of the nineteenth century, it became clear that much of northern Europe and the northern United States as well as most of Canada had been covered by great ice sheets during the so-called Ice Age. It also became evident that even areas not covered by ice had been affected because of the changes in climate and the redistribution of large amounts of water.

We now know that the last of the great North American ice sheets melted away from Canada less than 10,000 years ago. In many places, however, till from that ice sheet overlies older tills, deposited by earlier glaciations. The older till is distinguishable from the newer till because the older till was deeply weathered during times of warmer climate between glacial episodes.

19.5 ASTROGEOLOGY

Mars on a Glacier

Meteorites are extraterrestrial rocks—fragments of material from space that have managed to penetrate Earth's atmosphere and land on Earth's surface. They are of interest not only to astronomers but to geologists, for they help us date Earth (chapter 8) and give us clues to what Earth's interior is like (see chapter 2) because many of the meteorites are thought to represent fragments of destroyed minor planets. Meteorites are rarely found; they usually do not look very different from Earth's rocks with which they are mixed.

In recent years, a bonanza of meteorites has been found on the Antarctic ice sheets. Over a thousand meteorites have been collected from one small area where the ice sheet abuts against the Transantarctic Mountains. The reason for this heavy concentration is that meteorites landing on the surface of the ice over a vast area have been incorporated into the glacier and transported to where wastage takes place. The process is illustrated in box figure 1.

A few of the meteorites are especially intriguing. Some almost certainly are rocks from the moon, while several others apparently came from Mars. Their chemistry and physical properties match what we would expect of a Martian rock. But how could a rock escape Mars and travel to Earth? Scientists suggest that a meteorite hit Mars with such force that fragments of that planet were launched into space. Eventually, some of the fragments reached Earth.

In 1997, researchers announced that they found what could be signs of former life on Mars in one of the meteorites collected 13 years earlier in Antarctica. The evidence included carbon-containing molecules that might have been produced by living organisms as well as microscopic blobs that could be fossil alien bacteria. But there are alternate explanations for each line of evidence, and a hot debate has ensued between scientists with opposing viewpoints. The ongoing space missions to Mars will be looking closely for evidence of primitive life on Mars, past as well as present.

Related Web Resource

Antarctic Meteorite Program

www-curator.jsc.nasa.gov/curator/antmet/program.htm

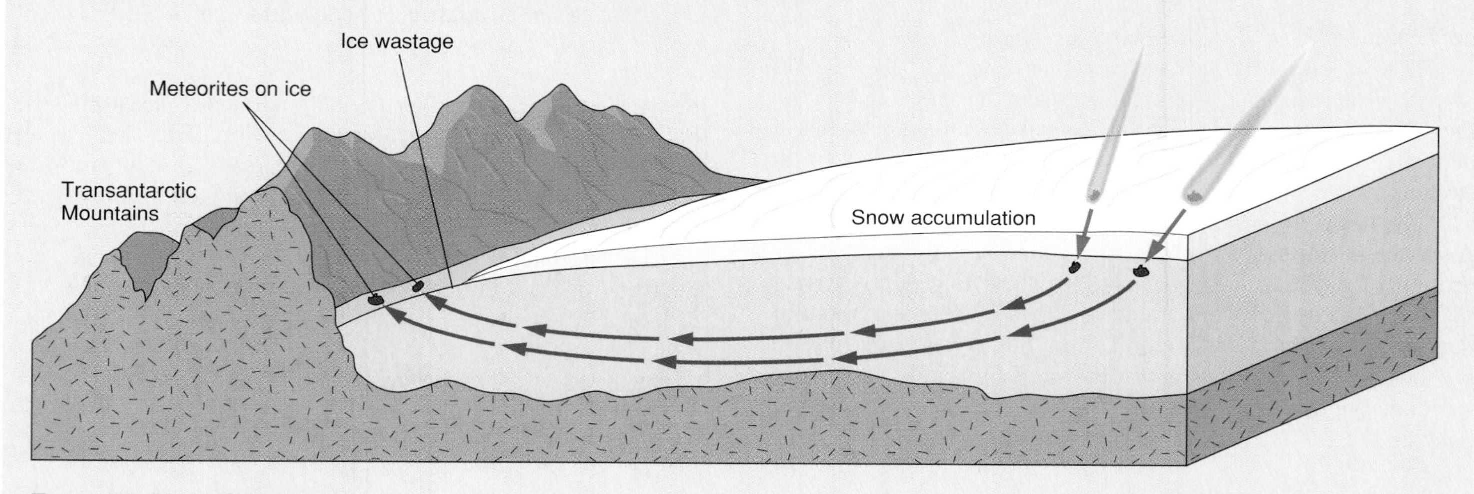

Box 19.5 — FIGURE 1

Diagram showing the way in which meteorites are concentrated in a narrow zone of wastage along the Transantarctic Mountains. Two meteorites are shown as well as the paths they would have taken from the time they hit the ice sheet until they reached the zone of wastage. The vertical scale is greatly exaggerated.

Source: Antarctic Journal of the United States

The Glacial Ages

Geologists can reconstruct with considerable accuracy the last episode of extensive glaciation, which covered large parts of North America and Europe and was at its peak about 18,000 years ago. There has not been enough time for weathering and erosion to alter significantly the effects of glaciation. Less evidence is preserved for each successively older glacial episode, because (1) weathering and erosion occurred during warm interglacial periods and (2) later ice sheets and valley glaciers overrode and obliterated many of the features of earlier glaciation. However, from piecing together the evidence, geologists can see that earlier glaciers covered approximately the same region as the more recent ones.

IN GREATER DEPTH 19.6

Causes of Glacial Ages

Go to the book's website at www.mhhe.com/plummer9e for a more in-depth presentation of the summary below.

The question of what caused the glacial ages has not been completely answered since the theory of glacial ages was accepted over a century ago. Only in the last few decades have climatologists thought they were beginning to provide acceptable answers.

The primary control on the Pleistocene glacial and interglacial episodes seems to be variations in Earth's orbit and inclination to the sun. The amount of heat from solar radiation received by any particular portion of Earth is related to the angle of the incoming sun's rays and, to a lesser degree, the distance to the sun. The angle of Earth's poles relative to the plane of Earth's orbit about the sun also changes periodically. Variations in orbital relationships and "wobble" of Earth's axis are largely responsible for glacial and interglacial episodes. These provide variations in incoming solar radiation cycles of 21,000, 41,000, and 100,000 years, as calculated by Milutin Milankovitch, a Serbian mathematician, in 1921. Proof of Milankovitch's cycles came from cores of deep sediment taken by oceanographic research ships. Deep sea sediment provides a fairly precise record of climatic variations over the past few hundred thousand years. The cycles of cooling and warming determined from the marine sediments closely match the times predicted by Milankovitch.

But the theory fails to explain the absence of glaciation over most of geologic time. Thus, one or more of the other mechanisms listed below (and described in the website) may have contributed to climate change resulting in glacial ages.

- **Changes in the atmosphere.** These changes include the amount of carbon dioxide in the atmosphere. Carbon dioxide has a "greenhouse effect," whereby the more of the gas in the atmosphere, the warmer the global climate. Large volcanic eruptions are known to lower temperature worldwide by placing SO_2 gas and fine dust in the high atmosphere. A series of large, volcanic eruptions might help trigger an ice age.
- **Changes in the positions of continents.** Plate tectonic movement of continents closer to the poles increases the likelihood of glaciation. Movement of northern hemisphere continents closer to the North Pole has placed landmasses in a position more favorable for glaciation.
- **Changes in circulation of sea water.** Landmasses block the worldwide free circulation of sea water, affecting which oceans are warmer than others.

Until a few years ago geologists thought the Pleistocene Epoch (see chapter 8) included all the glacial ages, but recent work indicates that worldwide climate changes necessary for northern continental glaciation probably began at least 3 million years ago, late in the Tertiary Period, at least a million years before the Pleistocene. Moreover, Antarctica has been glaciated for at least 20 million years.

Earth has undergone episodic changes in climate during the last 2 to 3 million years. Actually, the climate changes necessary for a glacial age to occur are not so great as one might imagine. During the height of a glacial age, the worldwide average of annual temperatures was probably only about 5°C cooler than at present. Some of the intervening interglacial periods were probably a bit warmer worldwide than present-day average annual temperatures.

Direct Effects of Past Glaciation in North America

Moving ice abraded vast areas of northern and eastern Canada during the growth of the North American ice sheets (figure 19.33). Most of the soil and sedimentary rock was scraped off and underlying crystalline bedrock was scoured. Many thousands of future lake basins were gouged out of the bedrock.

The directions of ice flow can be determined from the orientation of striations and grooves in the bedrock. The largest ice sheet (the Laurentide Ice Sheet) moved outward from the general area now occupied by Hudson Bay, which is where the ice sheet was thickest. The present generally barren surface of the Hudson Bay area contrasts markedly with the Great Plains surface of southern Canada and northern United States, where vast amounts of till were deposited.

Most of the till was deposited as ground moraine, which, along with outwash deposits, has partially weathered to yield excellent soil for agriculture. Rock flour that originally washed out of ice sheets has been redistributed by wind, as *loess,* over large parts of the Midwest and eastern Washington to contribute to especially good agricultural land (see chapter 18). In many areas along the southern boundaries of land covered by ground moraines, broad and complex end moraines extend for many kilometers, indicating that the ice margin must have been close to stationary for a long time (figure 19.34). Numerous drumlins are preserved in areas such as Ontario, New England, and upstate New York. New York's Long Island is made of terminal and recessional moraines and outwash deposits. Erratics there come from metamorphic rock in New England. Cape Cod in Massachusetts was also formed from moraines.

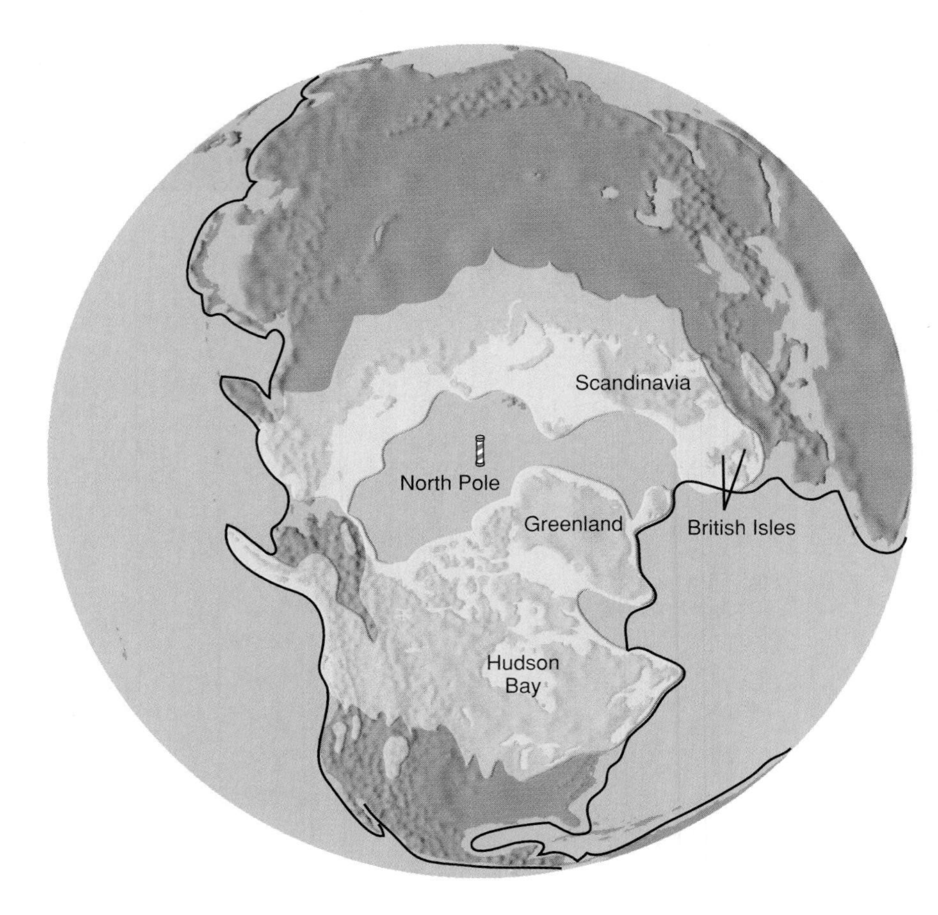

FIGURE 19.33

Maximum glaciation during the Pleistocene ice ages in the northern hemisphere. Note that glaciation extended beyond present continental shore lines. This is because sea level was lower than at present. Also shown are the shorelines for the unglaciated portions of North America.

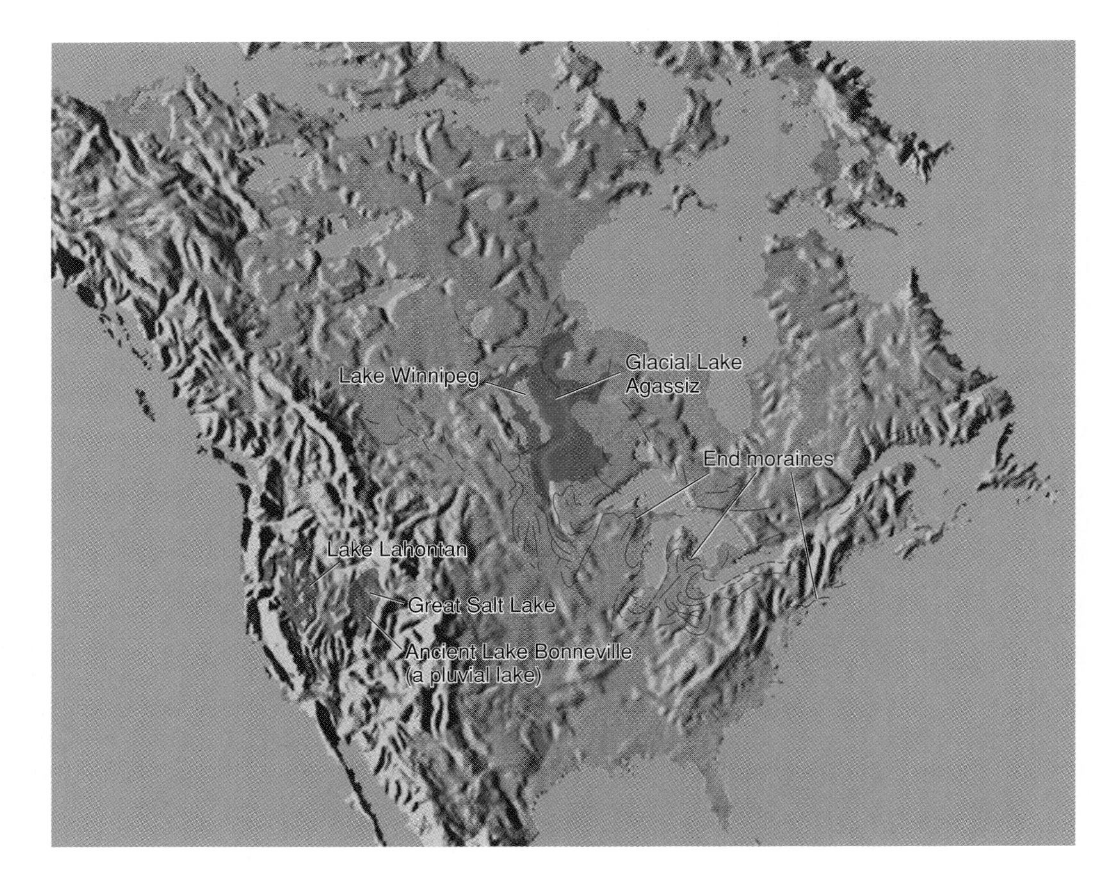

FIGURE 19.34

End moraines in the contiguous United States and Canada (shown by red lines), Glacial Lake Agassiz and pluvial lakes in the western United States (purple).

After C. S. Denny, U.S. Geological Survey, and the Geological Map of North America, Geological Society of America, and The Geological Survey of Canada

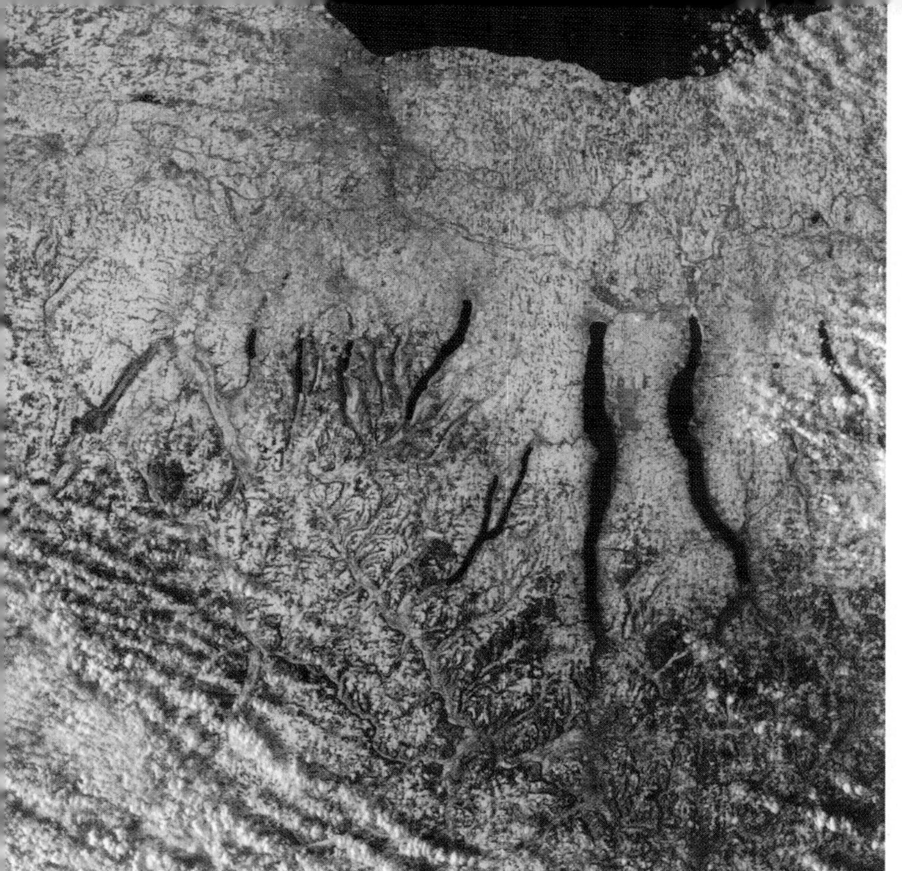

FIGURE 19.35
Satellite image of Finger Lakes in New York.
Photo courtesy of NASA

Glaciers have a tremendous capability for forming lakes through both erosion and deposition. Most states and provinces that were glaciated have thousands of lakes. By contrast, Virginia, which was not glaciated, has only three natural lakes. Minnesota bills itself as "the land of ten thousand lakes." Most of those lakes are kettle lakes. The Finger Lakes in New York (figure 19.35) are in long, north-south glacially modified valleys that are dammed by recessional moraines at their southern ends. The Great Lakes are, at least in part, a legacy of continental glaciation. Former stream valleys were widened by the ice sheet eroding weak layers of sedimentary rock into the present lake basins. End moraines border the Great Lakes, as shown in figure 19.34. Large regions of Manitoba, Saskatchewan, North Dakota, and Minnesota were covered by ice-dammed lakes (Lake Agassiz is the name for the largest of these). The former lake beds are now rich farmland.

Alpine glaciation was much more extensive throughout the world during the glacial ages than it is now. For example, small glaciers in the Rocky Mountains that now barely extend beyond their cirques were then valley glaciers 10, 50, or 100 kilometers in length. Yosemite Valley, which is no longer glaciated, was filled by a glacier about a kilometer thick. Its terminus was at an elevation of about 1,300 meters above sea level. Furthermore, cirques and other features typical of valley glaciers can be found in regions that at present have no glaciers, such as the northern Appalachians—notably in the White Mountains of New Hampshire.

Indirect Effects of Past Glaciation

As the last continental ice sheet wasted away, what effects did the tremendous volume of meltwater have on American rivers? Rivers that now contain only a trickle of water were huge in the glacial ages. Other river courses were blocked by the ice sheet or clogged with morainal debris. Large dry stream channels have been found that were preglacial tributaries to the Mississippi and other river systems.

Pluvial Lakes

During the glacial ages the climate in North America, even beyond the glaciated parts, was more humid than it is now. Most of the presently arid regions of the western United States had moderate rainfall, as traces or remnants of numerous lakes indicate. These **pluvial lakes** (formed in a period of abundant rainfall) once existed in Utah, Nevada, and eastern California (figure 19.34). Some may have been fed by meltwater from mountain glaciers, but most were simply the result of a wetter climate.

Great Salt Lake in Utah is but a small remnant of a much larger body of fresh water called Lake Bonneville, which, at its maximum size, was nearly as large as Lake Michigan is today. Ancient beaches and wave-cut terraces on hillsides indicate the depth and extent of ancient Lake Bonneville. As the climate became more arid, lake levels lowered, outlets were cut off, and the water became salty, eventually leaving behind the Bonneville salt flats and the present very saline Great Salt Lake (see figure 14.24).

Even Death Valley in California—now the driest and hottest place in the United States—was occupied by a deep lake during the Pleistocene. The salt flats that were left when this lake dried include rare boron salts that were mined during the pioneer days of the American West.

Lowering of Sea Level

All of the water for the great glaciers had to come from somewhere. The water was "borrowed" from the oceans, such that sea level worldwide was lower than it is today—at least 130 meters lower, according to scientific estimates.

Recall that if today's ice sheets were to melt, sea level worldwide would rise by over 60 meters and shorelines would be considerably further inland. It's important to realize that our present shorelines are not fixed and are very much controlled by climate changes. We should also realize that we are still in a cooler than usual (relative to most of Earth's history) time, perhaps the lingering effects of the last ice age.

What is the evidence for lower sea level? Stream channels have been charted in the present continental shelves, the gently inclined, now submerged edges of the continents (described in the chapter about the sea floor). These submerged channels are continuations of today's major rivers and had to have been above sea level for stream erosion to take place. Bones and

19.7 IN GREATER DEPTH

The Channeled Scablands

In chapter 10 we described how the Columbia plateau in the Pacific Northwest (see figure 10.27) was built by a series of successive lava floods. The northeastern part of the plateau features a unique landscape, known as the channeled scablands, where the basalt bedrock has been carved into a series of large, interweaving valleys. From the air, the pattern looks like that of a giant braided stream. The channels, however, which range up to 30 kilometers wide and are usually 15 to 30 meters deep, are mostly dry.

The scablands are believed to have been carved by gigantic floods of water. Huge ripples in gravel bars (box figure 1) support this idea. To create these ripples, a flood would have to be about 10 times the combined discharge of all the world's rivers. This is much larger than any flood in recorded history.

What seems to have occurred is that, during the ice ages, a lobe of the ice sheet extended southward into northern Washington, Montana, and Idaho, blocking the head of the valley occupied by the Clark Fork River. The ice provided a natural dam for what is now known as Glacial Lake Missoula. Lake Missoula drowned a system of valleys in western Montana that extended hundreds of kilometers into the Rocky Mountains.

Ice is not ideal for building dams. Upon failure of the glacial dam, the contents of Lake Missoula became the torrential flood that scoured the Columbia plateau. There were dozens of giant floods. Advancing ice from Canada would reestablish the dam, only to be destroyed after the reservoir refilled.

Related Web Resource
Ice Age Flood Homesite
www.uidaho.edu/igs/iafi/iafihome.html

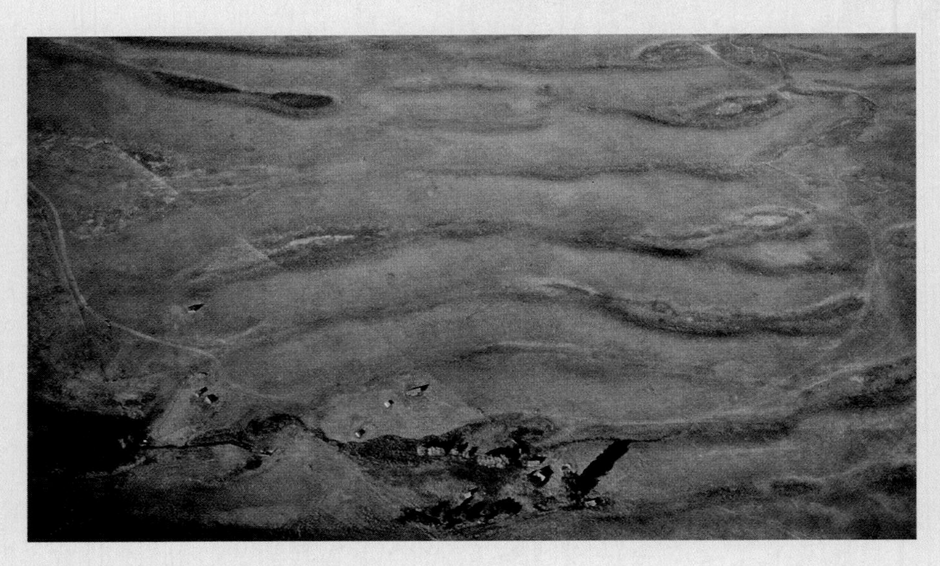

Box 19.7 — FIGURE 1
Giant ripples of gravel from the draining of Lake Missoula, Montana. Note farm buildings for scale.
Photo by P. Weiss, U.S. Geological Survey

teeth from now-extinct mammoths and mastodons have been dredged up from the Atlantic continental shelf, indicating that these relatives of elephants roamed over what must have been dry land at the time.

A **fiord** (also spelled fjord) is a coastal inlet that is a drowned glacially carved valley (figure 19.36). Fiords are common along the mountainous coastlines of Alaska, British Columbia, Chile, New Zealand, and Norway. Fiords are evidence that valleys eroded by past glaciers were later partly submerged by rising sea level.

Crustal Rebound

The weight of an ice sheet several thousand meters thick depresses the crust of Earth much as the weight of a person depresses a mattress. A land surface bearing the weight of a continental ice sheet may be depressed several hundred meters.

Once the glacier is gone, the land begins to rebound slowly to its previous height (see chapter 2 and figures 2.13 and 2.14). Uplifted and tilted shorelines along lakes are an indication of this process. The Great Lakes region is still rebounding as the crust slowly adjusts to the removal of the last ice sheet.

Evidence for Older Glaciation

Throughout most of geologic time, the climate has been warmer and more uniform than it is today. We think that the late Cenozoic Era is unusual because of the periodic fluctuations of climate and the widespread glaciations. However, glacial ages are not restricted to the late Cenozoic.

Evidence of older glaciation comes from rocks called tillites. A **tillite** is lithified till. Unsorted rock particles, including angular, striated, and faceted boulders, have been consolidated into a sedimentary rock. In some places, tillite layers overlie surfaces of older rock that have been polished and striated. Tillites of the late Paleozoic and tillites representing a minor part of the late Precambrian crop out in parts of the southern continents. (The striated surface in Australia shown in figure 19.13, is overlain in places by late Paleozoic tillite.)

The oldest glaciation for which we have evidence appears to have taken place in what is now Ontario around 2.3 billion years ago.

There is growing support for the idea that a late Precambrian ice age was so extensive that the surfaces of the world's oceans were frozen. Although the concept was first proposed in the early 20th century, scientists in the 1990s began taking it seriously and called it the *snowball Earth hypothesis*. Evidence for the hypothesis includes tillites that must, at the time, have been deposited near the equator. The hypothesis proposes that the extreme cold was due to the sun being weaker at the time and the absence of carbon dioxide and other greenhouse gases in the atmosphere.

Paleozoic glaciation provides strong support for plate tectonics. The late Paleozoic tillites in the southern continents (South Africa, Australia, Antarctica, South America) indicate that these landmasses were once joined (see chapter 4 on plate tectonics). Directions of striations show that an ice sheet flowed onto South America from what is now the South Atlantic Ocean. Because an ice sheet can build up only on land, it is reasonable to conclude that the former ice sheet was centered on the ancient supercontinent before the breakup and drift that formed the present continents.

FIGURE 19.36
Cruising in a fiord in Alaska. Note the glacier at the head of the fiord.
Photo by C. C. Plummer

Summary

A *glacier* is a large, perennial mass of ice that forms on land and moves under its own weight. A glacier can form wherever more snow accumulates than is lost. *Ice sheets* and *valley glaciers* are the two most important types of glaciers. Glaciers move downward from where the most snow accumulates toward where the most ice is wasted.

A glacier moves both by basal sliding and by internal flow. The upper portion of a glacier tends to remain rigid and is carried along by the ice moving beneath it.

Glaciers advance and recede in response to changes in climate. A receding glacier has a *negative budget* and an advancing one has a *positive budget*. A glacier's budget for the year can be determined by noting the relative position of the *equilibrium line*.

Snow recrystallizes into firn, which eventually becomes converted to glacier ice. Glacier ice is lost (or ablated) by melting, by breaking off as icebergs, and by direct evaporation of the ice into the air.

A glacier erodes by plucking and the grinding action of the rock it carries. The grinding produces rock flour and faceted and polished rock fragments. Bedrock over which a glacier moves is generally polished, striated, and grooved.

A mountain area showing the erosional effects of alpine glaciation possesses relatively straight valleys with U-shaped cross profiles. The floor of a glacial valley usually has a *cirque* at its head and descends as a series of rock steps. Small *rock-basin lakes* are commonly found along the steps and in cirques.

A *hanging valley* indicates that a smaller tributary joined the main glacier. A *horn* is a peak between several cirques. *Arêtes* usually separate adjacent glacial valleys.

A glacier deposits unsorted rock debris or *till*, which contrasts sharply with the sorted and layered deposits of glacial *outwash*. Till forms various types of *moraines*.

Fine silt and clay may settle as *varves* in a lake in front of a glacier, each pair of layers representing a year's accumulation.

Multiple till deposits and other glacial features indicate several major episodes of glaciation during the late Cenozoic Era. During each of these episodes, large ice sheets covered most of northern Europe and northern North America, and glaciation in mountain areas of the world was much more extensive than at present. At the peak of glaciation about a third of Earth's land surface was glaciated (in contrast to the 10% of the land surface presently under glaciers). Warmer climates prevailed during interglacial episodes.

The glacial ages also affected regions never covered by ice. Because of wetter climate in the past, large lakes formed in now-arid regions of the United States. Sea level was considerably lower.

Glacial ages also occurred in the more distant geologic past, as indicated by late Paleozoic and Precambrian tillites.

Terms to Remember

Testing Your Knowledge

Use the questions below to prepare for exams based on this chapter.

1. How do erosional landscapes that formed beneath glaciers differ from those that developed in rock exposed above glaciers?
2. How do features caused by stream erosion differ from features caused by glacial erosion?
3. How does material deposited by glaciers differ from material deposited by streams?
4. Why is the North Pole not glaciated?
5. How do arêtes, cirques, and horns form?
6. How does the glacial budget control the migration of the snow line?
7. How do recessional moraines differ from terminal moraines?
8. Alpine glaciation
   a. is found in mountainous regions
   b. exists where a large part of a continent is covered by glacial ice
   c. is a type of glacier
   d. none of the above
9. Continental glaciation
   a. is found in mountainous regions
   b. exists where a large part of a continent is covered by glacial ice
   c. is a glacier found in the subtropics of continents
   d. none of the above
10. At present about ____% of the land surface of the earth is covered by glaciers.
    a. 1/2 b. 1
    c. 2 d. 10
    e. 33 f. 50
11. Which is not a type of glacier?
    a. valley glacier b. ice sheet
    c. ice cap d. sea ice
12. The boundary between the zone of accumulation and the zone of ablation of a glacier is called the
    a. firn b. equilibrium line
    c. ablation zone d. moraine
13. The ice caps on Mars are composed mostly of frozen
    a. water b. carbon dioxide
    c. nitrogen d. helium
14. Recently geologists have been drilling through ice sheets for clues about
    a. ancient mammals b. astronomical events
    c. extinctions d. past climates
15. Glacially carved valleys are usually ____ shaped.
    a. V b. U
    c. Y d. all of the above
16. Which is not a type of moraine?
    a. medial b. end
    c. terminal d. recessional
    e. ground f. esker
17. The last episode of extensive glaciation in North America was at its peak about ____ years ago.
    a. 2,000 b. 5,000
    c. 10,000 d. 18,000
18. How fast does the central part of a valley glacier move compared to the sides of the glacier?
    a. faster b. slower
    c. at the same rate
19. During the Ice Ages, much of Nevada, Utah, and eastern California were covered by
    a. ice b. huge lakes
    c. deserts d. the sea

Expanding Your Knowledge

1. How might a warming trend cause increased glaciation?
2. How do, or do not, the Pleistocene glacial ages fit in with the principle of uniformitarianism?
3. Is ice within a glacier a mineral? Is a glacier a rock?
4. Could a rock that looks like a tillite have been formed by any agent other than glaciation?
5. What is the likelihood of a future glacial age? What effect might human activity have on causing or preventing a glacial age?

Exploring Web Resources

www.mhhe.com/plummer9e

Visit our Online Learning Center for some great animations of glacier development and movement. Check your answers for the Testing Your Knowledge section and try some additional interactive quizzing to further your understanding. This site also provides you with direct links to the sites listed below.

http://dir.yahoo.com/science/earth_sciences/geology_and_geophysics/glaciology/

Glaciers and Glaciology—list of sites. This site provides links and descriptions of numerous icy websites.

www.glacier.rice.edu/

Glacier. Explore Antarctica on Rice University's site. Go to "Ice." This site offers many topics you can click on to get information that expands on the information covered in this book. Examples are "How do Glaciers Move," "How do Glaciers Change the Land," "What Causes Ice Ages."

www.crevassezone.org/

Glacier movement studies on the Juneau Icefield, Alaska. Go to "Photo Gallery" to view photos of glacial features and other aspects of the project.

www.museum.state.il.us/exhibits/ice_ages/

Ice Ages. Illinois State Museum's virtual ice ages exhibit. The site features a tape clip showing the retreat of glaciers during the last ice age. You can download the video clip by going to:

www.museum.state.il.us/exhibits/ice_ages/laurentide_deglaciation.html

www-nsidc.colorado.edu/NSIDC/EDUCATION/

National Snow and Ice Data Center's Education Resources Site. General information on snow and ice. You can link to pages on glaciers, avalanches, icebergs.

Animations

This chapter includes the following animations available on our Online Learning Center at www.mhhe.com/plummer9e.

19.3 Ice Sheet
19.6 Valley Glacier
19.8 Movement of a Glacier
19.9 Crevasses on a Glacier

CHAPTER

20
Waves, Beaches, and Coasts

Chapters 13 and 16 through 19 have dealt with the sculpturing of the land by mass wasting, streams, ground water, glaciers, and wind. Water waves are another agent of erosion, transportation, and deposition of sediment. Along the shores of oceans and lakes, waves break against the land, building it up in some places and tearing it down in others.

The energy of the waves comes from the wind. This energy is used to a large extent in eroding and transporting sediment along the shoreline. Understanding how waves travel and move sediment can help you see how easily the balance of supply, transportation, and deposition of beach sediment can be disturbed. Such disturbances can be natural or human-made, and the changes that result often destroy beachfront homes and block harbors with sand.

Beaches have been called "rivers of sand" because breaking waves, as they sort and transport sediment, tend to move sand parallel to the shoreline. In this chapter we look at how beaches are formed and also examine the influence of wave action on such coastal features as sea cliffs, barrier islands, and terraces.

Opposite: A wave-eroded coast with sea stacks form the "Twelve Apostles" near Victoria, Australia.
Photo © Chad Ehlers/Stone

FIGURE 20.1

These beach homes in North Carolina were built too close to the ocean and are being destroyed as sand is removed from the beach by storm waves.
Photo © Bruce Roberts/Photo Researchers, Inc.

If you spend a week at the shore during the summer, you may not notice any great change in the appearance of the beach while you are there. Even if you spend the whole summer at the seaside, nothing much seems to happen to the beach during those months. Tides rise and fall every day and waves strike the shore, but the sand that you walk on one day looks very much like the sand that you walked on the previous day. The shape of the beach does not appear to change, nor does the sand seem to move very much.

On most beaches, however, the sand is moving, in some cases quite rapidly. The beach looks the same from day to day only because new sand is being supplied at about the same rate that old sand is being removed.

Where is the sand going? Some sand is carried out to deep water. Some is piled up and stored high on the beach. But on most shores much more of the sand moves along parallel to the beach in relatively shallow water. In this way, loose sand grains travel hundreds of feet per day along some coasts, especially those subject to strong waves.

On some beaches, sand is being removed faster than it is being replenished. When this happens, beaches become narrower and less attractive for swimming. Where erosion is severe, buildings close to the beach can be undermined and destroyed by waves as the beach disappears (figure 20.1). The sand moved from the beach may be redeposited in inconvenient places, such as across the mouth of a harbor, where it must be dredged out periodically. Because moving sand can create many problems for people in coastal towns and cities, it is important to understand something of how and why the sand moves.

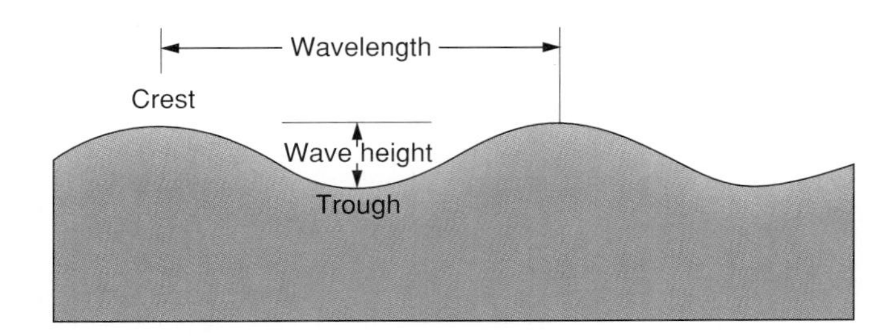

FIGURE 20.2

Wave height is the vertical distance between the wave crest and the wave trough. Wavelength is the horizontal distance between two crests.

WATER WAVES

The energy that moves sand along a beach comes from the wind-driven water waves that break upon the shore. As wind blows over the surface of an ocean or a lake, some of the wind's energy is transferred to the water surface, forming the waves that move through the water. The height of waves (and their length and speed) are controlled by the wind speed, the length of time that the wind blows, and the distance that the wind blows over the water (*fetch*).

Wave shapes can vary. Short, choppy *seas* in and near a storm create a confused sea surface, often with considerable white foam as strong winds blow the tops off waves. Long, rolling *swells* form a regular series of similar-sized waves on shores that may be thousands of kilometers from the storms that generated the waves. (Summer surfing waves in southern California can be generated by large storms north of Australia in the southern hemisphere winter.) When waves break against the shore as *surf,* a large portion of their energy is spent moving sand along the beach.

The height of waves is the key factor in determining wave energy. **Wave height** is the vertical distance between the **crest,** which is the high point of a wave, and the **trough,** which is the low point (figure 20.2). In the open ocean, normal waves have heights of about 0.3 to 5 meters, although during violent storms, including hurricanes, waves can be more than 15 meters high. The highest wind wave ever measured was 34 meters by the anxious crew of a ship in the north Pacific in 1933. (The highest tsunami ever measured, caused by a submarine earthquake rather than wind, was 85 meters, in the Ryukyu island chain south of Japan in 1971. See chapter 7.)

Wavelength is the horizontal distance between two wave crests (or two troughs). Most ocean wind waves are between 40 to 400 meters in length, and move at speeds of 25 to 90 kilometers per hour (15 to 55 miles per hour) in deep water.

The movement of water in a wave is like the movement of wheat in a field when wind blows across it. You can see the ripple caused by wind blowing across a wheat field, but the wheat does not pile up at the end of the field. Each stalk of wheat bends over when the wind strikes it and then returns to its original position. A particle of water moves in an *orbit,* a nearly circular path, as the wave passes (figure 20.3); the particle returns to its original posi-

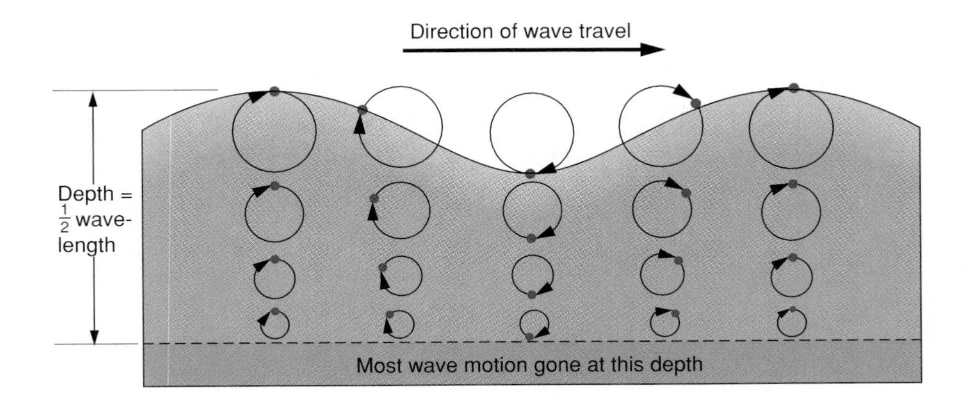

FIGURE 20.3

Orbital motion of water in waves dies out with depth. At the surface the diameter of the orbits is equal to the wave height.

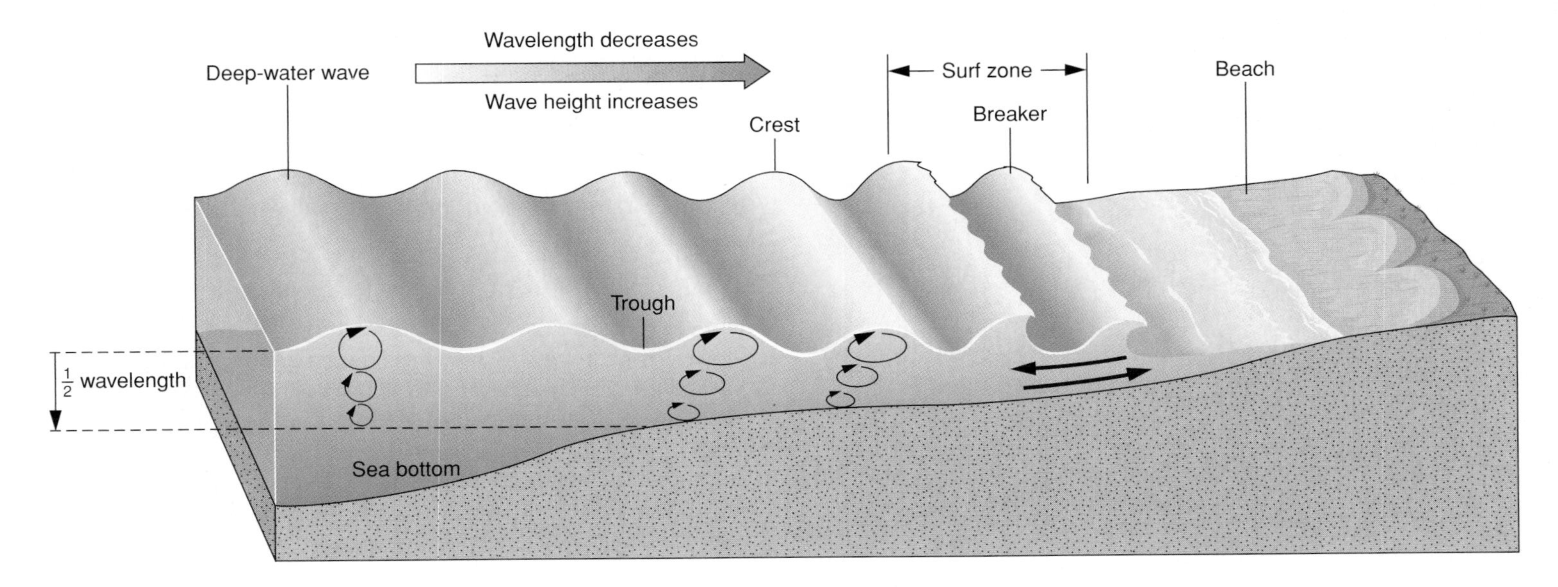

FIGURE 20.4

As a deep-water wave approaches shore, it begins to "feel" the sea bottom and slow down. Circular water orbits flatten and the wave peaks and breaks. In the foamy surf zone, water moves back and forth rather than in orbits.

tion after the wave has passed. In deep water, when a wave moves across the water surface, energy moves with the wave; but the water, like the wheat, does not advance with the wave.

At the surface, the diameter of the orbital path of a water particle is equal to the height of the wave (figure 20.3). Below the surface, the orbits decrease in size until the motion is essentially gone at a depth equal to half the wavelength. This is why a submarine can cruise in deep, calm water beneath surface ships that are being tossed by the orbital motion of large waves.

Surf

As waves move from deep water to shallow water near shore, they begin to be affected by the ocean bottom. A wave first begins to "feel bottom" at the level of lowest orbital motion—that is, when the depth to the bottom equals half the wavelength. For example, a wave 150 meters long will begin to be influenced by the bottom at a water depth of 75 meters.

In shallow water the presence of the bottom interferes with the circular orbits, which flatten into ovals (figure 20.4). The waves slow down and their length decreases. Meanwhile, the sloping bottom wedges the moving water upward, increasing the wave height. Because the height is increasing while the length is decreasing, the waves become steeper and steeper until they break. A **breaker** is a wave that has become so steep that the crest of the wave topples forward, moving faster than the main body of the wave. The breaker then advances as a turbulent, often foamy, mass. Breakers collectively are called **surf.** Water in the surf zone has lost its orbital motion and moves back and forth, alternating between onshore and offshore flow.

A

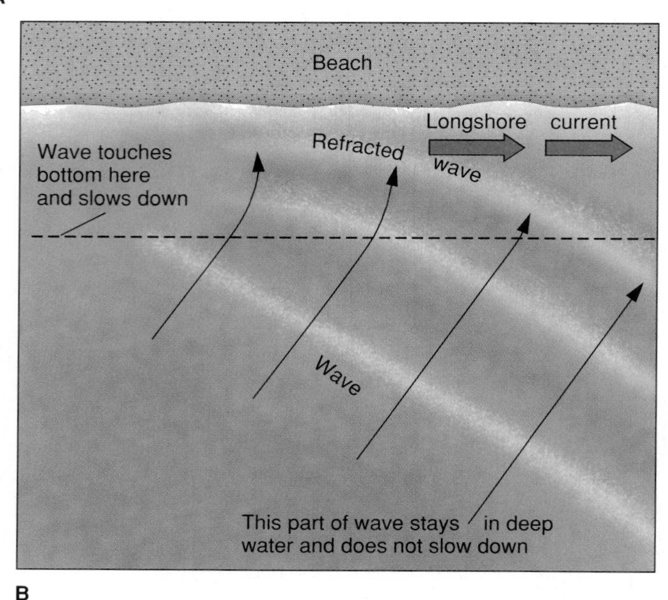

B

FIGURE 20.5

(*A*) These waves are arriving at an angle to the shoreline. They break progressively along the shore, from the upper right of the photo to the lower left. (*B*) Wave refraction changes the wave direction, bending the wave so it becomes more parallel to shore. The angled approach of waves to shore sets up a longshore current parallel to the shoreline.

NEARSHORE CIRCULATION

Wave Refraction

Most waves do not come straight into shore. A wave crest usually arrives at an angle to the shoreline (figure 20.5*A*). One end of the wave breaks first, and then the rest breaks progressively along the shore.

This angled approach of a wave toward shore can change the direction of wave travel. One end of the wave reaches shallow water first. This end of the wave "feels bottom" and slows down while the rest of the wave continues at its deep-water speed (figure 20.5*B*). As more and more of the wave comes into contact with the bottom, more of the wave slows down. As the wave slows progressively along its length, the wave crest changes direction and becomes more nearly parallel to the shoreline. This bending of waves is called **wave refraction** (figure 20.5*B*).

Longshore Currents

Although most wave crests become nearly parallel to shore as they are refracted, waves do not generally strike *exactly* parallel to shore. Even after refraction, a small angle remains between the wave crest and the shoreline. As a result, the water in the wave is pushed both *up* the beach toward land and *along* the beach parallel to shore.

Each wave that arrives at an angle to the shore pushes more water parallel to the shoreline. Eventually a moving mass of water called a **longshore current** develops parallel to the shoreline (figure 20.6). The width of the longshore current is about equal to the width of the surf zone. The seaward edge of the current is the outer edge of the surf zone, where waves are just beginning to break; the landward edge is the shoreline. A longshore current can be very strong, particularly when the waves are large. Such a current can carry swimmers hundreds of yards parallel to shore before they are aware that they are being swept along. It is these longshore currents that transport most of the beach sand parallel to shore.

Rip Currents

Rip currents are narrow currents that flow straight out to sea in the surf zone, returning water seaward that breaking waves have pushed ashore (figure 20.6). Rip currents travel at the water surface and die out with depth. They pulsate in strength, flowing most rapidly just after a set of large waves has carried a large amount of water onto shore. Rip currents can be important transporters of sediment, as they carry fine-grained sediment out of the surf zone into deep water.

As a single wave comes toward shore, its height varies from place to place. Rip currents tend to develop locally where wave height is low. Rip currents that are fixed in position are apt to be found over channels on the sea floor, because depressions on the bottom reduce wave height. Complex wave interactions can also lower wave height, and rip currents that form because of wave interactions tend to shift position along the shore. Such shifting rip currents are usually spaced at regular intervals along the beach.

Rip currents are fed by water within the surf zone. They flow rapidly out through the surf zone and then die out quickly. Where waves are nearly parallel to a shoreline, longshore feeder currents of equal strength develop in the surf zone on either side of a rip current (figure 20.6*A*). Where waves strike the shore at an angle and set up a strong, unidirectional, longshore current, a rip current is fed from one side by the longshore current, which increases in strength as it nears the rip current (figure 20.6*B*). Rip currents are also found alongside points of land and engineered structures such as jetties and piers, which can deflect longshore currents seaward.

You can easily learn to spot rip currents at a beach. Look for discoloration in the water where sediment is being picked up in the surf zone and moved seaward (figure 20.6*C*). Another sign

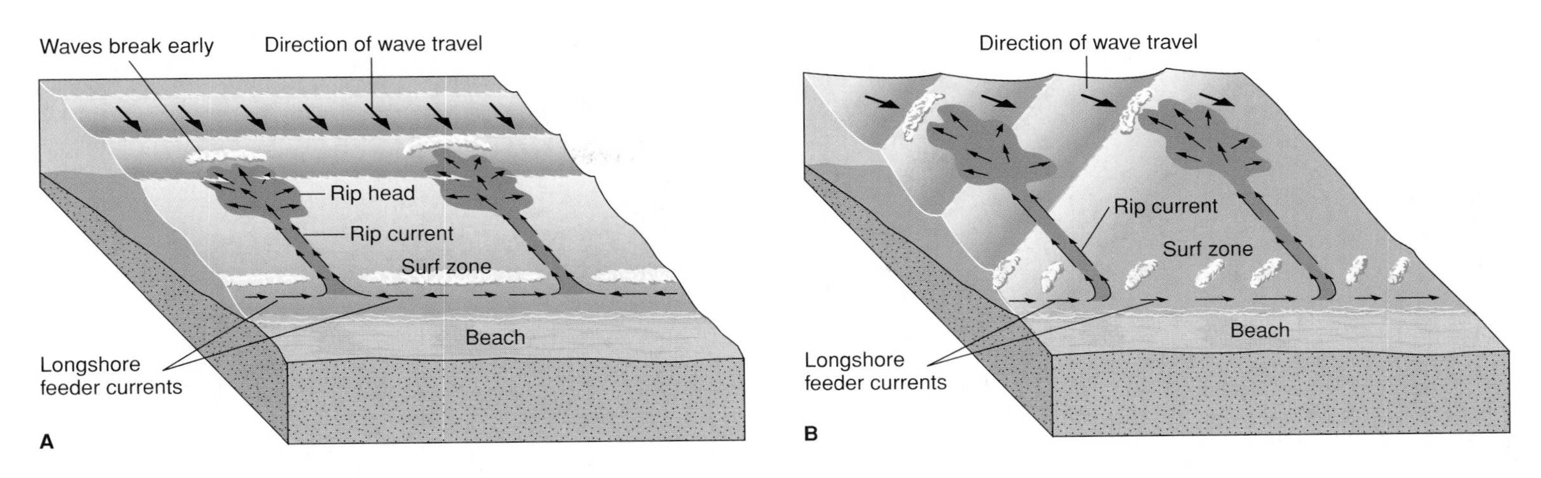

FIGURE 20.6

Rip currents and their feeder currents can develop regardless of the angle of approach of waves. (*A*) Waves approach parallel to shore; feeder currents on both sides of rip currents. (*B*) Waves approach at an angle to the shore; feeder current on only one side of rip current. (*C*) Rip currents carry dirty water and foam seaward; they can cause incoming waves to break early.

Photo *C* © Sanford Berry/Visuals Unlimited

is incoming waves breaking early within a rip current as they meet the opposing flow. The diffuse heads of rip currents outside the surf zone may be marked at the edge with foam lines. Even on very calm days rips can often be identified by subtle changes in the water surface, such as a different pattern of water ripples or light reflection off the water.

Getting caught in a rip current and being carried out to sea can panic an inexperienced swimmer—even though the trip will stop some distance beyond the surf zone as the rip dies out. A swimmer frightened by being carried away from land and into breaking waves can grow exhausted fighting the current to get back to shore. The thing to remember is that rip currents are narrow. Therefore, you can get out of a rip easily by swimming *parallel* to the beach instead of struggling against the current.

Surfers, on the other hand, often look for rip currents and paddle intentionally into them to get a quick ride out into the high breakers.

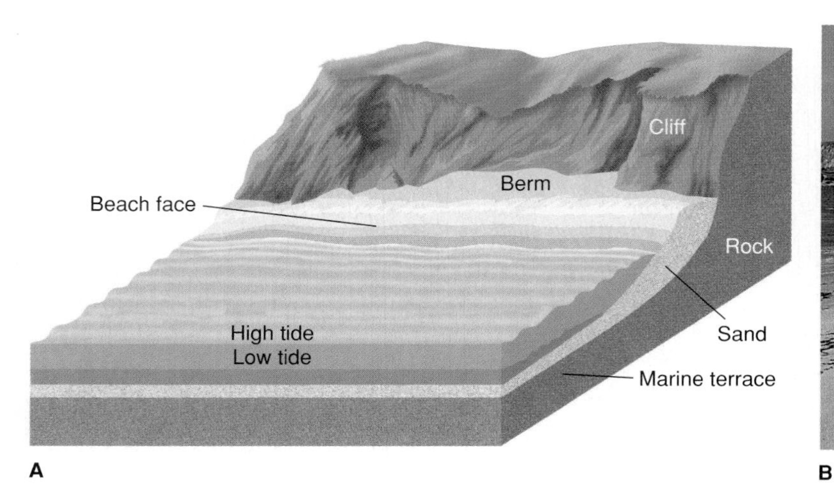

FIGURE 20.7
(*A*) Parts of a beach. (*B*) The beach face (on the left) and berm (on the right) on a northern California beach.
Photo *B* by Diane Carlson

BEACHES

A **beach** is a strip of sediment (usually sand or gravel) that extends from the low-water line inland to a cliff or a zone of permanent vegetation. Waves break on beaches, and rising and falling tides may regularly change the amount of beach sediment that is exposed above water (figure 20.7*A*).

The steepest part of a beach is the **beach face,** which is the section exposed to wave action, particularly at high tide. Offshore from the beach face there is usually a **marine terrace,** a broad, gently sloping platform that may be exposed at low tide if the shore has significant tidal action. Marine terraces may be *wave-built* terraces constructed of sediment carried away from the shore by waves, or they may be *wave-cut* rock benches or platforms, perhaps thinly covered with a layer of sediment.

The upper part of the beach, landward of the usual high-water line is the **berm,** a wave-deposited sediment platform that is flat or slopes slightly landward (figure 20.7*B*). It is usually dry, being covered by waves only during severe storms.

Beach sediment is usually sand, typically quartz-rich because of quartz's resistance to chemical weathering. Heavy metallic minerals ("black sands") can also be concentrated on some beaches as less dense minerals such as quartz and feldspar are carried away by waves or wind (titanium-bearing sands are mined on some beaches in Florida and Australia). Tropical beaches may be made of bioclastic carbonate grains from offshore corals, algae, and shells. Some Hawaiian beaches are made of sand-sized fragments of basalt. Gravel beaches are found on coasts attacked by the high energy of large waves (*shingle* is a regional name for disk-shaped gravel). Gravel beaches have a steeper face slope than sand beaches.

In seasonal climates, beaches often go through a summer-winter cycle (figure 20.8). This is due to the greater frequency of storms with strong winds during the winter months, which tend to produce high waves with short wavelengths. These high energy waves tend to crash onshore and erode sand from the beach face and narrow the berm. Offshore, in less turbulent water, the sand settles to the bottom and builds an underwater sandbar (parallel to the beach) that serves as a "storage facility" for the next summer's sand supply. The following summer, or during calmer weather, lower energy waves with long wavelengths break over the sandbar and gradually push the sand back onto the beach face to widen the berm. Each season the beach changes in shape until it comes into equilibrium with the prevailing wave type.

Many winter beaches can be dangerous because of high waves and narrowed beaches. Several beaches along the Pacific coast of the United States are nearly free of accidents in the summer, when they are heavily used, but are regularly marked by drownings in the winter as beachwalkers are swept off narrower beaches out to sea by large storm waves.

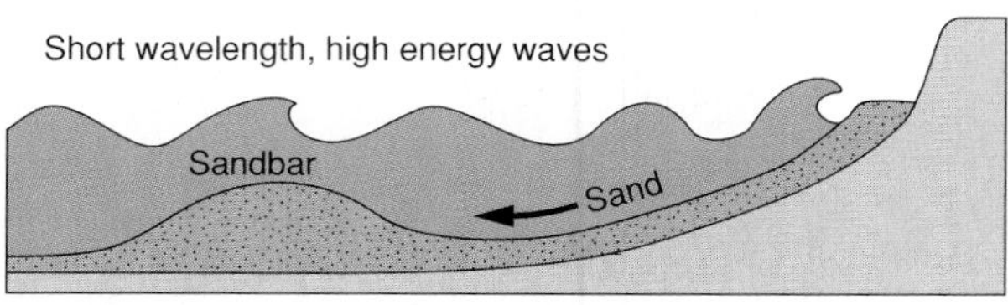

A Winter beach

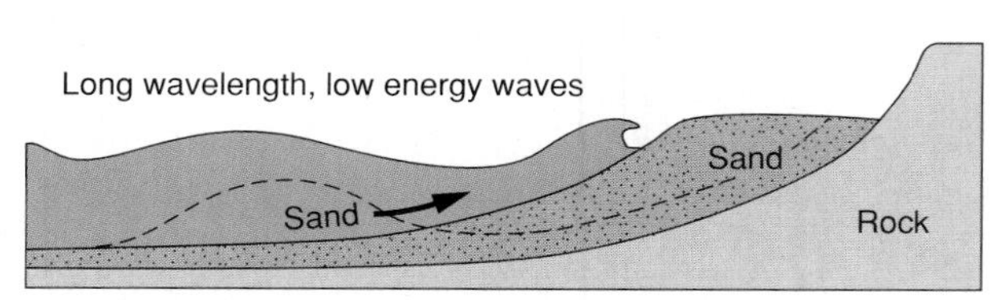

B Summer beach

FIGURE 20.8
Seasonal cycle of a beach caused by differing wave types. (*A*) Narrow winter beach. Waves may break once on the winter sandbar, then re-form and break again on the beach face. (*B*) Wide summer beach.

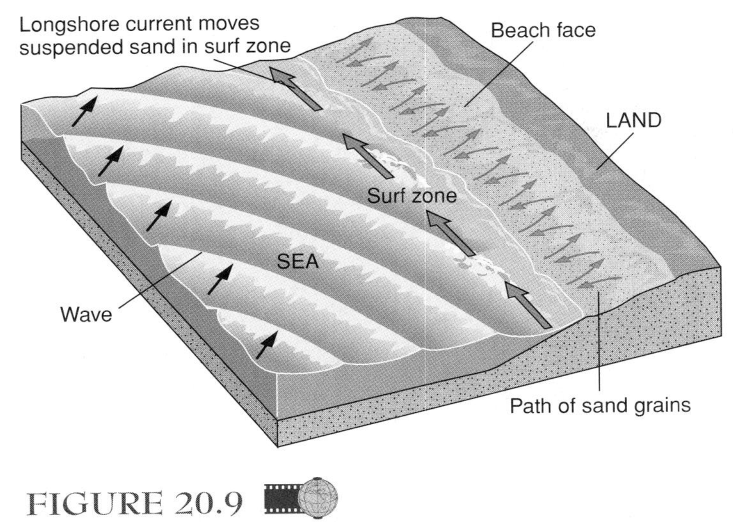

FIGURE 20.9

Longshore drift of sand on the beach face and by a longshore current within the surf zone.

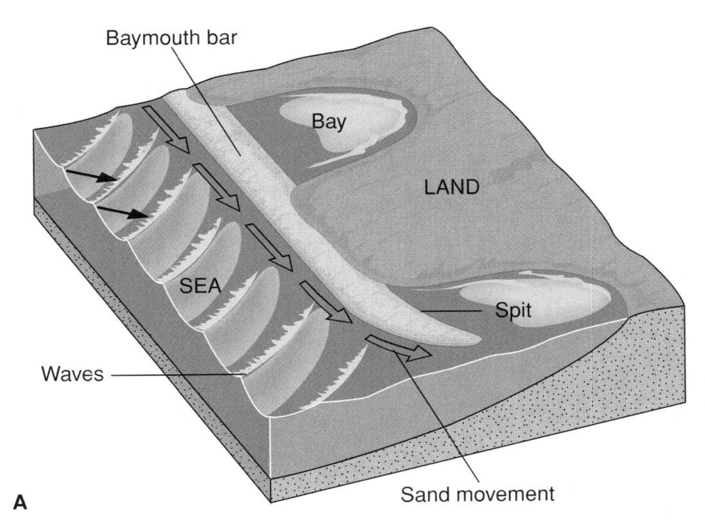

FIGURE 20.10

(*A*) Longshore drift of sand can form spits and baymouth bars. (*B*) Curved spit near Victoria, British Columbia. (*C*) A baymouth bar has sealed off this bay from the ocean as sand migrated across the mouth of the Russian River in northern California.

Photo *B* by D. A. Rahm, courtesy Rahm Memorial Collection, Western Washington University. Photo *C* by Diane Carlson

LONGSHORE DRIFT OF SEDIMENT

Longshore drift is the movement of sediment parallel to shore when waves strike the shoreline at an angle. Figure 20.9 shows the two ways in which this movement of sediment (usually sand) occurs. Some longshore drift takes place directly on the beach face when waves wash up on land. A wave washing up on the beach at an angle tends to wash sand along at the same angle. After the wave has washed up as far as it can go, the water returns to the sea by running down the beach face by the shortest possible route, that is, straight downhill to the shoreline, not back along the oblique route it came up. (Wave run-up is known as *swash,* the return as *backwash.*) The net effect of this motion is to move the sand in a series of arcs along the beach face.

Much more sand is moved by longshore transport in the surf zone, where waves are breaking into foam. The turbulence of the breakers erodes sand from the sea bottom and keeps it suspended. Even a weak longshore current can move the suspended sand parallel to the shoreline. The sand in the longshore current moves in the same direction as the sand drift on the beach face (figure 20.9).

Vast amounts of sand can be moved by longshore transport. The U.S. Army Corps of Engineers estimates that 436,000 cubic meters of sand per year are moved northward by waves at Sandy Hook, New Jersey, and 1,000,000 cubic meters of sand per year are moved southward at Santa Monica, California.

Eventually the sand that has moved along the shore by these processes is deposited. Sediment may build up off a point of land to form a **spit,** a fingerlike ridge of sediment that extends out into open water (figures 20.10*A,B*). A **baymouth bar,** a ridge of sediment that cuts a bay off from the ocean, is formed by sediment migrating across what was earlier an open bay (figures 20.10*A,C*). Off the western coast of the United States, a considerable amount of drifting sand is carried into the heads of underwater canyons, where the sediments slide down into deep, quiet water.

A striking, but rare, feature formed by longshore drift is a **tombolo,** a bar of sediment connecting a former island to the mainland. As shown in figure 20.11, waves are refracted around an island in such a way that they tend to converge behind the island. The waves sweep sand along the mainland (and from the island) and deposit it at this zone of convergence, forming a bar that grows outward from the mainland and eventually connects to the island.

Human Interference with Sand Drift

Several engineered features can interrupt the flow of sand along a beach (figure 20.12). *Jetties,* for example, are rock walls designed to protect the entrance of a harbor from sediment deposition and storm waves. Usually built in pairs, they protrude above the surface of the water. Figure 20.12*A* shows how sand piles up against one jetty while the beach next to the other, deprived of a sand supply, erodes back into the shore.

Groins are sometimes built in an attempt to protect beaches that are losing sand from longshore drifting. These short walls are built perpendicular to shore to trap moving sand and widen a beach (figure 20.12*B*).

Sand deposition also occurs when a stretch of shore is protected from wave action by a *breakwater,* an offshore structure built to absorb the force of large breaking waves and provide quiet water near shore. When the city of Santa Monica in California built a rock breakwater parallel to the shore to create a protected small-boat anchorage, the lessening of wave action on the shore behind the breakwater allowed sand to build up there (figure 20.12*C*), threatening eventually to fill in the anchorage. The city had to buy a dredge to remove the sand from the protected area and redeposit it farther along the shore where the waves could resume moving sediment.

A beach attempts to come into equilibrium with the waves that strike it. The type and amount of sediment, the position of the sediment, and especially the movement of the sediment, adjust to the incoming wave energy. Whenever human activity interferes with sand drift or wave action, the beach responds by changing its configuration, usually through erosion or deposition in a nearby part of the beach.

Sources of Sand on Beaches

Some beach sand comes from the erosion of local rock, such as points of land or cliffs nearby. On a few beaches replenishment comes from sand stored outside the surf zone in the deeper water offshore. Bioclastic carbonate beaches are formed from the remains of marine organisms offshore. But the greater part of the sand on most beaches comes from river sediment brought down to the ocean. Waves pick up this sediment and move it along the beach by longshore drift.

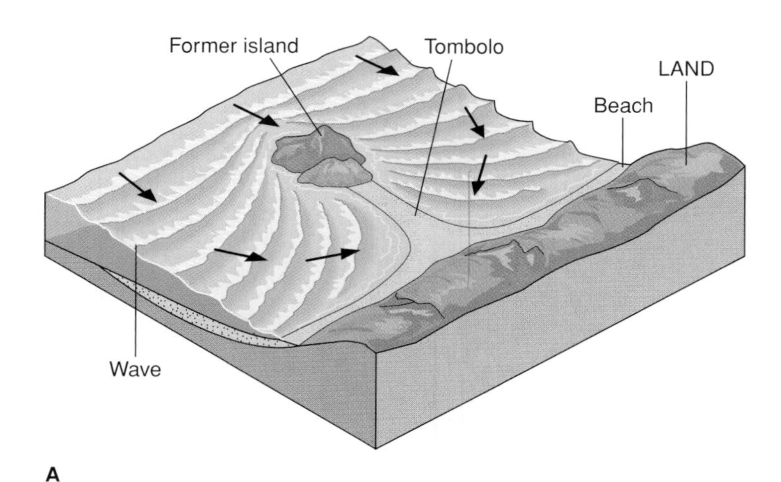

B

FIGURE 20.11

(*A*) Formation of a tombolo. Wave refraction around an island interrupts the longshore current and creates a sandbar that connects the island with the mainland. (*B*) A tombolo has connected this rock, once an island, to the shore. Note the waves bending around the two sides of the rock. Near Santa Cruz, California.

What happens to a beach if all the rivers contributing sand to it are dammed? Although damming a river may be desirable for many reasons (flood control, power generation, water supply, recreation), when a river is dammed, its sediment load no longer reaches the sea (see box 16.1). The sand that supplied the beach in the past now comes to rest in the quiet waters of the reservoir behind the dam. Longshore drift, however, continues to remove sand from beaches even though little new sand is being supplied, and the result is a net loss of sand from beaches. Beaches without a sand supply eventually disappear. To prevent this, some coastal communities have set up expensive programs of building pipelines or draining reservoirs and trucking the trapped sand down to the beaches.

A

B

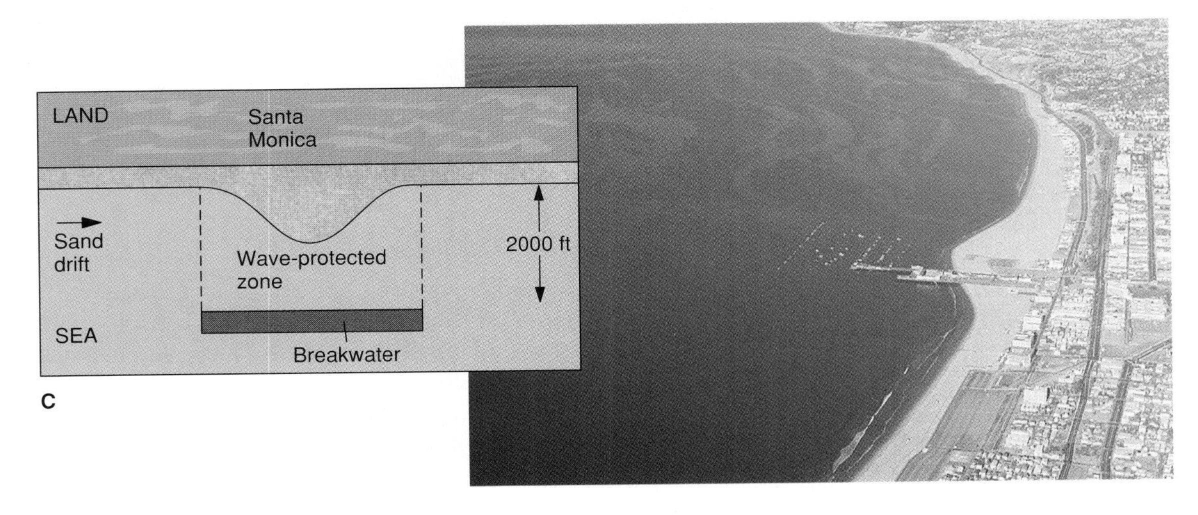

C

FIGURE 20.12

Sand piles up against obstructions and in areas deprived of wave energy. (*A*) Jetties at Manasquan Inlet, New Jersey. Sand drift to the right has piled sand against the left jetty and removed sand near the right jetty. (*B*) Groins at Ocean City, New Jersey. Sand drift is to the right. (*C*) Breakwater at Santa Monica, California, has caused deposition of sand in the wave-protected zone.

Photos *A* and *B* by S. Jeffress Williams; photo *C* © John Shelton

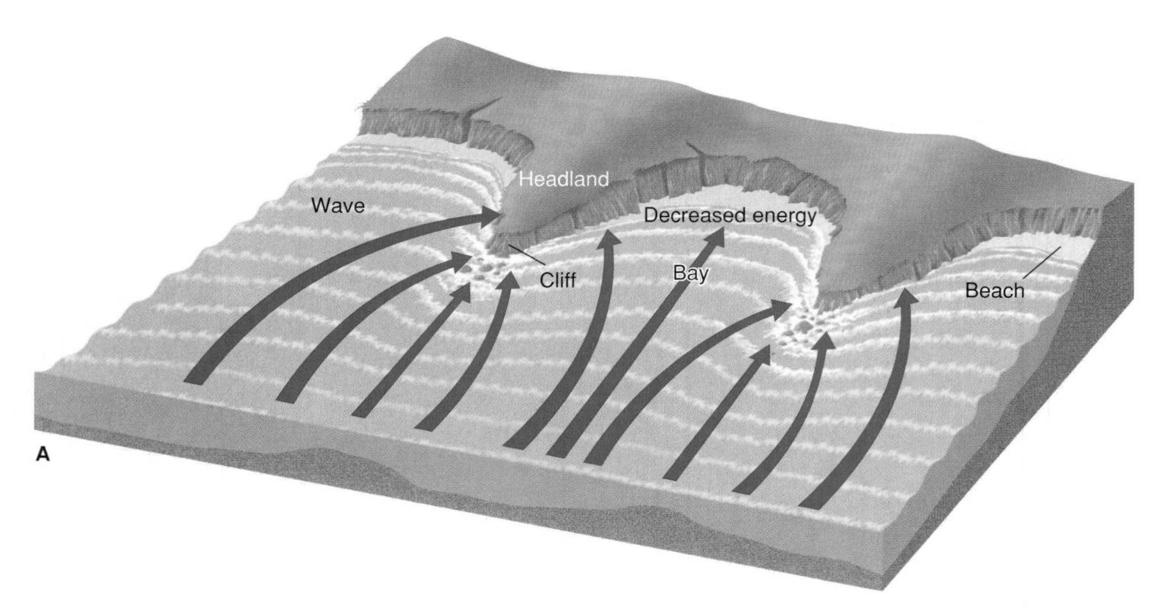

FIGURE 20.13

(*A*) Wave refraction on an irregular coast. Arrows show transport direction of wave energy, concentrated on headlands, spread out in bays. (*B*) Wave refraction around a point of land, Rincon Point, California. Note waves at right center have been bent 90 degrees from their original direction (parallel to the bottom of the photo).

Photo by Frank M. Hanna

COASTS AND COASTAL FEATURES

A beach is just a small part of the **coast,** which is all the land near the sea, including the beach and a strip of land inland from it. Coasts can be rocky, mountainous, and cliffed, as in northern New England and on the Pacific shore of North America; or they can be broad, gently sloping plains, as along much of the southeastern United States. Wave erosion and deposition can greatly modify coasts from their original shapes. Many coasts have been drowned during the past 15,000 years by the rise in sea level caused by the melting of the Pleistocene glaciers (see chapter 19). Other coasts have been lifted up by tectonic forces at a rate greater than the rise in sea level so that sea-floor features are now exposed on dry land.

Erosional Coasts

A great many steep, rocky coasts have been visibly changed by wave erosion. Soluble rocks such as limestone dissolve as waves wash against them, and more durable rocks such as granite are fractured by the enormous pressures caused by waves slamming into rock (wave impact pressures have been measured as high as 60 metric tons per square meter).

An irregular coast with bays separated by rocky **headlands** (points of land) can be gradually straightened by wave action. Because wave refraction bends waves approaching such a coast until they are nearly parallel to shore, most of the waves' energy is concentrated on the headlands, while the bays receive smaller, diverging waves (figure 20.13). Rocky cliffs form from wave erosion on the headlands. The eroded material is deposited in the quieter water of nearby bays, forming broad

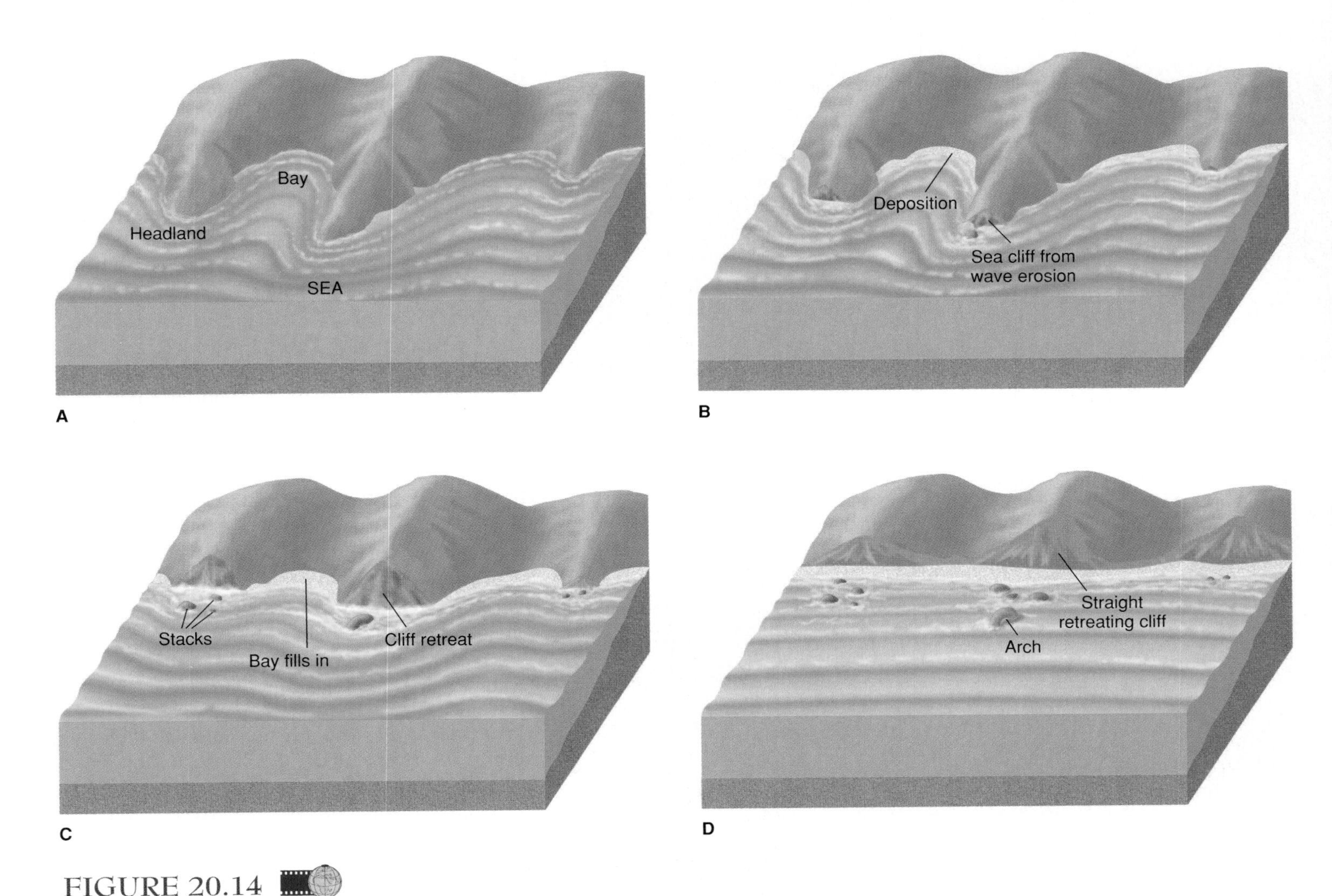

FIGURE 20.14

Coastal straightening of an irregular coastline by wave erosion of headlands and wave deposition of sediment in bays. Continued erosion produces a straight, retreating cliff.

beaches. **Coastal straightening** of an irregular shore gradually takes place through wave erosion of headlands and wave deposition in bays (figure 20.14).

Wave erosion of headlands produces **sea cliffs,** steep slopes that retreat inland by mass wasting as wave erosion undercuts them (figure 20.15). At the base of sea cliffs are sometimes found *sea caves,* cavities eroded by wave action along zones of weakness in the cliff rock. As headlands on irregular coasts are eroded landward, sea cliffs enlarge until the entire coast is marked by a retreating cliff (figure 20.14). On some exposed coasts the rate of cliff retreat can be quite rapid, particularly if the rock is weakly consolidated. Some sea cliffs north of San Diego, California, and at Cape Cod National Seashore in Massachusetts are retreating at an average rate of 1 meter per year. Because sea-cliff erosion in weak rock is often in the form of large, infrequent slumps (see chapter 13), some portions of these coasts may retreat 10 to 30 meters in a single storm. Some of these cliffs have "ocean-view" homes and hotels at their very edges. Sea cliffs in hard, durable rock such as granite and schist retreat much more slowly.

Seawalls may be constructed along the base of retreating cliffs to prevent wave erosion (figure 20.15). Seawalls of giant pieces of broken stone (riprap) or concrete tetrahedrons are designed to absorb wave energy rather than allow it to erode cliff rock. Vertical or concave seawalls of concrete, such as the one in Galveston, Texas (see box 20.1 figure 3), are designed to reflect wave energy seaward rather than allow it to impact the shore. Some of the reflected energy, however, is focused at the base of the seawall, which eventually undermines it and causes the seawall to collapse. Reflection of waves from a seawall also increases the amount of wave energy just offshore, often increasing the amount of sand erosion offshore. Thus a seawall designed to protect a sea cliff (and the buildings at its edge) may in some cases destroy a

FIGURE 20.15

Retreating wave-cut cliff, north of Bodega Bay, Sonoma County, California. A concrete seawall has been built at the base of the cliff to slow wave erosion and help protect the cliff-edge homes. Note fragments of wave-destroyed structures near the seawall. Seawalls usually increase the erosion of sand beaches.

sand beach at the base of the cliff. Seawalls are difficult and expensive to build and maintain, and they may destroy beaches, but political pressure to build more of them will increase as the sea level rises in the future.

Wave erosion produces other distinctive features in association with sea cliffs. A **wave-cut platform** (or *terrace*) is a horizontal bench of rock formed beneath the surf zone as a coast retreats by wave erosion (figure 20.16). The platform widens as the sea cliffs retreat. The depth of water above a wave-cut platform is generally 6 meters or less, coinciding with the depth at which turbulent breakers actively erode the sea bottom. **Stacks** are erosional remnants of headlands left behind as the coast retreats inland (figure 20.17). They form small, rocky islands off retreating coasts, often directly off headlands (figure 20.14). **Arches** (or *sea arches*) are bridges of rock left above openings eroded in headlands or stacks by waves. The openings are eroded in spots where the rock is weaker than normal, perhaps because of closely spaced fractures.

Depositional Coasts

Many coasts are gently sloping plains and show few effects of wave erosion. Such coasts are found along most of the Atlantic Ocean and Gulf of Mexico shores of the United States. These coasts are primarily shaped by sediment deposition, particularly by longshore drift of sand.

Coasts such as these are often marked by **barrier islands**—ridges of sand that parallel the shoreline and extend above sea level (figure 20.18). These barrier islands may have formed from sand eroded by waves from deeper water offshore, or they may be greatly elongated sand spits formed by longshore drift. The slowly rising sea level associated with the melting of the Pleistocene glaciers may have been a factor in their development. A protected lagoon separates barrier islands from the mainland. Because the lagoon is protected from waves, it provides a quiet waterway for boats. A series of such lagoons stretches almost continuously from New York to Florida, and many also exist along the Gulf Coast, forming an important route for barge traffic. As tides rise and fall, strong

A

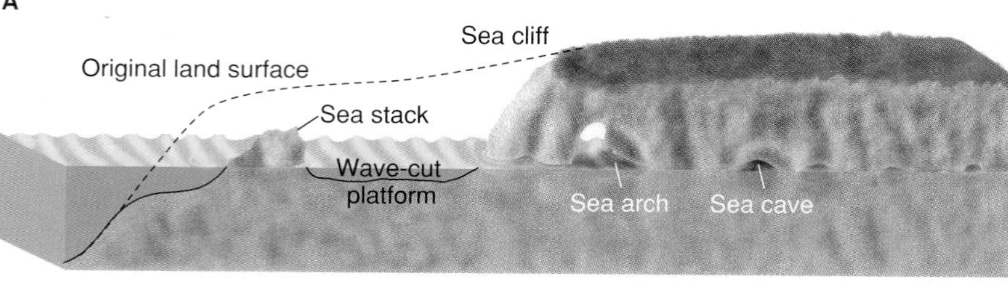

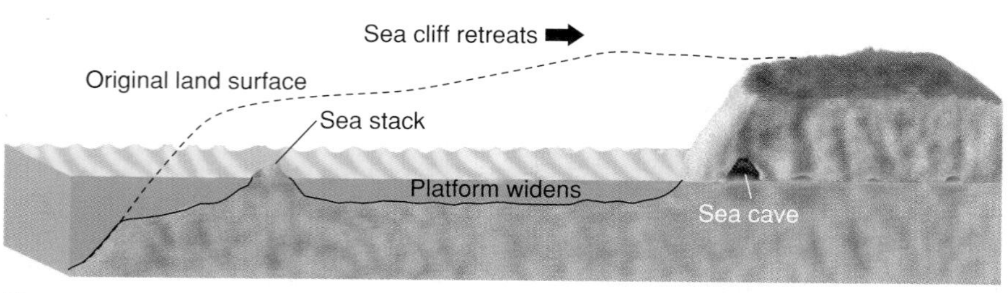

B

FIGURE 20.16

(*A*) A wave-cut platform (the wide, horizontal bench of dark rock at the base of the cliffs) is exposed at low tide. La Jolla, California. (*B*) A wave-cut platform widens as a sea cliff retreats.

FIGURE 20.17

Stacks and an arch mark old headland positions on this retreating, wave-eroded coast. Northern California.

tidal currents may wash in and out of gaps between barrier islands, distributing sand in submerged *tidal deltas* both landward and seaward of the gaps.

Some barrier islands along the Atlantic and Gulf coasts are densely populated. Atlantic City (New Jersey), Ocean City (Maryland), Miami Beach (Florida), and Galveston (Texas) are examples of cities built largely on barrier islands. In some of these cities, houses, luxury hotels, and condominiums are clustered near the edge of the sea; many are built upon the loose sand of the island (figure 20.19). These developed areas are vulnerable to late-summer hurricanes that sooner or later bring huge storm waves onto these coasts, eroding the sand and undermining the building foundations at the water's edge.

Nonmarine deposition may also shape a coast. Rapid sedimentation in *deltas* by rivers can build a coast seaward (see figure 16.30). *Glacial deposition* can form shoreline features. Several islands off the New England coast were glacially deposited; Long Island, New York, formed from the deposition of a recessional and end moraine.

Drowned Coasts

Drowned (or *submergent*) coasts are common because sea level has been rising worldwide for the past 15,000 years. During the glacial ages of the Pleistocene, sea level was 130 meters below its present level. The shallow sea floor near the continents was then dry land, and rivers flowed across it, cutting valleys. As the great ice sheets melted, sea level began to rise, drowning the river valleys. These drowned river mouths, called **estuaries,** mark many coasts today (figure 20.20). They extend inland as long arms of the sea. Fresh water from rivers

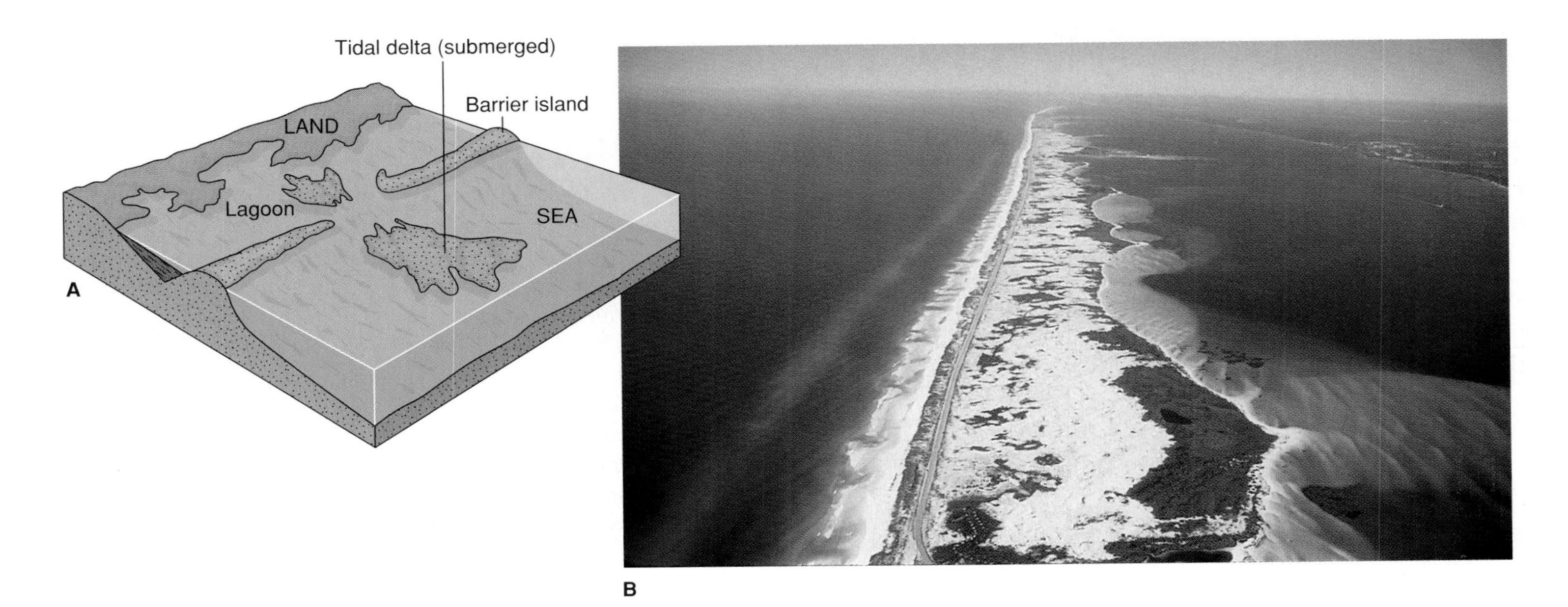

FIGURE 20.18

(*A*) A barrier island on a gently sloping coast. A lagoon separates the barrier island from the mainland, and tidal currents flowing in and out of gaps in the barrier island deposit sediment as submerged tidal deltas. (*B*) A barrier island near Pensacola, Florida. Open ocean to left, lagoon to right, mainland Florida on far right. Light-colored lobes of sand within lagoon were eroded from the barrier island by hurricane waves.

Photo by Frank M. Hanna

FIGURE 20.19

Hotels built upon the loose sand of a barrier island, Miami Beach, Florida. The lagoon and mainland Florida are visible at the upper left.
Photo by Florida Division of Tourism

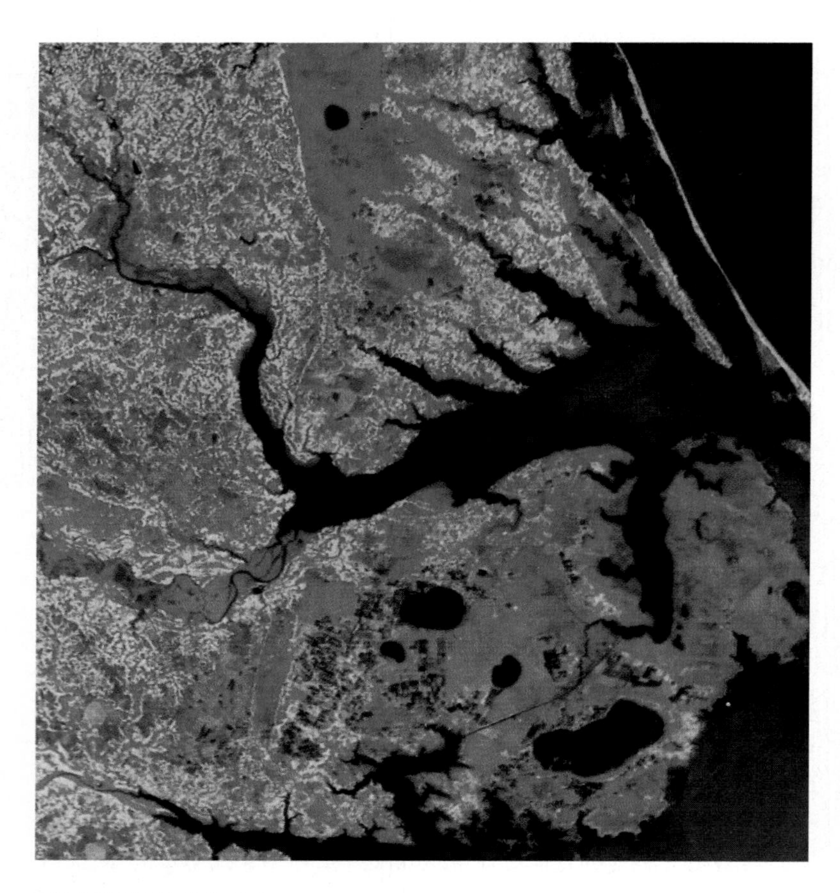

FIGURE 20.20

Landsat satellite photo of estuaries, Albemarle and Pamlico Sounds, North Carolina. Barrier islands are visible in upper right. Infrared image shows vegetation as red.
Photo by NASA

mixes with the sea water to make most estuaries brackish. The quiet, protected environment of estuaries makes them very rich in marine life, particularly the larval forms of numerous species. Unfortunately, cities and factories built on many estuaries to take advantage of quiet harbors are severely polluting the water and the sediment of the estuaries. The poor circulation that characterizes most estuaries hinders the flushing away of this pollution, and estuary shellfish are sometimes not safe to eat as a result.

Drowned coasts may be marked by **fiords,** glacially cut valleys flooded by rising sea level (see figure 19.36). They form in the same way as estuaries, except they were cut by glacial ice rather than rivers during low sea-level stands.

Uplifted Coasts

Uplifted (or *emergent*) coasts have been elevated by deep-seated tectonic forces. The land has risen faster than sea level, so parts of the old sea floor are now dry land.

Marine terraces form just offshore from the beach face, as described earlier in this chapter. These terraces can be wave-cut platforms caused by erosion of rock associated with cliff retreat, or they can be wave-built terraces caused by deposition of sediment. If the shore is elevated by tectonic uplift, these flat surfaces will become visible as *uplifted marine terraces* (figure 20.21). They formed below the ocean surface but are visible now because of uplift. The tectonically unstable Pacific coast of the United States and Canada has many areas marked by uplifted terraces, along with the erosional coast features described earlier.

FIGURE 20.21
Uplifted marine terrace, northern California. The flat land surface at the top of the sea cliff was eroded by wave action, then raised above sea level by tectonic uplift. The rock knob on the terrace was once a stack.

Coasts Shaped by Organisms

The growth of coral and algal *reefs* offshore can shape the character of a coast. The reefs act as a barrier to strong waves, protecting the shoreline from most wave erosion (see figure 3.22). Carbonate sediments blanket the sea floor on both sides of a reef and usually form a carbonate sand beach on land (see figure 14.18). Southernmost Florida has a coast of this type.

Branching *mangrove roots* dominate many parts of the southeastern United States coast. The roots dam wave and current action, creating a quiet environment that provides a haven for the larval forms of many marine organisms and that may trap fine-grained sediment. Mangroves also deposit layers of organic peat on low-lying coasts.

20.1 ENVIRONMENTAL GEOLOGY

THE EFFECTS OF RISING SEA LEVEL

LONG TERM

Sea level has risen about 130 meters in the past 15,000 years as the Pleistocene glaciers melted, adding water to the oceans. During this period, sea level initially rose at the rapid rate of 1.3 meters per century, but the rate of rise gradually slowed down, so that for the past 3,000 years the rise has only been about 4 cm per century. Since about 1930, however, the rate of sea-level rise has increased six-fold, to 24 cm per century along the Atlantic and Gulf coasts (the reasons for this increase are not clear). When sea level rose this rapidly a few thousand years ago, the coasts were rapidly eroded. If all the glacial ice on Earth were to melt, sea level would rise 60 meters, drowning many coastal areas.

On low-lying coasts a rise in sea level causes ocean waves (particularly storm waves) to extend much farther inland than before, flooding the land (box figure 1*A*). Barrier islands migrate landward. In the United States from New Jersey south to Florida, and along the Gulf of Mexico to southern Texas, the coastal land is very flat. A small rise in sea level can send seawater many kilometers inland.

On steep coasts a rise in sea level can accelerate the erosion of coastal cliffs, destroying oceanfront property (box figure 1*B*).

Some scientists predict that the rise in sea level may accelerate in the next century as a result of global warming by the "greenhouse effect." Burning of coal and oil has increased the amount of carbon dioxide in the atmosphere by more than 10% in this century. The carbon dioxide and other gases may trap more of the sun's energy on earth, warming the air and the sea. Warm air would accelerate the melting of glaciers on Greenland and Antarctica. Warm seawater expands. Together these two effects could raise sea level.

Predictions on the amount of sea level rise by the year 2100 vary widely. Early predictions of a rise of 1 to 2 meters have recently been revised downward to .3 to .6 meter as computer models were refined (by considering cloud cover, for example). A rise of even 1 meter on a low-lying coast could destroy thousands of beachfront houses and hotels.

Global warming is an emotional and controversial subject. The recent rise in atmospheric carbon dioxide is indisputable, but temperature data, while convincing some scientists that Earth is warming, indicate to other scientists that Earth is not changing in temperature, or is even cooling. The possibility of future warming is widely accepted, but evidence of warming to date is debatable. The debate over global warming does not change the fact that glaciers are now melting (as they have for 15,000 years) and sea level is now rising. The scientific argument is about whether the present rise of 24 centimeters per century will continue or will accelerate.

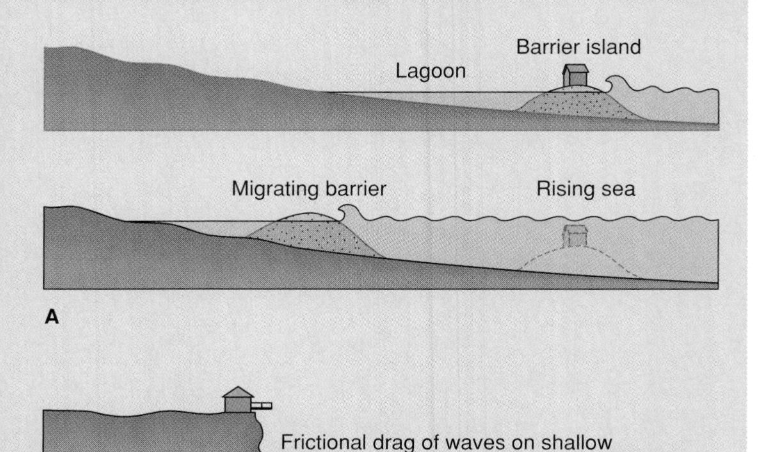

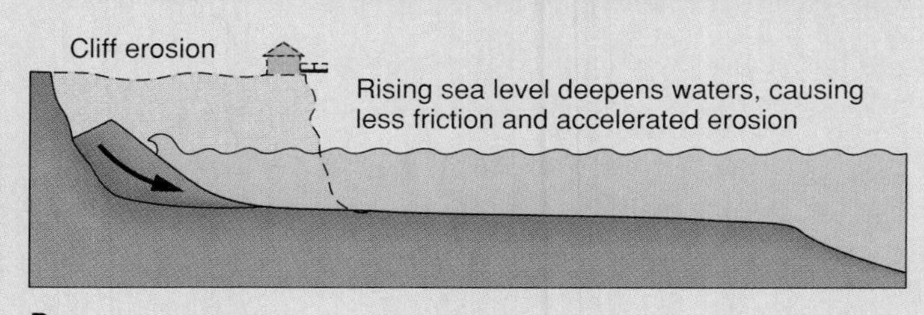

BOX 20.1 — FIGURE 1

Rising sea level can cause erosion of both gentle and steep coasts, leading to destruction of buildings.

SHORT TERM

Hurricanes, which are common along the Atlantic and Gulf coasts, tend to cause short-term rises in sea level called *storm surges.* Hurricanes consist of strong winds rotating counterclockwise (in the northern hemisphere) around a region of very low air pressure. They may be 500 kilometers in diameter and

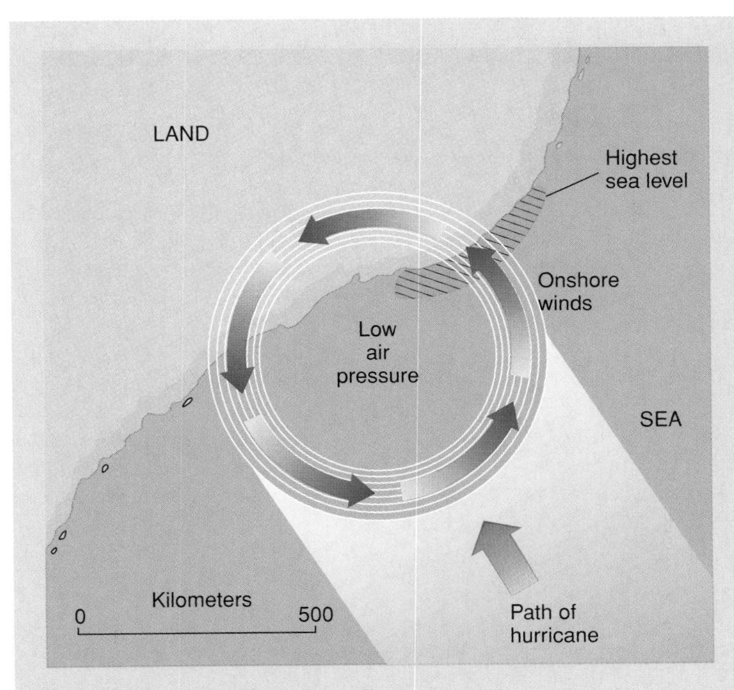

Box 20.1 — FIGURE 2

Strong onshore winds in a hurricane pile water against the shore, forming a storm surge (high sea level) that may cause severe flooding on a low-lying coast. Damage is worst at high tide.

Box 20.1 — FIGURE 3

Seventeen-foot seawall of concrete and granite blocks at Galveston, Texas, built to protect the barrier island from storm waves and storm surges. Note the lack of a sand beach seaward of the seawall.

contain winds up to 300 kilometers (180 miles) per hour. The low air pressure in the center of a hurricane allows the sea surface to rise, forming a broad dome, and this rise in sea level is accentuated by strong onshore winds, which pile water against shore (box figure 2). Storm surges can easily raise sea level 5 meters and are most devastating at high tide. A storm surge 8 meters high struck Galveston, Texas, in 1900, completely covering the barrier island that Galveston is built upon. High storm waves on top of the high sea level destroyed countless buildings, and 6,000 people died, many by drowning. Hurricane-tracking programs cut the death tolls of storm surges today by providing advance warning to coastal communities. In September 1989, hurricane Hugo hit South Carolina with 220 kilometer (135 mile per hour) winds and a 5-meter storm surge north of Charleston, causing $10 billion in damage. Because more than 500,000 people were evacuated along the low-lying coast, however, the death toll was only 29.

Careful planning for the use of coastal land in your lifetime will be necessary to lessen destruction caused by storm surges and long-term sea-level rise. Developed land at the ocean's edge may face a grim choice in the future—abandonment of existing buildings or expensive "armoring" of the coast with structures such as seawalls in hopes of protecting the buildings from wave damage (box figure 3).

Further Reading

Williams, S. J., K. Dodd, and K. K. Gohn. 1990. *Coasts in crisis.* U.S. Geological Survey Circular 1075. Online version can be found at http://pubs.usgs.gov/circular/c1075/

Summary

Wind blowing over the sea surface forms waves, which transfer some of the wind's energy to shorelines. Orbital water motion extends to a depth equal to half the wavelength.

As a wave moves into shallow water, the ocean bottom flattens the orbital motion and causes the wave to slow and peak up, eventually forming a *breaker* whose crest topples forward. The turbulence of *surf* is an important agent of sediment erosion and transportation.

Wave refraction bends wave crests and makes them more parallel to shore. Few waves actually become parallel to the shore, and so *longshore currents* develop in the surf zone. *Rip currents* carry water seaward from the surf zone.

A beach consists of a *berm, beach face,* and *marine terrace.* Summer beaches have a wide berm and a smooth offshore profile. Winter beaches are narrow, with offshore bars.

Longshore drift of sand is caused by the waves hitting the beach face at an angle and also by longshore currents.

Deposition of sand that is drifting along the shore can form *spits* and *baymouth bars.* Drifting sand may also be deposited against jetties or groins or inside breakwaters.

Rivers supply most sand to beaches, although local erosion may also contribute sediment. If the river supply of sand is cut off by dams, the beaches gradually disappear.

Coasts may be erosional or depositional, drowned or uplifted, or shaped by organisms such as corals and mangroves.

Coastal straightening by waves is caused by headland erosion and by deposition within bays.

A coast retreating under wave erosion can be marked by *sea cliffs,* a *wave-cut platform, stacks,* and *arches.*

Waves can form *barrier islands* off gently sloping coasts. River and glacial deposition can also shape coasts.

Drowned coasts are marked by *estuaries* and *fiords. Uplifted marine terraces* characterize coasts that have risen faster than the recent rise in sea level.

Terms to Remember

arch (sea arch) 510
barrier island 510
baymouth bar 505
beach 504
beach face 504
berm 504
breaker 501
coast 508
coastal straightening 509
crest (of wave) 500
estuary 511
fiord 512
headland 508
longshore current 502
longshore drift 505
marine terrace 504
rip current 502
sea cliff 509
spit 505
stack 510
surf 501
tombolo 506
trough (of wave) 500
wave-cut platform 510
wave height 500
wavelength 500
wave refraction 502

Testing Your Knowledge

Use the questions below to prepare for exams based on this chapter.

1. Show in a sketch how longshore drift of sand can form a baymouth bar.
2. In a sketch, show how and why sand moves along a beach face when waves approach a beach at an angle.
3. How are summer beaches different from winter beaches? Discuss the reasons for these differences.
4. What would happen to the beaches of most coasts if all the rivers flowing to the sea were dammed? Why?
5. What does the presence of an estuary imply about the recent geologic history of a region?
6. Describe how waves can straighten an irregular coastline.
7. Describe the transition of deep-water waves into surf.
8. Show in a sketch the refraction of waves approaching a straight coast at an angle. Explain why refraction occurs.
9. What is a longshore current? Why does it occur?
10. What is a rip current? Why does it occur? How do you get out of a rip current?
11. The path a water particle makes as a wave passes in deep water is best described as
    a. elliptical
    b. orbital
    c. spherical
    d. linear
12. The easiest method of escaping a rip current is to
    a. swim toward shore
    b. swim parallel to the shore
    c. swim away from the shore
13. Why is beach sediment typically quartz-rich sand?
    a. other minerals are not deposited on beaches
    b. quartz is the only mineral that can be sand-sized
    c. quartz is resistant to chemical weathering
    d. none of the above

14. Longshore drift is
    a. the movement of sediment parallel to shore when waves strike the shoreline at an angle
    b. a type of rip current
    c. a type of tide
    d. the movement of waves
15. Which structure would interfere with longshore drift?
    a. jetties b. groins
    c. breakwaters d. all of the above
16. What is the most common source of sand on beaches?
    a. sand from river sediment brought down to the ocean
    b. land next to the beach
    c. offshore sediments
17. Which would characterize an erosional coast?
    a. headlands b. sea cliffs
    c. stacks d. arches
    e. all of the above
18. Which would characterize a depositional coast?
    a. headlands b. sea cliffs
    c. stacks d. arches
    e. barrier islands
19. A glacial valley drowned by rising sea level is
    a. a fiord b. an estuary
    c. a tombolo d. a headland
20. The surf zone is
    a. the region in which waves break
    b. water less than one half wavelength in depth
    c. where the longshore current flows
    d. all of the above
21. The storm surge of a hurricane is
    a. the highest winds
    b. the tallest waves
    c. the dome of high water in the center of the hurricane
    d. the area of high pressure within the storm

Expanding Your Knowledge

1. Sea level would rise by about 60 meters if *all* the glacial ice on Earth melted. How many U.S. cities would this affect?
2. What happens to a coast if its offshore reef dies?
3. Is a beach a good place to mine sand for construction? Explain your answer carefully.
4. The seaward tip of a headland may be the most rapidly eroding locality on a coast, yet may also be the most expensive building site on a coast. Why is this so?

Exploring Web Resources

www.mhhe.com/plummer9e
Visit our Online Learning Center to check your answers for the Testing Your Knowledge section and to watch some great animations created to better your understanding of waves, beaches, and coastal shift. Click on the direct links to go to the other websites listed below.

http://marine.usgs.gov/
Web page for the *Coastal and Marine Geology Program* of the U.S. Geological Survey contains information about numerous geologic studies of U.S. coastal areas.

http://pubs.usgs.gov/circular/c1075/
Online version of *Coasts in crisis.* U.S. Geological Survey Circular 1075.

http://woodshole.er.usgs.gov/
Web page for the U.S. Geological Survey *Wood's Hole Field Center* for coastal and marine research contains information and data from ongoing scientific projects.

www-ccs.ucsd.edu/
Web page for the *Center for Coastal Studies at the Scripps Institute of Oceanography* provides information about their research and access to data collected from various coastal studies projects.

www.esdim.noaa.gov/ocean_page.html
Oceans Web page from *National Oceanic and Atmospheric Administration (NOAA)* provides numerous links to oceanography research projects and data.

Animations

This chapter includes the following animations available on our Online Learning Center at www.mhhe.com/plummer9e.

20.4 Orbital Motion in Shallow Water
20.8 Seasonal Cycle of a Beach
20.9 Longshore Drift of Sand
20.10A Spits and Baymouth Bars
20.14 Coastal Straightening

CHAPTER

21
Geologic Resources

Throughout this book, we have mentioned human use of Earth materials, most of which are nonrenewable. Our purpose in this chapter is to survey briefly some important geologic resources of economic value.

We first look at energy resources to see which ones might help replace our disappearing supplies of oil. Then we discuss metals and their relation to igneous rocks and plate tectonics and conclude with nonmetallic resources such as sand and gravel.

Opposite: Copper mine at San Zavier, Arizona
Photo © Jim Hark/Peter Arnold, Inc.

Nearly every manufactured object requires geologic resources. An automobile contains substantial amounts of iron, chromium, manganese, nickel, platinum, tin, copper, lead, and aluminum in its body and engine and quartz sand in its window glass. It consumes petroleum in several forms—as fuel and lubricants, as synthetic rubber for tires, and as plastic for electrical parts, upholstery, and steering wheels. Dozens of other geologic resources go into automobiles, from tungsten in light bulb filaments to sulfur in battery acid.

People are beginning to realize how heavily dependent on geologic resources they are. Some have tried to limit their consumption of resources or at least the rate at which their consumption *increases*. But it is impossible to stop consuming geologic resources. Think about some of the objects near you as you are reading this chapter. Your shirt may be partly polyester, which is made from petroleum, as are the plastic buttons. If you're wearing jeans, they may be made of fabric that is 100% cotton, but many jeans are made of shrinkproof fabrics that blend cotton with petroleum-based synthetic fibers. The brass zipper is made of copper and zinc. Some brands of jeans have pocket rivets made of copper. To make the leather tags on the back of some jeans, either aluminum or chromium was used during tanning. The fabric dye almost certainly came from petroleum. A pencil uses many geologic resources (figure 21.1) and a radio many more. Computer chips require silicon from quartz sand. More than 40% of gold that is mined is used for computers. Eyeglasses are made of quartz sand and petroleum. Dental fillings are made of mercury, silver, and other metals. Cassettes and videotapes are made of vinyl, a petroleum derivative, and iron or chromium particles from metal mines. Every one of us uses geologic resources and therefore is indirectly responsible for the existence of mines and oil wells.

Most of our energy comes from petroleum. As the world's population increases, as standards of living rise, and as new technology produces more desirable goods, our energy needs increase. If you are living in a home that is moderately old, it is likely that there are not enough electrical outlets and you have multiple adapters on each to plug in all your appliances and gadgets. Americans, in particular, take cheap and plentiful energy for granted. When prices rise abruptly, they expect politicians to find a fix. Few understand that, once used, petroleum can't be recycled or grown back like a vegetable crop.

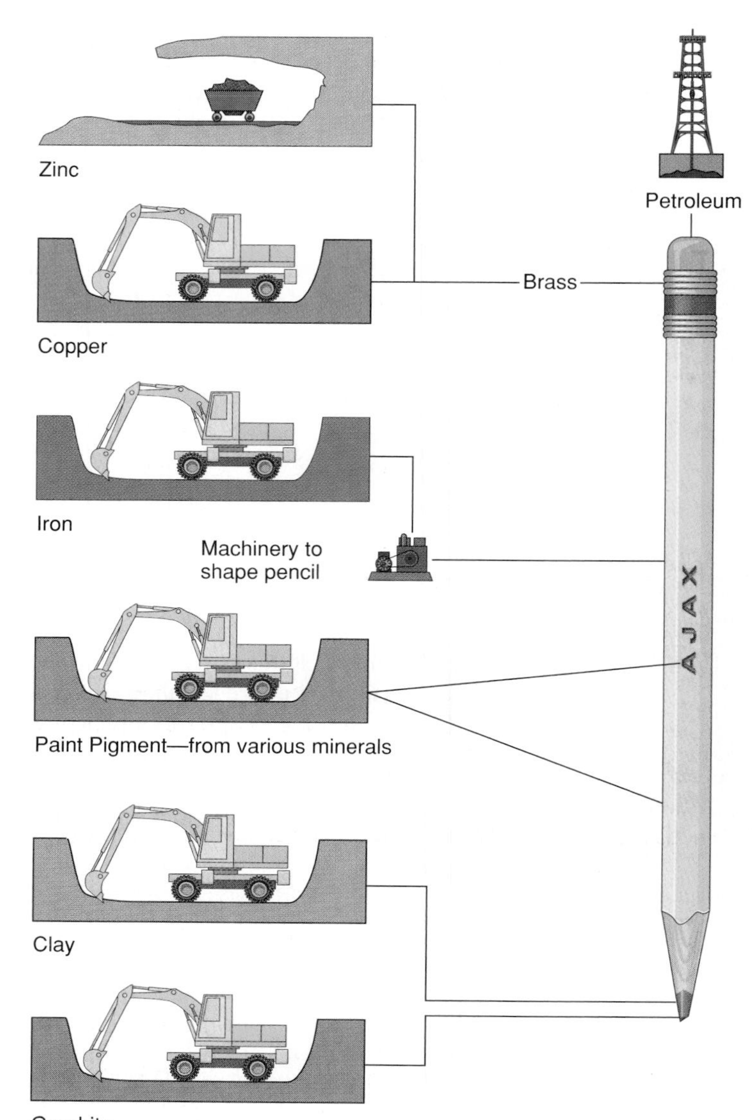

FIGURE 21.1
Mineral resources necessary to make a wooden pencil.

TYPES OF RESOURCES

Geologic resources are valuable materials of geologic origin that can be extracted from Earth. There are three main categories of geologic resources:

1. Energy resources—petroleum (oil and natural gas), coal, uranium, and a few others, such as geothermal resources.
2. Metals—iron, copper, aluminum, lead, zinc, gold, silver, and many more.
3. Nonmetallic resources—sand and gravel, building stone, limestone (for cement), sulfur, gems, gypsum, fertilizers, and many more. Ground water (chapter 17) should also be regarded as an important geologic resource.

Geologic resources are sometimes called *mineral resources,* but the term, though widely used, is not accurate. Oil, natural gas, and coal, for example, are not formed of solid, crystalline minerals.

A very important point about geologic resources is that they are **nonrenewable resources,** except for ground water. They form so slowly that, at the present rapid rates of consumption, they can easily become exhausted. Some Earth resources, such as food or timber, usually can be produced as fast as they are consumed. These are *renewable resources.* But petroleum, iron, lead, uranium, sulfur, sand and gravel, and essentially all other geologic resources are being used at rates far greater than the rates at which new deposits form. In some areas even ground water is being consumed faster than it is being replenished. This means that eventually these resources are going to either run out or be priced so high because of their scarcity that their use will drop to insignificant levels. Recycling, new discoveries, and substitutes can help prolong the life of some resources, but it is likely that within your lifetime some of the resources in common use today will be scarce. This need not mean that civilization will come to a standstill, but it does mean that we must find inexpensive substitutes for these resources. Otherwise, skyrocketing costs will cause a drop in nearly everyone's material standard of living.

Resources and Reserves

Two different terms are used to describe the amount of geologic material not yet extracted from Earth. **Resources** is a broad term used for the total amount of a geologic material in all deposits, discovered and undiscovered. It includes both the deposits that can be economically extracted under present conditions and those that may be extracted economically in the future (figure 21.2). It is very difficult to estimate resources because educated guesses must be made about the existence and sizes of deposits yet undiscovered as well as about what type of deposit might someday be economical to extract. **Reserves** are a small part of resources. They are the *discovered* deposits that can be extracted economically and legally under present conditions. That is, they are the short-term supply of a geologic material.

Once *resources* have been carefully estimated, the amount should not change from year to year, for an estimate of resources is basically an estimate of the total future supply. Estimates of *reserves,* however, change all the time. The extraction and use of a substance lowers reserves. New discoveries add to reserves, for part of the definition of reserves is that we have to know the deposit is there. (*Reserves* are like your bank balance, rising and falling with time; *resources* are like your future, lifetime income.)

A deposit also has to be profitable to extract, and many factors determine whether a profit can be made. The costs of extraction, including workers' salaries and the fuel used to run equipment, are important. New inventions that make extraction cheaper can increase reserves. The final price a company can receive for its product is also important. When prices are low, reserves are naturally low because few deposits can be extracted profitably at that price. When prices rise, more deposits become economical, and reserves increase, even without new discoveries. Reducing the amount of taxes that extraction companies pay can also increase reserves.

Changes in laws can affect reserves. Large areas of government-owned land are off-limits to mining and drilling, so any geologic materials under these areas are not legally extractable and cannot be included in reserves. Opening more land to extraction can therefore increase reserves.

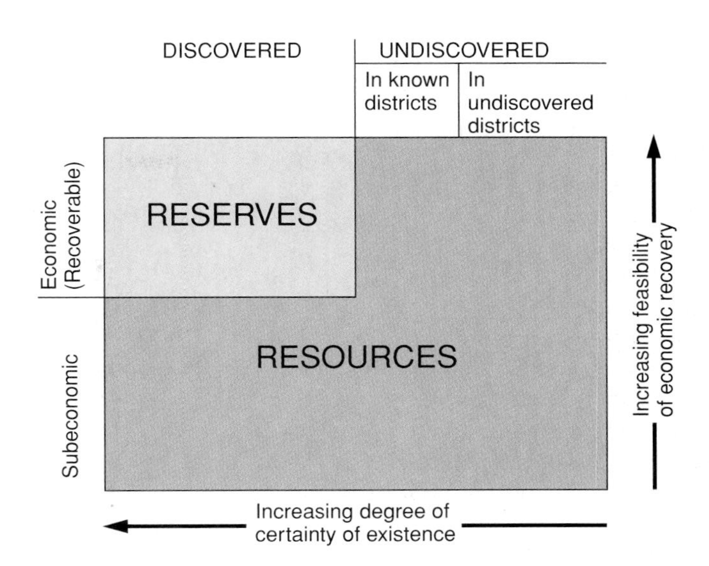

FIGURE 21.2
The difference between reserves and resources. Reserves (discovered deposits that are economically and legally extractable) form a small part of total resources (discovered and undiscovered deposits that are economical to extract now or may be economical in the future).

ENERGY USE

The United States consumes huge amounts of energy. By 1990 the United States was consuming ten times the amount of energy that it used in 1900. In 1973 and 1978 came steep price rises for petroleum. The need for conservation and fuel-efficient buildings and transportation became more and more apparent. From 1979 to 1985 total United States energy use declined and so did its dependence on petroleum (particularly imported oil). Following a precipitous drop in oil prices at the beginning of 1986, however, there was a sharp increase in both United States energy use and its reliance on imported oil. By the time energy prices again rose sharply in 2000 and 2001, the lessons of the 1970s had been forgotten and gas-guzzling vehicles were being sold in record numbers.

The sources of United States energy have been changing in recent years in response to changing prices.

	1975	1985	1993	1997	1999
Oil	46%	42%	40%	36%	41%
Natural gas	28	24	25	31	24
Coal	18	24	23	21	24
Nuclear	3	5.5	8	7	9
Hydroelectric	5	4.5	4	4	3

Note that oil and natural gas account for two-thirds of the United States energy supply; the fossil fuels (coal, oil, and gas) provide almost 90% of our energy.

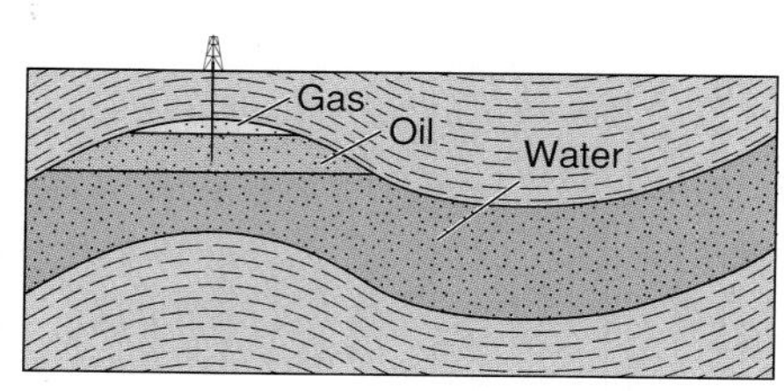

A Anticline

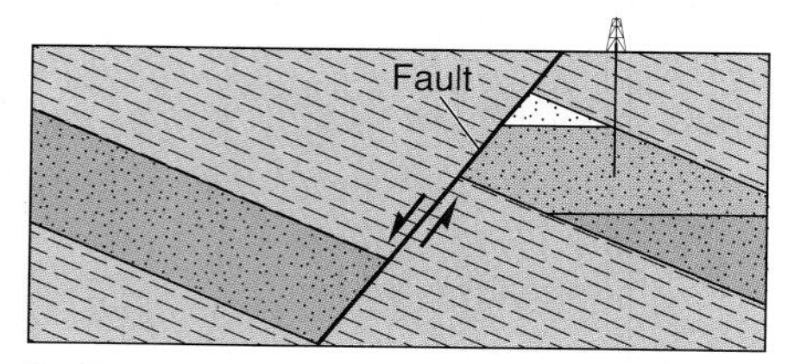

B Normal fault

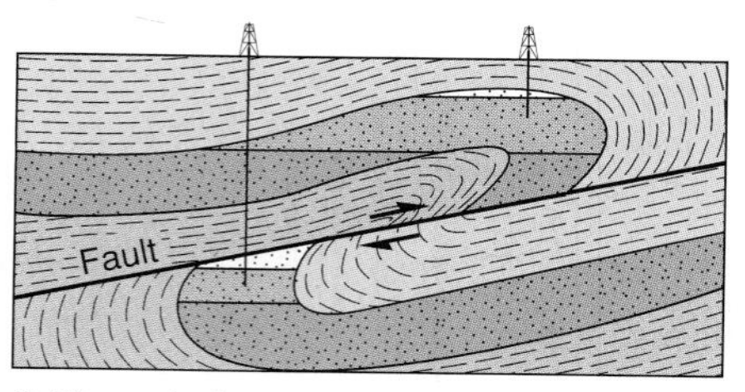

C Thrust fault

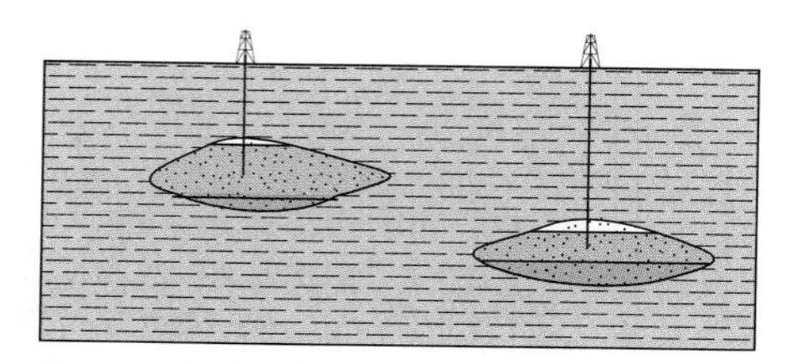

D Sandstone lenses

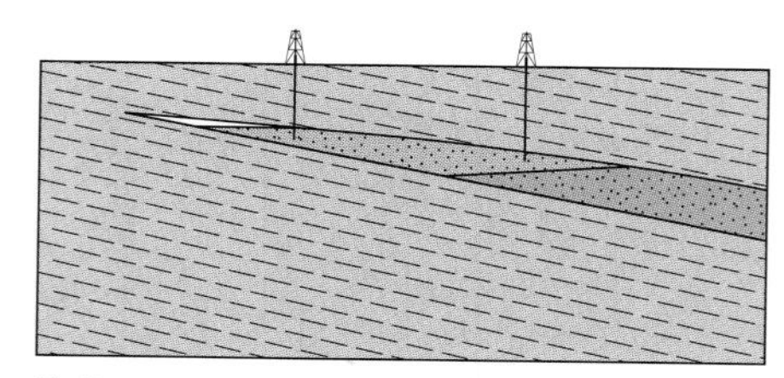

E Sandstone pinchout

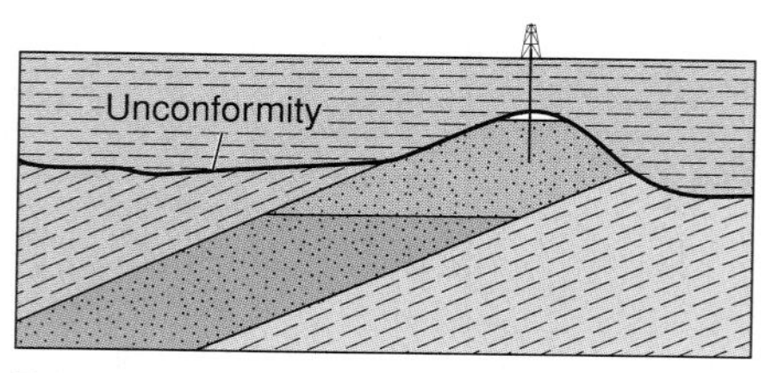

F Unconformity

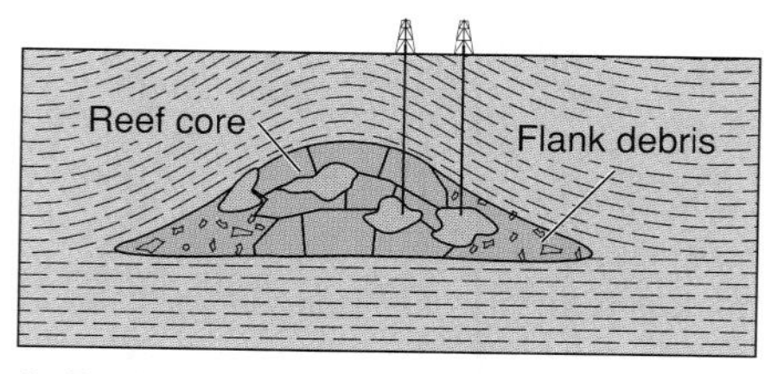

G Reef (a small "patch" reef)

FIGURE 21.3

Traps for oil and gas. (*A–C*) are structural traps. (*D–G*) are stratigraphic traps. Oil floats on water, and is often capped by natural gas.

OIL AND NATURAL GAS

Within the petroleum industry, **petroleum** is a broad term that includes both crude oil and natural gas. That will be our usage. (In common usage *petroleum* is synonymous with crude oil; many geologists also use it this way.)

Crude oil is a liquid mixture of naturally occurring hydrocarbons (compounds containing hydrogen and carbon), which can be distilled to yield a great variety of products.

Natural gas is a *gaseous* mixture of naturally occurring hydrocarbons. Its origin and occurrence closely parallel that of oil. Many wells that recover oil also recover natural gas, although either may exist alone.

The Occurrence of Oil and Gas

The origin of oil and gas in sediment and sedimentary rocks was described in chapter 14.

Oil pools are valuable underground accumulations of oil. They are found only where four specific conditions occur together: (1) a **source rock** (such as shale) containing organic matter that is converted to petroleum by burial and other post-depositional changes; (2) a **reservoir rock** (usually sandstone or limestone) that is sufficiently porous and permeable to store and transmit the petroleum; (3) an **oil trap,** a set of conditions to hold petroleum in a reservoir rock and prevent its escape by migration; and (4) deep enough burial (or *thermal maturity*) to "cook" the oil and gas out of the organic matter. All four conditions must be met; if one of the four is missing, the rock will hold no oil or gas.

Natural gas requires the same conditions as oil for accumulation. Gas can exist at greater depth than oil, however, and variations in source rock, depth of burial, and thermal history of the organic matter probably control whether gas, oil, or both accumulate in a region.

Figure 21.3 depicts several types of *structural traps* for oil and gas. *Anticlines* (described in chapter 6) are the most common oil traps. Where oil and water occur together in folded sandstone beds, the oil droplets, being less dense than water, rise within the permeable sandstones toward the top of the fold. There the oil may be trapped by impermeable shale overlying the sandstone reservoir rocks. Because natural gas is less dense than oil, the gas collects in a pocket, under fairly high pressure, on the top of the oil.

Faults may create oil traps when permeable reservoir rocks break and slide next to impermeable rocks. Thrust faults are often associated with folds because both are caused by compression. The backarc thrust belts on the landward side of both the Cordilleran and Appalachian mountain belts have been intensively explored for oil and gas.

A *stratigraphic trap* is a result of natural sedimentation rather than folding or faulting. It may be a lens of sandstone within a larger bed of shale. Another such trap is the narrow edge of a gently dipping sandstone layer where it pinches out within a shale sequence. Oil can collect under *unconformities* if shale seals off a reservoir rock. Limestone *reefs* can form a variety of traps. The core of a reef is usually full of large openings formed by the irregular growth of coral and algae. Oil can collect both in a reef core and in the dipping beds of wave-broken debris on the reef flanks. *Salt domes* create a variety of traps (see box 6.1).

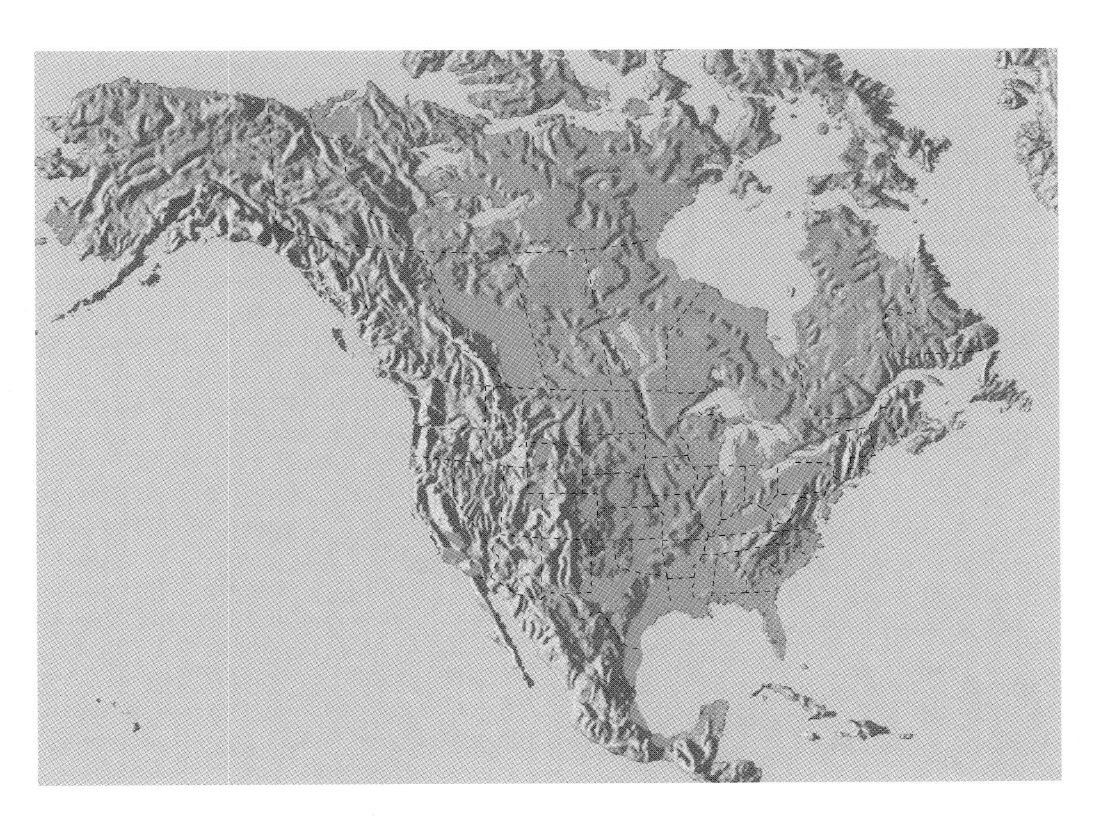

FIGURE 21.4
Major oil fields in North America. The amount of oil in a field is not necessarily related to its areal extent on a map. It is also governed by the vertical "thickness" of the oil pools in a field. The fields with the most oil are in Alaska and east Texas.
From U.S. Geological Survey and other sources

Oil fields are regions underlain by one or more oil pools. Figure 21.4 shows the location of most of the major North American oil fields. The two largest oil fields within the United States are in eastern Texas and on Alaska's North Slope. Most of the world's remaining oil lies in giant fields in the Middle East (especially in Saudi Arabia and Kuwait), Russia and Azerbaijan, Venezuela, and Mexico.

Recovering the Oil

When an oil pool or field has been discovered, wells are drilled into the ground. Permanent derricks used to be built to handle the long sections of drilling pipe. Now portable drilling rigs are set up to drill (figure 21.5) and are then dismantled and moved. When the well reaches a pool, oil usually rises up the well because of its density difference with water or because of the pressure of the expanding gas cap above the oil. Although this rise of oil is almost always carefully controlled today, spouts of oil, or *gushers,* were common in the past. Gas pressure gradually dies out, and oil is then pumped from the wells. Water or steam may be pumped down adjacent wells to help push the oil out. At a refinery the crude oil from underground is separated into natural gas, gasoline, kerosene, lubricating oil, fuel oil, grease, asphalt, and paraffin. *Petrochemicals,* manufactured from petroleum, include dyes, fertilizers, medicines, synthetic

FIGURE 21.5
Drilling rig on Alaska's North Slope.
Photo from Bureau of Mines, U.S. Department of the Interior

rubber, explosives, perfumes, paints, saccharin, solvents, synthetic fibers, and plastics used for such varied products as compact discs, cassettes, floor tile, and garbage bags.

As oil becomes increasingly difficult to find, the search for it is extended into more hostile environments. The development of the oil field on the North Slope of Alaska and the construction of the Alaska pipeline are examples of the great expense and difficulty involved in new oil discoveries. Offshore platforms extend the search for oil to the ocean's continental shelves and even beyond. More than one-quarter of the world's oil and almost one-fifth of the world's natural gas comes from offshore, even though offshore drilling is six to seven times as expensive as drilling on land.

Environmental Effects

Getting petroleum out of the ground and to the consumer can create environmental problems anywhere along the line. Salty *brine,* a by-product of most oil wells, can pollute surface water. Pipelines carrying oil can be broken by faulting, landsliding, or acts of war, causing an *oil spill.* Tanker spillage from collisions or groundings (such as the *Exxon Valdez* off Alaska in 1989) can create oil slicks at sea. Offshore platforms may also lose oil (such as the Santa Barbara blowout off southern California in 1969). Oil slicks can drift ashore, fouling the beaches. Natural oil seeps can also cause pollution. Tar-like lumps collect on some California beaches as a result of natural seepage of petroleum from the ground. At one place in Alaska, oil seeps build a crude oil lake behind ice during the winter; in the summer, when the ice melts, the oil flows out to a nearby bay.

Subsidence of the ground can occur as oil is removed. The Wilmington field near Long Beach, California, has subsided 9 meters in 50 years; dikes have had to be built to prevent sea water from flooding the area. The refining and burning of petroleum and its products can cause *air pollution.* Advancing technology and strict laws are helping control some of these adverse environmental effects.

How Much Oil Do We Have Left?

The World Situation

The world's oil *reserves* are estimated to be 1,000 to 1,100 billion barrels (a barrel contains 42 gallons). Total *resources,* of course, are much larger than reserves. One estimate is 3,000 billion barrels (including reserves).

Since the world's present consumption of oil is 24 billion barrels per year, known *reserves* will last about 40 to 45 years. Total *resources* may last well over 100 years. The world is not running rapidly out of oil; the United States *is,* however, as we will see in the next section.

There is far more oil underground than is listed in the resources estimate; some oil cannot be recovered. The oil may be in a pool too small, or lie too far from market, to justify the expense of drilling. Oil lies under regions where drilling is forbidden, such as national parks or other public lands. Even using the best extraction techniques, only about 30% to 40% of the oil in a given pool can be brought to the surface. The rest is far too difficult to extract and has to remain underground.

The Outlook in the United States

United States consumption of oil is 6.3 billion barrels of oil per year (with 6% of the world's population, the United States uses 29% of the world's oil production each year). Sixty-five percent of U.S. consumption is used for transportation. United States *reserves* are currently about 25 billion barrels. The United States imports 51% of the oil it uses; domestic production is 3.1 billion barrels per year. At current production rates, there is an 8-year supply in reserves (less if the production rate goes up).

Because the reserve estimate changes each year, an estimate of *resources* is more important for planning purposes. Such estimates have fluctuated wildly in the past 25 years. In recent years the estimate of resources has been going down rapidly as promising drill sites have proved to be dry. The current resource estimate is 74 billion barrels (including reserves). This means that at the current production rate there is a 24-year supply of oil left, counting both reserves and resources. The United States *is* truly running out of oil, and we will need alternate fuels and/or more imported oil in the future.

The situation becomes more serious if the temporary ban on most new offshore drilling, enacted after the *Exxon Valdez* spill, becomes permanent. About 16 billion barrels of the 74-billion total are in undiscovered offshore oil. If this oil is placed off-limits, the country's oil resources will run out in 19 years at the current production rate.

Past estimates of *natural gas* have been similarly gloomy, largely because estimators assumed that gas discoveries are linked to oil discoveries. In the past, gas has been found in a fixed ratio to oil, so some estimators assume that if you are running out of oil, you are also running out of gas.

The United States produces more than 18 trillion cubic feet (TCF) of natural gas per year. Reserve estimates are 173 TCF, with total resources perhaps 570 TCF. By these estimates, natural gas will be gone in the United States in 30 years.

Decontrol of natural gas prices, however, led to renewed interest in gas and revised estimates of resources. The new estimates are based on unconventional gas sources, not linked to oil. There may be 300 to 500 TCF of gas within coal beds, for example (the gas is a hazard to coal miners because it is toxic and tends to explode). Since 1995, two areas have begun to be important producers of coal bed gas, the San Juan Basin of northwestern New Mexico and the Warrior Basin of Alabama. Tightly compacted sandstones ("tight sands") in the western United States may hold 200 to 800 TCF. Devonian shales rich in organic matter in the eastern United States may hold 500 to 600 TCF. Gas can exist at greater depth than oil, so it is likely that deep drilling (beyond 5 kilometers) in sedimentary basins can produce substantial amounts of gas. Gas is also dissolved in hot salty water within highly fractured rock along the Gulf Coast ("geopressured zones") that may contain as much as 3,000 TCF of gas. Truly staggering amounts of gas may be tied up as gas hydrates, a peculiar solid that forms from natural gas and water in two unusual environments: (1) on the deep sea floor and (2) under permafrost. Recent estimates of worldwide gas in hydrate form range from 100,000s to 1,000,000s of TCF. All these estimates are for total gas in place, not for recoverable reserves, which would be far lower. Some geologists discount all these estimates as wildly optimistic and misleading. All these unconventional gas sources are getting serious attention today, however, and although some difficult extraction problems remain unsolved, there is at least cautious optimism that the United States may not be as short of gas as was once thought.

HEAVY CRUDE AND OIL SANDS

Heavy crude is dense, viscous petroleum. It may flow into a well, but its rate of flow is too slow to be economical. As a result, heavy crude is left out of reserve and resource estimates of less viscous "light oil," or regular oil. Heavy crude can be made to flow faster by injecting steam or solvents down wells, and if it can be recovered, it can be refined into gasoline and many other products just as light oil is. Most California oil is heavy crude.

Oil sands (or *tar sands*) are asphalt-cemented sand or sandstone deposits. The asphalt is solid, so oil sands are often mined rather than drilled into, although the techniques for reducing the viscosity of heavy crude often work on oil sands as well.

The origin of heavy crude and oil sands is uncertain. They may form from regular oil if the lighter components are lost by evaporation or other processes. Oil sands and asphalt seeps at Earth's surface (such as the Rancho La Brea Tar Pits in Los Angeles) probably formed from evaporating oil. But some heavy crudes and oil sands are found as far as 4,000 meters underground. Most of them have much higher concentrations of sulfur and metals, such as nickel and vanadium, than does regular oil. These facts suggest that heavy crude and oil sands may have a somewhat different origin than light oil.

The best-known oil sand deposit is the Athabasca Oil Sand in northern Alberta, Canada (figure 21.6). The deposit, currently being strip-mined, contains 900 billion barrels of oil, of which at least 80 billion barrels are recoverable. About 10% of Canada's oil production comes from these oil sands. Venezuela has even more oil sand than Canada. The United States has more than 100 billion barrels of heavy crude and oil in oil sands, including about 30 billion in the form of oil sands in Utah. Half of it may be ultimately recoverable. This means that heavy crude and oil sands may supply as much oil in the future as our light oil reserves.

OIL SHALE

Oil shale is a black or brown shale with a high content of solid organic matter from which oil may be extracted by distillation. The best-known oil shale in the United States is the Green River Formation, which covers more than 40,000 square kilometers in Colorado, Wyoming, and Utah, with deposits up to 650 meters thick (figures 21.6 and 21.7). The oil shale, which includes numerous fossils of fish skeletons, formed from mud deposited on the bottom of large, shallow Eocene lakes. The organic matter came from algae and other organisms that lived in the lakes.

The Green River Formation includes more than 400 billion barrels of oil in rich beds that yield over 25 gallons of oil per ton of rock. Another 1,400 billion barrels of oil occur in lower-grade beds yielding 10 to 25 gallons per ton. An estimated 300 to 600 billion barrels may be recoverable.

Relatively low-grade oil shales in Montana contain another 180 billion barrels of recoverable oil in shale that should be economical to mine because of its high content of vanadium, nickel, and zinc. Therefore, oil shale can supply potentially vast amounts of oil in the future as our liquid petroleum runs out.

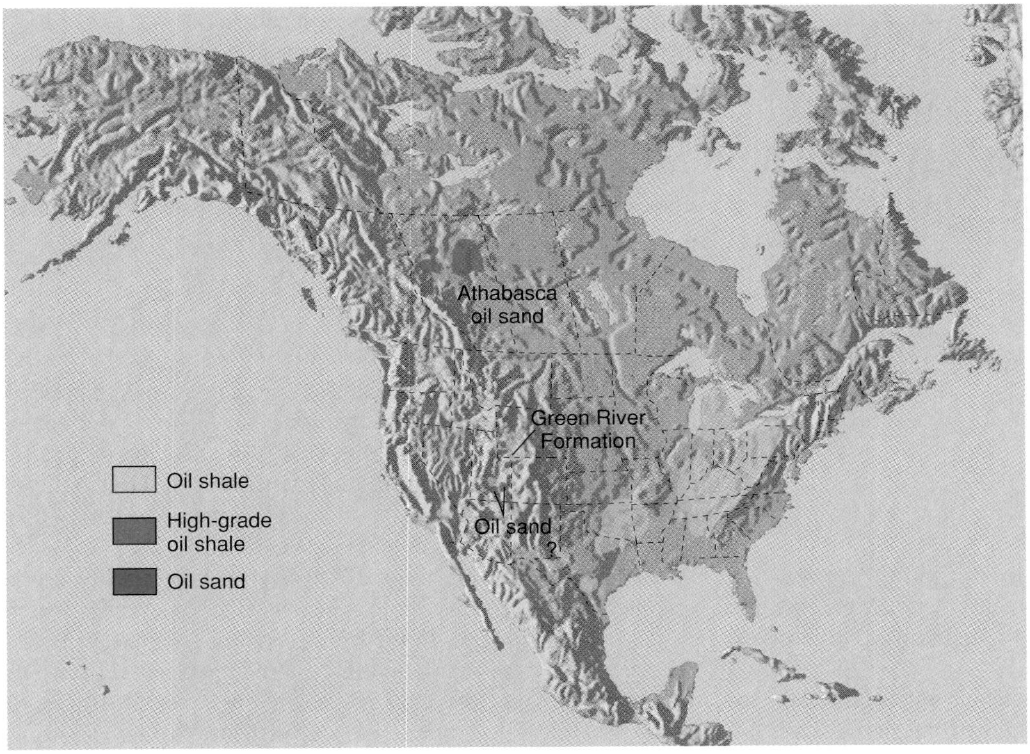

FIGURE 21.6
Distribution of major deposits of oil sand and oil shale in the United States and Canada.
From U.S. Geological Survey and other sources

21.1

ENVIRONMENTAL GEOLOGY

FLAMMABLE ICE. METHANE HYDRATE DEPOSITS—SOLUTION TO ENERGY SHORTAGE OR MAJOR CONTRIBUTOR TO GLOBAL WARMING?

Methane hydrates are an unusual mixture of ice and gas in which methane (one of the gasses in natural gas) is trapped in ice crystals. These are found in extreme environments, notably permafrost in polar regions and in the deep ocean floor. If lit, a piece of methane hydrate ice will burn with a red flame. The amount of methane hydrate in the ocean floors is staggering. Although estimates of methane hydrate resources vary, it appears that there is at least twice the amount of potential fuel tied up in methane hydrates as in all petroleum and coal combined.

Commercially exploiting methane hydrate deposits presents formidable challenges. Most of the deposits are in lenses frozen in sediment at deep ocean floors, beneath a kilometer or more of water. There, methane hydrate is stable because of the cold and high pressure. If pressure is reduced or the substance heated, it becomes unstable and the methane escapes. Mining anything at this depth is difficult, but trying to remove the icy substance from the sediment and get it to the surface without losing the methane is an extreme technologic problem.

Methane hydrate could significantly exasperate global warming. Methane, like carbon dioxide, is a greenhouse gas. However, unlike CO_2, methane will only stay in the atmosphere for around 10 years. But, methane reacts with oxygen in the atmosphere and produces CO_2, which remains in the atmosphere indefinitely. Significant volumes of methane could be released if methane hydrate sediments are disrupted by submarine landslides or other means. They could also be released if the oceans warm even slightly.

Related Web Resource

Scientific American: Gas Blasts

www.sciam.com/explorations/1999/122099hydrate/index.html

Additional Reading

E. Suess, G. Bohrmann, J. Greinert and E. Lausch, 1999, *Flammable Ice.* Scientific American, Nov. 1999, p. 76–83.

FIGURE 21.7

Cliffs of oil shale near Rifle, Colorado.
Photo by William W. Atkinson

A few distillation plants extract shale oil, but the current low price for oil makes shale oil uneconomical. A price increase for oil may make large-scale production of shale oil feasible in the future.

The mining of oil shale can create environmental problems. There is a space problem, for during distillation the shale expands. Spent shale could be piled in valleys and compacted, but land reclamation would be troublesome. A great amount of water is required, both for distillation and for reclamation, and water supply is always a problem in the arid West. New processing techniques that extract the oil in place without bringing the shale to the surface may eventually help solve some of the problems and lower the water requirements. It is possible to burn fractured oil shale in large underground excavations. The heat separates most of the oil from the rock; the oil can be collected as a liquid. (The fires, however, would be hard to control and would affect ground-water levels.) Another proposal involves heating the shale with radio waves or microwaves to separate the liquid oil from the rock, but to use either technique, a substantial amount of shale must first be removed.

COAL

Coal, as described in chapter 14, is a sedimentary rock that forms from the compaction of plant material that has not completely decayed. After oil and natural gas, coal is our third major energy resource. It provided 90% of the energy used in the United States in 1900 but provides only 21% today. Coal use has increased as petroleum becomes scarcer and more expensive.

About 88% of the present use of coal in the United States is for generating electricity. Coal is also used to make *coke,* which is used in steel making. In the future, coal may be used instead of petroleum in the manufacture of some chemicals. *Coal gas* and *coal oil* made from coal are reasonable approximations of natural gas and oil and can be used for some of the same purposes, although they are still very expensive to produce. Coal can also be powdered and mixed with water to form a *slurry,* which can be transported through pipelines and burned as a liquid fuel.

table 21.1 Varieties (Ranks) of Coal

	Color	Water Content (%)	Other Volatiles (%)	Fixed Carbon[2] (%)	Approximate Heat Value (BTUs of heat per pound of dry coal)
Peat[1]	Brown	75	10	15	Varies
Lignite	Brown to brownish-black	45	25	30	7,000
Subbituminous coal	Black	25	35	40	10,000
Bituminous coal (soft coal)	Black	5 to 15	20 to 30	50 to 75	12,000 to 15,000
Anthracite (hard coal)	Black	5	5	90	14,000

1. Peat is not a coal.
2. "Fixed carbon" means solid combustible material left after water, volatiles, and ash (noncombustible solids) are removed.

A

B

FIGURE 21.8

(*A*) A layer of peat being cut and dried for fuel, island of Mull, Scotland. Coal often forms from peat. (*B*) A bed of subbituminous coal near Gillette, Wyoming.

Varieties of Coal

Table 21.1 shows the common varieties (ranks) of coal. *Peat,* a mat of unconsolidated plant material, is not coal but probably represents the initial stage of coal development (figure 21.8*A*). When dry, it can be burned as a fuel (peat fires used to dry malted barley give Scotch whisky its smoky flavor). With compaction, peat can become *lignite* (*brown coal*), which may still contain visible pieces of wood. Lignite is soft and often crumbles as it dries in air. It is subject to spontaneous combustion as it oxidizes in air, and this somewhat limits its use as a fuel. *Subbituminous coal* and *bituminous coal* (*soft coal*) are black and often banded with layers of different plant material (figure 21.8*B*). They are dusty to handle, ignite readily, and burn with a smoky flame. *Anthracite* (*hard coal*) is actually a metamorphic rock, generally formed only under the regional compression associated with folding. It is hard to ignite but is dust-free and smokeless.

Occurrence of Coal

Coal occurs in beds that range in thickness from a few centimeters to 30 meters or more. If the beds are deeply buried, underground mines are dug to extract the coal (figure 21.9). If the coal beds are close to the land surface, the coal is mined in a **strip mine,** in

FIGURE 21.9
An underground coal mine.
Photo by Larry Lee/West Light

FIGURE 21.10
Coal strip mine near Gillette, Wyoming. The upper half of a 30-meter-thick bed of subbituminous coal is shown here, overlain by 3 to 10 meters of overburden.

which the overburden is removed to expose coal at the surface (figure 21.10). When a strip of coal has been uncovered and removed, the resulting trench is filled in with the overburden from the adjoining strip.

Figure 21.11 shows the coalfields of the "lower 48" states. We will discuss three major regions: the Appalachian fields, the interior fields, and the far western fields.

The major coal-producing fields in the United States are the *Appalachian fields,* which stretch from Pennsylvania to Alabama and contain extensive beds of bituminous coal. The coals are mostly of Pennsylvanian age; a few are Mississippian and Permian. The coal beds, which thin westward, were included in the late Paleozoic folding and faulting of the Appalachian orogeny, so they are strongly deformed in the eastern part of the belt. Steeply dipping coal beds here are mostly mined underground. The folds are gentler to the west, where the coals can be extracted by both underground and strip mining. Northeastern Pennsylvania has some anthracite that resulted from intense folding.

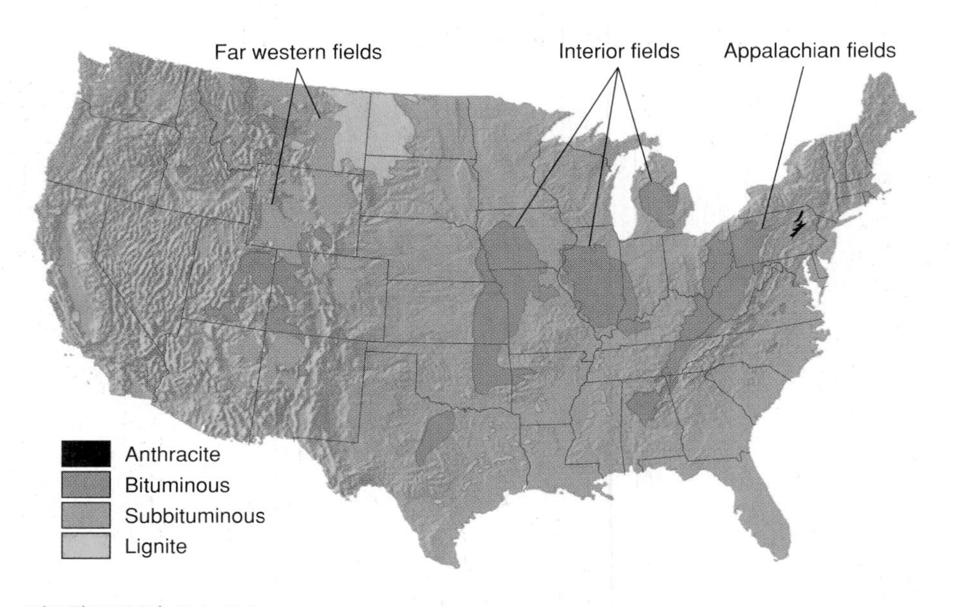

FIGURE 21.11
Coal fields of the United States. Alaska also has coal.
From U.S. Geological Survey

There are twenty-five to fifty coal beds over most of this region; each bed generally is 2 meters or less in thickness, although some are locally thicker. The coal lies within repeated sequences of sandstone, shale, and limestone, which indicate alternating continental and marine conditions. This implies a low-lying environment near the sea, such as lagoons, large deltas, and swampy coastal plains similar to those that exist in present-day Florida, Georgia, and South Carolina.

The *interior fields* extend from Michigan through Illinois to Texas and are extensions of the Appalachian rocks westward onto the continental interior. In Michigan and Illinois the rocks are preserved in large basins; the fields from Iowa to Texas are mostly horizontal. The coals are usually strip-mined, particularly near major industrial centers around the basin's edge in Illinois, Indiana, and Kentucky.

The *far western fields* extend from New Mexico northward through the Rocky Mountains to Montana and the Great Plains of North Dakota (and also into Canada). These coal beds are thicker and younger than eastern coals. They are generally of Cretaceous or Eocene age and are up to 30 meters thick. The coals occur in large basins and are generally of low rank, being either lignite or subbituminous coal, although some good-quality bituminous coals occur in Colorado and Utah. Many thick beds are very close to the surface and are strip-mined, although underground mining is common in some states. Western coals are attractive fuels, presently in very high demand, because they typically have less sulfur than eastern coals (sulfur compounds can pollute the air when coal burns).

Environmental Effects

Extracting and using coal creates environmental problems. The presence of a mine usually lowers the local water table as ground water is pumped out of the mine. The drainage out of the mines tends to be highly acidic, polluting surface streams and water supplies. In the past, strip mines have been refilled as barren, unsightly piles, but new techniques of recontouring the overburden and restoring the topsoil and vegetation help reclaim mined-out areas for other uses. When coal is burned, ash and sulfur gases can pollute air, but most of the harmful components can be removed with existing technology. When coal is burned, its carbon becomes carbon dioxide. CO_2 is a "greenhouse gas" and can contribute to global warming. Solving environmental problems associated with coal, of course, raises the immediate cost of extraction and the price to the consumer. Long-term costs may be *decreased,* however; preventing environmental damage is usually less expensive than cleaning it up.

Reserves and Resources

Coal production in the United States is about 1 billion tons per year (the United States uses 0.9 billion tons and exports the rest). Sixty percent of the production is from surface strip mines. Although Kentucky, West Virginia, and Pennsylvania still produce almost half the coal, the rapid development of the huge beds of low-sulfur western coal may soon cause the Northern Rockies states to surpass the Appalachian states.

Recoverable *reserves* in the United States are about 265 billion tons (40% of this is in Montana and Wyoming). As you can see, there are centuries of coal left at the present rate of production.

Coal *resources* within 1,000 meters of the surface are 1,700 billion tons. A greater amount is estimated to lie deeper, so the total United States resources are an impressive 3,900 billion tons (much of which, of course, is not presently usable). Even as we step up extraction for the export market, it seems obvious that our coal supplies will last a very long time. The United States has been called the Saudi Arabia of coal (there are also huge deposits in China, Australia, Ukraine, and Russia).

URANIUM

The metal *uranium,* which powers nuclear reactors, occurs as *pitchblende,* a black uranium oxide found in hydrothermal veins, or, much more commonly in the United States, as yellow *carnotite,* a complex hydrated oxide found as incrustations in sedimentary rocks. Ground water easily transports oxidized uranium, which is highly soluble. Organic matter reduces uranium, making it relatively insoluble, so uranium precipitates in association with organic matter.

Most of the easily recoverable uranium in the United States is found in sandstone in New Mexico, Utah, Colorado, and Wyoming, some of it in and near petrified wood. In the 1950s uranium boom, western prospectors looked for petrified logs and checked them with Geiger counters. Some individual logs contained tens of thousands of dollars worth of uranium. Some petrified logs have so much uranium that it would be dangerous to keep them as souvenirs. Most of the uranium, however, is in sandstone channels that contain plant fragments.

Organic phosphorite deposits of marine origin in Idaho and Florida also contain uranium. The uranium is not very concentrated, but the deposits are so large that overall they contain a substantial amount of uranium. The black Devonian shales of the eastern United States also contain uranium. These shales are really low-grade oil shales (figure 21.6), and they contain large amounts of natural gas, as you have seen. Uranium may be recovered from phosphorites or shales as a by-product or from another mining operation.

Uranium is used in nuclear reactors to produce electricity and in nuclear weapons and some naval craft. At present, nuclear reactors produce 5% of the energy needs of the United States. In France, 75% of its electricity is produced by nuclear power. Use of nuclear power may not rise appreciably in the future, depending on public acceptance. Nuclear plants produce long-lived waste products that remain dangerous for centuries, and no suitable storage site yet exists, although Yucca Mountain in Nevada is being intensively studied as a possible nuclear waste repository (see chapter 17). Accidents at the Three Mile Island reactor in Pennsylvania in 1979 and the Chernobyl reactor near Kiev in Ukraine in 1986 caused a major rethinking of the desirability of nuclear power. In 1999 the United States had 104 operational reactors. This total is down from 119 in 1991 and compares with 236 built or planned before the Three Mile Island accident. There have been no new plants ordered since 1978. The retreat from nuclear power may change as a result of the 2001 electricity crisis that began in California, where electricity supplies became inadequate ("rolling blackouts" became part of the language) and prices skyrocketed. Concerns about air pollution and global warming from fossil fuels also may cause Americans to rethink their opposition to nuclear power.

All of the United States's nuclear reactors use ^{235}U, which as we saw in the geologic time chapter, is much less abundant than ^{238}U (only 0.7% of uranium is ^{235}U). Half of France's nuclear reactors are breeder reactors, in which some of the neutrons given off during a chain reaction of ^{235}U bombard nuclei of ^{238}U, converting them into plutonium. Plutonium, like ^{235}U, is also capable of nuclear chain reactions. Thus, breeder reactors can greatly extend the amount of uranium that can be used as fuel.

Recoverable reserves of nearly 500,000 tons of uranium oxide in western sandstones seem adequate to power the existing reactors in the United States well into the century. Phosphorites and black shales contain millions of tons of low-grade uranium-oxide deposits as resources.

ALTERNATIVE SOURCES OF ENERGY

Several other sources may contribute enough energy in the future to help reduce the expected demand for oil, natural gas, coal, and uranium. *Hydroelectric power* contributes about 4% of the energy needs of the United States. Electricity is generated by turbines turned by water falling from dammed reservoirs. Hydroelectric power will probably not increase because most suitable rivers in the United States have already been dammed. Public pressure is growing to preserve most of the remaining undammed rivers in their wild state. *Geothermal power* (chapter 17) may contribute substantially to our power needs, particularly if successful techniques are developed for tapping the heat of areas not marked by surface hot springs. Currently, geothermal power contributes only 0.2% of our energy use. *Solar power* and *wind power* may contribute to our needs in the future, particularly if improved methods of storing the energy are devised. At the present, a great effort is being made to improve the technology for collecting solar energy. More exotic forms of generating energy include harnessing *tidal power, wave power, ocean current power,* and the energy represented by *vertical temperature differences in the sea.* The burning of *hydrogen from the dissociation of water* also may be developed; however, major technological problems are slowing the adoption of this method.

One great attraction of many of these sources of energy is that they are *renewable.* We do not use up sunlight, wind, or tides by harnessing their power. Widespread use of renewable resources in the 21st century may lessen the world's demand for fossil fuels.

METALS AND ORES

The successful search for metals depends on finding **ores,** which are naturally occurring materials that can be profitably mined. It is important to recognize that the local concentration of a metal must be greater (usually much greater) than its average crustal abundance to be a potential ore body. Metals must be concentrated in a particular place in a large enough amount to be viable ore bodies. Take gold in sea water. You could become fabulously wealthy if you could extract a fraction of the gold in sea water, because there are over 10^{11} troy ounces of gold—around 52 trillion dollars worth in the world's oceans. But the concentration is 4 grams per 1 million tons of water. It would cost you far more to remove that gold than you could sell it for.

Whether or not a mineral (or rock) is considered a metal ore depends on its chemical composition, the percentage of extractable metal, and the market value of the metal. The mineral hematite (Fe_2O_3), for example, is usually a good *iron ore* because it contains 70% iron by weight; this high percentage is profitable to extract at current prices for iron. Limonite ($Fe_2O_3 \bullet nH_2O$) contains less iron than hematite and hence is not as extensively mined. Even a mineral containing a high percentage of metal is not described as an *ore* if the metal is too difficult to extract or the site is too far from a market; profit is part of what defines an ore. As the prices of metals and the energy used to extract them fluctuate, so do the potential profits from minerals. Some of the common ore minerals are listed in table 21.2.

table 21.2 Common Ore Minerals

Metal	Ore Mineral	Composition
Aluminum	Bauxite (a mineral mixture)	$Al_2O_3 \bullet nH_2O$
Chromium	Chromite	$FeCr_2O_4$
Copper	Native copper	Cu
	Chalcocite	Cu_2S
	Chalcopyrite	$CuFeS_2$
Gold	Native gold	Au
Iron	Hematite	Fe_2O_3
	Magnetite	Fe_3O_4
Lead	Galena	PbS
Manganese	Pyrolusite	MnO_2
Mercury	Cinnabar	HgS
Nickel	Pentlandite	(Fe, Ni)S
Silver	Native silver	Ag
	Argentite	Ag_2S
Tin	Cassiterite	SnO_2
Uranium	Pitchblende	U_3O_8
	Carnotite	$K(UO_2)_2(VO_4)_2 \bullet 3H_2O$
Zinc	Sphalerite	ZnS

ORIGIN OF METALLIC ORE DEPOSITS

Table 21.3 summarizes most of the ways ore deposits form. Several of the processes have been discussed in earlier chapters.

table 21.3 Some Ways Ore Deposits Form

Type of Ore Deposit	Some Metals Found in This Type of Ore Deposit
Crystal settling within cooling magma	Chromium, platinum, iron
Hydrothermal deposits (contact metamorphism, hydrothermal veins, disseminated deposits, hot-spring deposits)	Copper, lead, zinc, gold, silver, iron, molybdenum, tungsten, tin, mercury, cobalt
Pegmatites	Lithium, rare metals
Chemical precipitation as sediment	Iron, manganese, copper
Placer deposits	Gold, tin, platinum, titanium
Concentration by weathering and ground water	Aluminum, nickel, copper, silver, uranium, iron, manganese, lead, tin, mercury

Ores Associated with Igneous Rocks

Crystal Settling

Crystal settling occurs as early-forming minerals crystallize and settle to the bottom of a cooling body of magma (figure 21.12). This process was described under differentiation in chapter 11. The metal chromium comes from chromite ore bod-

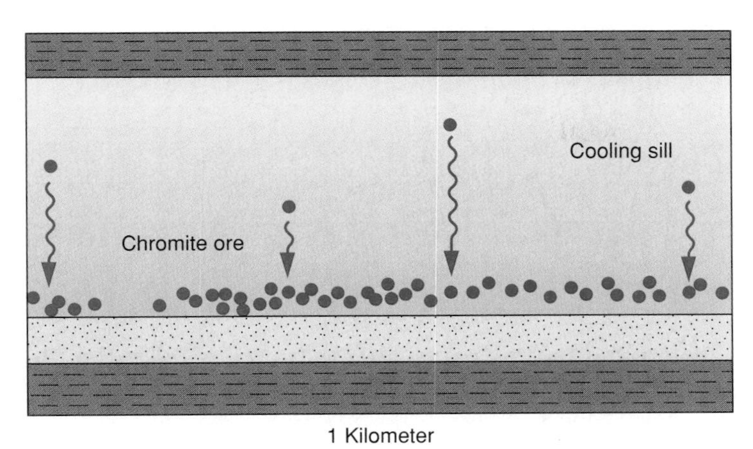

FIGURE 21.12

Early-forming minerals such as chromite may settle through magma to collect in layers near the bottom of a cooling sill.

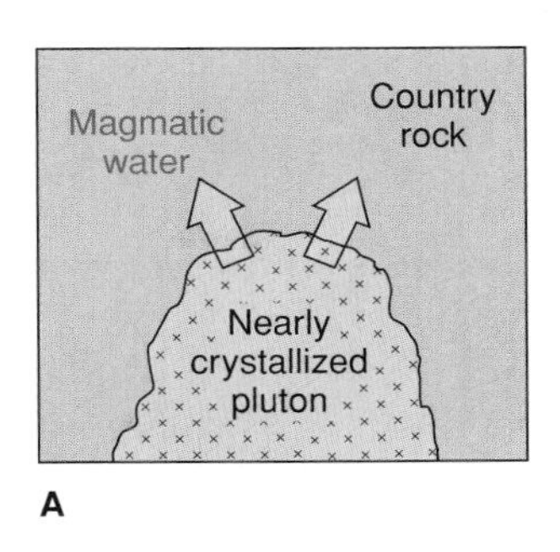

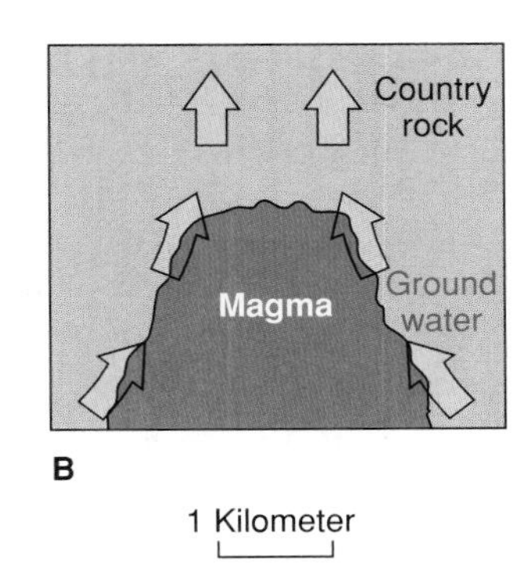

FIGURE 21.13

Two possible origins of hydrothermal fluids. (*A*) Residually concentrated magmatic water moves into country rock when magma is nearly all crystallized. (*B*) Ground water becomes heated by magma (or by a cooling solid pluton), and a convective circulation is set up.

ies near the base of sills and other intrusions. Most of the world's chromium and platinum come from a single intrusion, the huge Bushveldt Complex in South Africa. In Montana, another huge Precambrian sill called the Stillwater Complex contains similar but lower grade deposits of these two metals.

Hydrothermal Fluids

Hydrothermal fluids, discussed in chapter 15, are the most important source of metallic ore deposits. The hot water and other fluids are part of the magma itself, injected into the surrounding country rock during the last stages of magma crystallization (figure 21.13*A*; see also figure 15.21). Atoms of metals such as copper and gold, which do not fit into the growing crystals of feldspar and other minerals in the cooling pluton, are concentrated residually in the remaining water-rich magma. Eventually a hot solution, rich in metals and silica (quartz is the lowest-temperature mineral on Bowen's reaction series), moves into the country rock to create ore deposits. Most hydrothermal ores are metallic sulfides, often mixed with milky quartz. The origin of the sulfur is widely debated.

A magma body or hot rock may heat ground water and cause convection circulation. This water may mix with water given off from solidifying magma or it may leech metals from solid rock and deposit metallic minerals elsewhere as the water cools. However the hydrothermal solutions form, they tend to create four general types of hydrothermal ore deposits: (1) contact metamorphic deposits, (2) hydrothermal veins, (3) disseminated deposits, and (4) hot-spring deposits.

Contact metamorphism can create ores of iron, tungsten, copper, lead, zinc, silver, and other metals in country rock. The country rock may be completely or partially removed and replaced by ore (figure 21.14*A*). This is particularly true of limestone beds, which react readily with hydrothermal solutions. (The metasomatic addition of ions to country rock is described in chapter 15.) The ore bodies can be quite large and very rich.

Hydrothermal veins are narrow ore bodies formed along joints and faults (figure 21.14*B*). They can extend great distances from their apparent plutonic sources. Some extend so far that it is questionable whether they are even associated with plutons. The fluids can precipitate ore (and quartz) within cavities along the fractures and may also replace the wall rock of the fractures with ore. Hydrothermal veins form most of the world's great deposits of lead, zinc, silver, gold, tungsten, tin, mercury, and to some extent, copper.

Hot solutions can also form *disseminated deposits* in which metallic sulfide ore minerals are distributed in very low concentration through large volumes of rock, both above and within a pluton (figure 21.14*C* and box 15.4). Most of the world's copper comes from disseminated deposits (also called *porphyry copper deposits* because the associated pluton is usually porphyritic). Along with the copper are deposited many other metals, such as lead, zinc, molybdenum, silver, and gold (and iron, though not in commercial quantities).

Where hot solutions rise to Earth's surface, *hot springs* form. Hot springs on land may contain large amounts of dissolved metals. Some California hot springs contain so much mercury that the water is unfit to drink. More impressive are hot springs on the sea floor (figure 21.14*D*), which can precipitate large mounds of metallic sulfides, sometimes in commercial quantities. We will look at submarine hot springs later in the chapter in connection with plate tectonics.

Pegmatites (box 11.1) are another type of ore deposit associated with igneous rocks. They may contain important concentrations of minerals containing lithium, beryllium, and other rare metals, as well as gemstones such as emeralds and sapphires.

Ores Formed by Surface Processes

Chemical precipitation in layers is the most common origin for ores of iron and manganese. A few copper ores form in this way too. Banded iron ores, usually composed of alternating layers of iron minerals and chert, formed as sedimentary rocks in many

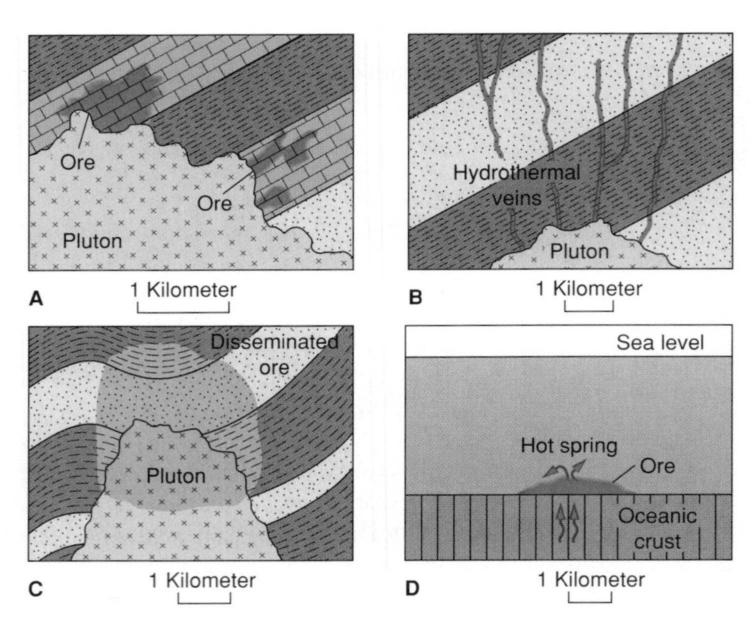

FIGURE 21.14

Hydrothermal ore deposits. (*A*) Contact metamorphism in which ore replaces limestone. (*B*) Ore emplaced in hydrothermal veins. (*C*) Disseminated ore within and above a pluton (porphyry copper deposits, for example). (*D*) Ore precipitated around a submarine hot spring (size of ore deposit exaggerated). (*E*) Hydrothermal veins of milky quartz in granite, northern California.

parts of the world during the Precambrian, apparently in shallow, water-filled basins (figure 21.15). Later folding, faulting, metamorphism, and solution have destroyed many of the original features of the ore, so the origin of the ore is difficult to interpret. The water may have been fresh or marine, and the iron may have come from volcanic activity or deep weathering of the surrounding continents. The alternating bands may have been created by some rhythmic variation in volcanic activity, river runoff, basin water circulation, growth of organisms, or some other factor. Since banded iron ores are all Precambrian, their origin might be connected to an ancient atmosphere or ocean chemically different from today's.

Placer deposits in which streams have concentrated heavy sediment grains in a river bar are described in chapter 16. Wave action can also form placers at beaches. Placers include gold nuggets and dust, native platinum, diamonds and other gemstones, and worn pebbles or sand grains composed of the heavy oxides of titanium and tin.

Ore deposits due to *concentration by weathering* were described in chapter 12. Aluminum (in bauxite) forms through weathering in tropical climates.

Another type of concentration by weathering is the *supergene enrichment* of disseminated ore deposits. Through supergene enrichment, low grade ores of 0.3% copper in rock can be enriched to a minable 1% copper. The major ore mineral in a disseminated copper deposit is chalcopyrite, a copper-iron sulfide containing about 35% copper. Near Earth's surface, downward-moving ground water can leach copper and sulfur from the ore, leaving the iron behind (figure 21.16). At or below the water table the dissolved copper can react with chalcopyrite in the lower part of the disseminated deposit, forming a richer ore mineral such as chalcocite, which is about 80% copper:

$3\,Cu^{++}$	+	$CuFeS_2$	→	$2\,Cu_2S$	+	Fe^{++}
Copper dissolved in ground water		Chalcopyrite		Chalcocite		Iron in solution

In this way copper is removed from the top of the deposit and added to the lower part. The ore below the water table may be several times richer than in the rest of the deposit (silver can move with the copper). The iron left behind at the surface forms a rusty cap called a *gossan,* or "iron hat," which is a visible clue to the ore below.

FIGURE 21.15

Banded iron ores of Precambrian age probably accumulated in shallow basins.

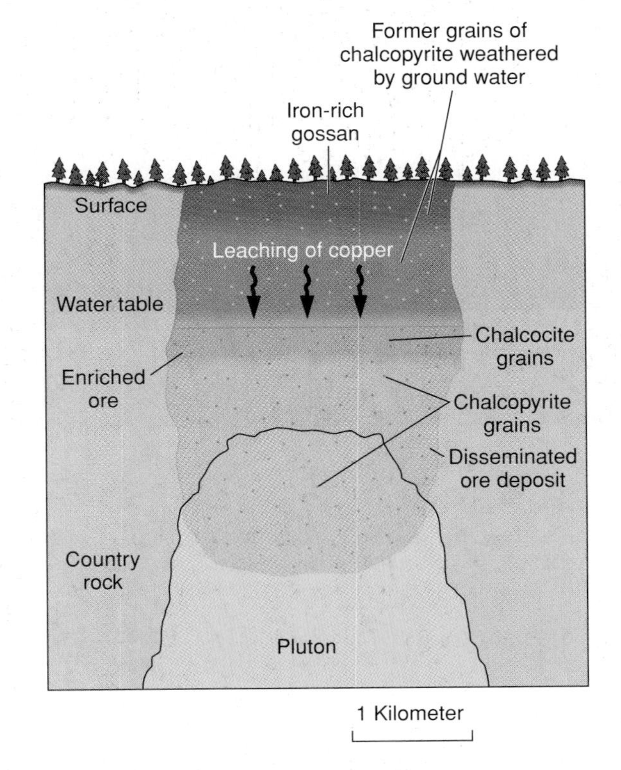

FIGURE 21.16

Supergene enrichment. Ground water leaches copper from upper part of disseminated deposit and precipitates it at or below the water table, forming rich ore.

METAL ORES AND PLATE TECTONICS

Plate tectonics has provided a model that is important to geologists searching for new metallic ore deposits. The relationships between plate tectonics and ore deposits were described in chapter 4. The following is a summary.

At *divergent* plate boundaries, metals form in rift valley hot springs due to the circulation of ground water, driven by magma. Nearly solid metal sulfide mounds precipitate from the hot water as it is cooled by sea water. The metals that accumulate in these mounds are predominantly iron, copper, and zinc, with smaller amounts of manganese, gold, and silver. Mining them is presently impractical because they are small and scattered on the sea floor.

Metallic ores that form at a divergent boundary are carried away from the ridge crest (figure 4.49) toward another plate boundary. Some *ophiolites* contain rich ore deposits that may have originally been metallic mounds. Ophiolites represent former crust and mantle created at spreading axes that has accreted onto continents (chapter 3). Some ophiolites also contain chromite ores within serpentinized ultramafic rock.

At *convergent* plate boundaries metallic ore deposits form in a variety of settings. Hot-spring deposits form on the flanks of andesitic volcanoes at *island arcs*. The ore sometimes collects as sedimentary layers in shallow basins above local magma bodies (figure 4.50). These usually contain more lead than the mounds that form along spreading axes. Rich *massive sulfide* ore deposits may have originally formed in this manner.

In western North America, metallic ore deposits tend to be distributed in broad belts that parallel past or present convergent plate boundaries. The type of ore deposit that forms seems to be related to the depth of subduction as described in chapter 4 (see figure 4.50). One model postulates that metallic minerals originally formed in a divergent plate boundary setting are carried down a subduction zone. Rock becomes hotter as it progresses down a subduction zone. The metals are remobilized and rise into the overlying continental crust. Metals that mobilized at relatively low temperatures rise closer to the plate boundary than those that mobilize at higher temperatures. Other metals may also be derived directly from the continental crust or underlying mantle. Rising magma and associated hot water circulation redistribute and concentrate the ores into veins and other hydrothermal deposits.

MINING

Mining can be carried out on Earth's surface or underground (figure 21.17). Two forms of surface mines are **strip mines,** used for mining some beds of coal, and **open-pit mines,** in which ore is exposed in a large excavation. Some geologists use *strip mine* and *open-pit mine* interchangeably; however, strip mines generally expose coal or another resource in a long band that is later filled in, while open-pit mines are roughly circular and are excavated to extract huge amounts of material. They are not usually refilled. **Placer mines** are surface mines in which valuable sediment grains are extracted from stream bar or beach deposits.

Environmental Effects

Some of the environmental problems associated with mining can be partially solved if care is taken. For example, in the past *waste rock* (or *tailings*) was routinely left in unsightly

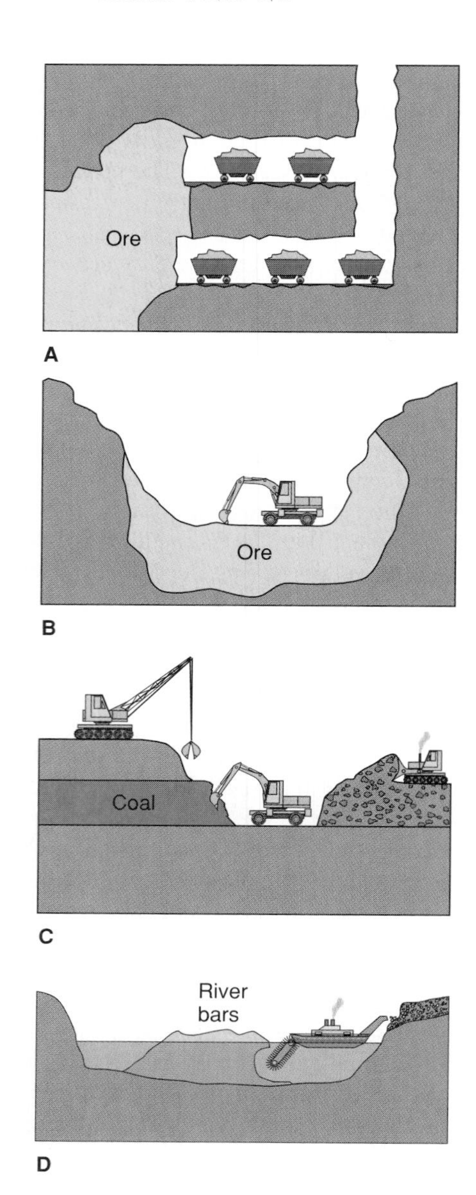

FIGURE 21.17

Types of mines: (*A*) Underground. (*B*) Open pit. (*C*) Strip. (*D*) Placer (being mined by a floating dredge).

FIGURE 21.18

(*A*) Waste rock piles from coal mining in the 1950s, Sweetwater County, Wyoming. (*B*) Subsidence caused by the collapse of an underground coal mine, Acme, Wyoming. Mining continued from 1900 to 1943 (photo taken in 1976).

Photo *A* by H. E. Maide, U.S. Geological Survey; photo *B* by C. R. Dunrud, U.S. Geological Survey

heaps and dumps (figure 21.18*A*). The excavations for strip mines and placer mines can be filled in with waste rock, leveled or graded, and then covered over with topsoil to restore the land to usable condition. In some cases crops can be grown on reclaimed land within 1 year after mining operations are completed. Open pits, being larger, are rarely filled in, for the filling cannot be done gradually while mining is in progress. Underground mines are sometimes back-filled with waste rock to prevent land *subsidence* after ore is removed. Figure 21.18*B* shows extensive subsidence caused by mine collapse.

One of the more difficult problems to deal with is *acid drainage* from mines, caused by ground water running or being pumped out of a mine. Sulfide ore minerals (table 21.2) and pyrite (FeS_2) are most often the source of the trouble. Ground water transports oxygen to the sulfides where they are oxidized to iron oxide and sulfuric acid. In some mines, expensive programs of holding and neutralizing drainage water in ponds or artificial wetlands prevent pollution of surface streams and harm to forests and wildlife. The worst problem is with long-abandoned mines that are still draining acid waters. Many of these may never be neutralized.

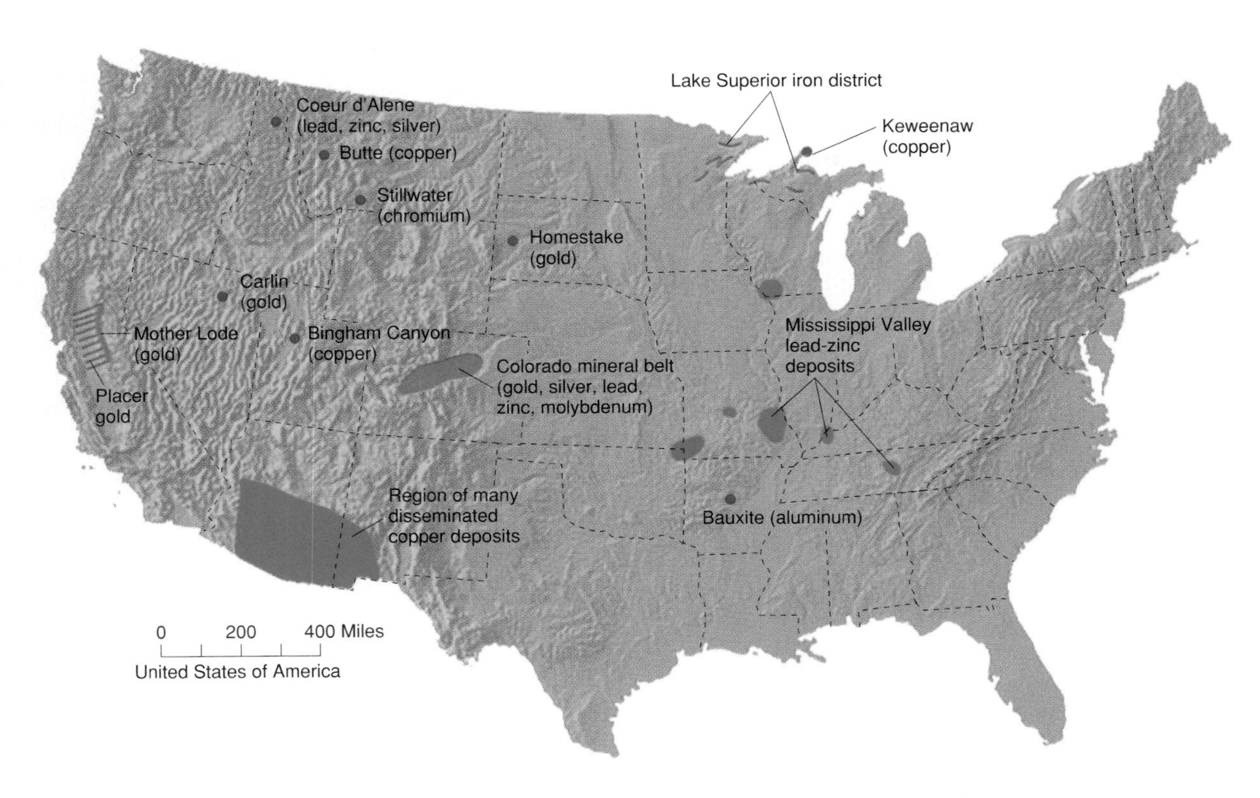

FIGURE 21.19
Some metallic ore localities mentioned in the text.

SOME IMPORTANT METALS

Iron

The basis for any modern, industrialized economy is *iron,* which is used to make steel. Iron and steel are used in a huge variety of products, from cast-iron frying pans to locomotives.

The major iron ore minerals are hematite and magnetite. Most of the iron in the United States comes from the region around Lake Superior (figure 21.19), most notably, Minnesota. The ores are banded iron ores of Precambrian age, typical of most iron ores in the world. Mining is done mostly by open-pit methods.

Copper

Less abundant is *copper,* another important metal for industry. More than half the copper used in the United States goes into electrical wire and equipment and one-third into the manufacture of brass, a copper-zinc alloy.

Most copper ores are sulfides. Chalcopyrite is the most important copper ore mineral. Some vein deposits of copper exist (as at Butte, Montana), but most major deposits are disseminated through large volumes of rock; so most copper mines are open pits (figure 21.20). Arizona, Utah, and a few other

FIGURE 21.20
Open-pit copper mine, Morenci, Arizona.

western states are the major producers in the United States (figure 21.19). The concentration of copper averages about 0.5% in most currently worked deposits; that is, 1 kg of copper is recovered for every 200 kg of rock processed.

Aluminum

Widely used in the United States, *aluminum* is consumed in the manufacture of beer and soft drink cans, airplanes, electrical cable, and many other products. The use of aluminum is increasing rapidly.

The ore of aluminum is bauxite, which forms under tropical weathering conditions. The United States has very little bauxite, so it imports 90% of its aluminum ore from tropical countries. The largest aluminum mine in the United States is in Arkansas (figure 21.19). Open-pit mining is the usual technique for extracting bauxite.

Lead

The main use of *lead* (79%) is in batteries. Substantial amounts are recycled, largely from automobile batteries.

The most important ore of lead is galena. Major deposits occur in Missouri, Idaho, Utah, and Colorado (figure 21.19). The Missouri deposits are mostly found in limestone beds; their precise origin is a matter of some controversy. The ore is mined both underground and from open pits. Deposits in Idaho occur mostly as veins and are usually mined underground.

Zinc

Widely used in industry, *zinc* is necessary for galvanizing and the manufacture of brass and other alloys.

The major zinc ore is sphalerite. As sphalerite usually is found closely associated with galena, most lead mines also extract zinc. Zinc occurs without lead in some areas, however.

Silver

Coins, tableware, jewelry, photographic film, and many other products are made of *silver.*

Silver, found as a native metal and in sulfide ores, is a common by-product of lead and copper mining. The lead-zinc mines of Idaho (the Coeur d'Alene district) are the largest silver producers in the United States (figure 21.19).

Gold

The rare and valuable metal *gold* is used in coins, jewelry, decoration, dentistry, electronics, and the space program. Gold bars are stored to back national currency, although this use is rapidly disappearing.

Gold is found as a native element in the form of nuggets and grains. In some parts of the world these are concentrated in placer deposits (California's Gold Rush of 1849 was triggered by discoveries of placer gold). Gold nuggets, flakes, and dust can be separated from the other sediments by (1) *panning;* (2) *sluice boxes* (figure 21.21), which catch the heavy gold on the bottom of a box as gravel is washed through it; (3) *hydraulic mining* (figure 21.22), which washes gold-bearing gravel from a hillside into a sluice box; or (4) floating *dredges* (figure 21.23), which separate gold from gravel aboard a large barge, piling the spent gravel behind. When gold is found in hydrothermal veins associated with milky quartz, as it is in parts of Colorado and in California's Mother Lode (figure 21.19), it is mined underground. The largest underground gold mine in the United States is the Homestake Mine in South Dakota, where finely disseminated gold is extracted from folded metamorphic rock. A large amount of disseminated gold is mined in open pits near Carlin, Nevada, and at several other localities in Nevada and California.

FIGURE 21.21
Sluice box used to separate gold from gravel, Alaska.
Photo by D. J. Miller, U.S. Geological Survey

FIGURE 21.22
Hydraulic mining for gold, Alaska.
Photo by T. L. Péwé, U.S. Geological Survey

FIGURE 21.23
A gold dredge separates gold from gravel.

The world has large reserves of iron and aluminum, moderate reserves of copper, lead, and zinc, and scanty reserves of gold and silver.

Other Metals

Many other metals are vital to our economy. *Chromium, nickel, cobalt, manganese, molybdenum, tungsten,* and *vanadium* are all important in the steel industry, particularly in the manufacture of specialty products such as stainless steel and armor plate. Most of these metals have other uses as well. *Tin* is used in solder and for plating steel in "tin cans." *Mercury* is used in thermometers, silent electrical switches, medical compounds, and batteries. *Magnesium* is used in aircraft. *Titanium,* as strong as steel but weighing half as much, is used in aircraft. *Platinum* is used in catalytic converters to clean automobile exhaust.

NONMETALLIC RESOURCES

Nonmetallic resources are Earth resources that are not mined to extract a metal or as a source of energy. Most rocks and minerals contain metals, but when nonmetallic resources are mined it is usually to use the rock (or mineral) as is (for example, using gravel and sand for construction projects), whereas metallic ores are processed to extract metal. With the exception of the gemstones such as diamonds and rubies, nonmetallic resources do not have the glamour of many metals or energy resources. Nonmetallic resources are generally inexpensive and are needed in large quantities (again, except for gemstones); however, their value exceeds that of all mined metals. The large demand and low unit price means that these resources are best taken from local sources. Transportation over long distances would add significantly to the cost.

Construction Materials

Sand and *gravel* are both needed for the manufacture of concrete for building and highway construction. Sand is also used in mortar, which holds bricks and cement blocks together. The demand for sand and gravel in the United States has more than doubled in the last 25 years. Sand dunes, river channel and bar deposits, glacial outwash, and beach deposits are common sources for sand and gravel. Cinder cones are mined for "gravel" in some areas. Sand and gravel are ordinarily mined in open pits (figure 21.24).

Stone refers to rock used in blocks to construct buildings or crushed to form roadbed. Most stone in buildings is limestone or granite, and most crushed stone is limestone. Huge quantities of stone are used each year in the United States. Stone is removed from open pits called *quarries* (figure 21.25).

Limestone has many uses other than building stone or crushed roadbed. Cement, used in concrete and mortar, is made from limestone. Pulverized limestone is in demand as a soil conditioner and is the principal ingredient of many chemical products.

Fertilizers and Evaporites

Fertilizers (phosphate, nitrate, and potassium compounds) are extremely important to agriculture today, so much so that they are one of the few nonmetallic resources transported across the sea. *Phosphate* is produced from phosphorite, a sedimentary rock

FIGURE 21.24
Sand and gravel pit in a glacial esker, near Saranac Lake, New York.
Courtesy Ward's Natural Science Est., Inc., Rochester, NY

formed by the accumulation and alteration of the remains of marine organisms. Major phosphate deposits in the United States are in Idaho, Wyoming, and Florida. *Nitrate* can form directly as an evaporite deposit but today is usually made from atmospheric nitrogen. *Potassium* compounds are often found as evaporites.

Rock salt is coarsely crystalline halite formed as an evaporite. Salt beds are mined underground in Ohio and Michigan; underground salt domes are mined in Texas and Louisiana. (Some salt is also extracted from sea water by evaporation.) Rock salt is used in many ways—de-icing roads in winter, preserving food, as table salt, and in manufacturing hydrochloric acid and sodium compounds for baking soda, soap, and other products. Rock salt is heavily used by industry.

Gypsum forms as an evaporite. Beds of gypsum are mined in many states, notably California, Michigan, Iowa, and Texas. Gypsum, the essential ingredient of plaster and wallboard (Sheetrock), is used mainly by the construction industry, although there are other uses.

Sulfur occurs in bright-yellow deposits of elemental sulfur. Most of its commercial production comes from the cap rock of salt domes. Sulfur is widely used in agriculture as a fungicide and fertilizer and by industry to manufacture sulfuric acid, matches, and many other products.

Other Nonmetallics

Gemstones (called *gems* when cut and polished) include precious stones such as diamonds, rubies, emeralds, and sapphires and semiprecious stones such as beryl, garnet, jade, spinel, topaz, turquoise, and zircon. Gems are used for jewelry, bearings, and abrasives (most are above 7 on Mohs' scale of hardness). Diamond drills and diamond saws are used to drill and cut rock. Old watches and other instruments often have hard gems at bearing points of friction ("17-jewel watches"). Gemstones are often found in pegmatites or in close association with other igneous intrusives. Some are recovered from placer deposits.

Asbestos is a fibrous variety of serpentine or chain silicate minerals. The fibers can be separated and woven into fireproof fabric used for firefighters' clothes and theater curtains. It is also used in manufacturing ceiling and sound insulation, shingles, and brake linings, although the use of asbestos is being rapidly curtailed because of concern about its connection with lung cancer (box 9.3). The United States produces little asbestos, mostly from belts of serpentinized ultramafic rocks in the Appalachians and the Pacific Coast states. Large amounts are mined in Canada, chiefly in Quebec. *Talc,* used in talcum powder and other products, is often found associated with asbestos (see box 5.1).

Other nonmetallic resources are important. *Mica* is used in electrical insulators. *Barite* ($BaSO_4$), because of its high specific gravity, is used to make heavy drilling mud to prevent oil gushers. *Borates* are boron-containing evaporites used in fiberglass, cleaning compounds, and ceramics. *Fluorite* (CaF_2) is used in toothpaste, Teflon finishes, and steel-smelting. *Clays* are used in ceramics, manufacturing paper, and as filters and absorbents. *Diatomite* is used in swimming pool filters and to filter out yeast in beer and wine. *Glass sand,* which is over 95% quartz, is the main component of glass. *Graphite* is used in foundries, lubricants, steel-making, batteries, and pencil "lead."

FIGURE 21.25

A limestone quarry in northern Illinois. The horizon marks the original land surface before the rock was removed.

SOME FUTURE TRENDS

Ocean mining, now rather uncommon, will increase in the near future. Mining of *manganese nodules* from the deep sea floor (see chapter 3) could provide the United States with substantial amounts of copper and nickel, together with far more manganese than United States industry can consume. The copper content of many nodules is four to five times higher than most deposits on land. *Metallic brines and deposits* of the Red Sea type may be a source of several metals in the future.

Several tools are of great help in mineral exploration on land. Highly sophisticated *geochemical tests* of soils and soil gases point to ore bodies underground. *Geophysical techniques* continue to be refined for resource exploration. Satellites and space-shuttle crews photograph Earth's surface in many different wavelengths of energy, and careful analysis of satellite imagery is proving to be of great help in prospecting. The economic returns from the satellite photography program will far outweigh its costs.

IN GREATER DEPTH 21.2

Substitutes, Recycling, and Conservation

Substitutes for many geologic resources now exist, and others will be found. Aluminum is replacing the more expensive copper in many electrical uses, particularly in transmission lines. Glass fibers also are replacing copper for telephone lines. Aluminum has largely replaced tin-coated steel for beverage containers. Cotton and wool use could increase, replacing polyester and other petroleum-based synthetic fibers in clothes.

Suitable substitutes, however, seem unlikely for some resources. Nothing yet developed can take the place of steel in bridges or mercury in thermometers, although electronic thermometers are being more widely used. Cobalt is vital for strong, permanent magnets. Although substitutes may help prolong the life of the supplies of some resources, they are not available for many others.

Recycling helps augment the supply of some resources such as aluminum, gold, silver, lead, plastics, and glass. No resource, however, receives even half its supply from recycling. Increased volunteer recycling on the part of the public and waste reclaiming of urban trash could increase these percentages. New ore will always be needed, however, because some uses of products prevent the material from being recyclable. A steel can that rusts away beside a road will never be recycled. The iron oxides are scattered in low percentage in the soil and can never be recovered. Many resources, such as petroleum and coal, are consumed during use and cannot be recycled. Some people argue against recycling when newly derived materials are cheaper. This is a shortsighted view and neglects the fact that there is a finite supply of all nonrenewable resources in Earth's crust. The more we conserve now, the longer our reserves will last. By squandering resources now we may be depriving future generations of resources.

Conservation of scarce resources is extremely important. The United States' use of oil from 1979 to 1985 declined as a result of conservation efforts such as the 55-mile-per-hour speed limit, more fuel-efficient automobiles, the upgrading of insulation in buildings, and the elimination of unneeded heating and lighting. But in 1986 oil use rose sharply as its price declined. Politically caused shortages and gluts of oil can obscure the fact that the supply of oil in the United States is severely limited. There will be difficult times ahead as the United States' domestic oil becomes scarcer; therefore, the need for conserving fuel should be repeatedly stressed. The price of gasoline will rise substantially as the United States becomes increasingly dependent on foreign oil. Conservation of metals, particularly those imported in large quantities, will become increasingly important. Smaller automobiles and more durable appliances can help conserve metals.

As the relation of plate tectonics to the distribution of resources such as metals and petroleum becomes clearer, selecting areas for exploring for these materials should become more sophisticated.

The Human Perspective

There is a tendency for humans to take a one-sided view toward geologic resources and the problems created by the extraction and transportation of those resources. The conflicts are between (1) maintaining or increasing our standard of living and raising the quality of life not only for ourselves, but for poverty-stricken people throughout the world; (2) maintaining the environment; and (3) making sure that we do not deprive future generations of the resources that sustain us.

The extreme position for each of the three concerns could be stated as follows: (1) Extreme exploiter: "Let's mine what we can and make ourselves as rich as possible now. Technological breakthroughs will provide for future generations. Damage to the environment due to extraction is insignificant (or it's where it doesn't bother me)." (2) Extreme environmentalist: "Any mine or oil well does environmental damage and therefore should not be permitted. We can maintain our lifestyles by recycling or by leading less technologically dependent lifestyles." (3) Extreme conservationist: "Let's not mine anything now because there are many future generations that need to rely on these resources."

You probably agree that none of the three extreme positions is reasonable. The middle ground among the three is where almost everyone thinks we should be. The challenge is deciding where in the middle ground. Should we lean toward more exploitation and away from environmental concerns in order that underdeveloped countries can raise their standards of living? Should we minimize mining and energy consumption so as to reduce any impact on air pollution or wildlife? Our hope is that we can at least understand the perspective of those who may disagree to strike a balance and try to deal with each case with enlightenment. Your understanding of geology is an important step in your being able to help resolve moral dilemmas that we face in which there is no ideal solution.

Summary

Geologic resources include energy resources, metals, and nonmetallic resources. All are nonrenewable, except ground water.

Reserves are known deposits that can be legally and economically recovered now—the short-term supply. *Resources* include reserves as well as other known and undiscovered deposits that may be economically extracted in the future.

Petroleum (*oil* and *natural gas*) supplies almost two-thirds of the energy used by the United States.

The occurrence of oil and natural gas is limited to regions having these conditions together: *source rocks, reservoir rocks, traps,* such as anticlines, faults, stratigraphic traps, and salt domes, and *thermal maturity* from burial.

Natural gas, *heavy crude, oil sand,* and *oil shale* may all help replace liquid petroleum in the future. Most of these resources are in western North America.

The United States has huge *coal* reserves, enough for centuries of use at the present rate. Coal, now used mostly for generating electricity, will probably be used more widely in the future as oil runs out.

The United States has ample *uranium* for its reactor program, mostly in sandstones in western states.

Metallic ores, which can be profitably mined, are often associated with igneous rocks, particularly their hydrothermal fluids, which can form *contact metamorphic deposits, hydrothermal veins, disseminated deposits,* and submarine *hot-spring deposits.* Iron occurs in sedimentary layers, and aluminum ores form from weathering.

Metallic ores form from hot springs at divergent plate boundaries, on the flanks of island arcs, and in belts on the edges of continents above subduction zones.

Ores are mined *underground* and also at Earth's surface in *strip mines, open-pit mines,* and *placer mines.*

Metals are vital to an industrial economy, particularly *iron* for steel production and *copper* for electrical equipment.

Nonmetallic resources such as *sand and gravel* and *limestone* for crushed rock and cement are used in huge quantities. *Fertilizers, rock salt, gypsum, sulfur,* and *clays* are also widely used.

Substitutes, recycling, and conservation can help cut consumption of some resources but will not eliminate the need for finding new deposits.

Deep-ocean mining and increasingly sophisticated exploration techniques will help supply some of our future resource needs.

Terms to Remember

crude oil 522
geologic resources 520
heavy crude 525
natural gas 522
nonrenewable resource 520
oil field 523
oil pool 522
oil sand 525
oil shale 525
oil trap 522
open-pit mine 533
ore 530
petroleum 522
placer mine 533
reserves 521
reservoir rock 522
resources 521
source rock 522
strip mine 527

Testing Your Knowledge

Use the questions below to prepare for exams based on this chapter.

1. Name the three major classes of geologic resources. Give four examples of each class.
2. Discuss the United States's supplies and potential use of natural gas, heavy crude, oil sand, oil shale, and uranium.
3. List in decreasing order of use the energy resources used in the United States. Discuss possible future trends in this ranking.
4. What geologic conditions are necessary for the accumulation of oil and natural gas?
5. Differentiate between *reserves* and *resources.* Can reserves be increased? Can resources be increased?
6. Compare oil reserves with coal reserves. What might this indicate for the future use of each?
7. Describe several ways in which ore deposits form. Which are the most important?
8. Describe four ways in which resources are mined.
9. Discuss the environmental effects of oil extraction and coal mining.
10. Discuss common uses for iron, copper, lead, zinc, and aluminum.
11. Describe the potential of substitutes, conservation, recycling, and deep-ocean mining for meeting an increasing need for geologic resources.
12. All of the following are nonrenewable resources except
    a. ground water b. oil
    c. coal d. iron
13. The major source of energy of the United States is
    a. natural gas b. coal
    c. oil d. nuclear power

14. Which is not a type of coal?
    a. peat b. lignite
    c. bituminous d. anthracite
15. Which metal would most likely be found in an ore deposit formed by crystal settling?
    a. copper b. gold
    c. silver d. chromium
16. Which metal would not be found in hydrothermal veins?
    a. aluminum b. lead
    c. zinc d. silver
    e. gold
17. Metal ore deposits have been found at all these tectonic settings except
    a. mid-oceanic ridges b. island arcs
    c. subduction zones d. mantle plumes
18. The main use of lead is in
    a. coins b. gasoline
    c. batteries d. pencils
19. What factors can increase reserves of Earth resources (choose all that apply)
    a. extraction of the resource b. new discoveries
    c. price controls d. new mining technology
20. The largest use of sand and gravel is
    a. glassmaking b. extraction of quartz
    c. construction d. ceramics
21. Oil accumulates when the following conditions are met (choose all that apply)
    a. source rock where oil forms b. permeable reservoir rock
    c. impermeable oil trap d. shallow burial
22. Coal forms
    a. by crystal settling
    b. through hydrothermal processes
    c. by compaction of plant material
    d. on the ocean floor

Expanding Your Knowledge

1. Many underdeveloped countries would like to have the standard of living enjoyed by the United States, which has 6% of the world population and uses 15% to 40% of the world production of many resources. As these countries become industrialized, what happens to the world demand for geologic resources? Where will these needed resources come from?
2. If driven 12,000 miles per year, how many more gallons of gasoline per year does a sports utility vehicle or pickup truck rated at 12 miles per gallon use than a minicompact car rated at 52 mpg? Over 5 years how much more does it cost to buy gasoline at $2 per gallon for the low-mileage car? At $5 per gallon (the price in many European countries)?

Exploring Web Resources

www.mhhe.com/plummer9e

Visit the Online Learning Center for the answers to the Testing Your Knowledge section, for more interactive quizzing, and for additional articles and media resources. Click on the direct links to go to the websites listed below.

www.microserve.net/~doug/

Mine Net. This is a gateway to mining information from around the world.

www.klws.com/gold/gold.html

Gold prospecting site. This site is mainly aimed for the amateur gold prospector, but contains facts and information on gold.

www.NRCan.gc.ca/

Natural Resources Canada. Use this site to get information on Canada's mineral and energy resources.

http://minerals.usgs.gov/

U.S. Geological Survey Mineral Resources Program. Provides current information on occurrence, quality, quantity, and availability of mineral resources.

www.eia.doe.gov/

U.S. Energy Information Administration. Provides data, analysis, and forecasts of energy and issues related to energy.

Appendix A

Identification of Minerals

Each mineral is identified by a unique set of physical or chemical properties. To determine some of these properties requires specialized equipment and techniques. Most common minerals, however, can be distinguished from one another by tests involving simple observations. Cleavage is an especially useful property. If cleavage is present, you should determine the number of cleavage directions, estimate the angles between cleavage directions, and note the quality of each direction of cleavage. Other easily performed tests and observations check for hardness (abbreviated H), luster, and color, and determining crystal form (if present). A simple chemical test can be made using dilute hydrochloric acid to see if the mineral effervesces.

The identification tables included here can be used to identify the most common minerals (the rock-forming minerals) and some of the most common ore minerals. For identifying less common minerals, refer to one of the books or websites on mineralogy listed at the end of chapter 9. Mineral identification takes practice, and you will probably want to verify your mineral identifications with a geology instructor.

Because the common rock-forming minerals are the ones you are most likely to encounter, we have included a simple key for identifying them. The key is based on first determining whether or not the mineral is harder than glass and then checking other properties that should lead to identification of the mineral. You should verify your identification by seeing whether other properties of your sample correspond to those listed for the mineral in table A.1.

Ore minerals are usually distinctive enough that a key is unnecessary. To identify an ore mineral, read through table A.2 and determine which set of properties best fits the unknown mineral.

Key for Identifying Common Rock-Forming Minerals

Determine whether a fresh surface of the mineral is harder or softer than glass. If you can scratch the mineral with a knife blade, the mineral is softer than glass.

I. Harder than glass—knife will not scratch mineral. (If softer than glass, go to II.)

  A. Determine if cleavage is present or absent (this may require careful examination). If cleavage is absent proceed to 1.; if cleavage is present, proceed to B.

    1. Vitreous luster
      a. Olive green or brown—*olivine*
      b. Reddish brown or in equidimensional crystals with twelve or more faces—*garnet*
      c. Usually light-colored or clear—*quartz*
    2. Metallic luster
      a. Bright yellow—*pyrite*
    3. Greasy or waxy luster
      a. Mottled green and black—*serpentine*

  B. Cleavage present. Determine the number of directions of cleavage in an individual grain or crystal.

    1. Two directions, good, at or near 90°—*feldspar*
      a. If striations are visible on cleavage surfaces—*plagioclase*
      b. If pink or salmon-colored—*potassium feldspar* (or *orthoclase*)
      c. If white or light gray without striations, it could be either type of feldspar
    2. Two directions, fair, at 90°
      a. Dark green to black—*pyroxene* (usually augite)
    3. Two directions, excellent, not near 90°
      a. Dark green to black—*amphibole* (usually hornblende)

II. Softer than glass—knife scratches mineral

  A. No cleavage detectable

    1. Earthy luster, in masses too fine to distinguish individual grains—*clay group* (for instance, *kaolinite*)

  B. Cleavage present

    1. One direction
      a. Perfect cleavage in flexible sheets—*mica:*
        Clear or white—*muscovite mica*
        Black or dark brown—*biotite mica*
    2. Three directions
      a. All three perfect and at 90° to each other (cubic cleavage)—*halite*
      b. All three perfect and not near 90° to each other:
        If effervesces in dilute acid—*calcite*
        If effervesces in dilute acid only after being pulverized—*dolomite*

table A.1 Diagnostic Properties of the Common Rock-Forming Minerals
Mineral groups are in uppercase letters.

Name (mineral groups shown in capitals)	Chemical Composition	Chemical Group	Diagnostic Properties	Other Properties
AMPHIBOLE (A mineral group in which *hornblende* is the most common member.)	$XSi_8O_{22}(OH)_2$ (*X* is a combination of Ca, Na, Fe, Mg, Al)	Chain silicate	2 good cleavage directions at 60° (120°) to each other.	H = 5–6 (barely scratches glass). Hornblende is dark green to black; tends to form in needle-like or elongate crystals; vitreous luster.
Augite (see Pyroxene)				
Biotite (see Mica)				
Calcite	$CaCO_3$	Carbonate	3 excellent cleavage directions, *not* at right angles (they define a rhombohedron). H = 3. Effervesces vigorously in weak acid.	Usually white, gray, or colorless; vitreous luster. Clear crystals show double refraction.
CLAY MINERALS (*Kaolinite* is a common example of this large mineral group.)	Compositions include $XSi_4O_{10}(OH)_8$ (*X* is Al, Mg, Fe, Ca, Na, K)	Sheet silicate	Generally microscopic crystals. Masses of clay minerals are softer than fingernail. Earthy luster. Clay-like smell when damp.	Seen as a chemical weathering product of feldspars and most other silicate minerals. A constituent of most soils.
Dolomite	$CaMg(CO_3)_2$	Carbonate	Identical to calcite (rhombohedral cleavage, H = 3) except effervesces in weak acid only when pulverized.	Usually white, gray, or colorless. Vitreous luster.
FELDSPAR (Most common group of minerals.) The group includes:	Framework	Framework silicates	H = 6 (scratch glass). 2 good cleavage directions at about 90° to each other.	Vitreous luster but surface may be weathered to clay, giving an earthy luster. Perfect crystal, shaped like an elongated box.
Potassium feldspar (orthoclase)	$KAlSi_3O_8$		White, pink, or salmon-colored.	Never has striations on cleavage surfaces.
Plagioclase (sodium and calcium feldspar)	Mixture of: $CaAl_2Si_2O_8$ and $NaAlSi_3O_8$		White, light to dark gray, rarely other colors. *May* have striations on cleavage surfaces.	Calcium-rich varieties generally a darker gray and may show an iridescent play of colors.
GARNET	$XAl_2Si_3O_{12}$ (*X* is a combination of Ca, Mg, Fe, Al, Mn)	Isolated silicate	No cleavage. Usually reddish brown. Tends to occur in perfect equidimensional crystals, usually 12 sided. H = 7.	Rarely yellow, green, or black. Usually found in metamorphic rocks. Vitreous luster.
Gypsum	$CaSO_4{\cdot}2H_2O$	Sulfate	H = 2. 1 good and 2 perfect cleavage directions. Vitreous or silky luster.	Clear, white, or pastel colors. Flexible cleavage fragments.

table A.1 Diagnostic Properties of the Common Rock-Forming Minerals Mineral groups are in uppercase letters. (*continued*)

Name (mineral groups shown in capitals)	Chemical Composition	Chemical Group	Diagnostic Properties	Other Properties
Halite	NaCl	Halide	3 excellent cleavage directions at 90° to each other (cubic). H = 2 1/2. Salty taste. Soluble in water.	Usually clear or white.
Hematite (*see* Ore mineral table)				
Hornblende (*see* Amphibole)				
Kaolinite (*see* Clay)				
MICA	K(X)($AlSi_3O_{10}$)$(OH)_2$	Sheet silicate	1 perfect cleavage direction (splits easily into flexible sheets).	H = 2–3. Vitreous luster.
The group includes:				
Biotite	(*X* is Mg, Fe, and Al)		Black or dark brown.	
Muscovite	(*X* is Al)		White or transparent.	
Olivine	X_2SiO_4 (*X* is Fe, Mg)	Isolated silicate	No cleavage. Generally olive green or brown. H = 6–7 (scratches glass). Vitreous luster.	Usually as small grains in mafic or ultramafic igneous rocks.
Orthoclase (*see* Feldspar)				
Plagioclase (*see* Feldspar)				
Pyrite ("fool's gold")	FeS_2	Sulfide	H = 6 (scratches glass). Bright, yellow, metallic luster. Black streak.	Commonly occurs as perfect crystals: cubes or crystals with five-sided faces. Weathers to brown.
PYROXENE (A mineral group; *Augite* is most common member.)	$XSiO_3$ (*X* is Fe, Mg, Al, Ca)	Chain silicate	2 fair cleavage directions at 90° to each other.	H = 6. Augite is dark green to black. Vitreous luster; usually stubby crystals.
Quartz	SiO_2	Framework silicate	H = 7. No cleavage. Vitreous luster. Does not weather to clay.	Almost any color but commonly white or clear. Good crystals have six-sided "column" with complex "pyramid" on top.
Serpentine	$Mg_6Si_4O_{10}(OH)_8$	Sheet silicate	Hardness variable but softer than glass. Mottled green and black. Greasy luster. Fractures along smooth curved surfaces.	Sometimes fibrous (asbestos).

table A.2 Diagnostic Properties of the Most Common Ore Minerals

Name	Chemical Composition	Diagnostic Properties	Other Properties
Azurite	$Cu_3(CO_3)_2(OH)_2$	Azure blue; effervesces in weak acid.	H = 3–4.
Bauxite	$Al_2O_3 \cdot nH_2O$	Earthy luster. A variety of clay. Generally pea-sized spheres included in a fine-grained mass.	
Bornite	Cu_3FeS_4	Metallic luster, tarnishes to iridescent purple color.	Gray streak; H = 3 (softer than glass).
Chalcopyrite	$CuFeS_2$	Metallic luster, brass-yellow. Softer than glass.	Black streak.
Cinnabar	HgS	Scarlet red, bright red streak.	Softer than glass. Generally an earthy luster.
Galena	PbS	Metallic luster, gray; 3 directions of cleavage at 90° (cubic). High specific gravity.	Softer than glass; gray streak.
Gold	Au	Metallic luster, yellow. H = 3 (softer than glass, can be pounded into thin sheets, easily deformed).	Yellow streak; high specific gravity.
Halite	$NaCl$	Salty taste; 3 cleavage directions at 90° (cubic).	Clear or white; easily soluble in water.
Hematite	Fe_2O_3	Red-brown streak.	Either in earthy reddish masses or in metallic, silver-colored flakes or crystals.
Limonite	$Fe_2O_3 \cdot nH_2O$	Earthy luster; yellow-brown streak.	Yellow to brown color; softer than glass.
Magnetite	Fe_3O_4	Metallic luster, black; magnetic.	Harder than glass; black streak.
Malachite	$Cu_2(CO_3)(OH)_2$	Bright-green color and streak.	Softer than glass; effervesces in weak acid.
Sphalerite	ZnS	Brown to yellow color; 6 directions of cleavage.	Lusterlike resin; yellow or cream-colored streak; softer than glass.
Talc	$Mg_3Si_4O_{10}(OH)_2$	White, gray, or green; softer than fingernail (H = 1).	Greasy feel.

Appendix B

Identification of Rocks

Igneous Rocks

Igneous rocks are classified on the basis of texture and composition. For some rocks, texture alone suffices for naming the rock. For most igneous rocks, composition as well as texture must be taken into account. Ideally, the mineral content of the rock should be used to determine composition; but for fine-grained igneous rocks, accurate identification of minerals may require a polarizing microscope or other special equipment. In the absence of such equipment, we rely on the color of fine-grained rocks and assume the color is indicative of the minerals present.

To identify a common igneous rock, use either table 11.1 or follow the key given below.

Key for Identifying Common Igneous Rocks

I. What is the texture of the rock?

A. Is it glassy (a very vitreous luster)? If so, it is *obsidian,* regardless of its chemical composition. Obsidian exhibits a pronounced conchoidal fracture.

B. Does it have a frothy appearance? If so, it is *pumice.* Pumice is light in weight and feels abrasive (it probably will float on water).

C. Does it have angular fragments of rock embedded in a volcanic-derived matrix? If so, it is a *volcanic breccia.* If the precise nature of the rock fragments and matrix can be identified, modifiers may be used; for instance, the rock may be an *andesite* breccia or a *rhyolite* breccia.

D. Is the rock composed of interlocking, very coarse-grained minerals? (The minerals should be more than 1 centimeter across.) If so, the rock is a *pegmatite.* Most pegmatites are mineralogically equivalent to granite, with feldspars and quartz being the predominant minerals.

E. Is the rock entirely coarse-grained? (That is, does it have an interlocking crystalline texture in which nearly all grains are more than 1 mm across?) If so, go to part II of this key.

F. Is the rock *entirely* fine-grained? (Are grains less than 1 mm across or too fine to distinguish with the naked eye?) If so, go to part III of this key.

G. Is the matrix fine-grained with some coarse-grained minerals visible in the rock? If so, go to part III and add the adjective *porphyritic* to the name of the rock.

II. Igneous rocks composed of interlocking coarse-grained minerals.

A. Is quartz present? If so, the rock is a *granite.*
Confirmation: Granite should be composed predominantly of feldspar—generally white, light gray, or pink (indicating high amounts of potassium or sodium in the feldspar). Rarely are there more than 20% ferromagnesian minerals in a granite.

B. Are quartz and feldspar absent? If so, the rock should be composed entirely of ferromagnesian minerals and is *ultramafic.*
Confirmation: Identify the minerals as being olivine or pyroxene (or less commonly, amphibole or biotite).

C. Does the rock have less than 50% feldspar and no quartz? If so, the rock should be a *gabbro.*
Confirmation: Most of the rock should be ferromagnesian minerals. Plagioclase can be medium or dark gray. There would be no pink feldspars.

D. Is the rock composed of 30% to 60% feldspar (and no quartz)? If so, the rock is a *diorite.*
Confirmation: Feldspar (plagioclase) is usually white to medium gray but never pink.

III. Igneous rocks that are fine-grained.

A. Can quartz be identified in the rock? If so, the rock is a *rhyolite.*

B. If the rock is too fine-grained for you to determine whether quartz is present but is white, light gray, pink, or pale green, the rock is most likely a *rhyolite.*

C. Is the rock composed predominantly of ferromagnesian minerals? If so, the rock is *basalt.*

D. If the rock is too fine-grained to identify ferromagnesian minerals but is black or dark gray, the rock is probably a *basalt.*

1. Does the rock have rounded holes in it? If so, it is a *vesicular basalt* or *scoria.*

E. Is the rock composed of roughly equal amounts of white or gray feldspar and ferromagnesian minerals (but no quartz)? If so, the rock is an *andesite.*
Confirmation: Most andesite is porphyritic, with numerous identifiable crystals of white or light gray feldspar and lesser amounts of hornblende crystals within the darker, fine-grained matrix. Andesite is usually medium to dark gray or green.

Sedimentary Rocks

The following key shows how sedimentary rocks are classified on the basis of texture and composition. The descriptions of the rocks in the main body of the text provide additional information, such as common rock colors. *Equipment* needed for identification of sedimentary rocks includes a bottle of dilute hydrochloric acid, a hand lens or magnifying glass, a millimeter scale, a glass plate for hardness tests, and a pocketknife or rock hammer.

Begin by testing the rock for carbonate minerals by applying a small amount of dilute hydrochloric acid to the surface of the rock.

1. The rock does not effervesce (fizz) in acid, or effervesces weakly, but when powdered by a knife or hammer, the powder effervesces strongly. If so, the rock is *dolomite.*
2. The rock does not effervesce at all, even when powdered, or effervesces only in some places, such as the cement between grains. Go to part I of this key.

3. The rock effervesces strongly. The rock is *limestone.* Go to part II of this key to determine limestone type.

I. With a hand lens or magnifying glass, determine if the rock has a clastic texture (grains cemented together) or a crystalline texture (visible, interlocking crystals).
   A. If clastic:
      1. Most grains are more than 2 mm in diameter.
         a. Angular grains *sedimentary breccia.*
         b. Rounded grains—*conglomerate.*
      2. Most grains are between 1/16 and 2 mm in diameter. Rock feels gritty to the fingers. *Sandstone.*
         a. More than 90% of the grains are quartz—*quartz sandstone.*
         b. More than 25% of the grains are feldspar—*arkose.*
         c. More than 25% of the grains are fine-grained rock fragments, such as shale, slate, and basalt—*lithic sandstone.*
         d. More than 15% of the rock is fine-grained matrix—*graywacke.*
      3. Rock is fine-grained (grains less than 1/16 mm in diameter). Feels smooth to fingers.
         a. Grains visible with a hand lens—*siltstone.*
         b. Grains too small to see, even with a hand lens.
            1. Rock is laminated, fissile—*shale.*
            2. Rock is unlayered, blocky—*mudstone.*
   B. If crystalline:
      1. Crystals fine to coarse, hardness of 2—*rock gypsum.*
      2. Coarse crystals that dissolve in water—*rock salt.*
   C. Hard to determine if clastic or crystalline:
      1. Very fine-grained, smooth to touch, conchoidal fracture, hardness of 6 (scratches glass), nonporous—*chert* (*flint* if dark)
      2. Very fine-grained, smooth to touch, breaks into flat chips—*shale.*
      3. Black or dark brown, readily broken, soils fingers—*coal.*

II. *Limestone* may be clastic or crystalline, fine- or coarse-grained, and may or may not contain visible fossils. Usually gray, tan, buff, or white. Some distinctive varieties are:
   A. *Bioclastic limestone*—clastic texture, grains are whole or broken fossils. Two relatively rare varieties are:
      1. *Coquina*—very coarse, recognizable shells, much open pore space.
      2. *Chalk*—very fine-grained, white or tan, soft and powdery.
   B. *Oolitic limestone*—grains are small spheres (less than 2 mm in diameter), all about the same size.
   C. *Travertine*—coarsely crystalline, no pore space, often contains different-colored layers (bands).

Metamorphic Rocks

The characteristics of a metamorphic rock are largely governed by (1) the composition of the parent rock and (2) the particular combination of temperature, confining pressure, and directed pressure. These factors cause different textures in rocks formed under different sets of conditions. For this reason, texture is usually the main basis for naming a metamorphic rock. Determining the composition (e.g., mineral content) is necessary for naming some rocks (e.g., *quartzite*), but for others, the minerals present are used as adjectives to describe the rock completely (e.g., *biotite* schist).

Metamorphic rocks are identified by determining first whether the rock has a *foliated* or *nonfoliated* texture.

Key for Identifying Metamorphic Rocks

I. If the rock is *nonfoliated,* then it is identified on the basis of its mineral content.
   A. Does the rock consist of mostly quartz? If so, the rock is a *quartzite.* A quartzite has a mosaic texture of interlocking grains of quartz and will easily scratch glass.
   B. Is the rock composed of interlocking coarse grains of calcite? If so, it is *marble.* (The individual grains should exhibit rhombohedral cleavage; the rock is softer than glass.)
   C. Is the rock a dense, dark mass of grains mostly too fine to identify with the naked eye? If so, it probably is a *hornfels.* A hornfels may have a few larger crystals of uncommon minerals enclosed in the fine-grained mass.

II. If the rock is *foliated,* determine the type of foliation and then, if possible, identify the minerals present.
   A. Is the rock very fine-grained and does it split into sheetlike slabs? If so, it is *slate.* Most slate is composed of extremely fine-grained sheet silicate minerals, and the rock has an earthy luster.
   B. Does the rock have a silky sheen but otherwise appear similar to slate? If so, it is a *phyllite.*
   C. Is the rock composed mostly of visible grains of platy or needlelike minerals that are approximately parallel to one another? If so, the rock is a *schist.* If the rock is composed mainly of mica, it is a *mica schist.* If it also contains garnet, it is called a *garnet-mica schist.* If hornblende is the predominant mineral, the rock is a *hornblende schist.* If talc prevails, it is a *talc schist* (sometimes called soapstone).
   D. Are dark and light minerals found in separate lenses or layers? If so, the rock is a *gneiss.* The light layers are composed of feldspars and perhaps quartz, whereas the darker layers commonly are formed of biotite, amphibole, or pyroxene. A gneiss may appear similar to granite or diorite but can be distinguished from the igneous rocks by the foliation.

Appendix C

The Elements Most Significant to Geology

table C.1

Atomic Number	Name	Symbol	Atomic Weight	Some Usual Charge of Ions
1	Hydrogen	H	1.0	+1
2	Helium	He	4.0	0 inert
3	Lithium	Li	6.9	+1
4	Beryllium	Be	9.0	+2
5	Boron	B	10.8	+3
6	Carbon	C	12.0	+4
7	Nitrogen	N	14.0	+5
8	Oxygen	O	16.0	−2
9	Fluorine	F	19.0	−1
10	Neon	Ne	20.2	0 inert
11	Sodium	Na	23.0	+1
12	Magnesium	Mg	24.3	+2
13	Aluminum	Al	27.0	+3
14	Silicon	Si	28.1	+4
15	Phosphorus	P	31.0	+5
16	Sulfur	S	32.1	−2
17	Chlorine	Cl	35.5	−1
18	Argon	Ar	39.9	0 inert
19	Potassium	K	39.1	+1
20	Calcium	Ca	40.1	+2
22	Titanium	Ti	47.9	+4
23	Vanadium	V	50.9	
24	Chromium	Cr	52.0	
25	Manganese	Mn	54.9	+4, +3
26	Iron	Fe	55.8	+2, +3
27	Cobalt	Co	58.9	
28	Nickel	Ni	58.7	+2
29	Copper	Cu	63.5	+2
30	Zinc	Zn	65.4	+2
33	Arsenic	As	74.9	+3
35	Bromine	Br	79.9	—
37	Rubidium	Rb	85.5	+1
38	Strontium	Sr	87.3	+2
40	Zirconium	Zr	91.2	—
42	Molybdenum	Mo	95.9	+4
47	Silver	Ag	107.9	+1
48	Cadmium	Cd	112.4	—
50	Tin	Sn	118.7	+4
51	Antimony	Sb	121.8	+3
52	Tellurium	Te	127.6	—
55	Cesium	Cs	132.9	—
56	Barium	Ba	137.4	+2
60	Neodymium	Nd	144	+3
62	Samarium	Sm	150	+3
74	Tungsten	W	183.9	—
78	Platinum	Pt	195.2	—
79	Gold	Au	197.0	—
80	Mercury	Hg	200.6	+2
82	Lead	Pb	207.2	+2
83	Bismuth	Bi	209.0	—
86	Radon	Rn	222	0 inert
88	Radium	Ra	226.1	
90	Thorium	Th	232.1	—
92	Uranium	U	238.1	—
94	Plutonium	Pu	239.0	—

Appendix D

Periodic Table of Elements

Legend: 6 / C / 12.01 — Atomic number / Symbol / Atomic weight

Representative elements; Noble gases; Transition metals; Metalloids; Lanthanides; Actinides

1 Group IA	2 IIA	3 IIIB	4 IVB	5 VB	6 VIB	7 VIIB	8 VIIIB	9 VIIIB	10 VIIIB	11 IB	12 IIB	13 IIIA	14 IVA	15 VA	16 VIA	17 VIIA	18 VIIIA
1 **H** 1.008																	2 **He** 4.00
3 **Li** 6.94	4 **Be** 9.01											5 **B** 10.81	6 **C** 12.01	7 **N** 14.01	8 **O** 16.00	9 **F** 19.00	10 **Ne** 20.18
11 **Na** 22.99	12 **Mg** 24.31											13 **Al** 26.98	14 **Si** 28.09	15 **P** 30.97	16 **S** 32.06	17 **Cl** 35.45	18 **Ar** 39.95
19 **K** 39.10	20 **Ca** 40.08	21 **Sc** 44.96	22 **Ti** 47.90	23 **V** 50.94	24 **Cr** 52.00	25 **Mn** 54.94	26 **Fe** 55.85	27 **Co** 58.93	28 **Ni** 58.69	29 **Cu** 63.54	30 **Zn** 65.37	31 **Ga** 69.72	32 **Ge** 72.61	33 **As** 74.92	34 **Se** 78.96	35 **Br** 79.91	36 **Kr** 83.80
37 **Rb** 85.47	38 **Sr** 87.62	39 **Y** 88.91	40 **Zr** 91.22	41 **Nb** 92.91	42 **Mo** 95.94	43 **Tc** (98)	44 **Ru** 101.07	45 **Rh** 102.90	46 **Pd** 106.42	47 **Ag** 107.87	48 **Cd** 112.41	49 **In** 114.82	50 **Sn** 118.69	51 **Sb** 121.75	52 **Te** 127.60	53 **I** 126.90	54 **Xe** 131.29
55 **Cs** 132.91	56 **Ba** 137.34	57 **La** 138.91	72 **Hf** 178.49	73 **Ta** 180.95	74 **W** 183.85	75 **Re** 186.21	76 **Os** 190.2	77 **Ir** 192.22	78 **Pt** 195.08	79 **Au** 196.97	80 **Hg** 200.59	81 **Tl** 204.37	82 **Pb** 207.19	83 **Bi** 208.98	84 **Po** (209)	85 **At** (210)	86 **Rn** (222)
87 **Fr** (223)	88 **Ra** 226.03	89 **Ac** 227.03	104 **Rf** (261)	105 **Ha** (262)	106 **Sg** (263)	107 **Ns** (262)	108 **Hs** (265)	109 **Mt** (266)									

Transition metals: groups 3–12

58 **Ce** 140.12	59 **Pr** 140.91	60 **Nd** 144.24	61 **Pm** (145)	62 **Sm** 150.36	63 **Eu** 151.96	64 **Gd** 157.25	65 **Tb** 158.92	66 **Dy** 162.50	67 **Ho** 164.93	68 **Er** 167.26	69 **Tm** 168.93	70 **Yb** 173.04	71 **Lu** 174.97
90 **Th** 232.04	91 **Pa** 231.04	92 **U** 238.03	93 **Np** (237)	94 **Pu** (244)	95 **Am** (243)	96 **Cm** (247)	97 **Bk** (247)	98 **Cf** (251)	99 **Es** (252)	100 **Fm** (257)	101 **Md** (258)	102 **No** (259)	103 **Lr** (260)

Elements with an atomic number greater than 92 are not naturally occurring.

Appendix E

Selected Conversion Factors

table E.1

	English Unit	Conversion Factor	Metric Unit	Conversion Factor	English Unit
Length and Distance	inch (in)	2.54	centimeters (cm)	0.4	inch (in)
	foot (ft)	0.3048	meter (m)	3.28	feet (ft)
	inch (in)	0.026	meter (m)	39.4	inches (in)
	mile, statute (mi)	1.61	kilometers (km)	0.62	mile (mi)
Area	square inch (in^2)	6.45	square centimeters (cm^2)	0.16	square inch (in^2)
	square foot (ft^2)	0.093	square meter (m^2)	10.8	square feet (ft^2)
	square mile (mi^2)	2.59	square kilometers (km^2)	0.39	square mile (mi^2)
	acre	0.4	hectare	2.47	acres
Volume	cubic inch (in^3)	16.4	cubic centimeters (cm^3)	0.06	cubic inch (in^3)
	cubic yard (yd^3)	0.76	cubic meter (m^3)	1.3	cubic yards (yd^3)
	cubic foot (ft^3)	0.0283	cubic meter (m^3)	35.5	cubic feet (ft^3)
	quart (qt)	0.95	liter	1.06	quarts (qt)
Weight	ounce (oz)	28.3	grams (g)	0.04	ounce (oz)
	pound (lb)	0.45	kilogram (kg)	2.2	pounds (lb)
	ton, short (2,000 lb)	907	kilograms (kg)	0.001	ton, short
	ton, short	0.91	ton, metric	1.1	ton, short
Temp.	degrees Fahrenheit (°F)	$-32° \times 5/9$	degrees Celsius (°C) (centigrade)	$\times 1.8 + 32°$	degrees Fahrenheit (°F)

Appendix F

Rock Symbols

Shown below are the rock symbols used in the text. In general, these symbols are used by all geologists, although they sometimes are modified slightly.

Breccia

Conglomerate

Sandstone

Rock salt

Granite

Any igneous rock

Shale

Limestone

Dolomite

Basalt flows

Folded schist or gneiss

Appendix G

Commonly Used Prefixes, Suffixes, and Roots

abyss deep (Greek)
alluvium deposited by flowing water (Latin)
anti- opposite (Greek)
archea (archaeo)- ancient (Greek)
astheno- weak, lack of strength (Greek)
ceno recent (Greek)
circum- about, around, round about (Latin)
clast broken (Greek)
-cline tilted, gradient (Greek)
de- lower, reduce, take away (Latin)
dis- separation, opposite of (Latin)
ex- out of, away from (Greek)
feld field (Swedish, German)
folium leaf (Latin)
geo- Earth (Greek)
glomero- cluster (Latin)
hydro- water (Greek)
iso- equal (Greek)
-lith stone or rock (Greek)
meso- middle (Greek)
meta- change (Greek)
-morph form, shape (Greek)
paleo- ancient (Greek)
ped- foot (Latin)
pelagic pertaining to the ocean (Greek)
petro- stone or rock (Greek)
phanero- visible, evident (Greek)
pheno- large, conspicuous ("to show" in Greek)
pluto- deep-seated (from Roman god of the underworld or infernal regions)
pre- before, in front (Latin)
proto- first, primary, primitive (Greek)
pyro- fire (Greek)
spar crystalline material (German)
-sphere ball (Greek)
stria small groove, streak, band (Latin)
sub- under, less than (Latin)
super- above, more than, in addition to (Latin)
syn- together, at the same time (Latin, Greek)
tecto- means building or constructing in Greek and Latin; in geology it means movement or structures caused by internal forces.
terra, terre pertaining to the Earth (Latin)
thermal pertaining to heat (Greek)
trans- over, beyond, through, across (Latin)
xeno- strange, foreign (Greek)
zoo, zoic- animal (Greek)

Glossary

A

aa A lava flow that solidifies with a spiny, rubbly surface.

ablation The loss of the glacial ice or snow by melting, evaporation, or breaking off into icebergs. (Also called *wastage*.)

abrasion The grinding away of rock by friction and impact during transportation.

absolute age Age given in years or some other unit of time. (Also known as *numerical age.*)

abyssal fan Great fan-shaped deposit of sediment on the deep-sea floor at the base of many submarine canyons.

abyssal plain Very flat sediment-covered region of the deep-sea floor, usually at the base of the continental rise.

accreted terrane Terrane that did not form at its present site on a continent.

accretionary wedge (subduction complex) A wedge of thrust-faulted and folded sediment scraped off a subducting plate by the overlying plate.

accumulation stage Stage in the evolution of major mountain belts characterized by the accumulation of great thicknesses (several kilometers) of sedimentary or volcanic rocks.

active continental margin A margin consisting of a continental shelf, a continental slope, and an oceanic trench.

actualism The principle that the same processes and natural laws that operated in the past are those we can actually observe or infer from observations as operating at present. Under present usage, *uniformitarianism* has the same meaning as actualism for most geologists.

advancing glacier Glacier with a positive budget, so that accumulation results in the lower edges being pushed outward and downward.

aftershock Small earthquake that follows a main shock.

A horizon The top layer of soil, characterized by the downward movement of water; also called *zone of leaching.*

alkali soil Soil containing such a great quantity of sodium salts precipitated by evaporating ground water that it is generally unfit for plant growth.

alluvial fan Large fan-shaped pile of sediment that usually forms where a stream's velocity decreases as it emerges from a narrow canyon onto a flat plain at the foot of a mountain range.

alpine glaciation Glaciation of a mountainous area.

amphibole group Mineral group in which all members are double-chain silicates.

amphibolite Amphibole (hornblende), plagioclase schist.

andesite Fine-grained igneous rock of intermediate composition. Up to half of the rock is plagioclase feldspar with the rest being ferromagnesian minerals.

angle of dip A vertical angle measured downward from the horizontal plane to an inclined plane.

angular Sharp-edged; lacking rounded edges or corners.

angular unconformity An unconformity in which younger strata overlie an erosion surface on tilted or folded layered rock.

anorthosite A crystalline rock composed almost entirely of calcium-rich plagioclase feldspar.

antecedent stream A stream that maintains its original course despite later deformation of the land.

anticline An arched fold in which the rock layers usually dip away from the axis of the fold.

aquifer A body of saturated rock or sediment through which water can move readily.

arch (sea arch) Bridge of rock left above an opening eroded in a headland by waves.

Archean Eon The oldest eon of Earth's history.

arête A sharp ridge that separates adjacent glacially carved valleys.

arid region An area with less than 25 cm of rain per year.

arkose A sandstone in which more than 25% of the grains are feldspar.

artesian aquifer *See* confined aquifer.

artesian well A well in which water rises above the aquifer.

artificial recharge Ground-water recharge increased by engineering techniques.

aseismic ridge Submarine ridge with which no earthquakes are associated.

ash (volcanic) Fine pyroclasts (less than 4 mm).

assimilation The process in which very hot magma melts country rock and assimilates the newly molten material.

asthenosphere A region of Earth's outer shell beneath the lithosphere. The asthenosphere is of indeterminate thickness and behaves plastically.

atoll A circular reef surrounding a deeper lagoon.

atom Smallest possible particle of an element that retains the properties of that element.

atomic mass number The total number of neutrons and protons in an atom.

atomic number The total number of protons in an atom.

atomic weight The sum of the weight of the subatomic particles in an average atom of an element, given in atomic mass units.

augite Mineral of the pyroxene group found in mafic igneous rocks.

aulacogen *See* failed rift.

aureole Zone of contact metamorphism adjacent to a pluton.

axial plane A plane containing all of the hinge lines of a fold.

axis *See* hinge line.

B

backarc spreading A type of sea-floor spreading that moves an island arc away from a continent, or tears an island arc in two, or splits the edge of a continent, in each case forming new sea floor.

backshore Upper part of the beach, landward of the high-water line.

bajada A broad, gently sloping, depositional surface formed at the base of a mountain range in a dry region by the coalescing of individual alluvial fans.

bar A ridge of sediment, usually sand or gravel, that has been deposited in the middle or along the banks of a stream by a decrease in stream velocity.

barchan A crescent-shaped dune with the horns of the crescent pointing downwind.

barrier island Ridge of sand paralleling the shoreline and extending above sea level.

barrier reef A reef separated from the shoreline by the deeper water of a lagoon.

basal sliding Movement in which the entire glacier slides along as a single body on its base over the underlying rock.

basalt A fine-grained, mafic, igneous rock composed predominantly of ferromagnesian minerals and with lesser amounts of calcium-rich plagioclase feldspar.

base level A theoretical downward limit for stream erosion of Earth's surface.

batholith A large discordant pluton with an outcropping area greater than 100 square kilometers.

bauxite The principal ore of aluminum; $Al_2O_3 \cdot nH_2O$.

baymouth bar A ridge of sediment that cuts a bay off from the ocean.

beach Strip of sediment, usually sand but sometimes pebbles, boulders, or mud, that extends from the low-water line inland to a cliff or zone of permanent vegetation.

beach face The section of the beach exposed to wave action.

bedding An arrangement of layers or beds of rock.

bedding plane A nearly flat surface separating two beds of sedimentary rock.

bed load Heavy or large sediment particles in a stream that travel near or on the stream bed.

bedrock Solid rock that underlies soil.

Benioff zone Distinct earthquake zone that begins at an oceanic trench and slopes landward and downward into Earth at an angle of about 30° to 60°.

bergschrund The crevasse that develops where a glacier is pulling away from a cirque wall.

berm Platform of wave-deposited sediment that is flat or slopes slightly landward.

B horizon A soil layer characterized by the accumulation of material leached downward from the A horizon above; also called *zone of accumulation.*

biochemical Precipitated by the action of organisms.

bioclastic limestone A limestone consisting of fragments of shells, corals, and algae.

biotite Iron/magnesium-bearing mica.

block Large angular pyroclast.

blowout A depression on the land surface caused by wind erosion.

body wave Seismic wave that travels through Earth's interior.

bomb Large spindle- or lens-shaped pyroclast.

bonding Attachment of an atom to one or more adjacent atoms.

bottomset bed A delta deposit formed from the finest silt and clay, which are carried far out to sea by river flow or by sediments sliding downhill on the sea floor.

boulder A sediment particle with a diameter greater than 256 mm.

Bowen's reaction series The sequence in which minerals crystallize from a cooling basaltic magma.

braided stream A stream that flows in a network of many interconnected rivulets around numerous bars.

breaker A wave that has become so steep that the crest of the wave topples forward, moving faster than the main body of the wave.

breakwater An offshore structure built to absorb the force of large breaking waves and provide quiet water near shore.

brittle strain Cracking or rupturing of a body under stress.

butte A narrow pinnacle of resistant rock with a flat top and very steep sides.

C

calcareous Containing calcium carbonate.

calcite Mineral with the formula $CaCO_3$.

caldera A volcanic depression much larger than the original crater.

capacity (of stream) The total load that a stream can carry.

capillary action The drawing of water upward into small openings as a result of surface tension.

capillary fringe A thin zone near the water table in which capillary action causes water to rise above the zone of saturation.

carbonaceous chondrite Stony meteorite containing chondrules and composed mostly of serpentine and large quantities of organic materials.

carbonic acid H_2CO_3, a weak acid common in rain and surface waters.

cave (cavern) Naturally formed underground chamber.

cement The solid material that precipitates in the pore space of sediments, binding the grains together to form solid rock.

cementation The chemical precipitation of material in the spaces between sediment grains, binding the grains together into a hard rock.

Cenozoic Era The most recent of the eras; followed the Mesozoic Era.

chain silicate structure Silicate structure in which two of each tetrahedron's oxygen ions are shared with adjacent tetrahedrons, resulting in a chain of tetrahedrons.

chalk A very fine-grained bioclastic limestone.

channel (*Mars*) Feature on the surface of the planet Mars that very closely resembles certain types of stream channels on Earth.

chaotic terrain (*Mars*) Patch of jumbled and broken angular slabs and blocks on the surface of Mars.

chemical sedimentary rock A rock composed of material precipitated directly from solution.

chemical weathering The decomposition of rock resulting from exposure to water and atmospheric gases.

chert A hard, compact, fine-grained sedimentary rock formed almost entirely of silica.

chill zone In an intrusion, the finer-grained rock adjacent to a contact with country rock.

chondrule Round silicate grain within some stony meteorites.

C horizon A soil layer composed of incompletely weathered parent material.

cinder (volcanic) Pyroclast approximately the size of a sand grain. Sometimes defined as between 4 and 32 millimeters in diameter.

cinder cone A volcano constructed of loose rock fragments ejected from a central vent.

circum-Pacific belt Major belt around the edge of the Pacific Ocean on which most composite volcanoes are located and where many earthquakes occur.

cirque A steep-sided, amphitheater-like hollow carved into a mountain at the head of a glacial valley.

clastic sedimentary rock A sedimentary rock composed of fragments of preexisting rock.

clastic texture An arrangement of rock fragments bound into a rigid network by cement.

clay Sediment composed of particles with diameter less than 1/256 mm.

clay mineral A hydrous aluminum-silicate that occurs as a platy grain of microscopic size with a sheet silicate structure.

clay mineral group Collective term for several clay minerals.

cleavage The ability of a mineral to break along preferred planes.

coal A sedimentary rock formed from the consolidation of plant material. It is rich in carbon, usually black, and burns readily.

coarse-grained rock Rock in which most of the grains are larger than 1 millimeter (igneous) or 2 millimeters (sedimentary).

coast The land near the sea, including the beach and a strip of land inland from the beach.

coastal straightening The gradual straightening of an irregular shoreline by wave erosion of headlands and wave deposition in bays.

cobble A sediment particle with a diameter of 64 to 256 mm.

column A dripstone feature formed when a stalactite growing downward and a stalagmite growing upward meet and join.

columnar structure Volcanic rock in parallel, usually vertical columns, mostly six-sided; also called *columnar jointing.*

comet Small object in space, no more than a few kilometers in diameter, composed of frozen methane, frozen ammonia, and water ice, with small solid particles and dust imbedded in the ices.

compaction A loss in overall volume and pore space of a rock as the particles are packed closer together by the weight of overlying material.

competence The largest particle that a stream can carry.

composite volcano (stratovolcano) A volcano constructed of alternating layers of pyroclastics and rock solidified from lava flows.

compressive directed pressure Directed pressure that tends to compress some portions of a body more than other portions.

compressive stress A stress due to a force pushing together on a body.

conchoidal fracture Curved fracture surfaces.

concordant Parallel to layering or earlier developed planar structures.

concretion Hard, rounded mass that develops when a considerable amount of cementing material precipitates locally in a rock, often around an organic nucleus.

cone of depression A depression of the water table formed around a well when water is pumped out; it is shaped like an inverted cone.

confined aquifer (artesian aquifer) An aquifer completely filled with pressurized water and separated from the land surface by a relatively impermeable confining bed, such as shale.

confining pressure Pressure applied equally on all surfaces of a body; also called *geostatic* or *lithostatic pressure.*

conglomerate A coarse-grained sedimentary rock (grains coarser than 2 mm) formed by the cementation of rounded gravel.

consolidation Any process that forms firm, coherent rock from sediment or from liquid.

contact Boundary surface between two different rock types or ages of rocks.

contact (thermal) metamorphism Metamorphism under conditions in which high temperature is the dominant factor.

continental crust The thick, granitic crust under continents.

continental drift A concept suggesting that continents move over Earth's surface.

continental glaciation The covering of a large region of a continent by a sheet of glacial ice.

continental rise A wedge of sediment that extends from the lower part of the continental slope to the deep sea floor.

continental shelf A submarine platform at the edge of a continent, inclined very gently seaward generally at an angle of less than 1°.

continental slope A relatively steep slope extending from a depth of 100 to 200 meters at the edge of the continental shelf down to oceanic depths.

contour current A bottom current that flows parallel to the slopes of the continental margin (along the contour rather than down the slope).

contour line A line on a topographic map connecting points of equal elevation.

convection (convection current) A very slow circulation of a substance driven by differences in temperature and density within that substance.

convergent plate boundary A boundary between two plates that are moving toward each other.

coquina A limestone consisting of coarse shells.

core The central zone of Earth.

correlation In geology, correlation usually means determining time equivalency of rock units. Rock units may be correlated within a region, a continent, and even between continents.

country rock Any rock that was older than and intruded by an igneous body.

covalent bonding Bonding due to the sharing of electrons by adjacent atoms.

crater (of a volcano) A basinlike depression over a vent at the summit of a volcanic cone.

craton Portion of a continent that has been structurally stable for a prolonged period of time.

creep Very slow, continuous downslope movement of soil or debris.

crest (of wave) The high point of a wave.

crevasse Open fissure in a glacier.

cross-bedding An arrangement of relatively thin layers of rock inclined at an angle to the more nearly horizontal bedding planes of the larger rock unit.

cross-cutting relationship A principle or law stating that a disrupted pattern is older than the cause of disruption.

cross section *See* geologic cross section.

crude oil A liquid mixture of naturally occurring hydrocarbons.

crust The outer layer of rock, forming a thin skin over Earth's surface.

crustal rebound The rise of Earth's crust after the removal of glacial ice.

crystal A homogeneous solid with an orderly internal atomic arrangement.

crystal form Arrangement of various faces on a crystal in a definite geometric relationship to one another.

crystalline Describing a substance in which the atoms are arranged in a regular, repeating, orderly pattern.

crystalline texture An arrangement of interlocking crystals.

crystallization Crystal development and growth.

crystal settling The process whereby the minerals that crystallize at a high temperature in a cooling magma move downward in the magma chamber because they are denser than the magma.

cuesta A ridge with a steep slope on one side and a gentle slope on the other side.

Curie point The temperature below which a material becomes magnetized.

D

daughter product The isotope produced by radioactive decay.

debris Any unconsolidated material at Earth's surface.

debris avalanche Very rapid and turbulent mass wasting of debris, air, and water.

debris fall A free-falling mass of debris.

debris flow Mass wasting in which motion is taking place throughout the moving mass (flow). The common varieties are earthflow, mudflow, and debris avalanche.

debris slide Rapid movement of debris as a coherent mass.

deflation The removal of clay, silt, and sand particles from the land surface by wind.

delamination *See* lithospheric delamination.

delta A body of sediment deposited at the mouth of a river when the river velocity decreases as it flows into a standing body of water.

dendritic pattern Drainage pattern of a river and its tributaries that resembles the branches of a tree or veins in a leaf.

density Weight per given volume of a substance.

deposition The settling or coming to rest of transported material.

depth of focus Distance between the focus and the epicenter of an earthquake.

desert A region with low precipitation (usually defined as less than 25 cm per year).

desertification The expansion of barren deserts into once-populated regions.

desert pavement A thin layer of closely packed gravel that protects the underlying sediment from deflation; also called *pebble armor.*

detachment fault Major fault in a mountain belt above which rocks have been intensely folded and faulted.

diapir Bodies of rock (e.g., rock salt) or magma that ascend within Earth's interior because they are less dense than the surrounding rock.

differential stress When pressures on a body are not of equal strength in all directions.

differential weathering Varying rates of weathering resulting from some rocks in an area being more resistant to weathering than others.

differentiation Separation of different ingredients from an originally homogeneous mixture.

dike A tabular, discordant intrusive structure.

diorite Coarse-grained igneous rock of intermediate composition. Up to half of the rock is plagioclase feldspar and the rest is ferromagnesian minerals.

dip *See* angle of dip, direction of dip.

dip-slip fault A fault in which movement is parallel to the dip of the fault surface.

directed pressure *See* differential stress.

direction of dip The compass direction in which the angle of dip is measured.

discharge In a stream, the volume of water that flows past a given point in a unit of time.

disconformity A surface that represents missing rock strata but beds above and below that surface are parallel to one another.

discordant Not parallel to any layering or parallel planes.

dissolved load The portion of the total sediment load in a stream that is carried in solution.

distributary Small shifting river channel that carries water away from the main river channel and distributes it over a delta's surface.

divergent plate boundary Boundary separating two plates moving away from each other.

divide Line dividing one drainage basin from another.

dolomite A sedimentary rock composed mostly of the mineral dolomite.

dolomitic marble Marble in which dolomite, rather than calcite, is the prevalent mineral.

dome *See* structural dome.

double refraction The splitting of light into two components when it passes through certain crystalline substances.

downcutting A valley-deepening process caused by erosion of a stream bed.

drainage basin Total area drained by a stream and its tributaries.

drainage pattern The arrangement in map view of a river and its tributaries.

drawdown The lowering of the water table near a pumped well.

dripstone Deposits of calcite (and, rarely, other minerals) built up by dripping water in caves.

drumlin A long, streamlined hill made of till.

ductile Capable of being molded and bent under stress.

ductile strain Strain in which a body is molded or bent under stress and does not return to its original shape after the stress is removed.

dust (volcanic) Finest-sized pyroclasts.

dynamic pressure *See* directed pressure.

E

earthflow Slow-to-rapid mass wasting in which debris moves downslope as a very viscous fluid.

earthquake A trembling or shaking of the ground caused by the sudden release of energy stored in the rocks beneath the surface.

earthy luster A luster giving a substance the appearance of unglazed pottery.

echo sounder An instrument used to measure and record the depth to the sea floor.

elastic limit The maximum amount of stress that can be applied to a body before it deforms in a permanent way by bending or breaking.

elastic rebound theory The sudden release of progressively stored strain in rocks results in movement along a fault.

elastic strain Strain in which a deformed body recovers its original shape after the stress is released.

electron A single, negative electric charge that contributes virtually no mass to an atom.

element A substance that cannot be broken down to other substances by ordinary chemical methods. Each atom of an element possesses the same number of protons.

emergent coast A coast in which land formerly under water has recently been placed above sea level, either by uplift of the land or by a drop in sea level.

end moraine A ridge of till piled up along the front edge of a glacier.

environment of deposition The location in which deposition occurs, usually marked by characteristic physical, chemical, or biological conditions.

eon The largest unit of geological time.

epicenter The point on Earth's surface directly above the focus of an earthquake.

epoch Each period of the standard geologic time scale is divided into epochs (e.g., Pleistocene Epoch of the Quaternary Period).

equilibrium Material is in equilibrium if it is adjusted to the physical and chemical conditions of its environment so that it does not change or alter with time.

equilibrium line An irregular line marking the highest level to which the winter snow cover on a glacier is lost during a melt season. (Also called *snow line.*)

era Major subdivision of the standard geologic time scale (e.g., Mesozoic Era).

erosion The physical removal of rock by an agent such as running water, glacial ice, or wind.

erratic An ice-transported boulder that does not derive from bedrock near its present site.

esker A long, sinuous ridge of sediment deposited by glacial meltwater.

estuary Drowned river mouth.

etch-pitted terrain (*Mars*) A terrain on the surface of Mars characterized by small pits.

evaporite Rock that forms from crystals precipitating during evaporation of water.

exfoliation The stripping of concentric rock slabs from the outer surface of a rock mass.

exfoliation dome A large, rounded landform developed in a massive rock, such as granite, by the process of exfoliation.

exotic terrane Terrane that did not form at its present site on a continent and traveled a great distance to get to its present site.

expansive clay Clay that increases in volume when water is added to it.

extension Strain involving an increase in length. Extension can cause crustal thinning and faulting.

extrusive rock Any igneous rock that forms at Earth's surface, whether it solidifies directly from a lava flow or is pyroclastic.

F

faceted A rock fragment with one or more flat surfaces caused by erosive action.

failed rift (aulacogen) An inactive, sediment-filled rift that forms above a mantle plume. The rift becomes inactive as two other rifts widen to form an ocean.

fall The situation in mass wasting that occurs when material free-falls or bounces down a cliff.

fault A fracture in bedrock along which movement has taken place.

fault-block mountain range A range created by uplift along normal or vertical faults.

faunal succession A principle or law stating that fossil species succeed one another in a definite and recognizable order; in general, fossils in progressively older rock show increasingly greater differences from species living at present.

feldspars Group of most common minerals of Earth's crust. All feldspars contain silicon, aluminum, and oxygen and may contain potassium, calcium, and sodium.

felsic rock Silica-rich igneous rock with a relatively high content of potassium and sodium.

ferromagnesian mineral Iron/magnesium-bearing mineral, such as augite, hornblende, olivine, or biotite.

fine-grained rock A rock in which most of the mineral grains are less than 1 millimeter across (igneous) or less than 1/16 mm (sedimentary).

fiord A coastal inlet that is a glacially carved valley, the base of which is submerged.

firn A compacted mass of granular snow, transitional between snow and glacier ice.

firn limit *See* equilibrium line.

fissility The ability of a rock to split into thin layers.

flank eruption An eruption in which lava erupts out of a vent on the side of a volcano.

flash flood Flood of very high discharge and short duration; sudden and local in extent.

floodplain A broad strip of land built up by sedimentation on either side of a stream channel.

flow A type of movement that implies that a descending mass is moving downslope as a viscous fluid.

flowstone Calcite precipitated by flowing water on cave walls and floors.

focus The point within Earth from which seismic waves originate in an earthquake.

fold Bend in layered bedrock.

fold and thrust belt A portion of a major mountain belt characterized by large thrust faults, stacked one upon another. Layered rock between the faults was folded when faulting was taking place.

fold axis *See* hinge line.

foliation Parallel alignment of textural and structural features of a rock.

footwall The underlying surface of an inclined fault plane.

foreland basin A sediment-filled basin on a continent, landward of a magmatic arc, and caused indirectly by ocean-continent convergence.

foreset bed A sediment layer in the main part of a delta, deposited at an angle to the horizontal.

foreshock Small earthquake that precedes a main shock.

foreshore The zone that is regularly covered and uncovered by the rise and fall of tides.

formation A body of rock of considerable thickness that has a recognizable unity or similarity making it distinguishable from adjacent rock units. Usually composed of one bed or several beds of sedimentary rock, although the term is also applied to units of metamorphic and igneous rock. A convenient unit for mapping, describing, or interpreting the geology of a region.

fossil Traces of plants or animals preserved in rock.

fossil assemblage Various different species of fossils in a rock.

fracture The way a substance breaks where not controlled by cleavage.

fracture zone Major line of weakness in Earth's crust that crosses the mid-oceanic ridge at approximately right angles.

fracturing Cracking or rupturing of a body under stress.

framework silicate structure Crystal structure in which all four oxygen ions of a silica tetrahedron are shared by adjacent ions.

fretted terrain (*Mars*) Flat lowland with some scattered high plateaus on the surface of Mars.

fringing reef A reef attached directly to shore. (*See* barrier reef.)

frost action Mechanical weathering of rock by freezing water.

frost heaving The lifting of rock or soil by the expansion of freezing water.

frost wedging A type of frost action in which the expansion of freezing water pries a rock apart.

gabbro A mafic, coarse-grained igneous rock composed predominantly of ferromagnesian minerals and with lesser amounts of calcium-rich plagioclase feldspar.

gaining stream A stream that receives water from the zone of saturation.

geode Partly hollow, globelike body found in limestone or other cavernous rock.

geologic cross section A representation of a portion of Earth in a vertical plane.

geologic map A map representing the geology of a given area.

geologic resources Valuable materials of geologic origin that can be extracted from Earth.

geology The scientific study of the planet Earth.

geophysics The application of physical laws and principles to a study of Earth.

geothermal energy Energy produced by harnessing naturally occurring steam and hot water.

geothermal gradient Rate of temperature increase associated with increasing depth beneath the surface of Earth (normally about 25°C/km).

geyser A type of hot spring that periodically erupts hot water and steam.

geyserite A deposit of silica that forms around many geysers and hot springs.

glacier A large, long-lasting mass of ice, formed on land by the compaction and recrystallization of snow, which moves because of its own weight.

glacier ice The mosaic of interlocking ice crystals that form a glacier.

glassy (or vitreous) luster A luster that gives a substance a glazed, porcelainlike appearance.

gneiss A metamorphic rock composed of light and dark layers or lenses.

gneissic The texture of a metamorphic rock in which minerals are separated into light and dark layers or lenses.

goethite The commonest mineral in the limonite group; $Fe_2O_3 \cdot nH_2O$.

Gondwanaland The southern part of *Pangaea* that formed South America, Africa, India, Australia, and Antarctica.

graben A down-dropped block bounded by normal fault.

graded bed A single bed with coarse grains at the bottom of the bed and progressively finer grains toward the top of the bed.

graded stream A stream that exhibits a delicate balance between its transporting capacity and the sediment load available to it.

granite A felsic, coarse-grained, intrusive igneous rock containing quartz and composed mostly of potassium- and sodium-rich feldspars.

granitization The process by which granite is created from other rock without a melt being involved.

gravel Rounded particles coarser than 2 mm in diameter.

gravity The force of attraction that two bodies exert on each other that is proportional to the product of their masses and inversely proportional to the square of the distance from the centers of the two bodies.

gravity anomaly A deviation from the average gravitational attraction between Earth and an object. *See* negative gravity anomaly, positive gravity anomaly.

gravity meter An instrument that measures the gravitational attraction between Earth and a mass within the instrument.

graywacke A sandstone with more than 15% fine-grained matrix between the sand grains.

groin Short wall built perpendicular to shore to trap moving sand and widen a beach.

ground moraine A blanket of till deposited by a glacier or released as glacier ice melted.

ground water The water that lies beneath the ground surface, filling the cracks, crevices, and pore space of rocks.

guyot Flat-topped seamount.

H

half-life The time it takes for a given amount of a radioactive isotope to be reduced by one-half.

hanging valley A smaller valley that terminates abruptly high above a main valley.

hanging wall The overlying surface of an inclined fault plane.

hardness The relative ease or difficulty with which a smooth surface of a mineral can be scratched; commonly measured by Mohs' scale.

headland Point of land along a coast.

headward erosion The lengthening of a valley in an uphill direction above its original source by gullying, mass wasting, and sheet erosion.

heat engine A device that converts heat energy into mechanical energy.

heat flow Gradual loss of heat (per unit of surface area) from Earth's interior out into space.

heavy crude Dense, viscous petroleum that flows slowly or not at all.

hematite A type of iron oxide that has a brick-red color when powdered; Fe_2O_3.

highland (*Moon*) A rugged region of the lunar surface representing an early period in lunar history when intense meteorite bombardment formed craters.

hinge line Line about which a fold appears to be hinged. Line of maximum curvature of a folded surface.

hinge plane *See* axial plane.

hogback A sharp-topped ridge formed by the erosion of steeply dipping beds.

horn A sharp peak formed where cirques cut back into a mountain on several sides.

hornblende Common amphibole frequently found in igneous and metamorphic rocks.

hornfels A fine-grained, unfoliated metamorphic rock.

horst An up-raised block bounded by normal faults.

hot spot An area of volcanic eruptions and high heat flow above a rising mantle plume.

hot spring Spring with a water temperature warmer than human body temperature.

hydraulic action The ability of water to pick up and move rock and sediment.

hydrologic cycle The movement of water and water vapor from the sea to the atmosphere, to the land, and back to the sea and atmosphere again.

hydrology The study of water's properties, circulation, and distribution.

hydrothermal metamorphism Alteration of a rock by hot water passing through it.

hydrothermal rock Rock deposited by precipitation of ions from solution in hot water.

hypocenter Synonym for the focus of an earthquake.

hypothesis A tentative theory.

I

iceberg Block of glacier-derived ice floating in water.

ice cap A glacier covering a relatively small area of land but not restricted to a valley.

icefall A chaotic jumble of crevasses that split glacier ice into pinnacles and blocks.

ice sheet A glacier covering a large area (more than 50,000 square kilometers) of land.

igneous rock A rock formed or apparently formed from solidification of magma.

incised meander A meander that retains its sinuous curves as it cuts vertically downward below the level at which it originally formed.

inclusion A fragment of rock that is distinct from the body of igneous rock in which it is enclosed.

inclusion, principle of Fragments included in a host rock are older than the host rock.

index fossil A fossil from a very short-lived species known to have existed during a specific period of geologic time.

intensity A measure of an earthquake's size by its effect on people and buildings.

intermediate rock Rock with a chemical content between felsic and mafic compositions.

intrusion (intrusive structure) A body of intrusive rock classified on the basis of size, shape, and relationship to surrounding rocks.

intrusive rock Rock that appears to have crystallized from magma emplaced in surrounding rock.

ion An electrically charged atom or group of atoms.

ionic bonding Bonding due to the attraction between positively charged ions and negatively charged ions.

iron meteorite A meteorite composed principally of iron-nickel alloy.

island arc A curved line of islands.

isoclinal fold A fold in which the limbs are parallel to one another.

isolated silicate structure Silicate minerals that are structured so that none of the oxygen atoms are shared by silica tetrahedrons.

isostasy The balance or equilibrium between adjacent blocks of crust resting on a plastic mantle.

isostatic adjustment Concept of vertical movement of sections of Earth's crust to achieve balance or equilibrium.

isotherm A line along which the temperature of rock (or other material) is the same.

isotopes Atoms (of the same element) that have different numbers of neutrons but the same number of protons.

isotopic dating Determining the age of a rock or mineral through its radioactive elements and decay products (previously and somewhat inaccurately called *radiometric or radioactive dating*).

J

jetty Rock wall protruding above sea level, designed to protect the entrance of a harbor from sediment deposition and storm waves; usually built in pairs.

joint A fracture or crack in bedrock along which essentially no displacement has occurred.

joint set Joints oriented in one direction approximately parallel to one another.

K

karst topography An area with many sinkholes and a cave system beneath the land surface and usually lacking a surface stream.

kettle A depression caused by the melting of a stagnant block of ice that was surrounded by sediment.

KREEP (*Moon*) A lunar basalt enriched in potassium (K), the rare Earth elements (REE), and phosphorus (P).

L

laccolith A concordant intrusive structure, similar to a sill, with the central portion thicker and domed upward. Laccoliths are not common and are not discussed in this textbook.

laminar flow Slow, smooth flow, with each drop of water traveling a smooth path parallel to its neighboring drops.

laminated terrain (*Mars*) Area where series of alternating light and dark layers can be seen on the surface of Mars.

lamination A thin layer in sedimentary rock (less than 1 centimeter thick).

landform A characteristically shaped feature of Earth's surface, such as a hill or a valley.

landslide The general term for a slowly to very rapidly descending mass of rock or debris.

lateral continuity Principle that states that an original sedimentary layer extends laterally until it tapers or thins at its edges.

lateral erosion Erosion and undercutting of stream banks caused by a stream swinging from side to side across its valley floor.

lateral moraine A low ridgelike pile of till along the side of a glacier.

laterite Highly leached soil that forms in regions of tropical climate with high temperatures and very abundant rainfall.

lava Magma on Earth's surface.

lava tube Tunnel-like cave within a lava flow. It forms during the late stages of solidification of a mafic lava flow.

left-lateral fault A strike-slip fault in which the block seen across the fault appears displaced to the left.

limb Portion of a fold shared by an anticline and a syncline.

limestone A sedimentary rock composed mostly of calcite.

limonite A type of iron oxide that is yellowish-brown when powdered; $Fe_2O_3 \bullet nH_2O$.

liquefaction A type of ground failure in which water-saturated sediment turns from a solid to a liquid as a result of shaking, often caused by an earthquake.

lithification The consolidation of sediment into sedimentary rock.

lithosphere The rigid outer shell of Earth, 70 to 125 or more kilometers thick.

lithospheric delamination The detachment of part of the mantle portion of the lithosphere beneath a mountain belt.

lithostatic pressure Confining pressure due to the weight of overlying rock.

loam Soil containing approximately equal amounts of sand, silt, and clay.

loess A fine-grained deposit of wind-blown dust.

longitudinal dune (seif) Large, symmetrical ridge of sand parallel to the wind direction.

longitudinal profile A line showing a stream's slope, drawn along the length of the stream as if it were viewed from the side.

longshore current A moving mass of water that develops parallel to a shoreline.

longshore drift Movement of sediment parallel to shore when waves strike a shoreline at an angle.

losing stream Stream that loses water to the zone of saturation.

Love waves A type of surface seismic wave that causes the ground to move side to side in a horizontal plane perpendicular to the direction the wave is traveling.

low-velocity zone Mantle zone at a depth of about 100 kilometers where seismic waves travel more slowly than in shallower layers of rock.

luster The quality and intensity of light reflected from the surface of a mineral.

M

mafic rock Silica-deficient igneous rock with a relatively high content of magnesium, iron, and calcium.

magma Molten rock, usually mostly silica. The liquid may contain dissolved gases as well as some solid minerals.

magmatic arc A line of batholiths or volcanoes. Generally the line, as seen from above, is curved.

magmatic underplating *See* underplating.

magnetic anomaly A deviation from the average strength of Earth's magnetic field. *See* negative magnetic anomaly, positive magnetic anomaly.

magnetic field Region of magnetic force that surrounds Earth.

magnetic pole An area where the strength of the magnetic field is greatest and where the magnetic lines of force appear to leave or enter Earth.

magnetic reversal A change in Earth's magnetic field between normal polarity and reversed polarity. In normal polarity the north magnetic pole, where magnetic lines of force enter Earth, lies near the geographic North Pole. In reversed polarity the south magnetic pole, where lines of force leave Earth, lies near the geographic North Pole (the magnetic poles have exchanged positions).

magnetite An iron oxide that is attracted to a magnet.

magnetometer An instrument that measures the strength of Earth's magnetic field.

magnitude A measure of the energy released during an earthquake.

major mountain belt A long chain (thousands of kilometers) of mountain ranges.

mantle A thick shell of rock that separates Earth's crust above from the core below.

mantle diapir A body of mantle rock, hotter than its surroundings, that ascends because it is less dense than the surrounding rock.

mantle plume Narrow column of hot mantle rock that rises and spreads radially outward.

marble A coarse-grained rock composed of interlocking calcite (or dolomite) crystals.

maria (*Moon*) Lava plains on Moon's surface (singular, *mare*).

marine terrace A broad, gently sloping platform that may be exposed at low tide.

mass wasting (or mass movement) Movement, caused by gravity, in which bedrock, rock debris, or soil moves downslope in bulk.

matrix Fine-grained material found in the pore space between larger sediment grains.

meander A pronounced sinuous curve along a stream's course.

meander cutoff A new, shorter channel across the narrow neck of a meander.

meander scar An abandoned meander filled with sediment and vegetation.

mechanical weathering The physical disintegration of rock into smaller pieces.

medial moraine A single long ridge of till on a glacier, formed by adjacent lateral moraines joining and being carried downglacier.

Mediterranean-Himalayan belt (Mediterranean belt) A major concentration of earthquakes and composite volcanoes that runs through the Mediterranean Sea, crosses the Mideast and the Himalaya, and passes through the East Indies.

melt Liquid rock resulting from melting in a laboratory.

Mercalli scale *See* modified Mercalli scale.

mesa A broad, flat-topped hill bounded by cliffs and capped with a resistant rock layer.

Mesozoic Era The era that followed the Paleozoic Era and preceded the Cenozoic Era.

metallic bonding Bonding, as in metals, whereby atoms are closely packed together and electrons move freely among atoms.

metallic luster Luster giving a substance the appearance of being made of metal.

metamorphic facies Metamorphic rocks that contain the same set of pressure or temperature sensitive minerals are regarded as belonging to the same facies, implying that they formed under broadly similar pressure and temperature conditions.

metamorphic rock A rock produced by metamorphism.

metamorphism The transformation of preexisting rock into texturally or mineralogically distinct new rock as a result of high temperature, high pressure, or both, but without the rock melting in the process.

metasomatism Metamorphism coupled with the introduction of ions from an external source.

meteor Fragment that passes through the earth's atmosphere, heated to incandescence by friction; sometimes incorrectly called "shooting" or "falling" stars.

meteorite Meteor that strikes Earth's surface.

meteoroid Small solid particles of stone and/or metal orbiting the sun.

mica group Group of minerals with a sheet silicate structure.

microcline (potassium) feldspar A feldspar with the formula $KAlSi_3O_8$.

mid-oceanic ridge A giant mountain range that lies under the ocean and extends around the world.

migmatite Mixed igneous and metamorphic rock.

mineral A naturally occurring, crystalline solid that has a specific chemical composition.

mineraloid A substance that is not crystalline but otherwise would be considered a mineral.

model In science, a model is an image—graphic, mathematical, or verbal—that is consistent with the known data.

modified Mercalli scale Scale expressing intensities of earthquakes (judged on amount of damage done) in Roman numerals ranging from I to XII.

Mohorovičić discontinuity The boundary separating the crust from the mantle beneath it (also called **Moho**).

Mohs' hardness scale Scale on which ten minerals are designated as standards of hardness.

molecule The smallest possible unit of a substance that has the properties of that substance.

moment magnitude An earthquake magnitude calculated from the strength of the rock, surface area of the fault rupture, and the amount of rock displacement along the fault.

monocline A local steeping in a gentle regional dip; a steplike fold in rock.

moraine A body of till either being carried on a glacier or left behind after a glacier has receded.

mountain range A group of closely spaced mountains or parallel ridges.

mud Term loosely used for silt and clay, usually wet.

mud crack Polygonal crack formed in very fine-grained sediment as it dries.

mudflow A flowing mixture of debris and water, usually moving down a channel.

mudstone A fine-grained sedimentary rock that lacks shale's laminations and fissility.

multi-ringed basin (*Moon*) Large lunar crater surrounded by a series of concentric rings with intervening lowlands.

muscovite Transparent or white mica that lacks iron and magnesium.

N

natural gas A gaseous mixture of naturally occurring hydrocarbons.

natural levee Low ridges of flood-deposited sediment formed on either side of a stream channel that thin away from the channel.

nebula A large volume of interstellar gas and dust.

negative gravity anomaly Less than normal gravitational attraction.

negative magnetic anomaly Less than average strength of Earth's magnetic field.

neutron A subatomic particle that contributes mass to an atom and is electrically neutral.

nonconformity An unconformity in which an erosion surface on plutonic or metamorphic rock has been covered by younger sedimentary or volcanic rock.

nonmetallic luster Luster that gives a substance the appearance of being made of something other than metal (e.g., glassy).

nonrenewable resource A resource that forms at extremely slow rates compared to its rate of consumption.

normal fault A fault in which the hanging-wall block moved down relative to the footwall block.

nucleus Protons and neutrons form the nucleus of an atom. Although the nucleus occupies an extremely tiny fraction of the volume of the entire atom, practically all the mass of the atom is concentrated in the nucleus.

numerical age Age given in years or some other unit of time.

oblique-slip fault A fault with both strike-slip and dip-slip components.

obsidian Volcanic glass.

oceanic crust The thin, basaltic crust under oceans.

oceanic trench A narrow, deep trough parallel to the edge of a continent or an island arc.

O horizon Dark-colored soil layer that is rich in organic material and forms just below surface vegetation.

oil *See* crude oil.

oil field An area underlain by one or more oil pools.

oil pool Underground accumulation of oil.

oil sand Asphalt-cemented sand deposit.

oil shale Shale with a high content of organic matter from which oil may be extracted by distillation.

oil trap A set of conditions that hold petroleum in a reservoir rock and prevent its escape by migration.

olivine A ferromagnesian mineral with the formula $(Fe, Mg)_2SiO_4$.

oolite (ooid) A small sphere of calcite precipitated from seawater.

oolitic limestone A limestone formed from oolites.

opal A mineraloid composed of silica and water.

open fold A fold with gently dipping limbs.

open-pit mine Mine in which ore is exposed at the surface in a large excavation.

ophiolite A distinctive rock sequence found in many mountain ranges on continents.

ore Naturally occurring material that can be profitably mined.

ore mineral A mineral of commercial value.

organic sedimentary rock Rock composed mostly of the remains of plants and animals.

original horizontality The deposition of most water-laid sediment in horizontal or near-horizontal layers that are essentially parallel to Earth's surface.

orogeny An episode of intense deformation of the rocks in a region, generally accompanied by metamorphism and plutonic activity.

orthoclase (potassium) feldspar A feldspar with the formula $KAlSi_3O_8$.

outcrop A surface exposure of bare rock, not covered by soil or vegetation.

outwash Material deposited by debris-laden meltwater from a glacier.

overburden The upper part of a sedimentary deposit. Its weight causes compaction of the lower part.

overturned fold A fold in which both limbs dip in the same direction.

oxbow lake A crescent-shaped lake occupying the abandoned channel of a stream meander that is isolated from the present channel by a meander cutoff and sedimentation.

P

pahoehoe A lava flow characterized by a ropy or billowy surface.

paired terraces *Stream terraces* (*see* definition) that occur at the same elevation on each side of a river.

paleomagnetism A study of ancient magnetic fields.

Paleozoic Era The era that followed the Precambrian and began with the appearance of complex life, as indicated by fossils.

Pangaea A supercontinent that broke apart 200 million years ago to form the present continents.

parabolic dune A deeply curved dune in a region of abundant sand. The horns point upwind and are often anchored by vegetation.

parent rock Original rock before being metamorphosed.

partial melting Melting of the components of a rock with the lowest melting temperatures.

passive continental margin A margin that includes a continental shelf, continental slope, and continental rise that generally extends down to an abyssal plain at a depth of about 5 kilometers.

paternoster lakes A series of rock-basin lakes carved by glacial erosion.

peat A brown, lightweight, unconsolidated or semi-consolidated deposit of plant remains.

pebble A sediment particle with a diameter of 2 to 64 mm.

pediment A gently sloping erosional surface cut into the solid rock of a mountain range in a dry region; usually covered with a thin veneer of gravel.

pegmatite Extremely coarse-grained igneous rock.

pelagic sediment Sediment made up of fine-grained clay and the skeletons of microscopic organisms that settle slowly down through the ocean water.

peneplain A nearly flat erosional surface presumably produced as mass wasting, sheet erosion, and stream erosion reduce a region almost to base level.

perched water table A water table separated from the main water table beneath it by a zone that is not saturated.

period Each era of the standard geologic time scale is subdivided into periods (e.g., the Cretaceous Period).

permafrost Ground that remains permanently frozen for many years.

permeability The capacity of a rock to transmit a fluid such as water or petroleum.

petrified wood A material that forms as the organic matter of buried wood is either filled in or replaced by inorganic silica carried in by ground water.

petroleum Crude oil and natural gas. (Some geologists use petroleum as a synonym for oil.)

Phanerozoic Eon of geologic time. Includes all time following the Precambrian.

phenocryst Any of the large crystals in porphyritic igneous rock.

phyllite A metamorphic rock in which clay minerals have recrystallized into microscopic micas, giving the rock a silky sheen.

physical continuity Being able to physically follow a rock unit between two places.

physical geology A large division of geology concerned with Earth materials, changes of the surface and interior of Earth, and the forces that cause those changes.

piezoelectricity Electrical current generated when pressure is applied to certain minerals.

pillow structure Rocks, generally basalt, formed in pillow-shaped masses fitting closely together; caused by underwater lava flows.

placer mine Surface mines in which valuable mineral grains are extracted from stream bar or beach deposits.

plagioclase feldspar A feldspar containing sodium and/or calcium in addition to aluminum, silicon, and oxygen.

plastic Capable of being molded and bent under stress.

plastic flow Movement within a glacier in which the ice is not fractured.

plate A large, mobile slab of rock making up part of Earth's surface.

plateau Broad, flat-topped area elevated above the surrounding land and bounded, at least in part, by cliffs.

plateau basalts Layers of basalt flows that have built up to great thicknesses.

plate tectonics A theory that Earth's surface is divided into a few large, thick plates that are slowly moving and changing in size. Intense geologic activity occurs at the plate boundaries.

playa A very flat surface underlain by hard, mud-cracked clay.

playa lake A shallow temporary lake (following a rainstorm) on a flat valley floor in a dry region.

Pleistocene Epoch An epoch of the Quaternary Period characterized by several glacial ages.

plunging fold A fold in which the hinge line (or axis) is not horizontal.

pluton An igneous body that crystallized deep underground.

plutonic rock Igneous rock formed at great depth.

pluvial lake A lake formed during an earlier time of abundant rainfall.

point bar A stream *bar* (*see* definition) deposited on the inside of a curve in the stream, where the water velocity is low.

polarity *See* magnetic reversal.

polar wandering An apparent movement of the Earth's poles.

pore space The total amount of space taken up by openings between sediment grains.

porosity The percentage of a rock's volume that is taken up by openings.

porphyritic rock An igneous rock in which large crystals are enclosed in a matrix (or ground mass) of much finer-grained minerals or obsidian.

positive gravity anomaly Greater than normal gravitational attraction.

positive magnetic anomaly Greater than average strength of the Earth's magnetic field.

potassium feldspar A feldspar with the formula $KAlSi_3O_8$.

pothole Depression eroded into the hard rock of a stream bed by the abrasive action of the stream's sediment load.

Precambrian The vast amount of time that preceded the Paleozoic Era.

Precambrian shield A complex of old Precambrian metamorphic and plutonic rocks exposed over a large area.

pressure release A significant type of mechanical weathering that causes rocks to crack when overburden is removed.

progressive metamorphism Metamorphism in which progressively greater pressure and temperature act on a rock type with increasing depth in Earth's crust.

Proterozoic Eon of Precambrian time.

proton A subatomic particle that contributes mass and a single positive electrical charge to an atom.

pumice A frothy volcanic glass.

P wave A compressional wave (seismic wave) in which rock vibrates parallel to the direction of wave propagation.

P-wave shadow zone The region on Earth's surface, 103° to 142° away from an earthquake epicenter, in which P waves from the earthquake are absent.

pyroclast Fragment of rock formed by volcanic explosion.

pyroclastic debris Rock fragments produced by volcanic explosion.

pyroclastic flow Turbulent mixture of pyroclastics and gases flowing down the flank of a volcano.

pyroxene group Mineral group, all members of which are single-chain silicates.

quartz Mineral with the formula SiO_2.

quartzite A rock composed of sand-sized grains of quartz that have been welded together during metamorphism.

quartz sandstone A sandstone in which more than 90% of the grains are quartz.

Quaternary Period The youngest geologic period; includes the present time.

R

radial pattern A drainage pattern in which streams diverge outward like spokes of a wheel.

radioactive decay The spontaneous nuclear disintegration of certain isotopes.

radioactivity The spontaneous nuclear disintegration of atoms of certain isotopes.

radon A radioactive gas produced by the radioactive decay of uranium.

rain shadow A region on the downwind side of mountains that has little or no rain because of the loss of moisture on the upwind side of the mountains.

rampart crater (*Mars*) Meteorite crater that is surrounded by material that appears to have flowed from the point of impact.

rayed crater (*Moon*) Crater with bright streaks radiating from it on the moon's surface.

Rayleigh waves A type of surface seismic wave that behaves like a rolling ocean wave and causes the ground to move in an elliptical path.

receding glacier A glacier with a negative budget, which causes the glacier to grow smaller as its edges melt back.

Recent (Holocene) Epoch The present epoch of the Quaternary Period.

recessional moraine An end moraine built during the retreat of a glacier.

recharge The addition of new water to an aquifer or to the zone of saturation.

reclamation Restoration of the land to usable condition after mining has ceased.

recrystallization The development of new crystals in a rock, often of the same composition as the original grains.

rectangular pattern A drainage pattern in which tributaries of a river change direction and join one another at right angles.

recumbent fold A fold overturned to such an extent that the limbs are essentially horizontal.

reef A resistant ridge of calcium carbonate formed on the sea floor by corals and coralline algae.

regional (dynamothermal) metamorphism Metamorphism that takes place at considerable depth underground.

regolith Loose, unconsolidated rock material resting on bedrock.

relative time The sequence in which events took place (not measured in time units).

relief The vertical distance between points on Earth's surface.

reserves The discovered deposits of a geologic material that are economically and legally feasible to recover under present circumstances.

reservoir rock A rock that is sufficiently porous and permeable to store and transmit petroleum.

residual clay Fine-grained particles left behind as insoluble residue when a limestone containing clay dissolves.

residual soil Soil that develops directly from weathering of the rock below.

resources The total amount of a geologic material in all its deposits, discovered and undiscovered (*see* reserves).

reverse fault A fault in which the hanging-wall block moved up relative to the footwall block.

rhyolite A fine-grained, felsic, igneous rock made up mostly of feldspar and quartz.

Richter scale A numerical scale of earthquake magnitudes.

ridge push The concept that oceanic plates diverge as a result of sliding down the sloping lithosphere-asthenosphere boundary.

rift valley A tensional valley bounded by normal faults. Rift valleys are found at diverging plate boundaries on continents and along the crest of the mid-oceanic ridge.

right-lateral fault A strike-slip fault in which the block seen across the fault appears displaced to the right.

rigid zone Upper part of a glacier in which there is no plastic flow.

rille (*Moon*) Elongate trenched or cracklike valley on the lunar surface.

rip current Narrow currents that flow straight out to sea in the surf zone, returning water seaward that has been pushed ashore by breaking waves.

ripple mark Any of the small ridges formed on sediment surfaces exposed to moving wind or water. The ridges form perpendicularly to the motion.

rock Naturally formed, consolidated material composed of grains of one or more minerals. (There are a few exceptions to this definition.)

rock avalanche A very rapidly moving, turbulent mass of broken-up bedrock.

rock-basin lake A lake occupying a depression caused by glacial erosion of bedrock.

rock cycle A theoretical concept relating tectonism, erosion, and various rock-forming processes to the common rock types.

rockfall Rock falling freely or bouncing down a cliff.

rock flour A powder of fine fragments of rock produced by glacial abrasion.

rock gypsum An evaporite composed of gypsum.

rock salt An evaporite composed of halite.

rockslide Rapid sliding of a mass of bedrock along an inclined surface of weakness.

rotational slide In mass wasting, movement along a curved surface in which the upper part moves vertically downward while the lower part moves outward. Also called a *slump*.

rounded knobs (glacial) Bedrock that is more resistant to glacial erosion stands out as rounded knobs, usually elongated parallel to the direction of glacier flow. These are also known as *roche moutonnées* (French for "rock sheep").

rounding The grinding away of sharp edges and corners of rock fragments during transportation.

rubble Angular sedimentary particles coarser than 2 mm in diameter.

S

saltation A mode of transport that carries sediment downcurrent in a series of short leaps or bounces.

sand Sediment composed of particles with a diameter between 1/16 mm and 2 mm.

sand dune A mound of loose sand grains heaped up by the wind.

sandstone A medium-grained sedimentary rock (grains between 1/16 mm and 2 mm) formed by the cementation of sand grains.

saturated zone A subsurface zone in which all rock openings are filled with water.

scale The relationship between distance on a map and the distance on the terrain being represented by that map.

schist A metamorphic rock characterized by coarse-grained minerals oriented approximately parallel.

schistose The texture of a rock in which visible platy or needle-shaped minerals have grown essentially parallel to each other under the influence of directed pressure.

scientific method A means of gaining knowledge through objective procedures.

scoria A basalt that is highly vesicular.

sea cave A cavity eroded by wave action at the base of a sea cliff.

sea cliff Steep slope that retreats inland by mass wasting as wave erosion undercuts it.

sea-floor spreading The concept that the ocean floor is moving away from the mid-oceanic ridge and across the deep ocean basin, to disappear beneath continents and island arcs.

seamount Conical mountain rising 1,000 meters or more above the sea floor.

seawall A wall constructed along the base of retreating cliffs to prevent wave erosion.

sediment Loose, solid particles that can originate by (1) weathering and erosion of preexisting rocks, (2) chemical precipitation from solution, usually in water, and (3) secretion by organisms.

sedimentary breccia A coarse-grained sedimentary rock (grains coarser than 2 mm) formed by the cementation of angular rubble.

sedimentary facies Significantly different rock types occupying laterally distinct parts of the same layered rock unit.

sedimentary rock Rock that has formed from (1) lithification of any type of sediment, (2) precipitation from solution, or (3) consolidation of the remains of plants or animals.

sedimentary structure A feature found within sedimentary rocks, usually formed during or shortly after deposition of the sediment and before lithification.

seismic gap A segment of a fault that has not experienced earthquakes for a long time; such gaps may be the site of large future quakes.

seismic profiler An instrument that measures and records the subbottom structure of the sea floor.

seismic reflection The return of part of the energy of seismic waves to Earth's surface after the waves bounce off a rock boundary.

seismic refraction The bending of seismic waves as they pass from one material to another.

seismic sea wave *See* tsunami.

seismic wave A wave of energy produced by an earthquake.

seismogram Paper record of earth vibration.

seismograph A seismometer with a recording device that produces a permanent record of Earth motion.

seismometer An instrument designed to detect seismic waves or Earth motion.

serpentine A magnesium silicate mineral. Most asbestos is a variety of serpentine.

shale A fine-grained sedimentary rock (grains finer than 1/16 mm in diameter) formed by the cementation of silt and clay (mud). Shale has thin layers (laminations) and an ability to split (fissility) into small chips.

shear force In mass wasting, the component of gravitational force that is parallel to an inclined surface.

shearing Movement in which parts of a body slide relative to one another and parallel to the forces being exerted.

shear strength In mass wasting, the resistance to movement or deformation of material.

shear stress Stress due to forces that tend to cause movement or strain parallel to the direction of the forces.

sheet erosion The removal of a thin layer of surface material, usually topsoil, by a flowing sheet of water.

sheet joints Cracks that develop parallel to the outer surface of a large mass of expanding rock, as pressure is released during unloading.

sheet silicate structure Crystal structure in which each silica tetrahedron shares three oxygen ions.

sheetwash Water flowing down a slope in a layer.

shield volcano Broad, gently sloping cone constructed of solidified lava flows.

silica A term used for oxygen plus silicon.

silicate A substance that contains silica as part of its chemical formula.

silica tetrahedron *See* silicon-oxygen tetrahedron.

silicic rock or magma Silica-rich igneous rock or magma with a relatively high content of potassium and sodium.

silicon-oxygen tetrahedron Four-sided, pyramidal object that visually represents the four oxygen atoms surrounding a silicon atom; the basic building block of silicate minerals. Also called a silica tetrahedron or a silicon tetrahedron.

sill A tabular intrusive structure concordant with the country rock.

silt Sediment composed of particles with a diameter of 1/256 to 1/16 mm.

siltstone A sedimentary rock consisting mostly of silt grains.

sinkhole A closed depression found on land surfaces underlain by limestone.

sinter A deposit of silica that forms around some hot springs and geysers.

slab pull The concept that subducting plates are pulled along by their dense leading edges.

slate A fine-grained rock that splits easily along flat, parallel planes.

slaty Describing a rock that splits easily along nearly flat and parallel planes.

slaty cleavage The ability of a rock to break along closely spaced parallel planes.

slide In mass wasting, movement of a relatively coherent descending mass along one or more well-defined surfaces.

slip face The steep, downwind slope of a dune; formed from loose, cascading sand that generally keeps the slope at the angle of repose (about 34°).

slump In mass wasting, movement along a curved surface in which the upper part moves vertically downward while the lower part moves outward. Also called a *rotational slide.*

snow line See *equilibrium line.*

soil A layer of weathered, unconsolidated material on top of bedrock; often also defined as containing organic matter and being capable of supporting plant growth.

soil horizon Any of the layers of soil that are distinguishable by characteristic physical or chemical properties.

solifluction Flow of water-saturated debris over impermeable material.

solution Usually slow but effective process of weathering and erosion in which rocks are dissolved by water.

sorting Process of selection and separation of sediment grains according to their grain size (or grain shape or specific gravity).

source area The locality that eroded to provide sediment to form a sedimentary rock.

source rock A rock containing organic matter that is converted to petroleum by burial and other postdepositional changes.

spatter cone A small, steep-sided cone built from lava spattering out of a vent.

specific gravity The ratio of the mass of a substance to the mass of an equal volume of water, determined at a specified temperature.

speleothem Dripstone deposit of calcite that precipitate from dripping water in caves.

spheroidally weathered boulder Boulder that has been rounded by weathering from an initial blocky shape.

spit A fingerlike ridge of sediment attached to land but extending out into open water.

spreading axis (or spreading center) The crest of the mid-oceanic ridge, where sea floor is moving away in opposite directions on either side.

spring A place where water flows naturally out of rock onto the land surface.

stable Describing a mineral that will not react with or convert to a new mineral or substance, given enough time.

stack A small rock island that is an erosional remnant of a headland left behind as a wave-eroded coast retreats inland.

stalactite Iciclelike pendant of dripstone formed on cave ceilings.

stalagmite Cone-shaped mass of dripstone formed on cave floors, generally directly below a stalactite.

standard geologic time scale A worldwide relative scale of geologic time divisions.

static pressure *See* confining pressure.

stock A small discordant pluton with an outcropping area of less than 100 square kilometers.

stony-iron meteorite A meteorite composed of silicate minerals and iron-nickel alloy in approximately equal amounts.

stony meteorite A meteorite made up mostly of plagioclase and iron-magnesium silicates.

stoping Upward movement of a body of magma by fracturing of overlying country rock. Magma engulfs the blocks of fractured country rock as it moves upward.

storm surge High sea level caused by the low pressure and high winds of hurricanes.

strain Change in size (volume) or shape of a body (or rock unit) in response to stress.

stratovolcano *See* composite volcano.

streak Color of a pulverized substance; a useful property for mineral identification.

stream A moving body of water, confined in a channel and running downhill under the influence of gravity.

stream capture *See* stream piracy.

stream channel A long, narrow depression, shaped and more or less filled by a stream.

stream discharge Volume of water that flows past a given point in a unit of time.

stream-dominated delta A delta with fingerlike distributaries formed by the dominance of stream sedimentation; also called a birdfoot delta.

stream gradient Downhill slope of a stream's bed or the water surface, if the stream is very large.

stream headwaters The upper part of a stream near the source.

stream mouth The place where the stream enters the sea, a large lake, or a larger stream.

stream piracy The natural diversion of the headwaters of one stream into the channel of another.

stream terrace Steplike landform found above a stream and its floodplain.

stream velocity The speed at which water in a stream travels.

stress A force acting on a body, or rock unit, that tends to change the size or shape of that body, or rock unit. Force per unit area within a body.

striations (1) On minerals, extremely straight, parallel lines; (2) Glacial—straight scratches in rock caused by abrasion by a moving glacier.

strike The compass direction of a line formed by the intersection of an inclined plane (such as a bedding plane) with a horizontal plane.

strike-slip fault A fault in which movement is parallel to the strike of the fault surface.

strip mine A mine in which the valuable material is exposed at the surface by removing a strip of overburden.

structural basin A structure in which the beds dip toward a central point.

structural dome A structure in which beds dip away from a central point.

structural geology The branch of geology concerned with the internal structure of bedrock and the shapes, arrangement, and interrelationships of rock units.

subduction The sliding of the sea floor beneath a continent or island arc.

subduction complex *See* accretionary wedge.

subduction zone Elongate region in which subduction takes place.

submarine canyon V-shaped valleys that run across the continental shelf and down the continental slope.

submergent coast A coast in which formerly dry land has been recently drowned, either by land subsidence or a rise in sea level.

subsidence Sinking or downwarping of a part of the Earth's surface.

superposed stream A river let down onto a buried geologic structure by erosion of overlying layers.

superposition A principle or law stating that within a sequence of undisturbed sedimentary rocks, the oldest layers are on the bottom, the youngest on the top.

surf Breaking waves.

surface wave A seismic wave that travels on Earth's surface.

suspect terrane A terrane that may not have formed at its present site.

suspended load Sediment in a stream that is light enough in weight to remain lifted indefinitely above the bottom by water turbulence.

S wave A seismic wave propagated by a shearing motion, which causes rock to vibrate perpendicular to the direction of wave propagation.

S-wave shadow zone The region on Earth's surface (at any distance more than 103° from an earthquake epicenter) in which S waves from the earthquake are absent.

swelling clay *See* expansive clay.

syncline A fold in which the layered rock usually dips toward an axis.

T

talus An accumulation of broken rock at the base of a cliff.

tarn *See* rock-basin lake.

tectite Small, rounded bits of glass formed from rock melting and being thrown into the air due to a meteorite impact.

tectonic forces Forces generated from within Earth that result in uplift, movement, or deformation of part of Earth's crust.

tectonostratigraphic terrane *See* terrane.

tensional stress A stress due to a force pulling away on a body.

tephra *See* pyroclastic debris.

terminal moraine An end moraine marking the farthest advance of a glacier.

terminus The lower edge of a glacier.

terrane (tectonostratigraphic terrane) A region in which the geology is markedly different from that in adjoining regions.

terrigenous sediment Land-derived sediment that has found its way to the sea floor.

theory An explanation for observed phenomena that has a high possibility of being true.

theory of glacial ages At times in the past, colder climates prevailed during which significantly more of the land surface of Earth was glaciated than at present.

thermal metamorphism *See* contact (thermal) metamorphism.

thrust fault A reverse fault in which the dip of the fault plane is at a low angle to horizontal.

tidal delta A submerged body of sediment formed by tidal currents passing through gaps in barrier islands.

"tidal wave" An incorrect name for a tsunami.

tide-dominated delta A delta formed by the reworking of sand by strong tides.

till Unsorted and unlayered rock debris carried by a glacier.

tillite Lithified till.

time-transgressive rock unit An apparently continuous rock layer in which different portions formed at different times.

tombolo A bar of marine sediment connecting a former island or stack to the mainland.

topographic map A map on which elevations are shown by means of contour lines.

topset bed In a delta, a nearly horizontal sediment bed of varying grain size formed by distributaries shifting across the delta surface.

traction Movement by rolling, sliding, or dragging of sediment fragments along a stream bottom.

transform fault The portion of a fracture zone between two offset segments of a mid-oceanic ridge crest.

transform plate boundary Boundary between two plates that are sliding past each other.

translational slide In mass wasting, movement of a descending mass along a plane approximately parallel to the slope of the surface.

transportation The movement of eroded particles by agents such as rivers, waves, glaciers, or wind.

transported soil Soil not formed from the local rock but from parent material brought in from some other region and deposited, usually by running water, wind, or glacial ice.

transverse dune A relatively straight, elongate dune oriented perpendicular to the wind.

travel-time curve A plot of seismic-wave arrival times against distance.

travertine A porous deposit of calcite that often forms around hot springs.

trellis pattern A drainage pattern consisting of parallel main streams with short tributaries meeting them at right angles.

trench *See* oceanic trench.

trench suction The concept that overlying plates move horizontally toward oceanic trenches as subducting plates sink at an angle steeper than their dip.

tributary Small stream flowing into a large stream, adding water to the large stream.

trough (of wave) The low point of a wave.

truncated spur Triangular facet where the lower end of a ridge has been eroded by glacial ice.

tsunami Huge ocean wave produced by displacement of the sea floor; also called seismic sea wave.

tufa A deposition of calcite that forms around a spring, lake, or percolating ground water.

tuff A rock formed from fine-grained pyroclastic particles (ash and dust).

turbidity current A flowing mass of sediment-laden water that is heavier than clear water and therefore flows downslope along the bottom of the sea or a lake.

turbulent flow Eddying, swirling flow in which water drops travel along erratically curved paths that cross the paths of neighboring drops.

U

ultramafic rock Rock composed entirely or almost entirely of ferromagnesian minerals.

unconfined aquifer A partially filled aquifer exposed to the land surface and marked by a rising and falling water table.

unconformity A surface that represents a break in the geologic record, with the rock unit immediately above it being considerably younger than the rock beneath.

unconsolidated In referring to sediment grains, loose, separate, or unattached to one another.

underplating The pooling of magmas at the base of the continental crust.

uniformitarianism Principle that geologic processes operating at present are the same processes that operated in the past. The principle is stated more succinctly as "The present is the key to the past." *Also, see* actualism.

unloading The removal of a great weight of rock.

unpaired terraces *Stream terraces* (*see* definition) that do not have the same elevation on opposite sides of a river.

unsaturated zone *See* vadose zone.

U-shaped valley Characteristic cross-profile of a valley carved by glacial erosion.

V

vadose zone A subsurface zone in which rock openings are generally unsaturated and filled partly with air and partly with water; above the saturated zone.

valley glacier A glacier confined to a valley. The ice flows from a higher to a lower elevation.

Van der Waal's bonds or forces Weak bonds in crystals.

varve Two thin layers of sediment, one dark and the other light in color, representing 1 year's deposition in a lake.

vent The opening in Earth's surface through which a volcanic eruption takes place.

ventifact Boulder, cobble, or pebble with flat surfaces caused by the abrasion of wind-blown sand.

vertical exaggeration An artificial steepening of slope angles on a topographic profile caused by using a vertical scale that differs from the horizontal scale.

vesicle A cavity in volcanic rock caused by gas in a lava.

viscosity Resistance to flow.

vitreous luster *See* glassy luster.

volcanic breccia Rock formed from large pieces of volcanic rock (cinders, blocks, bombs).

volcanic dome A steep-sided, dome- or spine-shaped mass of volcanic rock formed from viscous lava that solidifies in or immediately above a volcanic vent.

volcanic neck An intrusive structure that apparently represents magma that solidified within the throat of a volcano.

volcanism Volcanic activity, including the eruption of lava and rock fragments and gas explosions.

volcano A hill or mountain constructed by the extrusion of lava or rock fragments from a vent.

W

wastage *See* ablation.

water table The upper surface of the zone of saturation.

wave crest *See* crest.

wave-cut platform A horizontal bench of rock formed beneath the surf zone as a coast retreats because of wave erosion.

wave-dominated delta A delta formed by the reworking of sand by wave action.

wave height The vertical distance between the crest (the high point of a wave) and the trough (the low point).

wavelength The horizontal distance between two wave crests (or two troughs).

wave refraction Change in direction of waves due to slowing as they enter shallow water.

wave trough *See* trough.

weathering The group of processes that change rock at or near Earth's surface.

welded tuff A rock composed of pyroclasts welded together.

well A hole, generally cylindrical and usually walled or lined with pipe, that is dug or drilled into the ground to penetrate an aquifer below the zone of saturation.

Wilson cycle The cycle of splitting of a continent, opening of an ocean basin, followed by closing of the basin and collision of the continents.

wind ripple Small, low ridge of sand produced by the saltation of windblown sand.

wrinkle ridge (*Moon*) Wrinkle on lunar maria surface.

X

xenolith Fragment of rock distinct from the igneous rock in which it is enclosed.

Z

zone of ablation That portion of a glacier in which ice is lost.

zone of accumulation (1) That portion of a glacier with a perennial snow cover; (2) *See* B horizon (a soil layer).

zone of leaching *See* A horizon (a soil layer).

zone of saturation *See* saturated zone.

zoning Orderly variation in the chemical composition within a single crystal.

Index

Note: Citations of figures and tables are denoted by an *f* or *t;* figures are cited only when they occur outside of related text citations.

A

B

C

D

E

F

G

N

O

P

Q

R

S

T

U